ADVANCES IN POWER AND ENERGY ENGINEERING

PROCEEDINGS OF THE 8TH ASIA-PACIFIC POWER AND ENERGY ENGINEERING CONFERENCE (APPEEC 2016), SUZHOU, CHINA, 15–17 APRIL 2016

Advances in Power and Energy Engineering

Editor

Yuanzhang Sun
School of Electrical Engineering, Wuhan University, Wuhan, China

CRC Press
Taylor & Francis Group
Boca Raton London New York

CRC Press is an imprint of the
Taylor & Francis Group, an **informa** business

A BALKEMA BOOK

Published by:
CRC Press/Balkema
P.O. Box 447, 2300 AK Leiden, The Netherlands
e-mail: Pub.NL@taylorandfrancis.com
www.crcpress.com – www.taylorandfrancis.com

First issued in paperback 2020

© 2016 by Taylor & Francis Group, LLC
CRC Press/Balkema is an imprint of the Taylor & Francis Group, an informa business

No claim to original U.S. Government works

Typeset by V Publishing Solutions Pvt Ltd., Chennai, India

ISBN 13: 978-0-367-73729-0 (pbk)
ISBN 13: 978-1-138-02846-3 (hbk)

Visit the Taylor & Francis Web site at
http://www.taylorandfrancis.com

and the CRC Press Web site at
http://www.crcpress.com

Table of contents

Preface   xi

Organization   xiii

New energy and technologies

Research on the DFIG wind generation comprehensive control   3
M.X. Shangguan & J.X. Zhao

Economic analysis of supporting policies for residential solar Photovoltaic in Massachusetts,
USA: A case from Boston   7
S. Qiu & S. Ghosh

Fuzzy control based self-adaptive illumination system   17
C.K. Zhang, Z.Y. Li, Q.L. Zhang, W.H. Zhou & F.Z. Yu

PSO-Hammerstein model identification on MGT-based LiBr absorption chiller system   23
D.C. Zhao, L. Pan & J.L. Zhang

ERGA-SVD method based decoupled controller design for Micro Gas Turbine Combined
Cooling Heating and Power system   29
Y.X. Liu, F. Zhang, J.L. Zhang, X. Zhou & J. Shen

Auto-tuning model predicting the control of Micro Gas Turbine Combined with Cooling
Heating and Power systems   35
X.F. Liang, X. Wu, Y.G. Li, J.L. Zhang & J. Shen

Research on the master-slave control method for the multi-inverters of the micro-grid   41
S.F. Wen & S.T. Wang

Coordinated economic dispatch of interconnected microgrids based on rolling
optimization   47
Y. Du & W. Pei

Research on a novel power control method of Distributed Generators in the micro grid   55
L. Peng, S. Su, Z.Y. Zhang, X.N. Lin & X.S. Li

Positive feedback voltage drift island detection based on droop control   63
T. Yan, Z.Z. Qu, Z.B. Liu & G.W. Zhu

Coordinated control strategy of AC/DC hybrid distribution system   69
X. Zhang, W. Deng, W. Pei, T. Yu & T.J. Pu

Optimal protection coordination based on dual directional over-current relays for ring
network distribution power systems   75
X.Y. Wang, X.G. Wang, D.H. Chen, S. Su & X.N. Lin

Analysis on water requirement of Concentrating Solar Power plant in western China   83
X.Y. Wang, L.Z. Zhu, L. Zhang & R.Y. Yu

Island operation control strategy for battery micro-grid based on RT-LAB   87
T. Yan, Z.Z. Qu, G.W. Zhu & Z.B. Liu

A review of the Energy Storage Systems applied in power grids   91
B. Han, S.F. Lu, L. Jiang & Y. Du

Modeling and simulation on regenerative braking of the electric vehicle
W. Shi, Y.Q. Zhang & G.Z. Song
97

Probabilistic power flow calculation of large scale Photovoltaic grid-connected
power generation system
H. Fan, L.H. Zuo & D.F. Wang
101

Study on the mechanism and mitigation of SSI phenomena in large-scale
wind farms with DFIGs
Y. Huang, X.T. Wang, H.J. Yu, K.M. Chen, G.L. Wang & Y.M. Huang
107

Research on the design of the power quality control equipment testing platform
Z.F. Liu, L.X. Wu, R.K. He, M. Ma & B.Y. Xu
115

Optimal operation strategy of power systems with renewable generations via coordinated
control of hydrogen consumption
W. Wang & S.R. Wang
121

DC bus current reconstruction of 3L NPC dual PWM converter based on feed-forward control
W.J. Wu, W.X. Wang & X.Y. Li
129

Control strategy of the hybrid three terminal HVDC system for wind power transmission
Y. Zhang, X.T. Wang, Y. Huang & Z.B. Wang
135

Review of solar thermal energy storage materials
C. Peng, M.Y. Li, Y.Y. Li, X.M. Chen & G.M. Yu
141

Research on estimation of Electric Vehicles real-time schedulable capacity
X.B. Ju, C.Y. He, X.W. Zhang, J.R. Xia, H.H. Sun, C. Wang & C. Zhang
153

Research on wind-storage system participating in frequency regulation for energy internet
*Q. Fan, W. Gu, Y. Xiao, C.H. Lin, X.K. Wen, J.G. Chen, M.M. Xu, Y.T. Xu,
Y.Y. Chen & Y. He*
159

Research on modeling and simulation of a kind of AC/DC power distribution device
P. Ye, Y.N. Hu, Z.F. Teng, L. Liu, Q. Gao, G.M. Zhang & Q. Liu
163

Power generation—conventional and renewable

Bioethanol production from the co-fermentation and co-culture fermentation
of glucose and xylose by *Saccharomyces cerevisiae* and *Pichia stipitis* ATCC 58785
in a stirred tank bioreactor
F.J. Shalsh, N.A. Ibrahim, M. Arifullah & A.S. Mero Hussin
171

Implication of crystallites nucleation in petroleum based transformer oil in electric
power system
L.L. Kaanagbara, H.I. Inyang, Jy. Wu, V.P. Lukic & D. Young
177

Research on pressure fluctuation of Francis turbine based on CFD
L. Fu, P.G. Kou & Y.G. Cheng
185

Dynamic behavior of synchronous generator during a sudden change in input torque
using MATLAB/SIMULINK
M.A. Hassan Ibraheem
193

On-load capacity regulation design of distribution transformer
L.C. Ma & F.Y. Chang
199

Numerical Study of acoustic characteristics of gas-liquid coaxial injectors
H.H. An & W.S. Nie
205

Low Voltage Ride Through characteristics of large-scale grid-connected
Photovoltaic power stations
W. Li, F. Wu, W.Y. Kong & Z.J. Meng
211

Optimal sizing of Photovoltaic and battery for residential house using genetic algorithms
J. Li & J.H. Braslavsky
217

Research on thermal and power characteristics of conventional solar cells
with auxiliary equipment 225
X. Guo, E.C. Xin, H.W. Yuan, J.X. Lv, L.M. Zhou & Z. Ma

Preliminary analysis of AP1000 PCCS and its enhanced performance 231
C. Li, L. Li & Y.J. Zhang

Application of the Particle Swarm Optimization method in rectangular flue gas ducts 237
Q.C. Niu, B. Yan, X.K. Fan, Y. Yuan & D.M. Wu

Software redevelopment for rectangular flue gas ducts based on ABAQUS 241
X.K. Fan, B. Yan, Q.C. Niu, Y. Yuan & D.M. Wu

Research on the calculation of reversed flow of coal slime pastes in long-distance
pipeline transportation 245
Y.Y. Liu, Y.B. Yang, J. Gao, X.D. Hao & M. Wu

Wind speed probability function in the coast of Rio Grande do Norte, Brazil 251
M.A. Aredes & M. Aredes

An optimal and real-time control strategy for energy storage system 257
L.B. Yang, Y.L. Gao, Y. Niu, F. Zhang & T. Du

A voltage control strategy of Wind Farms Cluster for Steady State Voltage
Stability improvement 263
S. Yang, W.S. Wang, C. Liu, Y.H. Huang, J. Wang & T.F. Guo

Dynamic security assessment method of wind power farm 269
Y.F. He, Y.X. Zhuo, X.N. Lin, Z.X. Wang & C.G. Tu

Study on dynamic model with feedback from flexible structure of wind turbine 275
H.T. Li, Z.J. Li & S. Shen

A new completing method of wind power data based on back propagation
neural network 283
S.Y. Ye, Q. Tang, H. Feng, Y. Lu & X.R. Wang

Parallel computing strategy of power grid risk early warning system considering
wind farm integration 289
W. Zhang, L.Y. Xu, Y.X. Zhuo, C. Chen, X. Zhou, Y.F. He

Development of Fast Reactor power plant and its economic analysis 295
L. Liu & S.P. Qi

Power system management

Seamless switching control method of micro grid based on master-slave configuration 303
L. Wang, K. Jiang, Z.X. Wang, B.F. Lu & C. Zhang

Influence of fluctuating load on distribution network voltage stability 309
Y.H. Huo, L.X. Zhang, Q.L. Zhang, S.Y. Li & J.T. Gui

Pole assignment design of the Power System Stabilizer by phase compensation method 317
C. Lv, T. Littler & W. Du

Stochastic stability of power system with asynchronous wind turbines 323
B. Yuan & J. Zong

A novel control strategy for UPQC applied to large scale industrial enterprises 327
S. Liu, Y. Liang, J.Q. Wang, Y. Liu, L.M. Tu, J. Xiao, Y.C. Wang & C.X. Mao

An active power regulation model for wind farms considering Static Voltage
Stability margin 335
Y. Qi, X.D. Wang, N.S. Chen, X.J. Ge, M.X. Liu & M.S. Wang

Study on low power wireless energy transmission technology 341
Z.X. Wang & Y.G. Wei

Analysis method of weak point for high voltage distribution network with high density distributed generation 347
H. Zhao, Z.J. Li & H.R. Yan

Research on off-line quantification method of transient stability 355
D. Liu, Q.Y. Liu, R.G. Zhao, Q.F. Liu & R.H. Liu

Review and prospect of power system mid-long-term load forecasting methods 363
L. Xing, D. Yang, Z. Jiang, N. Wu & X. Zhao

Parameter identification of equivalent model of hydro generator units group based on PSO method with distributed computing structure 371
X. Xia & W. Ni

An improved calculation method for direct transient stability assessment with detailed generator model 377
Y. Wang, H.S. Sun, X.J. Pan, Y.P. Xu & J.H. Xi

Application of wavelet mutation detection technology in the ultrasonic localization of Partial Discharge in the transformer 385
Y.W. Dong, Y.Y. Liu, Q. Zou & L.X. Shi

Measurement accuracy analysis of active power under effects of reactive power compensation 391
S.Q. Li & B.L. Lei

Trip fault study of 500 kV common-tower double-circuit transmission lines based on induced voltage 395
A.W. Yu, W. Tang & B.C. Feng

Thermal fault diagnosis on power transformer with Grey Relational Analysis 401
L.F. Li, S.C. Deng, X. Chen & J. Wang

Study of time series clustering based on FCM 405
X. Guo, H.B. Yuan, J.X. Lv, M. Ling, L.M. Zhou & Z. Ma

Prediction of closing spring's energy-storing state-based on circuit breaker's running characteristic parameters 411
T. Shi, B.W. Guo, S.Q. Bai, W.J. Zhou, W. Dong & P.F. Li

State assessment of circuit breaker's spring based on fuzzy comprehensive evaluation method 417
X.Q. Xu, B.W. Guo, J. Yuan, W.J. Zhou, R.F. Wei & S.Y. Zhao

Real-time degradation assessment system for hydropower equipment based on the streaming data processing 423
W.P. Tan & J. Xiao

Study on the localization technology of reflected pulse in Partial Discharge detection of HV power cables 429
T. Dong, H. Zhang, B. Zhang, G.K. Yu, X.R. Li, T.S. Hu & D.Z. Xu

Design and realization of the monitoring system of a DC micro-grid experiment and research platform 435
C.N. Song, F. Li, X.F. Lin & B. Liu

The research and implementation of a new grounding resistance on-line monitoring method of tower transmission system 441
B. Zhang, J.J. Song, J.B. Zhang, W. Li & M. Xu

An online coordinated optimal control strategy of a wind/photovoltaic/storage hybrid system 447
D. Yang, P. Sun, W. Ci, L. Xing & T. Zheng

Dynamic Optimal Power Flow and analysis of the active distribution network 453
W.B. Li, S. Yang, F.M. Meng, Z. Jiang, D.D. Zhang & J.Y. Jiang

Mixed measurement-based power system state estimation with measurement correlation 461
Z.G. Lu, S.H. Yang, S. Yang & J. Wang

Research on undervoltage problems and coordination strategies of prevention and control in isolated receiving power grid
Z.H. Li, L.H. Wu, Q. Wang, X. Chen & Y.X. Yu
467

Impact of VSC–HVDC integration on the rotor angle stability of power systems
T.Z. Pan & X.S. Tang
473

Analysis of the reliability of multi-circuit-on-same-tower power transmission systems
Y.J. Sun, Y.H. Zhang, Y.C. Zhang, X.H. Qin & Q.Y. Zhou
479

A bus voltage automatic optimization and adjustment method based on AIMMS software
P. Xu, F. Li, Y. Wang, F. Shi & A.A. Ni
487

Research on Demand Side Management
M.L. Dong, D.H. You, J. Hu, G. Wang, C. Long, F. Zhang, Z. He & L. Dai
493

An integrated decision-making method for condition assessment of power transformer
L.J. Sun, H.W. Yuan, B.Y. Zhang, W.J. Zuo, Z. Ma & Y.W. Shang
501

Determination of harmonic responsibilities under the change of background harmonic impedance
J. Chen, T.L. Zang, L. Fu & Z.Y. He
507

Research and application of power flow diagram automatic generation method based on logical geographic information
W.C. Fang, S.Y. Liang, W. Zhu, R. Hu, T.F. Kang, Y.G. Li & R.P. Zhang
513

The design of an automatic batch test system for power quality monitoring equipment
R.S. Qin, Z. Xu, X. Wang & J.W. Geng
519

Transmission line optimization on wind farm considering the effect of wind speed
Z.F. Jiang, C. Ma, X.G. He & Y.L. Zhong
525

Optimal allocation of STATCOM using improved Harmony Search algorithm
X.Q. Xu, T. Zhang & Y.Q. Liu
533

The harmonic model and its application to traction power supply system considering high-pass filter
D.D. Li, F.L. Zhou, Q. Liu, P. Zhu & H. Yu
539

A comprehensive accessibility evaluation method for power network planning
J.Y. Xu, J. Yan, C. Wei, L. Qiao, Y.X. Zhuo, Z.C. Wang, X. Zhou, C. Chen & X.N. Lin
545

A novel ADP-based method for SCUC with power flow constraints
D.L. Long
551

Multi-source generation optimal dispatching strategy with significant renewable energy penetration under non-market condition
S. Ma, X.F. Li, Y.P. Xu, Y.F. Wang & J.T. Zhang
557

Research on the differentiated construction mode of smart distribution network
Z. Huang, S.G. Liu, Y.H. Liu, W. Zhang, W. Liu, L.M. Zhou & Y.M. Hou
565

Effect analysis of Distributed Generations in the Active Distribution Network
H. Fan, H. Hui, Y.M. Hou, W. Liu, Z. Huang & J. Su
571

Constructive Heuristic Algorithm for integrated Generation and Transmission Network Expansion Planning
E.B. Cedeno
577

The electrical properties of polyester materials as insulators
K.B. Ewiss, L.S. Nasrat & R.M. Sharkawy
583

Power transmission and distribution

Testing and analysis of 220 kV transformer's abnormal noise
W. Jiang, X.S. Lan & Y.Q. Zhou
591

Critical line identification of the power system considering network structure
and operating state   597
Z.Y. Qu, Q. Li, J.H. Du & Y.X. Ren

Simulation and experimental study of a state-of-the-art M-STATCOM   605
S.A. Kamran, J. Muñoz & Y. Chen

DC voltage characteristic analysis of MTDC distribution network   611
Z.Y. Zhao, B.S. Su & D. Xie

Synchronverter to damp multiple electromechanical oscillations   617
E.L. van Emmerik, B.W. França, A.R. Castro, G.F. Gontijo, D.S. Oliveira & M. Aredes

Research on power stability of hybrid system with UHVDC Hierarchical
Connection to AC grid   623
Y. Tang, B. Chen, L.L. Zhu, J.C. Pi & C.G. Wang

Electrical breakdown of air gap between conductor and tower covered
by insulation sheath   631
J. Chen, Z. Fan, Z.C. Zhou, Y. Liu, Y.L. Lu & W. Liang

Impulse characteristics of grounding devices of transmission tower based
on a simulation experiment in a two-layer soil   635
L.S. Xiao, Q. Li, Z.Q. Rao, X. Yang & J.R. Huang

Investigation on fault diagnosis of substation grounding grid using distribution
of ground potential   641
W.D. Qu, D. Huang, M. Yang, W.R. Si, Z.B. Xu, X. Guo & F.H. Wang

Study on impulse characteristics of substation grounding grid with experimental analysis   647
Z.X. Lu, R. Zhou, M.B. Wu, W.R. Si, X. Guo, Z.B. Xu & F.H. Wang

One high sensitive residual current protection apparatus   651
F. Du, W. Chen, Y. Zhuo & M. Anheuser

The arc model of single phase grounding fault and EMTP/ATP simulation research   655
Y. Yang & C.M. Li

Study on electric field distribution of 1000 kV AC double-circuit transmission
line Y-type insulator string   661
F. Huo, T. Xu, X.H. Fan, M.T. Wei & C.P. Huang

Study on the power grid structural scheme of the Ultra-High Voltage power delivery
system used in large-scale thermal power bases   667
*P. Ji, J.X. Liu, H.T. Yang, W. Tang, Y.T. Song, C. Zheng, L.N. Zhang, X.F. Hu,
A. Wang & J.J. Wang*

Test research of live working of 1000 kV UHV substation   673
B. Xiao, K. Liu, T. Wu, T. Liu, Y. Peng, Z.M. Su, P. Tang & X.L. Lei

Smart grid technologies

Application of disconnecting circuit breaker in new generation smart substation   681
J.B. Li, R. Hu, J. Chen, Y.X. Wen & J.G. Yang

Research of medium voltage flexible DC device locating   685
Y.Y. Cui, T. Wei, W. Liu & F.X. Hui

Comparative analysis on the economy of coal transportation and electricity transmission   691
K. Zheng, F.Q. Zhang & Y. Wang

Comprehensive evaluation of regional grid smart level based on key technology   697
Q.M. Zhao, W.J. Qi, X.J. Li, H.J. Jia, Y.L. Liu, H. Huang & L. Liu

Author index   707

Advances in Power and Energy Engineering – Sun (Ed.)
© *2016 Taylor & Francis Group, London, ISBN 978-1-138-02846-3*

Preface

It is our great pleasure to present the proceedings of the 8th Asia–Pacific Power and Energy Engineering Conference (APPEEC 2016), held from April 15 to 17, 2016 in Suzhou, China. We would like to take this opportunity to express our sincere gratitude and appreciation to all the authors and participants for their support of this conference.

With the development of new techniques and theories to deal with expanding challenges facing the energy and power industry, there are many new research approaches to be validated, probing questions to be asked and fresh ideas to be discussed. Besides, more and more scientists all around the world are dedicating themselves to such an interdisciplinary area, accumulating a lot of interesting results. The last seven APPEECs have done a lot of work in promoting the information exchange in power and energy engineering in the Asia-Pacific region attracting a broad array of participants

We are proud to see that the previous APPEEC conferences were succesfull in providing an ideal platform for them to exchange their exciting findings, to stimulate the further development of Power and Energy Engineering, and to enhance its impacts on various areas of energy and power. We have reasons to believe that the APPEEC 2016 will do even better in this regard.

On behalf of the organizing committee of APPEEC 2016, we would like to take this opportunity to express our gratitude to the conference's sponsors: Wuhan University and the 1000 Think Tank, and the conference's co-organizers: Southeast University and Tianjin University.

Our appreciation and gratitude are also extended to all the papers' reviewers and the Conference Organization Committee members. It is impossible to hold such a grand conference without their help and support.

The papers collected in "Advances in Power and Energy Engineering: Proceedings of APPEEC 2016" provide the detailed results of some oral presentations that will be of value to the readership.

Editor
Prof. Yuanzhang Sun
Wuhan University, China
2016

Advances in Power and Energy Engineering – Sun (Ed.)
© 2016 Taylor & Francis Group, London, ISBN 978-1-138-02846-3

Organization

This volume contains the Proceedings of the 8th Asia-Pacific Power and Energy Engineering Conference in Suzhou, China, held 15th to 17th April 2016. APPEEC16 has been organized by Wuhan University, Southeast University, Tianjin University and 1000 Thinktank.

TECHNICAL PROGRAMME COMMITTEE

General Chair

Prof. Yuanzhang Sun, *Wuhan University, China*

Prof. René Wamkeue, *University of Quebec, Canada*
Prof. M. Ramamoorty, *Equipment Manufacturing Company, India*
Dr. Önder Turan, *Anadolu University, Turkey*
Dr. Meng Ni, *The Hong Kong Polytechnic University, Hong Kong (China)*
Prof. Pierluigi Siano, *University of Salerno, Italy*
Prof. Young Nam Chun, *Chosun University, Korea (South)*
Prof. Almoataz Y. Abdelaziz, *Ain Shams University, Egypt*
Prof. Yi Ding, *Technical University of Denmark, Denmark*
Dr. Kua-anan Techato, *Prince of Songkla University, Thailand*
Prof. Ahmed Faheem Zobaa, *Brunel University, UK*
Dr. Chien-Hung Yeh, *Feng Chia University, Taiwan (China)*
Prof. Tao Lin, *Wuhan University, China*
Dr. Eng. Ahmed Kadhim Hussein, *University of Babylon, Iraq*
Vahid Jabbari, *University of Texas at El Paso, USA*
Dr. Sanjeevikumar Padmanaban, *Ohm Technologies, India*
Prof. Xusheng Chen, *Seattle University, USA*

New energy and technologies

Advances in Power and Energy Engineering – Sun (Ed.)
© 2016 Taylor & Francis Group, London, ISBN 978-1-138-02846-3

Research on the DFIG wind generation comprehensive control

M.X. Shangguan & J.X. Zhao

Shanxi Electric Power Company of State Grid, Shanxi, Taiyuan, China

ABSTRACT: The Doubly Fed Induction Generator (DFIG) active power and reactive power models, oscillating with twice the grid frequency, are analyzed under the unbalanced grid voltage conditions. Based on the deduced mathematical expressions of the DFIG stator and Grid Side Converter (GSC), two different targets for the Rotor Side Converter (RSC) and GSC are researched, respectively, to enhance the capability of the DFIG wind power generation system. Considering the associated control ability of the RSC and GSC fully, this paper proposes an unbalanced control scheme in the positive—and negative—synchronous reference frames, respectively. According to the characteristics of the GSC that comes under unbalanced grid voltage conditions, a novel power compensation method for converter voltage is presented for the purpose of eliminating the impacts on grid-side current control loop due to the direct voltage pulsation. The simulation results show that the proposed comprehensive strategy significantly enhances the wind system operation performance and improves the DFIG asymmetrical fault ride through its capability.

Keywords: DFIG; Unbalance; Control

1 INTRODUCTION

Generally, the traditional excitation control strategy is based on the ideal grid voltage. However, the unbalanced operation of the power grid is more often than not the balanced operation, and there are always some unbalanced cases in the voltage (ZHAO Yang et al. 2009, HU Jiabin et al. 2006). When there is three-phase unbalanced in the power grid, it will cause unbalanced heat in the winds of stator and rotor, and the pulsation of torque for the DFIG, and the oscillation of the output power (Miguel Castilla et al. 2010, QIAO Wei et al. 2008).

The paper deduced a new expression of the second pulsating power for the DFIG system's stator and grid side under unbalance grid voltage. And the paper presented a new kind of united control method for the grid side and rotor side converters. The theory analysis and simulation result tested the correct and availability of the new control method.

2 MODEL AND CONTROL

2.1 Power of stator of DFIG

The stator side of DFIG adopts generator routine, and the rotor side of DFIG adopts motor routine. When the grid voltage is unbalanced, the output power of the stator of DFIG is:

$$S = p_s + jq_s = \frac{3}{2}\mathbf{U}_{dqs}^+(\mathbf{I}_{dqs}^+)^*$$
$$= \frac{3}{2L_s}\mathbf{U}_{dqs}^+(L_m\mathbf{I}_{dqr}^+ - \psi_{dqs}^+)^* \qquad (1)$$

where S = apparent power of the stator; p_s = active power of the stator; q_s = reactive power of the stator; U_{dqs}^+ = voltage vector of the stator side; I_{dqs}^+ = current vector of the stator; I_{dqs}^+ = current vector of the rotor; Ψ_{dqs}^+ = magnetic linkage of the stator and "*" = conjugate. And for convenience, the superscript "+,-" means positive sequence and negative sequence synchronous revolution coordinate system, and the subscript "+,-" means positive sequence component and negative sequence component.

In the condition of grid unbalance voltage, there are negative sequence components in both the voltage and current of the stator and rotor, which leads to fluctuation of the active power and reactive power. After referring to the document (WANG Hongsheng et al. 2010, XU Lie 2008, HU Jiabing et al. 2010, HU Jiabing et al. 2008) giving the definitions of the active power and reactive power of stator as:

$$\begin{cases} p_s = p_{s0} + p_{sc2}\cos(2\omega_1 t) + p_{ss2}\sin(2\omega_1 t) \\ q_s = q_{s0} + q_{sc2}\cos(2\omega_1 t) + q_{ss2}\sin(2\omega_1 t) \end{cases} \qquad (2)$$

P_{s0} means the direct current component of the active power of stator; q_{s0} means the direct current component of the stator reactive power; p_{ss2} means twice frequency sine component of the stator active power; q_{ss2} means twice frequency sine component of the stator reactive power; p_{sc2} means twice frequency cosine component of the stator active power; q_{sc2} means twice frequency cosine component of the stator reactive power. All of these parameters mentioned above are all constant in the unbalanced condition. Also, ω_l is the frequency of the electric network.

We can calculate the matrix expression of the second pulsation active and reactive power of the stator, which is shown in formula (3). The subscripts "s" and "r" mean the windings of the stator and rotor of the DFIG. The subscripts "d" and "q" mean the d axis and q axis of the synchronic rotating coordinate system.

$$
\begin{pmatrix} p_{ss2} \\ p_{sc2} \\ q_{ss2} \\ q_{sc2} \end{pmatrix} = \frac{3L_m}{2L_s} \begin{pmatrix} u_{qs-}^- & -u_{ds-}^- & -u_{qs+}^+ & u_{ds+}^+ \\ u_{ds-}^- & u_{qs-}^- & u_{ds+}^+ & u_{qs+}^+ \\ -u_{ds-}^- & -u_{qs-}^- & u_{ds+}^+ & u_{qs+}^+ \\ u_{qs-}^- & -u_{ds-}^- & u_{qs+}^+ & -u_{ds+}^+ \end{pmatrix} \begin{pmatrix} i_{dr+}^+ \\ i_{qr+}^+ \\ i_{dr-}^- \\ i_{qr-}^- \end{pmatrix}
$$
$$
- \frac{3}{2L_s\omega_1} \begin{pmatrix} -u_{ds-}^- & -u_{qs-}^- & u_{ds+}^+ & u_{qs+}^+ \\ u_{qs-}^- & -u_{ds-}^- & u_{qs+}^+ & -u_{ds+}^+ \\ u_{qs-}^- & -u_{ds-}^- & -u_{qs+}^+ & u_{ds+}^+ \\ u_{ds-}^- & u_{qs-}^- & u_{ds+}^+ & u_{qs+}^+ \end{pmatrix} \begin{pmatrix} u_{ds+}^+ \\ u_{qs+}^+ \\ u_{ds-}^- \\ u_{qs-}^- \end{pmatrix}
$$
(3)

2.2 Comprehensive control of the inverters

According to the formula (3), because of the exit of negative-sequence component in the case of unbalanced grid voltage, there are two frequency pulsation components in the active and reactive power of the inverters for both the rotor side and grid side. The twice frequency of the pulsation power component will give some effects on the operation of DFIG, the two inverters system and the capacitor in the DC link.

It is necessary to remove the effect of the negative-sequence voltage using the double inverters to improve the fault ride through capability of the DFIG system.

From formula (3) we can see that in order to remove the pulsation output power of the stator, the positive and negative sequence currents are the controlled variables for the inverter of the rotor side. Formula (3) is thus simplified as:

$$\mathbf{P} = \mathbf{AI} + \mathbf{U} \qquad (4)$$

The determinant of coefficient matrix is:

$$|\mathbf{A}| = -4[(u_{qs-}^-)^2 + (u_{ds-}^-)^2] \times [(u_{qs+}^+)^2 + (u_{ds+}^+)^2] \qquad (5)$$

When there is the negative sequence voltage component, the determinant of coefficient matrix is not equal to zero. In theory, there is the unique solution $\mathbf{I} = (i_{dr0+}^+, i_{qr0+}^+, i_{dr0-}^-, i_{qr0-}^-)^T$ to remove the active and reactive second pulsation power. Remarkably, it is impossible that the reference value of positive sequence current component which is according to the maximum wind energy tracking and reactive power optimized, is equal to the solution $\mathbf{I} = (i_{dr0+}^+, i_{qr0+}^+, i_{dr0-}^-, i_{qr0-}^-)^T$. Thus, there are only two negative sequence currents as the unbalanced control variables in one control strategy, and one of the control strategies can only remove two of the four pulsation powers.

The comprehensive control of removing the pulsation of reactive power is in the formula (3), if $q_{ss2} = q_{sc2} = 0$, we can get the reference value of positive sequence current is:

$$
\begin{cases} i_{dr-}^- = k_{dd} i_{dr+}^+ + k_{qq} i_{qr+}^+ \\ i_{qr-}^- = k_{qq} i_{dr+}^+ - k_{dd} i_{qr+}^+ \end{cases} \qquad (6)
$$

where
$$
\begin{cases} k_{dd} = \dfrac{u_{ds-}^- u_{ds+}^+ - u_{qs-}^- u_{qs+}^+}{{u_{qs+}^+}^2 + {u_{ds+}^+}^2} \\ k_{qq} = \dfrac{u_{qs+}^+ u_{ds-}^- + u_{ds+}^+ u_{qs-}^-}{{u_{qs+}^+}^2 + {u_{ds+}^+}^2} \end{cases}
$$

The reference value mentioned above can remove both the reactive pulsation power and the two frequency pulsations of the DFIG electromagnetism torque (ZHENG Jianwen et al. 2009, XU Lie 2008, HU Jiabing et al. 2010, HU Jiabing et al. 2008).

In the formula (3), if $p_{ss2} = p_{sc2} = 0$, we can get the reference value of negative sequence current as:

$$
\begin{cases} i_{dr-}^- = -\dfrac{2}{L_m\omega_1} u_{qs-}^- - k_{dd} i_{dr+}^+ - k_{qq} i_{qr+}^+ \\ i_{qr-}^- = \dfrac{2}{L_m\omega_1} u_{ds-}^- - k_{qq} i_{dr+}^+ + k_{dd} i_{qr+}^+ \end{cases} \qquad (7)
$$

The paper designed as "Comprehensive power compensation" based on the grid side voltage, which can remove the effect of the second pulsation DC voltage to the controller. It is seen in the dotted line frame in Figure 1. The trap filter is added into the feedback of DC voltage to filter the second pulsation disturbance of active power current and remove the adjust error of the PI controller.

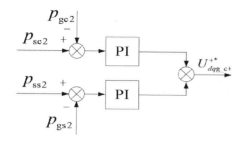

Figure 1. Power compensation structure.

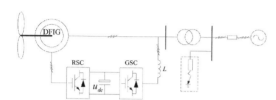

Figure 2. DFIG simulation model.

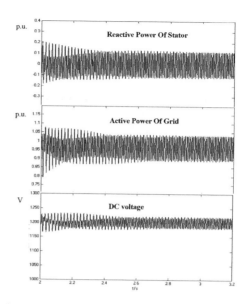

Figure 3. Simulated results without unbalanced control.

The principle of the comprehensive power compensation is shown in Figure 1, in which $U_{dqg\_c+}^{+*}$ is the voltage of power compensation. The method of the power compensation can remove the effect to the control of the grid side secondary disturbance current caused by the adjust function of the DC voltage. On the other hand, it can also remove the effect to design of the DC voltage control loop caused by the grid side converter power secondary pulsation component.

3 SIMULATION

In order to verify the availability of the comprehensive control strategy under the unbalanced grid voltage, the simulation model is constructed. The parameters of the simulation model are in appendix B. The degree of unbalanced of grid voltage is 5% in 2 s, and the rated wind speed is 12 m/s.

The simulation system is shown in Figure 2.

3.1 Without unbalanced control

When the electric network is unbalanced, there is a negative sequence component in the electric network. Under the traditional exciter control strategy, without unbalanced control, all of the physical quantities DFIG wind power generation system appears to be pulsating or oscillating waveforms of reactive power of stator, and the active power and DC voltage are shown in Figure 3.

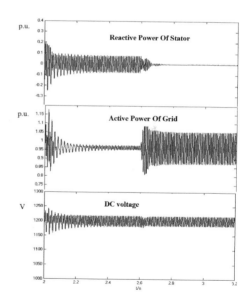

Figure 4. Simulated results of rotor side converter control.

3.2 With unbalanced control

In order to verify the availability of the Comprehensive control objective for the rotor side converter, the control strategy is different in different time sections. The strategy of suppressing the pulsation active power of stator is adopted during 2–2.625s,

5

and the strategy of suppressing the pulsation reactive power of stator is adopted during 2.625–3.3s.

In Figure 4, waveforms of reactive power of stator and the active power and DC voltage are shown, under the case of unbalanced grid voltage, the rotor side converter realized the setting control objective. During 2–2.625s, while the twice frequency oscillation of the stator output active power is removed, the twice frequency components of the stator output reactive power and electromagnetism torque are not be suppressed. Similarly, during 2.625–3.3s, while the twice frequency components of the stator output reactive power is removed, the twice frequency oscillation of the stator output active power is not suppressed.

4 CONCLUSION

The paper deduced a new mathematical model of the DFIG generation system under the unbalanced grid voltage. And the paper presented the comprehensive unbalanced control strategy of DFIG system based on positive and reverse direction synchronism rotate coordinate system. It designed a compensated power arithmetic based on voltage being aimed at the twice fluctuated indirect voltage being used in the DFIG system under the unbalanced grid voltage. Through the simulation, the comprehensive power control strategy improved the operation property of the DFIG system under unbalanced grid voltage, and improved the fault ride through capability for the DFIG system.

REFERENCES

HU Jiabing, SUN Dan, HE Yikang, et al. Modeling and control of DFIG wind energy generation system under grid voltage dip. *Automation of Electric Power Systems, 2006, 30(8):21–26.*

HU Jiabing, HE Yikang. Modeling and control of grid-connected voltage-source converters under generalized unbalanced operation conditions. *IEEE Trans on Energy Conversion, 2008, 23(3): 903–913.*

HU Jiabing, HE Yikang, WANG Hongsheng, et al. Proportional-resonant current control scheme for rotor-side converter of doubly-fed induction generators under unbalanced network voltage conditions. *Proceedings of the CSEE, 2010, 30(6): 48–56.*

HU Jiabing, HE Yikang, WANG Hongsheng, et al. Coordinated control of grid—and rotor—side converters of doubly-fed induction generator under unbalanced network voltage conditions. *Proceedings of the CSEE, 2010, 30(9): 97–104.*

Miguel Castilla, Jaume Miret, Jose Matas, et al. Direct rotor current-mode control improves the transient response of doubly fed induction generator-based wind turbines. *IEEE Trans on Energy Conversion, 2010, 25(3): 722–731.*

QIAO Wei, HARLEY R G. Improved control of DFIG wind turbines for operation with unbalanced network voltages//*Proceedings of 2008 IEEE Industry Applications Society Annual Meeting, 2008, Edmonton, Canada: 7p.*

WANG Hongsheng, ZHANG Wei, HU Jiabing, et al. A control strategy for doubly-fed induction generator wind turbines under asymmetrical grid voltage conditions caused by faults. *Automation of Electric Power Systems, 2010, 34(4): 97–102.*

XU Lie. Enhanced control and operation of DFIG-based wind farms during network unbalance. *IEEE Trans on Energy Conversion, 2008, 23(4): 1073–1081.*

XU Lie. Coordinated control of DFIG's rotor and grid side converters during network unbalance. *IEEE Trans on Power Electronics, 2008, 23(3): 1041–1049.*

ZHAO Yang, ZOU Xudong, DING Wenfang. Excitation control strategy of grid-connecting DFIG under unbalanced grid voltage. *Engineering Journal of Wuhan University, 2009, 42(1): 110–113.*

ZHENG Jianwen, LI Yongdong, CHAI Jianyun, et al. Research on control strategy for doubly-fed generation system under unbalanced voltage condition. *Automation of Electric Power Systems, 2009, 33(15): 89–93.*

Advances in Power and Energy Engineering – Sun (Ed.)
© *2016 Taylor & Francis Group, London, ISBN 978-1-138-02846-3*

Economic analysis of supporting policies for residential solar Photovoltaic in Massachusetts, USA: A case from Boston

S. Qiu
Newhuadu Business School, Minjiang University, Fuzhou, Fujian, China

S. Ghosh
Department of Accounting and Finance, Northeastern State University, Tahlequah, OK, USA

ABSTRACT: The development of solar Photovoltaic (PV) has been very rapid and successful in Massachusetts (MA) over recent years. This paper makes an economic analysis of supporting policies for residential solar PV in MA using a case study from Boston metro area. It finds that the net present value, the payback period and the internal rate of return of the project investigated are $49,474.56, 6 years and 19.0% respectively, which compare favorably with those from past studies on United States, Europe and China. Furthermore, a scenario analysis of seven policies affecting the profitability of solar PV investment and a sensitivity analysis of the profitability responding to changes of four key parameters are conducted. Finally, a conclusion is drawn that the rapid development of residential PV in MA is mainly driven by high profitability from the incentives, especially Solar Renewable Energy Certificate, and the substantial decrease in installed prices of PV.

1 INTRODUCTION

The state of Massachusetts (MA), considered as the most energy efficient state in the United States of America (USA), has attached great importance to the development of solar Photovoltaic (PV) over the past decade. By carrying out a series of effective supporting policies, the state's goal to achieve 400 MW_p of solar power installations was met in 2013, four years earlier than the scheduled date. Then, an aggressive new goal was set of 1,600 MW_p by 2020[i]. By May 7, 2015, MA has installed more than 841 MW_p of solar electricity, bringing the state more than halfway to the new goal of 1,600 MW_p[ii]. Cumulative PV capacity in MA has increased by a factor of 38.6 to 714.7 MW_p from 18.53 MW_p over 2009–2014[iii]. In contrast, over the same time period, cumulative PV capacity in USA has increased by a factor of only 10.8 to 18.3 GW_p from 1.7 GW_p, and that in the world has increased by a factor of only 7.6 to 177 GW_p from 23.2 GW_p. Still, the development of PV in MA compares favorably against that of countries like Germany and Japan, both of which have been pioneers in the solar PV market. Over 2009–2014, Germany and Japan have seen an installed capacity growth just by a factor of 3.9 and 9.0 respectively.[iv] Evidently, the development and adoption of PV technology has been very rapid and successful in MA so far. However, very few studies in the empirical literature on solar PV have investigated the role of major state and regional incentive policies in the rapid development of PV in MA. Exceptions include some recent studies by Flynn et al. (2010), Wiser et al. (2010), Burns et al. (2012) and Bird et al. (2014), who primarily based their conclusions upon Solar Renewable Energy Certificate (SREC) in MA.

Several researchers have performed comparative economic analysis on residential PV investments. In a comparative analysis of 17 western European countries, Dusonchet et al. (2010a) demonstrated

[i]The Official Website of the Executive Office of Energy and Environmental Affairs. http://www.mass.gov/eea/energy-utilities-clean-tech/renewable-energy/solar/, accessed on 05-18-2014.
[ii]The Official Website of the Executive Office of Energy and Environmental Affairs. http://www.mass.gov/eea/pr-2015/solar-milestone-for-massachusetts.html, accessed on 07-18-2015.
[iii]The Official Website of the Executive Office of Energy and Environmental Affairs. http://www.mass.gov/eea/grants-and-tech-assistance/guidance-technical-assistance/agencies-and-divisions/doer/renewable-energy-snapshot.html, accessed on 07-18-2015.

[iv]REN21. GLOBAL STATUS REPORT (2010–2015). http://www.ren21.net/resources/publications/, accessed on 07-18-2015.

the high profitability and low payback period for solar PV within a lifetime of 25 years, from state level supporting policies in Greece, Italy, and France. Along the same lines, the authors showed that the Czech Republic, Slovakia and Bulgaria, have the highest internal rate of return and lowest payback periods for small scale PV within the same lifetime, among the Eastern European countries (Dusonchet et al. 2010b). Within the United States (U.S.), Denholm et al. (2009) found that MA is one of the few states in the nation where a classical residential PV investment (10 kW$_p$) can operate at the breakeven point within a lifetime of 30 years. Burns et al. (2012) found that the Internal Rate of Return (IRR) of a classical residential PV investment (4 kW$_p$) in MA over 15 years is only 6.54%. Utilizing data from SREC implementation in seven states, they concluded that only three of them—New Jersey, MA and Delaware show profitable investment within the first 15 years, with the rest showing negative NPV from solar PV installation. On the other hand, Bird et al. (2014) investigating the evolution and functioning of SREC markets in USA, found that MA is only one among three states with project capacity greater than 250 kW$_p$, while the solar PV capacity is expected to increase to 400 MW$_p$ by 2020 in the state.

However, it is not certain from the above studies as to whether the rate of return and/or the lower payback period have been responsible for the rapid development of PV in the state of MA. Though they acknowledge the success of solar PV development in MA to some extent, they fail to provide insights into the collective role of federal and state incentive policies for renewables, especially solar energy in recent years. In the absence of high levels of solar irradiation in the state, the supporting policies are likely to generate economic incentives for adopting solar PV that may far outweigh the initial costs of installation. So, in this study, it is hypothesized that the various incentive policies encouraging solar PV installation in residential units, in particular, have been an instrumental factor behind this rapid development in solar energy in the state of MA. The study thus, has two major objectives:

- Determine at a micro level the economic effects in terms of net present value and returns from federal and state supporting policies for residential solar PV installations in the state of MA;
- Assess the relative importance of each policy or each key factor in determining the economic returns from a residential PV installation within the time period of the study.

With these underlying objectives, the study attempts to make an economic analysis of supporting polices for residential solar PV investments in MA, using a unique case study from the Greater Boston Area.

This research contributes to the existing literature on renewable energy policies and economic incentives on two fronts. First, it attempts to combine all existing federal and state supporting policies for solar PV to calculate the net present value, the rate of return and the payback period for a residential PV installation in MA. Previous studies have either based their conclusions by focusing primarily upon one policy (Burns et al. 2012, Bird et al. 2014) or have performed the economic analysis for a number of countries together (Dusonchet et al. 2010a, b). Secondly, by targeting the residential sector in MA which has seen a steep decline in net electricity consumption in the last decade while the state witnessed higher levels of electricity generation through renewable sources[v], the study provides compelling ground for further research on the economic viability of solar PV supporting policies in the state, and other regions across the country.

The following section provides an overview of all the supporting policies for residential PV development in MA. In Section 3, the profile of a case study in the Greater Boston Area, where adoption of solar PV has seen a substantial increase in recent years, is presented. In Section 4, the methodology for economic analysis of residential PV projects is outlined, taking into account the supporting policies currently provided in MA. In Section 5, the results are presented, some additional scenarios are depicted by sequentially changing the implementation of each supporting policy, and a sensitivity analysis is conducted by varying some of the major parameters affecting the outcomes. Finally, Section 6 draws a conclusion from the analysis.

2 SUPPORTING POLICIES FOR RESIDENTIAL PV IN MA

2.1 *Federal incentive: Residential Renewable Energy Tax Credit (RRETC)*

The federal incentive, RRETC was initially established by *The Energy Policy Act of 2005,* and came into force on January 1, 2006. *The Energy*

---

[v]U.S. Energy Information Administration. Retail sales of electricity to ultimate customers. http://www.eia.gov/electricity/data.cfm#sales, accessed on 09-28-2015. The Official Website of the Executive Office of Energy and Environmental Affairs. http://www.mass.gov/eea/pr-2015/solar-milestone-for-massachusetts.html, accessed on 07-18-2015.

Improvement and Extension Act of 2008 extended the tax credit to December 31, 2016. The two laws provide that a taxpayer may claim a credit of 30% of qualified expenditures for a PV system that serves a dwelling unit located in the U.S. and that is owned and used as a residence by the taxpayer. Expenditures include labor costs for on-site preparation, assembly or original system installation, and for piping or wiring to interconnect a system to the home. If the federal tax credit exceeds tax liability, the excess amount may be carried forward to the succeeding taxable year until 2016.[vi]

2.2 State incentives

2.2.1 Residential Renewable Energy Income Tax Credit (RREITC)

MA allows a 15% credit—up to $1,000—against the state income tax for the net expenditure of a PV system installed on an individual's primary residence. The net expenditure is defined as the total of the purchase price for a PV system and installation costs, less any federal tax credits and rebates or grants received from the U.S. Department of Housing and Urban Development. If the credit amount is greater than a resident's income tax liability, the excess credit amount may be carried forward to the next succeeding year for up to three years.[vii]

2.2.2 Solar Renewable Energy Certificate (SREC)

SRECs represent the renewable attributes of solar generation, bundled in minimum denominations of one MWh of production. MA's Solar Carve-Out program stipulates that all electric suppliers must use SRECs to demonstrate compliance with their solar Renewables Portfolio Standard (RPS) requirements. SRECs can be bought and sold in credit markets or can be sold to utilities through long-term contracts (City of Boston, 2010). The price of SRECs is determined primarily by market availability, although the MA Department of Energy Resources (DOER) has created a certain amount of market stability by establishing a state Solar Credit Clearinghouse Auction, (where prices are fixed at $300/MWh minus a 5% administrative fee, for a total of $285/MWh), as well as the Solar Alternative Compliance Payment for the state RPS. The Solar Credit Clearinghouse will only be utilized if or when SREC generators cannot sell their SRECs on the open market. Only solar electric facilities with a capacity less than 6 MW_p built after January 1, 2008, may be qualified to generate SRECs. SRECs are generated on or after January 1, 2010, since that is the date the Solar Carve-Out program took effect. Facilities that received funding prior to the effective date of the Solar Carve-Out from the MA Renewable Energy Trust or the MA Clean Energy Center (MassCEC), or received more than 67% of project funding from *the American Recovery and Reinvestment Act of 2009*, are ineligible. A solar PV system is eligible for generating SRECs for 10 years.[viii]

2.2.3 Commonwealth Solar II (CSII) Rebates

The CSII program offered by MassCEC, provides rebates for the installation of PV systems at residential, commercial, industrial, institutional and public facilities. The Host Customer must be a customer of a MA electric distribution utility that collects the Renewable Energy Systems Benefit Charge from its customers and deposits those funds into MassCEC Renewable Energy Trust Fund (Massachusetts Clean Energy Center, 2014). CSII rebates are available to electricity customers served by the following MA investor-owned electric utilities—Fitchburg Gas and Electric Light, National Grid, NSTAR Electric and Western Massachusetts Electric, as well as certain municipal lighting plant utilities. The program provides direct rebates for residential and small commercial PV systems between 1 kW_p and 15 kW_p, and the rebates were based on the first 10 kW_p only until 2012 while they have been provided on the first 5 kW_p only after 2012 (City of Boston, 2010). It offers incentives on a first-come-first-served basis. Funding is released in "blocks" every quarter. All rebate applications must be approved before the project installation begins. Rebate amounts are based on the total PV system size per building, regardless of the number of electric meters in use and certain other characteristics of the project. The proposed CSII rebate levels for residential and commercial PV systems are:

- Base incentive: $0.40/watt
- Adder (additional incentive) for MA company components: $0.05/watt
- Adder for moderate home value: $0.40/watt (applicable to residential projects only), or

[vi]Database of State Incentive for Renewables & Efficiency (DSIRE). 12-11-2012. http://www.dsireusa.org/incentives/incentive.cfm?Incentive_Code = US37F&re = 1&ee = 0, accessed on 05-18-2014.

[vii]DSIRE. 07-06-2013. http://www.dsireusa.org/incentives/incentive.cfm?Incentive_Code = MA06F&re = 1&ee = 0, accessed on 05-18-2014.

[viii]DSIRE. 03-01-2013. http://www.dsireusa.org/incentives/incentive.cfm?Incentive_Code = MA98F&re = 1&ee = 0, accessed on 05-18-2014.

- Adder for moderate income: $0.40/watt (applicable to residential projects only)
- Natural Disaster Relief Adder: $1.00/watt.[ix]

2.2.4 *Net Metering*

NSTAR in MA provides PV system owners with the ability to export excess electricity back into the electric distribution grid. This benefit, known as Net Metering, allows system owners to spin their electricity meter backwards, effectively crediting their billing account for electricity produced but not consumed on site. The NSTAR Net Metering tariff allows system owners to carry forward their monthly excess generation to future months and to use those credits to offset future electricity consumption. Excess generation is credited at or near the building's retail electricity rate (City of Boston, 2010).

2.2.5 *Renewable Energy Property Tax Exemption (REPTE)*

MA law provides that solar-energy systems used as a primary or auxiliary power system for the purpose of heating or otherwise supplying the energy needs of taxable property, are exempt from local property tax for a 20-year period. However, this incentive applies only to the value added to a property by an eligible system, according to the MA DOER.[x] It does not constitute an exemption for the full amount of the property tax bill, where the property tax rate is at 1.04%[xi].

2.2.6 *Renewable Energy Equipment Sales Tax Exemption (REESTE)*

MA law exempts the state's sales tax from equipment directly relating to any solar powered system, which is being utilized as a primary or auxiliary power system for the purpose of heating or otherwise supplying the energy needs of an individual's principal residence in the state[xii]. The average tax rate is 6.25%[xiii].

[ix]DSIRE. 05-22-2014. http://www.dsireusa.org/incentives/incentive.cfm?Incentive_Code = MA71F&re = 1&ee = 0, accessed on 05-25-2014. The rebate levels may vary from quarter to quarter and the specific levels and qualification requirements are released by the Program Manual every quarter. http://www.masscec.com/programs/common-wealth-solar-ii, accessed on 05-25-2014.

[x]DSIRE. 07-20-2013. http://www.dsireusa.org/incentives/incentive.cfm?Incentive_Code = MA01F&re = 1&ee = 0, accessed on 05-25-2014.

[xi]Tax-rates.org. http://www.tax- rates.org/taxtables/propert y-tax-by-state, accessed on 06-28-2014.

[xii]DSIRE. 12-17-2012. http://www.dsireusa.org/incentives/incetive.cfm?Incentive_Code = MA05F&re = 1&ee = 0, accessed on 2014-05-25.

[xiii]Tax-rates.org. http://www.tax-rates.org/taxtables/sales-tax-by-state, accessed on 06-28-2014.

3 A RESIDENTIAL PV SYSTEM FROM BOSTON

The previous section described briefly all of the major solar PV incentive policies currently endorsed by the state of MA. In what follows, a residential PV system in the Greater Boston Area is taken up as a case study for assessing the economic and financial feasibility of a solar PV in the light of these incentive policies. Boston serves as an appropriate case study because of two reasons. One, it is the primary metro area of Massachusetts with around two thirds of its population residing in this region. Two, the Renew Boston municipal program (City of Boston, 2011) was specifically designed to encourage energy efficiency in buildings, with 25% of the expenditure on renewable energy incentives allocated for residential customers.

The residential PV system in question is installed on the south rooftop of a dwelling located in Jamaica Plain, in Boston, MA. This dwelling is a three-story building (Fig. 1) consisting of three separate apartments. Its retired owner lives on the second floor with his wife in employment, his son at university and his daughter in high school. The two apartments on the first and the third floor are for rent. The family's annual income is about $95,000. NSTAR is this dwelling's electricity supplier.[xiv]

This is a grid-connected system with a installed capacity of 11.04 kW$_p$. Its major components include 48 solar panels made by SunPower, and one inverter manufactured by SMA America, and the whole system was installed by Sunlight Solar Energy. The project was initiated on September 8, 2011 (applications made to MassCEC on the same date) and the construction was completed in

Figure 1. A residential PV system in Boston.

[xiv]As a tenant in the household for almost 1 year, the first author had access to data sources from the owner regarding details about the economic and financial aspects of the PV installation in the building.

December, 2011. The business operation started on January 27, 2012, with a warranty period of five years. In SREC, this project joined the consortium of SRECTrade, inc., and its qualification certificate was obtained on March 15, 2012. The effective duration to generate SRECs started on the business operation start date. On October 30, 2012, six solar panels were blown off by gale and the system was inoperative until the maintenance was completed. The total generation from the solar PV amounted to 11,308.7 kWh from January 27, 2012 to October 30, 2012, and 13,584.8 kWh from January 9, 2013 to December 31, 2013.

Total installation costs of this project were $46,487.16, with an average cost of $4.21 per watt. The homeowner's investment on this project was $38,487.16, of which $15,000 came from his own capital and $23,487.16 was loaned from a bank.[xv]

4 METHODOLOGY FOR ECONOMIC ANALYSIS

4.1 *Theoretical background*

Determination of the economic viability of a PV investment generally includes the following four major components—Calculation of Cash Flows (CF) from the investment or the net flow of cash benefits over the time period under consideration; the Net Present Value (NPV) of the investment defined as the present value of all future benefits from the investment discounted at an appropriately chosen rate; the Payback Period (PBP) or the length of time for the cumulative benefits to cover the initial costs of investment; and finally, the IRR which is the rate at which the net present value of the investment equals zero or the discounted benefits just equal the costs of investment (Burns et al. 2012, Dusonchet et al. 2010a, b, Ziuku et al. 2012).

The cash flows are affected by a number of factors, including average local energy price, solar radiation, PV price, and various policies including the federal and state-level incentives as described above. The cash flows are calculated as the sum of all the profits and costs in any year t using the following formula (Burns et al. 2012):

$$CF_t^* = T_{fed} + T_{state} + R_{state} + c_{kWh,t} * E_t$$
$$+ F_t * E_{SREC,t} - \mu * I_0 - \eta * I_0 \quad (1)$$

where T_{fed} = federal RRETC; T_{state} = RREITC in MA; R_{state} = CSII rebates in MA; $c_{kwh,\,t}$ = retail price of residential electricity per kWh in year t; E_t = generated energy in kWh in year t; F_t = SREC value in year t; $E_{SREC,t}$ = SREC production in year t; I_0 = initial costs of a PV system, or initial investment; μ = coefficient used to evaluate annual maintenance and management costs; η = coefficient used to evaluate annual insurance cost. Both μ and η are estimated as a percentage of initial costs.

These cash flows may be annualized using the classical expression:

$$CF_t = \frac{CF_t^*}{(1+i)^t} \quad (2)$$

where i = discount rate or cost of capital.

NPV, PBP and IRR may be determined as follows:

$$NPV = \sum_{t=0}^{N} \frac{CF_t^*}{(1+i)^t} - I_0 \quad (3)$$

$$\sum_{t=0}^{PBP} \frac{CF_t^*}{(1+i)^t} - I_0 = 0 \quad (4)$$

$$\sum_{t=0}^{N} \frac{CF_t^*}{(1+IRR)^t} - I_0 = 0 \quad (5)$$

In the expressions (3) and (5), N = lifetime of a PV investment.

4.2 *Operational assumptions*

The following assumptions largely drawn from recent studies on PV installation are used for the economic analysis.

- The lifetime of this project is taken to be 25 years (Burns et al. 2012, Dusonchet et al. 2010a, b).
- The homeowner's family federal adjusted gross income and taxable income are assumed to be $95,000 and $71,700 respectively. So the family's federal income tax liability amounts to $9,985[xvi], which is smaller than their RRETC. As a result, the tax credit is offset in the first and second year.
- 15% of the net expenditure of the PV system amounts to $4,881.15, which is larger than the maximum amount of RREITC, $1,000. The family's state income tax liability amounts to $3,943.5, according to the family taxable income and the state income tax rate of 5.5%. So the

[xv]The loan was offered with free interest for 2 years by a bank, thanks to the homeowner's excellent credit history.

[xvi]Tax Rate Schedule Y-1 for 2012 is applied.

RREITC for this project is offset completely in the first year.

- As the family's gross income is lower than $95,200, it falls into the category of moderate income families, thereby entitling it to an additional CSII rebate of $0.40/watt for moderate income residential projects. However, the major components of the solar PV are not manufactured by MA companies. As a result, this project can obtain a rebate of $8,000 in the CSII program once the construction is complete.
- The energy production is assumed to be 13,584.8 kWh in 2013, the second year of the investment, and the annual decrease in energy production is assumed to be 0.8% of the total energy production since 2013 (Dusonchet et al. 2010a, b). The loss of energy production owing to maintenance is not considered.
- As the residential electricity price was $0.1658 per kWh in 2012 and the price of electricity production was $0.0799 per kWh, the price of electricity transmission was 0.0859 per kWh. The growth rate of residential electricity price is assumed to 3%, based on trends in the U.S. residential electricity prices over 2001–2013.[xvii]
- The discount rate is assumed to be 3% (Burns et al. 2012, Dusonchet et al. 2010a, b).
- The family's electricity consumption per year is estimated to be 3,600 kWh. So, in accordance with the rules of Net Metering, the surplus electricity produced by the PV system may be transferred for electricity consumption by the tenants at the retail price of residential electricity.
- The production of SRECs should be counted in round numbers according to the regulations of SREC. The SRECs are assumed to be traded annually at the price of auction in the MA Solar Credit Clearinghouse. The effective length of time for generation of SRECs is assumed to be 2012–2021, although the end date extends to January 16, 2022.
- The value of the PV property is assumed to follow a linear depreciation rate within a period of 20 years.
- The equipment cost is the difference between the initial costs and the installation labor cost and profit. The installation labor cost and profit are assumed to be $1,050/kW$_p$.[xviii] So the equipment

cost is $34,895.15. Hence, the PV equipment is entitled to a sales tax exemption amounting to $2,180.95, according to the REESTE once its purchase is complete.

- Both the annual maintenance and management costs and the annual insurance cost are assumed to be 0.5% of the initial costs (Burns et al. 2012).

5 RESULTS AND ANALYSIS

5.1 Results

We obtain the results following the incentives for residential PV in MA as described in Section 2 and using the formulae from (1) to (5), and the operational assumptions. The cumulative discounted cash flows, the initial costs and the NPV from our case over the lifetime are depicted in Figure 2, and the NPV, PBP and IRR are found to be $49,474.56, 6 years and 19.0% respectively. By contrast, PBP and IRR of a classical residential PV system (10 kW$_p$) in Greece were found to be 8 years and 12.92% respectively (Dusonchet et al. 2010a), which comprises the best outcome among the western European countries, while PBP and IRR of a classical residential PV system (5 kW$_p$) in the Czech Republic were 11 years and 8.57% respectively (Dusonchet et al. 2010b), which is the best outcome among the eastern European countries. In China, currently PBP of a distributed residential PV system is generally about 10 years, after a subsidy policy with ¥0.42 per kWh within 20 years took effect in 2013.[xix] So for this project, PBP is fairly short and IRR is quite high.

5.2 Scenario analysis

This section presents eight different scenarios, using a continuation or expiration of the seven supporting policies as independent variables and NPV, PBP, IRR as dependent variables, in order to reveal how a policy has an impact on the profitability of the PV project. The scenarios are presented in Table 1. The base scenario assumes that all the seven supporting policies are continued, while for the other seven scenarios, it is assumed that one of the seven supporting policies does not hold while the others remain valid. ΔNPV in Table 1 is actually an incentive benefit, which is the present value of additional or incremental

[xvii]U.S. Energy Information Administration. Average retail price of electricity to ultimate customers. http://www.eia.gov/ electricity/data.cfm#sales, accessed on 08-28-2014. The residential electricity price refers to the one offered by NSTAR while the price of electricity production is the average price in MA.
[xviii]The Solar Energy Bible. http://energybible.com/solar_energy/typical_costs.html, accessed on 06-28-2014.

[xix]China New Energy Website. Newly installed PV capacity maybe increases by a factor of 12 in 2014, 02-21-2014. http://www.chinastock.com.cn/yhwz_about.do?method Call = getDetailInfo&docId = 4005249, accessed on 09-19-2014.

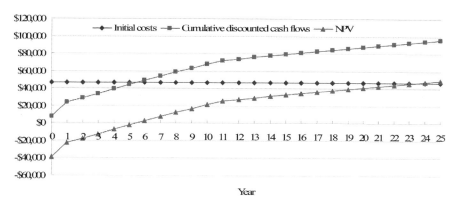

Figure 2. The cumulative discounted cash flows, initial costs and NPV of the residential PV project in MA.

Table 1. Scenario analysis on the profitability of the residential PV project in MA.

	Scenarios							
Variables	Base	1	2	3	4	5	6	7
Federal incentive								
RRETC	Cont.*	Expired	Cont.	Cont.	Cont.	Cont.	Cont.	Cont.
State incentives								
RREITC	Cont.	Cont.	Expired	Cont.	Cont.	Cont.	Cont.	Cont.
SREC	Cont.	Cont.	Cont.	Expired	Cont.	Cont.	Cont.	Cont.
CSII	Cont.	Cont.	Cont.	Cont.	Expired	Cont.	Cont.	Cont.
Net Metering	Cont.	Cont.	Cont.	Cont.	Cont.	Expired	Cont.	Cont.
REPTE	Cont.	Cont.	Cont.	Cont.	Cont.	Cont.	Expired	Cont.
REESTE	Cont.	Cont.	Cont.	Cont.	Cont.	Cont.	Cont.	Expired
NPV ($)	49,474.56	36,046.62	48,503.69	19,055.07	41,474.56	30,659.30	45,706.58	47,293.61
PBP (yr)	6	9	6	14	8	7	6	6
IRR (%)	19.0	11.3	18.2	8.3	13.8	14.6	17.6	17.3
ΔNPV ($)**	–	13,427.94	970.87	30,419.49	8,000.00	18,815.26	3,767.98	2,180.95
ΔPBP (yr)**	–	–3	0	–8	–2	–1	0	0
ΔIRR (%)**	–	7.7	0.8	10.7	5.2	4.4	1.4	1.7

*Cont. is an abbreviation for continued. **ΔNPV, ΔPBP or ΔIRR is the difference of NPV, PBP or IRR between the base scenario and one of the seven scenarios respectively.

income generated or cost avoided by an incentive policy for a PV investment. Similarly, ΔPBP or ΔIRR is the additional impact an incentive policy has on PBP or IRR respectively. Total incentive benefits, the sum of all of the seven incentive benefits, amount to $77,582.48, and total incentive income, the sum of the five incentives except the costs avoided by REPTE and REESTE, comes up to $71,633.56. Total income from this project is $104,056.60, including $32,423.04 earned from electricity generation, after the incentive benefit from Net Metering is deducted. Obviously, total income from this project is dominated by total incentive income, and the profitability depends mainly on the total incentive benefits.

As far as the effect of a sole incentive is concerned, results show that the effect of SREC is undoubtedly the biggest among the seven policies, with the ΔNPV, ΔPBP and ΔIRR in Scenario 3 being the largest among the seven scenarios. If SREC were expired, the profitability of a PV investment in MA would be too low to attract most investors. This resonates with the conclusion reached in Burns et al. (2012), where MA came second to New Jersey in terms of NPV and IRR of solar PVs receiving SREC incentives in eight states of the U.S. The effect of RRETC, Net Metering or CSII is also quite large, implying that if any of these polices were to be expired, the profitability of a PV investment in MA would be reduced greatly.

But it probably could still attract part of investment. The effect of REPTE, REESTE or RRE-ITC is found to be very small. This suggests that the profitability of a PV investment in MA is not affected much in the absence of these policies.

5.3 Sensitivity analysis

In order to ensure that the above outcomes are robust to small changes in four important parameters guiding the success of solar PV investment—initial costs, residential electricity price, price of SRECs and discount rate, we vary those parameters. As shown in the panel A of Figure 3, the NPV and IRR from this study are highly sensitive to a decrease in initial costs of installation. If the initial costs decline by 10 percentage points, the NPV and IRR are found to increase by over $4,000 and 4 percentage points respectively. This indicates that the profitability of residential PV installations in MA would improve drastically, as installed prices of PV have dropped by more than 50% over 2009–2014 in USA (Barbose et al. 2015). The NPV and IRR are mildly sensitive to the change of residential electricity price or price of SRECs (panel B and C). If the price of SRECs or the residential electricity price changes respectively by 10 percentage points, the NPV would respond to change by $3,042 or $5,124 respectively, and the IRR to change by around 1.1 percentage points. This implies that even if a similar project in MA has a lower price of SRECs or residential electricity, its profitability would keep the extent to which it is high enough to attract investors. Finally, while the NPV is found to be greatly influenced by the change of discount rate, the IRR is almost insensitive to it (panel D). With a discount rate as high as 7%, the NPV amounts to $26,693 which is equivalent to 57.4% of the initial costs, implying that the investment is still attractive.

5.4 Discussion

RRETC is expected to expire at the end of 2016. If this becomes true, the high profitability of residential PV investments would be reduced largely in MA. In addition, as the first phase of the Solar Carve-Out program ended in April 2014, the second phase of the Solar Carve-Out was launched immediately to continue supporting new solar photovoltaic installations, until 1,600 MW of capacity is installed across the entire state of MA. Obviously, after the goal of 1,600 MW of capacity has been met, whether or not the Solar Carve-Out will continue, remains uncertain. So it is expected and also apparent that many people are seizing the opportunity to install PV systems recently when these two policies are still in force. This has

A
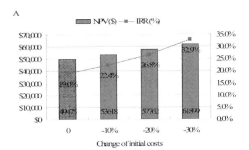

B
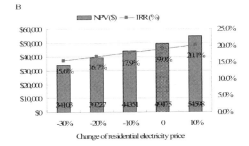

C
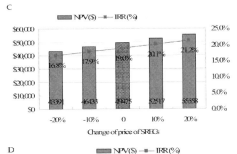

D
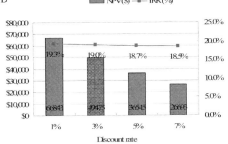

Figure 3. Sensitivity of NPV and IRR to initial costs, residential electricity price, price of SRECs and discount rate.

undoubtedly contributed to the fast development of solar PV in MA.

The PV system investigated in this study was installed in 2011. After 2012, the rebates in CSII were cut down gradually, until the program was successfully phased out in January, 2015. However, with the cost of installation of solar PV going down to a large extent, the profitability of PV investments in MA is supposed to be even better at present than before. So there seems to be a continuing trend of the rapid development of PV within the state.

Given these situations, the results from this study seem encouraging and confirm findings in past studies where Massachusetts has often been shown to have played a leading role in solar PV projects in the last decade. The assumptions and values of various parameters are conservatively and prudently drawn from previous studies, so it will not be an oversight to draw the above conclusions based on these values.

6 CONCLUSION

The analysis above allows us to make a conclusion that the rapid development of residential PV in MA over the past years is mainly driven by high profitability from the incentives, especially SREC, and the substantial decrease in installed prices of PV. It implies that high profitability of PV investments is vital for a region or country to formulate policies to push PV development, at least on the early stage.

ACKNOWLEDGMENTS

This research is financially supported by National Natural Science Foundation of China (Grant no. 71273052), and Minjiang University (Grant no. YSY13016).

REFERENCES

Barbose, G., Darghouth, N. 2015. Tracking the Sun VIII. Lawrence Berkely National Laboratory (LBNL).

Bird, L., Heeter, J., Kreycik, C. 2014. Solar Renewable Energy Certificate (SREC) Markets: Status and Trends. National Renewable Energy Laboratory (NREL).

Burns, J.E., Kang, J.S. 2012. Comparative economic analysis of supporting policies for residential solar PV in the United States: Solar Renewable Energy Credit (SREC) potential. *Energy Policy* 44: 217–225.

City of Boston. 2010. Solar Boston Permitting Guide: A resource for building owners and solar installers. http://www.cityofboston.gov/Images_Documents/ Solar%20Boston%20Permitting%20Guide%20 NEW%20 Sept%202011_tcm3–27989.pdf, accessed on 05–20–2014.

City of Boston. 2011. A Climate of Progress: City of Boston Climate Action Plan Update 2011. http:// www.cityofboston.gov/Images_Documents/A%20 Climate%20of%20Progress%20-%20CAP%20 Update%202011_tcm3–25020.pdf. accessed on 9-28-2015.

Denholm, P., Margolis, R.M., Ong, S. 2009. Break-Even Cost for Residential Photovoltaics in the United States: Key Drivers and Sensitivities, National Renewable Energy Laboratory (NREL).

Dusonchet, L., Telaretti, E. 2010a. Economic analysis of different supporting policies for the production of electrical energy by solar photovoltaics in western European Union countries. *Energy Policy* 38: 3297–3308.

Dusonchet, L., Telaretti, E. 2010b. Economic analysis of different supporting policies for the production of electrical energy by solar photovoltaics in eastern European Union countries. *Energy Policy* 38: 4011–4020.

Flynn, H., Breger, D., Belden, A., et al. 2010. System Dynamics Modeling of the Massachusetts SREC Market. *Sustainability* 2: 2746–2761.

Massachusetts Clean Energy Center. 2014. Commonwealth Solar II Photovoltaic Rebate Program Manual. Solicitation No. 2014 CSII-Version 18.0. http://www. masscec.com/programs/commonwealth-solar-ii, accessed on 05–25–2014.

Wiser, R., Barbose, G., Holt, E. 2010. Supporting Solar Power in Renewables Portfolio Standards: Experience from the United States. Lawrence Berkeley National Laboratory (LBNL).

Ziuku, S., Meyer, E.L. 2012. Economic viability of a residential building integrated photovoltaic generator in South Africa. *International Journal of Energy and Environment (IJEE)* 3 (6): 905–914.

Advances in Power and Energy Engineering – Sun (Ed.)
© 2016 Taylor & Francis Group, London, ISBN 978-1-138-02846-3

Fuzzy control based self-adaptive illumination system

C.K. Zhang
College of Energy and Electrical Engineering, Hohai University, Nanjing, Jiangsu, China

Z.Y. Li
College of Environment, Hohai University, Nanjing, Jiangsu, China

Q.L. Zhang, W.H. Zhou & F.Z. Yu
College of Energy and Electrical Engineering, Hohai University, Nanjing, Jiangsu, China

ABSTRACT: Smart LED illumination system can maintain a comfortable level of indoor illumination by changing the quantity of Luminous flux of LED lamp automatically based on the environment needed. This paper proposes a new control strategy for the Self-Adaptive illuminating system. SISO fuzzy control is used to avoid the durative actions of the system when the indoor illumination is close to the system operating value. It is manifested by experimental prototype, that the control scheme is correct and effective.

1 INTRODUCTION

Nowadays, the waste in illumination is very serious, especially in the traditional illuminating system (Huang 2014). The traditional illuminating system cannot adjust the indoor luminance, which not only wastes a lot of power but also brings inconvenience to users.

LED Self-Adaptive illuminating systems have been highly developed these years (Zhang 2015). The new self-adaptive illuminating system is economical, practical and worth promoted because it not only saves power but also brings convenience to users. However, this system will cause the durative actions of the system when the indoor illumination is close to the system operating value.

To reduce the durative action time, this paper proposes a control strategy for the Self-Adaptive illuminating system. In this paper, the Fuzzy Control method is used to make system operate based on the action hysteresis curve to avoid the system durative actions when the indoor illumination conditions are close to the system operating value (Li 2014). Through this control strategy, the new self-adaptive system can avoid getting into an infinite loop with LED lamp flash.

2 TRADITIONAL SELF-ADAPTIVE ILLUMINATION SYSTEM

When the system finds that the indoor illumination is getting dim (Gao 2011) because of sunset or curtains closed, the system will recalculate illumination after a time delay, which is defined before using, to avoid the influence of short-time changes of the indoor luminance. After the delayed time, the self-adaptive illuminating system calculates out the number of the LED lamp lighted based on current indoor illumination and the level users need (Niculescu 1998). Then the system lights the LED lamp required. For a typical Self-Adaptive illuminating system, its operation follow is: when the indoor illumination is slightly below the operating value of the chosen illuminating level, the system turns on one more LED lamp to make the indoor illumination reach the system level. When the indoor illumination is slightly above the system operating value, the system discharges unneeded LED lamp to make the indoor illumination slightly

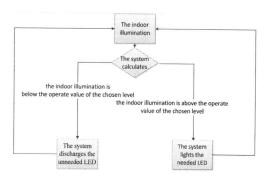

Figure 1. The durative actions of the system.

below the chosen illuminating level. The main problem for the current Self-Adaptive illuminating system is that the system may keep doing these durative actions nonstop (turning on and turning off the lights alternatively), which results a dead loop flash. Figure 1 shows the durative actions of the traditional system.

3 FUZZY CONTROL BASED SELF-ADAPTIVE ILLUMINATION SYSTEM

3.1 Control flow chart

After users choosing the illuminating level they need, the SCM compares the indoor illumination with the needed illumination (Li 2003). When the SCM finds that the indoor illumination cannot maintain users' demand, the SCM calculates out the number of the required lighting LED based on the indoor illumination and the level that users need. Then the system lights the LED needed. The SCM is used to control the ON or OFF of the LED which works under nominal voltage to save power (Li 2011). In addition, the time delay is used to avoid the influence of short-time changes of the indoor luminance. Figure 2 shows the control flow of the system.

3.2 Fuzzy control strategy

It is shown that the relationship between the illumination of LED and the indoor illumination is about linear by actual experiment.

The Adaptive regulator is adopted to make the SISO fuzzy control adapt to the Self-Adaptive

illuminating system better (Li 2006). The block diagram of the SISO fuzzy control is shown in Figure 3.

The self-adaptive illuminating system based on fuzzy control uses adaptive regulator and SISO fuzzy control (Han 2003). The single input is the deviation between the needed indoor illumination and the real-time indoor illumination. In addition, for the reason that the fuzzy control may cause a result of turning on or off a LED nonstop, the domain of the input and the output are divided symmetrically. In this paper, the domain of the input was set as [−1,1], the relative fuzzy subset is {N,Z,P}; and the domain of the output was set as [−1,1], its fuzzy subset is {N,Z,P}. (N means that the deviation is negative; Z means no deviation; P means that the deviation is positive;) 'E' is the fuzzy volume of the division. After the fuzzification, if 'E' is '−1' (N), make the output 'U' be '1' (P); if 'E' is '1' (P), make the output 'U' be '−1' (N); if 'E' is '0' (Z), make the system lights one more LED; 'U = −1' means the system discharges one more LED; 'U = 0' means no action; Table 1 shows the fuzzy control query table.

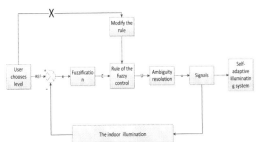

Annotation: X—the chosen level; REF—accurate indoor illumination the users' need; e—accurate volume of the deviation; E—fuzzy volume of the deviation; U—fuzzy volume of output; u—the volume of the U after ambiguity resolution.

Figure 3. The block diagram of the single-input single-output fuzzy control.

Table 1. The fuzzy control query table.

	Level 0 & Level 1			Level 2			Level 3			Level 4		
E*	−1	0	1	−1	0	1	−1	0	1	−1	0	1
U*	1	0	−1	1	0	−1	1	0	−1	1	0	−1
C*	70			75			80			75		

*The deviation: E—fuzzy volume of the deviation; U—fuzzy volume of the output; C—a parameter changed by the levels to modify the rule.

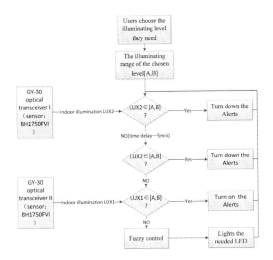

Figure 2. Control flow of the system.

$$E = < \frac{REFx - lx}{C} >\qquad(1)$$

where '< >' = Integer; U = –E; C = a parameter changed by the levels to modify the rule; REFx = the accurate needed illumination of level X; lx = the real-time indoor illumination.

We make the system working as a hysteresis curve to avoid the durative action of the system, and the width of the hysteresis is decided by parameters (C). In addition, the rule of the fuzzy control differs because the range and the need of levels are different. So, parameters (C) of the SISO fuzzy control was set based on the testing system which has five illuminating levels (level 0 to level 4).

4 EXPERIMENTAL CHECK

To validate the effectiveness of the proposed algorithm, a simulation environment is established, as is shown in Figure 4. This model simulated the characteristics of the traditional Self-Adaptive illuminating systems including the durative actions (dead loop flash). Under the durative action working condition, the proposed fuzzy control is used to make the system stable.

4.1 *Adaptive regulator of the single-output system*

The illumination for one LED of this testing system is 70 ± 10 lux. Five illumination levels 0–4 were set according to the Illumination Standard of International Commission on Illumination (S008-2001) (Yuan 2003) (Yu 2004). Where level 0

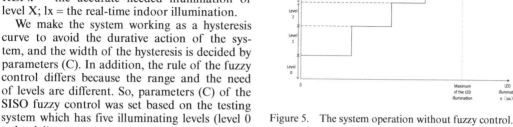

Figure 5. The system operation without fuzzy control.

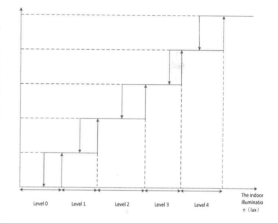

Figure 6. The system operation through fuzzy control.

indicates that the indoor illumination needed is less than 70 lux. Level 1 indicates that the indoor illumination needed is between 165 lux and 225 lux. Level 2 indicates that the indoor illumination needed is between 225 lux and 330 lux. Level 3 indicates that the indoor illumination needed is between 330 lux and 434 lux. Level 4 indicates that the indoor illumination needed is over 434 lux.

When the indoor illumination is close to the system operating value, the SISO fuzzy control is applied to make the system operate based on the hysteresis curve. The system operation without fuzzy control and the system operation under fuzzy control are compared and shown in Figure 5 and Figure 6, respectively. The experimental data is shown in the Table 2.

Figure 4. Simulation environment.

Table 2. Experimental data.

System change	Data				
	Initial indoor luminance (lux)	Needed luminance (lux)	Switching action time (second)	Indoor luminance of stability (lux)	The situation of system
0–1 Level (without fuzzy control)	34	157–181	10.20 s System stable	169	System stable
	75	157–181	5.76 s 2 lamps on 8.16 s 1 lamp on 10.04 s 3 lamps on	LED flash	Error: system gets into a dead loop
	104	157–181	3.33 s 2 lamps on 4.36 s 1 lamp on 5.89 s 3 lamps on	LED flash	Error: system gets into a dead loop
0–1 Level (with fuzzy control)	43	165–225	1.96 s System stable	157	System stable
	76	165–225	5.68 s System stable	190	System stable
	105	165–225	2.19 s System stable	217	System stable
0–2 Level (without fuzzy control)	45	181–285	2.63 s System stable	188	System stable
	100	181–285	2.40 s System stable	276	System stable
	160	181–285	2.14 s System stable	270	System stable
0–2 Level (with fuzzy control)	54	225–330	1.14 s System stable	192	System stable
	108	225–330	1.98 s System stable	248	System stable
	158	225–330	1.80 s System stable	300	System stable
0–3 Level (without fuzzy control)	60	285–364	2.53 s System operate 13.65 s System stable	358	System stable
	274	285–364	2.53 s 1 lamp on 4.51 s 0 lamp on 6.89 s 1 lamp on	LED flash	Error: system gets into a dead loop
0–3 Level (with fuzzy control)	47	330–434	1.43 s System stable	294	System stable
	275	330–434	1.58 s 4 lamps on 4.43 s 3 lamps on 5.89 s 4 lamps on	LED flash	Error: system gets into a dead loop
0–4 Level (without fuzzy control)	68	364+	2.95 s System stable	389	System stable
	330	364+	9.56 s 1 lamp on 12.54 s 0 lamp on 13.99 s 1 lamp on	LED flash	Error: system gets into a dead loop
0–4 Level (with fuzzy control)	55	434+	1.79 s System stable	375	System stable
	319	434+	1.51 s System operate 5.12 s System stable	464	System stable

5 CONCLUSIONS

Based on the comparison of experimental data, LED self-adaptive illuminating system is capable to adjust the indoor luminance. When the fuzzy control is not applied, the system will get into an infinite loop with LED flash when the indoor illumination is close to the operating value of the system.

After the SISO fuzzy control method is applied to the system, there were no durative actions exist, which demonstrate that the proposed method is effective to reduce system durative actions and keep the whole system stable.

Unlike the illumination fixed system, the new self-adaptive illuminating system is more intelligent and has a multi-illumination level which is convenient to the users. In self-adaptive illuminating system, all the users have to do is to choose the illuminating level they like and the system adjusts the indoor luminance to meet the user requirements. In addition, by taking the new SISO fuzzy control, the system gets more stable and avoids LED flashing.

REFERENCES

Chengkai Zhang. A New Adaptive Illumination System Based on STC89C5. manuscript submitted for publication.

Dazhong Li & Weiwei Ning & Weiqiang Ni & Wei Li. 2006. A Fuzzy Model Free Adaptive Control Method for SISO Systems, *Industry Control and Applications*, 06:23–25.

Huai Li & Yifei Chen. 2008. Study on Indoor Illumination Control Based on Fuzzy Neural Network, *Building electricity*, 07:27–30.

Jinkang Huang. 2014. The Design and Realization of Indoor Illumination Control Based on MCU, *Heilongjiang Science*, 03:37.

Junfeng Han & Yuhui Li. 2003. Fuzzy Control Technology. Chongqing, *Chongqing University Press*, 2003: 8–56.

Lie Li. 2011. The Research of Tunnel Lighting Intelligent Control Strategy for Multi-level Luminance Based on Grey Theory, *China Illuminating Engineering Journal*, 04:7–10.

Lihua Yu & Hong Luo & Deyue Chen. 2004. The Review of Intelligent Illumination Control, Shanghai Illumination Institution, *China Yangtze River Lighting Technology Forum Proceedings. Shanghai Illumination Insititution*: 4.

Niculescu S.I. & Verriest E.I. & et al. 1998. Stability and Robust Stability of Time-delay Systems: A Guided Tour. Stability and Control of Time-delay Systems, *Lecture Notes in Control and Information Sciences*, London: Springer-Verlag, 228:1–71.

Qiao Yuan. 2003. The Illumination Standard of International Commission on Illumination (S008-2001)—indoor workplace lighting (Continued), *China Illuminating Engineering Journal*, 14(1):55–62. DOI:10.3969/j.issn.1004-440X.2003.01.013.

Ying Gao & Jianbo Yang & Rong Wang. 2011. Study on Control Strategy of LED Energy Saving Lighting Based On LonWorks, *Electric Drive*, 04:52–55+59.

Yunjiang Li & Huiming Peng & Bo Xu. 2003. ComParison and Research on Some Methods of Luminance Calculation *of China Three Gorges Univ. (Natural Sciences)*, 01:30–32.

Advances in Power and Energy Engineering – Sun (Ed.)
© *2016 Taylor & Francis Group, London, ISBN 978-1-138-02846-3*

PSO-Hammerstein model identification on MGT-based LiBr absorption chiller system

D.C. Zhao, L. Pan & J.L. Zhang
School of Energy and Environment, Southeast University, China

ABSTRACT: Micro Gas Turbine (MGT)-based LiBr absorption chiller system is one of the most widely used distributed energy systems. Modeling on the system is difficult because of its complicated coupled inner chemical loops with nonlinear dynamics. A novel Particle Swarm Optimization (PSO) based Hammerstein model is proposed for control design and simulation of the system. PSO algorithm is used to identify the parameters of the Hammerstein model. Identification results indicate that the proposed Hammerstein model can approximate the nonlinear dynamics of the MGT-based LiBr absorption chiller system with high prediction accuracy. This modeling approach is also efficient for reproducing simulation system and for controller design due to its block-oriented structure.

1 INTRODUCTION

1.1 Distributed energy system

The Distributed Energy System (DES), which is usually composed of Micro Gas Turbine (MGT) and chiller system, receives great attention because of its advantages in energy saving, environmental protection and safety of the energy supply (Alanne & Saari, 2006, Sö derman & Pettersson, 2006, Duan & Yang, 2006). MGT-based LiBr absorption chiller system is one of the most widely used distributed energy systems. It provides distributed energy with multiple utility including cooling, heating and power supply with high quality. Therefore it is a multivariable and multi-disturbance system which brings challenges to control system design with modeling issues. A MGT-based LiBr absorption chiller system is hard to be modeled in First Principle because the chemical loops in it are complicated. Also the founded mathematic model has too many formulas to be used directly in controller design and simulation. Considering its high nonlinearity and time-varying characteristics, we use a kind of block-oriented identified model called Hammerstein model for describing MGT-based LiBr absorption chiller system. The system is modeled in apparent structure consisting of static-nonlinear block and linear dynamic block which is efficient for describing nonlinear dynamics, for reproducing simulation system and

for controller design such as predictive controller (Cervantes et al. 2003).

1.2 Hammerstein model

Block-oriented Hammerstein model, which includes a nonlinear static block followed by a linear dynamic block, is usually applied to model the static and dynamic characteristics of nonlinear systems. A large number of researches have been carried out on Hammerstein modeling and its related control of nonlinear systems (Dolanc & Strmcnik, 2006, Narendra & Gallman, 1966, Liu & Bai, 2007). Comparing to the linear model which can only approximate a system around a given operating point (Gómez & Baeyens, 2004). Hammerstein model can represent dynamic characteristics in full range in only one structured model. Thus Hammerstein model is more concise for control design and more convenient for model reconstruction than other multi-model approaches. It has been successfully used to represent nonlinear systems in many practical applications within the area of biomedical engineering, process engineering, and thermal engineering. A lot of methods have been proposed for Hammerstein model identification from given input and output data. These methods are different from each other for their structure of the static nonlinearity block. Polynomial functions (Jose, 2004), cubic spline functions (Dempsey & Westeick, 2004), neural networks (Janczak, 2004), and least squares support vector machines (Goethals, Pelckmans & Suykens, 2005) have been used in known structures. Particle Swarm Optimization

This work was supported by the National Natural Science Foundation of China under Grant 51576040.

(PSO) algorithm is easy to implement and with few parameters to adjust, which makes it be used widely to obtain the optimal solution. Thus PSO algorithm can be used to approach the static nonlinearity block. In this paper, a PSO based Hammerstein model is identified to describe the dynamic characteristic of the mechanism model of the MGT based LiBr absorption chiller system. The paper is organized as follows. Section 2 introduces a mathematic model of MGT-based LiBr absorption chiller system which we have built and will be taken as the system to be modeled in Hammerstein form; Section 3 presents our work of identifying PSO Hammerstein model on this system. Section 4 concludes the paper.

2 DESCRIPTION AND ANALYSIS OF SYSTEM

2.1 The process flow

A MGT-Based LiBr absorption chiller system is a typical complicated nonlinear dynamic process of energy converting (Zhang, Shen & Ge, 2013). Figure 1 illustrates the physical structure of the system. The Combined Cooling, Heating And Power (CCHP) or trigeneration system consists mainly of an 80 kW recuperator-adjustable micro gas turbine and a 425 kW double-effect LiBr absorption chiller. MGT converts the primary energy from fossil fuels to mechanical energy of the rotor and thermal energy of the flue gas. It acts as the prime mover of the entire system. An electrical generator converts mechanical energy into electricity energy. The MGT exhaust gas is the heating medium to drive

the chiller. An absorption chiller is a machine that, driven by heat, produces chilled water for ventilation air cooling. The absorption chiller requires two working fluids, a refrigerant and a sorbent solution of the refrigerant. In a LiBr absorption chiller, the refrigerant is water the sorbent is LiBr solution. In the absorption chiller cycle the water refrigerant undergoes a phase change in the condenser and evaporator while the sorbent solution undergoes a change in concentration in the absorber and evaporator. In the generator, the water-LiBr solution is heated by the flue gas and generates refrigerant vapor. Then the vapor is condensed in the condenser. Low operating temperature and pressure are required in the evaporator to vaporize refrigerant which can absorb heat from the chilled water. In the absorber, the cooling-water temperature determines the composition of the sorbent solution so that it absorbs the refrigerant vapor at the pressure determined by the evaporator as required. The high pressure generator refrigerant vapor valve assigns the proportion of the heating and cooling.

2.2 Simulation of the system

A simulation of the system is created using Simulink in Matlab in our previous work (Zhang, Shen & Ge, 2013). The simulation model method is of modularity and mechanism. This simulation model is multivariable with three inputs and three outputs. The inputs of the system are as follow: the fuel flow (G_f), the recuperator valve opening ratio (X_φ) and the high pressure generator refrigerant vapor valve opening ratio (μ_{hgr}).The three outputs includes the rotor speed of the MGT(n), the exit

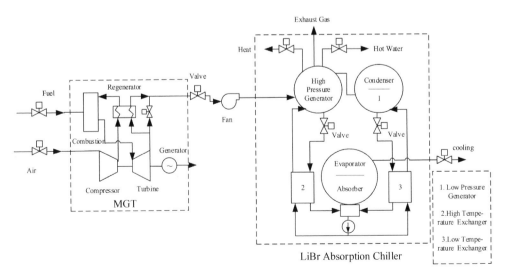

Figure 1.   Schematic diagram of the system.

temperature of the hot water (T_{hw}) and the exit temperature of the chilled water (T_{cw}), respectively correspond to the power, heating and cooling load.

We make some simulations to manifest the nonlinear dynamics and disturbances of the system in Figures 2–4. The open-loop step response for step change in the fuel flow (G_f) from rated amount (0.0114 kg/s) to its 90% at 800 s is shown in Figure 2. Likewise, Figures 3 and 4 show the open-loop step responses for step changes in X_φ and μ_{hgr} respectively. X_φ changes from 0% to 10% at 800 s and μ_{hgr} changes from 18% to 10% at 800 s.

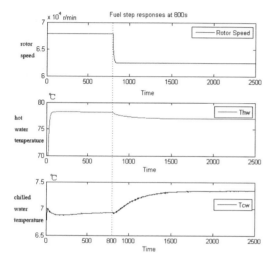

Figure 2.   Step response for G_f.

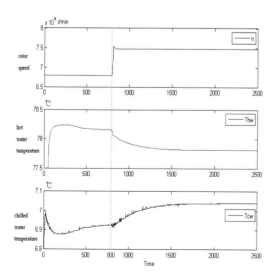

Figure 3.   Step response for X_φ.

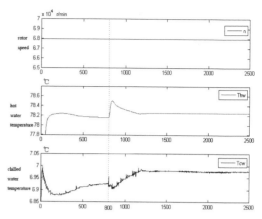

Figure 4.   Step response for μ_{hgr}.

It can be seen from the figures that the disturbances of G_f and X_φ have significant effects on all the outputs while the disturbance of μ_{hgr} only affects T_{hw} and T_{cw}. Comparing to T_{hw} and T_{cw}, n responds quicker for the inputs. Namely, MGT responds much quicker than the chiller. For the multivariable system, the nonlinearity is apparent from the figures, especially from the changes of T_{cw}. Because the system of mathematic model describes nonlinear dynamics of the system, it can be chosen as a realistic example for identifying Hammerstein model of MGT-based LiBr absorption chiller system.

3   PSO-BASED HAMMERSTEIN MODEL

3.1   Hammerstein model structure

The Hammerstein model consists of a series connection of a memoryless nonlinearity block f, and linear dynamics block G, where $y(k)$ is the output vector, $u(k)$ is the input vector, and $x(k)$ represents the transformed input variable, as shown in Figure 5.

The input-output relationship is then given by

$$A(q^{-1})y(k) = B(q^{-1})x(k) \tag{1}$$
$$x(k) = f(u(k)) \tag{2}$$

where:

$$A(q^{-1}) = 1 + a_1 q^{-1} + \cdots + a_n q^{-n} \tag{3}$$
$$B(q^{-1}) = b_1 q^{-1} + \cdots + b_m q^{-m} \tag{4}$$

The dynamic linear part of the model is described by linear autoregression.

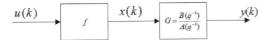

Figure 5.   Hammerstein model.

q^{-1} is the delay operator. It is assumed that the roots for characteristic equation

$$z^n + a_1 z^{n-1} + a_2 z^{n-2} + \cdots + a_n = 0 \qquad (5)$$

are all in the unit circle.

The model can be rewrite as follows:

$$y_t = \sum_{i=1}^{n} a_i y_{t-i} + \sum_{j=1}^{m} b_j f(u_{t-j}) + e_t \qquad (6)$$

with $u_t, y_t \in R$, $t \in Z$ and $\{(u_t, y_t)\}$ a set of input and output measurements. The equation error e_t is assumed to be white. m and n denote the order of the numerator and denominator in the transfer function of the linear model.

In order to apply PSO function to identify the model parameters, the nonlinear static function f should be approximated by a p-order polynomial function described by Equation (7).

$$x(k) = \sum_{i=0}^{p} r_i u^i(k) \qquad (7)$$

Parameter vector θ is introduced:

$$\theta = (a_1, a_2, \cdots a_n, b_1, b_2, \cdots b_m, r_1, \cdots r_p)^T \qquad (8)$$

Then the purpose of the identification is to estimate θ with the given inputs and outputs data. The estimation of θ is described by Equation (9).

$$\hat{\theta} = (\hat{a}_1, \hat{a}_2, \cdots \hat{a}_n, \hat{b}_1, \hat{b}_2, \cdots \hat{b}_m, \hat{r}_1, \cdots \hat{r}_p) \qquad (9)$$

The error of the estimation can be evaluated by the criterion function as Equation (10).

$$J(k) = \sum_{i=0}^{s} \beta(k, i) \left[y(k-i) - \hat{y}(k-i) \right]^2 \qquad (10)$$

where s is the width of the identification window, $\beta(k, i)$ is the weighting factor, $\hat{y}(k)$ is the output of the estimation model.

Equation (10) is an optimization problem. Its extremum can be solved by the PSO algorithm. Therefore the parameter of the estimation is gained.

3.2   PSO algorithm

PSO algorithm was inspired by social behavior of bird flocking. It can search the optimal solutions of complex space through cooperation and competition among individuals in group. PSO algorithm can retain personal and global optimal information. PSO algorithm is easy to implement and there are few parameters to adjust.

Steps of using PSO algorithm to identify system parameters are like follow:

1. Set the parameters. The parameters include the population size, space dimension, the maximum number of iterations, the inertia factor and the accelerating factor. At the same time, initialize the particles' position and set the initial speed;
2. Calculate the objective function. Find out current individual and global best value;
3. Update the particles' position and velocity;
4. Judge whether if the maximum number of iteration is reached. If it is true, the calculation ends. We get the optimal solution of the parameters which need to identify. Otherwise, return to step 2.
5. When the optimization process is finished, we obtain the estimates of parameters.

3.3   Hammerstein model identification

In this paper, we identify the Hammerstein model of the fuel flow tunnel by PSO from the mathematical model of the system which has been built up in Matlab to generate the training and testing data. The fuel flow affects all the outputs apparently so we choose the tunnel of fuel flow to rotor speed to present the method for simplicity. Firstly, a random magnitude sequence of the fuel flow is used as the excitation signal to the simulation, as shown in Figure 6. Then the response of the rotor speed can

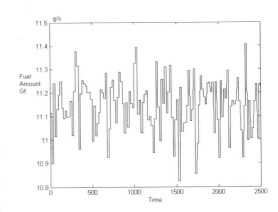

Figure 6.   Input signal of the fuel flow.

be obtained from mathematical model. A set of 2500 input-output samples are collected and normalized, the first 1500 data are used to identify the Hammerstein model and the remaining 1000 data are used for model validation.

Finally, the linear dynamic parameters of the model and the nonlinearity of the model can be obtained. Training results are shown in Figure 7. Output errors which described in Equation (11) are shown in Figure 8.

$$error = (\hat{y} - y)/y \tag{11}$$

The prediction results are shown in Figure 9. Output errors which described by Equation (11) are shown in Figure 10.

Prediction results indicate that the PSO based Hammerstein model has high prediction accuracy for the mechanism simulation of the MGT based LiBr absorption chiller system, and it can approximate the dynamic behavior of the system efficiently.

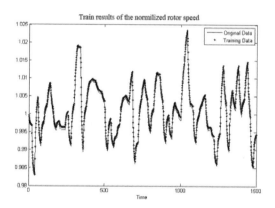

Figure 7.   Results of the Hammerstein model training.

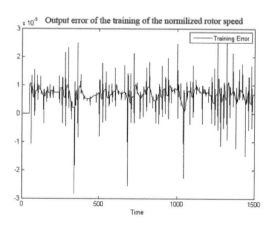

Figure 8.   Output errors of the training.

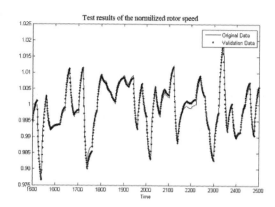

Figure 9.   Results of the Hammerstein model validation.

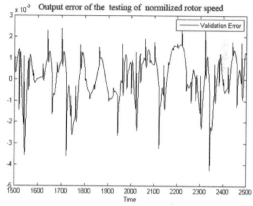

Figure 10.   Output errors of the testing.

Finally, we obtain the linear dynamic parameters of the model denoted by the coefficients of $A(q^{-1})$, $B(q^{-1})$ and the static nonlinearity parameters denoted by the coefficients of the polynomial or $(r_1, r_2, \ldots r_n)$. The results are as follow:

$$\frac{B(q^{-1})}{A(q^{-1})} = \frac{0.0569q^{-1} + 0.0624}{0.3342q^{-2} + 0.5465q^{-1} + 1} \tag{12}$$

$$(r_0, r_1, r_2) = (0.0819, 0.0223, 0.0955) \tag{13}$$

The PSO-based Hammerstein model identification for other tunnels can be obtained similarly.

4   CONCLUSION

MGT based LiBr absorption chiller system is complex and of high nonlinearity. A PSO based Hammerstein model of the MGT based LiBr absorption chiller system is reported to facilitate valid control strategy design and analysis. The identification of the dynamic linear model parameters and the static nonlinearity is achieved by PSO based Hammerstein model identification algorithm. Simulation results indicate that the PSO based Hammerstein model has high prediction accuracy. In the future, based on the Hammerstein model, some control scheme studies such as model predictive control can be developed based on the Hammerstein model.

REFERENCES

Alanne K, Saari A. 2006. Distributed energy generation and sustainable development [J]. *Renewable and Sustainable Energy Reviews*, 10: 539–558.

Cervantes A.L., Agamennoni O.E., Figueroa J.L. 2003. A Nonlinear Model Predictive Control System Based on Wiener Piecewise Linear Models, *Journal of Process Control*, Vol. 13, No. 7, 655–666.

Dempsey E.J., Westeick D.T. 2004. Identification of Hammerstein Models with Cubic Spline Nonlinearities, *IEEE Transactions on Biomedical Engineering*, Vol. 51, No. 2, 237–245.

Dolanc G., Strmcnik S. 2005. Identification of Nonlinear Systems Using a Piecewise-linear Hammerstein Model Block-oriented Nonlinear Models, *Systems & Control Letters*, Vol. 54, No. 2, 145–158.

Goethals I., Pelckmans K., Suykens J.A.K., et al, 2005. Identification of MIMO Hammerstein Models Using Least Squares Support Vector Machines, *Automatica*, Vol. 41, No. 7, 1263–1272.

Gómez J.C., Baeyens E. 2004. Identification of Block-oriented Nonlinear Systems Using Orthonormal Bases, *Journal of Process Control*, Vol. 14, No. 6, 685–697.

Janczak A. 2004. Neural Network Approach for Identificatio of Hammerstein Systems, *International Journal of Control*. Vol. 76, No. 17, 1749–1766.

Jose Vieira. 2004. Parameter estimation of non-linear systems with Hammerstein models using neuro-fuzzy and polynomial approximation approaches [J]. *IEEE International Conference*. Vol. 2: 849–854.

Junli Zhang, Jiong Shen, Bin Ge. 2013. Demand response with MGT-CCHP based on MLD model [C]. *The International Conference on Electrical Engineering 2013*, Fujian, Xiamen, China, July 14–17.

Liqiang Duan, Yongping Yang. 2006. Theoretical study on integration mechanism of distributed energy system [J]. *The 19th ECOS conference*, 1285–1293.

Liu Y., Bai E.W. 2007. Iterative identification of Hammerstein systems [J]. *Automatica*, 43(2): 346–354.

Narendra K.S., Gallman P.G. 1966. An iterative method for the identification of nonlinear systems using a Hammerstein model [J] *IEEE Transactions on Automatic Control*, 11(3): 546–550.

Sö derman J., Pettersson F. 2006. Structural and operational optimization of distributed energy systems [J]. *Applied Thermal Engineering*, 26: 1400–1408.

Advances in Power and Energy Engineering – Sun (Ed.)
© 2016 Taylor & Francis Group, London, ISBN 978-1-138-02846-3

ERGA-SVD method based decoupled controller design for Micro Gas Turbine Combined Cooling Heating and Power system

Y.X. Liu, F. Zhang, J.L. Zhang, X. Zhou & J. Shen
Key Laboratory of Energy Thermal Conversion and Control of Ministry of Education (Southeast University), Nanjing, Jiangsu Province, China

ABSTRACT: The Micro Gas Turbine Combined Cooling Heating and Power (MGT-CCHP) system, which is designed for generating electricity as well as cold/hot water, is more and more used on many occasions. To achieve outstanding control performance of the MGT-CCHP system, a decoupled control strategy has been proposed in this paper. Effective Relative Gain Array (ERGA) is firstly used to select the best pairing of input and output variables. Then a decoupled controller for the system is designed based on Singular Value Decomposition (SVD). The simulation results of the coordination control system show the superiority of the proposed method over traditional PID.

1 INTRODUCTION

With the development of the distributed energy technology, the MGT-CCHP system attracts increasing attention due to its functionality and convenience. Various literatures can be found on the research on the static characteristic of the MGT-CCHP system (Yang, 2010; Wang, 2011; Li, 2011; Jayasekara, 2012). However, few papers discuss the control strategy and optimization method.

The MGT-CCHP system that is designed for generating electricity as well as cold/hot water is comprised of a Micro Gas Turbine (MGT) and a Lithium Bromide Absorption Chiller (LBAC) subsystem (Li, 2011). The micro gas turbine generates electricity and its exhaust gas enters into the chiller for producing cold water and hot water. According to the open-loop response experiments of MGT-CCHP system (Zhang, 2010), it can be observed that the disturbance of the high pressure cryogen steam valve has little impact on the turbine speed. The micro gas turbine and lithium bromide absorption chiller were distributed in the upstream and downstream on the cascade of energy (Ming, 2011). Thus, the MGT-CCHP system can be simplified to two cascaded subsystems. With the purpose of efficient control in the complex industrial production process, decoupling design is often introduced in an engineering application. The common decoupling methods include cascade decoupling, SVD decoupling, adaptive decoupling, intelligent decoupling, fuzzy decoupling, and so on. A PID controller (Zhang, 2015) is designed on the basis of a multivariable frequency domain method for the MGT-CCHP system. However, the demerits

of PID, which include long regulating time, strong coupling, and great overshoot are not eliminated completely (Mohammadpour, 2009).

Due to the complex relationship of multivariate in this system, the correlation degree of input and output variables and the reasonable pairing (Moezzi, 2008) must be determined firstly for efficient control. This paper employs ERGA (Zhang, 2012), which introduces the concept of bandwidth based on the Relative Gain Array (RGA), which can make a more comprehensive description for the relationship of the multivariable system.

In light of the effectivity of ERGA (He, 2013) and in determining the pairing and the convenience of SVD (Tao, 2011) in calculations, this paper raises a decoupling method, which combines ERGA with SVD. The method judges the pairing methods according to ERGA and decouples the control object by SVD (Yam, 2014). The PID controllers in each loop can be designed independently.

The rest of this paper is organized as follows: Section 2 describes the model of the controlled object; Section 3 discusses the design of the decoupled controller based on ERGA and SVD; Section 4 shows the result of the simulation of the coordinated control system. Finally, the conclusion of the paper is outlined in section 5.

2 THE CONTROL OBJECT OF MGT-CCHP SYSTEM

2.1 The structure of MGT-CCHP system

The flow chart of the MGT-CCHP system is shown in Figure 1. The system which includes a micro gas

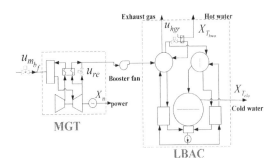

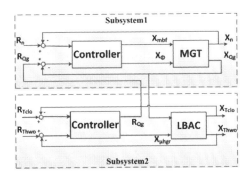

Figure 1. The flow chart of the MGT-CCHP system.

Figure 3. The coordinated control system of MGT-CCHP.

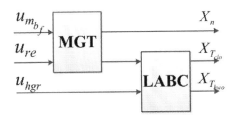

Figure 2. The simplified control block.

turbine and a lithium bromide absorption chiller can produce power, cold water, and hot water at the same time (Jayasekara, 2012). The micro gas turbine generates electricity and its exhaust gas enters into the chiller for producing cold water and hot water.

The simplified control block is shown in Figure 2.

Where u_{mbf} = fuel amount, kg/s; u_{re} = regenerative valve position, %; u_{hgr} = high pressure cryogen steam valve position, %; X_n = turbine speed, rad/min; X_{Tclo} = cold water temperature, and °C; and X_{Thwo} = hot water temperature, °C.

2.2 Decomposition models

The coordinate control system of MGT-CCHP is shown in Figure 3. In paper (Zhang, 2015), it points out that the MGT system has contact with the LBAC system through the exhaust gas temperature X_{Qg}. The MGT system is mainly used to allocate the power generation and the exhaust heat reasonably, while the LBAC system distributes the cold load and the hot load according to the exhaust heat.

Where, R_n = the setting value of the turbine speed, R_{Qg} = the setting value of the exhaust gas temperature, R_{Tclo} = the setting value of the cold water, and R_{Thwo} = the setting value of the hot water.

For the MGT system, the LBAC system provides the setting value of the exhaust temperature. The effect of the LBAC system can be neglected during the process of parameter design of the

MGT system. On the contrary, for the LBAC system the MGT system acts as the manipulate variable. Since the inertia of the MGT system is far less than that of LBAC system, controller of LBAC system can be designed independently. It is proved that under certain conditions, the MGT system is equivalent to a fast servo system, which implies that the reference of the gas temperature can be followed by the real value, i.e., $X_{Qg} \approx R_{Qg}$ (Zhang, 2015).

The problem of the coordinate control of MGT-CCHP can be converted into two independent control problems, namely the independent close-loop control problems of the MGT system and the LBAC system.

3 ERGA-SVD METHOD BASED DECOUPLED CONTROLLER DESIGN

In this section, the best variable paring of the MGT system and the LBAC system are determined based on the ERGA criterion. Then, the decoupled controllers of two subsystems are designed, respectively. All the parameters are calculated at the operating condition of 100% rated load.

3.1 Variables pairing process

In order to simplify the calculation, the order of transfer function model is reduced. The gain array and bandwidth are obtained according to the order of reduced models. Then, the ERGA can be calculated and the best pairings are determined.

1. Order reduction of models
   The transfer function of two subsystems' (Zhang, 2015) order is reduced. The higher-order inertia object is approximately equal to a one-order inertia object with pure delay. The model of the MGT system and the LBAC system with order reduction is shown in Equation 1 and Equation 2.

$$\begin{bmatrix} X_n \\ X_{TMex} \end{bmatrix} = \begin{bmatrix} \dfrac{0.7235}{1+15s}e^{-s} & \dfrac{0.129}{1+17s}e^{-s} \\ \dfrac{0.375}{1+0.5s}e^{-s} & \dfrac{-0.0156}{1+0.1s} \end{bmatrix} \begin{bmatrix} X_{mbf} \\ X_{\varphi} \end{bmatrix} \quad (1)$$

$$\begin{bmatrix} X_{Tclo} \\ X_{Thwo} \end{bmatrix} = \begin{bmatrix} \dfrac{-0.497}{1+252s}e^{-s} & \dfrac{-0.223}{1+17s}e^{-s} \\ \dfrac{0.48}{1+113s}e^{-s} & \dfrac{-0.215}{1+15s}e^{-3s} \end{bmatrix} \begin{bmatrix} X_{Qg} \\ X_{uhgr} \end{bmatrix} \quad (2)$$

2. ERGA of the MGT system

According to the ERGA calculation method (Zhang, 2012), the gain array, bandwidth array, and ERGA are calculated as following:

$$G_1(0) = \begin{bmatrix} 0.7235 & 0.129 \\ 0.375 & -0.0156 \end{bmatrix} \quad (3)$$

$$\Omega_1(0) = \begin{bmatrix} \dfrac{1}{15} & \dfrac{1}{17} \\ 2 & 10 \end{bmatrix} \quad (4)$$

$$E_1 = G_1(0) \otimes \Omega_1 = \begin{bmatrix} 0.0482 & 0.0076 \\ 0.7500 & -0.1560 \end{bmatrix} \quad (5)$$

$$\Phi_1 = E_1 \otimes E_1^{-T} = \begin{bmatrix} 0.5694 & 0.4306 \\ 0.4306 & 0.5694 \end{bmatrix} \quad (6)$$

where $G_1(0)$ = gain array; Ω_1 = bandwidth array; E_1 = effective gain array; and Φ_1 = effective relative gain array.

According to the ERGA pairing criterion, all pairing elements must be positive and approach one another as closely as possible (Zhang, 2012). The result indicates that the diagonal pair is the best pairing method of the MGT system.

3. ERGA of the LBAC system

In the same way, the gain array, bandwidth array, and ERGA are calculated as following:

$$G_2(0) = \begin{bmatrix} -0.497 & -0.223 \\ 0.48 & -0.215 \end{bmatrix} \quad (7)$$

$$\Omega_2(0) = \begin{bmatrix} \dfrac{1}{252} & \dfrac{1}{150} \\ \dfrac{1}{113} & \dfrac{1}{15} \end{bmatrix} \quad (8)$$

$$E_2 = G_2(0) \otimes \Omega_2 = \begin{bmatrix} -0.0020 & -0.0015 \\ 0.0042 & -0.0143 \end{bmatrix} \quad (9)$$

$$\Phi_2 = E_2 \otimes E_2^{-T} = \begin{bmatrix} 0.8168 & 0.1832 \\ 0.1832 & 0.8168 \end{bmatrix} \quad (10)$$

where $G_2(0)$ = gain array, Ω_2 = bandwidth array, E_2 = effective gain array, and Φ_2 = effective relative gain array.

Similarly, the result indicates that the diagonal pair is the best pairing method of the LBAC system.

3.2 *SVD decoupled controller of the MGT system design*

The gain array of the MGT system is decomposed with the approach of SVD. Then the PID controller is designed for each loop in the decoupled system independently. The close-loop control structure of the MGT system is shown in Figure 4.

The SVD approach is (Gauthier, 2009) as below:

$$U^T U = I \quad (11)$$

$$V^T V = I \quad (12)$$

$$\Sigma = U^T \left(U\Sigma V^T V \right) \quad (13)$$

where U = left singular array; V = right singular array; and Σ = dominant principal diagonal array.

Then, the gain array of the MGT system is decomposed by SVD.

$$U_1 = \begin{bmatrix} -0.8932 & -0.4497 \\ -0.4497 & 0.8932 \end{bmatrix} \quad (14)$$

$$V_1 = \begin{bmatrix} -0.9913 & 0.1316 \\ -0.1316 & -0.9913 \end{bmatrix} \quad (15)$$

$$\Sigma_1 = \begin{bmatrix} 0.8220 & 0 \\ 0 & 0.0726 \end{bmatrix} \quad (16)$$

where U_1 = left singular array; V_1 = right singular array; and Σ_1 = dominant principal diagonal array.

The decoupled system can be assumed to have the diagonal dominance, so the controllers of the two loops can be designed independently. The MGT system requires small overshoot, fast regulation process and zero steady state error. In light of the relatively small inertia of the MGT system, PI controller is used. The PI controllers of two loops can be obtained by adjusting the regulation time and the overshoot of the response curve.

$$PI_n = 8 + \frac{0.2}{s} \quad (17)$$

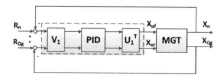

Figure 4. The close-loop control structure of the MGT system.

$$PI_{Qg} = 1 + \frac{12}{s} \tag{18}$$

where, PI_n = the controller of the turbine speed after decoupling; PI_{Qg} = the controller of the exhaust gas temperature after decoupling.

The parameters before decoupling are given for comparison.

$$PI'_n = 0.9 + \frac{0.2}{S} \tag{19}$$

$$PI'_{Qg} = -0.1 + \frac{-0.2}{S} \tag{20}$$

where PI'_n = the controller of the turbine speed before decoupling; and PI'_{Qg} = the controller of the exhaust gas temperature before decoupling.

Under the above controller parameters, the close loop simulation experiments where the setting value of the turbine speed decreases to 10% and the setting value of the exhaust gas temperature drops to 10% are shown in Figures 5–6. The dotted line is the response curve before decoupling and the solid line is the response curve after decoupling.

As can be seen in the Figure 5, the response of the turbine speed responds faster. Besides, the overshoot of the exhaust gas temperature reduce a lot after decoupling. In Figure 6, the coupling is significantly diminished and the exhaust gas temperature responds more rapidly.

3.3 SVD decoupled controller of the LBAC system design

In the LBAC system, the hot load accounts for small proportion. As a result, the high pressure cryogen steam valve is only used as an auxiliary adjustment. The regulation of cold load is realized by adjusting the hot source parameters while the hot load is regulated by adjusting the high pressure cryogen steam valve.

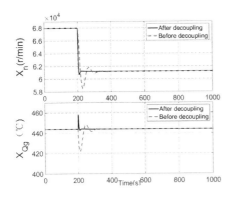

Figure 5. Decrease of setting value for turbine speed.

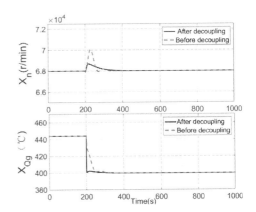

Figure 6. Decrease of setting value for the exhaust gas temperature.

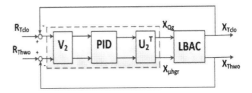

Figure 7. The close-loop control structure of the LBAC system.

Since, the reference of the gas temperature R_{Qg} can be followed by the real value X_{Qg} as mentioned in section 2.2, the close-loop control structure of the LBAC system can be designed as seen in Figure 7.

In the same way, the gain array of the LBAC system is decomposed by SVD.

$$U_2 = \begin{bmatrix} -0.7255 & 0.6882 \\ 0.6882 & 0.7255 \end{bmatrix} \tag{21}$$

$$V_2 = \begin{bmatrix} 0.9998 & 0.0200 \\ 0.0200 & -0.9998 \end{bmatrix} \tag{22}$$

$$\Sigma_2 = \begin{bmatrix} 0.6911 & 0 \\ 0 & 0.3095 \end{bmatrix} \tag{23}$$

where U_2 = left singular array; V_2 = right singular array; and Σ_2 = dominant principal diagonal array.

The PI controllers of two loops can be tuned by adjusting the regulation time and the overshoot of the response curve.

$$PID_{Tclo} = 2.5 + \frac{0.02}{s} + 50s \tag{24}$$

$$PI_{Thwo} = 5 + \frac{1}{s} \tag{25}$$

where PID_{Tclo} = the controller of the cold water temperature after decoupling; and PI_{Thwo} = the controller the hot water temperature after decoupling.

$$PID'_{Tclo} = -2.8 + \frac{-0.02}{s} - 140\,s \qquad (26)$$

$$PI'_{Thwo} = -3 + \frac{-0.8}{s} \qquad (27)$$

where PID_{Tclo}' = the controller of the cold water temperature before decoupling; and PI_{Thwo}' = the controller of the hot water temperature before decoupling.

Under the above controller parameters, the close loop simulation experiments where the setting values of the cold water temperature decreases to 10% and the setting value of the hot water temperature drops to 10% are shown in Figures 8–9. The dotted line is the response curve before decoupling and the solid line is the response curve after decoupling.

It can be seen in the Figure 8 that the coupling of two loops are weakened and the overshoot become smaller after eliminating coupling. From Figure 9, the responses of the cold water temperature and the hot water become faster and the overshot is limited within reasonable range.

4 SIMULATION OF THE COORDINATED CONTROL SYSTEM

According to the results of the MGT and the LBAC control system design, the coordinated control system structure can be achieved as shown in the Figure 10. The step experiments for the setting value of the turbine speed, cold water temperature, and hot water temperature are performed.

Case 1: step of the setting value of the turbine speed
The response curves of different parameters are shown in Figure 11. With a 10% decrease of the setting value of the turbine speed, the turbine speed can follow the change of setting value rapidly due to the relatively small inertia of the MGT system. There is almost no change in the high pressure cryogen steam valve position and the regenerative valve position, avoiding large disturbance on the parameters of the LBAC system.

Case 2: step of the setting value of the cold temperature
The response curves of different parameters are shown in Figure 12. With a 10% decrease of the setting value of the cold water temperature, even though the system is not completely decoupled, each controlled variable can be steadied on account

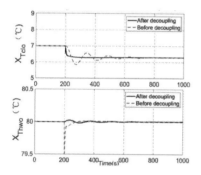

Figure 8. Decrease of setting value for the cold water temperature.

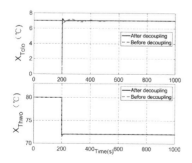

Figure 9. Decrease of setting value for the hot water temperature.

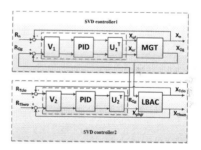

Figure 10. The coordinated control system structure.

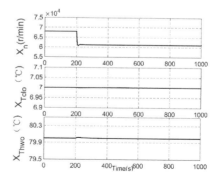

Figure 11. Decrease of setting value for the turbine speed.

33

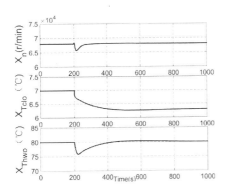

Figure 12. Decrease of setting value for the cold water temperature.

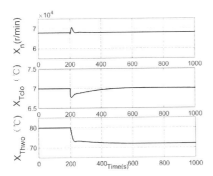

Figure 13. Decrease of setting value for the hot water temperature.

of SVD decoupled controller. At the same time, the overshoot is limited within an allowable range.

Case 3: step of the setting value of the hot temperature

The response curves of different parameters are shown in Figure 13. With a 10% decrease of the setting value of the hot water temperature, the cold water temperate takes more time to reach stability because the LBAC system has a relatively large inertia. The position variation of the high pressure cryogen steam valve makes no difference on the turbine speed in the original system. After adding the singular decoupled controller, the position variation of the high pressure cryogen steam valve has a very small impact on the turbine speed. Then, the disturbance can be eliminated quickly and the overshoot is small.

The results indicate that the decoupled system eliminates the main coupling and achieves better control effect. Also, the ERGA-SVD based decoupled control method is feasible to meet the requirements of the cooling heating and power for MGT-CCHP system.

5 CONCLUSION

Combining the validity of ERGA in pairing the convenience of the SVD in decoupling, this paper proposes a decoupling control method for MGT-CCHP system. The best pairing of input and output variables are selected by ERGA and the decoupled controller for each system is designed based on SVD. The simulation results of the coordination control system show that the method is beneficial to the reduction of regulating time, the weakening of coupling, and the decrease of overshoot. The system after decoupling obtains satisfying and dynamic characteristics.

REFERENCES

Gauthier, G. 2009 Terminal Iterative Learning Control design with singular value decomposition decoupling for thermoforming ovens. *American Control Conference, 2009*. ACC '09: 1640–1645.

He Guang. 2013. Interactions analysis in the maglev bogie with decentralized controllers using an effective relative gain array measure, *Control and Automation (ICCA)*: 1070–1075.

Jayasekara, S. 2012. A review on optimization strategies of combined cooling heating and power generation. *Information and Automation for Sustainability (ICIAfS)*: 302–307.

Li Zhengyi. 2011. Optimization and Analysis of Operation Strategies for Combined Cooling, Heating and Power System. *Power and Energy Engineering Conference (APPEEC)*: 1–4.

Ming Gang. 2011. Experimental studies on a CCHP system based on a micro-turbine. *Mechanic Automation and Control Engineering (MACE)*: 2159–2162.

Moezzi, M.-A. 2008. A novel automatic method for multivariable process pairing and control. *India Conference. Annual IEEE* (1): 262–267.

Mohammadpour, J. 2009. LPV decoupling for multivariable control system design. *American Control Conference*: 3112–3117.

Tao, K.M. 2011. A simple tunable method for Profile Control—Least-squares configuration. *American Control Conference (ACC)*: 4538–4539.

Wang Wei. 2011. Comparative Analysis of CCHP Systems Based on Different Gas Turbine Cycles Applied in North of China. *Power and Energy Engineering Conference (APPEEC)*: 1–4.

Yam, C.M. 2004. A SVD based controller of UPFC for power flow control. *Electric Power Systems Research* (70): 76–84.

Yang Wansheng. 2010. Analysis on Energy Saving for a Cool, Heat and Power Cogeneration System with a Micro Gas Turbine. *Power and Energy Engineering Conference (APPEEC)*: 1–5.

Zhang Junli. 2015. Integrated Optimization and Dynamic Characteristic Research On Micro turbine-based Cooling, Heating and Power System. *Ph.D thesis*. Southeast University.

Zhang, Yufei. 2012. ERGA pairing criterion for multivariable processes with large time delay. *Journal of Southeast University (Natural Science Edition)* (42): 880–885.

Advances in Power and Energy Engineering – Sun (Ed.)
© 2016 Taylor & Francis Group, London, ISBN 978-1-138-02846-3

Auto-tuning model predicting the control of Micro Gas Turbine Combined with Cooling Heating and Power systems

X.F. Liang, X. Wu, Y.G. Li, J.L. Zhang & J. Shen*
*Key Laboratory of Energy Thermal Conversion and Control of Ministry of Education,
School of Energy and Environment, Southeast University, Nanjing, China*

ABSTRACT: To overcome the control difficulties of the Micro Gas Turbine Combined with Cooling, Heating, and Power (MGT-CCHP) systems such as immeasurable disturbances, strong coupling among multi-variables and large-inertial behavior, an off-set free Model Predictive Control (MPC) strategy based on an incremental state space model is proposed in this paper, and a novel auto-tuning method to obtain output suppression factors of the MPC controller is presented. The simulation results for an 80 kw MGT-CCHP simulator show the advantage and effectiveness of the proposed MPC.

1 INTRODUCTION

The application of Combined Cooling, Heating, and Power (CCHP) technology has been thought of as the most advantageous branch of distributed generation recently (Xu, J. et al. 2010, Dong, L. 2009, Arashnia, I. 2015, Wu, D. W. & Wang, R. Z. 2006). Among various CCHP techniques, the Micro Gas Turbine-based Combined Cooling, Heating, and Power system (MGT-CCHP) is promising, owing to its advances in providing clean, reliable, and high-quality distributed energy. Although, the dynamic characteristics and modeling of the MGT-CCHP system have been studied extensively in the past decades (Gao, P. Li. et al. 2014, Wang, J.L. et al. 2014, Kong, X. et al. 2010, Rey, G. et al. 2015, Anvari, S. et al. 2015), the control approaches for the MGT-CCHP system still remain in the stage of conventional PID control.

Yang, J. (2009) proposed that PID control loops are designed for the MGT-CCHP system and the control parameters are well-tuned by trial and error at the given operating point. However, when the set-point changes in a wide range, strong oscillation occurs for the output variable, which may reduce the efficiency of the plant and threaten the safety of many devices. Zhang J. (2015) proposed a decoupling PID algorithm based on the frequency design methods, which results in a smaller overshot and faster response compared with the conventional PID loops. However, since the dynamic of the MGT-CCHP system has properties such as a large inertia, strong coupling, and unknown disturbances, the PID approaches that are devised on the basis of separate single-input single-output loops are no longer sufficient in meeting the performance specifications.

Given these reasons, we propose to use the MPC to achieve an integrated multi-variable control of the MGT-CCHP system. To overcome the effect of unknown disturbances and modeling mismatches, an incremental state-space model is employed, and an integral action is added into the predictive controller to achieve an offset-free tracking of the output variables. Moreover, a novel method for auto-tuning output suppression factors of the MPC is proposed, which can eliminate the parameters setting issues and the resulting performance degradation.

The proposed MPC approach is applied to an 80 kW MGT-CCHP system simulator. The remainder of this paper is organized as follows: the MGT-CCHP system is presented in Section 2, with a short introduction to the model identification of the MGT-CCHP system. The proposed MPC strategy and parameter tuning method are discussed in detail in Section 3. Simulation results are given in Section 4, starting with a brief introduction to the PI control system, which is used for the comparison and conclusion, which are drawn in Section 5.

2 THE MGT-CCHP SYSTEM DESCRIPTION

The simulator used in this paper describes the dynamics of an MGT-CCHP system, which consists of an 80 kW recuperator-adjustable micro turbine and a 425 kW double-effect LiBr-H_2O

*Corresponding author
This work was supported in parts by the Doctoral Fund of the Ministry of Education of China under Grant 20130092110061.

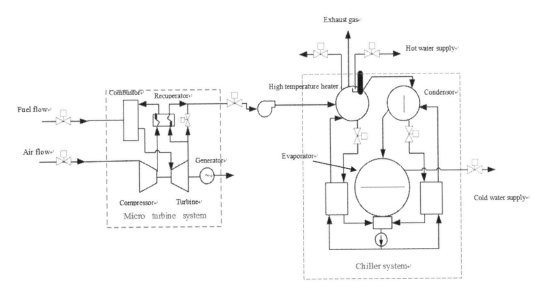

Figure 1. Diagram of MGT-CCHP system.

absorption chiller. The simulator is built from first-principle approach in the MATLAB environment (Zhang, J. 2015). The schematic diagram of the system is shown in Figure 1.

The working principle of the MGT-CCHP system could be generalized as follows. The compressed air is preheated in the recuperator by the exhaust gas from gas turbine and then piped into the combustor, where the mixed fuel gas and air is burned, producing flue gases with high pressure and temperature. The flue gases expand and drive a micro gas turbine to produce electrical energy and then it gets sent into the chiller as the heat source of cryogen after heating the compressed air. The cryogen heats the water in the high temperature heater for hot water supply and then it is chilled in the condenser. After leaving the condenser, the condensed cryogen absorbs the heat from the water in the evaporator for refrigeration and is fed into the high temperature heater for another cycle. Following this method, the MGT-CCHP system can provide cooling, heating, and electrical power to a building or a district simultaneously.

For the MGT-CCHP system used in this paper, the output variables are turbine rotation speed ω (r/min), cold water temperature TC (°C), and hot water temperature TH (°C); the input variables are fuel valve opening g (dimensionless), regenerative heat load φ (dimensionless), and cryogen valve opening μ (dimensionless).

To provide a suited model for the MPC design, prediction error approach (Ljung, L. 1998) is utilized and the transfer function matrix model of the

MGT-CCHP is identified using the input–output data. The model is then converted into the state space form because of its advances in the multivariable systems.

3 MPC CONTROL SYSTEM DESIGN

3.1 Incremental state space model

Suppose the following discrete state space model can be used to present the multivariable MGT-CCHP system:

$$x_m(k+1) = A_m x_m(k) + B_m u(k) \tag{1}$$

$$y(k) = C_m x_m(k) + D_m u(k) \tag{2}$$

where $x_m \in R^n$ = state vector; $u \in R^l$ = manipulated vector; $y \in R^t$ = output vector, and A_m, B_m, C_m, D_m are the system matrices, respectively.

Let:

$$\Delta x_m(k+1) = x_m(k+1) - x_m(k) \tag{3}$$

$$\Delta u(k) = u(k) - u(k-1) \tag{4}$$

we have:

$$\Delta x_m(k+1) = A_m \Delta x_m(k) + B_m \Delta u(k) \tag{5}$$

$$y(k+1) - y(k) = C_m A_m \Delta x_m(k) \\ + (C_m B_m + D_m)\,\Delta u(k) \tag{6}$$

Thus, we can have:

$$\begin{bmatrix} \Delta x_m(k+1) \\ y(k+1) \end{bmatrix} = \begin{bmatrix} A_m & O \\ C_m A_m & I \end{bmatrix} \begin{bmatrix} \Delta x_m(k) \\ y(k) \end{bmatrix}$$
$$+ \begin{bmatrix} B_m \\ C_m B_m + D_m \end{bmatrix} \Delta u(k) \qquad (7)$$

$$y(k) = \begin{bmatrix} O & I \end{bmatrix} \begin{bmatrix} \Delta x_m(k) \\ y(k) \end{bmatrix} \qquad (8)$$

where O is zero matrix, I is unit matrix.

Equation (8) and (9) can be used to predict the future output of the system, and the multistep ahead prediction equation is given in Section 3.3.

3.2 State estimation

Generally, for the identified state space model, (7) and (8), the state variables do not have physical meanings and, thus, cannot be measured directly. Therefore, linear discrete kalman filter (Brian D.O.A. & John B.M. 2005) is utilized here to estimate their value.

The main procedure of the algorithm is given as follows:

Prior state estimation:

$$x_{k|k-1} = A x_{k-1|k-1} + B \Delta u(k-1) \qquad (9)$$

Prior covariance matrix estimation:

$$P_{k|k-1} = A P_{k-1|k-1} A^T + Q_k \qquad (10)$$

Innovation:

$$\tilde{y} = y_m - C x_{k|k-1} \qquad (11)$$

Covariance innovation:

$$S_k = C P_{k|k-1} C^T + R_k \qquad (12)$$

Optimal kalman gain:

$$K = P_{k|k-1} C^T S_k^{-1} \qquad (13)$$

Updated state estimation:

$$x_{k|k} = x_{k|k-1} + K\tilde{y} \qquad (14)$$

Updated covariance estimation:

$$P_{k|k} = (I - KC) P_{k|k-1} \qquad (15)$$

where $x_{k|k}$ is the estimated state, Q_k is the covariance matrix of process noise, R_k is the covariance matrix of measurement noise, and y_m is the output measurement.

3.3 Control move calculation

Suppose the state estimated by kalman filter is $x_{k|k}$, then denote $y(k + m \mid k)$ ($m = 1, 2, ..., N_p$) as m step ahead prediction at time k. Multistep ahead prediction equation could be deduced by stacking up the model (8) and (9) for m steps.

The prediction equation matrix is:

$$Y = F x_{k|k}(k) + G \Delta U \qquad (16)$$

where

$$Y = \begin{bmatrix} y(k+1|k) & y(k+2|k) & ... & y(k+N_p|k) \end{bmatrix}^T$$
$$\Delta U = \begin{bmatrix} \Delta u(k) & \Delta u(k+1) & ... & \Delta u(k+N_c-1) \end{bmatrix}^T$$

$$F = \begin{bmatrix} CA \\ CA^2 \\ \vdots \\ CA^{N_p} \end{bmatrix}$$

$$G = \begin{bmatrix} CB & O & O & \cdots & O \\ CAB & CB & O & \cdots & O \\ CA^2B & CAB & CB & \cdots & O \\ \vdots & \vdots & \vdots & \vdots & \vdots \\ CA^{N_p-1}B & CA^{N_p-2}B & \cdots & \cdots & CA^{N_p-N_c}B \end{bmatrix},$$

and where N_p = prediction horizon and N_c = control horizon.

Control move can then be calculated by minimizing the quadratic objective function (17):

$$J = (R - Y)^T W^Y (R - Y) + \Delta U^T W^U \Delta U \qquad (17)$$

where

$$W^Y = \begin{bmatrix} W^y & O & \cdots & O \\ O & \ddots & \vdots & \vdots \\ \vdots & \cdots & W^y & O \\ O & \cdots & O & W^y \end{bmatrix}_{N_p \times N_p}$$

$$W^U = \begin{bmatrix} W^y & O & \cdots & O \\ O & W^y & \vdots & \vdots \\ \vdots & \cdots & \ddots & O \\ O & \cdots & O & W^y \end{bmatrix}_{N_c \times N_c}$$

$$W^y = \begin{bmatrix} r_{y1} & O & \cdots & O \\ O & r_{y2} & \vdots & \vdots \\ \vdots & \cdots & \ddots & O \\ O & \cdots & O & r_{yn} \end{bmatrix}$$

$$W^u = \begin{bmatrix} r_{u1} & O & \cdots & O \\ O & r_{u2} & \vdots & \vdots \\ \vdots & \cdots & \ddots & O \\ O & \cdots & O & r_{um} \end{bmatrix}$$

$(r_{y1}, r_{y2}, \ldots, r_{yn})$ are the output suppression factors and $(r_{u1}, r_{u2}, \ldots, r_{un})$ are the input suppression factors.

The analytical solution of (18) is:

$$\Delta U = (G^T G + W^U)^{-1} G^T W^Y (R - Fx_{k|k}(k)) \qquad (18)$$

3.4 *Auto-tuning of the output suppression factors*

Even though the MPC is effective, a set of unsuited parameters could lead to unsatisfactory control performance. Therefore, it is of great importance to study the tuning method of MPC controller. The MPC parameters to be tuned are prediction horizon, control horizon, and suppression factors. In general, the prediction horizon should be long enough to cover the major dynamics of the system. Control horizon is usually selected between 2~5 steps. The suppression factors are chosen by trial and error, which is a time consuming and unscientific exercise. For this reason, this paper presents a novel auto-tuning method to adjust the output suppression factors.

In Gous, G.Z. & de Vaal, P.L. (2012), manipulated variable overshot is used as a tuning metric to adjust the input suppression factors W^u, however, too much suppression on the manipulated variables leads to slower control move and longer transient time for the control system. To overcome this issue, this paper chooses the output variable overshot as a tuning metric.

Suppose the initial controller parameters are set as follows:

Sampling time: $T_s = 1$ s;
Prediction horizon: $N_p = 100$ s;
Control horizon: $N_c = 2$ s;
Output suppression factors: $r_\omega = r_{TH} = r_{TC} = 1$; and
Input suppression factors: $r_\sigma = r_\varphi = r_\mu = 1$.

Denote the setting point of ω, TC, TH as $R\omega$, RTC and RTH. The following experiment is then performed in a closed loop condition.

1. Step $R\omega$ and set RTC, RTH equal to zero. Then record the overshot δ of TC and TH. δ is defined as $\delta = \max|y - y_r|$;
2. Step RTC and RTH, respectively, and record the overshot of the other two outputs as described above.

During the experiment, we note that while stepping RTH, the values of δ for the other two outputs is much bigger than stepping $R\omega$ or RTC, which means the closed loop coupling effect of TH

is stronger than ω and TC. Therefore, if TH can have a smooth first order dynamic without overshot or oscillation, the dynamics of $R\omega$ and RTC are less disturbed by the strong coupling effect.

Therefore, we firstly determine r_{TH} through (19) and (20).

$$r_{TH} = 1 + \sigma \qquad (19)$$

$$\sigma = \begin{cases} 0 & \delta_{TH} < 0 \\ 2\delta_{TH} & 0 < \delta_{TH} < 50 \\ 100 & \delta_{TH} > 100 \end{cases} \qquad (20)$$

where $\delta_{TH} = 420|\Delta RTH| - 330|\Delta RTC|$ is a relationship extract during the experiment and ΔRTH and ΔRTC are the variations of RTH and RTC, respectively.

For the MGT-CCHP system used in this paper, the MPC control performance is already satisfactory after r_{TH} is tuned by (19) and (20), thus, we keep r_ω and r_{TC} equal to 1, which is their initial value.

4 SIMULATION RESULTS

In this section, the proposed auto-tuning MPC controller, fixed-parameter MPC controller and decoupling PI controller are compared through simulation, starting with a short introduction to the PI control system.

The decoupling PI controller is designed based on frequency domain methods (Zhang, J. 2015) and the schematic diagram of the PI control system is shown in Figure 2.

The parameter of PI 1 is:

$$K_1 = \begin{bmatrix} 5 & 0 \\ 0 & -0.5 \end{bmatrix} \begin{bmatrix} \dfrac{s+0.03}{s} & 0 \\ 0 & \dfrac{s+0.01}{s} \end{bmatrix}$$

$$\begin{bmatrix} 0.9299 & -0.8355 \\ -2.4814 & -7.848 \end{bmatrix}$$

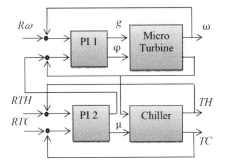

Figure 2. Diagram of decoupling PID control strategy (Zhang, J. 2015).

The parameter of PI 2 is:

$$K_2 = \begin{bmatrix} -0.3 & 0 \\ 0 & 5 \end{bmatrix} \begin{bmatrix} \dfrac{s+0.05}{s} & 0 \\ 0 & \dfrac{s+0.05}{s} \end{bmatrix} \begin{bmatrix} 1 & 1 \\ 1 & -1 \end{bmatrix}$$

Suppose that at initial time, the MGT-CCHP system works at 100% load operating point in steady state, then at $t = 100$ s, 300 s, and 1000 s, the set-points of rotation speed, cold water temperature, and hot water temperature drops by 10%, respectively. At 2000s, an immeasurable output disturbance $d = \{3.4 \times 103$ r/min; 0.35°C; 4°C$\}$, is added to the three outputs correspondingly.

Simulation results in Figures 3 and 4 show that the proposed auto-tuning MPC control has

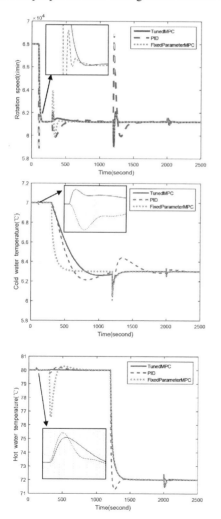

Figure 3. Performance of the MGT-CCHP system: Output variables.

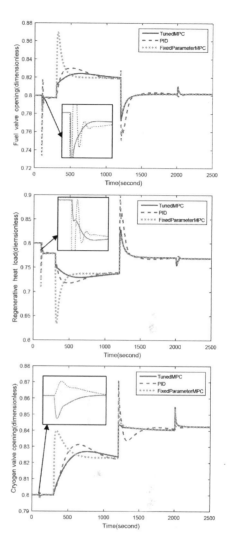

Figure 4. Performance of the MGT-CCHP system: Manipulated variables.

a much better control performance, which drives the outputs to the set points quickly and smoothly without any set-free. The PID controller can also track the references change, however, it has a repetitive oscillation and serious overshot, which may reduce the efficiency and threaten the safety of the plant.

For the fixed parameter MPC, we set $r_\omega = r_{TH} = r_{TC} = 1$, and $r_\sigma = r_\varphi = r_\mu = 1$, which could eliminate the overshot of ω when $R\omega$ steps. However, because the parameters are fixed, when RTC or RTH step-up, control performance degrades seriously. The simulation results also show that owing to the advantages of incremental model offset-free tracking of the output-variables can be achieved for the both MPCs, even in the case of strong unknown disturbances.

5 CONCLUSION

In order to solve the control problems of MGT-CCHP system, an auto-tuning MPC control strategy is proposed in this paper using incremental state-space model. The auto-tuning approach is devised for the output suppression factors to further improve the performance of MPC and reduce the complexity of manual parameters setting. The simulation results show that a satisfactory off-set free tracking control of the MGT-CCHP system can be achieved by the proposed MPC.

REFERENCES

Anvari, S., Taghavifar, H., Saray, R.K., Khalilarya, S. & Jafarmadar, S. 2015. Implementation of ANN on CCHP system to predict trigeneration performance with consideration of various operative factors. *Energy Conversion and Management* 101:503–514.

Arashnia, I., Najafi, G., Ghobadian, B., Yusaf, T. & Mamat R, Kettner M. 2015. Development of Microscale Biomass-fuelled CHP System Using Stirling Engine. *Energy Procedia* 75:1108–1113.

Brian D.O.A. & John B.M. 2005. *Optimal filtering*. Dover Publications Inc: New York.

Dong, L., Liu, H. & Riffat, S. 2009. Development of small-scale and micro-scale biomass-fuelled CHP systems: A literature review. *Applied Thermal Engineering* 29(11–12):2119–2126.

Gao, P., Li, W., Cheng, Y., Tong, Y. Dai, Y. & Wang, R. 2014. Thermodynamic performance assessment of CCHP system driven by different composition gas. *Applied Energy* 136:599–610.

Gous, G.Z. & de Vaal, P.L. 2012. Using MV overshoot as a tuning metric in choosing DMC move suppression values. *ISA Transactions* 51(5):657–664.

Ljung, L. 1998. *System identification: theory the user.* Prentice Hall: Upper Saddle River.

Kong, X., Wang, R., Li, Y. & Wu, J. 2010. Performance research of a micro-CCHP system with adsorption chiller. *Journal of Shanghai Jiaotong University (Science)* 15(6):671–675.

Rey, G., Ulloa, C., Cacabelos, A. & Barragáns, B. 2015. Performance analysis, model development and validation with experimental data of an ICE-based micro-CCHP system. *Applied Thermal Engineering* 76:233–244.

Wu, D.W. & Wang, R.Z. 2006. Combined cooling, heating and power: A review. *Progress in Energy and Combustion Science* 32(5–6):459–495.

Wang, J.L., Wu, J.Y. & Zheng, C.Y. 2014. Simulation and evaluation of a CCHP system with exhaust gas deep-recovery and thermoelectric generator. *Energy Conversion and Management* 86:992–1000.

Xu, J., Sui, J., Li, B. & Yang, M. 2010. Research, development and the prospect of combined cooling, heating, and power systems. *Energy* 35(11):4361–4367.

Yang, J. 2009. Integration and performance simulation study of combined cooling heating and power system based on micro turbine: Shanghai Jiao Tong University.

Zhang, J. 2015. Integrated optimization and dynamic characteristic research on micro-turbine-based cooling, heating and power system: Southeast University.

Advances in Power and Energy Engineering – Sun (Ed.)
© 2016 Taylor & Francis Group, London, ISBN 978-1-138-02846-3

Research on the master-slave control method for the multi-inverters of the micro-grid

S.F. Wen & S.T. Wang

Faculty of Electric Power, Inner Mongolia University of Technology, Hohhot, China

ABSTRACT: Micro-grid is a technology to solve distributed generations that are connecting the utility grid. This paper presents a master-slave control method for multi-inverters of micro-grid. In the grid-connected mode, all the DGs adopt PQ control to export reference active and reactive power. In the island mode, a DG is set to be the master source and run in the V/f mode to provide a stable voltage and frequency. To realize the smooth transition between the grid-connected mode and island mode, the current reference compensation algorithm is proposed. The simulation results show that this proposed control strategy can ensure a power balance and keep the bus voltage and frequency stable. Meanwhile, a smooth transition during micro-grid mode switching is realized.

1 INTRODUCTION

With the increasing consumption of traditional coal, oil, gas, and other non-renewable energy, the energy crisis and environmental problem is getting worse. Distributed Generation (DG) technologies using clean and renewable energy in power supply systems have been widely used. In order to make full use of their advantage and reduce the impact of large-scale distributed generations on a power system, CERTS (Consortium for Electric Reliability Technology Solutions) proposed the concept of micro-grid (CHEN Jie et al. 2014). The inverters play a vital role in micro-grid system because most of DGs are interfaced by the micro-grids, and so micro-grid control is the main control of the inverters. Currently, the control strategies of inverter include the active power-reactive power (PQ) control, voltage/frequency (V/f) control, and Droop control (P. Kanakasabapathy et al. 2014).

Micro-grid operates in two typical modes: grid-connected mode and island mode (WANG Chen-shan et al. 2012). In normal circumstances, the micro-grid operates in grid-connected mode and absorbs power from the utility grid or injects energy to the utility grid if necessary. When the utility grid has failures, micro-grid disconnects automatically from utility grid quickly and operates in the island mode in order to provide an uninterruptible power supply for the important local load. The coordination control methods of micro-grid can be mainly divided into master-slave control and peer-to-peer control (QIU Lin et al. 2014). All DGs in micro-grid adopt droop control under peer-to-peer control, which could ensure reasonable power assignment among DGs, but it could not ensure stable frequency and voltage. Besides, the prerequisite of droop control is that the line impedance is inductive, while the line impedance is resistive in the low-voltage micro-grid, so the master-slave control strategy has been widely adopted in the micro-grid system (CHEN Xin et al. 2013). That is, all DGs in micro-grid adopt PQ control in grid-connected mode, and one of DGs is changed to V/f control mode in the islanding mode in order to keep voltage and frequency stable. In this process, the smooth switching between grid-connected modes and islanding mode is important to micro-grid stable operations. Aiming at this problem, domestic and foreign scholars have been widely studied. WU Meirong et al. (2015) designed a constrained second Order Sliding Mode (SOSM) control scheme. Michele Cucuxxella et al. (2015) proposed a logical switch control and phase-locked loop control method. WU Shunyu et al. (2014) increased two PI controllers for current-loop. AI Xin et al. (2015) adopted the state following method.

In this paper, the master-slave control method is adopted. To minimize the transient current and voltage associated with the micro-grid operation mode transitions, a current compensation method has been developed. The simulation results demonstrate the efficiency and reliability of the proposed method.

2 MASTER-SLAVE CONTROLLED MICRO-GRID STRUCTURE

An AC micro-grid structure is used in this paper, that is, all DGs are connected to an AC bus by using three phase voltage source inverters. The master-slave

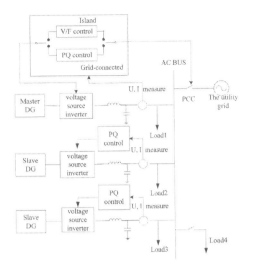

Figure 1. Schematic diagram of the micro-grid.

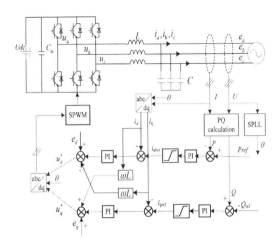

Figure 2. PQ control strategy diagram.

controlled micro-grid structure is shown in Figure 1. All DGs are assumed to be of dc voltage source. The micro-grid is transferred between two modes by using a static switch at PCC. When the switch is closed, the micro-grid operates in the grid-connected mode and all the DGs adopt PQ control. When the power generated by the DGs is greater than the load demand, the utility grid can also absorb power from DGs. When the switch is open, the island mode of the micro-grid is applied. For a master-slave controlled system, a DG is set to be the master source and it runs in the V/f mode to provide a reference voltage and frequency, however, other DGs are slave sources, and still run in the PQ model to provide a constant output. In this transformation, the master DG needs to be transferred between PQ control and V/f control. So the transition control methods mainly aim at the control of the master DG.

3 CONTROL STRATEGIES FOR DG INVERTERS

As discussed above, when using the master-slave control mode, the control strategies of DG inverters are PQ control and V/f control.

3.1 PQ control

In the grid-connected mode, the load power fluctuations, frequency, and voltage disturbances are supported by the utility grid. PQ control means that DG is controlled to give maximum output of power or specified power according to actual condition. The topology and PQ control strategy of DG inverter is shown in Figure 2. As shown in the figure, the whole control strategy can be split into two loops: the inner current loop and the outer power loop (XU shaohua & LI Jianlin 2013).

The outer power loop uses the grid voltage and grid-connected current. The power measured value is obtained by PQ calculation block, and then it is compared with the reference value P_{ref}, Q_{ref} respectively. The error signals are applied to PI controllers in order to obtain i_{dref} and i_{qref} of the inner current loop. The grid-connected current is measured and transferred to dq rotating frame in order to get i_d and i_q. The i_d and i_q are compared with reference quantities i_{dref} and i_{qref} to obtain error signals. The errors are also applied to PI controllers of the inner loop.

Neglecting the influence of filer capacitors, the current differential equation of the inverter model can be expressed as below:

$$\begin{cases} u_d^* = e_d - \omega L i_q + L\dfrac{di_d}{dt} \\ u_q^* = e_q + \omega L i_d + L\dfrac{di_q}{dt} \end{cases} \tag{1}$$

Then the voltage equation with PI controller can be written as below:

$$\begin{cases} u_d^* = e_d - \omega L i_q + (k_p + \dfrac{k_i}{s})(i_{dref} - i_d) \\ u_q^* = e_q - \omega L i_d + (k_p + \dfrac{k_i}{s})(i_{qref} - i_q) \end{cases} \tag{2}$$

The outputs of the inner current loop consists of PI controller outputs, the decoupling components of current and voltage feed forward compensation parts to the utility grid. The variables u_d^* and u_q^* are inversely transformed to the three phase abc frame and given to the pulse generator to generate

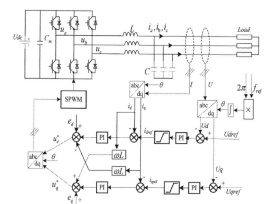

Figure 3. V/f control strategy diagram.

pulse for the inverter through the SPWM method. The reference phase θ of coordinate transformation is got by the SPLL (software phase locked loop). The PQ control strategy can ensure that the inverter output is corresponding power according to the power setpoint.

3.2 V/f control

When the micro-grid operates in the island mode, the master inverter needs to be transferred to V/f control in order to provide a strong voltage and frequency stability for the local load, and will adjust its output power according to the load demand. The topology and V/f control strategy of DG inverter is as shown in Figure 3. As shown in the figure, the whole control strategy can be split into two loops: the inner current loop and the outer voltage loop.

The outer voltage loop uses the load voltage, and the d and q axis components (U_d, U_q) are obtained by Park coordinate transformation. The U_d and U_q are compared with the reference value U_{dref}, U_{qref} respectively, and the error signals are applied to PI controllers. The outputs of the voltage loop control are used as the inner current references, that is, i_{dref} and i_{qref}. And the current control detail is the same as the control technique of PQ control mentioned above and is not described here. However, the required phase angle θ is obtained by f_{ref}. The f_{ref} is the internal assigned frequency reference of the micro-grid operating in island mode.

4 SMOOTH TRANSITION CONTROL METHOD

During the transition process from grid-connected mode to island mode, the master DG control strategy is switched between PQ control

and V/f control. As discussed above, the two control modes have the same current loop controls and different outer control loops. Due to the different control objectives of the outer control loop, the outputs are not synchronous, and a sudden change of the inner current references exists, and result in a great transient oscillation. To reduce transient fluctuation, a smooth transferring control method based on the current reference compensation is proposed.

In Figure 1, when micro-grid is grid-connecting, all DGs are operated under PQ control mode at the beginning. At this stage, power control system works to maintain the power to the reference value. The reference active power of all DGs is assumed to be P_0, the total active power for the load consumption is P_L, and the utility grid active power is P_g (P_g is positive when the grid absorb power, on the contrary, P_g is negative). Considering three parallel inverters neglecting the power loss, the relationship of P_0, P_L and P_g can be expressed as:

$$3P_0 = P_L + P_g \tag{3}$$

And then the power of the master DG can be expressed as:

$$P_0 = P_L + P_g - 2P_0 \tag{4}$$

The current reference value generated from PI controller of the outer power control loop can be calculated based on power, as seen below in the steady-state (take d-axis as an example):

$$i_{dref-PQ} = k \times P_0 \tag{5}$$

where k = scaling factor.

When the master inverter is changed from PQ control mode to V/f control mode, the utility grid active power P_g changes into 0, so the power value of the master DG can be expressed as:

$$P_1 = P_L - 2P_0 \tag{6}$$

If the slave inverters and loads are to maintain the original power value in this transformation process, the power of master inverter is also expressed as below, according to power equilibrium:

$$P_1 = P_0 - P_g \tag{7}$$

The current reference value generated from PI controller of the outer voltage control loop is calculated based on the power in the steady-state:

$$i_{dref-Vf} = k \times P_1 = k \times (P_0 - P_g) \neq i_{dref-PQ} \tag{8}$$

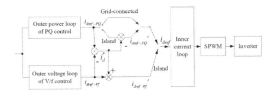

Figure 4. Structure of smooth transition control.

The inconsistency between the two references causes a step change and an undesirable transient after the operation mode transition. In order to remove the transient, the reference current compensation i_d^* is adopted. The current reference can be compensated in the following equation:

$$i_{dref-V'f} = i_{dref-Vf} + i_d^* = i_{dref-PQ} \quad (9)$$

where

$$i_d^* = i_{dref-PQ} - i_{dref-Vf} \quad (10)$$

Therefore, the V/f controller still follows the output state of PQ controller to ensure a smooth operation mode transition. The smooth transition control method is shown in Figure 4.

When the micro-grid reconnects to the utility, the master inverter goes through a transition from V/f control to PQ control. For the same reason, an undesired transient occurs. In order to ensure that the current references are consistent in the switching moment from V/f control to PQ control, the current reference of PQ control adds a compensation value during the island operation:

$$i_{dref-PQ'} = i_{dref-PQ} - i_d^* = i_{dref-Vf} \quad (11)$$

Once the smooth transition is obtained, the set-point of inner current loop can be adjusted automatically according to the outer loop.

5 SIMULATION RESULT AND ANALYSIS

In order to verify the effectiveness and feasibility of the proposed method, a micro-grid system (Fig. 1) is built based on MATLAB/SIMULINK. The main simulation parameters of the system are summarized in Table 1.

5.1 Simulation of grid-connected/island mode

At the beginning, the micro-grid operates in the grid-connected mode, both the master and the slave DGs adopt PQ control. At $t = 0.6$s, the micro-grid gets transferred to the island mode and reconnects to the utility grid at $t = 1.2$s. Figure 5 shows the

results for the transition between the grid-connected and the island mode.

In Figure 5, both the master and the slave DGs inverters export 100 kW during the grid-connected

Table 1. Parameters of the micro-grid system.

Parameters	Value
Rating power of DG	100 kW
DC-Link voltage	800 V
Utility grid voltage (phase to phase)	380 V
Utility grid frequency	50 Hz
Load 1, Load 2, Load 3	80 kW, 80 kW, 80 kW
Load 4	30 kW
Inverter switching frequency	20 kHz
Inverter filter inductance	3.5 mH
Inverter filter capacitance	20 μF
Current loop K_p, K_i	0.5, 20
Voltage loop K_p, K_i	50, 500
Power loop K_p, K_i	5, 50

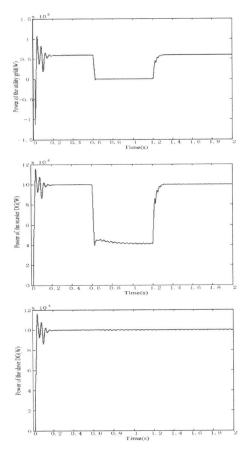

Figure 5. Continued

44

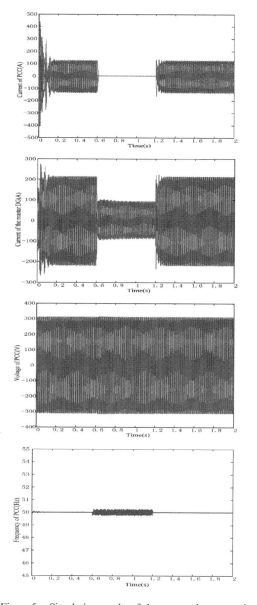

When the micro-grid reconnects to the utility grid at $t = 1.2$s, all state variables increase to its previous value. During the whole process, the AC bus voltage and frequency of the micro-grid system gets maintained constantly.

5.2 Simulation of throwing/cutting off load

During the island mode, a 30 kW load is introduced into the system at $t = 1$s and is taken out at $t = 1.5$s. The power and current change of the master DG for the additional load are shown in Figure 6. The output power of the master DG changes to 70 kW at 1–1.5s.

From the simulation results, a smooth transition is realized by using the proposed compensation control algorithm. The current of the master DG does not have any inrush, and the AC bus voltage and micro-grid frequency is kept stable.

Figure 5. Simulation results of the master-slave control.

mode, so the total micro-grid power is 300 kW. The load consumption within the micro-grid is 240 kW (Load4 is not switched on), so the utility grid absorbs 60 kW from the micro-grid. After getting transferred to island mode at $t = 0.6$s, the power and current of the utility grid decrease to 0 because there is no power exchange between the utility grid and micro-grid. The output power of the two slave DGs still stays at 100 kW, so the master DG adjusts its output power from 100 kW to 40 kW according to the load demand, the current of the master inverter decreases accordingly.

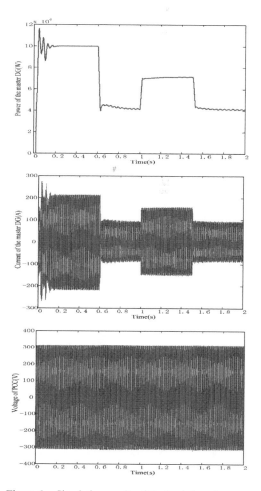

Figure 6. Simulation results of the load changing.

6 CONCLUSION

This paper studies the master-slave control method for multi-inverters of micro-grid and designs the control strategies of DG inverters including PQ control and V/f control. Aiming at the transition problem between the grid-connected mode and the island mode, a simple and effective approach based on the current reference compensation is proposed. Finally, the simulation model of the micro-grid system is built under MATLAB/Simulinik in order to test the control performance and the transition control method. The simulation results in different conditions show that this proposed control strategy can ensure power balance and reasonable power assignment, and keep the AC bus voltage and micro-grid frequency connected and stable during the switching of the micro-grid mode and load changing. Meanwhile, the stable operation and smooth transition of the micro-grid systems are guaranteed.

REFERENCES

Ai Xin, Deng Yuhui & Li Jingying 2015. Master-slave control strategy for distributed generation microgrid. *Journal of North China Electric Power University*, 42(1):1–6.

Chen Jie, Chen Xin, Feng Zhiyang, Gong Chunying & Yan Yangguang 2014. A control strategy of seamless transfer between grid-connected and islanding operation for microgrid. *Proceedings of the CSEE*, 34(19):3089–3097.

Chen Xin, Wang Yanhong & Wang Yuncheng 2013. A novel seamless transferring control method for microgrid based on master-slave configuration. *IEEE Energy Conversion Congress and Exposition(ECCE), Melbourne, Australia, 3–6 June 2013*, pp.351–357.

Michele Cucuzzella, Gian Paolo Incremona & Antonella Ferrara 2015. Master-slave second order sliding mode control for microgrids, *American Control Conference, Chicago, USA, 1–3 July 2015*, pp.5188–5193.

P.Kanakasabapathy & Vishnu Vardhan Rao.I 2014. Control strategy for inverter based micro-grid. *Power and Energy Systems: Towards Sustainable Energy(PESTSE), Bangalore, India, 13–15 March 2014*, pp.1–6.

Qiu Lin, Xu Lie, Zheng Zedong, Li Yongdong & Zheng Zhixue 2014. Control method of microgrid seamless switching. *Transactions of China Electrotechnical Society*, 29(2):171–176.

Wang Chenshan, Li Xialin, Guo LI & Li Yunwei 2012. A seamless operation mode transition control strategy for a microgrid based on master-slave control, *SCIENCE CHINA Technological Sciences*, 55(6):1644–1654.

Wu Meirong, Tao Shun & Xiao Xiangning 2015. Study on control method of smooth switchover for micro-grid based on master-slave control. *Modern Electric Power*, 32(1):1–7.

Wu Shunyu, Liu Shirong & Chen Xueting 2014. An intentionally seamless transfer strategy between grid-connected and islanding operation in micro-grid. *the 17th International Conference on Electrical Machines and Systems, Hangzhou,China, 22–25 October 2014*, pp.302–307.

Xu Shaohua & Li Jianlin 2013. Grid-connected/island operation control strategy for photovoltaic/battery micro-grid. *Proceedings of CSEE*, 33(34):25–33.

Advances in Power and Energy Engineering – Sun (Ed.)
© 2016 Taylor & Francis Group, London, ISBN 978-1-138-02846-3

Coordinated economic dispatch of interconnected microgrids based on rolling optimization

Y. Du
Institute of Electrical Engineering, Chinese Academy of Sciences, Beijing, China
University of Chinese Academy of Sciences, Beijing, China

W. Pei
Institute of Electrical Engineering, Chinese Academy of Sciences, Beijing, China

ABSTRACT: In this paper, a coordinated economic dispatch model of interconnected microgrids with multiple distributed energy resources is proposed, which allows for power exchange among microgrids to realize energy complementation and cascade utilization. In order to take into account the negative effects of uncertain surroundings, a rolling optimization strategy is further put forward to dynamically modulate real-time microgrid operation. The proposed model is tested through simulated scenarios and is further compared with the case of single microgrid operation. Numerical results indicate that both scheduling flexibility and systematic economy are enhanced to a great extent via microgrid interconnection.

1 INTRODUCTION

World-wide energy and environmental crisis has tremendously contributed to the integration of various Distributed Energy Resources (DERs) into the current power system at a distribution network level, for the sake of energy efficiency and economy. Microgrid has long and widely been acknowledged as the most effective way to aggregate all kinds of DERs into the utility grid to support system operation (Wang & Li, 2010). Fruitful research achievements have been gained in literature regarding control strategies (Olivares et al., 2014), modeling (Xu et al., 2015), planning and scheduling (Farzan et al., 2015) of multi-energy microgrid system.

Most recently, within the ongoing trend of energy internet (Rifkin, 2011) and the grand paradigm of smart grid, interconnection of microgrids with diverse Distributed Generators (DGs), Renewable Energy Sources (RES), and flexible demand has become the new heated topic in academic field, since networking microgrids has been proved to be favorable to system operation and reliability (Fathi & Bevrani, 2013) by realizing energy complementation and cascade utilization. Mainstream focus falls on intelligent energy management of microgrid cluster: Wang et al. (2015a,b) explored both coordinated and distributed manner of energy management of networked microgrids in scenarios with stochastic RES generation, and gave a detailed analysis of interaction between microgrids and the upper-level distribution network. However, the performance of networked microgrids was studied for only one time interval, and inter-temporal constraints and continuous operation were not fully modelled. Ouammi et al. (2015) presented a model predictive-based power flow control in a cluster of smart microgrids to maximize global profits. Nevertheless, only renewable energy and storage units were considered in the microgrid system, while more complicated distributed generators (i.e. micro turbines) and dispatchable load (i.e. deferrable and curtailable load) were not thoroughly described.

It's based on the past gains and insufficiencies that this paper brings out a coordinated economic dispatch model of microgrids in real-time environment to excavate the potential benefits of microgrid interconnection. The microgrid system encompasses diversified energy sources to meet both electric and thermal demand. Elastic residential load is also enclosed as demand response source to promote economic operation. Furthermore, rolling optimization strategy is adopted over the whole dispatch horizon to strengthen the robustness of microgrid schedule against real-time uncertainties.

The rest of this paper is organized as follows. Section II illustrates the mathematical model of interconnected microgrids. Section III outlines the concept and procedure of rolling optimization. Simulation results and numerical analysis are demonstrated in Section IV. Finally, Section V concludes the paper.

2 SYSTEM MODELING

2.1 *Composition of multi-energy microgrid system*

According to real-time scenarios, three types of microgrids are discussed in this paper:

1. residential area: consists of photovoltaic (PV) solar panel, wind turbines, Energy Storage System (ESS), dispatchable/fixed electric load and fixed thermal load;
2. commercial area: consists of PV solar panel, wind turbines, energy storage system, and fixed electric load;
3. industrial area: consists of CHPs, thermal energy storage, boilers and thermal load.

Figure 1 shows the interconnection of microgrids and their link with the upper-level distribution network.

The typical load profiles of residential, commercial and industrial areas are depicted in Figure 2 and Figure 3. It's noticeable that the peaks and valleys of the load profiles of different microgrids locate at different time intervals. Hence possible power exchange among the three interconnected areas could happen during those staggered peak and valley periods to realize energy complementation and increase operation economy at both individual and global scale.

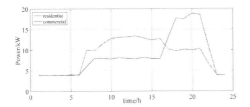

Figure 1. Network configuration of interconnected microgrids.

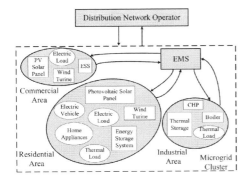

Figure 2. Electric load profiles.

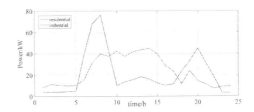

Figure 3. Thermal load profiles.

2.2 *Mathematical formulation*

1. Objective function:

At time step i, the EMS of the interconnected microgrids calculates the following objective to figure out the optimal power schedule of each microgrid member:

$$
\min c(i) = \sum_{t=i}^{i+N-1} \lambda_{RT}(t) \cdot P_{grid,M}(t)
$$
$$
+ \sum_{m \in M} \left(\sum_{t=i}^{i+N-1} \sum_{c=1}^{N_{CHP}} C_{chp}(P_{c,m}(t)) \right.
$$
$$
+ \sum_{t=i}^{i+N-1} \sum_{b=1}^{N_B} C_b(P_{b,m}(t))
$$
$$
\left. + \sum_{t=i}^{i+N-1} \sum_{e=1}^{N_{EV}} \beta_{EV} \cdot d_{e,m}(t) \right) \qquad (1)
$$

In equation (1), the first term is the cost of power exchange with the utility grid, where $\lambda_{RT}(t)$ = real-time price at interval t; $P_{grid,M}(t)$ = the amount of power exchange; N = the length of dispatch horizon; and M = microgrid cluster. The next three terms are generation cost of CHPs, boilers and compensation for V2G discharge, where β_{EV} = discharge price of EV. The cost of RES generators is assumed to be zero. Generation costs of CHPs and boiler are calculated as follows:

$$
C_{chp}(P_{c,m}(t)) = C_C \cdot \frac{P_{c,m}(t)}{\eta_{chp}} \qquad (2)
$$

$$
C_b(P_{b,m}(t)) = C_B \cdot \frac{P_{b,m}(t)}{\eta_b} \qquad (3)
$$

where C_C = fuel cost of CHP unit; $P_{c,m}(t)$ = CHP generation; η_{chp} = CHP efficiency; C_B = fuel cost of boiler; $P_{b,m}(t)$ = boiler generation; and η_b = boiler efficiency.

2. Operation constraints:

1. CHP:

$$
-R_c \le P_{c,m}(t) - P_{c,m}(t-1) \le R_c \qquad (4)
$$

$$P_c^{\min} \le P_{c,m}(t) \le P_c^{\max}$$
$$\forall m \in M, \ t = i, ..., i+N-1 \qquad (5)$$

Equation (4) is the ramping limit of CHP unit, equation (5) puts upper and lower boundary on CHP generation.

2. Boiler:

Boiler generation is constrained by equation (6):

$$P_b^{\min} \le P_{b,m}(t) \le P_b^{\max}$$
$$\forall m \in M, \ t = i, ..., i+N-1 \qquad (6)$$

3. Thermal storage:

$$S_{th,m}(t) = S_{th,m}(t-1) + \Delta t \left(P_{th,m}^c(t)\eta_{th} - P_{th,m}^d(t)/\eta_{th} \right) \qquad (7)$$

$$P_{th,m}^c(t) \le P_{th}^{c\,\max}, \ P_{th,m}^d(t) \le P_{th}^{d\,\max} \qquad (8)$$

$$S_{th,\min} \le S_{th,m}(t) \le S_{th,\max} \qquad (9)$$

$$S_{th,m}(N) = S_{th,m}(0) \qquad (10)$$

Equation (7) is the energy conservation constraint of thermal storage, where $S_{th,m}(t)$ = energy level of thermal storage; $P_{th,m}^c(t)$ = the charging thermal power; $P_{th,m}^d(t)$ = the discharging thermal power; η_{th} = charge/discharge efficiency; and Δt = the length of time interval. Equations (8)–(9) limit charging/discharging rate of thermal storage, and the stored energy level. Equation (10) implies that energy conservation returns to the initial level at the end of the dispatch horizon.

4. Energy storage system:

$$SOC_m(t) = \eta_{ess}SOC_m(t-1) + \Delta t P_{ess,m}(t) \qquad (11)$$

$$P_{ess}^{\min} \le P_{ess,m}(t) \le P_{ess}^{\max} \qquad (12)$$

$$SOC_{\min} \le SOC_m(t) \le SOC_{\max} \qquad (13)$$

$$SOC_m(N) \ge SOC_m(0) \qquad (14)$$

Similar to thermal storage constraints, energy storage system is also limited by energy conservation constraint, charging/discharging rate, storage capacity and initial state of charge, as is shown in equations (11)–(14), where $SOC_m(t)$ = the stored energy level; $P_{ess,m}(t)$ = the charging/discharging electric power; and η_{ess} = charge/discharge efficiency.

5. Flexible load:

The flexible load discussed in this paper mainly refers to home appliances in residential area. Appliances are categorized into three types according to their operation modes (Tushar et al., 2015):

a. *hard load*: some home appliances have strict scheduling requirement, and remain operational for a certain amount of time:

$$\sum_{h=t}^{t+T_m^I-1} P_{L,m}^I(h) = L_m^I su_m^I(t) \qquad (15)$$

$$P_{L,m}^I(h) = u_m^I(h)X_m^I, \ T_m^I = L_m^I/X_m^I$$

$$\sum_{t\in\chi I} su_m^I(t) = 1 \qquad (16)$$

$$u_m^I(h), \ su_m^I(t) \in \{0,1\}$$

The continuous operation of the hard load is defined by equation (15), where $P_{L,m}^I(h)$ = the actual power of the load; L_m^I = total energy consumption; $su_m^I(t)$ = start-up index; $u_m^I(h)$ = on/off status index; and X_m^I = the rated power.

(16) limits the start-up times, where χ^I = the allowed operation interval of the hard load.

b. *soft load*: within the given time slot, the appliance could be flexibly scheduled, as long as the total consumption reaches the preset amount:

$$\sum_{t\in\chi^{II}} P_{L,m}^{II}(t) = L_m^{II} \qquad (17)$$

$$P_{L,m}^{II}(t) = u_m^{II}(t)X_m^{II}, \ t \in \chi^{II}, \ u_m^{II}(t) \in \{0,1\} \qquad (18)$$

where $P_{L,m}^{II}(t)$ = the actual power of the soft load; L_m^{II} = total energy consumption; $u_m^{II}(t)$ = on/off status index; X_m^{II} = the rated power; and χ^{II} = the allowed operation interval of the soft load.

c. *electric vehicle (EV)*: EVs are categorized as the third type of appliance, for they have a stochastic amount of operation periods with randomly distributed arrival/departure time:

$$c_{EV,m}(t) \le \gamma_m^{III}(t)X_{m,\max}^{III}, \ d_{EV,m}(t) \le u_m^{III}(t)X_{m,\max}^{III} \qquad (19)$$

$$L_{m,\min}^{III} \le L_{m,init}^{III} + c_{EV,m}(t)\eta_c - d_{EV,m}(t)\eta_d \le L_{m,\max}^{III} \qquad (20)$$

$$L_{m,init}^{III} + \sum_{t=t_m^a}^{t_m^d} (c_{EV,m}(t)\eta_c - d_{EV,m}(t)\eta_d) \ge L_m^{III} \qquad (21)$$

$$u_m^{III}(t) + \gamma_m^{III}(t) < 2 \qquad (22)$$

$$u_m^{III}(t), \ \gamma_m^{III}(t) \in \{0,1\}, \ t \in \left[t_m^a, t_m^d\right]$$
$$u_m^{III}(t), \ \gamma_m^{III}(t) = 0, \ t \notin \left[t_m^a, t_m^d\right] \qquad (23)$$

Equations (19)–(20) limit charging/discharging rate of electric vehicle, and EV energy level, where $c_{EV,m}(t)$ = charging power; γ_m^{III} = charge index; $d_{EV,m}(t)$ = discharging power; μ_m^{III} =

discharge index; $X_{m,max}^{III}$ = the maximum charge/discharge rate; η_c = charge efficiency; η_d = discharge efficiency; $L_{m,init}^{III}$ = the initial energy level of EV; $L_{m,min}^{III}$ = the minimum energy level of EV; and $L_{m,max}^{III}$ = EV capacity.

Equation (21) indicates that EV must be charged to a target energy level before leaving, where L_m^{III} = the target energy level; t_m^a = the arrival time of EV; and t_m^d = the departure time of EV.

Equation (22) implies that EV cannot be charged and discharged at the same time. Equation (23) stipulates that V2G or G2V is only possible after EV arrives at home or before EV departs.

6. Power balance constraint:

$$
\sum_{m \in M} \left(P_{wnd,m}(t) + P_{pv,m}(t) + P_{c,m}(t) + P_{grid,m}(t) + d_{EV,m} \right)
$$
$$
- \sum_{m \in M} \left(P_{ess,m}(t) + P_{L,m}^{I}(t) + P_{L,m}^{II}(t) + c_{EV,m} \right)
$$
$$
- \sum_{m \in M} \left(P_{rload,m}(t) + P_{cload,m}(t) \right) = 0
$$

(24)

where $P_{wnd,m}(t)$ = wind power generation; $P_{pv,m}(t)$ = the electric power produced by PV solar panel; $P_{rload}(t)$ = residential electric load; and $P_{cload}(t)$ = commercial electric load.

7. Thermal balance constraint:

$$
\sum_{m \in M} \left(P_{tsp,m}(t) + P_{th,m}^{d}(t) + r_c\, P_{c,m}(t) + P_{b,m}(t) \right)
$$
$$
- \sum_{m \in M} \left(P_{th,m}^{c}(t) + P_{rload,m}^{t}(t) + P_{iload,m}^{t}(t) \right) = 0 \quad (25)
$$

where $P_{tsp,m}(t)$ = the thermal power produced by PV solar panel; r_c = heat/power ratio of CHP; $P_{rload,m}^{t}(t)$ = the residential thermal load; and $P_{iload,m}^{t}(t)$ = industrial thermal load.

3 ROLLING OPTIMIZATION STRATEGY

As can be seen from literature, microgrid schedule is usually carried out one day ahead based on the prediction of future wind speed, solar radiation, consumer demand and other uncertain factors. Distributed energy resources within the microgrid operate according to the day-ahead schedule in the next day. Given that this schedule relies highly on forecast precision, while current short-term prediction measures of renewable energy sources are still below expectation, a rolling optimization strategy is introduced in this paper to optimize real-time operation of interconnected microgrids, which can be described as the following steps:

1. At time interval i, sample the current weather condition, system demand, price signals and their forecasted information over the next $N-1$ intervals. In this case, N is set to be 24 hours and the length of each time interval is 1 hour;
2. Based on the sampled information, microgrid EMS decides the optimal set points of all units over the operation window [i, $i+N-1$];
3. The set point of the current interval is implemented, and system status is updated accordingly; the solutions for other time intervals are discarded;
4. Move to the next time interval and repeat the above steps until reaching the end of dispatch horizon.

A rolling optimization strategy makes generation plans for the future 24 hours in a row, and allows for timely revision of the plan by constantly recalculating the optimal schedule according to real-time conditions, hence compensates for power deviations caused by forecast errors of uncertainties and achieves robust optimization.

4 CASE STUDY

4.1 Simulation setup

In this paper, interconnection of three microgrids and their coordinated economic dispatch are simulated for 24 hours based on real-time scenarios. Parameters of DER units and flexible load are listed in Table 1.

Residential area is assumed to be composed of 20 households, all of which own the above home appliance I and II. It's further assumed that there are 20 EVs in the residential area, and their initial states are randomly generated between 20%–50% of the total capacity. The arrival and departure time of EVs are assumed to follow truncated Gaussian distribution (Vagropoulos & Bakirtzis, 2013), with a mean value of 19h and 7h, respectively, and a standard deviation of 2 hours. Real-time weather data and price information are taken from Jager & Andreas (1996) and PJM website (2015). The forecast error of wind speed is assumed to follow a normal distribution with zero mean and the standard deviation of 0.05. The model of photovoltaic solar panel is learnt from Zakharchenko et al. (2004). Electric and thermal load data is taken from Zhang et al. (2013).

4.2 Numerical analysis

Simulation of real-time economic dispatch of interconnected microgrids is carried out for a dispatch horizon of 24 hours, and is further compared with the case of single microgrid operation.

Table 1. Microgrid parameters.

Micro turbine

No.	P_c^{max} (kW)	η_{chp} (%)	Heat/power ratio	R_c (kW)	C_C ($/MWh)	$P_c(0)$ (kW)
1	20	29	2.1	10	42	0
2	20	29	2.1	10		

Boiler

No.	P_b^{max} (kW)	η_b (%)	C_B ($/MWh)
1	24	85	42

Thermal storage

$S_{th,max}$ (kWh)	$S_{th,min}$ (kWh)	η_{th}(%)	$P_{th}^{cmax}/P_{th}^{cmax}$ (kW)	$S_{th}(0)$ (kWh)
200	40	60	200	100

Energy storage system

SOC$_{max}$ (kWh)	SOC$_{min}$ (kWh)	P_{ess}^{max} (kW)	P_{ess}^{min} (kW)	η_{ess}	Δt	SOC(0) (kWh)
100	10	50	−50	0.95	1	50

Home appliance I				Home appliance II		Electric vehicle	
Type	Electric oven	Light bulb	TV	Type	Washing machine	$L_{max\,x}^{III}$ (kWh)	24
X^I (kW)	2	0.5	0.12	X^{II} (kW)	0.8	X_{max}^{III} (kW)	3
T^I (h)	2	4	3	T^{II} (h)	2	η_c/η_d	0.85
χ^I (h)	8–22	18–24	18–24	χ^{II} (h)	8–22	β_{EV} ($/kWh)	0.08

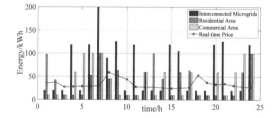

Figure 4. Energy level of ESS.

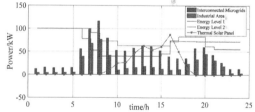

Figure 5. Thermal demand and thermal conservation.

Figure 4 demonstrates the dynamic changes of energy level of energy storage system in two cases. It can be discovered that the energy level of ESS has a wider range of variation from 10 kWh to 200 kWh when microgrids are interconnected, which can be explained as follows: because of a doubled ESS capacity and a real-time tracking of price signals, microgrids could store more energy at price valley periods (i.e. 4h, 7h) and discharge more at price peak periods (i.e. 8h, 10h) to fully avail the electricity at low price to reduce cost, therefore exhibit in-depth charge and discharge.

Figure 5 elucidates the function of thermal storage in coordinating thermal generation. The bars and solid lines in the figure stand for hourly thermal demand and energy level of thermal storage, respectively. Energy level 1 refers to the case of interconnected microgrids, and energy level 2 refers to the case of single microgrid operation. The energy level displays a wider range of variation in the former case, since the interconnection of microgrids results in a higher amount of thermal demand, and further a deeper discharge to satisfy the increased thermal requirement; in addition,

Table 2. Thermal generation and cost.

	Interconnected microgrids	Residential area	Industrial area
Thermal generation (kW)	670.7147	398.0359	336.6848 (imported)
Thermal cost ($)	40.182	26.6684	17.3532

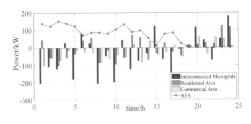

Figure 6. Power exchange with the distribution network.

Table 3. Power exchange and operation cost.

	Interconnected microgrids	Residential area	Commercial area	Industrial area
Power exchange (kW)	−865.9347	−31.4298	−530.2079	–
Operation cost ($)	−16.5915	20.741	−21.6781	17.3532

because photovoltaic solar panels are installed in both residential area and industrial area, thermal storage absorbs more solar energy at time intervals with rich solar radiation (i.e. 16h) in the case of interconnected microgrids, and shows an obvious leap in energy level.

Total thermal generation of CHPs and boilers and the associated cost is reported in Table 2. When operating separately, thermal storage supplies energy for the industrial area, while the residential area has to rely on solar panel generation and thermal power import for thermal energy. In this case, the price of imported thermal power is assumed to be 67$/kWh, which corresponds with natural gas price and its heat value. On the other hand, in-depth charge/discharge of thermal storage in the case of microgrid interconnection results in less imported thermal power, and reduces the total thermal cost by 8.7%.

Figure 6 compares the power exchange with the upper distribution network in the two cases. The negative value indicates the power exported to the utility grid and vice versa. As can be seen from the figure, compared with the latter, since the interconnected microgrids have a greater many wind turbines and PVs, they export more power to the utility grid at time intervals with higher RES generation (i.e. 1h, 5h) to make extra profits, and only import power at time intervals with insufficient RES generation (i.e. 6h, 15h), therefore makes full use of the cheap renewable energy.

The total amount of power exchange and operation cost over the whole dispatch horizon is shown in Table 3. A negative cost indicates income. Because there's no electric load residing in industrial area, the power exchange of industrial area is not listed. Via coordinated control of a larger number of available DERs and demand resources, the total income is doubled in the case of microgrid interconnection, which firmly substantiates its advantage of economic effectiveness over single microgrid operation.

In summary, interconnection of microgrids triggers the maximization of renewable energy use, temporal power transfer and conservation with the assistance of greater storage capacity, and heightens overall power dispatch flexibility and operational economy, therefore holds considerable feasibility in real-time application.

5 CONCLUSION

In this paper, the coordinated economic dispatch of microgrids are modelled and simulated to explore the potential advantages of microgrid interconnection in real-time operation. A rolling optimization approach is further presented to determine interaction among dispersed microgrids and the bulk system. Simulation results and comparisons with single microgrid operation reveal that interconnected microgrids possess more flexibility of sharing electric and thermal power with the assistance of enlarged energy storage capacity and increased DER units. As a consequence, system-wide economic benefit is greatly improved via their mutual support and cooperation.

ACKNOWLEDGEMENTS

The authors would like to thank the National Natural Science Foundation of China under Grant 51277170, 51377152 for financial support.

REFERENCES

Farzan, F., Jafari, M.A., Masiello, R., Lu, Y. 2015. Toward Optimal Day-Ahead Scheduling and Operation Control of Microgrids Under Uncertainty. *IEEE Transactions on Smart Grid* 6(2): 499–507.

Fathi, M., Bevrani, H. 2013. Adaptive energy consumption scheduling for connected microgrids under demand uncertainty. *IEEE Transactions on Power Delivery* 28(3): 1576–1583.

Jager, D., Andreas, A. 1996. NREL National Wind Technology Center (NWTC): M2 Tower, Boulder, Colorado (Data); NREL Report No. DA-5500–56489. Website:http://dx.doi.org/10.5439/1052222/ (Last accessed: Aug. 20, 2015).

Olivares, D.E., Mehrizi-Sani, A., Etemadi A.H., et al. 2014. Trends in microgrid control. *IEEE Transactions on Smart Grid* 5(4): 1905–1919.

Ouammi, A., Dagdougui, H., Dessaint, L., Sacile, R. 2015. Coordinated Model Predictive-Based Power Flows Control in a Cooperative Network of Smart Microgrids. *IEEE Transactions on Smart Grid* 6(5): 2233–2244.

PJM website. Website: http://pjm com/markets-and-operations aspx/ (Last accessed: Aug. 20, 2015).

Rifkin, J. 2011. *The third industrial revolution: how lateral power is transforming energy, the economy, and the world*. New York: Palgrave MacMillan.

Tushar, M.H.K., Assi, C., Maier, M. 2015. Distributed Real-Time Electricity Allocation Mechanism for Large Residential Microgrid. *IEEE Transactions on Smart Grid* 6(3): 1353–1363.

Vagropoulos, S., Bakirtzis, A.G. 2013. Optimal bidding strategy for electric vehicle aggregators in electricity markets. *IEEE Transactions on Power Systems* 28(4): 4031–4041.

Wang, C. & Li, P. 2010. Development and challenges of distributed generation, the micro-grid and smart distribution system. *Automation of electric power systems* 34(2): 10–14.

Wang, Z., Chen, B., Wang, J., Begovic, M.M., Chen C. 2015a. Coordinated energy management of networked microgrids in distribution systems. *IEEE Transactions on Smart Grid* 6(1): 45–53.

Wang, Z., Chen, B., Wang, J., Kim, J. 2015b. Decentralized Energy Management System for Networked Microgrids in Grid-Connected and Islanded Modes. *IEEE Transactions on Smart Grid*, in press.

Xu, X., Jia, H., Wang, D., Yu, D., Chiang, H. 2015. Hierarchical energy management system for multi-source multi-product microgrids. *Renewable Energy* 78: 621–630.

Zakharchenko, R., Licea-Jiménez, L., Pérez-García, S., et al. 2004. Photovoltaic solar panel for a hybrid PV/thermal system. *Solar Energy Materials and Solar Cells* 82(1): 253–261.

Zhang, D., Samsatli, N.J., Hawkes, A.D., et al. 2013. Fair electricity transfer price and unit capacity selection for microgrids. *Energy Economics* 36: 581–593.

Advances in Power and Energy Engineering – Sun (Ed.)
© 2016 Taylor & Francis Group, London, ISBN 978-1-138-02846-3

Research on a novel power control method of Distributed Generators in the micro grid

L. Peng, S. Su, Z.Y. Zhang & X.N. Lin
State Key Laboratory of Advanced Electromagnetic Engineering and Technology,
Huazhong University of Science and Technology, Wuhan, China

X.S. Li
College of Electrical Engineering and New Energy, China Three Gorges University, Yichang, China

ABSTRACT: When the Distributed Generators (DG) connect to the large power grid, the stability of the power output and the effects on the grid become quite important. The conventional power control methods show many drawbacks while facing this problem. This paper proposes a nonlinear control method that can control the power output of DG in micro grid. The nonlinear problem is transferred to a normal linear problem by using a specific nonlinear transformation. Through this method, we get final control over the whole nonlinear system. To prove the superiorities of this control method, we simulate examples on the electromagnetic simulating platform PSCAD/EMTDC.

1 INTRODUCTION

In recent years, as there was development in the Distributed Generation (DG), wide attentions were paid to researches of small capacity distributed micro-source control. Small capacity distributed micro-source in micro grid mainly includes: energy storage unit (super capacity, flywheel energy storage, and superconducting magnetic energy storage), micro gas turbine, wind turbine, storage battery, photovoltaic source, and other kinds of micro sources. Part of them needs power electronic interfaces to form some kind scale of micro grids and connect with large power system at PCC nodes. The problem of the control is very important because of the requirements of self-stability and effects on the stability of large power system.

For the control problem of distributed micro-sources, Ming Z. & Shi X. (2008) pointed out a way of micro-sources control, which can adapt to different operation modes: micro grid can adopt PQ control method in an island-mode, while it can adopt V/f control method while connecting in a power grid. The drawback of these methods is that the transient of switching in different operation modes is not smooth, and it may cause a large electromagnetic transient process and overshoot, which will produce impulse on frequency, voltage, and capacity of system and has an influence on the globe security and stability of power system. Wang J. & Li X. (2005) pointed out a method of decentralized control based on the local information of

micro-sources. Karel B. & Bruno B. proposed master/slave control method from the perspective of operation management of micro-grids. Chen D. & Zhu G. (2010) put forward a view that we can adopt constant power PQ control to sources of high intermittency such as wind power and solar power and adopt V/f control to stable sources based on different characteristics of micro-sources. It also gives the corresponding design methods of the various controllers. Control methods aimed at distributed micro-sources are linearization control methods. Its core idea is to stabilize the system by PID feedback regulations when small disturbance occurs at the equilibrium point in the power system. This control method does not meet the requirement of stability in the large power grid. Once the system working point changes in wide range, the system could lose control because parameters in PID controller are not adaptive.

The paper puts forward a non-linear control method aiming at distributing micro-sources in micro girds which is based on a state. It builds an affine nonlinear model of three-phase voltage source PWM converter in a rotating DQ frame coordinate and proves that the model satisfies the condition of feedback of the exact linearization. It gets linear model by appropriate co-ordinate transformation and realizes the decoupling control of active current and reactive current. We simulated examples on electromagnetic simulating platform PSCAD/EMTDC and proved the superiority of this control method.

2 MODELING

Micro power unit circuit as a reference is shown in Figure 1. In the picture, L_f and C_f are the filter elements of the inverter output, which can avoid the influence of the voltage and current harmonics signal on the quality of output power. The short circuit $R_1 + jX_1$ between bus 1 and bus 2 control the distribution of load flow by adjusting the output current of the inverter.

2.1 Inverter state space modeling

The state space average model of the three-phase voltage type VSR converter is established firstly, and the circuit topology structure is shown in Figure 2.

Using switch functions to give an accurate description of the switching process of VSR. The notation is introduced as follows.

$$s_k = \begin{cases} 1\ upper\ bridge\ arm\ breakover \\ 0\ lower\ bridge\ arm\ breakover \end{cases} k = a, b, c \quad (1)$$

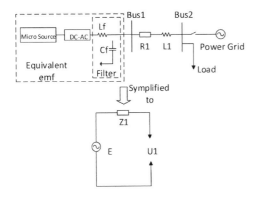

Figure 1. Circuit of DG section.

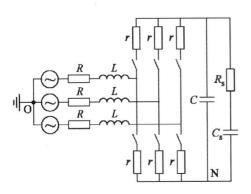

Figure 2. Topology of converter.

A three-phase VSR circuit equation by using KVL law:

$$\begin{cases} L\dfrac{di_a}{dt} + Ri_a = e_a - (V_{aN} + V_{NO}) \\ L\dfrac{di_b}{dt} + Ri_b = e_b - (V_{bN} + V_{NO}) \\ L\dfrac{di_c}{dt} + Ri_c = e_c - (V_{cN} + V_{NO}) \end{cases} \quad (2)$$

In addition to that, the application of KCL on cathode of the capacitor at DC side of VSR has the following relations.

$$C\frac{dV_{dc}}{dt} = i_a S_a + i_b S_b + i_c S_c - \frac{V_{dc} - e_L}{R_L} \quad (3)$$

Introducing state variable $X = (i_a\, i_b\, i_c\, V_{dc})^T$ among them there are three-phase currents i_a, i_b, and i_c on the AC side of the converter. V_{dc} is the capacitor voltage on DC side of the converter. The general mathematical model of three-phase VSR is:

$$ZX = AX + BE \quad (4)$$

where,

$$Z = diag(L, L, L, C) \quad (5)$$
$$B = diag(1, 1, 1, 1/R_L) \quad (6)$$
$$E = (e_a, e_b, e_c, e_L)^T \quad (7)$$

$$A = \begin{bmatrix} -R & 0 & 0 & -\left(S_a - \dfrac{1}{3}\sum\limits_{k=a,b,c} S_k\right) \\ 0 & -R & 0 & -\left(S_b - \dfrac{1}{3}\sum\limits_{k=a,b,c} S_k\right) \\ 0 & 0 & -R & -\left(S_c - \dfrac{1}{3}\sum\limits_{k=a,b,c} S_k\right) \\ S_a & S_b & S_c & -\dfrac{1}{R_L} \end{bmatrix} \quad (8)$$

Applying Park transformation on the above model, we obtain the state equation of the three phases VSR under the DQ synchronous rotation axis.

$$\frac{d}{dt}\begin{pmatrix} i_d \\ i_q \\ V_{dc} \end{pmatrix} = \begin{pmatrix} -\dfrac{R}{L} & \omega & \dfrac{m_d}{L} \\ -\omega & -\dfrac{R}{L} & \dfrac{m_q}{L} \\ -\dfrac{m_d}{C} & -\dfrac{m_q}{C} & 0 \end{pmatrix}\begin{pmatrix} i_d \\ i_q \\ V_{dc} \end{pmatrix} + \begin{pmatrix} -\dfrac{e_d}{L} \\ -\dfrac{e_q}{L} \\ 0 \end{pmatrix} \quad (9)$$

where m_d = modulation coefficient for d coordinate axis; m_q = modulation coefficient for q coordinate axis; e_d = grid voltage for d coordinate axis; e_q = grid voltage for q coordinate axis; i_d = active current; and i_q = reactive current in the VSR injection system under the rotating coordinate system. After the filter, the size of the injected current determines the power flow between the inverter and the grid in the case of stable voltage.

2.2 VSR control method

2.2.1 PQ control

The micro power supply as well as the inverter and the filter element are equivalent to an electric potential E (neglecting its internal impedance). The outlet voltage value equals to bus2 voltage U_2. Then it is connected to bus1 by a short circuit L_1. The power flow direction is bus 2 to bus 1. The distributed micro power supplies (EI-DG) with the inverter as the interface, together with distribution network, make up a single machine-infinite system. The phase diagram between each electrical parameter is shown in Figure 3.

U_1, U_2 meet the following relationship:

$$\dot{U}_2 = \dot{U}_1 + j\omega L_1 \dot{I}_{abc} \tag{10}$$

We can get the d and q axis component of U_2 by applying DQ decomposition on it. Among them they use the bus 1 voltage as d axis in order to reduce controlled variables, so that $U_{1q} = 0$, and $U_{1d} = U_1$. Changing the line current I_d and I_q can change U_{2d} and U_{2q}, thus, changing the amplitude of U_2 and its power angle between U_1, realizing effective control of the setting power. The reference current tracks the real-time current of the line, while the PI controller measures the current approaching the reference value. Eventually, the voltage equation of the three-phase inverter can be written as:

$$\begin{cases} U_{2d} = U_1 + j\omega L_1(I_q + \Delta I_q) \\ U_{2q} = j\omega L_1(I_d + \Delta I_d) \end{cases} \tag{11}$$

The structure of the PQ control is shown in Figure 4.

2.2.2 Affine nonlinear control

By using nonlinear state feedback and proper coordinate transformation, a nonlinear system can be accurately linearized under a certain condition. The exact state feedback can guarantee the stability and dynamic quality of the control system.

Select state variable $X = (x_1, x_2) = (i_d, i_q)$, input variable $u = (u_1, u_2)^T = (m_d, m_q)^T$, output variables

$$\begin{cases} y_1 = h_1(x(t)) = i_d \\ y_2 = h_2(x(t)) = i_q \end{cases}$$

We obtain an affine nonlinear model of three-phase VSR with two inputs and two outputs.

$$\begin{cases} \dot{x} = f(x) + g_1(x(t))u_1 + g_2(x(t))u_2 \\ y_1 = h_1(x(t)) \\ y_2 = h_2(x(t)) \end{cases} \tag{12}$$

where,

$$f(x) = \begin{bmatrix} -\dfrac{R}{L}x_1 + \omega x_2 - \dfrac{e_d}{L} \\ -\omega x_1 - \dfrac{R}{L}x_2 - \dfrac{e_q}{L} \end{bmatrix}$$

$$g_1(x(t)) = \begin{bmatrix} \dfrac{V_{dc}}{L}, 0 \end{bmatrix}^T$$

$$g_2(x(t)) = \begin{bmatrix} 0, \dfrac{V_{dc}}{L} \end{bmatrix}^T$$

The nonlinear system with double inputs and double outputs show the nonlinear feature x_1 and x_2, but the linear relationship to the control variable u_1, u_2.

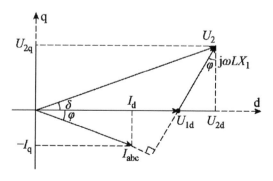

Figure 3.   Vector of DG.

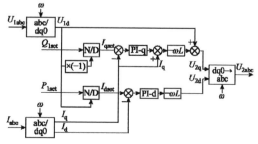

Figure 4.   Control logic of PID.

The lie derivative of output function $h_1(x)$, $h_2(x)$ along the vector field $h_1(x)$, $h_2(x)$ is shown as follows:

$$\begin{cases} L_f h_1(x) = -\dfrac{R}{L} x_1 + \omega x_2 - \dfrac{e_d}{L} \\ L_f h_2(x) = -\dfrac{R}{L} x_2 - \omega x_1 - \dfrac{e_q}{L} \end{cases} \tag{13}$$

For each output function, the nonlinear system is related as $r_1 = 1$ and $r_2 = 1$, the relationship degree of the system is $r = \{r_1\ r_2\} = \{11\}$ and the total system relationship degree is $r = r_1 + r_2 = 2$.

At the same time, there is the following relationship:

$$\begin{cases} L_{g_1} h_1(x) = -\dfrac{V_{dc}}{L} \\ L_{g_1} h_2(x) = 0 \\ L_{g_2} h_1(x) = 0 \\ L_{g_2} h_2(x) = \dfrac{V_{dc}}{L} \end{cases} \tag{14}$$

Simultaneously satisfy:

$$\begin{cases} ad_f g_1(x) = \dfrac{\partial g_1(x)}{\partial x} f(x) - \dfrac{\partial f(x)}{\partial x} g_1(x) = \begin{pmatrix} \dfrac{RV_{dc}}{L^2} \\ -\dfrac{\omega V_{dc}}{L} \end{pmatrix} \\ ad_f g_2(x) = \dfrac{\partial g_2(x)}{\partial x} f(x) - \dfrac{\partial f(x)}{\partial x} g_2(x) = \begin{pmatrix} \dfrac{\omega V_{dc}}{L} \\ \dfrac{RV_{dc}}{L^2} \end{pmatrix} \end{cases}$$

Therefore, the augmented matrix $G = (g_1(x), g_2(x), ad_f g_1(x), ad_f g_2(x))^T$ is of rank 2, equals to the number of state variables. At the same time, it proves that the rank of the vector field $(g_1(x), g_2(x), ad_f g_1(x), ad_f g_2(x))^T$ is still 2. The vector field is right. So there is a set of output functions that makes the relative order of the system be able to define itself while the total order $r(r = 2)$ is equal to the order $n(n = 2)$ of the system. In a word, the system can realize the exact linearization.

According to the standard linearization of this type of affine nonlinear system, the nonlinear coordinate change of the following form is selected.

$$z = \begin{pmatrix} z_1 \\ z_2 \end{pmatrix} = \Phi(x) = \begin{pmatrix} L_f^{r_1-1} h_1(x) \\ L_f^{r_2-1} h_2(x) \end{pmatrix} = \begin{pmatrix} h_1(x) \\ h_2(x) \end{pmatrix} \tag{15}$$

In order to achieve the decoupling effect of the system after accurate linearization, the linear decoupling system is expected to be shown as follows:

$$\begin{cases} \dot{z}_1 = -k_1 z_1 + k_1 v_1 \\ \dot{z}_2 = -k_2 z_2 + k_2 v_2 \end{cases} \tag{16}$$

Or written in matrix form:

$$\dot{Z} = AZ + BV$$

where,

$$A = \begin{pmatrix} -k_1 & 0 \\ 0 & -k_2 \end{pmatrix} \quad B = \begin{pmatrix} k_1 & 0 \\ 0 & k_2 \end{pmatrix}$$

In the formula, k_1, k_2 stand for the feedback control coefficient, v_1, v_2 are the feedback control variables in the linearized system. The excepted feedback control law of nonlinear system can be obtained by these simultaneous equations.

$$\begin{pmatrix} u_1 \\ u_2 \end{pmatrix} = \dfrac{L}{V_{dc}} \left(\begin{pmatrix} k_1 v_1 \\ k_2 v_2 \end{pmatrix} - \begin{pmatrix} k_1 x_1 - \dfrac{R}{L} x_1 + \omega x_2 - \dfrac{e_d}{L} \\ k_2 v_2 - \dfrac{R}{L} x_2 + \omega x_1 - \dfrac{e_q}{L} \end{pmatrix} \right)$$

According to the optimal control strategy of the linear system, the optimal control vector satisfies:

$$\begin{cases} V = -k^* Z \\ k^* = R^{-1} B^T P^* \\ A^T P^* + P^* A - P^* B R^{-1} B^T P^* + Q = 0 \end{cases} \tag{17}$$

Without loss of generality, let

$$Q = \begin{pmatrix} -1 & 0 \\ 0 & -1 \end{pmatrix}, \text{ we get } P^* = \begin{pmatrix} -1/k_1 & 0 \\ 0 & -1/k_2 \end{pmatrix}$$

By solving the Riccati equation, we can obtain the control law of linear system and nonlinear control system as below.

$$\begin{cases} V = \begin{pmatrix} 1 & 0 \\ 0 & 1 \end{pmatrix} Z \\ U = \begin{pmatrix} u_1 \\ u_2 \end{pmatrix} = \dfrac{L}{V_{dc}} \begin{pmatrix} \dfrac{R}{L} x_1 - \omega x_2 + \dfrac{e_d}{L} \\ \dfrac{R}{L} x_2 + \omega x_1 + \dfrac{e_q}{L} \end{pmatrix} \end{cases} \tag{18}$$

2.2.3 Robust verification of control law

As the exact linearization of the nonlinear system needs an assured mathematical model and precise parameters, once the model or parameters have small changes, the exact linearization condition may not be established, the effect of the design of the control law can also be reduced. The robust verification of the control law based on exact linearization and state feedback method is as follows.

58

The affine nonlinear model of three-phase VSR with two inputs and two outputs without disturbing their form is shown in type (12). Now, considering the parameters or model changes, the system can be described in disturbance form:

$$\begin{cases} \dot{X} = F(X(t)) + G_1(X(t))W + G_2(X(t))U \\ Z = H(X(t)) + K(X(t))U \end{cases} \quad (19)$$

where,

$$F(X(t)) = \begin{bmatrix} -\dfrac{R}{L}x_1 + \omega x_2 - \dfrac{e_d}{L} \\ -\dfrac{R}{L}x_2 - \omega x_1 - \dfrac{e_q}{L} \end{bmatrix}$$

$$G_1(X(t)) = (g_{11}, g_{12}) = \dfrac{V_{dc}}{L}\begin{bmatrix} 1 & 0 \\ 0 & 1 \end{bmatrix}$$

$$G_2(X(t)) = (g_{21}, g_{22}) = \dfrac{V_{dc}}{L}\begin{bmatrix} 1 & 0 \\ 0 & 1 \end{bmatrix}$$

$$H(X(t)) = (x_1, x_2, 0, 0)^T$$

$$K(X(t)) = \begin{pmatrix} 0 & 0 & 1 & 0 \\ 0 & 0 & 0 & 1 \end{pmatrix}^T$$

The control law, which does state feedback based on the system affine nonlinear model, is shown as type (18). The control law of the type (18) is denoted as U^{**}.

Establishing augmented functional $\bar{J}$ according to the system model:

$$\bar{J} = \int_0^T \left\{ \|Z\|^2 - \gamma^2\|W\|^2 + \Lambda^T\left[F(X(t)) + G_1(X(t))W \right.\right. $$
$$\left.\left. + G_2(X(t))U - \dot{X}\right] \right\}dt \quad (20)$$

Establishing Hamilton function $H(X, \Lambda, W, U)$ based on the integrand function established by augmented functional $\bar{J}$

$$H(X, \Lambda, W, U) = \|Z\|^2 - \gamma^2\|W\|^2 + \Lambda^T(F(X(t)) $$
$$+ G_1(X(t))W + G_2(X(t))U) \quad (21)$$

Assuming there is a first order differentiable function $V(x)$, replacing the Lagrange constant in Hamilton function by its gradient vector and further assuming that when $X \neq 0$, $V(x) > 0$ and $V(0) = 0$; a saddle point of a Hamilton function is obtained by solving the variation problem with constraints, as shown in (22) and (23).

$$W^* = \dfrac{1}{2\gamma^2}G_1^T(X(t))V_x \quad (22)$$

$$U^* = -R^{-1}(X(t))\left(\dfrac{1}{2}G_2^T(X(t))V_x^T \right.$$
$$\left. + K^T(X(t))H(X(t))\right) \quad (23)$$

In the formula, V_x is the non-negative solution of HJI inequality.

HJI inequality for three-phase VSR is shown in (24):

$$\left(\dfrac{1}{2\gamma}\dfrac{V_{dc}}{L}\right)^2\begin{pmatrix} 1 & 0 \\ 0 & 1 \end{pmatrix}V_x^2 + \begin{pmatrix} -\dfrac{R}{L}x_1 + \omega x_2 - \dfrac{e_d}{L} \\ -\dfrac{R}{L}x_2 - \omega x_1 - \dfrac{e_q}{L} \end{pmatrix}V_x$$
$$+ (x_1^2 + x_2^2) \leq 0 \quad (24)$$

The matrix inequality described above is a set of differential inequality equations, as shown in formula:

$$\begin{cases} \left(\dfrac{1}{2\gamma}\dfrac{V_{dc}}{L}\right)^2 V_x^2 + \left(-\dfrac{R}{L}x_1 + \omega x_2 - \dfrac{e_d}{L}\right)V_x + (x_1^2 + x_2^2) \leq 0 \\ \left(\dfrac{1}{2\gamma}\dfrac{V_{dc}}{L}\right)^2 V_x^2 + \left(-\dfrac{R}{L}x_2 - \omega x_1 - \dfrac{e_q}{L}\right)V_x + (x_1^2 + x_2^2) \leq 0 \end{cases} \quad (25)$$

And there is the following relationship:

$$H(X, \Lambda, W, U) = H^*(X, \Lambda, W^*, U^*) $$
$$+ \|U - U^*\|_R^2 - \gamma^2\|W - W^*\|^2 \quad (26)$$

Take the control law designed in section 2.2.2 into equation (26), get:

$$H(X, \Lambda, W, U^{**}) = H^*(X, \Lambda, W, U^{**}) $$
$$+ \|U^{**} - U^*\|_R^2 - \gamma^2\|W - W^*\|^2 \quad (27)$$

As $H^*(X, \Lambda, W^*, U^*) \leq 0$, we get

$$H(X, \Lambda, W, U^{**}) \leq \|U^{**} - U^*\|_R^2 - \gamma^2\|W - W^*\|^2 \quad (28)$$

In particular, by variable transformation method shown in equation (15), the nonlinear state equation (19) with disturbance form is transformed into a linear state equation as follows:

$$\dot{Z} = AZ + B_2V + \dfrac{\partial\Phi(x)}{\partial x}G_1(x)W \quad (29)$$

$$V = \begin{pmatrix} v_1 \\ v_2 \end{pmatrix} = \begin{pmatrix} L_f^{r_1}h_1(x) + L_{g_{21}}L_f^{r_1-1}h_1(x) \\ L_f^{r_2}h_2(x) + L_{g_{22}}L_f^{r_2-1}h_2(x) \end{pmatrix}$$

Letting

$$\overline{W} = \frac{\partial \Phi(x)}{\partial x} G_1(x)W = \frac{V_{dc}}{L}\begin{pmatrix} 1 & 0 \\ 0 & 1 \end{pmatrix}$$

The system can be written in a standard Brunovsky form.

$$\begin{cases} \dot{Z} = AZ + B_1W + B_2V \\ Y = CZ \end{cases} \quad (30)$$

In the formula, $A = $ zeros $(2, 2)$, $B_1 = B_2 = C = I$ $(2, 2)$.

Thus, the Riccati inequality for linear systems is:

$$A^T P + PA + \frac{1}{\gamma^2} PB_1B_1^T P - PB_2B_2^T P + C^TC < 0 \quad (31)$$

In order to achieve the comparison of the proposed control method and the verification of the system's robustness, a robust control law with one specific parameter is firstly formed; then it's compared with the proposed method. To this end, take γ as the set value, here you may wish to take $\gamma = 2$ to obtain a nonnegative solution, which meets the Riccati inequality:

$$P^* = \begin{pmatrix} \dfrac{2}{\sqrt{3}} & 0 \\ 0 & \dfrac{2}{\sqrt{3}} \end{pmatrix} \quad (32)$$

Based on the formula (31), the robust control law V^* of the linear system is shown in formula (33).

$$V^* = \begin{pmatrix} v_1^* \\ v_2^* \end{pmatrix} = -B_2^T P^* Z = -\begin{pmatrix} \dfrac{2}{\sqrt{3}} & 0 \\ 0 & \dfrac{2}{\sqrt{3}} \end{pmatrix}\begin{pmatrix} z_1 \\ z_2 \end{pmatrix} \quad (33)$$

The worst possible interference W^*

$$W^* = \begin{pmatrix} w_1^* \\ w_2^* \end{pmatrix} = \frac{1}{\gamma^2} B_1^T P^* Z = \begin{pmatrix} \dfrac{\sqrt{3}}{6} & 0 \\ 0 & \dfrac{\sqrt{3}}{6} \end{pmatrix}\begin{pmatrix} z_1 \\ z_2 \end{pmatrix} \quad (34)$$

The optimal robust control law:

$$U^* = \begin{pmatrix} u_1^* \\ u_2^* \end{pmatrix} = \frac{L}{V_{dc}}\begin{pmatrix} \left(\dfrac{R}{L} + \dfrac{2}{\sqrt{3}}\right)x_1 - \omega x_2 + \dfrac{e_d}{L} \\ \dfrac{R}{L}x_2 + \left(\omega - \dfrac{2}{\sqrt{3}}\right)x_1 + \dfrac{e_q}{L} \end{pmatrix} \quad (35)$$

Table 1. The effect of different disturbance on the system robustness.

Degree of disturbance	Robust indicator γ'	Controllers' robustness
$0.9W^*$	40	Unsatisfactory
$0.8W^*$	20	Acceptable
$0.7W^*$	13	Acceptable
$0.5W^*$	8	Ideal
$0.4W^*$	7	Ideal

Assuming the turbulence of different degrees applied, which is as shown in the following table can be described by formula (25), which also represents control effectiveness in robust by γ' presented under different disturbance. The results are shown in Table 1.

From the table, with the decrease of the disturbance degree, the smaller the γ', the better robustness of the proposed control method. It can be seen that the control law designed by this paper has robustness and can restrain a certain degree of disturbance and parameter variation. Optimal robust control and the proposed affine nonlinear control both belong to the category of the nonlinear control. Compared with the optimal robust control, the robustness of the algorithm proposed in this paper is weaker.

However, the optimal robust control mainly concerns the impact of the most serious disturbance and system parameters change. The optimal robust control design is relatively conservative in order to achieve the effective control of the most unfavorable situation, which means paying greater control price in larger conventional operating range. Meanwhile, the power supply of PQ nodes in the micro grid has more stable external operational mode, and the internal parameters seldom fluctuate greatly. Therefore, considering the control effect, cost of the controller and the adaptive range, the nonlinear control method proposed in this paper can meet the requirements of the accurate and stable power control.

3 SIMULATION ANALYSIS

The simulation models based on PSCAD/EMTDC electromagnetic transient simulation platform are built. Owing to the limited space and the complexity of the control link, the paper doesn't expend. EI interface in DG micro network is connected to the distribution network with low voltage level. For the sake of simplicity, we didn't add the link where DG connects to the network through the transformer. The system structure is shown in Figure 1.

A comparative simulation study of the different control strategies of the distributed micro power supply is carried out. As the distributed power supply in the micro grid is often clean and uses renewable energy or the energy storage unit, it should make the DG produce as much of active power as possible while producing less reactive power, in order to achieve full power factor operation.

Example 1

When system capacity reaches its shortfall and needs to start the standby DG in micro network rapidly to prevent the expanding imbalance of the micro network capacity, which may lead to collapse in the system and loss of the supporting role of the micro grid. The power output of the DG is observed to see if it can reach the needed level in time. Figure 5 and Figure 6, respectively, stands for the active and reactive power output curve following orders under classical PID control and nonlinear control.

The example sets the active power and reactive power of the DG as:

$$P_{ref} = 5.0 \text{ kW}, Q_{ref} = 0$$

The simulation results show that the classical PID control has a proportional component in controlling the unit, so the adjustment in the case of the fluctuation of the equilibrium point of a stable state may generate the overshoot phenomenon. Compared with this, the nonlinear control using accurate linearization can commendably eliminate the overshoot of the power output, and limit the overshoot of the current, which comes from the inverter to the outlet bus, to an ideal level, thus, eliminating the possibility of the DG exit owing to a large DG export current. It highlights the superiority of the nonlinear control.

Example 2

In the case of micro grid and distribution network of low voltage level increase of DG output at t = 3.0 s, the micro grid DG will send redundant power to the export bus, readjust its own output to support the important load inside the micro grid and sensitize load which needs to be powered uninterrupted. At the same time, it can be active in listing distributed power in micro grid on the scheduling power supply category. In this case, a comparative analysis of the traditional PID control and the exact linearized nonlinear control is carried out. In case of a change in the power instruction value, Figure 7 and Figure 8 show curve of power supply tracking output, respectively.

Setting $P_{ref} = 15.0 \text{ kW}, Q_{ref} = 0$.

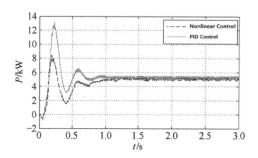

Figure 5. Comparison of active power output of DG by PID control and the nonlinear control.

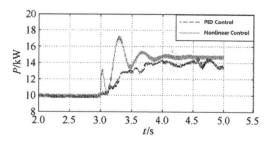

Figure 7. Comparison of active power output increase of DG by PID control and the nonlinear control.

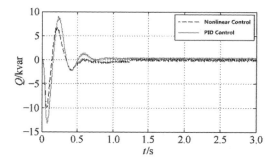

Figure 6. Comparison of reactive power output of DG by PID control and the nonlinear control.

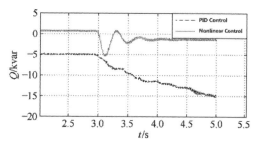

Figure 8. Comparison of reactive power output increase of DG by PID control and the nonlinear control.

The simulation results show that each DG needs to renew its output when a part of the distributed power supply needs to be involved in the energy management scheduling of the distribution network. When it comes to maintaining the local sensitive load, compared to the classical PID control, the nonlinear control leads to shorter electromagnetic transient process of power and current, smaller fluctuation range, and a more effective suppressed overshoot. Through the simulation, it is not difficult to find that when DG output instruction value changes PI parameters in controller's current loop lose control effect because of saturation, while the reactive power increases. According to those points, the proposed nonlinear control method does have superiority.

4 CONCLUSION

A nonlinear power control method for the distributed power supply in micro grid is proposed in this paper. The results show that when it comes to power control of distributed power in micro grid, the nonlinear control can effectively reduce the overshoot caused by the regulated quantity when compared with the traditional PID control. While changing the DG output is needed. Compared with traditional PID control method, nonlinear control method is more effective with a shorter transition process. Besides, it overcomes the difficulty in resetting parameters, which is typical of the traditional PID controller. Finally, the simulation results demonstrate the effectiveness and adaptability of the proposed method.

ACKNOWLEDGMENT

This work was supported in part by the National Natural Science Foundation of China (51277110), and in part by the National Key Basic Research Program of China (973 Program) (2012CB215100), and in part by the Research Fund for the Doctoral Program of Higher Education of China (20110142110055), and in part by Natural Science Foundation of Hubei Province (2012FFA075).

REFERENCES

Chen, D. 2010. Power transmission characteristics of low voltage microgrids. Transactions of China Electrotechnical Society 25(7): 117–122.

Karel, B.& Bruno, B. 2007. A voltage and frequency droop control method for parallel inverters. IEEE Transactions on Power Electronics 22(4): 1107–1115.

Le, J. & Xie, Y. 2010. Reactor magnetoelectric equivalent diagram based on gyrator model and its application. Electric Power Automation Equipment 30(2): 81–85.

Li, S. 2009. A discussion on the integrated operation mode of micro grid. Journal of Taiyuan University of Technology 40(2): 184–187.

Ming Z. & Shi X. 2008. PI regulator and parameter design of voltage PWM rectifier. Electric Power Science and Engineering 24(9): 19–23.

Peças, L.J. & Tomé, S.J. 2003. Management of micro grids. JIEE Conference, Bilbao, 28–29 October.

Pecas, L.J. & Moreira, C.L. 2006. Defining control strategies for microgrids islanded operation. IEEE Transactions on Power Systems 21(2): 916–924.

Wang, C. & Xiao, Z. 2008. Synthetical control and analysis of micro grid. Automation of Electric Power Systems 32(7):98–103.

Wang, J. & Li, X. 2005. Power system research on distributed generation penetration. Automation of Electric Power Systems 29(24):90–97.

Wu, Q.H. & Jiang L. 2001. Survey on nonlinear control theory and its application power systems. Automation of Electric Power Systems 2(3): 1–10.

Yan, C. & Sun, Y. 1993. Nonlinear controller design of SVC by exact linearization method. Journal of Tsinghua University 33(1): 18–24.

Zhao, Z. & Li, H. 2003. PI regulator and parameter design of PWM rectifier. Journal of North China Electric Power University 30(4): 34–37.

Advances in Power and Energy Engineering – Sun (Ed.)
© 2016 Taylor & Francis Group, London, ISBN 978-1-138-02846-3

Positive feedback voltage drift island detection based on droop control

T. Yan & Z.Z. Qu
China Electric Power Research Institute, Beijing, China

Z.B. Liu & G.W. Zhu
North China Electric Power University, Beijing, China

ABSTRACT: Isolated island detection is an important technology in the distributed generation system. In order to solve the problem that the detection efficiency and power quality can not be taken into account in the traditional island effect detection, this paper proposes an improved method of detecting the island detection based on the droop control principle. Through the simulation of the grid connected system of two inverters, the traditional droop control, the proposed method, and the improvement of the network voltage simulation waveforms are compared. Through the experiment, we can find that the improved detection method can change the detection time by adjusting the coefficient K, so as to improve the efficiency of the isolated island detection, and it will not affect the power quality.

Keywords: island detection; droop control; positive feedback; voltage shift

1 INTRODUCTION

Island effect refers to the grid occurring off-grid for some reason (such as regular maintenance or grid failure) when the normal power supply, and the protection devices start up because the user side of the grid connected power generation system has not detected the relevant fault in time, thus, forming a distributed power system and its load group is still running the state, which is called the island power supply system.

– Jia Shiping et al. 2004.

The non planned island can be harmful to the power network and people, so the island detection technology is becoming the focus of domestic and foreign scholars on the grid system.

2 RESEARCH STATUS OF ISLANDING DETECTION

At present, there are mainly three methods for the detection of the island effect: Switch state detection, Passive detection and Active detection.

– Yin Zhifeng et al. 2013, Zhang Shi et al. 2010, Liu Fangrui et al. 2008, Zheng Le et al. 1014.

The Switch state detection method is to detect the switch signals through wireless communication technology, with the advantages of high detection efficiency. But the transmission of the signal will interfere with other carrier communication, that is a high economic investment and it is not widely used. The Passive detection method is to determine whether there is an island effect, when power grids get off, by detecting the change in the situation of the voltage phase, the change situation of harmonics, Amplitude change, and Frequency change. It is easy to achieve and not affect the power quality. However, when output power of DG is close to or equal to the local load power, the method will be lost. So there is a large detection blind area. Active detection method is by injecting a certain disturbance to the system, perturbation motion will not be detected in smooth operation; once the power grid is off, these disturbances will accumulate over the limit, thus, the island effect will be detected. Active islanding detection methods are mainly harmonic current injection method, negative sequence voltage positive feedback method, voltage positive feedback, reactive current perturbation method, active current perturbation method, THD positive feedback method, active frequency offset method, slip frequency drift method, active phase shift method, and other methods.

– Zhang Qi et al. 2011, Guo Xiaoqiang et al. 2009, Liu Furong et al. 2012, Liu Wenhua 2011, Yang Tao et al. 2012, Yang Qiuxia et al. 2012.

Active islanding detection method has the advantages of high efficiency and small blind spot detection, but it will affect the power quality in the normal operation of the power grid, and it is not suitable for the places that require a higher power quality.

Since the communication without coordination between units, it will be able to balance the power of multiple distributed power, droop control has been widely applied in the relevant field of new energy power generation.

– Liang Jiangang et al. 2014, Fan Yuanliang et al. 2012.

In this paper, it proposed a voltage islanding detection algorithm and its improvement strategy based on the under droop control mechanism, it has no effect on the power quality in the grid during normal operation of multiple distributed power supplies in parallel apply. Through simulation, we verify the feasibility of this method.

3 THE PRINCIPLE DROOP CONTROL

Droop control is a control mode of distributed power, which is similar to conventional generators that have a droop characteristic of frequency. The basic principle can be expressed as follows:

$$\begin{cases} U = U^* + m(Q^* - Q) \\ f = f^* + n(P^* - P) \end{cases} \quad (1)$$

where U = the reference voltage; U* = the basic voltage of the grid; Q* = rating reactive power of the distributed power; Q = real reactive power of the distributed power; f = the reference frequency; f* = frequency of the distributed power output voltage; P* = rating active power of the distributed power; P = real active power of the distributed power; m = the control effect of distributed power output voltage; and n = the control effect of distributed power output voltage's frequency.

4 ISLANDING DETECTION ALGORITHM AND ITS IMPROVEMENT

4.1 *Islanding detection algorithm*

According to the relationship between voltage and reactive power in droop control, a method of islanding detection is proposed. When the grid is normal, make U* equal to Uo, where Uo is the fundamental voltage of the grid connection point. The fundamental voltage varies little due to the clamping effect of grid. When unplanned islanding occurs, due to the disappearance of the clamping effects, the fundamental voltage of converters shifts to one direction, an islanding is thus detected when the frequency reference exceeds a limit.

When faced with the need for proactive islanding operation, the droop controller is no longer used Uo as the reference voltage, but will set it a fixed value, to ensure that the system voltage stability.

In this paper, the detection of the voltage amplitude of the dot is detected, the basic principle of which can be described in Figure 1. Figure 1 is the improved droop control reactive power loop. It can be seen from the principle diagram that different from the traditional control strategy, the input voltage instruction follows the output voltage of the inverter, which can be understood as a quasi positive feedback system.

In conjunction with Figure 1, we can get this:

$$Uo(k) = U^*(k) + m\Delta Q = Uo(k-1) + m\Delta Q \quad (2)$$

After the change of reactive power grid off, grid ΔQ inverter side will affect the size of the output voltage. Ideally, the system can achieve zero error control, the inverter output voltage Uo(k) = U*(k) + mQ = Uo(k–1) + mQ. When the reactive power change is zero, the output voltage of the inverter will not change, which is the area of the detection of the island.

4.2 *Improved islanding detection algorithm*

Through the above analysis we can draw the following conclusions, when the Distributed reactive power output be similar to Consume reactive load, positive feedback inverter output voltage will be reduced, which would spend more time on islanding detection. For this reason, we add the increase coefficient K to strengthen its positive feedback effect on the basis of islanding detection, thus, reducing inspection time.

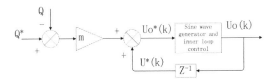

Figure 1. Schematic diagram of islanding detection.

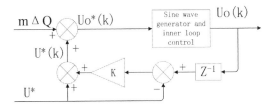

Figure 2. Improved method of islanding detection.

Figure 2 Islanding detection method schematic which strengthens the role of positive feedback. The output voltage feedback section in Figure 1 are improved.

Assuming this is an ideal control system, from the schematic improved we can get the output voltage expression after the off-grid power. As can be seen, the intensity of the feedback coefficient K may change intensity feedback inverter side voltage. According to Figure 2, to obtain the output voltage expression is as follows:

$$U_o(k) = U^*(k) + m\Delta Q$$
$$= (U_o(k-1) - U^*)K + U^* + m\Delta Q \qquad (3)$$

According to analysis of formula (3), we know that the detection effect is affected by the value of feedback strength factor K, specific content as follows:

1. K < 0, it's a negative feedback, which can't be detected;
2. K = 0, it will degrade into conventional droop control;
3. 0 < K < 1, it will be weaker than the voltage value of the positive feedback effect, which is not conducive to reduce detection time;
4. K = 1, the detection method has the equal effect; and
5. K > 1, it can strengthen the positive feedback effect which can shorten the detection time.

All the analysis shows that the larger the K value is, the faster detection speed will be, however detection speed can not be unlimited, because the controller is limited in response speed. And if the K value is too large, it will make the grid due to errors of voltage fluctuations and voltage detection introduces a large deviation during normal operation. This is not conducive to the stable operation of the system. Thus, the setting of the K value depends on the specific circumstances.

In addition, it requires us to notice that the RMS value of the voltage can not keep a constant value in the actual power grid in stable state. So, the voltage value of U* in Figure 2 and Type 3 can be updated at any time (such as 1 min) in the program, which is replaced with a new value when the voltage is detected within the normal range of the standard grid. Otherwise, it remains to be a constant value. In the power grid, there will be some disturbances which will cause the voltage to change and affect the criterion of the isolated island detection. The increase of the voltage will generate some errors. This can improve the sensitivity, but reduce the reliability. To solve this problem, the delay method is introduced to improve the reliability. The impact of power system disturbances on

the electrical parameters is a very short duration, which is usually in millisecond, and that can be quickly restored but the time will be more longer when island effect occurs. GBT19939-2005 requires the islanding effects must be detected within 2 seconds, if not, the detection is considered failed. If during the delay can the parameters restored to the normal range, it is determined that the island effects is false. Therefore, according to this feature, we can make the detection of electrical parameters exceeding the limits through the delay, then regard this as an island efforts when them not restore.

5 BOUNDARY CONDITION FOR SUCCESSFUL TESTING

According to (3) can be obtained:

$$U_o(z) = [m\Delta Q + (1-K)U^*]/(1-Kz^{-1}) \qquad (4)$$

When K = 1, this is the transfer function of pre-improved detection methods. The characteristic equation of the system is: z = 1. Unstable condition:

$$K > 1 \qquad (5)$$

Therefore, only when K > 1 can it make the inverter output voltage unstable, so as to provide conditions for the detection of the island state. With the increase of K value, the system is not easy to be stable, and the voltage island detection time is shorter. But the K value can not be infinite, but also according to the actual situation.

6 SIMULATION

The grid model of the energy storage system with 2 sets of droop control is built. Schematic diagram of the simulation model is shown in Figure 3. The capacity of single storage system is 720 kV·A, the effective value of the electric network line voltage is 380 V, fundamental frequency fs = 50 Hz, Rated voltage of DC link Ud = 700 V, LCL filter L1 = 300 μH,

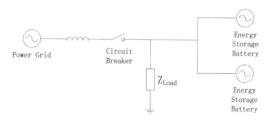

Figure 3. Schematic diagram of simulation model.

L2 = 180 μH, Cd = 190 μF, Rd = 0.1 Ω, droop control coefficient m = 3 × 10⁻⁶, n = 1 × 10⁻⁴. We set the island when t = 1 s. Using the traditional droop control, the proposed method and the improvement method, we do simulation of the situation that blind spots in the isolated island detection.

Two energy storage systems using the same droop control, almost have the same output waveform, so only one energy storage system of the output waveform is given.

Figures 4 and 5, respectively, show load reactive power consumption is less than and greater than the inverter rated output reactive power (reactive power load consumption is close to rating), using the traditional droop control algorithm, the proposed detection method and improved method of simulation comparison chart.

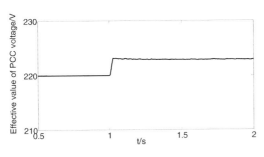

（a）The simulation of traditional droop control

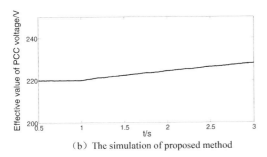

（b）The simulation of proposed method

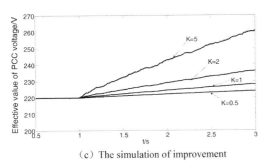

（c）The simulation of improvement

Figure 4.　The local load reactive power consumption is less than the inverter rated reactive power.

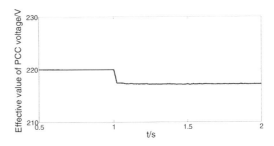

（a）The simulation of traditional droop control

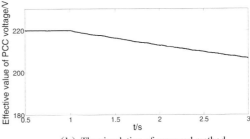

（b）The simulation of proposed method

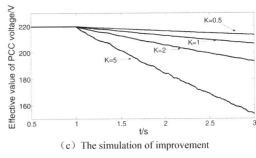

（c）The simulation of improvement

Figure 5.　The local load reactive power consumption is greater than the inverter rated reactive power.

The simulation results are shown in Figure 4 when the reactive power consumed by local loads is less than the nominal output of inverters. As is shown in Figure 4, the quiescent working point of RMS voltage would shift upward a little bit when traditional droop control algorithm is adopted under this circumstance and stabilize at some certain point within the limitation thus ensuring the smooth switching of grid connection and grid disconnection, which is the advantage of traditional droop control. But islanding detecting is also unavailable. Common islands could be detected when islanding detecting algorithm is adopted while the detecting time of islands where the load power consumption is near the inverter nominal output is longer. When improved islanding detecting algorithm is applied, the amplitude of phase voltage shifts faster as the feedback strength coefficient increases after island takes place so islanding detecting would be finished in a shorter time. The result is the same as when traditional algorithm is adopted when K = 1. Island could not be detected when k = 0.5 because k surpasses the limitation in equation (5).

In the case of reactive power consumption of local load is greater than the rated output power of the inverter, the simulation diagram as shown in Figure 5, it can be seen that the voltage RMS offset direction is the opposite of Figure 4, improved detection methods significantly reduced detection time, improve the detection efficiency.

In addition, when the grid runs the active island for regular detection reason, just set the feedback intensity coefficient K to 0, then that is the traditional droop control strategy, can realize smooth transition and off-grid.

7 CONCLUSION

In order to solve the problem that the detection efficiency and power quality can not be taken into account in the traditional island effect detection, this paper proposes a new method of detecting the island detection strategy based on droop control principle. In this method, the voltage value of PCC point voltage is given as a given value for the normal operation of the power network. Through simulation, it can be found that the detection time can be changed by adjusting the coefficient of K in the isolated island, and it will not affect the quality of the electric energy when the grid is connected.

ACKNOWLEDGEMENTS

This work was supported in part by the China State Grid Corp research project "Energy storage system grid connected/off characteristics and detection technology study".

REFERENCES

Fan Yuanliang et al. 2012. Generator control design based on droop control architecture for microgrid. *Electric Power Automation Equipment* 32(6): 125–130.

Guo Xiaoqiang & Wu Weiyang 2009. Non-devastating islanding detection for microgrids without non detection zone. *Proceedings of the CSEE* 29(25): 7–12.

Jia Shiping & Zheng Xiaoping 2004. Development of measuring methods and their implementing technologies for harmonic of power systems. *Electrical Switches*: 33–38.

Liang Jiangang et al. 2014. Improved grid-connection operation of microgrid converter based on droop control. *Electric Power Automation Equipment* 34(4): 59–65.

Liu Furong et al. 2012. Boundary conditions of voltage shift techniques for islanding detection. *Transactions of China Electrotechnical Society* 27(3): 247–251.

Liu Fangrui et al. 2008. Parameter optimization of active frequency drift with positive feedback islanding detection method. *Advanced Technology of Electric Engineering and Energy* 27(3): 22–25.

Liu Wenhua et al. 2011. Instantaneous current control and islanding detection for a three-phase grid connected photovoltaic generation system. *Journal of Tsinghua University: Science and Technology* 51(3): 345–350.

Yang Tao et al. 2012. A novel Islanding Detection Method Based On Positive Feedback Reactive Current And Frequency. *Automation of Electric Power Systems* 36(14): 193–199.

Yang Qiuxia et al. 2012. Modeling and analysis of current-disturbance based islanding detection for three-phase photovoltaic grid-connected inverters. *Automation of Electric Power Systems* 36(4): 45–49.

Yin Zhifeng et al. 2013. A new method for islanding detection of grid-connected photovoltaic inverter. *Power System Protection and Control* 41(22): 117–121.

Zhang Shi & Zhang Ruiyou 2010. Adaptive microgrid detection methods for PV grid-connected converters. *Power System Protection and Control* 38(21): 136–140.

Zhang Qi & Sun Xiangdong 2011. Even Harmonic Current Disturbing Method For Islanding Detection In The Distributed Power Generation Systems. *Transactions of China Electrotechnical Society* 26(7): 112–119.

Zheng Le et al. 2014. Research on Planning and Operation Model for Energy Storage System to Optimize Wind Power Integration. *Proceedings of the CSEE* 16: 2533–2543.

Advances in Power and Energy Engineering – Sun (Ed.)
© 2016 Taylor & Francis Group, London, ISBN 978-1-138-02846-3

Coordinated control strategy of AC/DC hybrid distribution system

X. Zhang, W. Deng & W. Pei
Institute of Electrical Engineering, Chinese Academy of Sciences, Haidian District, Beijing, China

T. Yu & T.J. Pu
China Electric Power Research Institute, Haidian District, Beijing, China

ABSTRACT: The control strategy of AC/DC hybrid system is one of the key technologies to ensure the safe and stable operation of the system. To solve this problem, this paper proposes a master slave control strategy for AC/DC hybrid system. According to the difference between the master station and the slave station, the control strategy is formulated, and the control block diagram and the detailed analysis are given. Considering the grid faults, the control strategy of converter station during the grid fault is given. The AC/DC hybrid system has the function of load transfer and AC/DC support by flexible switching VF control, PQ control and constant DC voltage control. Finally the validity and feasibility of the proposed method are verified by simulation results.

1 INTRODUCTION

As the energy crisis and environmental pollution are becoming more and more serious, the technology of grid connected power generation based on renewable energy is one of the effective measures to solve above problems. With the rapid development of grid connected PV and wind power technologies, the cumulative installed capacity of photovoltaic and wind power generation system have reached to 95.81 GW and 28.05 GW at the end of 2014 (Jiang et al. 2015). At the same time, a large number of renewable energy distributed generation system accessed to the traditional power system also presents new technical challenges.

Distributed generation unit with high penetration rate is the development trend of the future, since the continuous increase of the distributed element, the energy storage device and the DC load in renewable energy, the DC power network is widely concerned. The DC power grid has the advantages of low energy conversion; high efficiency; low cost; simple structure; no need to take into account the frequency, phase and reactive power compensation device (Zhang et al. 2015), (Justo et al. 2015). Although the DC micro grid has a unique advantage, however, due to the perfect infrastructure of the AC power grid, long-term existence of AC power supply and load, and other reasons make it difficult to replace the AC power grid, so it is a more effective way to use AC/DC hybrid power system. As the AC and DC link exists at the same time, it can be integrated with distributed unit,

energy storage device and load cell in the DC and AC bus independently, and then can get the advantages of reducing the energy conversion process, the cost of the system and so on. (Guerrero et al. 2013a, b).

AC/DC hybrid system usually include two or multiple converter stations as AC/DC interconnection device, the control strategy of interconnection device is key technology to ensure safe and stable operation of system. Concerning this issue, many scholars propose different control strategies. DC voltage control method of AC/DC hybrid system is necessary precondition to keep overall system stable, DC voltage control and schedule strategies of VSC-MTDC are proposed in (Xu et al. 2011), there are three operation modes depending on different schedule strategies, and the dc voltage coordinated control strategies are given under the three operation modes. The distributed dc voltage control strategy of VSC-MTDC is proposed in (Dierckxsens et al. 2012), the analysis results show that the system can obtain good transient performance when balance node adopt PI to control dc bus voltage and dc voltage droop control is applied for other converters. An adaptive droop control method is proposed for AC/DC hybrid system integrated electric vehicle (Pei et al. 2015), the capacity of converter stations can be determined by mesh adaptive direct search method, the line power of the transformer can be shared depending on the capacity of converter stations by adaptive droop control, more electric vehicles can be integrated and the utilization of transformer can be enhanced. The design method of MTDC system capacity and the control strategy are given in (Gavriluta et al. 2015a), transient influence degree of different converter stations capacity is analyzed under load power change conditions, the transient influence degree can be decreased by reasonable design of droop coefficients. The hierarchical control strategy of AC/DC MTDC system is proposed in (Gavriluta et al. 2015b), the optimal power flow can be achieved by adjusting the operation point based on secondary layer control. The hierarchical control strategy of AC/DC hybrid system is proposed in (Egea-Alvarez et al. 2015), the three-layer control strategy targets are analyzed, the design scheme of controller parameters is given. The adaptive droop control strategy based on minimum loss of dc system is proposed for AC/DC hybrid system, the adaptive droop coefficients can be determined by short circuit analysis and optimal design (Khazaei et al. 2015).

Previous research on the control strategy of AC/DC hybrid system mainly concentrated on DC voltage control and converter power sharing control. In this paper, a master-slave control strategy for AC/DC hybrid systems is proposed, different control strategies are developed according to the control target of master-slave converter, at the same time considering the grid faults, the corresponding control methods of converter station are given, the flexible switching of the three control schemes is capable of achieving continuous load current supply and mutual support between AC and DC systems, and improving the stability of the system. Finally, the simulation platform is built by simulation software and the methods are verified by simulation.

2 STRUCTURE AND CONTROL STRATEGY OF HYBRID SYSTEM

2.1 System structure

The structure of the three-terminal AC/DC hybrid system is shown in Figure 1.

There are three converter stations, and different converter stations are connected by the DC bus. The DC bus can integrate the distributed generation units, energy storage devices and the DC loads. The DC bus voltage can keep stable by the control of flexible DC interconnection devices. At the same time, the control method of the converter stations can be changed to support AC voltage and frequency, and ensure uninterrupted power supply for primary AC load during the AC system failure.

2.2 Control strategies of master converter station

The control strategies of converter stations mainly include: the master-slave control and droop control.

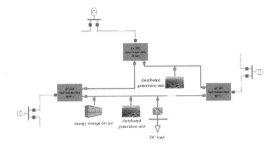

Figure 1. The system structure of AC/DC hybrid system.

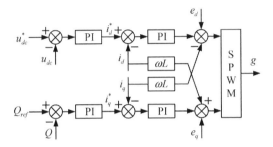

Figure 2. The control strategy of master converter station.

The paper adopts the master-slave control method, and the control structure of master-slave control is shown in Figure 2. The master converter station adopts the constant dc voltage and reactive power control strategy. First, the reference values of dc voltage and reactive power minus the actual values of dc voltage and reactive power can get outer-loop error signal, and the errors are sent into PI regulator to obtain the reference value of d-axis and q-axis current. The decoupling and feedforward control methods are applied to eliminate the d-q axis coupling and the impact of grid voltage disturbance. Finally, the three-phase modulated signals are sent into SPWM module to generate drive pulse.

2.3 Control strategies of slave converter station

During the grid operation, the PCC voltage is clamped by the grid voltage, and the output active power and reactive power of the inverter can be expressed as:

$$P = \frac{3}{2}\left(e_d i_d + e_q i_q\right) \tag{1}$$

$$Q = \frac{3}{2}\left(e_q i_d - e_d i_q\right) \tag{2}$$

where P = output active power of the inverter; Q = output reactive power of the inverter; $e_d = d$ axis component of PCC voltage; $e_q = q$ axis component of PCC voltage; $i_d = d$ axis component of the inverter output current; and $i_q = q$ axis component of the inverter output current.

When the d axis grid voltage is synchronized to the PCC voltage via a PLL so that $e_q = 0$, and substitute it into (1) and (2):

$$P = \frac{3}{2} e_d i_d \qquad (3)$$

$$Q = -\frac{3}{2} e_d i_q \qquad (4)$$

The reference values of active and reactive current can be calculated by (3) and (4):

$$i_{dref} = \frac{2P_{ref}}{3e_d} \qquad (5)$$

$$i_{qref} = -\frac{2Q_{ref}}{3e_d} \qquad (6)$$

where P_{ref} = reference value of inverter output active power; Q_{ref} = reference value of inverter output reactive power; $i_{dref} = d$ axis reference value of output current; and $i_{qref} = q$ axis reference value of output current.

The control target of slave converter station is to control active power and reactive power using the upper layer dispatching commands, and the control structure is shown in Figure 3. First, the reference values of active power and reactive power are obtained by upper layer dispatching center, substitute them into (5) and (6) can obtain current reference values. Comparing with control structures of Figures 2 and 3, the current internal loop is the same, and need not to be repeated here.

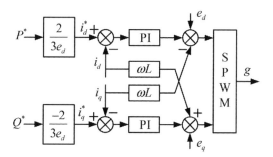

Figure 3. The control strategy of slave converter station.

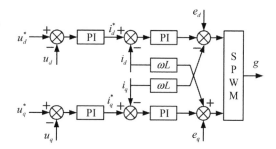

Figure 4. The control strategies of converter station under grid fault.

2.4 The control strategy of converter station during the AC system fault

The control strategies for master and slave converter station have been given under normal grid voltage conditions. However, in actual operation the grid is often affected by external disturbance, and the influence of short-circuit fault is the biggest. When short-circuit fault occurs in grid, the control methods of converter station usually are changed to support AC voltage and frequency and keep uninterrupted power supply of primary load. In order to deal with the problem, the constant voltage constant frequency control method is applied for converter station, the control structure can be shown in Figure 4.

As shown in Figure 4, the voltage error signals can be obtained by d-q axis reference values of PCC voltage minus d-q axis actual values of PCC voltage, and then sent voltage error signals into PI regulator to obtain d-q axis reference values of current internal loop, the closed-loop current control is the same as in Figure 2, and need not to be repeated here.

3 SIMULATION RESEARCH

In order to verify effectiveness and feasibility of proposed control method, the simulation platform of three-terminal AC/DC hybrid system is built by Matlab/Simulink simulation software in the paper, and the system structure is shown in Figure 5. The RMS value of grid line voltage is 10 kV; the Y/Δ transformer is applied; and the dc bus voltage is 10 kV. The converter station 1 is master converter station, and the others are slave converter station.

3.1 Simulation results of normal grid-connected operation

Figure 6 shows the simulation results of the three-terminal AC/DC hybrid system under normal operation conditions. According to the simulation

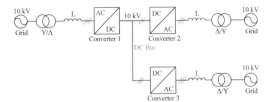

Figure 5. The simulation structure of three-terminal AC/DC hybrid system.

results, the DC bus voltage is controlled by the converter station 1 under initial condition, the constant active power and reactive power control method is applied for converter station 2 and 3, the reference value of active power and reactive power are 1 MW and 0 MVar for converter station 2, the reference value of active power and reactive power are 2 MW and 0 MVar for converter station 3. The actual active power and reactive power can track the reference value of active power and reactive power as shown in Figure 6(d) and (e). At t = 0.2 s, the active power reference values of converter station 2 and converter station 3 are changed to –1 MW and –2 MW, respectively. Then dc voltage increases instantaneously due to power flow reversal, and the dc voltage can keep constant by the control of the master converter station, and active power balanced again. The converter station 1 absorbs the active power of 3 MW, and the power factor is –1 since the phase position of grid voltage and current are opposite, as shown in Figure 6(f). At t = 0.4 s, the active power reference values of converter station 2 and converter station 3 are changed to 1 MW and 2 MW, respectively. Then dc voltage decreases instantaneously due to power flow reversal. In a similar way, the dc voltage can keep constant under active power balance conditions, and the output active power of converter station 1 is 3 MW, and then the power factor is 1 since the phase position of grid voltage and current are the same. The simulation results show that the dc bus voltage can be effectively controlled by master converter station under power reference change conditions, and the active power and reactive power can track power reference values for slave converter station, the system has good dynamic and steady-state performance.

3.2 *Converter station 2 grid-side faults*

The simulation results of AC/DC hybrid system under grid-side faults occurred in converter station 2 are shown in Figure 7. According to the simulation results, at t = 0.3 s, grid fault occurs in the converter station 2, the PQ control is switched to constant

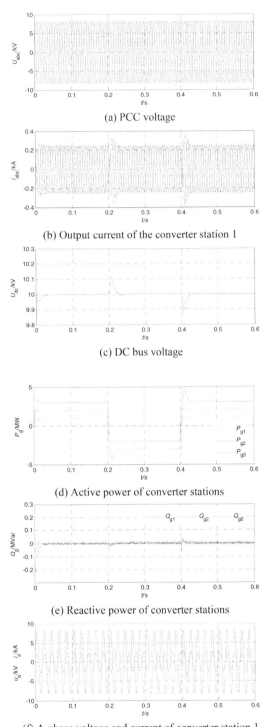

(a) PCC voltage

(b) Output current of the converter station 1

(c) DC bus voltage

(d) Active power of converter stations

(e) Reactive power of converter stations

(f) A phase voltage and current of converter station 1

Figure 6. The simulation results of AC/DC hybrid system under normal conditions.

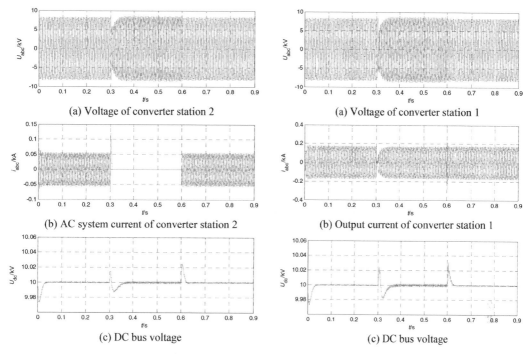

(a) Voltage of converter station 2

(b) AC system current of converter station 2

(c) DC bus voltage

Figure 7. The simulation results of AC/DC hybrid system under converter station 2 grid-side faults.

(a) Voltage of converter station 1

(b) Output current of converter station 1

(c) DC bus voltage

Figure 8. The simulation results of AC/DC hybrid system under converter station 1 grid-side faults.

voltage constant frequency control for converter station 2 to ensure uninterrupted power supply for primary AC load, and then the PCC voltage can keep constant, as shown in Figure 7(a). At t = 0.6 s, grid fault of converter station 2 is cleared, the constant voltage constant frequency control is switched to PQ control for converter station 2 to control active power and reactive power, the feasibility and effectiveness of the proposed control strategies are verified by the simulation results.

3.3 Converter station 1 grid-side faults

The simulation results of AC/DC hybrid system under grid-side faults occurred in converter station 1 are shown in Figure 8. According to the simulation results, at t = 0.3 s, grid-side fault occurs in the converter station 1, the constant DC voltage and reactive power control is switched to constant voltage constant frequency control for converter station 1 to ensure uninterrupted power supply for primary load. At the same time, the PQ control is switched to constant dc voltage and reactive power control for converter station 2 to keep dc bus voltage constant, and achieve DC network stable. At t = 0.6 s, the grid faults is cleared, the control strategy of each converter station returns to the initial state by the conversion of control mode.

4 CONCLUSIONS

In this paper, the coordinated control method for AC/DC hybrid system is proposed. Firstly, the control strategies of master converter station and the slave converter station are given under normal and grid fault conditions. The control structures of different control strategies are also given. The simulation platform of three-terminal AC/DC hybrid system is built by Matlab/Simulink simulation software. The simulation researchs are carried out under normal and grid fault conditions, respectively. The simulation results show that the system can keep stable under power reference values change and power flow reversal conditions and can achieve uninterrupted power supply for primary loads and AC-DC support control by the conversion of control mode

ACKNOWLEDGMENT

This work was supported in part by the National High Technology Research and Development

Program (863 Program) under Grant 2015AA050102 and in part by the National Natural Science Foundation of China under Grant 51407177.

REFERENCES

Dierckxsens, C., Srivastava, K., Reza, M., Cole, S., Beerten, J., Belmans, R. 2012. A distributed DC voltage control method for VSC MTDC systems. *Electric Power Systems Research* 82: 54–58.

Egea-Alvarez, A., Beerten, J., Hertem, D.V., Gomis-Bellmunt, O. 2015. Hierarchical power control of multiterminal HVDC grids. *Electric Power Systems Research* 121: 207–215.

Gavriluta, C., Candela, I., Citro, C., Luna, A., Rodriguez, P. 2015a. Design considerations for primary control in multi-terminal VSC-HVDC grids. *Electric Power Systems Research* 122: 33–41.

Gavriluta, C., Candela, I., Luna, A., Gomez-Exposito, A., Rodriguez, P. 2015b. Hierarchical Control of HV-MTDC Systems With Droop-Based Primary and OPF-Based Secondary. *IEEE Transactions on Smart Grid* 6(3): 1502–1510.

Guerrero, J.M., Chandorkar, M., Lee, T.L., Loh, P.C. 2013a. Advanced Control Architectures for Intelligent Microgrids—Part I: Decentralized and Hierarchical Control. *IEEE Transactions on Industrial Electronics* 60(4): 1254–1262.

Guerrero, J.M., Loh, P.C., Lee, T.L., Chandorkar, M. 2013b. Advanced Control Architectures for Intelligent Microgrids—Part II: Power Quality, Energy Storage, and AC/DC Microgrids. *IEEE Transactions on Industrial Electronics* 60(4): 1263–1270.

Jiang, L.P., Wang, C.X., Huang, Y.H., Pei, Z.Y., Xin, S.X., Wang, W.S., Ma, S., Brown, T. 2015. Growth in Wind and Sun: Integrating Variable Generation in China. *IEEE Power and Energy Magazine* 13(6): 40–49.

Justo, J.J., Mwasilu, F., Lee, J., Jung, J.W. 2013. AC-microgrids versus DC-microgrids with distributed energy resources: A review. *Renewable and Sustainable Energy Reviews* 24: 387–405.

Khazaei, J., Miao, Z.X., Piyasinghe, L.S., Fan, L.L. 2015. Minimizing DC system loss in multi-terminal HVDC systems through adaptive droop control. *Electric Power Systems Research* 126: 78–86.

Pei, W., Deng, W., Zhang, X., Qu, H., Sheng, K. 2015. Potential of Using Multiterminal LVDC to Improve Plug-In Electric Vehicle Integration in an Existing Distribution Network. *IEEE Transactions on Industrial Electronics* 62(5): 3101–3111.

Xu, L., Yao, L. 2011. DC voltage control and power dispatch of a multi-terminal HVDC system for integrating large offshore wind farms. *IET Renewable Power Generation* 5(3): 223–233.

Zhang, X., Pei, W., Deng, W., Du Y., Qi, Z.P., Dong, Z.M. 2015. Emerging smart grid technology for mitigating global warming. *International Journal of Energy Research* 39: 1742–1756.

Advances in Power and Energy Engineering – Sun (Ed.)
© 2016 Taylor & Francis Group, London, ISBN 978-1-138-02846-3

Optimal protection coordination based on dual directional over-current relays for ring network distribution power systems

X.Y. Wang & X.G. Wang
China Electric Power Research Institute, Haidian District, Beijing, China

D.H. Chen, S. Su & X.N. Lin
*State Key Laboratory of Advanced Electromagnetic Engineering and Technology,
Huazhong University of Science and Technology, Wuhan, Hubei Province, China*

ABSTRACT: In terms of Distributed Generation (DG), it is momentous to ensure a fast and reliable protection system for the looped network to avoid involuntary DG disconnection during fault conditions. In this paper, dual setting directional over-current relays are put forward to protect meshed distribution systems with DG. Compared to the conventional approach, which is using a directional relay of one-setting, the new protection coordination scheme counts on the dual setting relays. Dual setting relays are equipped with two inverse time-current characteristics whose settings will depend on the fault direction. The protection coordination problem for the dual setting directional relay is elaborated as a nonlinear programming problem where the objective is to minimize the overall time of operation of relays during primary and backup operation. The proposed scheme is applied to the power distribution network, which is equipped with synchronous and inverter-based DG. The results show that the new protection coordination scheme with the dual setting relay can considerably reduce the overall relay operating time, making it an attractive option for distribution systems with DG.

1 INTRODUCTION

Relay protection is vital for any power system, and protection coordination is an important aspect of the design of the protection system. It must ensure that the system can be fast, reliable and have a selection of the fault zone to be cut off. For the transmission system with an interconnected protection system, the direction of the relay is very popular for its high cost and favor. Due to the high permeability of distributed power supply, people pay close attention to the smart grid, and the distribution system is gradually transforming from the traditional radiation pattern to the ring network.

Generally speaking, the integration of distributed power supply will have different effects on the distribution system (Gao Fei-ling & Cai Jin-din 2008). The influence of distributed generation on the protection system is mainly determined by the type of distributed power supply and the structure of the protected system (Radiation pattern or ring network). In Literature (N. Nimpitiwan, G.T. Heydt, R. Ayyanar, and S. Surya-narayanan 2007), it is pointed out that the distributed generation of synchronous distributed power will create higher fault current and have worse effect than the distributed power which has contravariance. As the fault current of the inverter type distributed power supply is usually 1 to 2 units, so its influence is very small.

The typical distribution system usually adopts the recloser fuse and over-current protection relay to protect the system. For such a system, inverter-based distributed generation's influence almost can be ignored (H.B. Funmilayo, J.A. Silva, and K.L. Butler-Purry 2012). On the contrary, in case of a synchronous distributed power supply, the fuse may be fused before the action of the recloser, thereby, affecting the energy saving strategy of the fuse. Similarly, the synchronous distributed power supply may cause the whole feeder to have no fault trip (Zhang Chao, J.I. Jianren, X.I.A. Xiang 2006). In order to solve this problem, it is necessary to replace the fuse or reset the recloser and the overcurrent relay (J.A. Silva, H.B. Funmilayo, and K.L. Butler-Purry 2007). It is pointed out in the literature (S. Conti and S. Nicorta 2009) that for the distribution system with distributed power supply, the new protection scheme is dependent on an automatic remote control system, which will self-initiate after the tripping off of the line protection device. Literature (Wang Cheng-shan, Wang Shou-xiang 2008; Yuan Chao, Wu Gang, Zeng Xiang-jun 2009; Jing Qiang 2012) pointed out the changes and changes in the concept of distribution protection, especially the continuous development of distributed power research. For the ring network power distribution system, after the fault was cut off, distributed power supply can continue to supply other lines. It is important for protecting the device to cut off the system quickly, even though the low voltage duration is very short, but it can improve the fault handling capability of the distributed power supply (I. Erlich, W. Winter, and A. Dittrich 2006). Due to the characteristics of fault current in the distribution system of the ring network, the bidirectional relay has a great advantage (H.H. Zeineldin, E.F. El-Saadany, and M.A. Salama 2005). In order to overcome the influence of the distributed power, the distributed power must be considered in the optimization of relay parameters (Wu Gang, Lu Yu-ping 2007; W.K.A. Najy, H.H. Zeineldin, and W. L. Woon 2013). There are many ways to optimize the setting, such as traditional and heuristic techniques that can be used to set the optimal parameters (TDS, I_p) to ensure the total time of the coordination between the device and the minimization of the relay action. Other protection coordination strategies can improve the performance of the protection system, especially the system performance with distributed power supply. For example, the use of directional overcurrent relays digital capability, through the use of different or improved relay set of parameters and characteristics (Zeng Qi, Li Xing-yuan 2007), or by the use of digital relay communication potential (M. Khederzadeh 2012). As it is pointed out in the literature (M. Dewadasa, A. Ghosh, and G. Ledwich 2011), for the power system, usually the fault current flows in two directions from the point of failure, so you can use the big advantage which is that two-way relays can make different responses in two directions. Literature (Li Rui, Li Yue, Jiang Zhen, Zhao Yannan, Xu Hao, Liu Hai Tao, Wang Peng 2013), describes the principle and criterion of this new directional element, and it is verified by simulation. It is proved that the directional element based on the combined sequence constituent can be used to judge the direction of fault signal and can achieve high sensitivity and reliability.

A new protection coordination scheme based on two-way relay is proposed in this paper. Each relay is equipped with two sets of two directions, i.e., two time settings and two starting current values. The implementation and application of the new scheme is aimed at the distribution system with synchronous distributed power supply system. The parameters of the relays are established in the nonlinear programming model.

2 THE NEW WAY TO CONFIGURE PROTECTION WITH BIDIRECTIONAL OVERCURRENT RELAY

2.1 New scheme of protection configuration based on overcurrent relay with bidirectional setting

According to the time—current characteristic of the relay, the relay will act when the constraint condition of the relay protection coordination scheme is satisfied, thus, the action time limit is optimized. The protection coordination scheme makes it clear that which relay is the standby relay in the same system. Figure 1 shows an example of a three-bus system with six directional overcurrent relays. The traditional way of coordinating the relay is shown in Table 1. For example, if a fault occurs at the A point, the R_1 acts as the main relay, and R_5 will act as a standby relay when it refuses to act as the main relay. Similarly, for the same fault location as the main relay will be responsible for the fault, and when it refuses to move, R_6 will play a role of backup protection action.

New scheme makes full use of the advantages of the bidirectional relay flexibility and coordination ability, when a line fails, relays will act in the positive direction and also the negative one, but corresponding to their respective characteristics of the relay.

The time characteristic of the bidirectional relay is given as Figure 2. When the fault current flows in the positive direction set before, the relay will act as the main protection, and the other one acts as the backup protection. Relay has two pairs of different settings, TDS_{fw}, I_{pfw} are the parameters of the main protection. $TDSrv$, $Iprv$ is the parameters of the backup protection. Figure 3 shows an example of

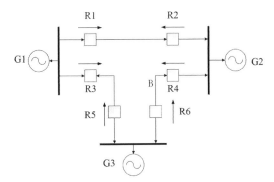

Figure 1.   Conventional directional overcurrent relays.

Table 1.   Primary/backup relay pairs based on conventional and proposed protection coordination schemes for a three-bus system.

Main relay	Backup relay in traditional scheme	Backup relay in new scheme
R1	R5	R3
R2	R6	R4
R3	R2	R1
R4	R1	R2
R5	R4	R6
R6	R3	R5

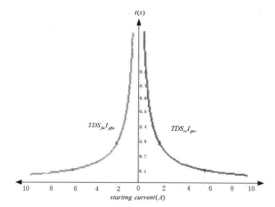

Figure 2. Time-current characteristics—the dual settings DOCR.

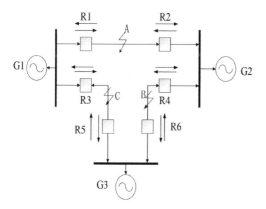

Figure 3. Protection with dual setting directional relays.

a three-bus system with six directional overcurrent relays. Each relay has two arrows to indicate the flow of the fault current, so as to indicate which of the relay can act. The bidirectional relay used by the new scheme is listed in Table 1. When the failure occurs at the point of A, R_3 will be used as the backup protection of R_1 and R_4 will be used as the backup protection of R_2. At the same fault point, when R_1 starts positive protection action, R_3 will start reverse protection action.

2.2 The way to specify the state of protection component

The fault location technology has been widely used and studied in the transmission network and distribution network. It has become a very mature technology, so it is assumed that the fault can be accurately located when the method of the configuration of the protection device is studied in this paper. It is specified that the configuration status of each protection device means a protection device plays the role of main protection or backup protection. For a given system, this paper uses the fault location to directly query the configuration state of the protection components, rather than the local current size and the current direction of the protection components to determine the protection configuration state. The configuration state of each protection device in Figure 3 is showed in Table 2.

In order to increase the fault tolerance capability of the protection device, that is, the system can still have the backup protection configuration when a protective device is out of operation because of its own power supply failure or its own hardware fault. In this paper, we assume that the failure of a large number of protection devices will not occur, and then we use an extended configuration method to confirm the configuration state of each protection device. This method is in fact to specify a reserve backup protection device for a given backup protection device on specific conditions. When a backup protection device fails to work, the reserve backup protection device automatically serves as backup protection and does its own work. Here, we give an example to illustrate this kind of mechanism. For example, as Table 3 shows, when fault happens at the point A, R3 is appointed to be the Backup Protection device (BP) and R5 is appointed to be the Reserve Backup Protection device (RBP). Then if R3 fails to work for some reason, R5 will upgrade to be the backup protection device immediately to take the place of R3 to execute the function of backup protection, in this way it can be avoided as the system loses its backup protection when it is under extreme conditions, and so the system fault tolerance is promoted.

Table 2. The configuration state of each protection device.

Fault location	Main protection	Backup protection
A	R1,R2	R3,R4
B	R4,R6	R2,R5
C	R3,R5	R1,R6

Table 3. The extended configuration for protection device.

Fault location	BP	RBP
A	R3	R5
	R4	R6
B	R2	R1
	R5	R3
C	R1	R2
	R5	R6

3 USE BIDIRECTIONAL RELAY TO SOLVE THE PROBLEMS OF PROTECTION COORDINATION

Action time of over current relay is in inverse ratio to its short circuit current, On the whole, it can be described as formula 1:

$$t_{ij} = TDS_i \frac{A}{\left(\dfrac{I_{scij}}{I_{pi}}\right)^B - 1} \tag{1}$$

where i = relay identifier; j = fault location identifier; constant parameters B and A are usually different, which is relative to the change of the type of the overcurrent relay, where the A is set to 0.14, and the B is set to 0.02. I_{scij} represents the current that flows through over current relay I_{pi} that represents the starting current.

As formula (1) shows, each directional overcurrent relay has a pair of main and backup action apparatus. The goal of optimization is to minimize the action time of relay protection device in the case of all protection coordination condition being fulfilling. The sum (T) of all objective functions can be described as formula (2):

$$T = \sum_{J=1}^{N} \left(\sum_{i=1}^{N} t_{fwij} + \sum_{k=1}^{N} t_{rvkj} \right) \quad \forall (i,k) \in \Omega \tag{2}$$

where Ω = collection of backup relays and the main relays; N = total number of relays; M = total number of all feeders, t_{fwij} = forward action time of relay I; and t_{rvkj} = reverse action time of rlay k in fault point j. Their linear relationship can be described as formula (3) (4):

$$t_{fwij} = TDS_{fwi} \frac{A}{\left(\dfrac{I_{scfwij}}{I_{pfwi}}\right)^B - 1} \tag{3}$$

$$t_{rvkj} = TDS_{rvk} \frac{A}{\left(\dfrac{I_{scrvkj}}{I_{prvk}}\right)^B - 1} \tag{4}$$

TDS_{fwi} and TDS_{rvk} are, respectively, the forward time setting value and reverse time setting value of relay i and relay k. I_{pfwi} and I_{prvk}, respectively, represent the forward and reverse current of relay i and relay k. As the current in the fault point flows through relay i, so it is marked as I_{scfwij}. Similarly, the current flowing through relay k is marked as I_{scrvkj}. Protection coordination problems can be solved on the condition that the following formula (5) must be met:

$$t_{rvkj} - t_{fwij} \geq CTI \qquad \forall i, k, j \tag{5}$$

CTI represents the minimum value of the action time difference between the main relay action and the standby relay. This time interval is usually between 0.2 and 0.5 seconds. In this paper, it is set to 0.3 seconds, in addition, the value of these values should be satisfied, just as formula (6), (7) shows:

$$I_{pi\_min} \leq I_{pfwi}, \qquad I_{prvi} \leq I_{pi\_max} \tag{6}$$

$$TDS_{i\_min} \leq TDS_{fwi}, \qquad TDS_{rvi} \leq TDS_{i\_max} \tag{7}$$

where I_{pi_min} = minimum value of the starting current of relay I; I_{pi_max} = maximum value of the starting current of relay I; TDS_{i_min} = minimum value of the setting time; and TDS_{i_max} = maximum value of the setting time.

Key variables that need to be optimized for protection coordination are TDS and I_p in two directions. The short circuit current is considered as the optimal parameter, but the location and capacity of the distributed power supply has a certain effect on the fault current level. The research in this paper has been done to analyze the fault current flowing through each relay before optimizing the relay setting. As it can be seen from formula (3) and (4), the action time of the relay is a nonlinear relationship between the starting current. Thus, the model is described as a nonlinear programming problem, an integrated master, backup protection model is considered to be the one most close to the protection scheme of coordination.

4 MODELING AND SIMULATION ANALYSIS

The two-way relay protection coordination scheme is tested in a modified partial system of IEEE30, as shown in Figure 4. This system has three substations; their voltage is 132/33 kV; their capacity is 50 MVA; and connecting points are 2812. The 33 kV side is equipped with 28 directional overcurrent relays, and each distributed power supply works according to the rated 5 MVA and the power factor is all unified. Distributed power generation unit's inlet wire system goes through a 480/33 kV boost transformer. Node is set in the middle of each line, so as to make three short circuit analyses.

In the conventional protection scheme, the system is equipped with 28 directional overcurrent relays, and each relay is equipped with a set of time and current setting. For example, if the fault node is F20, then the R17 and R18 are used as the

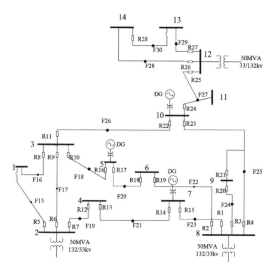

Figure 4. Modified distribution portion of the IEEE 30-bus system.

Table 4. Optimal relay TDS and Ip settings considering the conventional protection coordination.

Relay	TDS (s)	I_p (p.u)	Relay	TDS (s)	I_p (p.u)
1	0.1	0.855	15	0.3513	0.0684
2	0.1	0.6811	16	0.2773	0.0605
3	0.1	1.0708	17	0.1	0.4648
4	0.1	0.1395	18	0.1	0.4558
5	0.1	0.0768	19	0.1	0.2412
6	0.1	0.6758	20	0.2888	0.0789
7	0.1	0.7312	21	0.5966	0.0166
8	0.1	0.0196	22	0.2243	0.0961
9	0.1	0.3229	23	0.4176	0.0627
10	0.1	0.5952	24	0.1	0.1661
11	0.1	0.1975	25	0.2836	0.1502
12	0.1	0.2967	26	0.1	0.2174
13	0.1	0.538	27	0.1	0.0622
14	0.1	0.5503	28	0.1	0.0367

main relay. In the traditional protection coordination scheme, the relay R_{10} is used as the standby for R17, and R2 as a backup for R18. And in the proposed new scheme, the system is equipped with 28 relays, each relay is equipped with two sets, a forward one and a reverse one. Similarly, when the fault point is still F20, the main relay is still R17 and R18, and R16 will replace R10 as R17's standby and R19 will replace R2 as R18's standby. MATLAB's minimal constrained nonlinear multivariable function is used to solve the optimization model, this built-in MATLAB function depends on the gradient (first-order optimality) for solving nonlinear constrained optimization problem. Several algorithms can be used to solve the nonlinear programming problem. For example, SQP algorithm is adopted to solve the protection coordination model.

In this section, for the loop network distribution system with the distributed power supply, optimization scheme for traditional relay protection setting and that for a new relay protection setting are put forward, the comparison between these two schemes is used to highlight the superiority of the new scheme.

4.1 Comparison between the new scheme and the traditional scheme of the system with distributed power

In order to evaluate the performance of the new scheme, the model and the solution of the traditional scheme are first to be carried out. Table 4 shows that when the failure occurs between the node F15 and the node F30, the optimized settings of the main relay and the standby relay TDS and I_p. The sum of the relay's action time "T" uses 63.27 seconds like the traditional protection scheme. Each of the main standby relay has to meet the protection coordination constraints which is that the minimum coordination time interval should be 0.3 second.

New protection coordination plans are still tested in the same system, just as it is said before, each of the aforementioned relay use two sets of protection settings and it will not be repeated here. Table 5 gives two sets of optimal parameters. As it can be seen from the Table, in addition to the four relays, labeled as 5827 and 28, the other relays have a set of positive protection and also one of negative protection. For these four relays, as the fault current can only flow from the positive direction (with a normally open switch), they are, thus, only equipped with a set of protections.

From the Table 4 and Table 5, we can see that in the new protection scheme, the starting current, especially forward current, is smaller than that in the traditional protection scheme. For example, the starting current of the R1 is 0.855 p.u. in the conventional scheme, and the startup current in the new protection scheme is 0.064 p.u. The reason is that in the traditional protection scheme, R1 is only equipped with a set of protection and it must meet not only the constraints as the main relay, but also the constraints as the standby relay for R14. What needs to be pointed out is that for the same targets R1 relay, when it comes with two sets of protection, whether as a main relay or backup relay, it has its corresponding protection, so as to play its performance more effectively. As it is mentioned earlier, the application of two-way

Table 5. Optimal relay dual settings with the proposed protection coordination scheme.

Relay	TDS_{fw} (s)	I_{pfw} (p.u)	TDS_{rv} (s)	I_{prv} (p.u)
1	0.1	0.0684	0.2038	0.0684
2	0.1	0.0929	0.1031	0.1182
3	0.1	0.0789	0.2619	0.0789
4	0.1	0.0856	0.1	0.0856
5	0.1	0.0767		
6	0.1	0.1815	0.1007	0.1815
7	0.1	0.0746	0.1816	0.0746
8	0.1	0.0196		
9	0.1	0.1815	0.153	0.1815
10	0.1	0.0605	0.2321	0.0605
11	0.1	0.0547	0.24	0.0547
12	0.1	0.0746	0.1843	0.1117
13	0.1	0.0395	0.1977	0.1106
14	0.1	0.0395	0.1773	0.102
15	0.1	0.0684	0.2008	0.1107
16	0.1	0.0605	0.1766	0.1011
17	0.1	0.0279	0.1708	0.1041
18	0.1	0.0279	0.1850	0.1137
19	0.1	0.0673	0.1820	0.1061
20	0.1	0.0789	0.1862	0.111
21	0.1	0.0166	0.1754	0.1489
22	0.1	0.0197	0.1443	0.523
23	0.1	0.0627	0.1761	0.0627
24	0.1	0.0243	0.2637	0.0243
25	0.1	0.0485	0.1	0.1661
26	0.1	0.2174	0.1	0.0716
27	0.1	0.0622		
28	0.1	0.0367		

Table 6. Sample of optimal primary and backup relays operating times considering Near/Far points faults.

	P = primary, b = backup the unit is second		
Fault point	p	b_1	b_2
Fault point F18 nearby (traditional scheme)	R10 0.5433	R6 0.8433	R22 1.0989
	R16 0.934	R18 1.245	–
Fault point F18 far (traditional scheme)	R10 0.8565	R6 1.1565	R22 1.5898
	R16 0.857	R18 0.974	–
Fault point F18 nearby (new scheme)	R10 0.1699	R9 0.4699	R11 0.6202
	R16 0.2501	R17 0.6146	–
Fault point F18 far (new scheme)	R10 0.2055	R9 0.6414	R11 0.8981
	R16 0.2076	R17 0.5076	–

Table 7. Effect of fault resistance on total relay operating time-conventional and proposed coordination scheme.

Resistance (Ω)	Traditional scheme	New scheme
0	63.27 s	32.047 s
0.01	63.2734s	32.0512s
0.05	63.2763s	32.0681s
0.1	63.2795s	32.09 s
0.2	63.2842s	32.1337s
0.5	63.2874s	32.2924s

relay will produce a new protection scheme, the new protection scheme makes all relay's operating time to reduce obviously, compared with the traditional protection scheme the time decreased to about 50%, so the new protection scheme can significantly reduce the sum of all relays' operating time in reticular distributed system with distributed power supply.

4.2 Comparison between the new scheme and the traditional scheme of the system with distributed power

In order to explore the performance of the proposed new scheme further, we then study the situation of the fault point being in the line near the end and the remote, respectively, and give a summary of the three cases to make a comprehensive comparison. Table 6 briefly shows action time of relay R18 in two kinds of fault locations when on condition that a synchronous distributed power supply with a capacity of 5 MVA has fault in node 5, 7, and 10. Capacity of 5 MVA synchronous distributed power link node 5, 7, and 10. We have derived, through the comparison of the traditional method and the new scheme, irrespective of the fault that occurring at the proximal end of the relay or the distal end of the relay. The new scheme can effectively reduce the action time. Under the same condition, as it is shown in Table 7, when the fault resistance is not the same, the new scheme once again has demonstrated its superiority.

5 CONCLUSION

In this paper, a new scheme is proposed, which is based on two-way relay, which is equipped with the protection of negative direction and also that of positive direction and in each direction the relay has isolated setting values. As a result, each relay is designed to have two pairs of protection, and a pair of protection is responsible for their own direction. (For example TDS_{fw} and I_{pfw} are responsible for positive protection), however, the new scheme based on the application of this relay is different from the schemes before, which is described as an optimization problem, and the two optimal parameters of each relay is appointed. In the previous simulation experiment, compared to the previous use of traditional directional overcurrent relay protection scheme, the new scheme's simulation results show absolute superiority, and the action time of the relay was reduced by approximately 50%. Two-way relays can achieve faster fault isolation, thus, increasing the ability of distributed power going through the fault. Although, this may bring about economic and technological profit, this kind of extra function has raised its own production cost compared with the traditional relay.

REFERENCES

Conti S. and Nicorta, S. "Procedures for fault location and isolation to solve protection selectivity problems in MV distribution networks with dispersed generation," Elect. Power Syst. Res., vol. 79, no. 1, pp. 57–64, Jan. 2009.

Dewadasa, M., Ghosh, A., and Ledwich, G. "Protection of distributed generation connected networks with coordination of overcurrent relays," in Proc. 37th IEEE Annu. Conf. Ind. Electron. Soc. (IECON), Melbourne, VIC, Australia, Nov. 2011, pp. 924–929.

Erlich, I., Winter, W., and Dittrich, A. "Advanced grid requirements for the integration of wind turbines into the German transmission system," in Proc. IEEE PES Gen. Meeting, Montreal, QC, Canada, 2006, pp. 1–6.

Funmilayo, H.B., Silva, J.A., and Butler-Purry, K.L. "Over-current protection for the IEEE 34-node radial test feeder," IEEE Trans. Power Del., vol. 22, no. 2, pp. 459–468, Apr. 2012.

Gao Fei-ling, Cai Jin-din. Analysis for distributed generation impacts on current protection in distribution networks[J]. Journal of electric power science and technology, 2008, 23(3): 58–61.

Jing Qiang. Distributed power failure characteristics analysis and Research on the principle of micro grid protection [D]. Tianjin: Tianjin University, 2012:14–75.

Khederzadeh, M. "Adaptive setting of protective relays in micro-grids in grid connected and autonomous operation," in Proc. 11th Int. Conf. Develop. Power Syst. Protect. (DPSP), Birmingham, U.K., 2012, pp. 1–4.

Li Rui, Li Yue, Jiang Zhen, Zhao Yannan, Xu Hao, Liu Hai Tao, Wang Peng. Proceedings of the CSEE, Vol.33 Supplement Dec. 30, 2013.

Najy, W.K.A., Zeineldin, H.H., and Woon, W.L. "Optimal protection coordination for microgrids with grid connected and islanded capability," IEEE Trans. Ind. Electron., vol. 60, no. 4, pp. 1668–1677, Apr. 2013.

Nimpitiwan, N., Heydt, G.T., Ayyanar, R. and Surya-narayanan, S. "Fault current contribution from synchronous machine and inverter based distributed generators," IEEE Trans. Power Del., vol. 22, no. 1, pp. 634–641, Jan. 2007.

Silva, J.A., Funmilayo, H.B., and Butler-Purry, K.L. "Impact of distributed generation on the IEEE 34-node radial test feeder with over-current protection," in Proc. 39th North Amer. Power Symp. (NAPS), Las Cruces, NM, USA, 2007, pp. 49–57.

Wang Cheng-shan, Wang Shou-xiang. Study on some key problems related to distributed generation system[J]. Automation of Electric Power System, 2008, 32(20): 1–4, 31.

Wu Gang, Lu Yu-ping, et al. Impact of Fault Current Limiter to the Performance of Relay Protection in Distributed Generation[J]. Jiangsu Electrical Engineering, 2007.

Yuan Chao, Wu Gang, Zeng Xiang-jun. Protection technology for distributed generation systems[J]. Power System Protection and Control, 2009, 37(2): 99–105.

Zeineldin, H.H., El-Saadany, E.F., and Salama, M.A. "Optimal coordination of directional overcurrent relays," in Proc. Power Eng. Soc. Gen. Meeting, 2005, pp. 1101–1106.

Zeng Qi, Li Xing-yuan, et al. Research of a New Fault Current Limiter Controlled by PWM[J]. North China Electric Power, 2007.

Zhang Chao, Ji Jianren, Xia Xiang. Effect of distributed generation on the feeder protection in distribution network[J]. Relay, 2006, 34(13): 9–12.

Advances in Power and Energy Engineering – Sun (Ed.)
© 2016 Taylor & Francis Group, London, ISBN 978-1-138-02846-3

Analysis on water requirement of Concentrating Solar Power plant in western China

X.Y. Wang, L.Z. Zhu, L. Zhang & R.Y. Yu
China Electric Power Research Institute, Nanjing, China

ABSTRACT: For constructing a Concentrating Solar Power (CSP) plant, water resource is an important factor that should be considered by designers and developers. As a potential site for the large-scale deployment of a CSP plant, the water requirements of the CSP plant in western China have been studied using the System Advisor Model (SAM) software. The results of the comparative analyses show that dry cooling is a good choice for the CSP plant to be reducing the water consumption. Taking a 50 MW CSP plant with 7.5 full load storage hours as an example, a CSP plant with dry cooling only needs a water intensity of 0.293 m³/MWh, which is about 12 times lower than the wet cooling CSP with a water intensity of 3.505 m³/MWh. Furthermore, among these influencing factors, electric power output is the key point for the water requirement, while dry bulb temperature is the key for the parasitic energy requirement in western China, due to the alpine region.

1 INTRODUCTION

As one of the sustainable energy sources, Concentrating Solar Power (CSP) has been attracting more and more interests. According to the method of solar energy collection, there are three available CSP technologies including parabolic trough, power tower, and parabolic dish (Zhang, H.L. et al. 2013). Among them, parabolic trough and power tower have been in commercialized operation, which adapt to development in large scale. The function principle is similar to the conventional coal power plant. With the help of large arrays of mirrors, Direct Normal Irradiation (DNI) is reflected and concentrated onto a receiver where it is converted to heat, then the heat is used to drive a traditional steam Rankine cycle for power generation.

Considering the particularity of site selection of the CSP plant, the regions with high values of DNI are normally located in arid and semi-arid regions like western China (Qu, H. et al. 2008) and the Middle Eastern and North African region (MENA) (Liqreina, A. & Qoaider, L. 2014), where water recourse is scarce and expensive. Water requirement of the CSP plant is receiving more attention (Uzgoren, E. & Timur, E. 2015, Colmenar-Santos, A. et al. 2014). Studies show that 90%~95% of the water is consumed during the cooling process (Cohen, G. et al. 2005), and how much water required mainly depends on the choice of cooling technology. Wet cooling and dry cooling are the common cooling technologies. For wet cooling, a separate circuit of water is used to dissipate the heat energy to the environment. And ambient air is used for dry cooling, and no water is consumed. In general, dry cooling is less efficient than wet cooling in dissipating heat, due to much lower specific heat capacity of air. This means that the reduction in water consumption from dry cooling is at the cost of performance penalty. In this work, we comparatively analyzed the water requirement of CSP plant in western China for both wet and dry cooling options, using System Advisor Model (SAM), as well as the performance influence factors.

2 METHODOLOGY

2.1 Description of SAM

SAM is chosen as the modeling software for this study, which is a performance and financial model based on Transient Systems Simulation (TRNSYS) (NREL. 2011). SAM has been widely used in the simulation of renewable energy applications and specifically in CSP system analysis. For CSP system, SAM characterizes many system components from the first principles of heat transfer and thermodynamics. The condenser and all of the associated heat rejection equipment are modeled outside of the power cycle in detail due to the complication of multiple possible heat rejections. The wet and dry cooling models are included in SAM, which models both the water use and performance impacts of the different cooling technologies.

2.2 Site selection

DNI map of China shows that the solar source is abundant in China, but greatly diverse in various areas. The annual direct normal solar radiation is lesser than 700 kWh/m^2 in the southeast and more than 2500 kWh/m^2 in west. High DNI levels, vast deserts, and low-cost components make western China the most promising destination for large-scale deployment of CSP technology, such as West Qinghai, Southeast Xinjiang, Northwest Gansu, North Ningxia, and West Tibet. The first 10 MW phase of SUPCON 50 MW tower plant in Delingha and Qinghai provinces has been operated successfully, which is the first commercialized CSP station in China. The 50 MW Qinghai Delingha parabolic trough plant developed by China General Nuclear Power Group (CGN) is also being in construction. In addition, a notification released by National Energy Administration in 2015 informed that a number of CSP demonstration projects are to be constructed. These bring brighter light to China's CSP industry. So this work selects one place in western China as the case study site, which represents an attractive destination for CSP development.

2.3 Plant specifications

The reference that the CSP plant chose for this study is configured to have the main features of Andasol-1 plant in Spain (Solar Millennium, A.G. 2008, Herrmann, U. et al. 2002). Andasol-1 is the first commercial parabolic trough power plant in Europe, which has a capacity of 50 MW. Besides a solar field and a conventional power plant section, Andasol-1 integrates a two-tank molten salt heat storage system, allowing the plant to operate at a full capacity for 7.5 hours at night or during overcast periods. Thus, the annual operating hours can be increased greatly, and the dispatchability can be improved. As a grid-friendly technology, CSP with storage is expected to play an important role in the development of renewable energy. Further specifications of the plant are shown in Table 1.

3 RESULT AND DISCUSSION

3.1 Solar energy resource characteristics

Annual DNI distribution is shown in Figure 1, which indicates that the case study site has a good solar energy resource (ca. 2007 kWh/m^{2}a) for CSP plant. DNI in the latter half of the year is better than that in the first half. The highest monthly DNI value appears in October.

3.2 Comparison of water requirement

Table 2 represents the monthly water requirement when adopting the wet cooling and dry cooling option. It can be clearly found that the CSP plant with dry cooling requires much less water than that with wet cooling. The amount of water consumption of wet cooling plant is approximately 440,960 m^3/a, whereas, the amount of dry cooling plant is 34,611 m^3/a, saving above 90%. It is well known that the water consumption at a Rankine steam CSP plant occurs during the steam cycle, mirror washing, and mostly the cooling process. As for the plant with dry cooling, it only consumes roughly the same amount of water as the wet cooling plant for steam make-up and mirror washing.

Table 1. Technical specifications of the reference CSP plant.

Description	Numerical value
Nominal capacity (MW)	50
Aperture area of solar field (m^2)	510,120
Number of SCAs (–)*	624
Number of loops (–)	156
Number of SCAs per loop (–)	4
Length of SCA (m)	144
Aperture width of SCA (m)	5.77
Number of mirrors (–)	209,664
Number of modules per SCA (–)	12
Number of HCEs (–)**	22,464
Heat-transfer fluid	Dowtherm A
Design solar field inlet temperature (°C)	293
Design solar field outlet temperature (°C)	393
Full load storage hours (h)	7.5
Storage media	Solar salt

*SCAs, Solar Collector Assemblies; **HCEs, Heat Collector Elements.

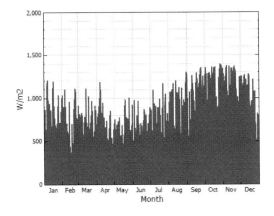

Figure 1. DNI of the case study site.

Table 2. Monthly water requirements of the wet cooled and dry cooled CSP plants.

Month	Wet cooled (m³)	Dry cooled (m³)	Saving (%)
January	16,272.6	2242.9	86.2
February	18,400.0	2319.8	87.4
March	27,701.1	2615.5	90.6
April	35,305.9	2852.3	91.9
May	38,976.8	2942.5	92.5
June	38,979.1	2931.9	92.5
July	47,111.3	3179.2	93.3
August	50,859.2	3302.2	93.5
September	53,444.2	3402.8	93.6
October	53,802.2	3436.4	93.6
November	38,513.3	2972.4	92.3
December	21,594.2	2413.4	88.8
Sum	440,959.9	34,611.3	92.2

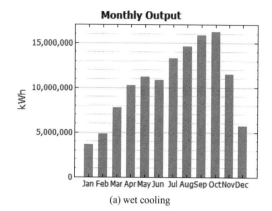

(a) wet cooling

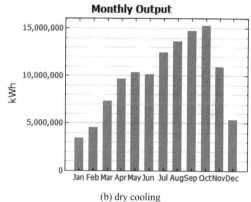

(b) dry cooling

Figure 2. Monthly electric power output of CSP plant with different cooling option: (a) wet cooling; (b) dry cooling.

Hence, dry cooling system is a potential alternative to wet cooling.

As shown in Table 2, for both wet and dry cooling, the peak of water requirement is observed in October, which is consistent with that of the DNI value. It might be due to a great amount of electricity generated for better solar source.

3.3 Influence factors analysis

3.3.1 Power output

Water requirement is a function of electricity produced in a CSP plant, so the amount of electric power output every month is studied, shown in Figure 2. Clearly, in spite of cooling options, similar monthly output profiles are presented. Compared to that of wet cooling, the power output of dry cooling has a slight decrease for every month. The annual power output is 125,796 MWh/a and 118,004 MWh/a for wet and dry cooling, respectively. This is in good agreement with the results of other researchers (EPRI. 2008).

To further describe the relationship between the water requirement and electric power output, we investigate the index "water intensity," which is defined as the water consumed per Megawatt-hour (MWh) of electricity produced (Carter N.T. & Campbell R.J. 2009).

As can be seen from Table 3, the water intensity of CSP plant using wet cooling is higher than that of other generation technologies, such as dry cooling CSP and wet cooling fossil thermal plant. As mentioned above, for dry cooling CSP, no water is needed during cooling process. However, for wet cooling fossil thermal plant, it may be because the water intensity reduces with the increase of installed unit. In general, the installed capacity of

Table 3. Water intensity of the power generation technology.

Technology	Water intensity (m³/MWh)
CSP-wet	3.505
CSP-dry	0.293
Fossil Thermal-wet	1.269–2.211
CSP-wet (ref)*	3.142–3.785
CSP-dry (ref)*	0.303–0.341

*The data is derived from report of others (Carter N.T. & Campbell R.J. 2009).

CSP unit is smaller (50 MW class) and different from the common installed capacity of thermal power generating units (600 MW/1000 MW).

3.3.2 Dry bulb temperature

Dry bulb temperature, also known as air temperature, is an important parameter for CSP plant, especially for dry cooling plant. On the one hand, dry bulb temperature determines whether the field freeze protection should be started or not, because the heat transfer fluid in the solar filed may cool to an unacceptably low temperature during times of extremely low temperature or extended shutdown. On the other hand, it limits efficiency of dry cooling. The higher the temperature is, the lower the efficiency will become. This eventually results in rising of water intensity. Herein, we make statistical analysis on the distribution of dry bulb temperature.

Figure 3 displays the scaled histogram of dry bulb temperature in the case study site. It can be observed that the dry bulb temperature ranges from –25 °C to 30 °C, mainly from –10 °C to 20 °C. And the number of hours of the temperature below zero is more than 3000 in one year. This suggests the case study site belonging to the alpine region. Effect of temperature on cooling efficiency is positive, so dry cooling option as a water-saving option is technically feasible in the study site. It is worth noting that the dry bulb temperature has an important influence on field freeze protection. In such circumstances, supplemental heat must be provided to maintain the temperature of the heat transfer fluid in the solar field at the minimum value. During operation of CSP plant, parasitic energy requirement of each section will be discussed in future works.

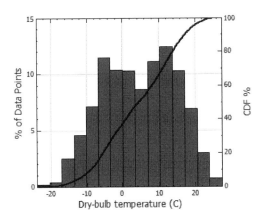

Figure 3. Scaled histogram of the dry bulb temperature in the case study site.

4 CONCLUSION

In conclusion, comparative analysis on water requirement of CSP plant in western China shows that dry cooling is technically feasible and is a good alternative to wet cooling for CSP deployment in water-constrained areas. The 50 MW CSP plant with 7.5 full load storage hours and dry cooling only has a water consumption of 34,611 m^3/a, reduced by 92.2% if compared to the wet cooling CSP. Although, the results are site specific, they can provide a reference to CSP plant development in other alpine regions.

REFERENCES

Carter, N.T. & Campbell, R.J. 2009. Water Issues of Concentrating Solar Power (CSP) Electricity in the U.S. Southwest.

Cohen, G. et al. 2005. Solar thermal parabolic trough electric power plants for electric utilities in California. *Solargenix Energy. Los Angeles, CA.*

Colmenar-Santos, A. et al. 2014. Water consumption in solar parabolic trough plants: review and analysis of the southern Spain case. *Renewable & Sustainable Energy Reviews* 34, 565–577.

EPRI. 2008. New Mexico Central Station Solar Power: Summary Report.

Herrmann, U. et al. 2002. The Andasol project—Workshop on Thermal Storage for Trough Power systems.

Liqreina, A. & Qoaider, L. 2014. Dry cooling of Concentrating Solar Power (CSP) plants, an economic competitive option for the desert regions of the MENA region. *Solar Energy* 103, 417–424.

NREL. 2011. Technical Manual for the SAM Physical Trough Model.

Qu, H. et al. 2008. Prospect of concentrating solar power in China-the sustainable future. *Renewable and Sustainable Energy Reviews* 12, 2505–2514.

Solar Millennium, A.G. 2008. The parabolic trough power plants Andasol 1 to 3.

Uzgoren, E. & Timur, E. 2015. A methodology to assess suitability of a site for small scale wet and dry CSP systems. *International Journal of Energy Research* 39 (8), 1094–1108.

Zhang, H.L. et al. 2013. Concentrated solar power plants: Review and design methodology. *Renewable and Sustainable Energy Reviews* 22, 466–481.

Advances in Power and Energy Engineering – Sun (Ed.)
© 2016 Taylor & Francis Group, London, ISBN 978-1-138-02846-3

Island operation control strategy for battery micro-grid based on RT-LAB

T. Yan & Z.Z. Qu
China Electric Power Research Institute, Nanjing, China

G.W. Zhu & Z.B. Liu
North China Electric Power University, Beijing, China

ABSTRACT: Micro-grid requires a stable voltage and frequency during island operation. For the variation in load and the fluctuation of PV and wind output, micro-grid needs Battery Energy Storage System (BESS) to balance the active and reactive power. This paper built an experiment platform of island battery Micro-Grid based on the RT-LAB, and the difference between the inductance current inner loop and the capacitive current inner loop was compared. Therefore the traditional v/f control was improved. Finally, this paper established the simulation model under Matlab/Simulink, and did hardware-in-the-loop simulation verification on the RT-LAB platform. The simulation and experimental results are provided to verify the effectiveness of the proposed control strategy.

1 INTRODUCTION

Micro-grid can be run from the grid-connected or island operation model. The latter has a distribution network as an infinite bus, and the control strategy is simple. Micro-grid requires a stable voltage and frequency during island operation. For the variation in load and the fluctuation of PV and wind output, micro-grid needs Battery Energy Storage System (BESS) to balance the active and reactive power.

The traditional control strategy for energy storage system uses v/f monocyclic control. The literature-Chen Zuo (2010) proposed traditional constant voltage constant frequency control in view of the LC filter type inverter. The output voltage monocyclic control method simplified the structure of controller and the control system, but the voltage harmonic content is much and the resistance to load current characteristics is poor; Literature-Huang Weihuang (2013), analyzed the closed loop controller design. For the current inner ring, the proportion controller is needed to improve the dynamic performance of the system and eliminating the static error can be done by resonance voltage outer loop controller. Literatures-(Yang Zilong 2010a,b, Sun Hai 2014) analyzed the advantage that by LC filter unit the BESS can be able to filter out harmonic, and stabilize voltage stability.

From all above, this paper built experiment platform of island battery Micro-Grid based on the RT-LAB, and the difference between the inductance current inner loop and the capacitive current inner loop was compared. Therefore the traditional v/f control was improved. Finally, this paper established the simulation model under matlab/simulink, and did hardware-in-the-loop simulation verification on the RT-LAB platform. The simulation experimental results verify the effectiveness of the proposed control strategy.

2 BUILDING EXPERIMENT PLATFORM ON RT-LAB

This paper studies off-grid operation control of the energy storage system based on the RT-LAB experiment platform. The main circuit including lithium battery energy storage, load and three-phase bridge uses the simulation software, and the embedded controller (DSP) makes up the control part. First the bidirectional converter simulation model is set up and run on the RT-LAB simulation system, using analog output control card to change the linear sampling ac and dc voltage and current of the converter into—3 v to 3 v level sampled by controller; And then the six-way PWM signal generated by the embedded controller through the digital input control card inputs RT-LAB simulation model of the target machine to control six IGBT tubes of the three-phase bridge, in which the function of rectifier and inverter is implemented-Guo Ruizhou (2013). Semi-physical real-time simulation system schematic diagram is shown in Figure 1.

In the diagram above, the battery model uses the RT-LAB standard lithium battery model, and the energy storage system uses three-phase bridge circuit and LC filter unit. In order to reduce the resonance of the system, the damping resistor in series with the shunt capacitance was adopted, so that the effect of peak suppression was achieved. In the Micro-grid, constant power and R-L-C load model was adopted.

Energy storage converter program flow of the embedded controller was shown in the following Figure 2. First frequency of 50 Hz was generated spontaneously, the outer ring voltage control section accepted voltage signal through the upper control unit and obtained the reference value of d and q axis by constant voltage control after coordinate change, and set expectations of active current and reactive current by PI adjustment, finally and through the current inner loop control, the SVPWM produces six-way PWM pulse. Program control block diagram of off-grid Energy storage system is shown in Figure 2.

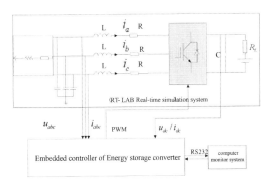

Figure 1. Hardware-in-the-loop simulation schematic.

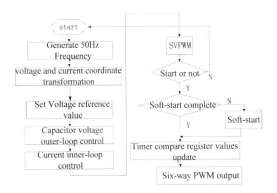

Figure 2. Program control diagram.

3 ENERGY STORAGE SYSTEM CONTROL

During energy storage system off-grid operation mode, the load requires a stable voltage and frequency. Load current resistance of traditional voltage monocyclic control is weak, the harmonic component in the load current is apt to cause the resonance of output voltage-Majumder R et al. 2011. The inductance current inner loop and the capacitive current inner two double loop control methods were analyzed, and the traditional constant voltage constant frequency control was improved. Voltage outer loop uses PI regulator. Because the current inner ring is mainly used to improve the system stability and dynamic performance, steady state error will not affect the precision of the outer ring-Park S H et al.2014, so the current inner ring uses the proportion. Two kinds of control mode block diagrams are shown in Figure 3.

In the Figure 3, $u_d^* u_q^*$ is d and q axis component of the reference voltage of capacitance voltage u_c, using voltage orientation. u_{cd} u_{cq} is d and q axis component of capacitance voltage u_c, i_{cd} i_{cq} is d and q axis component of capacitance current i_c. i_{id} i_{iq} is d and q axis component of inductance current i_i, u_{id} u_{iq} is d and q axis component of inverter voltage u_i. Simplified control ring structure as shown in Figure 4 (d axis for example).

Where K_{pwm} is equivalent gain of PWM, the value is 1. τ is PWM control parameter considering small inertia characteristics. Generally it is 0.5 times value of switch cycle. T_s is the system switching cycle. i_{od} is load current d axis component for energy storage systems, expression of the output voltage is obtained by Figure 4, and is shown as follows.

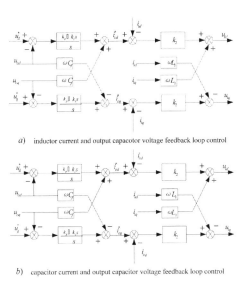

a) inductor current and output capacitor voltage feedback loop control

b) capacitor current and output capacitor voltage feedback loop control

Figure 3. Two control mode block diagram.

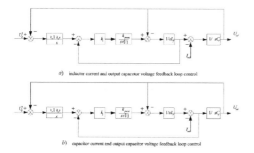

a) inductor current and output capacotor voltage feedback loop control

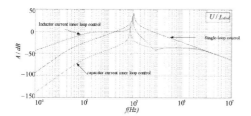

b) capacitor current and output capacitor voltage feedback loop control

Figure 4. The simplified system structure diagram.

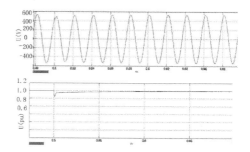

Figure 5. U_{cd}/I_{od} bode curve of three control mode.

Capacitance voltage outer loop inductance current inner 2-ring control:

$$U_{cd} = \frac{k_2(k_p s + k_i)}{\left(\begin{array}{c} C_f L_f \tau s^4 + C_f L_f (1+k_2)s^3 + \tau s^2 \\ + (k_2 k_p + 1)s + k_2 k_i \end{array} \right)} U_{dref}$$
$$- \frac{L_f \tau s^3 + L_f s^2 + k_2 s}{\left(\begin{array}{c} C_f L_f \tau s^4 + C_f L_f (1+k_2)s^3 + \tau s^2 \\ + (k_2 k_p + 1)s + k_2 k_i \end{array} \right)} I_{od}$$

(1)

Capacitance voltage outer ring capacitance current inner 2-ring control

$$U_{cd} = \frac{k_2(k_p s + k_i)}{\left(\begin{array}{c} C_f L_f \tau s^4 + C_f L_f (1+k_2)s^3 + \tau s^2 \\ + (k_2 k_p + 1)s + k_2 k_i \end{array} \right)} U_{dref}$$
$$- \frac{L_f \tau s^3 + L_f s^2}{\left(\begin{array}{c} C_f L_f \tau s^4 + C_f L_f (1+k_2)s^3 + \tau s^2 \\ + (k_2 k_p + 1)s + k_2 k_i \end{array} \right)} I_{od}$$

(2)

The expression consists of two parts. By the expression (1) and (2), when the inductance and capacitance current is applied to make the inner control, the voltage command tracking capability of the system is the same. The latter part of the expression can represent the system ability to resist load current changes, defined as U_{cd}/I_{od}. Amplitude frequency characteristics of three kinds of control method U_{cd}/I_{od} are showed in Figure 5.

By the above diagram, at low frequency stage, the amplitude frequency numerical characteristics of the capacitive current inner loop control is small, thus it has good resistance to load current performance and inhibit the impact on the output voltage effectively. At middle frequency stage, the resistance to load current performance of traditional loop control is poorer. At high frequency stage, the resistance to load current performance is the same. From the above, capacitor voltage outer ring and capacitive current inner loop control scheme has better resistance to load current performance and stabilizes the voltage in the permitted range.

4 SIMULATION

The energy storage system of 150 kVA were built up based on matlab/simulink, and the resistor is in series with the shunt capacitance, specific parameters are shown in Table 1.

The simulation process is as follows: at t = 0 s, energy storage system started up, and set up the voltage and frequency of the micro-grid. At t = 0.5 s, the load increased by 50 kW, the simulation results are shown in Figure 6.

By the Figure 6, under the inductance current inner loop control, the d axis component of output voltage (d axis overlaps the three-phase voltage

Table 1. Energy storage parameter.

The inverter capacity S	150 kVA
DC voltage	700 V
Filter L_f	5 mH
Filter C_f	150 uF
Damping resistance R	0.5 Ω
Switching frequency f_s	5000 Hz
Control parameter k_2	5
Control parameter k_p	5
Control parameter k_i	100
Phase voltage reference V_{ref}	311 V
Frequency f	50 Hz

Figure 6. Energy storage system output voltage (inductance current inner loop).

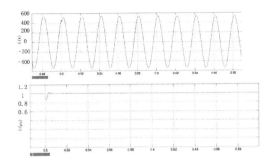

Figure 7. Energy storage system output voltage (capacitor current inner loop).

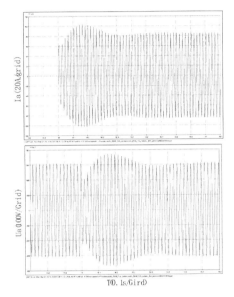

Figure 8. Energy storage system output current voltage waveform.

vector, on behalf of the output voltage amplitude) jumped in a large amplitude when the load increased, and needed a long time to restore the voltage rating. From Figure 7, under the capacitive current inner loop control, the energy storage system in microgrid can start quickly, and established the voltage and frequency. In the process of load change, it was able to recover voltage rating quickly.

5 EXPERIMENT ON RT-LAB

On RT-LAB experiment platform, 100 kVA energy storage system adopted the advanced v/f control. At t = 0 s, energy storage system start up. At t = 15 s, the load increased by 10 kW. The experimental results are shown in Figure 8.

The first image of Figure 8 shows that energy storage system provided stable current quickly after increasing load; the second image of Figure 8 is the voltage waveform. In the process of increasing load, the output voltage fluctuated in permissible range.

6 CONCLUSION

During island operation, the energy storage system establishes the stable voltage and frequency. This paper improved the traditional v/f control, using capacitive current inner loop control. The control strategy has better resistance to the load current performance. It is simulated under matlab/simulink, and to verify the effectiveness of the proposed control strategy finally on the RT-LAB experimental platform.

ACKNOWLEDGEMENTS

This work was supported in part by the China State Grid Corp research project "Energy storage system grid connected/off characteristics and detection technology study".

REFERENCES

Chen Zhuo. 2010. The Constant Voltage Constant Frequency Control Strategy for Three-phase SPWM Inverter. *Electrical Engineering*, 2010, 12: 24–26+38.

Guo Ruizhou. 2013. Research on the control mechanism of battery energy storage converter and construction of its hardware-in-the-loop simulation platform. Beijing: North China Electric Power University.

Huang Weihuang. 2013. Generalized Analysis and Design of Closed-loop Control Strategies for LC-filtered Voltage Source Inverters. *Automation of Electric Power Systems*, 2013, 37(19): 110–115.

Majumder R., Chakrabarti S., Ledwich G., et al. 2011. Control of battery storage to improve voltage profile in autonomous microgrid. *Power and Energy Society General Meeting*, IEEE. IEEE, 2011: 1–8.

Park S.H., Choi J.Y. & Won D.J. 2014. Cooperative control between the distributed energy resources in AC/DC hybrid microgrid. *Innovative Smart Grid Technologies Conference (ISGT)*, IEEE PES. IEEE, 2014: 1–5.

Sun Hai. 2014. Operating Characteristics Analysis and Control Strategy Research of Microgrid Stability Controller. Beijing: North China Electric Power University.

Wang Y., Tan K.T., So P.L. 2013. Coordinated control of battery energy storage system in a microgrid. *Power and Energy Engineering Conference* (APPEEC), IEEE PES Asia-Pacific. IEEE, 2013: 1–6.

Yang zi-long. 2010. Design of Three-phase Inverter System with Double Mode of Grid-connection and Stand-alone. *Power Electronics*, 2010(1): 14–16.

Advances in Power and Energy Engineering – Sun (Ed.)
© 2016 Taylor & Francis Group, London, ISBN 978-1-138-02846-3

A review of the Energy Storage Systems applied in power grids

B. Han & S.F. Lu
Xi'an Jiaotong-Liverpool University, Suzhou, China

L. Jiang
Liverpool University, Liverpool, UK

Y. Du
Xi'an Jiaotong-Liverpool University, Suzhou, China

ABSTRACT: Energy Storage Systems (ESS) play a crucial role in modern power systems owing to the integration of renewable energy sources. The utilization of ESS becomes a necessary component in the solution of electricity grids operation issues like stability, power quality, and systems balancing. In this review paper, based on their own intrinsic characteristics, various ESS technologies will be discussed including the working principle, application status, and development prospects for the future. Furthermore, advantages and disadvantages of ESS will be evaluated from the viewpoint of power systems. In particular, this paper will concentrate on the application of Battery Energy Storage Systems (BESS) technologies in distribution systems. The BESS management and control algorithms will be examined based on the effectiveness of BESS in power quality, peak load shaving, system balancing, and frequency control.

1 BACKGROUND

In recent years, faced by the growth of global electricity consumption, environmental problems and fossil fuel limitation issues, the penetration of renewable energy sources in the power generation sector are rapidly increasing, which reduces the dependency on traditional sources and, thus, decreases waste gas and CO_2 emissions. But there are also some disadvantages of renewable energy. One of the significant drawbacks of renewable sources is intermittency. For example, solar and wind power are uncertain, varying, and environmentally dependent. With a large penetration of renewable sources, various economics and stability challenges of power grids may immerge, such as power generation and consumption imbalance, power quality, and congestion management problems.

Among the solutions in solving a series of power system problems, one of the effective and promising approaches is to use Energy Storage Systems (ESS). ESS performs as a bank, which can convert electrical energy from the power grid to a storable form of energy and the stored energy can be converted back into electrical energy to power grids when needed (Chen *et al.* 2009). Different types of ESS have different characteristics, and thus perform differently to improve the performance of power systems (Miller *et al.* 2010). The main functions of ESS include:

1. Peak load shaving;
2. Power quality improvement;
3. Voltage/frequency control;
4. Congestion management.

2 CATEGORY OF ENERGY STORAGE SYSTEMS

Various types of ESS can be categorized by several criteria from different viewpoints. Generally, ESS can be grouped into three categories based on discharge duration and storage capacity. The classification is shown in Table 1.

Normally, the form of energy stored is the most widely used method for ESS classification, based on which ESS can be categorized into four groups: (1) Electrical; (2) Mechanical; (3) Chemical; and (4) Thermal. It is shown in Table 2. The detailed characteristics of each ESS will be discussed in Sections 2.1–2.4.

2.1 *Pumped Hydroelectric Storage (PHS) systems*

Pumped hydroelectric storage is a widely implemented storage system, which also has the long history in ESS technologies. As a mature technology, PHS takes almost 99% worldwide storage capacity and it is a vital part of future power grids. Figure 1

Table 1. ESS Classification 1.

Power density systems		Energy density systems
Short-term (seconds to minutes)	Long-term (minutes to hours)	Real-long term (hours to days)
Supercapacitor	Battery Systems	Pumped Hydroelectric Systems
Super Magnetic Energy Storage	Fuel Cell	Caged Air Energy Storage

Table 2. ESS Classification 2.

Electrical	Mechanical	Chemical	Thermal
Capacitor	Flywheel	Lead-acid Battery	Sensible Heat Storage
Supercapacitor	Pumped Hydro Systems	Nickel-metal Hydride Battery	Latent Storage
Super Magnetic Energy Storage	Caged Air Energy Storage	Lithium-ion Battery Fuel Cell	

Figure 1. Energy storage capacity in the US in 2011(Luo et al. 2015).

shows the ESS capacity in the US in 2011. Existing PHS units have various capacities ranging from 1MW to 5000MW, with an efficiency of 70%–85% and more than 40 years life time. In 1890s, PHS was firstly used in Switzerland and Italy and the first large scale commercial application is Rocky River PHS in USA, implemented in 1929. At present, PHS units are installed all around the world with around 21 GW in China, 27 GW in Japan, around 32 GW in EU and19.5 GW in US. The total capacity of PHS takes 3% of global generation capacity (Chen *et al*. 2009). In China, the Fengning PHS station started to construct based on the key project of "*12th five-year plan for Development*" with the capacity of 3.6 MkW (Li *et al*. 2015).

The structure of PHS consists of an upper reservoir, a lower reservoir, pump/turbine, and a motor/generator. It allows the conversion between electrical energy and water potential energy. During off-peak hours, the water is pumped from the lower reservoir to the upper reservoir. Water was stored at uphill for temporary potential storage. During peak hours, the stored water is released from the upper elevation, through the turbine/generator and finally stored in the lower reservoir. During this process, the gravitational potential energy is converted back to the electrical energy (Faias *et al*.) (Verma 2013).

In the past, PHS was built purely for electricity production. That is supplying the electricity

to the power systems in the day time (demand higher than supply) and pumping in the night time (demand lower than supply). Nowadays, PHS is also used for water management. Furthermore, the future PHS development will consider adopting Underground Pumped Hydro (UPH) systems, which overcomes the location limitation of traditional PHS (Pickard 2011).

2.2 *Compressed Air Energy Storage (CAES) systems*

Besides PHS, CAES is another type of commercial storage technology that can provide very large energy storage capacity with over 100MW capacity for a single unit. CAES works are based on conventional gas turbine generation. During off-peak hours, electricity energy is used to compress air, and then the compressed air is stored in a high pressure in underground caverns or over-ground tanks (air tight space) with a pressure 4.0 MPa–4.8 MPa. When the generation cannot meet the demands, the stored high-pressure air is released and is heated by a heat source. Finally, the gas turbine generates energy from the high temperature and high pressure air.

Similarly with PHS, CAES has the characteristics of large capacity, long storage period, and high efficiency. Typically, CAES capacity ranges from 50MW to 300MW, storage period can be more than one year with very small losses, and efficiency ranges from 70% to 89% (Zhang *et al*. 2012).

Two large scale CAES units exist in the world for commercial purposes at present. In 1978, the first CAES systems Huntorf was put into operation in Germany, with 290MW output power, located at around 600 m underground with a carven of 310,000 m³. The maximum pressure of the systems can reach to 10 MPa. This CAES plant was integrated into the power grids for more than

5000 times. So far, it shows excellent performance with 90%–99% starting and running reliability. The other CAES plant McIntosh was under operation since 1991 in Alabama, USA. This unit is located 450 m underground and the compressed pressure reaches up to 7.5 MPa, with a rated output power of 110MW. In 1992, The McIntosh consumed 46745MWh for compressing the air, and generated 39255MWh net electrical energy (Zhang et al. 2012).

There are several CAES under construction in countries like Japan and South Africa.

2.3 Flywheel Energy Storage (FES) systems

Flywheel has a long history of 6000 years and it was used as potter's wheel for earliest application. In recent times, the most common application of flywheel is found in internal combustion engine (Daoud et al. 2012). During operation, FES becomes kinetic energy storage systems, which stores electrical energy in the angular momentum of spinning mass (Chen et al. 2009). The stored energy is proportional to the square of rotational speed ω^2 and flywheel's momentum inertia I is calculated by Equation (1).

$$E_{stored} = 0.5 \times I \times (\omega_{max}^2 - \omega_{min}^2) \qquad (1)$$

The flywheel is composed of five basic components: (1) flywheel to rotate at a high velocity which stores huge kinetic energy; (2) group of bearings to provide mechanical support to the flywheel; (3) containment systems to provide a vacuum environment for reducing air resistance and minimizing the external disturbances and thus to guarantee that flywheel systems have high energy conversion efficiency typically in the range of 90%–95%; (4) motor/generator to implement the conversion between electrical energy and kinetic energy; and (5) power electronics converters to control the operation of flywheel increase or decrease rotation velocity (generate or store electricity) based on the demand.

Existing flywheel systems can be categorized into two groups: low speed systems (conventional metal rotor systems) normally having an operation speed below 6×10^3 rpm and high speed composite systems with an operation speed of typically around 10^5 rpm. The flywheel can be fully charged within a time scale of minutes, while normally battery systems take hours for charging.

2.4 Battery Energy Storage Systems (BESS)

The first battery was invented around 1800 in Italy, which consists of zinc and copper in the salt sink separated by a cardboard. The first rechargeable/secondary battery (lead-acid battery) was invented in 1859 by a French physicist. Battery stores electricity in a form of chemical energy. Chemical reactions occur inside the battery generating a flow of electrons between two electrodes through the external circuit. If external voltage is applied on two electrodes, the reaction is reversed and the battery will be charged.

There are various types of battery systems such as: lead-acid battery, nickel cadmium (NiCd) battery, sodium sulfur (NaS) battery, sodium nickel chloride (ZEBRA) battery, and lithium-ion (Li-ion) battery. Li-ion battery is a new developing technology, which has huge potential in the future.

2.4.1 Lead-acid battery

As a widely used rechargeable device with a long application history, lead-acid battery is installed all around the world, due to its advantage of low cost, technical maturity, abundant materials, and large scale manufacture. The largest one is 40MWh installed in China, and California (Chen et al. 2009). The drawbacks are low energy density (50–90 Wh/L), sensitivity to room temperature and relatively low recycling counts (up to around 2000), low energy density (50–90 Wh/L) and poisonous materials used.

Currently, there are several advanced lead-acid batteries developed, which have fast response time close to flywheels and supercapacitors (Luo et al. 2015).

2.4.2 Lithium-ion battery

The structure of lithium-ion battery consists of graphite carbon anode and lithiated metal oxide ($LiCoO_2$, $LiNiO_2$, $LiMn_2O_4$, and $LiFePO_4$) cathode. During charging period, Li atoms in cathode become ions and migrate to the carbon anode. Then ions are combined with external electrons in anode. During discharging periods, this process is reversed, and the chemical reaction occurs at two electrodes which generates electron flow in the external circuit.

Li-ion battery has characteristics of high energy density (300–400 kWh/m^3, 130 kWh/ton), power density (150–315 W/kg) and fast response time (around 20 milliseconds). In addition, it has relatively high conversion efficiency between 85%–95%, long life for 10 to 15 years, 2000 to 3000 cycling counts and 80% depth of discharge.

Li-ion battery has the characteristics of high power density & energy density, which is developing widely in transportation and other small scale applications. Table 3 shows characteristics of different ESS discussed in Section 2.

Table 3. ESS Characteristics.

	Maturity	Duration time	Capacity	Power	Efficiency	Life times
PHS	Mature	4–10h	5400–14000MWh	100–5000MW	70–85%	40–60 years
CAES	Commercial	6–20h	250MWh	100–300MW	70–89%	20–40 years
Flywheel	Demonstration	15 s–30 min	5MWh	5–10 kW	80–90%	15 years
Lead-acid	Commercial	1 min–3h	1 kW–50MW	100 kW–150MW	70–80%	5–15 years
Li-ion	Demonstration	Hours	80 kW	100 kW–100MW	90–94%	5–15 years

3 ENERGY STORAGE SYSTEMS IN POWER SYSTEMS

In 21st century, the global electric power systems are facing a dramatic revolution, due to the growth of load demand and the large integration of renewable sources. This situation raises challenges of security and flexibility to power systems. One of the effective solutions of the issues is to use energy storage systems, which can be utilized in each part of power systems. Typically, the traditional power systems structure can be considered consisting of five parts: (1) energy sources; (2) generation; (3) transmission; (4) distribution; and (5) customer. Figure 2 shows that ESS can be used in each part of power systems. Table 4 shows the advantages and disadvantages of ESS applied in power systems

3.1 *ESS in power generation*

PHS and CAES units are referred as the bulk energy storage systems. They are particularly suitable for energy management in the power generation side of the system in a scale of over 100 MW since these two systems have long lifetimes and unlimited cycle stability. PHS and CAES have good performance in balancing demand and supply, load leveling and spinning reserve when cooperate with renewable sources (Chatzivasileiadi 2012). In many countries, PHS is used to integrate with wind power generation systems forming a hybrid ESS systems named as the Wind Hydro Pumped Storage Systems.

Due to the characteristics of large capacity, one of drawbacks of PHS and CAES systems is its negative impact on environment. Construction of PHS demands a large area of lands. Moreover, PHS needs two large reservoirs and dams. Long construction time (>10 years) and high initial construction cost (hundreds to thousands of millions of dollars) are also two constraints of the wide application of PHS.

Similarly, CAES has geographic and environmental constraints for its real wide application. In general cases, CAES is suitable for the power plants near rock mines and salt caverns.

3.2 *ESS in power transmission*

Traditional transmission operates in a single direction where the electrical power is delivered from power plants to customers. Due to the penetration of distributed generators, the system turns into a bidirectional transmission (Kondoh 2000), which needs to cooperate with ESS systems for the purpose of systems stability. Supercapacitor, flywheel and BESS units are designed to install in the transmission and distribution systems since these systems have the characteristics of high power density and fast response speed.

Due to the rapidly increasing load demand, the transmission lines need to be upgraded urgently. Especially during peak hours, the load demand is beyond the maximum capacity of transmission line. Long construction period (normally 10 years to 15 years) (Masaud 2010) and high cost are the two main constraints of transmission line upgrade. Another problem is that the peak demand lasts for a short time, thus, most of time the transmission line capacity will not be fully used leading to a certain waste. Faced by these problems, ESS can be installed across the transmission line. During off-peak hours ESS can be charged; and during peak hours, electrical energy will be sent directly to the load without going through transmission line and, thus, decreases the transmission line congestion level (Suberuaet et al. 2014).

3.3 *ESS in power distribution*

In distribution networks, BESS have its special advantages of modularity, low space occupation, moveable and easy construction. Lead-acid battery and sodium batteries have a large power rating and relative long operation time. Such battery units in transmission and distribution systems can be used for load peak shaving.

Li-ion battery with a high power density and short response time would be especially suitable for power systems frequency regulation. The BESS could control the real power and thus regulate the frequency. BESS can help to reduce the absolute value of Area Control Error (ACE) in Load Frequency Control (LFC), and improve

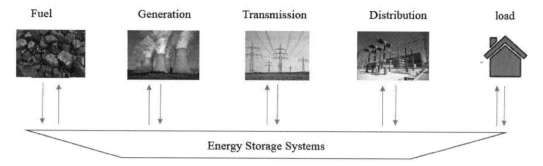

Figure 2. Energy Storage Systems in Power Systems (Chen *et al.* 2009).

Table 4. Advantages, disadvantages and applications of ESS.

	Advantages	Drawbacks	Applications
PHS and CAES	High power, large capacity, low cost	Geographical conditions, long construction period	Peak load shaving, frequency control, black starting, reserve supply
Lead-acid	Low cost, good reliability, high security	Low power & energy density, short life period	Power quality, reserve supply, black start, UPS/EPS
Li-ion	Large capacity, high power/energy density, high conversion efficiency	High cost, low security, short life period, small scale	Smooth renewable sources output, peak shaving, power quality

the systems stability (Chen, 2015). BESS also can cooperate with supercapacitor and work as hybrid energy systems (HES), which can satisfy various system application requirements (Feng *et al.* 2014).

4 CONCLUSION

In this paper, different ESS and their applications in power grids have been reviewed. ESS will play a vital rule in future power grids. ESS not only can improve economics benefits of power grids in increasing the renewable usage rate, reducing transmission line congestion, but also enhance stability of power grids like peak shaving and valley-filling. The utilization of various types of ESS is a necessity due to the large penetration of renewable energy and distributed generators.

Based on discharge duration classification, power density systems (fast response speed) could be used to regulate frequency deviation (LFC) and energy density systems (slow response speed) refer to be used for spinning reserve.

HES is a promising system combining the advantage of different ESS, such as BESS and supercapacitor that can fit different applications in future power grids.

REFERENCES

Chatzivasileiadi, K. 2012. Electrical energy storage technologies and the built environment. International renewable energy storage conference, Berlin, 12–14 November 2012.

Chen, H. Cong, T.N. Yang, W. Tan, C.Q. Li, Y.L. and Ding, Y.L. 2009. Progress in electrical energy storage system: A critical review. Progress in Natural Science 19: 291–312.

Chen, S.X. Zhang, T. Gooi, H.B Masiello, R.D. and Katzenstein, W. 2015. Penetration Rate and Effectiveness Studies of Aggregated BESS for Frequency Regulation. IEEE Transactions on Smart Grid, Vol:PP, Issue: 99.

Daoud, M.I. Abdel-Khalik, A.S. Massoud, A. Ahmed, S. and Abbasy, N.H. 2012. On The Development of Flywheel Storage Systems for Power System Applications: A Survey. International Conference on Electrical Machines: 2119–2125.

Faias, S. Santos, P. Sousa, J and Castro. An Overview on Short and Long-Term Response Energy Storage Devices for Power Systems Applications. Available: www.icrepq.com

Feng, X. Gooi, H.B. and Chen, S.X. 2014. Hybrid Energy Storage with Multimode Fuzzy Power Allocator for PV Systems. IEEE Transactions on sustainable energy, Vol.5, No.2.

Kondoh, K Ishii, I. Yamaguchi, H Murata, A. Otani, K. Sakuta, K. Higuchi, N. Sekine, S. and Kamimoto, M. 2000. Electrical energy storage systems for energy networks. Energy Conversion & Management, 41: 1863–1874.

Li, Y. Li, Y.B. Ji, P.F. and Yang, J. 2015. Development of energy storage industry in China: A technical and economic point of review. Renewable and Sustainable Energy Reviews 49: 805–812.

Luo, X. Wang, J.H. Dooner, M. and Clarke, J.2015. Overview of current development in electrical energy storage technologies and the application potential in power system operation. Applied Energy 137: 511–536.

Masaud, T.M. Lee, K. and Sen, P.K. 2010. An overview of energy storage technologies in electric power systems: what is the future. IEEE North American Power Symposium (NAPS): 1–6.

Miller, N. Manz, D. Roedel, J. Marken, P. and Kronbeck, E. 2010. Utility Scale Battery Energy Storage Systems. IEEE Power and Energy Society General Meeting: 1–7.

Pickard, W.F. 2011. The History, Present State, and Future Prospects of Underground Pumped Hydro for Massive Energy Storage. Proceedings of the IEEE 100 (2): 473–483.

Sebastian, R. and Alzola, R.P. 2012. Flywheel energy storage systems: Review and simulation for an isolated wind power system. Renewable and Sustainable Energy Reviews 16: 6803–6813.

Suberua, M.Y. Mustafa, M.W. and Bashirb, N. 2014. Energy storage systems for renewable energy power sector integration and mitigation of intermittency. Renewable and Sustainable Energy Reviews 35: 499–514

Verma, H. Gambhir, J. and Goyal, S. 2013. Energy Storage: A Review. International Journal of Innovative Technology and Exploring Engineering 3.

Zhang, X.J. Chen, H.S. Liu, J.C. Li, W. and Tan, C.Q. 2012. Research progress in compressed air energy storage system: A review. Energy Storage Science and Technology1(1): 291–312.

Advances in Power and Energy Engineering – Sun (Ed.)
© 2016 Taylor & Francis Group, London, ISBN 978-1-138-02846-3

Modeling and simulation on regenerative braking of the electric vehicle

W. Shi, Y.Q. Zhang & G.Z. Song
College of Environmental and Energy Engineering, Beijing University of Technology, Beijing, China

ABSTRACT: The deficiency of the endurance mileage has become the main obstacle to the development of electric vehicles because of the limited battery specific energy. To extend the mileage of electric vehicles, the energy management strategy based on recovering the braking energy is often used. In this paper, a simulation model using real vehicle experimental data is built and calibrated by the dynamic performance test data. On the basis of distribution of braking torque, the energy management strategy is implemented to fulfill the regenerative braking. Comparing with the original vehicle, the state of charge of the battery increases to 33.2% from 32% after ten-kilometer real running conditions and the per hundred kilometers consumption decreases to 10.61 kW·h from 13.24 kW·h, the electric energy consumption decreases by 19.3% accordingly.

1 INTRODUCTION

Electric vehicles have become the important development direction of the auto industry depending on the advantages of non-pollution and high energy efficiency (Hawkins et al. 2012). Restricted by the battery technology, it may cost expensive price for electric vehicles to reach the endurance mileage of the internal combustion engine vehicles (Nashed et al. 2012). To a certain extent, a reasonable energy control strategy can make up this defect. Braking energy consumption accounts for a high proportion of the total energy consumption in the travelling process. (Li et al. 2010) In city conditions, the percentage of braking energy is almost 47% of total energy consumptions (Zhang. 2012). An effective braking energy recovery strategy is necessary to improve fuel economy of electric cars.

This paper firstly establishes an electric vehicle model on the basis of a real electromobile. Then the dynamic performance is used to calibrate the model; the working condition is defined by experimental data. Finally, a new energy management strategy is presented and added to the model. The above procedures aim to verify the applicability of the strategy. The simulation results show that the fuel economy of electric car improves.

2 ELECTRIC VEHICLE MODEL AND ROAD CONDITION MODEL

2.1 Main parameters

The main parameters of the test electric vehicle are shown in Table 1.

The electric vehicle is equipped with lithium-ion batteries which have the energy of 26.5 KW·h. The motor is permanent-magnet synchronous motor which can speed up to 9000 RPM. The theoretical driving distance is 160 KM. The maximum gradeability is above 20% and the velocity can reach 120 km/h.

2.2 Vehicle model in CRUISE

According to the components parameters, the electric vehicle model is established in CRUISE. The model includes vehicle module, power system module, driven module, control system module and monitoring module. After the mechanical connection and electrical connection are finished (Gmbh. 2008), the accomplished model is shown in Figure 1.

2.3 Calibration of the model

In order to verify the reliability of the model, we simulate the dynamic performance of the electric vehicle. The result is shown in Table 2.

Table 1. Major parameters of the electric vehicle.

Parameters	Value
Curb weight/kg	1370
Length· width· height/mm	$3998 \times 1720 \times 1503$
Wheel span/mm	1460/1445
Battery capacity/Ah	66
Maximum torque of the motor/ N· m	144
Max speed of the motor/rpm	9000
Peak power/kW	45

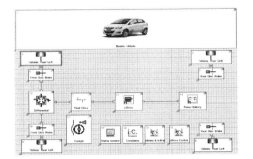

Figure 1. The electric vehicle model in CRUISE.

Table 2. The result of the simulation.

Parameters	Value
Constant velocity 60 km/h, drive distance/km	167.5
Maximum gradeability	23.35%
Max velocity/km · h⁻¹	122

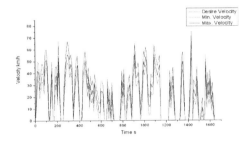

Figure 2. Velocity model in cycle run.

The simulation results prove that the model can meet the requirement of the test vehicle.

2.4 *The establishment of road condition model*

After the Vehicle Model is built, it's also necessary to set up the velocity profile. According to the experimental data of the test vehicle, Figure 2 shows that the target velocity is established in the Cycle Run task in CRUISE. And the threshold values are defined to restrict the actual velocity (maximum and minimum velocity is defined to reduce deviation).

The operating condition is actual driving condition. It sustains for 1690 Seconds. During the time the vehicle travels 10.037 kilometres, and the max velocity is 67 kilometres per hour.

3 BRAKING ENERGY RECOVERY STRATEGY

Braking energy recovery is a kind of unique technology for electric vehicle. In the process of automobile braking, a part of the braking energy can be transformed to electrical energy by the electric generator. The energy will be absorbed by the battery. So the efficiency can be improved. Accordingly the drive distance will increase. This process is accomplished by the Energy Management Strategy (Xie et al. 2010).

The quantity of the regeneration energy depends on the proportion of the electric braking. Under the premise of ensuring running safety and braking efficiency, raising the proportion of regenerative braking can improve the effect of the braking energy feedback. In order to maximize the use of regenerative braking, the control strategy is programmed by the following considerations:

1. If the desired braking torque is less than the maximum of regenerative braking torque, the braking torque is provided by regenerative braking.
2. If the regenerative braking torque cannot meet the torque requirement, then the mechanical braking will provide the rest part.
3. The torque threshold is regulated to ensure the automobile safety. If the desired braking torque exceeds the threshold value, the car is in a state of emergency braking then the braking torque will be provided by the mechanical braking.
4. In order to prevent the battery overcharge, the regeneration will be cut off when the SOC level is over 80%.

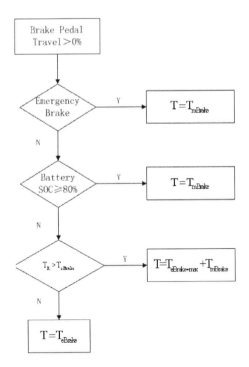

Figure 3. Flow chart of control strategy.

From the above, the flow chart of control strategy is shown in Figure 3. The strategy is implemented by the control module. Connect the module with data bus connection to finish the improved model.

4 ANALYSIS OF SIMULATION RESULT

Simulation study is fulfilled based on the constructions of vehicle model and the strategy, Figure 4 shows the simulation velocity is almost consistent with the actual velocity. Which proves that the braking performance won't be decreased after the control strategy adds to the model and guarantees the reliability of the simulation results.

The curve of the battery state of charge and current is shown in Figure 5 after a working cycle.

As shown in Figure 6, the initial SOC of the power battery is 38%. At the end of the simulation, it reduces to 33.2%. The battery state of charge decreases by 4.8%. In the actual test data, the SOC variation range is from 38% to 32%. The battery state of charge decreases by 6%. It can be seen that the electric quantity is saved by 1.2%. And the per hundred kilometers consumption decreases to 10.61 kW·h from 13.24 kW·h. The electric consumption decreases by 19.3%.

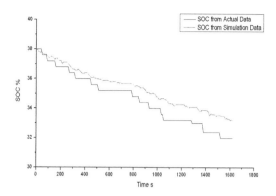

Figure 6. Comparison of experiment and simulation SOC.

5 CONCLUSIONS

In this paper, a simulation model based on an actual electric vehicle is built. The correctness of the model is verified. And a control strategy based on braking torque distribution is presented. The result shows the strategy for electric vehicle dynamic performance has a certain improvement and the fuel economy of the vehicle increases by 19.3%.

The simulation has not considered the battery charging efficiency and the influence of high-current charging to the battery. In further research, the mentioned points will be taken into account. The charging efficiency may affect the quantity of the regeneration energy and the maximum charging current should be limited to prevent the damages to the battery.

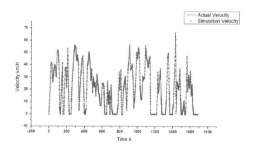

Figure 4. Comparison of actual and simulation velocity.

REFERENCES

Avl List GmbH. AVLCRUISE Users Guide [K]. 2008.
Hawkins T.R. et al. 2012. Environmental impacts of hybrid and electric vehicles-a review, The International Journal of Life Cycle Assessment, vol. 17, no. 8, 2012, pp. 997–1014.
Li Ke et al. 2010. Comparative Study of Induction Motor Efficiency Optimization Control Strategy for Electric Vehicle [C]. Proceedings of the 8th World Congress on Intelligent Control and Automation, 2010.
Nashed M.N., Wahsh S., Galal H. & Dakrory T., (2012). Application of Fuzzy Logic Controller for Development of Control Strategy in PHEV, Computer Technology and Application, vol. 3, 2012, pp. 1–7.
Xie Xing, Zhou su, Wang Tinghong, et al. A Simulation on Energy Management Strategy for the Power System of a Fuel Cell/Battery HEV Based on Cruise/SIMULINK [J]. China Automotive Engineering. 2010, 32(5), 373–378. (In Chinese).
Zhang Peibin, Research and Simulation of Electric Vehicle regenerative braking control [D], Wuhan University of Technology, Wuhan. (In Chinese).

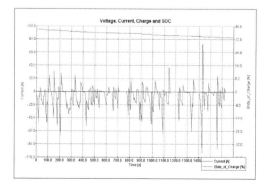

Figure 5. Current and SOC changing with time.

Advances in Power and Energy Engineering – Sun (Ed.)
© 2016 Taylor & Francis Group, London, ISBN 978-1-138-02846-3

Probabilistic power flow calculation of large scale Photovoltaic grid-connected power generation system

H. Fan & L.H. Zuo
College of Electric Engineering, Shanghai University of Electric Power, Shanghai, China

D.F. Wang
QingHai Province Key Laboratory of Photovoltaic Grid Connected Power Generation Technology (State Grid Qinghai Electric Power Company), Xining, Qinghai Province, China

ABSTRACT: This paper proposes an output model of the large-scale Photovoltaic (PV) power plants based on the condition of the light intensity on a typical day. This model takes the impact of environmental factors on photovoltaic cells into account. Under the condition of the deterministic grid system, according to the model we can calculate the probability power flow of comprising large-scale PV power plant system by Monte-Carlo simulation method and deterministic power flow method. This method not only is applicable to the transmission grid, but also applies to the distribution network, and the system can be estimated maximum capacity of photovoltaic power plants based on the sample's failure rate of power flow calculation. The paper verifies this method by use of two programs of the improved IEEE-14 nodes study.

1 INTRODUCTION

In recent years, photovoltaic power generation technology has made enormous progress and the generating cost has also greatly reduced with the national policy supports for new energy power generation. However, the regional abandonment photovoltaic phenomenon has occurred due to the great uncertainty of photovoltaic power generation and the high cost of power generation subsidy. However, the decrease of the cost of PV modules and the improvement of the efficiency of power generation make the energy structure in the western regions of China change, and then the construction of large-scale photovoltaic power station is accelerated (Lu-guang, Yan et al. 2010 & Lu-guang, Yan et al. 2011).

The randomness of photovoltaic power generation can cause that the large scale photovoltaic power station has a profound impact on active frequency, reactive power voltage, rotor angle stability, small disturbance stability, quality of electric energy and other characteristics of the power grid when the large scale photovoltaic power station is connected to the power grid (Ming, Ding et al. 2001). The operating mechanism of large capacity grid connected photovoltaic power station is described in Ref. (Ming, Ding et al. 2014), and the problems in operation of large capacity power station are proposed, including the multi

peak characteristics, temperature rise effect and hot spot effect of the combined PV array. In the assumption that the input and output variables of the photovoltaic power station are only the basic components of the photovoltaic power station, the Ref. (Zhengming, Zhao et al. 2011) proposes a model of the energy conversion and output of the power station by using the method of phase volume and the controlled source method. A mathematical model for simulating the output power of a photovoltaic power station is presented in Ref. (Jing, Li et al. 2008), which uses the solar radiation and the latitude and longitude to simulate the illumination intensity of a certain region. However, the complexity of the illumination condition is considered as a problem (Mei-qin, Mao et al. 2005). Improving the output efficiency of large-scale photovoltaic power plant depends on the excellent Maximum Power Point Tracking (MPPT) control strategy due to the constant change of environmental temperature and light intensity (Xiaojin, Wu et al. 2011, Yong, Guo et al. 2009).

Probabilistic power flow is an important basis for solving the problems caused by large scale uncertain factors. Three methods to solve the probabilistic power flow are described in Ref. (Dongran, Liu et al. 2011): the simulation method, the analytic method and the approximate method. Simulation method is widely used due to its highest accuracy of computational results, including Monte Carlo

Sampling method is widely used in the field of transmission and distribution network due to the highest accuracy (Yu, Liu et al. 2014, Chan, Chen et al. 2015, Chenshan, Wang et al. 2005, Yubing, Duan et al. 2011). In Ref. (Ming, Ding et al. 2012), a method of calculating the probability power flow of distribution network with the Monte-Carlo method is proposed, which considers the factors of the weather transformation. A new extended Quasi Monte Carlo method based on nonparametric kernel density estimation is proposed, and the selection method of convergence criterion which the error can be transformed into controllable quantity and improve accuracy is also proposed in Ref. (Sidun, Fang et al. 2015).

This paper presents a mathematical model for calculating the output of large-scale photovoltaic power plant based on the variation of typical daylight intensity and ambient temperature, and uses the Monte-Carlo sampling method to simulate the probability of the system.

2 PHOTOVOLTAIC POWER GENERATION MODEL ESTABLISHMENT

2.1 U-I mathematical model

In actual engineering calculation, the typical daylight intensity data of a certain area can be obtained from the monitoring station. The principle of photovoltaic cells is the photovoltaic effect that the surface of the photovoltaic cells can produce current and voltage when light shines on them. Through the physical structure of the photovoltaic cells, the U-I model of the typical grid connected photovoltaic power system is shown in Equation 1:

$$I = I_{sc}\left\{1 - C_1\left[\exp\left(\frac{U}{C_2 U_{oc}}\right) - 1\right]\right\} \quad (1)$$

where C_1, C_2 are the Intermediate variables, I_{sc} = short-circuit current; U_{oc} = open circuit voltage. The parameters I_{sc}, U_{oc} and the intermediate variables C_1, C_2 of the U-I model need to be corrected according to the light intensity and the environmental temperature, because the intensity of the sunlight is constantly changing, the coefficient of the modified equation is shown in Equations 2, 3:

$$C_1 = \left(1 - \frac{I_m}{I_{sc}}\right)\exp\left(-\frac{U_m}{C_2 U_{oc}}\right) \quad (2)$$

$$C_2 = \left(\frac{U_m}{U_{oc}} - 1\right)\left[\ln\left(1 - \frac{I_m}{I_{sc}}\right)\right]^{-1} \quad (3)$$

where I_m = maximum power point current; U_m = maximum power point voltage. I_{sc}, U_{oc}, I_m, U_m are related to changes in illumination and environment temperature, and the modified process of the photovoltaic cell technology parameters under different illumination intensities and ambient temperatures is shown in the following Equations 4~10:

$$k = \frac{S}{S_{ref}} \quad (4)$$

$$T_a(t) = \frac{1}{2}\left[T_{max} + T_{min} + (T_{max} - T_{min})\cos\frac{2\pi(t - t_p)}{24}\right] \quad (5)$$

$$\Delta T = T_a(t) - T_{ref} \quad (6)$$

$$I'_m = I_m k\,(1 + a\Delta T + bS) \quad (7)$$

$$I'_{sc} = I_{sc} k\,(1 + a\Delta T + bS) \quad (8)$$

$$U'_m = U_m(k + c)(1 - d\Delta T - eS) \quad (9)$$

$$U'_{oc} = U_{oc}(k + c)(1 - d\Delta T - eS) \quad (10)$$

where I_{sc}, U_{oc}, I_m, U_m are photovoltaic cell technology parameters, I_{sc}', U_{oc}', I_m', U_m' are the correction values of I_{sc}, U_{oc}, I_m, U_m under different conditions; a, b, c, d, e are constants, which is a = 0.0025/°C; b = 7.5 × e^{-5} m²/W; c = 0.5; d = 0.0028/°C; e = 8.4 × e^{-5} m²/W; S_{ref} = standard light intensity which is 1000 W/m²; S = the real-time light intensity on a typical day of the month; T_{ref} = reference temperature which is 25°C; T_a = the temperature of a day; T_{max}, T_{min} are the temperature maximum and minimum; t_p = the moment of the highest temperature of the day, which is generally considered to be 14:00.

2.2 Mean and variance of output power of photovoltaic power station

Due to the uncertainty of illumination intensity and the technical characteristics of photovoltaic cells, the MPPT control algorithm is needed to determine the maximum power point P and the corresponding voltage U on the P-U curve in order to obtain the maximum output power. In this paper, we use the typical U-I characteristics to solve the maximum power point of the algorithm directly. The equation is shown in the following Equations 11~13:

$$P = UI \quad (11)$$

$$\frac{dP}{dU} = I + U\frac{dI}{dU} \quad (12)$$

$$\frac{dP}{dU} = 0 \quad (13)$$

where Equation 13 is the maximum power point discrimination conditions, P = the theoretical maximum power of photovoltaic power generation system under the condition of P_{max} at the moment.

As the photovoltaic battery only produces DC voltage and current, the large scale PV power station need to use the inverter to make the inversion, which is the DC/AC transform. The output efficiency of PV modules is greatly influenced by the environmental temperature. These influences can not be ignored in the actual calculation. Assuming that the output power of the large scale PV grid connected PV system considering the MPPT element and the inverter efficiency is P_{pv}, the calculation formula is as Equations 14, 15 is shown:

$$P_{pv} = \eta P_{max} \eta_{mppt} \eta_{inv} \tag{14}$$

$$\eta = \eta_0 \left[1 - \varepsilon(T - T_{ref}) \right] \tag{15}$$

where η_0 = reference temperature of the PV module efficiency; ε = PV module temperature coefficient, generally taken as 0.003~0.005; η = efficiency of PV modules; η_{mppt} = efficiency of MPPT control components; η_{inv} = efficiency of the inverter. In addition, if the $P_{min} \leq P_{min}$ (P_{min} is the threshold value of PV grid connected power generation system), then the PV grid connected system is closed, the system output is 0. According to the Equations 14, 15, the maximum output power of the PV power station in each light intensity acquisition point can be obtained after obtaining the typical daily light data, and then the mean value of the output of Photovoltaic (PV) is obtained.

3 PROBABILISTIC POWER FLOW CALCULATION USING MONTE-CARLO SIMULATION

The core of probabilistic load flow calculation of Monte-Carlo simulation method is based on the load and power of the probability parameters, expectation and variance. Then for a large number of samples, a large number of qualified sampling samples are obtained under some constraints, and the expectation and variance of the node voltage and line power are obtained.

3.1 *Monte-Carlo sampling*

The load point and power point of the system in the grid and the system are required before the probability power flow calculation. Assuming that the power system network node number is n, the loads are recorded as L_i (i = one or several of the 1, 2, ..., n, the nodes without loads are denoted as

$L_m = 0$, m = the non-load point number). Source is recorded as S_j (j = one or several of the 1, 2, ..., n, the nodes without source are denoted as $S_f = 0$, f = the non-source point number). The probability density of the load and power supply is in accord with normal distribution. According to the load and power of the expectations and the average variance of Monte-Carlo sampling, the number of samples is k, the random vector of each load and power is:

$$L_i = \begin{bmatrix} L_{1i} & L_{2i} & L_{3i} & \cdots & L_{ki} \end{bmatrix}^T$$

$$S_j = \begin{bmatrix} S_{1j} & S_{2j} & S_{3j} & \cdots & S_{kj} \end{bmatrix}^T$$

where the elements of the vector are the random number generated by Monte-Carlo sampling. In addition, the power supply of the random number to be bound, because the power has its own output on the lower limit. Assuming that the output of the power supply is S_{jmax}, the lower limit is S_{jmin}, if the power supply is limited, it needs to be corrected, and the correction method is shown in Equations 16, 17:

$$S_{lj} = \min(S_{j\max}, S_{lj}) \tag{16}$$

$$S_{lj} = \max(S_{j\min}, S_{lj}) \tag{17}$$

where l = 1, 2, ..., k; S_{lj} should meet the above correction equation simultaneously. The final sample matrix is:

$$L = \begin{bmatrix} L_1 & L_2 & L_3 & \cdots & L_n \end{bmatrix}$$

$$S = \begin{bmatrix} S_1 & S_2 & S_3 & \cdots & S_n \end{bmatrix}$$

where the matrix $L = k \times n$ order matrix; the matrix $S = k \times n$ order matrix.

3.2 *Seeking the probability of the tide*

In the process of calculating probability power flow, the simulation of the sampling samples data could make the power flow calculation not convergent, so we need to consider the node voltage constraints, and then according to the Newton-Raphson algorithm to calculate the probability power flow of power system. The specific steps of calculating probability power flow are as follows:

Step 1: According to the Monte-Carlo simulation method, k groups of sampling sample matrixes which consist of injection power of deterministic load nodes and power points are generated;

Step 2: Using the Newton-Raphson algorithm and k groups of sampling sample matrixes of the step 1 calculates power flow, namely, k times power flow calculation, considering the node voltage constraints in the iterative processes.

Step 3: The mean and standard deviation of each node voltage and current of each branch can be calculated by Step 2 generated k′ results of the power flow calculation (k′ ≤ k, consider the case of no convergence power flow because of node voltage constraints. If power flow calculation does not converge, the results are not included in the summary matrix). Figure 1 is the probability power flow calculation flowchart considering large-scale photovoltaic power generation grid power system.

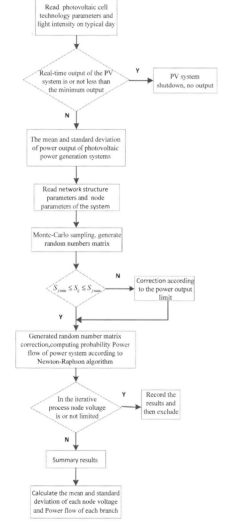

Figure 1. The probabilistic power flow calculation flow chart of large scale PV grid connected generation system.

4 STOCHASTIC POWER FLOW CALCULATION EXAMPLE

In this paper, a numerical example is used to improve the IEEE-14 node system, as shown in Figure 2, the nodes 1, 2 are the access points for large scale photovoltaic power stations, and the nodes 3, 4, 6 are the access points of the units, respectively. The data of the 14 nodes is shown in Table 1, and the data is the standard value, especially (the reference power is 100 MVA). The technical parameters of photovoltaic cells used in large scale photovoltaic power station are I_{sc} = 3.846 A, U_{oc} = 21.1 V, I_m = 3.5 A, U_m = 17.1 V, T_{max} = 40°C, T_{min} = 20°C.

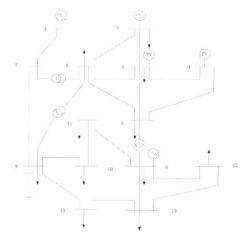

Figure 2. Grid structure of a 14 node system.

Table 1. The expected and average variance of the power supply of the case.

Node number	Generator active expectation	Mean square deviation	Reactive power of generator	Mean square deviation
1	PV	PV	0	0
2	PV	PV	0.424	0.03
3	0	0	0.2339	0.03
4	0	0	0.1333	0.03
5	0	0	0	0
6	0.85	0.03	0.12	0.03
7	0	0	0	0
8	0	0	0	0
9	0	0	0	0
10	0	0	0	0
11	0	0	0	0
12	0	0	0	0
13	0	0	0	0
14	0	0	0	0

Table 2. The results of the calculation of the Scheme 1 and 2.

Scheme 1		Scheme 2	
Node voltage expectation	Mean square deviation	Node voltage expectation	Mean square deviation
1.03E+00	4.43E-04	1.04E+00	5.03E-04
1.04E+00	3.24E-13	1.05E+00	3.24E-13
1.01E+00	1.48E-13	1.01E+00	1.47E-13
1.02E+00	2.32E-03	1.02E+00	2.34E-03
1.02E+00	2.04E-03	1.03E+00	2.06E-03
1.01E+00	1.48E-13	1.01E+00	1.47E-13
1.00E+00	0.00E+00	1.00E+00	0.00E+00
1.05E+00	3.94E-13	1.05E+00	3.94E-13
9.98E-01	6.08E-03	9.99E-01	6.09E-03
9.89E-01	1.08E-02	9.89E-01	1.09E-02
9.92E-01	1.02E-02	9.92E-01	1.04E-02
9.92E-01	8.70E-03	9.92E-01	8.66E-03
9.89E-01	8.45E-03	9.89E-01	8.33E-03
9.72E-01	1.66E-02	9.74E-01	1.62E-02
Scheme 1 sampling rate 0.01%		Scheme 2 sampling rate 0.03%	

In this example, the number of Monte-Carlo is set to 20,000 times. According to different photovoltaic power plant capacities, two schemes are adopted to calculate.

Scheme 1: probabilistic power flow calculation using node 1, 2 whose capacities are 10 MW.
Scheme 2: probabilistic power flow calculation using node 1, 2 whose capacities are 13 MW.

The results are shown in Table 2.

The computing results of the two programs are compared to get that increasing the capacity of photovoltaic power system will increase the node voltage deviation which the node voltage is more and more dispersed. At the same time, the failure rate of the calculation results increases considering the node voltage constraints because of the increasing capacity of the photovoltaic power station. With the increasing capacity of grid connected photovoltaic power plant, the number of non convergence of the sampling sample will be added. Therefore, considering the reliability of power system planning, the model and calculation method can be used in the calculation of large scale PV power station in power system.

5 CONCLUSIONS

This paper presents an output model of the large-scale Photovoltaic (PV) power plants based on light intensity which is suitable for large-scale photovoltaic power output model. The model can modify the photovoltaic cells automatically based on the technical parameters of ambient temperature and light intensity and look for the maximum output power through the MPPT control strategy. After determining the network structure, calculating the power flow through the Monte-Carlo simulation method and the deterministic power flow calculation can be implemented.

In this paper, the feasibility of the proposed method is verified through a practical example. The method is not only suitable for probabilistic load flow calculation of transmission network, but also probabilistic load flow calculation of power distribution networks. This method can also be used to estimate the ability to accept power of PV by qualified rate contrast sample, so it is very convenient in engineering application.

ACKNOWLEDGEMENT

Qing Hai Province Key Laboratory of Photovoltaic grid connected power generation technology (No. 2014-Z-Y34A). This paper is supported by National Natural Science Foundation of China (No. 51307104).

REFERENCES

Chan, Chen et al. 2015. Probabilistic Load Flow of Distribution Network Considering Correlated Photovoltaic Power Output. *Automation of Electric Power Systems* 39(9):41–47.
Chenshan, Wang et al. 2005. Probabilistic Power Flow Containing Distributed Generation Distribution System. *Automation of Electric Power Systems* 29(24):39–44.
Dongran, Liu et al. 2011. A Review on Models for Photovoltaic Generation System. *Power System Technology* 35(8):47–52.
Jing, Li et al. 2008. Dynamic Modeling and Simulation of the Grid-connected PV Power Station. *Automation of Electric Power Systems* 32(24):83–87.
Lu-guang, Yan et al. 2010. A proposal for planning and constructing a national integrated energy base combined with large-scale photo-voltaic power and hydropower in Qinghai province. *Advanced Technology of Electrical Engineering and Energy* 29(4):1–9.
Lu-guang, Yan et al. 2011. A roposal for planning and constructing a national integrated energy base combined with large-scale photo-voltaic power and hydropower in Qinghai province (Continuation). *Advanced Technology of Electrical Engineering and Energy* 30(1):8–11.
Mei-qin, Mao et al. 2005. Simulation of large-scale photovoltaic grid-connected systems. *Journal of HEFEI Universityof Technology* 28(9):1069–1072.

Ming, Ding et al. 2001. Probabilistic Load Flow Analysis Based on Monte-Carlo Simulation. *Power System Technology* 25(11):10–15.

Ming, Ding et al. 2012. Three-phase Probabilistic Power Flow Calculation in Distribution Systems with Multiple Unsymmetrical Grid-connected Photovoltaic Systems. *Automation of Electric Power Systems* 36(16):47–52.

Ming, Ding et al. 2014. A Review on the Effect of Large-scale PV Generation on Power Systems. *Proceedings of the CSEE* 34(1):1–14.

Sidun, Fang et al. 2015. An Extended Quasi Monte Carlo Probabilistic Load Flow Method Based on Non-parametric Kernel Density. Automation of Electric Power Systems 39(7):21–27.

Xiaojin, Wu et al. 2011. Study of Output Characteristics of PV Array Under Complicated Illumination Environment. *Proceedings of the CSEE* 31(Z):162–167.

Yong, Guo et al. 2009. Research on Maximum Power Point Tracking Method for Photovoltaic System. *Power Electronics* 43(11):21–23.

Yu, Liu et al. 2014. Review on Algorithms for Probabilistic Load Flow in Power System. *Automation of Electric Power Systems* 38(23):127–135.

Yubing, Duan et al. 2011. Probabilistic Power Flow Calculation in Microgrid Based on Monte-Carlo Simulation. *Transactions of China Electrotechnical Society* 26(Z):274–278.

Zhengming, Zhao et al. 2011. Overview of large-scale Grid-connected Photovoltaic Power Plants. *Automation of Electric Power Systems* 35(12):101–107.

Advances in Power and Energy Engineering – Sun (Ed.)
© 2016 Taylor & Francis Group, London, ISBN 978-1-138-02846-3

Study on the mechanism and mitigation of SSI phenomena in large-scale wind farms with DFIGs

Y. Huang & X.T. Wang
Shanghai Jiaotong University, Shanghai, China

H.J. Yu, K.M. Chen, G.L. Wang & Y.M. Huang
Shanghai Electric Wind Power Equipment Co. Ltd., Shanghai, China

ABSTRACT: The wind farms with Double Fed Induction Generators (DFIGs) that operate near series compensated AC transmission lines are susceptive to un-damping Sub-Synchronous Oscillating (SSO). This paper establishes the equivalent model of the practical power system and simulates in PSCAD/ EMTDC to analyze the Sub-Synchronous Interactions (SSI) phenomenon. By means of the particular settings of the models in PSCAD/EMTEDC, the power system is able to be divided into an electrical part, a mechanical part, and a control part. Through individual simulation analysis, the impact of the series capacitor, converter control, and drive chain of wind turbines on SSI are analyzed, and the results show that SSI in wind farms occur because of the interaction between series capacitors and Rotor Side Control (RSC). Then, the influence of SSI on the electrical signals in RSC is investigated, and the practical strategy of mitigating SSI by adding filters in RSC is presented, and the methods of designing filters are proposed. Finally, the effectiveness and generality of this method is proved through simulation.

1 INTRODUCTION

Integrating wind power is the future of the development of Chinese power system. In the *"Wind power development '12th Five Year Plan',"* the China National Energy Administration mentioned that there will be several Million kilowatt class wind power bases built in west and north of China during the "12th Five Year Plan" period (Liu 2012). However, the load centers are located in the east and south of China. Therefore, the efficiency of wind power highly relays on the high-rating and long-distance transmission.

It has long been understood that series compensation can effectively increase the capacity of transmission lines and is being widely used in long distance transmission. However, the use of series compensated transmission lines will introduce the risk of Sub-Synchronic Resonance (SSR) between generators and the transmission system, and the mechanism in thermal power transmission had been studied (Xu et al. 1999). In recent years, the mechanism of SSR in wind turbines with doubly fed induction generators (DFIGs) topology with a series compensated transmission lines have started to attract some attentions (Varma et al. 2008, Fan et al. 2010, Li et al. 2013, Zhang, et al. 2013, Varma et al. 2013). In October 2009, an unplanned outage in the Electric Reliability Council of Texas (ERCOT) introduced a new phenomenon: Sub-Synchronous Control Interaction (SSCI). After this accident, many researches had been conducted on this phenomenon (Adams et al. 2012, Badrzadeh et al. 2012). Reference (Irwin et al. 2011) summarized all the mechanisms of SSR in DFIG and put the concept of Sub-Synchronic Interactions (SSI). It pointed out that SSI involving the exchange of energy between a generator and a transmission system at the AC frequencies below the system nominal frequency. They include SSR, SSTI, and SSCI.

During the last two years, many wind farms located in northern China occurred SSI accidents several times, and caused many DFIGs' unplanned outage. The resonant frequency is about 6~8 Hz and the compensation level seen from the wind farms is very small, which is very different from what happened in Texas. So far, some researches have been published. Reference (Dong et al. 2014) uses eigenvalue analysis and time-domain simulation to research the SSI characteristics of wind farms under all operation regions of DFIG and finds out that with the increase of the wind speed, the area of stable operation region of DFIG is gradually increasing. By means of eigenvalue analysis, reference (Wang et al. 2015) points out that the SSI of wind farms is a special electrical resonance and mainly effected by the wind speed, number of generators and control parameters of DFIGs. Then an equivalent circuit is deduced to analyze the mechanism of SSI impacts from the aforementioned factors. Reference (Don et al. 2015) obtains the predominant factors influencing the SSI of wind farms through the eigenvalue sensitivity analysis and proposes that decreasing the parameters of PI control in the inner current loop of Rotor Side Control (RSC) would reduce the risk of SSI. However, changing those parameters has the risk of impacting the control property of the controller and the coordination of the protector.

This paper establishes the equivalent model of the practical power system and simulates in PSCAD/EMTDC to analyze the SSI phenomenon. By means of the particular settings of the models in PSCAD/EMTEDC, the power system is able to be divided into electrical part, mechanical part, and control part. Through individual simulation analysis, the impact of the series capacitor, converter control and drive chain of wind turbines on SSI are analyzed. Then, the impact of SSI on the electrical signals in RSC is investigated, and the practical strategy of mitigating SSI by adding filters in RSC is presented, and the methods of designing filters are proposed. Finally, the effectiveness and generality of this method are proved through simulation.

2 SIMULATION SYSTEM FOR SSI ANALYSIS

2.1 Equivalent model of the large scale wind farms of DFIGs

The wind farms located in the north of China which suffer SSI have two key features: 1) Most of the WTGs are DFIGs with the rated power 1.25 MW and the rated voltage 690 V; 2) There are series compensation devices on the 500 kv transmission lines and the compensation level seen from the wind farms is very small. To analyze this accident precisely, the system is modeled on the PSCAD/EMTDC since: this software makes sure of the non-linear characteristic of the system; this separation simulation analysis of mechanical, electrical and control property can be operated on the basis of some particular settings in the models of PSCAD/EMTDC.

First, the DFIG is modeled through three parts: generator, converter, and wind turbine. The generator can be modeled directly by the induction machine model in the master library of PSCAD/EMTDC, which can be chosen as wound rotor induction machine. The converter can be modeled by the 6 pulse rectifier inverter circuit; the power electronic switches are modeled in the master library.

The wind farms should be modeled by a large number of identical low-rating WTGs, which are connected to one collection bus. This system is equivalent to the one with an aggregated generator when the number of WTGs is given (Fernández, et al. 2008).

Suppose that all the generators in this area are DFIGs of the same type; then, the operating states of the same type of devices are same when they are in working state. The complex system can be simplified to an equivalent system shown in Figure 1.

2.2 Strategy of converter control

The convert control of DFIG can be divided into RSC and GSC. The RSC controls the active power and inactive power, while the GSC maintains the DC bus voltage and generator terminal voltage as the rated value (Song, et al. 2012). The control strategies are shown in Figure 2 and Figure 3.

After finishing all parts of the wind farms, the simulation is taken where the wind speed is set as 6 m/s and the compensation level seen from the

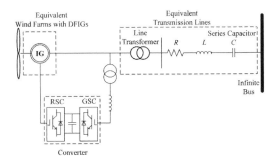

Figure 1. Equivalent system of wind farms with DFIGs used for SSI analysis.

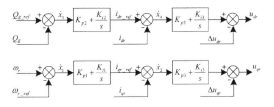

Figure 2. Control strategy of RSC.

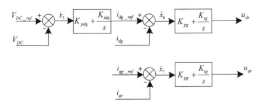

Figure 3. Control strategy of GSC.

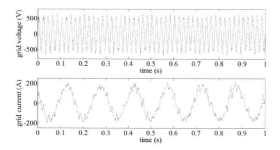

Figure 4. Simulation results of grid voltage and current while SSI occurs.

wind farm is 6.67%. The grid voltage and current are shown in Figure 4.

From the results above, it is easy to find that under this wind speed, although the compensation level seen from the wind farm is very low, the SSI still occurs; the grid current was distorted sharply when SSI occured. It is obvious that there is a sub-synchronic component in the grid current, and the frequency is around 6 Hz; comparing to current, the grid voltage is not affected seriously by SSI; the voltage only contains a 50 Hz component. This is because the frequency of SSI is very low, and the corresponding line impedance is not large enough to make a large distortion on the grid voltage.

3 SIMULATION ANALYSIS OF THE MECHANISM OF SSI

In order to analyze the mechanism of SSI phenomenon shown in Figure 4, the separation simulation analysis is used. The method of this analysis is to separate one part of the system by the means of particular settings in the models of PSCAD/EMTDC and simulate, then find the differences between the before and after separation. Since it is not easy to find the inner law through time domain simulation, the Prony analysis is involved.

For the system in this paper, the potential factors that will impact SSI are contained in the series compensation devices, the converter control, and the drive chain of wind turbine. In order to find the law, every time one part of system is separated, the operating situations in different wind speed (5 m/s, 6 m/s, 7 m/s) are simulated. During the simulation, the operation of separating acts after the system is stable, and the series compensation devices are put in 10 s later, which is taken as a disturbance of the system.

Before the separation simulation starts, the results of considering all the factors need to be obtained. Figure 5 shows the waveforms of the

A-phase grid current when the wind speed is 5 m/s, 6 m/s, and 7 m/s. Table 1 shows the corresponding results of Prony analysis.

From the results in Figure 5 and Table 1, it is easy to notice that the gird currents emerge as sub-synchronous components after series compensation devices are switched in; with the increase of wind speed, the degree of the compensation is reduced, which means the degree of SSI is reduced as the wind speed increases.

3.1 Impact of the series compensation devices on SSI

To separate the impact of series compensation, we can bypass the series capacitors on the 500 kV transmission lines. The simulation results after separating the impact of series compensation are shown in Figure 6, and the corresponding Prony analysis results are shown in Table 2.

Comparing the waveforms in Figures 5, 6, and the results in Tables 1, 2, it is obvious that when series capacitors are bypassed, the grid current doesn't contain the sub-synchronic component and the frequency only has the 50 Hz component, which means that the system is stable and no SSI occurs without series compensation. In other words, the series compensation devices have significant impact on SSI phenomenon.

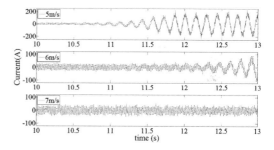

Figure 5. Simulation results of A-phase grid current while considering all the factors.

Table 1. Results of Prony analysis while considering all the factors.

Wind speed	Frequency	Real part of Eigenvalue	Damping
5 m/s	50.0 Hz	0.00	0.00
	6.6 Hz	2.89	−6.91
6 m/s	50.0 Hz	0.00	0.00
	6.6 Hz	1.02	−2.38
7 m/s	50.0 Hz	0.00	0.00
	6.6 Hz	−0.91	2.22

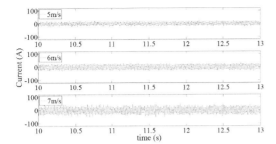

Figure 6. Simulation results of A-phase grid current without series compensation on the transmission line.

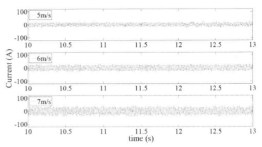

Figure 7. Simulation results of A-phase grid current without the impact of RSC.

Table 2. Results of Prony analysis without series compensation on the transmission line.

Wind speed	Frequency	Real part of Eigenvalue	Damping
5 m/s	50.0/Hz	0.00	0.00
6 m/s	50.0/Hz	0.00	0.00
7 m/s	50.0/Hz	0.00	0.00

Table 3. Results of Prony analysis without the effect of RSC.

Wind speed	Frequency	Real part of Eigenvalue	Damping
5 m/s	50.0 Hz	0.00	0.00
	6.8 Hz	0.51	−1.19
6 m/s	50.0 Hz	0.00	0.00
	6.6 Hz	−0.26	0.63
7 m/s	50.0 Hz	0.00	0.00
	6.5 Hz	−2.66	6.54

3.2 *Impact of the converter control on SSI*

The converter control contains two parts, the RSC and GSC. Both of them have the risk to impact SSI.

To separate the effect of RSC means to separate the influence of electrical signals in RSC. To achieve this, the input of each PI should be held as the value when the system is stable, so the output of RSC could be constant. The simulation results after separating the impact of RSC are shown in Figure 7, and the corresponding Prony analysis results are shown in Table 3

Comparing the above results with Figure 5 and Table 1, it is easy to find that when the outputs of RSC are constant, the SSI still occurs, but the divergence speed is much slower, and the corresponding real part of eigenvalues are smaller. This means SSI occurs whether the RSC is involved or not, but the divergence speed is impacted by the RSC. Therefore, the impact of RSC on SSI is not as big as the series compensation devices, but RSC can obviously impact the degree of SSI.

The method of separating GSC is similar to the RSC and the simulation results are shown in Figure 8, while the corresponding Prony analysis results are shown in Table 4.

Based on the results shown above, SSI still occurs even when GSC's outputs are constant. Comparing those results with the results in Figure 5 and Table 1, it is clear that the degree of SSI decreases when GSC's outputs are constant, but the level of decreasing is much smaller if compared with the

results in Figure 7 and Table 3. In other words, the SSI still occurs without GSC's impact and the impact of GSC is much smaller while comparing with RSC.

3.3 *Impact of wind turbine's drive chain on SSI*

The method of separating the impact of the turbine's drive chain is to lock the rotor speed input signal in the induction machine and set the reference of rotor speed as a constant. The simulation results after separating drive chain are shown in Figure 9 and the corresponding Prony analysis results are shown in Table 5.

From the results shown above, it is obvious that SSI still occurs when the impact of drive chain is avoided. Comparing the results with Figure 5 and Table 1, it is easy to find that there are few differences between them. Therefore, the drive chain nearly has no effect on SSI.

In conclusion, the SSI phenomenon happening in wind farms is mainly caused by the interaction between the series compensation devices and RSC, where the series compensation is the key of SSI that occurs and the RSC impacts the degree of SSI significantly. The impact of GSC and drive train of wind turbine is much smaller compared with those of series compensation and RSC.

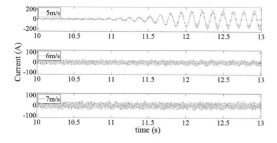

Figure 8. Simulation results of A-phase grid current without the impact of GSC.

Table 4. Results of Prony analysis without the impact of GSC.

Wind speed	Frequency	Real part of Eigenvalue	Damping
5 m/s	50.0 Hz	0.00	0.00
	6.8 Hz	0.51	−1.19
6 m/s	50.0 Hz	0.00	0.00
	6.6 Hz	−0.26	0.63
7 m/s	50.0 Hz	0.00	0.00
	6.5 Hz	−2.66	6.54

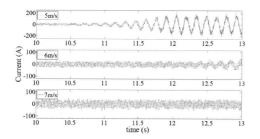

Figure 9. Simulation results of A-phase grid current without the impact of drive chain of wind turbines.

Table 5. Results of Prony analysis without the impact of drive chain of wind turbines.

Wind speed	Frequency	Real part of Eigenvalue	Damping
5 m/s	50.0 Hz	0.00	0.00
	6.5 Hz	2.21	−5.40
6 m/s	50.0 Hz	0.00	0.00
	6.6 Hz	1.34	−3.28
7 m/s	50.0 Hz	0.00	0.00
	6.7 Hz	−0.90	2.12

4 PRACTICAL MITIGATION STRATEGIES FOR THE SSI IN THE WIND FARMS WITH DFIGs

Based on the simulation analysis above, changing the structure of transmission lines and improving the control strategy of RSC are the primary methods to mitigate the SSI phenomenon in the wind farms with DFIGs.

The most useful way to change the structure of transmission lines is switching off the series compensation devices. However, the compensation level directly impacts the capacity of transmission lines. If the compensation level is too small, the system would not be stable. While improving the control strategy of RSC is cheaper and easy to achieve, and the effect of mitigating SSI is obvious. Therefore, the method of advancing the control strategy of RSC is more economical and practical.

4.1 *The Impact of SSI on the RSC*

From the control strategy of RSC shown in Figure 2, there are four intermediate variables (x_1, x_2, x_3 and x_4) and three electrical signals (Q_g, i_{dr} and i_{qr}). The frequency analysis on those variables is taken and the results are shown in Table 6.

Based on the results above, it is easy to find that when the wind speed is between 5~7 m/s, all the seven variables only contain a direct component when the system is stable and obtains a particular component with the frequency near 43.4 Hz when the SSI occurs, which happens to be the compensation of the sub-synchronous frequency, and the frequency changes a little when the wind speed changes.

Therefore, when the SSI occurs, a particular component will appear in the control channel whose frequency is the compensation of the sub-synchronous frequency. If a control strategy could reduce this particular component, block the formation loop of SSI phenomenon, SSI will be prevented in the beginning. The filter just satisfies this purpose.

4.2 *The method of filter design*

When the system is stable, the frequency in the control channel of RSC is 0, and when SSI occurs, the frequency changes to near 43.4 Hz. Hence, the

Table 6. Results of frequency analysis on intermediate variables in RSC.

Wind speed	Variable name	Frequency	
		No SSI	SSI
5 m/s	x_1~x_4	0.0 Hz	43.3 Hz
	Q_g, i_{dr}, i_{qr}	0.0 Hz	43.3 Hz
6 m/s	x_1~x_4	0.0 Hz	43.4 Hz
	Q_g, i_{dr}, i_{qr}	0.0 Hz	43.4 Hz
7 m/s	x_1~x_4	0.0 Hz	43.4 Hz
	Q_g, i_{dr}, i_{qr}	0.0 Hz	43.4 Hz

function of the filter should satisfy the following request: avoiding affecting the low frequency component and depressing the particular high frequency component. The low-pass filter just has this ability. Moreover, using low-pass filter benefits the system's robust stableness.

On the basis of the data in Table 6, there are two potential control strategies. The first one is depressing the particular component at x_2 and x_4, since these two are the final places that are impacted by the electrical signals. The block of this strategy is shown in Figure 10, where T presents the time constant, n is the number of order. The other one is depressing the particular component at x_1 and x_3, and filtering out the particular component in i_{dr} and i_{qr}. Through this combination, the entire particular component could be depressed. Figure 11 hows this strategy, where T_1, T_2 presents the time constant, n_1, n_2 presents the number of order.

To decide the value of parameters in each filter, some rules should be obeyed: 1) the degree of damping on the frequency around 40 Hz should be larger than 20 dB; 2) the decrease degree of phase should be smaller than 30°. Through calculation, the values of parameters in each filter are $T_1 = T_2 = T = 0.10$, $n_1 = n_2 = n = 1$.

4.3 The effectiveness and generality of the mitigation strategy

To find out which strategy is more effective and has less negative impact on the property of system control, the simulation has been made where wind speed is 6 m/s. The grid A-phase current in time domain and frequency domain are shown in Figure 12.

From the results shown in Figure 12, both strategies are able to prevent SSI. However, the impact of these strategies on the system is different. For the first strategy; although, the subsynchronous component is mitigated, there are some new components whose frequency is about 30 Hz. For the second one, it is easy to notice that the grid current mainly contains 50 Hz component after the series compensation is switched in. Therefore, the second strategy mitigates the SSI phenomenon and has less negative impact on the property of system control, which is better than the first one.

Furthermore, to find the generality of the second strategy, some simulations have been made. The results are shown in Figure 13.

Figure 13 shows the situation of grid current after the series compensation is switched in at 10 s while the wind speed is 5 m/s, 6 m/s, 7 m/s. Comparing Figures 5 and 13, it is obvious that using this strategy will mitigate SSI in different wind speed.

In conclusion, the second strategy is effective and universal for mitigating SSI phenomenon.

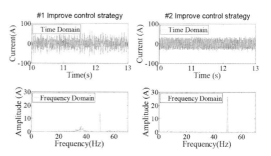

Figure 12. Simulation results of both improved control strategies in time domain and frequency domain.

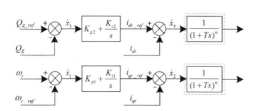

Figure 10. #1 improved control strategy of RSC.

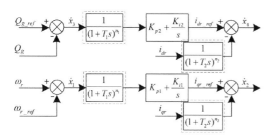

Figure 11. #2 improved control strategy of RSC.

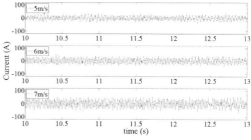

Figure 13. Simulation results of A-phase grid current with advanced control strategy of RSC in different wind speeds.

5 CONCLUSION

This paper establishes the equivalent model of the practical system and simulates in PSCAD/EMTDC to demonstrate SSI phenomenon. Analysis results indicate that SSI occurs even when the equivalent transmission system compensation level seen from the wind farms is only 6.67%. By means of the particular settings in the models of PSCAD/EMTEDC, the power system is divided into electrical part, mechanical part, and control part. Through individual simulation analysis, the impact of the series capacitor, converter control, and drive chain of wind turbines on SSI are analyzed and the results show that SSI in wind farms occurs because of the interaction between series capacitors and Rotor Side Control (RSC). Then, the influence of SSI on the electrical signals in RSC is investigated, study shows that there would be a particular component appearing in the control channel of RSC whose frequency is the compensation of the subsynchronous frequency. With those results, the practical strategy of mitigating SSI by adding filters in RSC is presented, and the two methods of designing filters are proposed. Finally, the effectiveness of the two methods is studied through several simulations.

REFERENCES

Adams J. et al. 2012, ERCOT experience with sub-synchronous control interaction and proposed remediation. 2012 Transmission and Distribution Conference and Exposition, IEEE PES, 7–10 May: 1–5.

Badrzadeh B. et al. 2012, A. Gole. Sub-synchronous Interaction in Wind Power Plants—Part II: An ERCOT Case Study. 2012 IEEE PES General Meeting, 22–26 July: 1–9.

Dong Xiaoliang et al. 2014, SSR characteristics of a wind farm connected to series-compensated transmission system under all operation region of DFIG. Power System Technology, 09:2429–2433.

Dong Xiaoliang et al. 2015, Impacting factors and stable area analysis of subsynchronous resonance in DFIG based wind farms connected to series-compensated power system[J]. Power System Technology, 01:189–193.

Fan Lingling et al. 2010, Modeling of DFIG-Based Wind Farms for SSR Analysis. IEEE Transactions on Power Delivery, 25(4): 2073–2082.

Fernández L.M. et al. 2008, Aggregated dynamic model for wind farms with doubly fed induction generator wind turbines[J]. Renewable Energy, 33(1): 129–140.

Irwin G.D. et al. 2011, Sub-synchronous control interactions between type 3 wind turbines and series compensated AC transmission systems. 2011 IEEE PES General Meeting, 24–29 July: 1–6.

Li Ran et al. 2013, Mechanism analysis on subsynchronous oscillation caused by grid-integration of double fed wind power generation system via series compensation. Power System Technology, 11: 3073–3079.

Liu Zhenya, 2012, Electric Power and Energy. Beijing: China Electric Power Press.

Song Yixu. 2012, Theory and control of wind-power generators. Beijing: China Machine Press.

Varma R.K. et al. 2008, Mitigation of Subsynchronous Resonance in a Series-Compensated Wind Farm Using FACTS Controllers. IEEE Transactions on Power Delivery, 23(3): 1645–1654.

Varma R.K. et al. 2013, SSR in Double-Cage Induction Generator-Based Wind Farm Connected to Series-Compensated Transmission Line. IEEE Transactions on Power Systems, 28(3): 2573–2583.

Wang L. et al. 2015, Investigation of SSR in Practical DFIG-Based Wind Farms Connected to a Series-Compensated Power System. Power Systems IEEE Transactions on, 30.

Xu Zheng, et al. 1999, Review on methods of analysis for subsynchronous oscillations of power systems. Power System Technology, 06:36–39.

Zhang Jian et al. 2013, Mechanism and characteristic study on sub-synchronous control interaction of a DFIG-based wind-power generator. Transactions of China Electrotechnical Society, 12:142–149+159.

Advances in Power and Energy Engineering – Sun (Ed.)
© 2016 Taylor & Francis Group, London, ISBN 978-1-138-02846-3

Research on the design of the power quality control equipment testing platform

Z.F. Liu
Electrical Power Research Institute of Guangdong Power Grid Corporation, Guangdong, China

L.X. Wu & R.K. He
Institute of Electrical Engineering, Chinese Academy of Sciences, Beijing, China

M. Ma & B.Y. Xu
Electrical Power Research Institute of Guangdong Power Grid Corporation, Guangdong, China

ABSTRACT: Nowadays, the power quality equipment not only lacks uniform and feasible assessment criteria, but has lots of difficulties building a simulation experiment environment. In view of the above problems, the cause and solution method of power quality problems are analyzed, and a design for a power quality control equipment testing platform is presented. According to the design scheme of power quality testing platform, the topology and parameter of the main circuit are discussed and the simulation model of the testing platform is built with PSIM software. The physical model is also built. Simulation and experimental results show that the scheme can achieve the test of the middle and low voltage power quality equipment.

1 INTRODUCTION

With the development of power electronic device and technology, power electronics, and nonlinear equipment has been widely used, which meanwhile has caused serious pollution to the grid and serious deterioration of power quality (Xie Xiaorong & Jiang Qirong (2006), Wang Zhaoan et al. 2006) on the other hand, with the slather use of electrical equipment which is complex, precise, sensitive to power quality, requirement of power quality, and reliability is higher. Nowadays, the power quality equipment, such as DVR (Dynamic Voltage Restorer)[3], active power filter (Yang Huayun & Ren Shiyan (2009)), D-STATCOM (Distribution Static Compensator) (Kumar, C. & Mishra, M. (2015), Hsieh, S. (2015)), TSC (Thyristor Switched Capacitor), UPQC (Unified Power Quality Controller), UPS (Uninterruptable Power Supply), SSTS (Solid State Transfer Switch) gained more and more popularity. But, the power quality equipment lacks of uniform and feasible assessment criteria. To establish a testing system has a very important significance for the fair use of power quality control equipment and improvement power quality.

In practice, depending on the actual power grid load model, the simulation testing system gains the value of every index of power quality to test whether the indicators meet the requirements of national standard, and makes an overall evaluation on the power quality; or to test the actual power grid power quality, get the system data files by measuring the instruments such as waveform recorder, and then analyze the data by the simulation test system to get the value of every index of power quality for ease of the offline test and analysis. And simulation test system of power quality can be the development foundation of the power quality test system based on a PC or the IPC machine.

At present; although, each country has made the relevant standard of power quality according to the actual situation which can judge whether the power quality is qualified. But the power quality control equipment test platform and related technology is not perfect enough. Further research is needed. This paper presents a scheme of comprehensive power quality test platform, which can realize the integrated test of low and middle voltage series compensation equipment and parallel compensation equipment. Part of the performance indicators are tested in real environment, while some indicators are tested by the equivalent circuit. Based on low voltage series compensation equipment test system, the simulation model and the physical experiment platform are built and run.

2 CAUSES, SOLUTIONS AND INDICATORS OF POWER QUALITY PROBLEMS

Construction of power quality control equipment testing platform, we first need to clear the reason for power quality problems and equipment required to solve the power quality (Hu Ming & Chen Yan (2000)); second, whether power quality control equipment meets the power quality indicators, the indicators generally includes voltage deviation, voltage fluctuation and flicker, three-phase imbalance degree, harmonic, voltage interruption, frequency deviation, the power supply reliability and the demand side management, and so on. The nature of power quality problems, characteristic index, causes, consequences, and solutions of the power system were summarized, as shown in Table 1.

At present, China has formulated and promulgated six national power quality standard: GBT 12325-2008 《Power Quality-Supply Voltage Deviation》, GBT 12326-2008 《Power Quality—Voltage fluctuations and flicker》, GBT 15543–2008 《Power Quality—Three-phase unbalanced》, GBT 15945–2008 《Power Quality—Power System Frequency Deviation》, GBT 24337–2009 《Power Quality—Harmonics between Utility Grid》, GBT 30137–2013 《Power Quality—Voltage Dip and short interruption》, GBT 14549–1993 《Power Quality—Utility Grid Harmonics》, GBT 18481–2001 《Power Quality—Temporary over-voltage and Transient over-voltage》 (Lin Haixue & Xu Jing (2005)).

3 DESIGN SCHEME OF TEST PLATFORM

3.1 Topology of the test platform

In this paper, the high and middle voltage large-capacity digital test system test is shown in Figure 1, and the system can also be used to research the low voltage system. The system is composed of the real-time digital simulation subsystem, dynamic physical simulation subsystems, monitoring unit, and so on. Main circuitry is composed of a composite power quality disturbance source, the physical device under test and programmable load. Real-time digital simulation subsystem includes a digital simulation unit, a digital equivalent of

Table 1. Causes, solutions, and indicators of power quality problems.

Type	Nature	Character	Cause	Consequence	Solution
Harmonic	Steady	Harmonic spectrum voltage, current waveform	Non-linear loads, fixed switching load	Overheating, Relay malfunction, equipment breakdown	Active, passive filter
Three-phase unbalanced	Steady	Unbalance factor	Asymmetric load	Overheating, Relay malfunction, communication interference	Static var compensation
Notch	Steady	Duration, amplitude	Speed Drive	Timer error, communication interference	Capacitors, Isolated inductor
Voltage flicker	Steady	Fluctuations in the amplitude, qualifying frequency, modulation frequency	Electric arc furnace, motor start	Servo motor running not normally	Static var compensation
Transient resonance	Transient	Waveform, peak, duration	Switch of line, load and capacitor banks	Equipment breakdown, Damage power electronic equipment	Filters, isolated transformer, lightning arrester
Pulse transient	Transient	Rise time, peak, duration	Lightning shock, Switch of inductive circuit	Equipment breakdown	Lightning arrester
Voltage swell and sag	Transient	Amplitude, duration, instantaneous value/time	Distal failure, Motor start	Equipment outage, sensitive load run improperly	DVR, UPS
Noise	Steady/transient	Amplitude, spectrum	Abnormal grounding, Solid switching load	Abnormal running of equipment controlled by Microprocessor	Properly ground filter

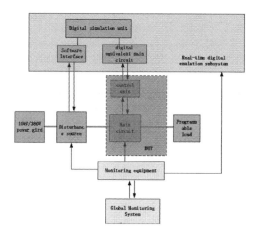

Figure 1. UUT digital test system.

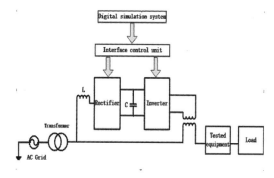

Figure 2. Programmable voltage quality disturbance system.

the controlled source, an output port, monitoring equipment, and so on. Dynamic physical analog subsystem includes high-power interface device being tested equipment, monitoring equipment.

3.2 Structure of programmable power quality disturbance system

Power quality control equipment mainly includes voltage quality control equipment DVR, and the current quality control equipment includes APF and others. The paper chooses DVR as a research program routine, research on the principle of DVR test equipment.

Major equipment of DVR test platform is a high-power electronic interface device, which works to generate the required voltage phase and amplitude at a particular time, and simulates grid working conditions, such as voltage swells, dips, and others, through superposition with grid voltage. Figure 2 shows diagram of programmable power quality disturbance system.

The basic idea of programmable voltage quality disturbance device is to insert a dynamic controlled voltage U_c in the subscriber line through a series transformer. By appropriate control methods, U_c can produce various disturbances, and thus make the normal voltage to produce the desired simulated power grid failures such as voltage drop, phase shift, phase unbalance, and harmonics. For controllable voltage quality disturbance device, the system mainly consists of a control unit, a series transformer, output filters, and other parts.

First, confirm the disturbance waveform to reach, then generate disturbance signal under the given disturbance policy by the control circuit; and then form the PWM drive signal through the PWM circuit, control the inverter power devices on and off through the drive circuit; and finally filter out

the high harmonics by the LC filter, resulting in the same disturbance voltage on the series transformer that comes along with the disturbance instruction to produce a variety of grid voltage distortions, to achieve the target of controllable voltage quality disturbance.

Power electronic interface device of the integrated test platform is controlled by a virtual digital power system, providing energy for the required power system environment of the tested equipment and to complete the required energy input and feedback. Power electronic interface device should have a fast response, quick-setting of waveform, and so on. Power non-linear load is controlled by the real-time digital simulation system, which can simulate all types of load characteristics and provide load test platform environment. Power electronic interface device shall be able to achieve energy exchange and feedback, so that the desired minimum energy consumption is throughout the experimental system. Power electronic interface device with full controller parts—IGBT and freewheeling diodes—mainly includes the rectifier, inverter circuit, and a control unit.

The DUT test system is consisted of a rectifier circuit connected to AC power, and the inverter circuit connected to constitute the basic circuit. To provide a stable DC power, the rectifier circuit uses an independent control method; using PWM rectifier control strategy to reduce the harmonic injection grid and to achieve unity power factor control with consideration of the impact of DC side load changes. Inverter unit requires fast trace port signals, using fast response deadbeat control technology combined with the platform overall control in order to realize the precise tracking of port signal.

Accuracy of the output disturbance voltage depends on the stability of the DC capacitor voltage, and the control range of DC capacitor voltage is restricted to the distribution of electrical parameters on the main circuit. Therefore, an appropriate set of primary circuit parameters is the basis for fast and stable response of the DC capacitor voltage.

4 PARAMETER DESIGN OF THE DISTURBANCE DEVICE MAIN CIRCUIT

4.1 Parameter design of inductor in rectifier unit AC side

To suppress harmonic pollution generated by the programmable voltage quality disturbance, AC side of the rectifier unit needs filter inductor in series. Large filter inductance can get a better effect of current filter, but may result in lower level of the rectified input voltage. Therefore, we design the filter inductor to meet the requirements of the rectified input voltage and current fluctuations as follows.

In order to improve the level of input voltage of the rectifier unit AC-side, increase control range of the DC capacitor voltage, and the drop on the filter inductor should be as small as possible. Set the load power of the programmable voltage power quality disturbances as PL, amplitude of the AC current is:

$$I_{am} = \frac{2P_L}{3E_{sam}\cos\varphi} \qquad (1)$$

where E_{sam}, I_{am} are AC power supply voltage and current amplitude, respectively; $\cos\varphi$ is input power factor. The inductor drop is not greater than 20% of the AC power supply voltage, shown as formula (2):

$$\omega L I_{am} \leq 0.2\, E_{sam} \qquad (2)$$

Binding of formula (1) and (2), to give an upper limit for the filter inductor:

$$L \leq \frac{3E^2_{sam}\cos\varphi}{10\omega P_L} \qquad (3)$$

In order to suppress the current overshoot, the filter inductor should not be chosen too small. In a switching period, trend of the filter inductor current is decided by two factors. First, excitation between AC power supply and inductor, which leads to the rising of the current; the second is the demagnetization between the voltage of the rectifier unit AC outlet and inductor, resulting in a lower current. To simplify the progress, the average duty cycle of the power switch is set to 1/2. In this case, the peak value of the AC current within the control cycle can be obtained from the voltage equation:

$$\Delta i_{a\,max} = \frac{1}{L}\left(\int_0^{Ta} E_{sam}dt - \int_0^{\frac{Ta}{2}}\frac{U_{dc}}{2}dt\right) \qquad (4)$$

Limit the $\Delta i_a\,max$, not more than 10% of rated current, and then use the formula (1) and the formula (4) to obtain a low limit expression of the filter inductor:

$$L \geq \frac{3(E_{sam} - 0.25u_{dc})E_{sam}T_s\cos\varphi}{2P_L} \qquad (5)$$

Setting the amplitude of the AC 311V, operating voltage of DC capacitor Udc = 700V, the control period is 250μs, power factor is 1, and load power is 60 kW. Combined with the formula (3) and (5), select the filter inductor L = 1.15 mH.

4.2 Parameter design of capacitor in rectifier unit AC side

Combined with the power fluctuation characteristics and requirements of capacitor voltage fluctuation, design the DC capacitor. When the disturbance voltage suddenly changes, the sudden change in the flow of power will lead to fluctuations in the DC capacitor voltage. DC voltage fluctuations can affect the accuracy of the output waveform disturbance. For DC capacitor:

$$\frac{du_{dc}}{dt} = \frac{i_s - i_L}{C} \qquad (6)$$

where, u_{dc} is DC capacitor voltage, i_s is the current flowing into the DC capacitor from rectifier unit, i_L is the current flowing into the inverter unit. As can be seen, the increase of the DC capacitor can suppress the fluctuation of the DC voltage, but the cost of the apparatus increases at the same time.

Programmable voltage power quality disturbances in the plant load change are the main cause of the DC capacitor voltage fluctuations. In an extreme case, the programmable voltage quality disturbance device change from absorption maximum load power P suddenly becomes regenerative power P, we hope the capacitor voltage fluctuations are limited to the permissible range in a control period. Since the delay of the control, inflow of power rectifier unit control can not mutate. At this time, the power supply should be much larger than the change in capacitance value of the maximum load power, take 10 times load power change value, there is:

$$u_{dc}C\frac{\Delta u_{dc}}{T_s} \geq 20P \qquad (7)$$

Set the maximum fluctuation range $\Delta u_{dc}\% = 10\%$, maximum load power is 60 kW.

According to equation (7), select DC capacitor, C = 14100 μF.

While maintaining the stability of the DC voltage, the power of the test equipment should not exceed the capacity programmable voltage power quality disturbances device. The over power of the test equipment will lead to fluctuations in the DC link capacitor voltage. The capacity of controllable programmable voltage quality disturbance device in this paper is 60 kW.

The control system bases on DSP TMS320F2812. Control interface and DSP communicate via RS–232. CPLD is mainly used for logic control; A / D is mainly used for data acquisition and conditioning; TMS320F2812 completes the implementation of some of the major detection and control algorithms, PWM is the pulse generator and a main circuit of each contactor control. The control circuit board is mainly used to achieve the signal acquisition, operation and control strategies achievement, while the main circuit board integrated power module, signal conditioning modules, optical trigger module and system protection module.

5 SIMULATION VERIFICATION

In order to verify the effectiveness and feasibility of power quality control equipment testing platform technology solution presented in this paper, we build a DVR model test platform in the PSIM. It simulates the DVR test equipment to produce 90° grid voltage drop compensation and 0° grid voltage drop compensation, respectively. The simulation results are shown in Figure 3 and Figure 4. It generates a compensation voltage of 40%, which is 40% of the voltage drop seen from the DVR network side.

Figure 3 is a voltage waveform generated by DVR testing equipment, Figure 4 is the grid

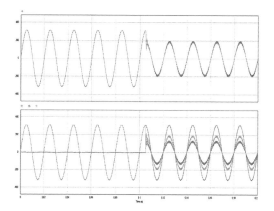

Figure 3.   Simulation of 90° voltage drop.

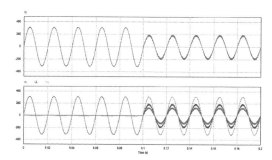

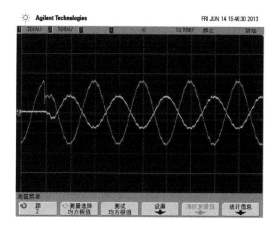

Figure 4.   Simulation of 0° voltage drop.

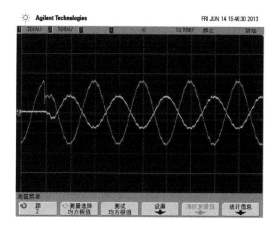

Figure 5.   Waveform.

voltage, compensation voltage generated by test equipment, voltage generated by test equipment. Figure 3 and Figure 4 shows that DVR test equipment in 90° and 0° are well compensated ideal voltage amplitudes; the angle and amplitude of the response are accurate. Simulation results show that the DVR testing equipment works properly.

6 EXPERIMENTAL VERIFICATION

In order to validate the feasibility of the proposed scheme, we set up a physical platform for the comprehensive power quality experiment. Test system consists of power supply, power distribution cabinet, voltage disturbance device, and DVR, which are in series from source to load. The test object is the independently developed storage DVR. The compensation for voltage drops to a depth of 50%. Response time compensation is less than 5 min, compensating time is 5 s, completing the performance test of the DVR—based on the platform.

In Figure 5, channel 1 is the output waveform of DVR response, channel 2 is the load voltage waveform. As can be seen from Figure 5, we produce a 50% instant grid voltage dip with the grid voltage disturbance device during 90°. After about 1 ms, DVR detects the grid voltage drop and begins to compensate, at around 3 ms the grid voltage is compensated to 100% to achieve full compensation. Compensation response time is less than 5 ms. The DVR equipment meets the technical specifications of the response.

7 CONCLUSION

Based on the proposed technical solutions for platform power quality testing, the test platform for critical parts of the structure are analyzed, the key parameters of the main circuit has been designed, we built a simulation model based on PSIM, also a physical model, simulation and experiment indicating that the test platform can achieve low-voltage series compensation devices in a variety of functional and performance tests, which proves that the technology is feasible.

REFERENCES

Hu Ming, Chen Yan. Survey of Power Quality and its Analysis Methods [J]. Power System Technology, 2000, 24(2): 36–38.

Hsieh, S., "Economic Evaluation of the Hybrid Enhancing Scheme With DSTATCOM and Active Power Curtailment for PV Penetration in Taipower Distribution Systems," Industry Applications, IEEE Transactions on, vol.51, no.3, pp.1953–1961, May-June 2015.

Kumar, C. Mishra, M. "Operation and Control of an Improved Performance Interactive DSTATCOM," Industrial Electronics, IEEE Transactions on, vol. PP, no.99, pp.1–1, 2015.

Lin Haixue, Xu Jing. Introduction of the National Electrical power system quality [J]. Shanghai Electric Power, 2005(3):221–227.

Wang Zhaoan, Yang Jun, Liu Jinjun. Harmonic Suppression and Reactive Power Compensation [M]. Beijing: China Machine Press, 2006.

Xie Xiaorong, Jiang Qirong. Flexible AC Transmission System: Principles and Applications [M]. Beijing: Tsinghua University Press, 2006.

Yang Huayun, Ren Shiyan. Optimal Design of Active Power Filter Based on Fundamental Magnetic Potential Self-balance and Harmonic Counteraction [J]. Electric Power Automation, 2009, 05:99–103.

Zhou Hui, Qi Zhi-ping. A Survey on Detection Algorithm and Restoring Strategy of Dynamic Voltage Restorer [J]. Proceedings of the CSEE, 30(6):23–29.

Advances in Power and Energy Engineering – Sun (Ed.)
© 2016 Taylor & Francis Group, London, ISBN 978-1-138-02846-3

Optimal operation strategy of power systems with renewable generations via coordinated control of hydrogen consumption

W. Wang
China-EU Institute for Clean and Renewable Energy, China

S.R. Wang
School of Electrical and Electronic Engineering, Huazhong University of Science and Technology, Wuhan, Hubei, China

ABSTRACT: Due to the stochastic nature of renewable energy resources, peak load shifting control by the energy storage system is needed to reach a high penetration level of renewable power generation into power grid. Using hydrogen as an energy storage carrier has many advantages, while the high cost of hydrogen storage is the main drawback. In this paper, a new optimal operation strategy is proposed, the main idea is to set electrolysis cells, fuel cells and hydrogen storage devices geographically closed to a hydrogen consumer. In this situation, hydrogen produced by surplus electricity can be directly delivered to the hydrogen consumer through gas pipes, thus the coordinated control of the hydrogen consumer's production schedule can be realized. In this study, the power balance model of a renewable energy system with hydrogen storage is built, and the minimum hydrogen storage volume needed for power balance is calculated through MATLAB simulation. The simulation results show that the needed hydrogen storage volume for power balance can be reduced via properly coordinated control of the cooperated hydrogen consumer's production schedule.

1 INTRODUCTION

Renewable Energy Systems (RES) are expected to address both the energy crisis and global warming issues. While power systems with high renewable penetration levels contain operational problems that reduce the economic value of renewable energy and present a barrier to the unlimited development of renewable energy. However, because other strategies cannot completely mitigate the random nature of renewable energy sources such as wind and solar, the large-scale use of renewable power will ultimately require the development of the Energy Storage System (ESS) (Beccali et al. 2013).

Using hydrogen as a storage medium for intermittent energy sources is a very interesting alternative in the long term. Hydrogen has a very high enthalpy of 120 MJ/kg, which is about three times that of gasoline (Yu Shuang et al. 2009). When hydrogen is used as a source of energy, only water and heat are produced as by-products with no carbon emission. Nowadays, the rapidly developing Proton Exchange Membrane Electrolysis Cell (PEMEC) and Proton Exchange Membrane Fuel Cell (PEMFC) are easy to operate and provide a fast start-up; the good dynamic performances make them suitable for discontinuous fluctuant

hydrogen production (Andrijinovits & Beldjajev 2012) (Magnus Korp'as & Terje Gjengedal 2006). Concerning the Hydrogen Storage System (HSS), due to the research advance of carbon fiber polymer materials, the working pressure of new type cylinders can be increased to as high as 70 MPa. The new light material under the pressure of 80 MPa is studied. In this condition, the density of hydrogen can reach as high as 36 kg per 1 m³ (Xu Wei et al. 2006). With these advances in technology, the development of the Renewable Energy System with Hydrogen Storage (RESHS) can be expected.

However, the total cost of the HSS is still too high, and acts as the main drawback of the whole RESHS's economic performance, so it is significant to reduce the needed storage volume. In order to reach this goal, a new optimal operation strategy is proposed. The main idea is setting electrolysis cells, fuel cells and hydrogen storage devices geographically closed to a hydrogen consumer. In this situation, hydrogen produced by surplus electricity can be directly delivered to the hydrogen consumer through gas pipes, thus the coordinated control of the hydrogen consumer's production schedule can be realized. The needed storage volume is expected to be reduced via properly coordinated control of the cooperated hydrogen consumer's production schedule.

In this paper, the power balance model of a RESHS was built first, and chapter 2 describes the details of this model. In order to verify our proposed strategy, three cases are studied in chapter 3. The simulation results are presented in chapter 4, based on these results the conclusion is drawn in chapter 5.

2 POWER BALANCE MODEL OF A RESHS

2.1 Models of system components

The formation of the RESHS with our proposed strategy is shown in Figure 1. As our study focuses on the hydrogen storage volume, the power control of PEMEC and PEMFC is the key part. These control processes are implemented through regulating the converters that link PEMEC and PEMFC to the DC bus. Moreover, since the converters' dynamic is considered to be much faster than that of the system, it was neglected and the converters are treated as devices providing a fixed output power that is determined by the controller of this system (Bilodeau & Agbossou 2006).

The input of the PEMEC sub-model is the input electric power P_{ec} and the output is the hydrogen production rate dV_{ec}. The PEMEC sub-model used in our research is chosen from the literature (Ozcan Atlam & Mohan Kolhe 2011). Similarly, the input of the PEMFC sub-model is the output electric power P_{fc} and the output is the hydrogen consumption rate dV_{fc}. The PEMFC sub-model

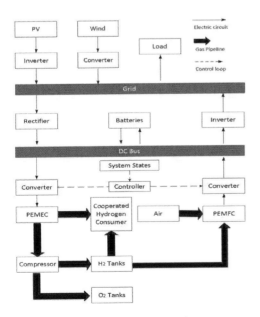

Figure 1.   RESHS with the new optimal strategy.

used in our research is chosen from the literature (Wang Caisheng et al. 2005).

Let us consider that the high-pressure HSS subsystem has two inputs and one output. The input and output hydrogen flow rates of HSS $\dot{V}_{in}$, $\dot{V}_{out}$ are these two inputs, which are equal to the hydrogen production rate of PEMEC dV_{ec} and hydrogen consumption rate of PEMFC dV_{fc} respectively. The storage volume of hydrogen V is the output. Although the actual high-pressure HSS has a compressor and the hydrogen is stored in a high-pressure tank, the equation used to show the relationship between $\dot{V}_{in}$, $\dot{V}_{out}$ and V assumed hydrogen is stored at normal pressure and temperature, which is given as follows:

$$\frac{dV}{dt} = \dot{V}_{in} - \dot{V}_{out} \tag{1}$$

The battery is a main component of RES, and generally plays the role of an energy buffer to handle current spikes and for short-term energy storage. For stationary applications, such as RESHS, the main parameters that determine the battery's performance are its internal resistance, the polarization effect and the long-term self-discharge rate (Kélouwani et al. 2005). This self-discharge rate is difficult to estimate and is itself subject to a number of factors, such as the operating temperature, the number of operation cycles, and the materials and technology used in its manufacture. In a properly controlled RESHS, the overcharge effect will not occur, and hence is not included in the model. Since the battery is connected in parallel with the DC bus and acts as an energy buffer, the current flowing into or out from the batteries is defined by

$$I_B = I_{pv} + I_{wind} + I_{fc} - I_{ec} - I_{load} \tag{2}$$

where I_{pv} is the photovoltaic array's current; I_{wind} is the wind turbine's current; I_{fc} is the fuel cell's current; I_{ec} is the electrolysis cell's current; and I_{load} is the load's current.

This current is positive when the batteries are charging and negative otherwise. Knowing the current, it is possible to deduce the voltage of the battery by:

$$U_B = (1 + at)U_0 + R_i I_B + K_i Q_R \tag{3}$$

where a is the self-discharge rate (Hz); U_0 is the open circuit voltage at time 0; R_i is the internal resistance; K_i is the polarization coefficient, and Q_R is the rate of accumulated charge.

The battery used in our research is chosen from the literature (Bilodeau & Agbossou 2006); a, Q_R and K_i are set as 0; R_i and U_0 are equal to 76 ohm and 48 kV respectively.

Equation (1) can be rewritten as

$$I_B = \frac{P_{pv} + P_{wind} + P_{fc} - P_{ec} - P_{load}}{U_B} \quad (4)$$

where P_{pv} is the power produced by the PV array; P_{wind} is the power produced by the wind turbine; P_{fc} is the output power of the fuel cell; P_{ec} is the input power of the electrolysis cell; and P_{load} is the power consumed by the load.

The total energy stored in the batteries is given by

$$E_B = E_0 + \frac{1}{3600} \int I_B dt \quad (5)$$

where E_0 is the battery's initial energy (Ah).

Finally, this energy can be expressed as a State-of-Charge (SOC) using the following equation:

$$SOC = \frac{E_B}{E_{max}} \times 100\% \quad (6)$$

Using the above equations, the model of the battery is created, and P_{pv}, P_{wind}, P_{fc}, P_{ec} and P_{load} are the inputs of this sub-model. On the other hand, the SOC of the battery is the output.

2.2 System control strategy

The main goal of our control strategy is to maintain the balance of power, mainly by properly and timely regulate the input power of the electrolysis cell P_{ec} and the output power of the fuel cell P_{fc}, the control of the battery's charge and discharge is the auxiliary method. Avoiding deep discharge and over charge of the battery is another mission. The desired behaviors, as described above, were implemented with the help of two input variables: the net power flow dP and the battery's SOC. The net power flow dP is simply the difference between the power provided by the sources and the power consumed by the loads. A detailed calculation is given as follows:

$$dP = P_{pv} + P_{wind} - P_{load} \quad (7)$$

This information is useful to determine whether there is excess power available or not and how much. The battery's SOC given by Equation (6) is used to prevent a deep discharge or an over charge of the battery and to know how much energy stored in the battery is available. The output variable of the controller is a power set point P^*. When the output is positive, the set point signal is sent to the converter and the fuel cell is started. On the other

hand, when the output is negative, the set point signal is sent to the converter and the electrolysis cell is started. P^* meets the following relations:

If P^* is positive: $P_{fc} = P^*$ and $P_{ec} = 0$; $\quad (8)$

If P^* is negative: $P_{ec} = P^*$ and $P_{fc} = 0$; $\quad (9)$

Based on Equations (7), (8) and (9), Equation (4) can be simplified as

$$I_B = \frac{dP + P^*}{U_B} \quad (10)$$

In our system, energy is mainly stored in the form of hydrogen fuel, while when the power unbalance between the power plant and the load is small, the value of dP fluctuates around 0, and the electrolysis cell or the fuel cell will be started and closed frequently. It is more efficient to use these devices for at least a few minutes at a time since they operate more efficiently at high temperatures, and their temperature rises only when they are in operation. Even if dP does not change between positive and negative frequently, on the condition that dP remains low, the efficiency of the electrolysis cell and the fuel cell will be very low, as they are constantly working at a low power operation point that is very inefficient.

So a threshold power P_{th} is set in this system; when the absolute value of dP is below the threshold, both the electrolysis cell and the fuel cell will not be started, which means that the power set point P^* is 0 and the battery now acts as the energy buffer. While when the absolute value of dP is beyond the threshold, first the surplus energy charges up the battery and raises the value of SOC to around 50%, in order to prepare for the next period when dP is below the threshold power. At this time, the value of P^* is given as

If SOC is high: $P^* = -dP - P_{th}$ $\quad (11)$

If SOC is low: $P^* = -dP + P_{th}$ $\quad (12)$

If SOC is moderate: $P^* = -dP$ $\quad (13)$

The boundaries of "low", "high" and "moderate" can be flexibly chosen; in our model, 60% and 40% are chosen as the upper and lower limits of "moderate".

2.3 Evaluating index

Similar to the SOC of the battery, the State of Hydrogen Tanks (SOHT) is used to describe the vacancy rate of the system's hydrogen storage tanks. SOHT can be simply written as

$$SOTH = \frac{\text{Volume of stored hydrogen}}{\text{Volume of total tanks}} \times 100\% \quad (14)$$

At the beginning of our simulation, SOTH is set as 50%, the volume of stored hydrogen can be calculated by Equation (1) and the total volume of tanks is properly assigned so that during our simulation, the value of SOTH will stay between 0% and 100%.

The difference value between the maximum value and the minimum value of SOTH can be regarded as the minimum needed volume of hydrogen storage, which is our greatest concern, so this difference value is chosen as the evaluating index to assess the effect of our new strategy.

3 CASE STUDY BASED ON MATLAB SIMULATION

In this chapter, the simulation model of the RESHS will be built based on Matlab/Simulink, and three cases will be studied:

Case 1: RESHS without coordinated hydrogen consumer.

Case 2: RESHS with coordinated hydrogen consumer while its production schedule does not match up to the power generation of RES;

Case 3: RESHS with coordinated hydrogen consumer and its production schedule is coordinated controlled that depends on the power generation of RES.

3.1 Simulation model

Figure 2 shows the overall simulation model. The aim of our simulation is an overview of the system's behavior over a long period of time. In order to achieve the goal, the system was simulated for 20000 minutes, which was about 2 weeks using an integration step of 60 s. While this value is high, it has no significant impact on the precision of the results, because there are no fast dynamics in the model (Bilodeau & Agbossou 2006). Let us assume that the simulation starts at day 1 at 2:30 hours and ends at day 14 at 24:00 hours, which lasts 20000 minutes totally). The model needs three inputs: wind

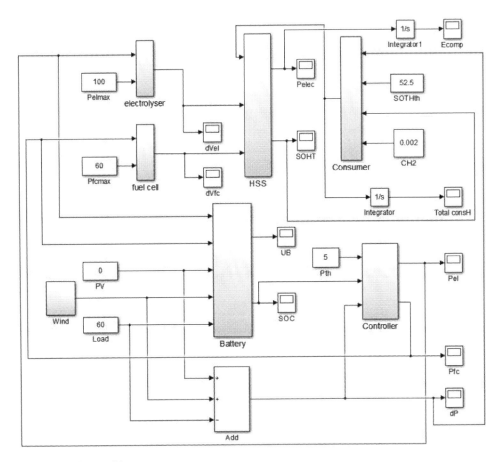

Figure 2.   Simulation model.

turbine power, Photovoltaic array (PV) power and load power. It is better to use the real data, while unfortunately only a set of real measurements of a 160 MW wind plant is acquired. So PV power is set as 0, and load power is assumed to remain as a constant value, given as 60 MW.

As the output range of wind power is from 0 to 160 MW, the value of dP can be as high as 100 MW and as low as −60 MW. Therefore, the maximum input power of the electrolysis cell and the maximum output power of the fuel cell are set as 100 MW and 60 MW. In our simulation, the threshold power P_{th} is set as 5 MW.

3.2 *Modification of simulation model for Case 2*

In the simulation of Case 2, a coordinated hydrogen consumer is added, so the simulation model needs some modification. Let us assume that the plant of the consumer starts every day from 9:00 to 17:00 hours, with a constant hydrogen consumption rate C_{H2}, which is assigned as 0.002 ML s⁻¹ (2 m³ s⁻¹). The production schedule of the hydrogen consumer has no relation with the status of the RESHS. If the hydrogen production rate of the RESHS is no less than C_{H2}, only the excessive part of hydrogen produced by the electrolysis cell is compressed and stored in tanks, and the rest of the part is directly delivered to the consumer's plant through pipelines. If the production rate is less than C_{H2}, hydrogen produced by the electrolysis cell is entirely delivered to the consumer and the insufficient part is provided from the storage tanks.

3.3 *Modification of simulation model for Case 3*

In Case 3, the production schedule of the hydrogen consumer can be controlled, which depends on the status of the RESHS. Hydrogen produced by electrolysis is better to be directly delivered to the consumer instead of being compressed into the tanks, so the consumer's plant is better to be scheduled to operate during the period when the net power flow dP is high as much as possible. On the other hand, the hydrogen stored in tanks is needed to balance the power of the whole RESHS when the power production of the renewable power plant is low, so it is important to ensure that there is always enough hydrogen being stored in the HSS.

In order to make sure that the SOHT is always high enough to meet the requirement of power balance, in our simulation, a threshold value SOHT$_{th}$ is given. When the SOHT is less than this threshold, the consumer's plant always keeps closed, no matter how large the value of dP is. Only if the SOHT is high enough and dP is positive can the consumer's plant start to work. If dP is higher than C_{H2}, the over part of hydrogen is compressed into

the tanks; on the contrary, if dP is lower than C_{H2}, the lack part is provided by the HSS.

Another key factor that we need to take into consideration is that the coordinated control should also ensure the total yield of the coordinated hydrogen consumer, or it will be difficult to persuade them to sign the agreement. In order to reach the same total consumption volume of hydrogen during the simulation, the value of SOHT$_{th}$ is assigned as 52.5%. The total consumption V_{con} can be simply calculated by

$$V_{con} = \int C_{H2}\, dt \tag{15}$$

4 RESULT ANALYSIS

All the three cases have no difference in the control strategy of power balance and the battery's charge and discharge, so the net power flow dP, the electrolysis cell's output power, the fuel cell's input power and the SOC of the battery are all the same in these three cases.

Figure 3 clearly shows that when the net power flow is positive in the system, the electrolysis cell starts to store hydrogen. On the other hand, the fuel cell is started when there is not enough energy

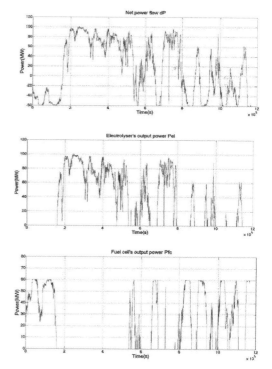

Figure 3. System's power flow.

125

available to power the load. Another characteristic illustrated in Figure 4 shows that the controller is able to keep the batteries' SOC not too high and not too low during these 2 weeks, thus helping to increase the batteries' lifetime.

The operation schedule of the cooperated hydrogen consumer during the simulation is shown in Figure 5. It can be clearly observed that in Case 2, the hydrogen consumption rate is regular, no matter how the states of the system change. On the other hand, in Case 3, hydrogen is mostly consumed during the period from 30000 s to 80000 s. From Figure 7, it is easy to find that SOTH is always beyond 52.5% during this period, and Figure 3 shows that in most time of this period, dP is positive. Therefore, the result is reasonable, and the control strategy for hydrogen consumer in Case 3 works.

Figure 6 shows the comparison between the accumulated hydrogen consumption of cooperated hydrogen consumer in Case 2 and Case 3, it

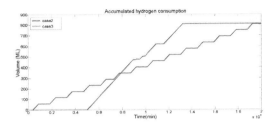

Figure 6. Comparison of accumulated hydrogen consumption in Cases 2 and 3.

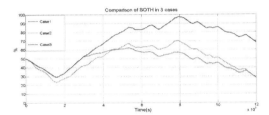

Figure 7. Comparison of SOTH in three cases.

Table 1. The comprehensive comparison of three cases.

Indicators	Case 1	Case 2	Case 3
Min of SOTH (%)	29.198	23.431	29.198
Max of SOTH (%)	97.431	70.353	56.947
Minimum needed hydrogen storage volume (SOTH) (%)	68.233	46.922	27.749
Total hydrogen consumption of hydrogen consumer (ML)*	None	806.52	810.55

*The volume of hydrogen is measured under room temperature (20°C) and normal pressure (1 atm).

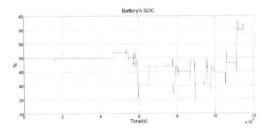

Figure 4. Battery's SOC.

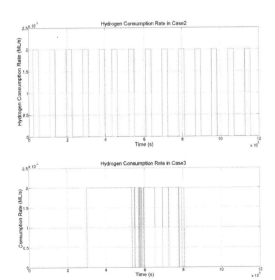

Figure 5. Hydrogen consumption rate in Case 2 and Case 3.

is easy to find out that this value of both cases is almost the same, which indicates that during these 2 weeks, the total yield of the cooperated hydrogen consumer in these two cases is almost the same. Specific numbers of the total hydrogen volume consumed in these two cases are presented in Table 1.

Figure 7 shows the comparison of SOTH in these three cases. For each case, the variation trend of SOTH is displayed as one wave, as mentioned before. The needed minimum storage volume can be simply calculated as the difference between the wave crest and the wave troughs. To make things more clearly, the detailed numerical results are summarized in Table 1.

From the results, it is obvious that in the new operation strategy, the required storage volume of hydrogen is greatly decreased. In Case 3, even more than half of the storage volume is saved when compared with Case 1.

5 CONCLUSION

In this study, the effect of our new strategy is examined in a power balance simulation model. The comparison of the simulation results between Case 1 and Case 2 clearly shows that our new strategy can reduce the needed storage volume of hydrogen in a RESHS. The simulation result of Case 3 indicates that with a proper control of the cooperated hydrogen consumption plant, the needed hydrogen storage volume can be further reduced.

With the reduction of the needed hydrogen storage volume in a HSS, this strategy can make hydrogen storage more economically acceptable, so as to provide a new approach to improve the penetration level of renewable power generation. This strategy can also provide a new clean way of large-scale hydrogen production instead of traditional thermo-chemical processes, which not only consume a large amount of fossil fuels but also have impacts on the environment.

ACKNOWLEDGMENT

The authors thank LIAO Shiwu and LI Jiaming, who are colleagues of our laboratory, for providing the real source power data of the 160 MW wind farm used in this research.

REFERENCES

Andrijinovits & Beldjajev. 2012. Techno-Economic Analysis of Hydrogen Buffers for Distributed Energy Systems. International Symposium on Power Electronics Electrical Drives Automation and Motion, IEEE, pp: 1401–1406.

Beccali et al. 2013. Method for size optimization of large wind–hydrogen systems with highpenetration on power grids. Applied Energy, 102: 533–544.

Bilodeau & Agbossou. 2006. Control analysis of renewable energy system with hydrogen storage for residential applications, Journal of Power Sources, 162: 757–764.

Kélouwani et al. 2005. Model for energy conversion in renewable energy system with hydrogen storage, Journal of Power Sources, 140: 392–399.

Magnus Korp'as & Terje Gjengedal. 2006. Opportunities for Hydrogen Storage in connection with Stochastic Distributed Generation, 9th International Conference on Probabilistic Methods Applied to Power Systems, IEEE, pp: 1–8.

Ozcan Atlam & Mohan Kolhe. 2011. Equivalent electrical model for a Proton Exchange Membrane (PEM) electrolyser, Energy Conversion and Management, 52: 2952–2957.

Wang Caisheng et al. 2005. Dynamic Models and Model Validation for PEM Fuel Cells Using Electrical Circuits. IEEE TRANSACTIONS ON ENERGY CONVERSION, pp: 442–451.

Xu Wei et al. 2006. Progress of Research on Hydrogen Storage, Progress in Chemistry, 18(2): 200–210.

Yu Shuang et al. 2009. A New Methodology for Designing Hydrogen Energy Storage in Wind Power Systems to Balance Generation and Demand Wind Engineering, Sustainable Power Generation and Supply Conference, IEEE, pp: 1–6.

Advances in Power and Energy Engineering – Sun (Ed.)
© 2016 Taylor & Francis Group, London, ISBN 978-1-138-02846-3

DC bus current reconstruction of 3L NPC dual PWM converter based on feed-forward control

W.J. Wu, W.X. Wang & X.Y. Li

Xi'an University of Technology, Xi'an, Shaanxi, China

ABSTRACT: Based on the mathematical model of the 3L PWM converter, the reconstitution of DC bus current was deduced based on FFT analysis in this paper. Then, the DC bus active power current was reconstructed and adopted into the control system of the 3L NPC dual PWM converter. Finally, the simulations of 3L back-to-back converter were completed, in which the active power current feed-forward control strategy was adopted. The DC bus active power current errors were studied based on different modulation depths, different mutation load powers and different load power factors. The simulation results indicate that the DC bus active power current reconstructed method was simple, effective and correct.

1 INTRODUCTION

3L (three level) NPC (neutral point clamped) dual PWM (Pulse width modulation) converter is widely used in wind power generation, variable speed drive, HVDC and other fields (Behera 2012, Chaves et al. 2011, Bueno et al. 2008) for its lower current harmonic distortion and double power flow. However, the inverter running status and the load carried changes will cause the DC bus voltage fluctuation, and have an impact on the safe operation of the system. Meanwhile, the DC bus current flowing into the inverter can reflect the operation status of the inverter in time, and it can be also used to improve the system's stability of current feed-forward control in the 3L dual PWM converter (Gu 2005, Zhang et al. 2011). But it is difficult to acquire the DC bus current well and truly because it is a high-frequency pulse current. So, the key research of current feed-forward control in the 3L dual PWM converter is to obtain the DC bus current accurately.

In a study by Hur et al. (2001), the current sensor was installed in the DC bus to get the DC bus current. This method is simple, but it will increase the stray inductance, and the filter parameters of high-frequency pulse DC bus current are difficult to be determined. In another study by Gu & Nam (2005), motor load information of the inverter was adopted to construct the DC bus current, which can reflect the motor load running state in time, but it is restricted by the control strategy, modulation method as well as the parameters of the motor. State observation was established to obtain the DC bus current in previous work (Li 2006a,b, Li et al. 2006). This method avoids the installation

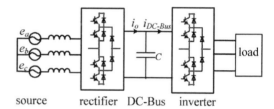

Figure 1. The 3L NPC dual PWM converter.

of the current sensor, but it depends on the rectifier output current and DC bus voltage; moreover, its algorithm is more complex and has a large number of calculation.

The DC bus current obtained by all of the above methods is rms (root-mean-square) current. As shown in Figure 1, the $i_{DC\text{-}Bus}$ flowing into the inverter is a high-frequency pulse current. However, its components and influence factor of $i_{DC\text{-}Bus}$ has not been reported yet. In this paper, $i_{DC\text{-}Bus}$ based on the 3L PWM converter working principle and the mathematical model is deduced. Furthermore, the DC bus active current is reconstructed based on FFT analysis, which provides a theoretical guidance for the DC bus current filter to be designed and the system stability to be improved.

2 THE RECONSTRUCTION AND ANALYSIS OF DC BUS CURRENT

2.1 DC bus current reconstruction

The high-frequency pulse DC bus current reconstruction is derived based on the working principle and the mathematical model of the 3L PWM converter (Bai et al. 2011; Li et al. 2010). Figure 2 (a) shows the topology of the 3L NPC PWM inverter and Figure 2 (b) shows the AC-side equivalent circuit of the inverter.

Assuming that the 3L inverter switching function is s_k, the expression of u_k can be written as follows:

$$u_k = \frac{1}{2} u_{dc} s_k \quad (k = a, b, c) \tag{1}$$

According to Figure 1, the differential equations can be expressed as

$$\begin{bmatrix} L\dfrac{di_a}{dt} \\ L\dfrac{di_b}{dt} \\ L\dfrac{di_c}{dt} \end{bmatrix} = \begin{bmatrix} u_a \\ u_b \\ u_c \end{bmatrix} + \begin{bmatrix} -R & 0 & 0 \\ 0 & -R & 0 \\ 0 & 0 & -R \end{bmatrix} \begin{bmatrix} i_a \\ i_b \\ i_c \end{bmatrix} \tag{2}$$

That is,

$$L\frac{di_k}{dt} + i_k R = u_k \tag{3}$$

where i_k is the constituted of transient and steady-state components according to Expression (3) and the first-order circuit response (Zhang 2007a, b). Its full solution is assumed as follows:

$$i_k = I \sin(\alpha_k - \varphi) + A_k e^{-\frac{t}{\tau}} \tag{4}$$

where time constant and the load phase angle are as follows:

$$\tau = \frac{L}{R} \quad \varphi = tg^{-1}\frac{\omega L}{R} \tag{5}$$

Here A_k is the undetermined coefficient. Substituting $i_k(0) = 0$ into (4) yields

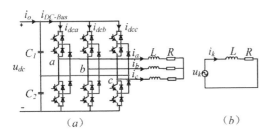

$$(a) \qquad\qquad (b)$$

Figure 2. 3L NPC PWM inverter topology and AC-side equivalent circuit.

$$i_k = I \sin(\alpha_k - \varphi) + Ie^{-\frac{t}{\tau}} \sin\left[\varphi + (k-1)120^0\right] \tag{6}$$

According to the high-frequency mathematical model of the 3L PWM inverter, $i_{DC\text{-}Bus}$ can be expressed as

$$\begin{cases} i_{DC\text{-}Bus} = i_{dca} + i_{dcb} + i_{dcc} \\ i_{dck} = \dfrac{1}{2} s_k (s_k + 1) i_k \end{cases} \tag{7}$$

The DC bus current $i_{DC\text{-}Bus}$ is high-frequency pulse signal associated with the PWM inverter switching function. The $i_{DC\text{-}Bus}$ may differ when the modulation strategy is different, and the specific analysis is given by

2.1.1 SPWM modulation strategy

$$\begin{cases} s_k = m_1 \sin \alpha_k \\ \alpha_k = \omega t - (k-1)120^0 \end{cases} \quad (k = a, b, c) \tag{8}$$

From (6)–(8), the DC bus current expression based on the SPWM modulation strategy can be obtained as follows:

$$i_{DC-Bus} = \frac{3}{4} Im_1 \cos\varphi - \frac{3}{8} m_1^2 \sin(3\omega t - \varphi)$$
$$- \frac{3}{4} e^{-\frac{t}{\tau}} Im_1 \cos(\omega t + \varphi) + \frac{3}{8} e^{-\frac{t}{\tau}} Im_1^2 \sin(2\omega t - \varphi) \tag{9}$$

2.1.2 SVPWM modulation strategy

For the value relationship of the modulation strategy between the SVPWM and SPWM, the switching function s_k can be expressed as

$$\begin{cases} s_k = m_1 \sin \alpha_k + m_2 \sin(3\omega t + \theta_2) \\ \alpha_k = \omega t - (k-1)120^0 \end{cases} \tag{10}$$

Applying (6), (7) and (10), the DC bus current expression based on the SVPWM modulation strategy can be obtained as follows:

$$i_{DC-Bus} = \sum_{k=1}^{3} \frac{1}{2} s_k (s_k + 1) i_k$$
$$= \frac{3}{4} Im_1 \cos\varphi - \frac{3}{8} Im_1^2 \sin(3\omega t - \varphi)$$
$$+ \frac{3}{2} Im_1 m_2 \cos\varphi \sin(3\omega t + \theta_2)$$
$$- \frac{3}{4} e^{-\frac{t}{\tau}} Im_1 \cos(\omega + \varphi) + \frac{3}{8} e^{-\frac{t}{\tau}} Im_1^2 \sin(2\omega t - \varphi)$$
$$- \frac{3}{2} e^{-\frac{t}{\tau}} Im_1 m_2 \cos(\omega t + \varphi) \sin(3\omega t + \theta_2) \tag{11}$$

2.2 Analysis of DC bus current reconstruction

We set f_1 as the fundamental wave frequency and f_s as the switching frequency. From (9) and (11), we know that the DC bus current expression is different when using different modulation methods. But the DC bus current is made up of transient and steady-state components in both expressions.

According to Equations (9) and (11) and the character of the PWM converter, when the circuit is in steady state, the high-frequency pulse DC bus current $i_{DC\text{-}Bus}$ is made up of DC and frequency of $3f_1$, nf_s and $nf_s \pm 3f_1$ components, which is irrelevant with the control strategy, modulation method, load properties and output voltage of the inverter.

Based on Figure 2(a), simulation studies were carried out on PSIM under the following conditions:

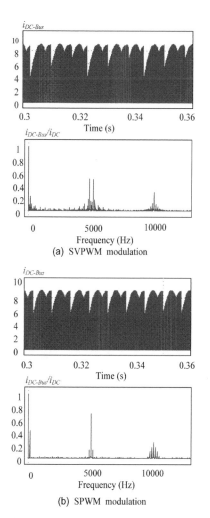

Figure 3. DC bus current with different modulating methods.

$u_{dc} = 600$ V, power factor $= 0.9$, modulation depth $= 0.8$, $C_1 = C_2 = 600$ uF and $f_s = 5000$ Hz.

The $i_{DC\text{-}Bus}$ waveform and their FFT analysis under different conditions are shown in Figure 3 while the circuit is in steady state. Apparently, the DC bus current and their harmonic components correspond to the above theoretical analysis.

3 THE RECONSTRUCTION AND ANALYSIS OF DC BUS ACTIVE POWER CURRENT

Most of the coordinated control strategy of the dual PWM converter is based on active power balance. From the algorithm analysis of part 2, we know that the DC bus current is a high-frequency pulse signal, which is comprised of DC and frequency of $3f_1$, nf_s and $nf_s \pm 3f_1$ components. But the component on behalf of the active power is unknown. The construction and analysis of DC bus active current based on the system active power conservation is represented below.

3.1 DC bus active power current reconstruction

3.1.1 The DC-side power of the 3L PWM inverter

$$p_{dc} = u_{dc} i_{DC-Bus} \tag{12}$$

While the SPWM modulation strategy is adopted, substituting (9) into (12) yields:

$$
\begin{aligned}
p_{dc} &= u_{dc} i_{DC-Bus} \\
&= \frac{3}{4} I u_{dc} m_1 \cos\varphi - \frac{3}{8} I u_{dc} m_1^2 \sin(3\omega t - \varphi) \\
&\quad - \frac{3}{4} e^{-\frac{t}{\tau}} I u_{dc} m_1 \cos(\omega t + \varphi) \\
&\quad + \frac{3}{8} e^{-\frac{t}{\tau}} I u_{dc} m_1^2 \sin(2\omega t - \varphi)
\end{aligned} \tag{13}
$$

While the SVPWM modulation strategy is adopted, substituting (11) into (12) yields:

$$
\begin{aligned}
p_{dc} &= u_{dc} i_{DC-Bus} \\
&= \frac{3}{4} I u_{dc} m_1 \cos\varphi - \frac{3}{8} I u_{dc} m_1^2 \sin(3\omega t - \varphi) \\
&\quad + \frac{3}{2} I u_{dc} m_1 m_2 \cos\varphi \sin(3\omega t + \theta_2) \\
&\quad - \frac{3}{4} e^{-\frac{t}{\tau}} I u_{dc} m_1 \cos(\omega t + \varphi) + \frac{3}{8} e^{-\frac{t}{\tau}} I u_{dc} m_1^2 \sin(2\omega t - \varphi) \\
&\quad - \frac{3}{2} e^{-\frac{t}{\tau}} I u_{dc} m_1 m_2 \cos(\omega t + \varphi) \sin(3\omega t + \theta_2)
\end{aligned} \tag{14}
$$

3.1.2 The AC-side active power of the 3L PWM inverter

$$p_{ac} = \sum_{k=1}^{3} u_k i_k = \sum_{k=1}^{3} \frac{1}{2} u_{dc} s_k i_k \tag{15}$$

While the SPWM modulation strategy is adopted, applying Equations (1), (6), (8) and (15), we obtain

$$
\begin{aligned}
p_{ac} &= \sum_{k=1}^{3} \left\{ \frac{1}{2} u_{dc} m_1 \sin \alpha_k \times \left[I \sin(\alpha_k - \varphi) \right. \right. \\
&\quad \left. \left. + I e^{-\frac{t}{\tau}} \sin\left(\varphi + (k-1)120^0 \right) \right] \right\} \\
&= \frac{3}{4} I u_{dc} m_1 \cos \varphi - \frac{3}{4} e^{-\frac{t}{\tau}} I u_{dc} m_1 \cos(\omega t + \varphi) \tag{16}
\end{aligned}
$$

While the SVPWM modulation strategy is adopted, applying Equations (1), (6), (10) and (15), we obtain

$$
\begin{aligned}
p_{ac} &= \sum_{k=1}^{3} \left\{ \frac{1}{2} u_{dc} \left[m_1 \sin \alpha_k + m_2 \sin(3\omega t + \theta_2) \right] \right. \\
&\quad \left. \times \left[I \sin(\alpha_k - \varphi) + I e^{-\frac{t}{\tau}} \sin\left(\varphi + (k-1)120^0 \right) \right] \right\} \\
&= \frac{3}{4} I u_{dc} m_1 \cos \varphi - \frac{3}{4} e^{-\frac{t}{\tau}} I u_{dc} m_1 \cos(\omega t + \varphi) \tag{17}
\end{aligned}
$$

Analyzing Equations (15) and (16), we reach a conclusion that no matter which modulation method is employed, the AC-side active power expression is uniform. The p_{ac} is a part of p_{dc}, which corresponds to the DC component and frequency of f_1 damping component of i_{DC-Bus}.

Therefore, DC bus active power current expression is reconstructed as follows:

$$
\begin{aligned}
i_{DC-Bus-act} &= \frac{3}{4} I m_1 \cos(\theta_1 + \varphi) \\
&\quad - \frac{3}{4} e^{-\frac{t}{\tau}} I m_1 \cos(\omega t + \theta_1 + \varphi) \tag{18}
\end{aligned}
$$

From the above analysis, we know that the $i_{DC-Bus-act}$ that is comprised of DC component and frequency of f_1 damping component accurately reflects the AC-side active power.

3.2 Analysis of DC bus active power current reconstruction

The actual DC-side active current is defined as

$$i_{real} = \frac{p_{ac}}{u_{dc}} \tag{19}$$

Its relative error between $i_{DC-Bus-act}$ and i_{real} is defined as

$$\varepsilon = \frac{i_{DC-Bus-act} - i_{real}}{i_{real}} \times 100\% \tag{20}$$

The simulating results of ε under the different situations are shown in Figure 4.

Here, m and PF denote the modulation depth and the load power factor, respectively.

The relationship between ε and load power factor at the modulation depth of 0.9 is shown in Figure 4(a). Although the load power factor changes from 0.1 pu to 2 pu, the relative errors are lower than 1%. The relative errors based on different modulation depths and load power factors are shown in Figure 4(b). Apparently, they are lower than 5%. These results demonstrate that the reconstructed DC bus active power current can accurately reflect the AC-side active power in this paper.

Defining the i_{dc} as the DC component of $i_{DC-Bus-act}$, the relative error between i_{dc} and i_{real} is given by

$$\varepsilon_1 = \frac{i_{dc} - i_{real}}{i_{real}} \times 100\% \tag{21}$$

The simulating results of ε_1 under the different situations are shown in Figure 5. In this figure, although there is variation in the modulation depth and the load power factor, the relative error

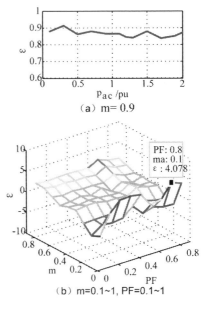

(a) m= 0.9

(b) m=0.1~1, PF=0.1~1

Figure 4. The relative error ε between $i_{DC-Bus-act}$ and i_{real} with different modulating depths as well as load power factors.

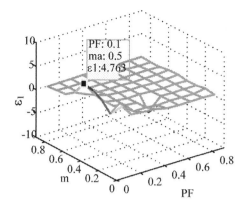

Figure 5. The relative error ε_1 between i_{dc} and i_{real} with different modulating depths as well as load power factors.

between i_{dc} and i_{real} is lower than 5%. In other words, the i_{dc} can reflect the AC-side active power under the condition of acceptable error. It is simpler to get the DC bus active current and to design the filter of the DC bus current.

4 THE FEED-FORWARD CONTROL OF THE 3L DUAL PWM CONVERTER

The DC component i_{dc} of reconstructed DC bus active power current is applied to the feed-forward control strategy based on the 3L dual PWM converter system. The simulation circuit is completed in PSIM, and the control block diagram is depicted in Figure 6. Dual-loop control strategy based on grid voltage orientation is used in the rectifier and the VVVF (variable voltage variable frequency) control strategy is adopted in the inverter.

Here, $\hat{u}_{dc}$, $\hat{u}$, $\hat{i}_{dc}$ and denote the DC bus reference voltage fluctuation, source voltage fluctuation and load current disturbance, respectively.

$$
\begin{cases}
G_{ref} = \dfrac{\Delta u_{dc}}{\Delta u_{ref}} = \dfrac{G_K Z_L G_U}{\Delta} \\[2mm]
G_v = \dfrac{\Delta u_{dc}}{\Delta u} = \dfrac{Z_L(G_D - G_u G_K)}{\Delta} \\[2mm]
Z_o = \dfrac{\Delta u_{dc}}{\Delta i_{load}} = \dfrac{-Z_L(1 - G_i G_K)}{\Delta} \\[2mm]
\Delta = 1 + Z_L(G_K G_U + G_Z)
\end{cases} \tag{22}
$$

Simulating parameters are as follows.

The power rating is 7.5 kW, the source voltage is 380 V, $C_1 = C_2 = 600$ uF, the switching frequency is 5 kHz, and the resistance and inductance of the inverter are 0.5 Ω and 5 mH, respectively.

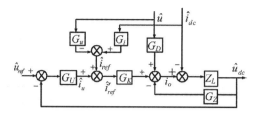

Figure 6. Feed-forward control based on DC component i_{dc} of reconstructed $i_{DC\text{-}Bus}$.

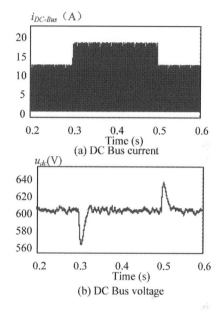

(a) DC Bus current

(b) DC Bus voltage

Figure 7. Feed-forward control strategy is not applied.

Figure 7(a) and Figure 7(b) show the DC bus current and DC bus voltage waveform when the feed-forward control strategy is not applied. When t = 0.3 s, the load power is suddenly increased by 150%, the DC bus voltage drops by 6% and the recovery time is 18 ms. When t = 0.5 s, load power is suddenly reduced by 150%, the DC bus voltage rises to 5% and the recovery time is 17 ms.

Figure 8 shows the system output waveform while the feed-forward control strategy based on the DC component of reconstructed DC bus active power current is adopted. As shown in Figure 8(a), the DC bus voltage drops by 0.67% and the recovery time is 5 ms when the load power suddenly increased by 150%, while DC bus voltage rises to 1.67% and the recovery time is 6 ms when the load power is suddenly reduced by 150%. Figure 8(b) shows the waveform of DC component i_{dc} of reconstructed DC bus active power current $i_{DC\text{-}Bus\text{-}act}$. It is evident that the AC-side active power can be reflected accurately.

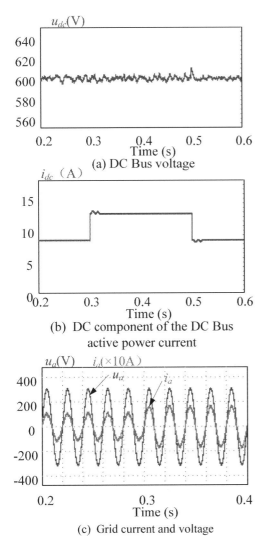

(a) DC Bus voltage

(b) DC component of the DC Bus active power current

(c) Grid current and voltage

Figure 8. Feed-forward control strategy is applied.

The grid voltage u_a synchronizes with the grid current i_a, as shown in Figure 8(c).

Comparing Figure 7(a) with Figure 8(a), it shows that the feed-forward control based on the reconstructed DC bus active current can effectively restrain the DC bus voltage fluctuation caused by the load disturbance of the inverter, and can improve the system's stability.

5 CONCLUSION

In this paper, the reconstructed method of DC bus active power current in the 3L NPC dual PWM converter is studied based on the FFT analysis. The DC

bus active power current is also obtained based on the active power balance. By algorithm analysis, we reach the conclusion that the DC component of the DC bus active power current can reflect the AC-side active power within acceptable error rates. The reconstruction method is applied in the feed-forward control based on the dual PWM converter system, in which the DC component of the DC bus active power current corresponds to the AC-side active power approximately. Simulation results reveal that it has a high precision and effectiveness. This method is simple and easy to be applied. It has not only lower system price, but also provides a theoretical guidance for the DC bus current filter to be designed in the three-level NPC converter.

REFERENCES

Bao dong Bai. et al. 2011. Study on the improved control strategy for PWM rectifier. Electrical Machines and Systems (ICEMS) vol., no., pp.1–4, 20–23.
Behera, R.K. 2012. Utility friendly three-level neutral point clamped converter-fed high-performance induction motor drive. IET Power Electronics 5(7): 1196–1203.
Emilio J. et al. 2008. Design of a Back-to-Back NPC Converter Interface for Wind Turbines With Squirrel-Cage Induction Generators. IEEE transactions on energy conversion 23(3):932–945.
Gu Bon gwan. 2002. A dc link capacitor minimization method through direct capacitor current control. IEEE Trans. on Power Elec 33(5):811–817.
Gu, B.-G. & Nam, K. 2005. A Theoretical minimum DC-link capacitance in PWM converter-inverter systems. Electric Power Applications, IEE Proceedings vol.152, no.1, pp.81–88, 7.
Hur, N. et al. 2001. A fast dynamic DC-link power-balancing scheme for a PWM converter-inverter system. Industrial Electronics, IEEE Transactions on, vol.48, no.4, pp.794–803.
Li Shi-jie & Li Yao-hua. 2006. Study of the Feed-forward Control Strategy based on the Luenberger Observer Power Electronics. 40(5):4–6.
Li Shi-jie. et al. 2006. Study the optimum feed-forward control strategy in back-to-back converter system. Proceedings of the CSEE 26(22):74–79.
Li Tao. et al 2010. The research on the macro-equivalence between SPWM and SVPWM. Proceedings of the CSEE30 (S1):178–184.
Miguel Chaves. et al. 2011. HVDC transmission systems Bipolar back-to-back diode clamped multilevel converter with fast optimum-predictive control and capacitor balancing strategy. Electric Power Systems Research 81:1436–1445.
Zhang Jiasheng & Zhang Lei. 2007. Research on the DC-side equivalent model of PWM inverters [J]. Proceedings of the CSEE 27(4):103–107.
Zhang Yingchao. et al. 2011. Integrated control scheme for three-level NPC based dual-PWM converter. Transactions of China Electrotechnical Society 26(11): 25–31.

Advances in Power and Energy Engineering – Sun (Ed.)
© 2016 Taylor & Francis Group, London, ISBN 978-1-138-02846-3

Control strategy of the hybrid three terminal HVDC system for wind power transmission

Y. Zhang, X.T. Wang & Y. Huang
School of Electronic Information and Electrical Engineering, Shanghai Jiao Tong University, Shanghai, China

Z.B. Wang
Renewable Energy Department, China Electric Power Research Institute, Qinghe District, Beijing, China

ABSTRACT: Hybrid three terminal HVDC system is suitable for wind power transmission in western China considering making full use of the existing LCC-HVDC lines and the stochastic and intermittent of wind power. A WF grid integration by MMC technology with LCC-based HVDC transmission system and the corresponding control strategies are proposed in this paper. The control objective is the power balance between three terminals. The LCC rectifier is operated on minimum conduction angle (α_{min}) control; The active power-frequency droop control and the AC bus voltage magnitude control are brought into the vector control of MMC; the LCC inverter is operated on improved current control or turn-off angle(γ) control. Simulations are carried out in PSCAD in order to illustrate the performance of the system under normal conditions and abnormal conditions. The results suggest that the proposed scheme is technically feasible.

1 INTRODUCTION

With the decrease of fossil energy and serious air pollution, the wind power, as a clean and renewable energy, has gotten a quick development and application over the past twenty years. In 2014, the newly added installed capacity of wind power is over 50 GW all over the world, in which China's newly added installed capacity of wind power is 23.3 GW, which is the most in the world (GWEC. 2015). China has planned to build 8 ten-million-kilowatts-class wind power bases at this stage, which are mostly concentrated in the Northeast, Northwest and North China. Especially in the Northwest, Most of the wind power should be transmitted to the central and eastern China where is the load center, because the local load is low (Xu Dong et al. 2015). When the large-scale wind farms are directly integrated into AC grid, the power quality problems and stability problems usually limit the transmission capability of the ac lines. The most viable option for transmission of wind power is High Voltage Direct Current (HVDC) (Cheng-hao Li et al. 2013). There have been several Line Commutated Converter (LCC) HVDC lines in China, which can be used to wind power transmission.

As a form of the renewable energy resources, the influences of wind power are different from the regular power after grid integration. If it is transmitted bundled with thermal conventional power which is generated near the wind power bases, frequency fluctuation will be caused according to the stochastic and intermittent of wind power because the local power system is weak. Thus, wind power should be operated in island.

Voltage Source Converter (VSC) has been applied to grid integration of offshore Wind Farm (WF) considering several advantages of VSC, such as the start-up capability, does not need an external commutation voltage for its operation; and has a relatively more compact size (Raymundo E. Torres-Olguin et al. 2012). Thus, the wind power can be connected to existing LCC-HVDC lines via VSC (Mai Huong Nguyen et al. 2012, Raymundo E. Torres-Olguin et al. 2012, Temesgen M. Haileselassie et al. 2011, Xia Chen et al. 2011).

Modular Multilevel Converter (MMC) has become one of the most concerned converter topologies in the field of VSC-HVDC system, The advantages of MMC-based HVDC system are very straightforward. The multilevel capability produces ac voltage waves with very low distortion, which eliminates most of the filtering. The modular structure provides the flexibility to scale the voltage and power level by adding more Submodules (SMs). Moreover, compared with the traditional VSC-HVDC systems, the MMC-based HVDC system can reduce losses significantly due to its low-frequency switching operation (Kaijian Ou et al. 201).

The feasibility of using a conventional LCC-HVDC transmission in combination with a VSC to integrate offshore wind farms has been investigated in paper (Raymundo E. Torres-Olguin et al. 2012). In paper (Xu Dong et al. 2015), a novel control strategy based on the Model Predictive Control (MPC) was proposed to control the VSC used for transmitting the wind power in the hybrid multi-terminal DC grid. A three-terminal LCC-HVDC system was proposed totransmit bulk inland wind power over thousands of kilometers, and control strategies for the individual converter and the whole system were designed in paper (Lin Weixing et al. 2012). All these previous works can be used as references for the proposed system structure and control strategies in this paper.

This paper is focused on investigating the feasibility of using a LCC-HVDC in combination with a MMC to integrate WF. Control strategies are presented based on the hybrid three-terminal HVDC system. The control objective is the power balance between three terminals. The LCC rectifier is operated on minimum conduction angle (α_{min}) control; the control strategy of MMC is vector control, the inner loop control is current control, and the outer loop control is power control, in which the active power-frequency droop control is brought into the active power control, and the AC bus voltage magnitude control is brought into the reactive power control; the LCC inverter is operated on current control or turn-off angle(γ) control. PSCAD/EMTDC numerical simulation cases are provided to illustrate the performance of the system under normal conditions, such as changes in wind speed, and abnormal conditions, such as ac faults in the MMC and the LCC rectifier.

2 CONTROL STRATEGY

The proposed configuration of the hybrid three terminal HVDC system is shown in Figure 1. LCC-HVDC is based on the well-known benchmark model proposed by the international council on large electric systems (CIGRE) (Manitoba HVDC Research Center. 2003). The conventional power plant and the load center are represented by The venin equivalent voltage sources with the

corresponding source impedance respectively, and the conventional power plant equivalent system is considered as a weak grid. Each LCC converter consists of two twelve-pulse bridges. Both ac filters, tuned to the characteristic harmonics, and reactive compensation are considered in each LCC. A large WF is connected to the LCC-HVDC through a MMC. To reduce the simulation time, an aggregated model has been used to represent the entire WF. The control objectives are that the power of conventional power plant is constant, and the power transmitted to load center changes along with the power of WF, which is $P_{d1} + P_{d2} - P_{loss} = P_{d3}$, P_{d1} is the power of LCC rectifier (rLCC), P_{d2} is the power of MMC, P_{d3} is the power of LCC inverter (iLCC), P_{loss} is the line loss power. The design of the controllers is divided into three sections: the LCC rectifier control, the MMC control, the LCC inverter control. Below, the control objectives and strategy for each system element are described.

2.1 LCC rectifier control

The control strategy of LCC rectifier is shown in Figure 2. I_{d1}^* is the dc current reference of LCC rectifier, which represents for the reference power of conventional power plant. I_{d1} is the dc current of LCC rectifier, and ΔI_m is the current margin. In steady state, $I_{d1}^* = I_{d1} + \Delta I_m$, there are always current error ΔI_m between I_{d1}^* and I_{d1}. The output of controller PI1 is saturated, which reaches its upper limit. The firing angle of LCC rectifier $\alpha_1 = \alpha_{min}$, LCC rectifier is operated on minimum firing angle state, and α_{min} is set based on I_{d1}^*.

2.2 MMC control

The widely used controller structure of MMC rectifier is shown in Figure 3. I_{d2} and U_{d2} are the dc current and dc voltage at MMC respectively. Active power and reactive power independent control are realized via the dq decoupling of MMC, in which active power is controlled by d-axis, and reactive power is controlled by q-axis. The outer loop power control provides dq current references i_{dref} and i_{qref} for the inner loop current control. Inner loop control makes the dq current follow its references by controlling

Figure 1. Schematic of hybrid three terminal HVDC system.

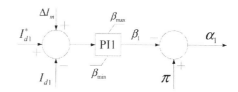

Figure 2. Schematic of LCC rectifier control.

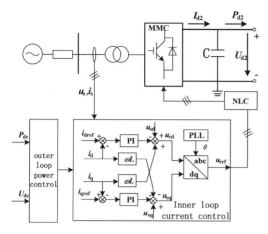

Figure 3.   Schematic of MMC control.

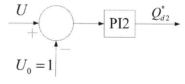

Figure 4.   Schematic of active power-frequency droop control.

the voltage output of converter u_{ref}. The Nearest Level Control (NLC) is used to generate the modulating signal of MMC.

2.2.1   Active power-frequency droop control

As we know, wind power has the features of stochastic and intermittent, and the power at MMC should follow the output power of WF. The transmission power can be regarded as a load of wind turbine generator. From the view of energy conservation, such as when the wind speed rises, the mechanical power of prime mover will rise, which makes the output power of wind turbine generator rises. If the transmission power of MMC does not follow output power of wind turbine generator rapidly, the frequency of AC bus will fluctuate, which is unsatisfactory. Thus, the frequency of AC bus can be brought into the active power control of MMC, it is expressed as

$$P_{d2}^* = P_g + K*(f - f_0) \tag{1}$$

where P_{d2}^* = active power reference of MMC; P_g = output power of wind turbine generator; $1/K$ = slope of active power-frequency droop curve; f = per unit frequency of AC bus; and $f_0 = 1$.

The structure of control block is shown in Figure 4.

2.2.2   AC bus voltage magnitude control

The stabilization of AC bus voltage magnitude is necessary for the normal operation of wind turbine generator. The control strategy of MMC is the decoupling control of active power and reactive power, and reactive power is closely related to voltage magnitude. Thus, the AC bus voltage magnitude can be brought into the reactive power control of MMC, which is

$$Q_{d2}^* = \left(K_p + \frac{K_i}{s}\right)(U - U_0) \tag{2}$$

where Q_{d2}^* = reactive power reference of MMC; U = per unit value of AC bus voltage magnitude; $U_0 = 1$; K_p = proportional action factor; and K_i = integration constant.

The structure of control block is shown in Figure 5.

2.3   LCC inverter control

The control strategies of LCC inverter are consist of the outer loop power control and the inner loop current control as shown in Figure 6. When the value of P_{d3}^* changes, the error between P_{d3}^* and P_{d3} leads to the current reference change though PI5, which is added to I_{d30}^*. The object of controlling P_{d3} can be achieved via the change of I_{d3}^*. P_{d3}^* and P_{d3} are the active power reference and the measured value of active power respectively; PI5 is the proportional plus integral controller; I_{d30}^* and I_{d3}^* are the initial value of current reference and current reference respectively. The other control strategies are the same with CIGRE HVDC Benchmark Model.

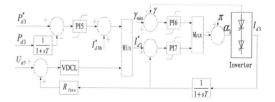

Figure 5.   Schematic of AC bus voltage magnitude control.

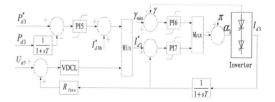

Figure 6.   Schematic of LCC inverter control.

3 CASE STUDY

For illustrating the performance of the proposed controllers, the HVDC system, shown in Figure 1, has been implemented using EMTDC/PSCAD. The system, which is rated at 1000 MW and 500 kV, is composed of two 12-pulse LCCs connected through a long dc cable of 200 km. At The rLCC side, the ac voltage is 345 kV at 50 Hz with a Short Circuit Ratio (SCR) of 2.5, and the rated power of conventional power plant is set to 600 MW. At the iLCC side, the ac voltage is 230 kV at 50 Hz with SCR of 2.5. The MMC is connected via a 40 km dc cable at the beginning of sending end of the other cable. The MMC is simulated using a fast numerical simulation model (Feng Yu et al. 2013). The WF is simulated using an aggregate model, i.e., a single unit of 300 MW. The dc line between LCCs is simulated using t-type equivalent circuits. The detailed parameters are shown in Table 1. The simulations were conducted under different conditions to investigate the operating characteristics of the pro-posed system. These conditions include variations in the wind speed, ac faults at the MMC and ac faults at the LCC rectifier.

3.1 *Response to change in the wind speed*

To study the response of the system to change in the wind speed, the WF is tested in various conditions as shown in Figure 7(a). The initial wind speed is 10 m/s. At 7 s, a wind speed gust occurs which increases amplitude 3 m/s with 2 s duration. Figure 7(b) shows the active power at the input of the MMC, at the input of the LCC rectifier and the output of the LCC inverter. The figures show that the power at rLCC remains 600 MW, the variation

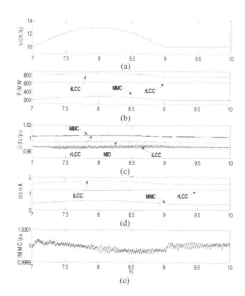

Figure 7. Proposed system response to change in the wind speed: (a) wind speed, (b) active powers, (c) dc voltages, (d) dc currents, (d) frequency at MMC ac bus, respectively.

of MMC power is from 200 MW to 300 MW, and the power fluctuations of the iLCC change along with the power of the MMC, in order to transmit all the power generated by the conventional power plant and WF. Figure 7(c) presents the dc voltages at the terminals of MMC, the rLCC, the iLCC, and the middle point of dc line. It is noteworthy that dc voltages are regulated satisfactorily to their references even with the power fluctuations at the MMC. The differences between these dc voltages represent the voltage drop on the dc lines. Figure 7(d) gives the dc currents at the MMC and the rLCC and iLCC, it can be seen again the consistency between the dc currents at the MMC and the iLCC. Figure 7(e) shows the frequency at MMC ac bus, the results suggest that the frequency is almost invariant with the function of power-frequency droop control.

3.2 *AC faults at the MMC*

Figure 8 shows the performance during three-phase to ground fault. The fault was applied at ac side of the MMC at 7 s and was cleared after 100 ms. Figure 8(a) shows that the power at rLCC increases during the fault, the power at iLCC changes along with the change of power at MMC (from the MMC), and power at MMC drops to negative at the beginning of fault. The power unbalance is reflected in the dc current of the system, as shown in Figure 8(c). Also Figure 8(c) shows that the dc current of the rLCC rises, and the rLCC provides most of the reverse dc current of MMC. As it can be observed

Table 1. Parameters of hybrid three terminal system.

Parameter	Value
Base power/MW	1000
Base dc voltage/kV	500
Base dc current/kA	2
Base ac voltages/kV	345 (sending end)
	230 (receiving end)
Base frequency/Hz	50
Rated power of conventional power plant/MW	600
Rated power of WF/MW	300
Resistance of dc line/Ω	5
Inductance of dc line/H	1.1936
Ground capacitance of dc line/μF	26
Resistance of WF dc line/Ω	1
Ground capacitance of MMC dc line/μF	150

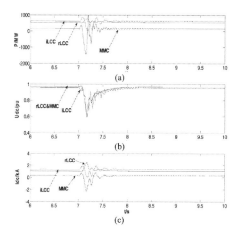

Figure 8. Proposed system response during three-phase to ground fault at MMC: (a) active powers,(b) dc voltages, (c) dc currents, respectively.

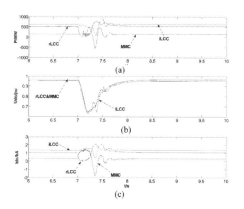

Figure 9. Proposed system response during three-phase to ground fault at LCC inverter: (a) active powers, (b) dc voltages, (c) dc currents, respectively.

in the Figure 8(b), there are drops in the dc voltages. The system is able to recover successfully in about 1 second after the fault occurs as seen in Figure 8.

3.3 AC faults at the LCC rectifier

Figure 9 shows the performance during three-phase to ground fault. The fault was applied at ac side of the LCC rectifier at 7 s and was cleared after 100 ms. Figure 9(a) shows that no power transfer during the fault from rLCC, and power fluctuations occur at MMC and iLCC. The dc current from rLCC is reduced to zero, as shows the Figure 9(c). Figure 9(b) shows that there are slight drops of the dc link voltages. The system is able to recover successfully in about 1 second after the fault occurs as seen in Figure 9.

4 CONCLUSION

Hybrid three terminal HVDC system is a structure which is suitable for the island transmission of wind power. A WF grid integration by MMC technology with LCC-based HVDC transmission system and the corresponding control strategies are proposed in this paper. The simulations were conducted under different conditions to investigate the operating characteristics of the proposed system. The results suggest that the proposed system have the abilities to track the variations in the wind speed and recovery from the faults.

REFERENCES

Cheng-hao Li et al. 2013, Offshore Wind Farms Integration and Frequency Support Control Utilizing Hybrid Multi-terminal HVDC Transmission. Industry Applications, IEEE Transactions on, 2014, 50(4): 2788–2797.

Feng Yu et al. 2013, Fast Voltage-Balancing Control and Fast Numerical Simulation Model for the Modular Multilevel Converter. Power Delivery, IEEE Transactions on, 2015, 30(1): 220–228.

GWEC. 2015, Global Installed Wind Power Capacity in 2014, http://www.gwec.net/global-figures/graphs/.

Kaijian Ou et al. 2014, MMC-HVDC Simulation and Testing Basedon Real-Time Digital Simulator and Physical Control System. Emerging and Selected Topics in Power Electronics, IEEE Journal of, 2014, 2(4): 1109–1116.

LIN Weixing et al. 2012, A Three Terminal HVDC System to Bundle Wind Farms With Conventional Power Plants. Power Systems, IEEE Transactions on, 2013, 28(3): 2292–2300.

Mai Huong Nguyen et al. 2012, Investigation on the impact of hybrid multi-terminal HVDC system combining LCC and VSC technologies using system identification. Universities Power Engineering Conference (AUPEC), 2012 22nd Australasian. IEEE, 2012: 1–6.

Manitoba HVDC Research Center. 2003, PSCAD/EMT-DCusers' manual. Winnipeg, Manitoba: Manitoba HVDC Research Center, 2003.

Raymundo E. Torres-Olguin et al. 2012, Offshore Wind Farm Grid Integration by VSC Technology With LCC-Based HVDC Transmission. Sustainable Energy, IEEE Transactions on, 2012, 3(4): 899–907.

Temesgen M. Haileselassie et al. 2011, Main Grid Frequency Support Strategy for VSC-HVDC Connected Wind Farms with Variable Speed Wind Turbines. PowerTech, 2011 IEEE Trondheim. IEEE, 2011: 1–6.

Xia Chen et al. 2011, Integrating Wind Farm to the Grid Using Hybrid Multiterminal HVDC Technology. Industry Applications, IEEE Transactions on, 2011, 47(2): 965–972.

Xu Dong et al. 2015, Model Predictive Control for VSC With Wind Farm Connected in the Hybrid Three-terminal DC Grid. Proceedings of the CSEE on, Vol. 35 No. 13 Jul. 5, 2015.

Advances in Power and Energy Engineering – Sun (Ed.)
© 2016 Taylor & Francis Group, London, ISBN 978-1-138-02846-3

Review of solar thermal energy storage materials

C. Peng, M.Y. Li & Y.Y. Li
School of Materials Science and Engineering, Wuhan University of Technology, Wuhan, P.R. China

X.M. Chen & G.M. Yu
College of Mechanical and Electrical Engineering, Huanggang Normal University, Huanggang, P.R. China

ABSTRACT: This paper focuses on the research and application of thermal storage materials. It reviewed the related research on phase change materials for solar thermal utilization. Thermal storage materials were classified according to working temperature. Both of the advantages and disadvantages of these materials were displayed in this paper. At the end of the paper, the further research and application about solar thermal storage materials were given.

1 INTRODUCTION

With the development of the economy of human society, the fossil fuels, such as coal, oil, natural gas and other resources, are increasingly insufficient. The production of energy has become more and more difficult to meet the needs of our human beings. The process of fossil fuel combustion is accompanied by the release of a large amount of toxic substances, so the pollution problem is becoming more and more serious. Hence, it is significant to look for renewable energy acting as alternative sources such as water, wind, nuclear, biomass and so on (Li et al., 2011). In terms of the renewable energy, solar energy is the most abundant eco-friendly resource with wide use. (Luo et al., 2005, Wu, 2006). As for solar thermal utilization, it is particularly important to store the solar energy effectively by energy storage materials because of the intermittent of solar radiation. In order to improve the utilization efficiency of solar energy, the important part of research focuses on new energy storage materials and technologies which are studied as hot subjects internationally (Farid et al., 2004). According to the heat storage modes, heat storage materials can be classified as follows: sensible heat storage materials, latent heat storage materials and the thermo chemical energy storage materials (Zalba et al., 2003). According to material properties, they can be classified to organic heat storage materials and inorganic one. According to operating temperature, it also can be classified to low (<80°C), medium (80°C~250°C) and high temperature (>250°C) heat storage material. In this paper, a variety of solar energy storage materials were introduced according to their operating temperatures.

2 LOW-TEMPERATURE THERMAL STORAGE MATERIALS

2.1 *Sensible thermal storage material*

Sensible thermal storage material stores thermal energy by its specific heat. The storing and releasing of thermal energy can be implemented through increasing and decreasing the temperature of the materials. Sensible thermal storage materials, the earliest studied and applied material, have many advantages such as long service life, low cost, relatively safe while it is working. However, the thermal storage density of sensible thermal storage materials is low. Meanwhile, this kind of material cannot maintain a constant temperature when it stores or releases thermal energy. Thus, a large volume of thermal storage device is needed to act as a container, which would limits application of this material to some extent. The familiar sensible thermal storage material includes water (Carlos, 2008), rocks, heat transfer oil (Montes et al., 2009), molten nitrate (Reddy, 2012, Peng et al., 2013), liquid sodium, high temperature concrete and so on.

Water and rock are the most common low-temperature sensible thermal storage materials. Water is clean, lowcost and eco-friendly, and it has a large specific heat capacity, so it can meet the most part of requirements for low temperature thermal storage. Meanwhile, water can be used as a Heat Transfer Fluid (HTF) in civil building heating system and household solar water heaters. Xiong Anhua et al. (Xiong & Dai, 2008) designed a low temperature solar heating system using water to store thermal energy, which has been put into use in Lhasa city in China in consideration of the abundant solar energy resources there.

Because of the popularity of household solar water heater, it becomes more important to improve the efficiency of heat exchanging. S. Jaisankar et al. (Jaisankar et al., 2011) reviewed many methods of increasing the efficiency of the solar water heater, indicating that an effective and energy saving method to strengthen the effect of heat transfer is using spiral ties located in the pipeline, which can lead to whirlpools of water.

The equipment using water as thermal storage material is usually huge due to its low thermal conductivity. Rock can be applied in underground solar thermal storage system because of its high thermal conductivity and abundant resource. But this thermal storage system is restricted greatly in application due to the geographic limitation and serious heat loss. Actually, rock has been applied in the solar thermal power stations at an early stage. For example, the Solar One tower thermal power station which was put into operation in 1982 in America has already applied oil and rocks in the thermal storage system (Yang & Li, 2012).

2.2 Latent thermal storage material

Phase Change Material (PCM) can store thermal energy in the form of latent heat. Compared with sensible thermal storage material, the main advantages of PCM are high thermal storage density and constant temperature in the process of heat charging and releasing, which enable the heat storage process to be better controlled. In the field of low-temperature application of thermal energy, many diagrams and tables of properties of PCMs have been given through loads of research. From the research, it is found that paraffin wax and hydrated salt are proper PCMs at low temperature (Ma & Wang, 1991, Zhang et al., 1996).

2.2.1 Inorganic hydrated salt

Inorganic hydrated salt is an important low temperature PCM. It has low and constant melting point (less than 70°C), high energy storage density (200~700 J/cm³), high latent heat, high thermal conductivity 0.5 W/(m·K), two times higher than the paraffin wax materials and low volume change rate. These properties make it suitable for the heat application at low temperature, such as building heating and water heater system. In organic hydrated salt was paid enough attention at an early age because of the wide varieties, low cost and abundant resource (Lin et al., 1987).

The major defects of hydrated salt are supercooling and phase separation in the thermal cycling process. Because of supercooling the material remains liquid when the temperature is lower than its freezing point. As a result, the stored heat energy cannot be released. Phase separation means some precipitations of salt are separated when it is applied. These defects have a great effect on the heat storage properties (Ding et al., 1996), so the research of inorganic hydrated salt focuses on these defects and the thermo stability in heat recycling. Generally, cold fingering and nucleating agent are used to solve supercooling. Cold fingering takes frozen part of the salt as nucleating agent which maintains the interior of the material cold. The subject which has the similar crystal structure with the solid salt is used as nucleating agent. Thickener, microstructure-changing agent, shaking and stirring are used to relieve the phase separation (Zhang, 2012).

Loads of research has been conducted to solve these defects. Duan Zhijun et al. (Duan et al., 2014) prepared a composite containing $CaCl_2 \cdot 6H_2O$ and expanded graphite whose tightness was enhanced by surfactant. This material has a high thermal conductivity and thermo stabiliy. Sheng Qiang et al. (Sheng et al., 2013) found foamy copper can improve the thermal conductivity and nucleating ability of $Ba(OH)_2 \cdot 8H_2O$, and this kind of copper makes the supercooling reduced by half. Biswas et al. (Biswas, 1977) mixed $Na_2SO_4 \cdot 10 H_2O$ with 3 wt% borax and some attapulgite (thickener), and it is found that it can reach 50% of the original heat storage properties after many heating and cooling cycles. Kenfack et al. (Kenfack & Bauer, 2014) prepared an inorganic hydrated salt for low temperature span, PYCO-PCM-1, which has a high thermo stabiliy, and the supercooling and phase separation has been solved basically. In addition, it has been proved that expanded graphite can improve the thermal conductivity effectively.

2.2.2 Organic paraffin

Paraffin is mixed by straight chain alkanes (formula: C_nH_{2n+2}), and it is a kind of organic phase change materials for thermal energy storage. Its melting point and latent heat will increase with the extension of the carbon chain in a certain range, but the melting point tends to be constant when the carbon chain reaches to a certain length (Zhang et al., 2006). The industrial mixed paraffin is generally used as PCM because of the higher price of the pure paraffin. The thermal physical properties of some paraffin are shown in Table 1 (Zhang, 2009)

Paraffin has a lower melting point and a higher latent heat. Compared with inorganic hydrated salt, paraffin nearly has no supercooling, phase separation, toxicity or corrosivity. Meanwhile, the resource of paraffin is abundant. Owing to these advantages, a lot of research has been conducted on paraffin. Sharma A. et al. (Sharma et al., 2002) conducted three hundreds of heat charge and discharge cycles each year for the same

142

Table 1. Thermal physical properties of paraffin.

Composition	Melting temperature [°C]	Heat of fusion [J/g]	Thermal conductivity [W/(m . K)]	Density [kg/cm³] Liquid	Solid
C16~C18	20~22	152			
C13~C24	22~24	189	0.21	760	900
C18	27.5/28	243.5/244	0.148 (40°C) 0.358 (25°C)	774	814
C16~C28	42~44	189	0.21	765	910
C20~C33	48~50	189	0.21	769	912
C22~C45	58~60	189	0.21	795	920
Paraffin wax	64	173.6/266	0.167 (63.5°C) 0.346 (33.6°C)	790	916
C21~C50	66~68	189	0.21	830	930

paraffin during 5 years, and it comes to the conclusion that paraffin has a better thermal physical property than fatty acids. Bascetincelik et al. (Bagetinelik et al., 1994) has successfully put the paraffin with the properties of melting point of 46°C~62°C and the latent heat of 192 J/g into the application of room temperature heating. In China, Liang Caihang et al. (Liang et al., 2004) developed a phase change mortar by adding about 20% microencapsulated paraffin, whose capacity of heat storage is equivalent to that of about 20 cm thick brick wall. Now the phase change mortar has been successfully used in building energy-saving projects in Germany.

However, the disadvantages of paraffin are obvious, such as the lower thermal conductivity, the lower density and so on. In recent years, the main research focuses on the modified paraffin through the technologies of adding relevant materials and composite to improve the thermal conductivity of paraffin and heat storage properties which would raise the applying value of paraffin. Some researchers (Qin et al., 2002, Sheng & Zhang, 2008) added the different polymer materials into some paraffin with different melting points, which showed that the heat storage capacity was improved significantly. Zhang-Tao et al. (Zhang & Yu, 2007) developed the modified paraffin with higher thermal conductivity by adding foamed aluminum and foamed copper. Mills et al. (Mills et al., 2006) made some use of the high thermal conductivity of expanded graphite to improve the thermal conductivity of paraffin. Li Min (Li, 2013) developed a kind of composite materials by adding nanographite into paraffin, and the result showed that the thermal conductivity of paraffin was increased. When the mass fraction of nanographite reached to 10%, the thermal conductivity of paraffin reached to 0.9362 W/(m·K).

2.2.3 *Organic fatty acid*

Fatty acid is also a kind of common organic PCM, which has the similar properties with paraffin. The formula of fatty acid is $C_nH_{2n}O_2$, and the melting point is related to the number of carbon atoms. Most of the fatty acids can be extracted from animals and plants, and have some advantages, such as high latent heat, relatively safe, non-flammable and so on (Jin, 2009). Fatty acids are easy to decompose and volatile. However, their performance is not very stable. Furthermore, the cost is higher than paraffin, so fatty acids are less practical in terms of application.

In order to solve the problems about the thermal physical property of single fatty acid, binary and multivariate mixed fatty acids are developed through the mix melting technique which can adjust the phase change temperature to make them more practical. Sarı et al. (Sarı et al., 2004, Sarı, 2005) developed different kinds of binary fatty acids with the phase changed temperature of 34°C~53°C and the latent heat of 165~185 J/g. After a long period of melting-freezing cycle test, the analysis of the thermal stability showed that the mixed PCM has an advantage in solar heating system.

In recent years, the study of paraffin materials focuses on the research of modified paraffin. The thermal conductivity of paraffin can be improved by the addition of relevant materials and composite technology, so that the thermal storage performance of the material can be improved. Common composite materials include expanded graphite, polymers, metal, etc. For example, Zhou Weibing et al. (Zhou et al., 2012) mixed stearic acid with expanded graphite. After adding 8% of expended graphite into the pure stearic acid, the thermal conductivity was increased from 0.18 W/(m·k) to 2.52 W/(m·k), but the latent heat was nearly unchanged. Fu Lujun et al. (Fu et al., 2013) developed a binary low-melting stearic acid

PCM through blending capric acid and lauric acid. Then when composited it with porous silica, he prepared a stable PCM with good thermo physical properties.

3 PCM AT MEDIUM AND HIGH TEMPERATURE

3.1 Sensible thermal storage material

3.1.1 Heat transfer oil

Based on the component, heat transfer oil can be classified into mineral heat transfer oil and synthesis heat transfer oil. Mineral oil can be gained from crude oil by distillation. It is non-toxic, plentiful, low-cost, and can be produced through a simple processing. Mineral oil can be used under 320°C. Synthetic oil can be mainly synthesized by some kinds of chemical products. Compared with mineral oil, it has more complicated production processes and has highcost. But its working temperature can reach up to 400°C, which is obviously higher than that of mineral oil. Because of the good thermal stability, synthesis oil has higher application value in the field of high temperature heat utilization. From the perspective of some commercial solar power stations heat transfer oil can be used as heat storage material as well as heat conduction material in the field of solar energy application. For example, SEGS I parabolic trough Concentrating Solar Power (CSP) plant in California, United States, use Caloria mineral oil as heat transfer and storage material. It used two-tank thermal storage system for heat storage which could store 120 MWh energy. The following SEGS power stations mostly applied synthesis oil as heat transfer and storage material that can be used at 400°C.

Heat transfer oil has a defect of thermal degradation whether it was applied as heat transfer or storage material. Heat transfer oil has coke formation tendency and high vapor pressure at high temperature, so it cannot be used at high temperature. The research of heat transfer oil mainly focused on the control of coke formation tendency to enhance application temperature.

For example Liu Tianyang et al. (Liu et al., 2014) developed a kind of heat transfer oil with high temperature resistance, which is obviously more outstanding than common heat transfer oil. Its boiling point is over 580°C, flash point over 280°C and ignition point over 300°C. It has low viscosity and high oxidation temperature.

3.1.2 Molten salt

Compared with heat transfer oil, the advantages of molten salt are that it has higher operating temperature, larger specific heat capacity, low vapor pressure, low viscosity, low flow resistance, high heat transfer and storage efficiency and non-toxic. As the same to heat transfer oil, it can be used both as heat transfer and storage material, which means the system is relatively safe. The research and application of molten salt has lasted for a long time (Kearney et al., 2003).

Eurelios, a tower solar power station built in Sicilia, Italy in 1981, used liquid nitrate which is called Hitec as heat storage material. In 1983, the power station CESA-1 in Spain was put into operation, which also used nitrite as heat storage material. In 1996, the solar power station, Solar Two, was accomplished in California, United States, which applied solar salt as heat transfer material. Three kinds of sensible heat storage material widely used are Solar Salt, Hitec and Hitec Xl respectively. Their thermo physical properties and mass proportions are shown in Table 2 (Kearney et al., 2003).

The Data show that the freezing points of these three kinds of mixed nitrate molten salt are higher than the environment temperature, which makes the operating temperature hard to control in thermal storage system. Molten salt tends to clog-pipes at low temperature as it is easy to crystallize, and tends to decompose at high temperature, which would have a serious impact on its usability due to the great changes of structures and thermal properties of decomposed molten salt. Molten salt has corrosivity, so studying on its compatibility with container is also important. Currently, research on molten salt focuses on reducing the freezing point through different mass composition, increasing

Table 2. Thermophysical properties and mass proportions of the three common molten salts used as sensible thermal storage material.

Name	Ingredient	Freezing point [°C]	Decomposition temperature [°C]	Density [Kg/m³]	Specific heat capacity [J/(Kg·h)]
Solar salt	40%KNO_3-60%$NaNO_3$	220	600	1899	1495
Hitec	40%$NaNO_2$-7%$NaNO_3$-53%KNO_3	142	535	1640	1560
HitecXL	48%$Ca(NO_3)_2$-45%KNO-7%$NaNO_3$	120	500	1992	1447

the decomposition temperature and improving the compatibility with containers, which would make these molten salts more suitable for practical application. Peng Qiang (Peng et al., 2009, 2010) improved composite heat storage performance by adding 5% additive to KNO_3-$NaNO_2$-$NaNO_3$ mixed molten salt, raising the operating temperature from 400°C~500°C to 550°C, improving the degradation of the original ternary molten salt. Cordaro et al. (Cordaro et al., 2011) researched the influence of different ion proportions on the freezing point and developed a new type of molten salt with freezing point reducing to about 70°C. Guillot S., et al. (Guillot et al., 2012) analyzed and compared the corrosion phenomena of carbonate, phosphate, sulfate, nitrate on ceramics using testing technologies like XRD and SEM. The result showed a well compatibility between nitrate and ceramic material, which provided a basic principle for the selection of packaging materials of molten salts.

3.1.3 *Concrete and ceramics*

The sensible heat storage materials described above are liquid materials which are generally used in low temperature field of solar thermal utilization. In order to acquire materials that can be used at higher temperature, solid sensible heat storage material has become the hot research subject. Common solid sensible heat storage materials are concrete, casting ceramic, sodium chloride, silica refractory bricks, refractory bricks, Magnesium Oxide, cast iron, etc. From an economic and practical point of view, concrete and casting ceramic have some obvious advantages at high temperature because of the lower price and better heat transfer performance, which makes them to be the two most widely studied solid sensible heat storage materials for high temperature now. In recent years, DLR (Tamme et al., 2005) (the German aerospace center), located in Stuttgart, Germany, indicated that concrete is suitable for thermal storage material because of the advantages including low cost, easy control, stable structure etc. In order to improve the related thermal physical properties of a normal concrete, they developed a new type of high temperature concrete which has a better performance in thermal storage. Table 3 shows are thermal physical property parameters of high temperatures concrete and casting ceramic (Laing et al., 2006) that DLR have researched (Wang et al., 2004, Liu et al., 2012).

Zhu Jiaoqun et al. (Zhu et al., 2007, 2009) conducted a lot of research on the concrete thermal storage materials. To solve the short coming of low thermal conductivity of concrete, they use aluminate cement as gelling agent and materials with high specific heat and thermal conductivity

Table 3. Two kinds of common solid sensible heat storage material properties.

Material	Casting ceramic	High temperature concrete
350°C density [Kg/m³]	3500	2750
350°C specific heat [J/kg·K]	866	916
350°C thermal conductivity [W/m·K]	1.35	1.0
350°C thermal expansion coefficient [10^{-6}/K]	11.8	9.3

(natural basalt stone, graphite, etc.) as aggregate to prepare a new type of concrete thermal storage material, and the mechanical performances were analyzed and compared with materials which contain different contents of additives.

3.2 *Latent thermal storage material*

3.2.1 *Inorganic salt*

High-temperature inorganic salt has high phase change temperature and operating temperature. It generally refers to the alkali metal or alkaline earth metal halide (e.g., chloride, fluoride salt), carbonate, sulfate, phosphate salts and their combination. High-temperature inorganic salt is quit proper to act as thermal storage material for application in CSP system because it has high operating temperature, no supercooling and phase separation, low vapor pressure, large latent heat of phase change and good thermal conduction performance. Every single high temperature inorganic salt has its own advantages and disadvantages. For example, carbonate has low price, high phase change latent heat, low corrosion resistance, but the viscosity is high at the molten state and some carbonates are easily decomposed when heated, which limit the scope of the application in the certain degree. Fluoride salt has a high melting point and phase change latent heat, and it is also very stable at the molten state, but its thermal conductivity is lower and it has large volume change in the liquid-solid phase change. Clearly, there are some restrictions for fluoride salt which acts as an ideal thermal storage material at high temperature. As a kind of cheap inorganic salt, chloride also has a very high melting point and phase change latent heat, but the corrosion of chloride is so strong that nearly all of the metal material can be corroded in wet conditions. Some thermal properties of high temperature inorganic salts are shown in Table 4 (Wang et al., 2004, Liu et al., 2012).

Because complex salts for general application have more significant advantages compared with

Table 4. Thermal properties of high-temperature inorganic salts.

Material	Phase change temperature [°C]	Latent heat of phase change [J/g]
Na_2SO_4	884	165
Na_2CO_3	858	275.7
$MgCl_2$	714	452
KOH	400	155
LiOH	471	879
LiF	849	1041
MgF_2	1263	938
KF	857	452

Table 5. Thermal properties of hybrid inorganic salts.

Mixed molten salt	Proportion	Phase change temperature [°C]	Latent heat [J/g]
NaF/MgF_2	75:25	832	650
LiF/MgF_2	67:33	746	947
Na_2CO_3/Li_2CO_3	56:44	496	368
Li_2CO_3/K_2CO_3	47:53	488	342
$NaCl/MgCl_2$	50:50	450	429
$NaCl/CaCl_2$	33:67	500	281
$NaF/KF/K_2CO_3$	17:21:62	520	274
$Li_2CO_3/Na_2CO_3/$ K_2CO_3	32:33:35	397	277

the single inorganic salt, a certain proportion of inorganic salts are usually mixed to prepare composite salt which has many advantages such as adjusting the melting temperature by a different mass ratio, extending the operating temperature, improving the energy storage density, reducing the volume change in the process of phase change. The mixture of low and high price inorganic salts can also reduce the cost without changing the heat capacity. In the study of hybrid inorganic salt, Liao Min et al. (Liao et al., 2008) prepared a composite salt by mixing Na_2CO_3 and K_2CO_3 in equal mass ratio with the static melting method. Based on this composite salt, 22.81 wt% NaCl was added into it. The results showed that the melting point declined by 133°C compared with the binary carbonate (Na_2CO_3 and K_2CO_3), while the latent heat increased by 1.9 fold and it remained stable at 850°C. Hu Baohua et al. (Hu et al., 2010) mixed NaCl and $CaCl_2$ in the same way. It was found that the melting point was 497°C and the latent heat was 86.85 J/g by using TGA/DTA, and the salt remained stable under 800°C. Some thermal properties of hybrid inorganic salts are shown in Table 4 (Zuo et al., 2006, Kenisarin, 2010, Wei et al., 2014).

The main disadvantage of inorganic salt is low heat conductivity coefficient (usually 0.2~0.8 W/m·k) which affects the efficiency of the heat storage directly in thermal storage system. Researchers focus on the modified inorganic salts in recent years. Modified inorganic salt can be prepared by mixing inorganic salt with high thermal conductivity materials. Generally, the modified salt has high latent heat of inorganic salts as well as the high thermal conductivity of additives. The main methods are compositing with metal matrix materials (foam nickel, copper, etc.), ceramic matrix materials and the expanded graphite etc. Qi Xianjin et al. (Qi et al., 2005) developed a composite material which was composited foamed nickel (metal matrix) with a series of high-temperature inorganic salts. The results show that the heat conductivity was 13.21~24.7 W/(m·K) at room temperature and the heat storage density was increased dramatically. The heat storage capacity of Composite ceramic made by Glück et al. (Glück et al., 1991) was greatly improved when added 20% Na_2SO_4 into ceramic matrix. Zhang Tao et al. (Zhang et al., 2010) found the heat conductivity of salt was increased by 38% after mixing with expanded graphite. Lopez et al. (Lopez et al., 2010) mixed the foam metal which has high thermal conductivity, expanded graphite and nitrate salt and found that the heat conductivity was largely increased.

3.2.2 *Alloy*

In terms of medium and high temperature PCM, the normal defect of molten slat is the low heat conductivity which would limits its practical application. Compared with salts, thermal conductivity of metals is tens or even hundreds of times higher. The metal material has more advantages in the medium and high temperature application because of its high latent heat, thermal storage density and good thermal stability. From the end of 1970s, Birchenal et al. (Birchenall and Riechman, 1980) started to study the possibility of storing thermal energy through the phase change of metals. He analyzed the thermal physical properties of

Table 6. Thermal properties of some common metals (Zhao et al., 2010, Farkas & Birchenall, 1985).

Metal	Melting temperature [°C]	Heat of fusion [J/g]
Al	660.2	395.4
Si	1414.0	1805.3
Cu	1083.0	203.5
Zn	419.5	103.1
Mg	651.0	376.8

Table 7. Thermal properties of some alloys (Farkas & Birchenall, 1985).

Alloys	Mass ratio	Melting temperature, [°C]	Heat of fusion [J/g]	Density, [g/cm³]	Specific heat, [J/(g/K)]	
					Solid	Liquid
Al-Si	87.5:12.5	577	515	2.25	1.49	
Cu-Si	80:20	803	197	6.6	0.5	
Al-Si-Cu	65:5:30	571	422	2.73	1.3	1.2
Al-Mg-Zn	59:35:6	443	310	2.38	1.63	1.46
Cu-Mg-Zn	25:60:15	452	254	2.8		
Cu-Mg-Zn	45:6:49	705	176	8.67	0.42	
Cu-Mg-Si	56:17:27	770	422	4.15	0.75	
Cu-Si-Zn	74:7:19	765	125	7.17		
Mg-Si-Zn	47:38:15	800	314			
Al-Cu-Mg-Zn	54:22:18:6	520	305	3.14	1.51	1.13

a series of alloy, especially for Al, Si, Cu, Mg and Zn. Thermal properties of common metals and alloys are presented in Tables 6 and 7 respectively. Gasanaliev (Gasanaliev, 2000) analyzed thermal properties of a series of metal alloys, and found that metal alloys are suitable for thermal storage at high temperature.

The alloy PCMs which are composed of metals with low melting point, such as Sn, Bi, Pb, Cd, In and Sb, have high energy storage density, high thermal conductivity and wide service temperature, so they can be used as latent medium-temperature heat storage material. The major part of research on alloys is about the melting point. Through the research on a series of binary alloys with low melting point, scholars have found the melting points of binary alloys comprised of Pb, Sn, Bi and In mainly distribute between 100 and 200°C (Li & Cheng, 2013). Abtew et al. (Abtew & Selvaduray, 2000) found the melting points of alloys based on Sn, Cu and Ag distribute between 200°C–300°C. El Daly et al. (El-Daly et al., 2009) analyzed the melting points of several binary alloys consist of Sn, Zn and Bi. They found their melting points show a decreasing trend with the increase of the mass content of Bi. In recent years, the research group of the author (Yu, 2012, Li et al., 2012b) also studied the thermal physical properties of some binary alloys and ternary alloys with low melting point. Thermal physical properties of these alloys are given in Table 8.

As a potential PCM, low-melting-point alloys are paid more attention due to its advantages such as large heat storage density and thermal coefficient (Ge et al., 2013). Because these alloys have a low phase change temperature, they are obviously dominant to be a Heat Transfer Fluid (HTF) (Friedman et al., 2013, Song et al., 2014). Compared to the HTF and molten salt, the liquid alloy has a higher

Table 8. Thermal physical properties of some alloys (Yu, 2012, Li et al., 2012b).

Alloy	Melting temperature [°C]	Heat of fusion, [J/g]
Sn-9Zn	198.1	65.8
Sn-58Bi	136.9	40.8
Sn-38Pb	180.9	37.7
Bi-45Pb	125.3	20.7
Bi-3.5Zn	251.4	50.3
40Sn-4Zn-56Bi	130.6	40.6
88Sn-9Zn-3Bi	196.1	63.4
86Sn-9Zn-5Bi	184.8	55.3
88Sn-9Zn-3Cu	208.1	54.3
86Sn-9Zn-5Cu	210.2	48.9

operating temperature and thermal conductivity, which means it has a greater value theoretically. The main problem for application of low-melting-point alloy is the poor compatibility with the container. Due to the serious corrosion of low-melting-point alloy, it is important to study the corrosion mechanism and look for suitable containers. Ilinčev et al. (Ilinčev, 2002) researched the liquid corrosion of Pb-Bi alloy with a series of different proportions, and the results showed that the corrosion became severe with the increasing mass of Bi.

As for the high-temperature phase change alloy materials, the majority research focused on the aluminum alloy first, and then the magnesium, copper and iron-based metal matrix alloys. Zhang Renyuan et al. (Zhang et al., 2006) studied the stability in thermal cycling test of Al-Si alloys with different mass content of Si, and the thermal properties of some alloys were measured. Zhang Yinping et al. (Zhang et al., 2006) measured the thermal properties of Al-Si binary alloy with different proportions, and the result of comparison showed that the Al-12Si has better thermal storage properties. Chen Xiaomin et al. (Cheng et al., 2010a, 2010b, 2010c, Li et al., 2012a) have studied the thermal physical properties of a variety of high-temperature alloys, such as Al-7Si, Al-33Cu and Al-7Si-4Cu, and designed the thermal storage system with aluminum alloy as the thermal storage materials. For the high-temperature waste heat recovery in industry, the application advantages of using Cu and Pb as the PCMs were analyzed by Maruoka et al. (Maruoka & Akiyama, 2003).

The most serious drawback of metal-based PCM is the corrosion with containers at high temperature, especially in the phase-change processing, so the study of packaging technologies of alloy and its compatibility with container are the hot research spots in recent years. A lot of comparative studies have been conducted on the compatibility between different kinds of metal alloys and containers such as stainless steel, heat-resistant steel and ceramic. Maruoka et al. (Maruoka & Akiyama, 2003) encapsulate the copper ball by adding a layer of nickel on rolling plat, and then he analyzed the thermal stability of the encapsulated material. Li Huipeng et al. (Li et al., 2009) compared the compatibility between the Al-Si alloy with silicon carbide ceramics and graphite materials respectively, and the result showed that the silicon carbide ceramics has better corrosion resistance. Scholars summarized the requirements of containers for alloy phase change materials, and they found out that container should have a certain thermal conductivity, corrosion resistance, toughness and good thermal isolation.

As a high-temperature latent heat storage material, metal-based PCM have a high thermal conductivity and high energy density, and it has great research value, but its practical application is not enough mainly because of the strong corrosion of liquid alloys. Thus, the next focus is to find a suitable container and developing packaging technology.

4 SENSIBLE-LATENT COMPOSITE HEAT STORAGE MATERIALS

Sensible or latent thermal storage materials were usually studied separately. However, the single thermal storage materials have many shortcomings both in performance and practical applications, and more researches began to focus on the sensible-latent composite heat storage technology. This composite material stores thermal energy through sensible heat and latent heat synthetically, and it can overcome some defects existing in each single material.

The common technologies studied in relative research are compositing inorganic salt and ceramic. Inorganic salt/ceramic composite thermal storage material can store thermal energy through the melting latent heat of inorganic salt and the sensible heat of inorganic salt/ceramic. As early as the late 1980s, a large number of German and American scholars conducted a series of studies relating to the complex recipe, preparation and performance, and applied some materials in the solar thermal storage system. In China, Zhang Renyuan et al. (Zhang, 2005) started to conduct research on inorganic salt/ceramic composite heat storage materials at an early age, analyzing and comparing its thermal physical properties and preparation process, which provided a reference value for other researchers. In recent years, some scholars focus on the composition of inorganic non-metallic material with high temperature and corrosion resistance (acts as substrate), PCM with large latent heat and some materials with high thermal conductivity. Leng Guanghui et al. (Leng et al., 2012) prepared andalusite honeycomb ceramics, and then encapsulated PCM into the ceramic substrate. This type of sensible-latent composite heat storage material has a large thermal storage density. For example, the thermal storage density could reach 987.7 kJ/kg, if it was encapsulated with 20 wt% of K_2SO_4, and it would reach to 796.4 kJ/kg if encapsulated with 16 wt% of NaCl. At the same time, the material has a good thermal stability.

5 CONCLUSION

Currently, research on thermal storage materials is abundant. However, in Concentrating Solar Power (CSP) system, only sensible heat storage materials are used widely. Although the latent heat storage material is one of the research hot spots in recent years for its superior thermal storage performance, it has not been promoted in large scale for the practical applications. Nonetheless, this type of thermal storage material has a well application prospect of solar energy. The research and application technology about the sensible heat storage material are relatively mature, but the problems, like how to reduce the solidifying point of the molten salt and how to improve the thermal conductivity and thermal stability of the concrete, etc., remain to be studied.

The latent heat storage material has a large heat storage density, and the temperature in heat exchanging process remains constant, but different types of PCMs have their own disadvantages in terms of thermal physical properties. For example, the under cooling and phase separation are main problems for inorganic hydrated salts and liquid corrosion for high-temperature inorganic salts and alloy. These problems limit their applications greatly. Therefore, the research should focus on improving the thermal physical properties, thermal stability and the compatibility between materials and the containers, etc.

In recent years, many research started to investigate sensible-latent composite heat storage technology as well as new high performance thermal storage system. This kind of composite material stores thermal energy through sensible heat and latent heat synthetically, and it will be the main part of research on thermal storage materials in the future.

ACKNOWLEDGEMENTS

This work was supported by the National Key Technology R&D Program of China (Grant No. 2012BAA05B05) and the National Natural Science Foundation of China (Grant No. 51206125).

REFERENCES

Abtew, M. & Selvaduray, G. 2000. Lead-free solders in microelectronics. *Materials Science and Engineering: R: Reports,* 27, 95–141.

Bagetinelik, A., Demirel, Y., Paksoy, H.O. & Oztfirk, H.H. 1994. Greenhouse heating with solar energy stored by phase change material. 15th National Agricultural Mechanisation Congress, Antalya, 20–22.

Birchenall, C.E. & Riechman, A.F. 1980. Heat storage in eutectic alloys. *Metallurgical and Materials Transactions A,* 11, 1415–1420.

Biswas, D.R. 1977. Thermal energy storage using sodium sulfate decahydrate and water. *Solar Energy,* 19, 99–100.

Carlos, M.S. 2008. An overview of GSP in Europe. *North Africa and the Middle. Madrid: CSP and CPV Finance and Investment Summit.*

Cheng, X.M., Zhang, S.K. & Wu, X.W. 2010c. Study on Mechanism of Phase Change Heat Storage in Al-Cu Alloy. *Hot Working Technology,* 21–23+26.

Cheng, X.M., He, G. & Wu, X.W. 2010a. Application and Research Progress of Aluminum-based Thermal Storage Materials in Solar Thermal Power. *Materials Review,* 139–143.

Cheng, X.M. Dong, J., Wu, X.W. & Gong, D.Q. 2010b. Thermal storage properties of high-temperature phase transformation on Al-Si-Cu-Mg-Zn alloys. *Heat Treatment of Metals,* 13–16.

Cordaro, J.G., Rubin, N.C. & Bradshaw, R.W. 2011. Multicomponent molten salt mixtures based on nitrate/nitrite anions. *Journal of Solar Energy Engineering,* 133, 011–014.

Ding, Y.M., Yan, L.C. & Xue, J.H. 1996. Nucleation of Salt-hydrates as the Thermal Energy Storage Material. *Chinese Journal of Chemical Physics,* 83–86.

Duan, Z.J., Zhang, H.Z., Sun, L.X., Cao, Z., Xu, F., Zou, Y.J., Chu, H.L., Qiu, S.J., Xiang, C.L. &

Zhou, H.Y. 2014. $CaC_{12} \cdot 6H_2O$/Expanded graphite composite as form-stable phase change materials for thermal energy storage. *Journal of Thermal Analysis and Calorimetry,* 115, 111–117.

El-Daly, A.A., Swilem, Y., Makled, M.H., El-Shaarawy, M.G. & Abdraboh, A.M. 2009. Thermal and mechanical properties of Sn–Zn–Bi lead-free solder alloys. *Journal of Alloys and Compounds,* 484, 134–142.

Farid, M.M., Khudhair, A.M., Razack, S.A.K. & Al-Hallaj, S. 2004. A review on phase change energy storage: materials and applications. *Energy conversion and management,* 45, 1597–1615.

Farkas, D. & Birchenall, C.E. 1985. New eutectic alloys and their heats of transformation. *Metallurgical Transactions A,* 16, 323–328.

Friedman, H., Reich, S., Popovitz-Biro, R., Von Huth, P., Halevy, I., Koltypin, Y., Gedanken, A. & Porat, Z. 2013. Micro-and nano-spheres of low melting point metals and alloys, formed by ultrasonic cavitation. *Ultrasonics sonochemistry,* 20, 432–444.

Fu, L.J., Dong, F.Q., Yang, Y.S. & He, P. 2013. Preparation and characterization of binary fatty acid/SiO_2 composite phase change energy storage materials [J]. *Journal of Functional Materials,* 4, 020.

Gasanaliev, A.M. & Gamataeva, B.Y. 2000. Heat-accumulating properties of melts. *Russian Chemical Reviews,* 69, 179.

Ge, H.S., Li, R.Y., Mei, S.F. & Liu, J. 2013. Low melting point liquid metal as a new class of phase change material: An emerging frontier in energy area. *Renewable and Sustainable Energy Reviews,* 21, 331–346.

Glück, A., Tamme, R., Kalfa, H. & Streuber, C. 1991. Investigation of high temperature storage materials in a technical scale test facility. *Solar energy materials,* 24, 240–248.

Guillot, S., Faik, A., Rakhmatullin, A., Lambert, J., Veron, E., Echegut, P., Bessada, C., Calvet, N. & Py, X. 2012. Corrosion effects between molten salts and thermal storage material for concentrated solar power plants. *Applied Energy,* 94, 174–181.

Hu, B.H., Ding, J., Wei, X.L., Peng, Q. & Liao, M. 2010. Test of thermal physics and analysis on thermal stability of high temperature molten salt. *Inorganic Chemicals Industry,* 1, 22–24.

Huang, R.Y. & Zhang, J. 2005. Research progress of salt/ceramic composite energy storage materials. *Materials Review,* 19, 106–108.

Ilinčev, G. 2002. Research results on the corrosion effects of liquid heavy metals Pb, Bi and Pb–Bi on structural materials with and without corrosion inhibitors. *Nuclear Engineering and Design,* 217, 167–177.

Jaisankar, S., Ananth, J., Thulasi, S., Jayasuthakar, S.T. & Sheeba, K.N. 2011. A comprehensive review on solar water heaters. *Renewable and Sustainable Energy Reviews,* 15, 3045–3050.

Jin, L. 2009. *Research on Preparation and Performance of Stearic Acid-PCM Microcapsule.* Postgraduate, Wuhan university of technology.

Kearney, D., Herrmann, U., Nava, P., Kelly, B., Mahoney, R., Pacheco, J., Cable, R., Potrovitza, N., Blake, D. & Price, H. 2003. Assessment of a molten salt heat transfer fluid in a parabolic trough solar field. *Journal of Solar Energy Engineering,* 125, 170–176.

Kenfack, F. & Bauer, M. 2014. Innovative Phase Change Material (PCM) for heat storage for industrial applications. *Energy Procedia*, 46, 310–316.

Kenisarin, M.M. 2010. High-temperature phase change materials for thermal energy storage. *Renewable and Sustainable Energy Reviews*, 14, 955–970.

Laing, D., Steinmann, W.D., Tamme, R. & Richter, C. 2006. Solid media thermal storage for parabolic trough power plants. *Solar energy*, 80, 1283–1289.

Leng, G.H., Wu, J.F. & Xu, X.H. 2012. Encapsulation of PCM in ceramic thermal energy storage materials. *Energy*, 1.

Li, H.P., Zhang, R.Y., Chen, X. & Liu, Z.J. 2009. The Study of the Container Used for Storing Al-Si Alloy for Heating [J]. *Journal of Guangdong University of Technology*, 2, 010.

Li, M. 2013. A nano-graphite/paraffin phase change material with high thermal conductivity. *Applied energy*, 106, 25–30.

Li, R.Z., Xi, F.Y., Yang, N. & Zou, J.S. 2011. The Analysis of World Energy Supply & Demand in 2010—An Explanation of BP Statistical Review of World Energy 2011. *Petroleum and Petrochemical Today*, 30–37+50.

Li, Y.Y., Cheng, X.M. & Yu, T.M. 2012b. Thermal cycling stability of Sn-Zn phase change heat storage alloy. *Special Casting & Nonferrous Alloys*, 7, 674–677.

Liang, C.H., Huang, X., Li, Y. & D, Y.H. 2004. Application of Phase Change Materials on Buildings. *Building Energy and Environment*, 23–26.

Liao, M., Ding, J., Wei, X.L., Yang, X.X. & Yang, J.P. 2008. Preparation and heat transfer and thermal storage property of high-temperature carbonate molten salt. *Inorganic Chemicals Industry*, 10, 004.

Li, Y.Y. & Cheng, X.M. 2013. Review on the low melting point alloys for thermal energy storage and heat transfer applications. *Energy Storage Science and Technology*, 189–198.

Li, Y.Y., Cheng, X.M., He, G. & Ye, P. 2012a. Investigation on Stability of Al-Cu-Mg-Zn Phase Change Heat Storage Alloy During Thermal Cycling. *Hot working technology*, 107–109.

Lin, W.X., Li, Y.C. & Lv, E.R. 1987. Review of Inorganic Hydrous Salt for Latent Heat Storage Material. *Journal of Yunnan Normal University(Natural Sciences Edition)*, 74–79.

Liu, M., Saman, W. & Bruno, F. 2012. Review on storage materials and thermal performance enhancement techniques for high temperature phase change thermal storage systems. *Renewable and Sustainable Energy Reviews*, 16, 2118–2132.

Liu, T.X., Cao, Z.W., Fan, X., Lan, Q., Cao, K., Huang, Y.W., Liu, T. & Yang, J. 2014. Synthesis of 1,3-dioctyl-1,1,3,3-tetraphenyl disiloxane with high thermal stability and flash point. *New Chemical Materials*, 42, 83–85.

Lopez, J., Acem, Z. & Del BarrIO, E.P. 2010. KNO 3/NaNO 3–Graphite materials for thermal energy storage at high temperature: Part II.–Phase transition properties. *Applied Thermal Engineering*, 30, 1586–1593.

Luo, Y.J., He, Z.N. & Wang, C.G. 2005. Solar energy utilization technology. Chemical Industry Press, Beijing.

Ma, X.G. W, Y.H. 1991. Research and Application of Latent Heat storage technology. *Energy Research and Utilization*, 13–17.

Maruoka, N. & Akiyama, T. 2003. Thermal Stress Analysis of PCM Encapsulation for Heat Recovery of High Temperature Waste Heat. *Journal of chemical engineering of Japan*, 36, 794–798.

Mills, A., Farid, M., Selman, J.R. & Al-Hallaj, S. 2006. Thermal conductivity enhancement of phase change materials using a graphite matrix. *Applied Thermal Engineering*, 26, 1652–1661.

Montes, M.J., Ab Nades, A., Martinez-Val, J.M. & Vald S.M. 2009. Solar multiple optimization for a solar-only thermal power plant, using oil as heat transfer fluid in the parabolic trough collectors. *Solar Energy*, 83, 2165–2176.

Peng, Q., Ding, J., Wei, X.L., Yang, J.P. & Yang, X.X. 2010. The preparation and properties of multi-component molten salts. *Applied Energy*, 87, 2812–2817.

Peng, Q., Yang, X.X., Wei, X.L., Yang, J.P., Ding, J. & Lu, J.F. 2013. New Molten Salt Heat Transfer Fluid for Solar Thermal Power Plant. 2, 496–499.

Peng, Q., Wei, X.L., Ding, J., Yang, J.P. & Yang, X.X 2009. Research on the Preparation and Properties of Mult-component Molten Salts. *Acta Energiae Solaris Sinica*, 1621–1626.

Qi, X.J., Wang, H., Wang, S.L. & He, F. 2005. Preparation and Research of Composite Heat Storage Material with Metal Ni and Molten Salts. *Industrial Heating*, 34, 8–10+18.

Qin, P.H., Yang, R. & Zhang, Y.P. 2002. Research on Preparation of Shape-stabilized Phase Change Material and the Thermal Properties. The biennial meeting of China's HVAC & R, 43, 833–835.

Reddy, R.G. Molten Salt Thermal Energy Storage Materials for Solar Power Generation. 2012 Beijing, China. 18.

Sarı, A. 2005. Eutectic mixtures of some fatty acids for low temperature solar heating applications: thermal properties and thermal reliability. *Applied Thermal Engineering*, 25, 2100–2107.

Sarı, A., Sarı, H. & Önal, A. 2004. Thermal properties and thermal reliability of eutectic mixtures of some fatty acids as latent heat storage materials. *Energy conversion and management*, 45, 365–376.

Sharma, A., Sharma, S.D. & Buddhi, D. 2002. Accelerated thermal cycle test of acetamide, stearic acid and paraffin wax for solar thermal latent heat storage applications. *Energy Conversion and Management*, 43, 1923–1930.

Sheng, Q.Q., Zhang, X.L. 2008. Research about the Wax Composite Phase Change Materials. *Refrigeration Air Conditioning and Electric Power Machinery*, 18–20+31.

Sheng, Q., Xing, Y.M. & Wang, Z. 2013. Experiment on heat transfer enhancement of barium hydroxide octahydrate phase change material. *Journal of Aerospace Power*, 1927–1932.

Song, S.H., Shen, W.D., Wang, J.L., Wang, S.C. & Xu, J.F. 2014. Experimental study on laminar convective heat transfer of microencapsulated phase change material slurry using liquid metal with low melting point as carrying fluid. *International Journal of Heat and Mass Transfer*, 73, 21–28.

Tamme, R., Steinmann, W.D. & Laing, D. Thermal energy storage technology for industrial process heat applications. ASME 2005 International Solar Energy Conference, 2005. American Society of Mechanical Engineers, 417–422.

Wang, S.L., Wang, H. & Qi, X.J. 2004. Research progress of latent thermal energy storage at high temperatures. *Energy Engineering,* 6, 002.

Wang, X., Liu, J., Zhang, Y., Di, H. & Jiang, Y. 2006. Experimental research on a kind of novel high temperature phase change storage heater. *Energy conversion and management,* 47, 2211–2222.

Wei, G.S., Xing, L.J., Du, X.Z. & Yang, Y.P 2014. Research Status and Selection of Phase Change Thermal Energy Storage Materials for CSP Systems. *Proceedings of the CSEE,* 325–335.

Wu, Z.J. 2006. New energy and renewable energy utilization. *Mechanical Industry Press, Beijing.*

Xiong, A.H. & Dai, Y.J. 2008. Research of Solar Hybrid Heating System in Lhasa. *Building Energy & Environment,* 56–58.

Yang, Y.L. & Li, Y. 2012. The technique of Solar Tower Power Plant and its Application Prospect in China. *Journal of Liuzhou Vocational and Technical College,* 43–47.

Yu, T.M. 2012. *Research on the Heat Storage Properties of Sn-Bi-Zn-Cu-Pb Alloys as Phase Change Materials in Medium Temperature.* Postgraduate, Wuhan university of technology.

Zalba, B., Marı N, J.M., Cabeza, L.F. & Mehling, H. 2003. Review on thermal energy storage with phase change: materials, heat transfer analysis and applications. *Applied thermal engineering,* 23, 251–283.

Zhang, R.Y., Sun, J.Q., Ke, X.F, & Zhou, X.X. 2006. Heat storage properties of Al-Si alloy. *Chinese Journal of Materials Research,* 156–160.

Zhang, R.Y. 2009. Phase change materials and phase change energy storage technology. Science Press, Beijing.

Zhang, T. & Yu, J.Z. 2007. Experiment of solid-liquid phase change in copper foam. *Journal of Beijing University of Aeronautics and Astronautics,* 1021–1024.

Zhang, T., Zeng, L. & Zhang, D. 2010. Improvement of thermal properties of hybrid inorganic salt phase change materials by expanded graphite and graphene. *Inorganic Chemicals Industry,* 5, 24–26.

Zhang, X.Y. 2012. Thermal Performance of Modified Paraffin Phase Change Heat Storage Materials.postgraduate, Beijing Forestry University.

Zhang, Y.P., Hu, H.P. & Kong, X.D. 1996. Phase change energy storage—theory and application. *Press of University of Science and Technology of China, Hefei.*

Zhang, Z.G., Wang, X.Z. & Fang, X.M. 2006. Structure and Thermal Properties of Composite Paraffin/ Expanded Graphite Phase-Change Material. *Journal of South China University of Technology (Natural Science Edition),* 1–5.

Zhao, C.Y., Lu, W. & Tian, Y. 2010. Heat transfer enhancement for thermal energy storage using metal foams embedded within Phase Change Materials (PCMs). *Solar Energy,* 84, 1402–1412.

Zhou, W.B., Zhang, L., Zhu, J.Q., Sun, Z. & Cheng, X. M. 2012. Kinetics Study of Phase Change on Stearic Acid/Expanded Graphite Composite as Heat Storage Material. *Wuhan Ligong Daxue Xuebao (Journal of Wuhan University of Technology),* 34, 9–13.

Zhu, J.Q., Tong, Y.Z., Zhou, W.B., Xie, B.L. & Ye, L.H. 2009. The properties research of high-temperature thermal storage concrete. *Energy Conservation,* 23–25+2.

Zhu, J.Q., Zhang, B. & Zhou, W.B. 2007. Preparation and researching of sensible heat storage material [J]. *Energy Conservation,* 4, 010.

Zuo, Y.Z., Ding, J. & Yang, X.X. 2006. Current status of thermal energy storage technologies used for concentrating solar power systems. *Chemical Industry and Engineering Progress,* 25, 995–1000+1030.

Advances in Power and Energy Engineering – Sun (Ed.)
© 2016 Taylor & Francis Group, London, ISBN 978-1-138-02846-3

Research on estimation of Electric Vehicles real-time schedulable capacity

X.B. Ju & C.Y. He
State Key Laboratory of Advanced Electromagnetic Engineering and Technology,
Huazhong University of Science and Technology, Wuhan, China

X.W. Zhang & J.R. Xia
China Electric Power Research Institute, Beijing, China

H.H. Sun
State Key Laboratory of Advanced Electromagnetic Engineering and Technology,
Huazhong University of Science and Technology, Wuhan, China

C. Wang
China Electric Power Research Institute, Beijing, China

C. Zhang
State Key Laboratory of Advanced Electromagnetic Engineering and Technology,
Huazhong University of Science and Technology, Wuhan, China

ABSTRACT: As an important part of smart grid, Electric Vehicles (EVs) can supply reservation to grid by regulating charging and discharging power, providing ancillary services such as peak regulation and frequency regulation etc. But the capacity is influenced by user demand, battery loss, etc. Firstly, the control model of EV involving in power dispatching is analyzed. Then an evaluation algorithm of EV real-time schedulable capacity is proposed. The algorithm considers the user demand, battery life and battery power constraints comprehensively. Finally, a scene of smart charging to stabilize the total load fluctuation is simulated, and the proposed algorithm is applied to calculate EV aggregator real-time schedulable capacity used for primary frequency regulation, secondary frequency regulation, and tertiary frequency regulation, based on required duration of schedulable capacity, which verifies the effectiveness of the algorithm.

1 INTRODUCTION

With the aggravation of energy crisis and environment pressure, energy-saving and emission-abating is gaining growing concerns. As a brand-new kind of transportation, EVs have various advantages over orthodox cars to alleviate this problem.

Large scale of EVs integration will have unneglected influence on the power grid (Ciwei, G., & Liang, Z. 2011) (Zhang, X. et al. 2015). The development of Vehicle-to-Grid (V2G) can improve safety and economy of power grid operation. EVs can provide virtual reserve capacity, peak-regulation service, and frequency control service for the grid (Wang, X. et al. 2013) (Mets, K. et al. 2010) (Lopes, J. et al. 2009). As a new kind of transportation, schedulable capacity of EVs is limited with restrictions of user demand and battery loss. In addition, power grid needs EVs schedulable capacity as reference, when considering EVs integration in unit

commitment and frequency regulation (Wang, C. et al. 2015). Therefore, EVs schedulable capacity is significant in EVs related research.

Up to now, there are some researches about EVs schedulable capacity. Reference (Kumar, K. et al. 2014) proposed a calculation method for V2G schedulable capacity to figure out maximum EVs discharging power, while user demand considered. In (Han, S. et al. 2011), a probability distribution algorithm for real-time V2G capacity estimation using dynamic EV scheduling is proposed. And the algorithm estimates the V2G capacity while ensuring the chargeability of the EVs. The EVs demand/supply model was established based on queuing theory, exhibiting stochastic characteristics of EVs (Chukwu, U.C., & Mahajan, S.M. 2011). According to a Monte Carlo simulation considering EVs stochastic mobility behavior, economy of EVs schedulable capacity is studied and capacity for first and secondary frequency regulation is evaluated in (Dallinger, D., Krampe, D., &

Wietschel, M. 2011). In reference (Weng, G. et al. 2011), the real-time evaluation models of V2G available capacity for each kind of battery group are proposed for target microgrid.

However, current literatures mainly focus on evaluations of EVs V2G schedulable capacity and ignore capacity provided by EVs as controllable load. In this paper, an algorithm for real-time V2G capacity estimation considering large scale EVs integration is proposed. This algorithm takes into account the main constraints such as user travelling demand, battery life and State of Charge (SOC). Firstly, under constraints of user demand and battery life, the upper-limit capacity of EVs regulating charging and discharging power is calculated. According to SOC restrictions, the lower limit of EVs regulating charging and discharging power can be obtained. Combing calculation results in each scheduling period, we can get instructions for peak regulation and frequency regulation.

2 EVALUATION OF EVS REAL-TIME SCHEDULABLE CAPACITY

Being load and power supply simultaneously, EVs in intelligent charging mode can accomplish optimal interaction with power grid by varying charging and discharging power (Dai, X. et al. 2013) (Li, M. et al. 2014). Therefore, for a EV at a certain time, if its charging and discharging power is within adjustable range, it has the ability to provide capacity for grid.

Regarding EVs as distributed power supply, when considering user demand, battery life and SOC constraints, the load capacity that EVs can increase equals to power generating capacity can be decreased, called EVs down-schedulable capacity; the load capacity that EVs can decrease equals to power generating capacity can be increased, called EVs up-schedulable capacity. Apparently, up-schedulable capacity is positive and down-schedulable capacity is negative. Details are shown in Figure 1.

In Figure 1, $P_{dischar,max}$: the maximum EVs discharging power; $P_{char,max}$: the maximum EVs charging power; $P_{EV,t}^i$: current EVs charging power. When $P_{EV,t}^i > 0$, EV is in discharging mode; when $P_{EV,t}^i < 0$, EV is in charging mode. Δt represents schedulable capacity requiring duration, and it varies in different application. For instance, the requiring duration in secondary frequency regulation is longer than in first frequency regulation. During Δt, $\overline{P_{\Delta t}^i}$ is the up limit of charging and discharging power and $P_{\Delta t}^i$ is the down limit. The difference between $P_{\Delta t}^i$ and $P_{EV,t}^i$ is EVs real-time up-schedulable capacity, the difference between $P_{EV,t}^i$ and $P_{\Delta t}^i$ is EVs real-time down-schedulable capacity. The EVs schedulable capacity evaluation method is discussed in the following part.

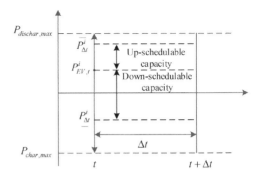

Figure 1. The schedulable capacity of EVs.

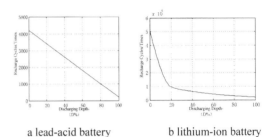

a lead-acid battery   b lithium-ion battery

Figure 2. The relationship between the EV battery discharge depth and the battery life.

2.1 Up-schedulable capacity evaluation

Whether EVs charging and discharging power has room to increase mainly depends on EVs user travelling demand and battery life. When user demand is satisfied, discharging depth is within permissible value, and current EVs power is less than the maximum discharging power, EVs charging power can be decreased or EVs can give more power back to grid.

1. User demand restriction
   When EVs leave, their State of Charge (SOC) need to reach user expected SOC—SOC_{need}^i. When $t = t + \Delta t$, the SOC restriction is as Equation (1):

$$SOC_{min,t+\Delta t}^i = SOC_{need}^i$$
$$- \frac{\left(-P_{char,max}\right) \times \eta_{char} \times \left(T_{dep,i} - (t+\Delta t)\right)}{C_i} \times 100\%$$

(1)

C_i: battery capacity; η: EVs charging and discharging efficiency; $T_{dep,i}$: leaving time of EVi.

2. Battery life restriction
   Battery life can be indicated by the number of recharge cycles. In Figure 2, there are two typical relationships between battery discharging

depth and recharging cycle life. The larger discharging depth is, the shorter the cycle life will be. Thus, to ensure battery life, intensified recharge cycle need to be avoid and discharging depth cannot be too large (Liu, L. et al. 2012). In this paper, we take lithium-ion battery as an example. In Figure 2, lithium-ion battery life is longer than lead-acid battery. Lithium-ion battery can survive 4358 cycles at the depth of discharge equaling to 60%. When the depth of discharge increases to 80%, lithium-ion battery can survive 3000 cycles. This could reach lithium-ion battery using life standards (Zhou, C. et al. 2011).

Compared to tertiary frequency regulation, first and secondary frequency regulations are more frequent. In order to provide enough schedulable capacity and reduce battery consumption, the depth of discharge in first frequency regulation is set to be 50%, the depth of discharge in secondary frequency regulation is set to be 60% and the depth of discharge in tertiary frequency regulation is set to be 80%.

Therefore, at $t + \Delta t$, SOC of EV must be larger than ($100\% - D_{max}$), shown in Equation (2).

$$SOC^i_{min,t+\Delta t} \geq 100\% - D_{max} \tag{2}$$

D_{max}: The maximum EVs depth of discharge.

The EVs real-time up-schedulable capacity in Δt is $\Delta P^i_{up} = \overline{P^i_{\Delta t}} - P^i_{EV,t} \cdot \overline{P^i_{\Delta t}}$ is shown in Equation (3).

$$\overline{P^i_{\Delta t}}$$

$$= \begin{cases} max\left(P_{char,max}, -\dfrac{\Delta SOC^i_t \times C_i}{\eta_{char} \times \Delta t}\right), & \Delta SOC^i_t > 0 \\ min\left(P_{dischar,max}, -\dfrac{\Delta SOC^i_t \times C_i}{\eta_{dischar} \times \Delta t}\right), & \Delta SOC^i_t < 0 \end{cases}$$

$$\tag{3}$$

$$\Delta SOC^i_t = SOC^i_{min,t+\Delta t} - SOC^i_t \tag{4}$$

ΔSOC^i_t is the difference between the minimum acceptable SOC and current SOC at $t + \Delta t$; η_{char} and $\eta_{dischar}$ are EVs charging and discharging efficiency. If $\overline{P^i_{\Delta t}} > 0$, $\overline{P^i_{\Delta t}}$ is the maximum acceptable discharging power; if $\overline{P^i_{\Delta t}} < 0$, $\overline{P^i_{\Delta t}}$ is the minimum acceptable charging power.

In Figure 1, when $\overline{P^i_{\Delta t}} > P^i_{EV,t}$, it indicates EV at current time can increase charging/discharging power. Thus, single EV real-time up-schedulable capacity is:

$$\Delta P^i_{up} = \overline{P^i_{\Delta t}} - P^i_{EV,t} \tag{5}$$

EVs group real-time up-schedulable capacity is:

$$P_{total,up} = \sum_{i=1}^{N} \Delta P^i_{up} \tag{6}$$

2.2 Down-schedulable capacity evaluation

Whether EVs charging and discharging power has room to decrease is mainly limited by maximum charging power of charging equipment and SOC restrictions. Only when charging power does not reach a limit and SOC is not full, EV is capable of increasing its charging power. Therefore, EVs can provide down-schedulable capacity for the power grid.

Considering SOC constraints:

$$SOC^i_{t+\Delta t} \leq 100\% \tag{7}$$

During Δt, the maximum capacity EVs can increase is:

$$\Delta SOC^i_t = 100\% - SOC^i_t \tag{8}$$

Thus, the maximum acceptable EV charging power is:

$$P^i_{\Delta t} = max\left(P_{char,max}, -\dfrac{\Delta SOC^i_t \times C_i}{\eta_{char} \times \Delta t}\right) \tag{9}$$

In Figure 1, when $\overline{P^i_{\Delta t}} < P^i_{EV,t}$, it indicates EV at current time can decrease charging/discharging power. Thus, single EV real-time down-schedulable capacity is:

$$\Delta P^i_{down} = \overline{P^i_{\Delta t}} - P^i_{EV,t} \tag{10}$$

EVs group real-time up-schedulable capacity is:

$$P_{total,down} = \sum_{i=1}^{N} \Delta P^i_{down} \tag{11}$$

Figure 3 is EV real time schedulable capacity evaluation algorithm flow chart. It proceeds as follows:

1. Collect each EV real time status information, including expected leaving time, current SOC, expected SOC, charging power battery capacity and so on;
2. Based on Equation (1) and Equation (2), calculate the minimum acceptable SOC at $t + \Delta t$. Based on Equation (8), calculate the maximum capacity EV can increase;
3. According to Equation (1) and Equation (2), figure out the up limit and down limit of EV charging/discharging power during Δt;

155

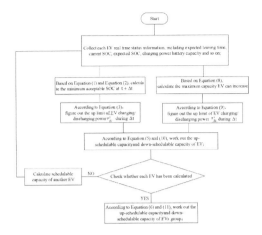

Figure 3. The calculation process of the real-time schedulable capacity of EVs.

4. According to Equation (5) and (10), work out the up-schedulable capacity and down-schedulable capacity of EV;
5. Check whether each EV has been calculated, if not, return to Step 1). Otherwise, send EVs group schedulable capacity as output.

3 CASE STUDY

3.1 *Scene simulation*

To verify the above algorithm, this paper simulates a EV charging/discharging scenario in one EV aggregator and get real-time information in 24 hours for each EV in this scenario. Based on these, evaluate the real-time schedulable capacity of this aggregator for one day. Aiming at stabilizing load fluctuation, take EV charging/discharging power as control variables and the specific model is as follows (Zhang, C. et al. 2014):

1. Objective function

$$min\sqrt{\sum_{t=1}^{T}\left(P_{load,t}-\sum_{i=1}^{N}P_{EV,t}^{i}-P_{average}\right)^{2}} \quad (12)$$

$P_{load,t}$ is the conventional load; N is number of EVs; T is the time nodes number in one day; $P_{EV,t}^{i}$ is control variable, i.e. EV charging/discharging power. When $P_{EV,t}^{i}>0$, EV is discharging; when $P_{EV,t}^{i}<0$, EV is charging; $P_{average}$ is average load:

$$P_{average}=\frac{1}{T}\sum_{t=1}^{T}\left(P_{load,t}-\sum_{i=1}^{N}P_{EV,t}^{i}\right) \quad (13)$$

2. The constraints
Charging/discharging constraints:

$$P_{char,max}\leq P_{EV,t}^{i}\leq P_{dischar,max} \quad (14)$$

At the leaving time, EVs should be capable of satisfying user travelling demand. Due to limited battery capacity, SOC of EV must be less than 100%.

$$SOC_{need}^{i}-SOC_{0}^{i}\leq\frac{\Delta E_{i}}{C_{i}}\times100\%\leq100\%-SOC_{0}^{i} \quad (15)$$

SOC_{0}^{i} is EV initial capacity; ΔE_{i} is EVs increasing capacity.

$$\Delta E_{i}=-\sum_{t=1}^{T}P_{EV,t}^{i}\times\Delta t\times\eta \quad (16)$$

We use MATLAB to establishing models and simulating. Assuming there are 2000 EVs paralleling into the grid, the maximum charging and discharging power are both for 5 kW, and the efficiency of charging and discharging is 95%. EVs user behavior is uncertain, like initial SOC, paralleling time, leaving time and other parameters. Statistical methods are applied to define their probability distributions for case simulation. According to Reference (Tian, L. et al. 2010), parameters are set as Table 1.

Figure 4 and Figure 5 show the influence EVs charging/discharging load has on conventional load and interaction with grid. The results shows 9:00–12:00 and 18:00–21:00 are load peak period. EV aggregator uses substantial energy of battery to feedback power to grid, which can reduce peak load and clip peak. However, in load valley period (13:00–15:00 and 23:00–6:00), most EVs are in charging mode for meeting user travelling demand. This can increase load in valley period and can help with valley regulation.

3.2 *Analysis of results*

Based on the simulation, we can get each EV real time status information, including expected leaving time, current SOC, expected SOC, charging power battery capacity and so on. According to the proposed algorithm for EV schedulable capacity evaluation, real time capacity that the aggregator can provide can be calculated. In the basis of different frequency regulation capacity, the lasting time and depth of discharge for first, secondary and tertiary frequency regulation are set as Table 2.

From Figure 6, we can know that EVs schedulable capacity varies with time and it depends on EVs

Table 1. EV parameters of the aggregator.

Period	Battery capacity (kWh)	Initial SOC (%)	Expected SOC	Paralleling time	Leaving time
Day	40	N(65,15)	85	N(9.3,0.15)	N(17.5,0.25)
Night		N(45,15)	90	N(18.5,1.2)	N(7.5,0.5)

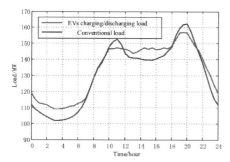

Figure 4. Influence of EV charging and discharging on the traditional load.

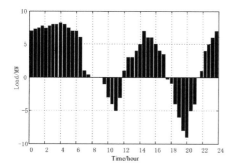

Figure 5. The energy interaction of EV aggregator and the power grid in one day.

Table 2. Parameters of frequency control.

Type	Lasting time Δt (h)	Depth of discharge (D%)
First frequency regulation	0.25	50
Second frequency regulation	0.5	60
Tertiary frequency regulation	2	80

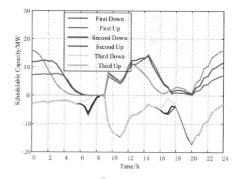

Figure 6. The real-time schedulable capacity of EV aggregator.

status and user demand. Compared to Figure 4, when EV is discharging, it can convert to charging mode. So its down-schedulable capacity is great. When EV is charging, it has the potential to give power back to grid. So its up-schedulable capacity is great. Besides, for different lasting time and depth of discharge, the capacity of different frequency regulation is variable, especially up-schedulable capacity.

In Figure 6, in the beginning period of paralleling into the grid (20:00–1:00), the acceptable depth of discharge is larger, and the capacity that EVs can provide is greater. Therefore, for schedulable capacity, the tertiary frequency regulation is the largest, and first frequency is the smallest. However, in the end period of paralleling into the grid (13:00–17:00 and 4:00–7:00), first frequency regulation is the largest, and the tertiary frequency regulation is

the smallest. Because many EVs have left and they are out of charging. The lasting time is too long to control EVs. The down-schedulable capacity only depends on leaving time and SOC, and is not related to lasting time and depth of charge. So their down-schedulable capacity is almost equal.

Figure 7 shows the depth of discharge influence schedulable capacity. Take schedulable discharge used for secondary frequency at 2:00 as example, we can see that depth of discharge has no influence on down-schedulable capacity. But up-schedulable capacity grows with the increment of depth of discharge. When the depth of discharge reaches 80%, the schedulable capacity is almost the maximum, and when the depth of discharge is 20%, the EV should not be allowed to participate discharge, and up-schedulable capacity is 0. So the depth of discharge should be determined accordingly, to guarantee both less loss of battery and fully use of EV schedulable capacity.

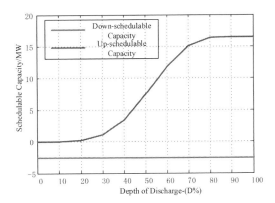

Figure 7. The relationship between the schedulable capacity the EV and battery discharge depth.

4 CONCLUSION

When EVs are applied in smart charging mode, it can realize interaction with gird by regulating charging and discharging power, and provide ancillary services such as peak regulation and frequency regulation etc. But the capacity is influenced by user demand and battery loss etc.

Firstly, an evaluation algorithm of EV real-time schedulable capacity is proposed. The algorithm considers the user demand, battery life and battery power constraints comprehensively.

Finally, a scene of smart charging to stabilize the total load fluctuation is simulated, and the proposed algorithm is applied to calculate EV aggregator real-time schedulable capacity used for primary frequency regulation, secondary frequency regulation, and tertiary frequency regulation. Results shows EVs schedulable capacity varies with lasting time and depth of discharging. And shorter lasting time and larger depth of discharge lead to larger schedulable capacity, which verifies the effectiveness of the algorithm. This evaluation algorithm can give instructions in peak regulation and frequency regulation.

REFERENCES

Chukwu, U.C., & Mahajan, S.M. 2011, July. V2G electric power capacity estimation and ancillary service market evaluation. In Power and Energy Society General Meeting, 2011 IEEE (pp. 1–8). IEEE.

Ciwei, G., & Liang, Z. 2011. A Survey of Influence of Electrics Vehicle Charging on Power Grid [J]. Power System Technology, 2, 128–130.

Dai X., Yuan Y., Fu Z., et al. 2013. Discharge strategy and economic benefits evaluation of electric vehicles in user side [J]. Proceeding of the CSU-EPSA, 06:55–61.

Dallinger, D., Krampe, D., & Wietschel, M. 2011. Vehicle-to-grid regulation reserves based on a dynamic simulation of mobility behavior. Smart Grid, IEEE Transactions on, 2(2), 302–313.

Han, S., Han, S., & Sezaki, K. 2011. Estimation of achievable power capacity from plug-in electric vehicles for V2G frequency regulation: Case studies for market participation. Smart Grid, IEEE Transactions on, 2(4), 632–641.

Kumar, K.N., Sivaneasan, B., Cheah, P.H., So, P.L., & Wang, D.Z.W. 2014. V2G capacity estimation using dynamic EV scheduling. Smart Grid, IEEE Transactions on, 5(2), 1051–1060.

Li M., Su X., Yan X., et al. 2014. New development of the coordinated charging and discharging control for plug-in electric vehicles [J]. Power System and Clean Energy, 06: 70–76+80.

Liu, L., Sohn, C., Balzer, G., Kessler, A., & Teufel, F. 2012, October. Economic assessment of Lithium-Ion batteries in terms of V2G utilisation. In Electrical and Power Engineering (EPE), 2012 International Conference and Exposition on (pp. 934–938). IEEE.

Lopes, J., Almeida, P.R., & Soares, F.J. (2009, June). Using vehicle-to-grid to maximize the integration of intermittent renewable energy resources in islanded electric grids. In Clean Electrical Power, 2009 International Conference on (pp. 290–295). IEEE.

Mets, K., Verschueren, T., Haerick, W., Develder, C., & De Turck, F. 2010, April. Optimizing smart energy control strategies for plug-in hybrid electric vehicle charging. In Network Operations and Management Symposium Workshops (NOMS Wksps), 2010 IEEE/IFIP (pp. 293–299). IEEE.

Tian L., Shi S., Jia Z. 2010. A statistical model for charging power demand of electric vehicles [J]. Power System Technology, 11:126–130.

Wang C., Wu K., Zhang X., et al. 2015. Unit commitment considering coordinated dispatch of large scale electric vehicles and wind power generation [J]. Power System Protection and Control, V43 (11):41–48.

Wang, X., Shao, C., Wang, X., & Du, C. 2013, January. Survey of electric vehicle charging load and dispatch control strategies. In Zhongguo Dianji Gongcheng Xuebao (Proceedings of the Chinese Society of Electrical Engineering) (Vol. 33, No. 1, pp. 1–10). Chinese Society for Electrical Engineering.

Weng G., Zhang Y., Qi J., et al. 2011. Evaluation for V2G available capacity of battery groups of electric vehicles as energy storage elements in microgrid [J]. Transactions of China Electrotechnical Society, 2(2):302–313.

Zhang C., Xu X., Sun H., Zhou X. 2014. Smart charging strategy of large-scale electric vehicles based on adaptive genetic algorithm [J]. Power System Protection and Control, 14: 19–24.

Zhang X., Wang X., Wang L., et al. 2015. Research on assessment methods of distribution network's ability of admitting electric vehicles [J]. Power System Protection and Control, V43 (12):14–20.

Zhou, C., Qian, K., Allan, M., & Zhou, W. (2011). Modeling of the cost of EV battery wear due to V2G application in power systems. Energy Conversion, IEEE Transactions on, 26(4), 1041–1050.

Advances in Power and Energy Engineering – Sun (Ed.)
© 2016 Taylor & Francis Group, London, ISBN 978-1-138-02846-3

Research on wind-storage system participating in frequency regulation for energy internet

Q. Fan, W. Gu, Y. Xiao, C.H. Lin, X.K. Wen, J.G. Chen, M.M. Xu & Y.T. Xu
Guizhou Power Grid Co. Ltd., Electric Power Research Institute, Guizhou, China

Y.Y. Chen & Y. He
Electrical Engineering College, Guizhou University, Guizhou, China

ABSTRACT: Energy internet is a complex network system based on renewable energy sources and distributed energy storage. With the rapid development of large-scale renewable energy sources, represented by wind power, the randomness, intermittence and instability of wind generation affect the stable operation of the power grid. An important way to participate in power dispatching is to fully make use of wind-storage system for joint frequency regulation. This paper summarized the characteristics of energy internet and researched the primary frequency regulation and secondary frequency regulation of the wind storage system. Finally, the development trend of the wind storage system participating in the frequency regulation for energy internet is proposed.

1 INTRODUCTION

The global energy development experiences Firewood Age, Coal Age, Oil-gas Age and Electrification Age. With huge reserves of global fossil energy sources, energy situation has become increasingly serious with large-scale development (Liu Z.Y. 2015). The global energy structural adjustment is strongly demanded along with the development of renewable energy sources. More and more renewable Distributed Generation (DG) has been applied in the power grid in recent years (Shen, Z. et al. 2014).

The concepts and technology framework of energy internet are proposed in the Project "Future Renewable Electric Energy Transmission and Management System" of National Science Foundation (NSF) in 2008, and the book Third Industrial Revolution (Refkin, J. 2012).

The renewable energy sources are main energy supplies of energy internet. The volatility and spatio-temporal randomness of energy requirement driven by customers causes uncertainty and disorder (Ci, S. et al. 2014). With the rapid development of large-scale renewable energy, represented by wind power, the randomness, intermittence and instability of wind generation affect the stable operation of the power grid.

Energy storage efficiently eliminates the volatility and randomness of energy flow. Meanwhile, energy storage devices have a huge potential for their application in the field of frequency regulation, not only rapidly but also accurately.

2 ENERGY INTERNET

2.1 Concepts of energy internet

Energy internet is a shared network with bidirectional exchange of energy and information flow. It consists of large numbers of DGs and distributed storage devices, based on the combination of power electronics technology, information technology, renewable energy technology and internet technology.

Its aim is to optimize energy structure and increase the efficiency of energy utilization (Zhou, H.M. et al. 2014). Basic Architecture and Components of Energy Internet is shown in Figure 1. It includes energy flow, information flow and traffic flow,

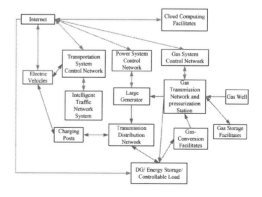

Figure 1. Basic architecture and components of energy internet.

represented by the red, green and blue arrows (Yu, S.H. et al. 2010). As can be seen from the figure, renewable energy and energy storage devices, located on the energy conversion side, play important roles in connecting sources and loads.

2.2 *Characteristics of energy internet*

Energy internet means that energy consumption will change from a local mode to a wide-area mode due to large amounts of distributed generations and devices. The characteristics are classified as follows:

1. **Diversity and wide-area distribution of energy sources**. Energy internet includes multiple power-generating approaches on the basis of DGs widely used. As DGs are distributed in a wide area, a network is required to collect, store and consume energy.
2. **Interconnectedness of energy sources**. A node of energy internet is constructed by a micro-distributed energy network. All the nodes make the energy internet to realize energy exchanges.
3. **Plug-and-play technology**. (Yu, S.H. et al. 2010). Traditional power system is supplied by remote centralized power units. But energy internet is not only to possess traditional power supply ways, but also to provide a public platform through the plug-and-play technology, in which DGs, intelligent devices and electric vehicles, achieve cutting-in and cutting-out by connecting intelligent agent terminals.
4. **Flexible energy storage**. Energy storage plays an important role in DGs and interconnection in the Third Industrial Revolution (Li, T.L. et al. 2015). Flexible bidirectional energy flow regulation by storage devices improves reliability and economics of the power system with large-scale DGs, whose function time ranges from seconds to hours.

3 RESEARCH STATUS OF WIND POWER PARTICIPATING IN FREQUENCY REGULATION

Power system frequency regulation includes inertial regulation, primary frequency regulation and secondary frequency regulation. It tends to select the generation units, which could rapidly start or stop and flexibly operate, to regulate frequency. Hydro power units have more advantages than thermal power units in traditional units. But it should not be deployed on the large scale because of its geographic restriction (Xiu, X.Q. et al. 2013).

Renewable energy, represented by wind power, has been rapidly developed and widely applied. China is the biggest and fastest wind power generation development country in the world. With large-scale and centralized development of wind power, it is more and more considered to be auxiliary of traditional generation. As a standard, AQSIQ (2011) requests that wind farm should participate in frequency regulation and peak regulation.

The virtual inertial control, droop control, pitch angle control and coordinate control are the main ways used in wind power participating in frequency regulation (Tang, X.S. et al. 2014). Most research aims for a single wind turbine or single wind farm. Given the energy internet, it needs to configure energy storage devices in a wind farm or a certain area, assisting wind power participating in frequency regulation.

4 RESEARCH ON WIND-STORAGE FREQUENCY REGULATION TECHNOLOGY FOR ENERGY INTERNET

Energy internet is a highly open energy system combined with the character of distribution and centralization. Energy storage devices realize spatio-temporal translation of energy flow, with a rapid response, on a large scale. The utilization efficiency significantly improves to the greatest extent by storage devices storing and releasing energy. Energy storage system is classified into different storage ways, as shown in Figure 2.

Energy storage system could efficiently suppress the power fluctuations in real time, improve wind power operation stability, eliminate the blind zone of frequency regulation, and increase the reserve capacity to improve the controllability and reliability of wind farm (Yan, G.G. et al. 2014). So large-scale wind-storage system meets the new requirements

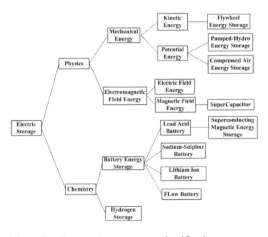

Figure 2. Energy storage system classification.

of energy internet. And it also improves economics to the greatest extent by properly controlling joint frequency regulation (Lv, J.F. et al. 2015).

Energy storage system, participating in primary frequency regulation and secondary frequency regulation, effectively maintains the grid frequency in the standard range. It would be an effective way to raise up wind power availability and reduce the requirements of reserve capacity. Energy storage system gradually applies in wind farms to meet different needs by changing the control strategy. Active power control or frequency regulation is only supported by energy storage devices in wind turbines or wind farms. It certainly leads to high cost and poor economy for redundant devices. Wind-storage frequency regulation mainly concentrates on primary frequency regulation and secondary frequency regulation.

4.1 Wind-storage primary frequency regulation

Primary frequency regulation is that power load frequency characteristics and generator governors prevent the frequency deviating from the standard. Wind-storage system makes use of its rapid response and stable operation ability to realize the inertial response and primary frequency regulation. Its frequency characteristics model is shown in Figure 3 (Miao, F.S. et al. 2015), and its transfer function equation is as follows:

$$G_{WE}(s) = \frac{\Delta P_W + \Delta P_E'}{\Delta f} = G_W(s) + G_E'(s) \qquad (1)$$

A control strategy based on storage technology is proposed to compensate the inertia of wind farms and coordinate energy exchange between the wind-storage system and the conventional grid. Frequency stability of the power system improves by rapid response frequency variation (Liu, J. et al. 2015, Tian, X.S. et al. 2015). An equivalent synchronous machine model, including wind turbine, storage battery and their inverters, is put forward to automatically regulate the output active power and keep the system frequency

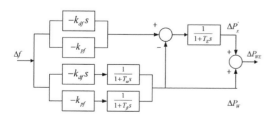

Figure 3. Frequency characteristic model of wind-storage system.

stability (Wu, Z.R. & Xu, X.N. 2015). A control strategy of the variable frequency regulating coefficient is presented for the load-frequency coordinated control-containing storage battery, with great advantages in primary frequency regulation (Lei, B. 2014). A wind turbine with a flywheel storage device maintains output active power generation and controls the wind turbine frequency (Thatte, A.A. et al. 2011).

A coordinated control strategy, combined with rotor speed control, pitch control and storage, is effectively used in a wide range of primary frequency regulation processes with a better economy and response (Tang, X.S. et al. 2012).

A double-fed wind turbine with a storage device suppresses wind power fluctuations using the storage device to regulate the output power. But it is still in the stage of laboratory research (Chen, Z.W. et al. 2014, Zhu, Y,K. & Chen, Y. 2013, Jia, J.C. et al. 2010).

4.2 Wind-storage secondary frequency regulation

Secondary frequency regulation is realized by Automatic Generation Control (AGC) basically [23]. Interconnected power system generally adopts tie line bias frequency control (TBC). AGC units receive real-time AGC signals from the dispatching center, automatically adjusting generated power to maintain system frequency rates and tie-line power plan values.

A subsection frequency regulation control strategy of the wind-storage system improves the wind power capacity factor by controlling the generation of active power (Li, P. et al. 2013). The Superconducting Magnetic Energy Storage (SMES) system with fuzzy normal forms controller participates in two-area frequency regulation (Li, L. et al. 2012). A control strategy is presented in which the wind-storage system and conventional units regulate high frequency and low frequency separately (Hu, Z.C. et al. 2014).

Model Predictive Control (MPC) method is applied to configure distributed or centralized energy storage capacity to regulate wind power frequency (Thatte, A.A. et al. 2011, Khalid, M. & Savkin, A.V. 2012). The optimal control strategy of the Battery Energy Storage System (BESS) based on MPC suppresses power fluctuations in real time. MPC provides a new way in research on the wind-storage system participating in power grid frequency regulation (Hong, H.S. et al. 2013).

5 CONCLUSION

The core of energy internet is renewable energy sources. With rapid development on a large scale, wind power is more and more considered to be auxiliary of traditional generation. Coordinated control with wind farms and energy storage system in interconnecting areas is a direction of the wind-storage system participating in power grid frequency regulation for energy internet. The MPC effectively uses wind power prediction information for AGC planning and responding rapidly. Many countries in the world conduct research on energy storage nowadays. So, how to solve the problems that volatility and randomness of wind power affect interconnection areas is becoming more and more urgent. New control strategies need to be studied and proposed to meet the needs of green, efficient, economic development of the power system.

ACKNOWLEDGMENT

This work was financially supported by the National Science and Technology Support Project of China (2013BAA02B02) and the Science and Technology Fund Project of Guizhou province (Qian Ke He J Zi [2011] 2059).

REFERENCES

Chen, Z.W., Zou, X.D. & Chen, Y.H. 2014. A Control Strategy for Double-fed Wind Power Generation System with Energy Storage. *Automation of Electric Power Systems* 38(18):1–4, 85.

Ci, S., Li, H.J. & Chen, X. 2014. The cornerstone of energy internet: research and practice of distributed energy storage technology. *Scientia Sinica* 44(6):762–773.

General Administration of Quality Supervision, Inspection and Quarantine of the People's Republic of China (AQSIQ). 2011. *GB/T 19963-2011 Technical rule for connecting wind farm to power system*. Beijing: China Electric Power Press.

Hong, H.S., Jiang, Q.Y. & Yan, Y.T. 2013. An Optimization Control Method of Battery Energy Storage System with Wind Power Fluctuations Smoothed in Real Time. *Automation of Electric Power Systems* 37:103–109.

Hu, Z.C., Xie, X. & Zhang, F. 2014. Research on Automatic Generation Control Strategy Incorporating Energy Storage Resources. *Proceedings of the CSEE* 34(29): 5080–5087.

Jia, J.C., Liu, J. & Zhang, Y.G. 2010. Power Control of DFIG Based Wind Power System Incorporated with Embedded Energy Storage. *Automation of Electric Power Systems* 34:80–84.

Khalid, M. & Savkin, A.V. 2012. An Optimal Operation of Wind Energy Storage System for Frequency Control based on Model Predictive Control. *Renewable Energy* 48(3):127–132.

Lei, B. 2014. *Research on Application of Battery Energy Storage System in Power System Frequency Regulation*. Huban: Hunan University.

Li, L., Zhang, J. & Jiang, Z.B. 2012. Frequency Regulation for a Power System with Wind Power and Battery energy storage. *Power System Technology (Power Con), 2012 IEEE International Conference on Auckland*.

Li, P., Huang, Y.H., & Xu, X.Y. 2013. Research of Frequency Control Strategy for Wind- PV-Storage Power Generation System. *East China Electric Power* 41(1):144–147.

Li, T.L., Tian, L.T. & Li, C.L. 2015. Key Technology of Energy Storage combined Grid-connected Distributed Renewable Energy. *Electrotechnical Application* 34(9):28–33.

Liu, J., Yao, W. & Wen, J.Y. 2015. A Wind Farm Virtual Inertia Compensation Strategy Based on Energy Storage System, *Proceedings of the CSEE* 35(7): 1596–1605.

Liu, Z.Y. 2015. *Global Energy Internet*. Beijing: China Electric Power Press.

Lv, J.F., Wu, L.L. & Sheng, S.Q. 2015. The Application Review of Energy Storage Technology Satisfying the Scheduling Requirement. *North China Electric Power* 3:1–7.

Miao, F.S., Tang, X.S. & Qi, Z.P. 2015. Analysis of Frequency Characteristics of Power System Based on Wind Farm-energy Storage Combined Frequency Regulation. *High Voltage Technology* 41(7): 2209–2216.

Refkin, J. 2012. *The Third Industrial Revolution*. Beijing: China CITIC Press.

Shen, Z., Zhou, J.H. & Yuan, Y.D. 2014. Development and Suggestion of the Energy-Internet. *Jiangsu Electrical Engineering* 33:81–84.

Tang, X.S., Miao, F.F. & Qi, Z.P. 2014. Survey on Frequency Control of Wind Power. *Proceedings of the CSEE* 34(25):4304–4314.

Tang, X.S., Miao, F.F. & Qi, Z.P. *A coordination control method of wind power and energy storage cluster*. China, 201210477712. 3.

Thatte, A.A., Zhang, F. & Xie, L. 2011. Coordination of Wind Farms and Flywheels for Energy Balancing and Frequency Regulation. *2011 IEEE Power & Energy Society General Meeting. Detroit, USA: IEEE Power & Energy Society* 1–7.

Tian, X.S., Wang, W.S. & Chi, Y.N. 2015. A New Coordinated Virtual Inertia Control Strategy for Wind Farm. *Automation of Electric Power Systems* 39(5): 22–26, 33.

Wu, Z.R. & Xu, X.N. 2015. The Grid- connected Control of Synchronverters in Direct- Driven Wind Turbine Generator with Energy Storage Battery. *Journal of Longdong University* 26(1):45–48.

Xiu, X.Q., Li, J.L. & Hui, D. 2013. Capacity Configuration and Economic Evaluation of Energy Storage System for Grid Peak Load Shifting. *Electric Power Construction* 34(2):1–5.

Yan, G.G., Feng, S. & Li, J.H. 2014. Review on combined wind power generation and energy storage systems. *Energy Storage Science & Technology* 3(4):297–301.

Yu, S.H., Sun, Y. & Niu, X.N. 2010. Energy Internet system based on distributed renewable energy generation. *Electric Power Automation Equipment* 30(5):104–108.

Zhou, H.M., Liu G.Y. & Liu, C.Q. 2014. Study on the Energy Internet Technology Framework. *China Electric Power* 47:140–144.

Zhu, Y,K. & Chen, Y. 2013. Operation and Control of a Novel Doubly- fed Induction Generator with Energy Storage. *East China Electric Power* 41(9):1851–1855.

Advances in Power and Energy Engineering – Sun (Ed.)
© *2016 Taylor & Francis Group, London, ISBN 978-1-138-02846-3*

Research on modeling and simulation of a kind of AC/DC power distribution device

P. Ye, Y.N. Hu, Z.F. Teng & L. Liu
Shenyang Institute of Engineering, Shenyang, Liaoning, P.R. China

Q. Gao, G.M. Zhang & Q. Liu
State Grid Liaoning Electric Power Research Institute, Shenyang, Liaoning, P.R. China

ABSTRACT: In this paper, a kind of AC/DC power distribution device was proposed, which was based on the back-to-back HVDC system. Its function was to undertake the integration of new energy sources and DC loads. The power transmission strategy and the DC voltage control method of the AC/DC power distribution device were designed and explored. The simulation model of the above system was set up with electric converters, i.e. wind power generation, photovoltaic and energy storage battery. Then, dynamic simulation of operation characteristics was researched and analyzed. The simulation results indicate that the proposed AC/DC distribution system can operate steadily; the DC bus voltage is stable; flexible energy conversion can be achieved in the system by means of power control.

1 INTRODUCTION

Due to the advantages of less pollution and flexible installation, the distributed power supply has developed rapidly in recent years (Begovi M. 2001, Lasseter R. 2009, Yu T. & Tong J.P. 2009). Along with the rapid increase of distributed power supply in the low-voltage distribution network, the application of distributed power supply becomes more and more complex. Traditional distribution network is unable to access and integrate those new energies. It is imperative to realize the "plug and play" for the multiple sources and charges in the power distribution system.

In this paper, a kind of AC/DC power distribution device was proposed, which was based on the back-to-back HVDC system. Its function was to undertake the integration of new energy sources and DC loads. The power transmission strategy and the DC voltage control method of the AC/DC power distribution device were designed and explored. A simulation model of the above system was set up with electric converters; wind power generation, photovoltaic and energy storage battery. And then dynamic simulation of operation characteristics was researched and analyzed. The simulation results indicate that the proposed AC/DC distribution system can operate steadily, the DC bus voltage is stable, and flexible energy conversion can be achieved in the system by means of power control.

2 SYSTEM STRUCTURE

As shown in Figure 1, the system structure is studied in this paper, and the dotted frame is an AC/DC hybrid power distribution device. The system mainly consists of two voltage source converters VSC1, VSC2 and a DC power grid. The DC side of VSC1, VSC2 is connected with the DC bus, and the AC side is connected with the distribution

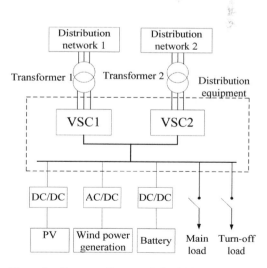

Figure 1. Structure diagram of the AC/DC power distribution system.

networks 1 and 2, respectively, and the DC grid is connected with distributed power supply such as wind power, PV, battery and DC loads. The key to the above system operation is the control of voltage source converters and DC/DC converters.

3 MICRO POWER MODEL

3.1 PV model

Solar cells can be equivalent to the circuit, as shown in Figure 2, consisting of the current source, the parallel resistor and the series resistor (Su J.H. 2001). In Figure 2, I is the output current of the solar cell, V is the solar cell output voltage, I_{sc} is the photocurrent, I_d is the diode current, and I_{sh} is the current shunt resistor.

3.2 Direct drive permanent magnet wind generator

In the *d-q* rotating coordinate system, the case of a three-phase symmetrical condition is considered, according to the motor model of the direct-drive permanent magnet wind turbine, taking the rotate coordinate system of the stator magnetic field as a reference (Yu M. et al. 2007, Lang Y.Q. et al. 2007), ignoring the stator electromagnetic transient situation. The equivalent circuit of the permanent magnet synchronous generator in the d-q synchronous rotating coordinate system is shown in Figure 3.

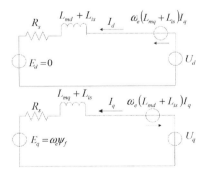

Figure 2.   PV equivalent circuit.

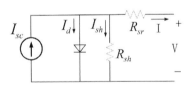

Figure 3.   Direct-drive permanent magnet synchronous fan d-q axis equivalent circuit diagram.

In the figure, U_d and U_q are the output d-axis and q-axis voltage components of the generator, ω_e is the electrical angular velocity, ψ_f is the perpetual flux, R_s is the stator resistance, L_{md} and L_{mq} are the stator inductance components in the d-axis and q-axis, and L_{is} is the leakage reactance.

3.3 Battery model

As shown in Figure 4, this paper adopts the universal storage battery equivalent circuit model, which uses the controlled voltage source in series with constant resistance (Li Y. et al. 2012, Hu Y.F. et al. 2008). The circuit structure is simple. Due to the nonlinear characteristic of the internal battery, it has a high fitting degree in the process of short-term dynamic simulation.

3.4 The model of Voltage Source Converter

The basic circuit topology of the Voltage Source Converter (VSC) has a two-level structure. In this paper, the structure of three-phase voltage source converters with two-level structures is adopted, as shown in Figure 5.

The model of the three-phase VSC in the *d-q* coordinate system is obtained through the Park transformation:

$$\begin{cases} L\dfrac{di_d}{dt} + Ri_d - \omega Li_q = U_{sd} - U_{cd} \\ L\dfrac{di_q}{dt} + Ri_q + \omega Li_d = U_{sq} - U_{cq} \\ C\dfrac{dU_d}{dt} = \dfrac{3}{2}\left(S_d i_d + S_q i_q\right) - i_L \end{cases} \qquad (1)$$

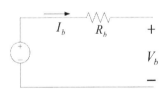

Figure 4.   General battery equivalent model.

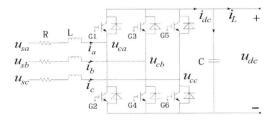

Figure 5.   Voltage source converter circuit topology.

164

where U_{sd} and U_{sq} are the bus voltage components of the d axis and q axis, connected with the large power grid; U_{cd}, U_{cq} are the VSC voltage components of the d axis and q axis at the AC side; and i_d, i_q are the VSC current components of the d axis and q axis.

4 SYSTEM CONTROL STRATEGY

4.1 Control objectives

In this paper, the AC/DC power distribution device is a smart power distribution hub that can realize the object of load transfer control and the integration of new energy. The biggest difference with the traditional back-to-back VSC-HVDC is that the DC bus between the two VSC relates to the wind, light, storage and other micro source and loads, which is composed of a DC micro grid. The control strategy should consider the following principles:

1. The DC bus voltage is stable.
2. The energy conversion between the buses is realized through the control of VSC and DC/DC converters.

4.2 Control strategy

As for the AC/DC power distribution system mentioned in this paper, when the system is operating, it mainly uses the large grid as the support of power and voltage. And it can achieve the energy conversion between the DC bus and the access of distributed power supply and loads by the control of VSC and DC/DC converters.

The control strategy of the system is as follows: VSC1 controls the DC line voltage and AC side reactive power. VSC2 controls the active and reactive power.

VSC1 is mainly used to control the stability of DC bus voltage. In accordance with the set value, VSC2 absorbs or emits power. The balance of the DC micro grid power is jointly maintained by VSC1 and VSC2. The voltage source converter adopts the vector control mode, in which the double closed-loop structure of the current loop and target outer loop is used. The rotating coordinate system and the three-phase grid voltage rotate synchronously, and the d axis coincides with the grid phase voltage vector. Then, the d axis component is the active current component, while the q axis component is the reactive current component.

Under the d-q coordinate system, the VSC mathematical model is shown in (1). Apparently, the d axis and q axis current components are associated with the component of the d-q voltage, with a serious coupling relationship. Therefore, decoupling

algorithm is introduced in the current inner loop control. The design of the inner current decoupling controller structure is shown in Figure 6.

The output of the decoupling controller can be expressed as follows:

$$\begin{cases} u_{cd} = -k\left(1+\dfrac{1}{\tau s}\right)\left(i_{dref}-i_d\right)+\omega L i_q + u_{sd} \\ u_{cq} = -k\left(1+\dfrac{1}{\tau s}\right)\left(i_{qref}-i_q\right)-\omega L i_d + u_{sq} \end{cases} \quad (2)$$

For voltage source converters, VSC 1 adopts the control mode of permanent DC voltage and AC side reactive power. The control block diagram is shown in Figure 7. VSC2 uses a fixed active and reactive power control mode. The control block diagram is shown in Figure 8. Phase-locked loop is

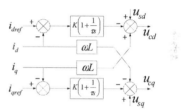

Figure 6. The inner ring current decoupling controller structure.

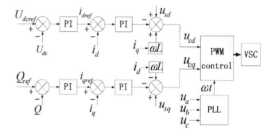

Figure 7. Constant DC voltage and reactive power control block diagram.

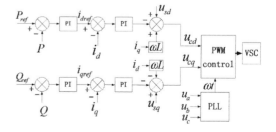

Figure 8. Constant active power and reactive power control diagram.

165

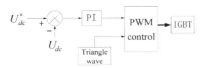

Figure 9. Buck regulator circuit control block.

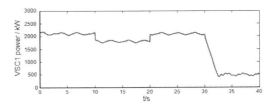

Figure 10. VSC1 power waveform.

used in the *dq* coordinate system to keep pace with the *abc* three-phase system.

In order to enhance the ability of controlling the DC bus voltage, Buck step-down chopper circuit is set before the photovoltaic power and wind power is connected to the DC bus. It can make the micro source undertake the task of maintaining the bus voltage stability. The Buck regulator voltage control circuit diagram is shown in Figure 9.

The DC bus standard voltage U^*_{DC} subtracts the real-time monitoring bus voltage U_{dc}. After the difference is fed back to the PI link and adjusted, it compares with the triangular wave and carries out PWM modulation. The output modulation signal control IGBT switch of Buck circuit is used to achieve the stable voltage control.

5 SYSTEM SIMULATION AND RESULT ANALYSIS

According to the system structure shown in Figure 1, the system model is simulated in PSCAD software. In the whole process of simulation, the complex source and the load accessing condition are set by changing the demand of load and the sudden access of the distributed power supply. Through observing the stability of the DC bus voltage under the disturbance and power conversion, the effectiveness of the system control strategy can be validated.

The simulation parameters can be set as follows: VSC2 inverter power is 1 MW and reactive power is 0 MVar. The constant temperature is 25°C. DC bus voltage is 800 V, the main load is 700 kW and the turning off load is 300 kW. In the initial state, the wind speed is 8 m/s, the illumination is W/m²; the load is 1000 kw when the photovoltaic power is not connected. The disturbance of the simulation is set as follows: at 10 s, 300 kW load is shut off, at 20 s, 300 kW load is reconnected to the DC bus; at 30 s, illumination is suddenly changed to 900 W/m² and the photovoltaic cell is connected to the DC grid; in 40 s simulation ends.

The system model is simulated in PSCAD. Figures 10–13 show the simulation waveform diagrams of the system. Here, the rectified power is positive in the power waveform diagram of VSC1, the inverter power is positive in the power waveform diagram of VSC2, and the absorbed power from VSC1 to VSC2 and to the DC

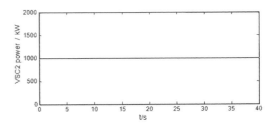

Figure 11. VSC2 power waveform.

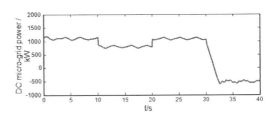

Figure 12. DC micro-grid power waveform.

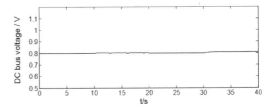

Figure 13. DC bus voltage waveform.

the DC bus is positive in the power waveform diagram of the DC micro-grids.

In the above situation of the waveform, the specific analysis is shown as follows:

0~10 s: the micro source power of the DC micro grid is less than the load power. The DC micro grid absorbs the power from the DC bus. VSC1 operates in the state of rectification, and the power flows from VSC1 to VSC2 and to the DC micro grid.

10~20 s: at 10 s, the load of 300 kW is cut off and the power absorption of the DC micro grid power is reduced, and the rectifying power of VSC1 is also reduced accordingly.

20~30 s: at 20 s, the 300 kW load is reconnected to the micro grid. The power of the DC micro grid and the power of VSC1 increase accordingly.

30~40 s: at 30 s, when the light is changed to 900 W/m², the photovoltaic power generation is connected to the network; the micro grid microsource power is higher than the load power; the excessive power feeds back to the DC bus; VSC1 reduces the power output.

Considering the simulation results, when the system is operating, it mainly uses the large grid as the energy and voltage support. At any time, the power exchange of the VSC1, VSC2 and DC micro power grids through DC bus is balanced from each other; the total power absorbed from the DC bus is equal to the total power injected to the DC bus; the flexible energy conversion is achieved in the system by means of power control. The system can quickly and stably track the set standard DC bus voltage. In the process of simulation, the DC bus voltage keeps stable under complex disturbance.

6 CONCLUSIONS

This paper proposed a kind of AC/DC power distribution device, which is based on back-to-back HVDC. The main results can be summarized as follows:

1. The basic principles, equivalent model and control model of the important components structure were analyzed. They included photovoltaic power generation, direct-drive permanent magnet wind turbine, batteries and voltage source converter.

2. A control strategy, which was based on keeping the stability of the DC bus voltage and the balance of the power, was designed to realize the energy conversion.

3. The model of the system was established in the PSCAD software and the control strategy was demonstrated by the simulation. It showed that the AC/DC distribution device can work smoothly and effectively.

REFERENCES

Begovi M., et al. Impact of renewable distributed generation on power systems [C]//Proceedings of the 34th Hawaii International Conference on System Sciences. Hawaii: IEEE, 2001:654–663.

Hu Y.F. et al. Modeling of Ni/MH battery for hybrid electric vehicle [J]. Battery Bimonthly, 2008, 38(4): 236–238.

Lang Y.Q. et al. Reactive Power Analysis and Control of Doubly Fed Induction Generator Wind Farm [J]. Proceedings of the CSEE, 2007, 27(9):77–82.

Lasseter R. The role of distributed energy resources in future electric power systems [EB/OL]. http://www.energy.wisc.edu/wpcontent/uploads/2006/04/lasseter Distributed Generation.pdf, 2009-02-09.

Li Y. et al. A Mathematical Model of Versatile Energy Storage System and Its Modeling by Power System Analysis Software Package [J]. Power System Technology, 2012, 36(1):51–56.

Su J.H. et al. Investigation on engineering analytical model of silicon solar cells [J]. Acta Energiae Solaris Sinica, 2001, 22(4):409–412.

Yin M. et al. Modeling and Control Strategies of Directly Driven Wind Turbine [J]. Power System Technology, 2007, 31(15):61–65.

Yu T., & Tong J.P. Modeling and simulation of the micro-turbine generation system [J]. Power System Protection and Control, 2009, 37(3):27–31.

Power generation—conventional and renewable

Advances in Power and Energy Engineering – Sun (Ed.)
© 2016 Taylor & Francis Group, London, ISBN 978-1-138-02846-3

Bioethanol production from the co-fermentation and co-culture fermentation of glucose and xylose by *Saccharomyces cerevisiae* and *Pichia stipitis* ATCC 58785 in a stirred tank bioreactor

F.J. Shalsh
Industrial Microbiology Department, Directorate of Agricultural Research/Ministry of Science and Technology, Formerly The Iraqi Atomic Energy Commission (IAEC), Baghdad, Iraq

N.A. Ibrahim & M. Arifullah
Faculty of Agro Based Industry, University Malaysia Kelantan, Kelantan, Malaysia

A.S. Mero Hussin
Faculty of Food Science and Technology, University Putra Malaysia, Selangor, Malaysia

ABSTRACT: This study investigated co-fermentation and co-culture fermentation techniques by using *Saccharomyces cerevisiae* and *Pichia stipitis* (ATCC 58785) in a stirred tank bioreactor to produce bioethanol. A medium containing 60.0 g/L glucose and 20 g/L xylose at 30 °C and pH 5 was agitated at 150 rpm to 300 rpm. Co-fermentation was achieved by *P.stipitis* in the medium containing the glucose and xylose mixture. *P.stipitis* fermented both sugars, produced 28.3 g/L ethanol, and consumed 60% of xylose in 12 h to 72 h at a utilization rate of 0.16 g/L h in the presence of glucose. The co-culture fermentation of *S. cerevisiae* and *P.stipitis* produced 30.12 g/L ethanol; the ethanol yield was 0.43 g/g after 48 h. The ethanol yield through the co-fermentation by *P.stipitis* at the same duration was 0.40 g/g. Therefore, co-culture fermentation can be economically applied to simultaneously utilize glucose and xylose and produce a high ethanol yield; indeed, the proposed technique can be used for industrial ethanol production.

1 INTRODUCTION

Lignocellulosic biomass is one of the major potential sources of commercial bioethanol because of its abundance, cost effectiveness, and non-competition with food resources (Khan & Dwivedi, 2012). However, the ability to release and ferment sugars from lignocellulosic biomass to ethanol is a technological limitation (Wan, Zhai, Wang, Yang, & Tian, 2012). The primary constituent of lignocellulosic biomass is cellulose, which accounts for 40% to 45% of lignocellulosic biomass; hemicellulose is the second most abundant component, which accounts for 20% to 30% of lignocellulosic biomass. Two monomer sugars, particularly glucose and xylose, should be efficiently converted to ethanol to establish an economical bioethanol production from lignocellulosic biomass (Chen, 2011). Although *S. cerevisiae* is generally used to ferment hexose sugars in biomass, this yeast cannot ferment pentoses. *P.stipitis* can also convert xylose to ethanol from lignocellulosic biomass because this species ferments glucose, xylose, and cellobiose with a high ethanol yield and a low co-product yield (Bellido et al., 2011). Co-cultures are industrially applied to

produce bioethanol. In such co-culture techniques, xylose- and glucose-fermenting microorganisms are consolidated to simultaneously ferment glucose and xylose (Wan et al., 2012). In these techniques, xylose- and glucose-fermenting microorganisms are initially selected to estimate their compatibility and ability to grow together; their co-fermentation efficiency is then evaluated. As such, the fermentation parameters of these two combined microorganisms should be compatible. For instance, the temperature and pH at which *S. cerevisiae* utilizes and converts glucose to ethanol are compatible with those of *P.stipitis*, a xylose-fermenting yeast. Therefore, the interactions between *P.stipitis* and *S. cerevisiae* can be employed in co-culture fermentation (Chen, 2011).

S. cerevisiae and *P.stipitis* present one of the most promising cost-effective challenges in bioethanol production. However, the simultaneous fermentation of glucose and xylose by co-cultures is associated with several difficulties (Laplace, Delgenes, Moletta, & Navarro, 1993), including the diauxic behavior of xylose-fermenting organisms, which cannot utilize xylose in the presence of glucose. Moreover, *P.stipitis* is more responsive to oxygen levels and other possible mass transfer limitations than *S. cerevisiae* (Karagöz & Özkan, 2014). Oxygen concentration is a critical factor affecting *P.stipitis* to ferment glucose and xylose to ethanol. High aeration increases biomass production but decreases ethanol yield. Under strict anaerobic conditions, *P.stipitis* cells cannot produce ethanol and cannot live for more than one generation (Papini, Nookaew, Uhlén, & Nielsen, 2012). Low oxygen levels, such as approximately 2 mmol L/h, are a critical factor stimulating xylose-fermenting yeasts to efficiently ethanol production from xylose because of the influence on nicotinamide adenine dinucleotide balance and cell viability. The microaerophilic condition and glucose preference of *P.stipitis* decrease its xylose fermentation; by contrast, *S. cerevisiae* does not require oxygen to utilize glucose (Karagöz & Özkan, 2014). Respiratory-deficient mutant *S. cerevisiae* strains are used to resolve this limitation related to oxygen requirements in a co-culture approach because respiratory-deficient *Saccharomyces* mutants can generate oxygen required by xylose-fermenting yeasts. A decrease in the oxygen requirements of aerobic yeast strains creates anaerobic conditions that enhance the growth of anaerobic or microaerophilic strains. This mode of microbial co-culture forfending from environmental effects (Delgenes, Escare, Laplace, Moletta, & Navarro, 1998). Furthermore, oxygen transfer must be achieved in the complete fermentation duration in different processes, such as those employed in approximately 93% of applications, including bioreactors with scheme agitation and air bubble dispersion in media. Studies on bioreactors are also advantageous because these bioreactors can be used to estimate the fermentation kinetics that cannot be well controlled in rotatory incubators; these bioreactors can also enhance the necessary conditions and resemble those applied to an industrial scale (Silva, Mussatto, Roberto, & Teixeira, 2011). Co-culture is a potential bioprocess if microorganisms can co-exist in the same environment and if each microorganism metabolizing its corresponding substrate is unaffected by the presence of the other microorganism. In bioethanol research, the co-culture fermentation of *S. cerevisiae* and *P.stipitis* has been extensively explored to perform a simple, one-batch process to ferment glucose and xylose by using lignocellulosic feedstock (Taniguchi & Tanaka, 2004). However, the co-culture process is characterized by technical drawbacks, such as low ethanol yields associated with different optimal oxygen transfer rates required by each organism. Despite the great importance of understanding the interactions between yeast strains and co-culture systems, very limited research has been performed because of the complicated nature of mixed culture systems. This study evaluated the different types of co-fermentation and co-culture fermentation of *S. cerevisiae* and *P.stipitis* to produce bioethanol.

2 MATERIALS AND METHODS

2.1 *Microorganisms and culture conditions*

Two yeast strains were used in this study. *S. cerevisiae* was obtained from the Industrial Biotechnology Research Laboratory (IBRL0) USM. *P.stipitis* (ATCC 58785) was purchased from American Type Culture Collection. *S. cerevisiae* was maintained at 4 °C in a medium containing 5 g/L yeast extract, 3 g/L peptone, 20 g/L glucose, and 20 g/L agar. *P.stipitis* was maintained at 4 °C in a medium containing 5 g/L yeast extract, 3 g/L peptone, 20 g/L xylose, and 20 g/L agar. *S. cerevisiae* was inoculated in an YPD medium containing 10 g/L yeast extract, 20 g/L peptone, and 20 g/L glucose. The inoculum culture was prepared by transferring a loop full of cells to 250 mL Erlenmeyer flask with 50 mL of the medium and the cells grown at 30 °C in a rotary shaker at 150 rpm for 24 h. The batch culture of *P.stipitis* was prepared in a 250 mL flask with 50 mL of the medium incubated at 30 °C and shaken at 170 rpm. The culture broth containing the YPD medium consisting of 30 g/L xylose, 5 g/L peptone, and 3 g/L yeast extract at pH 5. The cells pellet were harvested through centrifugation at 3000 rpm for 10 min and washed with 0.9% (w/v) sterile sodium chloride to reach the desired cell concentration, which was 5 g/L of dry weight. The cells were centrifuged again twice and the pelleted cells were inoculated in the fermentation media.

2.2 *Fermentation*

2.2.1 *Fermentation media*
Fermentation was performed using the medium containing 5.0 g/L of yeast extract, 5.0 g/L of peptone, 5.0 g/L of KH_2PO_4, 0.2 g/L of $(NH_4)_2SO_4$, and 0.4 g/L of $MgSO_4 \cdot 7H_2O$. The xylose and glucose mixture was used as carbon sources in the fermentation experiments; pH was adjusted to 5.

2.2.2 *Co-fermentation of glucose or xylose by using P.stipitis in a mono-culture*
Co-fermentation was conducted using a cell concentration at OD 600 of 15 and at 30 °C, pH 5, and 150 rpm. The initial cell concentration was 5 g/L dry cell weight.

2.2.3 *Co-culture fermentation of the glucose and xylose mixture by using S. cerevisiae and P.stipitis*
S. cerevisiae (7 g/L dry cell weight) was inoculated and incubated anaerobically at 30 °C for 30 h and was agitated at 150 rpm of the *P.stipitis* inoculum

(5 g/L dry cell weight) was then added. Fermentation was continuously performed at 30 °C for 48 h before the samples were agitated at 300 rpm. The samples were withdrawn at 12 h intervals and were centrifuged for 15 min.

2.2.4   *Fermentation in a stirred tank bioreactor*

Batch fermentation was conducted using two vessels of bioreactors (BIOSTAT C Q 1000, Sartorius BBI Systems GmbH, and Germany). A working volume of 0.75 L was used. The first fermenter vessel 1 was the co-fermentation type, which was inoculated with 10% (v/v) *P.stipitis* inoculum at pH 5.0. The vessel was incubated at 30 °C and then agitated at a stirrer rate of 300 rpm. The second vessel fermenter 4 was the co-culture fermentation type, which was inoculated with 7% (v/v) *S. cerevisiae* inoculum and 5% *P.stipitis* at pH 5.0. The vessel was incubated at 30 °C for 24 h, agitated at 150 rpm, and fermented for 48 h as shown in Figure 1.

2.3   *Analytical methods*

Cell growth was monitored turbidometrically by measuring absorbance at 600 nm, and this factor was correlated with the calibration curve (dry weight × OD). Glucose and xylose were analyzed using a Schimadzu LC-6 A high-performance liquid chromatograph (Kyoto, Japan) and detected by a refractive index detector RI-1530 and a Cosmosil

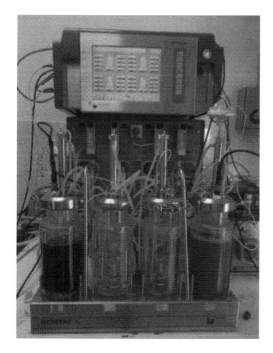

Figure 1.   Bioreactors BIOSTAT C Q 1000.

packed column sugar-D (4.6 ID × 250 mm). This procedure was conducted with BORWIN chromatography software under the following conditions: mobile phase of 80% acetonitrile and flow rate of 0.7 mL/min. The samples were filtered through 0.2 μm syringe filters before 20 μL was injected. Standard glucose (D-(+)-Glucose SUPELCO, USA) and standard xylose (D-(+)-Xylose SUPELCO, USA) were used. The serial concentrations of glucose and xylose standards were prepared on the basis of a five-point calibration (0, 1.25, 2.5, 5, and 10 g/L) with R2 > 0.999. The retention time of glucose and xylose were 9.292 and 6.825, respectively.

2.4   *Estimation of ethanol content through gas chromatography*

Ethanol production was estimated using a gas chromatograph equipped with a flame ionization detector and data acquisition system composed of IRIS 32 computer software. A SUPELCOWAX-10 Capillary Column (30 m × 0.25 mm × 0.25 μm) was installed. Oven temperature was maintained at 80 °C. The injector and detector temperatures were 270 and 250 °C, respectively. The flow rate of helium used as a carrier gas was 1.0 mL/min and the oven temperature was set at 70 °C for 10 min; the split mode was set at 5. The injection sample volume was 0.5 μL. The program run time and the ethanol retention time were 6 min and approximately 2.658 min, respectively. The area of standard ethanol was 3.71750e5. The ethanol percentage was calculated using the following equations:

$$
\begin{aligned}
\text{Concentration of ethanol} \\
= (\text{Volume of standard ethanol} \\
\times \text{Area of unknown sample}) \\
/(\text{Area of standard ethanol})
\end{aligned} \tag{1}
$$

$$
\begin{aligned}
\text{Ethanol\%} = 100 - ((\text{Volume of standard ethanol} \\
\times \text{Concentration of ethanol}) \\
/(\text{Area of standard ethanol}) \times 100)
\end{aligned} \tag{2}
$$

Ethanol yield (YP/S, g/g) was determined as the ratio between ethanol concentration and substrate (glucose + xylose) consumed. Ethanol productivity (QP, g/Lh) was calculated as the ratio between the ethanol concentration (g/L) and the fermentation time (h). The efficiency of sugar conversion (η, %) was calculated as the ratio of YP/S (g/g) to the theoretical value (0.51 g/g) (Hahn-Hägerdal, Jeppsson, Skoog, & Prior, 1994).

3   RESULTS AND DISCUSSION

3.1   *Glucose fermentation by S. cerevisiae*

Although *S. cerevisiae* is the most commonly used microorganism for the fermentation of hexose sugars present in biomass; however, *S. cerevisiae* does not use xylose as a carbon source, which is the second major sugar component of biomass (Matsushika & Sawayama, 2008). Both xylose reductase and xylitol dehydrogenase, which catalyze the first two steps of xylose metabolism in yeasts, are present in *S. cerevisiae*; despite this phenomenon, the rate of *S. cerevisiae* is fivefold to tenfold lower than that of xylose-fermenting yeasts. Although *S. cerevisiae* can not grow in a xylose-containing medium and does not produce ethanol, this species produces limited amounts of xylitol. In addition, xylose transport is less dynamic than glucose transport. Xylose is transported four times slower than glucose under aerobic conditions and twice as slow under anaerobic conditions (Kötter, Amore, Hollenberg, & Ciriacy, 1990). In our study, *S. cerevisiae* rapidly fermented sugar and produced a high ethanol yield under static conditions at 30 °C. Glucose was completely consumed by *S. cerevisiae* after 24 h of fermentation, and the resulting ethanol concentration was 24.64 g/L, with ethanol yield and productivity of 0.41 g/g and 1.02 g/h, respectively as illustrated in Figure 2.

3.2 Glucose and xylose co-fermentation by P.stipitis

P.stipitis has been extensively used to produce ethanol from biomass because of its capacity to ferment xylose, glucose, galactose, and mannose, cellobiose, and xylan oligomer sugars (Basso, De Amorim, De Oliveira, & Lopes, 2008). In the xylose consumption by *P.stipitis*, the lag phase covers 10 h to 12 h, whereas lag phase in glucose fermentation was not remarkable. Glucose was completely utilized within 24 h at a concentration of 57.86 g/L and a consumption rate of 3.07 g/L h with an ethanol concentration of 15.32 g/L. After glucose was completely consumed, *P.stipitis* used xylose and produced 28.45 g/L ethanol (Fig. 3). Furthermore, 60% of xylose was consumed from 12 h to 72 h of fermentation at a utilization rate

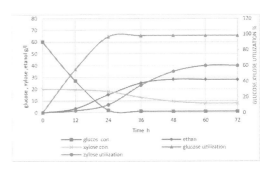

Figure 3. Co-fermentation of glucose and xylose by *P.stipitis*.

of 0.16 g/h. The expression of glucose on the utilization of xylose was observed in the existence of glucose (Ji et al., 2009). This behavior may be in accordance with the glucose preference of *P.stipitis* even under microaerophilic conditions (Hanly & Henson, 2013). The maximum ethanol yield was 0.40 g/g at 48 h because of the ethanol yield from xylose fermentation by *P.stipitis*. *P.stipitis* can utilize ethanol as a carbon source. Therefore, ethanol concentration slightly decreased, although xylose was utilized. The reaction was terminated at 28.33 g/L ethanol. Approximately 8 g/L xylose remained intact because of the low ethanol tolerance of *P.stipitis* (Delgenes et al., 1998). Delegnes et al. (1998) indicated that the ethanol inhibition of *P.stipitis* occurs at 30 g/L ethanol. The rapid ethanol production from glucose via the co-culture approach promoted the possibility of repressed xylose fermentation by ethanol. Glucose and xylose co-fermentation may be enhanced by decreasing the effect of ethanol through the selection of more ethanol-tolerant strains or by connecting the fermentation system to the ethanol removal scheme. The microaerophilic nature and glucose preference of *P.stipitis* reduces its xylose fermentation rate.

3.3 Glucose and xylose consumption by S. cerevisiae and P.stipitis co-culture

Glucose was consumed rapidly by *S. cerevisiae* and *P.stipitis*. The glucose consumption rate (5.68 g/h) was much higher than that in the presence of *P.stipitis* or *S. cerevisiae* alone. By contrast, 10.12 g/L xylose was consumed from 12 h to 72 h. After 72 h of co-culture fermentation, the glucose consumption ratio, xylose consumption ratio, and ethanol yield were 100%, 50.2%, and 28.2 g/L, respectively (Fig. 4). The result shows that the maximum ethanol production was 30.12 g/L with an ethanol productivity rate of 0.43 g/g at 48 h. (Chen, 2011) found that the co-culture fermentation of *P.stipitis* and *S. cerevisiae* yields a higher

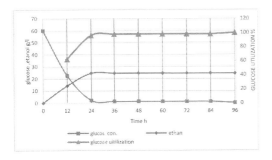

Figure 2. Glucose fermentation by *S.cerevisiae*.

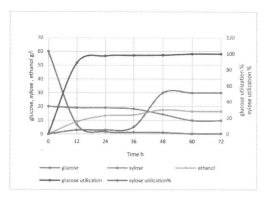

Figure 4. Co-culture of *S. cerevisiae* and *P.stipitis* of mixed sugars.

final ethanol concentration than the culture with either *P.stipitis* or *S. cerevisiae* alone. The co-culture fermentation of these two yeasts from the hydrolysate of date palm fronds yields 0.45 g/g in 66 h of incubation with a fermentation efficiency of 88.6% (Ali, Aziz, Syahadah, Kelvin, & Arifin). The result is consistent with a previous finding, which demonstrated that the ethanol yield from the co-culture fermentation of *S. cerevisiae* (local) and *P.stipitis* is much higher than that in the culture of *P.stipitis* alone when oil palm plantation is used to produce ethanol (Norhazimah & Faizal, 2014). In a study of the production of ethanol from corn stover through different fermentation strategies, the ethanol productivity of *S. cerevisiae* and *P.stipitis* co-culture is lower than that of the co-fermentation by *P.stipitis* (135.4 g/1000 g) (Q. Chu et al., 2014). This finding is possible because the oxygen uptake required for xylose fermentation facilitates the high-cell biomass production of *S. cerevisiae* by conversion glucose into the cell biomass rather than into ethanol (B.C. Chu & Lee, 2007) in co-culture. An increase in glucose contusion and ethanol production in the presence of *P.stipitis* may be related to the fermentation preference of *P.stipitis*, which is known as a Crabtree-negative and microaerophilic organism, in contrast to *S. cerevisiae*, which is a Crabtree-positive organism. Crabtree negative of *P.stipitis* exhibits predominantly respiratory metabolism even at high glucose concentrations; the beginning of the fermentation of *P.stipitis* does not rely on sugar concentration, and this yeast requires oxygen-limiting conditions to undergo fermentation (Papini et al., 2012). *P.stipitis* may start fermentation when glucose concentration or oxygen concentration is decreased by *S. cerevisiae* activity. In the first stage of the co-culture fermentation by *S. cerevisiae* and *P.stipitis*, ethanol concentration was high at a low agitation rate of 150 rpm for glucose utilization. In

the second stage of xylose fermentation, *P.stipitis* exhibits a high activity at a high agitation rate. *P.stipitis* produced a high ethanol yield at 300 rpm stirring that could be caused by the effect of agitation in xylose fermentation. This effect can increase the surface area in contact with air and thus decrease the anaerobic condition favorable to the strain. In both co-culture and co-fermentation, the yeast preferred glucose over xylose; as such, a relatively high yield of ethanol could be produced before xylose was used, and this factor may negatively influence xylose fermentation. Although *S. cerevisiae* cannot convert xylose to xylulose, *S. cerevisiae*can ferment xylulose (Richard, Toivari, & Penttilä, 2000). Therefore, the use of *P.stipitis*/*S. cerevisiae* co-culture is likely advantageous for optimal xylose fermentation. By contrast, *P.stipitis* can convert both glucose and xylose to ethanol. The competition between *P.stipitis* and *S. cerevisiae* for glucose may be an important factor contributing to theenhancement of ethanol production.

4 CONCLUSIONS

In this study, the proposed method is a practical application of ethanol production in a bioreactor by using S. cerevisiae and *P.stipitis*. Medium composition, inoculum times of each yeast, and substrate addition should be carefully considered for the successful co-culture fermentation of ethanol production. The ethanol production was higher in the culture containing the two yeast species. Our results supported previous co-culture findings demonstrating that *P.stipitis* and *S. cerevisiae* co-culture exhibits a higher substrate utilization rate of glucose than the mono-culture of either of this species. Although *S. cerevisiae* cannot convert xylose to xylulose, this yeast can ferment xylulose. By contrast, *P.stipitis* cells can utilize both glucose and xylose to ethanol. The ethanol production was limited because *P.stipitis* could tolerate approximately 30 g/L ethanol. The competition between *S. cerevisiae* and *P.stipitis* for glucose may be an important factor contributing to the enhancement of ethanol production. Oxygen concentration is also a relevant factor causing *P.stipitis* to ferment glucose and xylose into ethanol. Therefore, the use of *S. cerevisiae* and *P.stipitis* co-culture is likely advantageous for optimal xylose fermentation because glucose is effectively fermented by *S. cerevisiae* and the remaining xylose can be fermented efficiently by *P.stipitis*. The main challenge related to co-culture fermentation is simultaneously determining and providing optimal environmental conditions for the two different strains. The maximum ethanol production was 30.12 g/L with an ethanol productivity of 0.43 g/g for 48 h of fermentation. As an efficient fermentation of glucose and xylose, the co-culture fermentation to produce ethanol from lignocellulosic biomass can improve ethanol yield, shorten fermentation time, and reduce process costs.

REFERENCES

Ali, I.W., Aziz, K.K., Syahadah, A.A., Kelvin, S., & Arifin, Z. Bioethanol Production from Acid Hydrolysates of Date Palm Fronds Using a Co-culture of Saccharomyces cerevisiae and Pichia stipitis.

Basso, L.C., De Amorim, H.V., De Oliveira, A.J., & Lopes, M.L. (2008). Yeast selection for fuel ethanol production in Brazil. *FEMS yeast research, 8*(7), 1155–1163.

Bellido, C., Bolado, S., Coca, M., Lucas, S., González-Benito, G., & García-Cubero, M.T. (2011). Effect of inhibitors formed during wheat straw pretreatment on ethanol fermentation by Pichia stipitis. *Bioresource technology, 102*(23), 10868–10874.

Chen, Y. (2011). Development and application of co-culture for ethanol production by co-fermentation of glucose and xylose: a systematic review. *Journal of industrial microbiology & biotechnology, 38*(5), 581–597.

Chu, B.C., & Lee, H. (2007). Genetic improvement of Saccharomyces cerevisiae for xylose fermentation. *Biotechnology advances, 25*(5), 425–441.

Chu, Q., Li, X., Yang, D., Xu, Y., Ouyang, J., Yu, S., & Yong, Q. (2014). Corn Stover Bioconversion by Green Liquor Pretreatment and a Selected Liquid Fermentation Strategy. *Bioresources, 9*(4), 7681–7695.

Delgenes, J., Escare, M., Laplace, J., Moletta, R., & Navarro, J. (1998). Biological production of industrial chemicals, ie xylitol and ethanol, from lignocelluloses by controlled mixed culture systems. *Industrial Crops and Products, 7*(2), 101–111.

Hahn-Hägerdal, B., Jeppsson, H., Skoog, K., & Prior, B. (1994). Biochemistry and physiology of xylose fermentation by yeasts. *Enzyme and Microbial Technology, 16*(11), 933–943.

Hanly, T.J., & Henson, M.A. (2013). Dynamic metabolic modeling of a microaerobic yeast co-culture: predicting and optimizing ethanol production from glucose/xylose mixtures. *Biotechnology for biofuels, 6*(1), 1–16.

Ji, X.-J., Huang, H., Du, J., Zhu, J.-G., Ren, L.-J., Li, S., & Nie, Z.-K. (2009). Development of an industrial medium for economical 2, 3-butanediol production through co-fermentation of glucose and xylose by Klebsiella oxytoca. *Bioresource technology, 100*(21), 5214–5218.

Karagöz, P., & Özkan, M. (2014). Ethanol production from wheat straw by Saccharomyces cerevisiae and Scheffersomyces stipitis co-culture in batch and continuous system. *Bioresource technology, 158*, 286–293.

Khan, Z., & Dwivedi, A.K. (2012). Open Access Review Article. *Planning, 2011*(12.85), 66.90.

Kötter, P., Amore, R., Hollenberg, C.P., & Ciriacy, M. (1990). Isolation and characterization of the Pichia stipitis xylitol dehydrogenase gene, XYL2, and construction of a xylose-utilizing Saccharomyces cerevisiae transformant. *Current genetics, 18*(6), 493–500.

Laplace, J., Delgenes, J., Moletta, R., & Navarro, J. (1993). Effects of culture conditions on the co-fermentation of a glucose and xylose mixture to ethanol by a mutant of Saccharomyces diastaticus associated with Pichia stipitis. *Applied microbiology and biotechnology, 39*(6), 760–763.

Matsushika, A., & Sawayama, S. (2008). Efficient bioethanol production from xylose by recombinant Saccharomyces cerevisiae requires high activity of xylose reductase and moderate xylulokinase activity. *Journal of bioscience and bioengineering, 106*(3), 306–309.

Norhazimah, A., & Faizal, C. (2014). Bioconversion of Oil Palm Trunks Sap to Bioethanol by Different Strains and Co-Cultures at Different Temperatures. *Journal of Medical and Bioengineering Vol, 3*(4).

Papini, M., Nookaew, I., Uhlén, M., & Nielsen, J. (2012). Scheffersomyces stipitis: a comparative systems biology study with the Crabtree positive yeast Saccharomyces cerevisiae. *Microb Cell Fact, 11*(1), 136–136.

Richard, P., Toivari, M.H., & Penttilä, M. (2000). The role of xylulokinase in Saccharomyces cerevisiae xylulose catabolism. *FEMS microbiology letters, 190*(1), 39–43.

Silva, J.P.A., Mussatto, S.I., Roberto, I.C., & Teixeira, J. (2011). Ethanol production from xylose by Pichia stipitis NRRL Y-7124 in a stirred tank bioreactor. *Brazilian Journal of Chemical Engineering, 28*(1), 151–156.

Taniguchi, M., & Tanaka, T. (2004). Clarification of interactions among microorganisms and development of co-culture system for production of useful substances *Recent Progress of Biochemical and Biomedical Engineering in Japan I* (pp. 35–62): Springer.

Wan, P., Zhai, D., Wang, Z., Yang, X., & Tian, S. (2012). Ethanol production from nondetoxified dilute-acid lignocellulosic hydrolysate by cocultures of Saccharomyces cerevisiae Y5 and Pichia stipitis CBS6054. *Biotechnology research international, 2012*.

Advances in Power and Energy Engineering – Sun (Ed.)
© 2016 Taylor & Francis Group, London, ISBN 978-1-138-02846-3

Implication of crystallites nucleation in petroleum based transformer oil in electric power system

Life L. Kaanagbara
GIEES and Department of Civil and Environmental Engineering, University of North Carolina, NC, USA

Hilary I. Inyang
Duke Energy Distinguished Professor and Director GIEES, CARC 236, University of North Carolina, NC, USA

Jy. Wu
Department of Civil and Environmental Engineering, University of North Carolina, Charlotte, NC, USA

V.P. Lukic
Department of Electrical and Computer Engineering, University of North Carolina, Charlotte, NC, USA

David Young
Department of Civil and Environmental Engineering, University of North Carolina, Charlotte, NC, USA

ABSTRACT: Petroleum-based insulating oils are essential components of electric power transformers. Their aging during the service life of transformers usually manifests as changes in physico-chemical characteristics. In some cases, impurities that may compromise the performance of transformers can exsolve from the oils. Herein, we report the discovery of an organic crystallite with the calculated formula $C_{15}H_{24}O$, during voltage breakdown of type II transformer oil at a constant temperature levels, within the 55 °C–70 °C range. This crystallite is a form of Butylated Hydroxytoluene (BHT) which is generally known to be an anti-oxidant. Studies using a scanning electron microscope with capacity for elemental composition determination and a metallurgical microscope show exsolved crystallites as a colonnade network of elongated fibers. It is conceivable that the presence of this exsolved material can shorten the effective service life of petroleum-based transformer oils in electric power transformers, thereby necessitating their frequent replacement to sustain efficient and uninterrupted power distribution systems.

1 INTRODUCTION

Insulating oil is a critical component of oil-based electric power transformers. It is designed to cool the system and maintain good operating efficiency (Franklin and Franklin, 1983; Karsai et al., 1987; Lobeiras and Sabau, 2000). Lamarre et al. (1987) and Polovick and Hydro (1998) indicate that in addition to its insulating function, the oil extinguishes arcs, transmits heat to the radiators, and dissolves gases generated as a result of oil degradation. Insulating oils for transformers play a significant role not only in protecting the winding of the core, but also in ensuring that the coil used in the transformers performs its functions to maintain transformer efficiency. Transformer oils are designed to withstand thermal, electrical, chemical, and mechanical stresses (Kamata and Kako, 1980; Paloniemi and Ab, 1981; Franklin and Franklin, 1983; Vincent and Crine, 1986; Karsai et al., 1987). As many countries

and regions posture to expand and sustain electric power supply systems within their economic development programs, the long term performance of power transformers of electric power distribution systems has become very important.

Aging of petroleum-based oil which is frequently used as transformer oils under operating conditions in the field can change oil composition. The rates of occurrence of aging processes along the path to eventual breakdown of oil depend on physico-chemical and electrical properties of the oil, and the magnitude of electrical and associated stresses imposed on it. According to Armstrong (1998), changes in composition and increase in conductivity are good indicators of the occurrence of aging but not the underlying mechanisms. Others as typified by Sabau (1998) have queried the existing standard procedures used in oil testing and have argued that in some cases, they do not provide adequate information on aging mechanisms.

For instance, standards for the detection of gas in insulating oil are silent on the true causes of gas evolution. Also, Lamarre et al. (1987) report that direct correlation between various parameters that affect aging is yet to be fully established. This implies that there is still much to be done for better understanding of aging processes in oil.

While aging mechanisms that derive from thermal, electrical, mechanical and chemical stresses can be tracked to evaluate oil resilience, the response of transformers need to be analyzed wholistically within an overall electric power distribution efficiency assessment. For instance, Pugh (1998) observes that aging of oil insulation may not be observed during short time periods, but the damage is cumulative and irreversible. The significance of this observation is that the performance of physical tests may not completely reveal the potential extent of oil aging, as well as the kinetics of by-product generation, although small initial indicators can build up to levels of concern. For instance, aging products are usually free radicals of compounds such as methane, propane, and propene, produced by other free radicals, hydroperoxy, termination and free carbon preserving reactions, which require chemical separation for reactions to occur (Paloniemi & Ab, 1981; Bae et al., 1990; Ruggeri et al., 1990; Sabau, 1998). As these processes and reactions manifest, they make transformer oils to perform the dual role of both insulator and electrolyte. Also, they cause the distortion of free energy levels of various oil constituents, leading to changes in chemical composition and physical morphology of oil insulation. The most significant effects of this on transformer oil are reductions in voltage and interfacial properties of the oil. Dominant oil deterioration processes such as decomposition, depolymerisation, photodegradation, redox reaction, acidification, gas dissolution, sludge accumulation, and hydrolysis depend on the existence of a favorable physico-chemical environment. As aged transformer oils undergo these processes, discharges can develop, leading to overheating that can in turn, soften and/ or thermally decompose the oil.

2 CHANGE IN OIL COMPOSITION

Apart from the fact that a transformer oil is initially a complex blend of more than 3000 hydrocarbon compounds (Lobeiras and Sabau, 2000), its physical and chemical deterioration can significantly alter the assemblage of chemical constituents in the oil. Aging oil degenerates into a mixture that contains particles such as sludge, gas, water, air, and metals. This change in composition also accelerates the breakage of long chain hydrocarbons into short ones through various processes. Ultimately, oil

durability decreases to a level that can negatively affect transformer performance. During the process of oxidation, some by-products generated are capable of attacking the oil inhibitors. Once the oil redox process is initiated, it immediately generates a series of reactions culminating in the formation of compounds of the form: ROH, RCOR, RCHO, RCOOR, and RCOOH (Paloniemi and Ab, 1981; Bae et al., 1990). Reactions that produce free radicals, hydroperoxy radicals, and associated reactions could lead to process propagation, process termination and free carbon preservation, (Paloniemi and Ab, 1981; Bae et al., 1990; Ruggeri et al., 1990; and Sabau, 1998) which all negatively impact transformer oil performance. For instance, oxidation by-products of hydroperoxides are usually short-lived but are extremely reactive and, are implicated in oil accelerated aging. The physical-chemical processes that drive the generation of products in transformer oils (Bae et al., 1990; Ruggeri et al., 1990; and Sabau, 1998) are summarized and presented as equations (1) through (4).

$$2RH \xrightarrow{O_2} 2R^o + H_2O$$
$$\text{(oil free radical reaction)} \qquad (1)$$

$$R_o + O_2 \rightarrow RO_2^o \text{ (oil hydroperoxy and}$$
$$\text{propagation reactions)} \qquad (2)$$

$$RO_2^o + RO_2^o \rightarrow ROH + RCOOR + O_2$$
$$\text{(oil termination reation)} \qquad (3)$$

$$C_2H_6(ethane) \rightarrow CH_4 + \underline{C} + H_2 \text{(oil free carbon}$$
$$\text{preservation reaction)} \qquad (4)$$

Hydrocarbon compounds are major constituents of insulating oil and are subject to degradation or transformation that can bring about changes of oil composition as described in the above reactions. As the resilience of the oil insulation is affected by changes in composition, age indicators such as formation of gases, sludges, and particles appear. It has been observed that significant changes in oil indexes or parameter values during or after oxidation occur (Franklin and Franklin, 1983).

2.1 *Particle formation*

A transformer oil system operates smoothly if the oil has good oxidation stability. As soon as there is a temperature change in the system (even at moderate levels), air dissolution or partial air exclusion, oxidation begins and gradually attacks the inhibitors that are usually present in the oil. This action produces insoluble sludges that affect the winding, cooling ducts, and possibly the porous insulation. Snow (1998) observes that the dissolution of air in oil is the progenitor of oxidation activity. This is in agreement with the views of other researchers, exemplified by Armstrong (1998), who indicated that oxidation occurs in oil during partial exclusion of air. Oil has the natural capacity to dissolve up to 10% air by volume (LeBlanc et al., 1993; Grant and Hydro, 1998; and Armstrong, 1998). This implies that a small volume of dissolved air coupled with temperature increase, can trigger oxidation processes. However, Torkos et al. (1984), Griffin (1998), Duval (1989), LeBlanc et al. (1993), Bengton (1996), and Chen and Lin (2001) observe that there is no consensus on the origin of gases that evolve in transformer oil. Many of these investigators believe that gases in oil evolve as a result of discharges of waxes and fluorescent materials in the oil or are attributable to minor deterioration of insulation that results in incipient faults, in the form of arcs or sparks arising from thermally induced breakdown of the insulation. These gases can be seen in oil as dissolved or evolved products and together with overheating, can lead to degradation of oil insulation as various compounds decompose. The presence of weak bonds can facilitate oil decomposition. As large molecules, especially those with weak internal bonds disintegrate gases such as H_2, CO, CO_2, CH_4, C_2H_2, C_2H_6 and C_3H_8 can be formed. Dissolved gases alone do not provide sufficient information for diagnosing or detecting transformer oil degradation (Griffin, 1998), because material dissolution can also evolve gases. The solubilities of N_2, CO, O_2, and H_2 are somewhat low in oil, whereas C_2H_2, C_2H_4, C_2H_6 and CO_2 are highly soluble. It should be noted that gas solubility in oil varies with temperature and pressure. The solubilities of N_2, CO, O_2, and H_2 have been found to increase with temperature, while CO_2, C_2H_2, C_2H_4, and C_2H_6 decrease in solubility with temperature; usually CH_4 solubility is insensitive to temperature (Baker, 1979 and Griffin, 1998).

The presence of water in transformer oil degrades the oil. This manifests as lowering of the voltage capacity of the oil. Water may exist in oil insulation in three ways: dissolved, finely distributed as an emulsion, and a roughly distributed form that is capable of producing droplets. The quantity of water present at any time depends on the process hydrolysis or moisture saturation of the oil. The undesirable effects of moisture in oil are enhanced by the presence of foreign particles such as long fibers. Fibers are oriented by the direction of electrical field in the water and constitute a conductive block between parts of transformer oil–filled chambers, thereby causing flashovers that may severely damage the transformer. Long fibers such as pieces of paper or yarn are produced as a result of water penetration or hydrolysis of paper insulation. The latter causes degradation and depolymerisation of furanic compounds (Franklin and Franklin, 1983).

2.2 Organic crystallite formation

Disintegration and recombination of a series of chemical compounds under changing thermal and electric fields in transformer oils can produce insoluble products in the form of crystallites. Being that hydrocarbons dominate the chemistry of petroleum-based transformer oils, it is conceivable that such crystallites are primarily of hydrocarbon origin. On the contrary, the exsolution of organic crystals has long been regarded as impossible by early researchers. This postulation is based on the observation that crystalline cross-linked polymers are abundant in inorganic materials but are absent in organic compounds. However, McCarthy and Calvin (1967) postulate that biotic and abiotic processes can produce porphyrin crystals. Similarly, Brodskii et al. (1975) identify long chain polymers of crystals recovered from high boiling fractions of crude oil as that of polycyclic and hydroaromatic hydrocarbons. According to Wang et al. (2004) and Wu et al. (2004), in recent research reveal that syntheses of hydrocarbon crystals are no longer constrained. As demonstrated by Wu et al. (2004), they particularly succeeded in their works that focused on creating porous crystallite covalent organic frameworks that already exist in some materials. These results however, build on earlier research (Pope and Swenberg, 1999, and Vodak et al., 2002). All these corroborate investigations that suggest that the strong covalent bonds of hydrocarbons can under favorable circumstances enable the formation of crystallites (Desiraju, 2002; Pestov, 2003). Such crystallites are impurities that contribute negatively to transformer oil durability and performance

2.3 Electrostatic application

During the passage of current in transformers, petroleum-based oils experience strong electrostatic forces at the microscopic scale irrespective of their covalent bonding. These forces hold together by attraction, positive nuclei and negative electron clouds. Thus force field is also responsible for the intermolecular attraction that comprises hydrogen bonding, dispersion and dipole—dipole orientation. Petroleum-based oil is an insulator by classification but at the same time, it stores both long term and short term charges that make it a material that possess electrical properties (Griffiths, 1999; Guru and Hiziroglu, 2004). At the ground state, which is a time frame of inertness due to charge immobility, it is electrically neutral.

A detailed analysis of various investigations implies a possibility of stable current existence in mineral oils, a property that relates to stored electrical potential, particularly as presented by Zaky and Hawley (1973), Dissado and Fothergill (1992),

Zakrevskii et al. (2003), and Theodosou et al. (2004). The complexity associated with these parameters of investigations can be viewed from interpretations adduced to them, which are primarily based on either classical or quantum perception. Mineral oils suffer immensely in this regard, as their electrostatic characteristics are often related to, assessed by or interpreted on the basis of conduction in solids or perhaps, fluid plasma. As it is the case, explanation of liquid insulation or conduction is sandwiched between existing knowledge of conduction in solid and fluid plasma. Therefore, it is the degree of conductivity and resistivity of mineral oils that pose contentions. For instance, it has been demonstrated that there exists static electrification in petroleum-based transformer oil due to liquid dielectric flow (Gross, 1969; Dissado and Fothergill, 1992; Lee et al., 1998) and that it can be quantified (Gross, 1969; Zaky and Hawley, 1973; Saville, 1997; Wahab et al., 2000; Guru and Hiziroglu, 2004).

A mathematical expression of energy and charges in oil and assuming steady state conditions, the increment of internal energy (∂U) in transformer oil is:

$$\partial U = \begin{bmatrix} energy\ loss\ by\ N\ charges \\ due\ to\ collision\ with\ distance \end{bmatrix} - \begin{bmatrix} energy\ gain\ by\ N\ charges \\ due\ to\ electric\ field\ with\ distance \end{bmatrix} = 0 \quad (5)$$

which results into equation (6).

$$\partial U = \Delta V \partial \varphi \quad (6)$$

But internal energy with respect to time, power, is equation (7).

$$\frac{\partial U}{\partial t} = VI \quad (7)$$

If equation (7) is substituted into equation (6) then, it is defined by equation (8).

$$\frac{\int \partial V}{\partial t} = \frac{-VI}{\partial \varphi} \quad (8)$$

From equation (8), the rate at which the voltage capacity transfers into internal energy is defined by equation (9) and when substituted into equation (8) with integration, it becomes equation (10),

$$k = -\frac{I}{\partial \varphi} \quad (9)$$

$$V_t = V_0 e^{-kt} \quad (10)$$

wherein V_t is the oil voltage capacity at any time (t) in the equation, K is the transformation rate, and V_o is the initial voltage capacity of the oil when time is zero. The V_0 of equation (10) forms the target of experimentation in which there arise formation of crystallites nucleation.

2.4 *Inhibitor reaction in transformer oil*

In order to stabilize critical characteristics of transformer oil to maintain its resistance to electrical, mechanical, chemical, and thermal stresses, various approaches have been adopted. The function of an inhibitor in oil is to preserve the oil stability or at least, to minimize the rate of its deterioration (Armstrong, 1998; Robertson, 1998; Polowick and Hydro, 1998). Transformer oils are supplemented with Polynuclear Aromatic Hydrocarbons (PAH) to prolong service life (Dovgopolyi et al., 1972; Pillai et al., 2005) or amended with antioxidants such as 2,6-Di-Tertiary-Butylphenol (DBP), 2,6-di-tertiary-butyl-4-methylphenol and 2,6-Di-Tertiary-Butyl-Para-Cresol (DBPC) to combat oil oxidation (Polowick and Hydro, 1998; Pugh, 1998, Jada et al., 2002). Inhibitors provide the insulating oil with the resilience to withstand decomposition, minimize gas bubble formation, maintain interfacial tension with negligible temporal variation, and moderate oxidation so as to deactivate radicals that are formed (Franklin and Franklin, 1983; Sabau, 1998; Bae et al., 1990).

Although the key role of an oxidation inhibitor in transformer oil is already emphasized, information on driving chemical reactions is even more critical. The chemical model by which anti-oxidant breaks the chain of oxidation by-products, stabilizes oil insulation properties, deactivates radicals in transformer oil, prevents particle precipitation even at low temperatures, and makes transformer oil a primary determinant of transformer service life is shown as equation (11)

$$(11)$$

3 MATERIAL AND METHODS

The transformer oil samples used in this research were collected from Duke Power Station in

Mooresville, North Carolina. Three different transformer oil samples of the type II variety were collected. Two of the samples (T14 and T24) were in service oils and (T44) was a new Duke Power oil. A fourth sample (T34) was Exxon oil bought from a local supplier. All the oils were of type II as shown in Figure 1. They exhibited differences in chemical composition based on the operating conditions of the transformers that contain them. The in-use oil samples were collected through a drained cork such that atmospheric interference was minimized. Oil samples for experiments were in 2–3 liters high temperature resistant containers loosely corked and placed in oven under a constant temperature and periodically taken for voltage breakdown measurement.

The DTS-100D Oil Tester was used for voltage breakdown as oil internal energy increases due to the application of an external magnetic field. Experiments were performed as stipulated in the International Electrochemical Commission (IEC) and American Society of Testing and Materials (ASTM) D1816-protocols for voltage breakdown of petroleum-based transformer oils. The voltage breakdowns of oil samples were determined periodically within the 0–1000 hours timeframe. Tests were performed at constant oven temperature selected within the range of 55 °C to 70 °C to simulate field conditions. The oil samples were carefully labeled to indicate when they were to be tested to avoid protocol errors.

In addition to taking photomicrographs of the organic crystallites that formed, an optical lens or metallurgical microscope (Meiji), with inbuilt INFAVIEW software, was used to observe the internal matrix of the crystals. Further nano-analyses of the oil crystallites were made using a scanning electron microscope (JSM—6460 LV). The latter equipment was very effective as regards the observation of primary and secondary folding of structures within the crystallites. The elemental composition of the crystallites was determined using scanning techniques.

4 ANALYSIS AND RESULTS OF DEVELOPED CRYSTALLITES

Crystallites of various geometrical configurations that formed on the internal walls of the fire-resistant container that was used in the voltage breakdown experiments are discussed herein. Oil samples that developed crystallites were analyzed prior to test chamber and characterized by neutralization number less than 6.5×10^{-1} (mg/KOH/g), interfacial tension less than 3×10^{-2} N/m, and oil quality index less than 280. Figure 2 shows the colonnade network structures of the crystallites. The crystallites have long structural backbones with projected branches that resemble fern leaves. Figure 2a and 2d show crystallites that are feather-like. Dense clusters appear in some areas (Figure 2 (a & b) while portions of Figure 2 (c & d) are fallow.

Analyses of structural forms obtained from the metallurgical microscope (Meiji) with its INFAVIEW software, show three primary sections: namely, a distinct internal matrix with elongated fibers, an amorphous region, and a nucleus that is engrossed with radiating fibers.

Figure 3 shows these structural forms. Also, the secondary electron micrographs obtained at different accelerated voltages and magnifications show primary and secondary folding as in Figure 4.

With no knowledge of the mineralogy of the crystallites, and baseline for comparison of signatures of the elements present, a standardless nano-analysis was considered appropriate for mineral compositional determination. Scanning of portions of the crystallites produced peaks on carbon and oxygen as shown in Figure 5. With this information, hydrogen is assumed to be present because the JSM—6460 LV measurement technique is incapable of detecting hydrogen. More so,

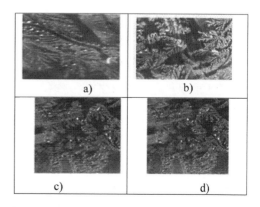

Figure 2. Photomicrographs of oil crystallite colonade network.

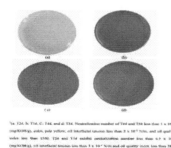

Figure 1. Photos of sampled oils.

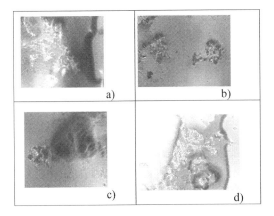

Figure 3. Optical micrographs of the internal structure of crystallite.

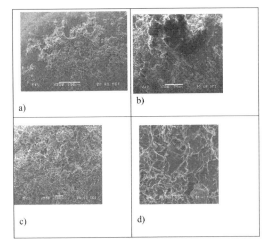

Figure 4. Electron micrographs of primary and secondary folding crystallites.

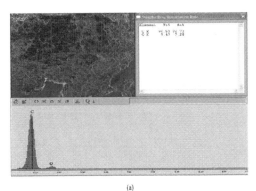

(a)

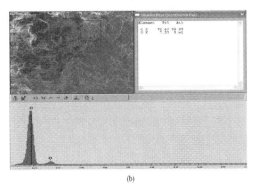

(b)

Figure 5 (a & b). Nano-analysis of composition variation.

hydrogen appears as the likely element with less atomic mass based on detailed compositions and analysis of transformer oil that is insensitive to JSM—6460 LV. Hydrogen is common with most compounds found in transformer composition (Duval, 1989 and Griffin, 1998). Therefore, the definitive appearance of carbon and oxygen is indicative of the organic origin of the exsolved crystallite. Such an origin justifies an assumption of the presence of hydrogen as well. This enables a rough estimate of the chemical formula of the crystallites. However, secondary electron micrographs presented as Figure 5 (a & b), show a compositional variation in the oil crystallites and allow for empirical formular computation. With the inclusion of hydrogen and taking into account

the compositional variation in locations reflected in Figures 5a and 5b, carbon wt% in the tripartite relationship (C, H, O) is 81.8 that depicts result for Figures 5a and 5b. Consequently, the empirical formular $C_{15}H_{24}O$ provides with a molecular weight of 220 g/mole, indicates that the crystallites is most likely BHT. It is known that the temperature range of oil aging experiments (55 °C–70 °C) is within the range of sublimation of Butylated Hydroxytoluence (BHT), an organic substance.

5 CONCLUSIONS

With the knowledge that the original pre-aging chemistry of the transformer oils tested is hydrocarbon dominated, and the observation of carbon and oxygen as major constituents of exsolved crystallites from the oil, it is concluded that the crystallites are of organic origin. It has been determined through calculations that the crystallites formed are primarily Butylated Hydroxytoluence (BHT). BHT is an anti-oxidant and is known to have a sublimation temperature in the range of 55 °C–70 °C, which approximates the temperature range at which these oil aging experiments were performed. Thus analysis throws light on the generation of impurities that can significantly affect the thermo-mechanical characterization of transformer oils, and hence the service life of electric power transformers. It may be necessary to device cooling systems for transformers such that sublimation temperatures of damaging substances are not attained. Some power transformers have radiator-oil cooling systems that are less effective in restraining or controlling cases of exsolution of substances in the transformer systems. However, the probability of exsolution of such substances can be deduced from initial oil composition, operational voltages and temperature of transformers, as well as the service life of the transformers.

ACKNOWLEDGEMENTS

This research was performed at the Global Institute for Energy and Environmental Studies (GIEES) and Civil Engineering Department of the University of North Carolina-Charlotte, USA using a grant awarded by Duke Energy Corporation. The co-authors are grateful to Duke Energy for this sponsorship. Duke Energy supports strategic research but does not necessarily concur with the methods and/or the results of research that it sponsors at grantee institutions.

REFERENCES

Armstrong, D.A. (1998). Partial discharges and gasing; a theoretical perspective. *ASAIT/CEA Technology Symposium,* Calgary, Alberta, Canada.

Bae, J.; Chon, Y. and Kang, Y. (1990). A new method for measuring the degradation level of transformer insulating oil in service. *IEEE, International Symposium on Electrical Insulation,* Toronto, Canada, 36–42.

Baker, A.E. (1979). Solubility of gases in transformer oil. *46th Annual International Conference Doble Clients,* Watertown, MA.

Bengton, C. (1996). Status and trends in transformer oil. *IEEE Transactions on Power Delivery,* 3: 1379–1384.

Brodskii, E.S.; Barabadze, S.S.; Siryuk, A.G.; Lukashenko, I.M.; Lebedevskaya, V.G. and Shnyrev, G.D. (1975). Investigation of crystals recovered from high boiling point petroleum. *Chemistry and Technology of Fuels and Oils.* 2: 143–147.

Chen, A. and Lin, C. (2001). Fuzzy approaches for fault diagnosis of transformers. *Fuzzy set and Systems,* 118: 139–151.

Desiraju, G.R. (2002). Cryptic crystallography. *Nature Materials,* 1: 77–79.

Dissado, L.A. and Fothergill, J.C. (1992). *Electrical Degradation and Breakdown in Polymers* Peter Peregrinus, London.

Dovgopolyi, E.E.; Eminov, E.A.; Lipshtein, R.A.; Kozlova, E.K. Efficiency of oxidation inhibitors for use in transformer oils. *Chem. & Tech. of Fuels and Oils,* 1972; 8: 908–911.

Duval, M. (1989). Dissolved gases analysis: it can save your transformer. *IEEE Electrical Insulation Magazine,* 6: 22–27.

Franklin, A.C., and Franklin, D.P. (1983). *The J & P Transformer Book.* Butterworth, London.

Grant, D. and Hydro, M. (1998). Current reclamation techniques for aged mineral insulating oils. *ASAIT/CEA Technology Symposium,* Calgary, Alberta, Canada.

Griffin, P.J. (1998). Criteria for the interpretation of data for dissolved gases in oil from Transformers (a review). *Electrical Insulation Oils, ASTM,* Philadelphia, 89–107.

Griffiths, D.J. (1999). *Introduction to Electrodynamic.* Prentice-Hall Inc, New Jersey.

Gross, M.J. (1969). Conditions for the static equilibrium and circulation of insulating liquids in electric fields. *Nature* 224: 763–766.

Guru, B.S. and Hiziroglu, H.R. (2004). *Electromagnetic Field Theory Fundamentals.* University Press, Cambridge.

Jada, A; Chaou, A.A.; Bertrand, Y. and Moreau, O. (2002). Adsorption and properties of silica with transformer insulating oils. *Fuel,* 81: 1227–1232.

Kamaka, Y. and Kako, Y. (1980). Flashover characteristics of extremely long gaps in transformer oil under non-uniform field conditions. *IEEE Trans. Elec. Insul.,* 1: 18–26.

Karsai, K., Kerenyi, D., and Kiss, L. (1987). *Studies in Electrical and Electronics Engineering 25.* Elsevier Science Publishing Company, Inc., New York.

Lamarre, C.; Crine, J.P. and Duval, M. (1987). Influence of oxidation on the electrical properties of inhibited naphthenic and paraffinic transformer oils. *IEEE, Trans. Elect. Insul.,* EI 22, No.1: 57–62.

LeBlanc, Y.; Gilbert, R.; Duval, M. and Hubert, J. (1993). Static headspace gaschromatographic determination of fault gases dissolved in transformer insulating oils. *Journal of Chromatography,* 633: 185–193.

Lee, M.J. and Nelson, J.K. (1998). Dielectric integrity associated with circulating insulating fluids. *IEEE Trans. Elect. Insul.* 23: 707–715.

Lobeiras, A. and Sabau, J. (2000). Particle counting of insoluble decay products in mineral insulating oils. *Symposium of The American Society for Testing and Materials (ASTM),* Toronto, ON.

McCarthy, E.D. and Calvin, M. (1967). Organic Geochemical Studies. I. Molecular Criteria for Hydrocarbon Genesis. *Nature* 216, 642–647.

Paloniemi, P., and Ab, Q.S. (1981). Theory of equalization of thermal ageing processes of Electrical insulating materials in thermal endurance tests 1: review of theoretical basis of test Methods and chemical and physical aspects of ageing. *IEEE, Trans. Elect. Insul.* I: 1–6.

Pestov, S. (2003). *Physical Properties of Liquid Crystals.* Springer, New York.

Pillai, I; Ritchie, L.; Heywood, R.; Wilson, G.; Pahlavanpour, B.; Setford, S. and Saini, S. Development of an improved analytical method for determination of carcinogenic polycyclic aromatic hydrocarbons in transformer oil. *Chromatography A,* 2005; I064: 205–212.

Polowick, G. and Hydro, B.C. (1998). Current techniques of mineralizing oxidation of oil and paper insulation in power transformers. *ASAIT/CEA Technology Symposium*, Calgary, Alberta, Canada.

Pope, M. and Swenberg, C.E. (1999). *Electronic Processes in Organic Crystals and Polymers*. Oxford University Press.

Pugh, D. (1998).The damaging effect of aged oils upon the reliability of power transformers. *ASAIT/CEA Technology Symposium*, Calgary, Alberta, Canada.

Ruggeri, B.; Tundo, P. and Tumiatti, W. (1990). Supported Liquid Phase Reactor (SLPR) for PCBs in oil decontamination. *Chem. Eng. Sci.*, 45: 2687–2694.

Sabau, J. (1998). An alternative interpretation of dissolved gas analysis and the stability of oil under the effect of high voltage fields. *ASAIT/CEA Technology Symposium*, Calgary, Alberta, Canada.

Saville, D.A. (1997). Electrohydrodynamics: The Taylor-Melcher leaky dielectric model. *Annu. Rev. Fluid Mech.* 29: 27–64.

Snow, G. (1998). An environmental friendly technique designed to arrest chemical decay process of oil-paper insulation. *ASAIT/CEA Technology Symposium, Calgary*, Alberta, Canada.

Torkos, K.; Borossay, J. and Szekely, A. (1984). Determination of fault gases in transformer oils. *Journal of Chromatography,* 286: 317–321.

Theodosou, K.; Vitellas, I.; Gialas, I. and Agoris, D.P. (2004). Polymer films degradation and breakdown in high voltage AC fields. *J. Elect. Engineering.* 55: 9–10.

Vincent, C. and Crine, J. (1986). Evaluation of in-service transformer oil condition from Various measurements. *IEEE Int. Symposium on Electrical Insulation*, Washington DC, USA.

Vodak, D.T.; Braun, M.; Lordanidis, L.; Plevert, J.; Stevens, M..; Beck, L.; Spence, J.C.H., O'Keeffe, M. and Yaghi, O.M. (2004). One step synthesis and structure of an oligo (spiro-orthocarbonate. *Am.chem. Soc.*, 124: 4942.

Wahab M.A.A.; Hamada, M.M.; Ismail, G. and Zeitoun, A.G. (2000). Novel modeling of residual operating time of transformer oil. *Alexanda Engineering Journal*, 39: 543–553.

Wang, G.; Sun, Y. and Yang, G. (2004). Synthesis and crystal structures of two new organically templated borates. *Solid State Chemistry*, 12: 4648–4654.

Wu, J.S.; Melcer, N.; Sharp, W.P.; O'Keeffe, M.; Spence, J.C.H. and Yaghi, O.M. (2004). Structural study of new hydrocarbon nano-crystals by energy-filtered Electron diffraction. *Ultramicroscopy*, 98: 145–150.

Zakrevskii, V.A.; Sudar, N.T.; Zaopo, A. and Dubitsky, Y.A. (2003). Mechanism of electrical degradation and breakdown of insulating polymers. *J. App. Phys.* 93: 2135–2139.

Zaky, A.A. and Hawley, R. (1973). *Conduction and Breakdown in Mineral Oil*. Peter Peregrinus, London.

184

Advances in Power and Energy Engineering – Sun (Ed.)
© 2016 Taylor & Francis Group, London, ISBN 978-1-138-02846-3

Research on pressure fluctuation of Francis turbine based on CFD

L. Fu & P.G. Kou
Hunan Electric Science Research Institute, Changsha, China

Y.G. Cheng
State Key Laboratory of Water Resources and Hydropower Engineering Science, Wuhan, China

ABSTRACT: The hydraulic turbine vibration is the most important factor influencing the safe operation of hydropower station. Because of the relatively wide vibration area, the regulating capacity of the Francis turbine has been greatly limited in the grid. In order to analyze the mechanisms of inducing vibration of the Francis turbine, the 3-D CFD model of Francis turbine has been established and analyzed using numerical simulation in this paper. And the results of CFD numerical simulation are compared with the data of field testing. Through the numerical simulation, the mechanisms of inducing vibration and the variation of the pressure fluctuation have been identified. And in this paper, the pressure fluctuation contours of Francis turbine has been drawn out by the calculation in 25 conditions and the study can guarantee the hydropower station's safe and stable operation.

1   INTRODUCTION

With a large number of great capacity hydropower units going into operation, the contradiction between regulating capacity of hydro turbine and grid dispatching is becoming more and more seriously. Hydropower units which have the advantages of flexible operation modes and superior regulation performance bear the tasks of peak load and frequency regulation in the grid, but the hydraulic vibration which exists in hydro turbine inherently and is difficult to be solved has limited the hydropower units' peak load and frequency regulation capacity, and it also affects the stability of the grid. One of the main sources of the hydraulic vibration is the pressure fluctuation which is investigated by the model experiment and the field test at present (Zhang F. et al. 2011, He C. & Wang Z. 2002). Because the number of field testing is limited, it is difficult to measure the pressure fluctuation and the hydraulic vibration under various operation conditions, and also because of the heavy workload of field test, it cannot measure the water flow state accurately in the whole flow channel system, so it is difficult to identify the mechanisms of inducing vibration by the field testing. And because of scale effect, the law of the pressure fluctuation which is obtained by model experiment is not always consistent with that of the prototype turbine (Tridon, S, Barre, S, Ciocan, G.D et al. 2010).

With the development of CFD technology, the research on pressure fluctuation of prototype hydro turbine by numerical simulation has been a hot issue which is cared by the both academic research and industrial applications (Yexiang, X, Zhengwei, W, Zongguo, Y et al. 2010, Jošt, D, Lipej, A. 2011, Liu, S, Shao, Q, Yang, J et al. 2005).

It becomes possible to calculate the complex flow accurately recently with the development of computer performance. Numerical simulation shows that it is much more economic and accurate comparing with traditional model experiment and filed testing methods. Choosing the suitable turbulence model aiming at the unsteady turbulent simulation in the hydro turbine is the problem which is cared by the most scholars. In the relevant literatures, such as large eddy simulation, Reynolds stress model based on Reynolds Navier-Stokes equation, k–ε series models, are used in simulation. At present, the research of CFD simulation aiming at the large prototype hydro turbine of Three Gorges and Wanjiazhai hydropower station has been carried out in some domestic universities and research institutions, in their studies the pressure fluctuation in hydro turbine was calculated and analyzed (Wang, Z, Zhou, L, He, C. 2005, Gui, Z, Tang, S, Pan, L. 2006). According to the results, although the CFD calculation could predict the frequency characteristic of the pressure fluctuation in draft tube accurately, but it could not predict the amplitude of the pressure fluctuation accurately. In their studies, the simulation did not combine with testing, which caused the deviations, and in the most of the studies, only the flow pattern and pressure fluctuation in one or two operating conditions were calculated, the variation of pressure fluctuation under the whole load conditions was not obtained, so it could not get the pressure fluctuation area of the prototype hydro turbine.

In this paper, the research on pressure fluctuation of Francis turbine based on CFD is carried out. The 3-D CFD hydro turbine model of the FengTan hydropower station has been established, and numerical simulation of the Francis turbine is carried out. And based on test data, model and method of numerical simulation are been improved. By the study the mechanisms of inducing vibration of the Francis turbine has been indentified and the variation of the pressure fluctuation in the Francis turbine under the whole load conditions has been analyzed.

2 INTRODUCTION OF EQUIPMENTS

The FengTan hydropower station has the total installed capacity of 800 MW. 4 hydro generator sets of 100 MW which designed and manufactured by Harbin Electrical Machinery were installed at the first construction of the hydropower station. The hydraulic turbine runner was introduced to China from Soviet Union in the 1950s. The main parameters of the plants are shown in Table 1.

3 MATHEMATICAL MODELS

3.1 N-S equations

The N-S equations can be obtained considering the continuity equation and momentum equations and ignoring the energy conservation equation.

Continuity equation:

$$\frac{\partial \rho}{\partial t} + div(\rho \vec{u}) = 0 \qquad (1)$$

Table 1. The main parameters of the plants.

Hydraulic turbine		Generator	
Type	HL702-LJ-410	Type	TS854/210–40
Rated capacity	103 MW	Rated capacity	100 MW
Design head	73 m	Rated voltage	13.8 kV
Range of head	57.5–86.2 m	Power factor	0.85
Rated flow	160 m³/s		
Rated speed	150 r/min		

Momentum equations:

$$\left. \begin{array}{l} \dfrac{\partial(\rho u)}{\partial t} + div(\rho u \vec{u}) = -\dfrac{\partial p}{\partial x} + div(\mu \cdot gradu) + S_u \\[2mm] \dfrac{\partial(\rho v)}{\partial t} + div(\rho v \vec{u}) = -\dfrac{\partial p}{\partial x} + div(\mu \cdot gradv) + S_v \\[2mm] \dfrac{\partial(\rho w)}{\partial t} + div(\rho w \vec{u}) = -\dfrac{\partial p}{\partial x} + div(\mu \cdot gradw) + S_w \end{array} \right\} \qquad (2)$$

where $\vec{u}$ is velocity of certain point in the flow field, u, v, w are the components of $\vec{u}$ in x, y, z directions, ρ is fluid density, μ is fluid viscosity, p is the pressure which acts on fluid micro element, $S_u = F_x + s_x$ $S_v = F_y + s_y$, $S_w = F_z + s_z$ are the generalized source terms, F_x, F_y, F_z are the body force which acts on fluid micro element, in general, the values of S_x, S_y, S_z are very small, and for incompressible fluid while the viscous is constant, $s_x = s_y = s_z = 0$.

3.2 Turbulence model

There are three methods of the turbulence simulation: Reynolds Averaged Navier-Stokes (RANS), Large Eddy Simulation (LES) and Direct Numerical Simulation (DNS).

RANS is the most widely used of the turbulence simulation methods. The basic idea of RANS is that: The instantaneous N-S equations is decomposed into time-averaged N-S equations and pulsate N-S equations. Instead of solving instantaneous N-S equations directly, only the time-averaged N-S equations are solved, and the pulsation value is showed in the time-averaged N-S equations by introducing certain model. Thus, not only the precision requirement of the simulation can be met, but also the computational work is reduced.

The Reynolds stress term which concerns the turbulence impulse value in N-S equations is unknown, it needs to model the Reynolds stress term, and make empirical and semi-empirical assumptions on the Reynolds stress term according to turbulence theory, test data and simulation result. It also needs to establish turbulence model equation or stress expression of the Reynolds stress term, and these can link the time-average value with the impulse value of the turbulence. There are two kinds of commonly used turbulence models according to the treatment methods of the Reynolds stress term, the eddy viscosity model and the Reynolds stress model. But the application of the Reynolds stress model is greatly limited, because of the large computational complexity of the Reynolds stress model, so application of the eddy viscosity model is much more extensive.

In this article, the RNG $k - \varepsilon$ model one of the eddy viscosity models is be used. The coefficients of this model is got by "reforming group" method which can embody the effects of the small scale by the viscosity term modified and movement in large scale, and the method can remove the movement in small scale from control equations systematically. The k and ε transport equations are as follows:

k equation:

$$\frac{\partial(\rho k)}{\partial t} + \frac{\partial(\rho k u_i)}{\partial x_i} = \frac{\partial}{\partial x_j}\left[\alpha_k \mu_{eff}\frac{\partial k}{\partial x_j}\right] + G_k - \rho\varepsilon \quad (3)$$

ε equation:

$$\frac{\partial(\rho\varepsilon)}{\partial t} + \frac{\partial(\rho\varepsilon u_i)}{\partial x_i} = \frac{\partial}{\partial x_j}\left[\alpha_k \mu_{eff}\frac{\partial\varepsilon}{\partial x_j}\right] + \frac{C_{1\varepsilon}^*\varepsilon}{k}G_k$$
$$- C_{2\varepsilon}\rho\frac{\varepsilon^2}{k} \quad (4)$$

where: $\mu_{eff} = \mu + \mu_t$, $\mu_t = \rho C_\mu\frac{k^2}{\varepsilon}$, $C\mu = 0.0845$, $\alpha_k = \alpha_\varepsilon = 1.39$; $C_{1\varepsilon}^* = C_{1\varepsilon} - \frac{\eta(1-\eta/\eta_0)}{1+\beta\eta^3}$, $C_{1\varepsilon} = 1.42$, $\eta = S\frac{k}{\varepsilon}$, $S = \sqrt{2S_{ij}S_{ij}}$, $S_{ij} = \frac{1}{2}\left[\frac{\partial u_i}{\partial x_j} + \frac{\partial u_j}{\partial x_i}\right]$, $\eta_0 = 4.377$, $\beta = 0.012$, $C_{2\varepsilon} = 1.68$.

4 NUMERICAL SIMULATION AND ANALYSIS

4.1 Calculation model

Taking the Francis turbine of HL702-LJ-410 as the object of calculation, wooden patterns is used to establish the model of runner blade as Figure 1 and Figure 2 shown. The 3-D shape of the whole turbine model as Figure 3, and the parameters as follows: the number of hydro turbine blades $zb = 14$, the number of guide vanes $zgv = 24$, and the number of stay vanes $zsv = 12$ (including baffle tongue in the nasal tip), the inlet diameter of spiral

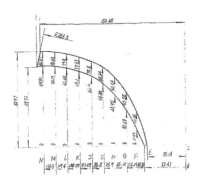

Figure 1. Wood patterns of the hydro turbine.

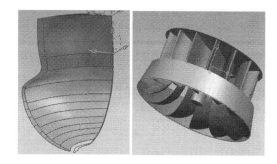

Figure 2. Simulation model of the blade and hydraulic turbine runner.

Figure 3. 3-D shape of the whole turbine simulation model.

Figure 4. Computational grid of the simulation model.

case $D_s = 5.6$ m, the inlet diameter of runner $D_1 = 4.1$ m, the guide vane height $H = 1008$ mm.

The computational grid based on hybrid grid is constructed as Figure 4, in the area of spiral case and runner, the tetrahedral mesh is used; in the area of guide vane, the wedge mesh is used; in the area of draft tube, the hybrid grid of tetrahedron and hexahedron is used. At last, 2300000 grids are used by the sensitivity analysis of the grid number.

4.2 Boundary conditions and treatment method

Boundary conditions: pressure import boundary condition is used at the inlet of spiral case; pressure export boundary condition is used at the outlet of draft tube. The pressure difference value of the inlet and outlet is the calculation head, and its value is adjusted with the different head condition. Slip boundary is used on the wall face.

The results of steady calculation are used as initialized condition of unsteady calculation. In the area of the runner, the sliding mesh is used in the unsteady calculation, the speed $n = 150$ rpm, time step is set as 0.004 s. In the calculation, the residual convergence target is set as 0.0001, the largest iteration number each time step is set as 30 times.

To analysis the pressure pulsation of Francis turbine under various operating conditions, in the unsteady calculation, 14 monitoring points of pressure pulsation are set in the turbine, and the pressure changing process of these points are recorded. The distribution of the monitoring points is as Figure 5 shown.

4.3 CFD calculation results

The calculation conditions include five different guide vane openings (a) and five different heads (H), with the total of 25 conditions. The guide vane openings are respectively a = 100 mm, a = 150 mm, a = 200 mm, a = 240 mm, a = 290 mm, and the heads are respectively maximum head H = 86.2 m, many years' average head H = 78.84 m, rated head H = 73 m, minimum head H = 57.5 m and the head H = 65 m.

The results of calculation are shown as Figure 6~9, the amplitude in the longitudinal coordinate Y represents of the peak peak value of pressure pulsation, and the abscissa coordinate X represents of the number of monitoring points.

As observed in Figure 6 and Figure 7, when the guide vane opening is at 150 mm in the head of 86.2 m and 78.84 m, the amplitude of pressure pulsation of every monitoring point is larger than other guide vane openings. With the decreasing of the head, the largest amplitude of pressure pulsation transfers to the guide vane opening of 200 mm. In the minimum head, at the 200 mm guide vane opening, the amplitude of pressure pulsation gets the largest. As observed in Figure 6~9, it also shows that the largest amplitude of pressure pulsation

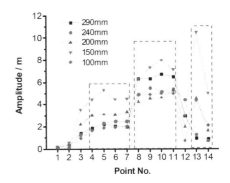

Figure 6. Pressure pulsation's calculation results of different openings at the head of 86.2 m.

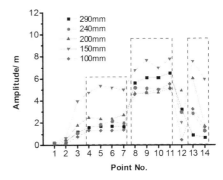

Figure 7. Pressure pulsation's calculation results of different openings at the head of 78.84 m.

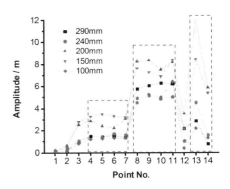

Figure 8. Pressure pulsation's calculation results of different openings at the head of 65 m.

occurred in the part load condition but not in full load and extremely low load conditions, this is consistent with the characteristic of the pressure pulsation in draft tube, and this also explains that the turbine is not suitable to operate at the 150 mm guide vane opening in the head which is higher than

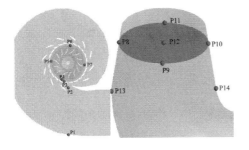

Figure 5. Monitoring points in the hydro turbine.

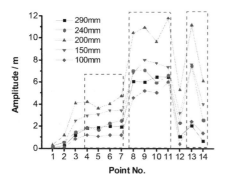

Figure 9. Pressure pulsation's calculation results of different openings at the head of 57.5 m.

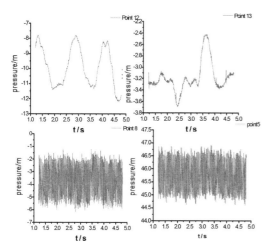

Figure 10. Typical monitoring points' time domain wave calculation results in the condition of a = 240 mm, H = 86.2 m.

rated head 73 m and 200 mm guide vane opening in the head which is lower than 73 m.

The time domain wave calculation results of some typical monitoring points in the condition of a = 240 mm, H = 86.2 m are shown as in Figure 10. From the results we can get that, the high frequency pressure pulsation which is induced by dynamic and static interference of guide vane and runner transmits to the upstream and downstream. On the transmission to the spiral case's inlet, the pulsation is enhanced at the side wall of the inlet of draft tube. And the high frequency pulsation decays rapidly at the center of the draft tube, while the main component of the pulsation is the low frequency which is induced by the vortex band in the draft tube. From the calculation of pressure pulsation along the way of the hydro turbine, we can get that

the high frequency pulsation which is induced by dynamic and static interference is superimposed on the low frequency pulsation which is induced by vortex band. The amplitude of the low frequency pulsation which is larger transmits to the upstream, and the amplitude is gradually reduced on the transmission.

The Figure 11 shows the calculation results of the vortex band in the draft tube of the typical two conditions.

1) When the guide vane opening is 150 mm, in the maximum head condition, high intensity helical vortex rope is induced in the draft tube. With the decreasing of the head, the flow separates and becomes complex, the vortex rope is not concentrated at the draft conic tube, but induced at the runner crown near the outlet of turbine blade, and it appears several vortex structures. And in the low flow rate condition, the vortex band is not integrated and slender, and the phenomenon of mutual interference and combination which is similar to the detached vortex of hydrodynamic form happens.

2) When the guide vane opening is 200 mm, in the maximum head condition, the velocity's tangential component of runner blade's outflow is very small, and the flow pattern is good, there is not helical vortex rope in draft tube, only detached vortex induced. With the decreasing of the head, high intensity helical vortex rope is induced in the draft tube, and the intensity enhanced, the

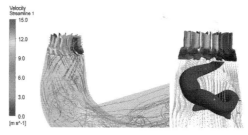

Condition d1 (a=150mm,H=86.2m)

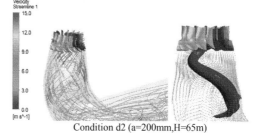

Condition d2 (a=200mm,H=65m)

Figure 11. Calculation results of the vortex band in the draft tube of the condition a = 150 mm H = 86.2 m and the condition a = 200 mm H = 65 m.

pitch increased with the decreasing of the head, the vortex rope even contact to the wall of draft tube, and it induces strong pressure pulsation.

5 RESULT COMPARISON BETWEEN CALCULATION AND TEST

The stability field test was performed in May 2014, the average head of the hydropower station was 68.96 m when the test. The pressure pulsation both at the spiral case inlet and draft tube inlet was measured by the high precision pressure sensors. The result comparison between calculation and test is shown as Table 2. In the table, A represents the amplitude of frequency mixing, A1 represents the amplitude of dominant frequency and f1 represents the value of dominant frequency.

From the comparison, the result of calculation is basically in accordance with the test data. On the pressure pulsation at the spiral case inlet, between the results of test and CFD, the difference of the amplitude of frequency mixing is less than 26%, the difference of the amplitude of dominant frequency is less than 21%, and the difference of

the dominant frequency is less than 10%. On the pressure pulsation at the draft tube inlet, between the results of test and CFD, the difference of the amplitude of frequency mixing is less than 17%, the difference of the amplitude of dominant frequency is less than 19%, and the difference of the dominant frequency is less than 12%.

In general, the difference between the calculation result and the test data of the spiral case inlet is larger than that of the draft tube. This is because pressure import boundary condition is used at the inlet of spiral case in the CFD calculation, and this method will probably reduce the pressure pulsation in the range of spiral case inlet.

6 ANALYSIS OF PRESSURE PULSATION CONTOURS IN DIFFERENT CONDITIONS

The CFD calculations include 25 conditions which have basically covered the normal operating range of the hydro turbine. So from the CFD calculation results we can draw the pressure pulsation contours in different conditions as the Figure 12~13 shown.

Table 2. The result comparison between calculation and test.

Conditions	Opening (mm)		Pressure pulsation at spiral case inlet						Pressure pulsation at draft tube inlet					
			A (kPa)		A1 (kPa)		f1 (Hz)		A (kPa)		A1 (kPa)		f1 (Hz)	
			Test	CFD	Test	CFD	Test	CFD	Test	CFD	Test	CFD	Test	CFD
1	87.58	Val	19.04	15.80	11.14	10.03	3.48	3.95	82.02	75.66	24.22	21.60	3.48	3.12
		Err (%)	(17.02)		(9.96)		(13.51)		(7.75)		(10.81)		(10.26)	
2	142.4	Val	22.71	16.82	4.86	3.89	3.82	3.95	69.71	65.18	11.82	10.23	1.17	1.04
		Err (%)	(25.94)		(19.96)		(3.40)		(6.50)		(13.46)		(11.45)	
3	200	Val	21.28	18.74	7.04	6.85	0.69	0.65	27.63	26.58	8.89	9.93	0.85	0.82
		Err (%)	(11.94)		(2.70)		(5.80)		(3.80)		(11.67)		(3.18)	
4	219.8	Val	23.23	19.51	7.59	6.24	2.44	2.24	24.87	29.15	4.84	3.96	0.34	0.37
		Err (%)	(16.02)		(17.79)		(8.20)		(17.21)		(18.13)		(8.88)	

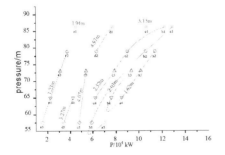

Figure 12. Pressure pulsation contours in different conditions of the blade free sections.

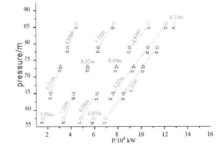

Figure 13. Pressure pulsation contours in different conditions of the draft tube inlet.

1. From the pressure pulsation contours of the blade free sections as shown in Figure 12, we can get that: the distribution of the pressure pulsation contours is close to that of the equal opening lines, the amplitude of the pressure pulsation is the largest in the range of 150 mm–200 mm openings. The pressure pulsation contours are almost continuous along the equal opening lines, and it explains that the pressure pulsation of the blade free sections is insensitive to the head.

2. From the pressure pulsation contours of the draft tube inlet as shown in Figure 13, we can get that: the distribution of the pressure pulsation contours is close to that of the equal opening lines, the amplitude of the pressure pulsation is the largest in the range of 150 mm–200 mm openings. The pressure pulsation contours are discontinuous along the equal opening lines, and it explains that the pressure pulsation of the draft tube inlet is sensitive to the head.

7 CONCLUSION

In this article, the characteristics of Francis turbine's pressure pulsation and flow pattern are studied using CFD method. The calculation conditions include 25 conditions which have basically covered the normal operating range of the hydro turbine, the generation mechanism of the pressure pulsation and the association between pressure pulsation and flow pattern are also analyzed. On the basis of the study, mainly have following conclusions:

1. Instability operation area of the Francis turbine can be obtained using CFD calculation. The instability operation area of the FengTan hydropower station turbine is at the guide vane openings of 150 mm–200 mm, and the area changes with the changes of head. From the study, the turbine of the FengTan hydropower station is not suitable to operate at the guide vane opening of 150 mm in the head which is higher than rated head and 200 mm in the head which is lower than 73 m.

2. There are mainly two kinds of pressure pulsation existing in the hydro turbine. The high frequency pressure pulsation which is induced by dynamic and static interference of guide vane and runner transmits to the upstream and downstream. And the high frequency pulsation is enhanced at the side wall of the inlet of draft tube and reduced rapidly at the center of the draft tube. The low frequency pressure pulsation which is induced by the vortex band in the draft tube transmits to the upstream, and the amplitude is gradually reduced on the transmission.

3. From the comparison, the result of calculation is basically in accord with the test data. And the difference between the calculation result and the test data of the spiral case inlet is larger than that of the draft tube, this is because pressure import boundary condition is used at the inlet of spiral case in the CFD calculation.

4. The pressure pulsation contours in different conditions can be drawn using the CFD calculation, and this can play an instructive role on the stable operation of hydro turbine. On the pressure pulsation curves of the FengTan hydropower station turbine, the amplitude of the pressure pulsation is the largest in the range of 150 mm-200 mm guide vane openings, this is also consistent with the result of the vortex band and pressure pulsation amplitude analysis.

REFERENCES

Gui, Z., Tang, S., Pan, L. 2006. Simulation and pressure pulsation prediction of unsteady flow in draft-tubes of Francis turbine. *Journal of China Institute of Water Resources and Hydropower Research* 1: 012.

He, C., Wang, Z. 2002. Experimental study on pressure surge in hydraulic turbine draft tube. *Chinese Journal of Mechanical Engineering* 38(11): 62–65.

Jošt, D., Lipej, A. 2011. Numerical prediction of non-cavitating and cavitating vortex rope in a Francis turbine draft tube. *Strojniški vestnik-Journal of Mechanical Engineering* 57(6): 445–456.

Liu, S., Shao, Q., Yang, J., et al. 2005. Unsteady turbulent simulation of prototype hydraulic turbine and analysis of pressure fluctuation in draft tube. *Journal of Hydroelectric Engineering* 24(1): 74–78.

Tridon, S., Barre, S., Ciocan, G.D., et al. 2010. Experimental analysis of the swirling flow in a Francis turbine draft tube: Focus on radial velocity component determination. *European Journal of Mechanics-B/Fluids* 29(4): 321–335.

Wang, Z., Zhou, L., He, C. 2005. Pressure oscillations in a hydraulic turbine draft tube. *Qinghua Daxue Xuebao/Journal of Tsinghua University (China)* 45(8): 1138–1141.

Yexiang, X., Zhengwei, W., Zongguo, Y., et al.2010. Numerical analysis of unsteady flow under high head operating conditions in Francis turbine. *Engineering Computations* 27(3): 365–386.

Zhang, F, Gao, Z.X., Pan, L.P, et al. 2011. Study on pressure fluctuation in Francis turbine draft tubes during partial load. *Journal of Hydraulic Engineering* 42(10): 1234–1238.

Advances in Power and Energy Engineering – Sun (Ed.)
© 2016 Taylor & Francis Group, London, ISBN 978-1-138-02846-3

Dynamic behavior of synchronous generator during a sudden change in input torque using MATLAB/SIMULINK

M.A. Hassan Ibraheem

Hebei University of Technology, Tianjin, China

ABSTRACT: The Synchronous generators are considered as the most important part in electrical power system. Hence, it is very instructive to observe and analyze the dynamic behavior of synchronous generators during normal operation or during disturbances. Thus, this paper aims to observe the dynamic behavior of synchronous generators during a sudden change in input Torque from no load to full load. So, a new proposed simulation technique was developed to simulate the synchronous generator characteristics in one simulation block using MATLAB SIMULINK Program. This technique was obtained using Subsystem and Masking procedure to collect all fundamental mathematical equations of synchronous generator in one subsystem block. This block is simple, easy to use and it completely describes the electrical characteristics of synchronous generator. Therefore the terminal components of synchronous generator such as currents, voltage, load angle, power, and speed were observed and analyzed at zero torque and following step change in torque occurrence.

1 INTRODUCTION

A three-phase synchronous generator which converts mechanical energy into electrical energy is the main power source of the power system (Kundur, P., 2004). Thus, to study the behavior of the power system, the characteristics of the synchronous generator must be investigated very carefully during normal or disturbance operation. This can be obtained by the many of tests one of them is a sudden to make step change in input torque from no load to full load. In this study, it was assumed that the machine has a constant voltage and frequency. In general, the equations which describe the performance of a synchronous generator are nonlinear, therefore can only be solved by computer simulation techniques (Anderson, P.M. & Fouad, A.A., 1977). The prime concern is to accommodate all differential equations, including stator transients which are usually neglected in traditional method to arrive at an accurate representation of machine dynamics (Ong, Chee-Mun, 1998). The SIMULINK Program provides a feature of encapsulating all synchronous generator modeling equations in one subsystem simulation block that can subsequently be used as a single element in SIMULINK library.

2 MODELING OF SYNCHRONOUS GENERATOR

The mathematical description or model developed in this paper is based on the concept of an ideal synchronous machine with two basic poles. The fields produced by the winding currents are assumed to be sinusoidally distributed around the air gap (Kundur, P., 2004). This assumption of sinusoidal field distribution ignores the space harmonics, which may has secondary effects on machine's behavior. It is also assumed that the stator slots cause no appreciable variation of any of the rotor winding inductances with rotor position. Also the saturation effect is not explicitly taken into account in this model. Figure 1 show the circuit involved in the analysis of a synchronous machine. The stator circuits consist of three-phase armature windings carrying alternative currents (i_a, i_b, i_c). The rotor circuit may have one physically identifiable field winding (f), additional windings

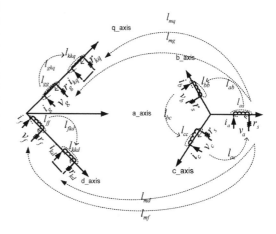

Figure 1.   Stator and rotor circuits of a synchronous machine.

are often used to represent the damper windings (k_d, k_q) and effects of current flow in the rotor iron (g) (Ong, Chee-Mun, 1998).

2.1 Stator circuit equations

The voltage equations for the three phases are

$$v_a = \frac{d\lambda_a}{dt} + r_s i_a = p\lambda_a + r_s i_a \tag{1}$$

$$v_b = p\lambda_b + r_s i_b \tag{2}$$

$$v_c = p\lambda_c + r_s i_c \tag{3}$$

where i_a, i_b, i_c = instantaneous stator currents in phases a, b, and c; $\lambda_a, \lambda_b, \lambda_c$ = Flux linkages in the stator windings; and r_s = Stator resistance.

Flux linkage in the stator windings at any instant is given by

$$\lambda_a = l_{aa} i_a + l_{ab} i_b + l_{ac} i_c + l_{mf_a} i_f + l_{md_a} i_{kd} + l_{mg_a} i_g + l_{mq_a} i_{kq} \tag{4}$$

$$\lambda_b = l_{ab} i_a + l_{bb} i_b + l_{bc} i_c + l_{mf_b} i_f + l_{md_b} i_{kd} + l_{mg_b} i_g + l_{mq_b} i_{kq} \tag{5}$$

$$\lambda_c = l_{ac} i_a + l_{bc} i_b + l_{cc} i_c + l_{mf_c} i_f + l_{md_c} i_{kd} + l_{mg_c} i_g + l_{mq_c} i_{kq} \tag{6}$$

where l_{aa}, l_{bb}, l_{cc} = self inductance of stator windings; l_{ab}, l_{bc}, l_{ac} = mutual inductance between stator winding; $l_{mf}, l_{mg}, l_{md}, l_{mq}$ = mutual inductance between stator and rotor windings; i_{kd}, i_{kq} = d, q axis damper currents; i_f = field current; and i_g = effects of current flow in the rotor iron.

2.2 Rotor circuit equations

The rotor circuit voltage equations are

$$v_f = p\lambda_f + r_f i_f \tag{7}$$

$$v_g = p\lambda_g + r_g i_g \tag{8}$$

$$0 = p\lambda_{kd} + r_{kd} i_{kd} \tag{9}$$

$$0 = p\lambda_{kq} + r_{kq} i_{kq} \tag{10}$$

where v_f, v_g = field voltages; λ_f, λ_g = rotor field flux linkages; $\lambda_{kd}, \lambda_{kq}$ = rotor damper flux linkages; r_f, r_g = d- and q axes filed winding resistance; and r_{kd}, r_{kq} = d- and q axes damper winding resistance.

$$\lambda_f = l_{ff} i_f + l_{fkd} i_{kd} + l_{mf} \left[i_a \sin\theta_r + i_b \sin\left(\theta_r - \frac{2\pi}{3}\right) + i_c \sin\left(\theta_r + \frac{2\pi}{3}\right) \right] \tag{11}$$

$$\lambda_{kd} = l_{kdf} i_f + l_{kkd} i_{kd} + l_{md} \left[i_a \sin\theta_r + i_b \sin\left(\theta_r - \frac{2\pi}{3}\right) + i_c \sin\left(\theta_r + \frac{2\pi}{3}\right) \right] \tag{12}$$

$$\lambda_g = l_{gg} i_g + l_{gkq} i_{kq} + l_{mg} \left[i_a \cos\theta_r + i_b \cos\left(\theta_r - \frac{2\pi}{3}\right) + i_c \cos\left(\theta_r + \frac{2\pi}{3}\right) \right] \tag{13}$$

$$\lambda_{kq} = l_{kqg} i_g + l_{kkq} i_{kq} + l_{mq} \left[i_a \cos\theta_r + i_b \cos\left(\theta_r - \frac{2\pi}{3}\right) + i_c \cos\left(\theta_r + \frac{2\pi}{3}\right) \right] \tag{14}$$

where l_{fkd}, l_{gkq} = mutual inductance between rotor windings; l_{ff}, l_{kkd} = d-axis self inductance of rotor windings; and l_{gg}, l_{kkq} = q-axis self inductance of rotor windings.

Equation 1 to Equation 14 are completely describe the electrical performance of a synchronous machine, but these equations contain inductance terms which vary with angle θ_r, which in turns varies with time (Hadi Saadat, P., 2002).

In order to solve this problem the axes of the rotor winding are already along the q- and d-axes, and the $qd0$ transformation need only to be applied for transform stator winding quantities into qd reference frame as shown in Figure 2. After that the following equations has been deduced.

2.3 Stator and rotor flux linkages

$$\lambda_q = l_{ls} i_q + l_{mq}(i_q + i_g^- + i_{kq}^-) \tag{15}$$

$$\lambda_d = l_{ls} i_d + l_{md}(i_d + i_f^- + i_{kd}^-) \tag{16}$$

$$\lambda_f^- = l_{lf}^- i_f^- + l_{md}(i_d + i_f^- + i_{kd}^-) \tag{17}$$

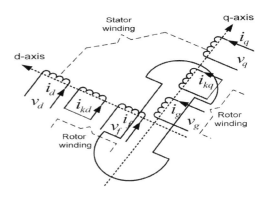

Figure 2. Stator and rotor winding in qd0 reference frame.

$$\lambda_{kd}^- = l_{lkd}^- i_{kd}^- + l_{md}(i_d + i_f^- + i_{kd}^-) \tag{18}$$

$$\lambda_g^- = l_{lg}^- i_g^- + l_{mq}(i_q + i_g^- + i_{kq}^-) \tag{19}$$

$$\lambda_{kq}^- = l_{lkq}^- i_{kq}^- + l_{mq}(i_q + i_g^- + i_{kq}^-) \tag{20}$$

where $i_f^-, i_g^-, i_{kd}^-, i_{kq}^-$ = rotor field and damper currents referring to stator; l_{ls} = stator leakage inductance; $l_{lf}^-, l_{lkd}^-, l_{lg}^-, l_{lkq}^-$ = rotor self inductances referring to stator; and $\lambda_f^-, \lambda_{kd}^-, \lambda_g^-, \lambda_{kq}^-$ = rotor fluxes referring to stator.

2.4 Stator and rotor voltage equations

$$v_q = r_s i_q + p\lambda_q + \lambda_d \omega_r \tag{21}$$

$$v_d = r_s i_d + p\lambda_d - \lambda_q \omega_r \tag{22}$$

$$v_f^- = r_f^- i_f^- + p\lambda_f^- \tag{23}$$

$$v_{kd}^- = r_{kd}^- i_{kd}^- + p\lambda_{kd}^- \tag{24}$$

$$v_g^- = r_g^- i_g^- + p\lambda_g^- \tag{25}$$

$$v_{kq}^- = r_{kq}^- i_{kq}^- + p\lambda_{kq}^- \tag{26}$$

where $r_f^-, r_{kd}^-, r_g^-, r_{kq}^-$ = rotor resistances referring to stator; $v_f^-, v_{kd}^-, v_g^-, v_{kq}^-$ = rotor voltages referring to stator; and ω_r = electrical rotor angular velocity in rad/sec.

From Equation 15 to Equation 26, the equivalent circuit to represent complete electrical characteristics of synchronous machine including the voltage equations can be developed as shown in Figure 3. In these equivalent circuits, voltages as well as flux linkages appear. Therefore, flux linkages are shown in terms of their time derivatives.

2.5 Steady-state operation

In order to better understand the operational behavior of synchronous generator, the equations and phase diagram must be examined under steady-state and no load conditions derived from calculation of initial values. Balanced steady state condition is assumed with the rotor rotating at synchronous speed, $\omega_r = \omega_e$ and field excitation held constant. Under steady-state condition, stator and rotor fluxes are constant which means that all time derivative terms $p\lambda_d, p\lambda_q, p\lambda_f, p\lambda_{kd}, p\lambda_g, p\lambda_{kq}$ equal zero and occur only in the transient case. Also under steady-state, stator currents (i_d, i_q) and field current (i_f) are constant, and all damper currents $(i_{kd}^-, i_{kq}^- = 0)$ with neglecting the iron currents flow in rotor i_g^-.

3 SIMULATION OF SYNCHRONOUS GENERATOR

The winding equations of synchronous machine model derived in previous section can be implemented in one simulation block to represent a synchronous generator that uses; Terminal voltage V_m and Mechanical torque T_{mech} as inputs to this model, and the Measured Quantities as the outputs that are shown in Figure 4(a). The Output measured quantities signal can be connected with Bus Selector Block to select the output signals

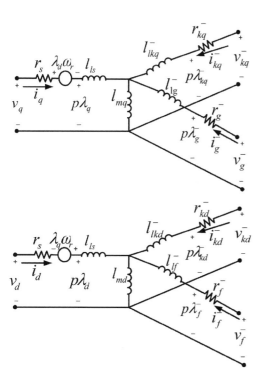

Figure 3.   Complete q- and d-axes equivalent circuits.

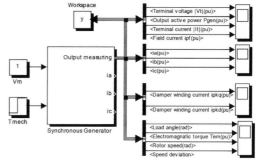

Figure 4.   (a) Synchronous Generator SIMULINK block subjected to step change in mechanical torque.

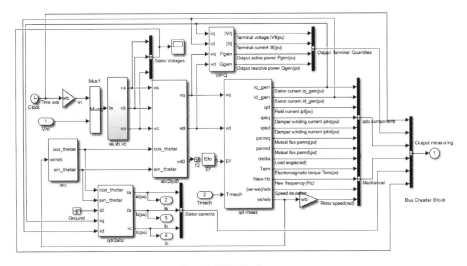

Figure 4.   (b) Inside Synchronous Generator SIMULINK block.

Table 1.   Synchronous generator manufacturer parameters.

S_{rated}	N_{rated}	x_{ls}	x_d	x'_d	x''_d	T'_{d0}	T''_{d0}	H
555 MVA	3600 $r.pm$	0.15 $p.u$	1.81 $p.u$	0.3 $p.u$	0.23 $p.u$	8 s	0.03 s	3.77 s
V_{rated}	$P.F$	r_s	x_q	x'_q	x''_q	T'_q0	T''_q0	D
24 KV	0.9	0.003 $p.u$	1.76 $p.u$	0.65 $p.u$	0.25 $p.u$	1.0 s	0.07 s	2 $p.u$

what has been wanted to measure i.e. terminal voltages, active power, terminal currents, field current and damper currents also to observe; load angle, electromagnetic torque rotor speed, and speed deviation. Bus Selector is also gives the ability of observing the dynamic behavior of these signals on Scope Block as shown in Figure 4(a).

The Synchronous Generator block in Figure 4(a) is further divided into other subsystem blocks which have been made to simulate the synchronous generator performance during sudden change in torque using MATLAB SIMULINK Program such as shown in Figure 4(b). These subsystems contain the blocks necessary to simulate the flux linkage dynamics as well as the algebraic equations relating flux linkages to currents and voltages, and the blocks implementing the qd0 transformation, and motion equation of torque. Table 1 shows the machine parameters data that was used in this paper (Kundur, P., 2004).

4   RESULT AND DISCUSSION

In this paper the performance of transient characteristics of synchronous generator has been examined by considering the response of applying a sudden step change in mechanical torque at the terminals with fixed voltage and frequency. This analysis gives the insight of understanding of physical concepts of all machine transient performance such as; terminal currents, field current, and damper currents, active power and all mechanical component.

The manufacturer parameters which in Table 1, can be easily inserted in the dialog box after pressing a double click on the main block of Figure 4(a) such as shown in Figure 5(a). Also a new method had been developed by inserting the initial value equations of steady-state operation, which were written as the M-file script inside this block using Mask Editor procedure known as "initialization" as shown in Figure 5(b). And in order to apply the step change in mechanical torque for example at t = 0 sec, just the value Step Input Block has been changed from 0 p.u to 1.

The model shown in Figure 4(a) can be run initially from no load at rated terminal voltage V_m = 1.0 p.u, and initial real and reactive power P = 0, Q = 0, respectively, with constant speed i.e. no governor action T_{mech} = 0 p.u. The assumption is made

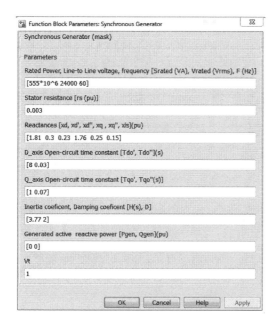

Figure 5(a).   Manufacturer parameters in dialog box.

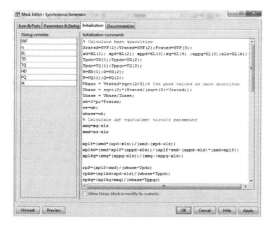

Figure 5(b).   M-file scripts inside block.

that the synchronous generator initially operate with no load i.e zero input torque with the excitation current held constant at the value that gives rated open circuit terminal voltage at synchronous speed as shown in Figure 6. Also as shown in Figure 6(b) at zero input torque, no generated power at terminal because no terminal currents appear at no-load as shown in Figure 7. As we know at steady state condition which $\omega_r = \omega_e$ there is no transients eddy currents will appear on the surface of the rotor damper winding, that means the damper currents i_{kq} and, i_{kd} equal zero in simulation curve as shown in Figure 8.

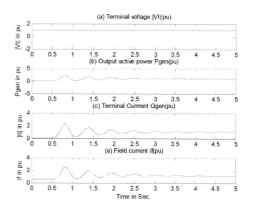

Figure 6.   Dynamic Behavior of; terminal voltage, real power generated, terminal current and field current during a sudden change in torque.

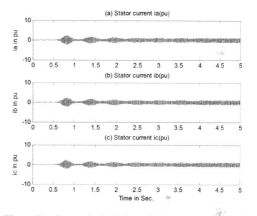

Figure 7.   Dynamic Behavior of; phases currents a, b, and c during a sudden change in torque.

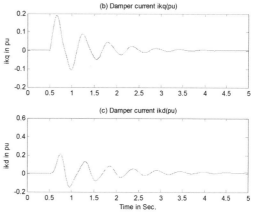

Figure 8.   Dynamic Behavior of; Q- and D-axis damper currents during a sudden change in torque.

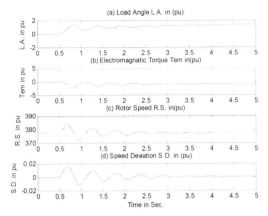

Figure 9. Dynamic Behavior of; load angle (delta), electromechanical torque (T_{em}), rotor speed (ω_r), and speed deviation during a sudden change in torque.

If the synchronous generator is subjected to a sudden increase in input torque from zero p.u (no torque) to 1.0 p.u (full torque), the rotor speed begins to increase immediately because of electrical torque at that time is still equal zero and that will causes the load angle to increase in accordance to Figure 9(a–c). Then the rotor speeds up until the accelerating torque on the rotor reach zero, this is acts in an approximately from 377 to 385 electrical radians per second. Also with accelerating torque is zero, the rotor is running above synchronous speed, hence load angle δ will continue to increases, and the electrical torque will increase in negative direction as shown in Figure 9(b).

The increase in electrical torque T_{em}, which is an increase in the output power, causes the rotor to speed down toward synchronous speed. However, when rotor speed is reached the synchronous speed, the speed deviation will become zero at a new steady state operating point as shown in Figure 9(d). Also the value of load angle δ has become larger than necessary to satisfy the input torque. Furthermore after the step change occurs, the damper currents for "direct and quadrature-axis" have been appeared in a transient waveform and continue until to reach the new steady state operating point at full load.

5 CONCLUSION

It is very useful to simulate, observe, and analyze the dynamic performance of the synchronous generator during normal operation or during disturbances, hence in this paper the dynamic behavior of synchronous generator was examined and investigated using MATLAB SIMULINK during a sudden "step change" in mechanical torque from no load to full load. So in general the terminal currents and power was increased transiently from zero to one per unit i.e. to full load power. Also the damper currents appear clearly in transient curve and then vanished to zero after sudden change in torque.

ACKNOWLEDGEMENT

Foremost and first of all, great thankfulness to Almighty God for granting me strength and ability to finish this paper. I would like to thank my parents, wife, and colleagues for their constant support and encouragement. I am indebted to my father, for his constant guidance, encouragement, strong support, and indispensable role in the success of this paper.

REFERENCES

Anderson, P.M. & Fouad, A.A., 1977, Power System Control and Stability Volume 1, Ephrata Since Press, pp. 83–206.
Hadi Saadat, P., 1993, Computational Aids in Control System Using MATLAB, New York, McGraw-Hill, Inc,.
Hadi Saadat, P., 2002, Power System Analysis, New York, McGraw-Hill Companies, Inc, pp. 314–347.
Kalsi, S. & Lipo, T.A., 1977, A Model Approach to the Transient Analysis of Synchronous Machines, New York, Hemisphere Publishing Corporation.
Kundur, P., 2004, Power System Stability and Control, New York, McGraw-Hill, In, pp. 45–198.
Louie, Kwok-Wai, 1995, Phase-Doman Synchronous Generator Model for Transient Simulation, Columbia, M.Sc. thesis, University of British.
Nanda, J. & Kothar M.L., Emerging Trends in Power System, Allied Publishers Limited.
Ong, Chee-Mun, 1998, Dynamic Simulation Of Electric Machinery Using Matlab Simulink, New Jersey 07458, Prentice-Hall, Inc, pp. 259–340.
Padiyar, K.R., Ltd 2005, HVDC Power Transmission System, New Age International (P).
Rebizant, Waldemar & Terzija Vlodimir, 2003, Asynchronous Generator Behavior after a Sudden Load Rejection, Bologna, 7.
The Math Works, 1990–1999, Using Simulink, Inc, United State.
The Math Works, 1999–2004, Learning Simulink, Inc, United State.

Advances in Power and Energy Engineering – Sun (Ed.)
© 2016 Taylor & Francis Group, London, ISBN 978-1-138-02846-3

On-load capacity regulation design of distribution transformer

L.C. Ma
Henan Mechanical and Electrical Engineering College, Xinxiang, China

F.Y. Chang
North China Electric Power University, Baoding, China

ABSTRACT: In order to reduce the loss of transformer in single-shift system factories and rural areas, a control project of on-load capacity regulation distribution transformer is designed, in which thyristors are used as a capacity regulation switch. The instantaneous overvoltage and overcurrent are simulated with MATLAB, and the results indicate that the project can effectively restrain instantaneous overvoltage and overcurrent and can remarkably save energy, so it has a practical application value.

1 INTRODUCTION

Comparing with the older products, the new S_9 and S_{11} transformers obviously decrease the load loss and the no-load loss. But in some cases, such as agricultural transformer, there are electricity peak season and electricity slack season. The peak season occupies about three months in a year and operates at full load. The slack season occupies the rest of the time and operates at a load factor of 20 percent. Besides, this situation also applies for 8-hour factory with a nearly full-load operation at work time and a low-load factor at spare time, respectively. Because the transformer underloads at a longer cycle time, decreasing underloaded depletion is of vital importance to save energy for this kind of transformer.

The generation of capacity regulating transformer is based on the above circumstances. Overloading operation makes the transformer work at full capacity, while underloading operation turns down the capacity of and thus decreasing the operational deterioration.

Current capacity regulating methods of the transformer are star-delta conversion and winding series-parallel connection (Wang, J.L. & Sheng, W.X. 2009 Yu, 1988). They all operate under the no-load situation and cannot achieve automation control, which limits the utilization and development of capacity regulating transformer. Therefore, accomplishing the on-load capacity regulating of transformer shows a realistic significance to the popularization, application and energy conservation of capacity regulating transformer.

2 THE CONTROL OF ON-LOAD CAPACITY REGULATING DISTRIBUTION TRANSFORMER

Principle of capacity regulating: The main core flux is reduced in underloading that lowers iron-loss and can effectively decrease no-load loss (Yao, Z.S. & Yao, L. 2006).

Method of capacity regulating: Each phase of the high and low voltages is equally divided into 11′, 22′, 33′, and 44′. The ratio of the number of turns is 3.5:1.5:1.5:3.5. Through changing the series-parallel system of coil 22′ and coil 33′ from part 2 and part 3, we can achieve the capacity regulating transformer. The controlling method of capacity regulating transformer is shown in Figure 1.

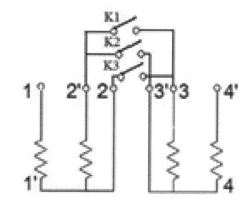

Figure 1. Capacity regulation capacity mode of the transformer.

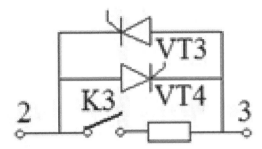

Figure 2.   Principle of regulation switch.

To illustrate any phase, when the load is heavy and the transformer adopts high capacity, coil 22′ and coil 33′ turn to the parallel connection from the series connection. Besides, K1 switches off while K2 and K3 switch on. When the load is light and the transformer adopts low capacity, coil 22′ and coil 33′ turn to the series connection from the parallel connection. Then, K2 and K3 switch off while K1 switches on.

Power switch of capacity regulating: As shown in Figure 2, thyristors VT3 and VT4 are adopted as control switches of instantaneous motion and contactor K3 as the stable switch. On the high-voltage side, each phase covers six thyristors and three phases reach 18 thyristors. On the low-voltage side, each phase covers six bidirectional thyristors and three phases reach 9. On both high- and low-voltage sides, three AC contactors are used as stable switches reciprocally.

3   THE ANALYSIS AND RESTRAINING OF INSTANTANEOUS OVERVOLTAGE AND OVERCURRENT

3.1   Simulation model

In order to analyze the characteristics of the instantaneous overvoltage and overcurrent of each capacity regulating switch, we use MATLAB to establish the Simulink simulation model, as shown in Figure 3.

1. Transformer model
   We adopt one phase of the multicoil transformer from the simpower system if $S_N = 250$ KVA, U_1/ $U_2 = 5774/220$ V, and high- and low-voltage winding are separated equally into four parts, which are coil 1(5), 2(6), 3(7), and 4(8) with the ratio of number of turns being 3.5:1.5:1.5:3.5, then the minimum value of ohmmeter in each part are, respectively, 0.00256, 0.0011, 0.0011 and 0.00256, and the minimum value of leakage resistance standard in each part are, respectively,

0.00728, 0.00312, 0.00312, and 0.00728. The minimum value of excitation resistance standard is, respectively, 50 and 50.

2. Switch model
   We adopt the ideal switch from the simpower system. If the internal resistance is 0.001 Ω and the initial state is 0 or 1, and then the buffer resistance and the buffer capacitor are 10 Ω and 1 μF, respectively. By changing three switches, the high- and low-voltage sides of the transformer will accomplish the series-parallel transition between 2 and 3 as well as 6 and 7.

3. Switches controlling module
   We use four switches controlling modules to control the switchover of six switches. By changing the parameters of the switches controlling module, we can change the timing of capacity regulation.

4. Battery module
   We adopt the alternating current power of 50 HZ with an amplification of 8164 V.

3.2   Results and analysis

1. As for certain situations that the series connection turns into the parallel connection and simultaneous motions during switches ($t_1 = t_2 = 0.225$ s, $\varphi_{u1} = 0$) in high- and low-voltage sides when the load is heavy as well as that the parallel connection turns into the series connection and simultaneous motion during switches ($t_1 = t_2 = 0.225$ s, $\varphi_{u1} = \pi/4$) in high- and low-voltage sides when the load is light, the conditions are simulated and the results are shown in Figures 4 and 5. The result indicates that at any time of simultaneous motions in high- and low-voltage sides, there is an overcurrent that is ten times larger than the general one if it switches over once. If it switches off twice, there is a serious overvoltage that is ten times or even hundred times larger than the normal one.

2. As for certain situations that the series connection turns into the parallel connection and asynchronous motions during switches ($t_1 = 0.2$ s, $t_2 = 0.20122$ s, $\varphi_{u1} = 0$, $\varphi_{i2} = 0$) in high- and low-voltage sides when the load is heavy as well as that the parallel connection turns into the series connection and asynchronous motions during switches ($t_1 = 0.2$ s, $t_2 = 0.2016$ s, $\varphi_{u1} = 0$, $\varphi_{i2} = 0$) in high- and low-voltage sides when the load is light, the conditions are simulated and the results are shown in Figures 6 and 7.

The result indicates that by controlling the motion movement of switches once and twice, we can effectively restrain overvoltage and overcurrent. When choosing the first motion movement of switches t_1 and making $\varphi_{u1} = 0$, we can effectively

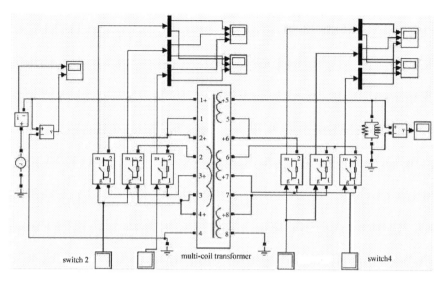

Figure 3.　Simulation model of on-load capacity regulation distribution transformer.

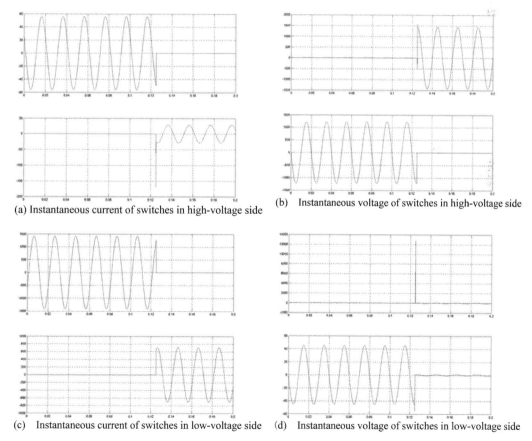

(a)　Instantaneous current of switches in high-voltage side

(b)　Instantaneous voltage of switches in high-voltage side

(c)　Instantaneous current of switches in low-voltage side

(d)　Instantaneous voltage of switches in low-voltage side

Figure 4.　Emulation wave during winding connection conversion from series to parallel ($t_1 = t_2 = 0.225$ s, $\varphi_{u1} = \pi/4$).

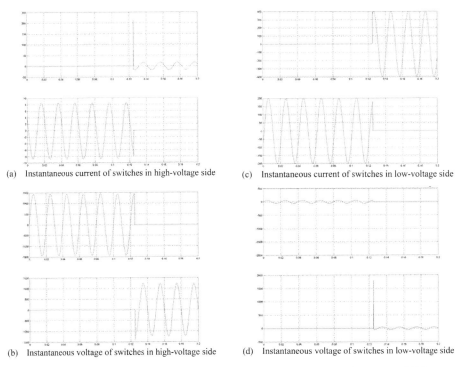

(a) Instantaneous current of switches in high-voltage side   (c) Instantaneous current of switches in low-voltage side

(b) Instantaneous voltage of switches in high-voltage side   (d) Instantaneous voltage of switches in low-voltage side

Figure 5.   Emulation wave during winding connection conversion from parallel to series ($t_1 = t_2 = 0.225$ s, $\varphi_{u1} = \pi/4$).

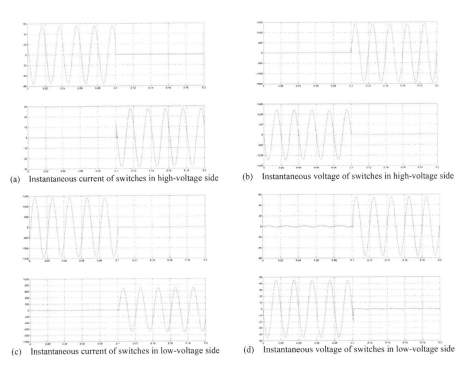

(a) Instantaneous current of switches in high-voltage side   (b) Instantaneous voltage of switches in high-voltage side

(c) Instantaneous current of switches in low-voltage side   (d) Instantaneous voltage of switches in low-voltage side

Figure 6.   Emulation wave during winding connection conversion from series to parallel ($t_1 = 0.2$ s, $t_2 = 0.20122$ s, $\varphi_{u1} = \varphi_{i2} = 0$).

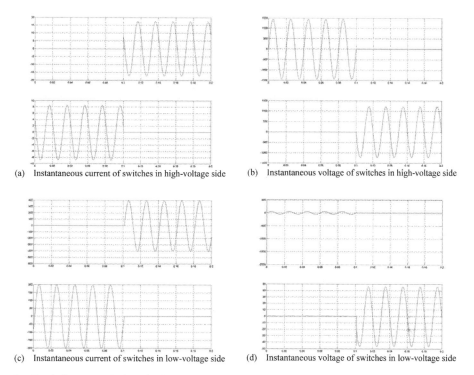

(a) Instantaneous current of switches in high-voltage side (b) Instantaneous voltage of switches in high-voltage side

(c) Instantaneous current of switches in low-voltage side (d) Instantaneous voltage of switches in low-voltage side

Figure 7. Emulation wave during winding connection conversion from parallel to series ($t_1 = 0.2$ s, $t_2 = 0.2016$ s, $\varphi_{u1} = \varphi_{i2} = 0$).

Table 1. Loss analysis and contrast with different series-parallel connection winding proportions under different loads for 315 KVA transformer.

Series-parallel winding proportion	Load factor	High-voltage winding number	Low-voltage winding number	Flux density (T)	Iron loss per unit weight (W/kg)	No-load loss (W)	Load loss (W)	Overall loss (W)	Energy conservation (W)
Parallel connection	100%	761	29	1.405	0.72	650	3420	4070	–
Parallel connection	20%	761	29	1.405	0.72	650	136.8	786.8	–
Series connection		1522	58	0.703	0.0527	49	547.2	596.2	190.6
3.5:1.5:1.5:3.5		1268	48	0.849	0.1841	166	342	508	278.8
Parallel connection	15%	761	29	1.405	0.72	650	77	727	–
Series connection		1522	58	0.703	0.0527	49	308	357	370
3.5:1.5:1.5:3.5		1268	48	0.849	0.1841	166	192	358	369

eliminate overcurrent once. The overcurrent and voltage peak endured by capacity regulating switches in the high-voltage side are lower than 20 A and 1500 V, respectively. When choose the second motion movement of switches t_2 and make $\varphi_{i2} = 0$, we can effectively eliminate overvoltage twice. The overcurrent and voltage peak endured by capacity regulating switches in the low-voltage side are lower than 400 A and 50 V, respectively. We can use thyristors as transient capacity regulating switches (Wang, Y.L. 2006).

4 THE ANALYSIS OF ENERGY CONSERVATION OF ON-LOAD CAPACITY REGULATING TRANSFORMER

To illustrate the 315 KVA transformers whose connection group is yyn0, the voltage ratio is set to range from 10 KV to 0.4 KV. If the primary winding and secondary winding of each phase is equally distributed into four parts, we adopt the winding series and parallel connections.

Different proportions of the series-parallel connection from the high- and low-voltage winding have different flux densities, iron loss per unit weight (Zhang 2007, Yin 2003) and depletion, as indicated in Table 1.

From the above table, we can see that to illustrate a junior transformer 315 KVA, 15 percent of the power depletion of the rated load decreases by about 370 W, and the power depletion of the rated load under 15 percent decreases more. According to the operational time of the rated load under 15 percent, which is 9 months, the overall year energy-conservation is 2431 KWh, which is about 2000 yuan. Therefore, we can save 20000 yuan in 10 years. The equipment costs 6000~7000 yuan, so it will only take 3~4 years to gain the cost.

5 CONCLUSION

This paper proposes the implementation of transformer on-load capacity regulating plan controlled by thyristors. It realizes automation control by accomplishing dynamic capacity regulating under on-load conditions, avoiding troubles of no-load artificial capacity regulating and saving labors and materials. This plan can effectively suppress instantaneous overvoltage and overcurrent from capacity regulating switches and is well dependent. Besides, the investment cost is low and the effects of energy conservation are significant. It possesses a practical application and promotion value.

REFERENCES

Wang, J.L. & Sheng, W.X. 2009. The simulation analysis of capacity regulating transformer. *Transformer* 46(7):19–23.
Wang, Y.L. 2006. *Power and Electron Technology*. Beijing: Electronic Industry Press.
Yao, Z.S. & Yao, L. 2006. *The Structure, Principle and Application of New Type Transformer*. Beijing: China Machine Press.
Yin, K.N. 2003. *The Design Principles of Transformer*. Beijing: China Electric Power Press.
Zhang, Z.B. 2007. *The Principles and Application of Transformer*. Beijing: Chemical Industry Press.

Advances in Power and Energy Engineering – Sun (Ed.)
© 2016 Taylor & Francis Group, London, ISBN 978-1-138-02846-3

Numerical Study of acoustic characteristics of gas-liquid coaxial injectors

H.H. An & W.S. Nie
Equipment Academy, Beijing, P.R. China

ABSTRACT: The acoustic characteristic of the liquid rocket combustor with a single injector is studied by adopting a method of linear acoustic analysis. The injector is simplified to a quarter-wave resonator and a half-wave resonator, respectively. Acoustic-damping effect of the injector on the first tangential mode of the chamber is evaluated under hot condition. It is found that the length and the inlet geometry of the injector can play a significant role in the acoustic damping capacity while they are chosen properly. While the inlet jet is choked, the optimum length of the injector to maximize damping capacity is near quarter of a full wavelength of the first longitudinal mode, which travels in the injector with the acoustic frequency intended for damping. On the contrary, this optimum length is half of a full wavelength of the first longitudinal mode while the inlet jet is unchoked.

1 INTRODUCTION

The high-frequency instable combustion of the liquid-propellant rocket engine is the result of interaction between the release of combustion heat and the acoustic mode in the combustion chamber (Harrje, D.T. & Reardon, F.H., 1972). It might cause the dramatic increase of partial temperature, damage the cooling layer of the inner wall of the combustion chamber, burn through the faceplate and the wall surface, and trigger unpredictable serious accidents. Instable combustion is extensive in solid and liquid rocket engines, combustion motors, gas turbines and the other heat engines. However, it is especially serious in liquid rocket engines. Although a large number of scholars have studied the instable generation mechanism, the effective methods for predicting control of the instable combustion have not been gained. Currently, baffles, acoustic cavities, acoustic liners and other supplementary devices are used to dissipate or absorb the acoustic energy to damp the instability. Research findings suggest that these devices have good damping capacity (Kim, S.K. et al., 2004) (Nie W.S. & Zhuang F.C., 1998). However, the installation of these devices increases the manufacturing difficulty, even impose a negative influence on the performance of the engine.

The acoustic damping capacity of the gas-liquid co-axial injector of staged combustion rocket engine has also drawn equal attention. Literature introduced an acoustic test for cold condition in the high-frequency instable performance research (Vigor Y. & William E., 1995)

(Haksoon K. & Chae H.S., 2006). Zhang M.Z. employed a simulated combustion chamber to conduct an experimental research on the acoustic characteristics of the gas-liquid co-axial injector (Zhang M.Z. & Zhang Z.T., 2007). Li L.F. used a simulated combustion chamber to study the influence of the gas-liquid co-axial injector's structure on the margin of the high-frequency instability combustion (Wang F., Li L.F. & Zhang G.T., 2012). Zhou J. adopted an oxy-hydrogen co-axial injector in an acoustic test, finding out that sharp squeal within certain scope of working parameters was appeared (Zhou J., & Hu X.P., 1996). Soo H.K. studied the effect of the injector's internal structure on the acoustic damping (Soo H.K., et al., 2014). It can be seen that the previous research work mainly employed a sub-scale combustion chamber for acoustic analysis. The acoustic characteristic of the full-scale engine's injector has not been studied nearly. This paper mainly studies an acoustic simplified method of the gas-liquid co-axial injector used in the full-scale liquid-oxygen/kerosene engine. Acoustic damping capacity under different working conditions was studied systematically.

2 ACOUSTIC CAVITY INJECTOR

The acoustic resonator is a straight pipeline which can influence the acoustic oscillation. Various types of the resonator are shown in Figure 1 (a–g), which can be divided into closed tubes and open tubes. The former are unilaterally end-closed (a-b), while the latter are bilaterally end-open (c-d). According to the

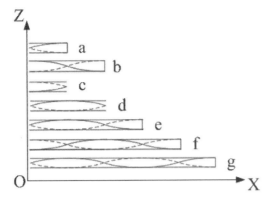

Figure 1. Acoustic resonator.

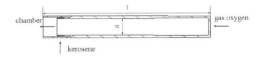

Figure 2. Gas liquid coaxial injector.

tube length and the boundary condition, the resonator can be divided into four typical kinds, namely quarter-wave closed tube (a), half-wave open tube (d), quarter-wave open tube (c), and half-wave closed tube (b). The former two kinds have the largest damping capability, while the damping capability of the latter two kinds is the smallest.

The staged combustion rocket engine adopts the pre-combustion chamber and the regenerative cooling technique. The injector structure of the main combustion chamber is shown in Figure 2. Gaseous oxygen flows through the inner passage of the injector and mixes with the liquid fuel injected through several peripheral holes. In the pre-combustion chamber, most oxygen and a little fuel will form high-temperature mixed gas injected into the main combustion chamber. On the other hand, the temperature of the liquid fuel will reach its critical temperature after going through the cooling channel and mixing with the co-axial high-temperature gaseous oxygen. Under these conditions, the physical properties of the oxidizer and the fuel being injected approach the properties of the dense gas. The whole injector channel is filled with gas. The shape changes of the injector are neglected because they are smaller than the feature size of the injector. Thus, the injector is simplified into a cylinder with equal internal diameter, which will not impose an essential influence on the injector's acoustic characteristics. In this way, the injector can be regarded as a resonator, also known as an acoustic cavity injector that might damp the acoustic oscillation of the combustion chamber. This can also avoid the installation of extra baffles, acoustic cavities and other auxiliary devices.

At the same time, both ends of the injector are connected by the oxygen manifold and the combustion chamber. The high-temperature oxygen in the oxygen manifold is injected through the right end of the injector. The flow injecting speed of the

end can be divided into the sound speed and the subsonic speed. When the jet speed is equal to the sound speed, the inlet is choked, which can be regarded as acoustic closed end. When the jet speed is less than the sound speed, it can be regarded as acoustic open-end. Both acoustic closed-end and open-end can conduct a total reflection of the acoustic wave. Based on the above analysis, the gas-liquid co-axial injector can be simplified into a quarter-wave resonator and a half-wave resonator. There are many injectors on the faceplate of a liquid rocket engine. This paper mainly studies the acoustic characteristics of the chamber with a single injector. The research results supply significant guiding data for the injector design.

3 NUMERICAL METHODS

3.1 *Governing equation*

For a homogeneous and non-dissipative medium, the conventional wave equation is

$$\nabla^2 p' - \frac{1}{c^2}\frac{\partial^2 p'}{\partial t^2} = 0 \qquad (1)$$

where p' = pressure fluctuation; c = the sound speed in the medium; and t = time.

$$\nabla^2 = \frac{\partial^2}{\partial x^2} + \frac{\partial^2}{\partial y^2} + \frac{\partial^2}{\partial z^2} \qquad (2)$$

The pressure fluctuation is expressed as follows:

$$p' = p(x, y, z) \cdot e^{jwt} \qquad (3)$$

Then the helmholtz equation is expressed in the form:

$$\nabla^2 p(x, y, z) - k^2 p(x, y, z) = 0 \qquad (4)$$

where $w = 2\pi f$ = the angular frequency; $k = w/c = 2\pi f/c$ = wavenumber.

3.2 *Geometric models*

The engine adopted in this paper is shown in Figure 3. The diameter of the combustion chamber

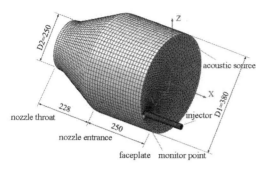

Figure 3. Combustor model.

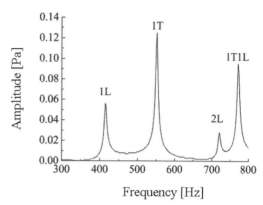

Figure 4. Acoustic responses without the injector.

and the throat part is 380 mm and 190 mm, respectively. The length from the injection panel to the entrance of the nozzle is 250 mm and to the throat is 478 mm, respectively. The half-angle of the convergence part is 30°. The number of computational grids is 196896. Through grid-dependency check of the numerical results, this grid system was suitable.

3.3 Boundary conditions

In actual reactive flows, the sonic condition is formed at the nozzle throat. Thus the throat can be considered to be an acoustically closed end which reflects all the incident waves. To consider the boundary condition at wall, boundary absorption coefficient is defined as 0.005, which resulted from the experiment (Kim, S.K. et al., 2004). The injector is set near the wall surface. As shown in Figure 3, the sound field is excited by a monopole sound source. The sound source position is opposite to that of the injector, and is close to the injection faceplate near the chamber wall. The monitoring point is located near the injector at the opposite side to the sound source. The influence of the injector on the acoustic characteristics can be evaluated by monitoring the acoustic amplitude changes near the injector.

4 RESULTS AND ANALYSIS

First of all, the calculation employs a quiescent air as the medium, with the temperature of 298 K, the density of 1.2 kg/m³ and the sound speed of 343 m/s. Under the condition of no injector, different sound sources are adopted for the sound response calculation. The acoustic modal shape of the acoustic field in the combustion chamber remains the same. Through the estimation of equations 5–7 and the spatial distribution of acoustic amplitude, the former two peak values are defined as the first order longitudinal and the first order tangential modal, as shown in Figure 4.

The calculation results coincide with the test results in literature 5. It shows that the calculation model built up is reasonable. At the same time, it is noted that the amplitude of the first order tangential modal is the largest, which coincides with the first order tangential oscillation being the most destructive.

The research into two kinds of acoustic cavity injector with different lengths and diameters under hot condition has been conducted. According to the engine thermodynamic parameters obtained by calculating flow inside the combustor, density and sound speed in the combustor are 12.299 kg/m³ and 1255.5 m/s, respectively. And density and sound speed in the injector are 103.3 kg/m³ and 471.19055 m/s, respectively.

$$f = \frac{c}{2}\sqrt{\left(\frac{q}{L}\right)^2 + \left(\frac{2\alpha}{D1}\right)^2} \qquad (5)$$

$$f_{1L} = f_{1T} = \frac{c_i}{2(l + \Delta l)} \qquad (6)$$

$$l = \frac{c_i}{2f} - \Delta l \qquad (7)$$

where c = the sound speed in the combustor; c_i = the sound speed in the injector; $D1$ = chamber diameter; L = chamber length; q = the counter of the longitudinal mode; f_{1L} = the frequency of the first longitudinal mode; f_{1T} = the frequency of the first tangential mode; l = injector length; and Δl = correction value (Chen X.H. et al., 2014).

Acoustic behavior in chamber with a single injector is investigated for hot condition over a wide range of injector length. The injector diameter is selected to be 21 mm. To evaluate acoustic damping capacity of the injector quantitatively, a parameter of damping coefficient as shown in

Figure 5, is introduced and evaluated by bandwidth method in the form (Dravnovsky, M.L., 2007):

$$\eta = \frac{f_2 - f_1}{f_{peak}} \qquad (8)$$

where f_{peak} = the frequency at which the peak response appears; f_1 and f_2 = the frequencies at which the pressure amplitude corresponds to $P/\sqrt{2}$ with $f_2 > f_1$.

4.1 Influence of the length of the close-end acoustic cavity injector on the acoustic characteristics

Through the acoustic response calculation, the minimum of acoustic amplitudes at the monitoring point is coincided with the maximum of acoustic damping coefficient. The coordinate of the corresponding points is 58 mm, 174 mm and 290 mm, respectively, as shown in Figures 6 and 7, which is corresponding to a quarter, three quarters and five quarters of the wavelength of the first order longitudinal modal in the close-end acoustic cavity injector. The damping capability reaches the maximum when the length of the close-end acoustic cavity injector is integer multiples of a quarter wavelength. It can be seen that with the increase of the counter of the mode, the damping capability decreases gradually. In order to maintain the maximum damping capability, the close-end acoustic cavity injector length should be the lowest counter, namely a quarter wavelength.

4.2 Influence of the area of the close-end acoustic cavity injector on the acoustic characteristics

The acoustic response of the five kinds of closed-end acoustic cavity injectors has been calculated, of which diameter is 0, 7 mm, 14 mm, 21 mm,

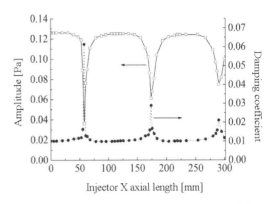

Figure 6. Acoustic amplitudes and damping coefficient as a function of injector length.

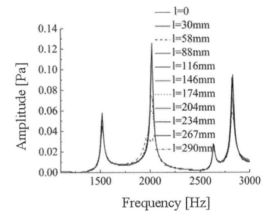

Figure 7. Acoustic responses with different injector length.

28 mm and 35 mm, respectively ("0" stands for the combustion chamber without injector). Results show that only the amplitude of the first order tangential modal changes, the others remain the same, as shown in Figure 8. The resonant frequency of all modals is not changed when the diameter is 0, 7 mm, 14 mm and 21 mm; while the acoustic pressure amplitude shows divergences when the diameter is 28 mm and 35 mm. In other words, the first-order tangential modal can be divided into two oscillation modals. This suggests that the diameter can affect the damping capacity of the injector. When the diameter is relatively small, the acoustic modal frequency is not changed. However, when the diameter exceeds certain range, the corresponding acoustic modal will be divided into two oscillation modals. The frequency will be located on two sides of the original modal. Because the acoustic pressure amplitude shows divergences when the

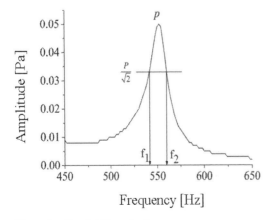

Figure 5. Bandwidth method.

208

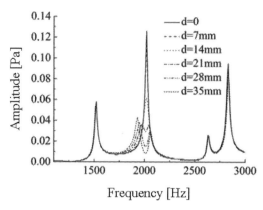

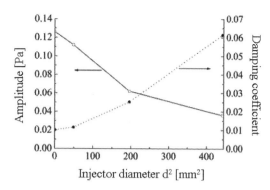

Figure 8. Acoustic amplitudes with different injector diameter.

Figure 9. Acoustic amplitudes and damping coefficient as a function of injector diameter.

diameter is 28 mm and 35 mm, it is unsuitable to evaluate the damping capacity by bandwidth method. Therefore, only four diameter conditions, namely 0, 7 mm, 14 mm and 21 mm, are adopted for analysis. From Figure 9, we can see that with the increase of the diameter, the acoustic pressure amplitude decreases, but the acoustic damping coefficient increases. The amplitude and the damping coefficient show a linear relationship with the square of the diameter.

4.3 Influence of the length of the open-end acoustic cavity injector on the acoustic characteristics

The open-end acoustic cavity injector is shown in Figure 1. The entrance end is in the oxygen cavity (d), which is an acoustic soft boundary. In order to simplify the boundary conditions in the entrance, the injector is extended by a quarter wavelength at the entrance as shown in Figure 1(e). The new entrance end is defined as the acoustic closed end. Under this condition, the wave node of the acoustic pressure is still at the original entrance end (e, f and g). In other words, the acoustic characteristics of the acoustic cavity injector remain the same. Here, this approach is defined as the "quarter-wave method." The longitudinal acoustic oscillation within the injector before and after the extension is shown in Figure 1(d and e). Through the acoustic response calculation, the minimum of acoustic pressure amplitudes at the monitoring point is coincide with the maximum of acoustic damping coefficient. The coordinate of the corresponding points is 116 mm and 234 mm, respectively, as shown in Figure 10, which is corresponding to a half wavelength and one wavelength of the first-order longitudinal modal in the open-end acoustic cavity injector. The damping

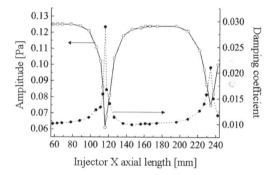

Figure 10. Acoustic amplitudes and damping coefficient as a function of injector length.

capability reaches the maximum when the length of the open-end acoustic cavity injector is integer multiples of a half wavelength. It can be seen that with the increase of the counter of the mode, the damping capability decreases gradually. In order to maintain the maximum damping capability, the open-end acoustic cavity injector length should be the lowest counter, namely a half wavelength.

4.4 Influence of the area of the open-end acoustic cavity injector on the acoustic characteristics

The acoustic response of the five kinds of open-end acoustic cavity injectors has been calculated, of which diameter is 0, 7 mm, 14 mm, 21 mm, 28 mm and 35 mm, respectively. Results show that the amplitude of the first order tangential modal changes, while the others remain the same, as shown in Figure 11. The resonant frequency of all modals is not changed when the diameter is 0, 7 mm, 14 mm and 21 mm; while the acoustic pressure amplitude shows divergences when the diameter is 28 mm and 35 mm. This suggests that the diameter can

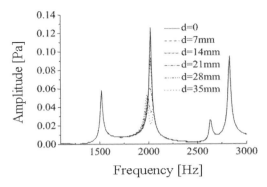

Figure 11. Acoustic amplitudes with different injector diameter.

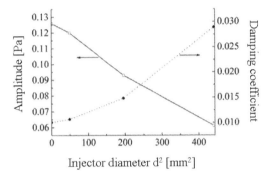

Figure 12. Acoustic amplitudes and damping coefficient as a function of injector diameter.

affect the damping capacity of the injector. When the diameter is relatively small, the acoustic modal frequency is not changed. However, when the diameter exceeds certain range, the corresponding acoustic modal will be divided into two oscillation modals. The frequency will be located on two sides of the original modal. Because the acoustic pressure amplitude shows divergences when the diameter is 28 mm and 35 mm, it is unsuitable to evaluate the damping capacity by bandwidth method. Therefore, only four diameter conditions, namely 0, 7 mm, 14 mm and 21 mm, are adopted for analysis. From Figure 12, we can see that with the increase of the diameter, the acoustic pressure amplitude decreases, but the acoustic damping coefficient increases. The amplitude and the damping coefficient show a linear relationship with the square of the diameter.

5 CONCLUSIONS

Acoustic behavior of the gas-liquid injector of the liquid rocket engine is investigated. Different acoustic characteristics of the injector under different entrance injection conditions are analyzed. Results suggest that the acoustic stability of the engine can be improved through reasonable injector design. When the entrance injection of the injector is choked, the optimal length of the injector should be a quarter wavelength of the first order longitudinal modal within the injector so as to achieve the maximum damping capability. Otherwise, the optimal length should be a half wavelength of the first order longitudinal modal to ensure the damping capability to be maximal. The damping capability of two kinds of the injector can increase along with the increase of the injector diameter. Based on the research findings in this paper, apart from considering the atomization performance of injector, designers should also pay attention to its acoustic damping performance.

REFERENCES

Chen X.H., Tian X.H. & SU L.Y., 2014. Theory of spacecraft propulsion [M]. Beijing: National Defense Industry Press.

Dravnovsky M.L., 2007. Combustion Instabilities in Liquid Rocket Engines: Testing and Development Practices in Russia, edited by V. Yang, Culick F.E.C. & Talley D.G., Vol. 221, Progress in Aeronautics and Astronautics, Washington, D.C.

Harrje D.T., Reardon F.H., 1972. Liquid Propellant Rocket Combustion Instability. Washington, D.C., NASA SP-194.

Haksoon K., Chae H.S., 2006. Experimental study of the role of gas-liquid scheme injector as an acoustic resonator in a combustion chamber. Journal of Mechanical Science and Technology, 6(20), 896–904.

Kim S.K., Kim H.J., Seol W.S. & Sohn C.H. July 2004. Acoustic stability analysis of liquid propellant rocket combustion chambers, AIAA Paper 2004-4142.

Nie W.S., Zhuang F.C., 1998. The study of characteristics of resonators used in liquid rocket engines combustion stabilities suppression. Journal of National University of Defense Technology. 20(2), 12–16.

Soo H.K., Young J.K., Chae H.S., 2014. Acoustic damping of gas-liquid scheme injectors with a recess in a subscale combustor. Journal of Mechanical Science and Technology. 28(9), 3813–3823.

Vigor Y., William E., 1995. Liquid rocket engine combustion instability. Volume 169, Progress in Aeronautics and Astronautics. The American institute of Aeronautics and Astronautics.

Wang F., Li L.F., Zhang, G.T., 2012. Experimental study on high frequency combustion instability with coaxial injector of staged combustion LOX/kerosene rocket engine. Journal of Astronautics, 2(33), 260–264.

Zhang M.Z., Zhang Z.T., 2007. Simulation test and application of chamber acoustic characteristics. Experimental Technology and Management, 24(8), 39–42.

Zhou J., Hu X.P., 1996. An experimental study on acoustic characteristics of gas-liquid coaxial injector of liquid rocket engine[J]. Journal of Propulsion Technology, 17(4), 37–41.

Advances in Power and Energy Engineering – Sun (Ed.)
© 2016 Taylor & Francis Group, London, ISBN 978-1-138-02846-3

Low Voltage Ride Through characteristics of large-scale grid-connected Photovoltaic power stations

W. Li, F. Wu & W.Y. Kong
College of Energy and Electrical Engineering, Hohai University, Nanjing, Jiangsu, China

Z.J. Meng
State Grid Electric Power Research Institute, Nanjing, Jiangsu, China

ABSTRACT: Recently, the capacity of Photovoltaic (PV) power stations integrated into the power grid has been increased significantly, whereas, the Low Voltage Ride Through (LVRT) capability of large-scale PV power station has not been fully investigated. Hence, a large-scale PV power station with LVRT model is built in this paper, which consists of three parts: the PV generation system, electrical control model and converter model. To address the voltage depression problem, the cap and ramp rate limit control that eliminate the current impact is proposed, and the reactive control is also considered. From the results of the electromechanical simulation on the provincial power grid PV systems, the active power recovery and the dynamic reactive power support of the proposed model have been discussed and analyzed.

1 INTRODUCTION

Recently, the power generated from the renewable energy sources has been increasing continuously due to the energy crisis and environment pollution concerns. The PV system with enormous and promising potentials is expected to be one of the major renewable energy conversion systems which can be utilized. Apart from small residential applications, the PV plants with a capacity of megawatts have been connected to the power network with high and medium voltage level (Eftekharnejad et al., 2013, Eltawil et al., 2010). Attributed to the large-scale penetration of PV power stations, the dynamic behavior of PV turbines during and after grid disturbances is the major concern for system operators (Tamimi et al., 2013, Piwko et al., 2010).

The isolation of large PV power generations probably results in the loss of active power and the decreasing grid frequency when the fault happens. Thus, the designed task for PV plants is to take LVRT technology in the case of voltage depressions. In this regard, the state grid has promulgated the newest LVRT requirements: The PV turbines should not trip off-line during the voltage dip or fault and need to provide enough reactive support until the grid recovers to the normal condition (Zhang et al., 2012). This practice is essential to overcome the shortcomings of the traditional way in which the grid must tolerate the loss of large-scale PV plants. What's more, it can reduce the

depression stresses of the electronics so as to highlight the widespread of large-scale PV plants.

The studies on LVRT characteristics of large-scale grid-connected PV power stations mainly concentrate on the optimization of converter controller (Ma et al., 2012) or the increase of the LVRT capacity (Islam et al., 2011, GE Energy, 2010). The dynamics of the PV power stations with and without reactive control were compared and it was concluded that the implementation of the reactive control can rise up the LVRT capacity (Islam et al., 2011). The capacity of LVRT was increased by changing the topology of the converters (GE Energy, 2010).

Although numerous studies about LVRT assessment in PV systems have carried out, the scale of the simulated PV power stations is too low to effectively demonstrate the LVRT characteristics of large-scale grid connected PV power station. Additionally, the LVRT control for large-scale PV plant was usually ignored in the previous studies.

In this paper, the model of the large-scale PV power station with LVRT is proposed, which comprised of three parts: the PV generation system, electrical control model and converter model. Moreover, under the voltage depression, the cap and the ramp rate limit control which further eliminate the current impact, and the reactive control is proposed. The model is carried out on the provincial power grid PV systems and the simulation results verify the effectiveness of the proposed model in term of the fundamental demands, the active power recovery and the dynamic reactive power support.

2 LVRT REQUIREMENTS FROM STATE GRID CODES

The LVRT capacity has been widely required in wind power generation system, then followed by few standards specially restricted on PV system, which means the operation of the renewable generation is changing from disconnection in case of any grid failure to a maximized active power delivery.

In 2012, State Grid Corporation of China (SGCC) released the newest standard: GB/T 19964-2012 Technical Rule for PV Power Station Connected to Power Grid which restricted on PV power plant power quality, LVRT, safety and protection, anti-islanding, general technical conditions, relations, metering and other technical standards (GB/T, 2012).

Particularly, the State Grid enterprise standards required that all medium and large scale PV converters in Chinese market should have the LVRT capacity. Otherwise the PV station cannot be allowed to be integrated to power systems. The requirements of the PV LVRT are divided into four parts.

2.1 *The connected point voltage requirements*

The PV power stations should guarantee a continuous operation connected with the grid if the connected point voltage is within the area above the voltage curve 1 in Figure 1. Otherwise the PV power station should be tripped from the power grid.

2.2 *Failure type and assessment voltage*

As shown in Table 1, the LVRT assessment voltage is evaluated by different failure types.

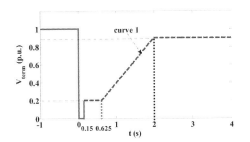

Figure 1.   PV power station LVRT requirements.

Table 1.   PV LVRT assessment voltage.

Failure type	Assessment voltage
Three-phase short circuit fault	Line voltage
Two-phase short circuit fault	Line voltage
Single-phase earthling-short fault	Phase voltage

Generally, the voltage depression caused by three-phase faults is much server than single-phase or two-phase faults. Thus, the three-phase fault is considered in this paper.

2.3 *The active power recovery*

For those PV power stations remaining connected to the grid during the failure of the power system, their active power should be restored as quickly as possible since the failure has been eliminated. Specifically, the active power of PV power stations should be restored to the primitive value with a power changing rate of at least 30% of the rated power per second.

2.4 *Dynamic reactive support capacity*

If the large-scale PV plant was connected to the grid by the 220 kV (or 330 kV) PV collection system and then boosted voltage to 500 kV (or 750 kV), the plant should have dynamic reactive support capacities:

- The react time of the dynamic reactive current ought not to be less than 30 ms upon the connected point voltage depression.
- When the fault is cleared and the voltage recovers to at least 0.9 p.u., the dynamic reactive current I_T should change in time as follows:

$$I_T \geq \begin{cases} 1.05 \times I_N & (U_T < 0.2) \\ 1.5 \times (0.9 - U_T)I_N & (0.2 \leq U_T \leq 0.9) \\ 0 & (U_T > 0.9) \end{cases} \quad (1)$$

where U_T is the PV connected point voltage in per unit system, and $I_N = S_N / \left(\sqrt{3} \times U_N \right)$.

3 LARGE-SCALE PHOTOVOLTAIC POWER STATION MODEL

3.1 *Photovoltaic system model*

The general model of utility scale PV mainly includes: the solar power model, the electrical control model and the converter model. The schematic of the model is represented in Figure 2.

A PV cell is a semiconductor diode which converts the incident sunlight into electrical energy. The equivalent circuit of a PV module is shown in Figure 3.

In practice, a PV module consists of a number of PV cells. The performance of the solar cell depends on the radiation and temperature strongly, which is different from the traditional power station. When the temperature is 25°C, the relationship between the illumination intensity and the maximum output power is depicted in Figure 4.

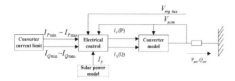

Figure 2.   Schematic of the model used to represent utility scale PVs.

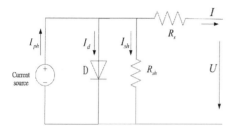

Figure 3.   The equivalent of the PV cell.

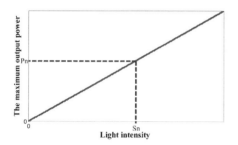

Figure 4.   Curve of illumination intensity and the maximum output power ($T = 25°C$).

As shown in Figure 4, when the temperature T is 25°C, the relationship between the illumination intensity and the maximum output power can be written as:

$$y = p1 \times x^4 + p2 \times x^3 + p3 \times x^2 + p4 \times x + p5 \qquad (2)$$

where x is the illumination intensity, y is the maximum output power of the PV cells, $p1, p2, p3, p4$ and $p5$ are coefficients, they are respectively: 0.01408, −0.0752, 0.1677, 0.8981 and −0.004697.

3.2   Converter model

The converter model is shown in Figure 5. The real and reactive current command signals are developed in the electrical control model described in Section 3.3. Also, the Low Voltage Power Logic (LVPL) is the key part of the converter model in this paper.

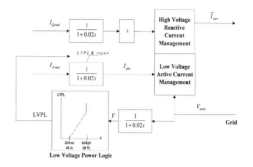

Figure 5.   GE converter model.

The LVPL of the converter reduces the system stress during and after sustained faults by limiting the real current command with both a cap and a ramp rate limit. Under normal operating conditions, the filtered terminal voltage is above a user-specified breakpoint (brkpt) and there is no upper limit. When the voltage falls below the breakpoint during a fault, a cap is calculated and applied. When the voltage is below the user-specified zero-crossing (zerox), the cap becomes zero. Meanwhile, the ramp rate limit (rrpwr) that is determined by the converter capacity is the key to the post-fault power recovery. During this recovery period, the voltage may exceed the breakpoint and the cap is removed. However, the real command rate of increase will be restricted by the ramp rate limit.

The recommendation of the settings for the low voltage power logic of the converter model is included in Table 2.

3.3   Electrical control model

The controller system is very important in the PV systems. Controllers are included in this model to control the active and reactive power as well as the converter current.

The input signal of the active model P_{ord} is initialized to match the PV generation power output from the power flow results. As shown in Figure 6, the real current command signal I_{Pcmd} is developed from this power order and the terminal voltage V_{term}.

The reactive control model contains two parts: reactive power supervision model and electrical control model, as is shown in Figure 7.

The proposed PV LVRT model is implemented on the electromechanical simulation software FASTEST environment. FASTEST is developed by NARI Group Corporation of SGCC in the FORTRAN programming language interface, and its run-way model includes the normal and breakdown condition. The input information contains the local light intensity, and the output contains the results of multiple variables correspondingly.

Table 2. LVRT model parameters.

Parameter name	Recommended values
brkpt	0.9 p.u.
zerox	0.4 p.u.
lvpll	1.1 p.u.
rrpwr	10%

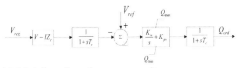

Figure 6. Modeling of active control.

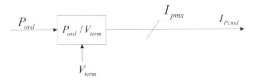

(a). Modeling of reactive power monitor.

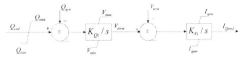

(b). Modeling of reactive power electrical control.

Figure 7. Modeling of reactive control.

4 SIMULATION RESULTS AND DISCUSSION

The provincial power grid, as the large-scale PV plant M with the capacity of 80 MW is taken as the simulation case in this paper. It typically operates under conditions of unity power factor to output the active power as much as possible. The local single line diagram of the PV plant M is shown in Figure 8.

The large-scale PV plant M is integrated to the grid through 330 kV PV collection system and then boosts voltage to 750 kV, which means this PV plant with a multi-megawatt capacity is connected to the high and medium voltage network. A three-phase permanent short circuit fault is applied in a line which belongs to B-C double lines, and cleared in 0.1 second.

To eliminate the influence of the active and reactive power due to the fluctuation of light intensity in short terms, the input data, i.e. light intensity that varies with time smoothly, is shown in Figure 9.

The dynamics of the terminal voltage V_{term} of the PV plant M during and after the fault are shown in Figure 10.

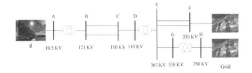

Figure 8. Local single line diagram of the PV plant M.

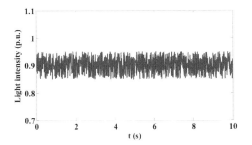

Figure 9. Performance of light intensity.

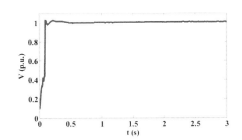

Figure 10. Performance of voltage in and after the fault.

The dynamics of the voltage depression can be separated into two stages. The first stage is from the normal condition to the faulty condition during the voltage drop. The other one is the PV plant recovering to the primitive value after the voltage stresses dismiss.

Figure 11 illustrates the comparison of the active current I_p injected into the grid in three different cases:

Case 1: the convert model with LVPL and rrpwr;
Case 2: the convert model with LVPL but without Srrpwr;
Case 3: the convert model without LVPL and rrpwr.

The node voltage V_{term} sustains a sudden change upon the voltage of the grid dropping as shown in Figure 10. Meanwhile, from Figure 11, it can be seen that the active current Ip rises up to 2 times than normal, which is so high that it destroys the electronics easily due to the flux conversation. Once the LVRT controlled with LVPL and rrpwr is applied, at the first stage of the voltage depression,

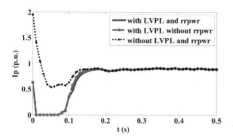

Figure 11. Comparison of I_p in three different cases.

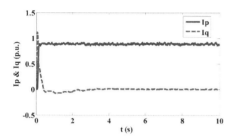

Figure 12. Performance of I_p & I_q with LVRT control in and after the disturbance.

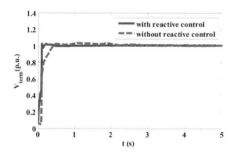

Figure 13. Comparison of V_{term} without & with var control.

V_{term} fluctuates around 0.3 p.u. and Ip is controlled in the rational range. Then at the next stage, the increasing rate of Ip is restricted by 10%, which further slows down the pressure of the grid. Meanwhile, the system recovers into the normal state at around 0.18 seconds.

Converter controllers are included in this paper to control the active and reactive power as well as current of the converter. In Figure 12, the performance of Ip and Iq with LVRT controlled during and after the disturbance is presented.

As shown above, once the fault is cleared, the active current Ip recovers to the normal condition as soon as possible. When the fault happens, the active current Iq is restrained to 1.1p.u. by the reactive current limiting controller. In 0.1 seconds, Iq comes back to 0 p.u.

As a consequence, after the grid fault occurred, Figure 13 would present the comparison of V_{term} with and without var control.

On the one hand, if without the reactive control, V_{term} will drop to 0 p.u. at the first stage. What is more, it will recover to the normal condition for a long time. On the other hand, the PV plant takes much less time for V_{term} with reactive control to recover to the normal condition, demonstrating that the reactive control can support enough reactive power for the voltage recovery.

5 CONCLUSIONS

In this paper, a model of large-scale PV power stations with LVRT is proposed. Upon the voltage depression, the cap and the ramp rate limit control and the reactive control further eliminate the current impact. Also, the scale of PV power station in the simulation is enough to effectively perform LVRT characteristics of large-scale grid connected PV power stations. Finally, the simulation results have verified the effectiveness and applicability of the proposed LVRT model for large-scale PV plants.

REFERENCES

Eftekharnejad, S. & Vittal, V. 2013. Impact of increased penetration of photovoltaic generation on power systems. *IEEE Trans. Power Syst* 28(2): 893–901.

Eltawil, M.A. & Zhao, Z.M. 2010. Grid-connected photovoltaic power systems: *Technical and potential problems—A review*. Renew. Sustain. Energy Rev 14(1): 112–129.

GB/T 19964-2012 Technical Rule for PV Power Station Connected to Power Grid.

GE Energy, 2010. Modeling of GE Solar Photovoltaic Plants for Grid Studies, Version 1.1, Schenectady, USA.

Islam, G.M.S. & Al-Durra, A. Low voltage ride through capability enhancement of grid connected large scale photovoltaic system. *In IECON 2011-37th Annual Conference on IEEE Industrial Electronics Society-IEEE Oceania*, Melbourne, Australia, Nov 7–10, 2011: 884–889.

Ma, L. & Liao, H. 2012. Analysis of Chinese Photovoltaic Generation System Low Voltage Ride Through Characters. *In 2012 IEEE 7th International Power Electronics and Motion Control Conference—ECCE Asia*, Harbin, China, June 2–5, 2012, 1178–1182.

Piwko, R. & Miller, N. 2010. Generator fault tolerance and grid codes. *IEEE Power Energy Mag* 8(2): 18–26.

Tamimi, B. & Canizares, C. 2013. System stability impact of large-scale and distributed solar photovoltaic generation: The Case of Ontario, Canada. *IEEE Trans. Sustain. Energy* 4(3): 680–688.

Zhang, J.J. & Ju, R.R. 2012. Latest Technical Specifications & Testing Procedures for Grid-connected Photovoltaic Power Station in China. *In 2012 International Conference on Intelligent Systems Design and Engineering Application*, Nanjing, China, 2012: 1435–1439.

Advances in Power and Energy Engineering – Sun (Ed.)
© 2016 Taylor & Francis Group, London, ISBN 978-1-138-02846-3

Optimal sizing of Photovoltaic and battery for residential house using genetic algorithms

J. Li
CSIRO Data61 Flagship, Sydney, Australia

J.H. Braslavsky
CSIRO Energy Flagship, Newcastle, Australia

ABSTRACT: A genetic algorithm based optimization methodology for solar PV generation and battery storage systems is presented. The purpose of the proposed methodology is to minimise the annual energy cost for a household by optimally sizing and operating the battery and PV system. Genetic algorithm optimises the battery size together with the corresponding charge and discharge cycles under a time-of-use electricity tariff. Three optimisation scenarios are considered: optimization of battery size for households with and without pre-existing PV generation, and optimisation of both battery and PV size simultaneously for a household without PV generation. For each scenario, the optimization is performed using realistic battery models for a commercial deep-cycle lead-acid battery system. The results show that optimal battery capacities and PV size can significantly save energy cost for the householder. However, it varies strongly with the electricity consumption profile of the household, and is also affected by electricity and battery prices.

1 INTRODUCTION

Photovoltaic (PV) generation of electricity is an important renewable energy resource. In some states in Australia, around one in four houses already boast rooftop solar PV systems (RenewEconomy 2015). The growth in PV capacity has an increasingly significant impact on the operation of national electricity networks, and the traditional business models of electricity generators, retailers and network operators.

The Virtual Power Station (VPS) concept proposed by CSIRO aims to explore the aggregation of a large number of geographically disperse, technically diverse, small-scale PV generation panels and associated grid-connected battery systems. A preliminary VPS study involving 20 sites was conducted by the CSIRO (Platt 2008, Ward 2008 and Li 2013), which led to a second stage VPS (VPS2) project funded by the Australian Renewable Energy Agency (ARENA VPS2). The VPS2 project aims to undertake pilot-scale testing of the methodologies developed to coordinate distributed loads, generation and storage. A key question addressed by the VPS is how such set of distributed residential power systems could be presented to the grid and/or electricity markets, as a more reliable, even dispatchable power entity. A related question concerns how much penetration

of distributed residential PV generation could be accommodated by existing distribution networks, and to what extent it needs to be accompanied by battery storage. Answering these questions leads to the analysis of the potential economic benefits of residential PV and energy storage systems, and the determination of the optimal PV and battery sizes for given individual household energy profiles.

Several research groups have carried out PV and battery size optimization studies for renewable energy systems. Borowy and Salameh (1996) developed an algorithm to optimize a PV array with a battery bank in a standalone hybrid PV/wind system. The model proposed was based on a long-term hourly solar irradiance and peak load demand data of the site chosen. The number of PV modules in the hybrid energy system was calculated and the cost of the PV modules and battery systems was taking into account in the same algorithm. A graphic construction technique was used to optimize the size of the PV/wind energy system by Markvart (1996), by considering monthly average solar and wind energy values. More recently, Lalwani et al. (2011) proposed an optimization technique following the Loss of Power Supply Probability (LPSP) model for a PV system taking reliability of the system into account. They demonstrated the utility of their model for different weather conditions to minimize the cost of the system under the specific load demand. Recently, Yang et al. (2011) presented the methodology for optimally sizing of battery banks and PV arrays for PV/wind hybrid System. The power supply reliability and the installed capacity of the system are taken into consideration for the size optimization. Very recently, Von Appen et al. (2015) introduced an approach based on a mixed integer linear program, which allows integrating the sizing and operation decision of PV and PV battery systems in one optimization problem and assessing their impact on key indicators of grid planning and operation.

The studies carried out by Borowy and Salameh (1996), Markvart (1996) and Yang et al. (2011) are based on hourly, daily, or monthly average and peak data of the chosen sites. These types of data lost their statistical probability, lack of considering electricity demand profiles for different sites. Lalwani et al. (2011) used simple battery charge/discharge actions, and did not take into account the variable electricity price through the day. In their algorithm the energy is stored in the battery when solar array produces more power than load demand or supplies power to load when the solar array produces less power than the load demand during cloudy or rainy days or at nights. In contrast, Von Appen et al. (2015) use yearly load energy consumption from all analysed households, which leads to a more general optimization for all households. However, for each individual household, the electricity consumption profile varies one by one. The optimized results may be hard to be met by individual customer requirements.

The present paper introduces a Genetic Algorithm (GA) methodology for the optimization of combined rooftop solar PV generation and energy storage systems using one year of real-time electricity demand data from households. The main objective is the minimization of the total energy cost of the household as a function of PV size and battery capacity. Individual household and battery/PV size are optimized simultaneously to achieve the most economic result. This optimization is formulated in terms of the cost of electricity, the cost and technical capabilities of PV and battery systems, and the energy usage profiles of real users in NSW, Australia. The optimisation considers variable electricity prices through the day in a Time-Of-Use (TOU) tariff split into peak, shoulder, and off peak. The GA technique used to perform the optimization does not only produce optimal battery and PV sizes, but also provides corresponding battery charge and discharge cycles under the considered TOU electricity tariff. The optimised annual energy costs include the cost of electricity, network charges, and the cost of the energy storage and PV system with investment payback periods of 5 and 10 years respectively.

Assumptions on the electricity market prices and battery selection/modelling are provided in Sections 2 and 3. The genetic algorithm technique is described in Section 4, and includes the details on algorithms for searching optimal battery size and the actions that could minimize the total system costs. The optimisation scenarios considered include optimization of battery size for households with and without pre-existing PV generation, and optimisation of both battery and PV size simultaneously for a household without PV generation. For each scenario, the optimization is performed using realistic battery models for a commercial deep-cycle lead-acid battery system and under a time-of-use electricity tariff. A series of experimental results are presented in Section 5 to show the optimization performance. Finally, conclusions are drawn in Section 6.

2 ELECTRICITY TARIFF STRUCTURE

AusGrid is the one of Australia's largest and most experienced electricity infrastructure company. Ausgrid TOU tariffs are adopted in our optimization algorithms. These tariffs include electricity supply price and solar electricity buy-back price. The electricity supply price has different values during the peak, shoulder and off peak time of use periods. The details are as follows.

- Peak: 2 pm–8 pm on working weekdays with price 51.128 C/ kWh
- Shoulder: 7 am–2 pm and 8 pm–10 pm working weekdays and 7 am–10 pm on weekends and public holidays with price 19.657 C/kWh
- Off Peak: All other times with price: 10.758 C/ kWh.

Solar electricity buy-back price is 6 c/kWh.

To minimize the system cost, the battery storage is generally considered an effective way. In this paper, we developed a GA based algorithm that optimizes the battery size together with the corresponding charge and discharge operations.

3 BATTERY SELECTION AND MODELLING

3.1 Battery selection

The cost of battery has dropped significantly in the last 5 years. This reduced price increased demand for batteries in PV systems.

There are a lot of different types of batteries in the market. For rooftop PV system, battery should have small size, low cost, deep discharge using most of its capacity, long lasting and free maintenance.

Lead-acid batteries satisfy these conditions. Lead-acid battery is a deep-cycle battery designed to use most of their capacity on each charge-discharge cycle with advantages of low cost, high capability, and free maintenance. Lead-acid batteries are perfect for applications that require either frequent cycling or renewable power storage. Common uses include running appliances when camping away from power, using with a solar panel for charging, running in a dual battery system in a vehicle, or providing safe power on a boat. Lead-acid battery is a great choice for our application.

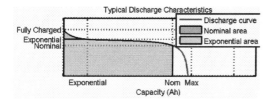

Figure 1. Typical charge/discharge characteristics of lead-acid battery.

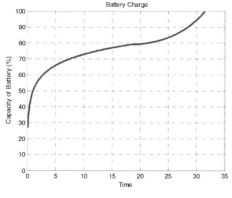

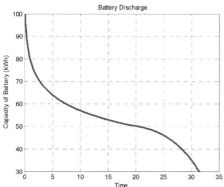

Figure 2. Modeled battery charge and discharge functions.

3.2 *Battery modeling*

Realistic battery models for a commercial deep-cycle lead-acid battery system are used in our optimization algorithms. It has typical characteristics of a lead-acid battery charge and discharge functions (Mathworks, Jackey 2007 & Spotnitz 2005) as shown in Figure 1.

A realistic battery model includes battery charge and discharge functions defined by three main elements: a charge/discharge curve, a nominal area and an exponential area. The modelled battery charge and discharge functions are shown in Figure 2. In our battery modelling, battery degradation and leakage have not been considered.

4 GENETIC ALGORITHM AND OPTIMIZATION DESIGN FOR OUR APPLICATION

A Genetic Algorithm (GA) is a robust search method requiring little information to search effectively in a large or poorly-understood search space. It is typically used to provide good approximate solutions to problems that cannot be solved easily using other techniques.

In our application, the optimization is based on one year data for each household. The given individual house energy profile changes year by year slightly. A near-optimal solution is sufficient for us. On the other hand, electricity market price varies during the day as introduced in section 2. Based on this price, it is too computationally-intensive to find an exact solution about total energy cost of the household as a function of PV size, battery capacity and battery charge/discharge actions, e.g. charge/discharge time and amount. Although GA is never guaranteed to find an optimal solution for any problem due to its random nature, it will often find a good solution if one exists. Therefore, GA is a good choice for our application. The brief introduction about GA and its design for our application will be given as below.

4.1 *Genetic algorithm*

The genetic algorithm (Rennard 2000) search method is inspired by natural selection and survival of the fittest in the biological world. GA is a type of "evolutionary algorithm". One common application of GAs is function optimization, where the goal is to find a set of parameter values that maximize a complex multi-parameter function. This complex multi-parameter function is called fitness function. It has been successfully applied to a lot of applications, such as optimization, building modelling (Wang 2006, Platt 2010 & Li 2004).

A colony of rule sets can be evolved for a number of generations, improving the performance of the colony. Techniques of fitness determination, selection, cross-over, reproduction, and mutation are applied to the rules and their chromosomal representation. Here GA is used to find optimized battery maximum capacity that could minimize the total system cost, i.e., maximise fitness function.

The first generation of parameters (initial population) can be either random or predetermined values. At each subsequent generation of parameters, every individual of the population must be evaluated to be able to distinguish between good and bad individuals. This is done by mapping the objective function to a "fitness function", which is a non-negative well-behaved measure of relative fitness of the parameters. The better the fitness of a given rule, the more likely it is to be selected. After the two parent rules are selected, each is represented by a "chromosomal" string and then combined to determine the chromosomes of the two resulting rules (offspring). These chromosomes are subjected to potential mutation, and are then converted back to their equivalent rule representation. The selected parents are then replaced in the colony by the offspring. This mechanism of natural selection is expected to eventually result in a population with a higher performance.

4.2 Genetic algorithm design for PV system

Since electricity market price varies during the day, a battery can be used to minimize system costs by charging and discharging during proper times. The installed battery capacity directly affects the energy consumption of the house, which determines the energy cost. In general, a larger battery capacity leads to a larger electricity supply saving because more discharge can be covered within the peak period of the electricity. On the other hand, the cost of the battery increases with rising battery capacity. Consequently, the optimal battery size from the perspective of household considers the combination costs of battery system investment, network charges and electricity supply. Here the GA is used to search for an optimal battery size to minimize overall costs.

The system cost components are listed as below.

- Cost to buy a battery (\$): β
- Number of battery needed: N
- Cost of other equipments of battery system: E
- Battery warranty time (year): Y
- Electricity price (\$/kW): Ω.
  Where Ω is variable electricity price with peak, off-peak and shoulder
- Electricity usage (kW): μ
- Electricity service fee: F

- Cost of electricity supply (\$): S = Ω * μ,
- Cost of battery electricity supply for charging battery (\$): B.

GA is designed not only to optimize battery size for households with and without local PV generations, but also to optimize both battery and PV size simultaneously for a household without PV generation. Total three optimisation scenarios will be introduced below.

4.2.1 GA design of battery optimization for the system without local PV

For the residential home without local PV system, GA tries to find out optimal battery size that minimizes the total system cost by charging battery during off-peak hours and using minimum supply electricity during peak hours through battery discharge.

The parameters for GA are defined as below.

- Battery cost per day:
  The battery cost measurement (\$/day) is defined as

$$\rho = \frac{N \times \beta + E}{365 + Y} \tag{1}$$

  where ρ is cost of battery per day (\$/day), β is cost to buy one battery (\$), N is number of battery needed for battery bank, E is cost of other equipments of storage system, Y is warranty time (year).

- Chromosome of GA:
  Chromosome is a set of parameters that the GA optimizes to maximize a complex multi-parameter function, namely, a fitness function. The chromosome for the system without PV is simply the battery size needed, i.e. the maximum battery capacity (kWh) C_{mx}.

- System cost per day:

$$\xi = S + B + \rho + F \tag{2}$$

  where S is the cost of electricity supply for the house; B is cost of electricity supply for charging battery, B = $\Omega \times \Delta C$, Where Ω is electricity price during battery charge, e.g. off-peak electricity price, ΔC is battery charge amount; ρ is cost of battery per day, F is electricity service fee, 82.5 c/day.

- Fitness function:
  The goal of GA is to find the optimal battery size, C_{mx}, that maximizes the fitness function. The fitness function is taken as the inverse of total system costs, expressed as

$$\Lambda = \frac{1}{\sum \xi} \tag{3}$$

4.2.2 GA design of battery optimization for the system with pre-existing PV

For the residential home with pre-existing PV system, GA tries to find out an optimal way that minimizes the total system cost by using minimum supply electricity during peak hours through battery discharge. The questions will include what size battery need to be bought, what conditions batteries need to be charged or discharged, and what condition electricity need to be sold to grid.

Figure 3 average energy consumption and generation of a house. From the sub-figure on the top we can see that power usage has two peaks. One is in the morning between 6:30–11 am (breakfast, morning tea), and another is later afternoon and over night between 3:30–11 pm (afternoon tea, dinner, shower, entertainment, etc.). The first peak is during electricity off-peak and shoulder price times. The second peak is during electricity shoulder and peak price times. Solar energy output is between 6 am and 6 pm. Sub-figure in the middle is average PV generation. We can see that peak solar generation is at around 12 pm, which is in electricity shoulder price period. Therefore, peak electricity supply price time is not matching with peak solar generation time. The difference between consumption and PV supply is shown in bottom sub-figure. We can see that the second peak electricity cost will be much more than the first one.

In order to minimize the total system cost, the best way is to reduce second electricity consumption peak through battery discharge. GA is used to realize this.

Apart from chromosome, all others are the same as section 4.2.1.

- Chromosome of GA:
  Battery size and battery capacities in percentage are defined for different electricity price zones. Total seven battery statuses are defined as chromosome.

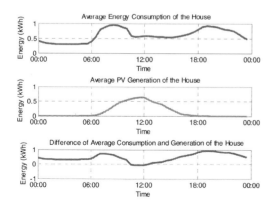

Figure 3. Average energy distribution.

- o C1—battery size, i.e. maximum battery capacity
- o C2—% battery capacity at 7 am in working days (beginning of 1st shoulder price)
- o C3—% battery capacity at 2 pm in working days (beginning of peak price)
- o C4—% battery capacity at 7 am in weekends (beginning of shoulder price)
- o C5—% battery capacity at 2 pm in weekends (end of shoulder price)
- o C6—% battery capacity at 8 pm in working days (end of peak price)
- o C7—battery capacity at 10 pm in working days (end of 2nd shoulder price).

- Actions:
  System actions include battery charge, battery discharge, use supply electricity or solar electricity and sell electricity to grid. These actions are determined by load consumption, solar generation and battery status defined in the chromosome. In each time zone, the system follows three rules as below.
  - o Solar energy: consumed by loads first, then charged by batteries or sold to grid.
  - o Supply energy: consumed by loads if solar energy and battery discharge energy is not enough, then charged by battery.
  - o Battery: charged if capacity is less than required percentage capacity, discharged for load consumption or sold to grid if capacity is more than requited percentage capacity.

4.2.3 GA design of both battery and PV optimization

The drawback of PV system is the high capital cost as compared to conventional energy sources. For the house without local PV generation, the first question will be what the optimal PV size and battery capacity should be installed. This algorithm will achieve optimization of PV and battery systems so that the size of PV and capacity of battery are optimally selected simultaneously.

The system actions follow the same rule as section 4.2.2. The functions defined by GA are described as follows.

- Solar panel cost per day:
  Solar panel cost measurement $/day is defined as equation (4)

$$\Psi = \frac{\alpha \times M + \Phi}{365 + Y} \tag{4}$$

where Ψ is cost of solar panel per day ($/day), α is cost to buy solar panel ($/kW), M is size of solar panel (kW), Φ is cost of other equipments of solar system ($), Y is warranty time (year).

- Chromosome of GA:
  One more chromosome C8, PV size, is added, other chromosomes are the same as section 4.2.2.
- System cost per day:

$$\xi = S + B + \rho + P + F \tag{5}$$

where S is the cost of electricity supply for loads; B is cost of electricity supply for charging battery, $B = \Omega \times \Delta C$, Where Ω is electricity price during battery charge, e.g. off-peak electricity price, ΔC is battery capacity difference during charge period; ρ is cost of battery per day, Ψ is cost of solar panel per day, F is electricity service fee, 82.5 c/day.

- Fitness function:
  The goal of GA is to find optimal battery and solar panel size, C_{mx} & P, that could maximize the fitness function. Fitness function will be the inverse of total system costs as expressed as equation (3).

5 EXPERIMENTAL RESULTS

Data is collected from different houses. Three parameters are used in our experiments: Time, General_Supply (total energy consumption of the house) and Gross_Generation (total energy generated by the PV system).

The battery with capacity 3.36 kWh was selected from the market. The battery system cost is $1539 with a 5-year warranty.

For the experiments below, the battery cost is assumed $500, while other ancillary equipment cost, such as the inverter, is assumed as $1039. The battery is assumed to be capable of being fully charged and discharged, without degradation or leakage.

Three optimization scenarios are considered: optimization of battery size for households with and without pre-existing PV generation, and optimization of both battery and PV size simultaneously for a household without PV generation. For each scenario, the optimization is performed using realistic battery models for a commercial deep-cycle lead-acid battery system as introduced in section 3.2. The results are introduced in each section below.

5.1 Battery size optimization for household without pre-existing PV generation

For the residential home without PV, the supply cost without battery is $7.8082 kWh per day. GA optimized battery size is 16.4867 kWh. Using this battery, charge battery during off-peak time and discharge battery during peak time of electricity price, the electricity supply cost is reduced to $4.0128 per day.

For the battery we selected, one battery size is 3.36 kWh. To get 16.4867 kWh, 5 batteries need to be bought. Batteries and system equipment costs are $3539. This battery has a 5-year warranty. Battery cost will be $1.9391 per day. So total cost for optimized system will be electricity supply cost and battery cost i.e., $5.9519 per day. Comparing with cost without battery, $7.8082, the saving by using battery is $1.8563 per day and $677.55 per year.

In order to identify the optimal battery capacity required under different battery prices, our battery sizing optimization algorithm is run for different battery prices. The relationship of optimal battery capacity to battery price is shown in Figure 4, in which the battery price varies between $50 and $300 per kWh. From Figure 4 we can see that as the battery price is becoming cheaper and cheaper, bigger battery capacity can be used and less electricity will supply for the system to realize more saving for the residential household.

5.2 Battery size optimization for household with pre-existing PV generation

For the case of a residential home with pre-existing PV, the net supply cost without battery is $6.3879 kWh per day. The GA optimized chromosome is [11.0081, 1, 0.997009, 0.956951, 0, 0, 0.0523882]. This means that optimized battery size is 11.0081 kWh. Battery is 100% full at 7 am in working days, 99.7% full at 2 pm in working days, 95.70% full at 7 am in weekends, 0% at 10 pm in weekends, 0% at 8 pm and 5% at 10 pm in working days. The optimization results show us that battery tries to completely discharge its capacity during peak price time in the working days and shoulder price time in weekends. This is quite reasonable.

Using this battery and its optimized status in different time zones, the electricity supply cost is reduced to $3.9804 per day.

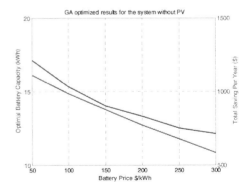

Figure 4. Battery price versus optimal battery capacity and total saving for the system without PV.

For the battery we selected, one battery size is 3.36 kWh. To get 11.0081 kWh, 4 batteries need to be bought. Batteries and system equipment costs are $3039. This battery has 5 year warranty. Battery cost will be $1.6652 per day. So total cost for optimized system will be electricity supply cost and battery cost i.e., $5.6456 per day. Comparing with cost without battery, $6.3879, the saving by using battery is $0.7423 per day and $270.9395 per year.

In order to identify the optimal battery capacity required under different battery prices, our battery sizing optimization algorithm is run for different battery prices. The relationship of optimal battery capacity to battery price is shown in Figure 5, in which the battery price varies between $50 and $300 per kWh. From Figure 5 we can see that as the battery price is becoming cheaper and cheaper, bigger battery capacity can be used and less electricity will supply for the system to realize more saving for the residential household.

5.3 Battery and PV size optimization for household without pre-existing PV generation

5.3.1 Solar panel cost
Based on the information on May 2015 (Solar-Quotes, 2015), the fully installed solar panels cost is as below.

- 3 kW: $5,000—$6,500 ($1667/kW ~)
- 4 kW: $6,500—$7,500 ($1625/kW ~)
- 5 kW: $7,500—$8,500 ($1500/kW ~)
- 10 kW: $13,000—$16,000 ($1300/kW ~ 1600/kW).

Since solar panel prices will be dropped significantly day by day, solar panel cost in our algorithm is chosen $1500/kWh.

5.3.2 Battery and PV optimization
Figure 6 is the average PV outputs from our experimental data. We suppose PV size is PV's maximum

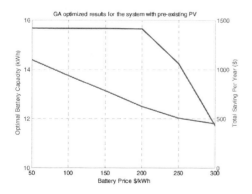

Figure 5. Battery price versus optimal battery capacity and total saving for the system with PV.

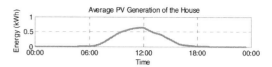

Figure 6. Average PV generation.

output, which is 1.023 kW in our database. In our algorithm, PV output will base on this PV size. PV output will be expressed in equation below.

$$O = \frac{P}{1.023}\zeta \qquad (6)$$

where O is PV output, P is optimized PV size in kW, ζ is PV generation in each day.

The optimized battery and solar panel are 6.1479 kWh and 3.8462 kW respectively. Using this battery and PV, the household will earn $1.5233 per day. For the battery we selected, one battery size is 3.36 kWh. To get 6.1479 kWh, 2 batteries need to be bought. Batteries and system equipment costs are $2039. This battery has 5 year warranty. Battery cost will be $1.117 per day. To install 3.8462 kW solar panel cost is $5769.3. This PV has 10 year warranty. PV cost will be $1.5806 per day. Total PV and battery cost is $2.6978 per day. So total cost for optimized system will be electricity supply cost and battery cost i.e., $1.1745 per day. Comparing with cost without battery and PV, 7.8082, the saving by using battery is $6.7337 per day and $2457.80 per year.

6 CONCLUSIONS

We have investigated the use of a GA to find optimized battery and PV sizes to minimize total energy cost for households. The results show that the optimal battery capacities for a household with an average daily consumption of 29 kWh per day without and with pre-existing PV generation are 16.5 kWh and 11 kWh respectively. The corresponding annual energy cost savings are $677 and $271. The optimal battery capacity and PV size for a household without PV generation are 6.1479 kWh and 3.8462 kW respectively. The annual energy cost saving is $2457.80. The results also show that the cheaper battery price, the larger battery capacity can be bought and therefore more energy cost savings may be achieved, which agrees with existing results (Weniger, 2013).

We have successfully developed GA based optimization algorithms for battery and PV size optimization for household to realize energy cost savings. The optimization does not only generate optimal battery and PV size but also provides corresponding battery charge/discharge operations under a time-of-use electricity tariff.

The current GA algorithm has been tested on single household. In order to quantify statistic performance more generally and optimal battery and PV size patterns, the proposed optimization algorithm will be applied to a large residential dataset that includes 7112 households. It should be noted that the battery model used in this paper still includes idealistic assumptions, such as no leakage and full charge/discharge capabilities. More realistic battery models will be considered in further studies in order to generate more realistic results.

ACKNOWLEDGEMENT

This research received funding from ARENA, the Australian Renewable Energy Agency.

The views expressed herein are not necessarily the views of the Australian Government, and the Australian Government does not accept responsibility for any information or advice contained herein.

REFERENCES

ARENA VPS2 http://arena.gov.au/project/virtual-power-station-2.

Borowy, B.S. Salameh, Z.M. 1996 "Methodology for optimally sizing the combination of a battery bank and PV array in a wind–PV hybrid system" *IEEE Trans Energy Convers, 11 (2) (1996), pp. 367–375.*

Jackey, R.A. 2007 "A Simple, Effective Lead-Acid Battery Modeling Process for Electrical System Component Selection", *The MathWorks, Inc. 2007-01-0778.*

Lalwani, M. Kothari, D.P. Singh, M. 2011 "Size optimization of stand-alone photovoltaic system under local weather conditions in India", *International Journal of Applied Engineering Research, Volume 1, No 4, 2011.*

Li, J. Guo, Y. & Poulton, G. 2004 "Critical Damage Reporting in Intelligent Sensor Networks" Proceedings, 17th Australian Joint Conference on Artificial Intelligence—AI04: Advances in Artificial Intelligence, 6th–10th December 2004, Cairns, Australia, vol. 3339, pp. 26–38.

Li, J. Guo, Y. Platt, G. & Ward, J.K. 2013 "Renewable Energy Aggregation with Intelligent Battery Controller" *Journal Renewable Energy Volume 59, November 2013, Pages 220–228.*

Markvart, T. 1996 "Sizing of hybrid photovoltaic–wind energy systems', Solar Energy, 57 (4) (1996), pp. 277–281.

MathWorks *http://au.mathworks.com/help/physmod/sps/powersys/ref/battery.html.*

Platt, G. Guo, Y. Li, J. & West, S. 2008 "The Virtual Power Station" *IEEE International Conference on Sustainable Energy Technologies (ICSET2008), Vols 1 and 2, Nov. 2008, pp. 526–530.*

Platt, G. Li, J. Li, R. Poulton, G. James, J & Wall, J. 2010 "Adaptive HVAC Zone Modeling for Sustainable Buildings", *Journal of Energy and Buildings, Vol. 42, Issue 4, April 2010, pp. 412–421.*

RenewEconomy 2015 *http://reneweconomy.com.au/2014/why-australian-households-are-desperate-for-battery-storage-13709.*

Rennard, J.P. 2000 "Introduction to genetic algorithms", *PhD thesis, 2000.*

SolarQuotes, 2015 *https://www.solarquotes.com.au/panels/cost/.*

Spotnitz, R.M. 2005 "Battery Modeling", *The Electrochemical Society Interface, Winter 2005.*

Von Appen, J.V., Braslavsky, J.H. Ward, J.K. & Braun. M. 2015, "Sizing and grid impact of PV battery systems—a comparative analysis for Australia and Germany", *In Proceedings of the 2015 International Symposium on Smart Electric Distributions Systems and Technologies (EDST), Vienna, Austria, September 2015.*

Wang, S.W. & Xu, X.H. 2006 "Simplified Building Model for Transient Thermal Performance Estimation Using GA-Based Parameter Identification", *International Journal of Thermal Science, V45(4), 2006, pp. 419–432.*

Ward, J.K. Platt, G. & Li, J. 2008 "The Virtual Power Station—Reliably Meeting Electricity Network Demands with Photovoltaics", *3rd International Solar Energy Society Conference, Sydney, Nov. 2008.*

Weniger, J. Tjaden, T. & Quaschning, V. 2013, "Sizing of residential PV battery systems," *Energy Procedia, vol. 46, pp. 78–87, 2014, 8th International Renewable Energy Storage Conference and Exhibition (IRES 2013).*

Yang, G. & Chen, M. 2010 "The Methodology for Size Optimization of the Photovoltaic/Wind Hybrid System", *Energy Sources, Part A, 32:1644–1650, 2010.*

Advances in Power and Energy Engineering – Sun (Ed.)
© 2016 Taylor & Francis Group, London, ISBN 978-1-138-02846-3

Research on thermal and power characteristics of conventional solar cells with auxiliary equipment

X. Guo, E.C. Xin, H.W. Yuan & J.X. Lv
School of Automation Science and Electrical Engineering, Beihang University, Beijing, China

L.M. Zhou & Z. Ma
China Electric Power Research Institute, Beijing, China

ABSTRACT: In order to maximize the efficiency of the photovoltaic power generation, a concentrating photovoltaic power generation system with the cooling system and adjustable concentration ratio was designed based on the research on the characteristics of conventional single crystal silicon. The test was conducted with or without the cooling system under the condenser condition. The results indicated that the conversion efficiency of conventional solar cells was 16.6% under the natural environment. Under the condenser condition but without the cooling system, the conversion efficiency was 17.4%. Under the condenser condition with the cooling system, the conversion efficiency was 23.7%. When the concentration multiple was within 50 times, the cooling device of the concentrating photovoltaic system can ensure that the temperature of the sun battery is below 60°C, making the battery work under a higher conversion efficiency.

1 INTRODUCTION

With the fossil energy exhausting, people's attention to the problem of environmental protection is increasing, and the issues looking for a clean alternative energy are becoming more and more urgent. Solar energy as a kind of renewable clean energy has sustainable use, thus it has a broad application prospect; moreover, photovoltaic (PV) power generation technology is gaining more attention of people (Ball, G.J. et al. 2002, Benner, J.P & Kazmerski, L. 1999, Green, M.A. 2002, Palz, W. 2000, Taylor, R. et al. 1996). Grid-connected photovoltaic power generation is one of the main forms of solar energy application, and also it is the main research direction for solar power. It has been indicated that the technology of grid-connected photovoltaic power generation has entered a new historical stage. From the international development trend of energy utilization, photovoltaic power generation will play the role of alternative sources in the electric power market, and grid-connected will be the only way for photovoltaic power generation in the power market. However, the photoelectric conversion efficiency of the solar cell would be affected in the work by illumination and temperature. So, how to further improve the conversion efficiency of the solar cell has become a key technical problem for researchers.

By using the method of concentrating, making the solar battery work several times and even under increased light intensity conditions, the output power of the solar battery on unit area can be improved. The cost of photovoltaic power generation can be greatly reduced, which has a very good application prospect (Whitfield, G.R. et al. 2002).

In this paper, based on the research on the characteristics of the conventional single crystal silicon photovoltaic battery output, a photovoltaic power generation system was developed with cooling function whose concentration ratio can be adjusted. The system can not only realize the efficient utilization of solar energy and reduce the cost but also improve the reliability of the system.

2 PHOTOVOLTAIC CELL CHARACTERISTICS

2.1 *Photovoltaic cell equivalent model*

When the light is constant, photo-generated current I_{ph} does not change with the working state of the photovoltaic battery. So, I_{ph} can be thought of as a constant current source in the equivalent circuit. After accessing the load R at both ends of the photovoltaic battery, both ends of the load will set up the voltage V when the photo-generated current flows through the load R. The current I_d, which moves in an opposite direction to that of the photo-generated current, will be generated after the load voltage reacts on the PN junction of the photovoltaic battery. In addition, due to the photovoltaic battery surface's electrode and the resistivity of the material itself, when the working current flows through the board, it will inevitably cause internal loss of the panels. Therefore, the series resistance R_s is introduced. The greater the series resistance is, the more the line loss is; thus, the output efficiency of the photovoltaic battery is lower. For the actual PV battery, series resistance is small, mostly between 10^{-3} ohm and several ohms. Moreover, due to the manufacturing process factors, the edge of the photovoltaic battery and metal electrodes may produce tiny cracks or scratches, which can give rise to electric leakage, making the photo-generated current originally flow through the load short circuit. So, a shunt resistance R_{sh} is introduced in the equivalent circuit. Relative to the series resistance, the parallel resistance is larger, more than 1 kΩ. The equivalent circuit of the solar photovoltaic battery is shown in Figure 1.

The solar photovoltaic battery equivalent circuit can be expressed as follows:

$$I = I_{ph} - I_d - I_{sh} \tag{1}$$

where I is the load current; I_{ph} is the photo-generated current, which is directly proportional to the sunlight intensity; I_d is the diode current; and I_{sh} is the solar PV battery leakage current. And I_d can be expressed as

$$I_d = I_0 \left\{ \exp\left[\frac{q(V + IRs)}{AKT} \right] - 1 \right\} \tag{2}$$

where I_0 is the reverse saturation current (generally speaking, its order of magnitudes is 10^{-4} A); q is the electron charge (1.6×10^{-19} C); K is the Boltzmann constant (1.38×10^{-23} J/K); T is the absolute temperature ($= t+273$ K); A is the PN junction ideal factor (value 1–2); R_{sh} is the photovoltaic battery parallel resistance; and R_s is the photovoltaic battery series resistance. And I_{sh} is given by

$$I_{sh} = \frac{V + IR_s}{R_{sh}} \tag{3}$$

According to (1) ~ (3), I can be expressed as

$$I = I_{ph} - I_0 \left\{ \exp\left[\frac{q(V + IR_s)}{AKT} \right] - 1 \right\} - \frac{V + IR_s}{R_{sh}} \tag{4}$$

2.2 Effect of temperature on the PV battery

The conventional single crystal silicon photovoltaic battery is used for testing. When the light intensity is 1 kW/m^2 and the temperature is 25°C, the battery conversion efficiency is from 16% to 24%. Other parameters are as follows: output power 100 W, open-circuit voltage 21 to 23 V, short-circuit current 5.4 to 5.6 A and output voltage 18 V.

For the solar battery, the diffusion coefficient of the carrier will raise with the increasing temperature, so the minority carrier diffusion length also increases slightly with the increase in temperature. As a result, the photo-generated current also improves with the increase in temperature (Radziemska, E. 2003, Masoum, M. et al. 2002). Figure 2 shows the output characteristic curve of the solar battery at different temperatures under the same sunshine intensity. The figure shows that the solar photovoltaic battery's open circuit voltage will drop, the short circuit current will increase and the maximum output power will decrease with the increase in temperature at the same sunshine intensity.

2.3 Effect of sunshine intensity on the PV battery

Sunshine intensity directly affects the size of the solar photovoltaic battery's output power (Bendel, C. & Wagner, A. 2003). Figure 3 shows the output characteristic curve of different sunshine intensities at room temperature and the output characteristic of the photovoltaic battery is nonlinear. From the characteristic curve, we can know that the PV battery's open circuit voltage is almost the same, the short circuit current increases and the maximum output power increases with the increase in sunlight intensity at the same temperature.

2.4 Photovoltaic cell conversion efficiency

When the temperature is constant, the battery conversion efficiency will be improved with the increase in the sunlight intensity. For a given output power, the conversion efficiency of the battery determines the required number of solar panels, so it is extremely important to make the battery conversion efficiency as high as possible.

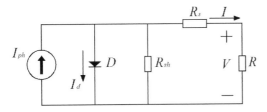

Figure 1. Solar photovoltaic battery equivalent circuit.

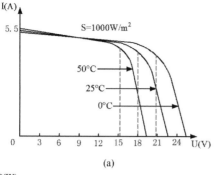

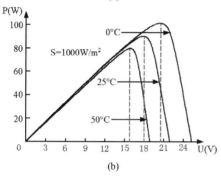

Figure 2. Output characteristic curve of different temperatures: (a) U-I and (b) U-P.

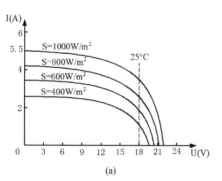

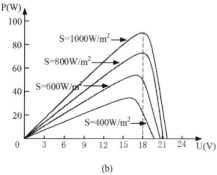

Figure 3. Output characteristic curve of different lighting values: (a) U-I and (b) U-P.

This conclusion provides a way to improve the efficiency of conversion: the addition of the condenser enhances the light intensity, thus reducing the use of the photovoltaic battery and the cost of photovoltaic power generation.

The ratio of the photovoltaic battery's output power and incident light power is called the efficiency of the solar battery. The efficiency η is given by

$$\eta = \frac{P_m}{AP_{in}} = \frac{I_m V_m}{AP_{in}} = \frac{(FF)I_{sh}V_{oc}}{AP_{in}} \quad (5)$$

where I_m is the PV array maximum current (A); V_m is the PV array maximum voltage (V); A is the effective area of the battery; P_{in} is the incident light power of unit area; and FF is the fill factor. And FF can be expressed as

$$FF = \frac{P_m}{V_{oc} \cdot I_{sc}} = \frac{V_m \cdot I_m}{V_{oc} \cdot I_{sc}} \quad (6)$$

Fill factor is an important parameter of photovoltaic battery performance. In general, $FF < 1$, the higher value indicates the higher photoelectric conversion efficiency. In order to improve the conversion efficiency of the solar battery, the short-circuit current I_{sc}, the open circuit voltage V_{oc} and the fill factor FF must be improved.

From the output characteristic of the photovoltaic battery, the solar battery efficiency will increase with the increase in light intensity, so the efficiency of the battery can be improved by increasing the illumination of unit area, i.e. using the condensing technology. However, the efficiency of the battery will decline with the increasing temperature, i.e. the solar battery conversion rate has a negative temperature coefficient. So, the use of the cooling measures will reduce the working temperature of the photovoltaic battery, making the concentration effect more prominent.

3 CONCENTRATING PHOTOVOLTAIC SYSTEM DESIGN

Because the incident light intensity is not up to the standard conditions (atmospheric quality 1.5, light intensity 1 kW/m² and environment temperature 25°C), the output power of conventional monocrystalline silicon photovoltaic cells is often less than the nominal value in practical applications. The condenser adopts a special lens that is larger than the area of solar panels, to ensure uniform sunlight. Concentration multiple can be controlled by the distance controller that can adjust the distance between the condenser and solar panels. When the natural light intensity is

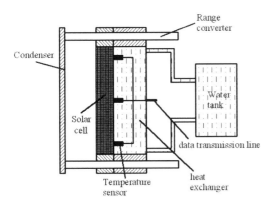

Figure 4. Schematic diagram of a concentrating photovoltaic system.

low, the condenser that increases the concentration multiple can make the PV battery work with higher efficiency (Takeo, W. et al. 2000). The concentration ratio is proportional to the size of the lens, but the large size of the lens has a large processing difficulty and high cost. Therefore, this method is only applicable for a low power condenser, with the concentration ratio ranging from 1 to 50. The rise in the operating temperature will lead to gains in efficiency. So, in order to ensure the efficiency of photovoltaic cells, the cooling system is used under the condenser condition. The concentrating photovoltaic system diagram is shown in Figure 4. The cooling system consists of two parts: water tank and heat exchanger. Water tank connects the water inlet and outlet of the heat exchanger by a plastic pipe, and an electric water pump is used to control the convective heat transfer rate. The working face of the heat exchanger, which works against the solar battery, can take the heat of the photovoltaic battery, realizing the goal of heat dissipation. The surface of the heat exchanger, which clings to the photovoltaic battery, has six holes with temperature sensors in it. Measurement data can be transmitted to the principal computer through the data line, realizing the real-time online monitoring of the solar battery temperature. By controlling the condenser multiples and the cooling water flow rate, the temperature of the PV battery can be controlled within a reasonable range.

4 PERFORMANCE TEST AND ANALYSIS

The test environment temperature is 20°C, and the cooling water temperature is 12°C. Figure 5 shows the changing rule of the photovoltaic battery temperature t in the process of the test with and without the cooling measures. It can be seen from the

diagram that without the cooling measures, the solar battery surface temperature increases quickly with the increase in the concentration multiple. After using the cooling measures, the surface temperature of the solar battery rises slowly with the increase in the concentration multiple.

In the actual concentrating photovoltaic system, there exists a specific concentration ratio. The conversion efficiency of the solar battery increases with increases in the concentration ratio when it is less than this value, and decreases when it is greater than this value. Figure 6 shows the relationship between the transformation efficiency and the concentration ratio in the process of the actual measurement. As it can be seen from the figure, the conversion efficiency of the solar battery increases with increases in the concentration ratio; when the concentration ratio is more than 40, the conversion efficiency of the solar battery reaches its limit value by 23.7%, and then the conversion efficiency begins to slow down as the concentration ratio increases. This shows that significant benefits can be obtained after cooling for the solar battery. Therefore, focusing on the conventional photovoltaic battery can increase the output power of the battery, so as to reduce the cost of photovoltaic power generation. The use of the cooling system

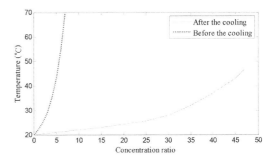

Figure 5. Photovoltaic battery temperature under different condensing ratios.

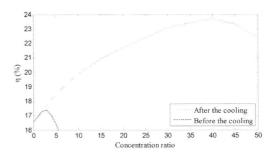

Figure 6. Photovoltaic battery efficiency under different condensing ratios.

can ensure that the solar battery has a relatively low temperature under the concentration condition, making the battery work efficiently.

5 CONCLUSION

Based on the research on the thermal and power characteristics of the conventional single crystal silicon photovoltaic battery under the condenser condition, it was found that the use of the condenser with the cooling system can not only improve the efficiency of the photovoltaic battery, but also improve the reliability of the photovoltaic power generation system. This is a feasible way to decrease the cost of solar photovoltaic power generation. The concentration ratio and the working temperature of the photovoltaic battery are the main factors that affect the output characteristics of the conventional single crystal silicon photovoltaic battery when it is used at a low concentration multiple. The use of the cooling system can ensure that the temperature of the solar battery falls below 60°C under the concentration condition, making the battery work with higher conversion efficiency.

ACKNOWLEDGMENT

The research was supported by the SGCC (State Grid Corporation of China) "Thousand Talents program" special support project (EPRIPDKJ (2014)2863).

REFERENCES

Ball, G.J. et al. 2002. Recent application and performance of large, grid-connected commercial PV systems. *Photovoltaic Specialists Conference, 2002 Conference Record of the Twenty-Ninth IEEE. 19–24 May 2002.* New Orleans, LA, USA.

Bendel, C. & Wagner A. 2003. Photovoltaic measurement relevant to the energy yield. *Proceedings of 3rd World Conference on Photovoltaic Energy Conversion. 18–18 May 2003.* Osaka, Japan.

Benner, J.P. & Kazmerski L.1999. Photovoltaics gaining greater visibility. *IEEE SPeetrum* 36(9):34–42.

Green, M.A. 2002. Third generation photovoltaics: comparative evaluation of advanced solar conversion options. *Photovoltaic Specialists Conference, Conference Record of the Twenty-Ninth IEEE. 19–24 May 2002.* New Orleans, LA, USA.

Masoum, M. et al. 2002. Theoretical and experimental analyses of photovoltaic systems with voltage and current-based maximum power-point tracking. *IEEE Transaction on Energy Conversion* 17(4):514–522.

Palz, W. 2000. P.V for the new century status and prospects for PV in Europe. *Renewable Energy World* 3(2):24.

Radziemska, E. 2003. The effect of temperature on the power drop in crystalicon solar cells. *Renewable Energy* 28(1):1–12.

Takeo, W. et al. 2000. Evaluation of Transmission Line Arresteis against Winter Lightning. *IEEE Transactions on Power Delivery* 15(2):684–701.

Taylor, R. et al. 1996. Opportunities and issues in international photovoltaic market development. *Photovoltaic Specialists Conference, Conference Record of the Twenty-Fifth IEEE. 13–17 May 1996.* Washington, DC.

Whitfield, G.R. et al. 2002. The development of small concentrating pv systems. *Photovoltaic Specialists Conference, 2002 Conference Record of the Twenty-Ninth IEEE. 19–24 May 2002.* New Orleans, LA, USA.

Advances in Power and Energy Engineering – Sun (Ed.)
© 2016 Taylor & Francis Group, London, ISBN 978-1-138-02846-3

Preliminary analysis of AP1000 PCCS and its enhanced performance

C. Li
Department of Thermal Engineering, Tsinghua University, Beijing, China

L. Li & Y.J. Zhang
Institute of Nuclear and New Energy Technology, Tsinghua University, Beijing, China

ABSTRACT: To analyze the advanced AP1000 nuclear power plant PCCS operation performance, a code was developed with the previous works adopted for the containment outside cooling. The one-volume model was used for the containment inner gas mixture heat transfer and the multi-volumes model was used for the containment wall conduction and for the containment outside cooling. The simulation results fitted well with those from WGOTHIC results and from Wang. By discussing the containment operation performances, the containment output performance was concluded. Finally, the scheme of suppression vessel was provided to improve the PCCS performance. The preliminary simulation shows that the scheme can effectively suppress the containment pressure and at the same time it improved the containment utilization efficiency and could also improve AP1000 competitiveness by reducing the construction investment.

1 INTRODUCTION

The containment functions have been extended up by restraining the radioactive materials within it to enhance nuclear power safety and it is also used as the last barrier of the effective defense system in depth. Loss of Coolant (LOCA) and main steam line break are two typical Pressurized Water Reactor accidents. The accident prevention and mitigation are two key features of the Generation III nuclear power plants and they could effectively prevent the reactor core vessel from melting. The Westinghouse AP1000 is the passive features (Schuz 2006, Lin 2008 & Westinghouse Electric Company 2008), typical of the Passive Containment Cooling System (PCCS), which depends upon the natural phenomena, such as gravity, buoyancy and condensation, to continually remove the reactor core decay heat out for days with no emergent power requirements. In March, 2011, the emergency power failure, to some extent, resulted in the Fukushima Daiichi severe nuclear accidents. This accident further indicates the importance and inherent safety of the passive features. Chinese has been developing the passive Generation III nuclear power plant since 2010.

Based on the PCCS heat removal techniques, the current work will firstly analyze the postulated LOCA accident scenario of AP1000 PCCS and obtain the pressure characteristics for the whole accident scenario. Accordingly, the preliminary analysis on the AP1000 PCCS operation characteristics is then discussed. An enhanced

PCCS heat removal method is then put forward and the postulated LOCA accident of AP1000 is used to preliminarily compare for the PCCS enhancement techniques.

2 AP1000 PCCS ACCIDENT SCENARIO

2.1 Code development

Rahim et al (2012) used the two-volume method simulated the AP1000 PCCS and they compared the results with those obtained from the Westinghouse. The WGOTHIC software was used for the AP1000 PCCS design and it has been seen to be valid from NRC for AP1000 PCCS safety analysis. Their results proved that the lumped parameter volume method was acceptable.

In order to effectively analyze the PCCS operation features, presently the code of one volume for containment inner free volume and multi-volumes for the outside water film cooling is developed. The lumped parameter volume method for the containment inner free volume is presently used for code development to predict the Double-Ended Cold Leg Guillotine (DECLG) LOCA. The whole PCCS operation includes scenarios of the inner condensation and broken pipe mass injection, the containment shell conduction and outside cooling. A schematic considering the AP1000 PCCS structure features and the important phenomena (PIRTs) is shown in Figure 1 and it is used for the analysis of the present code development.

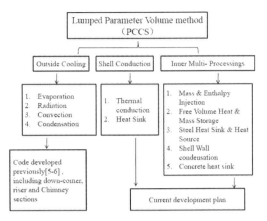

Figure 1. Code development schematics.

For the outside cooling, previous reports have given the code development specification by Li (2013, 2015). It simulated the PCCS outside down-comer, riser and chimney heat and mass transfer among air, water film and the containment shell. The mixed convection evaporation there used the Churchill relation, which was from Yang & Tao (2007) and was composed of the McaDAMS correlation (MacDAms 1985) and the Colburn correlation (Colburn 1964). Its total sub-volume number was 1950, which was larger than those of the WGOTHIC for the outside cooling and so it takes its advantages to achieve simulating the condensation around the chimney structures.

For the containment shell conduction and containment inner heat sinks, the transient heat transfer shows the same phenomenon. Thus, similar numerical method is used for both of them. Yu (1998) listed the main heat sink and heat source. To simplify the numerical calculation, these steel heat sinks were classified into two plates of the equivalent and fictitious thicknesses with the total volumes and the total heat transfer surfaces of being the same as the AP1000 PCCS. As the numerical code was used to analyze the enhanced heat removal technique and it was not used for safety analysis to accurately supporting the engineering design, those irregular heat sinks were simplified to the plate structure model (Rahim 2012 & Yu 1998).

For transient conduction heat transfer, if Bi number is less than 0.1, the lumped parameter conduction can be used accurately. However, for the AP1000 containment, the heat sinks, such as the crane and the containment shell, are such thick that the equivalent heat transfer coefficient is relatively large and the lumped parameter method cannot be directly used for the transient conduction. Thus, the total thickness here were split into

10 even sub-volumes, so that the temperature difference within every sub-volume was small enough and its average temperature could indicate the temperature feature of this sub-volume. Thus, the lumped parameter method was valid for the steel conductor every sub-volume. The conservation of energy equation within every sub-volume was:

$$\int \rho C \delta_{steel} dt = \int \left(q_{in} - q_{out}\right) d\tau \tag{1}$$

Thus, the lumped parameter temperature, t, in the above equation was used for the heat flux by heat sink effect and the heat flux from one sub-volume to the next sub-volume. Figure 2 showed the schematic of the steel wall conduction. The heat from the steel plate or shell surface entered sub-volume i and this sub-volume would be heat up and at the same time it could heat up the sub-volume j. if Bi was less than 0.1, the local temperature, $t_{i,1}$ an $t_{i,2}$ were approximately the same as t_i.

The heat flux entering sub-volume i was:

$$q_{in} = h_{conv}\left(t_{bulk} - t_i\right) \tag{2}$$

The heat flux leaving sub-volume i from the fictitious boundary between sub-volume i and sub-volume j was seen as:

$$q_{out} = \lambda \frac{t_i - t_j}{\delta_{sub}} \tag{3}$$

If the sub-volume included the steel surface wall, the Uchida correlation (Yu 1998) was used for the surface heat flux. The heat transfer coefficient was:

$$h_{Uchida} = 450.5\left(\frac{\rho_a}{\rho_g}\right)^{0.8} \tag{4}$$

Equation (1) can also be used for the concrete heat transfer. However, as the concrete thermal conductivity was small enough and it was relatively thick, its thickness was split into 50 sub-volumes.

The perfect gas law was used for the containment inner gas mixture. The gas mixture conservation of

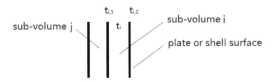

Figure 2. Sub-volume schematic for steel plate.

energy equation within the containment free volume was:

$$\Delta E = \int Q d\tau + \int m_{in}\left(h + \frac{1}{2}u^2 + gz\right)_{in}$$

$$- \int \sum_{i=1,2,3} m\left(h + \frac{1}{2}u^2 + gz\right) d\tau \qquad (5)$$

The first term in the right hand of Equation (5) was the heat source, the transient heat transferred into the free volume from the hot Accumulator, the hot steam generator and the hot pipes. The second term was the total enthalpy from the broken pipe for DECLG LOCA (Westinghouse Electric Company 2008). The last term was the condensation heat transfer on the surface of the containment, the steel plate and water tank. The subscript 1, 2 and 3 in Equation (5) are the containment inner wall surface condensation enthalpy, the steel plate surface condensation enthalpy and the concrete surface condensation enthalpy.

Assuming that there is no air leakage from the containment inside to the airside, the total mass and the density of the dry air in the containment is then constant. For the severe accident of DECLG LOCA, experimental results for AP1000 showed that the gas and the steam injected in the containment could mix well (Lin 2008). Thus, as the perfect gas, the dry air partial pressure in the free volume could be expressed as the function of temperature for the whole accident scenario:

$$p_{air} = p_0 \frac{t + 273.15}{t_0 + 273.15} \qquad (6)$$

However, the total mass of the steam within the containment changes all the time during the entire accident scenario. Thus, the conservation of steam mass equation was:

$$\frac{dm}{d\tau} = m_{in} - \sum_{i=1,2,3} m \qquad (7)$$

The AP1000 containment maximum operation pressure is around 0.5 MPa, so the steam partial pressure is less than 0.4 MPa, and the steam compressibility factor is among 0.95–1. Thus, the steam can also be seen as a perfect gas. The steam partial pressure is:

$$p_{steam} = \frac{mR(t + 273.15)}{V} \qquad (8)$$

The above equations are used for the current code development. The specific parameters for AP1000 PCCS are listed in Table 1.

Table 1. Geometry & initial conditions (Rahim 2012).

Setting	Description	Value (unit)
V	Containment volume	58969.067 (m²)
$A_{out.cont}$	Outer active area of containment	5934.8567 (m²)
$A_{in.cont.}$	Inner active area of containment	5922.1553 (m²)
$A_{conc.}$	Active area of concrete shield	6552.1346 (m²)
$A_{AirBafle}$	Active area of air baffle	4258.6595 (m²)
A_{gap1}	Active area of gap1	6183.729 (m²)
A_{gap2}	Active area of gap2 (riser)	5984.7455 (m²)
δ_{cont}	Containment thickness	0.0444 (m)
δ_{gap2}	Riser thickness	0.6561 (m)
$\delta_{AirBaffle}$	Air baffle thickness	0.015 (m)
δ_{gap1}	Down-comer thickness	0.6561 (m)
$\delta_{cont.}$	Concrete shield thickness	0.9144 (m)
$d_{in.cont}$	Internal radius of containment	19.812 (m)
$d_{in.conc}$	Internal concrete shield radius	21.132 (m)
t_0	Initial containment temperature	50 (°C)
RH	Initial relative humidity	48 (%)
p_0	Initial containment pressure	0.1082 (MPa)
$p_{ColdLeg}$	Cold leg pressure	15.9268 (MPa)
$t_{ColdLeg}$	Cold leg temperature	280.66 (°C)
m_{air}	Inlet air mass flow rate	788.05 (kg/s)
h	Active height of heat transfer	47.8209 (m)

The AP1000 accident scenario could be divided into four stages. They are the blow-down, re-flood, pressure peak and the long term depressurizing. There are different mass and enthalpy injection from the broken pipe for these stages. Their effects on the containment pressure here are treated as the mass and enthalpy inlet boundary conditions. The AP1000 PCCS injection curves for the mass and the enthalpy were are given by Rahim et al (2012) and Wang (2013), respectively.

The current code was developed by the Fortran 90. It was used to predict the AP1000 PCCS DECLG LOCA accident scenario and at the same time to obtain the operation pressure features. Its results were compared with those from Westinghouse for AP1000 and from Wang (2013). The comparisons were shown in Figure 3. Generally, the current results also showed the similar two pressure peaks. It indicated that the present simplified code could reasonably predict the containment accident scenario.

There were two pressure peaks with the latter peak being higher. They respectively appeared at around 30 s and at 1800–2000 s. The first pressure peak for the current code was a little later than that by WGOTHIC software (Westinghouse Electric Company 2008, Gavrilas 1996) and it is almost the same as those from Wang (2013). It is possible for the

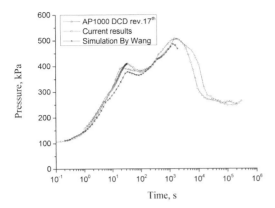

Figure 3. Comparisons with the WGOTHIC & Wang for DECLG LOCA.

differences that the present simplified plate model for heat sink causes some distortions of certain heat sink, as the heat sinks with different thicknesses and with different shapes have different response time for the transient heat transfer. Yu (1998) listed the complex heat sinks in the containment. Their thicknesses are from 2.3 mm to 76.2 mm. However, the current work was not for PCCS safety analysis, it is valid to predict the gas mixture pressure behavior within the containment free volume.

2.2 *Performance and analysis*

It can be concluded from the above analysis that the containment heat removal is mainly by means of the inner free volume storage and the condensation of both the containment wall and inner cold structures. By analyzing the effects of different heat removal ways on the whole scenario pressure, it could be concluded that the functions of the containment inner free volume storage is crucial for the initial stages and only it could store large energy and mass transiently. In contrast, the heat removal by surface condensation heat transfer need much more time. However, the containment storage capacity is limited by its design operation pressure. As is shown in Figure 3, the second peak appears at around 1800–2000 s. Before and after this time, the storage effect gradually disappears and the design operation pressure is meaningless as the local pressure is much lower that the designed one. Thus, there are mainly two factors which affect the PCCS heat removal performances. They are the peak pressure and the condensation heat transfer characteristics, with former restricting the nuclear power plant design pressure and construction investment.

Although the composed scheme of the inner free volume storage and the wall condensation

can conservatively restrain the containment inner gas pressure, the containment potential performance is not explored totally. The first example is that only a little time interval of around 1000 s operates around its peak pressure ranges. For all the accident evolution except this time interval, the PCCS operating pressure is very low. It means that the designed PCCS of relatively large volume can efficiently output only for this time interval. Thus, the optimized heat removal enhancement method is encouraged to make the inner free volume storage capacity and the wall condensation capacity match each other, so that the wall condensation and evaporation heat removal can be efficient. The second example is that for the same operation pressure, the condensation heat removal capacity is influenced by the non-condensable gas. The gas is the atmosphere air, which has the same density for the entire accident. If the non-condensable gas concentration becomes low enough, the condensation heat removal can be enhanced (Chun, Kim & Park 1996) too.

3 ENHANCED TECHNIQUE AND DISCUSSIONS

The discussions above show that the present PCCS performance on the core decay heat removal is not the optimized one. The current objective is to reduce the containment design pressure to make it have a lower economic investment. Based on the above two example, a method of the suppression vessel measure is proposed. The vessel, shown in Figure 4, contains both gas and the liquid phases and it is at the atmosphere pressure before accidents. The gas phase is the moisture air and the liquid part is the cold water. Their volume ratio should be calculated to make sure that the vessel can effectively operate. For severe accidents, the pressure signal or temperature signal in the containment triggers

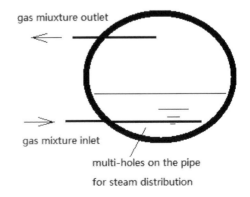

Figure 4. The conceptual suppression vessel schematic.

the vessel inlet pipe, then the steam and air mixture enter the liquid phase space due to the pressure difference between the containment and the vessel. The gas mixture is redistributed at the end of the pipe so that they can have a large surface-volume ratio when they meet the cold water in the vessel. As a result, most of the steam is transformed into liquid water by direct condensation. The condensation latent heat heats up the water temperature. At the same time, the gas and the residual steam go into the gas space and they heat up the air space by direct mixing. As a result, the pressure in the vessel also increase until the direct condensation suppression effects disappear and it reaches the similar pressure with the containment. When the containment pressure drops later and the pressure will be larger than the vessel. A trigger signal is used to release the air via the pipe exit and they return the containment. The valve closes down when the pressure difference is small enough. Thus, the vessel can suppress the containment pressure and at the same time it has the self-adaptation function for any non-DBA accidents. Besides, during the vessel operation, much air within the containment flows into the vessel and is stored and compressed within the air space of the vessel. As a result, the air concentration in the containment becomes lower and at the same time heat removal capacity of the containment wall condensation is enhanced.

For the preliminary scheme, a vessel of 4000 m³ is used with the air-water volume ratio of 20. The trigger pressure is of 440 kPa and then the suppression rapidly takes in lots of gas and steam mixture from the containment free volume. The Chun correlation (Norman 2007) is used to predict the direct condensation heat and mass transfer. The heat and mass transfer for the gas phase or the liquid phase within the vessel is similar with that of the containment free volume, so the lumped parameter volume method is also valid for vessel heat removal simulation. Finally, the direct condensation within the vessel is developed and is added to the current numerical code. Numerical calculation is applied for the new simulation. The results are compared with those from AP1000.

It is shown in Figure 5 for the scheme that the second containment peak pressure becomes significantly low and the maximum operation pressure can reduce by around 10%, which may bring an obvious investment reduction. At the same time, the time interval for high operation pressure becomes longer.

Besides, about 30% air in the containment is compressed into the vessel. According to Equation (4), the containment wall condensation heat transfer is enhanced. Li (2013, 2015) had made the numerical simulation on the outside cooling. They concluded that the outside water film evaporation

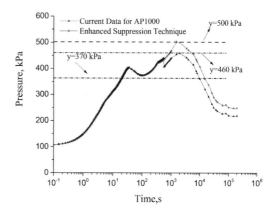

Figure 5. Pressure comparisons for the enhanced and the prototype ones.

capacity is higher than that of the containment inner wall condensation. The possible reason is that the condensation is a natural convection mass transfer while the water film evaporation is the forced one. Thus, the main resistance of the core decay heat removal is on the condensation side. The enhanced condensation will improved the containment heat removal capacity.

4 CONCLUSIONS

To master the passive containment feature, a code was developed, based on the former code on the outside cooling by Li (2013) & Li (2015). The containment cooling characteristics were then obtained and the pressure variation within the containment free volume was obtained. By comparing the results with those from Westinghouse and from Wang (2013), it can be seen that the one volume method for the containment free volume simulation is acceptable. Some little differences with Westinghouse results may be of the simplified heat sink classification. Besides, results from Wang (2013) shows the similar differences with those from Westinghouse results.

The analysis on the PCCS indicates that it cannot output well as only around 1000s is next to the maximum operation pressure. The design pressure is invalid except this interval. Thus, a preliminary scheme by water suppression is provide and a 4000 m³ vessel is added to the PCCS system with the trigger pressure of 440 kPa. The results show that the new scheme could suppress the containment pressure of around 10%, which can greatly reduce the PCCS material investment and engineering difficulties and at the same time improve the containment utilization efficiency for nuclear power accident.

Similar scheme can be also worth further analyzing. For example, the whole containment free volume is split into two similar containments with one for the primary loop arrangements as the current AP1000 and with another being used for the suppression. Thus, the containment will be not as high as skyscrapers. Besides, further studies on the economical efficiency of the improved scheme are needed.

REFERENCES

Chun, M.H. Kim, Y.S & Park, J.W. 1996. An investigation of direct condensation of steam jet in sub-cooled water, International Communications of Heat and Mass Transfer 23(7): 947–958.

Colburn, A.P. 1964. A Method of Correlating Forced Convection Heat Transfer Data and a Comparison with Fluid Friction, International Journal of Heat Mass Transfer 7(12): 1359–1384.

Gavrilas, M. Hejzlar P. & Todreas, N.E., et al. 1996. Gothic code evaluation of alternative passive containment cooling features, Nuclear Engineering and Design 166(3): 427–442.

Li, C., Fan, H.R. & Zhao, R.C. 2013. The outside transport phenomena of the passive containment cooling system, Proceedings of the ASME 2013 Summer Heat Transfer Conference, Minneapolis, July 14–19.

Li, C., Fan, P.C., & Fan, H.R. 2015. Effects of the upcomer width on the PCCS heat removal, Proceedings of the 23th International Conference on Nuclear Engineering, Chiba, May 17–21.

Lin, C.G. 2008. An Advanced passive plant AP1000, Atomic Energy Press. Beijing: 195–197.

Mcadams, W.H. 1985. Heat Transmission (18th), McGraw—Hill, Singapore, 1985, pp: 65–175.

Norman, T.L. 2007. Steam-air mixture condensation in a subcooled water pool, Ph.D. Thesis, Purdue University Graduate School. West Lafayette: 61–62.

Rahim, F.C. Rahgoshay, M. & Mousavian, S.K. 2012. A study of large break LOCA in the AP1000 reactor containment, Progress in Nuclear Energy 54: 132–137.

Schulz, T.L. 2006. Westinghous AP1000 advanced passive pant, Nuclear Engineering and Design 236: 1547–1557.

Wang, Y. 2013. Preliminary Study for the Passive Containment Cooling System Analysis of the Advanced PWR, Energy Procedia 39: 240–247.

Westinghouse Electric Company, 2008. Design of structure, components, equipment and systems, AP1000TM Desgin Control Document (Rev.17) 2: 55–59.

Yang, S.M. & Tao, W.Q. 2007. Heat transfer (4th edition), High Education Press. Beijing: 5–10, 369–370.

Yu, J.Y. 1998. Three dimensional research of advanced pressurized water reactor passive containment cooling system, Doctor thesis, Tsinghua University: 75–77.

Advances in Power and Energy Engineering – Sun (Ed.)
© 2016 Taylor & Francis Group, London, ISBN 978-1-138-02846-3

Application of the Particle Swarm Optimization method in rectangular flue gas ducts

Q.C. Niu, B. Yan & X.K. Fan
College of Aerospace Engineering, Chongqing University, Chongqing, China

Y. Yuan & D.M. Wu
Southwest Electric Power Design Institute, Power Engineering Consulting Group Corporation, Chengdu, China

ABSTRACT: Rectangular flue gas duct, composed by pipe, reinforcement rib and internal overarm support, is of highly importance for thermal power planet. At present, the arrangement of reinforcement rib and internal overarm support in huge flue gas duct is mostly based on technical regulations, not on optimization analysis. To optimize the rectangular flue gas duct, an optimization plugin of finite element analysis software ABAQUS is developed by combining Particle Swarm Optimization (PSO) algorithm with ABAQUS using Python language. This plugin is designed for modeling, computing and optimizing the rectangular flue gas duct. By counting the lightest weight of structure as the goal and the requirements of strength, stiffness and dynamic characteristics as constraint conditions, the optimal number and the distribution of transverse and longitudinal reinforcement ribs as well as the internal overarm supports can be calculated. Numerical examples show that the optimization plugin is valid and stable.

1 INTRODUCTION

Proposed by Eberhart and Kennedy in 1995 (Eberhart, R.C. 1995), the Particle Swarm Optimization (PSO) algorithm was designed for solving optimization problem by simulating the behavior of the flock, considering that there is much similarity between flock behavior rule and optimization problem. The PSO algorithm is particularly applicable to large-scale optimization problem (Ellips, M. et al. 2010), for the characteristics of inherent parallelism and distributed procession it remains. With high efficiency, good convergence and simplicity in calculation, the PSO algorithm is widely used in various fields.

Rectangular flue gas duct, bearing a wide range of loads such as wind load and internal pressure, is made of pipe, reinforcement rib, bracket, hanger, internal overarm support and compensator. Either the waste of material or accident may occur if the design of the system is inappropriate. The design of flue gas duct, constituted by throttle control device, platform escalator and pipe arrangement, offering a high demand of stiffness, strength, endurance and stability for pipe, part, bracket and hanger (Yuan, J.Y. 2011). However, because the restriction of design calculation method, the component of pipe set may be unreasonable and consequently wasting material or causing accident. At present, the arrangement of reinforcement rib

and internal overarm support in huge flue gas duct is mostly based on technical regulations, not on optimization analysis. Therefore, there is much significance in optimization of reinforcement rib and internal overarm support.

This paper combines PSO algorithm with finite element software ABAQUS, using the Python language. The optimization plugin is derived from the redevelopment function of ABAQUS (Cheng, L. et al. 2009). By regarding the lightest weight of structure as the goal and the requirements of strength, stiffness and dynamic characteristics as constraint conditions, the optimal number and the distribution of transverse and longitudinal reinforcement ribs as well as the internal overarm supports can be calculated.

2 PARTICLE SWARM OPTIMIZATION ALGORITHM

2.1 *The standard particle swarm algorithm*

The PSO algorithm is one kind of intelligent optimization algorithm. It simulates the process of a flock of birds looking for food. Each bird is a particle in the PSO. They constantly change their position and speed flying in the air until find food.

The PSO algorithm regards each individual as a particle without quality and volume in D dimension. Moreover, this particle can fly at a certain speed, which is decided by individual experience and group experience (Lin, Q. et al 2015). The speed equation of standard particle swarm optimization algorithm is as follow:

$$v_j(t+1) = w * v_j(t) + c_1 * r_1 * (p_j(t) - x_j(t)) \\ + c_2 * r_2 * (p_g(t) - x_j(t)) \quad (1)$$

where $v_j(t)$ = speed of particle j at the t generation; $x_j(t)$ = position of particle j at the t generation; w = inertia weight; $c_1 = c_2$ = learning factors; $r_1 = r_2$ = random numbers of uniformly distributed between 0 and 1; $p_j(t)$ = optimal position of the individual history of the particle j; $p_g(t)$ = optimal position of group history.

Correspondingly, the position equation of standard particle swarm optimization algorithm is as follows:

$$x_j(t+1) = v_j(t+1) + x_j(t) \quad (2)$$

2.2 Constriction coefficient

The learning factors c_1 and c_2, reflecting the exchange of information among particles, determine the particle's experience information and the influence of other particle's experience information on this particle's movement (Poli, R. et al. 2007). Bigger c_1 will make the particles wander around the local area while bigger c_2 will urge the particles to converge to a local minimum.

In order to control the flying speed of particles effectively, the algorithm should achieve the effective balance between global exploration and local development. Clerc constructed the PSO algorithm for introducing the constriction coefficients. The speed updates formula as follows:

$$v_j(t+1) = \varphi\{v_j(t) + c_1 * r_1 * [p_j - x_j(t)] \\ + c_2 * r_2 * [p_g - x_j(t)]\} \quad (3)$$

$$\varphi = \frac{2}{\left|2 - C - \sqrt{C^2 - 4*C}\right|}, \quad C = c_1 + c_2 \quad (4)$$

To ensure the smooth solution of the algorithm, the constant $c_1 + c_2$ must be greater than 4. Generally speaking, the constant $c_1 = c_1 = 2.05$ when the constant C is 4.1 and the constriction coefficient φ is 0.729. This formula is equivalent to the standard PSO speed formula when $c_1 = c_1 = 1.4962$ and $w = 0.7298$.

2.3 The discrete particle swarm algorithm

The basic PSO and its improvement algorithm are mainly designed for finding the optimal value of the function in the continuous domain. Experiments proved the algorithm to be an effective tool. However, in practice, a great many problems are arisen in discrete space like scheduling, path. Generally speaking, the standard PSO can't be used for discrete problem because the obtained solution by general PSO cannot guarantee that it must be in the discrete space. Therefore, the algorithm lost its advantages. In order to solve the disadvantages in discrete problems, a rounded particle swarm algorithm has been proposed.

This rounded method is simple and effective in solving discrete problems, which does not change any part of PSO. Just in calculating the fitness function, it transforms the particle's position from the real space to the integer space at first. In addition, the way of conversion is also very simple, for it directly takes the integer part of the particle position.

3 PIPELINE OPTIMIZATION PLUGIN DESIGN

3.1 The application of PSO in rectangular flue gas duct

The rectangular flue gas duct is mainly made up of pipe, reinforcement rib, bracket and hanger, internal overarm support and a variety of compensators. Pipes are divided into straight pipe, curved pipe and all kinds of special-shaped connection pipe (Hu, W.Y. 2013). This paper focuses on the optimization of reinforcement rib and internal overarm support on straight pipe.

The commercial finite element analysis software ABAQUS is compiled by Python language. Using Python, the GUI can be developed in ABAQUS. In this paper, an optimization plugin is developed through the redevelopment function of ABAQUS. The interface of the plugin is shown in Figure 1.

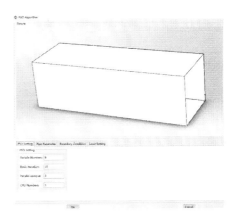

Figure 1. The interface of the optimization plugin.

In this paper, optimization plugin has been written for rectangular flue gas ducts by Python through combining the commercial finite element analysis software ABAQUS and PSO optimization algorithm.

In the optimization process, the optimization variables are composed of seven variables. They are the arrangement number of transverse reinforcement rib, the type of transverse reinforcement rib, the arrangement number of longitudinal reinforcement rib, the type of longitudinal reinforcement rib, the arrangement number of transverse internal overarm support, the support arrangement number of longitudinal internal overarm and the type of internal overarm support. Users can choose some of the variables or all of the variables to optimize. The constraint condition is the condition for the stress, displacement and frequency to satisfy the user's inputs. The optimization objective is to make the pipe the lightest weight.

Mathematical model of straight pipe is as follows:

$$min\ f(x) = W_h + W_z + W_n \tag{5}$$

$$S_{max} < S' \tag{6}$$

$$f > f' \tag{7}$$

$$d_{max} < d' \tag{8}$$

where W_h = weight of transverse reinforcement rib; W_z = weight of longitudinal reinforcement rib arrangement; W_n = weight of internal overarm support; S_{max} = maximum stress; f = frequency; D_{max} = maximum deflection.

The implementation of the optimization algorithm shows in Figure 2.

3.2 Example analysis

In this paper, the optimization plugin is used to optimize the rectangular flue gas ducts. The duct is under the action of gravity, internal pressure, wind load, ash load, and heat insulation layer. The initial input parameters are shown in Tables 1 and 2. Results of optimization are shown in Table 3.

As shown in Table 3, under the condition of satisfying the stress, deflection and frequency constraint, the engineering calculation need the total weight of reinforcement rib and internal overarm supports 7001.5 kg; however, the PSO optimization algorithm just need 4800.9 kg. The material for rectangular flue gas duct is decreased by 31.4% according to the optimization result, comparing to that according to the engineering calculation.

Figure 3 shows the stress, frequency and displacement distribution of ducts under the action of the above loads after optimization.

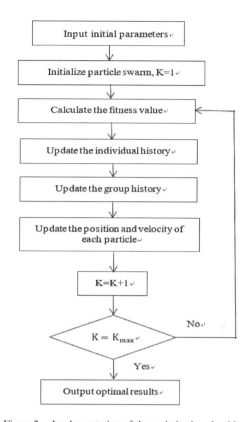

Figure 2. Implementation of the optimization algorithm.

Table 1. Initial geometric parameters.

Parameters	Value
Pipe length (mm)	20000
Pipe width (mm)	7000
Pipe height (mm)	6000
Thickness (mm)	8
Bracket 1 position (mm)	5000
Bracket 1 type	Fixed
Bracket 2 position (mm)	15000
Bracket 2 type	Fixed
Hanger 1 position (mm)	10000
Hanger 1 type	Fixed

Table 2. Load parameters.

Load	Value
Gravity (mm/s²)	9800
Internal pressure (kPa)	−4
Wind load (kPa)	0.5
Ash load (kPa)	7.5
Heat insulation layer (kPa)	0.24

239

Table 3. Optimization results.

Variable name	Engineering calculation results	PSO
Transverse reinforcement rib distance (mm)	1000	1250
Transverse reinforcement rib type	P-I10	P-C12_6
Internal overarm support number on side	1	
Internal overarm support number on top and button	2	
Internal overarm support type	Ø57 × 3	
Weight (kg)	7001.5	4800.9

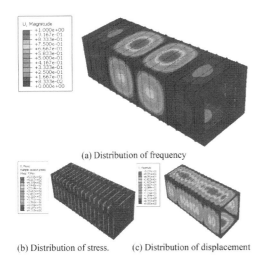

(a) Distribution of frequency

(b) Distribution of stress.　　(c) Distribution of displacement

Figure 3. Distribution of stress, frequency and displacement.

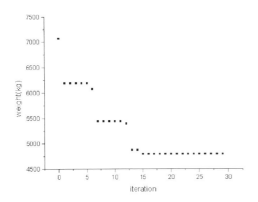

Figure 4. Optimal value varying with the number of iterations.

The optimal values of the iterative optimization process are shown in Figure 4.

Figure 4 indicates that when the number of iterations is 15 times, with the increase of the number of iterations, the optimal solution tends to be stable. At this point, the optimal solution converges to the theoretical solution.

4 CONCLUSIONS

To solve the discrete optimization problem with constrained condition, rounding method and the PSO algorithm with constriction are adopted. The PSO algorithm is embedded into the commercial finite element analysis software ABAQUS with Python language. The redeveloped plugin package can be applied in rectangular flue gas duct without error. The optimal number and the distribution of transverse and longitudinal reinforcement ribs as well as the internal overarm supports can be calculated with iterative solution. This plugin package is designed for the design and analysis of rectangular flue gas duct.

REFERENCES

Cheng, L. et al. 2009. Second development of ABAQUS based on the scripting interfaces. *Modern Machinery* (2): 58–59.

Eberhart, R.C. & Kennedy, J. 1995. A new optimizer-using particle swarm theory. *In Proceedings of the sixth international symposium on micro machine and human science*: 39–43.

Ellips, M. & Davoud, S. 2010. Multi-objective robot motion planning using a particle swarm optimization model. *Journal of Zhejiang University-Science C (Computers & Electronics)* (08): 607–619.

Hu, W.Y. 2013. Software redevelopment for finite element modeling and computation of circular flue gas ducts based on ABAQUS. Chongqing University.

Lin, Q. et al. 2015. A novel multi-objective particle swarm optimization with multiple search strategies. *European Journal of Operational Research* 247(3): 732–744.

Poli, R. et al. 2007. Particle swarm optimization. *Swarm Intell*: 33–57.

Yuan, J.T. 2011. Finite element analysis method for the design of air & circular flue gas duct in thermal power plant. Chongqing University.

Advances in Power and Energy Engineering – Sun (Ed.)
© 2016 Taylor & Francis Group, London, ISBN 978-1-138-02846-3

Software redevelopment for rectangular flue gas ducts based on ABAQUS

X.K. Fan, B. Yan & Q.C. Niu
College of Aerospace Engineering, Chongqing University, Chongqing, China

Y. Yuan & D.M. Wu
Southwest Electric Power Design Institute, Power Engineering Consulting Group Corporation, Chengdu, China

ABSTRACT: Rectangular flue gas ducts of thermal power plants are usually affected by complex external loads. An unreasonable design may lead to immeasurable economic loss and engineering accidents. The commercial software based on the finite element method has been developed to a higher level, which is one of the most important means to solve this problem. For modeling and analysis of the flue gas ducts system, software is redeveloped based on ABAQUS and Python language. Finite element models of the ducts can be established quickly by using the redeveloped software, which means that strength, stiffness and stability analysis of the ducts system can be carried out easily. Our implemented project is trying to offer a new way of calculation and analysis for engineers and thus to meet the practical application for the purpose of low-cost and efficiency in the construction of thermal power plants.

1 INTRODUCTION

With the rapid development of economy and the growing demand for electricity in China, the size of the thermal power plant is also getting larger, leading to the increase in the section area of the rectangular flue gas ducts. Flue gas ducts have been regarded as the core part of the thermal power plant. Currently, the technical specification for the design of thermal power plant flue gas ducts is very simple for the simplification of the model and calculation analysis (Zhang, Q. et al. 2011). The technical specification cannot calculate the overall analysis of the ducts system. The above situations may inevitably cause not only the unreasonable design of the duct system, but also the waste of processed materials and even security issues (Lian, C.W. 2006). The practical application shows that the technical specifications cannot meet the design requirements of the modern thermal power plant, while the design of rectangular flue gas ducts is the key issues for the safety operation of the power plant. So, modeling and analysis of the flue gas ducts system play a key role in the design phase.

As a result of the complexity of the ducts structure and the variability of the load factor, the design process of the ducts system becomes considerably complex. In order to improve design efficiency (Li, X.C & Cen, E.F. 2007), it becomes an urgent necessity to develop a targeted software system to achieve a satisfied goal of rapid modeling, analysis and design of the flue gas ducts system. The commercial software based on the finite element method is developed to a higher level. Software is redeveloped based on the current advanced universal finite element software ABAQUS platform. The object-oriented Python language and parametric modeling methods are used in the redeveloped software, which will be a very effective tool for engineers to calculate and analyze during the design phase.

2 FRAME OF THE SOFTWARE

The purpose of redeveloped software systems is to reasonably design rectangular flue gas ducts for the thermal power plant, and the software is convenient for the engineering designer for modeling and analysis quickly and accurately. At the same time, static, dynamic and stability analysis of flue gas ducts can be carried out. The software uses the ABAQUS/CAE powerful open interface function and the standard graphical user interface. The plug-in modules suitable for the rectangular flue gas ducts system are redeveloped based on the basic module of ABAQUS (Cheng, L. et al. 2009). Basic frame of the redeveloped software based on the design method and user requirements is shown in Figure 1.

"Ducts Component Module" is used to create the main components of flue gas ducts, such as duct, reinforcing ribs, supporting steel bars, supports, hangers and a variety of compensators. It is required

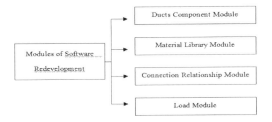

Figure 1. Basic modules of redevelopment software.

to analyze the geometric characteristic parameters of each component and use the simple way to orientate the spatial position of each component.

"Material Library Module" is used to manage the mechanical parameters of materials and profiles of section steel, and it is convenient to directly recall and edit materials parameters for designers.

"Connection Relationship Module" is used to define the connection relationship between the various components, including bind connection, contact and compensator connection relationship.

"Load Module" is used to define the loading surface and direction on the component, and calculate the size of various loads according to the relevant power facilities design specification.

3 METHOD OF MODELING

Based on the interfaces and parametric modeling methods of the ABAQUS /CAE software Plug-in program, finite element modeling of the duct system can be divided into geometric modeling, material parameters, grid cell division, defining boundary conditions and load applying part.

Rectangular flue gas ducts system is mainly composed of pipes, supports and hangers, reinforcing ribs, supporting steel bars and a variety of compensators. According to the needs, engineering designers select only the ducts component menu from the ABAQUS/CAE Graphical User Interface (GUI), and enter the geometric feature parameters, the overall coordinate of the reference point and rotation angle of each component, and modeling and assembly will be achieved automatically.

In the structure design of flue gas ducts, the structural steel Q235, Q345, #10, #20, are usually used for material, and the physical and mechanical properties of materials are mainly Young's modulus, Poisson's ratio, density and thermal expansion coefficient. In general, the component's material is in a linear elastic state. Reinforcing ribs welded on the outer surface of the ducts are mainly made of section steel, such as T-section steel, L-section steel, I-section steel, H-section steel and channel. For the negative pressure duct, in order to prevent instability,

sometimes we also need to set up supporting steel bars; generally, hot-rolled seamless steel tubes are used for supporting steel bars. Because types of section steel are more, it is necessary to establish a database of the material and section steel, so that designers can quickly recall and modify material parameters according to their own demand.

Connection relationships of duct system components are mainly welding, sliding contact and compensation connection. Tie function of ABAQUS can be used to define the binding connection relationship. The duct in the heating system will lead to flexible deformation, in order to reduce the heat stress caused due to the structure. On the one hand, a compensator, such as metal compensator, is simulated by setting the connection unit and axial stiffness. On the other hand, sliding contact between the bottom of the bracket and the limit base can be simulated by setting the contact pair and the friction coefficient (Guo, X. et al. 2013).

Shell element, spatial beam element, three-dimensional (3D) solid element and spring element are applied to the discrete finite element model of flue gas ducts (Hu, W.Y. 2013). Generally, the ducts are welded by the steel plate, thickness is relatively thin, and can be simulated by the shell element. Reinforcing ribs and stand bars are discretized by the spatial beam element because their section dimensions are very small relative to the length. The limit base bearing compression deformation should be discretized by the 3D solid element, and the compensator can be simulated by the spring element. Meshing is very important in the calculation, and the program will automatically separate element by using the ABAQUS original meshing module.

In practical applications, the flue gas ducts subject to various complex loads should consider the weight of the load, internal pressure, wind load, snow load, ash load, insulation and static seismic load in the design phase (Yuan, J.T. 2011). For each specific load in this redeveloped software, engineering designer only needs to enter the corresponding parameter and selects the appropriate load acting plane from the drop-down menu, and then programs automatically to calculate the load size and loads to the corresponding surface.

4 ANALYSIS AND RESULTS

The finite element model of this application case is established according to the rectangular flue gas ducts CAD design graphics of a thermal power plant, which includes a straight duct, elbow duct, compensation, damper, rectangular reinforcing ribs, horizontal rectangular duct limit bracket, and rectangular limit base and other components, as shown in Figure 2.

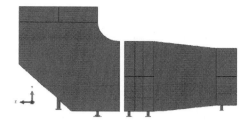

Figure 2. The finite element model of a flue gas duct.

Each component material of this model is selected in Q235, the thickness of the pipe is 12 mm, and reinforcing ribs are made of channel steel. Tie is defined as connection relationships, and the bottom of each bracket is fixed. The finite element model is divided into 127832 elements.

4.1 Mode analysis

In order to avoid the resonance phenomenon, a modal analysis is carried out to calculate the first 10 steps of natural frequencies and modes. The natural frequencies are listed in Table 1, and the corresponding first four steps of modes are shown in Figure 3.

It can be seen from Figure 3, the lateral stiffness of the duct section is small. When the frequency of the vibration is low, the duct easily leads to large deformation. The natural frequency of the duct system should be improved, and it is valid to increase the number of reinforcing ribs or select the proper section of steel models. If the requirements are still not met, addition to supporting steel bars should be suggested.

4.2 Buckling analysis

Working pressure of flue gas ducts is generally negative, which is the reason why buckling analysis is carried out. The first 10 buckling modes are calculated and corresponding critical buckling pressures are required. Table 2 lists the first 10 corresponding buckling eigenvalues. Figure 4 shows the first and third buckling modes.

Based on the buckling analysis results, the size of the minimum buckling pressure of the duct is 0.019 MPa and the working pressure of the ducts is 0.0084 MPa, so it will be stable under the negative pressure.

4.3 Static analysis

The static response of the model under combined loads is calculated to analyze its security. The load considered for static analysis is given in Table 3.

Figure 5 shows the stress distribution under the action of the above loads. It can be clearly seen

Table 1.   Natural frequencies of the first 10 modes.

Mode	Frequency
1	1.062
2	3.051
3	3.228
4	3.231
5	3.265
6	3.322
7	3.412
8	3.517
9	3.868
10	4.196

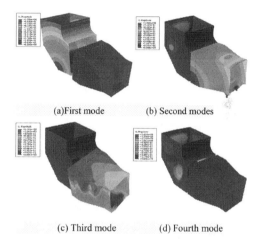

(a)First mode   (b) Second modes

(c) Third mode   (d) Fourth mode

Figure 3.   The first four modes.

Table 2.   Eigenvalues of the first 10 buckling modes.

Buckling mode	Eigenvalue
1	0.019
2	0.025
3	0.026
4	0.028
5	0.029
6	0.033
7	0.035
8	0.036
9	0.039
10	0.046

that the maximum stress is 207 MPa, and the place of the maximum stress is located at the connection surface of the support and the pipe. In addition, the corners of the rectangular duct are prone to the stress concentration phenomenon.

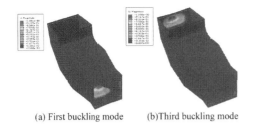

(a) First buckling mode   (b)Third buckling mode

Figure 4.   Buckling modes.

Table 3.   Loads on the model.

Load type	Value
Gravity acceleration (m/s^2)	9.806
Ash load (KPa)	0.840
Snow load (KPa)	0.300
Wind load (KPa)	0.350
Insulation layer pressure (KPa)	0.240
Internal pressure (KPa)	6.400

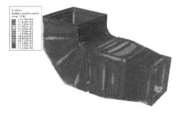

(a) Stress distribution of ducts system.

(b) Stress distribution of supports.   (c) Stress distribution of reinforcement ribs.

Figure 5.   Stress distribution of the ducts system under combined loads.

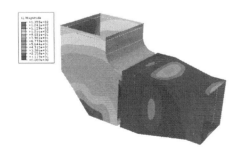

Figure 6.   Displacement distribution of the ducts system under combined loads.

Figure 6 shows the displacement distribution of the ducts system, in which the maximum displacement is 135.5 mm. The position of the maximum displacement occurs in the lateral of the elbow, whose stiffness is weak. The result is consistent with the result of the buckling analysis.

5   CONCLUSIONS

This paper introduces the frame and module of the redevelopment software, and studies the parametric finite element modeling method of the ducts system. Based on ABAQUS and Python language, software is redeveloped for modeling and analysis of the flue gas ducts system. Engineering designer only needs to input a few model parameters, and the finite element model of the ducts system will be quickly established. And so it will be much easier and more effective for engineers to carry out the relevant calculation analysis. In addition, according to the design drawing of a thermal power plant rectangular duct, the finite element model of the ducts is established by using this redevelopment software. Static response analysis of the ducts was completed under the action of gravity, ash load, snow load, wind load, insulation layer pressure and internal pressure. Besides, the results of modal and buckling analyses are reasonable and conformed to the practical situation. The validity of the software is verified.

REFERENCES

Cheng, L. & L, H.B. (2009). Second development of ABAQUS based on the scripting interfaces. *Modern Machinery* (2): 58–59.

Guo, X., Guan, Z., Liu, S., Chen, P., Liu, J., & Wang, X., et al. (2013). Application of abaqus secondary development based on python in simulation of laminate repair. *Computer Aided Engineering*.

Hu, W.Y. (2013). Software redevelopment for finite element modeling and computation of circular flue gas ducts based on ABAQUS. Chongqing University.

Li, X.C & Cen, E.F. (2007). Rectangular air & flue gas ducts CAD system of thermal power plant. *Energy Engineering* 5(2): 22–25.

Lian, C.W., Wang, Z.Y., Chuan-Jun, D.U., & Sun, J.X. (2006). Application of second-developed abaqus post-process on numerical simulation of plastic forming. *Forging & Stamping Technology* 31(4): 111–114.

Yuan, J.T. (2011). Finite element analysis method for the design of air & circular flue gas duct in thermal power plant. Chongqing University.

Zhang, Q., M, Y. & L, S.C. (2011). Method and application of second-development ABAQUS based on Python. *Ship Electronic Engineering* 38(2): 131–134.

Advances in Power and Energy Engineering – Sun (Ed.)
© 2016 Taylor & Francis Group, London, ISBN 978-1-138-02846-3

Research on the calculation of reversed flow of coal slime pastes in long-distance pipeline transportation

Y.Y. Liu, Y.B. Yang, J. Gao, X.D. Hao & M. Wu
School of Mechanical Electronic and Information Engineering, China University of Mining and Technology, Beijing, China

ABSTRACT: With the rapid growth of coal slime pastes, higher requirements were put forward on its transportation and disposal system, and the application of long-distance large-diameter engineering pipelines is increasing. However, this also brought up two technical problems, the reversed flow of pastes and a strong vibration to the pipeline system. In this paper, we studied the transient flow of coal slime pastes during the switching process of pump cylinders based on transient flow theory. The flow rate of pastes in the transient process was calculated. By calculating time integral of flow rate when it is negative, the reversed flow rate was obtained. A comparison between calculated and actual reversed flow rate was made. Results show that, the method based on the theory of transient flow can calculate reversed flow rate of pastes in long-distance pipeline effectively, which is significant for further research on how to reduce the reversed flow and vibration.

1 INTRODUCTION

Coal slime is a two-phase solid-liquid waste, which is produced during the washing process and has a high concentration, viscosity, and a paste-like form (Hao et al. 2009). Conventional disposal methods, such as stockpiling on the ground and burying under the ground, have caused serious environmental problems. With the rapid expansion of the industry, the production of coal slime pastes increased greatly. It's estimated that the production of coal slime pastes has reached more than 100 million tons/year. In fact, coal slime pastes still have great value in use. Generating elasticity through burning coal slime can save a lot of coals. Comprehensive utilization of resources is an effective way to solve the two core issues of rational use of resources and environmental pollution in sustainable development. As a clean and efficient delivery method, pipeline transportation of coal slime pastes is significant in improving the comprehensive utilization rate of resources.

Currently, the application of engineering pipelines delivering coal slime pastes in short distance for combustion and electricity generation is quite mature. With the rapid growth of coal slime pastes and pithead power plant, the needs of long-distance pipelines with high flow rate become increasingly urgent. However, two technical problems arose when the transportation distance exceeded 1000 m and the flow rate reaches 60 m³/h. First, the back flow of coal slime pastes in the transportation process results in a very low actual flow rate and even leads to delivery failures. Second, the strong

vibration in the delivering process reduces the reliability of transportation equipment greatly. Both of these two issues are closely related to the elastic deformation of coal slime pastes during long-distance pipeline transportation, so it is necessary to analyze the transient flow of coal slime pastes during the switching process of pump cylinders.

At present, researches on coal slime pastes mainly focus on the rheological properties (Zhao et al. 2006) and transportation resistance (Qu et al. 2009). With the emergence of vibration and reversed flow in long-distance transportation, Hao conducted a series of researches on the measurement of bulk modulus of elasticity (Hao et al. 2013, Hao et al. 2014). The bulk modulus of elasticity of coal slime pastes were tested based on the method of definition and the method of wave velocity, respectively. As to the research on the transient flow of water in pipelines, it has matured over the years and has complete researching methods, which can be referred to in this case. In 1897, the Soviet Union hydraulics expert Joukowsky conducted lots of water hammer experiments using the Moscow water pipeline systems and deduced the relation between liquid pressure rise Δp and speed decrease Δv. Allievi did numerical calculations of his water hammer equations which were established in 1913 and applied the mathematical methods and graphical methods to the water hammer analysis. In the 1960s, the United States hydrodynamics expert Professor Streeter V.L. lectured the calculation of water hammer using the Method of Characteristics (MOC). Professor Jin did research on the separation of water column for over 10 years, summarized the theory of water hammer prevention, and made full use of its computerization calculation method (Jin et al. 2004).

This paper established the differential equations of coal slime pastes in long-distance pipeline transportation based on transient flow theory. The variations of pressure and flow rate in the switching process of pump cylinders were calculated using the MOC. By calculating the time integral of flow rate when the flow rate is negative, the reversed flow rate can be obtained. Calculating the reversed flow rate accurately lays a foundation for reducing reversed flow and pipeline vibration.

2 MECHANISM OF REVERSED FLOW

Take a long-distance pipeline system delivering coal slime pastes for example, the pump has two cylinders. In the pumping process, two cylinder pistons driven by hydraulic force suck and discharge coal slime alternately. As a result, the coal slime was delivered into the pipeline gradually. The diagram of pump and pipeline transporting coal slime pastes is as shown in Figure 1.

The coal slime pastes will be compressed in the pipeline under the pumping pressure as the pastes are compressible. The compressibility of pastes becomes even more obvious in long-distance pipeline. As can be known from the operating principle, the pumping pressure is down to 0 momentarily when the pump is switching between two cylinders. As a result, the compressed pastes will have some elastic recovery and back flows into the feed, which is called reversed flow.

3 THE BASIC THEORY OF TRANSIENT FLOW

3.1 Basic theory of transient flow

In a pipeline filled with pressured water, due to the momentum transfer caused by dramatic changes in velocity, a series of abrupt alternating pressure changes in the pipeline was caused. This phenomenon is called water hammer, also known as fluid transients.

The basic theory of water hammer can be divided into the rigid water column theory and the elastic water column theory. The main difference between the two theories is that the elastic water column theory considers the elasticity of the pipe wall and the compressibility of water under changing pressure. This study was based on the elastic water column theory.

3.2 Differential equation of motion of transient flow

As shown in Figure 2, the two parallel planes B and C are perpendicular to the axis of the pipeline. The distance between B and C is dx. Take the small fraction of pastes between plane B and C as an isolator, to analyze the forces acting on it.

Take an instant of water hammer as example, the force analysis is carried out under the consumption that the small fraction of pastes is being compressed. The forces in the X direction are as follows:

$$\gamma A(H-z) - \gamma\left(A + \frac{\partial A}{\partial x}dx\right)\left(H - z + \frac{\partial H}{\partial x}dx + \sin\alpha dx\right)$$
$$+ \gamma\left(A + \frac{dx}{2}\frac{\partial A}{\partial x}\right)dx\sin\alpha$$
$$+ \gamma\left(H - z + \frac{\partial H}{\partial x}\frac{dx}{2} + \frac{dx}{2}\sin\alpha\right)\frac{\partial A}{\partial x}dx - \pi D dx\,\tau_0$$

The physical meanings of each item are as follows:

First item, total pressure acting on plane B; Second item, total pressure acting on plane C; Third item, the component of gravity of the pastes in the X direction; Fourth item, the pressure of pipe wall in the X direction caused by the swelling of the cross section; Fifth item: the force of friction caused by the wall; and γ is unit weight of pastes.

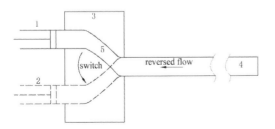

Figure 1. Schematic diagram of pump and pipeline.
1, pumping cylinder 1; 2, pumping cylinder 2; 3, feed bin; 4, pipeline; and 5, S pipe.

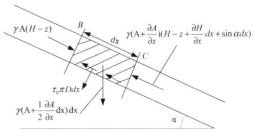

Figure 2. The force analysis of pastes in the pipe.

Expand the equation and omit the terms with higher order, the following equation can be obtained according to Newton's second law.

$$-dx\left(\gamma A\frac{\partial H}{\partial x}+\pi D\tau_0\right)=\frac{\gamma A dx}{g}\frac{dV}{dt} \tag{1}$$

Expand item dV/dt by the definition of total differential and substitute the hydraulic slope of coal slime pastes (Wang et al. 2008) into Equation (1), and the differential equation of motion of transient flow can be obtained as Equation (2).

$$\frac{\partial H}{\partial x}+\frac{1}{g}\frac{\partial V}{\partial t}+\frac{V}{g}\frac{\partial V}{\partial x}+\frac{f}{D}\frac{V|V|}{2g}\frac{\rho_m}{\rho}=0 \tag{2}$$

3.3 Differential equation of continuity of transient flow

The small fraction of water is continuous under such condition: in the process of the whole water hammer, the whole space of the small fraction of pipe is filled with water. According to the law of conservation of mass, the mass difference between water that flows into and out the small fraction dx, should be equal to water mass increased by the expansion of pipe wall and the compression of water.

Set A as the cross section of plane B, V as the velocity, ρ as the density of water, then ΔM_1, ΔM_2 can be obtained.

$$\Delta M_1 = \rho A V dt$$
$$-\left(\rho+\frac{\partial \rho}{\partial x}dx\right)\left(A+\frac{\partial A}{\partial x}dx\right)\left(V+\frac{\partial V}{\partial x}dx\right)dt$$
$$\Delta M_2 = \frac{d}{dt}(\rho A dx)dt = dxdt\left(\rho\frac{dA}{dt}+A\frac{d\rho}{dt}\right)$$

From the above analysis, we know $\Delta M_1 = \Delta M_2$, then the differential equation of continuity of transient flow is as follows:

$$\frac{\partial H}{\partial t}+V\left(\frac{\partial H}{\partial x}+\sin\alpha\right)+\frac{a^2}{g}\frac{\partial V}{\partial x}=0 \tag{3}$$

where, a = spread speed of water hammer wave (Hao et al. 2013) and α = angle between the pipeline and the horizontal line.

3.4 Finite difference equation

To use computerization method is used to solve the basic differential equation of transient flow. The basic equations need to be converted to finite differential equations by MOC to facilitate computation process.

Equation (2) and Equation (3) are a pair of quasi linear hyperbolic partial differential equations. The MOC is used to transform the two partial differential equations into ordinary differential equation forms, the results are Equations (4)–(7).

$$\frac{dH}{dt}+\frac{a}{g}\frac{dV}{dt}+V\sin\alpha+\frac{af}{2gD}\frac{\rho_m}{\rho}V|V|=0 \tag{4}$$

$$\frac{dx}{dt}=V+a \tag{5}$$

$$\frac{dH}{dt}-\frac{a}{g}\frac{dV}{dt}+V\sin\alpha-\frac{af}{2gD}\frac{\rho_m}{\rho}V|V|=0 \tag{6}$$

$$\frac{dx}{dt}=V-a \tag{7}$$

The above are the characteristics equations of transient flow. There meanings can be illustrated in Figure 3.

As shown in Figure 3, integrate differential Equation (4) from point A to point P along the characteristic line $C+$, which equals to Equation (5). Then substitute flow rate Q for velocity V, where, $Q = VA$. Equation (8) can be obtained as follows:

$$H_P - H_A + \frac{a}{gA}(Q_P - Q_A) + \frac{Q_A(t_P - t_A)}{A}\sin\alpha$$
$$+ \frac{af\rho_m}{2gDA^2\rho}Q_A|Q_A|(t_P - t_A) = 0 \tag{8}$$

In the same way, integrate differential Equation (6) from point B to point P along the characteristic line C, which equals to Equation (7). Equation (9) can be obtained, too.

$$H_P - H_B + \frac{a}{gA}(Q_P - Q_B) + \frac{Q_B(t_P - t_B)}{A}\sin\alpha$$
$$+ \frac{af\rho_m}{2gDA^2\rho}Q_B|Q_B|(t_P - t_B) = 0 \tag{9}$$

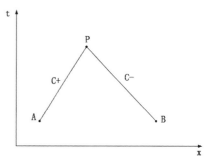

Figure 3.   The meaning of the characteristics equations.

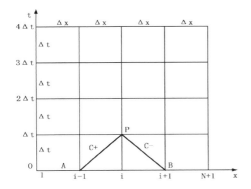

Figure 4. Rectangular grid of the simplified differential equation.

Equations (8) and (9) are called the finite differential equations.

Items in the equations with subscript A and B denote the initial value at the beginning time t_A and t_B of calculation and t_P is the ending time of calculation. So in the two equations above, only H_P and Q_P are unknown, which can be calculated easily by solving the equations.

Use the rectangular grid of x–t coordinate to describe the calculation process of finite differential equations. As shown in Figure 4, the pipeline is divided into N sections and the length of each section is Δx. Use i to denote the section number and each calculation time is Δt, where $\Delta t = \Delta x/a$.

In Figure 4, all the diagonal lines are characteristics line. The computation starts from time 0. In the first period of computation time (from $t = 0$ to $t = \Delta t$), parameters of node A and node B are known. Parameters of node P can be calculated by solving Equations (8) and (9). At the end of the first period of computation time (time $t = \Delta t$), the parameters of all the nodes P will be obtained. The parameters of boundary nodes can be decided according to the boundary conditions. Then the second period of computation time starts (from $t = \Delta t$ to $t = 2\Delta t$), and so on, until the parameters of all the nodes in the rectangular grid are obtained.

4 CALCULATION OF REVERSED FLOW

4.1 Calculation of reversed flow rate

Take a pipeline system transporting coal slime pastes for example, the processes of calculating reversed flow rate are shown as below. The parameters of the pipelines are, the diameter is 293 mm, length of the pipeline is 900 m, and the height of the pipe at the end is 80 m. To simplify the calculation

process, the pipeline is simplified to a linear one, as shown in Figure 5.

When the pump output is 50%, the initial pressure of the outlet is 2.5 MPa and the switching time t between two cylinders is 0.72 s. The outlet pressure of the pump drops to 0 during the period of time t. Pressure at the outlet of the pipeline is considered as 0 throughout the calculation. The relationship between the water head (H) and the distance (x) from the pump outlet is as follows:

$$H = -0.0964\,x + 166.67.$$

For nodes at both end of the pipeline, their flow rates can be calculated using Equations (8) and (9), respectively, as the water head at both nodes are known according to the above analysis. The water head at both ends in the entire computation time period are already known, so the flow rates of the two nodes can be calculated. At the same time, the initial pressure and flow rates of the whole pipeline are known. Following the calculation steps already introduced in Section 3.4 using Equation (8) and Equation (9), the water head and flow rates at all the nodes of the pipeline in the whole calculation time period can be obtained.

The water head and flow rate of the pump outlet in the computation time period are as shown in Figures 6 and 7, respectively. The water head at nodes with a distance of 221, 271, and 321 m from the pump outlet are shown in Figure 6. When the pressure at the pump outlet dropped, the pressure at other nodes also had a certain degree of decline.

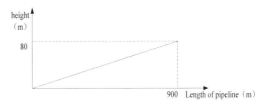

Figure 5. Schematic diagram of pipeline.

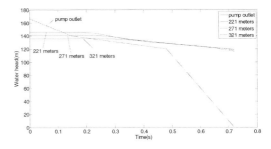

Figure 6. Pressure at the pump outlet.

248

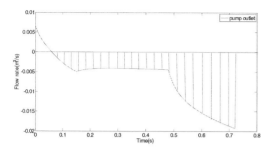

Figure 7. Flow rate at pump outlet.

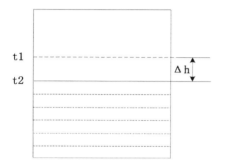

Figure 8. Height changes of material in the silo.

In Figure 7, the reversed flow rate can be obtained by calculating time integral of flow rate when flow rate is negative. The shaded part is reversed flow rate, as shown in Figure 7.

It is calculated that the reversed flow rate is 1.51 m³/h when the pump output is 50%.

4.2 The actual reversed flow rate

According to the height changes of material in the silo, the actual flow rate can be calculated as the silo parameters are already known. Theoretical flow rate is proportional to the pump output, so it can be calculated easily. Then the actual reversed flow rate is the difference between the theoretical and the actual flow rates. As shown in Figure 8, the height of the material has reduced Δh from t_1 to t_2. Here, Δh is 0.16 m, Δt is 37 min. It is calculated that the actual reversed flow rate is 2 m³/h. The result is very close to that of theoretical calculation based on the transient flow, and the error is 24.5%. It shows that this method can be used in the preliminary calculation of reversed flow rate.

5 CONCLUSION

In this paper, we studied the transient flow of coal slime pastes in pipeline transportation based on the transient flow theory. The basic differential equations of transient flow of coal slime pastes are established. The MOC was employed to simplify and to solve differential equations. The pressure and flow rates at every nodes of the pipeline are calculated. By calculating the time integral of flow rate when flow rate is negative, the reversed flow rate can be obtained. In comparison with the actual reversed flow rate, results show that the method can calculate the reversed flow rate of pipeline system effectively.

Research on transient process of coal slime pastes, when the pump switches between two cylinders and calculating the reversed flow rate accurately, is significant in solving the problems of reversed flow and pipeline vibration. As can be seen from the equations, the length of the pipeline and the friction coefficient are key factors, which affect the amount of pastes backflow and are also critical in reducing the calculation error and improving the calculation accuracy.

ACKNOWLEDGEMENT

This project is supported by the funds from National Natural Science Foundation of China (No. 51075389 and 51406106).

REFERENCES

Hao, X.D., Zhang, P.C., Guo, J.C. 2009. Experimental-study on pressure loss feature of coals lurry pipe linetran sportation. Coal Engineering: 86–88.

Hao, X.D., Li, N., Jia, X.K. 2013. Measurement of bulk modulus of elasticity of dense pastes and its effects on flow rate in long pipeline. Minerals Engineering: 145–153.

Hao, X.D., Liu, Y.Y., Li, N. 2014. Measurement of bulk modulus of coal slime pastes and it's application in long pipelines delivering pastes. China Coal: 84–88.

Jin, Z., Jiang, N.C., Wang, X.H. 2004. Water hammer caused by stop pumping and it's protection. China Building Industry press.

Qu, Y.Y., Zou, D.C., Liu, M. 2009. Experiment Research on Bent Tube Resistance Loss Based on Hydraulic Pipeline Rheology TestRig. Coal Science and Technology: 69–70, 87.

Wang, X. 2008. Numerical and Experimental Studies on Resistance Characteristics of Pipeline Transport High Concentration and Viscous Materials. China University of Mining and Technology, Beijing.

Zhao, C.Z., Zhou, H.Q., Qu Q.D. 2006. Experimental study on rheology performances of paste backfilling slurry. Coal Science and Technology: 54–56.

Advances in Power and Energy Engineering – Sun (Ed.)
© 2016 Taylor & Francis Group, London, ISBN 978-1-138-02846-3

Wind speed probability function in the coast of Rio Grande do Norte, Brazil

M.A. Aredes & M. Aredes
Federal University of Rio de Janeiro, Rio de Janeiro, Brazil

ABSTRACT: As is well known, the operating performances of wind farms depend essentially on wind speed distribution, since it has a cubic relation with the potential power of the site. Many studies have been performed in order to determine the best probability function for properly describing wind speed. These studies emphasize that the most probabilistic model adopted in wind studies is the Weibull distribution. Due to its flexibility, it can represent a large range of probabilistic functions. In this paper, wind speed data from the coast of Rio Grande do Norte (northeast of Brazil) are analyzed using the Weibull distribution function.

1 INTRODUCTION

The production cost of wind power is considered high compared to other sources, but on the other hand, wind is considered the cleanest energy source of the planet. In a time where sustainable alternatives are increasingly necessary, wind energy has become a path to attempt conserving natural resources and consuming conscientiously. In addition, it can be an alternative to the Brazilian energy source, which primarily uses power from hydropower. Although hydropower is clean, the use of our hydroelectric potential causes more environmental impact and is more susceptible to crises, as it depends on the amount of rain, which is subject to great variation.

In Brazil, most of the wind farms are concentrated in the Northeast and South. However, most of the country has the potential to generate this kind of energy. In this paper, the potential wind power in the state of Rio Grande do Norte was analyzed. This state receives regular winds in much of its territory, which do not have abrupt changes in frequency and speed.

The analyzed wind speed data originates from a region in Rio Grande do Norte where there are three wind power plants in operation: Alegria I, Alegria II, and Miassaba 3; as shown in Figure 1.

Figure 1. Wind farm locations.

2 WIND SPEED DISTRIBUTION

2.1 *Weibull distribution*

In probability theory and statistics, the Weibull distribution is a continuous probability distribution. It was named after Waloddi Weibull, who described it in detail in 1951 as a statistical distribution function that could be used in a wide range of problems (Weibull 1951). After that, the Weibull distribution became one of the most widely used lifetime distributions in reliability engineering. Moreover, it is a versatile distribution that can take on the characteristics of other types of distributions, based on the value of the shape parameter β. The Weibull probability distribution function with two parameters is given by:

$$f(x) = \frac{\beta}{\eta} \left(\frac{x}{\eta} \right)^{\beta-1} \exp\left(-\left(\frac{x}{\eta} \right)^{\beta} \right) \qquad (1)$$

where β is the shape parameter and η is the scale parameter.

In the study of wind speed, this function can be understood as the probability of wind speed 'x' during any time interval chosen. Since wind speed behavior is repetitive over the years, it is normally used in a period of 1 year. By definition of probability function, the probability of the wind speed between zero and infinity during a chosen period is unity, i.e.,:

$$\int_0^\infty f(x)dx = 1 \tag{2}$$

As mentioned before, a range of probabilistic distributions can be found by varying the shape parameter β. Figure 2 shows the Weibull distribution versus wind speed. The curve for $\beta = 1$ has a heavy bias to the left, where most days are windless ($x = 0$). This kind of distribution is simply an exponential function and the Weibull probability function is given by:

$$f(x) = \frac{1}{\eta} \exp\left(-\frac{x}{\eta}\right) \tag{3}$$

The second curve, when $\beta = 2$, is a Rayleigh distribution which is a typical wind distribution found at most sites given by Equation (4). In this distribution, more days have lower winds than the mean speed, while few days have high wind.

$$f(x) = \frac{2x}{\eta^2} \exp\left(-\frac{x^2}{\eta^2}\right) \tag{4}$$

For $\beta \geq 3$ the curve approaches to a normal distribution, where some days have high wind, and equal number of days have low wind.

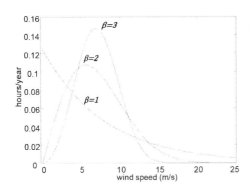

Figure 2.   Weibull probability distribution function with scale parameter $\eta = 10$ and shape parameters $\beta = 1, 2,$ and 3.

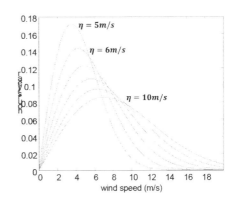

Figure 3.   Weibull probability distribution with shape parameter $\beta = 2$ and the scale parameters ranging from 5 to 10 m/s.

Figure 3 shows the distribution curves corresponding to $\beta = 2$ (Rayleigh distribution) with different values of η ranging from 5 to 10 m/s. For greater values of η, the curves shift right to higher wind speeds. In other words, the higher the scale parameter η, the more number of days have high winds. Since this shifts the distribution of hours at a higher speed scale, it is called the scale parameter and has the same unit of wind speed.

For most sites, de shape parameter (β) varies from 1.5 to 2.5, and the scale parameter varies from 5 m/s to 10 m/s (Patel 1999).

2.2   Estimation of Weibull distribution parameters

2.2.1   Cumulative Weibull distribution function
By definition, the Weibull distribution density function is given by Equation (1), and the cumulative Weibull distribution function can be obtained by:

$$F(X) = f(x \leq X) = \int_0^X f(x)dx \tag{5}$$

$$F(X) = 1 - \exp\left(-\left(\frac{X}{\eta}\right)^\beta\right) \tag{6}$$

2.2.2   Least square method
The least square method is used to estimate the parameters in a formula when modeling an experiment of a phenomenon, (Engelhardt 1975, Mann et al. 1974, Miller s.d.). When using the least square method, the sum of the squares of the deviations e, which is defined as below, should be minimized.

$$e = \sum_{i=1}^n w_i^2 [y_i - g(x_i)]^2 \tag{7}$$

where x_i is the wind speed, y_i is the probability of wind speed rank, w_i is a weight value of the plot and n is the number of data plot.

The Linear Least Square Method (LLSM) is a computational approach to fitting a mathematical or statistical model to data. It is commonly applied in engineering and mathematics problems that are often not thought of as estimation problems. The LLSM is a special case of the least square method with a formula which consists of some linear functions, and it is easy to use. Furthermore, in the more special case that the formula is a line, the LLSM is much easier. The Weibull distribution function is a nonlinear function, but the cumulative Weibull distribution function (Equation 7) is transformed to a linear function as follows:

$$F(X) = 1 - \exp\left(-\left(\frac{X}{\eta}\right)^\beta\right) \qquad (8)$$

$$1 - F(X) = \exp\left(-\left(\frac{X}{\eta}\right)^\beta\right) \qquad (9)$$

$$\ln(1 - F(X)) = \left(\frac{X}{\eta}\right)^\beta \qquad (10)$$

$$\ln\left(\frac{1}{1 - F(X)}\right) = \left(\frac{X}{\eta}\right)^\beta \qquad (11)$$

$$\ln\left(\ln\left(\frac{1}{1 - F(X)}\right)\right) = \beta\ln(X) - \beta\ln(\eta) \qquad (12)$$

Equation (12) can be written as:

$$Y = bX + a \qquad (13)$$

where,

$$Y = \ln\left(\ln\left(\frac{1}{1 - F(X)}\right)\right) \qquad (14)$$

$$X = \ln(X) \qquad (15)$$

$$b = \beta \qquad (16)$$

$$\alpha = \beta\ln(\eta) \qquad (17)$$

By linear regression formula:

$$b = \frac{n\sum_{i=1}^{n} X_i Y_i - \sum_{i=1}^{n} X_i \sum_{i=1}^{n} Y_i}{n\sum_{i=1}^{n} X_i^2 - \left(\sum_{i=1}^{n} X_i\right)^2} \qquad (18)$$

$$a = \frac{\sum_{i=1}^{n} X_i^2 \sum_{i=1}^{n} Y_i - \sum_{i=1}^{n} X_i \sum_{i=1}^{n} X_i Y_i}{n\sum_{i=1}^{n} X_i^2 - \left(\sum_{i=1}^{n} X_i\right)^2} \qquad (19)$$

2.3 Wind power distribution

For a given wind speed v_w, the available power that can be extracted by the rotor blade is given by:

$$P_w = \frac{1}{2}\rho A v_w^3 \qquad (20)$$

where ρ is the air density and A is the area covered by the rotor blades.

The wind power density distribution, which represents the distribution of wind energy at a given wind speed is given by Equation (21), where $f(v_w)$ is the Weibull probability function of the wind speed v_w.

$$e_w = \frac{1}{2}\rho v_w^3 f(v_w) \qquad (21)$$

Therefore, the total wind power density is:

$$E_w = \int_0^\infty e_w(v_w)\,dv \qquad (22)$$

$$E_w = \frac{1}{2}\rho\int_0^\infty v_w^3 f(v_w)\,dv \qquad (23)$$

Considering the Weibull probability function, the wind power density is expressed by Equation (24).

$$E_w = \frac{1}{2}\rho c^3 \Gamma\left(\frac{k+3}{k}\right) \qquad (24)$$

where Γ is the gamma function (Equation 25).

$$\Gamma(x) = \int_0^\infty e^{-u} u^{x-1}\,du \qquad (25)$$

2.3.1 Wind power curve

A typical power curve of a wind turbine is shown in Figure 4. The operational speed range is between the cut-in speed and the cut-out speed. The cut-in speed is the wind speed at which the turbine begins to generate power. The cut-out speed is chosen to protect the turbine and structure from high loads. In the speed control mode, the wind turbine is controlled for the maximum power extraction. In addition, in the power limitation control mode, the captured wind power is turbine rated power and/or the generator speed is rated.

Thus, the output power can be expressed by:

$$p(v_w) = \begin{cases} 0 & v_w < vcut_{in} \text{ or } v_w > vcut_{out} \\ \frac{1}{2}\rho A C_p v_w^3, & vcut_{in} < v_w < v_{rated} \\ P_r, & v_{rated} < v_w < vcut_{out} \end{cases} \qquad (26)$$

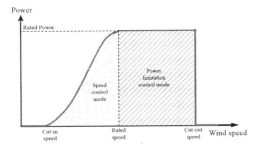

Figure 4.   Typical power curve.

where C_p is the turbine performance coefficient. The theoretical maximum value of C_p is 0.59; it is determined by aerodynamic laws and thus may change from one wind turbine type to another. In practical designs, the maximum achievable C_p is below 0.5 for high-speed turbines and between 0.2 and 0.4 for slow speed turbines.

Therefore, the generated power of a wind turbine depends not only on the controls of the generator and the turbine aerodynamic and mechanical systems but also the wind speed probability.

3   WIND FARMS IN RIO GRANDE DO NORTE

As mentioned before, there are three wind farms located at the analyzed wind data area. The wind farm Alegria is located at Guamaré city in the state of Rio Grande do Norte. The wind farm consists of two units, Alegria I and Alegria II. Alegria I unit consists of 31 wind turbines with total capacity of 51.15 MW, while in Alegria II, 61 wind turbines were installed with total capacity of 100.65 MW. The 92 wind turbines of the complex were manufactured by Danish company Vestas, model V82. Miassaba 3 is near the city of Macau. The project comprises 41 wind turbines model ECO 86 ALSTOM, and total power of 68.47 MW. The specifications and operating data of the wind turbines are given in Table 1, and the power curves can be seen in Figures 5 and 6.

4   RESULTS

Wind speed data from 2010 to 2012 from the coast of Rio Grande do Norte were analyzed. The time interval of wind speed measurements is 10 min. A graphic with the monthly average is given in Figure 7.

The LLSM was performed to estimate the shape and scalar parameters of the Weibull distribution for the given wind data. The values of

Table 1.   Wind turbines specifications and operating data.

Wind turbine	Rated power (MW)	Cut-in wind speed (m/s)	Cut-out wind speed (m/s)	Power control system
Vestas V82	1.65	3.5	20	Fixed-speed, active-stall control
Alstom wind ECO 86	1.67	3	25	Variable speed, pitch control

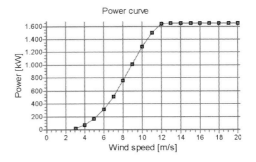

Figure 5.   Vestas V82 power curve.

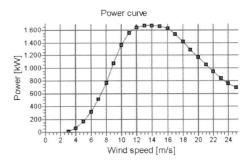

Figure 6.   Alstom wind ECO 86 power curve.

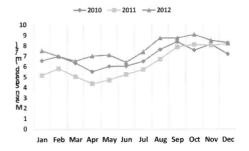

Figure 7.   Monthly average wind speed.

the parameters for each year are given separately in Table 2. The resulting distribution curves are shown in Figure 8.

Considering all data, the parameters of the Weibull distribution of the wind speed are: $\beta = 2.7221$ and $\eta = 7.8136$ m/s. The probability density function curve is shown in Figure 9.

For the wind speed distribution function found in Figure 9, the probabilities that the wind turbines are offline, in speed control mode, and in power limitation control mode is given in Table 3.

For the location of interest, the wind turbines operate in the speed control mode most of the time. The probability that the wind turbine generates rated power in power limitation control mode is very low. The probability that the turbine is disconnected from the grid due to a high wind above the cut-out speed is almost zero.

Table 2. Weibull distribution function parameters.

	2010	2011	2012
η	77,463	71,026	87,360
β	29,317	25,061	29,832

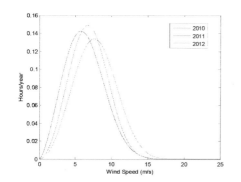

Figure 8. Weibull distribution of 2010, 2011, and 2012.

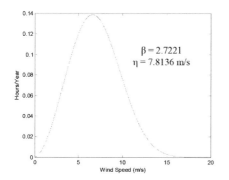

Figure 9. Weibull distribution.

Table 3. Probabilities that the wind turbines are offline, in speed control mode, and in power limitation control mode.

Wind farm	Wind turbine	Off line (%)	Speed control mode (%)	Power limitation control mode (%)
Alegria I and II	Vestas V82	7.12	88.87	4.01
Miassaba 3	Alstom wind ECO 86	10.63	87.54	1.83

5 CONCLUSION

The great wind potential in Rio Grande do Norte, coupled with the search for renewable energy sources to diversify the national energy matrix, is attracting investments in the energy sector.

As can be seen at the Weibull probability function of the wind speed, the area of interest receives regular winds, which do not have abrupt changes in frequency and speed. Therefore, the wind turbines operating in wind farms operate in the speed control mode in most of the time. In this way, the maximum power is extracted from the wind and the probability that the wind turbines waste energy in power limitation control mode is very low.

REFERENCES

Borges, C., Leite, A. & Falcao, D., 2007. *Probabilistic Wind Farms Generation Model for Reliability Studies Applied to Brazilian Sites*. Tampa, Florida, s.n.

Chiodo, E. & Lauria, D., 2009. *Analytical Study of Different Probability Distributions for Wind Speed Related To Power Statistics*. Capri, s.n.

Engelhardt, M., 1975. *On simple estimation of the parameters of the Weibull or extreme-value distribution*. s.l.:American Statistical Association.

Mann, N.R., Schafer, R.E. & Singpurwalla, N.D., 1974. *Methods for statistical analysis of reliability and life data*. New York: John Wiley and Sons.

Miller, S.J., s.d. *The Method of Least Squares*. Mathematics Department, Brown University: s.n.

Patel, M., 1999. *Wind and Solar power systems*. s.l.: CRC Press LLC.

Weibull, W., 1951. *A Statistical Sistribution Sunction of Side Applicability*. Stockholm, s.n.

Advances in Power and Energy Engineering – Sun (Ed.)
© *2016 Taylor & Francis Group, London, ISBN 978-1-138-02846-3*

An optimal and real-time control strategy for energy storage system

L.B. Yang
State Grid Qinghai Electric Power Research Institute, Xining, Qinghai, P.R. China

Y.L. Gao
State Grid Shandong Electric Power Company, Jinan, Shandong, P.R. China

Y. Niu & F. Zhang
Department of Electrical Engineering, Shandong University, Jinan, Shandong, P.R. China

T. Du
Shandong University Electric Power Technology Co. Ltd., Jinan, P.R. China

ABSTRACT: This paper focuses on the control strategy for Battery Energy Storage System (BESS), which is integrated within substation to mitigate the fluctuations of the aggregated wind power from clustered wind farms. The main objective of the control strategy is to encourage BESS take part in peak shaving besides fluctuation smoothing when the State of Charge (SOC) varies in a narrow range. Thus the control target aims to simultaneous effectively smooth the fluctuations and achieve economic profits from peak shaving. Case studies based on actual wind power data are provided, and the results show this method can achieve significant peak shaving profits and keep BESS in good operation condition.

1 INTRODUCTION

In recent years, the power grid has faced great challenges due to the intermittency and randomness of wind power (J.X. Feng et al. 2013). The energy storage system has fast response capacity which makes it possible to integrate a battery energy storage station with a large wind farm to smooth the fluctuations (M. Ding et al. 2013).

The researches of distributed energy storage and BESS have been studied extensively in the literature. The energy storage systems in engineering project mainly focus on the wind power fluctuations (Teleke S et al. 2010, Teleke S et al 2009 & J. Chen et al. 2013). The coordination of ESS and wind farms can improve the quality of wind power generation, and profit its penetration level (G.N. Bao et al. 2012 & D.Y. Chen et al. 2013). Many researchers have been focusing on ESS operation issues. In (Yoshimoto K et al. 2006), the proposed control strategy aims to increase the renewable energy penetration level and minimize the energy production curtailments by energy buffering. Some control strategies in (T. Han et al. 2010 & Q. Cui et al. 2013) pay attention to the lifetime of storage system by considering the operating constrains, such as SOC. A study in (Ted K.A. Brekken et al. 2011) determines the lowest-cost flow-battery storage system, and the cost function is defined as the cost sum of power capacity and energy capacity. Another research in (M. Ding et al. 2011) considers the cost of storage unit and power conversion unit as the main cost of the storage system. In fact, beside fluctuation smoothing, BESS can also play great roles in other areas such as peak load shaving, frequency regulation, and emergency support (M. Ding et al. 2012).

Based on the presented literatures, different business forms of BESS are discussed in this paper. The control strategy in this research aims to shave the peak load and smooth the fluctuation of wind power simultaneously, during which the profits from peak shaving can be obtained. Furthermore, the profits distribution and cost allocation methods are also discussed. Simulation result verifies the effectiveness of the proposed control strategy of BESS.

2 THEORETICAL ANALYSIS

2.1 Smoothing effect of clustered wind farms

Theoretical analysis and actual operation data of wind power demonstrate that the smoothing effect exists during clustered wind farms, which is essentially caused by the delay and filter effect of wind (Y.H. Liu. 2013 & D.Y. Yu et al. 2011). This effect can decrease the fluctuation itself, and BESS for wind farm clusters is prone to show better performance in fluctuation smoothing compared with distributed energy storage.

2.2 Operation modes of BESS

BESS with lower capacity is always allocated in micro-grid and distributed networks [8], which is designed to improve the power supply reliability or for some economic purposes. Meanwhile, BESS with high capacity are integrated with large-scale intermittency power with aim to mitigate power fluctuation. However, due to the seasonal features of wind power, in mild wind season only partial BESS is needed to smooth the fluctuations, in such a case BESS is capable to participate in peak load shaving. In this research, the BESS will be arranged to smoothing the wind power fluctuations and meanwhile shaving the peak load to achieve extra benefits.

In consideration of the operation modes of BESS, BESS will be difficult to deal with the frequency regulation as it needs stochastic dispatch of the BESS, so it is difficult for the size-limited BESS to guarantee the smoothing effect of wind power fluctuation, and meanwhile have enough energy for frequency regulation at any time. Comparatively, peak shaving is quite suitable for spare BESS. When BESS has spare capacity if the wind power fluctuations are mild, then BESS can engage in peak shaving for benefits, and when BESS is difficult to address the fluctuation issues in wild wind season, then BESS will quit peak shaving. So the seasonal feature of wind power provides the possibility of such an operation mode.

3 OPERATION RULES OF BESS

The mentioned operation mode including peak load shaving and fluctuation smoothing is reasonable since the seasonal difference of wind farm clusters power output result in capacity redundancy which can be utilized in certain time period. Besides, profits are generated due to the fact that valley load price is lower than peak load price. In other word, BESS can store energy during valley load period and supply energy to grid during peak load period which can bring benefit for both BESS and power grid.

3.1 Charging and discharging strategy

The main purpose of charging and discharging strategy is to gain operation profits and meanwhile guarantee the smoothing effect of power fluctuation. Considering this, the spare capacity should be fully utilized to get additional profits by consulting the difference between peak load price and valley load price.

In the process of power fluctuation smoothing, if the wind power $P(t)$ exceeds the expected output power $P_{ref}(t)$, BESS will charge. If $P(t) < P_{ref}(t)$, BESS will be in discharging state. If $P(t) = P_{ref}(t)$, BESS stays in floating charging state. During the

procedure, when the charging power rate, the State Of Charge (*SOC*) or the change rate of charging power exceed the limited values, which are depicted in (1–3), abandoned power will appear.

$$P(t) - P_{ref}(t) > P_{max\text{-}cha} \tag{1}$$

$$SOC(t) > SOC_{max} \tag{2}$$

$$\Delta P_{cha}(t) > \Delta P_{max\text{-}cha} \tag{3}$$

where $P_{max\text{-}cha}$ represents the maximum charging power, SOC_{max} is the maximum limit of *SOC* which is set as 1.0 in this paper. $\Delta P_{cha}(t) = P(t) - P(t-1)$ denotes the variation rate of charging power and $\Delta P_{max\text{-}cha}$ is the upper limit of $\Delta P_{cha}(t)$. Likewise, when the discharging and the corresponding variance rate exceed the limited values, and the *SOC* is less than the lower bound, as shown in (4–6), smoothing effect will be affected because the BESS is shortage of discharge power and energy to smooth the fluctuations effectively.

$$|P(t) - P_{ref}(t)| > P_{max\text{-}discha} \tag{4}$$

$$SOC(t) < SOC_{min} \tag{5}$$

$$\Delta P_{discha}(t) > \Delta P_{max\text{-}discha} \tag{6}$$

where $P_{max\text{-}discha}$ denotes the maximum value of discharging power. SOC_{min} represents the lower bound of *SOC*. $\Delta P_{discha}(t)$ is the variance rate of discharging power and $\Delta P_{max\text{-}discha}$ is set as the maximization limit.

When BESS participates in peak load regulation, it is assumed that the charge and discharge power for peak shaving are constant, which are expressed as:

$$E_{cha} = P_{cha} \cdot \Delta t_{cha} \tag{7}$$

$$E_{discha} = P_{discha} \cdot \Delta t_{discha} \tag{8}$$

where E_{cha} is the stored energy and E_{discha} is the released energy. P_{cha} and P_{discha} are the charging and discharging power respectively. Δt_{cha} and Δt_{discha} denote the time duration of charging and discharging interval.

It should be noted that when BESS participates in peak load regulation, the utilizaion of capacity will increase the risk of the wind power curtailment and power shortage for fluctuation smoothing.

3.2 Optimal control strategy for operation modes

Benefits from the participation of BESS in peak load regulation, and on the other hand the operating cost will increase due to the utilization of energy capacity. For this reason, maximizing the operation profits is set as the optimization objective to get the tradeoff between the profits of peak load shaving and the increased operation cost. Specifically, for each day $[t_{pi}, t_{pl}]$ and $[t_{vi}, t_{vl}]$ represent the peak load period and the valley load period. $P_p(t)$ and $P_v(t)$ are the discharging power rate and charging power for peak load shaving. r_p and r_v are price of selling power to the grid and purchasing power from the power plant, respectively. Δt is the sampling period. So, the profits are expressed as follows:

$$G = \sum_{t=t_{pi}}^{t_{pl}} r_p \cdot P_p(t) \cdot \Delta t - \sum_{t=t_{vi}}^{t_{vl}} r_v \cdot P_v(t) \cdot \Delta t \qquad (9)$$

When BESS operates in the time interval $t_{pi} \leq t \leq t_{pl}$:

$$\begin{cases} SOC_{st1}(t) = SOC(t_{pi}) + \dfrac{1}{V} \cdot \displaystyle\sum_{t=t_{pi}}^{t_{pl}} P_p(t) \cdot \Delta t \\[4mm] \qquad\qquad + \dfrac{1}{V} \cdot \displaystyle\sum_{t=t_{pi}}^{t_{pl}} P_s(t) \cdot \Delta t \\[4mm] C_{st1} = C_{aba}(SOC_{st1}) + C_{lack}(SOC_{st1}) \\[2mm] \qquad\quad + C_{outline}(SOC_{st1}) \end{cases} \qquad (10)$$

where $SOC_{st1}(t) = SOC$ in the time interval $t_{pi} \leq t \leq t_{pl}$, $SOC(t_{pi})$ = the initial value of SOC in this time interval. $P_s(t)$ = the power for smooth fluctuation. C_{st1} represents the operating cost. It can be seen from (10) that the equation for C_{st1} varies according to SOC. In this time interval BESS will shave the peak load and meanwhile smooth the wind power fluctuations.

When BESS operates in the time interval $t_{pl} \leq t \leq t_{vi}$:

$$\begin{cases} SOC_{st2}(t) = SOC(t_{pl}) + \dfrac{1}{V} \cdot \displaystyle\sum_{t=t_{pl}}^{t_{vi}} P_s(t) \cdot \Delta t \\[4mm] C_{st2} = C_{aba}(SOC_{st2}) + C_{lack}(SOC_{st2}) \\[2mm] \qquad\quad + C_{outline}(SOC_{st2}) \end{cases} \qquad (11)$$

During this time period, BESS is only used to smooth power fluctuations.

When BESS operates in the time interval $t_{vi} \leq t \leq t_{vl}$:

$$\begin{cases} SOC_{st3}(t) = SOC(t_{vi}) - \dfrac{1}{V} \cdot \displaystyle\sum_{t=t_{vi}}^{t_{vl}} P_v(t) \cdot \Delta t \\[4mm] \qquad\qquad + \dfrac{1}{V} \cdot \displaystyle\sum_{t=t_{vi}}^{t_{vl}} P_s(t) \cdot \Delta t \\[4mm] C_{st3} = C_{aba}(SOC_{st3}) + C_{lack}(SOC_{st3}) \\[2mm] \qquad\quad + C_{outline}(SOC_{st3}) \end{cases} \qquad (12)$$

BESS in this time interval has the functions of storing power in valley load period and fluctuation smoothing.

When BESS operates in the time interval $t_{vl} \leq t \leq t_{pi}$:

$$\begin{cases} SOC_{st2}(t) = SOC(t_{pl}) + \dfrac{1}{V} \cdot \displaystyle\sum_{t=t_{pl}}^{t_{vi}} P_s(t) \cdot \Delta t \\[4mm] C_{st2} = C_{aba}(SOC_{st2}) + C_{lack}(SOC_{st2}) \\[2mm] \qquad\quad + C_{outline}(SOC_{st2}) \end{cases} \qquad (13)$$

During this period, BESS is mainly used to mitigate power fluctuation.

Based on the analysis above, the daily operation profits are described in the following equation:

$$Q_d = G - \left[\sum_{i=1}^{4} C_{sti} - \sum_{i=1}^{4} C'_{sti} \right] \qquad (14)$$

where C'_{sti} = the operating cost when BESS only participates in power fluctuation mitigation. Thus, maximizing the operation profits Q_d is the optimization objective in which $P_p(t)$, $P_v(t)$ and $P_s(t)$ are the optimization variables. When $Q_d > 0$, it demonstrates that the participation in peak load shaving can bring more benefits than cost to BESS. However, when the risk cost incurred in fluctuation mitigation exceeds the profits brought by peak load regulation, Q_d will drop to zero, which means no participation in peak load regulation.

Solutions for the optimization target are based on the following assumptions in this research, in which the parameter varies in different regions but will not affect the reasonability of this method:

1. $[t_{pi}, t_{pl}]$ and $[t_{vi}, t_{vl}]$ are various due to the regional and seasonal differences. In this paper, by consulting actual data of Shandong Power Grid, $[t_{pi}, t_{pl}]$ and $[t_{vi}, t_{vl}]$ are set to be [19:00, 22:00] and [2:00, 5:00] respectively. Meanwhile, in order to simplify the problem, $P_p(t)$ and $P_v(t)$ are regarded as constant.

2. $P_p(t)$ and $P_v(t)$ are supposed to have some influence on the planned wind power output during the time period $[t_{pi}, t_{pl}]$ and $[t_{vi}, t_{vl}]$. To solve this, the optimized result will be reported to the grid control center within the pre-scheduling-time level (i.e., 2 hours), which can provide response time for the operators.

4 PROFITS DISTRIBUTION AND COST ALLOCATION

4.1 Cost allocation

The allocation scheme is established according to the interconnection of profits and responsibility and is rooted at the difference of the potential profits. In terms of the running cost C_{run}, the equipartition scheme is adopted considering the unity and the proportionality of the benefit in wind farm clusters. The construction cost C_{cons} is proportionally allocated according to the installed capacity of each wind farm and reasons are as follows:

1. The climate and geography conditions of each wind farm in a wind farm clusters are similar. The power output of wind farms are in approximate linearity with the installed capacity.
2. The wind farms are geographically close to each other, which can weaken the effect of real-time change of wind speed. The fluctuation characteristics have a certain relationship with the installed capacity.

4.2 Profits distribution

The operation profits consist of the potential profits stemming from fluctuation smoothing and the direct benefits of participating in peak load shaving. The potential profits are already contained in the feed-in tariff contract made by power generation enterprise and the power grid.

The operation profits G generated from the value promotion of the energy stored in valley load period. So G is proportional distributed according to total power generation during the time period $[t_{vi}, t_{vl}]$ of each wind farm.

5 CASE STUDY

In this paper, the optimization model has been verified with the simulation of $c++$ programming language in *visual studio*2010, and then computed with the optimization software *CPLEX*. The actual wind power data is collected from the clustered wind farms located in southeast coast of China. There are three wind farms which are geographically close to each other in this case. Their installed capacities are 45 *MW*, 40 *MW* and 65 *MW* respectively.

According to the optimal strategy for operation mode, it can be seen than in December when the wind power fluctuates most, the participation in peak load regulation for BESS would aggravate the condition of *SOC*. Moreover, the calculation results indicate that the profits from peak load regulation cannot balance the increased operating costs. Therefore, the strategy that all energy capacity is utilized to smooth power fluctuation is reasonable and economical. But in July when the wind is mild with few dramatic fluctuations, BESS can perfectly smooth the wind power fluctuations with partial capacity. So with spare capacity BESS is able to participate peak load shaving. Wind power data in a typical day of 24 hours is chosen to show

the operation condition of the BESS. The corresponding profits are shown in Table 1, and the power smoothing curve and *SOC* conditions are shown in Figure 1.

The construction cost of the unit capacity (*MWh*) is chosen as the reference value, which is assumed to be $5.9 \times 10^5\$$/MWh. The benefits in Table 1 and 2 are converted to the unit values based on the reference value.

The results in Table 1 show that the operation profits from peak shaving are considerable. According to the profits distribution and cost allocation strategy, each wind farm gains different profits. Besides, it can be seen from Figure 1 that the fluctuations can be perfectly smoothed at most of the time. However, due to the utilization of capacity for peak load regulation, during some intervals the fluctuation cannot be effectively smoothed. Even though, it is still acceptable to meet the smoothing requirement.

The operation condition of BESS is another concern in this control strategy. As shown in Figure 2, SOC' denotes the curve of SOC with no participation in peak load shaving. By comparing the two curves of SOC, it is easy to find out that the SOC keeps at high level after the BESS charging power for peak load shaving, and obviously this will increase the risk of out-of-limit of SOC. Similarly, after the BESS discharges power for peak load shaving, the SOC runs at a low level.

Table 1. Calculation results.

Item	Profit Q_d (10^{-3})
BESS	8.54
1# wind farm	2.41
2# wind farm	2.58
3# wind farm	3.55

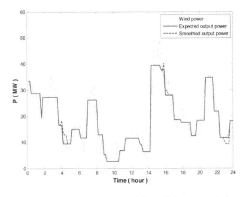

Figure 1. Smoothing curve with peak load regulation.

Table 2. Monthly benefits.

Month	1	2	3	4	5	6
$Q_d (10^{-3})$	0	0	69.3	107.4	122.1	166.8
Month	7	8	9	10	11	12
$Q_d (10^{-3})$	247.2	232.7	208.6	126.3	65.7	3.1

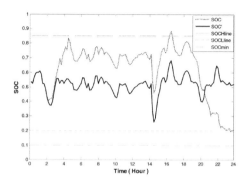

Figure 2. Curve of SOC with peak load regulation.

However, by the objective function in this study, the optimal control strategy can achieve the trade-off between the profits of peak load shaving and the increased operationcost. The operation profits for each month in a whole year are shown in Table 2, which further verifies the reasonability of the proposed control strategy.

It can be seen from the analysis above that the optimal control strategy for BESS can be verified by the actual power data in wind farm. Based on this, peak load shaving is quite suitable for BESS because of the seasonal variation of wind power output. The simulation with actual wind power data demonstrates that the optimal control strategy can generate considerable profits.

6 CONCLUSION

The control strategy for BESS is presented according to the characteristic of wind power output, and then the optimal operation mode for BESS which aims to obtain the maximum operating benefit is proposed. Moreover, the specific scheme for cost allocation and profits distribution is given in this study. The actual wind power data is calculated to prove the proposed method. The optimization results indicate that the operation mode can bring considerable profits to BESS, and this verifies the efficiency and feasibility of the proposed optimal model.

ACKNOWLEDGEMENT

This work was partially supported by QingHai Province Key Laboratory of Photovoltaic Grid Connected Power Generation Technology under Project No. 2014-Z-Y34A.

REFERENCES

J.X. Feng et al. 2013. A Method for the Optimization Calculation of Wind Farm Energy Storage Capacity. Automation of Electric Power Systems, 37(1):90–95.

M. Ding et al. 2013. A Real-time Smoothing Control Strategy With SOC Adjustment Function of Storage Systems. Proceedings of the CSEE, 33(1):22–30.

Teleke S et al. 2010. Rule-based Control of Battery Energy Storage for Dispatching Intermittent Renewable Sources. IEEE Trans on Sustainable Energy, 25(3):117–124.

Teleke S et al. 2009. Control Strategies for Battery Energy Storage for Wind Farm Dispatching. IEEE Trans. on Energy Conversion, 24(3):725–732.

J. Chen et al. 2013. Optimal Sizing for Stand-alone Microgrid Considering Different Control Strategies. Automation of Electric Power Systems, 37(11):1–6.

G.N. Bao et al. 2012. Load Shift Real-time Optimization Strategy of Battery Storage System Based on Dynamic Programming. Automation of Electric Power Systems, 36(12):11–16.

D.Y. Chen et al. 2013. Development of Energy Storage in Frequency Regulation Market of United States and Its Enlightenment. Automation of Electric Power Systems, 37(11):9–13.

Yoshimoto K et al. 2006. New control method for regulating state of charge of a battery in hybrid wind power/battery storage system. Power Systems Conference and Exposition, PSEC: 1244–1251.

T. Han et al. 2010. Optimized Scheme of Energy Storage Capacity for Grid-connected Large-scale Wind Farm. Power System Technology, 34(1):169–173.

Q. Cui et al. 2013. Study on Linkage Electricity Price and Benefit Analysis Considering Energy Storage Station Operation in Market Environment. Proceedings of the CSEE, 33(13):62–68.

Ted K.A. Brekken et al. 2011. Optimal Energy Storage Sizing and Control for Wind Power Application. IEEE Trans on Sustainable Energy, 2(1):69–77.

M. Ding et al. 2011. Modeling and Comparative Study on Multiple Battery Energy Storage System. Automation of Electric Power Systems, 35(15):34–39.

M. Ding et al. 2012. Static Function of the Battery Energy Storage System. Transactions of China Electrotechnical Society, 27(10):242–248.

Y.H. Liu. 2013. Analysis and Application of Wind Farm Output Smoothing Effect. Power System Technology, 37(4):987–991.

D.Y. Yu et al. 2011. Study on the Profiling of China's Regional Wind Power Fluctuation Using GEOS-5 Data Assimilation System of National Aeronautics and Space Administration of America. Automation of Electric Power Systems, 35(5):77–81.

Advances in Power and Energy Engineering – Sun (Ed.)
© 2016 Taylor & Francis Group, London, ISBN 978-1-138-02846-3

A voltage control strategy of Wind Farms Cluster for Steady State Voltage Stability improvement

S. Yang, W.S. Wang, C. Liu & Y.H. Huang
China Electric Power Research Institute, Haidian District, Beijing, China

J. Wang
Electric Power Planning and Engineering Institute, Xicheng District, Beijing, China

T.F. Guo
State Grid Handan Electric Power Supply Company, Handan, Hebei, China

ABSTRACT: This paper presents a voltage control strategy for improving the Steady State Voltage Stability (SSVS) of the large scale Wind Farms Cluster (WFC) integrated system. This control strategy employs the local L-index to evaluate the voltage stability of the WFC. Aiming at ensuring more wind power can be safely send out from the WFC, the SSVS margin, the voltage security margin and the dynamic reactive power reserve are considered as the multi-objective functions of this control strategy. Besides binding the voltage of the point of interconnection of the WFC to the reference value given by the higher level control, the voltage control strategy can also coordinate different kinds of reactive power sources to regulate the voltage profile of all wind farms. The simulation results of a typical WFC integrated system in Northwest China validate the proposed model and strategy.

1 INTRODUCTION

With the high speed development of wind power generation in the last two decades, the wind power installed capacity in China has been more than 114.609 GW by the end of 2014. However, most of the large-scale wind power bases under construction or planning are located in the regions far away from the load center. These regions where Wind Farms Cluster (WFC) integrated generally possess the weakly connected grid structure. The reactive power and voltage issues caused by the wind power fluctuation have become the prominent challenges for the large scale development of the wind energy (M.J. Hossain, et al, 2012; E. Vittal, et al, 2008).

In order to improve the reactive power and voltage problem caused by wind power integration, some Transmission System Operators (TSOs) in different countries have formulated the grid codes to require the wind farms having the voltage regulation capability to smooth the fluctuations of the Point of Common Coupling (PCC). But the existing control strategies of the wind farm only aims at the reactive power balance of each single wind farm, they cannot satisfy the comprehensive voltage stability needs of the WFC in whole. Therefore, the study on the voltage control of the WFC has great practical significance.

In previous works, many scholars have investigated in reactive power and voltage control of wind farms integrated system. A method using the reduced Jacobian matrix to assess the reactive power and voltage supporting ability of the integrated wind farm is proposed in SHAO Yixiang (2009). A coordinated voltage control strategy which regards the wind farms and interconnection substations as a whole system is proposed in CHEN Huifen (2010). A two-tier and multi-stage voltage coordination control model which considers different time scale of the fast response and slow response reactive power sources is proposed in Amir Ahmidi (2012). Some simple cooperating distribution methods of reactive power control among wind farms in the wind power integrated regions are proposed in WANG Qi (2012) from different points of view.

The above strategies mainly focus on the features of the reactive power compensation equipment in the wind farm and the central substation. Actually, the WFC is the sending system of the clean energy, the TSOs are more concerned about how to increase the SSVS margin and send out more wind power safely from the WFC by optimizing the reactive power distribution. Therefore, the voltage control of the WFC should take maintaining adequate SSVS as an important control goal. In this paper, a new voltage control strategy of the WFC considering the SSVS improvement is proposed. This control strategy considers the SSVS margin, the voltage security margin and the dynamic reactive power reserve as the optimization goals, and the Interior Point Filter Algorithm (IPFA) is used to solve such a multi-objective optimization model.

This paper is organized as follows. The voltage control framework and voltage stability assessment of the WFC are respectively described in Section 2 and Section 3. The mathematical model and solution algorithm are respectively described in Section 4 and Section 5. The simulation results are presented in Section 6. Finally, the conclusions are drawn in Section 7.

2 WFC VOLTAGE CONTROL FRAMEWORK

The WFC voltage control is based on different kinds of reactive power sources in the wind farm and the central substation, such as the On Load Tap Changer (OLTC), the shunt capacitor/reactor, the Static Var Compensator (SVC) and the variable speed wind turbine generators. On the one hand, the voltage control of the WFC should effectively employ these reactive power sources to quickly respond the voltage of Point of Interconnection (POI) of the WFC to the reference value given by the higher level control. On the other hand, the control strategy should achieve other goals like voltage stability, voltage security and reactive power reserve. Based on the hierarchical voltage control, the typical system structure of the WFC voltage control is shown in Figure 1. Here, V_{POI} is the voltage of the POI after control, V_{PCC} and T_{WF} are respectively the real time voltage of the PCC and the OLTC tap of the wind farm, Q_{SUB} and T_{SUB} are respectively the real time reactive power output

and OLTC tap of the central substations, the variables with the superscript ref are the reference values given by the higher level control.

3 WFC VOLTAGE STABILITY ASSESSMENT

Many indexes are proposed for voltage stability assessment in conventional power system, such as the eigenvalue index, the sensitivity index and the load margin index. On the one hand, some of these indexes have to evaluate the singularity of the Jacobian matrix at the critical operation point, the calculation process is complex and time-consuming. On the other hand, the indexes deviated from the conventional power system without wind power integration may hardly be utilized in the system only integrated with large scale wind power.

Based on the above considerations, the local voltage stability L-index proposed by Kessel (1986) is employed to assess the SSVS of the WFC in this work. In this paper, the applicability of the L-index is proved firstly and then considered as one of the multi-objective functions in the control strategy.

The topological structure of the generalized WFC system with n-wind farms is shown in Figure 2.

The network equations in terms of the node admittance matrix can be simply written as

$$\dot{I}_{node} = \dot{Y}_{node}\dot{V}_{node} \qquad (1)$$

For computing the voltage stability index value, all the nodes of the WFC are divided into two categories that only involve infinite power grid node (N_s) and the other nodes (N_D) as follows

$$\begin{bmatrix} I_S \\ I_D \end{bmatrix} = \begin{bmatrix} Y_{SS} & Y_{SD} \\ Y_{DS} & Y_{DD} \end{bmatrix} \begin{bmatrix} U_S \\ U_D \end{bmatrix} \qquad (2)$$

where, Y_{SS}, Y_{SD}, Y_{DS}, Y_{DD} are the sub-matrices of the full order node admittance matrix Y_{node}.

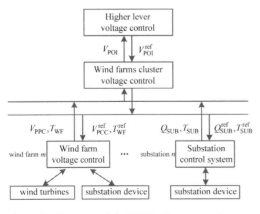

Figure 1. Structure of the WFC voltage control.

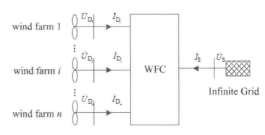

Figure 2. Topological structure of generalized WFC system.

Equation (2) can be transformed as follows

$$\begin{bmatrix} U_{\mathrm{D}} \\ I_{\mathrm{S}} \end{bmatrix} = \begin{bmatrix} Y_{\mathrm{DD}}^{-1} & -Y_{\mathrm{DD}}^{-1}Y_{\mathrm{DS}} \\ Y_{\mathrm{SD}}Y_{\mathrm{DD}}^{-1} & Y_{\mathrm{SS}} - Y_{\mathrm{SD}}Y_{\mathrm{DD}}^{-1}Y_{\mathrm{DS}} \end{bmatrix} \begin{bmatrix} I_{\mathrm{D}} \\ U_{\mathrm{S}} \end{bmatrix} \quad (3)$$

Equation (3) can be further transformed as

$$U_{\mathrm{D}} = ZI_{\mathrm{D}} + FU_{\mathrm{S}} \quad (4)$$

where $Z = -Y_{\mathrm{DD}}^{-1}$, $F = -Y_{\mathrm{DD}}^{-1}Y_{\mathrm{DS}}$.
For each node wind farm integrated, there is

$$\dot{U}_{\mathrm{D}_i} = \sum_{j=1}^{N_{\mathrm{D}}} Z_{ij}\dot{I}_{\mathrm{D}_j} + \dot{F}_i \dot{U}_{\mathrm{S}} \quad (5)$$

where Z_{ij} = mutual impedance between node i and j; $\dot{F}_i$ = i-th component in matrix F.
By introducing the power correction shown as follow

$$\dot{S}_{\mathrm{D}_i}^{\mathrm{corr}} = \dot{U}_{\mathrm{D}_i} \sum_{j \neq i}^{N_{\mathrm{D}}} \frac{Z_{ij}^*}{Z_{ii}^*} \frac{\dot{S}_{\mathrm{D}_j}}{\dot{U}_{\mathrm{D}_j}} \quad (6)$$

then Equation (5) can be further transformed as

$$U_{\mathrm{D}_i}^2 - \dot{E}_{\mathrm{S}_i}\dot{U}_{\mathrm{D}_i}^* = \left(\dot{S}_{\mathrm{D}_i} + \dot{S}_{\mathrm{D}_i}^{\mathrm{corr}}\right)^* Z_{ii} \quad (7)$$

According to the original definition of the local voltage stability L-index, the L-index which is suitable for the WFC can be described as follows

$$L_i = \frac{\left|\left(\dot{S}_{\mathrm{D}_i} + \dot{S}_{\mathrm{D}_i}^{\mathrm{corr}}\right)Z_{ii}^*\right|}{U_{\mathrm{D}_i}^2} = \left|1 - \frac{\dot{E}_{\mathrm{S}_i}}{\dot{U}_{\mathrm{D}_i}}\right|$$

$$= \left|1 - \frac{F_i \cdot U_{\mathrm{S}}}{U_{\mathrm{D}_i}} \angle \delta_i - \theta_i\right| \quad (8)$$

where, L_i = L-index voltage stability indicator for node i; δ_i = angle of i-th component of vector F; V_{D_i} = magnitude of i-th component of vector V_{D}; θ_i = angle of i-th component of vector V_{D}.
When $L < 1$, it means the voltage stability margin of the WFC is adequate; when $L = 1$, it means the status of the WFC is critical stability; when $L > 1$, the system is instability.

The comparison of the P-V curves and the P-L curves in the simulation system mentioned in Section 6 when the wind power increase gradually is shown in Figure 3. It can be seen that the L-index can monotonically reflect the SSVS of the WFC, and the variation trend is consistent with the PV curve.

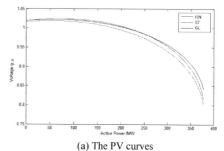

(a) The PV curves

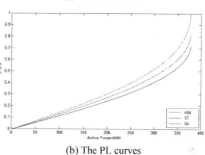

(b) The PL curves

Figure 3. Comparison of the PV and PL curves of the WFC.

4 MATHEMATICAL MODEL

4.1 Objective function

In this paper, the objective function of the WFC voltage control mathematical model is consisted of the SSVS margin, the voltage security margin and the dynamic reactive power reserve. Therefore, the multi-objective function of the WFC voltage control can be expressed as

$$\min \; f(x) = w_L F_L + w_V F_V + w_Q F_Q \quad (9)$$

where, F_L, F_V and F_Q are the SSVS margin, the voltage security margin and the dynamic reactive power reserve of the control strategy; w_L, w_V and w_Q are respectively the weight coefficients of the three functions.

4.1.1 Typography for references

In order to send out more wind power safely from the WFC, it is necessary to maintain adequate voltage stability margin under both normal and contingency conditions. From the above discussion, it is easy to know that keeping the L-index value of all nodes away from 1 and closer to 0 indicates an improved voltage stability margin. Therefore, the objective function for the WFC voltage stability margin is

$$\min \; F_L = \sum_{i \in N_{\mathrm{D}}} L_i^2 \quad (10)$$

4.1.2 *Voltage security margin*

In order to avoid cascading trip-off of wind turbine generators caused by the voltage steep rise/drop under contingency cases, it is necessary to keep the buses in WFC maintaining the adequate voltage security margin. Therefore, the objective function for WFC voltage security margin is

$$\min\ F_V = \sum_{i \in N_D} \left(V_{D_i} - V_{D_i}^{\mathrm{ref}} \right)^2 \tag{11}$$

where, V_{D_i} = real time voltage of i-th bus of N_D; $V_{D_i}^{\mathrm{ref}}$ = voltage reference of i-th bus of N_D.

4.1.3 *Reactive power reserve*

Dynamic reactive power reserve is an important issue which has a significant effect on reliable operation of the WFC. It can be said that adequate dynamic reactive power reserve in the WFC should be available to prevent voltage drop after any contingency. Therefore, the objective function for WFC dynamic reactive power reserve is

$$\min\ F_Q = \sum_{i \in N_C} \left(\frac{Q_{C_i}}{Q_{C_{i\max}} - Q_{C_{i\min}}} \right)^2 \tag{12}$$

where, N_C = the set of buses which have dynamic reactive power output capability; Q_{C_i} = reactive power output of i-th bus; $Q_{C_{i\max}}$ = upper bound of reactive power output; $Q_{C_{i\min}}$ = lower bound of reactive power output.

4.2 *Constraints*

The basic function of voltage control of WFC is effectively using these reactive power resources to quickly respond the voltage of point of interconnection of WFC to the reference value given by the higher level control. Therefore, constraint for WFC POI voltage deviation is

$$\left| V_{\mathrm{POI}} - V_{\mathrm{POI}}^{\mathrm{ref}} \right| \leq V_{\mathrm{err}} \tag{13}$$

where, V_{err} = the acceptable deviation.

Besides the POI voltage deviation, the constraints of WFC voltage control also contain load flow equations

$$P_i - V_i \sum_{j=1}^{N_S} V_j \left(G_{ij} \cos \theta_{ij} + B_{ij} \sin \theta_{ij} \right) = 0 \quad i \in N_N \tag{14}$$

$$Q_i - V_i \sum_{j=1}^{N_S} V_j \left(G_{ij} \sin \theta_{ij} - B_{ij} \cos \theta_{ij} \right) = 0 \quad i \in N_N \tag{15}$$

and variable parameter bounds

$$\begin{cases} V_{i\min} \leq V_i \leq V_{i\max} & i \in N_N \\ T_{i\min} \leq T_i \leq T_{i\max} & i \in N_T \\ Q_{C_{i\min}} \leq Q_{C_i} \leq Q_{C_{i\max}} & i \in N_C \end{cases} \tag{16}$$

Here, N_N = the set of all nodes of the WFC; N_T = the set of all OLTC in the WFC.

5 INTERIOR POINT FILTER ALGORITHM

The IPFA is an optimization algorithm for solving large scale non-linear problems. It has been warmly concerned because of possessing faster calculation speed, better convergence and good robustness (Wachter Andreas, et al., 2001; Yang Shuo, et al., 2011). The IPFA is based on a primal dual interior point algorithm with a filter set which can be updated after the iteration. And the filter will be augmented after every iterative calculation, ensuring that the iteration point x_{k+1} cannot return to the neighborhood of the previous iteration point x_k. If the iteration point x_{k+1} does not satisfy the filter rule, the backtracking line-search procedure will be used to decrease the trial step sizes unless the iteration point x_{k+1} does not meet the filter rule. Over all, these procedures ensure that the algorithm cannot cycle between two points that alternatively decrease the constraint violation and the barrier objective function.

In this paper, the mathematic model of the WFC voltage control is solved by the IPFA. The discrete variables are adjusted continuously at first, then the discrete variables calculation results are fixed on the nearest rounding value. In the end, the optimization problem is calculated again and the discrete optimal solution will be found later. The steps followed for the implementation of the algorithm are described as follows:

Step 1: Get the real time operating data of each wind farm and substation from the WFC monitoring and control platform, and initialize the starting point to IPFA.

Step 2: Check convergence for the overall problem. If the accuracy of convergence is satisfied, then STOP. Otherwise go to Step 3.

Step 3: Compute the search directions and the step sizes. If the iteration point x_{k+1} does not satisfy the filter rule, accept the trial point x_{k+1} and then augment the filter. Otherwise, use the backtracking line-search procedure to decrease the trial step sizes unless the iteration point x_{k+1} does not meet the filter rule. Complete this iteration and go back to Step 2.

6 CASE STUDY

A real WFC system in Northwest China is employed to verify the validity and performance of the proposed strategy. The structure of the WFC is shown in Figure 4. There are 12 wind farms connected to 4 cluster substations, and the total installed capability of the WFC is 990 MW. The installed capability and the initial wind power output in simulation of each wind farm is shown in Table 1.

The higher side node of substation A is considered as the POI of the WFC. The dynamic compensations equipped in each wind farm equal the 15% of its installed capability. All the wind turbine generators participate the voltage control, and the power factor can be adjusted from −0.98~+0.95. The OLTC in each wind farm substation can be controlled from 0.9~1.1 p.u. by 17 steps.

Three modes of different voltage control strategy are utilized to compare the control effect.

1. Mode I: There is no WFC voltage control, the PCC voltage of each wind farm is controlled within 0.97~1.07 p.u. by the reactive power sources in their own substation.
2. Mode II: The aim of the WFC voltage control is to equipoise the PCC voltage security margin for each wind farm. In the simulation, there is $V_D^{ref} = 1.0$ p.u.

3. Mode III: The WFC voltage control strategy proposed in this paper is used. In the simulation, there are $V_{err} = 0.002$ p.u.

Use the POI voltage in Mode I when the voltage of infinite grid sustains 0.97 p.u. as the POI voltage reference in Mode II and the Mode III. Figure 5 shows the voltage and L-index values of the PCC of each wind farm in different control modes. The statistical results in the three different control modes are contrasted in Table 2.

As indicated from Figure 5 and Table 2, in Mode I, the wind farm cannot coordinated with each other because of lacking the WFC voltage control, thus the node voltage profile in the WFC is lower and L-index value is higher, and the dynamic reactive power reserve of the WFC is the most inadequate as well. In Mode II, the node voltage profile is better because the WFC voltage control aims at enhancing the voltage security of all the PCC nodes. But the L-index doesn't become better because it isn't one of the multi-objective functions of the WFC voltage control. By considering the multi-objective functions of F_L, F_V and F_Q, the node voltage profile, the node L-index value and the dynamic reactive power reserve in Mode III are more satisfied than that in the Mode I and Mode II.

In order to validate the effect of sending out more wind power safely from the WFC, the PV

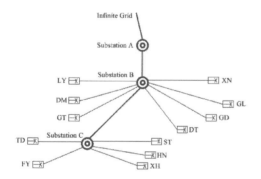

Figure 4. Structure of simulation system.

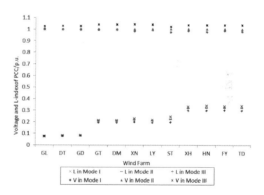

Figure 5. Voltage and L-index of PCC in different control modes.

Table 1. Installed capability and initial output of different wind farms (MW).

Wind farm	Installed capability	Initial Output	Wind farm	Installed capability	Initial Output
GL	49.5	12.3	LY	198	103.4
DT	49.5	10.9	ST	99	36.14
GD	49.5	13.4	XH	49.5	6.8
GT	49.5	12.3	HN	49.5	11.9
DM	198	138.6	FY	49.5	11.2
XN	99	65.3	TD	49.5	7.4

Table 2. Contrast of results of the three control modes.

	Mode I	Mode II	Mode III
L_{max}	0.805153	0.811076	0.741701
V_{min}	0.923136	0.973388	0.100968
Q_{max}	0.502341	0.627950	0.220559
F_L	17.719749	18.323493	15.366426
F_V	0.060161	0.012203	0.067877
F_Q	4.558761	5.726680	0.085365

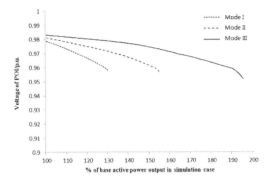

Figure 6. PV curve of the POI in different control modes.

curve which considers the voltage control capability of the wind farm is utilized. The active power output of each wind farm is increasing by 1% of the initial output of the simulation case until it reaches the installed capacity. The wind farm keeps its PCC voltage unchanged until the reactive power is inadequate. The PV curve of the POI in different control mode is shown in Figure 6.

As shown in Figure 6, the voltage collapse of the WFC occurs respectively at 130% (means 558.532 MW), 155% (means 665.942 MW) and 195% (means 837.798 MW) of the initial output in the three different control modes. By an overall decrease in L-index value of all buses of the WFC, the voltage control strategy proposed in this paper can maintain a sufficient voltage stability margin and ensure more wind power can be sent out safely from the WFC.

7 CONCLUSIONS

A voltage control strategy of the WFC which considers improving the SSVS is proposed in this paper. The local L-index is employed to evaluate the SSVS margin of the WFC. In this voltage control strategy, the voltage stability margin, the voltage security margin and the dynamic reactive power reserve are all considered as the optimization goals, and the IPFA is used to solve such a multi-objective optimization model. The simulation results show that the coordinated control between different kinds of reactive power sources

in wind farms and substations can be achieved. Also, it verifies that more wind power can be sent out safely from the WFC by improving the SSVS according to the control strategy.

ACKNOWLEDGEMENTS

This work has been supported financially by the National Natural Science Foundation of China (No. 51207145), the National Basic Research Program of China (973 Program) (2012CB215105), the Science and Technology Project of SGCC (No. NY71-14-035).

REFERENCES

Amir Ahmidi, Xavier Guillaud, Yvon Besanger, et al. 2012. A multilevel approach for optimal participating of wind farms at reactive power balancing in transmission power system. *IEEE Systems Journal*, 6(2): 260–269.

Chen Huifen, Qiao Ying, Lu Zongxiang, et al. 2010. Study on reactive power and voltage coordinated control strategy of wind farm group. *Automation of Electric Power Systems*, 34(18): 78–83.

Hossain, M.J., Hemanshu R.P. & M.A. Mahmud, et al. 2012. Investigation of the impacts of large-scale wind power penetration on the angle and voltage stability of power systems. *IEEE Systems Journal*, 6(1): 76–84.

Kessel P., Glavitsch H. 1986. Estimating the voltage stability of a power system. *IEEE Transactions on Power Delivery*, 11(3): 346–354.

Shao Yixiang, Chen Ning, Zhu Lingzhi, et al. 2009. Reduced jacobian matrix based method to assess local static reactive power/voltage supporting ability of interconnected wind farm. *Power System Technology*, 33(2): 14–19.

Vittal, E., A. Keane, M. O'Malley. 2008. Varying penetration ratios of wind turbine technology for voltage and frequency stability. *Proceeding of IEEE PES General Meeting*, Pittsburgh, PA.

Wachter Andreas, Lorenz T Bieglery. 2001. *Line Search Filter Methods for Nonlinear Programming: Motivation and Global Convergence*. Yorktown Heights: USA IBM T.J. Watson Research Center.

Wang Qi, Yuan Yue, Chen Ning, et al. 2012. The cooperating distribution methods of reactive power in multi-wind farms integrated region. *Power System Protection and Control*, 40(24): 76–83.

Yang Shuo, Zhou Jingyang, Li Qiang, et al. 2011. An interior point filter algorithm for reactive power optimization. *Power System Protection and Control*, 39(18): 14–19.

Advances in Power and Energy Engineering – Sun (Ed.)
© 2016 Taylor & Francis Group, London, ISBN 978-1-138-02846-3

Dynamic security assessment method of wind power farm

Y.F. He, Y.X. Zhuo, X.N. Lin & Z.X. Wang
*State Key Laboratory of Advanced Electromagnetic Engineering and Technology,
Huazhong University of Science and Technology, Wuhan, Hubei, China*

C.G. Tu
Center China Grid Company Limited, China

ABSTRACT: The random fluctuation of wind power output will cause voltage fluctuation randomly in the wind farm integration and even lead to system crash. Therefore, it is necessary to search an all-round and accurate evaluating method of voltage fluctuation. According to the characteristic of the voltage random distribution, this paper focused on the overall and partial voltage fluctuation and constructs four voltage evaluation indexes, including a voltage fluctuant index, a skewness index, maintenance coefficients, and dispersion coefficient. The bus indexes and system indexes are defined respectively. These proposed indexes can well reflect the random distribution of the voltage in the network with wind power integration. The paper develops the key risk indicator based on the analysis of a number of numerical examples, which is used for classifying voltage safety into five categories. This appraisal procedure can completely implement the real-time monitoring of voltage fluctuation in integration wind farm.

1 INTRODUCTION

Due to the lack of energy supply and the scarcity of global oil, coal, and natural gas resources, how to utilize new energy efficiently has become an important topic of research institutions and has been widely noticed (Thiringer et al. 2001). As a kind of clean and renewable energy, wind contains tremendous power. If we can efficiently covert wind power into electrical power, it's completely possible to meet a major proportion of power demand in the world (Abouzahr & Ramakumar 1991, Wang et al. 2004). However, the fact is that wind farm has never been taken into extensive operation because of randomness and uncontrollability in the operation of wind power plant, which result in power system accidents and unnecessary losses (Han & Cao 2004).

The most crucial problem the grid-connected wind farm faces now is the system crashes resulting from voltage fluctuations. In this paper, the safety of voltage in wind farm was selected as the major object of study. This paper analyzes the factors that contributing to the voltage fluctuations of wind farm integration, and gets a series of statistical data by constructing models of multi-machine wind farms in DIgSILENT software, and simulating the running states of wind farms under different conditions. By setting key risk index of safe operation of wind farms, we analyzed approaches and models of evaluations on wind farm dynamic operation.

2 MODELING AND SIMULATING OF WIND FARM

2.1 Doubly Fed Induction Generator (DFIG) based of DIgSILENT

Figure 1 is DFIG based on DIgSILENT. The model of DFIG is composed of wind speed, prime mover, double-fed machine, and control system. The above modules mentioned are implemented in the dynamic control of double-fed machine (Rather et al. 2013).

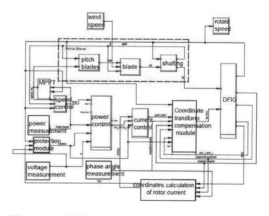

Figure 1. DFIG based on DIgSILENT.

2.2 Equivalent modeling of wind farm of DFIG

The current typical connection modes of wind farm is as follows: terminal voltage of fan is 6.9 kV, the generator terminal package transformer increases the number of volts to 35 kV, then the voltage increases to 220 kV by booster substations. Static VAR Compensator (SVC) is set on 35 kV bus in the wind farm to compensate reactive power loss (Zhuo et al. 2014).

The equivalent modeling of wind farms is built on DIgSILENT, as shown in Figure 2. Wake effect is considered in the simulation system, therefore, the results are more credible. The wind filed G1, G2, G3, and G4 are made up of 30 DFIG, each of which is 2 MW. The influx lines of 35 kV are 3 km. The total installed capacity of wind fan is 300 MW. Because the number of electrical equipment can be set in DIgSILENT, there is no need for equivalent calculation of wind fan, transformer, and transmission line.

The equivalent modeling of wind farm as shown in Figure 2 is connected to the electrical system which is shown in Figure 3 by Bus 6. The whole simulation model is finished.

2.3 Simulating voltage waveform of wind farm

We can get voltage waveforms in different operation conditions by changing the number of double-fed machines of wind farm in G1, G2, G3, and G4. The wave figures are shown in Figures 4–7. The bottom half of each figure represents the voltage of bus (MV) inside of wind farm, the top half represents the bus (BUS 6) voltage waveform when wind farm is connected to power system.

It can be inferred from the simulation that the system would appear unstable condition when the output of wind farm reaches a critical state, the number of generators reaches saturation, or the overload input power. And the increasing

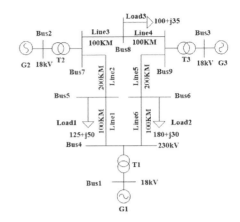

Figure 3. Electrical system accessed to wind farm.

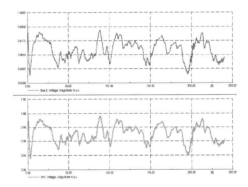

Figure 4. The numbers of double-fed machines in G1, G2, G3, and G4 are set to 30.

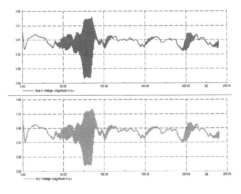

Figure 5. The number of double-fed machines in G1, G2, G3, and G4 are set to 45, 45, 45, and 44.

of capacity will intensify the unstable condition, which causes great damage to utility. Therefore, the paper puts forward four kinds of criterion and real-time warning. And it evaluates the safety condition of voltage fluctuation in wind farm.

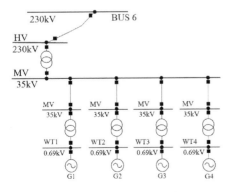

Figure 2. Equivalent modeling of wind farms.

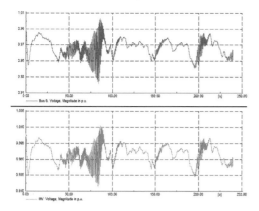

Figure 6. The number of double-fed machines in G1, G2, G3, and G4 are set to 45.

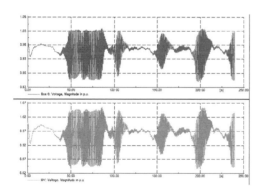

Figure 7. The number of double-fed machines in G1, G2, G3, and G4 are set to 48, 45, 45, and 45.

3 EVALUATION SYSTEM VOLTAGE FLUCTUATION

The operating process of a wind farm is complex with large amount of parameters. It may put threat to the secure operation of the electric power system due to its intermittent feature, complicated control and lack of the capability of frequency adjustment, etc (Wei et al. 2013). In order to achieve the function of real-time monitoring in integration wind farm, we set relevant indexes and criterions for real time assessment of voltage security to ensure the stability of the power system and sends out forewarning in time in case of failure, which allows staff to take remedial action to avoid accidents.

It's a gradual process from small voltage fluctuation to power grid failure. We can make forewarning more effective by setting the value of t which is based on the actual situation (t is a time segment).

3.1 Voltage Fluctuant Index (VFI)

Standard deviation is a measure that is used to quantify the amount of variation or dispersion of a set of data values. VFI is defined as standard deviation of voltage, and it is an accurate description of overall volatility of bus voltage, which is defined by:

$$\text{VFI} = \sqrt{\frac{1}{m-1}\sum_{i=1}^{m}\left(U_i - \overline{U}\right)^2} \tag{1}$$

where, U_i = observations of voltage at time i; U = the average voltage in time frame of t; and m = sample size of voltage in time frame of t.

3.2 Voltage Skewness Index (VSI)

Frequently, skewness coefficient is a measure of the asymmetry of the probability distribution of a real-valued random variable about its mean. The influence on raising or lowering bus voltage caused by wind power integration is described by asymmetric degree of bus voltage, which is defined as VSI:

$$\text{VSI} = \frac{\sum_{i=1}^{m}\left(U_i - U_0\right)^3}{(m-1)(m-1)\sigma^3} \tag{2}$$

where, U_0 = bus voltage before wind power connecting to the grid and σ = standard deviation of voltage.

Although VFI describes the overall volatility of bus voltage after wind power connecting to the grid, it can not present the increase or decrease in bus voltage after DFIG being integrated to power system. The introduction of VSI can solve this very well. It provides reference for rid planning and dispatching opera, and helps staff switch VAR compensator to safeguard power system safer when it's necessary.

3.3 Voltage maintenance coefficients

In order to express the change of bus voltage before and after the wind power is connected to the grid, this paper introduces Voltage Maintenance Coefficients (VMC) which are derived from deviation factor to describe the capacity of bus voltage to maintain the level before wind power connecting to the grid.

$$\text{VMC} = \frac{C(\Omega)}{m}$$
$$\Omega = \left\{U_i \middle| |U_i - U_0| \leq \alpha U_0 \right\} \tag{3}$$

where, $C(\Omega)$ denotes how many objects are contained in that set and α is the threshold value of voltage fluctuating range which is set to 10% according to national standards (Wei et al. 2013).

VMC provides a key support for decision-making on control of power system voltage. There is stronger randomness in the system voltage when wind power connecting to the grid. The related schedule and control decision can no longer depend on the criteria of certainty, so probabilistic and decision analyses are necessary. VMC presented in this article can accurately reflect the probability that bus voltage stays constant after grid is connected.

3.4 Voltage Dispersion Coefficient (VDC)

For some random variables, the relations between the value and mean value of volatility have practical implications. In order to denote the trend of voltage fluctuation in wind farm, this paper introduces dispersion coefficient of mean value to describe the fluctuation of random variable. As a supplement for VFI, voltage dispersion coefficient can give more accurate evaluation on system safety.

Supposed the voltage u of integration wind farm is random variable, its mean value is E_u, standard deviation is σ_u. VDC is given by:

$$VDC = \frac{\sigma_u}{E_u} \qquad (4)$$

VFI describes the discrete distance of u, while VDC denotes the discrete level of voltage value and non-zero mean.

4 VOLTAGE FLUCTUATION ANALYSIS BASED ON MATLAB

This paper introduces the operation data of wind fan model from DIgSILENT to MATLAB (Wu & Ding et al. 2009), and operates the program to get corresponding results, which are the index of wind farm.

After the wind farm is connected to the power system, the sum of positive and negative deviation of absolute value of voltage of the grid is less than 10% of rated voltage (GB/T 12326-2008). The operating condition of wind farm can be rated as safe, stable, critical, unstable, and collapsed based on the standard above. According to the analysis data in MATLAB and classifications of wind farm operating condition, we can divide voltage fluctuation safety grade into five levels (A, B, C, D, and E). Specific evaluation standards are shown in Table 1.

When VFI gets higher, it indicates that bus voltage in the power system is more discrete and the

Table 1. Wind farm voltage fluctuation safety grade evaluation based on index criterion.

	Voltage fluctuant index	Voltage dispersion coefficient	Voltage skewness index
A	<0.0063	<0.0064	1
B	0.0063–0.0097	0.0064–0.0146	1
C	0.0097–0.0408	0.0146–0.0421	0.8526–1
D	0.0408–0.0727	0.0421–0.0770	0.7246–0.8562
E	>0.0727	>0.0770	<0.7246

fluctuation range increases, which have adverse effect on the stability of power system voltage and imperil the power quality. VFI describes discrete distance of the value of voltage, while VDC describes the degree of discretization of value of voltage on nonzero mean. The means of judgment are the same as above and it could be a supplement to judge the safety grade of system.

VMC reflects the probability when bus voltage maintains the value before wind farm connect to the grid. High index represents high probability of maintaining at the value and vice versa.

VSC can not measure the safety of voltage fluctuation, but using it as an index, we can adjust the operation of wind farm. Positive value indicates the increase in voltage is more serious than the decrease in voltage after paralleling in, which is shown in the probability density curve of voltage distribution having a longer tail on the right side. Negative value is reverse. When VSC is zero, it represents the increase in voltage is about the same as the decrease in voltage after paralleling in, which is shown in the probability density curve of voltage distribution is bilateral symmetry.

5 IMPLEMENTATION OF REAL-TIME WARNING OF WIND FARM

The index judgment set in this paper is different from each other on the focus of safety evaluation of voltage fluctuation in wind farm. It will cause misleading results and error because each index can judge safety condition of wind farm independently. In order to achieve higher accuracy of real-time warning, we should analyze further after confirming the value interval and corresponding grades of VFI, VDC, and VMC.

To avoid the occurrence of safety incidents in greater extent, the paper adopts conservative approaches to conclude dynamic safety evaluation as shown in Figure 8. Suppose the priority order of safety grades is E > D > C > B > A, when different safety grades are given by VFI, VDC, and

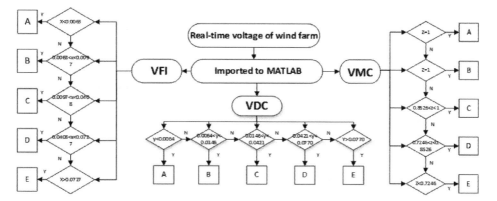

Figure 8. Dynamic safety evaluation system of wind farm.

VMC respectively, we select the highest priority as the safety grades of wind farm. For example: we obtain B level from VFI and obtain C level from VDC and obtain D level from VMC, according to the priority order of D > C > B, the safety grades of wind farm is D.

6 CONCLUSION

This paper begins with the feature of voltage random distribution, discussing the safety assessments about integration wind farm relates to voltage fluctuation with the method of statistics. It introduces standard deviation, skewness coefficient, and deviation factor and demonstrates the scientific nature, effective and easy operation of the evaluation system. The wake effect of wind farm is considered in the simulation model, so the results are credibility.

ACKNOWLEDGEMENT

This work is sponsored by the National Basic Research Program of China (2012CB215100), and in part by the National Natural Science Foundation of China (51277082), in part by National Natural Science Youth Foundation of China (51207064), in part by Natural Science Foundation of Hubei Province (2012FFA075), and in part by Huang He Ying Cai funding of Wuhan.

REFERENCES

Abouzahr, I & Ramakumar, R. 1991. An approach to assess the performance of utility-interactive wind electric conversion systems. *IEEE Trans on Energy Conversion* 6(4): 627–638.

Han Zhenxiang & Cao Yijia. 2004. Power system security and its prevention. *Power System Technology* 28(9): 1–6.

Rather Z.H, et al. 2013. Dynamic Security Assessment of Danish Power System Based on Decision Trees: Today and Tomorrow. *PowerTech, 2013 IEEE Grenoble*: 1–6.

Thiringer Torbjörn, et al. 2001. Power quality impact of a sea located hybrid wind park. *IEEE Trans on Energy Conversion* 16(2): 123–127.

Wang Zhiqun, et al. 2004. Impacts of distributed generation on distribution system voltage profile. *Automation of Electric Power Systems* 28(16): 56–60.

Wei Ran, et al. 2013. Research on the Factors Influencing Power Quality of Distribution Networks Integrated with Wind Turbines. *Power System and Clean Energy* 29(11): 96–107.

Wu Yichun & Ding Ming. 2009. Simulation Study on Voltage Fluctuations and Flicker Caused by Wind Farms. *Power System Technology* 33(20): 125–130.

Zhuo Yixin, et al. 2014. The Establishment of Wind Farm Dynamic Co-Simulation Platform and Influence Analysis of Wind Conditions. *Transactions of China Electrotechnical Society* 29(1): 356–364.

Advances in Power and Energy Engineering – Sun (Ed.)
© *2016 Taylor & Francis Group, London, ISBN 978-1-138-02846-3*

Study on dynamic model with feedback from flexible structure of wind turbine

H.T. Li & Z.J. Li
Beijing Electric Power Research Institute, Beijing, China

S. Shen
Shandong Ai Pu Zhi Xin Amorphous-Alloy Transformer Co. Ltd., Jinan, Shandong, China

ABSTRACT: The dynamic inflow model in yaw condition was constructed based on the theory of the Blade Element Momentum (BEM) revised model and the Pitt–Peter model. When taking large deformations of the flexible blade during the working process and the feedback of flexible structural deformation on the dynamic inflow model into consideration, the gradient mechanisms of induced factor and structural dynamic response on the wind turbine were obtained under different circumstances, such as in transition of rotor speed and blade pitch angel. Dynamic induced factor and structural dynamic response of one 5-MW wind turbine was figured out. All these calculations were achieved by employing MATLAB programming. The calculated results of several models reveal that it is necessary to consider the feedback of flexible structures.

1 INTRODUCTION

As wind turbine technology advances, the capacity of single wind turbines has been increasing rapidly, with rotor diameters soaring from 20 m in the early days to the current 160-plus m. In order to make full use of the wind resource and reduce the cost of wind energy, enlarging the capacity of single wind turbines has been the inevitable solution for the development of modern wind turbines. With the development of large scale wind turbines, the blades become larger and more flexible. As a result, the flexible deformation of the blades has to be taken into account. Thus, the aero-elastic model, based on the small transmogrification assumption, is no longer suitable for simulating more uncertain factors caused by the increase in wind turbines' flexibility.

The nonlinear aero-elastic analysis that allows for blade flexibility has become a research focus. A number of analytical methods, which have their own characteristics and are relatively mature in the scope of their application, have been developed (Xu et al. 2015). However, all these existing methods did not consider the influence of wind turbines' flexibility on dynamic aerodynamic models and take the corresponding changes of aerodynamic loads as feedback to aero-elastic analysis. With the increase of wind turbine capacity and the need to meet the special requirements of offshore wind field environment, seeking a full-system aero-elastic analytical method for the overall unit, which considers both the fluid-solid coupling effect of wind, wave, and unit structure, and the effect of the unit's nonlinear large flexible deformation has become an important direction of research. Based on Blade Element Momentum (BEM) and Pitt–Peters modified models, this study comprehensively considered various factors that influence wind turbines' aerodynamic performance and took flexible blade deformations, which were mainly in three degrees of freedom including flap-wise displacement, edgewise displacement, and twist angle, as feedback to the dynamic inflow model. It presents an exact description of dynamic changes of induced velocity and structural dynamic response with the change of each parameter and realizes an accurate analysis of dynamic inflow effect of the flexible rotor as well as structural dynamic analysis.

2 BASIC THEORY

2.1 BEM modified model

Considering the correction factors, such as blade tip loss, rotor shaft loss, and wind shear, on the basis of BEM theory, a coordinate system of wind turbine was established (Shen et al. 2013). After performing coordinate transformation and taking yaw into consideration, the expressions for calculating inflow angle and inducing factor can be written as:

$$\phi = \tan^{-1}\{(\sin\gamma * \cos\eta\cos\beta_0 + \cos\gamma * \sin\beta_0\sin\psi \\ - \sin\gamma * \sin\beta_0\sin\eta\cos\psi)U(1-a)/[-\Omega ra'\cos\beta_0 \\ + U(1-a)(\cos\gamma * \cos\psi + \sin\gamma * \sin\psi\sin\eta)]\} \tag{1}$$

$$a = \frac{Bc(C_L\cos\phi + C_D\sin\phi)\cos\beta_0}{8\pi rF\sin^2\phi + Bc(C_L\cos\phi + C_D\sin\phi)\cos\beta_0} \tag{2}$$

$$a' = \frac{Bc(C_L\sin\phi - C_D\sin\phi)}{8\pi rF\sin\phi\cos\phi\cos\beta_0 - Bc(C_L\sin\phi - C_D\sin\phi)} \tag{3}$$

The angle of attack is:

$$\alpha = \phi - \theta \tag{4}$$

$$\gamma * = \frac{\pi}{2} - \gamma \tag{5}$$

where θ = the sum of blade angle and twist angle of the section; β_0 = cone angle (initial value of blade incidence); ψ = azimuth angle; η = tilt angle of rotor shaft; and γ = yaw angel.

The relative velocity of airflow can be written as:

$$W_{X_P} = U(1-a)(\cos\gamma * \cos\psi + \sin\gamma * \sin\psi\sin\eta) \\ - \Omega ra'\cos\beta_0 \tag{6}$$

$$W_{Y_P} = U(1-a)(\sin\gamma * \cos\eta\cos\beta_0 \\ + \cos\gamma * \sin\beta_0\sin\psi - \sin\gamma * \sin\beta_0\sin\eta\cos\psi) \tag{7}$$

$$W = \sqrt{W_{X_P}^2 + W_{Y_P}^2} \tag{8}$$

According to the BEM theory, the normal force and tangential force on the blade element can be written as:

$$dF_{Y_P} = \frac{1}{2}\rho W^2 c(C_L\cos\phi + C_D\sin\phi)dr \tag{9}$$

$$dF_{X_P} = \frac{1}{2}\rho W^2 c(C_L\sin\phi - C_D\cos\phi)dr \tag{10}$$

Thus, the thrust and torsion on the blade element can be written as:

$$dT = dF_{Y_P}\cos\beta_0 \tag{11}$$

$$dQ = dF_{X_P}\cos\beta_0 \tag{12}$$

The torsional moment on the blade section can be written as:

$$M_{aero} = \frac{1}{2}\rho W^2 c(eC_L\cos\alpha + eC_D\sin\alpha + cC_M) \tag{13}$$

The force acting on the tower top can be written as (Liu 2006):

$$F_X = F_{X_P}\cos\psi + F_{Y_P}\sin\psi\sin\beta_0 \tag{14}$$

$$F_Y = F_{Y_P}(\cos\beta_0\cos\eta - \sin\eta\sin\beta_0\cos\psi) \\ + F_{X_P}\sin\eta\sin\psi \tag{15}$$

$$F_Z = F_{Y_P}(\cos\beta_0\sin\eta + \cos\eta\sin\beta_0\cos\psi) \\ - F_{X_P}\cos\eta\sin\psi \tag{16}$$

Finally, the yawing moment and heeling moment can be written, respectively, as:

$$M_z = (-r\sin\beta_0\cos\eta - r\cos\beta_0\cos\psi\sin\eta \\ - L\cos\eta)F_X - (r\cos\beta_0\sin\psi)F_Y \tag{17}$$

$$M_x = (-r\sin\beta_0\sin\eta + r\cos\beta_0\cos\psi\cos\eta \\ - L\sin\eta)F_Y - (-r\sin\beta_0\cos\eta \\ - r\cos\beta_0\cos\psi\sin\eta - L\cos\eta)F_Z \tag{18}$$

2.2 Pitt–Peters modified model

Starting by using the acceleration potential method, Pitt and Peters deduced the expression of added mass at the time of dynamic inflow on the basis of helicopter rotor theory and established the analytical model for dynamic inflow of wind turbines by applying the expression to the BEM theory.

By using acceleration potential method to calculate the pressure on the rotation plane of rotor, the added mass to the rotor can be expressed as $\frac{8}{3}\rho R^3$. Meanwhile, a tiny circle was taken from the rotation plane of the rotor and analyzed to combine with the BEM theory. Then, the corrected thrust coefficient can be expressed as Equation (19) (Shen 2013):

$$C_T = 4a(1-a) + \frac{8r}{\pi U}\dot{a} \tag{19}$$

Compared with the original BEM theory, the expression has an additional time-varying term of induced velocity. Thus, it reflects the influence of changes in the axial induction factor on thrust coefficient, namely considering that the changes of incident velocity are subject to the thrust. When calculating the axial velocity inducing factor, the differential equation can be used to replace the thrust coefficient expression in the BEM theory. By integrating the differential equation on every time step, the time-dependent incoming flow value at every section of every blade can be obtained as below:

$$\left[4a(1-a) + \frac{8r\dot{a}}{\pi U}\right]\rho\pi U^2 rFdr \\ = \frac{1}{2}\rho W^2(C_L\cos\phi + C_D\sin\phi)Bcdr \tag{20}$$

Equation (19) can be simplified as:

$$f(a) = A_a a^2 + B_a a + C_a = 0 \tag{21}$$

276

As shown in Equation (21), the axial induction factor, a, is the solution of a standard quadratic equation with one unknown, but the equation coefficients are changing with a. Using Newton iteration method, the iterative formula of a can be written as: (Liu et al. 2009),

$$a_{n+1} = [(UC_N Bc\Delta t + 8\pi r UF \sin^2 \phi \Delta t)a_n^2 \\ - (UC_N Bc\Delta t + 16Fa_{front}r^2 \sin^2 \phi)]/ \\ [2(UC_N Bc\Delta t + 8\pi r UF \sin^2 \phi \Delta t)a_n \\ - (2UC_N Bc\Delta t + 8\pi r UF \sin^2 \phi \Delta t \\ + 16Fr^2 \sin^2 \phi)] \tag{22}$$

$$C_N = C_L \cos\phi + C_D \sin\phi \tag{23}$$

where C_N = normal force coefficient.

The tangential velocity induced factor can be calculated by using the original method. The tangential and axial velocity inducing factors influence each other, which mean the result of the former is dependent on the calculation of the latter. Thus, the tangential velocity inducing factor can also reflect the influence of dynamic inflow.

3 FEEDBACK FROM FLEXIBLE STRUCTURES

This paper focuses on studying the flexible rotor by feeding deformations of flexible structures back to the dynamic inflow model. The feedback is carried out mainly to reflect the influence of section deformations in three degrees of freedom, including flap-wise displacement, edgewise displacement, and twist angle, on the blades' dynamic aerodynamic performance. The blades are simplified to non-uniform cantilevers. Finite element discretization is done to the blades by using two-node beam elements. The entire blade's equations of motion with multi-degree of freedom can be written as: (Chen et al. 2008),

$$\begin{cases} M\ddot{x}(t) + C\dot{x}(t) + Kx(t) = P(t) \\ M\ddot{\theta}(t) + C\dot{\theta}(t) + K\theta(t) = P(t) \end{cases} \tag{24}$$

where M = mass matrix; C = damping matrix; K = stiffness matrix; $P(t)$ = load element array; $x(t)$ = flap-wise/edgewise deformation; and $\theta(t)$ = torsional deformation.

With respect to the flap-wise and edgewise deformations (Clough & Penzien 1993),

$$M_e = \frac{\rho_e l A_e}{420} \begin{bmatrix} 156 & 54 & 22l & -13l \\ 54 & 156 & 13l & -22l \\ 22l & 13l & 4l^2 & -3l^2 \\ -13l & -22l & -3l & 4l^2 \end{bmatrix} \tag{25}$$

$$K_e = \frac{2E_e I_e}{l^3} \begin{bmatrix} 6 & -6 & 3l & 3l \\ -6 & 6 & -3l & -3l \\ 3l & -3l & 2l^2 & l^2 \\ 3l & -3l & l^2 & 2l^2 \end{bmatrix} \\ + \frac{N_e}{30l} \begin{bmatrix} 36 & -36 & 3l & 3l \\ -36 & 36 & -3l & -3l \\ 3l & -3l & 4l^2 & -l^2 \\ 3l & -3l & -l^2 & 4l^2 \end{bmatrix} \tag{26}$$

$$P_e(t) = \begin{bmatrix} (7p_1 + 3p_2)l/20 \\ (3p_1 + 2p_2)l/60 \\ (3p_1 + 7p_2)l/20 \\ -(2p_1 + 3p_2)l^2/60 \end{bmatrix} \tag{27}$$

where ρ_e = element density; l = element length; A_e = element area; E_e = modulus of elasticity; I_e = cross-sectional moment of inertia; and N_e = axial force acting on the elements.

For the torsional deformation (Clough & Penzien 1993),

$$M_e = \begin{bmatrix} \dfrac{\rho_e J_e}{3} & -\dfrac{\rho_e J_e}{6} \\ -\dfrac{\rho_e J_e}{6} & \dfrac{\rho_e J_e}{3} \end{bmatrix} \tag{28}$$

$$K_e = \begin{bmatrix} \dfrac{G_e J_e}{l} & -\dfrac{G_e J_e}{l} \\ -\dfrac{G_e J_e}{l} & \dfrac{G_e J_e}{l} \end{bmatrix} \tag{29}$$

where $J_e = \iint r^2 dA$ is the polar moment of inertia of the beam elements and G_e = shear modulus of the beam elements.

$$P_e(t) = \begin{bmatrix} (2m_1 + m_2)l/6 \\ (m_1 + 2m_2)l/6 \end{bmatrix} \tag{30}$$

Rayleigh damping model is used as the damping matrix (Xu 2001)

$$C = \alpha M + \beta K \tag{31}$$

$$\begin{bmatrix} \alpha \\ \beta \end{bmatrix} = \frac{2\omega_m \omega_n}{\omega_n^2 - \omega_m^2} \begin{bmatrix} \omega_n & -\omega_m \\ -1/\omega_n & 1/\omega_m \end{bmatrix} \begin{bmatrix} \xi_m \\ \xi_n \end{bmatrix} \tag{32}$$

where α and β = constants; ω_m = base frequency of the multi-degree-of-freedom system; ω_n is chosen from high-order modes that contribute to dynamic response significantly. As generally assumed, $\xi_m = \xi_n = \xi$.

After taking into consideration the deformations of flexible blades, the flap-wise and edgewise acting forces and the torsional moment on the blade elements can be written, respectively, as:

$$F_{flap} = F_{Y_P} + F_c \sin \beta_0 + F_g \sin \eta \sin \beta_0 \qquad (33)$$

$$F_{edge} = F_{X_p} - F_c \cos \beta_0 \sin \sigma \\ + F_g \cos \eta \sin \psi \cos \sigma \qquad (34)$$

$$M = M_{aero} + F_c \sin \beta_0 e_{em} \qquad (35)$$

where $F_c = mr\omega^2 \cos \beta_0$ = centrifugal force; e_{em} = the effective distance between the stiffness center of the section and the center of mass; and F_g = gravity.

The load on the blade is calculated by using Equations (33), (34), and (35). Putting the result into the equations of motion with multi-degree of freedom (Equation (24)), the corresponding deformations, including the variation of twist angle of the section ($\Delta\theta$), the variation of flap-wise angle ($\Delta\beta$), and edgewise angle (σ), are calculated; $\Delta\theta$ is fed back to θ and $\Delta\beta$ is fed back to β_0.

The above mentioned dynamic inflow model and the feedback from the structures constitute the complete computational model.

4 CALCULATION EXAMPLES

This paper took into consideration the gradually changing mechanism of induced velocity of large-scale wind turbines when their blades were distorted, and took the parameters of a 5-MW wind turbine as calculation parameters (Jonkman et al. 2009). By considering the feedback from the flexible structure, it studied the changes of induced velocity and dynamic response in the changing processes of pitch angle and rotor speed, respectively. On the premise of taking wind shear into account, it studied the wind turbine's flap-wise blade tip speed (i.e., out-of-plan speed), edgewise blade tip speed (i.e., in-plane speed), twist angle, and the time history of the normal force, tangential force, and torsion at blade root. All the four computational models are as follows:

Model 1: steady-state model with feedback from the flexible structure
Model 2: dynamic model with feedback from the flexible structure
Model 3: dynamic model without feedback
Model 4: steady-state model without feedback

4.1 Changing pitch angle

The changing coursed by the pitch angle is as follows. At the time of 40 s, the pitch angle was changed from 0° to 6° rapidly. It stayed unchanged till the time of 80 s and then rapidly changed back to 0°. The simulation time was 120 s. The operat-

ing wind speed was 11.4 m/s. The BEM modified model, dynamic inflow model, and dynamic inflow model with feedback from the flexible structure were used to calculate the axial induction factor at $r = 32.25$ m.

It can be seen from Figure 1 that: when using the BEM modified model to calculate the factor, the axial induced velocity changed instantaneously, while there was an obvious hysteresis of the axial induction factor and a transitional period for the value to reach the level corresponding to the pitch angle step when using the dynamic inflow model and the dynamic inflow model with feedback from the flexible structure. When the pitch angle stayed unchanged, the induction factor obtained by the models was basically the same, with its amplitude only changing with the azimuth. In actual situations, the induction factor does not change sharply, and the fact indicates that the dynamic inflow model reflects the dynamic changing process of the induction factor in a more reasonable way. The value of the induction factor obtained by using the dynamic inflow model with feedback from the flexible structure was slightly bigger than the original value, presenting the same changing trend on the whole. The feedback includes the variation of twist angle of the section ($\Delta\theta$) and the variation of flap-wise angle ($\Delta\beta$); $\Delta\theta$ was fed back to θ and $\Delta\beta$ was fed back to β_0. In actual situations, the direction of the torsion is related to the twist center of the blade element, center of gravity, and aerodynamic center. This paper assumes that the torsion makes aerofoil rotate clockwise; namely, it makes the aerofoil nose up and reduces the twist angle of the section. The reduction would lead to the increase in the angle of attack, which would increase the torsion in return. When the torsional rigidity is appropriate, no divergence will occur, and $\Delta\beta$ will lead to the reduction of the pitch angle and the increase of the attack angle as

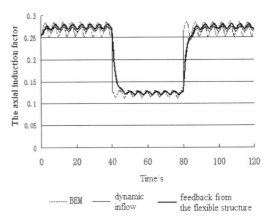

Figure 1. Axial induction factor at $r = 32.25$ m for step on the pitch angle.

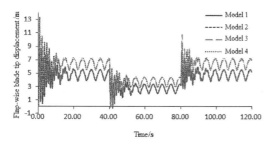

Figure 2. Time history of flap-wise blade tip displacement.

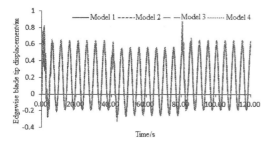

Figure 3. Time history of edgewise blade tip displacement.

well. Thus, after repeated iteration and calculation, the value of the induction factor would be slightly bigger than the original value, as shown in Figure 1.

By comparing Model 1 with Model 2, and Model 3 with Model 4 (as shown in Figure 2), we can see that the flap-wise blade tip displacement obtained by the dynamic models and steady-state models was the same when the pitch angle was in the steady state (both 0° and 6°). When the pitch angle was changed back to 0° at the time of 80 s, we can see that the flap-wise displacement obtained by the dynamic models fluctuated more greatly. The comparison between Model 2 and Model 3 shows that the flap-wise blade tip displacement obtained by the model with feedback from the flexible structure reduced significantly, decreasing by 23.04% on average.

By comparing Model 1, Model 2, Model 3, and Model 4 (as shown in Figure 3), we can see that the values obtained by the four models were close to each other, and the pitch angle had little influence on the edgewise blade tip displacement in all the models. When the pitch angle was changed back to 0° at the time of 80 s, the values obtained by Model 2 and Model 3 were significantly higher than that obtained by the steady-state models. The comparison between Model 2 and Model 3 shows that the edgewise blade tip displacement obtained by the model with feedback from the flexible structure reduced slightly, decreasing by 6.76% on average.

By comparing Model 1 with Model 2, and Model 3 with Model 4 (as shown in Figure 4), we can see that the twist angle of blade tip section obtained by the dynamic models and steady-state models was same when the pitch angle was in the steady states (both 0° and 6°). When the pitch angle was changed back to 0° at the time of 80 s, we can see that the twist angle of blade tip section obtained by the dynamic models fluctuated more greatly, presenting an obvious overshoot. The overshoots in Model 2 and Model 3 were, respectively, 13.68% and 16.32% higher than that in Model 1 and Model 4. The comparison between Model 2 and Model 3 shows that the twist angle of the blade tip section obtained by the model with feedback from the flexible structure increased by 29.30% on average.

By comparing Model 1 with Model 2, and Model 3 with Model 4 (as shown in Figure 5), we can see that the range of the torsion at the blade root obtained by the dynamic models was smaller than that obtained by the steady-state models. When the pitch angle was changed back to 0° at the time of 80 s, there were overshoots in Model 2 and Model 3, which were, respectively, 30.66% and 33.68% higher than that in Model 1 and Model 4. The comparison between Model 2 and Model 3 shows that the torsion at the blade root obtained by the model with feedback from the flexible structure increased by 18.41% on average.

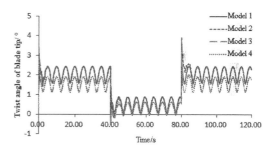

Figure 4. Time history of the twist angle of the blade tip.

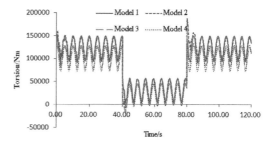

Figure 5. Time history of torsion at the root of the blade.

The changing course of the rotation speed is as follows. At the time of 40 s, the rotation speed was changed from the rated speed (1.26 rad/s) to 0.8 rad/s, then stayed unchanged till the time of 80 s, and then rapidly changed back to 1.26 rad/s. The simulation time was 120 s. The operating wind was at a constant speed of 11.4 m/s. The BEM modified model, dynamic inflow model, and dynamic inflow model with feedback from the flexible structure were used to calculate the axial induction factor at $r = 36.35$ m.

As shown in Figure 6, the general trends of the induction factor curves obtained by the three models under the condition of changing the rotation speed were the same as that under the condition of changing the pitch angle. When using the BEM modified model to calculate the factor, the axial induced velocity changed instantaneously, while there was an obvious hysteresis of the axial induction factor and a transitional period for the value to reach the level corresponding to the rotation speed step when using the dynamic inflow model and the dynamic inflow model with feedback from the flexible structure. When the rotation speed stayed unchanged, the induction factor obtained by the models was basically the same, whose amplitude only changing with the azimuth. The above difference indicates that the dynamic inflow model with feedback from the flexible structure can better reflect the dynamic changing process of the induction factor.

By comparing Model 1 with Model 2, and Model 3 with Model 4 (as shown in Figure 7), we can see that the flap-wise blade tip displacement obtained by the dynamic models and steady-state models was same when the rotation speed was in the steady state (both 1.26 rad/s and 0.8 rad/s). When the rotation speed was changed back to 1.26 rad/s at the time of 80 s, we can see that the

flap-wise displacement obtained by the dynamic models fluctuated more greatly. The comparison between Model 2 and Model 3 shows that the flap-wise blade tip displacement obtained by the model with feedback from the flexible structure reduced significantly, decreasing by 22.47% on average.

By comparing Model 1, Model 2, Model 3, and Model 4 (as shown in Figure 8), we can see that the values obtained by the four models were close to each other, and the rotation speed had little influence on the edgewise blade tip displacement in all the models. When the rotation speed was changed back to 1.26 rad/s at the time of 80 s, the values obtained by Model 2 and Model 3 were higher than that obtained by the steady-state models. It can be clearly seen from the figure that the periods of the sine waves became longer when the rotation speed was at a smaller value. The comparison between Model 2 and Model 3 shows that the edgewise blade tip displacement obtained by the model with feedback from the flexible structure reduced slightly, decreasing by 9.27% on average.

By comparing Model 1 with Model 2, and Model 3 with Model 4 (as shown in Figure 9), we can see that the twist angle of the blade tip section obtained by the dynamic models and steady-state models was same when the rotation speed was in the steady state (both 1.26 rad/s and 0.8 rad/s). When the rotation speed was changed back to 1.26 rad/s at the time of 80 s, we can see that the twist angle of blade tip section obtained by the dynamic

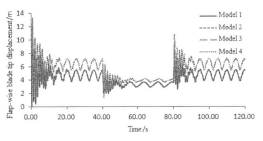

Figure 7. Time history of flap-wise blade tip displacement.

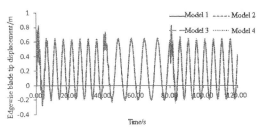

Figure 8. Time history of edgewise blade tip displacement.

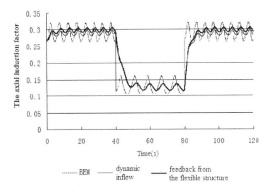

Figure 6. Axial induction factor at $r = 36.35$ m for the step on the rotor speed.

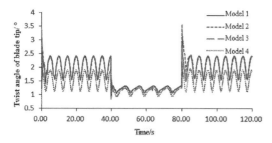

Figure 9. Time history of the twist angle of the blade tip.

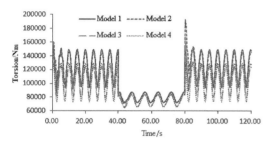

Figure 10. Time history of torsion at the blade root.

models fluctuated more greatly, presenting an obvious overshoot. The overshoots in Model 2 and Model 3 were, respectively, 15.64% and 20.33% higher than that in Model 1 and Model 4. The comparison between Model 2 and Model 3 shows that the twist angle of the blade tip section obtained by the model with feedback from the flexible structure increased by 29.28% on average.

By comparing Model 1 with Model 2, and Model 3 with Model 4 (as shown in Figure 10), we can see that the range of the torsion at the blade root obtained by the dynamic models was smaller than that obtained by the steady-state models. When the rotation speed was changed back to 1.26 rad/s at the time of 80 s, there were overshoots in Model 2 and Model 3, which were, respectively, 18.49% and 25.71% higher than that in Model 1 and Model 4. The comparison between Model 2 and Model 3 shows that the torsion at the blade root obtained by the model with feedback from the flexible structure increased by 22.10% on average.

5 CONCLUSION

Taking the BEM model and Pitt–Peters model as the theoretical basis, this study established the dynamic inflow model with structural deformations of the flexible rotor as the feedback and analyzed

the gradually changing mechanism of the induced velocity flow field in the changing processes of pitch angle and rotor speed respectively. By comparing different models, the following conclusions were obtained.

After adding the feedback from the flexible structure to the model: (1) The flap-wise blade tip displacement reduces significantly; this study assumes that the aerodynamic center coincides with the stiffness center, and under this condition, the calculated pitch moment of aerofoil is negative, thus positioning the aerofoil nose down and increasing the twist angle of the section; as a result, the aerodynamic load is reduced before stalling and hence the flap-wise blade tip displacement reduces. (2) The edgewise blade tip displacement changes slightly because the edgewise stress is mainly from the gravitational load of the blades (though the aerodynamic load has a certain influence, it is much smaller than the gravitational load); the edgewise blade tip displacement is barely changed. (3) The twist angle of blade tip section increases greatly, and the feedback of the twist angle of the section has little influence on it. The main reason for the increase is that the flap-wise angle increases the torsion on the blade. The torsion mainly consists of four parts: the pitch moment formed by lift force and resistance force around the torsional center, the pitch moment formed by the aerofoil itself around the stiffness center, and the pitch moments formed by centrifugal force and the flap-wise component of gravitational force; in addition, the flap-wise angle has great influence on centrifugal force and the flap-wise component of gravitational force. (4) The torsion at the blade root increases significantly; its changing trend is consistent with that of the twist angle of the blade tip section. The feedback of the twist angle of the section has a certain influence, but the main reason for the increase is that the absolute value of the torsion increases after adding the variation of the flap-wise angle. In addition, the influence of the flap-wise angle is much greater than that of the twist angle of the section, and the variation mainly arises from centrifugal force and the flap-wise component of gravitational force.

REFERENCES

Chen, Y. et al. 2008. Application of Dynamic Wake Model in Computing Aerodynamic Performance of Horizontal Axis Wind Turbine. Acta Energiae Solaris Sinica 29(10): 1297–1302.
Clough, R.W. & Penzien, J. 1993. Dynamics of structures. New York: McGraw Hill.
Jonkman, J. et al. 2009. Definition of a 5-MW reference wind turbine for offshore system development. Technical Report NREL/TP-500–38060.

Liu, J. 2006. Analysis of Full-system Loads of Wind Turbine and Integrated Development of Software for Optimal Design. Shantou University.

Liu, X. et al. 2009. Transient Aerodynamic Load Prediction Model for Horizontal Axis Wind Turbines Based on Dynamic Inflow Theory. Acta Energiae Solaris Sinica 30(4): 412–419.

Shen, S. 2013. Dynamic Aerodynamic Characteristics and Structural Dynamic Response of Flexible Wind Wheel. Shantou University.

Shen, S. et al. 2013. Research of dynamic inflow effect on flexible turbine rotor. Acta Aerodynamica Sinica 31(3): 401–406.

Xu L. et al. 2015. The random response analysis of flexible blade of wind turbine based on nonlinear aeroelastic coupling. Journal of Vibration and shock 34(10): 20–27.

Xu, R. 2001. Finite Element Method for Structural Analysis and MATLAB Program Design. Beijing: China Communications Press.

Advances in Power and Energy Engineering – Sun (Ed.)
© 2016 Taylor & Francis Group, London, ISBN 978-1-138-02846-3

A new completing method of wind power data based on back propagation neural network

S.Y. Ye, Q. Tang & H. Feng
State Grid Sichuan Economic Research Institute, Chengdu, China

Y. Lu & X.R. Wang
School of Electrical Engineering, Southwest Jiaotong University, Chengdu, China

ABSTRACT: The authenticity and integrality of historical wind power output data of a wind farm are the basis of wind power research. Therefore, the research of wind power data completion arouses increasing interest. This paper proposes a new wind power data completing method using Back Propagation Neural Network (BPNN), adapted for data loss of long time scales. Power data of the other seven wind turbines are selected as the inputs of the BPNN, based on the results of the Pearson Correlation Coefficient (PCC) analysis. Comparison studies are implemented to test the effectiveness of the proposed method by using the data obtained from an actual wind farm located in the southwest of China. The results obtained from the proposed method for 48 h, 96 h, and 192 h show a significant improvement compared with the adjacent wind turbine method.

1 INTRODUCTION

With the rapid development of wind power, randomness, volatility, and intermittency of wind power have had a significant influence on power grid scheduling and operation. At present, some research highlights are all based on wind farm historical power data, e.g., wind power output characteristics analysis (Banakar et al. 2008, Chen et al. 2010, Li et al. 2010, Xiao et al. 2010), wind power forecasting (Jps et al. 2011, Fan et al. 2008, Wu et al. 2011, Xiao et al. 2015), assessment of the influence of wind power on power grids and its control strategy (Yu et al. 2011), wind power curve calculation and correction (Shen et al. 2009), effects caused by wind power on power system, etc.

More data will develop rapidly and play a unique and significant role throughout each procedure of the electrical power industry (Ye 2015 & Zhu 2015). There are high possibilities of abnormal data and data loss in procedures of historical wind power data collection, measurement, delivery, and transformation caused by a system breakdown or disturbance. On the other hand, wind power loss due to the limitation of a grid's absorptive capacity also leads to unnatural wind power operation data (Lu 2013). Concentrated and mass wind power data loss increases the difficulty of analysis, causes errors in results, and affects the efficiency of research work at the same time. Overall, it has become more significant to reconstruct and polish the data series in order to decrease the influence on wind power research (Ma et al. 2013).

According to Q/GDW 588-2011, the previous wind power data should be utilized to replace the missing data. The most persistent method is the lowest data completing method (Brown et al. 1984), which is usually utilized to assess other data completing methods as a criterion. Adjacent wind turbine methods can complete historical data in long time scales when having an acquaintance with the spatial distribution of wind turbines. Quan and Cao (2007) considered that the average interpolation method gives an index to the sustainability and monotonicity of the missing data characteristics. Also, based on wind speed data, it's easy to get wind power output according to standard power curve. Yang et al. (2014) proposed the Adaptive Neuro-Fuzzy Inference system (ANFIS) method to complete the missing data and optimizes the results. A phenomenon known as correlation is that wind power peak and valley changes of adjacent wind turbines or wind farms power output reach unanimity in time. Some commonly used mathematical methods, such as the non-adjacent average completion method, recursive non-adjacent average completion method (Wang et al. 2009), etc., do not take the time series characteristics of wind power data and the correlation of wind power output of wind farms or wind turbines into consideration, which lead to inappropriate results. Zhang et al. (2014) proposed two methods to reconstruct historical wind power data based on the wind power output correlation of wind farms and data characteristics of the wind farm itself, and applied them into an actual project.

The traditional data completing methods mentioned above are mostly applied in short time scales with appropriate results. However, it's difficult to obtain decent results when it comes to wind power data completing problems in long time scales or without historical wind speed data. In this context, to solve wind power data loss in long time scales and take wind power output correlation of wind turbines into account in order to get decent results, this paper proposes a new method using Back Propagation Neural Network (BPNN) to complete historical wind power operation data. The proposed method is tested with the wind power data of eight wind turbines from an actual wind farm located in southwest of China from January 1st to December 31st, 2012, with 15-min intervals. This paper shows better results obtained for 48 h, 96 h, and 192 h compared with the adjacent wind turbine method, which shows that the proposed method is accurate and reliable.

2 WIND POWER CORRELATION

2.1 Pearson Correlation Coefficient (PCC)

Pearson correlation analysis is a statistical method to quantitatively measure correlativity between variables complied to normal distribution, where the correlation coefficient r reflects linear correlation degree of the two variables. The bigger the absolute value of r is, the stronger the correlation will be. The two variables have negative correlativity when r is between -1 and 0 as well as positive correlativity when r is between 0 and 1. Let X and Y be two zero-mean, real-valued random variables (Jacob et al. 2008). The PCC is defined as:

$$r = \frac{N\sum XY - \sum X \sum Y}{\sqrt{\left(N\sum X^2 - \left(\sum X\right)^2\right)\left(N\sum Y^2 - \left(\sum Y\right)^2\right)}} \quad (1)$$

where N = sample amount of X and Y and PCC = the ratio of covariance and product of standard deviation of the two variables, which is a standardization covariance without dimension.

2.2 Wind power correlation of wind turbines of a wind farm

Before completing the missing historical wind power output data, the correlation coefficients of should first be analyzed. The wind power data series approximately comply to normal distribution

Table 1. Wind power output correlation coefficients of eight wind turbines.

	1#	2#	3#	4#	5#	6#	7#	8#
1#	1.00	0.96	0.89	0.87	0.88	0.87	0.88	0.79
2#	0.96	1.00	0.91	0.88	0.89	0.87	0.89	0.81
3#	0.89	0.91	1.00	0.93	0.91	0.89	0.90	0.84
4#	0.87	0.88	0.93	1.00	0.91	0.89	0.94	0.85
5#	0.88	0.89	0.91	0.91	1.00	0.97	0.91	0.82
6#	0.87	0.87	0.89	0.89	0.97	1.00	0.88	0.80
7#	0.88	0.89	0.90	0.94	0.91	0.88	1.00	0.88
8#	0.79	0.81	0.84	0.85	0.82	0.80	0.88	1.00

in practical application so that PCC analysis is utilized. Table 1 shows wind power output correlation coefficients of the eight wind turbines.

It can be seen from the above data that these eight wind turbines have a strong correlation two by two, and the differences between these coefficients are not too distinct, where the minimum is 0.79 and the maximum is 0.97. Because of the strong correlation between each other, performances of these eight wind turbines' wind power output time series tend to be in accordance in peak and valley, i.e., wind power fluctuation characteristics of each wind turbine can help analyze the others' as a reliable reference.

3 BPNN

BPNN is noted for its strong nonlinear mapping ability and excellent learning ability among numerous fields, which is a multi-input, multi-output system typically consisting of an input layer, a hidden layer, and an output layer. Each layer employs several neurons which are connected to the neurons in the adjacent layer with different weights (Ashraf et al. 2012). The architecture of a typical BPNN is shown in Figure 1. The input of a neuron in hidden layer is calculated by the following form:

$$h_j = \sum_{i=1}^{n} x_i W_{ij} + B_j \quad (2)$$

where hj = input of the hidden layer neuron j; n = the amount of neurons in the input layer; xi = the input to the hidden layer neuron j from the input layer neuron i; and Wij = the weight between them; Bj = bias of the hidden layer neuron j.

In this paper, the sigmoid tansig function is utilized as the excitation function in the hidden layer, which is described as:

$$f(x) = \frac{2}{1 - e^{-2x}} - 1 \quad (3)$$

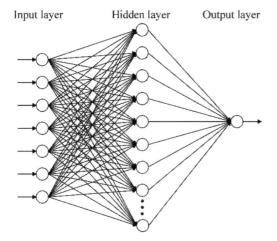

Figure 1. Architecture of the BPNN.

The network's different weights and biases basically decide the network's ability to complete the missing wind power output data accurately. In each iteration procedure to update parameters, the weights and biases are adjusted to minimize the error function which depends on the optimization algorithm where the gradient decline method is commonly used. The interconnectivity among nodes of BPNN makes it possible to learn the relationship between input–output pairs during the training phase. The network then obtains the wind power completing data from the test set using the trained network.

In this paper, historical wind power data of seven wind turbines are utilized as inputs and the corresponding wind power of the other one wind turbine is output.

4 CASE STUDY

4.1 *Accuracy measures*

To conduct the statistical analysis of error, two accuracy measures called Mean–Absolute–Percent Error (MAPE) and Root–Mean–Square Error (RMSE) are utilized to assess the completing performance according to Q/GDW 588-2011, which are defined as:

$$MAPE = \sum_{t=1}^{N} \frac{\left| P_t - P_t^e \right|}{N \cdot Cap} \qquad (4)$$

$$RMSE = \frac{\sqrt{\sum_{t=1}^{N} (P_t - P_t^e)^2}}{\sqrt{N} \cdot Cap} \qquad (5)$$

where Pt = real wind power data at t moment; Pte = wind power estimation value at t moment; and Cap = single wind turbine total capacity; and N = the sample amount.

4.2 *Data completing results*

In this paper, the proposed new method, using BPNN with 9000 points as the training dataset and 192 points (48 h), 384 points (96 h), and 768 points (192 h) as the testing dataset respectively, is applied to complete the missing historical wind power data of one wind turbine.

Figures 2, 3, and 4 separately show the 48 h, 96 h, and 192 h completing results of the 4# wind turbine compared with results of the adjacent wind turbine method. Meanwhile, the relevant performance comparisons are recorded in Tables 2, 3, 4 separately.

It can be seen from Figure 2 that the proposed method has a much better testing performance in 48 h than that of the adjacent wind turbine method

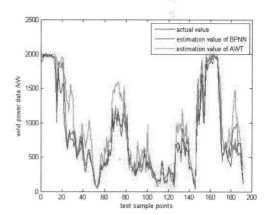

Figure 2. Testing results at 48 h.

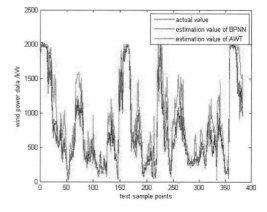

Figure 3. Testing results at 96 h.

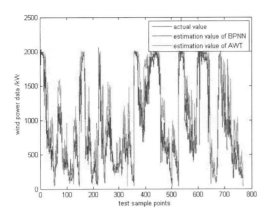

Figure 4. Testing results at 192 h.

Table 4. Comparison of 192 h testing performance of BPNN and the adjacent wind turbine method.

Wind turbine	MAPE (%)		RMSE (%)	
	BPNN	AWT	BPNN	AWT
1#	5.92	6.65	9.48	13.24
2#	6.26	10.70	10.97	16.93
3#	6.66	7.57	9.75	10.90
4#	4.81	10.05	7.05	14.21
5#	4.20	4.33	5.89	6.33
6#	4.36	7.59	6.01	11.21
7#	4.50	12.03	6.31	19.01
8#	9.62	12.03	15.32	19.01
Average	5.80	8.87	8.85	13.86

Table 2. Comparison of 48 h testing performance of BPNN and the adjacent wind turbine method.

Wind turbine	MAPE (%)		RMSE (%)	
	BPNN	AWT	BPNN	AWT
1#	5.39	5.60	9.85	10.23
2#	5.18	8.88	7.67	13.83
3#	5.57	6.66	8.62	9.77
4#	4.55	9.61	6.41	13.10
5#	4.21	4.52	5.88	6.24
6#	4.44	8.57	6.03	11.73
7#	4.48	7.57	6.64	11.49
8#	7.4	7.57	10.27	11.49
Average	5.15	7.37	7.67	10.99

AWT, adjacent wind turbine method.

Table 3. Comparison of 96 h testing performance of BPNN and the adjacent wind turbine method.

Wind turbine	MAPE (%)		RMSE (%)	
	BPNN	AWT	BPNN	AWT
1#	5.31	6.31	9.27	9.80
2#	4.89	8.11	7.60	13.24
3#	5.20	6.71	8.05	9.62
4#	4.08	10.08	5.78	13.95
5#	4.73	5.80	6.42	8.96
6#	4.42	8.07	6.08	11.49
7#	4.14	7.05	5.70	10.94
8#	6.41	7.05	8.91	10.94
Average	4.90	7.40	7.23	11.12

in all eight testing groups. In Table 2, the 5# wind turbine has the minimum MAPE value and RMSE value while 8# wind turbine has the maximum MAPE value and RMSE value, which are basically due to the wind power output correlation with the other wind turbines.

Figure 3 apparently reflects that the proposed method has a better testing performance in 96 h than that of the adjacent wind turbine method. Table 3 shows that 4# wind turbine has the minimum MAPE value but 7# wind turbine has the minimum RMSE value while 8# wind turbine has the maximum MAPE value but 1# wind turbine has the maximum RMSE value. Both the average MAPE and RMSE values are smaller than those of 48 h testing.

Also, Figure 4 reflects that the proposed method has better testing performance in 192 h. In Table 4, 5# wind turbine has the minimum MAPE value and RMSE value, while 8# wind turbine has the maximum MAPE value and RMSE value. Meanwhile, both of the average MAPE value and RMSE value are bigger than those of 48 h testing performance separately, which shows that performance of the proposed BPNN method does not fluctuate obviously owing to its not being sensitive to long time scales' changes.

5 CONCLUSION

This paper proposes a BPNN based method to complete historical wind power data loss in long time scales. Main conclusions are described as follows:

a. The proposed new method is not sensitive with long time scales' changes, i.e., results of 48 h, 96 h, and 192 h demonstrate that the accuracy will not fluctuate obviously.
b. The results show that the proposed data completing method presents a significant improvement for all three time scales compared with the adjacent wind turbine method.

REFERENCES

Ashraf, U.H. et al. 2012. A new strategy for predicting short-term wind speed using soft computing models. Renewable and Sustainable Energy Reviews, 16: 4563–4573.

Banakar, H.L. et al. 2008. Impacts of wind power minute-to-minute variations on power system operation [J]. IEEE Trans. on Power Systems, 23(1):150–160.

Brown, B.G. et al. 1984. Time series models to simulate and forecast wind speed and wind power [J]. Journal of Climate and Applied Meteorology, 23(8): 1184–1195.

Chen, P. et al. 2010. ARIMA-based time series model of stochastic wind power generation [J]. IEEE Trans. on Power Systems, 25(2):667–676.

Fan, G.F. et al. 2008. Wind power prediction based on artificial neural network [J]. Proceedings of the CSEE, 28(34):118–123.

Jacob, B. et al. 2008. On the Importance of the Pearson Correlation Codfficient in Noise Reduction. IEEE Trans. Audio, Speech, Lang. Process., 16(4): 757–765.

Jps, C.A. et al. 2011. Hybrid Wavelet-PSO-ANFIS approach for short-term wind power forecasting in Portugal [J]. IEEE Trans on Sustain Energy, 2(1):50–59.

Li, J.Q. et al. 2010. Research on statistical modeling of large-scale wind farms output fluctuations in different spacial and temporal scales [J]. Power System Protection and Control, 40(19):7–13.

Lu, Q. et al. 2013. A new evaluation method for wind power curtailment based on analysis of system regulation capability [J]. Power System Technology, 37(7): 1–8.

Ma, S.L. et al. 2013. Algorithm Research on Polishing the Mass Missing Data of Wind Power Based on Regression Model [J]. Power System and Clean Energy, 29(9): 74–80.

Quan, S.Y. & Cao, Y. 2007. Interpolation method research and application [J]. Science & Technology Information.

Wu, X.M. et al. 2011. Short-term wind power forecast based on the radial basis function neural network [J]. Power System Protection and Control, 39(15): 80–83.

Shen, X.H. et al. 2009. Analysis and modification method for wind turbine power curve [J]. Applied Energy Technology.

Wang, L. et al. 2009. Regression forecast and abnormal data detection based on support vector regression [J]. Proceedings of the CSEE.

Xiao, C.Y. et al. 2010. Power characteristics of Jiuquan wind power base [J]. Automation of Electric Power Systems, 34(17):64–67.

Xiao, L. et al. 2015. Combined forecasting models for wind energy forecasting: A case study in China [J]. Renewable and Sustainable Energy Reviews, 44:271–288.

Yang, M. et al. 2014. Data Completing of Missing Wind Power Data Based on Adaptive Neuro-fuzzy Inference System [J]. Automation of Electric Power Systems, 38(19): 16–22.

Yu, P. et al. 2011. Research on the method based on hybrid energy storage system for balancing fluctuant wind power [J]. Power System Protection and Control, 39(24): 35–40.

Zhang, D.Y. et al. 2014. Reconstruction Method of Active Power Historical Operation Data for Wind Farm [J]. Automation of Electric Power Systems, 38(5): 14–18.

Zhu, Q.W. & Ye, L. 2015. Methods for elimination and reconstruction of abnormal power data in wind farms [J]. Power System Protection and Control, 43(3): 38–45.

Advances in Power and Energy Engineering – Sun (Ed.)
© 2016 Taylor & Francis Group, London, ISBN 978-1-138-02846-3

Parallel computing strategy of power grid risk early warning system considering wind farm integration

W. Zhang & L.Y. Xu
China-EU Institute for Clean and Renewable Energy, Wuhan, Hubei, China

Y.X. Zhuo, C. Chen, X. Zhou & Y.F. He
School of Electrical and Electronic Engineering, Huazhong University of Science and Technology, Wuhan, Hubei, China

ABSTRACT: With the rapid growth of electric power grid and the access of wind farms, it is of great significance to establish a warning system to give an overall picture of the power grid in the least time. In this paper, first, an early risk warning system is established on the basis of stochastic power flow calculation and the Monte Carlo sampling method. Then, several scenarios are carried out to study the accelerating effect of different calculation processes in the system's parallel computing algorithm. To be specific, in the case study, the first scenario series focusses on the sampling process and grid information load method. The second scenario series concentrates on task allocation strategy. Finally, different data exchange methods are analyzed in the third scenario series. The results of the scenarios show that the best speed-up ratio can be achieved by using balanced task allocation, integrated sampling, and data exchange strategy.

1 INTRODUCTION

In recent years, the rapid development of new power plants and new transmission lines lead to the expansion of the power system. Meanwhile, the renewable energy technology based on wind power and photovoltaic technology has gradually attracted people's attention (Abouzahr & Ramakumar 1991, Wang et al. 2004). Considering this situation, an early warning system is required to evaluate the operational status of the power grid (Han & Cao 2004).

The concept of stochastic power flow has been studied and improved in power grid since 1974 (Borkowska 1974, Miao et al. 2007, Patra & Misra 1993). This method uses the probability theory to describe the uncertainty, and establishes the corresponding mathematical model, algorithm, and practical application (Silva & Arienti 1990).

Monte Carlo simulation is the most typical simulation method used in stochastic power flow calculation (Allan et al. 1981). The basic idea of Monte Carlo method is to solve the random problem by using statistical laws (Ding et al. 2001, Yu et al. 2009).

In this paper, first, an early warning system is established. Then several parallel computing scenarios are conducted to speed up the algorithm and analyze the different influences on

sub-calculation process. Specifically speaking, the calculation process is divided into several sub calculation modules. Namely, these are sampling, load of grid information, task allocation, power flow calculation, results collection, indexes calculation, and plotting. Consequently, three scenarios are designed to analyze the accelerating effect of different sub modules.

In Scenarios A.1–A.3, the main difference is whether sampling and grid information load are conducted in a holistic way. In Scenarios B.1–B.3, task allocation strategies are studied by using different numbers of cores in CPU on indexes calculation. While in Scenarios C.1–C.4, sub calculation modules are sent to different cores to study the effect of data exchange methods in parallel algorithm.

2 MODEL AND EVALUATION INDEXES OF THE EARLY WARNING SYSTEM

2.1 Models of system components

The two-status probability model is utilized to describe the probability model of generators and transmission lines. In a single period, the two-status model is that there are only two states: operation state and fault status (Yu 2007). If the status is operation state, the generator gives out rated power or the transmission line is working. Otherwise, the total power output is zero and the transmission line is broken. For continuous electric loads, the normal distribution is generally used to demonstrate the uncertainty of the load (Liu et al. 2011). The wind farm output probability model should be built on the basis of output forecast of wind power (Bludszuweit et al. 2008, Usaola 2009, 2010). So beta distribution is selected as it can simulate the changeable peak of wind farms.

2.2 Calculation process of the early warning system

The early warning method of the power system is mainly based on the existing power system static security assessment method.

As can be seen in Figure 1, the algorithm is divided into several sub calculation modules. Then in the following sections, different modules are conducted by using different task allocation strategy, data exchange strategy, and so on.

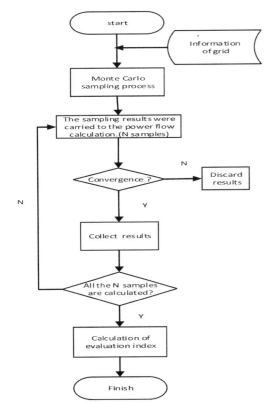

Figure 1. Calculation process of risk early warning system.

2.3 Evaluating indexes

The evaluation indexes for parallel algorithm are speedup ratio and theoretical speedup ratio. The speedup ratio is the evaluation index for the performance of overall program. In contrast, theoretical speedup ratio only gives the efficiency of a certain part of the parallelized program.

Formula of speedup ratio (Wang 2009):

$$S_p = \frac{t_1}{t_n} \qquad (1)$$

Where t_1 = calculation time of serial computing and t_n = parallel computing time using n cores.

Formula of theoretical speedup ratio (Akhter & Roberts 2006):

$$S_P' = \frac{1}{1gs + \frac{1-s}{n}} \qquad (2)$$

Where s = proportion of the serial part in the algorithm and n = numbers of the core in CPU.

3 PARALLEL COMPUTING SCENARIOS

3.1 Scenarios A.1–A.3 on different sampling and grid information load method

First, two concepts are explained: the client and lab. Client refers to the MATLAB serial program operation interface. Lab refers to each core of CPU, for instance, lab1 represents core 1. In the MATLAB software, through the built-in program compiler, the balanced tasks allocation and communication tasks are conducted automatically on parallel algorithms.

In Scenarios A.1–A.3, the automatic task allocation models, the effects of sampling process, and information reload strategy are considered. The power flow calculation program is written in parallel mode, and the evaluation indexes calculation and plotting parts are written in serial mode.

In Scenario A.1, the sampling process is completed for the client. Then, in each calculation loop in labs, client sends the sampling data to labs, and the other structural parameters' reloads are in each lab.

The only difference between Scenario A.2 and A.1 is the load type of structural parameters.

In Scenario A.3, the structural parameters are: first load to client; then, before each power flow calculation in the labs, the structural parameters are sent from the client and the sampling process in the lab is conducted only one time; and after that, the power flow calculation is conducted in each lab. The detailed information can be seen in the Table 1.

3.2 *Scenarios B.1–B.3 on different task allocation methods*

Compared to the power flow calculation, the workload of the evaluation indexes calculation is relatively small. In the Scenarios B.1–B.3, task allocation and communication strategy are determined by means of manual programming. The evaluation indexes are as above.

In Scenarios B.1–B.3, the sampling and parallel power flow calculation process are the same as scenario A.1. So these parts are not included in the Table 2. Yet there are some differences. In Scenario B.1, the task is allocated by MATLAB's compiler. In Scenario B.2, the evaluation indexes of node and branch are assigned to lab1 and lab2 respectively. In Scenario B.3, four indexes are allocated to four labs separately. As the workload of indexes are different, Scenario B.2 and B.3 can be classified as unbalanced allocation strategy. Specific difference is as follows:

3.3 *Scenarios C.1–C.4 on different data exchange methods*

In Scenarios B.1–B.3, the evaluation index calculation part is written in parallel algorithm. In Scenarios C.1–C.4, the power flow calculation part is parallelized. The main differences are the task allocation and communication strategy.

In Scenario C.1, all the tasks are allocated to four labs equally. There is no communication between the labs. After all the calculation is done, the results are sent to the client. Then the plotting process is carried out in the client.

In Scenario C.2, in each loop, the sampling process is done in lab 1. Then the sampling results and power grid information are sent to lab 2, 3, and 4. After the power flow calculation is done in each core, lab1 collects the results. Then the evaluation indexes computation is finished in lab 1. Then the plotting process is carried out for the client.

The only difference between Scenario C.2 and C.3 is that in C.3 the power flow results are entirely sent to lab 1 in after all the N samples are done.

Table 1. Difference of calculation in Scenarios A.1–A.3.

Sub calculation module	Scenario	A.1 C*	A.1 L**	A.2 C	A.2 L	A.3 C	A.3 L
Sampling	Holistic	•		•			
	In loop						•
Load of grid Information	Holistic			•	•		
	In loop	•					
Power flow calculation	Serial						
	Parallel		•		•		•
Indexes calculation and plotting	Serial	•					
	Parallel				•		•

C* means the calculation is conducted in client. L** means the calculation is conducted in lab.

Table 2. Difference of calculation in Scenarios B.1–B.3.

Sub calculation module	Scenario	B.1 C	B.1 L	B.2 C	B.2 L	B.3 C	B.3 L
Task allocation	Balanced	•					
	Unbalanced			•		•	
	Auto	•					
	Manual			•		•	
Indexes calculation	Lab usage	1–4		1–2		1–4	
	Serial						
	Parallel		•		•		•
Plotting	Serial	•		•		•	
	Parallel						

Table 3. Difference of calculation in Scenarios C.1–C.4.

Sub calculation module	Scenario	C.1 C	C.1 L	C.2 C	C.2 L	C.3 C	C.3 L	C.4 C	C.4 L
Sampling	Serial				•		•		•
	Parallel		•						
	Holistic								•
	In loop		•		•		•		
	Lab usage	1–4		1		1		1	
Load of grid information	Holistic		•						•
	In loop				•		•		
	Lab usage	1–4		1		1		1	
Task allocation	Balanced	•							
	Unbalanced			•		•		•	
	Auto								
	Manual	•		•		•		•	
	Lab usage	1–4		1		1		1	
Power flow calculation	Serial								
	Parallel		•		•		•		•
	Lab usage	1–4		2–4		2–4		2–4	
Results collection	Holistic	•					•		•
	In loop				•				
	Lab usage			1		1		1	
Indexes calculation	Lab usage	1–4		1		1		1	
Plotting	Serial	•		•		•		•	
	Parallel								

In Scenario C.4, N times' sampling process is conducted in lab 1, then the sampling results and structure information are sent to lab 2, 3, and 4 in its entirety. After the power flow calculation is finished in each lab, the results are sent to lab 1. Then the evaluation indexes computation is finished in lab 1 and the plotting process is carried out for the client.

The detailed difference can be found in Table 3.

4 RESULTS AND ANALYSIS OF DIFFERENT SCENARIOS

4.1 Case study

The system is calculated on the basis of IEEE 39-Bus power system. At bus 30 and 39, two 150-MW conventional generator units are exchange by two wind farms which have the equivalent rated output power. The calculation is carried out on a four-core Intel PC.

4.2 Results of Scenarios A.1–Scenario C.4

For different scenarios, the calculation results are listed in plotted in Figure 3.

It can be seen clearly that Scenario B.2 gets the minimum operational time. Scenario A.2 and B.3 are almost on the same level. The running times of Scenario C are higher than that of Scenario A and B, which means that the manual program has a lower efficiency in task allocation process compared to MATLAB's automatic assignment. Yet except scenario C.2 and C.3, other scenarios spend less time than the serial program. This means that

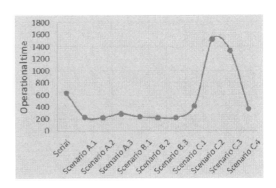

Figure 3. Operational time of different scenarios.

the parallelized programs do make a difference. Moreover, other conclusions can be deducted by analyzing the results among similar scenario series.

The operational time of Scenario A.1, A.2 is less than that of Scenario A.3. This means that the sampling process of the grid's operational status can be more efficiently conducted if it's finished entirely for the client rather than the carrying it out in the labs. The time difference between scenario A.1 and A.2 is caused by the information reload strategy. It is obvious that reloading for the client costs less time.

The calculation results of scenario B are on the same level as Scenario A.1 and A.2. Scenario B.1 runs a little slower than Scenario B.2 and B.3 because the MATLAB's complier takes more time to allocate tasks to each labs. In other words, for tasks which get a small workload of calculation, it usually takes more time in the task allocation process than the calculation itself. The small difference in Scenario B.2 and B.3 is caused by the communication strategy. In a single time, there is only one order in the control bus of the CPU. In other words, even the communication bus gets enough hardware sources; there is only one message between lab and lab at a single time. However, it is not a problem when the labs communicate with the client. This can be perfectly verified by Scenario B.2 and B.3. Before the last plotting process, the indexes should be collected from the labs by client. In Scenario B.2, the indexes calculation is allocated to two labs. So during the process of sending the indexes to client in a lab, the other lab has to wait. While in scenario B.3, the indexes calculation is allocated to four labs. So the other three labs have to wait during the communication process. The latter strategy costs more time. In Scenario B series, the influence of communication strategy is not as big as in Scenario C series because the size of transferred data is much smaller.

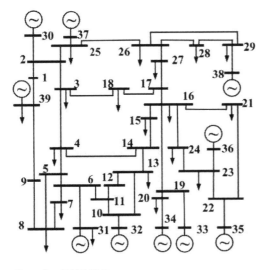

Figure 2. IEEE 39-Bus power system.

Then acceleration effect of the algorithms can be found in Figure 4. The speedup ratio (Sp) reflects the overall acceleration effect of the algorithm, while the theoretical speedup ratio (Sp') reflects the acceleration effect of the parallelized part in the algorithm.

It can be seen that in Scenarios C.2 and C.3, though they get the lowest speedup ratio, their theoretical speedup ratios are higher than others. This means some serial part in the program is perfectly parallelized; yet their overall efficiency is low. So in the real case, it is usually suggested that only the computing intensive part of the program should be parallelized.

4.3 Analysis of influence in different data exchange method

It is clearly demonstrated in Scenario C series that different data exchange strategies can really make a difference in parallel computing.

In Scenario C.1, the parallelized calculations in each lab are totally independent of each other. Therefore, there are no data exchanges among the labs. Moreover, no waiting time exists during the communication between the labs and the client. However, in Scenarios C.2–C.4, data exchange does happen. The exchanged data can be divided into three types: first, sampling data from lab1 to other labs; second, power flow results collected by lab 1 from other labs; and third, results of evaluation indexes from lab 1 to client. Generally speaking, the idle time is mainly caused by the first two kinds. The size of exchanged data in Scenarios C.2–C.4 is plotted in Figure 5. The results are obtained from a six-core CPU. Although the number of cores and total running times are different, the plotting can demonstrate the conclusion as well.

It can be seen that the totally transferred data is almost the same about 64 Mb. The size of data sent from lab 1 to other labs is much smaller than those sent from other labs to lab 1. Since the size of data is the same, what is the reason for operational time difference? Figure 6 can tell.

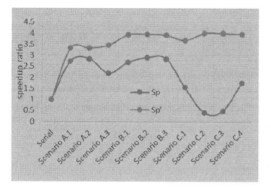

Figure 4. Speedup ratio of different scenarios.

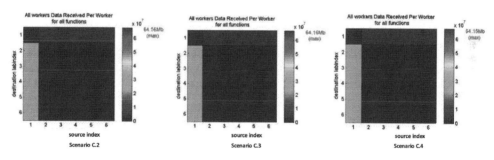

Figure 5. Data exchange size in Scenarios C.2–C.4.

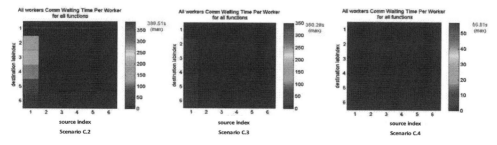

Figure 6. Waiting time of each lab.

In Scenarios C.2 and C.3, it takes much more time to complete the calculation. In Scenario C.2, there is data exchange in each loop. So building the communication pathway and exchanging data takes the majority part of the whole time. We can also see that during the communication process between two labs, the other labs have to wait. This causes the low efficiency of the algorithm. In Scenario C.3, by modifying the strategy of data being sent back to labs, the results of the power flow are sent to lab 1 as in entirety after N loops. We can see that after this modification, the total time decreases from 1521s to 1333s in a four-core CPU. Similarly, in a six-core CPU, the time plunges to 350 s from 388 s. Furthermore, in Scenario C.4, the sampling results are sent to other labs as a whole. Therefore, Scenario C.4 costs the least time. Also, we can see that there is no extra waiting time in the calculation process. Moreover, we can see there is no waiting time when the power flow results are sent back to lab 1. The reason is that during the power flow calculation process, there is enough time for the communication.

5 CONCLUSION

The parallel computing strategy can largely shorten the running time of an evaluation system, which is a big step toward the online application of current evaluation systems. By using balanced task allocation and integrated data exchange communication strategy, the efficiency of algorithm can be improved. With the development in IT technology and adoption of high-performance computers in grids, the parallel algorithm can be introduced to almost any control and analysis systems in the grid.

ACKNOWLEDGMENT

This work is sponsored by the National Basic Research Program of China (2012CB215100), in part by the National Natural Science Foundation of China (51277082), in part by National Natural Science Youth Foundation of China (51207064), in part by Natural Science Foundation of Hubei Province (2012FFA075), and in part by Huang He Ying Cai funding of Wuhan.

REFERENCES

Abouzahr, I & Ramakumar, R. 1991. An approach to assess the performance of utility-interactive wind electric conversion systems. *IEEE Trans on Energy Conversion* 6(4): 627–638.

Akhter Shameem & Roberts Jason. 2006. *Multi-Core Programming*. Hillsboro: Intel Press.

Allan R.N, et al. 1981. Evaluation Methods and Accuracy in Probabilistic Load Flow Solutions. *IEEE Transactions on Power Apparatus and Systems* 100: 2539–2546.

Bludszuweit Hans, et al. 2008. Statistical analysis of wind power forecast error. *IEEE Transactions on Power Systems* 23(3): 983–991.

Borkowska, B. 1974. Probabilistic Load Flow. *IEEE Transactions on Power Apparatus and Systems*, 93:7 52–759.

Ding Ming, et al. 2001. Probabilistic load flow analysis based on Monte Carlo simulation. *Power System Technology* 25(11): 10–14 & 22.

Han Zhenxiang & Cao Yijia. 2004. Power system security and its prevention. *Power System Technology* 28(9): 1–6.

Liu Yifang, et al. 2011. Probabilistic Load Flow Algorithm Considering Static Security Risk of the Power System. *Proceedings of the Chinese Society for Electrical Engineering* 31(1): 59–64.

Miao, L, et al. 2007. A Probabilistic Load Flow Method Considering Transmission Network Contingency. *In Proc. 2007 IEEE Power Engineering Society General Meeting* 2: 1–6.

Patra, S. & Misra R.B. 1993. Probabilistic Load Flow Solution Using Method of Moments. *In Proc. 2nd International Conference on Advances in Power System Control, Operation & Management* 2: 922–934 & 991.

Silva, A.M.L.d. & Arienti, V.L. 1990. Probabilistic Load Flow by a Multi-linear Simulation Algorithm. *IEE Proceedings Generation, Transmission and Distribution* 137: 276–282.

Usaola Julio. 2009. Probabilistic load flow with production uncertainty using cumulants and Cornish-Fisher expansion. *Electrical Power and Energy Systems* 31: 474–481.

Usaola Julio. 2010. Probabilistic load flow with correlated wind power injections. *Electrical Power Systems Research* 80(5): 528–536.

Wang Shunxu. 2009. Implementation and Analyses of Parallel Computation on Multi-core Computers. *Journal of Huaihai Institute of Technology (Natural Science Edition)* 18(3): 30–33.

Wang Zhiqun, et al. 2004. Impacts of distributed generation on distribution system voltage profile. *Automation of Electric Power Systems* 28(16): 56–60.

Yu H, et al. 2009. Probabilistic load flow evaluation with hybrid Latin hypercube sampling and cholesky decomposition. *IEEE Transactions on Power Systems* 24(2): 661–667.

Yu Juan. 2007. *New Models and Algorithms of Optimal Reactive Power Flow and Applications in Voltage Stability Risk Assessment*. Chongqing: Chongqing University.

Advances in Power and Energy Engineering – Sun (Ed.)
© 2016 Taylor & Francis Group, London, ISBN 978-1-138-02846-3

Development of Fast Reactor power plant and its economic analysis

L. Liu & S.P. Qi
China Institute of Atomic Energy, Beijing, China

ABSTRACT: This paper introduces the technology of Fast Reactors (FRs) and investigates its development all over the world, including China. Subsequently, the investments of FRs are researched, and the economic analysis is elaborated.

1 INTRODUCTION

Human beings have been facing the problem of environment pollution and lack of energy since this century. Nuclear energy, which is near-zero-carbon and with high energy, has become an important energy choice in many countries, including China.

Since the 'scientific development view' requires to establish an efficient, economic, and optimal nuclear fuel cycle system (National Development and Reform Commission 2007), it is important and pressing to make nuclear energy sustainable development. As we known that Fast Reactor (FR) is the crux of China's nuclear developing route, which is "thermal reactor—FR—fusion reactor." It can realize closed fuel cycle, improve the utilization of uranium resources, reduce nuclear waste, and achieve sustainable development of nuclear energy (Xu).

However, as the decontrolling of power market, economics become the core competitiveness for a power plant. Nuclear power plants including FR are facing economics challenge.

2 THE TECHNOLOGY OF FR

2.1 *Principles*

The nuclear reactors refer to devices in which controlled nuclear fission chain reactions induced by neutrons can be maintained, meanwhile, the FRs refer to reactors in which the average neutron energy is between 0.08 MeV and 0.1 MeV. Comparing with fossil energy, nuclear energy has many advantages, such as consuming less fuels and exhausting no greenhouse gases.

Nuclear power plants that have been or are being constructed are mostly the Pressurized-Water Reactors (PWRs). In comparison to PWR, FR has two advantages: breeding and transmutation. Breeding means FR can increase the utilization

coefficient of uranium resources. Transmutation means FR can decrease nuclear waste. FR is an important part of the closed nuclear fuel cycle (Su et al. 1990). The Sodium-cooled FR (SFR) is the main candidate of the Generation IV advanced nuclear systems. Figures 1 and 2 show a simple schematic of the nuclear fission chain reactions and comparison between nuclear energy and fossil energy, respectively.

2.2 *Structural composition*

SFR is usually loaded with oxide fuels or metal fuels. The coolant of SFR is liquid sodium, and the

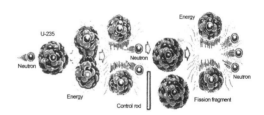

Figure 1. A simple schematic of the nuclear fission chain reactions.

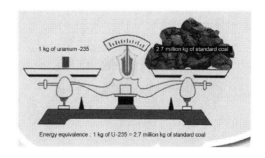

Figure 2. Comparison of nuclear energy and fossil energy.

main heat transport system is comprised of three circuits: sodium–sodium–water. According to the layout of primary system, SFR can be divided into two types: pool type and loop type. The main equipments of both type SFRs contain reactor core, primary pump, intermediate heat exchanger, steam generator and so on. The main equipments of primary circuit for the pool type SFR are located in the reactor vessel while the majority of the loop type SFR primary equipments except the core are located outside the reactor vessel.

Figures 3, 4, and 5 show schematics of pool type SFR, loop type SFR, and comparisons between them, respectively.

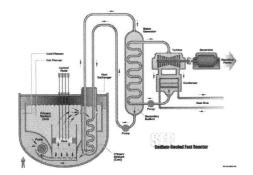

Figure 3.   Pool type SFR.

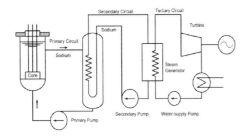

Figure 4.   Loop type SFR.

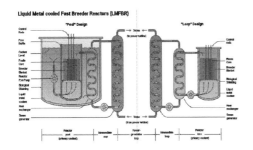

Figure 5.   Comparisons of the between the pool type SFR and the loop type SFR.

3   FR DEVELOPMENT IN COUNTRIES BESIDES CHINA

American scientists presented the concept of "breed reactor." The first FR was constructed in America. Table 1 shows the FRs that have been constructed in America (Cochran 2010).

Russia has begun to develop FR since the middle of 20th century. Table 2 shows the FRs that have been constructed in Russia (Cochran 2010).

France has 60 years experience of liquid cooling FR research and design, and operation. Moreover, France is the only country that has constructed and operated large-scale commercial FRs. Table 3 shows the FRs that were constructed in France (Crette 1998, Schneider 2009).

There are two FRs that have been constructed in Japan by now, which were named Joyo and Monju, respectively. The thermal power of Joyo is 50 MWt and the structure is loop type. Its first criticality is in the year of 1977. Monju was begun to be designed at the same time with Joyo, but the construction was delayed and finally reached the first criticality in the year of 1994. Monju is a loop type FR loading MOX fuel and the electric power is 280 MW (Kondo 1998).

In order to meet their country's nuclear plan and Global Nuclear Energy Partnership (GNEP) plan, Japan is now recombining FR research and development, and planning to construct a demonstration FR named JSFR. The electric power of JSFR is 1500 MW and the structure is loop type. JSFR will be fueled with MOX. Japan plan to develop commercial FR by the year of 2050.

There are also two FRs that have been constructed in India, which were named FBTR and PFBR. Table 4 shows the parameters of FBTR and PFBR (Rodriguez & Bhoje 1998).

4   FR DEVELOPMENT IN CHINA

FR has been deployed in the 1960s in China. Scientists focused on the research of physics, thermotics, structural material, and sodium process at the beginning and small size experimental facilities were constructed. In the 1980s, FR project was planned in the nuclear energy theme of "863" project. In 1995, China Experimental FR (CEFR) was approved by the government. The thermal power of CEFR is 65 MW; the net electrical power is 20 MW; the structure is pool type and starting fuel is UO_2. CEFR was begun to be constructed in May 2000. The main power house was capped in August 2008. On 21 July 2010, the first criticality was achieved and finally it was combined to the power grid on 21 July 2011.

Table 1. FRs in America.

Reactor	Thermal/electric power (MW)	Type	Coolant	Fuel	Criticality year
Clementine (E*)	0.025/0	Loop	Silver	Pu	1946
EBR-I (E)	1.2/0.2	Loop	NaK	U	1951
LAMPRE (E)	1.0/0	Loop	Na	Pu	1961
EBR-II (E)	62.5/20	Pool	Na	U–Zr	Dry-type 1961 Wet-type 1963
Fermi (E)	200/61	Loop	Na	U	1963
SEFOR (E)	20/0	Loop	Na	UO_2–PuO_2	1969
FFTF (E)	400/0	Loop	Na	UO_2–PuO_2	1980
CRBR (P)	975/380	Loop	Na	UO_2–PuO_2	1983 suspended
ALMR (P/D)	840/303	Pool	Na	U–Pu–Zr	1994 suspended

*(E), experiment; (P), prototype; (D), demonstration.

Table 2. FRs in Russia.

Reactor	Thermal/electric power (MW)	Type	Coolant	Fuel	Criticality year
BR-10 (E)	8/0	Loop	Na	UN (PuO_2)	1958
BOR-60 (E)	60/12	Loop	Na	PuO_2–UO_2	1968
BN-350 (P)	700/130	Loop	Na	UO_2	1972
BN-600 (P)	1470/600	Pool	Na	UO_2–PuO_2 (UO_2)	1980
BN-800 (D)	2100/800	Pool	Na	UO_2–PuO_2	2014

Table 3. FRs in France.

Reactor	Location	Operation year	Criticality year	Time combined to grid	Type	Power	Fuel
Rhapsody	Cadarache	1967–1983	1967	–	Loop	40 MWt	Pu-239:31.5 kg U-235: 79.5 kg
Phoenix	Marcoule	1973–2009	1973	1973	Pool	563 MWt	931 kg Pu (Pu-239 degree of enrichment 77%)
Super–Phoenix	Vier	1985–1998	1985	1986	Pool	1240 MWe	5780 kg Pu (Pu-239: 4054 kg)

Table 4. FRs in India.

Reactor	Thermal/electric power (MW)	Type	Coolant	Fuel	Criticality year
FBTR (E)	40/13	Loop	Na	PuC–UC	1985
PFBR (P)	1250/500	Pool	Na	PuO_2–UO_2	2015.9 (plan)

As the second step of China's development of FR, a demonstration FR is now under researching and designing. The electric power will be 600 MW. The project will contribute to closed fuel cycle theme and assist to achieve the long-term energy strategy in China.

5 ECONOMIC ANALYSIS OF FR

As we mentioned above that economics is the core competitiveness of a power plant, it is necessary to make research on the economics of FR power plant. For this purpose, the investments of FRs are investigated and shown in Table 5 (Cheng 2004).

The construction of CEFR project was finished in 2011. The investment of CEFR is **hundred million yuan, amount to **hundred million dollars. By converting the data in Table 5 into the year of 2011, Table 6 shows the comparison of investments of FRs in different countries. (The investment of CEFR is not shown for confidentiality. Although, it's about on the same investment level with other experimental FRs.)

According to the calculation of investment and the research results, factors like construction period, technology maturity, design function, policy support, and economic market in different countries or regions can lead influence to investment of FR power plants. The specific investments of Fermi and EBR-I diverge from the others for different reasons.

Fermi is the first experimental FR to generate electricity, which was supported by American government, and 25 power companies, 4 manufactures, and 4 engineering companies. Investments of government and the support companies were not included in the table data (Cheng 2004). That is the reason why its specific investment is lower.

Specific investment of EBR-I is very high because of the low electric power. Actually, specific investment will reduce as the electric power increasing in an index curve.

Comparing with PWR or fossil power, the investment of FR is higher. But when assessing economics of power plant, more factors should be considered besides investment. According to International Atomic Energy Agency's methodology for sustainability assessment of nuclear in the area of economics (International Atomic Energy Agency 2014) and the idea of comprehensive evaluation (Guo 2007), three criterions and four indicators are proposed to assess economics of FR, which are shown in Table 7.

For CR1 "cost competitiveness," comparisons of nuclear and Natural Gas Combined Cycle power-generation (NGCC), coal-fired power generation (PCSE), and Integrated Gasification Combined Cycle power generation (IGCC) are shown in Figure 6 (Zhang et al. 2009).

We can see from Figure 6 that in a low principle, nuclear cost is lower than all the other energies when capacity factor is in the range of 55–100%. The nuclear cost researched in Figure 6 maybe PWRs. The cost of FR is higher than the mature

Table 5. Investments of FRs.

Country	Reactor	Thermal/electric power (MW)	Finished/started year	Investment (10 thousands $)	Specific investment ($/kW)
India	FBTR	40/13	1972/1985(C*)	6250	4808
England	DFR	60/15	1954/1962	3653	2435
America	EBR-II	62.5/20	1958/1964	3522	1761
Germany	KNK-II	58/20	1975/1977	6341	3171
America	FFTF	400/–	1970/1980	61500	
America	EBR-I	1.2/0.2	–/1951	280	14,000
America	Fermi	200/61	1956/1966	2959	485
France	Rapsodie	40/–	1962/1967	120	
Korea	Kalimer	330/110	–	146 300	13 300

*C, criticality year.

Table 6. Comparison of FRs' investments.

Reactor	Thermal power (MW)	Electric power (MW)	Investment (10 thousands $)	Specific investment ($/kW)
FBTR	40	13	39,496	30,382
DFR	60	15	32,723	21,815
EBR-II	62.5	20	30,154	15,077
KNK-II	58	20	36,338	18,169
FFTF	400	–	297,775	
EBR-I	1.2	0.2	2781	139,071
Fermi	200	61	25,540	4187
Kalimer	330	110	238,875	21,716
CEFR	65	20	**	**

298

Table 7. Economics assessment of FR.

Criterion	Indicator (IN) and Acceptance Limit (AL)
CR 1 Cost competitiveness	IN1 cost of energy AL1 $C_{FR} \leq k * C_A$ (CN = cost of fast rector, and CA = cost of energy form alternative source; factor k is usually ≥ 1 and is based on strategic considerations.)
CR 2 Attractiveness of investment	IN2 financial figures of merit. Figures of merit for investing in FR are comparable with or better than those for competing energy technologies.
CR3 Social beneficial	IN3.1 benefit for environment Technology does not bring environmental harm and reduce environment harm. IN3.2 sustainable development Technology should be sustainable development.

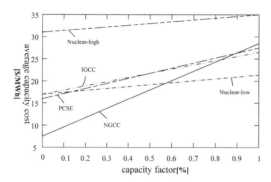

Figure 6. Average capacity cost among different energies.

PWR now. But construction cost of FR can be reduced by scale economy, then it will be economic comparable with PWR, so as to other energies.

For CR2 "attractiveness of investment," the financial figures of merit of FR can be comparable with PWR. (Although in this situation the sales price of FR is higher than mature PWR, but can be comparable with solar electricity plant.) The financial figures of merit and grid purchase price are based on the demonstration FR's economic assessment, but for confidentiality reason, the data will not be quoted here.

For CR3 "social beneficial," IN3.1 "benefit for environment," FR can transmute long-lived elements and reduce their radiation to environment. For IN3.2 "sustainable development," with the characteristic of breeding, FR can increase the utilization coefficient of uranium resources to more than 60% (XU), which can be seen as sustainable development technology.

6 CONCLUSION

FR is an innovative nuclear reactor. Many experimental FRs, prototype FRs and one commercial FR have been constructed by now and many countries have planned to develop FR, including China.

FR is not cost competitive with coal power plant, gas power plant and mature PWR at present. But FR is certainly beneficial to environment and can lead nuclear energy sustainable developing. Besides, with FR developing, the cost can be reduced by scale economy and technology maturity. And the financial figures of merit of FR can be comparable with PWR at a high price level.

REFERENCES

Cheng, Ping 2004. Investment Comparison among CEFR and Other Experimental Fast Reactors. Policy: China nuclear industry magazine Publishing House.

Cochran, Thomas B. et al. 2010. Fast Breeder Reactor Programs: History and Status. Research Report 8 International Panel on Fissile Materials: 63–96.

Crette, J.P. 1998. Review of the Western European Breeder Programs. Energy, Vol. 23, No. 7/8: 581–591.

Guo, Yajun 2007. Comprehensive Evaluation Method and Its Aplyin. Beijing: Science press.

International Atomic Energy Agency 2014. INPRO Methodology for Sustainability Assessment of Nuclear Energy Systems: Economics. Vienna.

Kondo, Shunsuke 1998. History and Perspective of Fast Breeder Reactor Development In Japan. Energy Vol. 23, No. 7/8, pp. 619–627.

Mycle, Schneider 2009. Fast Breeder Reactors in France. Science and Global Security, 17:36–53.

National Development and Reform Commission (NDRC) 2007. Long-term planning for nuclear power (2005–2020). Beijing.

Rodriguez, P. & Bhoje, S.B. 1998. The FBR Program in India. Energy Vol. 23, No. 7/8, pp. 629–636.

Su, Zhuting et al. 1990. Sodium-Cooled Fast Breeder Reactors. Beijing: Atomic Energy Press.

Xu, Mi. Fast Reactor-Recent Status and Prospects. China Academic Journal Electronic Publishing House. Vol. 20, No. 5:91–93.

Zhang, Yanfu et al. 20009. The Capacity Assignment Research among Different Kinds of Power Generators for Long-term Energy Resource Planning. APPEEC 2009.

Power system management

Advances in Power and Energy Engineering – Sun (Ed.)
© 2016 Taylor & Francis Group, London, ISBN 978-1-138-02846-3

Seamless switching control method of micro grid based on master-slave configuration

L. Wang
Department of Electrical Engineering, Shanghai Jiao Tong University, Shanghai, China

K. Jiang
Shanghai Municipal Electric Power Co., Shanghai, China

Z.X. Wang
Department of Electrical Engineering, Shanghai Jiao Tong University, Shanghai, China

B.F. Lu
Jiaxing Renewable Power Electrical Co. Ltd., Jiaxing, Zhejiang Province, China

C. Zhang
Shanghai Najie Complete Electrical Co. Ltd., Shanghai, China

ABSTRACT: The control strategies of micro-grid based on master-slave configuration in grid-connected and stand-alone operation modes were introduced in this paper. The mode switching process between grid-connected and stand-alone operation was also discussed. A kind of pre-synchronizing control was proposed to eliminate difference of phase angles, which can effectively avoid the complex calculation of voltage phase angle and decrease the current impact during the switching process from stand-alone operation to grid-connected operation mode. Then, a new seamless switching control method was proposed, which delayed the movement of static switch and avoided power shortage during the switching process from grid-connected operation to stand-alone operation mode. The current reference compensation control was also used, which can reduce the transient effects. Finally, MATLAB simulation verifies the correctness and validity of the model and controls strategies.

Keywords: micro grid; master-slave configuration; seamless switching; grid-connected operation; stand-alone operation; pre-synchronizing control

1 INTRODUCTION

With the energy crisis and environmental problems becoming increasingly prominent, Distributed Generation (DG) obtained extensive research and application (Blaabjerg 2006, Ault 2003). Micro grid is a kind of new form of distributed power supply access, which connects various distributed generations, energy storage devices, loads and control devices with the electricity grid (Pogaku 2007, Yuan 2011, Azevedo 2011). This power supply mode has greatly increased the power supply reliability of the load side and improved the efficiency of distributed power supply.

Micro grid usually has two kinds of operating mode, grid-connected mode and standalone mode. Under normal circumstances, micro power grid operates in grid-connected mode. When a grid failure occurs, the Static Transfer Switch (STS) immediately breaks away, and micro grid transfers into the standalone mode, which guarantee the power supply (Lasseter, 2004). When the micro grid switches between two kinds of operation modes, power deficiency or controller state mutations may happens, which can cause the system to produce larger impact current and voltage transient oscillation, affecting the micro grid reliability and power quality. So the seamless switching control method of micro grid is of great significance.

Based on the characteristics when micro grid runs in different modes, seamless switching control strategies between two modes are respectively presented in this paper. The simulation results show that the method can effectively restrain the impact current and voltage oscillation when micro grid switches between two operation modes.

2 SYSTEM MODELING AND ANALYSIS

At present, the control strategies of micro grid can be mainly classified as master-slave control and peer-to-peer control. Owing to simple control structure and high reliability, most of micro grid demonstration project using master-slave control currently (Chen 2013, Wang 2012). Master-slave control usually specifies an energy converter or diesel generators as the master converter. The master converter adopts PQ control method to provide power for local load when micro grid runs in the grid-connected mode, and adopts V/F control method to maintain the network voltage and frequency when micro grid runs in the standalone mode. While randomness of energy inverters, such as photovoltaic and wind power inverters, work as slave inverters, which always adopt PQ control method in order to maximize the energy utilization. Figure 1 illustrates the micro grid structure using master-slave control method.

The micro grid operation state is decided by the static transfer switch. That is to say, when the STS is closed, micro grid runs in grid-connected state, or it is in standalone state when STS is broke. The control method of master inverter is determined by the switch Sc. The master controller adopts PQ control when Sc contacts point 1, or it adopts V/F control when Sc contacts point 2, which ensures the voltage stability of PCC point. Figure 2 shows the inverter control system block diagram.

L filter model in the dq coordinate system can be expressed as in (1).

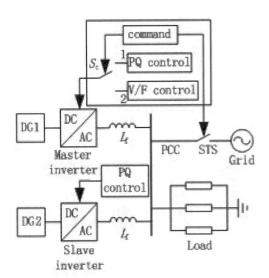

Figure 1. Master-slave controlled micro grid configuration.

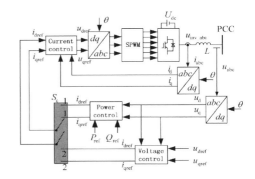

Figure 2. Control diagram of master converter.

$$
\begin{cases}
u_{\text{inv\_d}} = Ri_{\text{d}} + L\dfrac{di_{\text{d}}}{dt} - \omega Li_{\text{q}} + u_{\text{d}} \\[2mm]
u_{\text{inv\_q}} = Ri_{\text{q}} + L\dfrac{di_{\text{q}}}{dt} + \omega Li_{\text{d}} + u_{\text{q}}
\end{cases}
\tag{1}
$$

Based on instantaneous power theory, the instantaneous active power p and reactive power q of inverter can be expressed as in (2).

$$
\begin{cases}
p = u_{\text{d}}i_{\text{d}} + u_{\text{q}}i_{\text{q}} \\
q = u_{\text{q}}i_{\text{d}} - u_{\text{d}}i_{\text{q}}
\end{cases}
\tag{2}
$$

The equation of PQ controller can be calculated by (3).

$$
\begin{cases}
i_{\text{dref}} = \dfrac{P_{\text{ref}}u_{\text{d}} + Q_{\text{ref}}u_{\text{q}}}{u_{\text{d}}^{2} + u_{\text{q}}^{2}} \\[3mm]
i_{\text{qref}} = \dfrac{P_{\text{ref}}u_{\text{q}} - Q_{\text{ref}}u_{\text{d}}}{u_{\text{d}}^{2} + u_{\text{q}}^{2}}
\end{cases}
\tag{3}
$$

Without considering the waveform distortion of grid voltage, the d-q axis components of grid voltage can be express in (4).

$$
\begin{cases}
e_{\text{d}} = \sqrt{3}E_{0} \\
e_{\text{q}} = 0
\end{cases}
\tag{4}
$$

So, equation (3) can be simplified to (5).

$$
\begin{cases}
i_{\text{dref}} = \dfrac{P_{\text{ref}}}{\sqrt{3}E_{0}} \\[3mm]
i_{\text{qref}} = -\dfrac{Q_{\text{ref}}}{\sqrt{3}E_{0}}
\end{cases}
\tag{5}
$$

In order to realize the decoupling control of d-q axis and eliminate the interference of the power grid voltage, the current loop usually adopts the

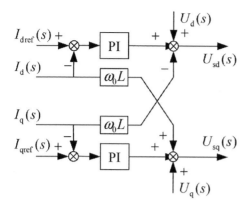

Figure 3.   Diagram of current decoupled control.

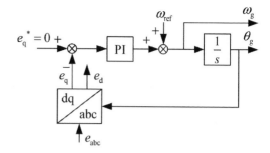

Figure 4.   Control block diagram of SRF-PLL.

voltage and current feed-forward control strategy, as shown in Figure 3.

When micro grid runs in grid-connected state, the master controller adopts voltage and current double closed loop control, the voltage controller equation is

$$\begin{cases} i_{dref} = \left(k_{pv} + \dfrac{k_{iv}}{s}\right)(u_{dref} - u_d) \\ i_{qref} = \left(k_{pv} + \dfrac{k_{iv}}{s}\right)(u_{qref} - u_q) \end{cases} \qquad (6)$$

When micro grid runs in standalone state, STS switches off, the reference voltage is set as (7).

$$\begin{cases} u_{dref} = \sqrt{3}E_0 \\ u_{qref} = 0 \end{cases} \qquad (7)$$

3   SEAMLESS SWITCHING CONTROL

3.1   Transition from stand-alone to grid-connected

When micro grid runs in the state of standalone, the phase theta in Park transformation is calculated by reference angular frequency, namely

$$\theta_{inv} = \int \omega_{ref} dt = 2\pi \int f_{ref} dt \qquad (8)$$

When micro grid runs in grid-connected state, the master inverter adopts PQ control, the phase theta in Park transformation is obtained by phase-locked loop, as shown in Figure 4.

In the process of state transition from standalone to grid-connected, the pre-synchronization control should be adopted to ensure the output voltage of master inverter is same with grid voltage in view of voltage amplitude value, frequency and phase.

Without considering the waveform distortion of grid voltage, the d-q axis components of PCC voltage can be express in (9).

$$\begin{cases} u_\alpha = \sqrt{3}U_0 \cos(\omega t + \theta_2) \\ u_\beta = \sqrt{3}U_0 \sin(\omega t + \theta_2) \end{cases} \qquad (9)$$

So

$$e_\beta u_\alpha - e_\alpha u_\beta = 3E_0{}^2 \sin(\theta_1 - \theta_2) \qquad (10)$$

Therefore, $e_\beta u_\alpha - e_\alpha u_\beta$ can be chose as control variables for the pre-synchronization controller, which can avoid the accurate calculation of voltage phase angle and eliminate the phase difference between output voltage of master inverter and grid voltage by original voltage signal directly. Based on the above analysis, this paper proposes a new method of pre-synchronization control, namely

$$\omega_{ref} = \left(k_{pw} + \dfrac{k_{iw}}{s}\right)(e_\beta u_\alpha - e_\alpha u_\beta) + 2\pi f_{ref} \qquad (11)$$

Before closing STS, the master inverter adopts pre-synchronization control, and the reference frequency of V/F control is no longer a fixed value, but by equation (13) is calculated.

3.2   Transition from grid-connected to stand-alone

When micro grid runs in grid-connected state, the grid phase voltage RMS E_0 can be considered to be constant, and the reference current can also be considered to be constant based on (5). But the reference current is calculated by the voltage PI controller when micro grid runs in standalone state. As a result, the reference current mutates in the sudden of grid operation state changes, which may cause the PCC point voltage drop, and even lead to a sustained oscillation or a substantial overshoot.

In order to transition seamlessly, the reference current is compensated. Namely, the reference current that calculated by the PQ controller is used as the initial value of the voltage PI controller when Sc switches from point 1 to 2, which can reduce the transient process.

Normally, when micro grid runs from grid-connected state to standalone state, the static switch STS and the control switch Sc switch at the same time. However, the local loads and the power of inverters usually keep constant at the switching moment, and power vacancy will appear, which has an effect on the PCC point voltage waveform. Therefore, this paper proposes that the switch control switch Sc is switched on before the static switch STS breaks away, which enables the master inverter to keep the output power stable when it changes from the PQ control to the V/F control.

4 VERIFICATION OF THE CONTROL STRATEGY

According to the micro grid structure shown in Figure 1, the simulation model is built in MATLAB to verify the control strategy. The master and slave inverter use the same simulation parameters, as shown in Table 1.

When micro grid runs in grid-connected state, the master inverter outputs 5 kW, the slave inverter outputs 15 kW, and the local load consumes 30 kW, the power grid outputs 10 kW. When micro grid runs in standalone state, the master inverter outputs 15 kW, the slave inverter outputs 15 kW, and the local load consumes 30 kW, the power grid outputs 0 kW.

The simulation procedure is as follows:

1. t = 0~0.05 s, micro grid runs in the standalone state, control switch Sc connects point 2. The master inverter adopts V/F control, the output voltage lags behind the grid voltage 60;
2. t = 0.05~0.15 s, the master inverter adopts pre synchronization control;
3. t = 0.15 s, the static switch STS is closed, the micro grid is switched to the grid connected state. Sc is switched to point 1, and the master inverter is transited to PQ control;
4. 0.2S = t, Sc is switched to point 2. 0.01 s later, STS is disconnected, and the micro grid transited to standalone state.

Simulation results are shown in Figures 5~7.

Table 1. Simulation parameters.

Parameters	Simulation	Parameters	Simulation
U_{dc}/V	800	k_{pv}	0.35
E_0/V	220	k_{iv}	210
f/Hz	50	k_{pi}	16.67
f_s/kHz	10	k_{ii}	33.33
L/mH	5	R/Ω	0.01

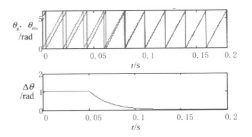

Figure 5. Simulation results of pre-synchronizing control.

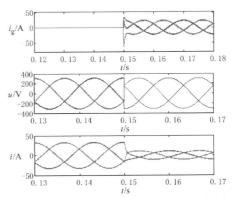

(a) Switching from stand-alone to grid-connected state directly.

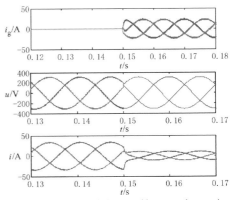

(b) Switching from stand-alone to grid-connected state using seamless switching control.

Figure 6. Simulation results of switching from stand-alone to grid-connected state.

Figure 5 shows the results for pre-synchronization control. The micro grid is in the state of standalone at first. When t = 0.05, the master inverter controller receives the grid connection signal, and adjust the V/F control reference frequency according to (13). The implementation of the pre-synchronization operation is completed in 0.1 s, and then STS switches to grid-connected state at 0.15 s. It can be

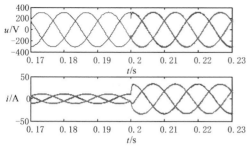

(a) Switching from grid-connected to stand-alone state directly.

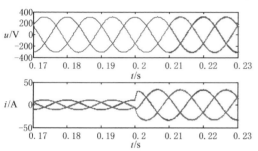

(b) Switching from grid-connected to stand-alone state using seamless switching control.

Figure 7. Simulation results of switching from grid-connected to stand-alone state.

seen from the simulation results, grid voltage phase can be tracked rapidly, achieving the purpose of smooth transition PCC voltage phase angle in the whole switching process.

Figure 6 shows the simulation results of the micro grid switches from standalone state to grid-connected state directly and using seamless switching control respectively. Due to the phase angle difference between PPC point voltage and grid voltage, big impact current will be produced if STS switched directly, which will caused great impact on the grid. In the simulation of switching directly, the voltage phase of PPC point voltage lags grid voltage 60°, and the maximum impact current is as high as 65.5 A, which is 2.7 times of the steady-state current. In comparison with Figure 6(a) and (b), it can be seen that the pre-synchronization control can reduce the impact current effectively and realize the seamless transition from standalone operation state to grid-connected operation state.

Figure 7 shows the simulation results of the micro grid switches from grid-connected state to standalone state directly and using seamless switching control respectively. If STS switches directly at 0.2 s, a vacancy of 10 kW appears at the instant of switching, resulting in a significant voltage drop of PCC voltage, as shown in Figure 7(a). By using the seamless switching control strategy, power vacancy can be avoided, and the PCC voltage amplitude is

stable during the switching process, realizing the seamless transition from grid-connected state to standalone state.

5 CONCLUSION

The control strategies of micro grid based on master-slave configuration in grid-connected and stand-alone operation modes were introduced in this paper. Aiming at the problems of power vacancy and controller state mutation in the transition between two running state, the seamless switching control strategy from standalone to grid-connected state and grid-connected to standalone state is put forward respectively. The simulation results show that the proposed pre-synchronous control method has good dynamic performance, and PCC voltage can tracks the grid voltage phase quickly. STS delay control in the process of switching from grid-connected state to standalone state is proposed in this paper to avoid the appearance of power vacancy. The control method can also inhibit the PCC voltage drop effectively.

REFERENCES

Ault G.W., McDonald J.R., & Burt G.M. Strategic analysis framework for evaluating distributed generation and utility strategies [J]. IEE Proceedings-Generation, Transmission and Distribution, 2003, 150(4): 475–481.

Azevedo G., Bradaschia F., Cavalcanti M.C., et al. Safe transient operation of microgrids based on master-slave configuration [C]//Energy Conversion Congress and Exposition (ECCE), 2011 IEEE. IEEE, 2011: 2191–2195.

Blaabjerg F., Teodorescu R., Liserre M., et al. Overview of control and grid synchronization for distributed power generation systems [J]. Industrial Electronics, IEEE Transactions on, 2006, 53(5): 1398–1409.

Chen X., Wang Y.H., Wang Y.C. A novel seamless transferring control method for microgrid based on master-slave configuration [C]//ECCE Asia Downunder (ECCE Asia), 2013 IEEE. IEEE, 2013: 351–357.

Lasseter R.H., Paigi P. Microgrid: a conceptual solution [C]//Power Electronics Specialists Conference, 2004. PESC 04. 2004 IEEE 35th Annual. IEEE, 2004, 6: 4285–4290.

Pogaku N., Prodanović M., Green T.C. Modeling, analysis and testing of autonomous operation of an inverter-based microgrid [J]. Power Electronics, IEEE Transactions on, 2007, 22(2): 613–625.

Wang C.S., Li X.L., Guo L., et al. A seamless operation mode transition control strategy for a microgrid based on master-slave control [J]. Science China Technological Sciences, 2012, 55(6): 1644–1654.

Yuan P., PeiQiang L., Xinran L., et al. Strategy of research and application for the microgrid coordinated control [C]//Advanced Power System Automation and Protection (APAP), 2011 International Conference on. IEEE, 2011, 2: 873–878.

Advances in Power and Energy Engineering – Sun (Ed.)
© 2016 Taylor & Francis Group, London, ISBN 978-1-138-02846-3

Influence of fluctuating load on distribution network voltage stability

Y.H. Huo, L.X. Zhang, Q.L. Zhang & S.Y. Li
China University of Petroleum, Qingdao, Shandong, China

J.T. Gui
Nanyang Power Supply Company, Nanyang, Henan, China

ABSTRACT: With the development of Distributed Generation (DG) and the decrease of load rate, the impact of fluctuating load on voltage quality is becoming more and more important. In this paper, the effect of voltage quality, line loss and transformer loss in local fluctuant 10 kV power network are studied, and the impact of load fluctuations on the power grid voltage quality is analyzed from two aspects: the value of load and the distance from loads to the power points. For the power grid with a small proportion of fluctuating loads, a method is designed to control the power grid by using Static Var Compensator (SVC). Simulation results prove that using SVC flexibly can improve voltage, reduce line loss and transformer loss obviously. It can not only improve voltage quality, but also "release" additional capacity and extend supply power distance.

1 INTRODUCTION

The 10 kV power grid has the characteristics of large load density, strong concentrated electricity consumption, and high requirement of power supply reliability for its single trunk line with many branch lines. There are a lot of questions such as high line losses, low power factor, insufficient transmission capacity and low voltage at the terminal of power lines. Accompanied the development of distributed power supply, the loads are power customer described as high randomness, strong fluctuations and uncontrollable output. Their developments propose a higher request of power quality (Kalla Ujjwal Kumar et al. 2014).

Thyristor Switched Capacitor (TSC), which is one of the most popular SVC devices, has characteristics of large capacity, low cost and no harmonic pollution (Jiaqiang Yang et al. 2014). Some scholar designed a dynamic var compensator by using TSC to correct power factor (Zhikang Shuai et al. 2009); some literature proposed to combine TSC and STATCOM to realize continuous compensation with large capacity (Wei Zhao et al. 2009). In the future, large-capacity TSC will be used in power grid sequentially.

This paper established a power grid with small proportion of fluctuating loads. TSC is applied with reasonable configured dynamic reactive compensation capacity. The relation among active power loss, node voltages and reactive compensation capacity are analyzed. A method of dispersed compensation is put forward for fluctuating load.

By comparing different influence of different compensation mode for the power grid, the advantage of this dispersed compensation method is demonstrated. The impact of load fluctuations on the power grid voltage quality is also analyzed. Simulation results prove that by using this method, power grid loss can be reduced; voltage quality and economic benefits can be improved effectively.

2 THE INTRODUCTION AND ANALYSES OF 10 KV POWER GRID

As shown in Figure 1, a 10 kV Power system model is used in this paper. This model has 73 nodes and 72 branches. This figure draws a major part only. In this grid, daily load curves of node 8 and node 40 fluctuate strongly, while daily load curves of node 12 and node 24 fluctuate less than node 8's and node 40's, as shown in Figure 2. Because the other nodes fluctuate slightly, only these 4 nodes' fluctuating loads are considered for the grid influence.

We select 4 instant moments: 4 a.m., 8 a.m., 12 a.m. and 4 p.m. In these moments, the instant loads of these 4 nodes are shown in Table 1. From Table 1 we can see that the loads of 8 a.m. and 4 p.m. are generally large while the loads of 4 a.m. are generally small. At 8 a.m. and 4 p.m., the loads reached the peak point while at 4 a.m., the loads fell to the low point. The proportions of these 4 fluctuating loads in all the loads are shown in Table 2. The total loads of these 4 loads account for 5%–12% of the grid's total loads, 10% on

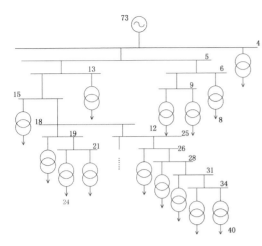

Figure 1.   10 kV grid structure.

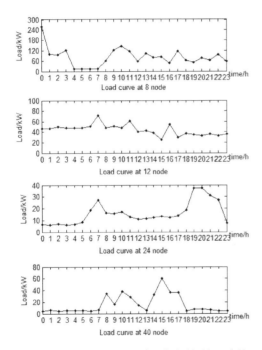

Figure 2.   Daily load curves of node 8, 12, 24, and 40.

Table 1.   4 nodes' loads in different time (kW).

Node time	8	12	24	40
4 a.m.	15.536	47.46	6.712	5.028
8 a.m.	61.104	47.85	16.184	34.119
12 a.m.	58.592	40.71	10.818	14.217
4 p.m.	47.2	53.86	12.444	35.922

Table 2.   Fluctuating loads' percentage in the overall loads.

Node time	8	12	24	40
4 a.m.	0.012841	0.038989	0.006252	0.004503
8 a.m.	0.050506	0.039309	0.015075	0.030554
12 a.m.	0.04843	0.033443	0.010077	0.012731
4 p.m.	0.039014	0.044246	0.011591	0.032168

average, which is a small percentage in the overall loads.

Maximum loads and minimum loads in daily load curves are shown in Table 3. The difference between them is large, and these curves fluctuate wildly. Therefore, the demand for dynamic reactive compensation becomes stronger.

During computation, set benchmark capacity to 200 kVA, set adjusting range of variable transformer's tap to 0.9–1.1, and set span to 0.0125. Take generator node 73 as balance node, and run Newton-Raphson power flow at different time to observe different node voltages.

2.1   Effect of fluctuating loads on power grid

At 4 instant times, these 4 nodes' voltages are as shown in Table 4. According to these load curves, loads are large and node voltages are small generally at 8 a.m., while loads are small and node voltages are large generally at 4 a.m.

In order to observe the impact of fluctuating loads, we change node 8's load in the designated area. Because there are lots of nodes in this system, we draw a figure instead of making a list to observe the differences between the voltages before changing load and the voltages after changing load, which are shown in Figure 3. X axis shows node, y axis shows voltage difference. We change load of node 8 three times and get three different lines. From the figure we can see the trends of these three lines are the same. The line of node 8 is higher than the others relatively. Though the other nodes' voltage differences are really small, they are not zero. Because of node 73 is balance node, its voltage standard value is 1, and thus only the voltage difference of node 73 is 0. We can easily find from the figure that changing a node's load can influence all the nodes in the grid, but influence the changed node's load largest.

2.2   Effect of transmission line length on power grid

In order to study the effect of transmission line length on power grid, we change the transmission

Table 3. 4 nodes' maximum\minimum load (kW).

Node load	8	12	24	40
P_{max}	260.672	71.55	37.478	60.426
P_{min}	14.136	24.88	6.002	4.173

Table 4. Node voltages in different time (p.u.).

Node time	8	12	24	40
4 a.m.	1.0264	1.0184	1.0177	1.0078
8 a.m.	1.0149	1.0179	1.0138	0.989
12 a.m.	1.0157	1.0198	1.016	1.0019
16 p.m.	1.0184	1.0164	1.0152	0.9879

Table 5. Effect of different transmission line length on grid voltage (p.u.).

Times node	2	4	8	16	32	64
8	0	0	0	0	1e-4	2e-4
12	1e-4	2e-4	4e-4	8e-4	1.7e-3	3.4e-3
24	0	0	0	0	0	1e-4
40	0	0	0	1e-4	1e-4	2e-4

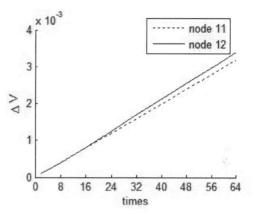

Figure 4. Effect of a node's transmission line length on ΔV.

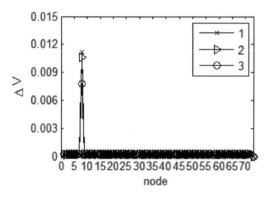

Figure 3. Effect of changing a node's load on the overall nodes.

line length of these 4 fluctuating nodes. These 4 transmission line length is scaled up simultaneously, the result of voltage differences are shown in Table 5. We can easily find that the voltage differences are growing with the increase of transmission line length. Though the differences are less obvious than the differences that fluctuating loads cause, they can't be ignored. From Figure 4 we find that there is an almost linear relationship between voltage difference and transmission line length.

For studying the effect to the other nodes when changing the transmission line length, we change transmission line length of 12 nodes only. After power flow calculation, only the voltages of node 11 and node 12 are changed. Node 11 connects node 12 to the main electric grid. ΔV is the difference between the original transmission line length's voltages and the changed transmission line length's voltages. The trends of ΔV are shown

in Figure 4, there is an almost linear relationship between ΔV and transmission line length.

Overall, the fluctuating loads and transmission line length influence the grid voltages. A node's fluctuating load can impact on the overall nodes' voltages while a node's transmission line length can effect on the voltages of area only near this node. Therefore, it's necessary to carry on the dynamic compensation according to the real situation.

3 THE DESIGN OF REACTIVE COMPENSATION

3.1 Fixed compensation

The principle of Fixed Compensation (FC) is adding capacitor on the compensatory node to make power factor of daily minimum load up to 0.95. Fixed compensation is used on fluctuating load node. The formula that calculates compensation capacity is as follow:

$$C_{fixed} = \frac{P_{min}(\tan\delta - \tan\delta')}{\omega U^2} \tag{1}$$

The reactive power need to be compensated is:

$$Q_C = P_{min}(\tan \delta - \tan \delta') \qquad (2)$$

where C_{fixed} = compensation capacity; Q_C = compensated reactive power; P_{min} = the node's minimum load; δ = power factor angle before compensation; δ' = power factor angle after fixed compensation; and U = node voltage (Feng Qian et al. 2009).

The fixed compensation capacity is shown in Table 6. Because the relevance between reactive power and capacitance is $Q = \omega CU^2$, when the capacitance is given, the reactive powers will fluctuated slightly along with voltage fluctuations. Thus Table 6 is reactive capacity at some point, and there's a little error on different time. After compensation, the nodes' voltages will increase, and Q will increase with it, which is a disadvantage of reactive compensation.

3.2 TSC reactive compensation

The circuit diagram of anode's TSC dynamic compensation is shown in Figure 5. It combines FC and 16 levels dynamic compensation (Qi Chen et al. 2006). The FC can compensate the node's minimum load, and TSC can be decided the level to compensate the other load due to the node's instant load.

The formula of FC compensation capacity is:

$$C_{FC} = \frac{Q_{min}}{\omega U^2} = \frac{P_{min} \tan \delta}{\omega U^2} \qquad (3)$$

where C_{FC} = fixed compensation capacity; P_{min} = node's minimum active power; and Q_{min} = node's minimum reactive power.

Table 6. Fluctuating loads' fixed compensation capacity (p.u.).

Node	8	12	24	40
Fixed compensation capacity	0.0688	0.1699	0.034	0.027

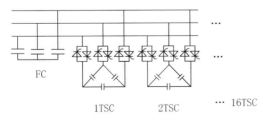

1TSC   2TSC   ··· 16TSC

Figure 5. Circuit diagram of TSC dynamic compensation.

Table 7. The dynamic compensation capacity of fluctuating loads (p.u.).

Node time	8	12	24	40
4 a.m.	0.094	0.3136	0.044	0.0342
8 a.m.	0.1943	0.3135	0.0872	0.2055
12 a.m.	0.1945	0.2636	0.0584	0.0917
4 p.m.	0.1953	0.3636	0.0584	0.2052

TSC dynamic compensation use 16 groups of equivalent capacitor, the capacitance of a group of TSC is:

$$C_{TSC} = \frac{Q_{max} - Q_{min}}{16\omega U^2} = \frac{(P_{max} - P_{min})\tan \delta}{16\omega U^2} \qquad (4)$$

where C_{TSC} = TSC compensation capacity; P_{max} = node's maximum active power; and Q_{max} = node's maximum reactive power.

Assume that this node's active load is P at one point, then the group of dynamic compensation capacitance that decide to be put into is:

$$k = \left(\frac{P(\tan \delta - \tan \delta')}{\omega U^2} - C_{FC} \right) \Big/ C_{TSC} \qquad (5)$$

where k = the group of TSC dynamic compensation which is put into circuit.

Because the k is an integer, dynamic compensation can't meet demand most of the time. If the power factor needed is 0.93, δ' can take 0.95. The instant compensation capacities are shown in Table 7, they means the capacity that one of these nodes is compensated at one point. In this table, nodes' capacities at 8 a.m. are generally larger because of their large load; nodes' capacities at 4 a.m. are lower because of its low load. The TSC device realizes flexibility on different load.

4 ECONOMIC BENEFIT ANALYZE

From the perspective of main operation equipment in power grid, power losses can be classified into transmission loss, transformer loss and other equipment losses. In the power grid operation, other devices such as capacitor, inductor and switch will also generate losses which are relatively small, therefore they can be ignored. Then the power losses can be considered the sum of transmission loss and transformer loss.

$$\Delta P_l = \sum \left(\frac{U_i - U_j}{R_{ij} + jX_{ij}} \right)^2 * R_{ij} \qquad (6)$$

$$\Delta P_l\% = \frac{\Delta P}{P_G} \tag{7}$$

$$\Delta P_T = \sum P - \Delta P_l \tag{8}$$

where ΔP_l = Transmission loss; $\Delta P_l\%$ = Line loss rate; ΔP_T = Transformer loss; U_i = the voltage of node i; U_j = the voltage of node j; $R_{ij} + jX_{ij}$ = branch's transmission impedance; P_G = power supply; and ΣP = vectorial sum of all nodes' power, supplied power is positive and consumed power is negative.

Nodes 24's and node 40's line losses at different time are shown in Figures 6 and 7 separately.

We can see from Figures 6 and 7 that the line losses after dynamic compensation are lower than the line loss after fixed compensation all the time. The line loss differences at 8 a.m. and 4 p.m. are large, thus the effect of dynamic compensation is more prominent when the fluctuating loads

are high. The line loss differences of node 40 are more obvious than node 24's, which states that dynamic compensation has more significant role for higher fluctuating loads. After TSC compensation, the reactive power which distribution network primary side transmits is reduced. Dynamic compensation can improve line loss rate 0.014% higher than fixed compensation. This impact is very considerable and has an important influence on developing economic benefit of power system.

In the same way, node 24's and node 40's transformer losses at different time can calculated, as shown in Figures 8 and 9.

Figures 8 and 9 showed that, although the differences between transformer losses after dynamic compensation and transformer losses after fixed compensation is not big, the largest difference between two types of compensation is about 0.0016, which can't be ignored for reducing the

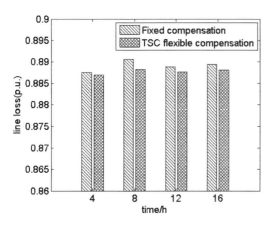

Figure 6. Node 24's line loss in different time.

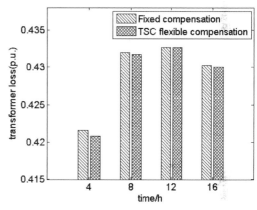

Figure 8. Node 24's transformer loss in different time.

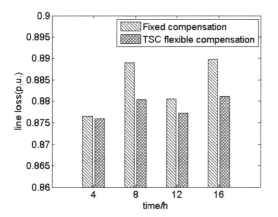

Figure 7. Node 40's line loss in different time.

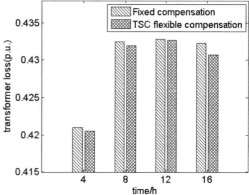

Figure 9. Node 40's transformer loss in different time.

power grid's loss. For reducing transformer loss, dynamic compensation has modest effect at low load, but the effect is obvious when the load is large. TSC dynamic compensation plays a great role in reducing the loss of transformers.

Therefore, the general idea of reactive power compensation in distribution network with fluctuating loads is: for a more stable load, fixed compensation should be used to guarantee power factor to reach the required value, which generally takes 0.93; For the fluctuating load, the combination of fixed compensation and TSC dynamic compensation should be used simultaneously: the fixed compensation can compensate daily minimum loads while the multiple levels TSC can compensate the rest part loads which are variable at different times. The TSC will put into high levels at the peak load and reduce them at low load automatically, which can meet the demand of the ideal conditions for the compensation capacitor instantly.

5 VOLTAGE QUALITY ANALYZE

Among those methods of improving voltage quality, making full use of reactive power compensation device such as parallel capacitor occupies an important position. On power system planning, grid structure should ensure the flexibility in operation, and reactive power planning in power distribution network should be done well. TSC dynamic reactive power compensation can not only improve the economic benefit, but also improve the power quality of the power grid (Linchuan Li et al. 1999). We calculate the power flow of 10 kV power grid and compare the voltages before and after compensation, as shown in Figures 10 and 11.

It can be seen from Figures 10 and 11 that the node voltages that is before compensation, after

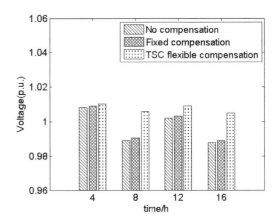

Figure 11. Node 40's voltages at different time in different compensation ways.

fixed compensation, after TSC flexible compensation are gradually rise, which means TSC flexible compensation is better than fixed compensation again.

By using the TSC reactive power compensation optimization, every node's secondary voltage is improved on a different level, and power losses are reduced and voltage quality is improved significantly. Due to the rapid response capability of TSC, it can quickly make reactive power compensation, support system voltage and prevent voltage collapse when the system voltage drops. After adding TSC, the fluctuation range is reduced, voltage quality is improved and losses caused by fluctuation loads are cut down greatly.

6 CONCLUSION

In this paper, the influence of fixed compensation and TSC dynamic compensation to distribution grid and the improvement of economic benefit are analyzed by applied fixed capacitor and TSC on fluctuating load. The main research achievements on this basis are as following: firstly, the impact of fluctuating load and transmission line length on power grid's voltage are analyzed. Secondly, fixed compensation and TSC dynamic compensation are proposed. On the basics of this, by comparing the effect of fixed compensation and TSC dynamic compensation for transmission line loss and transformer loss, the advantage of TSC dynamic compensation for distribution grid with fluctuating load is demonstrated. Finally, by analyzing economic benefit, nodes' voltage which is before compensation, after fixed compensation and after TSC dynamic compensation are compared and the effect of TSC on the voltage quality is analyzed. Studies and

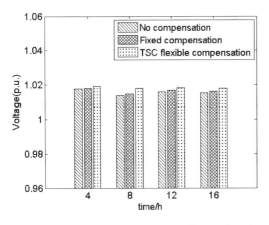

Figure 10. Node 24's voltages at different time in different compensation ways.

simulation results prove that TSC dynamic reactive power compensation improves voltage stability of each node in the network greatly, reduces reactive power flow and network loss and has a profound impact on the distribution network.

ACKNOWLEDGEMENTS

This study was supported by National Science Foundation of China (51207170) and Fundamental Research Funds for the Central Universities (14CX02172A).

REFERENCES

Feng Qian, Jianchao Zheng, Guangfu Tang & Zhiyuan He. New Approach to Determine Capacity of Dynamic Reactive Power Compensation Using Economic Voltage Difference. *Proceedings of the CSEE*. Jan. 2009. vol. 29, no. 1, pp. 1–6.

Jiaqiang Yang, Shilan Chen, Jie Zhu & Zheng Zeng. Control strategy designing of hybrid compensator based on APF and TSC. *Electric Machines and Control*. Jan. 2014. vol. 18, no. 1, pp. 11–18.

Kalla Ujjwal Kumar, Singh Bhim, Murthy S.S. Normalised adaptive linear element-based control of single-phase self excited induction generator feeding fluctuating loads. *Power Electronics, IET*. Aug. 2014. vol. 7, Iss. 8, pp. 2151–2160.

Linchuan Li, Jianyong Wang, Liyi Chen & Wennan Song. Optimal Reactive Power Planning of Electrical Power System. *Proceedings of the CSEE*. Feb. 1999. vol. 19, no. 2, pp. 66–69.

Qi Chen, Xiongwei Chen & Yupu Liu. Emulation and design of local low-voltage reactive power compensation in jerky and wavy load. *Electric Power Automation Equipment*. Apr. 2006. vol. 26, pp. 56–59.

Wei Zhao, An Luo, Jie Tang & Xia Deng. Hybrid Var Compensator Based on the Coordinated Operation of STATCOM and TSC. *Proceedings of the CSEE*. Jul. 2009. vol. 29, no. 19, pp. 92–98.

Zhikang Shuai, An Luo, Shen Z. John, Wenji Zhu, Zhipeng lv & Chuanping Wu. A Dynamic Hybrid Var Compensator and a Two-Level Collaborative Optimization Compensation Method. *Power Electronics, IEEE Transactions*. Sep. 2009. vol. 24, no. 9, pp. 2091–2100.

Advances in Power and Energy Engineering – Sun (Ed.)
© 2016 Taylor & Francis Group, London, ISBN 978-1-138-02846-3

Pole assignment design of the Power System Stabilizer by phase compensation method

C. Lv & T. Littler
Queen's University of Belfast, Northern Ireland, UK

W. Du
North China Electric Power University, Beijing, China

ABSTRACT: The small-signal angular stability problem caused by insufficient damping of low-frequency oscillations threatens the security and integrity of power grids. An effective way of suppressing oscillations has been through auxiliary controllers, which form Power System Stabilizers (PSSs), installed alongside generator excitation system to provide assisted damping. General PSS designs include the phase compensation and pole assignment methods. One accurate approach in this paper is proposed to set parameters of a PSS to achieve pole alignment using the phase compensation method. The required damping contribution of the PSS is calculated firstly from the design target obtained from the pole assignment method. Parameters of the PSS are set which satisfy the required damping contribution using the phase compensation method. Thus, oscillation modes are assigned accurately to the required position. This paper demonstrates that the proposed PSS method can effectively suppress identified low frequency oscillations and limit the impact on small-signal stability.

1 INTRODUCTION

Nowadays the small-signal angular stability problem caused by insufficient damping of low-frequency oscillations threatens the security and integrity of power grids when the power system is developing towards large or even super large interconnected grids. The function of a PSS is to extend the angular stability limits of a power system by providing supplemental damping to the oscillation of synchronous machines rotors through the generator excitation system. This damping is provided by the electric torque which is applied to the rotor. Furthermore, it is in phase with the speed variation (F.P. Dmello 1969a, b, C. Concordia & W.G. Heffron 1952a, b, R.A. Phillips & IEEE Committee Report 1981).

General PSS designs include the phase compensation and pole assignment methods. The phase compensation method is an approach based on the physical significance of the classical control theory which is easily understood and implemented in the field, and it has been widely used in the engineering practice (E.V. Larsen & D.A. Swann (1981)). The basic structure of a PSS is a 'lead-lag' transfer function. The output signal of the PSS is a voltage signal which is added as an input signal to the exciter. Since the PSS must produce a component of electric torque in phase with the speed deviation,

phase lead blocks circuits are used to compensate for the lag between the PSS output and the control action, the electric torque (E.V. Larsen & D.A. Swann 1981). The PSS gain is an important factor as the damping provided by the PSS increases in proportion to the gain of the PSS up to a certain critical gain value. The stabilizer frequency characteristic is adjusted by varying the time constants. It should be noted that the stabilizer transfer function includes the effect of both the input signal transducer and other filtering required to attenuate the stabilizer gain at generator shaft torsional frequencies. These effects must be considered in addition to the transfer function for which the stabilizer must compensate (E.V. Larsen & D.A. Swann 1981).

Pole assignment is another method for designing a state feedback controller on the basis of Decentralized Modal Control (DMC) principle. The stabilizer parameters are computed directly by solving a set of algebraic equations obtained from the exact assignment of several eigenvalues associated with the poorly damped electromechanical modes to preselected locations (C.L. Chen & Y.Y. Hsu 1987). Hence, it is more efficiently than the iterative algorithms used in previous work (Y.N. Yu 1983).

In this paper, the theoretical foundation of Damping Torque Analysis (DTA) is introduced and applied to a Single-Machine Infinite-Bus

(SMIB) power system to examine the effect of excitation control on power system small-signal stability under the complex frequency domain. One accurate approach is proposed to set parameters of a PSS to achieve pole assignment using the phase compensation method. The required damping contribution of the PSS is calculated firstly from the design target of the pole assignment method. Then, parameters of the PSS are formed which satisfy the required damping contribution using the phase compensation method. Thus, the oscillation modes are assigned accurately to the required position. The paper demonstrates that the proposed PSS method can effectively suppress identified low frequency oscillations and limit its impact on small-signal stability.

2 MODELLING OF POWER SYSTEM INSTALLED WITH PSS

2.1 *The Phillips-Heffron model of Single-Machine Infinite-Bus power system installed with PSS*

The Phillips-Heffron model of a Single-Machine Infinite-Bus (SMIB) power system has been successfully used for the analysis and design of the excitation system and Power System Stabiliser (PSS) for decades.

$$\begin{cases} \Delta\dot{\delta} = \omega_o \Delta\omega \\ \Delta\dot{\omega} = \dfrac{1}{M}(-K_1\Delta\delta - K_2\Delta E_q' - D\Delta\omega) \\ \Delta\dot{E}_q' = \dfrac{1}{T_{do}'}(-K_3\Delta E_q' - K_4\Delta\delta + \Delta E_{fd}') \\ \Delta\dot{E}_{fd}' = -\dfrac{1}{T_A}\Delta E_{fd}' - \dfrac{K_A}{T_A}(K_5\Delta\delta + K_6\Delta E_q' - \Delta u_{pss}) \end{cases}$$
(1)

Equation (1) is the so-called Phillips-Heffron model of the single-machine infinite-bus power system, which can be shown by Figure 1.

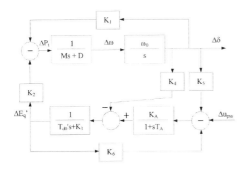

Figure 1. Phillips-Heffron model of a SMIB power system.

The Phillips-Heffron model can be written in the form of state-space representation:

$$\dot{X} = AX + b\Delta u_{pss}.$$
(2)

where

$$X = \begin{bmatrix} \Delta\delta \\ \Delta\omega \\ \Delta E_q' \\ \Delta E_{fd}' \end{bmatrix},$$

$$A = \begin{bmatrix} 0 & \omega_o & 0 & 0 \\ -\dfrac{K_1}{M} & -\dfrac{D}{M} & -\dfrac{K_2}{M} & 0 \\ -\dfrac{K_4}{T_{do}'} & 0 & -\dfrac{K_3}{T_{do}'} & \dfrac{1}{T_{do}'} \\ -\dfrac{K_A K_5}{T_A} & 0 & -\dfrac{K_A K_6}{T_A} & -\dfrac{1}{T_A} \end{bmatrix}, b = \begin{bmatrix} 0 \\ 0 \\ 0 \\ -\dfrac{K_A}{T_A} \end{bmatrix}$$

The oscillation model is $\lambda_g = \xi_g \pm j\omega_g$ in power system without a PSS.

2.2 *The principle of damping torque analysis in complex frequency domain*

The theoretical foundation of damping torque analysis is the classical control theory under complex frequency domain. From Figure 1 without considering the impact of power system stabilizer ($\Delta u_{pss} = 0$), it can be seen:

$$(Ms^2 + Ds + \omega_0 K_1)\Delta\delta(s) = -\omega_0\Delta T(s).$$
(3)

$$\Delta T(s) = F_{delta}(s)\Delta\delta(s).$$
(4)

where $F_{delta}(s)$ is transfer function from $\Delta\delta(s)$ to $\Delta T(s)$.

Combine above two equations, it can have:

$$[Ms^2 + Ds + \omega_0 K_1 + \omega_0 F_{delta}(s)]\Delta\delta(s) = 0.$$
(5)

Then the characteristic equation of the system can be obtained:

$$Ms^2 + Ds + \omega_0 K_1 + \omega_0 F_{delta}(s) = 0.$$
(6)

The solution of Equation (6) is the eigenvalues of matrix A, and only one pair of conjugate eigenvalues which is closely related with the electromechanical oscillation loop is defined as electromechanical oscillation mode, denoted by $\lambda_s = \xi_s \pm j\omega_s$. The real part of the electromechanical oscillation mode determines the damping of low frequency oscillation that is the oscillation stability of the system.

For $\lambda_s = \xi_s + j\omega_s$, in the complex frequency domain, the equations are rewritten as:

$$M\lambda_s^2 + D\lambda_s + \omega_0 K_1 + \omega_0 F_{delta}(\lambda_s) = 0. \quad (7)$$

Equation (4) becomes:

$$\Delta T(\lambda_s) = F_{delta}(\lambda_s)\Delta\delta(\lambda_s). \quad (8)$$

And from the first equation in Equation (1), it can have:

$$\Delta\omega(\lambda_s) = \frac{\xi_s + j\omega_s}{\omega_0}\Delta\delta(\lambda_s)$$
$$= \frac{\xi_s}{\omega_0}\Delta\delta(\lambda_s) + j\frac{\omega_s}{\omega_0}\Delta\delta(\lambda_s). \quad (9)$$

Assuming that:

$$\Delta T(\lambda_s) = T_s\Delta\delta(\lambda_s) + T_d\Delta\omega(\lambda_s). \quad (10)$$

From Equation (8), Equation (9) and Equation (10), it can be obtained that:

$$F_{delta}(\lambda_s)\Delta\delta(\lambda_s) = T_s\Delta\delta(\lambda_s) + T_d\frac{\xi_s}{\omega_0}\Delta\delta(\lambda_s)$$
$$+ jT_d\frac{\omega_s}{\omega_0}\Delta\delta(\lambda_s) \quad (11)$$

Thus

$$F_{delta}(\lambda_s) = T_s + T_d\frac{\xi_s}{\omega_0} + jT_d\frac{\omega_s}{\omega_0}. \quad (12)$$

where

$$\begin{cases} T_d = \dfrac{\omega_0}{\omega_s}\mathrm{Im}[F_{delta}(\lambda_s)] \\ T_s = \mathrm{Re}[F_{delta}(\lambda_s)] - \dfrac{T_d\xi_s}{\omega_0} \end{cases} \quad (13)$$

By substituting Equation (10) into Equation (1), it can have:

$$(M\lambda_s^2 + D\lambda_s + \omega_0 K_1)\Delta\delta(\lambda_s)$$
$$= -\omega_0 T_s\Delta\delta(\lambda_s) - \omega_0 T_d\Delta\omega(\lambda_s) \quad (14)$$
$$= -\omega_0 T_s\Delta\delta(\lambda_s) - T_d\lambda_s\Delta\delta(\lambda_s).$$

And then

$$M\lambda_s^2 + (D + T_d)\lambda_s + \omega_0 K_1 + \omega_0 T_s = 0. \quad (15)$$

The solution of Equation (15) is:

$$\xi_s = -\frac{D + T_d}{2M}. \quad (16)$$

2.3 Design PSS of pole assignment by phase compensation method

The oscillation mode without PSS is $\lambda_g = \xi_g + j\omega_g$, which is moved to $\lambda_s = \xi_s + j\omega_s$ in power system installed PSS based on the pole assignment method. Thus λ_s is the closed-loop oscillation mode in power system.

If the PSS takes the rotor speed of the generator as the feedback signal, the transfer function will be $G_{pss}(s)$.

The electric torque contribution from the PSS to the electromechanical oscillation loop of the generator is:

$$\Delta T_{pss} = F_{pss}(\lambda_s)G_{pss}(\lambda_s)\Delta\omega$$
$$= F_{pss}(\lambda_s)G_{pss}(\lambda_s)\frac{\lambda_s}{\omega_0}\Delta\delta$$

From Equation (6), it can have:

$$M\lambda_s^2 + D\lambda_s + \omega_0 K_1 + \omega_0 F_{delta}(\lambda_s)$$
$$+ \omega_0 F_{pss}(\lambda_s)G_{pss}(\lambda_s)\frac{\lambda_s}{\omega_0} = 0$$

From Equation (13) T_d and T_s can be calculated:

$$M\lambda_s^2 + (D + T_d)\lambda_s + \omega_0 K_1 + \omega_0 T_s$$
$$+ F_{pss}(\lambda_s)G_{pss}(\lambda_s)\lambda_s = 0 \quad (17)$$

Design the parameters of PSS, that is:

$$F_{pss}(\lambda_s)G_{pss}(\lambda_s) = D_{pss}. \quad (18)$$

From Equation (17) and Equation (18), it can have:

$$M\lambda_s^2 + (D + T_d + D_{pss})\lambda_s + \omega_0 K_1 + \omega_0 T_s = 0. \quad (19)$$

The solution of above equation is:

$$\xi_s = -\frac{D + T_d + D_{pss}}{2M}. \quad (20)$$

D_{pss} will be obtained from this solution.

In order to have the maximum efficiency from PSS design, ideally the PSS provides only damping torque, that is:

$$\Delta T_{pss} = D_{pss}\Delta\omega. \quad (21)$$

319

where D_{pss} is the coefficient of the damping torque.

Design of the PSS based on the phase compensation method should satisfy that:

$$D_{pss} = F_{pss}(j\omega_g)G_{pss}(j\omega_g). \qquad (22)$$

According to Equation (22), design of the PSS should set the phase of PSS, to be equal to the minus phase of the forward path, that $-\angle G_{pss}(j\omega_g)$.

The phase of the PSS should be set as the negative value of the phase of the forward path which is the $\angle G_{pss}(j\omega_g)$. in the design of the PSS.

If it is denoted that:

$$F_{pss}(j\omega_g) = |F_{pss}| \angle\phi, \quad G_{pss}(j\omega_g) = |G_{pss}| \angle\gamma. \qquad (23)$$

The phase compensation method requires:

$$\begin{cases} T_{pssd} = |F_{pss}G_{pss}|\cos(\phi+\gamma) = D_{pss} \\ T_{psss} = |F_{pss}G_{pss}|\sin(\phi+\gamma) = 0 \end{cases}. \qquad (24)$$

This can be achieved by setting:

$$\gamma = -\phi, \quad |G_{pss}| = \frac{D_{pss}}{|F_{pss}|}. \qquad (25)$$

Often the PSS is constructed to be a lead-lag network circuit with the main part of the transfer function to be:

$$\begin{aligned} G_{pss}(s) &= K_{pss}\frac{(1+sT_2)}{(1+sT_1)}\frac{(1+sT_4)}{(1+sT_3)} \\ &= K_{pss1}\frac{(1+sT_2)}{(1+sT_1)}K_{pss2}\frac{(1+sT_4)}{(1+sT_3)}. \end{aligned} \qquad (26)$$

where $K_{pss} = K_{pss1}K_{pss2}$. The parameters of PSS can be set to satisfy the conditions in phase compensation method for the PSS to provide a positive damping torque:

$$\begin{cases} K_{pss1}\dfrac{(1+j\omega_g T_2)}{(1+j\omega_g T_1)} = \dfrac{D_{pss}}{|F_{pss}|}\angle-\dfrac{\phi}{2} \\ K_{pss2}\dfrac{(1+j\omega_g T_4)}{(1+j\omega_g T_3)} = 1.0\angle-\dfrac{\phi}{2} \end{cases}. \qquad (27)$$

The state equation of the SMIB system integrated with a PSS is:

$$\begin{bmatrix} \Delta\dot{\delta} \\ \Delta\dot{\omega} \\ \Delta\dot{E}_q{}' \\ \Delta\dot{E}_{fd}{}' \end{bmatrix} = \begin{bmatrix} 0 & \omega_o & 0 & 0 \\ -\dfrac{K_1}{M} & -\dfrac{D}{M} & -\dfrac{K_2}{M} & 0 \\ -\dfrac{K_4}{T_{do}{}'} & 0 & -\dfrac{K_3}{T_{do}{}'} & \dfrac{1}{T_{do}{}'} \\ -\dfrac{K_A K_5}{T_A} & 0 & -\dfrac{K_A K_6}{T_A} & -\dfrac{1}{T_A} \end{bmatrix}$$

$$\times \begin{bmatrix} \Delta\delta \\ \Delta\omega \\ \Delta E_q{}' \\ \Delta E_{fd}{}' \end{bmatrix} + \begin{bmatrix} 0 \\ 0 \\ 0 \\ -\dfrac{K_A}{T_A} \end{bmatrix} \Delta u_{pss}.$$

$$\Delta u_{pss} = G_{pss}(s)\Delta\omega = K_{pss1}\frac{(1+sT_2)}{(1+sT_1)}K_{pss2}\frac{(1+sT_4)}{(1+sT_3)}\Delta\omega \qquad (28)$$

And

$$\Delta x_1 = K_{pss2}\frac{(1+sT_4)}{(1+sT_3)}\Delta\omega. \qquad (29)$$

$$\Delta u_{pss} = K_{pss1}\frac{(1+sT_2)}{(1+sT_1)}\Delta x_1. \qquad (30)$$

Thus, the closed-loop state equations of the power system are:

$$\begin{bmatrix} \Delta\dot{\delta} \\ \Delta\dot{\omega} \\ \Delta\dot{E}_q{}' \\ \Delta\dot{E}_{fd}{}' \\ \Delta\dot{x}_1 \\ \Delta\dot{u}_{pss} \end{bmatrix} = A_c \begin{bmatrix} \Delta\delta \\ \Delta\omega \\ \Delta E_q{}' \\ \Delta E_{fd}{}' \\ \Delta x_1 \\ \Delta u_{pss} \end{bmatrix}. \qquad (31)$$

where A_c is the closed-loop matrix of power system installed with a PSS, and the oscillation mode is moved to $\lambda_s = \xi_s + j\omega_s$ in power system installed with the PSS.

3 CASE STUDY

Table 1 gives the oscillation modes of power system with/without PSS installed.

Table 1. Oscillation mode with/without PSS.

	Oscillation mode
Without PSS	$-0.0785 + j7.0176$
With PSS	$-0.7100 + j6.9826$

320

The oscillation mode without a PSS installed is $\lambda_g = \xi_g + j\omega_g$, which is moved to $\lambda_s = \xi_s + j\omega_s$ in power system installed with PSS.

From Equation (13), it can have:

$$\begin{cases} T_d = \dfrac{\omega_0}{\omega_s} \text{Im}[F_{delta}(\lambda_s)] = 1.2175 \\ T_s = \text{Re}[F_{delta}(\lambda_s)] - \dfrac{T_d \xi_s}{\omega_0} = -0.0140 \end{cases}$$

From Equation (20), it can have:

$$\xi_s = -\frac{D + T_d + D_{pss}}{2M} = -0.7100. \quad (32)$$

Thus, $D_{pss} = 8.7244$.

According to Equation (27) of designing the parameters of PSS, the transfer function is:

$$G_{pss}(s) = K_{pss1} \frac{(1 + sT_2)}{(1 + sT_1)} K_{pss2} \frac{(1 + sT_4)}{(1 + sT_3)}$$

where $T_1 = 0.09s$, $T_3 = 0.09s$

For

$$G_{pss}(\lambda_s) = \frac{D_{pss}}{F_{pss}(\lambda_s)} = 2.1542 + j7.2392$$

If

$$\begin{cases} K_{pss1} \dfrac{(1 + \lambda_s T_2)}{(1 + j\lambda_s T_1)} = 7.5529\angle 36.7142° \\ K_{pss2} \dfrac{(1 + \lambda_s T_4)}{(1 + \lambda_s T_3)} = 1.0\angle 36.7142° \end{cases}$$

The parameters of PSS can be set as follows:

$K_{pss} = K_{pss1} K_{pss2} = 1.3582$,
$T_2 = 0.3710s$, $T_4 = 0.3710s$

From Equation (28) and Equation (31), it can have the eigenvalues of matrix A_c as:

$\lambda_1 = -19.8998$, $\bar{\lambda}_{2,3} = -0.7100 \pm j6.9826$,
$\bar{\lambda}_{4,5} = -7.5356 \pm j6.8701$, $\lambda_6 = -5.9221$

Thus, the oscillation mode is successfully moved to:

$$\bar{\lambda}_s = \xi_s + j\omega_s = -0.7100 \pm j6.9826.$$

Figure 2 shows the simulation result of oscillation modes of the power system (with/without

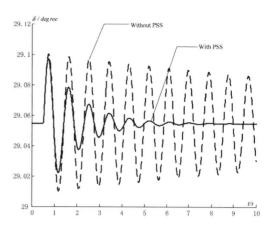

Figure 2. Simulation result of oscillation mode in power system.

PSS). It can be seen that the damping torque is positive in power system installed with the PSS. Hence, the stabilizers can effectively suppress identified low frequency oscillations and limit its impact on small-signal stability.

4 CONCLUSION

This paper proposes an accurate method of PSS design to achieve pole assignment using the phase compensation method. A linearized Phillips-Heffron model of a power system installed with a power system stabilizer is established. On the basis of damping torque analysis principle, the required damping coefficient is calculated by close-loop design of the pole assignment method. Then, parameters of the PSS are formed which satisfy the required damping contribution condition using the phase compensation method. As a consequence, the electromechanical oscillation modes are assigned accurately to the required position. The proposed PSS design method can effectively suppress identified low frequency oscillation and limit its impact on small-signal stability in power system.

ACKNOWLEDGEMENTS

The author would like to acknowledge the financial support of the Science Bridge Project funded by the UKRC and the Queen's University of Belfast, UK, and the National Basic Research Program of China (973Program) (2012CB215204).

REFERENCES

Chen, C.L., Y.Y. Hsu. 1987. Coordinated synthesis of multi-machine power system stabilizer using an efficient decentralized modal control algorithm. *IEEE Trans. Power Syst.*, 2(3): 543–550.

Dmello F.P., & C. Concordia. 1969. Concept of synchronous machine stability as affected by excitation control. *IEEE Trans. Power App. Syst.*, 88(4): 316–329.

Heffron, W.G., R.A. Phillips. 1952. Effect of modern amplidyne voltage regulators on under excited operation of large turbine generators. *AIEE Trans. Power App Syst.*, 71: 692–697.

IEEE Committee Report. 1981. Excitation system models for power system stability studies. *IEEE Trans. Power App. Syst.*, 100(1): 494–509.

Larsen, E.V., D.A. Swann. 1981. Applying power system stabilizers Part I: General concepts. *IEEE Trans. Power App. Syst.*, 100(6): 3017–3024.

Larsen, E.V., D.A. Swann. 1981. Applying power system stabilizers Part II: Performance objectives and tuning concepts. *IEEE Trans. Power App. Syst.*, 100(6): 3025–3033.

Larsen, E.V., D.A. Swann. 1981. Applying power system stabilizers Part III: Practical considerations. *IEEE Trans. Power App. Syst.*, 100(6): 3034–3046.

Yu. Y.N., 1983. Electric power system dynamics. *Academic Press*, New York.

APPENDIX

Parameters of example single-machine infinite-bus power system (in per unit except indicated):

Generator

$$X_d = 1.18, \ X_q = 1.0, \ X_{ad} = 1.0, \ X_f = 1.13,$$

$$X_d' = X_d - \frac{X_{ad}^2}{X_f} = 0.2951,$$

$$M = 7, \ D = 0, \ T_{d0}' = 5.044s$$

Transmission line

$$X_t = 0.3$$

AVR

$$K_A = 50, \ T_A = 0.05$$

Steady-state operating point

$$P_{t0} = 0.5, \ V_{t0} = 1.05, \ V_{b0} = 1.0.$$

Advances in Power and Energy Engineering – Sun (Ed.)
© 2016 Taylor & Francis Group, London, ISBN 978-1-138-02846-3

Stochastic stability of power system with asynchronous wind turbines

B. Yuan
State Grid Energy Research Institute, Beijing, China

J. Zong
State Grid Jibei Electric Power Co. Ltd., Research Institute, Beijing, China

ABSTRACT: This paper presents an approach to analyze the impact of random state matrix coefficients brought by wind power on power system stability. The system is modeled as a time-variant system and its stability is analyzed through stochastic Lyapunov function based on stochastic differential equation theory. A simple sufficient condition is obtained through mathematical deduction to ensure the mean-square stability of such a system. The numerical simulation result shows the conciseness and effectiveness of the proposed method.

1 INTRODUCTION

The stochastic nature of wind power has brought significant impact on power system stability, especially small signal stability. The dynamic systems are driven not only by operator's control inputs, but also random disturbances which cannot be modeled deterministically. The traditional approach to power system small signal stability is based on ordinary differential equations (Su et al. 2012, Du & Su 2011, Wang et al. 2011). However, such a deterministic analysis does not provide a realistic evaluation of system performance where the intermittency and variability of energy production associated with wind power that needs to be reflected and accurately modeled in system stability and performance assessments (Milano & Zarate 2013).

In this paper, the system with random variations is modeled by a set of stochastic differential equations with time-varying interval parameters. The stochastic mean-square stability for such a system is studied. Applying the decomposition technique of time-varying interval matrix and stochastic Lyapunov function based on Ito stochastic differential equation, a simple sufficient condition is obtained to ensure the stochastic mean-square stability of time-varying interval stochastic linear systems. A numerical simulation shows the effectiveness of the presented results.

2 PRELIMINARY

In this section, we will first give the definition of stochastic mean-square stability. Considering a stochastic system with time-varying interval parameters such as.

$$dX(t) = A_t X(t)dt + CX(t)dW(t), X(t_0) = X_0 \quad (1)$$

where $A_t = (a_{ij}(t))_{n \times n}, a_{ij}(t) \in [\underline{a}_{ij}(t), \bar{a}_{ij}(t)], \underline{a}_{ij}(t), \bar{a}_{ij}(t)$ are both bounded real function; $X(t) \in R^n$; $W(t)$ is a Wiener process; $E|X_0|^2 < \infty$. Let $e_{ij} = \inf_{t \geq 0} \underline{a}_{ij}(t), f_{ij} = \sup_{t \geq 0} \bar{a}_{ij}(t)$, and $E = (e_{ij})_{n \times n}, F = (f_{ij})_{n \times n}$; Then we define $D[\underline{A}(t), \bar{A}(t)]$ as:

$$D[\underline{A}(t), \bar{A}(t)] = \{A_t = (a_{ij}(t))_{n \times n}: \underline{a}_{ij}(t) \leq a_{ij}(t)$$
$$\leq \bar{a}_{ij}(t), i, j = 1, 2, ..., n\}$$

Then system (1) is called to be **stochastic mean-square stable** if $\forall A_t \in D[\underline{A}(t), \bar{A}(t)], t \geq t_0$ and initial condition $X(t_0)$, the solution $X(t)$ satisfies $\lim_{t \to \infty} E[X(t)X^T(t)] = 0$.

3 STOCHASTIC STABILITY ANALYSIS

In this section we are going to analyze the stochastic mean-square stability of a system like (1). According to Ito formula [5], the time-varying system (1) can be rewritten as a deterministic system.

$$\dot{M}_x(t) = A_t M_x(t) + M_x(t)A_t^T + CM_x(t)C^T, \quad (2)$$
$$M_x(t_0) = E\left(X_0 X_0^T\right)$$

where $M_x(t)$ is the mean square limit of $X(t)$. Apparently, the stochastic mean square stability of (1) can be equivalent to the mean stability of (2). Now we will try to deduce the mean square stability criterion for (2). First, we have

$$Vec(\dot{M}_x(t)) = (I_n \odot A_t + A_t \odot I_n + C \odot C)Vec(M_x(t))$$

$$(3)$$

where $Vec(A)$ is the column straightening of A, $\odot$ represents the Kronecker product of matrix. Let $Z = Vec(M_x(t))$, we have

$$\dot{Z} = NZ + (I_n \odot \Delta A_t + \Delta A_t \odot I_n)Z \qquad (4)$$

where $N = I_n \odot E + E \odot I_n + C \odot C$.

Here we need to choose a suitable Lyapunov function to evaluate its stability. The typically used function $V(Z(t),t) = Z T Z$ is adopted. Note that the Lyapunov function is deterministic and not stochastic. System (3) will be mean square stable if

$$\dot{V}(Z(t),t)\big|_{(3)} < 0$$

From (4) we have

$$\begin{aligned} \dot{V}(Z(t),t)\big|_{(3)} &= 2Z^T \dot{Z} = Z^T(N + N^T)Z \\ &\quad + 2Z^T(I_n \odot \Delta A_t + \Delta A_t \odot I_n)Z \\ &\leq Z^T(N + N^T)Z \\ &\quad + 2\big(\|I_n \odot \Delta A_t + \Delta A_t \odot I_n\|_2\big)Z^T Z \quad (5) \end{aligned}$$

From the property of matrix spectral norm, we can get

$$\begin{aligned} \|I_n \odot \Delta A_t + \Delta A_t \odot I_n\|_2 &= 2\max_i|\lambda_i(\Delta \dot{A}_t)| \\ &= 2\rho(\Delta A_t) \leq 2\rho(F - E) \end{aligned} \qquad (6)$$

where $\rho(A)$ is the spectral radius of matrix A, $\lambda(A)$ is the eigenvalue of matrix A.

Substitute (6) into (5), we have

$$\begin{aligned} \dot{V}(Z(t),t)\big|_{(3)} &\leq Z^T(N + N^T)Z + 2(2\rho(F - E))Z^T Z \\ &= Z^T(N + N^T + 2\mu I_{nn})Z < 0 \end{aligned}$$

$$(7)$$

Then we have **a sufficient condition to ensure the stochastic mean-square stability**:

Remark 1: System (1) is stochastic mean-square stable if $N + N^T + 2\mu I_{nn} < 0$; where $N = I_n \odot E + E \odot I_n + C \odot C$; $\mu = 2\rho(F - E)$; I_{nn} is a $n \times n$ unit matrix.

4 DYNAMIC MODEL OF POWER SYSTEM WITH RANDOM VARIATIONS BASED ON ITO SDE

In this section, based on the Ito integral and Ito SDE, a novel model is proposed to model the

dynamic characteristics of asynchronous wind turbines with stochastic mechanical power input.[6]

The asynchronous wind turbine is described as a The venin equivalent voltage source E′ behind the impedance $RS + jX'$ as shown in Figure 1. Assuming the generator is in a steady-state with the initial values of $\underline{E}'_0, s_0, \underline{I}_{s0}, P_0, U_0$, when small disturbance occurs, the system model is linearized around the equibrium point. The mechanical power input is modeled as a Wiener process; The second-order infinitesimal $\Delta \underline{E}' \Delta s$ and $\Delta \underline{E}' \Delta \underline{E}'^*$ are neglected; the the state equations can be expressed as a SDE with time-varying interval parameters:

$$\frac{d}{dt}\begin{bmatrix} \Delta E'_r \\ \Delta E'_m \\ \Delta s \end{bmatrix} = \begin{bmatrix} -K_7 & \omega_s s_0 + K_8 & \omega_s E'_{r0} \\ -\omega_s s_0 - K_8 & -K_7 & -\omega_s E'_{m0} \\ -\dfrac{K_9}{h} & -\dfrac{K_{10}}{h} & 0 \end{bmatrix}\begin{bmatrix} \Delta E'_r \\ \Delta E'_m \\ \Delta s \end{bmatrix}$$

$$+ \begin{bmatrix} 0 \\ 0 \\ 1/h \end{bmatrix}\vartheta B(t)$$

$$(8)$$

where E'_r and E'_m are the real and imaginary part of $\underline{E}'$.

$$K_7 + jK_8 = \frac{1}{T'_0}\left(1 + \frac{j(X_0 - X')}{R_S + jX' + \underline{Z}_{eq}}\right) \qquad (9)$$

$$K_9 = -G + \text{Re}\{\underline{I}_S(t)\} \qquad (10)$$

$$K_{10} = -B + \text{Im}\{\underline{I}_S(t)\} \qquad (11)$$

$$G + jB = \frac{E'_0}{R_S - jX' + \underline{Z}_{eq}^*}, h = 2H(1 - s_0) \qquad (12)$$

It can be seen that (8) is a stochastic differential equation with time-varying interval parameters.

Consider a two-machine infinite bus power system with a synchronous generator and an asynchronous wind turbine generator as shown in Figure 2.

The classic model of synchronous generator is deployed, then the system stochastic state equation can be expressed as

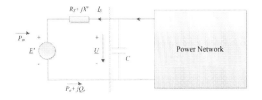

Figure 1. Dynamic model of the asynchronous wind turbine.

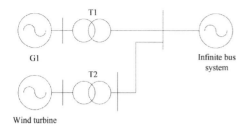

Figure 2. Two-machine infinite bus equivalent power system.

$$d\Delta X(t) = A_t \Delta X(t)dt + KdB_t \qquad (13)$$

where A_t is a time-varying matrix. Then we can know the stochastic mean-square stability of (13) by using *Remark 1*.

5 CASE STUDY

To validate the effectiveness of the proposed theory, simulation was carried out on the two-machine infinite bus power system as shown in Figure 2. The stochastic power system stability was analyzed both through the theoretical result in *Remark 1* and numerically using the Euler-Maruyama method (Oksendal 1994).

The parameters of the synchronous and asynchronous generators are drawn from other papers (Pidre 2003, Kundur 1994). Through calculation, we have

$$E = \begin{bmatrix} 0 & 342 & 0 & & & \\ -0.513 & 0 & -0.179 & & 0 & \\ -0.531 & 0 & -0.1905 & & & \\ & & & -16.392 & 3.454 & -67.387 \\ & 0 & & 3.454 & -16.392 & -198.32 \\ & & & 0.0905 & 0.6232 & -16.392 \end{bmatrix}$$

$$F = \begin{bmatrix} 0 & 342 & 0 & & & \\ -0.513 & 0 & -0.179 & & 0 & \\ -0.531 & 0 & -0.1905 & & & \\ & & & -15.287 & 4.122 & -66.266 \\ & 0 & & 3.96 & -15.287 & -179.42 \\ & & & 0.1205 & 0.7102 & -15.287 \end{bmatrix}$$

Then

$$F\text{-}E = \begin{bmatrix} 0 & 342 & 0 & & & \\ -0.513 & 0 & -0.179 & & 0 & \\ -0.531 & 0 & -0.1905 & & & \\ & & & -15.287 & 4.122 & -66.266 \\ & 0 & & 3.96 & -15.287 & -179.42 \\ & & & 0.1205 & 0.7102 & -15.287 \end{bmatrix}$$

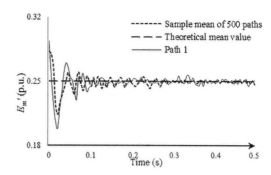

Figure 3. Response of asynchronous wind turbine under small disturbance.

From *Remark 1*, $N + N^T + 2\mu I_{nn} = -4.9356 < 0$, it can be seen that the system is stochastic mean-square stable.

To validate the result, simulation was also carried out using Euler-Maruyama method. The system response after a small disturbance occurs is shown in Figure 3.

We can see that when $t \rightarrow +\infty$, the fluctuation range is small. Therefore the system is stochastic mean-square stable, which is consistent with the theoretical result.

6 CONCLUSION

This paper develops an approach to analyze the impact of stochastic wind power generations on the small signal stability of a power system. The system is modeled as a set of stochastic differential equations with time-varying interval parameters. The definition of stochastic mean-square stability is presented and the system stochastic stability is explored through stochastic Lyapunov stability analysis. A simple sufficient condition is obtained to ensure the stochastic mean-square stability of such a system. Numerical simulations have validated the effectiveness of the proposed method.

REFERENCES

Du, W. & Su, B.Q. 2011. Effect of stochastic variation of grid-connected wind generation on power system small-signal probabilistic stability. Proceedings of the CSEE 31(12):7–11.

Kundur, P. 1994. Power System Stability and Control. New York: Mcgrwa-Hill: 33–34.

Milano, F. & Zarate, R. 2013. A systematic method to model power systems as stochastic differential algebraic equations. IEEE Trans. Power Systems 28(1):4537–4544.

Oksendal, B. 1994. Stochastic Differential Equations-An Introduction with Applications, 5th Edition. New York: Springer-Verlag Heidelberg: 167–168.

Pidre, J.C. 2003. Probabilistic model for mechanical power fluctuations in asynchronous wind parks. IEEE Trans. Power Systems 18(2):761–768.

Su, B.Q. & Du, W. 2012. Probabilistic analysis of small signal stability of large-scale power systems as affected by penetration of wind generation. IEEE Trans. Power Systems 27(1):762–770.

Wang, C. & Ni, Y.X. et al. 2010. Modeling analysis in power system small signal stability considering uncertainty of wind generation. Proceedings of the IEEE PES General Meeting 12(1):12–19.

Advances in Power and Energy Engineering – Sun (Ed.)
© 2016 Taylor & Francis Group, London, ISBN 978-1-138-02846-3

A novel control strategy for UPQC applied to large scale industrial enterprises

S. Liu
Beijing Sifang Automation Co. Ltd., Beijing, China

Y. Liang
School of Electrical and Electronic Engineering, Huazhong University of Science and Technology, Wuhan, China

J.Q. Wang, Y. Liu & L.M. Tu
Beijing Sifang Automation Co. Ltd., Beijing, China

J. Xiao, Y.C. Wang & C.X. Mao
School of Electrical and Electronic Engineering, Huazhong University of Science and Technology, Wuhan, China

ABSTRACT: This paper analyzes the characteristic of power supply and loads in large scale industrial enterprises. Three kinds of reasons qualitatively contribute to its power quality problems. They are imbalance of power supply or loads, impact of reactive or active power, harmonic of current. These factors will cause the load side PCC point voltage fluctuation and the source side current distortion. A power quality problem management scheme based on direct and sequence component independent control strategy for UPQC is proposed. The paper designs the detection and the control system in detail. A simulation model according to which is also built in Matlab/Simulink to test its performance. Simulation results show that the proposed scheme can guarantee the load side PCC point voltage stability and the supply side input current balance even in the condition of imbalance and nonlinear, which is a comprehensive scheme for managing power quality of large scale industrial enterprises.

1 INTRODUCTION

Power supply's reliability and quality is a precondition for safe, stable and efficient production of the large industrial enterprises. However, more and more complex power grid structure, nonlinear loads and impact loads, not only lead to the unbalanced power grid, voltage flicker, lower unity power factor and more other problems (Tu et al. 2008), but even extend to short circuit or open circuit failure, resulting in power outages. On the other hand, with the rapid increase of complexity and precision loads which are even more sensitive to power quality, even if the slight power problem can also interrupt production and bring a heavy loss to enterprises.

In view of the large industrial enterprises' actual working conditions, the paper analyzes the influence on load side and source side under the condition of unbalanced source voltage or unbalanced loads, operating impact loads and nonlinear loads. Results indicate that the actual working conditions of large industrial enterprises will not only bring about the load side voltage fluctuation of PCC point but the source side current distortion. The UPQC based

on direct control strategy (Tan et al. 2006) seems to be a suitable scheme. However, the conventional direct control strategy has problem to cope with unbalanced or nonlinear condition (M. & V. 2014, Peng et al. 2006), that is, getting either complicated control algorithm or poor control effect (Tan 2007). A novel scheme based on direct and sequence component independent control strategy for UPQC is proposed so as to synthetically manage power quality of large industrial enterprises. The paper designs the detection and control system in detail. A simulation model is also built according to this control strategy in Matlab/Simulink to verify its performance. Simulation results show that the proposed scheme can guarantee the voltage of load side PCC point stability, the input current of source side balance and the unity power factor 1.

2 POWER QUALITY IN LARGE INDUSTRY ENTERPRISES

The reason led to power quality problems in large industrial enterprises can be roughly divided into two categories. One is caused by source side like the short or open circuit fault occurred in power supply, the other is caused by load side like the operating impact or nonlinear loads.

2.1 Reasons lead to power quality problems

It is relatively simple to analyze the source side one. Because the power supply voltage harmonic is usually controlled within 5%, only symmetrical or asymmetrical voltage drop occurred with power failure contributes to power quality problem. As for load side, the paper takes electric arc furnace, large rolling mill and large induction motor as an example to analyze in detail.

Electric arc furnace uses electric arc produced by electrodes to smelt metal. Three aspects of reasons lead to power quality problem (Deepak & Ramesh 2014, Pedro et al. 2008). During melting period, three-phase current is changing sharply, resulting in serious fluctuation in active and reactive power. During the electrode regulation period, there are frequent short or open circuit between electrode and electrode or electrode and the furnace wall, resulting in serious imbalance of three-phase current. During the refining period, on account of the boiling metal, the equivalent arc length occurs periodical or quasi-periodical random fluctuation, leading to three-phase current serious distortion.

Large rolling mill, driven by cycloconverter system, operates on the main drive to complete steel rolling. It mainly causes two aspects of problems (Bao et al. 2010, He et al. 2011). While the mill is just rolling steel, owing to the big trigger angle of thyristor converter, the unity power factor decreases sharply, which makes for violently impacting on reactive power. Besides, cycloconverter system is usually constituted of three or twelve phase-bridge circuits based on thyristor. Because of its nonlinear, it produces harmonic current leading to power grid voltage distortion.

Large induction motor is the main load forms in large industrial enterprises (Petr et al. 2014). During its full-voltage start period, the starting current is 5 to 10 times higher as the steady-state one, which can suddenly cause voltage drop as it flows through the system impedance.

According to the analysis above, three kinds of reasons contribute qualitatively to power quality problem. They are imbalance of power supply or loads, impact of reactive or active power, harmonic current, respectively. The paper collectively refers to them as "flash power".

2.2 Flash power's influence on PCC point and source side

Simplified system is shown in Figure 1.

The system consists PCC point, power supply and loads. Where, $S_B = 100\ kVA$, $U_B = 381\ V$ are reference power and reference voltage, respectively. These three kinds of flash power are reflected in time axis $t = 0\sim5s$, successively. Then we use simulation tool to analyze.

2.2.1 The simplified flash power
① Imbalance of power supply

Imbalance of power supply in large industrial enterprises mainly presents as unbalanced sag or swell of voltage. Reflect it in time axis $t = 1\sim2s$ as shown in Table 1.
② Imbalance of loads

Imbalance of load mainly presents as asymmetric operation or fault. Reflect it in time axis $t = 2\sim3s$ as shown in Table 2.
③ Impact of reactive or active power

Impact of reactive or active power (Zhang et al. 2008) is reflected in time axis $t = 3\sim4s$ as shown in Table 3.
④ Nonlinear load

Nonlinear loads in large industrial enterprises is simplified into uncontrolled rectifier here. Resistive load $R = 2.75\ p.u.$ is in dc-side. Reflect it in time axis $t = 4\sim4.5s$.

2.2.2 Influence results of flash power on PCC point voltage and source side current
Use MATLAB/Simulink to analyze the influence of flash power on PCC point voltage and source side current.

① Voltage of PCC point

Figure 2 shows that both unbalanced power supply and impact loads constitute the primary influential factors. Meanwhile, unbalanced power supply is a factor as a direct result of PCC point voltage fluctuation but impact loads is a indirect one, which does so via system reactance.
② Source current

Separate source current of $t = 1\sim4s$ into positive, negative and zero sequence components and express them in dq coordinates, as shown in Figure 3, while source current of $t = 4\sim4.5s$ is expressed in ABC coordinate, as shown in Figure 4.

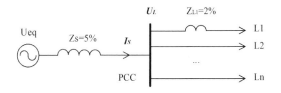

Figure 1. Simplified system.

Table 1. Imbalance of power, reflected in time axis $t = 1\text{-}2s$.

Time/s	1~1.2	1.2~1.4	1.4~1.5	1.5~1.7	1.7~1.9	1.9~2
Phase	A	BC	C	ABC	ABC	ABC
Flicker/ pu	0.9	0.85	1.1	1.15	0.9	1

Table 2. Imbalance of loads, reflected in time axis $t = 2\text{-}3s$.

Time/s		2~2.2	2.2~2.4	2.4~2.6	2.6~2.8	2.8~3
Asym type		A	BC	B	AC	ABC
Phase imp/pu		1	1	1	1	1
PF		0.62	0.62	0.62	0.62	0.62

Table 3. Impact of reactive or active power, reflected in time axis $t = 3\text{-}4s$.

Time/s	3~3.2	3.2~3.3	3.3~3.7	3.7~3.8	3.8~3.9	3.9~4
Fluct/pu	0.74	0.93	1.25	0.92	2.46	1.57
PF	0.45	0.45	0.53	0.61	0.47	0.53

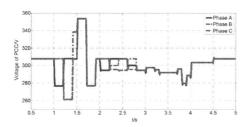

Figure 2. Voltage of PCC point.

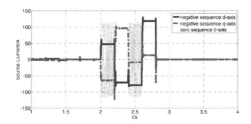

Figure 3. Source current of $t = 1\text{-}4s$.

The discussion above indicates that the flash power in large industrial enterprises in terms of its source can be divided into source-side and load-side. In terms of its formation can be divided into the imbalance of power supply or loads, impact

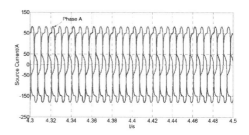

Figure 4. Source current of $t = 4\text{-}4.5s$.

of reactive or active power and harmonic current. When flash power is operating, not only leads to PCC point voltage fluctuation but imbalance and distortion of the source current. Hence, solving the power quality problems in large scale industrial enterprises is supposed to meet the above requirements. The scheme proposed should be a relatively comprehensive management solution.

3 UPQC SCHEME BASED ON DIRECT AND SEQUENCE COMPONENT INDEPENDENT CONTROL STRATEGY

Unified power quality controller UPQC has prominent advantages over synthetically managing power quality problems in both source-side and load-side. In general, series part of UPQC runs for voltage source, while shunt part runs for current source. The so-called indirect control method has significant computational cost in detection and control link (Mehdi & Saeed 2007). In addition, important load's continuous working needs to be supported by the shunt part during a power grid blackout, which requires switching control strategies, increasing the complexity of control algorithm. Direct control strategy is developed in (Li et al. 2007, Li 2006) to solve the problems above. Power quality problems in large industrial enterprises are even special as its PCC point voltage fluctuation and imbalance or distortion of source current according to the analysis before. However, the conventional direct control strategy is not appropriate for imbalance or distortion condition (Zhu et al. 2007) that is, getting either complicated control algorithm or poor control effect. A novel scheme based on direct and sequence component independent control strategy for UPQC is proposed so as to synthetically manage power quality of large industrial enterprises.

3.1 Mathematical model of UPQC

The topological structure of UPQC based on novel control strategy and its connected relation in system is illustrated in Figure 5.

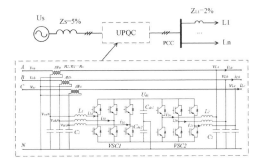

Figure 5. Topological structure of UPQC.

UPQC contains series part VSC1 and shunt part VSC2. On the basis of Figure 5. Mathematical model of UPQC in dq coordinate system is established as follows.

3.1.1 For VSC1

$$
\begin{cases}
\dfrac{di_{1d}}{dt} = \omega i_{1q} - \dfrac{R_1}{L_1}i_{1d} + \dfrac{v_{cd}}{L_1} - \dfrac{v_{dc}}{2L_1}d_{1d} \\[2mm]
\dfrac{di_{1q}}{dt} = -\omega i_{1d} - \dfrac{R_1}{L_1}i_{1q} + \dfrac{v_{cq}}{L_1} - \dfrac{v_{dc}}{2L_1}d_{1q} \\[2mm]
\dfrac{di_{1o}}{dt} = -\dfrac{R_1}{L_1}i_{1o} + \dfrac{v_{co}}{L_1} - \dfrac{v_{dc}}{2L_1}d_{1o} \\[2mm]
\dfrac{dv_{dc}}{dt} = \dfrac{d_{1d}}{C_{dc}}i_{1d} + \dfrac{d_{1q}}{C_{dc}}i_{1q} + \dfrac{1+d_{1o}}{C_{dc}}i_{1o} \\[2mm]
\qquad\quad -\dfrac{v_{dc}}{C_{dc}R_b} - \dfrac{2}{C_{dc}}i_{dc2} + \dfrac{2E_b}{C_{dc}R_b}
\end{cases}
\tag{1}
$$

3.1.2 For VSC2

$$
\begin{cases}
\dfrac{di_{2d}}{dt} = \omega i_{2q} - \dfrac{v_{Ld}}{L_2} + \dfrac{v_{dc}}{L_2}d_{2d} \\[2mm]
\dfrac{di_{2q}}{dt} = -\omega i_{2d} - \dfrac{v_{Ld}}{L_2} + \dfrac{v_{dc}}{L_2}d_{2q} \\[2mm]
\dfrac{di_{2o}}{dt} = -\dfrac{v_{Lo}}{L_2} + \dfrac{v_{dc}}{L_2}d_{2o} \\[2mm]
\dfrac{dv_{Ld}}{dt} = \omega v_{Lq} + \dfrac{i_{2d}}{C_2} - \dfrac{i_{Ld}}{C_2} \\[2mm]
\dfrac{dv_{Lq}}{dt} = -\omega v_{Ld} + \dfrac{i_{2q}}{C_2} - \dfrac{i_{Lq}}{C_2} \\[2mm]
\dfrac{dv_{Lo}}{dt} = \dfrac{i_{2o}}{C_2} - \dfrac{i_{Lo}}{C_2}
\end{cases}
\tag{2}
$$

where, d_k = derived from adopting state space averaging method is an approximation of switch function S_k. E_b, R_b = equivalent voltage and internal resistance of dc-link, respectively.

3.2 Separation of sequence components

There are positive, negative and zero sequence components when the system is unbalanced (Temporarily not consider harmonic). Electric quantity in the system can be expressed by:

$$
Sys = e^{j\omega t}X_{dq}^P + e^{-j\omega t}X_{dq}^N + x_0
\tag{3}
$$

where, $X_{dq}^P = x_d^P + x_q^P$ = positive dq-axis sequence components in synchronous rotating frame. $X_{dq}^N = x_d^N + x_q^N$ = negative dq-axis sequence components in reversal rotating frame. x_0 = zero sequence component in the system. According to equation (3), there is:

$$
x_\alpha + jx_\beta + x_0 = e^{j\omega t}X_{dq}^P + e^{-j\omega t}X_{dq}^N + x_0
\tag{4}
$$

where, x_α, x_β, x_0 = electric quantity in $\alpha\beta0$ coordinate.

Multiply equation (4) both sides by $e^{j\omega t}$ and $e^{-j\omega t}$. Positive and negative dq-axis sequence components can be consequentially calculated as:

$$
\begin{cases}
x_d^P = (x_\alpha \cos\omega t + x_\beta \sin\omega t) \\
\qquad -(x_d^N \cos 2\omega t + x_q^N \sin 2\omega t) \\
x_q^P = (x_\beta \cos\omega t - x_\alpha \sin\omega t) \\
\qquad -(x_q^N \cos 2\omega t - x_d^N \sin 2\omega t)
\end{cases}
\tag{5}
$$

$$
\begin{cases}
x_d^N = (x_\alpha \cos\omega t - x_\beta \sin\omega t) \\
\qquad -(x_d^P \cos 2\omega t - x_q^P \sin 2\omega t) \\
x_q^N = (x_\alpha \sin\omega t + x_\beta \cos\omega t) \\
\qquad -(x_q^P \cos 2\omega t + x_d^P \sin 2\omega t)
\end{cases}
\tag{6}
$$

It indicates that we can derive X_{dq}^P, X_{dq}^N from eliminating negative sequence component appearing as double frequency in synchronous rotating frame and eliminating positive sequence component appearing as double frequency in reversal rotating frame, respectively. Therefore, positive and negative sequence components can be extracted by notch filter, while zero sequence component is obtained though Park transformation.

3.3 Design of control system

3.3.1 Control system design of series part

The purpose of series part is controlling source current to be three-phase symmetrical, sinusoidal and power factor of source to be 1 in consideration that the power grid voltage is unbalanced or flickering. Based on positive, negative and zero sequence components obtained in detection part, the paper design corresponding three sequence controllers.

① Positive sequence controller

According to series part's mathematical model in *dq* coordinate system and positive sequence information obtained from detection part. When adopting the PI controller, positive sequence controller equation can be addressed as:

$$\begin{cases} v_{1d}^P = v_{cd}^P + \omega L_1 i_{1q}^P - [k_{ip}(i_{1d}^{P*} - i_{1d}^P) \\ \quad + k_{ii}\int(i_{1d}^{P*} - i_{1d}^P)], v_{1d}^P = d_{1d}^P \dfrac{v_{dc}}{2} \\ v_{1q}^P = v_{cq}^P - \omega L_1 i_{1d}^P - [k_{ip}(i_{1q}^{P*} - i_{1q}^P) \\ \quad + k_{ii}\int(i_{1q}^{P*} - i_{1q}^P)], v_{1q}^P = d_{1q}^P \dfrac{v_{dc}}{2} \end{cases} \quad (7)$$

where, $i_{1d}^{P*}, i_{1q}^{P'} = d, q$ axis positive sequence components reference signal of series converter input current. Meanwhile, $i_{1d}^{P*} =$ calculated by $i_{1d}^{P*} = \Delta I_1 + I_{1c}^*$. Where, $\Delta I_1 =$ an output of dc-link voltage regulator, which compensates UPQC for its active loss. $I_{1d}^* =$ input active current of source side, which is used for active power consumption of load. $i_{1q}^{P*} =$ the input reactive current of source side calculated by: $i_{1q}^{P*} = 0$.

② Negative sequence controller

Obviously, UPQC belongs to "static element". The series converter's mathematical model in the synchronous rotating coordinate is also appropriate for negative sequence controller in reversal rotating coordinate. Hence, the equation of series converter negative sequence controller can be described as:

$$\begin{cases} v_{1d}^N = v_{cd}^N + \omega L_1 i_{1q}^N - [k_{ip}(i_{1d}^{N*} - i_{1d}^N) \\ \quad + k_{ii}\int(i_{1d}^{N*} - i_{1d}^N)], v_{1d}^N = d_{1d}^N \dfrac{v_{dc}}{2} \\ v_{1q}^N = v_{cq}^N - \omega L_1 i_{1d}^N - [k_{ip}(i_{1q}^{N*} - i_{1q}^N) \\ \quad + k_{ii}\int(i_{1q}^{N*} - i_{1q}^N)], v_{1q}^N = d_{1q}^N \dfrac{v_{dc}}{2} \end{cases} \quad (8)$$

where, $i_{1d}^{N*}, i_{1q}^{N*} = d, q$ axis negative sequence components reference signal of series converter input current. To eliminate the influence of the imbalance grid voltage on control system, $i_{1d}^{N*}, i_{1q}^{N*} =$ calculated by: $i_{1d}^{N*} = 0, i_{1q}^{N*} = 0$.

③ Zero sequence controller

The zero sequence component in the synchronous rotating coordinate is independent of *dq*-axis. Therefore, its controller design is different from the positive or negative sequence one. Theoretically, only design simple PI controller can eliminate the zero sequence component appearing with the imbalance in power grid. As aforementioned, the simple PI control for zero sequence controller can be described as:

$$v_{1o} = v_{co} - [k_{ip}(i_{1o}^* - i_{1o}) + k_{ii}\int(i_{1o}^* - i_{1o})], v_{1o} = d_{1o}\dfrac{v_{dc}}{2} \quad (9)$$

where, $i_{1o}^* =$ the zero sequence component reference signal of source side input current. As the same analysis mentioned above, $i_{1o}^* =$ calculated by: $i_{1o}^* = 0$.

But notice that zero sequence in the synchronous rotating coordinate has the same frequency with fundamental wave. Simple PI controller designed above can't achieve tracking without steady-state error. Therefore, we add a resonant controller in parallel on the basis of PI controller, taking advantage of its infinite gain to the specific frequency of alternating signals to realize zero steady-state error tracking. The resonant controller's transfer function is:

$$G_r(s) = \dfrac{2k_r s}{s^2 + \omega_0^2} \quad (10)$$

where, $\omega_0 =$ the desired eliminating angular frequency of zero sequence component.

3.3.2 *Control system design of shunt part*

The purpose of shunt part is controlling PCC point voltage to be three-phase symmetrical, sinusoidal under the circumstances of existing unbalanced load and nonlinear load. Because the design for shunt part is similar to series part, a brief description is as follows.

① Positive sequence controller

$$\begin{cases} v_{2d}^P = \left[k_{ip}(i_{2d}^{P*} - i_{2d}^P) + k_{ii}\int(i_{2d}^{P*} - i_{2d}^P)\right] \\ \quad + v_{Ld}^P - \omega L_2 i_{2q}^P \\ i_{2d}^{P*} = \left[k_{vp}(v_{Ld}^{P*} - v_{Ld}^P) \\ \quad + k_{vi}\int(v_{Ld}^{P*} - v_{Ld}^P)\right] + i_{Ld}^P - \omega C_2 v_{Lq}^P \end{cases}, v_{2d}^P = d_{2d}^P v_{dc} \quad (11)$$

$$\begin{cases} v_{2q}^P = \left[k_{ip}(i_{2q}^{P*} - i_{2q}^P) + k_{ii}\int(i_{2q}^{P*} - i_{2q}^P)\right] \\ \quad + v_{Lq}^P + \omega L_2 i_{2d}^P \\ i_{2q}^{P*} = \left[k_{vp}(v_{Lq}^{P*} - v_{Lq}^P) \\ \quad + k_{vi}\int(v_{Lq}^{P*} - v_{Lq}^P)\right] + i_{Lq}^P + \omega C_2 v_{Ld}^P \end{cases}, v_{2q}^P = d_{2q}^P v_{dc} \quad (12)$$

where, $v_{Ld}^{P*} = 377.5 = d$-axis positive sequence component reference signal of shunt converter output voltage. $v_{Lq}^{P*} = 0 = q$-axis positive sequence component reference signal of shunt converter output voltage.

331

②. Negative sequence controller

$$
\begin{cases}
v_{2d}^N = \left[k_{ip}(i_{2d}^{N\,*} - i_{2d}^N) + k_{ii}\int(i_{2d}^{N\,*} - i_{2d}^N) \right] \\
\quad + v_{Ld}^N - \omega L_2 i_{2q}^N \\
i_{2d}^{N\,*} = \left[k_{vp}(v_{Ld}^{N\,*} - v_{Ld}^N) \right. \\
\quad \left. + k_{vi}\int(v_{Ld}^{N\,*} - v_{Ld}^N) \right] + i_{Ld}^N - \omega C_2 v_{Lq}^N
\end{cases}, \; v_{2d}^N = d_{2d}^N v_{dc}
$$

(13)

$$
\begin{cases}
v_{2q}^N = \left[k_{ip}(i_{2q}^{N\,*} - i_{2q}^N) + k_{ii}\int(i_{2q}^{N\,*} - i_{2q}^N) \right] \\
\quad + v_{Lq}^N + \omega L_2 i_{2d}^N \\
i_{2q}^{N\,*} = \left[k_{vp}(v_{Lq}^{N\,*} - v_{Lq}^N) \right. \\
\quad \left. + k_{vi}\int(v_{Lq}^{N\,*} - v_{Lq}^N) \right] + i_{Lq}^N + \omega C_2 v_{Ld}^N
\end{cases}, \; v_{2q}^N = d_{2q}^N v_{dc}
$$

(14)

where, $v_{Ld}^{N\,*} = 0 = d$-axis negative sequence component reference signal of shunt converter output voltage. $v_{Lq}^{N\,*} = 0 = q$-axis negative one.

③. Zero sequence controller

$$
\begin{cases}
v_{2o} = \left[k_{ip}(i_{2o}^* - i_{2o}) \right. \\
\quad \left. + k_{ii}\int(i_{2o}^* - i_{2o}) \right] + v_{Lo} \\
i_{2o}^* = \left[k_{vp}(v_{Lo}^* - v_{Lo}) \right. \\
\quad \left. + k_{vi}\int(v_{Lo}^* - v_{Lo}) \right] + i_{Lo}
\end{cases}, \; v_{2o} = d_{2o} v_{dc}
$$

(15)

where, $v_{Lo}^* = 0 =$ zero sequence component reference signal of shunt converter output voltage.

4 SIMULATION RESULTS

According to Figure 1, a simulation model is built in Matlab/Simulink, which adopts control strategies proposed in this paper. We verify the performance of this UPQC for solving load side PCC point voltage fluctuation and the source side current distortion under the circumstances of several flash power. Similarly, the system consists PCC point, power supply and load. Where, $S_B = 100 \; kVA$, $U_B = 381 \; V$ are reference power and reference voltage, respectively.

UPQC's main circuit parameters are listed in Table 4.

4.1 Flash power 1: imbalance of gird voltage

That is, $V_{SA} = 0.85$, $V_{SB} = V_{SC} = 1$, $Z_L = 1.86 + j2.35$.

4.2 Flash power 2: existing impact load

That is, $V_{SA} = V_{SB} = V_{SC} = 1$. Put impact load into operation at $t = 1s$ then quit operation at $t = 1.1s$. Meanwhile, impact current is 8 times large as the initial state one.

Table 4. Circuit parameters of the UPQC.

Circuit Parameters of UPQC	Value
Inductance of series part:L1/mH	5.8
Capacitance of series part:C1/uF	5
Inductance of shunt part:L2/mH	3.5
Capacitance of shunt part:C2/uF	250
Ratio of series transformer:Ns	3.5
Voltage of dc-link:Vdc1 = Vdc2/V	500
Capacitance of dc-link:Cdc1 = Cdc2/uF	24000
Switching frequency:fc/Hz	9000

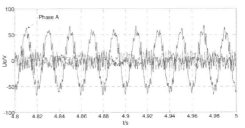

(a)Grid voltage drop compensation by series part

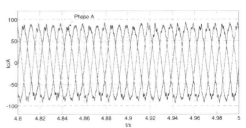

(b)Reactive current compensation by shunt part

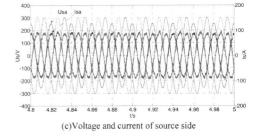

(c)Voltage and current of source side

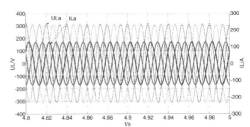

(d)Voltage of PCC point and current of load side

Figure 6. Simulation results of flash power 1.

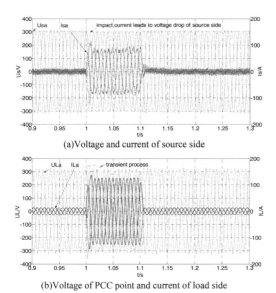

(a)Voltage and current of source side

(b)Voltage of PCC point and current of load side

Figure 7.   Simulation results of flash power 2.

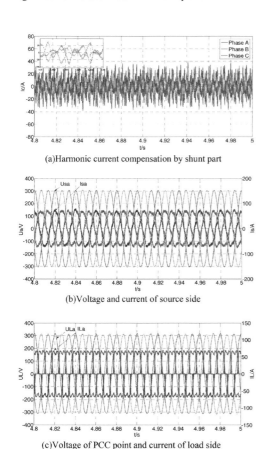

(a)Harmonic current compensation by shunt part

(b)Voltage and current of source side

(c)Voltage of PCC point and current of load side

Figure 8.   Simulation results of flash power 3.

4.3   Flash power 3: existing nonlinear load

That is, $V_{SA} = V_{SB} = V_{SC} = 1$. Load is three-phase uncontrolled rectifier, resistance in dc-link is pure resistive, $R = 0.36$.

Figures 6~8 illustrate the control effect of this UPQC for solving load side PCC point voltage fluctuation and the source side current distortion under the circumstances of several flash power. It demonstrates that the proposed scheme can guarantee the load side of PCC voltage stability, source side current balance and the unity power factor 1. More importantly, the results also prove that the control strategy based on direct and sequence component independent is so effective that can be especially applied in imbalance and nonlinear condition of the large industrial enterprises.

5   CONCLUSIONS

The actual working conditions of large industrial enterprises will not only bring about the load side PCC point voltage fluctuation but the power supply side current distortion. A novel scheme based on direct and sequence component independent control strategy for UPQC is proposed so as to synthetically manage power quality of large industrial enterprises. The paper designs the detection and control system of it in detail. A simulation model is also built in Matlab/Simulink to test its performance. Simulation results show that the proposed scheme can guarantee the load side of PCC voltage stability, supply side input current balance and the unity power factor 1. The novel control strategy for UPQC is a comprehensive scheme for managing power quality of large scale industrial enterprises.

REFERENCES

Bao Ming-hui, Xia Jun, Yin Xiao-gen & Dong Man-ling. 2010. Harmonic measurements and analysis in a modern steel manufacturing facility. *Power and Energy Society General Meeting, 1–6.*

Deepak C. Bhonsle & Ramesh B. Kelkar. 2014. New time domain electric arc furnace model for power quality study. In Kurukshetra, *IEEE India International Conf. Power Electronics, 1–6.*

He Chun, Zhao Gang & Li Tao. 2011. The improving study for scheme about high order harmonics produced by AC/DC/AC frequency alterable mill and application. *Power Electronics: 40–42.*

Li Peng, Bai Qian & Li Gengyin. 2006. Coordinated control Strategy for UPQC and its verification. In Montreal, *Power Engineering Society General Meeting, 2–8.*

Li Xun, Duan Shan-xu, Kang Yong & Chen Jian. 2007. Control Strategies for three-phase four-wire UPQC. *Automation of Electric Power System: 56–60.*

Li Xun. 2006. Analysis and control of unified power quality conditioner. Ph, D. Dissertation, Huazhong University of Science & Technology, Wuhan, China.

Mehdi Forghani & Saeed Afsharnia. 2007. Online wavelet transform-based control strategy for UPQC control system. *IEEE Transaction on Power Delivery: 481–491*.

Pedro E., Ferando & Gustavo A. 2008. Power quality measurements in a steel industry with electric arc furnaces. *Power Engineering Society General Meeting—Conversation and Delivery of Electrical Energy in the 21st Century, 1–8*.

Petr Orság, Stanislav Kocma & Jan Otýpka. 2014. Impact of mains power quality on operation characteristics of induction motor. In Krakow, *14th International Conf. Environment and Electrical Engineering, 328–385*.

Tan Zhi-li, Li Xun & Chen Jian. 2006. Direct control strategy of UPQC based on p-q-r theory. *Electric Power Automation Equipment: 12–17*.

Tan Zhi-li. 2007. Study of control strategy of three-phase four-wire UPQC used for compensating unbalance and distortion. Ph,D. Dissertation, Huazhong University of Science & Technology, Wuhan, China.

Tu Chun-ming, Pan Hong-bin, Tang Jie & Luo An. 2008. Design and Application of Power Quality Compensation System for Enterprise Distribution Network. *Power System Technology: 10–13*.

Qasim, M., & Khadkikar, V. 2014. ADALINE based control strategy for three-phase three wire UPQC system. In Bucharest, *16th IEEE International Conf. Harmonics and Quality of Power, 587–590*.

Zhang Ding-hua, Gui Wei-hua, Wang Wei-an & Liu Lian-gen. 2008. Comprehensive compensation system combining reactive power compensation and harmonic suppression for large-scale electric arc-furnace. *Power System Technology: 23–29*.

Zhu Peng-cheng, Li Xun, Kang Yong & Chen jian. 2007. Study of control strategy for a unified power quality conditioner. *Proceeding of the CSEE: 67–73*.

334

Advances in Power and Energy Engineering – Sun (Ed.)
© *2016 Taylor & Francis Group, London, ISBN 978-1-138-02846-3*

An active power regulation model for wind farms considering Static Voltage Stability margin

Y. Qi & X.D. Wang
State Grid Tianjin Electric Power Research Institute, Tianjin, China

N.S. Chen & X.J. Ge
China Electric Power Research Institute, Beijing, China

M.X. Liu
State Grid Tianjin Economic Research Institute, Tianjin, China

M.S. Wang
Key Laboratory of Smart Grid of Ministry of Education, Tianjin University, Tianjin, China

ABSTRACT: Considering the Static Voltage Stability (SVS) and economic operation of the power system, an active power regulation model for generators is developed for the Unit Commitment (UC) of the power system, integrated with a large scale of wind farms. It can be used for the economic dispatch of traditional generators to compensate the variation of wind farms. The Continuous Power Flow (CPF) method is utilized to determine the SVS margin of the power system. A Particle Swarm Optimization (PSO) algorithm is used to obtain the optimal regulation scheme for the active power of dispatchable generators. A modified IEEE 30-bus system integrated with wind farms is used to verify the proposed optimization model. The simulation results show that the traditional generators can respond to the variation of wind power with relatively high SVS margins and low generating costs with the developed active power optimal regulation model, which can increase the utilization of wind power to some extent.

1 INTRODUCTION

As one of the important renewable energy sources, wind power has the most mature technology and developed rapidly in recent years. Wind power is intermittent and non-schedulable, while providing clean energy. In addition, wind farms will absorb a considerable amount of reactive power from the system, while sending out active power. Due to the close correlation between voltage stability and reactive power, large-scale wind farms have a serious influence on the voltage stability of the power system.

In China, the cumulative installed capacity of wind power had reached 114 GW by 2014 (Chinese Wind Energy Association, 2015). But the uncertain and intermittent characteristics of wind power will increase the difficulty in Unit Commitment (UC) of the power system. Therefore, dispatchable generators, energy storage devices, or controllable loads will have to respond to the power variation of wind farms, to support the stable, and secure operation of the power system (Jiang et al. 2013, Wang et al. 2015).

This work was supported in part by the Research project of SGCC on the Control Technology for Energy Forecast and Coordination of Smart Grid Park. Several studies have been carried out to discuss active power regulation with the penetration of wind power. A real-time dispatching strategy of traditional generators is developed to regulate active power variation with the integration of wind power with large capacity (Xu et al. 2011). Active power regulation of wind farms is accomplished by controlling heat pumps and traditional generators, but the cost of generation and demand response has not been evaluated (Miao et al. 2012). The local boundaries of voltage stability regions in wind farms are obtained with the approach proposed in (Mu et al. 2010), which can help the active power regulation of generators. In (Mu et al. 2015), a directional control method considering the Static Voltage Stability (SVS) is developed to study the relationship between the generators and interface flow. However, the generating cost is not considered and the optimal regulation for the generators is not given in this study.

In this paper, the framework of an optimal regulation model responding to the variation of wind power is first introduced. Then the Continuous Power Flow (CPF) method is used to obtain the SVS margin of power system. Then the optimization model for active power regulation with the Particle Swarm Optimization (PSO) algorithm is developed. Then a test case of IEEE 30-bus system integrated with large scale of wind farms is utilized to verify the proposed active power regulation model.

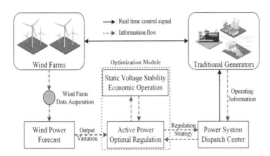

Figure 1. The framework of the active power regulation model.

2 FRAMEWORK OF THE REGULATION MODEL

The framework of the active power regulation model for UC of power system with wind farms is shown in Figure 1.

The power system dispatch center monitors the SVS margin of power system by using CPF method.

The active power regulation model is utilized in this paper to regulate the traditional generators for compensating the power output variation of wind farms based on the forecasted and real-time output of wind power.

From the power system dispatch center, the operating information of the generating system is obtained. Then combining the obtained information with the active power regulation model, the regulation scheme is determined considering both the SVS margin (from the CPF) and economic operation of power system.

3 SVS OF THE POWER SYSTEM WITH WIND FARMS

When the load demand of the conventional power system exceeds its transfer limit, voltage instability/collapse usually happens at the heavy-duty points of the receiving end. However, when large wind farms connect to the power grid, e.g., at the receiving end, the transmission mode of power system may be changed. Besides, wind farms usually absorb considerable amounts of reactive power from the system while sending out active power. Due to the close correlation between voltage stability and reactive power, large-scale wind farms have serious influence on the voltage stability.

Traditionally, the generation of each participating generator is regulated at a predefined rate or according to their spinning reserves for the active power regulation. However, it will be of significant importance to determine the optimum generation pattern, which can provide the best solution in

terms of the SVS margin. This will be beneficial for today's highly interconnected power system when a feasibility of scheduling generation is required to meet the SVS criteria.

$S_{GL} = S_G \cup S_L$ is used to represent the power injection space of power system, where $S_G = (P_{Gc} \cup Q_{Gc}) \cup (P_{Gw} \cup Q_{Gw})$ corresponds to the power injection vectors of the conventional generators and wind farms, respectively. $S_L = P_L \cup Q_L$ represents the load injection vector. Once S_{GL} is determined, the only stable operation state of power system x is obtained by the power flow Equation (1).

$$\begin{aligned} f(x, S_{GL}) &= 0 \\ g(x, S_{GL}) &\leq 0 \end{aligned} \tag{1}$$

where $f(\cdot)$ = power flow equation; $g(\cdot)$ = system operating constraint equation. When x yielded by Equation (1) also satisfies Equation (2), the system is said to be in voltage stability. While, if x also satisfies Equation (3), the system is at the critical status of voltage stability.

$$\det(f_x) \neq 0 \tag{2}$$

$$\det(f_x) = 0 \tag{3}$$

where f_x = Jacobian matrix of the power flow equation.

The generation and load pattern of S_{GL} can significantly impact the value of the SVS margin. In the proposed regulation model, the S_G is utilized as an independent parameter to regulate the power for compensating the variation of wind farms because the generation pattern is easier to control by the operators than by load. S_L is obtained by load forecasting and considered as a known element.

In the conventional power system, when load demand approaches or even exceeds its limit, the voltage instability or collapse happens at the heavy-duty point. The CPF algorithm can converge at the critical point of power flow, and it can be used to evaluate the SVS of power system (Rong et al. 2010). In order to analyze the voltage stability, a continuous parameter λ is introduced in the CPF, which indicates the SVS margin of power system. The CPF equation is given by Equation (4).

$$\begin{cases} P_{Gi}\left(1 + \lambda k_{Gi}\right) - P_{Li}\left(1 + \lambda k_{PLi}\right) \\ \quad -U_i \sum_{j \in i} U_j \left(G_{ij} \cos\theta_{ij} + B_{ij} \sin\theta_{ij}\right) \\ \quad = 0, \ i = 1, 2, \cdots, n \\ Q_{Gi} - Q_{Li}\left(1 + \lambda k_{QLi}\right) \\ \quad -U_i \sum_{j \in i} U_j \left(G_{ij} \sin\theta_{ij} - B_{ij} \cos\theta_{ij}\right) \\ \quad = 0, \ i = 1, 2, \cdots, n \end{cases} \tag{4}$$

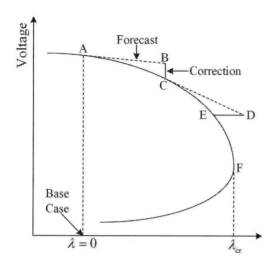

Figure 2. The P–V curve based on the SVS margin.

$$\min f = \lambda_{cr} \cdot \eta(\lambda)$$
$$+ \sum_{i=1}^{m} \left[a_i \cdot \left(P_{Gi0} + \Delta P_{Gi} \right)^2 + b_i \cdot \left(P_{Gi0} + \Delta P_{Gi} \right) + c_i \right],$$

$$(5)$$

where $\eta(\lambda)$ = a penalty function shown in Equation (6) and α = the penalty coefficient; m = the number of dispatchable generators; P_{Gi0} = the initial power output of generator i and ΔP_{Gi} = the power output variation of generator i; and a_i, b_i, and c_i = the coefficients of generating cost of generator i.

$$\eta(\lambda) = \begin{cases} \alpha \cdot \cos\left(\dfrac{\lambda}{\lambda_{cr}} \dfrac{\pi}{2} \right), & 0 < \lambda < \lambda_{cr} \\ 0 & , \lambda \geq \lambda_{cr} \end{cases} \qquad (6)$$

The constraints are subject to Equation (7).

$$\begin{cases} \sum_{i=1}^{m} \Delta P_{Gi} - \sum_{j=1}^{s} \Delta P_{Wj} = 0 \\ P_{Gi,\min} \leq P_{Gi0} + \Delta P_{Gi} \leq P_{Gi,\max} \end{cases} \qquad (7)$$

where s = the number of the integrated wind farms; ΔP_{Wj} = the power output variation of wind farm j; $\Delta P_{Gi,\min}$ and $\Delta P_{Gi,\max}$ = the lower and upper output limit of generator i; and ΔP_{Gi} = the control variables in this optimization.

In this optimization model, the power variation caused by wind farms will be compensated by the dispatchable generators in the power system. The PSO algorithm is used to obtain the minimum value of Equation (5) for active power optimal regulation of generators, which will contribute to the stable and economic operation of the power system.

where P_{Gi} = the active power injection of the generator at bus i; Q_G = the reactive power injection of the generator at bus i; P_{Li} = the active load power injection at bus i; Q_{Li} = the reactive load power injection at bus i; k_{Gi} = the active power growth direction of the generator at bus i; k_{PLi} = the active power growth direction of the load at bus i; k_{QLi} = the reactive power growth direction of the load at bus i; U_i = the voltage amplitude of bus i; θ_{ij} = the angle between bus i and j; $(G + jB)$ = the node admittance matrix J; and n = the bus number.

With the gradual growth of λ in Equation (1), the CPF starts from the base case A ($\lambda = 0$) to the next point C with the forecast AB and correction BC as the $P–V$ curve shown in Figure 2. Thus, the critical power flow point F and the SVS margin of λ_{cr} are obtained with a fixed load increasing and generation dispatch model (Chiang et al. 1995, 2009). The CPF method can be utilized to evaluate the SVS margin of power system with different regulation of generators.

4 THE ACTIVE POWER REGULATION MODEL

The SVS margin and the economic operation are the main factors considered in the active power optimal regulation model. The regulation model is a multi-objective issue consisting of the SVS margin (from the CPF) and the generating cost in order to get the optimal regulation scheme for generators with relatively low cost and high SVS of power system. The objective function of the optimization model is given by Equation (5):

5 CASE STUDY AND RESULTS

In this section, a modified IEEE 30-bus system (6 generator buses, 20 load buses, 1 wind farm, 41 branches, and 4 transformers) as shown in Figure 3 is used to verify the effectiveness of the proposed active power regulation model. Bus 1 is the slack bus, and a wind farm G_{W25} with an installed capacity of 100 MW is integrated at bus 25. We assume all the loads in the system are increased to 165%, and only G_{22}, G_{23}, and G_{27} (generators at bus 22, 23, and 27) are dispatchable. The power output ranges and generating cost coefficients of the three generators are given in Table 1.

To verify the effectiveness of the active power optimal model, it is defined that $\lambda_{cr} = 2.5$ and $\alpha = 1000$. It is assumed that the wind power output

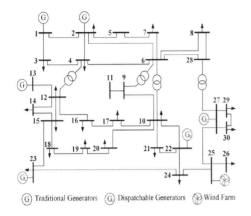

Figure 3. The IEEE 30-bus system.

Table 1. Power output range and generating cost coefficients.

G	P_{Gi} (MW)	$P_{Gi,min}$ (MW)	$P_{Gi,max}$ (MW)	a_i ($/ (MW²·hr))	b_i ($/ (MW·hr))	c_i ($/hr)
22	21.59	0	80	0.01750	1.75	0
23	19.20	0	80	0.00834	3.25	0
27	26.91	0	80	0.02500	3.00	0

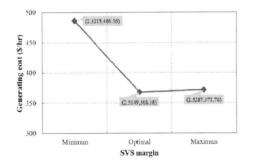

Figure 4. Generating cost under minimum, optimal, and maximum SVS margins.

of G_{W25} decreases from 100 MW to 0 MW. Based on the proposed optimization model, the active power optimal regulation for G_{22}, G_{23}, and G_{27} are obtained. A change to the objective function (only considering the value of λ_{cr}) gives the generating cost under minimum and maximum SVS margins as shown in Figure 4. For comparison, some defined regulation strategies for generators are also given in Table 2. It is clear that the SVS margin λ_{cr} is higher than 2.5 and the generating cost is relatively low with the active power optimal regulation.

6 CONCLUSION

In this paper, an active power optimal regulation model is proposed for generators' dispatching with the power variation of wind farms. The CPF method is used to evaluate the SVS of power system. The optimization model is a multi-objective issue with the SVS and generating cost, and a PSO algorithm is used to acquire the optimal regulation for generators. With the optimal regulation strategy, it is easy to see that the generators will operate at a low generating cost with relatively high SVS of power system. This optimal regulation model will contribute to the active power regulation for UC of power system, which will operate more stably and economically.

ACKNOWLEDGMENT

This work is supported in part by the research project of SGCC on Multi Energy interconnection and Management Techniques for Smart City (SGTJDK00DWJS1500100), and the research project of SGCC (Tianjin) on the Energy Forecast and Coordination Control Techniques of Smart Grid Park (SGTJDK00DWJS1500032).

Table 2. Optimal regulation and some dispatch strategies.

Dispatch strategy	P_{G22} (MW)	ΔP_{G22} (MW)	P_{G22} (MW)	ΔP_{G22} (MW)	P_{G27} (MW)	ΔP_{G27} (MW)	λ_{cr}	Generating cost ($/h)
1	54.92	33.33	52.53	33.33	60.24	33.33	2.4584	614.07
2	71.59	50.00	69.20	50.00	26.91	0.00	2.5087	578.64
3	71.59	50.00	19.20	0.00	76.91	50.00	2.4338	659.06
4	21.59	0.00	69.20	50.00	76.91	50.00	2.3277	689.39
5	80.00	58.41	60.79	41.59	26.91	0.00	2.5254	579.22
6	80.00	58.41	19.20	0.00	68.50	41.59	2.4713	640.28
7	60.79	39.20	80.00	60.80	26.91	0.00	2.4776	583.26
8	21.59	0.00	80.00	60.80	66.11	39.20	2.3469	666.91
9	68.50	46.91	19.20	0.00	80.00	53.09	2.4184	667.46
10	21.59	0.00	66.11	46.91	80.00	53.09	2.3218	697.25
Optimal	74.39	52.80	66.28	47.08	27.03	0.12	2.5149	578.43

REFERENCES

Chiang, H.D. & Jin, L.C. 2009. Matthew Varghese and so on. Linear and nonlinear methods for contingency analysis in on-line voltage security assessments. *Proceedings of 2009 IEEE Power and Energy Society General Meeting*, Calgary, AB, Canada, 1–6.

Chiang, H.D. et al. 1995. CPFLOW: a practical tool for tracing power system steady-state stationary behavior due to load and generation variations. *IEEE Transactions on Power System*, 10(2): 623–634.

Chinese Wind Energy Association (CWEA), 2015, *The installed capacity of wind power in China by 2014 (in Chinese)*.

Jiang, Q.Y. & Hong, H.S. 2013. Wavelet-based capacity configuration and coordinated control of hybrid energy storage system for smoothing out wind power fluctuations. *IEEE Transactions on Power Systems*, 28(2): 176–184.

Miao W.W. et al. 2012. Active power regulation of wind power systems through demand response. *Science China Technological Science*, 55(6): 1667–1676.

Mu, Y.F. & Jia, H.J. 2010. An approach to determine the local boundaries of voltage stability region with wind farms in power injection space. *Science China Technological Science*, 53(12): 3232–3240.

Mu, Y.F. et al. 2015. A directional control method for interface flow considering SVS. *International Journal of Electrical Power and Energy Systems*, 64: 176–184.

Rong, Y.J. & Zheng, X.H. 2010. Power system static voltage stability analysis including large-scale wind farms and system weak nodes identification. *Proceedings of 2010 International Conference on Power System Technology*, Chengdu, China, 2747–2753.

Wang, M.S. & Mu, Y.F. 2015. A preventive control strategy for SVS based on an efficient power plant model of electric vehicles. *Journal of Modern Power Systems and Clean Energy*, 3(1): 1363–1372.

Xu R. & Gao Z.H. 2011. Real-time dispatching and coordinated control of large capacity wind power integration. *Proceedings of 2011 International Conference on Advanced Power System Automation and Protection*, Beijing, China, 49–53.

Advances in Power and Energy Engineering – Sun (Ed.)
© 2016 Taylor & Francis Group, London, ISBN 978-1-138-02846-3

Study on low power wireless energy transmission technology

Z.X. Wang
School of Mechanical and Electrical Engineering, Heilongjiang University, Harbin, China
School of Electrical Engineering and Automation, Harbin Institute of Technology, Harbin, China

Y.G. Wei
School of Mechanical and Electrical Engineering, Heilongjiang University, Harbin, China

ABSTRACT: In this paper, the wireless power transmission devices, coupled with the magnetic field resonance technology, have been researched and designed. At first, the circuits of the transmitter and the receiver have been designed. Herein, the transmission is mainly made up of the control circuit and E class inverter circuit. Where the receiving device is composed mainly of a single-phase full-bridge rectifier circuit. Also, the emission and receiving coils have been designed with the involute type and the hollow spiral type, respectively. Then, the design of the circuit has been verified by the simulation software, and the device parameters have been corrected to make sure the system achieves the most excellent performance. Finally, the platform for experiment has been built to debug and test the system. The results show that the designed system could not only realize the wireless transmission of the electricity, but also have a simple structure and high overall efficiency.

1 INTRODUCTION

In 1889, Nikola Tesla proposed the radio theory which resulted as an emerging power transmission technology, and the development of the wireless energy transmission technology took place. (Jiang Lei-lei et al. 2012) After several years, the scientists have put a lot of efforts in researching, but there were very few researches in this aspect. Until 2007, Professor Marin Soljacic in MIT and his team used his magnetic coupling resonance technology to carry out the experiment, pointing at a 60-watt light bulb away from 2.13 meters. (Xie Wenyan & Lin Subin, 2014) Thereafter, the wireless energy transmission technology has begun to enter the public view, but also set off a hot research. (B. L. Cannon et al. 2009)

In 2005, Tao Fa made use of the wireless power transmission technology in implantable biological sensing device. (Tao Fa, 2005) To make the reception terminal voltage stable and keep the reference power source fixed, the transmitter was used in the form of class E amplifier circuit, the receiver device was used as a differential amplifier. This study verifies the feasibility of the wireless power transmission in a biomedical field. In 2011, Bai Yang, who proposed a radio capable of transmitting, that was based on an ultrasonic and this transducer had been designed through the use of mechanical equivalent and analogy. (Bai Yang et al. 2011) The simulation of the transducer conducted harmonic

response in Ansys. Finally, the feasibility of wireless power transmission has been verified by the experimental platform. In 2013, Zhao Zhengming, who carried out the research of the magnetically coupled resonant wireless power transmission, which was focused on the current and future development along with elaborate trend. (Zhao Zhengming et al. 2013)

The above text shows that most researchers tend to use magnetically coupled resonant wireless power transmission system for research and design. (Jiao Jun-ting & Yang Rong-hua, 2013) Therefore, on the basis of the previous work, the principle of the magnetic field resonance coupling for wireless power transmission systems has been studied further. (L. Chen et al. 2013) The transmitter and the receiver have been designed among the different structural dimensions of the transmitter and receiver coils, to debug and experiment.

2 SYSTEM HARDWARE DESIGN

The wireless power transmission coupled with the electromagnetic resonator is mainly composed of the high-frequency inverter circuit, the transmitter coil, the receiving coil, the rectifier filter circuit, and the voltage regulator circuit, as shown in Figure 1. The principle based on the electromagnetic resonant coupling wireless power transmission technology is that the input direct current is

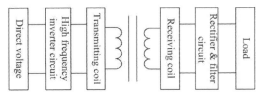

Figure 1. The design scheme of the wireless energy transmission with the electromagnetic resonant coupling.

Figure 2. Involute coils.

Figure 3. Hollow spiral coil.

transformed into the high frequency alternating current to the coils by the high-frequency inverter circuit, which is converted to a high-frequency magnetic field by the coil inductance and capacitance of the resonant coupling shock. Then, the receiving coils get the high-frequency magnetic field to the rectifier and the filter circuit into the direct current. At last, the direct current is supplied to the load.

Whether the high-frequency inverter circuit resonant couple with the two coils which is based on the above scheme to be good or not, will directly impact on the transmission performance of the wireless power transmission apparatus. For the technical specifications of this paper, the preliminary design of the various components of the wireless power transmission is as follows:

1. The control circuit takes LTC6900 chip to generate 5us PWM, to ensure that the circuit could operate at a constant frequency.
2. The inverter circuit takes E class inverter to work at a high frequency mode driven by PWM reverse into sine wave AC, to ensure that the energy conversion efficiency is approximately 100%.
3. In order to improve the energy transmission efficiency of the device, the rectifier receiving device could take the full bridge rectifier.
4. Respectively in comparison to the choice of the transmitting and the receiving coils, several common coils of involute, hollow spiral, tesla, and ring type were compared and analyzed.

The transmitting coil takes the involute coil with low resistance, as shown in Figure 2, which not only could reduce the skin effect at high frequency, but could also improve the transmission efficiency.

The receiving coil only receives the electromagnetic waves to convert it into the electrical energy. The hollow spiral coil is far more better and suitable to receive the magnetic signal than the involute coil, as shown in Figure 3.

The hollow spiral coil inductor is:

$$L = (0.08 \times D^2 \times N^2) \div (W \div D + 0.44) \qquad (1)$$

where L = inductance of a single-layer hollow coil, and the unit is micro-henry (uH); D = average diameter of the coil, and the unit is centimeters (cm); W = width of the coil, and the unit is centimeters (cm); and N = number of turns of the coil.

Because the signal got from the receiving coil system is of high frequency magnetic field, for which it is necessary to consider the resistance, capacitance influence, and the efficiency of the receiver coil. Taking this into consideration, the coil is designed as 4-turn coils and the radius is of 20 cm. This will not only reduce the multi-turn coil to bring the huge resistance, high capacitance and other issues, but will also solve that the coil turns were too little to bring to low inductance.

3 SYSTEM SIMULATION

In the wireless power transmission apparatus, compared with the system receiving device, the design of the transmitting device is more important, especially in the design of the inverter circuit. (Q. Yuan et al. 2010) This part analyzes mainly the simulation of the inverter circuit by Protues and Multiaim software.

The simulation circuit of the transmitter is shown in Figure 4, wherein, L4 is the load (transmitting coil), C1 is resonant capacitor. In order to facilitate the simulation, the driving circuit of the emitting device is simplified. The simulation circuit of the inverter in class E is shown in Figure 5. Setting of different frequency is done, to observe

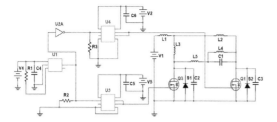

Figure 4.   The simulation circuit of the transmitter.

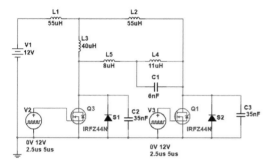

Figure 5.   The simulation circuit of class E inverter.

the output waveform of the simulation. The result shows that, when the driving signal switches Q1 and Q3 are not the same, the output waveform of the load is sinusoidal to meet the functional requirements. Therefore, the inversed design in class E meets the requirements of the electromagnetic resonant coupling wireless power transmission.

4   SYSTEM TEST AND ANALYSIS

4.1   *The construction of the experimental platform*

To facilitate the test, a simple experimental platform is built, as shown in Figure 6, and the main material is wooden. The scale of the experimental bench is marked to read data easily. The transmitting coil is fixed in the experimental stage, and the receiving coil can be moved in the experimental stage, to measure the reception terminal voltage and other parameters at the different distances. And, the distance could be read between the transmitter coil and the receiving coil directly in the experimental stage. (T. Imura & Y. Hori 2011).

The platform testing tools, which are mainly made of Tektronix oscilloscope and the multimeter, are used to measure the waveform, the voltage, the input current of the transmitting device and the load current of the receiving device.

4.2   *The emitting device testing and analysis*

The transmission test system includes two PWM output waveform tests and E class inverter test. Set PWM frequency of 500 kHz, two measured PWM waveforms show in Figure 7, where, channel 1 is the output waveform of LTC6900, channel 2 is the inverter output waveform. Figure 7 shows that PWM cycle is 2 V, and the frequency is 500 kHz. It is consistent with the preset value. The output high voltages are 5 V and 5.5 V, to meet the input requirements of UCC37321.

The test of E class inverter is the voltage and the waveform of the transmitter coil. In the 500 kHz PWM case, the waveform of the coil ends is shown in Figure 8. Seen from the figure, the frequency of the transmit coil is 538.8 kHz, which is approximately equal to 500 kHz, and the voltage is 33.4 V. The waveform is a sine wave, in which exists a small amount of harmonics.

4.3   *The receiving device testing and analysis*

The test of the receiving apparatus is the rectifier. Figure 9 (a) shows the input waveform before the rectifier, which is the voltage waveform of the receiving coil. Figure 9 (b) shows the output waveform after the rectifier. From the output waveform, the ripple of the output voltage is very little.

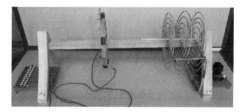

Figure 6.   The test platform of the wireless energy transmission system.

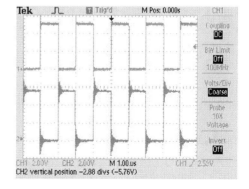

Figure 7.   PWM waveform.

343

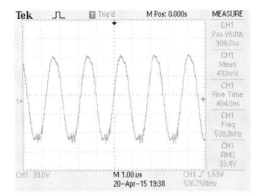

Figure 8.   The waveform of the transmitter coil.

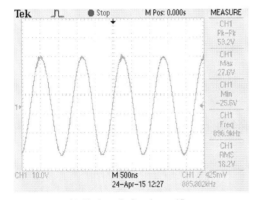

(a) The input before the rectifier.

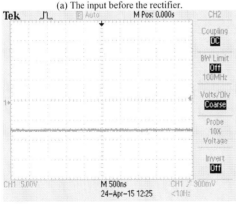

(b) The output after the rectifier and the filter.

Figure 9.   The waveform of the rectifier.

4.4   *The overall system test and analysis*

4.4.1   *Transmitting frequency effect on the system*
This part mainly measures the impact of the resonance frequency to the system. The load power level at the receiving end has been tested by adjusting the different resonance frequencies. That is, to ensure the output voltage and turns unchanged and adjusts to the sending and receiving of resonant coupling capacitance device. It sets different frequencies to ensure that the same efficiency of the system between LC resonant frequency and PWM frequency is maximized. The process-specific experiment is that the receiving end after the full-bridge rectifier ensures that it always lit three lamps 1 W LED lights. The test data results are shown in Table 1.

When three LED lights are lit, Table 1 shows the distance is not increased with the increasing frequency. When the frequency is 947 kHz, the distance between the three LED lit is farthest. At this time, the efficiency of the system achieves maximum.

4.4.2   *Receiving coil voltage parameter test*
In order to analyze the relationship between the voltage and the distance of the wireless power transmission, it is necessary to test the voltage variation of the receiver coils. During the test, the frequency needs to be 947 kHz, and the number of the load of 1 W LED lamp bead needs to be ten. The peak of the distance and voltage (PK–PK), the Maximum Voltage (MAX), the Minimum Voltage (MIN), the effective value of the voltage (RMS) have all been measured, as shown in Table 2. It shows that the voltage is reduced with the increase in the distance. When the distance is 10 cm, the indicator is maximum.

4.4.3   *The efficiency of the system test*
The measurement of the system efficiency mainly depends on the numbers of lighting the lamps. Ensure that the PWM frequency and the number tends to be same, the maximum distance which lit the different number of 1 W lamp has been measured. When the distance is 10 cm, the input current is of 1.32 A, and the input voltage is 12 V.

The frequency is kept to be 947 kHz in the test process. The transmit coil is the hollow coil in volute type. The different numbers of small lamps are lit after the rectifier, whose test results shown in Table 3, where, the numbers of LED lighting lamp could be the same as the system output power at this distance approximately.

Table 3 shows, the numbers of the small lamps which could be lit gradually decrease along with

Table 1.   The influence of frequency on the system.

Frequency (kHz)	500	700	900	947
Distance (cm)	34	36	38	43

Table 2. The influence of the voltage on the receiving coil distance.

Distance (cm)	PK–PK (V)	MAX (V)	MIN (V)	RMS (V)
10	43.6	22.8	−21.4	14.7
14	42.8	21.6	−20.0	14.1
18	40.2	21.6	−19.2	13.8
20	41.6	21.2	−19.2	13.6
24	41.2	20.8	−19.4	13.9
28	37.4	20.2	−18.2	13.1
30	36.2	19.2	−17.6	12.7
34	32.6	17.6	−15.2	10.6
38	26.6	14.4	−12.2	8.36
40	24.9	13.2	−11.8	8.25
44	22.4	11.8	−10.6	7.40

Table 3. The test of distance of the lighting lamp numbers.

Number of light	1	3	5	7	9	10	20
Distance (cm)	64	43	35	28	24	23	20

the increasing distance between the receiving coil and the transmitter coil, which means the efficiency decreases. When the distance is 20 cm, 20 W LED could be lit. When the distance is 64 cm, 1 W LED could be lit.

5 CONCLUSION

In this paper, the method of small power wireless energy transmission has been proposed and designed on the basis of magnetic field resonance coupling. The purpose of energy transfer could be achieved by the oscillation circuit generating the same resonant frequency. This device uses class E amplifier for transmitting and getting high-frequency sinusoidal AC circuit. The receiving device which is using the full bridge rectifier filter circuit converts AC to DC power supply. The methods of the combination of hardware and software have been verified to the feasibility of the circuit of the design.

The factors that affected the power transmission efficiency have been analyzed by the experimental results. The test results show that the system in the operating frequency of 947 kHz, could light 20 W LED at 20 cm, and could light 1 W little lamp at 64 cm.

The circuit of the designed system is of simple and stable performance, who realizes the wireless transmission of the electrical energy. However, the research on other factors that affected the efficiency of the transmission is still pending and not analyzed enough, it will be later studied further.

ACKNOWLEDGMENT

This work has been partly supported by the National Natural Science Foundation of China (Grant No. 51307045).

REFERENCES

Bai Yang, Huang Xueliang, Zou Yuwei, Ding Xiaochen. Study of Wireless Power Transfer System Based on Ultrasonic [J]. Piezoelectrics and Acoustooptics, 2011, 33(2).

Cannon, B.L. J.F. Hoburg, D.D. Stancil, et al. Magnetic Resonant Coupling As a Potential Means for Wireless Power Transfer to Multiple Small Receivers [J]. Power Electronics IEEE Transactions on, 2009, 24(7): 1819–1825.

Chen, L. S. Liu, Y.C. Zhou, et al. An Optimizable Circuit Structure for High-Efficiency Wireless Power Transfer [J]. Industrial Electronics IEEE Transactions on, 2013, 60(1): 339–349.

Imura, T. Y. Hori. Maximizing Air Gap and Efficiency of Magnetic Resonant Coupling for Wireless Power Transfer Using Equivalent Circuit and Neumann Formula [J]. Industrial Electronics IEEE Transactions on, 2011, 58(10): 4746–4752.

Jiang Lei-lei, Chen Qian-hong, Mao Lang. Optimization of 12-pulse autotransformer rectifier units and design of filter inductor [J]. Advanced Technology of Electrical Engineering and Energy, 2012, 31(3).

Jiao Jun-ting, Yang Rong-hua. Design Method Comparison of the Cross-section Reinforcement for Reinforced Concrete under Bending between China and America [J]. Journal of Xiamen Unversity of Technology, 2013, 21(4).

Tao Fa. Study on a new wireless power transmission used to the embedded telemetry instruments [D]. Nanjing University of Aeronautics and Astronautics, 2005. (In Chinese).

Xie Wenyan, Lin Subin. Analysis and Optimization of Wireless Power Transmission Magnetic Coupling Structures [J]. Electrical Engineering, 2014, (9).

Yuan, Q. Q. Chen, L. Li, et al. Numerical Analysis on Transmission Efficiency of Evanescent Resonant Coupling Wireless Power Transfer System [J]. Antennas & Propagation IEEE Transactions on, 2010, 58(5): 1751–1758.

Zhao Zhengming, Zhang Yiming, Chen Kainan. New Progress of Magnetically-coupled Resonant Wireless Power Transfer Technology [J]. Proceedings of the CSEE, 2013, 33(3).

Advances in Power and Energy Engineering – Sun (Ed.)
© 2016 Taylor & Francis Group, London, ISBN 978-1-138-02846-3

Analysis method of weak point for high voltage distribution network with high density distributed generation

H. Zhao & Z.J. Li
Beijing Electrical Power Research Institute, Beijing, China

H.R. Yan
China Agricultural University, Beijing, China

ABSTRACT: With the implementation of China's distributed generation incentive policies and deepening of the power system reforms, more users will install and use distributed generation, especially the photovoltaic power generation, which will help to reduce environmental pollution and electricity costs. But this will cause the distribution network, which is partially connected with high density distributed generation, to trigger random two-way fluctuations of power flow in the distribution network. As a result, the voltage stability issue arises. In order to identify the voltage weak nodes and meet the requirements of further security, the problem of voltage stability caused by the distributed generation and load random fluctuation is evaluated in this paper, using the method of probabilistic power flow. In this paper, the IEEE30 node is used to verify the proposed method, and the effectiveness of the method is being demonstrated.

1 INTRODUCTION

In recent years, the utilization of distributed generation system, which uses solar, wind, and other renewable energy sources as primary energy, is used more and more widely (Zhou et al. 2014). In Beijing, most of the distributed generation is dominated by photovoltaic power generation (Zhao et al. 2012), and there are also tri-generation system connected to the system (Wang et al. 2013). So it is necessary to study about the voltage violation evaluation method of the distribution network, which contains renewable energy to determine the weak link in distribution network. (Zhang et al. 2013) And then adopt some security corrective measures, to set those monitored nodes voltage to the operating range or expected values (Wang et al. 2015). It has important significance for guiding the distribution network working in stability and secured state (Xu et al. 2010).

The problem of power system caused uncertainty by intermittent power supply and load, which is a challenge to traditional deterministic analysis method (Zhu et al. 2013): 1. If the method is used to calculate the system power flow in every possible state, the calculation amount can be very large, so it is neither necessary nor realistic. 2. Deterministic analysis method is even less powerful when facing with the immerging of uncertainties, this method probably may not be able to find out the potential security problems on time. In other words, the

extreme operation conditions which were considered in the traditional deterministic analysis may not be the severest condition in the real operation system. 3Deterministic analyses would only give us the binary information of the node voltage like 'off-limits' and 'on-limits', 'stable,' and 'unstable.' It lacks statistical information, such as the system stability probability and risk index when the system is randomly changed. So it is not conducive to power the department of power grid planning and decision making (Kang et al. 2010).

In view of the above, all kinds of defects in the deterministic analysis method are measured, in order to be able to analyze the voltage and power flow of the power system more scientifically and provide effective information for the safe operation of the power system, probabilistic safety analysis criteria should be developed (Dong et al. 2009). The probabilistic method is a breakthrough of traditional deterministic analysis method, the randomness of power system is considered out of which (Qin et al. 2015), the analysis result is better than using deterministic method. In the traditional deterministic voltage violation evaluation method, there are some defects in considering the random and uncertain factors of the solar power generation and load (Zhu et al. 2013); therefore, the voltage violation evaluation method based on the probability has become a valuable and urgent research topics. It is important for the power department to provide the users with appropriate services.

At the same time, it has important significance for the access of distributed generation and the safe operation of the system. Thus, the report will focus on the computational method for calculating the random distribution of voltage with distributed generation, the evaluation index and the calculation method of the voltage violation probability, and then the weak nodes identification method for distribution network based on probabilistic power flow is formed.

2 METHOD OF PROBABILISTIC POWER FLOW CALCULATION BASED ON SEMI INVARIANTS AND GRAM-CHARLIER SERIES

2.1 Linearized model for probabilistic power flow calculation

The matrix form of power system power flow equations is:

$$W = G(X) \tag{1}$$

where W = the column vector of active and reactive power injected into the joint; X = a state vector composed of node voltage amplitude and phase angle; and $G(X)$ = the column vector of the output power function of the nodes represented by the state vector. In the probabilistic power flow model, the injected power and the state variables can be expressed as:

$$\begin{aligned} W &= W_0 + \Delta W \\ X &= X_0 + \Delta X \end{aligned} \tag{2}$$

where W_0 = the column vectors of the W; X_0 = the mean value of X; ΔW = the column vectors of the disturbance random variable of W; and ΔX = the column vectors of the disturbance random variable of X.

In the probabilistic power flow calculation, it can be derived from Equation (1) that the mean value of the node voltage random variable can be obtained by means of the mean value of the random variable of the injected node.

$$W_0 = G(X_0) \tag{3}$$

Equation (1) can be expressed as the tailor series in the base as,

$$W_0 + \Delta W = G(X_0 + \Delta X) \approx G(X_0) + J_0 \Delta X \tag{4}$$

where J_0 = the Jacobin matrix at the average points.

$$\Delta X = J_0^{-1} \Delta W = S_0 \Delta W \tag{5}$$

$\Delta S = J_0^{-1}$ is a sensitivity matrix

The Equation (5) is called the linear model of stochastic power flow calculation.

2.2 Probabilistic power flow calculation method

The assumption of probabilistic power flow is that the random input variables are independent of each other, the relationship between the order semi invariants of node voltage disturbance random variables ΔX and v order semi invariants of Random variable of the injected power of the node ΔX is:

$$\begin{bmatrix} \gamma_{v,\Delta X,1} \\ \gamma_{v,\Delta X,2} \\ \vdots \\ \gamma_{v,\Delta X,N} \end{bmatrix} = \begin{bmatrix} S_{12}^v & S_{12}^v & \cdots & S_{1N}^v \\ S_{12}^v & S_{12}^v & \cdots & S_{2N}^v \\ \vdots & \vdots & \cdots & \vdots \\ S_{N1}^v & S_{N2}^v & \cdots & S_{NN}^v \end{bmatrix} \begin{bmatrix} \gamma_{v,\Delta W,1} \\ \gamma_{v,\Delta W,2} \\ \vdots \\ \gamma_{v,\Delta W,N} \end{bmatrix} \tag{6}$$

$\gamma_{v,\Delta X,j}$ and $\gamma_{v,\Delta W,j}$ are respectively the voltage disturbance random variable and v order semi invariants of the injected power disturbance random variable of node j. S_{ij} is the corresponding element in the sensitivity matrix in the linear power flow model. N is the number of nodes in the system.

The probability density function of node voltage amplitude state variable is:

$$f(U_j) = \frac{1}{\sigma_{Uj}} \left[\varphi\left(\frac{U_j - U_{j0}}{\sigma_{Uj}}\right) + \frac{c_1}{1!} \varphi'\left(\frac{U_j - U_{j0}}{\sigma_{Uj}}\right) + \cdots \right.$$
$$\left. + \frac{c_k}{k!} \varphi^{(k)}\left(\frac{U_j - U_{j0}}{\sigma_{Uj}}\right) + \cdots \right] \tag{7}$$

where U_{j0} = the mean of the voltage amplitude of the node j; σ_{Uj} = the standard deviation of the voltage amplitude of the node j; $\varphi(.)$ = the probability density function of the standard normal distribution; and c_k is the coefficient K of Gram–Charlier series.

3 STOCHASTIC MODELS OF PHOTOVOLTAIC POWER AND LOAD

In probabilistic power flow calculation containing photovoltaic power, of the node injection power includes the stochastic model of photovoltaic power output and load. This study uses a truncated normal distribution stochastic load model and the stochastic model of photovoltaic power distribution obtained directly by the probability of the output irradiance.

3.1 Photovoltaic generation output power stochastic model

Solar photovoltaic power generation system consists of solar cell matrix, controllers, and inverters, of which the solar cell is the foundation and core of photovoltaic power generation system, its output power is closely related to the light intensity.

According to the average irradiance u and standard deviation σ in a certain period of time, the shape parameter of the Beta distribution can be obtained:

$$\alpha = \mu\left[\frac{\mu(1-\mu)}{\sigma^2} - 1\right] \tag{8}$$

$$\beta = (1-\mu)\left[\frac{\mu(1-\mu)}{\sigma^2} - 1\right] \tag{9}$$

Because of the randomness of the sunlight intensity, the output power of solar photovoltaic generation system is also random; the distribution density function of solar cell matrix output power is also distributed as Beta distribution.

$$f(P_{solar}) = \frac{\Gamma(\alpha+\beta)}{R_{solar}\,\Gamma(\alpha)\Gamma(\beta)}\left(\frac{P_{solar}}{R_{solar}}\right)^{\alpha-1}\left(1-\frac{P_{solar}}{R_{solar}}\right)^{\beta-1} \tag{10}$$

3.2 Load stochastic model

The uncertainty of load in electric power system is a factor that must be considered in the analysis of the voltage violation probability. In this paper, the truncated normal distribution is used to reflect the uncertainty of the load. This model describes the probability density function of the real part (active power) and imaginary part (reactive power) of load disturbance random variable.

$$f_{Load}(P;\mu_P,\sigma_P,P_L,P_H) = \begin{cases} \dfrac{\varphi\left[\dfrac{P-\mu_P}{\sigma_P}\right]}{\sigma_P\left\{\Phi\left[\dfrac{P_H-\mu_P}{\sigma_P}\right] - \Phi\left[\dfrac{P_L-\mu_P}{\sigma_P}\right]\right\}}, & P_L \le P \le P_H \\ 0 \end{cases} \tag{11}$$

$$f_{Load}(Q;\mu_Q,\sigma_Q,Q_L,Q_H) = \begin{cases} \dfrac{\varphi\left[\dfrac{Q-\mu_Q}{\sigma_Q}\right]}{\sigma_Q\left\{\Phi\left[\dfrac{Q_H-\mu_Q}{\sigma_Q}\right] - \Phi\left[\dfrac{Q_L-\mu_Q}{\sigma_Q}\right]\right\}}, & Q_L \le Q \le Q_H \\ 0 \end{cases} \tag{12}$$

where μ_P = the mean of the active load forecasting; σ_P = the standard deviation of the active load forecasting; μ_Q = the mean of the reactive load forecasting; σ_Q = the standard deviation of the reactive load forecasting; P_H = the upper limits of active load forecasting; P_L = the lower limits of active load forecasting; Q_H = the upper limits of reactive load forecasting; Q_L = the lower limits of reactive load forecasting; $\varphi(.)$ = the probability density function of standard normal distribution random variable; and $\Phi(.)$ = the probability distribution function of standard normal distribution random variable.

4 VOLTAGE VIOLATION PROBABILITY EVALUATION OF DISTRIBUTION NETWORK WITH PHOTOVOLTAIC GENERATION

The evaluation index of the voltage violation probability of the distribution network is divided into two categories: node type index and system type index. The node type index focuses on the voltage violation of each node, and the system type index focuses on the voltage violation of overall or the specific range (area node group).

4.1 The voltage violation probability evaluation index of node type

The voltage violation probability evaluation that indexes the node type are: the probability for the node voltage over the upper limitation, the probability for the node voltage under the lower limitation, and the node voltage violation probability.

The probability distribution density function $f(U_i)$ of the voltage amplitude U_i of any P_Q node i can be obtained by probabilistic power flow calculation.

Set O_{UiH} is a random variable U_i over the upper limitation event, O_{UiL} is U_i under the lower limitation event, and O_{UiL} is U_i over limitation event, then the probability of over the upper limitation, under the lower limitation and over limitation of node i is:

$$Pr(O_{UiH}) = Pr\{U_i > U_H\} = \int_{U_H}^{+\infty} f(U_i)dU_i \qquad (13)$$

$$Pr(O_{UiL}) = Pr\{U_i < U_L\} = \int_{-\infty}^{U_L} f(U_i)dU_i \qquad (14)$$

$$Pr(O_{Ui}) = Pr(O_{UiH}) + Pr(O_{UiL}) \qquad (15)$$

where $Pr(O_{UiH})$ = the probability for the voltage over the upper limitation of node i; $Pr(O_{UiL})$ = the probability for the voltage under the lower limitation of node i; and $Pr(_{OUi})$ = the probability for voltage violation of node i; U_H = the upper limits of the voltage qualified; and U_L = the lower limits of the voltage qualified.

4.2 The voltage violation probability evaluation index of the system and node group

When the voltage of all PQ nodes in the system is 1, it define voltage violation probability of the system probability to 1; and when the voltage of all PQ nodes in the system is 0, it define voltage violation probability of the system probability to 0; therefore, the index for voltage violation probability of the system is:

$$Pr(O_U) = \sum_{i \in \mathbf{D}_{PQ}} \frac{1}{N_{PQ}} Pr(O_{Ui}) \qquad (16)$$

$$Pr(O_{UH}) = \sum_{i \in \mathbf{D}_{PQ}} \frac{1}{N_{PQ}} Pr(O_{UiH}) \qquad (17)$$

$$Pr(O_{UL}) = \sum_{i \in \mathbf{D}_{PQ}} \frac{1}{N_{PQ}} Pr(O_{UiL}) \qquad (18)$$

where $Pr(O_U)$ = the evaluation index for the voltage violation of the system; $Pr(O_{UH})$ = the evaluation index for voltage over the upper limitation of the system; $Pr(O_{UL})$ = the evaluation index for voltage under the lower limitation of the system; N_{PQ} = the number of PQ nodes of the system; and D_{PQ} = a PQ node set.

5 EXAMPLES

To verify the probabilistic power flow analysis method for distribution network with photovoltaic generation and voltage violation probabilistic assessment method, this study has made the appropriate changes on IEEE30 node system, and the photovoltaic generation system is set up to work in low light period, stronger light period and the strongest light period, and then simulation calculation and analysis are carried out. The simulation results are divided into two groups, which are the probabilistic power flow calculation based on the

Gram–Charlier series and the rationality verification for voltage violation probabilistic evaluation method of the system and the node group.

5.1 Basic system information

IEEE30 node system contains 41 branches, 6 power supply nodes, and 21 load nodes. The power nodes are 1, 2, 5, 8, 11, and 13, where the node 1 is the balance node, the node 2, 5, 8, 11, and 13 are PV nodes. It is normally accepted in the physical principle that when the output power is small, the distance between the generator and the electricity is longer, the probability for the node voltage with heavy load under the lower limitation is higher; when the output power is large, the distance between the generator and electricity is shorter, the probability for the node voltage with light load over the upper limitation is higher. In order to examine the node voltage violation probability calculated by probabilistic power flow is in accordance with this principle, and analysis evaluation index about the voltage violation probability of the system and node group, the following changes are made to the IEEE30 node system: 1. the power supply nodes 8 and 13 in the original system are changed to photovoltaic generation; 2. the load of nodes 12, 14, and 16 reduced to 0.4 times the original, the load of other nodes increased to 1.2 times the original. In the actual operation of the system, the load condition is very likely to occur.

The system topology is shown in Figure 1

Set two photovoltaic power generation field work in the low light period, stronger light period, and the strongest light period, capacity of photovoltaic power generation system are

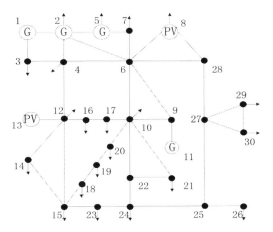

Figure 1. IEEE30 node system with photovoltaic generation.

20 MWp (node 8) and 30 MWp (node 13). Reactive power output of photovoltaic power generation $Q_{PV} = 0.5P_{PV}$.

The load power obeys the truncated normal distribution, the expected value is the given load value, the standard deviation is 5% of the expected value, The truncated lower bound is 0, and truncated upper bound is 1.5 times the expected value. Respectively the upper and lower limits of the voltage are 1.05 p.u. and 0.95 p.u..

5.2 Calculation results and analysis

5.2.1 Using probabilistic power flow method to calculate the voltage violation probability

In the low light period, stronger light period and the strongest light period, the three kinds of different light intensity conditions, the voltage violation probability of node 30 and node 14 are calculated by using the probabilistic power flow method based on Gram–Charlier series in the weak light time, the strong light time, and the strong light time. Results are listed in Table 1.

As can be seen from Table 1, in low light periods, the probability for the voltage under the lower limitation of the 30 node is 78.35%. In fact, because the output power of photovoltaic generation is proportional to the light intensity, the power output of photovoltaic power generation is very small in the low light period, so that the probability for the voltage under the lower limitation of node 30, which is far from the power supply has heavy load, and is bound to be very large. In the strongest light period, the probability for the voltage over the upper limitation of the node 14 is 59.24%. In the bright time, the probability for the high output power of the photovoltaic generation is very large, and thus the probability for the voltage over the upper limitation of node 14 can be very large from the physical principle.

The experimental results show that the probabilistic power flow calculation method based on Gram–Charlier series can objectively describe the random characteristics of the grid with photovoltaic generation.

5.2.2 Verification and calculation of the voltage violation probability evaluation method of the system and node group

In the period of stronger light and the strongest light, the probability for the voltage over the upper limitation of the nodes in another part is increased, these nodes are concentrated in the power supply and their loads are lighter, the probability of node 12 in the period of stronger light and the strongest light reached 90.01% and 91.28%, and node 14 in the strong light period reached 59.24%.

In the system, these nodes, the voltage violation probability of which is higher, are clearly divided into two groups. Node 14 and node 12 are in a group, the voltage violation probability is larger in the period of stronger light and the strongest light. Node 26, node 29, and node 30 are in a group, the voltage violation probability is larger in the period of low light. That is, in the low light period, the node group which is far from the power supply and the heavy load is greatly influenced by the photovoltaic power supply. In the period of stronger light and the strongest light, the node group which is near the power supply and with lighter load is greatly influenced by the photovoltaic power supply.

The evaluation index of system voltage violation probability in three kinds of different light intensity conditions and two node groups that are concerned are shown in Table 2.

As can be seen in Table 2, although the evaluation index of system voltage violation probability is not high (in the low light period is 7.38%,

Table 1. Typical node voltage violation probability in three different light intensities.

Node	Evaluation index of node voltage violation probability	Node voltage violation probability (%)		
		Low light period	Stronger light period	The strongest light period
14	The probability for the voltage over the upper limitation	0.004	8.92	59.24
	The probability for the voltage under the lower limitation	0	0	0
	Voltage violation probability	0.004	8.92	59.24
30	The probability for the voltage over the upper limitation	0	0	0
	The probability for the voltage under the lower limitation	78.35	5.65	4.85
	Voltage violation probability	78.35	5.65	4.85

Table 2. The evaluation index of voltage violation probability in the three kinds of different light intensity conditions.

The evaluation index of voltage violation probability (%)		Low light period	Stronger light period	The strongest light period
Evaluation index of system	Over the upper limitation	0.19	4.12	6.27
	Under the lower limitation	7.19	0.29	0.25
	Voltage violation	7.38	4.41	6.52
Node group (12, 14)	Over the upper limitation	2.26	49.47	75.26
	Under the lower limitation	0	0	0
	Voltage violation	2.26	49.47	75.26
Node group (26, 29, and 30)	Over the upper limitation	0	0	0
	Under the lower limitation	57.55	2.28	1.99
	Voltage violation	57.55	2.28	1.99

mainly composed of the index of under the lower limitation, and in the period of stronger light and the strongest light are 4.41% and 6.52% respectively, mainly composed of the index of over the upper limitation), the calculation results are in accord with the actual situation of the operation of the power system with photovoltaic generation. The calculation results of voltage violation probability of node group are more obvious, the probability for the voltage over the upper limitation of the node group including node 12 and node 14 in the period of stronger light and the strongest light is as high as 49.47% and 75.26% respectively, the probability for the voltage under the lower limitation of the node group including node 26, node 29 and node 30 in the low light period is as high as 57.55%.

The experimental results show that the proposed method can quantitatively identify the weak nodes or weak areas (node group) of the distribution network with photovoltaic generation.

6 CONCLUSION

In this paper we put forward the evaluation index and calculation method about the voltage violation probability of distribution network with photovoltaic power supply. This method obtains the probability distribution of each node voltage by using the probabilistic power flow calculation based on Gram–Charlier series and semi-invariants, and then gets the voltage violation probability of each node; finally, the index of system and regional node group voltage violation probability can be obtained. It is showed that probabilistic power flow calculation can give an objective description of the random characteristics of distribution network with photovoltaic generation; the voltage violation evaluation method can not

only give the probability for the voltage over the upper limitation, the voltage under the lower limitation, and voltage violation, but also get more significant indexes from evaluation calculation of different concerned nodes in different light intensity. Thus, it can be able to quantitatively identify the weak nodes or weak areas (node group) of the distribution network with photovoltaic generation.

REFERENCES

Dong Lei, Cheng Weidong & Yang Yihan. 2009. Probabilistic Load Flow Calculation for Power Grid Containing Wind Farms. *Power System Technology* 16: 87–91.

Kang Longyun, Guo Hongxia, Wu Jie & Chen Sizhe. 2010. Characteristics of Distributed Generation System and Related Research Issues Caused by Connecting It to Power System. *Power System Technology*, 11: 43–47.

Qin Wenping, Ren Chen, Han Xiaoqing, Wang Peng & Liu Zhijuan. 2015. Power System Voltage Stability Risk Assessment Considering the Limit of Load Fluctuation. *Proceedings of CSEE* 16: 4102–4111.

Wang Chengshan, Hong Bowen, Guo Li, Zhang Deju & Liu Wenjian. 2013. A General Modeling Method for Optimal Dispatch of Combined Cooling, Heating and Power Microgrid. *Proceedings of CSEE* 31: 26–33+3.

Wang Jing, Jiang Xiaoliang, Yang Zhuo, Guan Chaojie, Guo Zhengyi & Hu Po. 2015. Penetration Capacity under Voltage Constraint and Evaluation Methodology of Voltage Fluctuation Caused by Centralized Grid Connection of Photovoltaic Power. *Power System Technology* 09: 2450–2457.

Xu Xiaoyan, Huang Yuehui, Liu Chun & Wang Weisheng. 2010. Influence of Distributed Photovoltaic Generation on Voltage in Distribution Network and Solution of Voltage beyond Limits. *Power System Technology* 10: 140–146.

Zhang Zhe, Li Gengyin & Wei Junqiang. 2013. Probabilistic Evaluation of Voltage Quality in Distribution Networks

Considering the Stochastic Characteristic of Distributed Generators. *Proceedings of CSEE* 13: 150–156.

Zhao He, Yu Xijuan & Wang Lin. 2012. Analysis of optimal economic operation of Combined Cooling, Heating and Power Microgrid. *Electrotechnical Application* 21: 60–64.

Zhou Xiaoxin, Lu Zongxiang, Liu Yingmei & Chen Shuyong. 2012. Development Models and Key Technologies of Future Grid in China. *Proceedings of CSEE* 29: 4999–5008.

Zhu Xingyang, Liu Wenxia, Zhang Jianhua, Qiu Wei & Huang Yufeng. 2013. Reviews on Power System Stochastic Load Flow and Its Applications in Safety Evaluation. *Transactions of China Electrotechnical Society* 10: 257–270.

Zhu Xingyang, Liu Wenxia & Zhang Jianhua. 2013. Probabilistic Load Flow Method Considering Large-scale Wind Power Integration. *Proceedings of CSEE* 07: 77–85+16.

Advances in Power and Energy Engineering – Sun (Ed.)
© 2016 Taylor & Francis Group, London, ISBN 978-1-138-02846-3

Research on off-line quantification method of transient stability

D. Liu
Ziyang Company of Sichuan Electrical Power Company, Ziyang, Sichuan Province, China

Q.Y. Liu & R.G. Zhao
School of Automation and Engineering, University of Electronic Science and Technology, Chengdu, China

Q.F. Liu
CHINAGUODIAN Sichuan Ashine Power Co. Ltd., Chengdu, China

R.H. Liu
Sichuan Electric Vocational and Technical College, China

ABSTRACT: Energy margin based on Lyapunov direct method has been considered as an important index to quantify the transient stability of power system. Especially its on-line application has been paid much attention. Based on the classical equations of rotor motion, the transient energy function model is first established in this paper. Then the converting relationship between the potential energy boundary surface and the stability region characteristics is described. Based on these, the second kick method and the interpolation method based on the property of potential energy boundary are combined to calculate the maximum of the potential energy. At last, the energy margin model is established based on the maximum of potential energy and the transient energy. The simulation results in three-machine and IEEE-30 bus system and the comparison with CUEP verify the effectiveness and feasibility of the proposed method, which can be used in off-line assessment of transient stability of power system.

Keywords: transient stability; energy margin; the second kick; stability region

1 INTRODUCTION

With the increased assessment requirements of the transient stability in power system, the research on assessment theory of transient stability become necessary in order to analyze the degree of stability and even the development trend quantitatively. In recent years, almost all the methods are mainly based on the construction of the quantitative index model to evaluate the degree of transient stability and its development trend. By far, a lot of results have been achieved. The research thoughts are mainly concentrated on the uncertainty analysis method and the deterministic analysis method.

The uncertainty analysis method is based on stochastic theory (Wang, W. 2009 & Wang, L.J. 2011) and probability distribution (Izzri, A.W.N. 2007 & Ye, S.Y. 2011). The deterministic analysis methods mainly include SVM based method (Chen, L. 2009 & Zhu, L. 2012), Regression method (Pai, M.A. 1989), Sensitivity method for searching multi pendulum transient stability constraints (Fouad, A.A. 1981a), Solutions for tracking method (Fouad, A.A. 1981b & Chiang, H.D.

1994) and energy function method (Vaahedi, E, 1996, Roger T.T. 1996, Zhao, X.Y. 2009, Wang, X.M. 2011, Chao. D 2012, Wang, F.Z. 2011). As a direct method of the important branches of the deterministic analysis method, the research is most active during 1960s–1990s. The results have laid a solid theoretical foundation for the practical application of the transient stability (Vaahedi, E. 1996 & Roger, T.T. 1996). By far, the energy function is still the main method for the study of transient stability. References (Zhao, X.Y. 2009 & Wang, X.M. 2011, Chao, D. 2012) have researched the mechanism and evaluation method of transient stability from different angles base on energy function. The energy function based on the tie line is constructed by Zhao, X.Y. (2009), where the energy is decomposed into oscillatory component and quasi-steady state. So the low frequency oscillation of the system caused by multiple disturbances is analyzed. Wang, X.M. et al (2011) constructed a dynamic and practical model, in which the equivalent single machine infinite bus system is obtained by using the rotor acceleration equation to construct the equivalent transient energy function, which is advantageous to simplify the operation. Aiming at the disadvantages of the PEBS (potential energy boundary) method of the branch potential energy boundary, Branch RIDGE method is proposed by Chao, D. (2012) for power system transient stability analysis and critical cut set detection. A method is proposed by Wang, F.Z. (2011) to construct the Lyapunov function, in which the transient stability of system of the polynomial vector field is treated by using the positive definite polynomial, the semi definite programming and the square decoupling.

In order to improve the computational efficiency, the idea of parallel computing is put forward. Vahid, J.M. (2012) have applied the S-stage 2S-order Symplectic Runge-Kutta-Nyström method to the calculation of power system transient stability, where the classic power system model is utilized and the skill in Jacobian matrix splitting and the relaxation of matrix inversion is used to propose a new parallel computation method for power system transient stability. A large scale power system transient simulation method based on multi image processing unit with multi parallel structure is proposed by Roger, T.T. (1996), where the robust effective transient relaxation based on parallel processing technique highlights the explicit integrals and Full Newton iteration and a linear solver based on sparse matrix is used to run the multi image processing unit.

The above research provides a theoretical support for the online application of energy function in the power system transient stability. However, under the new situation of the development of power system, the online assessment of transient stability based on energy function requires a new challenge to the existing theoretical method. Faster and more convenient method is still the focus of power operation personnel.

In this paper, the transient energy function model is established based on the classical equations of rotor motion and the equations of power balance. Then, based on the conversion relationship between kinetic energy and potential energy of power system and the basic characteristics of the stable region, the maximum of the potential energy is obtained by using the second disturbance method and the interpolation method. Finally, the evaluation index of grid transient stability is constructed based on the maximum of potential energy. The simulation results in three-machine and IEEE-30 bus system and the comparison with CUEP verify the effectiveness and feasibility of the proposed method.

2 THE CONSTRUCTION OF ENERGY FUNCTION MODEL

Figure 1 shows the connection of Single-machine—infinite system. When the system is disturbed, the

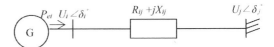

Figure 1. Single-machine infinite system.

equations of rotor motion of the generator are shown in (1) and (2).

$$\dot{\delta}_i = \omega_i \tag{1}$$

$$M_i \ddot{\delta}_i = P_{mi} - P_{ei} \tag{2}$$

As is shown in (1), δ_i and ω_i are respectively the rotor angle and speed of generator. P_{mi} and P_{ei} are respectively the mechanical power and electromagnetic power of the generator. M_i is the inertia time constant.

$$f_i(\delta_i) = P_{mi} - P_{ei} \tag{3}$$

According to Lyapunov direct method, the energy function expression of the generator by (1) and (3) is as follows

$$E_i = \frac{1}{2} M_i \omega_i^2 - \int_{\delta_i^s}^{\delta_i} f_i(\delta) d\delta \tag{4}$$

Among them,

$$E_{kei} = \frac{1}{2} M_i \omega_i^2 \tag{5}$$

$$E_{pei} = -\int_{\delta_i^s}^{\delta_i} f_i(\delta) d\delta \tag{6}$$

In a multi machine system, it is the superposition of kinetic energy and potential energy, expressed as

$$E_{tot} = \sum_{i=1}^{n} E_i = \sum_{i=1}^{n} E_{kei} + \sum_{i=1}^{n} E_{pei} \tag{7}$$

The first of the formula (4) and (7) represents the kinetic energy of the system, and the second represents the potential energy of the system. In this paper, we study on the method of obtaining the maximum of potential energy. The classical energy function is used to get the general conclusion. Therefore, the formula (4) and (7) have no more detailed information about the operation parameters of the system, such as active load, reactive load and voltage amplitude.

The stable equilibrium point is (δ_s, ω_s) after the fault. At this point, $E_{tot}(\delta_s, \omega_s) = 0$. The derivative of the energy function to the time is

$$E_m(UEP1) = E_{Pe2}\_UEP1 - E_{Pe_1} - UEP1 = 0 \tag{8}$$

In practical systems, the derivative of the total energy to time t is

$$\frac{dE_{tot}}{dt} = \sum_{i=1}^{n}(-D\omega^2 + 0) \leq 0 \qquad (9)$$

Among them, D is the damping constant. The formula (9) shows that the formula (4) meets the conditions of Lyapunov direct method in practice. It can be used to analyze the stability of power system. The second of the formula (4) and the formula (7) have the path-dependent problem. There is little impact on the accuracy of the result by using the trapezoidal method to approximate the calculation.

3 CONSTRUCTION OF TRANSIENT STABILITY OFF-LINE QUANTITATIVE INDEX BASED ON THE SECOND DISTURBANCE METHOD

To ensure that the first disturbance and the second disturbance will return to the same Stable Equilibrium Point (SEP) and the trajectory of the second disturbance is a large scale asymptotic stability. The second disturbance will be as close as possible to the unstable equilibrium point of the stable boundary, which is still in the area of attraction of the stable equilibrium. Based on this conclusion, the first disturbance is small disturbance and the small amplitude increase of the load is done. The second disturbance is to be applied to large disturbance, and the three-phase short circuit fault is adopted. Using such a two disturbance model, the simulation results has been obtained as Figure 2 (Madeleine, G.C. 2005).

In Figure 2, the two unstable equilibrium points on the stability region boundary are $UEP1$ and $UEP2$ respectively. The first disturbance and the second disturbance trajectories are respectively close to these two unstable equilibrium points. Corresponding to these two unstable equilibrium points, the potential energy extremum ($E_{Pe1\_}$ $UEP1$, $E_{Pe2\_}UEP1$, $E_{Pe1\_}UEP2$, $E_{Pe2\_}UEP2$) will be shown. When the first disturbance and the second disturbance approach the same unstable equilibrium point, the extreme value of potential energy is regard as the energy margin, which is expressed as:

$$E_m(UEP1) = E_{Pe2\_}UEP1 - E_{Pe_1}\_UEP1 \qquad (10)$$

$$E_m(UEP2) = E_{Pe_2}\_UEP2 - E_{Pe_1}\_UEP2 \qquad (11)$$

As is shown in (10) and (11), $E_m(UEP1)$ and $E_m(UEP2)$ are respectively the energy margin of unstable equilibrium point $UEP2$ and $UEP1$. Due to the influence of the simulation or the accuracy of the calculation, there is a certain gap between the results of (10) and (11). From the perspective of ensuring security, the minimum among the two results are taken. So the final energy margin is

$$E_m = \min(E_m(UEP1), E_m(UEP2)) \qquad (12)$$

The formula (12) measures the distance between the operating point of the current system and the stable boundary, the larger the distance, the more stable the system is. When $E_m = 0$, the system has been on the verge of the critical state of instability. In the transient process of the system, the stability trend of the system is related to the kinetic energy of the injection system on fault. If the kinetic energy of the injection system is too much, the grid structure of current system can not be fully absorbed, and the system is easy to lose.

4 THE CONCRETE REALIZATION OF THE SECOND DISTURBANCE METHOD

4.1 The basic principle of the second disturbance method

According to the existing literature, it is not necessary to find a complete stability boundary. To find out the stability margin of the system, only the corresponding CUEP (Controlling Unstable Equilibrium Point) of motion trajectory in a certain state is needed to obtain. According to CUEP, the potential energy limit ($E_{pe\_}UEP$) is obtained and determined whether the moving trajectory through the stability boundary.

Because the computation of CUEP takes a long time, here the two perturbation methods are combined with in the request of the $E_{pe\_}UEP$ to

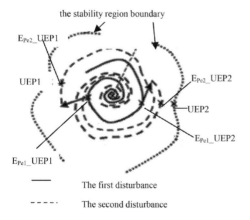

Figure 2. Second—disturbance based energy margin computation.

save calculation time. The realization process of the second disturbance method is shown in Figure 3.

In Figure 3, the time period $0–t_{\_start}$ is the duration of the first disturbance. At the $t_{\_start}$ moment, the minimum point of the kinetic energy of the system is $E_{ke(t\_start)}$. At the $t_{\_start+}$ moment, a big enough disturbance is applied to the system for the second time to make the system unstable. And the kinetic energy of the system increases. The moment of kinetic energy trajectory through the second largest kinetic energy is $t_{\_clear}$, which is the fault limiting removal time. The corresponding kinetic energy is $E_{ke(t\_clear)}$. When the system is fully restored, the corresponding time is t_{s2}, the corresponding kinetic energy is $E_{ke(ts2)}$ and the corresponding potential energy is $E_{pe(ts2)}$.

The potential energy limit $(E_{pe\_}UEP)$ is the potential energy of the unstable equilibrium point on the potential energy boundary, which is the maximum value of the potential energy after all the kinetic energy is converted. The traditional method is the CUEP method (Dong, C. 2012), its calculation is very time consuming. In order to reduce the calculation, the interpolation method is used to obtain the final results in the calculation of $E_{pe\_}UEP$.

4.2 The calculation of maximum of potential energy based on interpolation method

The limit value of the potential energy on the stability boundary is the maximum of potential energy of the corresponding UEP. Therefore, the change rate of potential energy with time at this point is 0. If the change rate of potential energy and the time of the two arbitrary points near UEP can be obtained, the time of the point can be calculated according to the interpolation method. Then the potential energy of the point can be calculated by the time of the point, which is the maximum of potential energy required.

In Figure 4, abscissa represents transient process and ordinate represents the variation of potential energy in transient process. In (6), because δ_i is a function of time, according to the potential energy of the critical moments of t_start and t_clear, the potential energy variation of $dE_{pe(t\_start)}$ and $dE_{pe(t\_clear)}$ at the time t_start and t_clear can

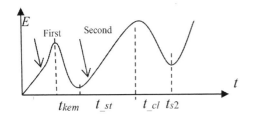

Figure 3. The curve of kinetic energy trajectory.

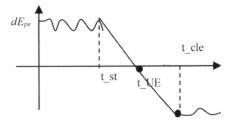

Figure 4. Applying the interpolation to compute maximum of potential energy.

be calculated. In the t_UEP moment, the potential energy boundary is the maximum of potential energy and its variation quantity is 0. The interpolation method is used as eqn. (13).

$$t\_UEP = t\_start - dE_{pe(t\_start)}$$
$$\cdot \frac{(t\_start - t\_clear)}{\left(dE_{pe(t\_start)} - dE_{pe(t\_clear)}\right)} \quad (13)$$

The basis of the idea is that the PEBS corresponds to the maximum of potential energy and the change rate of the potential energy at the crossing point of PEBS is 0. In (13), t_start is the moment of the first disturbance and t_clear is the moment when the kinetic energy reaches the maximum after the second disturbance occurs. They can be obtained according to the simulation curve of kinetic energy. Considering $dE_{pe(t\_start)}$ and $dE_{pe(t\_clear)}$, t_UEP can be calculated and $E_{pe}(t\_UEP)$ can be obtained at the moment according to (14).

$$E_{pe}(t\_UEP) = \sum_{i=1}^{n} \int_{t\_start}^{t\_UEP} \frac{\partial E_{pei}}{\partial t} dt \quad (14)$$

On the one hand, the calculation of (14) avoids the calculation of the kinetic energy and potential energy at the moment of t_{s2}. On the other hand, it avoids the direct solution of CUEP or the escape point of potential energy boundary. It can reduce calculation to a certain extent. The disadvantage is that the need to save the running state of the t_start time and the off-line implementation of the second disturbance. It increases the computation time to a certain extent.

5 EXAMPLE VERIFICATION

5.1 The simulation analysis of three-machine system

The three-machine system is shown in Figure 5.

In Figure 5, G_1, G_2 and G_3 are reference generator, they have a larger inertia and fast response

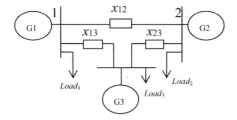

Figure 5.   Three-machine system.

controller, and both G_1 and G_2 have a fast excitation control system. There are different sizes of load at each generator bus bar. x_{12}, x_{23} and x_{13} is line reactance, and $x_{12} = x_{23} = x_{13} = 0.30$.

1. The analysis of change of system trajectories

   The first disturbance is three phase short circuit fault, which occurs at the midpoint of the branch 1–2, after the system was stabilized by 50 s, then last 0.5 seconds clear. The second disturbance occurs at the moment of 70 s, then last 0.5 seconds clear. The change curve of the system trajectory from the start to the stability (0 s–40 s) is shown in Figure 6(a). In Figure 6(a), sep1, sep2 and sep3 are respectively the initial stable equilibrium point of G_1, G_2 and G_3, and the corresponding coordinate are respectively (0.99, 54.37°), (0.99, 55.78°) and (0.99, 42.13°). From Figure 6(a), the three generators reached the stable equilibrium point at a very fast speed after starting.

   Figure 6(b) indicates that the first disturbance occurs at the 50 s moment, and shows the trajectory (50 s–60 s) from the failure of the system to achieve a stable equilibrium point. In Figure 6(b), sep1, sep2 and sep3 are respectively the stable equilibrium point of G_1, G_2 and G_3 after the first disturbance, and the corresponding coordinate are respectively (0.978, 54.67°), (0.978, 55.90°) and (0.978, 42.15°). Compared with the stable equilibrium point in Figure 6(a), there is a very small movement in both the angular velocity of the equilibrium point and the rotor angle. It shows that the three-phase short circuit fault has a great influence on the whole transient stability of the system, and cause the generator can not return to the initial stable point.

   Figure 6(c) indicates that the second disturbance (three phase short circuit fault, which occurs at the midpoint of the branch 1–2) occurs at the 70 s moment, and shows the trajectory curve (70 s–150 s) from the occurrence of disturbance to the removal of disturbance and the final stability. In Figure 6(c), sep1, sep2 and sep3 are respectively the stable equilibrium point of G_1, G_2 and G_3 after the second

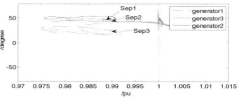

(a)the stability trajectory of three machine system and the initial stable equilibrium point

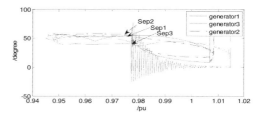

(b)The stability trajectory of three machine system and the initial stable equilibrium point after the first disturbance

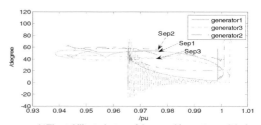

(c)The stability trajectory of three machine system and the initial stable equilibrium point after the second disturbance

Figure 6.   (a) The stability trajectory of three machine system and the initial stable equilibrium point. (b) The stability trajectory of three machine system and the initial stable equilibrium point after the first disturbance. (c) The stability trajectory of three machine system and the initial stable equilibrium point after the second disturbance.

disturbance, and the corresponding coordinate are respectively (0.978, 54.66°), (0.978, 55.88°) and (0.978, 42.14°). Compared with Figure 6(b), there is almost no movement in the position of the equilibrium point of three generators. Only from the change of the equilibrium point, the second three-phase short circuit fault has no effect on the stability of the system.

2. The calculation of the kinetic energy and the potential energy in the fault

   According to the trajectory of the two disturbances and the computational model of the kinetic energy and the potential energy, the change curves of the kinetic energy and the potential energy of the system can be obtained in the process of the two disturbances, which are shown in Figure 7 and Figure 8.

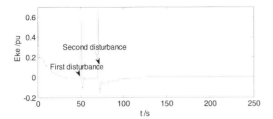

Figure 7. The kinetic energy curve of three-machine system.

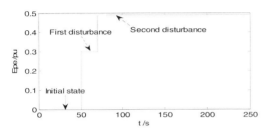

Figure 8. The energy of three-machine system corresponding to different faults.

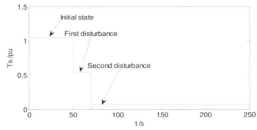

Figure 9. The energy margin curve during the two disturbance.

Table 1. The comparison of computing time and energy margin among CUEP method and the PEBS—based interpolation method.

	CUEP method	This method
Simulation time(s)	9.33	4.05
Energy margin computing time(s)	0.63	0.62
Total computing time(s)	9.96	4.67

In the system, the kinetic energy and the potential energy are conserved at any moment. From Figure 7 and Figure 8, the conservation of the kinetic energy and the potential energy is verified. In addition, the two disturbances are equivalent to the system injected with two kinetic energy injection. In the case of neglecting the other kinetic energy consumption of the system, the two injection of the kinetic energy will eventually be transformed into different forms of potential energy in the system. In Fig.8, the injected kinetic energy of the first disturbance is converted into potential energy completely, and no longer changes after reaching the steady state. After the second disturbance is injected into the kinetic energy, the potential energy step up, with the change of time presents the trend of potential superposition. Once all the kinetic energy is converted into the potential energy, the potential energy of the system will remain unchanged.

3. The results of the energy margin

Figure 9 shows the change of energy margin of the system in the initial state and after suffered two disturbances. The energy margin in the initial state of the system is near 1, and after the first disturbance, the energy margin dropped to about 0.5. The energy margin dropped to about 0.1 after the second disturbance, it is very small and the system has already begun to weaken. If there is a disturbance again, the system will be

on the verge of instability. In the 1) section, the stable equilibrium point of the first disturbance and the stability of the second disturbance are almost identical. The results of transient stability analysis based on the calculation of the energy margin are compared. The idea that whether the stability of the system is changed by judging the motion of the stable operating point is likely to cause errors in transient stability judgment.

5.2 The simulation analysis of IEEE-30 bus system

In IEEE-30 bus system, the proposed method in this paper and the CUEP method are simulated and compared with the results. In the use of CUEP method, the trajectory of the fault is integrated, and the positive intersection of the boundary of the stable region (CUEP) is found. Then the corresponding CUEP (δ) value is substituted into the equation (7), and the potential maximum of the corresponding point is calculated. Finally, the value is substituted into equation (11) to obtain the energy margin of three-phase short circuit fault occurs at different points in IEEE-30 bus system. The comparison of the computing time and results of two methods are respectively shown in Table 1 and Table 2.

As is shown in Table 1, the computing time of CUEP method is significantly longer than that in this paper. The reason is that the use of the fault path integral to obtain the CUEP needs to

Table 2. The comparison among the CUEP and the PEBS—based interpolation method in IEEE-30 bus system.

| Fault line | This method | CUEP | | This method | CUEP method | |
	First disturbance	Second disturbance	The error	First disturbance	Second disturbance	The error
1–2	0.423	0.435	2.76%	0.247	0.254	2.76%
2–5	0.417	0.421	0.95%	0.274	0.282	2.84%
5–7	0.578	0.592	2.36%	0.354	0.375	5.60%
12–16	0.701	0.719	2.50%	0.435	0.457	4.81%
9–21	0.759	0.780	2.69%	0.466	0.497	6.24%
18–19	0.898	0.912	1.54%	0.552	0.573	3.66%
15–23	0.869	0.893	2.69%	0.523	0.541	3.33%
29–30	0.953	0.977	2.46%	0.899	0.921	2.39%

consume a long time. The method of this paper does not need to carry out the integration of the fault trajectory, and time is consumed mainly in the implementation of the two disturbances. In the calculation of energy margin, the CUEP method is basically the same as that of the proposed method in this paper. This is because the calculation process of the energy margin is basically the same and the amount of calculation is also the same. From simulation to margin calculation, the CUEP method takes more time than the proposed method of this paper 5.29 seconds. So it is proved that the proposed method can improve the calculation speed, and the on-line application of transient energy function is close to a step further.

Table 2 shows the longitudinal comparison of the calculation results of the first disturbance and the second disturbance. When the three-phase short circuit fault occurs at the branch near the generator or the branch directly connected to the generator, the lower the stability margin, the branch 1–2, 2–5 and 5–7 compared to the branch 12–16, 9–21, 18–19 and 15–23, the stability margin decreased a lot after the first disturbance and the second disturbance. So these branches have important significance to maintain the stability of the system. The stability margin of the branch 29–30 is the largest after the first disturbance and the second disturbance, and having minimal impact on the stability of the system. From the horizontal comparison of Table 2, the energy margin of CUEP method is significantly larger than the calculation results of the proposed method in this paper. The reason is that in the process of obtaining the CUEP, multiple integrals and the selection of the step size will lead to large errors. In the process of the second disturbance, the determination of stable equilibrium point and CUEP of the second fault restoration will be because of a large number of calculations to increase the error. In addition, the selection of

the location of the fault has a certain effect on the accuracy of the results. As is shown in Table 2, the calculation errors of different fault locations are not the same. When the branch 9–21 is disconnected, the error is 6.24%, and the maximum is reached.

6 THE CONCLUSIONS

Combined with the transient energy function model, the interpolation method of potential energy boundary based on the second disturbance method is analyzed, and the quantitative index of transient stability of the system is derived. The comparative analysis of the CUEP method and the simulation results of the three machine system and the IEEE-30 bus system show that the interpolation method of potential energy boundary based on the second disturbance method can be avoided to obtain the unstable equilibrium point of the stable boundary directly. The disadvantage is that the state parameters of system stability after the first failure are needed to preserve, which depends on the occurrence of the second disturbance. The potential energy boundary based on the second disturbance method is directly affected by the duration of the second disturbance, so there are differences in the potential energy boundary value for different duration. To some extent, it affects the accuracy of the results. Compared to the CUEP method, the calculation of the quantitative index of transient stability based on the second disturbance method is improved and it can be adapted to different operating conditions. However, the limitation is that it can only be used for the transient stability offline analysis. The next step is to further study the construction of the online evaluation index of the power system transient stability based on the energy margin.

REFERENCES

Chen, L., Min, Y., Xu, F. & Wang, K.P. 2009. A Continuation-Based Method to Compute the Relevant Unstable Equilibrium Points for Power System Transient Stability Analysis, *IEEE Transactions on Power Systems*, 24(1): 165–172.

Chiang, H.D., Wu, F.F. and Varaiya, P.P. 1994. A BCU Method for Direct Analysis of Power System Transient Stability, *IEEE Transactions on Power System*, 9(3): 1194–1208.

Dong, C., Liu, D.C., Liao, Q.F., Wang, B. 2012. Research on Low Frequency oscillation in Power Grid and Location of Disturbance Source based on Energy Function, *Power System Technology*, 36(8): 175–181.

Fouad, A.A., Stanton, S.E. 1981. Transient Stability of A Multi-Machine Power System Part II: Critical Transient Energy, *IEEE Transactions on Power Apparatus and Systems*, PAS-100(7): 3417–3424.

Fouad, A.A., Stanton. S.E. 1981. Transient Stability of A Multi-Machine Power System Part I: Investigation of System Trajectories, *IEEE Transactions on Power Apparatus and Systems*, PAS-100(7): 3408–3416.

Izzri, A.W.N., Mohamed, A., Yahya, I. 2007. A New Method of Transient Stability Assessment in Power Systems Using LS-SVM[C]. *The 5th Student Conference on Research and Development –SCOReD*, 11–12 December 2007, Malaysia, 1–6.

Madeleine, G.C., Liu, C.C., Taoka, H.H.H. 2005. Energy-Based Stability Margin Computation Incorporating Effects of ULTCs, *IEEE Transactions on Power Systems*, 20(2): 843–851.

Pai, M.A. 1989. Energy Function Analysis for Power System Stability,. *Kluwer Academic Publisher*.

Roger, T.T, Vittal. V., Kliemann, W. 1996. An improved technique to determine the controlling unstable equilibrium point in a power system, *IEEE Transactions on Circuits and Systems-I: Fundamental Theory and Applications*, 43(4): 313–322.

Roger, T.T., Vitta. V., Kliemann, W. 1996. An Improved Technique to Determine the Controlling Unstable Equilibrium Point in a Power System, *IEEE Transactions on Circuits and Systems-I: Fundamental Theory and Applications*, 43(4): 313–322.

Vaahedi, E., Mansour. Y., Chang, A.Y., Corns, B., Tse. E.K. 1996. Enhanced "Second Kick" Methods for On-line Dynamic Security Assessment, *IEEE Transactions on Power Systems*, 1996, 11(4): 1976–1982.

Vahid, J.M., Zhou, Z.Y., Venkata, D.V. 2012. Large-scale transient stability simulation of electrical power systems on parallel GPUs, *IEEE Transactions on Parallel and Distributed Systems*, 23(7): 1255–1266.

Wan, W., Mao, A.J., Zhang, L.Z., Yang, X., Yuan, J. 2009. Risk Assessment of Power System Transient Security Under Market Condition. *Proceedings of the CSEE*, 29(1): 68–73.

Wang, F.Z., He, Y.F. 2011. A parallel computational method for power system transient stability based on Symplectic Runge-Kutta-Nyström method, *Power System Technology*, 35(4): 40–45.

Wang, L.J., Wang, G. 2011. Probabilistic Assessment of Transient Stability Based on Dynamic Security Region and Edge worth Series. *Proceedings of the CSEE*, 31(1): 52–58.

Wang, X.M, Liu, D.C., Wu, J., Huang, Y.Y., Zhao, H.S. 2011. Energy Function-Based Power System Transient Stability Analysis, *Power System Technology*, 35(8): 114–118.

Ye, S.Y., Wang, X.Y., Liu, Z.G. 2011. Power System Transient Stability Assessment based on Support Vector Machine Incremental Learning Method, *Automation of Electric Power Systems*, 2011, 35(11): 15–19.

Zhao, X.Y., Jia, Y.J. 2009. Transient Stability Research Considering flux decay dynamics, *Electric Power Automation Equipment*, 29(10): 50–54.

Zhu, L., Xu, M., Chen, Y.S., Cai, Z.X., Ni, Y.X. 2012. A Method for Improving Power System Transient Performance Using Estimator Resetting Based on Tracking Models, *Automation of Electric Power Systems*, 36(14): 1–5, 22.

362

Advances in Power and Energy Engineering – Sun (Ed.)
© 2016 Taylor & Francis Group, London, ISBN 978-1-138-02846-3

Review and prospect of power system mid-long-term load forecasting methods

L. Xing, D. Yang, Z. Jiang, N. Wu & X. Zhao
State Grid Shandong Electric Power Research Institute, Jinan, China

ABSTRACT: Mid-long-term load forecasting plays an important role in guiding development and fuel plans of an electric power system. The main task of it is forecasting the temporal and spatial distribution of the power load for the next few months, years, or even longer periods. In order to make full use of existing research findings, the characteristics of mid-long-term load forecasting, andpresented forecasting methods are discussed, and the solved problems are analyzed. With the development of an active distribution network, and big data of the electric power user side and spatial load forecasting, the directions for future researches on mid-long-term load forecasting are prospected.

1 INTRODUCTION

Load forecasting is a process of estimating the load of a certain period of time in the future, using certain prediction models and methods, based on historical load data and its related influencing factors, predicted results of future economic conditions, social development, and climate conditions. The main task of it is forecasting the temporal and spatial distribution of power load (He 2013). First of all, in terms of load forecasting content, the total load forecasting should be distinguished from spatial load forecasting (Kang et al. 2004). Total load forecasting includes forecasting of future electricity (power) demand, future power (energy) consumption, and load curve.

Load forecasting is a key task for effective operation and planning of power systems. Such forecasts are sometimes categorized as short-, mid—and long-term forecasts, depending on the time horizon. Usually, mid-long-term demands forecasting, correspondes to the forecast horizon from several months to several years ahead. This is an integral process in scheduling the construction of new generation facilities and in the development of transmission and distribution systems (Fan 2010, Hyndman 2010).

To date in literatures, short-term load forecasting has attracted substantial attention due to its importance for power system control, unit commitment, economic dispatch, and electricity markets. On the other hand, the mid-long-term forecasting has not received as much attention, despite their value for system planning and budget allocation, and few papers present practical approaches verified through field implementations at utilities (Fan 2010, Hyndman 2010, Hong et al. 2014).

Due to the time horizon of mid-long-term load forecasting is long, and it is affected by influence of various complex factors, such as politics, economic, and climate, which result in that forecasting accuracy cannot meet ideal demand (Li et al. 2011, Qian et al. 2007, Wu et al. 2012). An overestimate of long-term electricity demand will result in substantial wastage of investment in the construction of excess power facilities, while an underestimate of demand will result in insufficient generation and unmet demands (Hyndman 2010 & Fan 2010). Hobbs's report pointed out to reduce the average absolute percentage error by 1.5%, and there would be an economic benefit improvement of 7.6 million USD a year (Hobbs et al. 1998). The power grid planning department should recognize the importance of mid-long-term load forecasting well, and neither overvalue or undervalue the future load, to realize a reasonable and green development of the power grid.

2 CLASSIFICATION OF MID-LONG-TERM LOAD FORECASTING METHODS

Mid-long-term load forecasting methods can be divided into experiential methods and mathematical model methods. The experiential methods refer to the methods that rely on judgment of experts or expert groups for load forecasting, without quantitative models (Ye et al. 2012).

Classified by used data, mathematical model methods include inherent law extrapolation

Table 1. A classification of mid-long-term load forecasting methods.

Experiential methods	Expert forecasting method Analogy method Subjective probability methods	
Mathematical model methods	Inherent law extrapolation methods	Growth curve extrapolation Linear regression Exponential smoothing method Grey system method
	Correlation analysis methods	Multiple regression Consumption of per unit output value of each industry method Electricity elasticity coefficient method Load density method
Mathematical model methods	Nonparametric models	Expert system Fuzzy forecasting Neural network forecasting method Evidence theory Data mining technology Support vector machine
	Parametric models	Grey system method Electricity elasticity coefficient method Time series method Regression analysis method

methods and correlation analysis methods. On the basis of historical load data, inherent law extrapolation methods launches the change rule of load through analysis of historical load data, and the forecast future load by extrapolating the change and development model of load. The correlation analysis methods combine load forecasting with various social and economic factors, and forecast load though finding and establishing relationships or data models between load development and other related factors that are affecting the change of load (He 2013 & Yuan et al. 2012).

The mathematical model methods can be divided into parametric models and nonparametric models, according to whether parameter calculation is needed or not. Parametric models can grasp the change regularity of mid-long-term load better, but lack adequate consideration of random variation component of load. With development of new theory, the nonparametric models have been applied gradually in load forecasting (Wu et al. 2012). The physical meaning of the nonparametric models is not clear, and the adjustment of the model is not easy when forecasting results are not satisfactory. There is a classification of mid-long-term load forecasting methods given in Table 1.

A single load forecasting method is generally based on one or several change rules of the load, and ignores the influence of other factors, and thus a single load forecasting method has its deficiencies. For multivariable and non-deterministic

mid-long-term load forecasting, it is not enough to use a single qualitative or quantitative forecasting method. The problems how to combine to a variety of methods reasonably, realize complementary advantages, and further improve the accuracy of load forecasting have become the focus (Wu et al. 2012). In this case, the combined forecasting method becomes a consensus. It is widely considered to be a forecasting strategy progress that from a single forecasting model to combined forecasting method (Kang et al. 2004).

3 INHERENT LAW EXTRAPOLATION METHOD

This paper will illustrate typical and commonly used inherent law extrapolation methods.

3.1 Growth curve extrapolation method

By using some kinds of function curves and historical statistical data fitting of load, the growth curve extrapolation methods establish forecasting models that can describe development process of load, and then forecast load by extrapolating this model (He 2013). The mainly used function curves include polynomial curve and exponential curve.

There is a model identification problem in growth curve extrapolation methods, which is usually solved by pre analysis and calculation of time series, and recognition and trend analysis of graphs. The two conditions for application of growth curve extrapolation methods are: a) the development process of the load is not jumping, but changing gradually; b) factors that determine the past development of load still decide the future development of load in a large extent. The growth curve extrapolation methods are generally used for a small number of forecast periods, and the forecasting results will get inadequate accuracy for a large number of forecast periods.

3.2 Grey system method

The grey system theory was first put forward by Professor Julong Deng that plays an important role in forecasting field. A grey system is an inadequate information system; one important characteristic of it is the uncertain factors of the system or uncertain relationships between factors. The grey system theory considers a random variable as a grey variable that changes within a certain range, and does not seek its statistical rule and probability distribution, but processes the original data. The grey system transfers desultory original data into regularly generated data, and forecasts load by establishing dynamic model for generate data (He 2013 & Liu 2003).

The grey theory model has advantages of a small sample number, convenient calculation and accurate forecasting. The power system is a typical grey system, so it is very suitable for power load forecasting. But the grey model itself has some limitations, the greater the discrete degree of data, the larger the grey scale, and the worse the forecasting accuracy.

The grey time series model GM (1, 1) is the most commonly used in power system mid-long-term load forecasting (He 2013).

In recent years, in view of defects and the shortcomings of the GM (1, 1) model, many improvement measures are put forward. The improvement methods can be summarized as following three categories:

1. Improvement of estimation method for model parameters;
2. Improvement of construction method for background value of models;
3. Improvement of initial conditions for model calculation (Li 2011 & Mao 2011).

Li & Mao (2011) presented an optimized GM (1, 1) model based on the bi-directional difference method, which adopted minimum sum of fitting error square of forward difference and backward difference modeling to estimate parameters, and initial values of GM (1, 1) prediction model were optimized by using the method of minimum sum of error square of fitting values and initial values. Wang & Wang (2013) proposed a variable weights

buffer grey model, integrating the variable weights buffer operator with the background value optimized GM (1, 1) model, to implement dynamic preprocessing of original load data. The improvement methods of GM (1, 1) model in Li & Mao (2011) and Wang & Wang (2013) are all of above class 1) and class 3).

The traditional GM (1, 1) model showed great error when it was used to forecast the "S" type sequence. To overcome this shortage, Zhang et al. (2011) put forward a discrete grey Verhulst model. The equal-dimension grey number addition forecasting method was used to improve the grey discrete Verhulst model. Zhou (2009) built an improved Verhulst model based on Least Square-Support Vector Machine (LS-SVM) algorithm and equal-dimension and new-information technique. Both of Zhang et al. (2011) and Zhou (2009) make forecasting accuracy been improved.

3.3 Notes for inherent law extrapolation methods

The model parameters of inherent law extrapolation methods are determined by historical data, and cannot change with different situations in future. It is difficult for these methods to use expert experience and opinion. Therefore, adaptively these forecasting models need to be strengthened (Tan et al. 2011). In addition, inherent law extrapolation methods could not reflect the comprehensive effect of various factors on the load. The grey model has stable forecasting results. The regression analysis model based on various mathematical curves is worth recommending (Kang et al. 2004).

4 CORRELATION ANALYSIS METHOD

4.1 Consumption of per unit output value of each industry method

The consumption of per unit output value of each industry method analyzes the three major industries of national economy statistically, and combines with the economic development and industrial structure adjustment conditions, to determine the power consumption of per unit output value in these three industries. And then according to the national economic and social development planning indicators, the forecast value of load in the planning period will be calculated out.

4.2 Electricity elasticity coefficient method

The electricity elasticity coefficient is a ratio of the average annual growth rate of electricity consumption and the average annual growth rate of Gross Domestic Product (GDP) for a certain period:

$$E = \frac{\bar{R}_e}{\bar{R}_g} \qquad (1)$$

where $\bar{R}_e$ = average annual growth rate of electricity consumption for a certain period; and $\bar{R}_g$ = average annual growth rate of GDP for a certain period. The electricity elasticity coefficient is an indicator that reflects the relationship between the development of electric power industry and the development of the national economy from the macro point of view (He 2013).

The forecasting steps are as bellows: determine or forecast $\hat{E}$ for a period of time in future by a certain method, and the average growth rate of GDP has been forecasted as $\bar{R}_g$, then the growth rate of electricity consumption can be calculated:

$$\bar{R}_e = \hat{E} \times \bar{R}_g \qquad (2)$$

Then the forecasted electricity of the tth period will be:

$$\hat{W}_t = W_o(1 + \bar{R}_e)^t \qquad (3)$$

where W_o = electricity demand of the base year in forecasting.

4.3 *Notes for correlation analysis methods*

Correlation analysis methods can reflect influences of various social and economic factors on load forecasting, and have been applied in power utilities. For example, the electricity elasticity coefficient method has been widely used as an effective method, because of its accurate reflection on the national economy (Kang et al. 2004). But correlation analysis methods usually need large amount of data, and some data are not easily acquired (Yuan et al. 2012). These methods are dependent on statistical data, and the effectiveness of the relevant data has big influence on the accuracy of forecasting. Compared with inherent law extrapolation methods, the implementation of correlation analysis methods is more difficult.

5 COMBINED FORECASTING METHOD

For multivariable and non-deterministic mid-long-term load forecasting, advantages of combined forecasting methods have been presented gradually, and a lot of literatures have studied on basis of these methods.

The grey system and artificial neural network had been selected as the basic models to be integrated in Wu et al. (2012), and the evidential theory was used here to mix results of various basic forecasting methods. Yuan et al. (2012) improved the traditional electric consumption elasticity coefficient method, and combining with grey prediction model and secondary moving average method, established a new integrated model for load forecasting, and used genetic algorithm to optimize the weight of each single forecasting model. Tan et al. (2011) proposed a combined forecasting method based on econometrics approach and system dynamics. Two kinds of nonparametric models were combined in Xia et al. (2012), and a heuristic radial basic function neural network load forecasting model was proposed. The orthogonal signal correction method was used to process primary data, and neural network extended training samples were constructed. The forecasting results of heuristic neural network forecasting model were corrected on the basis of rough set theory to further improve the forecasted accuracy. A new combined forecasting model was put forward based on support vector machines and Markov chain theory in Chen & Su (2012). A new load forecasting procedure was described in Ghelardoni et al. (2013), which exploited the empirical mode decomposition method had to disaggregate a time series into two sets of components, respectively describing the trend and the local oscillations of the energy consumption values. These sets were then used for training support vector regression models.

There are two key issues in combined forecasting methods, one is how to determine single forecasting methods used, and the other one is how to determine the weight of each single forecasting method (Yuan et al. 2012). The accuracy of combination forecast methods tend to rely heavily on forecasting accuracy of each single forecasting model, and the high correlation between forecasting results of single forecasting methods will significantly reduce accuracy of a combined forecasting method (Liu 2012). There are two combination methods, permanent weights and variable weights. Combined forecasting methods with variable weights get good adaptability, but are also based on error theories, and are difficult to reflect the validity of the model (Yin et al. 2015).

Wen et al. (2011) proposed two indicators to calculate the weight of each single forecasting model. One referred to errors between each single-model predictions and the actual load, and the other one referred to the errors between annual growth rate of each forecast value and annual growth rate of GDP. The objective entropy method and subjective G1 method were adopted to determine the relative importance of the two indicators and to get the weight of each single model. A new combination model based on induced ordered weighted geometry averaging operator and weighted

Markov chain was proposed in Mao et al. (2010a). According to the level of accuracy, this model assigned the weight to each individual method to achieve the correlation between weight coefficient and fitting accuracy in any time point. But the calculation of weights is still based on historical data, and this model is applicable when data is continuous, and the forecasting error will be big when data changes suddenly. In Li et al. (2009), three index quantities, i.e., the total index quantity, the increasing index quantity and the index growth rate, were proposed. The presented comprehensive model selected the optimal forecasting model for each index quantity, then, by using radial basic function neural network, the forecasted results from the three optimal models were fitted. Weights calculation was instead by radial basic function neural network algorithm.

6 PROSPECT OF MID-LONG-TERM LOAD FORECASTING

6.1 Active distributed net

The active distributed net was put forward originally by the C6.11 team of distributed net and distributed generation special committee of the 2008 international council on large electric systems. The active distributed net is defined as a distribution system that manages power flow by flexible network topology, to control and manage local distributed energy actively (Zhong et al. 2014).

From the perspective of whether loads in active distributed net are controllable, they can be classified into friendly load and unfriendly load. Friendly load refers to load can be controlled completely, and can adjust actively according to power grid scheduling and load guiding mechanism, playing a useful role for safe and economic operation of distributed net. This is known as the reflection of demand side response in active distributed net (Zhong et al. 2014). It will affect analysis methods for mid-long-term load forecasting. The mid-long-term load forecasting technology should take technical features of the active distributed net into consideration.

Active demand had been considered in mid-long-term load forecasting in Garulli et al. (2015), based on grey-box models where the seasonal component of the load was extracted by a suitable preprocessing and active demand was considered as an exogenous input to a linear transfer function model. The effect of electric vehicle charging load should also be considered in mid-long-term load forecasting. There are literatures specializing in electric vehicle charging load at present. A forecasting method for electric vehicle charging load

considering the spatial and temporal distribution was proposed in Zhang et al. (2014), based on driving and parking characteristics of private cars. The Monte Carlo simulation method was adopted to simulate electric vehicle parking, driving and charging behaviors at different time and different places, for the forecasting of the spatial and temporal distribution characteristics of electric vehicle charging load.

6.2 Electric power big data and cloud computing

Information technology special committee of the Chinese Society of Electrical Engineering issued White Paper on Development of Chinese Electric Power Big Data in 2013. In this paper, it states the characteristics, value for society, prospects and technical challenges of electric power big. Big data of electric power user side are mainly from various kinds of sensors and smart home appliances. The coverage rate of intelligent electric meter has reached 40.5% at the end of January in 2013 (Wang 2015 & Sun 2015).

The cloud computing is a kind of basic platform and efficient supporting technology for management solution of big data. At present, the big data and cloud computing technology have been applied in access of new energy, safety state assessment of wind turbines and disaster warning of power grid, and will also play an important role in power system load forecasting. Professor Victor of the Oxford University Networking Academy has pointed out in his book "Age of Big Data": in age of big data, the correlation analysis will shine, and it can capture today and forecast the future by finding associated content in a phenomenon (Peng et al. 2015). Under the background of increasing large data of load statistical index in power system and temperature, it is of great significance to extract correlation characteristics between data effectively for planning and operation of power system (Ma et al. 2015). The big data will promote application of the earth space technology and the weather forecasting data in the smart grid, and improve the load forecasting accuracy and new energy generation (Zhang et al. 2015).

6.3 Spatial power load forecasting

Spatial power load forecasting refers to forecasting of the amount and distribution of the future electric load in a power supply area. However, traditional load forecasting methods only forecast the amount of the future load, and do not give a more precise distribution of load. Forecasting of power load distribution first appeared in studies abroad in the mid of 1930s, and domestic studies of spatial power load forecasting started relatively later. As the management of the power system is transferring from extensive to intensive, traditional load forecasting methods cannot meet the requirements of power system planning. Therefore, spatial power load forecasting has become one of the hottest researches (Xiao et al. 2013). For example, Xiao et al. (2014) analyzed abnormal data of cellular load, calculated the probability spectrum curve, and obtained the maximal cellular load suitable for spatial load forecasting by limitation and constraints of the probability spectrum curve.

In a power system covering a large geographical area, a single model for load forecasting of the entire area sometimes cannot guarantee satisfactory forecasting accuracy. One of the major reasons is the load diversity, usually caused by weather diversity throughout the area (Fan et al. 2009). Combination of total load forecasting and spatial load forecasting is feasible and effective for solution, and to generate accurate load forecasting results for large area power system. But, the spatial load forecasting could not substitute for total load forecasting, for the latter playing an irreplaceable guiding role in long-term development planning of the power industry.

After introductions of load coincidence factors and thought of multi-regional load forecasting, results of spatial load forecasting can be transformed into total load forecasting.

6.4 *Data preprocessing*

The historical load data is the basis of mid-long-term load forecasting, and the data preprocessing has a significant effect on the load forecasting accuracy.

Influence on the number of historical data periods on forecasting results will be different for different forecasting methods, and characteristics of forecasting methods should be considered in selection of period number. Generally speaking, it is advisable to select the number of historical data periods as 5–10 years for mid-long-term load forecasting (Kang et al. 2004). A principle of "Paying more attention to near historical periods" is also often mentioned, which indicates that the trend of physical quantities in the future depends more on their development law in near historical periods, and correlation between future development trend, and long-term historical data is weaker. The principle of "Paying more attention to the near historical periods" is easily to be achieved in mid-long-term load forecasting, and a commonly used method is weighted parameter estimation (Kang et al. 2004).

The abnormal historical data and historical data missing seriously affect the accuracy and effectiveness of load forecasting (Mao et al. 2010b). Identifying abnormal data correctly and filling missing data are important for load forecasting. Many scholars have turned their attention to the load forecasting data preprocessing technology. A method for missing data filling, based on both T^2 ellipse map to identify abnormal data and least square support vector machine was proposed in Mao et al. (2010b). In order to deal with the data uncertainty problem in mid-long term load forecasting. Monte Carlo method and interval arithmetic were introduced in Zheng et al. (2011), analyzing the effect of data uncertainty on prediction results strictly. Mao et al. (2009) presented an integrated method, which combined partial least square method with the improved orthogonal signal correction. This method removed the useless orthogonal information between the independent variable X and dependent variable Y effectively. The correlation between X and Y was strengthened and the explanatory ability of model's component under the condition of limited components was improved.

6.5 *Forecasting error evaluation*

Load forecasting accuracy is an ex post measurable indicator for the deviation degree between forecasting results and actual load, which is an assessment method for electricity competent authority promoting load forecasting accuracy. Forecasting error evaluation can evaluate different load forecasting methods objectively, analyze the causes of forecasting error, and select the optimal forecasting method, to further improve the forecasting accuracy and economic benefit.

At present, there were few studies on mid-long-term forecasting error evaluation technology and ex post forecasting accuracy checking. Wang & Xu (2012) comparatively analyzed forecasting accuracy evaluation methods and evaluation indicators. Load forecasting error evaluation is the foundation for development of adaptive forecasting methods, and can improve mid-long-term power load forecasting methods. Special attention should be paid to forecasting error evaluation.

6.6 *Probabilistic load forecasting*

Power load forecasting forecasts future load with historical and present load, which is beyond control and with uncertainty and conditionality (He et al. 2013). Although forecasting is a stochastic problem by nature, most utilities today are still developing and using forecasts point instead of probabilistic forecasts (Hong et al. 2014). From the perspective of confidence interval analysis in probability theory, a trumpet shaped ribbon area can be obtained, referring a variation range of forecasting data (Kang et al. 2004).

At present, probabilistic forecasting methods have been applied to daily load forecasting (Chen et al. 2011). In a long-term context, planners must adopt a probabilistic view of potential peak demand levels (Hyndman 2010 & Fan 2010). Hong et al. (2014) proposed a modern approach that took advantage of hourly information to create more accurate and defensible forecasts, and modernized predictive modeling, weather normalization, and probabilistic forecasting. He et al. (2013) proposed a probability density forecasting method using radical basis function neural network quantile regression based on the existed researches on combination forecasting and probability interval prediction. A method to predict medium-term probability density of power load based neural network quantile regression was proposed in He et al. (2015). Hyndman & Fan (2010) proposed a comprehensive methodology to forecast the density of annual and weekly peak electricity demand up to ten years ahead.

Density forecasts (providing estimates of the full probability distributions of the possible future values of the demand) are more helpful than point forecasts, and are necessary for utilities to evaluate and hedge the financial risk accrued by demand variability and forecasting uncertainty.

7 CONCLUSION

In this paper, the mid-long-term load forecasting methods are summarized and classified. Typical methods and characteristics of inherent law extrapolation methods, correlation analysis methods and combined forecasting methods are illustrated. The main issues of the future researches on this field are discussed. It is pointed out that active distributed net, electric power big data and cloud computing, data preprocessing, forecasting error evaluation, and probabilistic load forecasting should be included in the research system of mid-long-term load forecasting.

REFERENCES

Chen, J. & Su, H. 2012. A forecasting model of medium/long term power load in combination of the support vector machine and Markov chain algorithms. *Southern Power System Technology* 6(1): 54–58.
Chen, X. et al. 2011. Short term probabilistic forecasting of the magnitude and timing of extreme load. *Proceedings of the CSEE* 31(22): 64–72.
Fan, S. et al. 2009. Multiregion load forecasting for system with large geographical area. *IEEE Transactions on Industry Applications* 45(4): 1452–1459.

Garulli, A. et al. 2015. Models and techniques for electric load forecasting in the presence of demand response. *IEEE Transactions on Control Systems Technology* 23(3):1087–1097.
Ghelardoni, L. et al. 2013. Energy load forecasting using empirical mode decomposition and support vector regression. *IEEE Transactions on Smart Grid* 4(1): 549–556.
He, H. 2013. *Electric power load forecasting and load management*. Beijing: China Electric Power Press.
He, Y. et al. 2013. A power load probability density forecasting method based on RBF neural network quantile regression. *Proceedings of the CSEE* 33(1): 93–98.
He, Y. et al. 2015. A method to predict probability density of medium-term power load considering temperature factor. *Power System Technology* 39(1): 176–181.
Hobbs, B.F. et al. 1998. Artificial neural networks for short-term energy forecasting: accuracy and economic value. *Neurocomputing* 23(1): 71–84.
Hong, T. et al. 2014. Long term probabilistic load forecasting and normalization with hourly information. *IEEE Transactions on smart grid* 5(1): 456–462.
Hyndman, R.J. & Fan, S. 2010. Density forecasting for long-term peak electricity demand. *IEEE Transactions on Power System* 25(2): 1142–1153.
Kang, C. et al. 2004. Review of power system load forecasting and its development. *Automation of Electric Power Systems* 28(17): 1–11.
Li, C. et al. 2009. A comprehensive model for long–and medium-term load forecasting based on analytic hierarchy process and radial basis function neural network. *Power System Technology* 33(2): 99–104.
Li, H. & Mao, W. 2011. An optimized GM(1, 1) model based on bi-directional difference method and its application in long-term power demand forecasting. *Power System Protection and Control* 39(13): 53–58.
Li, J. et al. 2011. Mid-long term load forecasting based on simulated annealing and SVM algorithm. *Proceedings of the CSEE* 31(16): 63–66.
Liu, D. 2012. A model for medium—and long-term power load forecasting based on error correction. *Power System Technology* 36(8): 243–247.
Liu, S. 2003. Emergence and development of grey theory and its frontier trend. *Chinese Journal of Management Science* (11): 29–32.
Ma, R. et al. 2015. Data mining on correlation feature of load characteristics statistical indexes considering temperature. *Proceedings of the CSEE* 35(1): 43–51.
Mao, L. et al. 2009. Medium and long term load forecasting based on orthogonal signal correction and partial least-squares regression. *Proceedings of the CSEE* 29(16): 82–88.
Mao, L. et al. 2010a. Theoretical study of combination model for medium and long term load forecasting. *Proceedings of the CSEE* 30(16): 53–59.
Mao, L. et al. 2010b. Abnormal data identification and missing data filling in medium-and long-term load forecasting. *Power System Technology* 34(7): 148–153.
Peng, X. et al. 2015. Key technologies of electric power big data and its application prospects in smart grid. *Proceedings of the CSEE* 35(3): 503–510.

Qian, W. et al. 2007. Short-term correlation and annual growth based mid-long term load forecasting. *Automation of Electric Power Systems* 31(11): 59–64.

Tan, Z. et al. 2011. A model integrating econometric approach with system dynamics for long-term load forecasting. *Power System Technology* 35(1): 186–190.

Wang, D. & Sun, Z. 2015. Big data analysis and parallel load forecasting of electric power user side. *Proceedings of the CSEE* 35(3): 527–537.

Wang, D. & Wang, B. 2013. Medium–and long-term load forecasting based on variable weights buffer grey model. *Power System Technology* 37(1): 168–171.

Wang, G. & Xu, Y. 2012. Research on accuracy assessment method for load forecast. *Guangdong Electric Power* 25(11): 39–42.

Wen, Q. et al. 2011. The application of combination forecasting model in medium-long term load forecasting based on load error and economic development trend. *Power System Protection and Control* 39(3): 58–61.

Wu, Y. et al. 2012. The medium and long-term load forecasting based on improved D-S evidential theory. *Transactions of China Electrotechnical Society* 27(8): 157–162.

Xia, F. et al. 2012. A model for medium and long term load forecasting based on rough set theory and heuristic radial basic function neural network. *Power System Protection and Control* 40(16): 21–26.

Xiao, B. et al. 2013. Review and prospect of the spatial load forecasting methods. *Proceedings of the CSEE* 33(25): 78–92.

Xiao, B. et al. 2014. A probability spectrum method for ascertaining maximal value of cellular load in spatial load forecasting. *Automation of Electric Power Systems* 38(21): 47–52.

Ye, Z. et al. 2012. Medium/long term load forecast of exponential smoothing method based on information and equal dimensional operators. *Power System Protection and Control* 40(18): 47–51.

Yin, X. et al. 2015. Bus load forecasting model selection and variable weights combination forecasting based on forecasting effectiveness and Markov chain-cloud model. *Electric Power Automation Equipment* 35(3): 114–119.

Yuan, T. et al. 2012. Study on the comprehensive model of mid-long term load forecasting. *Power System Protection and Control* 40(14): 144–151.

Zhang, C. et al. 2011. Middle and long term power load forecasting based on grey discrete Verhulst model's theory. *Power System Protection and Control* 41(4): 45–49.

Zhang, D. et al. 2015. Research on development strategy for smart grid big data. *Proceedings of the CSEE* 35(1): 2–12.

Zhang, H. et al. 2014. A prediction method for electric vehicle charging load considering spatial and temporal distribution. *Automation of Electric Power Systems* 38(1): 13–20.

Zheng, Z. et al. 2011. Medium and long term load forecasting considering data uncertainty. *Power System Protection and Control* 39(7): 123–132.

Zhong, Q. et al. 2014. Load and power forecasting in active distribution network planning. *Proceedings of the CSEE* 34(19): 3050–3056.

Zhou, D. 2009. Application of improved gray Verhulst model in middle and long term load forecasting. *Power System Technology* 33(18): 124–127.

Advances in Power and Energy Engineering – Sun (Ed.)
© 2016 Taylor & Francis Group, London, ISBN 978-1-138-02846-3

Parameter identification of equivalent model of hydro generator units group based on PSO method with distributed computing structure

Xin Xia & Wei Ni
Faculty of Automation, Huaiyin Institute of Technology, Huaian, China

ABSTRACT: Dynamical equivalent can reduce the computing effort of the electrical power system. The characteristics of the hydro generator and the time taken for identification methods are without proper consideration in traditional dynamical equivalent methods. In this paper, a five-order generator model is used as the equivalent model of hydro generator group, and a novel objective function is also proposed. Then, a PSO identification method with distributed computing structure has been proposed to reduce the identification time. Finally, experiments have been utilized to verify the proposed method. The results indicate that the dynamic response of the equivalent model approaches the original system very well, and the proposed method can reduce the identification time significantly.

1 INTRODUCTION

With the expansion of the scale of the electrical power system, thousands of synchronous generators should be considered in transient stability studies and other routine power engineering studies which leads to the problem of massive computation. The main motivation of this work is to build an equivalent generator model of a group of hydro generator units. The aim of this work is to simplify the electrical power system effectively.

Traditional dynamical equivalence methods (Chen 2004, Podmore 1978, Jiang 2010) can be divided into three parts: First, dynamic recognition of coherent unit groups;second, network simplification. Finally, parameter aggregation of coherent unit is in groups. Because the generator parameters are not constant throughout the useful life, some parameters have changed because of repairing and aging. So the parameters obtained by manufacturer data sheets are not accurate which leads to the parameter aggregation method that cannot get accurate models of synchronous generators.

Some researchers have proposed the estimation for parameters of synchronous generators online instead of parameter aggregation (Miah 1998, Machowski 1985, Kyriakides 2005, Ju 2007 & 2008, Ghahremani 2011, Lu 2009). Online parameters identification can ensure the accuracy request. Elias Kyriakides (Kyriakides 2005) estimated the parameters of round rotor synchronous generators including magnetic saturation online. Ju Pin (Ju 2007 & 2008) used the data of WAMS (Wide Area Measurement System) to estimate the equivalent generator and load model of an electrical power system area. Ghahremani (Ghahremani

2011) has used the experimental online data to estimate the third-order model parameter of the synchronous generator. These researches make a great contribution to the development of electrical power system analysis. But hydro generators are salient pole generator; the third-order model is not suitable. So there is an urgent need to develop a new equivalent model which can describe the characteristic of hydro generator unit.

As the operating condition of hydro generator units changes frequently, the equivalent model is changing through time. Traditional identification methods cannot meet the real time demand. Real time identification methods have become a research topic, and have already attracted plenty of attention.

In this paper, first, a five-order synchronous generator is used as an equivalent model of hydro generator units group.. And a novel objective function is also proposed at the same time. Second, a new identification method with PSO and distributed computing has been proposed. The method has great efficiencies in parameters identification. And last, this method is employed on a typical region. The results indicate that the proposed method can get accurate equivalent model, and the identification time has been reduced significantly.

2 IDENTIFICATION OF EQUIVALENT GENERATOR MODEL BASED ON WAMS

As the expansion of the size of electrical power system, transient calculations encounter the computational bottleneck;equivalent methods can simplify the structure of electrical power system, which can greatly reduce the calculation effort. Equivalent methods are mainly divided into two categories: parameter aggregation and online parameter identification. As generator parameters are in general not constant throughout the useful life, parameter aggregation method cannot get the accurate models. With the WAMS widely applied on the electrical power system, a lot of data have been collected which makes parameter identification possible. Many researchers have indicated that online parameter identification method can get the equivalent model which can greatly suit the transient progress in the electrical power system.

Most online parameter identification methods set the equivalent generator model as a three-order model. Although, this method can suit most thermal power generators and have simple calculation, but three-order model is not suitable for hydro generators which have salient pole. In this paper, a five-order synchronous generator is used as an equivalent model of hydro generators, as follows:

$$\begin{cases} T'_{d0}\dfrac{dE'_q}{dt} = E_{fd} - [E'_q + (x_d - x'_d)I_d] \\ T''_{d0}\dfrac{dE''_q}{dt} = -E''_q - (x'_d - x''_d)I_d + E'_q + T''_{d0}\dfrac{dE'_q}{dt} \\ T''_{q0}\dfrac{dE''_d}{dt} = -E''_d + (x_q - x''_q)I_q \\ u_q = E''_q - x''_d I_d , \ \ u_d = -E''_d + x''_q I_q \end{cases} \quad (1)$$

$$\begin{cases} \dfrac{d\delta}{dt} = (\omega - 1)2\pi f \\ T_J\dfrac{d\omega}{dt} = P_m - P_e - D(\omega - \omega_0) \end{cases} \quad (2)$$

Equation 1 is the electromagnetic equations of generator, where T'_{d0} = open-circuit time constant of d axis; T''_{d0} = open circuit sub transient time constant of d axis; T''_{q0} = open-circuit sub transient time constant of q axis; x_d = reactance of d axis; x_q = reactance of q axis; x'_d = transient reactance of d axis; x'_q = transient reactance of q axis; x''_d = sub transient reactance of d axis; x''_q = sub transient reactance of q axis; E'_q = transient voltage of q axis; E''_q = sub transient voltage of q axis; E''_d = sub transient voltage of d axis; E_{fd} = the exciting voltage; I_d = current of d axis; I_q = current of q axis; u_d = voltage of d axis; u_q a = voltage of q axis.

Equation 2 is the rotor equations of generator, where δ = the rotor angle in rad; ω = the electrical speed in per unit; P_m = the input mechanical

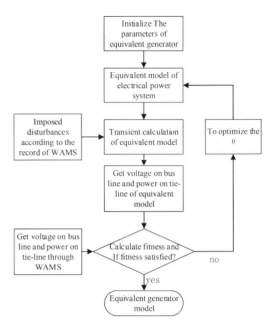

Figure 1. Equivalent generator model identification progress.

torque; P_e = the output electric; D = the damper coefficient; T_J = constant of inertia time.

From Equation 1 and Equation 2, δ, ω, E'_q, E''_q, E''_d are state variables. T'_{d0}, T''_{d0}, T''_{q0}, x_d, x_q, x'_d, x'_q, x''_d, x''_q, T_J are unknown variables. The unknown variables are set as:

$$\theta = [X_d, X'_d, X''_d, T'_{d0}, T''_{d0}, X_q, X'_q, X''_q, T''_{q0}, T_J] \quad (3)$$

In this paper, considering that different unknown variables compositions may result in the different system outputs: the voltage on bus line and power on tie-line. The basic idea of this parameter identification progress is to find a group of θ whose output is mostly approached to the data of WAMS. The identification progress is proposed as shown in Figure 1. The fitness is defined as Equation 4:

$$fitness = \frac{1}{N}\left(\sum_{i=1}^{N}(V_i - V'_i)^2 + \sum_{i=1}^{N}(P_i - P'_i)^2\right) \quad (4)$$

where V_i = the ith voltage amplitude sample of the tie-line of the original system, V'_i = the ith voltage amplitude sample of the tie-line of the identification system, P_i = the ith electrical power sample of the tie-line of the original system, P'_i = the ith sample electrical power of the tie-line of the identification system.

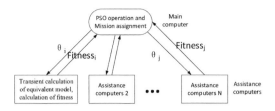

Figure 2. The structure and progress of distributed computing system.

3 A PSO IDENTIFICATION METHOD WITH DISTRIBUTED COMPUTING STRUCTURE

The Particle Swarm Optimization (PSO) (Kennedy 1995) is proposed by Kennedy and Eberhart in 1995. This method has the advantage of high computational efficiency, good convergence property and easy to implement. It has already attracted plenty of attention in the research of the electrical power system (Mohaghegi 2005).

Although, the PSO has high computational efficiency, but the transient calculation of equivalent model needs a lot of time which limit the efficiency of the identification progress. Distributed computing is proposed in recent year, and it is an effective method for solving complex large-scale engineering computation problems (Lu 2009).

In this paper, a PSO identification method based on distributed computing structure is proposed. A computer is set as the main computer, and it operates the PSO method, which doesn't contain the calculation of fitness. At the same time, the main computer controls the mission assignment. The other computers are set as assistant computers, and their functions are transient calculations of equivalent model, calculation of fitness and data communication based on TCP/IP.

The structure of distributed computing is shown in Figure 2.

4 EXPERIMENTAL STUDY

An experimental study has been utilized to test the proposed method. The experiment is based on the IEEE10g39 region as shown in Figure 3. The equivalent system is equated to a simple equivalent generator as shown in Figure 4. The structure of the equivalent generator is set as the model which is introduced before.

The parameters of transient simulation are set as: a three-phase short circuit fault, which happens in bus 25 at 0.5s. The simulation time is 5s. The simulation step is set as 0.05s.

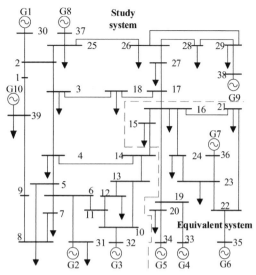

Figure 3. The IEEE10g39 region.

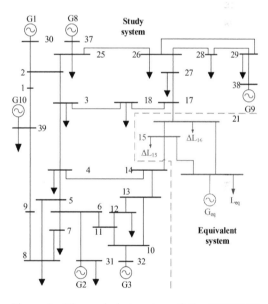

Figure 4. The equivalent system of the IEEE10g39 region.

The distributed computing system is organized by six computers. And the parameters of the PSO method are set as Table 1.

First, the proposed method is compared to the traditional identification method of one computer. The identification time is shown in Table 2. It indicated that the distributed computing system can reduce the identification time. The results also point out that as the increase number of distributed computer, the computing time will reduce significant.

Table 1. The parameters of the PSO method.

Popsize	Inertia weight	c_1	c_2	Max average error
100	0.7 and 0.2	1.49	1.49	0.01(pu)

Table 2. The identification time of the different methods.

	One computer	Distributed computing system
Average identification time (20 times) (s)	415	167

Table 3. The parameters of the equivalent generator.

Parameters	X_d	X_d'	X_d''	X_q	X_q'
Value (pu)	3.500	0.797	0.199	3.100	0.800
Parameters	X_q''	T_{d0}'	T_{d0}''	T_{q0}''	T_J
Value (pu)	0.200	6.256	0.047	0.035	5.656

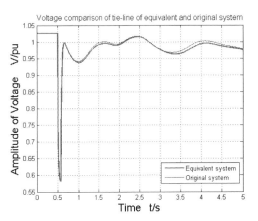

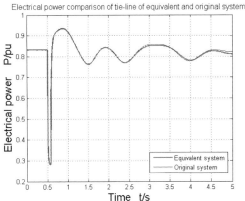

Figure 5. The dynamic response comparison of the original system and the equivalent system.

The parameter of the equivalent generator, which is identified by the proposed method, is shown in Table 3.

Then, the transient response of tie-line is utilized to verify the correctness of the proposed method. The voltage and electrical power of tie-line of the original system and the equivalent system is shown in Figure 5.

The results show that the dynamic response of the equivalent system approaches the original system very well; and indicates that the proposed method had a strong effect in reducing the power systems.

5 VERIFICATION OF ACTUAL EXAMPLE

In order to verify the proposed method, an actual example is employed. The hydro generator unit group of Enshi of China is set as the modeling object. The wiring diagram is shown as Figure 6.

In this example, the group of hydro generator units is equated to one simple hydro generator. And the data of WAMS is employed to identify the model parameters. The result is shown as Table 4.

In order to verify the veracity of the equivalent model, the equivalent model is used to simulate the dynamic response of the power system. The voltage magnitude and active power of the original

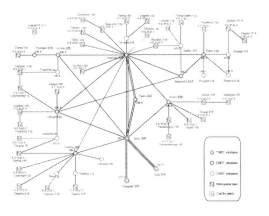

Figure 6. The wiring diagram of hydro generator units group of Enshi.

Table 4. The parameters of the equivalent generator.

Parameters	X_d	X_d'	X_d''	X_q	X_q'
Value (pu)	0.897	0.237	0.187	0.750	0.235
Parameters	X_q''	T_{d0}'	T_{d0}''	T_{q0}''	T_J
Value (pu)	0.187	3.120	0.025	0.089	6.66

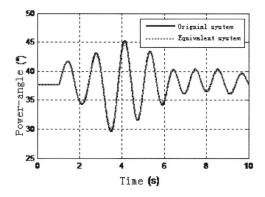

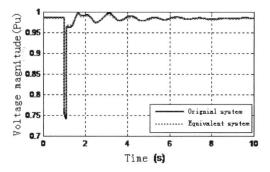

Figure 7. The dynamic response comparison of the original system and the equivalent system.

system and the equivalent system are compared in Figure 7.

From Figure 7, it can be seen that the dynamic response of the equivalent system approaches the original system well. The equivalent system can give a good description of the dynamic characteristic of the original system. At the same time, the model of the power system has great reduction which is useful to transient stability studies and other routine power engineering studies.

So the actual example also indicates that the proposed method is useful.

6 CONCLUSION AND DISCUSSION

Traditional dynamical equivalent methods donot consider the characters of hydro generator, and the identification method cost a lot of time. In this paper, a five-order generator model is used as the equivalent model of hydro generator group. Then, a PSO identification method with distributed computing structure has been proposed to reduce the identification time. And at last, the experiments

have been done. Hence, the results show the proposed method is useful.

ACKNOWLEDGMENT

This work is supported by National Science Foundation for Distinguished Young Scholars of China (61203056). Platform Construction Project of Technology Innovation of Huaian China (HAP201432).

REFERENCES

Chen E, Mackay D S. 2004. Effects of distribution-based parameter aggregation on a spatially distributed agricultural nonpoint source pollution model. *Journal of Hydrology* 295(1): 211–224.

Podmore R. 1978. Identification of Coherent Generators for Dynamic Equivalents. *IEEE Transactions on Power Apparatus & Systems* 97(4): 1344–1354.

Jiang H, Jin S, Wang C. 2010. Parameter-Based Data Aggregation for Statistical Information Extraction in Wireless Sensor Networks. V *IEEE Transactions on ehicular Technology* 59(8): 3992–4001.

Miah, A.M. 1998. Simple dynamic equivalent for fast online transient stability assessment. *IEE Proceedings—Generation, Transmission and Distribution* 145(1): 49–55.

Machowski, J. 1985. Dynamic equivalents for transient stability studies of electrical power systems. *International Journal of Electrical Power & Energy Systems* 7(85): 215–224.

Kyriakides, Elias, G.T. Heydt, and V. Vittal. 2005. Online Parameter Estimation of Round Rotor Synchronous Generators Including Magnetic Saturation. *IEEE Transactions on Energy Conversion* 20: 529–537.

Ping, J.U., et al. 2007. Identification Approach to Dynamic Equivalents of the Power System Interconnected with Three Areas. *Proceedings of the Csee* 13(13): 29–34.

P. Ju, and D. Ma. 2008. *Load model of power system*. Beijing: China Electric Power Press.

Ghahremani, E., and I. Kamwa. 2011. Online state estimation of a synchronous generator using unscented Kalman filter from phasor measurement units. *IEEE Transactions on Energy Conversion* 26(4): 1099–1108.

Lin, L., et al. 2009. A distributed hierarchical structure optimization algorithm based poly-particle swarm for reconfiguration of distribution network. *Power System Protection & Control* 37: 56–60.

Kennedy, James. 1995. Particle swarm optimization. *IEEE International Conference on Neural Networks* 4: 1942–1948.

Mohaghegi, S., et al. 2005. A comparison of PSO and backpropagation for training RBF neural networks for identification of a power system with STATCOM. *Swarm Intelligence Symposium SIS 2005. Proceedings 2005*: 381–384.

Advances in Power and Energy Engineering – Sun (Ed.)
© 2016 Taylor & Francis Group, London, ISBN 978-1-138-02846-3

An improved calculation method for direct transient stability assessment with detailed generator model

Y. Wang & H.S. Sun
State Key Laboratory of Advanced Electromagnetic Engineering and Technology, Huazhong University of Science and Technology, Wuhan, China

X.J. Pan, Y.P. Xu & J.H. Xi
Central China Branch of State Grid Corporation of China, Wuhan, China

ABSTRACT: In direct methods for transient stability analysis, the system transient energy at the end of the disturbance is compared with critical energy and the difference is used as an indication of stability. The transient energy function with a detailed model is always a tough question for direct method research. When the generator model is represented with its Thevenin equivalent circuit in detail, the expression of the transient energy function is similar to that of the classical model. Theoretical analysis and simulations show that generator equivalent internal impedance has a great influence on the computational results of critical energy. A computational method of the generator equivalent internal impedance is proposed and then an improved calculation method for the direct transient stability assessment with a detailed generator model has been developed. Case studies on a three-machine nine-bus system and Central-China power grid equivalent system are given to verify the effectiveness of the presented method in improving the precision.

1 INTRODUCTION

The issue of transient stability of power system has always been, and remains, a major concern for researchers and operators. At the present, there are mainly two methods for analyzing transient stability, one is time-domain simulation and the other is the direct method (Fouad, A.A. & Vittal, V. 1991) (Ni, Y.X. et al. 2002) (Liu, S. & Wang, J. 1996) (Liu, S. 1998). Transient Energy Function (TEF) is a representative of direct method. One of the major drawbacks of TEF is that the model is too simple. The model used initially in TEF is the classical model which is too simple to provide assessment results with high accuracy. How to construct the transient energy function which contains a more detailed element model has been a tough question in the direct method. Network structure preserving model (Bergen, A.R. & Hill, D.J. 1981) (Narasimhamurthi, N. & Musavi, M.T. 1984) (Tsolas, N. et al. 1985) is proposed to describe the power system accurately, but the traditional method of transient stability assessment is difficult to be applied to that model (Yin, M.H. & Zou, Y. 2003).

In reference (Fouad, A.A. et al. 1989), effect of excitation control is incorporated in direct transient stability assessment. When a synchronous generator with exciter is assumed to be represented with Thevenin equivalent circuit, expression of transient energy function is similar with that for classical model and state variables other than power angle in the TEF are treated as parameters. The investigation is primarily concerned with the effects of the internal voltage variations on the system transient energies while calculation method of equivalent internal impedance of generator is relatively rough. The author discovered that the equivalent impedance of the generator has great influence on the accuracy of the transient stability assessment. So calculation method of generator's equivalent impedance needs further discussion.

The structure of the paper is as follows. Section II gives an introduction of procedure for direct transient stability assessment with excitation control proposed in (Fouad, A.A. et al. 1989). In Section III, discussion is made on calculation methods of generator equivalent internal impedance, then a more accurate calculation method is proposed and also an improved calculation method for direct transient stability assessment with detailed generator model is derived. The results of two examples given in Section VI verify the effectiveness of presented method in improving accuracy.

2 DIRECT TRANSIENT STABILITY ASSESSMENT WITH EXCITATION CONTROL

TEF determines stability directly by comparing transient energy at the end of the disturbance and critical value of potential energy. The difference between them is called energy margin and used as an indication of stability. Generally, critical energy is considered at unstable control equilibrium point. When effects of exciters are considered, the derivatives of some state variables are not zero at the point where the system potential energy reaches its maximum. That point is referred to as peak point instead of control unstable equilibrium point (Fouad, A.A. et al. 1989). The critical potential energy is defined as the potential energy at the peak point. There are two key points in direct transient stability assessment, one is energy function form and the other is accurate determination of desired peak point. Both are closely related to power system model used in analysis.

2.1 System model

With the generator model by the so-called two-axis model and the excitation system represented by an equivalent one gain, one time constant, and one limiter model using the "noise-equivalent bandwidth" techniques (Brown, R.G. & Hwang, P.Y. 1992), the generator-exciter equations in the Center Of the Inertial (COI) frame are given by:

$$T'_{d0i}\dot{E}'_{qi} = E_{fdi} - E'_{qi}\left(x_d - x'_d\right)I_{di} \tag{1}$$

$$-T'_{q0i}\dot{E}'_{di} = E'_{di} - \left(x_q - x'_q\right)I_{qi} \tag{2}$$

$$T_{Ei}\dot{E}'_{fdi} = -E'_{fdi} + K_A\left(V_{refi} - V_{ti}\right) \tag{3}$$

$$M_i\dot{\tilde{\omega}}_i = P_{ai} = P_{mi} - P_{ei} - \frac{M_i}{M_T}P_{COI} \tag{4}$$

$$\dot{\theta}_i = \tilde{\omega}_i, i = 1,2,\cdots,n \tag{5}$$

where n = number of generators in the system; T'_{d0i}, T'_{q0i} = open circuit direct and quadrature axes time constants, respectively; x_{di}, x_{qi} = direct axis synchronous and transient reactance; x'_{di}, x'_{qi} = quadrature axis synchronous and transient reactance; K_{Ai}, T_{Ei} = equivalent gain and time constant of exciter; E_{fdi}, E'_{fdi} = stator emf corresponding to field voltage, after and before the limiter, respectively; E'_{qi}, E'_{di} = direct and quadrature axes stator emf corresponding to rotor flux components; I_{di}, I_{qi} = direct and quadrature axes stator currents;

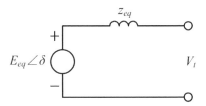

Figure 1. Synchronous generator equivalent model.

V_{ti} = generator terminal voltage; V_{refi} = exciter reference voltage; P_{mi} = generator's mechanical power; P_{ei} = generator's electrical power; P_{ai} = generator's accelerate power relative to COI; $\tilde{\omega}_i$ = generator angular-velocity with respect to COI; θ_i = power-angle with respect to COI; M_i = inertia time constant; $P_{COI} = \sum_{i=1}^{n}\left(P_{mi} - P_{ei}\right)$; and $M_T = \sum_{i=1}^{n}M_i$.

When generator is represented by two-axis model, its Thevenin equivalent circuit is shown in Figure 1. Z_{eq} is the transient reactance and E_{eq} is the voltage behind it. The circuit form is the same with classical model. Transient reactance and the internal voltage are considered constants for classical model while for detailed generator model, the voltage E_{eq} has two components varying with different time constants and z_{eq} is neither a constant.

With generator represented by the circuit shown in Figure 1 and loads represented by constant impedances, the fixed nodes are eliminated and the resulting network contains only the internal generator nodes. Network admittance matrix is denoted as $\mathbf{Y}_G$. The electrical power output of machine i is given by:

$$P_{ei} = G_{ii}\alpha_{ii} + \sum_{\substack{j=1\\j\neq i}}^{n}G_{ij}\left(\alpha_{ij}\cos\theta_{ij} + \beta_{ij}\sin\theta_{ij}\right)$$

$$+ \sum_{\substack{j=1\\j\neq i}}^{n}B_{ij}\left(\alpha_{ij}\sin\theta_{ij} - \beta_{ij}\cos\theta_{ij}\right) \tag{6}$$

where θ_{ij} = angle difference between node i and j; G_{ii}, B_{ii} = real and imaginary part of self-admittance for node i, respectively; and G_{ij} and B_{ij} = real and imaginary part of transfer admittance between node i and j, respectively.

$$\begin{cases}\alpha_{ii} = E'^2_{di} + E'^2_{qi}\\ \alpha_{ij} = E'_{di}E'_{dj} + E'_{qi}E'_{qj}, \beta_{ij} = E'_{di}E'_{qj} - E'_{qi}E'_{dj}\end{cases} \tag{7}$$

2.2 Energy function formulation

Based on the following assumptions:

1. A linear trajectory approximation is used to evaluate the path dependent terms.

2. The equivalent internal impedance of generator is regarded as constant.
3. Generator transient voltages are treated as constants, an average between the value at clearing and at the peak point, namely, $E'_d = \left(E'^{cl}_d + E'^{p}_d\right)/2$ and $E'_q = \left(E'^{cl}_q + E'^{p}_q\right)/2$.

System potential energy is given by (Fouad, A.A. et al. 1989):

$$V_{PE} = \sum_{i=1}^{n}\left(-P_{mi} + \alpha_{ii}G_{ii}\right)\left(\theta_i - \theta_i^{s2}\right) + \sum_{i=1}^{n-1}\sum_{j=i+1}^{n} B_{ij}$$
$$\left[\alpha_{ij}\left(\cos\theta_{ij}^{s2} - \cos\theta_{ij}\right) - \beta_{ij}\left(\sin\theta_{ij} - \sin\theta_{ij}^{s2}\right)\right]$$
$$+ \sum_{i=1}^{n-1}\sum_{j=i+1}^{n} \frac{\theta_i + \theta_j - \theta_i^{s2} - \theta_j^{s2}}{\theta_{ij} - \theta_{ij}^{s2}}G_{ij}$$
$$\left[\alpha_{ij}\left(\sin\theta_{ij} - \sin\theta_{ij}^{s2}\right) - \beta_{ij}\left(\cos\theta_{ij} - \cos\theta_{ij}^{s2}\right)\right] \quad (8)$$

where θ_{ij}^{s2} = angle difference between node i and j at postfault stable equilibrium point.

Corrected kinetic energy (Fouad, A.A. & Kruempel, K.C. 1981), which contributes to system separation, is given by

$$V_{KEcorr} = \frac{1}{2}M_{eq}\omega_{eq}^2 \quad (9)$$

where $M_{eq} = \frac{M_{cr}\cdot M_{sys}}{M_{cr}+M_{sys}}$; $\omega_{eq} = \omega_{cr} - \omega_{sys}$; M_{cr}, ω_{cr} = inertial constant and angular speed of critical generators; and M_{sys}, ω_{sys} = inertial constant and angular speed of the rest of generators.

And the transient energy becomes

$$V = V_{KEcorr} + V_{PE} \quad (10)$$

2.3 Peak point estimation

2.3.1 Approximate function for Eq′(θ)
As the generator excitation control mainly affects the q axis transient emf, it is assumed that varies exponentially between conditions at fault clearing and those at the limiting conditions:

$$E'_q(\theta) = E'^{cr}_q - \left(E'^{cr}_q - E'^{cl}_q\right)e^{\frac{\theta - \theta^{cl}}{\Delta\theta}} \quad (11)$$

where E'^{cl}_q = q axis transient emf at fault clearing; and E'^{cr}_q = q axis transient emf at the limiting conditions.

$$\Delta\theta = \left(E'^{cr}_q - E'^{cl}_q\right)/\left[\frac{dE'_q}{d\theta}\right]_{cl} \quad (12)$$

Since limiting condition is unknown, a simplified method is to use the value of q axis transient emf at the so-called "corner point" instead of that

at the limiting condition. The corner point (CRN) is obtained by solving:

$$\dot{E}'_{di} = 0, \dot{E}_{qi} = 0, \dot{E}'_{fdi} = 0, \dot{\bar{\omega}} = 0 \quad (13)$$

2.3.2 Peak point estimation by using approximate Eq′(θ)
The peak point is determined by the following two-step procedure: First, an approximate peak point, called ray point, is obtained by maximizing the potential energy on the ray from the post-disturbance stable equilibrium point to the corner point in angular space. Any point on the ray emanating from and passing can be denoted as:

$$\theta^{ray} = \theta^{s2} + \alpha\left(\theta^{CRN} - \theta^{s2}\right) \quad (14)$$

where α = optimal factor, superscript ray and CRN denote, respectively, conditions at ray point and corner point.

Substitute Equation (14) into Equation (8) and the corresponding first order differential equation of Equation (8) is

$$\frac{dV_{PE}}{d\alpha} \approx \sum_{i=1}^{n} P_{ai}\left[E'_q\left(\theta^{ay}\right), E'^{ray}_d, \theta^{ay}\right]\left(\theta_i^{CRN} - \theta_i^{s2}\right) = 0 \quad (15)$$

where $E'^{ray}_d \approx \left(E'^{s2}_d + E'^{CRN}_d\right)/2$ and expression of E'_q is given in Equation (11).

Hence, the problem of determining the point of maximum potential energy along the ray is a one-dimensional problem with respect to variable α and can be solved with conditions at corner point as the initial iteration values.

In the second step the exact peak point is determined by solving the set of equations with conditions at ray point as the initial iteration values.

$$P_{ai}\left[E'_q\left(\theta^p\right), E'^{p}_d, \theta^p\right] = 0, \quad i = 1, 2, \cdots, n \quad (16)$$

where $E'^{ray}_d \approx \left(E'^{s2}_d + E'^{CRN}_d\right)/2$ and expression for E'_q is the same as (11).

3 GENERATOR INTERNAL IMPEDANCE

3.1 Computation of generator internal impedance

According to Thevenin equivalent circuit shown in Figure 1, E_{eq} can be expressed as:

$$E_{eq} \triangleq U_d + jU_q + z_{eq}\left(I_d + jI_q\right) \quad (17)$$

where U_d= the component of the stator terminal voltage on axis d; and U_q = the component on axis q.

Form Equation (17), z_{eq} can be written as:

$$z_{eq} = \frac{E_{eq} - \left(u_d + ju_q\right)}{I_d + jI_q} \qquad (18)$$

As four-order model is used for generators, following relationship can be established

$$\begin{cases} E_{eq} = E_d' + jE_q' \\ E_d' = U_d - x_q'I_q + rI_d \\ E_q' = U_q + x_d'I_d + rI_q \end{cases} \qquad (19)$$

Substitute Equation (19) into Equation (17), we can get

$$z_{eq}^{(1)} = r + \frac{x_d' - x_q'}{I_{eq}^2}I_dI_q + j\frac{x_d'I_d^2 + x_q'I_q^2}{I_{eq}^2} \qquad (20)$$

When higher order generator model is used, calculation formula for generator internal impedance can be deduced using similar method.

In (Fouad, A.A. & Vittal, V. 1991), it is suggested that generator internal impedance for the postfault network is calculated by the following formula:

$$z_{eq}^{(2)} = r + j\frac{1}{2}\left(x_d' + x_q'\right) \qquad (21)$$

On comparing Equation (20) with Equation (21) the difference is

$$z_{eq}^{(1)} - z_{eq}^{(2)} = \frac{I_dI_q + j\left(I_d^2 - I_q^2\right)}{I_{eq}^2}\left(x_d' - x_q'\right) \qquad (22)$$

As can be known from the Equation (22), Equation (21) coincides with Equation (20) only when $x_d' = x_q'$. For hydro units, reactance parameters usually satisfy $x_d' > x_q''$ (parameters relative to quadrature axis subtransient process are denoted as E_q'', x_q'' and T_{q0}''). Transient parameters for thermal units usually satisfy $x_q' \geq x_d'$ (Kundur, P. 1994). Thus, calculating results of the two methods are different in most cases. The equivalent internal impedance affects reduced admittance matrix especially self-admittance of corresponding generator bus. Since generator internal impedances account for large proportion of the total network impedance (He, Y.Z. & Wen, Z.Y. 2002), effects of its calculating results on transient stability assessment cannot be neglected. A more accurate computation method of generator

internal impedance in direct transient stability method is necessary.

3.2 Improved direct method

As shown in Equation (20), equivalent internal impedance not only depends on machine parameters such as r, x_d' and x_q', it is also related to operating variables. In the transient process after fault clearing, both stator current and power angle are time-varying. Thus generator internal impedance is not constant, as mentioned in 2.1. In this paper, generator equivalent internal impedance is calculated by Equation (20) using conditions at peak point. Then an improved calculation method for direct transient stability assessment is developed. The procedure is outlined as follows.

1. Determine exact corner point

Conditions at corner point still satisfy Equation (13). Since peak point is unknown, generator equivalent impedance can't be regarded as constant. Generators are modeled in detail and network is reduced to generator terminal bus, not internal bus. Generator electrical power is calculated

$$P_{ei} = u_{di}I_{di} + u_{qi}I_{qi}, \quad i = 1,2,\cdots,n \qquad (23)$$

Generator terminal voltage and current satisfy the following nodal admittance equation:

$$\mathbf{I}_x + j\mathbf{I}_y = \mathbf{Y}_N\left(\mathbf{U}_x + j\mathbf{U}_y\right) \qquad (24)$$

where $\mathbf{Y}_N$ = network admittance matrix reduced to generator's terminal buses; $\mathbf{U}_x$, $\mathbf{U}_y$ = column vectors of terminal voltages in network reference frame; and $\mathbf{I}_x$, $\mathbf{I}_y$ = column vectors of generator currents in network reference frame.

Generator terminal voltages and currents are converted to the network reference frame from generator reference frame, using the relation

$$\begin{pmatrix} A_x \\ A_y \end{pmatrix} = \begin{pmatrix} \sin\theta & \cos\theta \\ -\cos\theta & \sin\theta \end{pmatrix}\begin{pmatrix} A_d \\ A_q \end{pmatrix} \qquad (25)$$

Solving simultaneous Equations (13) and (23)~(25), we can obtain the exact corner point.

2. Calculate generator equivalent impedance

Generator currents at the corner point are substituted into Equation (20) to get the value of generator equivalent impedance. Then, network is reduced to generator internal bus by supposing generator internal impedance as constant. Next steps of peak pint estimation are the same as introduced in 2.3.2.

Table 2.   Generator internal impedance at peak point.

4   VALIDATION STUDIES

4.1   *IEEE three-Machine nine-bus system*

Taking IEEE three-Machine nine-bus system based on PSCAD software with network topology and parameters presented in (Anderson, P.M. & Fouad, A.A. 2003). Synchronous generators are represented by four-order model with exciter and loads are represented by constant impedance. Setting short-circuits faults on different buses, corresponding Critical Clearing Time (CCT) is shown in Table 1. Superscript (1), (2) and (3) represent the method described in reference (Fouad, A.A. et al. 1989), method presented in this paper and time simulation, respectively.

As illustrated in Table 1, CCT using method proposed by Fouad, A.A. et al. 1989 is always smaller than time simulation results. That is to say, transient stability assessment is conservative. Data obtained with the improved method presented in this paper is more approximate to time simulation results.

Taking short circuit fault at Bus 5 as an example, stator current and internal impedance of G1~G3 after fault clearing are shown in Figure 2.

Table 2.   Generator internal impedance at peak point.

No	$z_{eq}^{(1)}$	$z_{eq}^{(2)}$	$z_{eq}^{(3)}$
G1	0.0304i	0.0006 + 0.0608i	−0.0151 + 0.0689i
G2	0.1583i	−0.0075 + 0.1205i	0.0310 + 0.1354i
G3	0.2157i	−0.0127 + 0.1837i	0.0260 + 0.1932i

In the first swing toward peak point from fault clearing, internal impedance of G2 and G3 decreased gradually. This is because in the process of system moving forward, critical generators had gradually deviated steady-state operating point, d component of stator current increased and q component decreased. According to Equation (20), internal impedance of G2 and G3 will decrease. For generator G1, both d and q components of stator current increased, thus, the internal impedance did not not change much compared to G2 and G3. The results indicate that for critical generators, the calculation value of generator internal impedance using Equation (21) is larger than the actual value at peak point, equivalent to lower the power limit of system. So the corresponding transient stability assessment is conservative. Table 2 compares generator internal impedance at peak point calculated by several methods. The meaning of superscript (1), (2), and (3) is the same with those in Table 1. We can know that the value of generator internal impedance using Equation (21) is smaller than time simulation results for hydro unit G1 while larger for thermal units G2 and G3. It is indicated that calculation values by the present method are more appropriate to simulation results.

4.2   *Case study for Central-China power grid*

According to plan of State Grid Corp. of China, Chongqing-Hubei HVDC Back-To-Back (BTB) project will be constructed around 2020, realizing asynchronous interconnection of southwest power system and Central China power grid. In order to facilitate research, equivalent network for Central-North China power grid equivalent system after operation of Chongqing-Hubei HVDC back-to-back project is developed (Liu, J. et al. 2014) and its topology is shown as Figure 3. A three-order model is used for a synchronous machine and exciter is 12-type regulator in PSASP. Theoretical derivation of electrical power of synchronous generator and system transient energy function when three-order model is used is presented in Appendix.

Chongqing-Hubei HVDC BTB project is transmitting power from Chongqing to Hubei, transmitted power of HVDC increases gradually by raising output of Sichuan generators while reducing output of Hubei generators while keeping

Table 1.   Comparisons of CCT of the nine-bus system.

Bus	Branch to be cut off	CCT[(1)]	CCT[(2)]	CCT[(3)]
		s	s	s
5	5–7	0.155	0.200	0.225
6	5–7	0.259	0.277	0.285
7	7–9	0.086	0.118	0.121
9	7–9	0.165	0.179	0.175

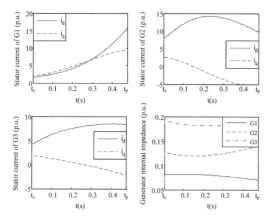

Figure 2.   Variations of stator current and internal impedance after fault clearing.

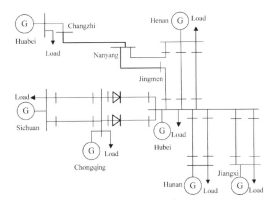

Figure 3. Structure diagram of Central-North China power grid equivalent system after operation of Chongqing-Hubei HVDC back-to-back project.

Table 3. Transient stability under Bi-polar block fault of Chongqing-Hubei HVDC BTB project.

Power MW	$\Delta V^{(1)}$	$\Delta V^{(2)}$	$\Delta V^{(3)}$	Simulation
1400	−74.6642	79.3509	0.6212	stable
1500	−79.3949	52.2593	0.4714	stable
1600	−79.9583	49.8093	0.0738	unstable
1700	−84.8342	46.3651	0.9548	unstable

transfer power of tie-line between Central China and North China power grid and transmission interface between other provinces in Central China unchanged. Compute energy margin ΔV under HVDC bi-polar block fault at different HVDC transmitted power, transient stability assessment results of direct methods are compared with results obtained with time simulation, as shown in Table 3. Let superscript (1), (2), and (3) denote method 1, method 2, and the method proposed in this paper, respectively. Method 1 and method 2 shall refer to reference (Fouad, A.A. et al. 1989) and formula for generator equivalent impedance is respectively $z_{eq}^{(1)} = r + j\frac{1}{2}(x_d' + x_q)$ and $z_{eq}^{(2)} = r + j\frac{1}{2}x_d'$. Formula in the presented method is $z_{eq}^{(3)} = \frac{(jx_d' + r)I_d + (jr - x_q')I_q}{I_d + jI_q}$.

As illustrated in Table 3, method 1 is on the conservative side while method 2 excesses the stable domain. Results using the presented method are in agreement with simulation results and misjudgment occurs only when power system is critically stable. For equivalent system, maximum transmitted power of HVDC is respectively 1550 MW and 1600 MW by using domain simulation and the presented method. For the original system, it is 1600 MW using time simulation. It can be seen that result of proposed method is very close to simulation result.

5 CONCLUSION

The expression of transient energy function for detailed generator model is similar to that for classical power system model when generator model is represented by the Thevenin equivalent circuit and the equivalent internal impedance is regarded as constant. Theoretical analysis and simulation results show that generator internal impedance is time-varying during the transient towards limiting conditions after fault clearing. This paper discusses the computation methods of generator equivalent internal impedance and deduces the exact computation formula based on the equivalent circuit. Value of equivalent internal impedance at corner point which is obtained on basis of detailed generator model is considered a good approximation of the value at peak point. Results of IEEE three-machine nine-bus system and Central China power grid show that the presented method can improve accuracy of transient stability assessment and reliability of security analysis to some extent.

APPENDIX

For power system containing n generators represented by three-order model, electrical power of synchronous generator k is given by:

$$
\begin{aligned}
P_{ek} = {}& E_{qk}'^2 G_{kk} + E_{qk}' \sum_{\substack{j=1 \\ j\neq k}}^{n} E_{qj}' \left(G_{kj}\cos\theta_{kj} + B_{kj}\sin\theta_{kj} \right) \\
& + \left(x_{qk} - x_{dk}' \right) \left(\sum_{j=1}^{n} E_{qj}' \left(G_{kj}\sin\theta_{kj} + B_{kj}\cos\theta_{kj} \right) \right) \\
& \cdot \left(\sum_{j=1}^{n} E_{qj}' \left(G_{kj}\cos\theta_{kj} + B_{kj}\sin\theta_{kj} \right) \right) - ri_k^2
\end{aligned}
\tag{26}
$$

System potential function is given by:

$$
\begin{aligned}
V_{PE} = {}& \int \sum_{k=1}^{n} \left(-P_{mk} + P_{ek} \right) d\theta_k \\
= {}& \int \sum_{k=1}^{n} \left(-P_{mk} + E_{qk}'^2 G_{kk} \right) d\theta_k \\
& + \sum_{k=1}^{n-1} \sum_{j=k+1}^{n} \int B_{kj} E_{qk}' E_{qj}' \sin\theta_{kj} d\theta_{kj} \\
& + \sum_{k=1}^{n-1} \sum_{j=k+1}^{n} \int G_{kj} E_{qk}' E_{qj}' \cos\theta_{kj} d\left(\theta_k + \theta_j \right) \\
& + \int \sum_{k=1}^{n} \left(\left(x_{qk} - x_{dk}' \right) i_{dk} i_{qk} - ri_k^2 \right) d\theta_k
\end{aligned}
\tag{27}
$$

Based on following assumptions:

1. Generator equivalent internal impedance is regarded as constant.
2. Generator transient emf is regarded as constant, an average of the value at fault clearing and that at the peak point, namely, $E'_d = (E'^{cl}_d + E'^{p}_d)/2$.
3. A linear trajectory approximation is used to evaluate the path dependent terms $\int B_{kj}E'^2_{qk} \sin\theta_{kj} d(\theta_k + \theta_j)$.
4. $\int \sum_{k=1}^{n}((x_{qk} - x'_{dk})i_{dk}i_{qk} - ri_k^2) \, d\theta_k$ is approximated

with the aid of initial and final points.

Critical potential energy can be expressed as:

$$V_{cr} = \sum_{k=1}^{n}\left(-P_{mk} + E'^2_{qk}G_{kk}\right)\left(\theta_k^p - \theta_k^{cl}\right)$$
$$+\sum_{k=1}^{n-1}\sum_{j=k+1}^{n} B_{kj}E'^2_{qk}\left(\cos\theta_{kj}^{cl} - \cos\theta_{kj}^p\right)$$
$$+\sum_{k=1}^{n-1}\sum_{j=k+1}^{n} \frac{\theta_k^p + \theta_j^p - \theta_k^{cl} - \theta_j^{cl}}{\theta_{kj}^p - \theta_{kj}^{cl}} G_{kj}E'^2_{qk}\left(\sin\theta_{kj}^p - \sin\theta\theta_{kj}^{cl}\right)$$
$$+\frac{1}{2}\left(x_{qk} - x'_{dk}\right)\left(i_{dk}^{cl}i_{qk}^{cl} + i_{dk}^p i_{qk}^p\right) - \frac{1}{2}r\left(i^{cl2} + i^{p2}\right)$$

(28)

REFERENCES

Anderson, P.M. & Fouad, A.A. 2003. *Power System Control and Stability*, 2nd ed. Piscataway, NJ: IEEE.

Bergen, A.R. & Hill, D.J. 1981. A structure preserving model for power system stability analysis. *IEEE Transactions on Power Apparatus and Systems* PAS-100: 25–35.

Brown, R.G. & Hwang, P.Y. 1992. *Introduction to Random Signal Analysis and Applied Kalman Filtering*. NewYork: John Wiley and Sons.

Fouad, A.A. & Kruempel, K.C. 1981. *Transient Stability Margin as a Tool for Dynamic Security Assessment*. USA: Electric Power Research Institute.

Fouad, A.A., Vittal, V., Ni, Y.X., Zein-Eldin, H.M. & Vaahdi, E. 1989. Direct transient stability assessment with excitation control. *IEEE Transactions on Power Systems* 4(1): 75–82.

Fouad, A.A. & Vittal, V. 1991. *Power system transient stability analysis using the transient energy function method*. Pearson Education.

He, Y.Z. & Wen, Z.Y. 2002. *Power System Analysis*. Wuhan: HuaZhong University of Science and Technology Press.

Kundur, P. 1994. *Power System Stability and Control*. New York: McGraw Hill.

Liu, S. & Wang, J. 1996. *Transient Energy Function of Power System*. Shanghai: Shanghai Jiao Tong University Press.

Liu, S. 1998. The recent progress of the transient energy function method. *Automation of Electric Power Systems*, vol. 22(9): 19–24.

Liu, J., Sun, H.S., Liu, Z.Q., Wen, J.Y. & Luo, C. 2014. A simplified equivalent model of Central-North China power grid, *Automation of Electric Power Systems* 38(6): 17–42.

Narasimhamurthi, N. & Musavi, M.T. 1984. A generalized energy function for transient stability analysis of power systems. *IEEE Transactions on Circuits and Systems* 31(7): 637–645.

Ni, Y.X., Chen, S.S. & Zhang, B.L. 2002. *Theory and Analysis of Power System Dynamics*. Beijing: Tsinghua University Press.

Tsolas, N., Arapostathis, A. & Varaiya, P.P. 1985. A structure preserving energy function for power system transient stability analysis. *IEEE Transactions on Circuits and Systems* 32(10): 1041–1049.

Yin, M.H. & Zou, Y. 2003. Controlling uep method in network structure preserving power system model, *Proceedings of the CSEE* 23(8): 32–37.

Advances in Power and Energy Engineering – Sun (Ed.)
© 2016 Taylor & Francis Group, London, ISBN 978-1-138-02846-3

Application of wavelet mutation detection technology in the ultrasonic localization of Partial Discharge in the transformer

Y.W. Dong, Y.Y. Liu & Q. Zou
Tianjin Electric Power Research Institute, Tianjin, China

L.X. Shi
Tianjin Chengxi Power Supply Branch Company, Tianjin, China

ABSTRACT: The use of ultrasonic technology to locate the partial discharge of the power transformer must obtain a correct time delay. Based on experimental research and detailed analysis, we found that there are multiple mutations in the ultrasonic signal generated by the partial discharge. The first mutation point is the starting time of the partial discharge, which should be marked. However, the traditional time-delay estimation methods usually use the peak signal mutation as a marker; as a result, the positioning accuracy is affected. Wavelet mutation detection technology can identify the first-time mutation of the signal. This article describes the principle of wavelet mutation detection technology and conducts tests to locate the partial discharge. The results indicated that the positioning accuracy of the method was improved significantly, and compared with traditional methods, the absolute error can be reduced by 1–2 cm.

1   INTRODUCTION

Partial Discharge (PD) is one of the main reasons that causes internal insulation deterioration of the power transformer. It is also an important manifestation and signal for deteriorated insulation. Online monitoring of transformer PD can determine the status of the transformer internal insulation timely and accurately. So, it plays an important role in preventing the occurrence of power transformer accidents. Ultrasound position method, which is a simple, strong anti-electromagnetic interference, and online testing, uses the time delay of the ultrasonic signal produced by PD to determine the location. So, it is widely used in the current transformer PD monitoring and diagnosis application of engineering practices. However, in practical applications, due to the influences of periodic ultrasonic, environmental noise and propagation path, the peak spectrum of ultrasonic wave may be expanded, resulting in the side peak spectrum. Traditional ultrasound method uses the time delay of the largest peak as the direct wave (Luo Richeng, Li Weiguo & Li Chengrong. 2005) signal start-up delay, which cannot be estimated accurately and thus will not give the desired positioning effect. Other cross-power phase methods include the adaptive method (Dong Zhi-feng, Wang Zeng-fu & Liu Qing-wen. 2002.), the generalized correlation method (Luo Yong-fen & Li Yan-ming. 2004.), the generalized spectrum, or high cumulative amount method. These methods are used to estimate the relative delay of the spec-tral peak position signal. Although they achieve relatively better results in some areas, the time-delay errors are still large. The wavelet mutation detection technology can identify the mutation point as the start-up time delay of the direct wave signal, so the first mutation point of the ultrasonic signal received by the sensor can correctly judge the time delay. This article describes a new transformer PD localization method based on wavelet mutation detection techniques. Using transformer PD test devices, we obtain PD ultrasonic signals and analyze their characteristics. Then, we describe the wavelet mutation detection technology principle and realization method. Finally, we conduct a test to verify the effectiveness of the positioning method.

2   ACQUISITION OF ULTRASOUND SIGNAL OF PD

2.1   *The principle and method of direction finding*

In the high-voltage laboratory, we set up the test platform mode of transformer PD, on which the needle-plate PD device was placed in the container filled with the transformer oil to simulate the PD inside the transformer. Then, we used a 3 × 3 square array ultrasound transducer coupled with butter and affixed to the outside wall of the transformer mode to acquire the ultrasonic signal of PD. The ultrasonic signal test platform for receiving the discharge produced is shown in Figure 1.

The major hardware and software devices of the test platform are described below:

2.1.1 Transformer model
To simulate the internal transformer PD, we used an analog box welded with a steel plate, with a size of 120 mm (length) × 100 mm (width) × 100 mm (H) and the cabinet chassis ground, which was equipped with the transformer oil.

2.1.2 Discharge model
In order to obtain the ultrasonic signal of PD, we selected the typical needle-plate discharge model. The needle part comprised one copper cylinder with a diameter of 10 mm at the bottom of the ground, which constitutes a conical tip needle. The plate part comprised a cylindrical copper with a diameter of about 30 mm and a height of about 10 mm. The needle and the plate spacing was about 5 mm apart and separated by an insulating cardboard, as shown in Figure 2. When the voltage was gradually increased, the needle tip first appeared at the corona and then the continuous discharge occurred.

2.1.3 The array sensor and its coupling position
The single sensor was made up of a piezoelectric wafer, acoustic, acoustic matching layer and other structures (e.g. electrode lead, housing) and other components. The ultrasonic array sensor was composed of a "square" flat array that consisted of nine (3 × 3) center sensors, with an array element spacing of 9 mm. The transmission line of the sensor array employed a high-quality shielded cable, with a length of 1.0 m, a core diameter of 1mm and an outer diameter of 1.8 mm, respectively. Both ends of the joint used Q7 head and BNC head.

The array sensor was coupled to the positive plate of the tank wall toward the discharge source to receive the direct wave ultrasonic signal.

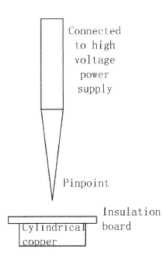

Figure 2. Needle-plate discharge schematic model.

2.1.4 Data gathering machine
In this paper, a 9-channel synchronous data acquisition card was used along with the model USB-1114 Opt10. The acquisition card with data acquisition and high-speed transmission is easy to connect, expand and use. The sampling frequency was 1 kHz–10 MHz and the sampling length was 1 k–512 k. Each channel was set to filter the amplifier with a high-pass filter of 20 kHz, 40 kHz, 80 kHz, an optional low-pass filter of 125 kHz, 250 kHz, 312.5 kHz, and the optional magnification channel of 1–256 times. There were external trigger, internal trigger and freedom trigger.

2.1.5 Software system
Experiment Lab-VIEW software system was used for the data acquisition and the storage system, enabling data acquisition, preservation, other amplification, and display functions. In this paper, Matlab software was also used to filter the raw data of the ultrasound signal and delay estimation process.

2.2 PD characteristics of ultrasonic signals

After the set-up of the test platform, the regulator was turned slowly until the voltage across the needle and the board reached the continuous discharge value, and then the data acquisition unit was opened to collect the signal in the event of discharge. The signal waveform obtained by using LabVIEW software is shown in Figure 3.

From the figure, we can see the waveform display of smooth undulating when the ultrasonic signal is not received. Once the ultrasonic signal is received, the waveform is displayed on the screen. The delay of the PD ultrasound direct wave signal

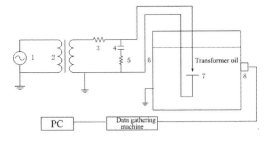

Figure 1. Transformer model PD test platform. 1. AC power supply and voltage regulator, 2. step-up transformer, 3. protection resistor, 4. coupling capacitor, 5. detects impedance, 6. transformer model, 7. needle-plate discharge model, 8. array sensor.

is about 650–750 µs. The ultrasonic signal of the transformer PD is similar to a spindle shape; after a brief start-up phase (usually about 10 µs), it reaches to a peak and then decays quickly, and the mutation will disappear. Start-up delay is the time difference between the occurrence of the time PD and the vibration of the time ultrasonic direct wave (i.e. PD ultrasonic signal reaches the sensor), which reflects the PD source ultrasonic signal propagation time to the sensor. Using the time delay, we can obtain the PD distance between the source and the sensor by multiplying the propagation velocity. Only the first point mutation of direct wave signals can obtain the start-up delay for positioning. However, the PD ultrasonic wave signal directly mixes with some random interference and noise signals, so it is difficult to distinguish the first point mutation. In this paper, the Fast Fourier Transform (FFT) filtering method and the wavelet denoising method were used for signal pretreatment. Figure 4 shows a comparison of the direct wave signal mutated portion of 600–800 µs before and after the pretreatment.

From Figure 4, we can see that the signal has a plurality of point mutations. The conventional

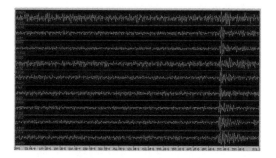

Figure 3. 3 × 3 plane array 9-channel direct wave ultrasonic signal of PD acquired by data acquisition unit.

time-delay estimation method estimates at the same time of the peak as the direct wave arrival time, which has some delay obviously. The wavelet transform mutation detection method is used to detect the time at which the first point mutation occurs. The resulting delay is closer to the start-up delay.

3 PRINCIPLE OF WAVELET SIGNAL MUTATION DETECTION

First, we obtain the mutant signal multiscale decomposition by using the wavelet multi-resolution analysis characteristic, and then analyze the decomposed signal to determine the mutation position of the signal. At higher sophisticated scale wavelet transform, the maximum value of the wavelet transform modulus and the signal discontinuity location is precise. The mutation decay rate depends on the signal at the point mutation of the Lipschitz index.

3.1 Signal mutation and wavelet transform

S. Mallat linked the local singularity and local maximum analog of the signal after wavelet transform with the wavelet transform modulus maximum decay rate at different scales to measure the local singularity signal (Lin Zhi-bin & Xu Bai-ling. 2004).

Wavelets are provided as a continuous and real function with decay:

$$|\psi(t)| \le K(1+|t|)^{-2-\varepsilon}(\varepsilon > 0) \tag{1}$$

Given that the function $f(t) \in L^2(R)$ is consistent on the interval I, there exists a constant c > 0, $\forall a,b \in I$, and the wavelet transform is satisfied:

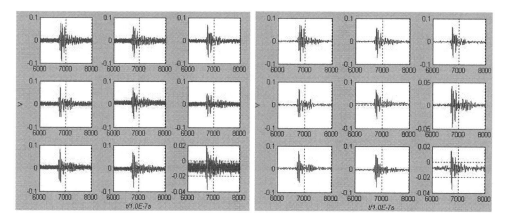

Figure 4. The original waveform and the waveform after the pretreatment.

387

$$|(W f)(a,b)| \le c a^{\alpha+1/2} \tag{2}$$

where α is the Lipschitz index $\alpha(-\varepsilon < \alpha \le 1)$.

Conversely, if the wavelet transform of $f(t) \in L^2(R)$ satisfies Equation (2), then f has a uniform Lipschitz index $\alpha(-\varepsilon < \alpha \le 1)$ on I.

When $b = t_0$, $|(W f)(a,b)|$ achieves the maximum value, being equal to Equation (2). The singular point of $f(t)$ is called t_0.

In the case of the binary wavelet transform, Equation (2) becomes

$$|(W f)(2^j,b)| \le c \times 2^{j(\alpha+1/2)} \tag{3}$$

In the signal and image processing, convolution type of wavelet transform is often used. We then set $f(t), \psi(t) \in L^2(R)$:

$$\psi_s(t) = \frac{1}{s} \psi\left(\frac{t}{s}\right), s > 0 \tag{4}$$

Convolution type of wavelet transform of $f(t)$ is given by

$$(W f)(s,b) = f * \psi_s(b) = \frac{1}{s} \int_{-\infty}^{+\infty} f(t) \psi\left(\frac{b-t}{s}\right) dt \tag{5}$$

Considering the wavelet transform of $f(t)$ as the convolution type, Equations (2) and (3) become:

$$|(W f)(s,b)| \le c s^{\alpha} \tag{6}$$

$$|(W f)(2^j,b)| \le c 2^{j\alpha} \tag{7}$$

3.2 Theory of signal mutation detection

Signal mutation detection first has the original signal polished from a different scale, and then detects the first or second derivative of the signal after finishing their extreme or crossing points. The polished function should be satisfied:

$$\int_{-\infty}^{+\infty} \theta(t) dt = 1 \tag{8}$$

$$\lim_{t \to \infty} \theta(t) = 0 \tag{9}$$

From Formula (8), $\hat{\theta}(t) = 1$, where $\theta(t)$ is a low-pass filter. The common polished function $\theta(t)$ can be selected from the Gauss function or B-spline function.

From the definition and the nature property of convolution, we can know the first derivative of wavelet transform $(W f)(s,t) = f * \psi_s^{(1)}(t)$ and $f * \theta_s(t)$ is proportional, and the number of the second derivative of $(W f)(s,t) = f * \psi_s^{(2)}(t)$ and $f * \theta_s(t)$ is proportional:

$$f * \psi_s^{(1)}(t) = f * \left(s \frac{d\theta_s}{dt}\right) = s \frac{d}{dt}(f * \theta_s(t)) \tag{10}$$

$$f * \psi_s^{(2)}(t) = f * \left(s^2 \frac{d^2\theta_s}{dt^2}\right) = s^2 \frac{d^2}{dt^2}(f * \theta_s(t)) \tag{11}$$

Convolution $f * \theta_s(t)$ expresses the function f after being polished as a signal, which intuitively means that the "corner" of f is ground into a smooth arc, so that f becomes a smooth function $f * \theta_s(t)$.

Thus, after choosing the smooth functions $\theta(t)$, mutation of the signal $f(t)$ can be detected and obtained by the wavelet transform and the modulus maximum value.

3.3 Error analysis

During the experiments, direction finding errors, noise, sensor differences, and the spacing and coupling of the array element errors were analyzed, and there was also a multi-path propagation of ultrasonic signals [9–11] that would generate errors.

4 WAVELET DETECTION OF PD ULTRASONIC SIGNAL DISCONTINUITY

To determine the time of the PD ultrasonic signal mutation point in the experiment, we use the db6 continuous wavelet transform, and then analyze and process the factor in order to determine the time point at which the mutation occurs. After filtering the noise pretreatment of PD array ultrasonic direct wave original signal for the aforementioned acquisition and storage, we use wavelets in scale 1–32 db6 from the signal of array element 5 to make continuous wavelet transform. The absolute value of the relative coefficients is shown in Figure 5.

Light areas shown in Figure 5 correspond to the event of the PD of about 700 μs. Thus, it can be concluded that point mutations are present in this region, which represents the received direct wave signal. Further analysis showed that the time at which the mutation points first appeared in this region is 65 μs, the calculated velocity is 1440 m/s, the PD source to the distance of the sensor is 936 mm, and the actual distance is 900 mm, with an error of 36 mm. By using the conventional delay estimation method, the delay crest is taken as 670 μs and the distance is calculated as 965 mm, with an error of 65 mm. Based on wavelet mutation detection, the PD ultrasound direct wave delay can be selected for positioning with an error smaller than the traditional method (about 30 mm).

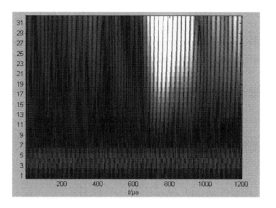

Figure 5. The relative absolute value of coefficients after continuous wavelet transform.

Table 1. Location results and errors.

Array element	Start-up delay of vibration	Calculation distance/ mm	Actual distance/ mm	Error/ mm
1	647	932	892	40
2	649	935	897	38
3	653	940	903	37
4	639	920	895	25
6	649	935	905	30
7	648	933	897	36
8	653	941	903	38
5	656	945	908	37

The results of the aforementioned processing of other array elements are summarized in Table 1. From Table 1, we can see that the sensor error is less than 40 mm based on the wavelet mutation detection method. Compared with the traditional method estimation error of more than 60 mm, this method obtains a more accurate positioning result.

5 CONCLUSION

Direct waveform characteristics of PD ultrasonic signal from the transformer model were analyzed in detail. It can be found that in the direct wave signal, there is a plurality of discontinuities. The first delay information at the point mutation can be used as its vibrating start time delay, while the wavelet transform that is used to analyze the position and strength of the signal singularity is a powerful tool for signal mutation detection. Using the wavelet multi-scale signal analysis with good judgment, we can accurately determine the position of the mutation points, and then obtain the spatial distance. The experimental results indicate that the error is smaller; thus, the theory and experiment have proved the effectiveness of this method.

REFERENCES

Dong Zhi-feng, Wang Zeng-fu & Liu Qing-wen. 2002. A Method for Eliminating False Location in Cross-localization, Journal of System Simulation, vol. 14, no. 4, pp. 527–530.

Lin Zhi-bin, Xu Bai-ling. 2004. Sound Source Localization Based on Microphone Array, Sound Technology, vol. 19, no. 5, pp. 19–23.

Luo Ri-cheng, Li Wei-guo & Li Cheng-rong. 2005. A High-Speed Algorithm to Remove Mendacious Points of Direction-Finding Cross-Localizaion in Multi-Target Passive Localization of Partial Discharge Source within Power Transformer, Power System Technology, vol. 29, no. 19, pp. 53–56.

Luo Yong-fen & Li Yan-ming. 2004. Simulation of PD Location Method in Oil Based on UHF and Ultra-sonic Phased Array Receiving Theory, Transactions of China Electrotechnical Society, vol. 19, no. 1, pp. 35–39.

Advances in Power and Energy Engineering – Sun (Ed.)
© 2016 Taylor & Francis Group, London, ISBN 978-1-138-02846-3

Measurement accuracy analysis of active power under effects of reactive power compensation

S.Q. Li
Wuxi Branch, Institute of Electrical Engineering, Chinese Academy of Sciences, China

B.L. Lei
Institute of Electrical Engineering, Chinese Academy of Sciences, China

ABSTRACT: The measurement of electricity data in factories is an important part of energy saving reconstruction. The active power is a key parameter to the energy consumption level analysis. Its accuracy is affected by many factors. Research on these factors would make the active power measurement accurate and efficient. A decrease of active power measurement data would not always mean the energy saving. Attention should be paid to inspect the precision and accuracy of each measurement. Restriction by the factories environment would lead some compromise on measuring method and equipment, which conceals greater uncertainty of the data. So the proper equipments and methods are essential for the active power measurement.

1 INTRODUCTION

The measurement of electricity data in factories is an important part of energy saving reconstruction. Through the measurement, data of production facilities is used to analyze the electricity energy consumption level. But, on most conditions, the power systems cannot break and the facilities cannot be shut down. The environment of factories always restricts the measurement. As a result, many different methods would be used on the measurement.

Power Quality Analyzer is a widely used high performance embedded instrument, and we used it to monitor the electricity data of factories. The power quality analyzer can measure phase to phase voltage up to 690 V. Current can be detected by current transformers connected to the power quality analyzer. Apparent power, active power, reactive power and power factor on the line would be calculated in the power quality analyzer (Cao 2012).

Split core current transformers and power quality analyzers are used to measure the electricity data from a steel rolling mill factory in this paper. The precision of inspection is of great significance on the credible and reliable of analysis results.

The active power is a key parameter to the energy consumption level analysis. Its accuracy is affected by many factors. Research on these factors would make the active power measurement accurate and efficient.

2 THE MEASUREMENT ACCURACY OF ACTIVE POWER (FEI 2014 AND TUMANSKI 2009)

2.1 Integration method of active power measuring

Regularly, integration method is used to measure active power. The active power is equal to the average value of the quadrature of the product of voltage and current. It can be described by Equation (1):

$$P = \frac{1}{T}\int_0^T u \cdot i \, dt \tag{1}$$

To most measurement equipment, the Equation (1) is changed to Equation (2) by discretization of voltage and current.

$$P = \frac{1}{N}\sum_{k=0}^{N-1} u_k \cdot i_k \tag{2}$$

On condition that facilities are working continuously and the energy consumption is steady, the RMS values of voltage and current keep stable. The measurement value is equivalent to Equation (3).

$$P = U \cdot I \cos\left(\theta_u - \theta_i\right) \tag{3}$$

2.2 Errors

The accuracy of current and voltage sensor influences on active power. Assuming the voltage, current and power factor angle remain errors: ΔU, ΔI, $\Delta \theta_u$, $\Delta \theta_i$.

By the function errors equation, the error of active power can be calculated by Equation (4).

$$\Delta P = \frac{\partial P}{\partial U} \Delta U + \frac{\partial P}{\partial I} \Delta I + \frac{\partial P}{\partial \theta_u} \Delta \theta_u + \frac{\partial P}{\partial \theta_i} \Delta \theta_i \quad (4)$$

But, in reality, ΔU, ΔI, $\Delta \theta_u$ and $\Delta \theta_i$ are not clear unless each sensor has been tested before used. The accuracy of active power derives from the parameters of sensor. The standard derivation of real power is:

$$\sigma_p^2 = \sum_{k=U,I,\theta_u,\theta_i} \left(\frac{\partial P}{\partial k}\right)^2 \sigma_k^2 \\ = a_u^2 \sigma_U^2 + a_i^2 \sigma_i^2 + a_{\theta_u}^2 \sigma_{\theta_u}^2 + a_{\theta_i}^2 \sigma_{\theta_i}^2 \quad (5)$$

where:

$$a_u = \frac{\partial P}{\partial U} = I \cdot \cos\left(\theta_u - \theta_i\right)$$

$$a_i = \frac{\partial P}{\partial I} = U \cdot \cos\left(\theta_u - \theta_i\right)$$

$$a_{\theta_u} = \frac{\partial P}{\partial \theta_u} = -U \cdot I \cdot \cos\left(\theta_u - \theta_i\right)$$

$$a_{\theta_i} = \frac{\partial P}{\partial \theta_i} = U \cdot I \cdot \cos\left(\theta_u - \theta_i\right)$$

2.3 Case study

For example, current transformer's ratio is 3000 A to 5 A, voltage is 400 V, and single-phase is measured. The accuracy of each session is shown as below:

- Current transformer precision level 0.5, phase precision level 5′ (Acrel 2013).
- Voltage precision ± 0.05% rdg ± 0.05% FS (PSL 2011), phase error is ignored.

The uncertainty of active power is calculated by Equation (6):

$$U_P = \sqrt{a_U^2 \cdot \Delta U^2 + a_I^2 \cdot \Delta I^2 + a_{\theta_i}^2 \cdot \Delta \theta_i^2} \quad (6)$$

And the uncertainty of current transformer and voltage measurement is:

- $\Delta I = \pm 0.5\% \times 3000\ A = \pm 15\ A$
- $\Delta U = \pm 0.05\% u = \pm 0.05\% \times 400\ V = \pm 4\ V$

Considering the regular situation, the power factor is assumed 0.8. When current is 2500 A, and voltage is 400 V, the transmission coefficient is:

$$a_U = I \cdot \cos\theta = 2000$$

$$a_I = U \cdot \cos\theta = 320$$

$$a_{\theta_i} = U \cdot I \cdot \cos\theta = 600000$$

In Equation (6),

$$a_U \cdot \Delta U = 2000 \times (\pm 4) = \pm 8000$$

$$a_I \cdot \Delta I = 320 \times (\pm 15) = \pm 4800$$

$$a_{\theta_i} \cdot \Delta \theta_I = \pm 872.7$$

Then:

$$U_P = \sqrt{8000^2 + 4800^2 + 872.7^2} = 9.37\ kW$$

The active power is 800 kW. From the calculation, the uncertainty of current and voltage is determined by the sensor itself and would not change a lot. The uncertainty of active power is mainly affected by the power factor.

The relationship between uncertainty of active power and the power factor can be described by Equation (7):

$$U_P = \sqrt{\begin{array}{c}(U\Delta I)^2 PF^2 + (I\Delta U)^2 PF^2 \\ + (UI)^2 \left(1 - PF^2\right) \Delta \theta_I^2 \end{array}} \quad (7)$$

With voltage 400 V, current 2500 A, the U_p changes with power factor is shown in Figure 1. And the relative error is shown in Figure 2.

In reality, the current changes with the power factor. The relationship between relative error and current can be described by Equation (8):

$$\eta\% = \sqrt{\frac{\Delta I^2}{I^2} + \frac{\Delta U^2}{U^2} + \frac{1 - PF^2}{PF^2} \Delta \theta^2} \quad (8)$$

From Figure 3, the $\eta\%$ is increasing as the power factor decrease under 0.05, while the $\eta\%$ does not change much when power factor is higher.

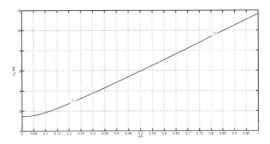

Figure 1. Errors of active power.

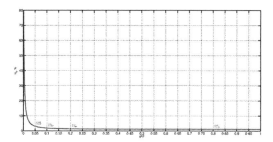

Figure 2. Relative error of active power.

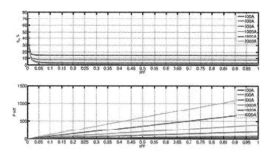

Figure 3. Relative errors of active power with current and PF.

From 0.1 to 1, the $\eta\%$ is determined by the current level, higher relative errors with lower current level.

3 EFFECTS OF REACTIVE POWER COMPENSATION

It is found that the active power measurement is different before and after the point of reactive power compensation in a steel rolling factory. To analyze the phenomenon, the errors between the point 1 and point 2 shown in Figure 4 is taken into consideration.

The error between point 1 and point 2 is:

$$U_P = \sqrt{3}I_1 \cdot PF_1 \cdot \Delta U_1 + \sqrt{3}U_1 \cdot PF_1 \cdot \Delta I_1 \\ + \sqrt{3}I_2 \cdot PF_2 \cdot \Delta U_2 + \sqrt{3}U_2 \cdot PF_2 \cdot \Delta I_2 \\ + \sqrt{3}U_1 I_1 \cdot \sin\theta_1 \cdot \Delta\theta_1 + \sqrt{3}U_2 I_2 \cdot \sin\theta_2 \cdot \Delta\theta_2$$

where:

$\Delta U_1 = \Delta U_2 = 660 \times 0.1\% = 0.66$ V

$\Delta I_1 = \Delta I_2 = 2000 \times 0.5\% = 10$ A

$\Delta\theta = 3 \times 10^{-3}$

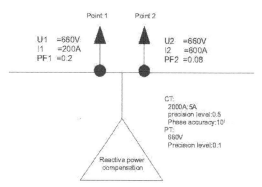

Figure 4. Measurement points of active power.

The $U_P = 6.02$ kW. Relative errors are 13.2% compared to point 1 and 10.9% compared to point 2. The active power of point 2 is larger than that of point 1 about 8 percentage, giving false appearance that the reactive power compensation have effects on energy saving.

But from analysis above, the measurement errors between two points would lead the phenomenon. The energy generating of reactive power compensation is fake.

From Figure 3, it is clear that when power factor is higher than 0.05, power factor has less influence on the uncertainty of active power measurement. The influence of the different power factor around reactive power compensation point could be ignored.

In the measurement, two same current sensors are used at two points. On two points, 200 A and 600 A current are not fit to the current sensors 2000 A.

But more adaptive current sensor cannot be used at this situation because much higher current is reached when facilities starting, often 1200 A to 1500 A. Current sensor with lower measuring range would be broken down. What's more, the reactive power increases rapidly, the time delay would let rapid reactive current to influence power system after compensation.

All these factors determined that 2000 A current sensors should be used for the measurement.

4 CONCLUSION

From the analysis above, it is known that a decrease of active power measurement data would not always mean the energy saving. Attention should be paid to inspect the precision and accuracy of each measurement.

Restriction by the factories environment would lead some compromise on measuring method and equipment, which conceals greater uncertainty of the data. So the proper equipments and methods are essential for the active power measurement.

REFERENCES

Acrel. 2013. Acrel Electrical Engineering Solution Manual.

Cao, W.D. et al. 2012. Research on Multiple signals monitor control circuit by one Power Quality Analyzer. APPEEC 2012.

Fei, Y.T. 2014. Error Theory and Data Processing. Beijing: China Machine Press.

PSL. 2011. Pqube Specification 2.0.

Tumanski, S. 2009. Principles of Electrical Measurement. Beijing: China Machine Press.

Advances in Power and Energy Engineering – Sun (Ed.)
© 2016 Taylor & Francis Group, London, ISBN 978-1-138-02846-3

Trip fault study of 500 kV common-tower double-circuit transmission lines based on induced voltage

A.W. Yu & W. Tang
Center China Grid Company Limited, Wuhan, Hubei, China

B.C. Feng
Beijing Kedong Electric Power Control System Co. Ltd., Beijing, China

ABSTRACT: 500 kV common-tower double-circuit transmission lines have a close distance between conductors and strong coupling. When faults occur in the transmission line, the sound phase has an electromagnetic-induced voltage compared with the fault phase. This paper proposes an induced voltage-based analytic strategy on the master station for the one-line tripping out of 500 kV common-tower double-circuit transmission lines. The fault property, the fault phases, the position of failure points and whether the fault has been removed can be decided based on our strategy. It can provide a reliable basis for the forced line charging.

1 INTRODUCTION

500 kV common-tower double-circuit transmission lines have great advantages, such as huge transmission capacity, small covering area and high economic value (Danhui Hu et al. 1927–1931). Its wide applications empower the significant change of electrical characteristics in the power network. Due to the close distance between 500 kV common-tower double-circuit transmission lines, the coupling on lines is strong and the induced voltage on the fault line is high. Therefore, we can get the real-time fault information on the fault property, the fault phase and the fault point location based on the characteristics of the induced voltage on the fault line (Aixiu You. 2000).

Current fault treatment strategies for 500 kV common-tower double-circuit transmission lines are mainly based on the logical reclosing on the substation, which means that the immediate reclosing is conducted when the fault phase trips out (Xin Qian et al. 2002). However, if the fault is the permanent fault or the electric fault on the fault point is not quenched during reclosing, the reclosing will fail due to the logical reclosing on the substation, and the remaining two phases on the one-line of the fault phase will also trip out.

In contrast to the above strategies, our work conducts a real-time surveillance for the statement of the fault line on the major station, and decides whether the fault is rescinding, the fault phase, and whether it can transmit forced power to the lines after the tripping out of lines by analyzing the

properties of the induced voltage after the one-line tripping out on 500 kV common-tower double-circuit transmission lines. Moreover, our work shows the fault information to the dispatcher, helping them to recover the normal running rapidly and improving the security, stability and economy at the operation level of the power network.

2 INDUCED VOLTAGE CALCULATION

When the one-line on 500 kV common-tower double-circuit transmission lines has a fault, certain induced voltages are produced on the fault line due to the coupling effect between lines (Xu Wei & Changyi Li. 1927–1931). The main factors influencing the induced voltages include running line voltage, load current, line length, distance between lines and lines that have transposition.

Our designed strategy depends on the following information: first, because the main study object in this paper is the common-tower double-circuit transmission lines' fault, the basic information about common-tower double-circuit transmission lines in the power network is needed. Second, as the main function that we want to achieve is the real-time surveillance, the real-time line information from the WAMS (Wide Area Measurement System) is needed.

Assuming that in 500 kV common-tower double-circuit transmission lines, the distance of each phase line, the distance of two loops and the height of lines are known and the lines are fully transposed, the matrices of self-capacitance and mutual capacitance of each phase, denoted as C, and self-inductance and mutual inductance, denoted as L, can be expressed as follows:

$$
C = \begin{bmatrix}
C_{aa} & C_{ab} & C_{ac} & C_{aA} & C_{aB} & C_{aC} \\
C_{ba} & C_{bb} & C_{bc} & C_{bA} & C_{bB} & C_{bC} \\
C_{ca} & C_{cb} & C_{cc} & C_{cA} & C_{cB} & C_{cC} \\
C_{Aa} & C_{Ab} & C_{Ac} & C_{AA} & C_{AB} & C_{AC} \\
C_{Ba} & C_{Bb} & C_{Bc} & C_{BA} & C_{BB} & C_{BC} \\
C_{Ca} & C_{Cb} & C_{Cc} & C_{CA} & C_{CB} & C_{CC}
\end{bmatrix} \quad (1)
$$

$$
L = \begin{bmatrix}
L_{aa} & L_{ab} & L_{ac} & L_{aA} & L_{aB} & L_{aC} \\
L_{ba} & L_{bb} & L_{bc} & L_{bA} & L_{bB} & L_{bC} \\
L_{ca} & L_{cb} & L_{cc} & L_{cA} & L_{cB} & L_{cC} \\
L_{Aa} & L_{Ab} & L_{Ac} & L_{AA} & L_{AB} & L_{AC} \\
L_{Ba} & L_{Bb} & L_{Bc} & L_{BA} & L_{BB} & L_{BC} \\
L_{Ca} & L_{Cb} & L_{Cc} & L_{CA} & L_{CB} & L_{CC}
\end{bmatrix} \quad (2)
$$

where A, B, C, a, b, c represent each phase of one-line and two-line, respectively. In this paper, we denote C as $\begin{bmatrix} C_1 & C_m \\ C_m & C_2 \end{bmatrix}$ and denote L as $\begin{bmatrix} l_1 & l_m \\ l_m & l_2 \end{bmatrix}$ for short.

Moreover, we combine the sound phase voltage obtained from the WAMS, denoted as $U_{abc} = [U_a\ U_b\ U_c]^T$, the induced voltage of each phase on the fault line can be obtained, denoted as $U_{ABC} = [U_A\ U_B\ U_C]^T$.

In this paper, we carry out our theory analysis on the following three different circumstances when lines trip out: 1) the one-line of common-tower double-circuit transmission lines has fault with both ends dangling; 2) the one-line of common-tower double-circuit transmission lines has fault with one end dangling and the other grounding, that is to say, it is grounding with one end; 3) the one-line of common-tower double-circuit transmission lines has fault with both ends grounding.

2.1 Both ends dangling

We assume that both of the two ends of the induced line are dangling. Therefore, the induced line has induced voltage, which is the coupling voltage. Moreover, we have the following function:

$$
\begin{bmatrix} C_1 & C_m \\ C_m & C_2 \end{bmatrix} \begin{bmatrix} U_{abc} \\ U_{ABC} \end{bmatrix} = \begin{bmatrix} Q_{abc} \\ 0 \end{bmatrix} \quad (3)
$$

where Q_{abc} is the matrix of the induced charge.

According to Formula (3), we have the following functions:

$$
C_1 U_{abc} + C_m U_{ABC} = Q_{abc} \quad (4)
$$

$$
C_m U_{abc} + C_2 U_{ABC} = 0 \quad (5)
$$

so

$$
U_{ABC} = -C_2^{-1} C_m U_{abc} \quad (6)
$$

Assuming the fact that the lines are transposed fully, and the corresponding fact that the fault line has a three-phase voltage that is mainly symmetrical, the induced line voltage is related to the voltage of the running line only.

2.2 One end grounding

We assume that the line is dangling with one end and the other end is grounding. Therefore, the current voltage difference is inductance coupling. If we denote ΔU_{ABC} as the voltage difference of two ends of the line, we have the following function:

$$
j\omega \begin{bmatrix} L_1 & L_m \\ L_m & L_2 \end{bmatrix} \begin{bmatrix} I_{abc} \\ 0 \end{bmatrix} = \begin{bmatrix} \Delta U_{abc} \\ \Delta U_{ABC} \end{bmatrix} \quad (7)
$$

According to Function (7), we have the following:

$$
j\omega L_1 I_{abc} = \Delta U_{abc} \quad (8)
$$

$$
j\omega L_m I_{abc} = \Delta U_{ABC} \quad (9)
$$

That is to say, the voltage of the fault line endpoint is given by

$$
U_{ABC} = j\omega L_m I_{abc} \quad (10)
$$

From the above functions, we know that the induced voltage on the fault line is proportional to the transmission power on the running line as well as to the line's length.

2.3 Both ends grounding

If the two ends of the line are grounding, the induced line has induced voltage caused by inductance. However, due to the fact that the two ends are grounding simultaneously, ΔU_{ABC} is nearly 0 kV.

2.4 Example of induced voltage calculation

We take one 500 kv common-tower double-circuit transmission line in the Central China Grid as an example. Its basic information of lines is as follows: the line's wire type is 4*LGJ500; the source voltage is 520 kv; the terminal voltage is 500 kv; the phase angle difference is 10°; the distance of two lines is 17 m; the phase space of the one-line is 10 m; and the height of the bottom line above ground is 15 m. The induced voltage under the

Table 1. Induced voltage under the three circumstances.

Phase voltage/kV	Phase A	Phase B	Phase C
Two ends dangling	29.9	30.7	29.1
One end grounding	1.26	1.25	1.22
Two ends grounding	0.14	0.13	0.14

above three circumstances, i.e. two ends dangling, one end grounding and two ends grounding, can be calculated by the above conditions. The calculated results are summarized in Table 1.

From Table 1, we can see that the induced voltage of the fault line is about 30kV when the two ends are dangling, the induced voltage of the fault line is about 1 kV when one end is grounding, and the induced voltage of the fault line is about 0 kV when the two ends are grounding. Therefore, we can judge the fault property, i.e. the permanent fault or the transient fault, according to the corresponding induced voltage under different circumstances when the line trips out. The specific judging method is presented in Section 3.

3   FAILURE ANALYSES

Faults treatment in the power network needs to be accurate and timely; thus, the dispatchers need to have the information on real-time power network running to recover equipments' power transmission (Wei Wu et al. 2002). Due to the fact that the percentage of the single-phase fault to the lines fault is higher than 90%, we mainly study the single-phase fault in the one-line of 500 kv common-tower double-circuit transmission lines.

3.1   Line's fault property determination

In this paper, faults need to be solved by different ways according to the different fault properties. When lines have single-phase faults, the three phases of the one-line tripping out and the other line is running normal. At such status, fault property can be distinguished into two types: the permanent fault or the transient fault. 1) If the fault is the transient fault, in which the three phases of the fault line are all dangling and electrostatic-induced voltage exists, the arc on fault point is extinguished, the three-phase voltage is mainly symmetrical and about 30 kV during the one-line tripping out, then the line fault is removed and the forced transmission for the line is permitted. 2) If the fault is the permanent fault, in which the fault phase is grounding and electromagnetic-induced voltage exists, the non-fault-phase is dangling and has electrostatic-induced voltage, thus the induced voltage of the non-fault phase is much higher than the fault phase, and the induced voltage of the fault phase is about 0kV during the one-line tripping out, then the fault phase and point are needed to be determined further, and the forced transmission for the line is permitted after the line fault is removed.

3.2   Line's fault area determination

3.2.1   Fault phase determination
When the fault is a transient fault, the fault can disappear instantaneously. Thus, the fault phase can be obtained according to the tripping signal of the protector when the fault occurs. When the fault is a permanent fault, the fault always exists. Thus, the fault phases can be judged by comparing the induced voltage at each phase. Specifically, when the induced voltage at one phase is about zero, the phase is the fault phase. Furthermore, it can be decided whether the fault has to be removed according to the testing for the induced voltage. That is to say, if the tested induced voltage of the fault phase is close to the other induced voltage, the fault is removed and the forced power transmission can be conducted; instead, if the fault is not removed, the forced power transmission cannot be conducted.

3.2.2   Determination of fault point location
Regardless of the transient fault or the permanent fault, the difference between the induced voltage tested from the fault point and the other location is evident. Therefore, we test and compare the induced voltage on different locations to decide on the location of the fault point.

3.3   Example of failure analyses

The fault circuit of 500 kv common-tower double-circuit transmission lines is shown in Figure 1, and the three-phase voltage in the one-line is given in Table 2.

From Table 2, we can see that the induced voltage on phase C is obviously low, about 0, while the induced voltage of phase A and phase B is obviously higher than that of phase C. Based on the above failure analyses, we can see that phase C is the fault phase, which has a transient fault, and the fault has not been removed.

In order to further determine the fault point location, we show the tested induced voltage on different locations on the line. From Table 3, we can see that the induced voltage of the 50% line point location is obviously different from that of the other locations. Therefore, we can determine that the fault point is the 50% of the line. The whole fault information of the fault line is shown in Figure 2.

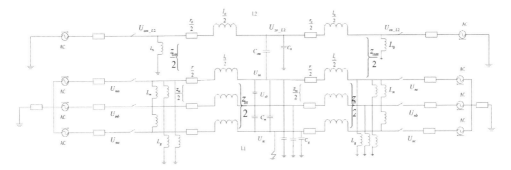

Figure 1. Fault in the phase C of 500 kv common-tower double-circuit transmission lines.

Table 2. Two side-induced voltage of the fault line after the one-line of 500 kv common-tower double-circuit transmission lines trips out.

Phase voltage/kV	Phase A	Phase B	Phase C
Source of line	32.10	27.44	0.15
Terminal of line	32.20	27.32	0.23

Table 3. Induced voltage of the tested points.

Percentage of the tested point to the whole line	0%	25%	50%	75%	100%
Induced voltage/kV	1.7	1.4	0.006	1.3	1.6

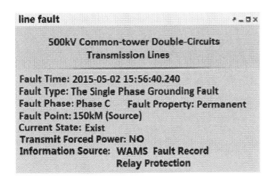

Figure 2. Failure analyses interface of the fault line.

4 CONCLUSION

In this paper, we conduct the failure analyses of 500 kV common-tower double-circuit transmission lines on the major station by calculating the induced voltage of fault lines. According to the property of the fault line-induced voltage, we analyze and get the real-time fault information on the fault line, and decide whether to transmit the forced power. Meanwhile, the fault information analyzed is shown on the dispatching platform and thus helps the dispatcher hold the real-time fault information. As we only study the situation of one-line of 500 kV common-tower double-circuit transmission lines, the inter-line fault is not analyzed in this paper. Therefore, future work will focus on the study of the inter-line fault.

ACKNOWLEDGMENTS

This work was supported by the Scientific and Technological Programme of Central China Grid (SGTYHT/13-JS-175).

REFERENCES

Aixiu You. 2000. Analysis and calculation of electromagnetic influence voltage on transmission line [J]. Shanxi Electric Power, 95(6): 14–16.

Danhui Hu et al. 1927–1931. Calculation of inductive voltage and current for 500 kV double circuit transmission lines on Single Tower. *High Voltage Engineering* 34(9).

Hua Zhao & Jiangjun Ruan. 2005. Calculation of the induction voltage on double circuit transmission lines [J]. Relay, 33(22): 37–41.

Wei Wu et al. 2002. Study on induced voltage and secondary arc current on Quanzhou2Jinjiang 500 kV double circuit transmission lines [J]. Fujian Power and Electrical Engineering, 22(4): 4–7.

Wenliang Zhang et al. 2002. A study on induced effect between the 500 kV Hong Long double circuit transmission lines with HV shunt reactor [J]. High Voltage Engineering, 28(2): 4–7.

Xin Qian et al. 2002. Influence of fault types and different transposition forms on coupling between two circuits on same tower [J]. Power System Technology, 26(10): 18–20, 57.

Xu Wei & Changyi Li. 1927–1931. Computation of induction current in 500 kV double circuit. *East China Electric Power* (3): 7–11.

Yanhua Han et al. Induced voltage and current on double circuits with same tower. *High Voltage Engineering* 33(1): 140–143.

Yi HU & Dingzhen Nie. 2000. Calculation of induced voltage and safety working condition on 500 kV double circuit transmission lines [J]. Electric Power, 33(6): 45–47.

Yi Hu et al. The research of safe distance for live working on 500 kV compact double circuit transmission line. *High Voltage Engineering* 29(8): 3–4.

Zhihong Guo et al. Study and measurement of induced voltage and current for 500 kV double-circuit line on same tower. *High Voltage Engineering* 32(5): 11–15.

Advances in Power and Energy Engineering – Sun (Ed.)
© 2016 Taylor & Francis Group, London, ISBN 978-1-138-02846-3

Thermal fault diagnosis on power transformer with Grey Relational Analysis

L.F. Li, S.C. Deng & X. Chen
Shenzhen Power Supply Co. Ltd., Shenzhen, Guangdong Province, China

J. Wang
Changjiang Institute of Survey, Wuhan, Hubei Province, China

ABSTRACT: Overheating fault is the most common fault of the transformer. It's very important for the transformer to identify and locate the overheating fault. DGA analysis method is useful in the identification of overheating fault, but it cannot locate the fault. This overheat can be caused by transformer core faults and conductive loop faults. When the overheating fault occurs, the temperature of transformer oil is lower than 1000 °C. The main components of transformer oil cracking gas are low molecular weight alkanes and alkenes at this temperature, so it is hard to tell whether the fault is conductive circuit overheating or non-conductive circuit overheating only with DGA data. At present, the identification of overheating fault location is based on the electrical test, such as DC resistance test, ratio test, no-load test, etc. The analysis of overheating fault position is largely dependent on the experience of the field staff. When the overheating fault exists at the conductive loops or magnetic circuit, the fault energy changes with current or voltage, and the oil chromatography data also changes accordingly. Based on the grey relational analysis theory, the correlation degree among oil spectrum data, voltage and current is analyzed quantitatively, and then the overheating is located. Firstly, construct gray sequences with Gray correlation analysis method. The sequences include the total hydrocarbon sequence, voltage sequence and current sequence. Secondly, calculate the correlation degree among the sequences. Lastly, compare the size of the correlation degrees. If the degree between total hydrocarbon and voltage is greater than that of current, the overheating fault is magnet loop overheating fault; if not, the fault is conductive circuit fault. This paper provides a method for the distinguishing of conductive circuit overheating and magnetic circuit overheating. The effectiveness of the method is verified by field examples.

1 INTRODUCTION

The thermal fault is a serious threat for the transformer. Most of these faults occur in the conducting circuit (such as tap switch, wire and connector) and nonconductive circuit (such as iron core eddy current and multi-point grounding). The others are caused by the local oil channel blockage or other reasons. In the Dissolved Gas Analysis, the three-ratio analysis method is a good method to judge the overheating, discharge fault, but not clear overheat fault is caused by the conductive loop is composed of magnetic circuit (Duval, M. 2002) (Duval, M. 2003).

At present, the common thermal fault is diagnosed by electrical test, switching operation, historical data, etc., but there are some defects such as power outages, low sensitivity. The A-V curve is just a qualitative description of transformer fault, rather than quantitative analysis. And it can be applied only in the natural running state. The numerical difference among current, voltage and total hydrocarbons is very large due to their units, and this makes fault diagnose difficult by A-V curve (Min-zhou, Y. 2003) (Yi-chang, L. 2011).

Grey Relational Analysis (GRA) was pioneered by Deng Julong in 1984 (Ju-long, D. 1984) (Ju-long, D. 1989). The method is commonly used in Asia. It is an impact evaluation model that measures the degree of similarity or difference between two sequences based on the grade of relation. In this paper, a novel Grey Relational Analysis (GRA) method is presented for the diagnosing thermal faults in power transformers through the analysis of total amount of hydrocarbon and A-V curve.

2 GREY RELATIONAL ANALYSIS METHOD

GRA is part of grey system theory, which is suitable for solving the complicated interrelationships between multiple factors and variables. In fact, GRA might be reckoned as a contrasting way, in wholeness, equipped with reference for contrasting. It is well known that the distance space found on numerical measure. However, which devoid of wholeness, it is only a partnering comparison. The set-point topology is characterized by neighborhood-ness, namely the wholeness, but devoid of numerical measure. Differing from the traditional mathematical analysis, GRA provides a simple scheme to analyze the series relationships or the system behavior, even if the given information is few. The grey relational analysis is used to determine the relationship between two sequences of stochastic data in a grey system. The procedure may bear some similarity to the pattern recognition technology. One sequence of data is called the "reference pattern" or "reference sequence," and the correlation of the other sequence (named "comparative pattern" or "comparative sequence") to the reference sequence is to be identified (Ju-long, D. 2007) (Chang-tian, L. 2006). The mathematics mode of grey relational analysis is shown as follows:

Suppose $\mathbf{x}_i$ be the reference pattern with n entries

$$\mathbf{x}_i = \{x_i(k)\} = \{x_i(1), x_i(2), ..., x_i(m)\}$$
$$k = 1, 2, ..., m \quad (1 \le k \le m) \tag{1}$$

and $\mathbf{x}_j$ be one of the patterns with p entries to be compared with $\mathbf{x}_i$ (each $\mathbf{x}_j$ has the same m number of criteria as $\mathbf{x}_i$). The $\mathbf{x}_j$ is written as

$$\mathbf{x}_j = \{x_j(k)\} = \{x_j(1), x_j(2), ..., x_j(m)\} \quad (1 \le k \le m) \tag{2}$$

Like general statistical analysis methods, grey relational analysis first calls for the appropriate normalization of raw data to remove anomalies. The raw data can be transformed into dimensionless forms either by initial-value processing or average-value processing. These two processing methods defined as follow:

$$\mathbf{x}_j^* = \{x_j^*(k)\} = \frac{x_j(k)}{x_j(1)} \tag{3}$$

$$\mathbf{x}_j^* = \{x_j^*(k)\} = \frac{x_j(k)}{\frac{1}{m}\sum_{k=1}^{m} x_j(k)} \tag{4}$$

The pattern of processing to use depends on the nature of data. Generally, average-value processing is applied to data series independent of time sequences, and maximum-value processing is more appropriate for data that vary with time. Normalize the sequences with average-value processing to ensure that all of them are in the same order, and the normalized sequences can be denoted as

$$\mathbf{x}_j^* = \{x_j^*(k)\} = \{x_j^*(1), x_j^*(2), ..., x_j^*(m)\} \quad (1 \le k \le m) \tag{5}$$

The set of the sequence $\mathbf{x}_j^*$ is generally the influencing factor to $\mathbf{x}_i$. The grey relational coefficient between the compared pattern $\mathbf{x}_j^*$, and the reference pattern $\mathbf{x}_i$ at the kth entry ($k = 1, 2, ..., m$) is defined

$$\xi_{jk}(k) = \xi_{jk}(x_i(k), x_j^*(k))$$
$$= \frac{\min\limits_{j}\min\limits_{k} |x_i(k) - x_j^*(k)| + \rho\max\limits_{j}\max\limits_{k} |x_i(k) - x_j^*(k)|}{|x_i(k) - x_j^*(k)| + \rho\max\limits_{j}\max\limits_{k} |x_i(k) - x_j^*(k)|} \tag{6}$$

where $|x_i(k) - x_j^*(k)|$ is the absolute difference of two comparative sequence, $\min\limits_{j}\min\limits_{k} |x_i(k) - x_j^*(k)|$ and $\max\limits_{j}\max\limits_{k} |x_i(k) - x_j^*(k)|$ are the minimum and maximum value of $|x_i(k) - x_j^*(k)|$ respectively, and ρ is the distinguish coefficient which its value is adjusted with the systematic actual need and defined in the range between 0 and 1. A value of 0.5 is used in most situations (Chang-tian, L. 2006). When $x_i(k)$ equals $x_j(k)$, the coefficient of grey relation is $\xi_{jk} = 1$. This indicates that $x_j(k)$ is highly related to $x_i(k)$. When $|x_i(k) - x_j^*(k)|$ is the maximum value over all k entries, ξ_{jk} reaches the minimum value over all k entries. The grey relational grade $\xi_{jk}(x_i, x_j)$ between the reference pattern $\mathbf{x}_i$ and the compared pattern $\mathbf{x}_j$ is taken as the average of ξ_{jk} over all k entries, when the coefficients of grey relation are equally important at all entries.

$$\gamma_{ij} = \gamma(x_i, x_j) = \frac{1}{m}\sum_{k=1}^{m} \xi_{ij}(k) \tag{7}$$

Grey relation grade γ_{ij} is the similarity indicator of the pattern $\mathbf{x}_i$ and the pattern $\mathbf{x}_j$. If, $\gamma_{ij} > \gamma_{il}(1 \le l \le p)$ then the patter $\mathbf{x}_j$ has characteristics closer to those of the reference pattern $\mathbf{x}_i$ than the pattern $\mathbf{x}_l$. A higher grey relational grade indicates that the compared sequence is the most similar to the reference sequence. However, there are unresolved problems in the above investigation. The main one is to put uniform information on each factor $\mathbf{x}_i$ for all sequences. Thus the information entropy approach is introduced.

3 APPLICATION STEPS OF GRA

When DGA data is given, the steps of grey relational analysis are shown as below.

Step 1: Building reference and comparative sequence.

With transformer oil data of total hydrocarbons $(C_1 + C_2)$ builds the reference sequence according to formula (1), namely:

$$\mathbf{x}_0 = \{x_0(k)\} = \{x_0(1), x_0(2), ..., x_0(m)\}$$
$$k = 1, 2, ..., m \qquad (8)$$

Then with transformer oil data of the average voltage, current builds comparative sequence respectively according to formula (2), namely:

$$\mathbf{x}_i = \{x_i(k)\} = \{x_i(1), x_i(2), ..., x_i(m)\} \quad i = 1, 2$$
$$(9)$$

where $i = 1$ is voltage, $i = 2$ is current.

Step 2: Standardized data

In analysis, due to different unit of the total hydrocarbon, voltage and current, data must carry on the standardization of data processing to ensure that all factors have equivalence and the properties of sequence. This paper uses the initial-value method processing data according to formula (3).

$$\mathbf{x}_0^* = \frac{x_0(k)}{x_0(1)} \qquad (10)$$

$$\mathbf{x}_i^* = \frac{x_i(k)}{x_i(1)} \qquad (11)$$

Step 3: Calculation of grey relational coefficient

According to formula (4) calculate the grey relation coefficient ξ_{0i}.

Step 4: Calculation of grey relational grade

According to formula (5) calculate the grey relation grade γ_{0i}.

Step 5: Ordering of the grey relation grade

Comparison of the value of the γ_{01} and γ_{02}, in order to determine the voltage, current of two factors, which one factor is more close to the development trend of the total hydrocarbon is more close, so as to determine the fault loop.

4 EXPERIMENTAL RESULTS

A transformer was put into operation in 2000, and its type is OSFPSL-120000/220. DGA data showed latent defects existed in this transformer in 2010. DGA data is shown in Table 1 (Yi-chang, L. 2011). Three-ratio code of the DGA data is 022, and it is thermal fault. Voltage and current data are shown in Table 2, the curve of the series data is shown in Figure 1.

According to formula (1)–(7), the results show as the following:

$$\gamma_{01} = 0.681$$

Table 1. DGA data (unit: uL/L).

Date	H_2	CH_4	C_2H_6	C_2H_4	C_2H_2	$C_1 + C_2$
2010.3.2	79.99	120.64	28.77	297.47	18.91	465.79
2010.3.6	77.54	127.70	25.48	267.67	19.91	440.76
2010.3.8	80.50	153.54	42.61	276.00	18.37	490.52
2010.3.10	76.01	114.67	40.57	276.00	15.23	446.47
2010.3.12	131.87	162.10	29.08	314.70	22.45	528.33
2010.3.14	107.22	143.88	31.10	290.68	21.83	487.49
2010.3.16	98.73	123.58	33.56	296.24	16.08	469.46
2010.3.19	109.59	133.96	39.30	361.00	23.43	557.69

Table 2. Voltage and current data (Unit: Voltage-kV, Current-A).

Date	V	A	Date	V	A
2010.3.2	131.5	226.5	2010.3.12	149.2	222.3
2010.3.6	93.4	225.2	2010.3.14	129.4	223.9
2010.3.8	131.0	225.3	2010.3.16	120.7	225
2010.3.10	141.2	223.5	2010.3.19	100.4	228.1

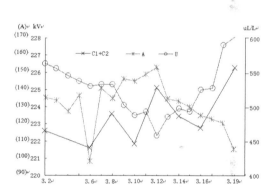

Figure 1. The curve of total hydrocarbons, voltage and current.

$$\gamma_{02} = 0.617$$

Because of the $\gamma_{01} > \gamma_{02}$, it shows that the total hydrocarbon is more large extent associated with the voltage, namely the development trend is more similar between them, known as magnetic circuit fault overheating. After the cover lifting, it founds that the overheating fault is caused by short circuit circulation between iron core and iron yoke clamp.

5 CONCLUSIONS

The total hydrocarbon, voltage and current of the transformer are interrelated and Interact, these factors have correlation. If we take the analysis time, total hydrocarbon, the voltage and the current relative value constitutes a multi-dimensional space coordinate system, although the values of these points is relative, but between these points is the distance absolute at the same energy excitation. Therefore, they are in the space reflects the distance relationship each other, which is composed of spatial distance, and total hydrocarbon, current, voltage value is not very stable, in different circumstances they are changing dynamically, namely a grey number, and thus meet the grey relational analysis.

The paper analysed the cause of transformer thermal fault using the grey relational analysis, it is a new exploration. The method is feasibility and validity from theory and example and has good application value, it is worth further research.

REFERENCES

Chang-tian, L. 2006. A Grey analysis of bank re-decreasing the required reserve ration. J. Grey Syst. 11 (2):119–132.

Duval, M. 2002. Using gas-in-oil standards to improve accuracy of dissolved gas analysis results and diagnoses, Electricity Today, 14(6):16–17.

Duval, M. 2003. New techniques for dissolved gas in oil analysis, IEEE Elect. Insul. Mag., 19(2):6–15.

Ju-long, D. 1984. The theory and methods of socio-economy grey system. Social sciences in China, (6):47–60.

Ju-long, D. 1989 Introduction Grey system theory, J. Grey Syst. 1:1–24.

Ju-long, D. 2007. The basic method of gray system. Wuhan: Hua Zhong science and engineering university publishing house.

Min-zhou, Y. 2008. Determination of over-heat fault of power transformer using method of amount of hydro carbon and A-V curve. High voltage engineering, 66(12):64–66.

Yi-chang, L. 2011. Transformer thermal fault diagnosis and its parts with chromatography. Transformer, 32(7):35–38.

Advances in Power and Energy Engineering – Sun (Ed.)
© 2016 Taylor & Francis Group, London, ISBN 978-1-138-02846-3

Study of time series clustering based on FCM

X. Guo, H.B. Yuan, J.X. Lv & M. Ling
School of Automation Science and Electrical Engineering, Beihang University, Beijing, China

L.M. Zhou & Z. Ma
China Electric Power Research Institute, Beijing, China

ABSTRACT: In this paper, Fuzzy Cognitive Map (FCM) is employed and discussed to analyze time series that represents fault dynamics. At first, the essential of FCM prototype is studied, and diversified data set that are produced from the prototype to construct candidate FCM model is proposed. The Particle Swarm Optimization (PSO) and Simulated Annealing (SA) learning algorithm is taken as training method based on computational intelligence. Secondly, analysis of representation for time series from real world application is carried out with the proposed FCM model to assess the quality of FCM design methodology and algorithmic performance. The dynamics of time series is depicted by fuzzy c-mean clustering, and the concept of the node in the candidate FCM model corresponds to the activation level of each cluster center. In the end, the parameters of node number in the candidate FCM model and steepness of activation function are taken into consideration in order to obtain the better result. The discussion and findings are offered and the results show the proposed method is desirable in the representation of time series both at numeric value level and linguistic level.

1 INTRODUCTION

Application analysis of time series representation is diversified in many fields, such as signal processing and fault management (Sapello, A. et al. 2010), stock market, business and power system (Yu, H. 2005, Chow, G.C. & Lawler, C.C. 2003, Chow, G.C. & Lawler, C.C. 2003). Comprehensive methods and algorithms have been proposed recently in order to exploit the dynamics of time series by representing the features that extracted from it. One of the approaches is knowledge representation regarding the domain of application to clarify the dynamic behavior of the time series. However, this approach lies in the disadvantage of domain knowledge restrictions, subjective inclination and overhead in dealing with large complex system. In this paper, FCM model is proposed to represent the dynamics of the time series using information clustering method and computational intelligence technique in order to develop semi-automated or automated design method to overcome the aforementioned drawbacks from expert based method. FCM was first proposed by Kosko (Kosko, B. 1986), which can be used to represent the relationships between the elements. The extension of FCM can be used to model complex systems in which a precise mathematical model is difficult to represent and can also provide an approach to knowledge representation by defining

concept according to application requirement. Due to the desirable properties of abstraction, flexibility and adaptability, the design of many different FCM models concerning real world application have been developed, especially in the case of failure analysis and reliability assessment (Lee, K. & Kim, S. 1999). However, little work is carried out to build FCM model for the representation and predication of time series.

In this study, we focus on developing the universal FCM model in an abstract way, and the up-to-date learning method is also employed to train the model so that the scheme of constructed architecture of the model can be suitable for analysis of time series representation. The features that imply the dynamic behavior of the time series is originated from clustered data set, and the treatment of data set with proper FCM model is carried out. In this way, the evolution of the time series is depicted by the state set in the FCM model. It is emphasized that the weights accounts for the cause-effect relationships among the nodes in a FCM model, the weights can be optimized by Particle Swarm Optimization (PSO) method or Simulated Annealing (SA) method. With the FCM model is proposed, we take a further step toward the performance assessment in order to capture and emulate the essence of numerical data set produced from real world time series. The main motivations and contributions of the study are as follows:

1. Comprehensive analysis of FCM model is carried out. Each node within the FCM model is regarded as input and output from current state to the next state respectively. Given the former data value, the successive data points can be reached according to the computation of the developed FCM topology, as a result, the value with regard to the node can be either numerical or linguistic, which can be in the form of specific data or the activation level of fuzzy set.
2. The formation of candidate FCM model is studied. The determination of proper weight matrix in the model is regarded as optimization objective problem and PSO is employed as learning method. The candidate FCM model is highly abstract in concept and is also easy to construct.
3. Attempt is made for analysis of time series representation based on proposed method. Amplitude and change of amplitude is clustered to create candidate FCM model. Diversified parameter configuration regarding candidate FCM model is studied as well.

The rest of the paper is organized as follows. In section 2, FCM design architecture as well as learning method is given. In section 3, method of time series representation is provided and the results are shown and discussed. Section 4 draws the conclusion.

2 VALIDATION OF FCM ARCHITECTURE

2.1 *Fundamentals and formation of FCM*

The structure of a FCM is based on graph topology and can be conveniently observed visually. Generally speaking, FCM is capable of describing a system by nodes and connections. In a FCM model, the nodes are also referred to as concepts, events, actions, goals or trends of the system, each node in the FCM stands for a state of the system at specific time. Each connection between two nodes also refers to weight, edge, arc or interaction of the nodes. The connection depicts the cause-effect relationship from one node to another. Detailed definition and implementation regarding nodes and connections are determined by different application aspect of system, the nodes and connections can be added or removed on demand as well.

A weight exists between two nodes can be indicated as positive causality, negative causality and neutral causality. In FCM, the weight takes the value in the interval [−1, 1], which describes the strength of the corresponding causality. In Figure 1, the weight W_{ij} indicates the cause-effect relationship from source node i to node j holds some conditions. The condition of $W_{ij} \in [-1, 0)$

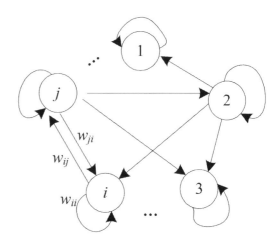

Figure 1. FCM model.

means the increase (decrease) in the value of node i can lead to the increase (decrease) in the value of node j, the condition of $W_{ij} \in (0, 1)$ means the increase (decrease) in the value of node i can lead to the decrease (increase) in the value of node j, whereas $W_{ij} = 0$ means the change in the value of node i does not effect the value of node j.

In general, the inference process in a FCM model is regarded as an iterative process, and feedback can be presented during the process as well. Once the initial state is determined in the FCM, the next state can be calculated and updated by applying the scalar product and activation function to each node synchronously. The values of each time state for all nodes are combined in order to form a discrete time system that can indicate cause-effect propagation in the value from the initial state to the final state. The iteration will continue until one of the following requirements on convergence is satisfied: equilibrium, limited cycle and chaotic behavior. The inference mechanism and state update of FCM can be described in the following formula:

$$v_i(t+1) = f\left(\sum_{j=1}^{n} v_i(t) \times w_{ij}(t)\right) \qquad (1)$$

where $i, j = 1, 2, \ldots, n$; $t = 1, 2, \ldots, k$; t denotes the iteration step, $v_i(t)$ denotes the state value of node, denotes threshold of activation function. Activation level of the nodes will change over iteration, which describes the dynamic behavior and characteristics of FCM. By comparison with neural network and other expert system, FCM can provide more flexible interaction and distinct meanings for the problem of real world application, and the weights and nodes can be configured and visually observed transparently (Tsadiras, A. 2008).

2.2 Design of candidate FCM model

The main objective for the development of candidate FCM model is to emulate a specific series of data set that are clustered from time series. To begin with, a FCM prototype with fully connected weights and self-feedback of each node is taken into consideration. In order to make it possible for further evaluation of both linguistic and numerical purpose, the feature that can be used to describe the trend of the time series is taken into account and will be presented to describe the behavior of the time series, which means the data values of the time series is not represented directly. Suppose input time series that is used for fault clustering $x(t)$, the amplitude and change of amplitude is then combined together to show the behavior. More specifically, we can use amplitude and change of amplitude to capture the dynamics of the time series, which is in the form of two dimensional data. The dynamics of time series can be described by information granules in the form of amplitude and change of amplitude $(x(t), \Delta x(t))$ using clustering method. Each data value in time series is given a numerical value of activation level with regard to the information granules. Therefore, for a given time series, the behavior of each data value centers around information granules, which are also regarded as the features of time series, and each data value has a degree of belonging to cluster center. The best way to form the information granules is to cluster time series. Fuzzy c-mean clustering is a useful way and can be used to divide data into different category. The cluster center will constitute the certain level of granules that imply the behavior of the original time series. In this way, the features with regard to information granules are extracted. The information granules are used. Each cluster is defined as a node in the FCM model, the concept of the node is described as the degree of belonging to the cluster.

2.3 Learning method and FCM training process

Training process is important in the design of FCM model, and the learning method plays vital rule during training process. Evolutionary strategy, particle swarm optimization and simulated annealing are dominate learning method. Having the data set available, the supervised learning method is taken into account. We are interested in the learning method of particle swarm optimization method for it is promising in tackling optimization problems. Apart from that, the activation function is crucial in manipulating the value of state vector. The proper selection of the activation function is dependant on some factors, for example, the nature of the problem, the required representation capabilities of the problem and the level of inference required by the case. Here, sigmoid function is generally used for it is suitable for qualitative and quantitative problems. The determination of the weights is made according to the objective function described as follows:

$$F_{obj} = \frac{1}{n(k-1)} \sum_{t=1}^{k} \sum_{i=1}^{n} \left\| f\left(\sum_{j=1}^{n} v_i(t) \times \overline{w}_{ij}(t) \right) - v_i(t+1) \right\|^2$$

(2)

The objective function is calculated for each iteration during optimization process, and the weights will be updated at the next iteration according to corresponding learning method. With particle swarm optimization is introduced (Clerc, M. & Kennedy, J. 2002), the swarm consisting of m particles is created initially, with each particle is the vector of $n \times n$ dimensions, and the dimension of each particle is corresponding to the number of cells in weight matrix W.

3 APPLICATION AND ANALYSIS

3.1 Data set and pre-processing

In this paper, a time series is introduced to demonstrate the proposed method. For simplicity, the time series under analysis comes from collection of electricity demand in New South Wales, Australia January, 2010. The data set is shown in Figure 2.

The values are recorded at an interval of 30 minutes. There are 48 data points altogether for one day of 24 hours. The procedure for analysis of representation begins with pre-processing of data set and setting the number of linguistic terms. Figure 3 shows the data set for one day, and the number of linguistic terms that is used to describe amplitude and change of amplitude is set to 2. In this way, the cluster center and its activation level matrix is obtained by using fuzzy c-mean

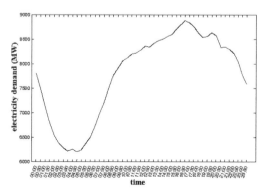

Figure 2.   Introduced time series.

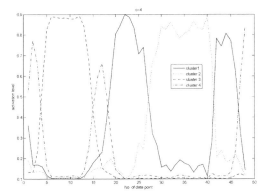

Figure 3. Cluster for data set from time series.

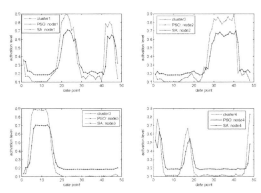

Figure 4. Representation of activation level for cluster ($\sigma = 1$).

clustering based on section 2. Simulation is performed regarding the process of FCM construction. In order to observe the influence of FCM parameters, the steepness of activation function is set to 1 and 6 separately for the same data points, the quality of representation from candidate FCM model is validated and assessed as well.

3.2 *Result and analysis*

In this section, the real world time series is taken to explore the proposed method, parameters of configuration are taken into account to observe the performance and quality of candidate FCM model as well. Figure 3 shows the activation level regarding data set, the number of linguistic terms is 2 labels for each of the two signals, i.e., amplitude and change of amplitude. The number of cluster is therefore set to 4, which is the same as the number of nodes in the candidate FCM model, after training process is completed, the candidate FCM model will be used subsequently to imitate the dynamics of the activation level as shown in Figure 3 over all data points.

In this respect, the parameters regarding candidate FCM model is initialized, the node number is set to 4, and the data points that are used to train the candidate FCM model is 48, the steepness of activation function is set to 1 and 6 respectively. The SA algorithm is also taken into consideration for comparison purpose. The red line denotes the original cluster, the blue line denotes the representation with PSO method, and the black line denotes the representation with SA method. Figures 4 and 5 show the result.

It can be seen in Figure 4 that the steepness has significant effect on the result of data representation, the performance is not desirable for each cluster in the case of $\sigma = 1$, although the trend of decrease and increase is depicted well, the quality

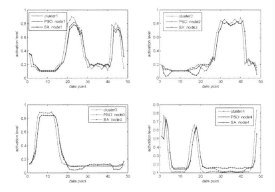

Figure 5. Representation of activation level for cluster ($\sigma = 6$).

is not good. In Figure 5, σ is set to 6, we can see that the quality of representation is improved. It is therefore concluded that the steepness of activation function is more sensitive in the design process of candidate FCM model. The result also shows that the data representation is advisable provided that the parameters are taken into consideration and the optimization performance is obtained.

4 CONCLUSIONS

In this paper, a design method based on FCM model is proposed for the representation of time series analysis. The up-to-data learning method is used to optimize the weight matrix, the steepness of activation function that affect FCM model is studied. The feature of the time series can be depicted by adjusting the parameters. It can be used for trend prediction and analysis for the application of real world time series.

ACKNOWLEDGMENTS

This research is supported by National Natural Science Foundation of China under Grant No. 61273165, Aero-Science Fund of China with Grant No. 2014ZD51047, SGCC (State Grid Corporation of China) "Thousand Talents program" special support project (EPRIPDKJ (2014) 2863) and SRF for ROCS, SEM.

REFERENCES

Chow, G.C. & Lawler, C.C. 2003. A Time-Series Analysis of the Shanghai and New York Stock Price Indices. *General Economics and Teaching* 4(1): 17–35.

Clerc, M. & Kennedy, J. 2002. The particle swarm—explosion, stability, and convergence in a multidimensional complex space. *IEEE Trans. Evolutionary Computation* 6(1): 58–73.

Kosko, B. 1986. Fuzzy cognitive maps. *International Journal of Man-Machine Studies* 24(1): 65–75.

Lee, K. & Kim, S. 1999. Fault diagnostic system based on fuzzy time cognitive map. *Transaction on Control, Automation and System Engineering* 6: 62–68.

Sapello, A. et al. 2010. Application of Time Series Analysis to Fault Management in MANETs. *International Conference on Network and Service Management. 25–29 October 2010*. Niagara Falls, Canada.

Tsadiras, A. 2008. Comparing the inference capabilities of binary, trivalent and sigmoid fuzzy cognitive maps. *Information Sciences* 178(20): 3880–3894.

Wang, J. 2011. Chaotic time series method combined with particle swarm optimization and trend adjustment for electricity demand forecasting. *Expert System with Applications* 38: 8419–8429.

Yu, H. 2005. A refined fuzzy time-series model for forecasting. *Physica A: Statistical Mechanics and its Applications* 346(3): 657–681.

Advances in Power and Energy Engineering – Sun (Ed.)
© *2016 Taylor & Francis Group, London, ISBN 978-1-138-02846-3*

Prediction of closing spring's energy-storing state-based on circuit breaker's running characteristic parameters

T. Shi
Kunming Power Supply Company of China Southern Power Grid Company, Kunming, China

B.W. Guo
School of Electrical Engineering, Wuhan University, Wuhan, China

S.Q. Bai
Kunming Power Supply Company of China Southern Power Grid Company, Kunming, China

W.J. Zhou
School of electrical engineering, Wuhan University, Wuhan, China

W. Dong
Kunming Power Supply Company of China Southern Power Grid Company, Kunming, China

P.F. Li
School of Electrical Engineering, Wuhan University, Wuhan, China

ABSTRACT: To achieve the diagnosis of the closing spring's energy-storing state in maintenance, the pressure sensor, the photoelectric encoder and the current transformer are used in the LW25–126 SF6 circuit breaker. The closing spring's energy-storing state and the running characteristic parameters can be obtained through experiments. The corresponding relationship between the characteristic parameters of the storing process and the energy-storing state is studied. The energy-storing pressure of the closing spring is predicted by the BP neural network. Based on the training results of the BP neural network and the analysis of experimental results, running characteristic parameters can effectively forecast the energy-storing condition of the closing spring. The result indicates that the energy-storing state of the closing spring can be effectively predicted by using the running characteristic parameters of the circuit breaker.

1   INTRODUCTION

The spring operating mechanism is widely used in the HV circuit breaker due to its advantages of small power capacity, simple mechanical structure and concise maintenance. The growth of the operational time, the manufacturing technology, material defects, environment and other factors probably cause faults such as loose mechanical connection, the fatigue and creep of energy-storing spring (Yuan Shun. 2001, Chen Baolun & Wen Yaning. 2010). It directly leads to insufficient or excessive storing energy (Chang guang. 2013). It exerts an extremely serious influence on the performance of the circuit breaker.

The maintenance of the energy-storing spring mainly concentrates on the spring whether there is corrosion and deformation and on the spring connection whether it is loose (The state grid corporation. 2005). But these methods of inspection

are not tested for the performance of the spring. The closing time, closing speed, opening time, opening speed, and other parameters are all tested for mechanical properties in the maintenance (GB3309–1989). The overall state of the circuit breaker can be judged by these parameters. But as the source of spring operating mechanism—energy-storing spring, the effective diagnostic method has not been put forward. In recent years, pressure sensors are added into the energy-storing spring for real-time monitor in the circuit breaker (Mei Fei, Mei Jun & Zheng Jianyong. 2013). This method can monitor the energy-storing spring state. But for the operating mechanism of the running spring, it is almost impossible for being equipped with the pressure sensor. At present, there are no diagnostic methods of the circuit breaker in the maintenance for the closing spring's energy-storing state.

In order to predict and diagnose the closing spring's energy-storing state, the LW25–126 SF6 circuit breaker is used as the tested object and its running parameters are used as the eigenvalue. The pressure sensor, the photoelectric encoder and the current transformer are respectively used to measure the pressure, closing curve and the current of the energy-storing motor. The mechanical characteristics of the parameters are extracted (Sun Laijun, Hu Xiaoguang & Ji Yanchao. 2007). The corresponding relationship between the running characteristic parameters and the closing spring's energy-storing state is established. This state is predicted and diagnosed by the BP neural network. The result indicates that the energy-storing state of the closing spring can be precisely predicted by not using the pressure sensor (Guo Qing, Xia Han & Han Wenwei. 2013).

2 BASIC PRINCIPLE

2.1 Running characteristic parameters of spring operation mechanism

The power of the spring operation mechanism is provided by the closing spring and the opening spring. The action of the circuit breaker includes three processes: closing process, opening process, and energy-storing process. When the opening signals are received, the opening spring releases energy and the static-dynamic contacts of operating mechanisms are driven to separate them. The closing process consists of two actions: first, the closing spring releases energy and completes the closing action; second, during the closing process, the part of the closing spring's energy transfers to the opening spring and it makes the opening spring complete the storing energy. It is the preparation for the next opening. After the completion of the closing action, the energy-storing motor starts running, the closing spring stores energy through the energy-storing transmission mechanism.

In this paper, the characteristic parameters of the closing process and the storing process are needed to study. In the maintenance of the circuit breaker, the relevant mechanical characteristic parameters include the closing time, the closing speed (maximum speed, average speed, and initial speed), the energy-storing electric current peak, the output power of the motor, and the effective working hours.

2.2 Energy-storing state characteristics of the closing spring

HV circuit breaker usually works in the open environment. With the increase in the working time, the phenomenon of the creep occurs on the spring under the long-term compression. The direct results of the creep are the decrease in the stiffness coefficient of the spring. It makes the spring's pressure decrease under the compression of the same length and also makes the storing energy decline.

The closing time of the HV circuit breaker is approximately in the range of 50 ms–120 ms. LW25–126 SF6 circuit breaker is used as an example for the experiment. It is normal when its closing time is within 90 ms–120 ms. In this period of closing time, the closing spring pressure declines from 30 kN to 20 kN. The great change in force results in a great vibration. This vibration leads to the loose of the screw and the nut in the closing spring. Thus, the preloading length of the closing spring may decrease or increase and the storing energy may also reduce or increase.

The increasing storing energy of the closing spring directly influences the closing speed of the circuit breaker. High closing speed will increase the vibration and friction of the circuit breaker. It can accelerate the aging of the contact. The service life of the circuit breaker is affected seriously. The decrease in energy-storing closing spring will cause the reduction in closing speed and reduce the ability of arc extinction. In summary, insufficient and excessive storing energy of the closing spring has a serious effect on the performance of the circuit breaker.

3 THE ACQUISITION OF RUNNING PARAMETERS

3.1 The acquisition of original data

LW25–126 SF6 circuit breaker is used for 110 kV. The closing spring is cylindrical spiral spring and adopts the storing method of compression. The parameters of the closing spring are listed in Table 1.

CCY-01 pressure sensor of the spiral spring is used to monitor the spring's pressure. The range of CCY-01 is 0–40000 N and test precision is less than 1%. The energy-storing pressure of the closing spring can be measured accurately. The E6B2-CWZ6C photoelectric encoder is used to monitor the speed of the main shaft. The closing curve of the circuit breaker is obtained by mathematical integral. In the CT20 operating mechanism, the rated current of the energy-storing motor is 4 A. The TR0108–2B current transformer is used and its range is 100 A.

The signals of three sensors are collected by MPS-140801 and uploaded to the host computer in time. The structure diagram of the hardware is shown in Figure 1.

Table 1. The parameters of the closing spring in the CT20 spring operation mechanism.

Number of coils	Wire diameter/mm	Inside diameter of the spring/mm	Free height/mm
9.5	28	167	590

Preloading height/mm	Pre-pressure/N	Energy-storing height/mm	Energy-storing pressure/N
420	18522	320	29400

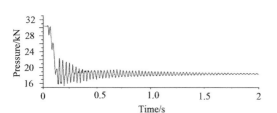

Figure 1. The structure diagram of the hardware.

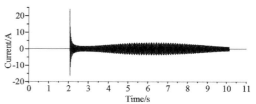

Figure 4. The motor current of storing energy.

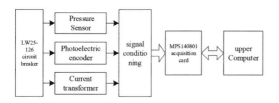

Figure 2. The pressure curve of the closing process.

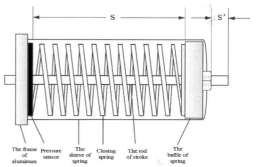

Figure 5. The regulated schematic diagram of the energy-storing state.

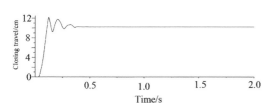

Figure 3. The closing stroke curve of the contact.

The pressure of the closing process and the closing curve is respectively shown in Figure 2 and Figure 3. The motor current curve in the energy-storing process is shown in Figure 4.

3.2 The extraction of running characteristic parameters

Before and after closing pressure (F_b and F_a) are, respectively, obtained through the pressure curve. They are used as the parameters of state indication in the closing spring. The closing curve of the contact is extracted to obtain the closing time T_c, the initial closing speed V_j, the maximum closing speed V_{max}, and the average closing speed V_{ave}. The energy-storing motor current is extracted to obtain the peak current I_p, the effective output power of the motor W, and the effective working time of the motor T_{ef}. These data are used as the diagnostic vector to diagnose the closing spring's energy-storing state.

LW25–126 SF6 circuit breaker with no load conditions is used as the experimental platform for test. Testing methods are shown in Figure 5.

The tail length of the rod S' is regulated to achieve the control of the length of closing spring S. The experiments have four groups, with experimental compressed lengths of 335 mm, 330 mm, 325 mm, and 320 mm, respectively. Each group of the experiments includes ten times and each running experiment includes closing energy, opening energy and storing energy. The running characteristic parameters are listed in Table 2.

Table 2. The running characteristic parameters in different states of storing energy.

Compressed length of storing energy L/mm	Closing time T_c/ms	Initial closing speed V_i/m/s	Maximum closing speed V_{max}/m/s	Average closing speed V_{ave}/m/s	Peak current I_p/A	Effective output power of motor W/kW	Effective working time of motor T_{ef}/s	Force before closing F_b/kN	Force after closing F_a/kN
335	95.73	2.38	2.38	1.76	4.11	3.575	7.764	27.45	16.38
330	93.16	2.44	2.45	1.81	4.26	3.642	8.047	28.17	17.02
325	90.84	2.49	2.57	1.86	4.31	3.701	8.081	28.88	17.60
320	90.25	2.56	2.58	1.89	4.44	3.778	8.077	29.36	18.00

4 DIAGNOSTIC ALGORITHM BASED ON THE BP NEURAL NETWORK

4.1 The design of the BP neural network

The BP (Back Propagation) neural network (Tan Kezhu, Chai Yuhua & Song Weixian. 2000) is a branch of the artificial neural network. It is composed of the input layer, the hidden layer, and the output layer. In order to realize or close the desired mapping relationship between the input and the output, error is passed back and network parameters are adjusted continuously by the methods of correcting error. This paper adopts the neural network's toolbox of MATLAB for predicting the pressure of the closing spring.

The characteristic parameters are closing time T_c, initial closing speed V_i, maximal closing speed V_{max}, average closing speed V_{ave}, motor effective output power W, and motor effective working time T_{ef}. These eigenvalues are used as the eigenvector. Before and after closing pressure are used as the output. The BP neural network is established. The number of the hidden layers is double the neurons of input based on the convention of the BP neural network. As indicated in Figure 6, the structure of the BP neural network has 7 input layers, 14 hidden layers, and 2 output layers.

4.2 The training and example of the BP neural network

The 40 experimental data are used as the samples to train the neural network. In the neural network's toolbox of MATLAB, any 20 data are randomly picked up from the 40 data. These data are used as original data for training network. The 10 random groups of data are used as verified data for verifying the network. The 10 random groups of data are used as testing data for testing network. The training results are shown in Figure 7. The iterative training number of the network is 13 times. The best testing performance is obtained after 7 iterations. Figure 8 shows the before and after closing pressure and the training output of the neural network in the 40 groups of the experiments.

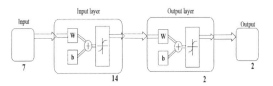

Figure 6. The structure of the BP neural network.

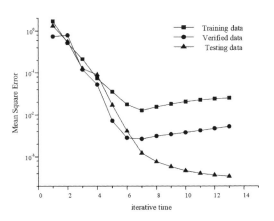

Figure 7. The training results of the BP neural network.

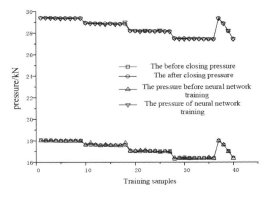

Figure 8. The comparison between the actual pressure and the trained outputs of the neural network.

414

Table 3. Running characteristic parameters of the verified experiments.

Compressed length of storing energy L/mm	Closing time T_c/ms	Initial closing speed V_j/m/s	Maximum closing speed V_{max}/m/s	Average closing speed V_{ave}/m/s	Peak current I_p/A	Effective output power of the motor W/kW	Effective working time of the motor T_{ef}/s
332	94.2	2.37	2.44	1.79	4.21	3.542	8.010
	94.2	2.35	2.41	1.76	4.22	3.642	8.047
322	90.91	2.50	2.54	1.87	4.37	3.692	8.054
	90.95	2.51	2.54	1.86	4.42	3.778	8.048

Table 4. The output vectors and diagnosis of the neural network.

Length of the compressed L/mm	Actual pressure/kN		Predicting pressure of the neural network/kN	
	Opening	Closing	Opening	Closing
332	28.09	16.75	28.16	17.01
	28.08	16.75	28.14	17.03
322	29.24	17.91	29.25	17.90
	29.20	17.82	29.24	17.91

As shown in Figure 8, the maximum error occurs in NO.18, whose error value is 170 N. The pressure of the error corresponding to the length of the spring is 1.4 mm. It can still determine that the length of the compressed spring is about 325 mm.

To verify the reliability of the training network, the compressed length of the spring is readjusted to 322 mm and 332 mm. Two experiments are conducted, and the running characteristic parameters are listed in Table 3.

The energy-storing pressure of the closing spring is predicted by sim function in the neural network's toolkit of MATLAB. The results are summarized in Table 4.

In Table 1, the spring's pressure, which is predicted by the neural network, is almost consistent with the actual pressure. However, the predicting pressure of the neural network is slightly larger than the actual pressure. To achieve a more accurate prediction, the types of the training parameters of the network need to be improved. A further in-depth study is required.

5 CONCLUSION

In this paper, the 7 running characteristic parameters of the circuit breaker are studied: closing time, initial closing speed, maximal closing speed, average closing speed, peak current of energy-storing motor, motor effective output, and motor effective working time. The relationship between these parameters and the energy-storing state of the closing spring is derived in this paper. The energy-storing pressure of the closing spring is predicted by the BP neural network. The result indicates that the energy-storing state of the closing spring can be effectively predicted by using only the running characteristic parameters of the circuit breaker without using the pressure sensor.

REFERENCES

Chang guang, Wang yi & Wang wei. 2013. Zero phase filter using vibration signal in frequency entropy of high voltage circuit breaker mechanical fault diagnosis. *Proceedings of the CSEE*, 33 (3): 155–162.

Chen Bao-lun & Wen Ya-ning. 2010. Circuit breaker spring operation mechanism is introduced. *High voltage electric equipment*, 46–48 (10): 75–80.

GB3309–1989. The mechanical test of high voltage switch equipment room temperature.

Guo Qing, Xia Han & Han Wen-wei. 2013. Based on the wavelet entropy and the BP neural network fault signal of motor research. *Instrument technique and sensor*, 1:96–99.

Mei Fei, Mei Jun & Zheng Jianyong. 2013. Development and Application of Distributed Multilayer On-line Monitoring System for High Voltage Vacuum Circuit Breaker. *Journal of electrical engineering & technology*, (8): 813–823.

Sun Lai-jun, Hu Xiao-guang & Ji Yan-chao. 2007. The improved wavelet packet—characteristics of entropy in the application of high voltage circuit breaker fault diagnosis. *Proceedings of the CSEE*, 7 (12): 103–108.

Tan Kezhu; Chai Yuhua; Song Weixian. Identification of diseases for soybean.

The state grid corporation. 2005. <The AC high voltage switch equipment operation management standard>.

Yuan Shun. 2001. High voltage circuit breaker spring operation mechanism. Beijing: *mechanical industry publishing house*.

Advances in Power and Energy Engineering – Sun (Ed.)
© 2016 Taylor & Francis Group, London, ISBN 978-1-138-02846-3

State assessment of circuit breaker's spring based on fuzzy comprehensive evaluation method

X.Q. Xu
Kunming Power Supply Company of China Southern Power Grid Company, Kunming, China

B.W. Guo
School of Electrical Engineering, Wuhan University, Wuhan, China

J. Yuan
Kunming Power Supply Company of China Southern Power Grid Company, Kunming, China

W.J. Zhou
School of Electrical Engineering, Wuhan University, Wuhan, China

R.F. Wei
Kunming Power Supply Company of China Southern Power Grid Company, Kunming, China

S.Y. Zhao
School of Electrical Engineering, Wuhan University, Wuhan, China

ABSTRACT: In order to evaluate the energy-storing state of the circuit breaker's opening and closing spring, the fuzzy comprehensive evaluation method is applied to state the assessment of the HV (High Voltage) circuit breaker's spring. The factor set is established by selecting characteristic parameters, which reflects the energy-storing state of the spring. The establishment of the evaluation grade set is based on the spring's condition. The evaluation weight set is established by the AHP (the Analytic Hierarchy Process). Fuzzy relationship between factors and evaluation grades is constructed, and fuzzy operators and result processing methods are set. The whole evaluation model is completed and used to evaluate the state of the circuit breaker's spring. The practical calculation example based on the LW25–126 circuit breaker shows that the model can objectively and effectively assess the energy-storing state of the HV circuit breaker's spring.

1 INTRODUCTION

The spring operating mechanism has the characteristics of reliability, concision and convenience. It is increasingly used in the HV circuit breaker. The spring's fault of the circuit breaker will lead to a large trouble, so its state assessment is of great significance (Chen Baolun & Wen Yaning. 2010, Shu Fuhua. 2007, Su Jufang, Zhang Hongjun & Wen Yaning. 2011).

At present, the research on the circuit breaker is mainly focused on the overall evaluation of the operating mechanism. In a study (Mei Fei, Mei Jun & Zheng Jianyong. 2013), the circuit breaker is judged by monitoring the overall performance of speed, time, synchronism and other parameters. In a previous study (Li Yu, Zhang Guogang & Geng Yingsan. 2007), fuzzy mathematics theory is applied to diagnose the whole HV circuit breaker, but the spring of the circuit breaker is not assessed by its model. In another study (Zeng Guo, Li Pengfei & Li Jun. 2015), the monitoring system of the closing spring is established, but the method of state evaluation is not formed. Those studies have achieved some results, but the effective method of state evaluation for the circuit breaker's spring has not been made yet.

The number of the circuit breaker spring's characteristic parameters is very large. The corresponding relationship between the state of energy-storing spring and these characteristics is fuzzy and complex. A fuzzy comprehensive evaluation method is used in the assessment of the spring. The evaluation model is established by analyzing characteristic parameters of the spring, and its effectiveness is verified in the LW25–126 circuit breaker.

2 FUZZY COMPREHENSIVE EVALUATION

Fuzzy comprehensive evaluation is a method that can make an overall assessment by synthesizing the attributes or factors of the objects that possess a variety of attributes. Their quality is influenced by many factors. The evaluation objects have characteristics of fuzziness and the relationship between factors, and the degree of influence also has characteristics of fuzziness. The theory of classical set for these objects cannot be defined accurately, because fuzzy objects are difficult to be simply described as "belong to" or "not belong to", and it can only be described through the degree of membership. Elements in the theory should not only be expressed in 0 or 1 for its membership, but rely on the real number between 0 and 1. The fuzzy evaluation makes the characteristic function flexible and elements for the set's membership extended to the value between (0, 1) (Chen S.W., Lin S.C. & Chang K.E. 2008, Zhang Hannie & He Zhengjia. 1992).

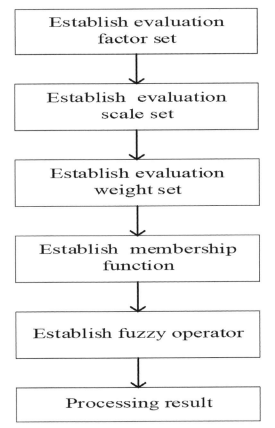

Figure 1. The flow chart of fuzzy comprehensive evaluation.

Fuzzy comprehensive evaluation is a method for state assessment by fuzzy transformation. The process is shown in Figure 1.

3 EVALUATION MODEL OF THE SPRING STATE

3.1 *The evaluation factor set*

HV circuit breaker has the advantages of long service time, low frequency of action and rapid movement. When the circuit breaker does not move, the spring is always in the long-term energy-storing compressed state. When the circuit breaker moves, the spring will release about 10 kN force within the tens of milliseconds. When the circuit breaker closes, the partial energy of the closing spring is used for closing, and an additional part of energy is used to store energy for the opening spring. After the closing spring releases energy, the energy-storing motor of the circuit breaker will start the energy storing for the closing spring immediately. There are many characteristic parameters of the spring's state in the circuit breaker. It mainly involves running parameters, force parameters and parameters of energy-storing motor. In addition, there are some other factors that need to be taken into consideration, such as working years and action numbers (Dai Wen-fang, Zhao Si-yang & Zeng Guo. 2015).

By considering the relationship between the action mechanism and the characteristic parameters, the second-order evaluation model of the spring's state is established. The evaluation model of the closing spring is used as an example, as given in Table 1.

3.2 *The evaluation scale set*

The level of the evaluation set is divided according to the actual condition of the evaluation objects. The accuracy of the evaluation and complexity of computation is considered comprehensively (Sun Minghua. 1994). The spring's state of the circuit breaker is divided into four grades: "good", "common", "maintenance" and "replacement". They correspond to the result set $V = \{V_1, V_2, V_3, V_4\}$.

Good (V_1): it means that the circuit breaker's spring is in good condition, and the basic characteristic of its state has an optimal value.

Common (V_2): it means that a certain degree of degradation has occurred in the circuit breaker's spring. But it is still in the normal state to provide the required power for the circuit breaker. There is no necessity to deal with the spring.

Maintenance (V_3): it means that the state of the spring is close to the safety threshold and results

The flow chart boxes read:

Establish evaluation factor set

Establish evaluation scale set

Establish evaluation weight set

Establish membership function

Establish fuzzy operator

Processing result

Table 1. Closing spring's model of the circuit breaker.

State of closing spring	Running parameter U_1	Closing time U_{11}
		Initial closing speed U_{12}
		Average closing speed U_{13}
		Maximum closing speed U_{14}
	Force parameter U_2	Maximum spring force U_{21}
		Minimum spring force U_{22}
	Energy-storing parameter U_3	Output power of motor U_{31}
		Maximum motor current U_{32}
		Working time of motor U_{33}
	Other parameter U_4	Cumulative actual number U_{41}
		Cumulative operation years U_{42}

in a high fault ratio. The spring needs to be maintained, and the problem of jam and debris needs to be checked. The spring should be paid more attention. The integrated performance of the circuit breaker still conforms to the operational requirements. There is no need for the spring to be replaced immediately.

Replacement (V_4): it means that the spring of the circuit breaker has severely exceeded safety thresholds. It may have developed cracks or stress relaxation. The overall performance of the circuit breaker will be seriously affected. The spring needs to be replaced immediately.

3.3 Evaluation weight set

Weight set can be regarded as the fuzzy relation between the evaluation objects and the evaluation factors. It is used to show the relative importance of assessment factors. The main method is the analytic hierarchy process. Analytic hierarchy process is used to obtain the comprehensive evaluation method of different feasible solutions by constructing the judgment matrix and calculating the weight. It provides a basis for choosing the optimal solution (Akira Notsu & Hirokazu Kawakami & Yuki Tezuka & Katsuhiro. 2013). The steps are as follows:

1. Determine the object and evaluation factor set.
2. A is the aim. U_i is evaluation factors. U_{ij} indicates the relative importance of U_i to U_j. The

discriminant matrix P is constructed. The values of U_{ij} are given in Table 2.
3. The eigenvalue and eigenvector are calculated. The discriminant matrix is obtained in step 2 and the maximum characteristic roots λ_{max} are calculated. λ_{max} corresponds to its eigenvectors $Q=[Q_1, Q_2,...Q_n]$. After the normalized processing, the importance of the evaluation factors is arranged, namely weight distribution. The weight distribution A is constituted after standardization ($A = \{a_1, a_2, ...a_n\}$, $a_1+a_2+...+a_n = 1$).
4. Inspection. A calculated in step 3 is the weight. Its consistency can be checked to verify the rationality of weight distribution by using the following formula:

$$CR = CI/RI \qquad (1)$$

where CR is the ratio of the judgment matrix's random consistency; CI is the index of the judgment for the matrix's consistency; RI is the index of the judgment for the matrix's average consistency:

$$CI = \frac{1}{n-1}\left(\lambda_{max} - n\right) \qquad (2)$$

The values of the RI that correspond to the judgment matrix are given in Table 3.

The values of the CR can be calculated based on the above data. When the CR is less than 0.01, the judgment matrix is regarded as having satisfied consistency. It indicates that the weight distribution is reasonable. Otherwise, the judgment matrix needs to be adjusted until a reasonable weight distribution is obtained (Guo Fengming & Cheng Jinhua. 1997).

Table 2. Judgment matrix's scale and implication.

Scale	Meaning
1	U_i and U_j are equally important
3	U_i is slightly more important than U_j
5	U_i is obviously more important than U_j
7	U_i is strongly more important than U_j
9	U_i is extremely more important than U_j
2, 4, 6, 8	Respectively indicate the intermediate value of 1–3, 3–5, 5–7 and 7–9

Table 3. The RI of the judgment matrix.

N	1	2	3	4	5	6	7	8	9
RI	0.00	0.00	0.58	0.90	1.12	1.24	1.32	1.41	1.45

3.4 Membership function

Membership function is established, which is used to indicate the fuzzy relationship between the state and the factors. For linear factors, the corresponding relationship between the actual change of signals and the performance's change of spring is linear. The relative original data are preprocessed, and then half trapezoid distribution function and trigonometric function are used as the membership function. For the nonlinear factor, the corresponding relationship between the actual change of signals and the performance's change of the spring is nonlinear. The signals are used as the input of normal distribution functions or power functions that are used as the input of the membership function. Thus, the results of membership functions corresponding to various levels are obtained (Wu Songtao, Hou Fenghua & Dai Feng. 2007).

3.5 Determination of the fuzzy operator

Every fuzzy evaluation model B is based on different synthesis conditions between the weight vectors A and the judgment matrix R. It is usually represented by M ($\overset{\cdot}{*}$, *). Among them, M represents the model and the symbol in parentheses represents the way of the synthetic operation. $\overset{\cdot}{*}$ = fuzzy 'and' operation. * = fuzzy 'or' operation. The synthetic method is the fuzzy operator (Hu Shikai. 2004).

The weighted average operator can be expressed as follows:

$$b_j = \sum_{i=1}^{m} a_i r_{ij}, j = 1, 2, \cdots, n \qquad (3)$$

where $\sum_{i=1}^{m} a_i = 1, b_j$ is the evaluation result; a_i is the weight parameter; r_{ij} is the factor's value.

This model not only takes into account the influence of all factors, but also retains all information of every single factor. Thus, this model conforms to the actual situation.

3.6 Process results

After the result sets are obtained, the results are processed according to a certain principle or method so as to obtain the final evaluation result. The set of the designed results has four levels. In addition, the levels mingle with each other. The results are processed based on the maximum membership principle, i.e. the evaluation results are equal to the evaluation factor v_i that corresponds to the largest evaluation index max (b_i) as follows (Tian Qinmo. 1990):

$$V = \left\{ v_i \middle| v_i \to max(b_j) \right\} \qquad (4)$$

4 EXAMPLE

In order to verify the effectiveness of the algorithm in the state evaluation of the spring, the springs of the LW25–126 circuit breaker in Kunming Power Supply Company are evaluated by the fuzzy comprehensive evaluation method. The circuit breaker is produced by Xikai Company. It has been on service since 2004. The cumulative actions are 248 times. Its main technical parameters are listed in Table 4. Through the preventive test on the circuit breaker, characteristic data of its springs are obtained. The data are summarized in Table 4.

4.1 Fuzzy relation matrix

Substituting the closing time into its membership functions, the following results can be obtained:

$$\mu_{v1} = 0.55 \ \mu_{v2} = 0.18 \ \mu_{v3} = 0 \ \mu_{v4} = 0$$

Table 4. Technical parameters and measured data of the LW25–126 circuit breaker's spring.

Technical parameters	Reference value	Measured value
Closing time/ms	70 ± 8	72
Opening time/ms	≤40	39
Initial closing speed/m/s	1.5 ± 0.5	1.6
Initial opening speed/m/s	3.0 ± 0.5	3.4
Average closing speed/m/s	2.0 ± 0.2	2.1
Average opening speed/m/s	4.0 ± 0.6	4.4
Maximum closing speed/m/s	2.5 ± 0.3	2.4
Maximum opening speed/m/s	5.0 ± 0.6	5.3
Maximum closing force/kN	29.4 ± 2.9	30.0
Maximum opening force/kN	17.6 ± 1.8	19.1
Minimum closing force/kN	18.5 ± 1.9	18.7
Minimum opening force/kN	7.9 ± 0.8	8.6
Maximum current of motor working/A	4.7 ± 0.3	4.5
Working time of motor/s	≤15	8.1
Running time/time	2000	248

The membership matrix of the closing time is $R_{11} = [0.55 \; 0.18 \; 0 \; 0]$.

The initiate closing speed, the average closing speed and the maximum closing speed are, respectively, substituted into their membership function. Their membership matrix is as follows:

$R_{12} = [0.65 \; 0.15 \; 0 \; 0]$, $R_{13} = [0.88 \; 0.47 \; 0 \; 0]$,
$R_{14} = [0.5 \; 0.3 \; 0 \; 0]$.

It is concluded that the membership matrix of running parameters R_1 is given by

$$R_1 = \begin{bmatrix} 0.55 & 0.18 & 0 & 0 \\ 0.65 & 0.15 & 0 & 0 \\ 0.88 & 0.47 & 0 & 0 \\ 0.5 & 0.3 & 0 & 0 \end{bmatrix}$$

In the same way, it is also concluded that the membership matrix of force parameters R_2 is given by

$$R_2 = \begin{bmatrix} 0.62 & 0.34 & 0 & 0 \\ 0.71 & 0.15 & 0 & 0 \end{bmatrix}$$

The membership matrix of the energy-storing parameters R_3 is given by

$$R_3 = \begin{bmatrix} 0.3 & 0.6 & 0.1 & 0 \\ 0.65 & 0.35 & 0 & 0 \\ 0.2 & 0.55 & 0.18 & 0 \end{bmatrix}$$

The membership matrix of the other parameters R_4 is given by

$$R_4 = \begin{bmatrix} 0.9 & 0.6 & 0 & 0 \\ 0.3 & 0.42 & 0.25 & 0 \end{bmatrix}$$

4.2 Calculation results

The determined weights A = {0.3 0.35 0.2 0.15}, A_1 = {0.556 0.278 0.103 0.063}, A_2 = {0.65 0.35}, A_3 = {0.582 0.309 0.109} and A_4 = {0.7 0.3} are substituted into type (3). The evaluation results at all levels for the closing spring can be obtained as follows:

B_1 = {0.609 0.209 0 0}, B_2 = {0.652 0.274 0 0}, B_3 = {0.397 0.517 0.078 0}, B_4 = {0.72 0.546 0.075 0}.

Finally, evaluation result of closing spring can be calculated as B = {0.598 0.344 0.027 0}.

According to the maximum membership principle, the closing spring of the circuit breaker is evaluated and its state belongs to the "good". It indicates that the performance of this closing spring is still at a good level. The actual measured data in the experiment have proved the conclusion. The state parameters of the closing spring in the circuit breaker basically have an optimal value.

Similarly, the energy-storing state assessment of the opening spring is derived. The result set B' = {0.226 0.443 0.079 0}. According to the maximum membership principle, its state belongs to the "common". It indicates that some deterioration has occurred in this opening spring, but its performance is still within a safe range. The actual measured data in the experiment have proved the conclusion. In the data of the opening spring, some state parameters have already deviated from the optimal value, but these parameters have not reached the safe threshold.

(a)

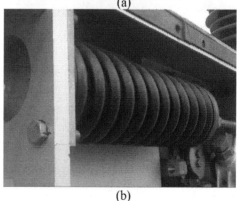

(b)

Figure 2. Springs of the LW25–126 circuit breaker. (a) Closing spring; (b) Opening spring.

4.3 Physical verification

In order to further verify the effectiveness of the proposed algorithm, the springs of the W25–126 circuit breaker are removed from the sleeve, as shown in Figure 2.

From Figure 2(a), it can be seen that the structure of the closing spring is complete. There are no cracks on its surface and the spring is not deformed. The surface is smooth without rust. From Figure 2(b), it can be seen that the structure of the opening closing is complete. There is no crack on its surface and the spring is not deformed. But there is some rust on its surface.

The above results can draw a conclusion that the actual state of the closing and opening springs and the calculated results are basically identical as revealed by the verifying experiment. The effectiveness of the evaluation model is further verified.

5 CONCLUSION

In this paper, the state assessment of the energy-storing spring is done based on the fuzzy comprehensive evaluation method. Running parameters, pressure parameters and other state parameters are selected to evaluate the performance of the circuit breaker's spring. According to the AHP, the weight distribution of evaluation factors is determined. The specific evaluation object is combined with the weight distribution to establish the membership function. Through the established evaluation model, the energy-storing spring of the LW25–126 circuit breaker is evaluated. Through the tested parameters and the example of the spring, the assessment results are verified. In this paper, the proposed algorithm is used to effectively evaluate the spring's state of the HV circuit breaker. It provides reference for the maintenance of the spring in the circuit breaker.

REFERENCES

Akira Notsu, Hirokazu Kawakami, Yuki Tezuka & Katsuhiro Honda. 2013. Integration of Information based on the Similarity in AHP. *Procedia Computer Science*.

Chen Bao-lun & Wen Ya-ning. 2010. Introduction of Spring Operating Mechanism for CB. *High Voltage Apparatus* 10(10):75–80.

Chen S.W., Lin S.C. & Chang K.E. 2008. Attributed concept maps: fuzzy integration and fuzzy matching. *IEEE Transactions on Systems, Man, and Cybernetics*, Part B: Cybernetics, Vol. 31 (5), pp. 842–852.

Dai Wen-fang, Zhao Si-yang & Zeng Guo, etc. 2015. Prediction of Closing Spring's Energy-Storing State Based on Circuit Breaker's Running Characteristic Parameters. *Technique and Sensor*, (10):12–16.

Guo Feng-ming & Cheng Jin-hua. 1997. Study on Inspection to Determine Unanimity of The Matrix in A Level Analysis Method and Deviation Correction Method. *China Geology & Mining Economic*, (7):19–24.

Hu Shi-kai, 2004. The study of fuzzy aggregation operators. *Chengdu*: Sichuan University.

Li Yu, Zhang Guo-gang & Geng Ying-san. 2007. The Condition Assessment Method for HV Circuit Breakers Based on Fuzzy Theory. *High Voltage Apparatus* (4):274–277.

Mei Fei, Mei Jun & Zheng Jianyong. 2013. Development and Application of Distributed Multilayer On-line Monitoring System for High Voltage Vacuum Circuit Breaker. *Journal of electrical engineering & technology* (8):813–823.

Shu Fu-hua. 2007. Closing Switch Spring Reliability Analysis & Improvement of High Voltage Circuit Breaker Operating Mechanisms *High Voltage Apparatus*, (5):368–370+373.

Su Ju-fang, Zhang Hong-jun & Wen Ya-ning, etc. 2011. Application of CT20 Spring Operation Mechanisms to 252 kV HV Circuit-breaker. *High Voltage Apparatus* (6):87–90.

Sun Ming-hua. 1994. Gradation's division and ranking of fuzzy comprehensive evaluation's results. *Journal of Beijing University of Technology* (1):72–79.

Tian Qin-mo. 1990. The Result Analysis of Fuzzy Comprehensive Evaluation. *Journal of Dalian Fisheries University* (1):61–65.

Wu Song-tao, Hou Feng-hua & Dai Feng. 2007. Linear Approximating Method in the Transacting Process of Nonlinear Standardization of Data. *Journal of Information Engineering University* (2):250–253.

Zeng Guo, Li Peng-fei & Li Jun, etc. 2015 State—monitoring of Circuit Breaker's Energy-storing Closing Spring. *Instrument Technique and Sensor* (4):65–68.

Zhang Han-nie & He Zheng-jia. 1992. Principle and application of Fuzzy Diagnosis. *Xi'An*: Xi'An Jiaotong University Press.

Advances in Power and Energy Engineering – Sun (Ed.)
© 2016 Taylor & Francis Group, London, ISBN 978-1-138-02846-3

Real-time degradation assessment system for hydropower equipment based on the streaming data processing

W.P. Tan & J. Xiao
HuNan Electric Research Institute of State Grid Corporation, Changsha, Hunan, China

ABSTRACT: This paper presents a new degradation assessment system for hydropower equipment based on the streaming data processing with the Storm framework. In the present study, the degradation assessment processes of hydropower equipment are discussed. The system was built with the cloud architecture based on Storm, including the following components: data acquisition web service, degradation assessment streaming computing model on the cloud server and the UI interface. The system also defines a multidimensional degradation model to assess the health degree.

1 INTRODUCTION

In recent years, with the development of the hydropower plant, the safety and stability operation of hydropower equipment has become increasingly important. To support critical decision-making processes such as maintenance replacement and system design, activities of performance degradation assessment and life prediction are of great importance to hydropower equipment. If the performance degradation could be detected at an early stage, the operational failure of hydropower equipment can be avoided and economic losses can be reduced (Lee, J et al. 2014). Moreover, the hydropower plant is usually built far away from the city, and experts could hardly analyze the equipment condition on the spot timely. Thus, it is of practical necessity to build an online monitoring system that can easily test the equipment condition without affecting its daily operation.

Numerous papers have reported on the condition monitoring and fault diagnosis of hydropower equipment. The main principal tool for condition monitoring is vibration analysis (Sinha, J.K. & Elbhbah, K. 2013). The vital fault information can be obtained from the vibration signals by some processing techniques. However, the infrared thermal image system, partial discharges system, Supervisory Control and Data Acquisition (SCADA) system and operation daily record should be acquired as the parameter of performance degradation assessment, which could make correct assessment result. The above mass data collected for the equipment remote monitoring require higher performance of data storage and data inquiry. Thus, a distributed cloud computing system is a matter of interest, which collects mass monitoring data and stores it securely in a cloud server. First, it can be accessed using any device such as a web browser or a mobile phone from anywhere. Users no longer need to install the software at their local machines. Second, the cost for cloud computing is greatly reduced based on the pay-per-usage scheme.

The traditional cloud computing system was designed and achieved based on Hadoop for equipment condition monitoring (Kondo, D. et al. 2009). The system consists of three parts: Hadoop cluster, storage client and inquiry client. Although this system has the advantages of mass storage and efficient inquiry, it is unsuitable for dealing with the real-time streaming data in condition monitoring of hydropower equipment (Xia, H., et al. 2013) (Fong, E.-M. & Chung, W.-Y. 2013).

Compared with the traditional batch processing architecture, the monitoring framework based on streaming data computing is superior in real-time monitoring and cluster extensibility.

Storm is a distributed, reliable, fault-tolerant system for processing streams of data. The work is delegated to different types of components that are each responsible for a simple specific processing task. In this paper, we proposed a streaming data processing framework for the real-time degradation assessment of hydropower equipment based on the Storm cloud computing system.

This paper is organized as follows. Section 2 details the Storm and degradation model. Section 3 introduces the real-time degradation assessment system. Section 4 details the engineering application. Finally, a brief conclusion is given in Section 5.

2 DEGRADATION MODEL

2.1 The stream data on condition monitoring

Storm cluster has two kinds of nodes: master node and worker nodes. Master node runs a daemon called Nimbus, which is responsible for distributing code around the cluster, assigning tasks to each worker node and monitoring for failures. Worker nodes run a daemon called Supervisor, which executes a portion of a topology. A topology in Storm runs across many worker nodes on different machines. The architecture of Storm is shown in Figure 1.

The work of Storm is delegated to different types of components that are each responsible for a simple specific processing task. The input stream of a Storm cluster is handled by a component called a spout. As shown in Figure 2, the spout passes the data to a component called a bolt, which transforms it in some way. A bolt either stores the data in some sort of storage or passes it to some other bolt. A Storm cluster could be considered as a chain of bolt components that each makes some kind of transformation on the data exposed by the spout.

2.2 Real-time degradation assessment model

There are two types of indicators that could be produced by hydropower equipment degradation inspections: (1) direct indicators, such as the crack

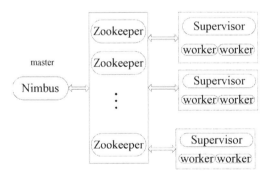

Figure 1. Architecture of Storm.

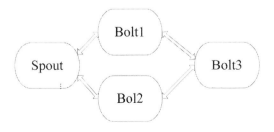

Figure 2. The topology model.

depth on a turbine blade, which directly relate to a failure mechanism; and (2) indirect indicators, such as the indicators extracted from vibration signals and oil analysis data, which can only partially reveal a failure mechanism. While direct indicators enable more precise references to equipment health conditions, they are often more difficult to obtain than indirect indicators (Gebraeel, N. & Pan, J. 2008).

According to the failure mechanism of hydropower equipment (Jin, G., et al, 2013), we designed a degradation assessment model combined with the direct indicators and indirect indicators, which could be calculated by the different types of data acquired from different sub-systems, such as online monitoring, productive management system, Supervisory Control and Data Acquisition (SCADA) system (Song, G., et al. 2008).

The degradation assessment model can be defined as follows:

$$L_{jz} = \frac{\sum_{n=1}^{N}(L_{bn}W_{bn})}{\sum_{n=1}^{N}(W_{bn})} \cdot 100\% \tag{1}$$

where L_{jz} is the degradation degree of the whole equipment; N is the number of components that is attended to be assessed; W_{bn} is the weight of the n-th component. The degradation degree of the n-th component could be calculated as follows:

$$L_b = \frac{\sum_{k=1}^{K}(L_{rk}W_{rk} + L_{tk}W_{tk})}{\sum_{k=1}^{K}(W_{rk} + W_{tk})} \cdot 100\% - L_f - L_x \tag{2}$$

where L_b is the degradation degree of each component; L_{rk} is the k-th degradation indicator based on the real-time monitoring; L_{tk} is the k-th degradation indicator based on the operation zone variation; W_{rk} is the weight of L_{rk}; W_{tk} is the weight of L_{tk}; L_f is the product problem, which is caused by the design problem or initial setup; and L_x is the degradation indicator by equipment inspection.

The degradation indicator based on the operation zone variation can be defined as follows:

$$L_t = \frac{W_{t\_w}\left(1-\left|\frac{\Delta P_r - \Delta P}{\Delta P}\right|\right) + W_{t\_c}\left(1-\left|\frac{P_{r\_ave} - P_{ave}}{P_{ave}}\right|\right)}{W_{t\_w} + W_{t\_c}} \tag{3}$$

Where P_{max}/P_{min} is the maximum/minimum value of the design operation zone; P_{r_max}/P_{r_min} is the maximum/minimum value of the real operation zone; ΔP is the width range of the design operation zone; ΔP_r is the width range of the real operation zone; P_{ave} is the center of the design operation zone, with $P_{ave} = (P_{max} + P_{min})/2; P_{r\_ave}$ is the center of the real operation zone, i.e. $P_{r\_ave} = (P_{r\_max} + P_{r\_min})/2$; and W_j is the weight of W_{j_c}, which is the degradation indicator based on the operation zone offset.

The degradation indicator based on the real-time monitoring can be defined as follows:

$$L_i = \begin{cases} L_{i\_linear} \\ L_{i\_e1} \\ L_{i\_e2} \end{cases}, \qquad (4)$$

where L_{i_linear} is the degradation model with a normal operation period, which can be defined as follows:

$$L_{i\_linear} = \begin{cases} 0 & V_r \geq V_{max} \\ 100 & V_r \leq V_{min} \\ 100 \times \dfrac{V_{max} - V_r}{V_{max} - V_{min}} & V_{min} \leq V_r \leq V_{max} \end{cases}, \quad (5)$$

where L_{i_e1} is the degradation model with an old period, which can be defined as follows:

$$L_{i\_e1} = \begin{cases} 0 & V_r \geq V_{max} \\ 100 & V_r \leq V_{min} \\ 100 \times (1 - e^{-k_1 \frac{V_{max} - V_r}{V_{max} - V_{min}}}) & V_{min} \leq V_r \leq V_{max} \end{cases},$$

$$\qquad (6)$$

where L_{i_e2} is the degradation model with an old period, which can be defined as follows:

$$L_{i\_e2} = \begin{cases} 0 & V_r \geq V_{max} \\ 100 & V_r \leq V_{min} \\ 100 \times e^{-k_2 \frac{V_r - V_{min}}{V_{max} - V_{min}}} & V_{min} \leq V_r \leq V_{max} \end{cases}, \quad (7)$$

where V_{max} is the extreme value of the indicator; V_{min} is the best value of the indicator; K_1, K_2 is the trend factor to control the degradation velocity; and V_r is the real monitoring data.

3 REAL-TIME DEGRADATION ASSESSMENT OF HYDROPOWER EQUIPMENT BASED ON THE STREAMING DATA PROCESSING

3.1 Brief introduction of monitoring data

With the development of hydropower plant construction and research, information technology

Table 1. The degradation parameters.

Parameter	Source	Data type
Vibration, run out, pressure fluctuation	Condition Monitoring	Numerical
Temperature, start-up and shutdown time	SCADA	Numerical
Operation zone	AGC	Numerical
Equipment defect	MIS	String
Maintenance log	MIS	String
Design flaw	Document	String
CAD	Document	dwg
Equipment inspection Log	Document	String

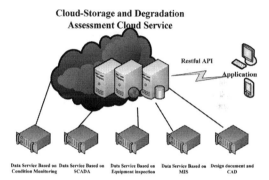

Figure 3. The architecture of the system.

of the hydropower plant are expanding in breadth and depth, and the process of state monitoring collection data is growing exponentially. In our system, we acquired the mass data from a multi-dimensional model. The type of data collected is listed in Table 1.

3.2 The proposed degradation assessment method based on Storm

In this paper, we proposed a novel degradation assessment method based on Storm. The architecture of the system is shown in Figure 3.

The assessment steps are described as follows:

Step 1: All the degradation data are acquired by different types of sub-system, which are described in Table.1. Moreover, all the data are wrapped as data service and the data acquisition application on the cloud server can collect the mass data in a common format such as XML and JSON.

Step 2: The cloud server collects the data and passes them to the degradation assessment system on Storm; a detailed assessment flowchart using Storm is shown in Figure 4.

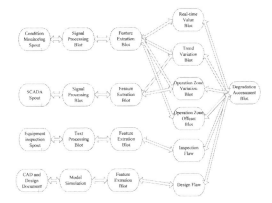

Figure 4. The flowchart of degradation on Storm.

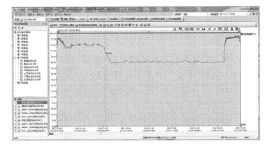

Figure 5. The main UI of the degradation system.

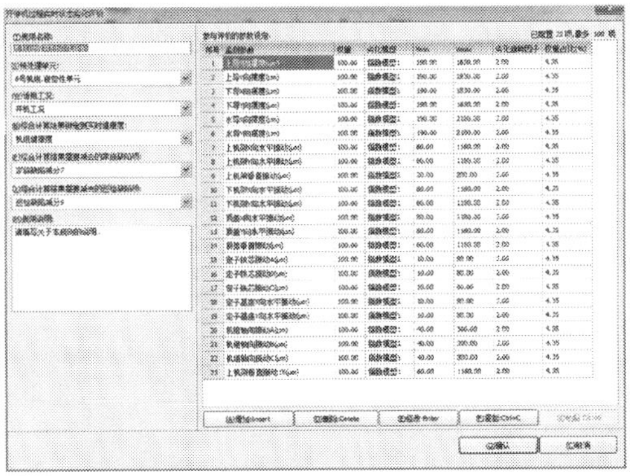

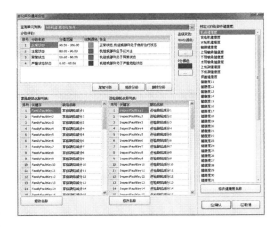

Figure 6. The channel setting of the degradation system.

Step 3: The health degree is calculated by the system and the result is coded as data service using RESTful API.

Step 4: Finally, the PC and smart phone can acquire the health degree by the RESTful API and manage the maintenance plan.

4 ENGINEERING APPLICATION

We applied our system in the DONGJIANG power plant, the FENGTAN power plant and the ZHEXI power plant, and the master server system based on Storm is located in the ChangSha, which linked the slave system by TCP/IP.

In the hydropower equipment degradation assessment working field, the degradation indicators are continuously acquired from the monitoring system. An abnormal situation is shown and the result of health degree calculation is obtained. The assessment message will be sent to the user interface, the main UI of which is shown in Figure 5. To illustrate the process of fault diagnosis, a degradation assessment case is given as follows.

Through the channel setting and indicator setting, we could acquire the degradation indicator from the monitoring system. This indicator could be acquired from various sources (e.g. condition monitoring system, SCADA system). As shown in Figure 6, all the degradation assessment rules could be self-defined such as the name, preprocess, operation zone and comments.

The real-time indicator can be set to one of the three types of degradation model. All the parameters of the single rule could be modified, such as weight, degradation model and degradation factor.

Figure 7 shows the time series of degradation degree index, which is queried from the degradation result history database.

Figure 7. The degradation degree time series.

Figure 8. The health degree of the degradation system.

Finally, the equipment condition can be set, as shown in Figure 8. A degradation degree is represented as 0–100, where 100 is the highest score representing the best condition of component health and 0 is the lowest score representing the worst condition. Four degree zones can be defined for the assessment result: normal condition, attention condition, abnormal condition and failure condition.

5 CONCLUSION

In this paper, the degradation assessment system for hydropower equipment based on the streaming data processing with the Storm framework is proposed. We introduced the cloud architecture of the system based on Storm and proposed the degradation model.

REFERENCES

Fong, E.-M. & Chung, W.-Y. 2013. Mobile cloud-computing-based healthcare service by noncontact ECG monitoring. Sensors, 13, 16451–16473.

Gebraeel, N. & Pan, J. 2008. Prognostic degradation models for computing and updating residual life distributions in a time-varying environment. Reliability, IEEE Transactions on, 57, 539–550.

Jin, G., Matthews, D.E. & Zhou, Z. 2013. A Bayesian framework for on-line degradation assessment and residual life prediction of secondary batteries inspacecraft. Reliability Engineering & System Safety, 113, 7–20.

Kondo, D., Javadi, B., Malecot, P., Cappello, F. & Anderson, D.P. Cost-benefit analysis of cloud computing versus desktop grids. Parallel & Distributed Processing, 2009. IPDPS 2009. IEEE International Symposium on, 2009. IEEE, 1–12.

Lee, J., Wu, F., Zhao, W., Ghaffari, M., Liao, L. & Siegel, D. 2014. Prognostics and health management design for rotary machinery systems—Reviews, methodology and applications. Mechanical Systems and Signal Processing, 42, 314–334.

Sinha, J.K. & Elbhbah, K. 2013. A future possibility of vibration based condition monitoring of rotating machines. Mechanical Systems and Signal Processing, 34, 231–240..

Song, G., He, Y., Chu, F. & Gu, Y. 2008. HYDES: A Web-based hydro turbine fault diagnosis system. Expert Systems with Applications, 34, 764–772.

Xia, H., Asif, I. & Zhao, X. 2013. Cloud-ECG for real time ECG monitoring and analysis. Computer methods and programs in biomedicine, 110, 253–259.

Advances in Power and Energy Engineering – Sun (Ed.)
© 2016 Taylor & Francis Group, London, ISBN 978-1-138-02846-3

Study on the localization technology of reflected pulse in Partial Discharge detection of HV power cables

T. Dong
Yunnan Electric Power Research Institute, Kunming, Yunnan Province, China

H. Zhang, B. Zhang, G.K. Yu, X.R. Li, T.S. Hu & D.Z. Xu
School of Electrical Engineering, Wuhan University, Wuhan, Hubei Province, China

ABSTRACT: Partial Discharge (PD) detection is an important method to assess the condition of the electrical insulation of HV power cables. The identification of reflected pulse is directly relevant to the inspection results and the PD source positioning accuracy in a PD measurement. But the reflected pulse that is automatically selected by the PD check system often suffers from a lack of precision, and needs correction by the operators based on their personal experiences. But there are still no specific criteria on how to determine the PD reflected pulse accurately. In this paper, a MATLAB/Simulink model is set up to simulate the PD pulse propagation in the power cable. A large number of PD measurements for different length preset fault cables and corresponding simulations prove that by using simulation software setup, the corresponding simulation model, the simulation results of the PD original pulse and the PD reflected pulse can help the operators to locate the true reflected pulse. In this way, it is possible to improve the positioning accuracy of the reflected wave.

1 INTRODUCTION

Partial discharge of the HV XLPE cable is the main reason to cause the aging of cable insulation and the main form of the early fault of power cable insulation, and it is also one of the main characteristics of cable insulation performance. This technology has a short time and will not cause damage to the cable, which is an ideal method for the completion of the HV power cable or the handover test after maintenance (Ying Qiliang 2001, Du Boxue 2010).

In a PD measure, the identification of the reflected pulse is directly relevant to the inspection results and the PD source positioning accuracy. But in fact, the reflected pulse that is automatically selected by the PD check system often suffers from a lack of precision in a large number of practical discharge measurements, and needs correction by the operators based on their personal experiences. But there are still no specific criteria on how to determine the PD reflected pulse accurately. In particular, when testing long lengths of power cables, after a long propagation time, the attenuation of the PD reflected pulse is more serious so that it is hard to locate. The detection accuracy and PD defect diagnosis ability not only depend on the accuracy of the detection equipment, measurement bandwidth and other technical indicators, but also on the attenuation of the PD pulse in the cable, which is also an important factor

(Guo Canxin 2009, Duan Naixi 2002). Therefore, in this paper, a MATLAB/Simulink model is set up to simulate the PD pulse propagation in the power cable, and a large number of PD measurements for different length preset fault cables are carried out by using the method of the Oscillating Wave Test (OWTS). Through the combination of simulation analysis and testing results, an auxiliary method is provided to improve the positioning accuracy of the PD reflected pulse.

2 THE INFLUENCE OF REFLECTED PULSE LOCALIZATION ACCURACY

The main electrical measurement methods for electric power cable PD detection include the capacitive coupling method, Ultra-High-Frequency (UHF) method and oscillating wave detection method, and almost all kinds of these partial discharge detection technologies use the Time Domain Reflectometry (TDR) analysis to locate the PD source. The rationale is that when the fault point is inspired in the test voltage, if at t_a the original pulse is detected and at t_b the reflected pulse is detected, the fault location can be calculated according to the time difference between t_a and t_b, and the wave propagation velocity, calculation principle and related waveforms shown in Figure 1.

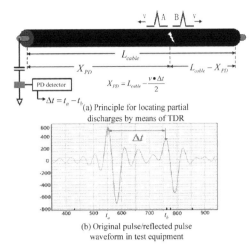

$$X_{PD} = L_{cable} - \frac{v \bullet \Delta t}{2}$$

$$\Delta t = t_a - t_b$$

(a) Principle for locating partial
discharges by means of TDR

(b) Original pulse/reflected pulse
waveform in test equipment

Figure 1. Principle for locating partial discharges by means of TDR and PD pulse waveform.

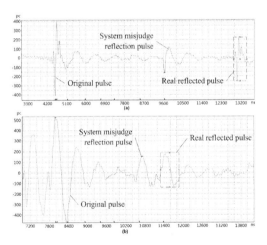

Figure 2. Some examples of the PD check system error locating the reflected pulse.

PD location:

$$X_{PD} = L_{cable} - v \bullet \Delta t / 2 \qquad (1)$$

where L_{cable} = length of tested cable; v = wave propagation velocity; and $\Delta t = t_a - t_b$. As shown in Figure 1 and formula 1, the location of the reflected pulse determines the value of Δt, and then directly determines the location accuracy of the PD source.

The automatic selection of the reflected pulse is based on the similarity between the original pulse and the reflected one. But due to the actual PD signal in the transmission process of cable suffer from attenuation, reflection, and various kinds of interference, the PD check system is often unable to accurately locate the reflected pulse. Some examples of the PD check system error locating the reflected pulse are shown in Figure 2.

In the actual measurement, the manual correction for the misjudged reflected pulse by the system can only rely on the operators' personal experiences. By using the MATLAB setup, the corresponding simulation model, the simulation results of PD original pulse and the PD reflected pulse can help the operators to locate the true reflected pulse.

3 SIMULATION MODEL OF PD SIGNAL PROPAGATION BASED ON SIMULINK

3.1 Partial discharge signal model

Some studies (Cheng Yuhong 2004, Ren Chenyan 2004, Edward Gulski 2006) carried out the theory, measurement and simulation analysis of the partial discharge signal, and the study shows that the partial discharge signal in a high voltage cable is a narrow pulse signal that contains high-frequency components. The most obvious characteristic is that there is a steep rise along much of high-frequency components. Some studies (Wang Nongyu 2014) established a physical model of partial discharge inside the cable and showed that the partial discharge waveform was in the form of exponential decay. Comprehensive domestic and foreign-related research literature on the common mathematical model for such a discharge pulse signal has four kinds of models: single exponential model; double exponential model; single exponential oscillation model; double exponential oscillation model (Hao Yanpeng 2000, Chia 2000). This paper adopts the double exponential decay model to fit the partial discharge waveform, the expression of which is given by

$$u_i(t) = V_i e^{-(t-t_i)/\tau_i} * 1(t - t_i) \qquad (2)$$

where $u_i(t)$ is the partial discharge pulse voltage; τ_i is the decay constant; and $1(t-t_i)$ is the unit step function. Taking the PD pulse rising edge of 150 ns, the PD pulse voltage amplitude is 7 mV. The simulation signal of the PD pulse is shown in Figure 3.

3.2 Power cable model

The frequency of the PD pulse is broadband, and the frequency range of the frequency distribution is from high frequency to ultra-frequency, while the tested cable length is usually no more than 10 km, so the transmission line length is not much larger than the wavelength of the PD pulse. Therefore,

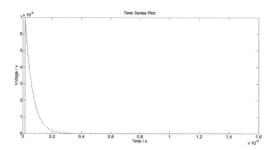

Figure 3. The simulation signal of the PD pulse.

Table 1. Sample cable parameters.

Parameter	Numerical
Conductor radius/mm	4.15
Inner semi-conductive layer thickness/mm	0.7
XLPE insulation thickness/mm	4.5
Outer semi-conductive layer thickness/mm	0.6
Ground screen thickness/mm	0.1
Outer sheath thickness/mm	1.76
Core conductivity/(S·m^{-1})	3.7×10^7
Ground screen conductivity/(S·m^{-1})	5.8×10^7
XLPE insulation relative dielectric constant	2.26

the distribution parameters model is suitable for the power cable model.

A study (Tang Zhong 2014) analyzed and compared three kinds of power cable distribution parameters model: LCC model in ATP/EMTP; Carson-Clem model and distribution parameters model in MATLAB, which is widely used at present, and confirmed the accuracy of MATLAB's distribution parameters model. MATLAB's distribution parameters model cannot change some parameters such as the radius of the cable conductor and insulation thickness, but it can change the equivalent inductance, equivalent resistance and equivalent capacitance. Another study (Xu Yingning 2009) described the conversion process and formulas.

The equivalent resistance of the cable per unit length is given by

$$R = \frac{4\rho}{\pi D_C^2} \left[1 + \alpha(\theta - 20)\right] k \, (\Omega/m) \qquad (3)$$

where ρ is the conductor resistivity; α is the temperature coefficient of the conductor; k is the metal resistivity increasing coefficient, approximately equal to 1; and θ is the test temperature.

The equivalent capacitance of the cable per unit length is as follows:

$$C_0 = \frac{2\pi\varepsilon\varepsilon_0}{\ln D_A/D_c} \, (F/m) \qquad (4)$$

where ε is the relative dielectric constant and ε_0 is the permittivity of vacuum.

The equivalent inductance of the cable per unit length is as follows:

$$L_0 = L_i + L_e = \frac{\mu_0}{8\pi} + \frac{\mu_0}{2\pi} \ln \frac{D_A}{D_C} \, (H/m) \qquad (5)$$

where L_i is the internal inductance; L_e is the external inductor; μ_o is the permeability; D_A is the cable shield layer diameter; and D_c is the core wire

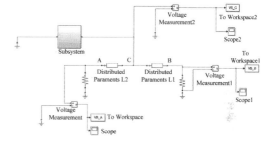

Figure 4. The partial discharge circuit model in MATLAB.

diameter. From the above three formulas, according to the specific parameters of the power cable, the power cable parameters can be converted to the distribution parameter in the MATLAB model. Cable specific parameters, taking YJV 8.7/10 kV high-voltage cross-linked polyethylene power cable as an example, are listed in Table 1.

Based on the above parameters, the equivalent inductance of the unit length of the power cable is 2.185×10^{-7} H/m, the equivalent inductance of the unit length of the power cable is 1.65×10^{-10} F/m, the equivalent resistance of the cable per unit length is 4.5×10^{-4} Ω/m, and the wave impedance is 36.4 Ω.

3.3 Partial discharge circuit model

Using MATLAB to build a power cable model AB, we injected the simulation PD signal at point C and observed the attenuation of the pulse waveform at points A and B. The partial discharge circuit model is shown in Figure 4.

4 TEST VERIFICATION

We take the partial discharge detection of 320 m preset fault cable by the OWTS, and the test result is shown in Figure 5.

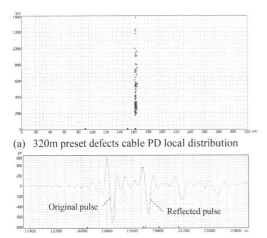

(a) 320m preset defects cable PD local distribution

(b) 320m preset defects cable PD waveform

Figure 5. The test result of the 320 m preset fault cable.

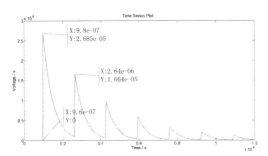

Figure 6. Simulation results, injection at point C and measurement at point A.

As shown in Figure 5, the PD source is at about 162 meters from the test port. As the preset discharge fault is very obvious and the measurement is carried out in the high pressure hall, the original pulse and the reflected pulse is clearly visible, as shown in Figure 5(b), with no obvious interference signal. As shown in Figure 5(b), the attenuation rate between the reflected pulse amplitude relative to the original pulse amplitude is 340/550 = 61.8%.

Placing the PD source location and cable parameter data into the partial discharge circuit model, the cable AC section is set to 162 meters and the cable CB section is set to 158 meters. The PD pulse waveforms obtained by the A port of the cable are shown in Figure 6.

As shown in Figure 6, the attenuation rate between the simulated reflected pulse amplitude relative to the simulated original pulse amplitude is 1.664/2.685 = 61.9%, and the simulated attenuation ratio is basically consistent with the measured data (61.8%).

Taking the XLPE cable electrical signal velocity as 194 $m/\mu s$, the time of the pulse signal transmitted from point C to point A is 162/194 = 0.835 μs, the simulation time of the pulse signal transmitted from point C to point A is 0.96 μs, minus the pulse injection start time is 0.14 μs, the simulation time is 0.84 μs, and the delay time difference between the simulation and the theory is also very small. The delay time between the original pulse and the reflected pulse is 1.59 μs in simulation, and the actual measurement delay time is 1.62 μs, as shown in Figure 5(b).

In order to apply the simulation model to the practical measurement of HV power cable partial discharge detection, we set the cable parameters of the model and the position of the PD source to the variable, and compile the simulation model to an executable program by the MATLAB compiler. Then, this method can be easily applied to detect the partial discharge of the HV power cable.

The actual measurement data and program simulation data for a 1400 meter-long HV power cable with a partial discharge fault are shown in Figure 7.

(a) 1400m preset defects cable PD waveform

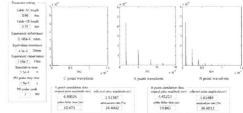

(b) 1400 meter fault cable software simulation data

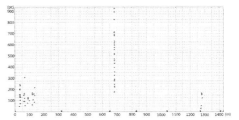

(c) Fault distribution map of the 1400 meter fault cable after reflected pulse correction

Figure 7. Actual measurement data and program simulation data for a 1400 meter-long HV power cable.

As shown in Figure 7(a), the amplitude attenuation rate of the reflected pulse A, which is automatically selected by the system, is 100/400 = 27.3%. If this reflected pulse is the real one, the corresponding PD source should be located at about 1000 meters from the test port. Substituting the corresponding parameters into the simulation model, the attenuation rate is calculated to be 41.2%, which is far away from the test result (27.3%). The calculated result in the simulation under this condition, with the amplitude attenuation rate between reflected pulse B and the original pulse being 150/440 = 34%, is shown in Figure 7(b). The attenuation rate is 34.4%, very close to 34%, and the simulation delay time is also basically consistent with the measured value, so reflected pulse B is the correct one. Figure 7(c) shows the fault distribution map of the 1400 meter-long fault cable after reflected pulse correction, as it shows the PD source at about 680 meters from the test port.

5 CONCLUSIONS

In a large number of practical discharge detections, the reflected pulse that is automatically selected by the PD check system often suffers from a lack of precision, while the operators can only calibrate it based on their personal experiences. In this paper, a MATLAB/Simulink model is set up to simulate the PD pulse propagation in the power cable. Taking the actual measurement and simulation of the HV power cables with different lengths, the simulation results are in agreement with the experimental results and theoretical analysis.

The above experimental and simulation research proved that the simulation results calculated by the corresponding simulation model can help the operators to locate the true reflected pulse. If the reflected pulse automatically selected by the system is true, the amplitude attenuation ratio of the reflected pulse calculated by the simulation model should be very close to the test result. In order to facilitate the application of this simulation model to the actual PD measurement, the simulation model is compiled into an executable program by the MATLAB compiler and used for the practical discharge detection of a 1400 meter-long HV power cable. The test result verifies the validity of this scheme in the PD measurement of the HV power cable.

REFERENCES

Chia, A.C. 2000. Novel Approach to Partial Discharge Signals Modeling in Dielectric Insulation Void Extension of Lumped Capacitance Model, International *Conference on Power System Technology*. Perth, Australia.

Cheng Yuhong. 2004. Study on the Fractal Characteristics of Ultra-wide Band Partial Discharge in Gas-insulated System With Typical Defect, *Proceedings of the CSEE* 24(8):99–102.

Du Boxue. 2010. Recent research status of techniques for power cables, *High Voltage Apparatus*: 100–104.

Duan Naixi. 2002. Experimental Study on The Propagation Characteristics of The PD Pulse in XLPE Power Cable, *High Voltage Apparatus* 38(4):16–18.

Edward Gulski. 2006. Advanced partial discharge diagnostic of MV power cable system using Oscillating Wave Test System, *IEEE Electrical Insulation in Magazine* 2(16):17–25.

Guo Canxin. 2009. Current Status of Partial Discharge Detection and Location Techniques in XLPE Power Cable, *High Voltage Apparatus*: 56–60.

Hao Yanpeng. 2000. A Double Exponential Function Fitting Algorithm for Impulse Parameter Evaluation, *High Voltage Apparatus* 26(3):23–25.

Ren Chenyan. 2004. Simulation Calculation of Partial Discharge in Single Void Using Variable Void Resistance Model, *Journal of Xi'an Jiaotong University*, 38(10):1019–1021.

Tang Zhong. 2001. Calculation and Research of Single-Core XLPE Power Cable Distribution Parameters, *East China Electric Power* 42(3):534–540.

Wang Nongyu. 2014. Model Research for OWTS Partial Discharge Test System, *Measurement and Detecting Technics* 36(4):99–102.

Xu Yingning. 2009. *Wire and Cable Guide*. Beijing: Machinery Industry Press.

Ying Qiliang. 2001. Development and Application of High Voltage and Extra-high Voltage XLPE Power Cable System in China, Electric Wire & Cable: 3–9.

Advances in Power and Energy Engineering – Sun (Ed.)
© 2016 Taylor & Francis Group, London, ISBN 978-1-138-02846-3

Design and realization of the monitoring system of a DC micro-grid experiment and research platform

C.N. Song, F. Li, X.F. Lin & B. Liu
School of Electrical Engineering, Guangxi University, Nanning, China

ABSTRACT: A DC micro-grid experiment and research platform is designed. This platform consists of a Photovoltaic System (PV), a Wind Power system (WP), a Li-Ion Battery Pack (LIB), a lead acid battery pack, an Ultra-Capacitor bank (UC), an electric vehicle, a controllable electric load, LED lamps and a central control unit all together. Multi-source and complementary experiment and research can be carried out, such as wind and light complement, wind, light and energy storage complement, and complement of different kinds of energy storage. Monitoring and control system is designed and implemented by the research requirements of the experiment platform, research on the communication technique and a DC micro-grid monitoring device.

1 INTRODUCTION

With the looming of global shortage of energy, distributed power generation has attracted the attention of many scholars in many countries (Ahmed T. Elsayed et al. 2015). However, distributed power generation also suffers from its weaknesses such as uncontrollability, intermittence and fluctuation, which may bring about a negative influence on the stability of the power system (Kumars Rouzbehi et al. 2014) (Kai Strunz et al. 2014). To address the related practical issues, some scholars have proposed a new type of network, i.e. the micro-grid (R. Noroozian et al. 2010). A DC micro-grid is a system that includes DC bus, distributed generators, energy storage device, converters, loads, monitoring device and protection device. Compared with the AC micro-grid, the operation of a DC micro-grid mainly depends on the voltage of the DC bus (Manuela Sechilariu et al. 2014). It is easier to realize the operation coordination and control. Distributed generators in a DC micro-grid are connected to the DC bus through converters that prevents reactive power loss and eddy-current loss in AC systems to enhance the efficiency greatly (Z.H. Jian et al. 2013). And the power supply is not interrupted by problems such as voltage imbalance, voltage flicker, and increased reliability and stability of the power system (Salomonsson D. et al. 2009). As a result, the DC micro-grid becomes a significant direction of research.

With respect to the DC micro-grid, it is most important to ensure the stability and security of the power system (Xiangyang Zhao et al. 2013). The monitoring and control system could achieve the operation coordination among all the units in the system, and optimize the performance of the complete system (Yuwu Liao et al. 2012). Before the realization of local and remote monitoring and control, an excellent human-computer operation interface is also needed. This paper establishes a multi-source and complementary DC micro-grid experiment and research platform. Monitoring and control system is designed and implemented by the research requirements of the experiment platform, research on the communication technique and a DC micro-grid monitoring device.

The monitoring system uses LabVIEW as the software development platform. Its intuitive programming interface is conducive to the follow-up of the system. And it reduces the cost of experimental research.

2 OVERALL DESIGN OF THE DC MICRO-GRID PLATFORM FOR EXPERIMENT AND RESEARCH

The platform of the DC micro-grid system consists of a PV system, a WP system, a LIB pack, a lead acid battery pack, a UC bank, an electric vehicle, controllable electric loads, LED lamps and a central control unit all together. Figure 1 shows its design framework.

2.1 Distributed generation

The adopted distributed generators are a PV generation system and a WP generation system.

The PV system can generate 3 kW power at most with 14 crystalline silicon panels of Type BP4175. Two parallels of seven-series panels provide a voltage

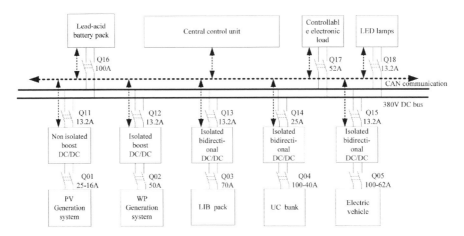

Figure 1.   Design framework of the DC micro-grid experiment platform.

of 247.8 V. By stepping up the output voltage of uni-directional DC/DC converters (rated power 5 KW) with the input voltage ranging from 200–320 V to 380 V, the goal of energy transfer is achieved. The DC/DC converter employs DSP28335 as the core processor to perform control functions.

The small-scale wind turbines include the transverse axis, longitudinal axis and Type H wind turbines with a rated power of 600 W, 400 W and 300 W. Considering the fluctuation of the wind power, an axial fan is used to simulate the variation of wind speed. With unidirectional DC/DC converters (rated power 1 KW) with the input voltage ranging from DC12 to 38 V, the output voltage is stepped up to 380 V.

2.2   Energy storage unit

Energy storage unit contains a lead acid battery pack, a UC bank and a LIB pack.

The lead acid battery pack (380 V) possesses 32 batteries in series, while the LIB pack (380 V) contains 22 batteries in series. By using the converter with a rated power of 5 kW and an input voltage of DC 50–80 V, bidirectional power delivery can be achieved.

The UC bank (480 V/165 F) is designed as 10 single capacitors in series. By using the converter with a rated power of 10 kW and an input voltage of DC100–250 V, bidirectional power delivery can be achieved.

2.3   Loads

A total of seven LED lamps in series and programmable DC electric load (Type chroma63804) with a maximum power of 4500 W are connected to the DC 380 V bus.

2.4   Electric vehicle

The electric vehicle can function as an energy storage device as well as electric load. It is a bidirectional power delivery platform based on a combined system of a Li-ion battery pack and super capacitors. AC electric dynamometer is utilized to simulate the driving conditions. Its structure is shown in Figure 2. The energy of the Li-ion battery pack is 7.2 kWh. Each of the 22 units in series is made at the level of 3.2 V/100 Ah. After the electric vehicle is integrated to the micro-grid, the Li-ion battery pack is charged to supply power to critical loads during emergency. By using the converter with a rated power of 5 kW and an input voltage of DC50–80 V, bidirectional power delivery can be achieved.

2.5   DC micro-grid protection system and sensors

The DC micro-grid protection system is located in the DC integration interface. It mainly consists of an electric energy meter, circuit breakers, CTs, PTs, protection output terminal strips and a micro-grid power supply bus. The integration interface connects the outward power network and the micro-grid via CTs and PTs. The adopted circuit breakers are of Schneider NSX100F, with overload delay protection and short circuit transient protection to distribute power and protect the transmission lines against faults such as overload and short circuit.

2.6   DC micro-grid monitoring system

Data exchange between the industrial control computer and the device is accomplished through the CAN network for the monitoring system. Thereafter, display and data exchange of the monitoring system interface based on LabVIEW can guarantee the control of subordinate computers.

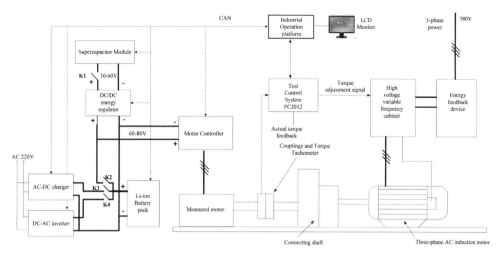

Figure 2. Diagram of the electric vehicle experiment platform.

3 MONITORING SYSTEM

This paper chooses an industrial personal computer, mainly aiming at control and driving, coordination and dispatch of the application program, interaction between tasks, interruption and memory system distribution and management of the hardware such as CAN card and data acquisition card on the PCI board.

3.1 *Realization of the monitoring system*

The operating micro-grid management system takes a PCI-1680 CAN communication card and two PCI-1762 data acquisition cards. The structure of the monitoring system is shown in Figure 1. The CAN communication card can transmit the current and voltage data of eight modules and the data acquisition cards can measure the state of each switch. Real-time data of the system are transmitted to the IPC via RS232, RS485 and CAN devices. DC/DC converter cabinet, lead acid battery pack, controllable electronic load and LED load communicate with the host computer through the CAN bus. DC/DC converter cabinet sends its working condition to the upper computer and accepts the control command of the upper computer through the CAN bus.

As required, CAN 2.0B extending data mode is used and the baud rate is 1Mbit/s. The network system sets communication nodes with communication capabilities including the Micro-Grid Control Unit (MGCU), PV generation (DC/DC1), WP generation (DC/DC2), LIB pack (DC/DC3), UC pack (DC/DC4) and electric vehicle (DC/DC5).

Micro-Grid Control Unit (MGCU) serves as the information exchange center that facilitates the data exchange between the nodes for the network system. Thanks to the embedded CAN controller, PCI-1680U retransmit automatically to realize bus arbitration and error detection, which decreases lost data and guarantees the reliabilities of the system. When the host computer sends operation commands and data to PCI-1680U, CAN ID and data field of messages are packed and generated to send the packed program of the data to DC/DC, in accordance with the command format of the CAN2.0B application layer protocol in this paper.

3.2 *Monitoring system function analysis*

For micro-grid operation, the monitoring system needs to gather the operational parameter and state information of every module for control to achieve the coordinated control and fault protection of each component. The structure of the monitoring system includes monitoring interface, control interface, power quality interface and event recording interface. There exists an automatic control and manual control for the monitoring system. Their functions are as follows:

1. Operational functions of the DC micro-grid monitoring system include setting operated authorization of users, remote control of each switch in the user interface, modification of operation mode and parameter, and delivering data of each module online.
2. Using gathering and recording functions, the operators can record the operation procedures, reasons and times of warning, check online history record that can be printed and gather the voltage and current of each module

for calculation of power and energy whose waveforms can be displayed in the monitoring interface.

3. Protection functions include active protection, overload warning and maximum operational current. If the parameter of modules exceeds the active protection value, corresponding modules would be disconnected from the DC bus. If it only exceeds the overload warning point, the modules would not perform the action while the warning light flickers and warning appears on the screen. Maximum operational current is preset in the interface. If the current set by the user is beyond the limit, the system would remind to reset the parameter so as to force the current within the safe range.

3.3 Data process and storage

3.3.1 Database

In this paper, standard Microsoft Access in Windows is chosen. Though LabVIEW cannot directly access the database, it can establish the connection with the database through ADO (Data Access Object), which is an external authoring interface. In this database management system, the database written with the ADO technique can take advantage of LabSQL to encapsulate the complex bottom layer ADO and SQL operation into a series of LabSQL sub VI for LabVIEW to directly call and use to finish the access of the database. Stored database is mainly used to store the real-time data and information of the experiment platform, including DC/DC1–5 data lists, event recording data lists and main data lists.

3.3.2 Power and energy calculation

The power and energy generated and absorbed by the micro-grid is the product of the voltage and current of every module. Output power is positive if the raising voltage mode is chosen, whereas the output power is negative if the lowering voltage mode is chosen. The overall energy can be obtained by integration of the total power of the micro-grid, which is as follows:

$$W_{out} = \int_0^t P_{out}(t)\, dt = \int_0^t U_{out}(t)\, I_{out}(t)\, dt \tag{1}$$

where P_{out} is the output power of the PV system and the wind power system. The total loss of energy for the micro-grid can be obtained by the integration of the power that the loads consume, which is as follows:

$$W_{in} = \int_0^t P_{in}(t)\, dt = \int_0^t U_{in}(t)\, I_{in}(t)\, dt \tag{2}$$

where P_{in} is the power of the LED load, electronic load and electric vehicle.

4 MONITORING PROTECTION

Protection functions include passive protection, active protection, overload warning and maximum operational current. Passive protection is controlled by DC/DC controllers. Other functions are carried out by monitoring software.

Over-current protection and overload protection have the ability of early warning and action. The logic structure is shown in Figure 3. The program

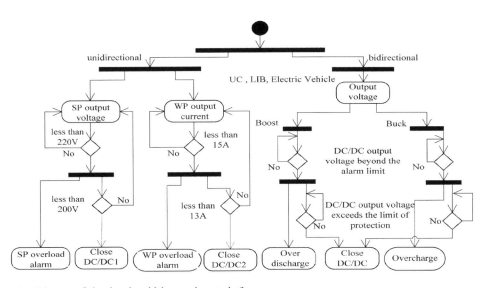

Figure 3. Diagram of the electric vehicle experiment platform.

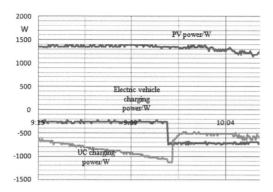

Figure 4. Power curve.

measures the output voltage or voltage of a single piece to judge whether to warn or perform the action. If the limit value of early warning is exceeded, the warning light will be ignited. If the limit value of action is exceeded, a trip command will be sent to DC/DC converters.

5 REALIZATION OF THE DC MICRO-GRID MONITORING SYSTEM

The switches Q01, Q11, Q04, Q14, Q05, Q15, Q16 and Q17 are closed to connect the PV system, UC, electric vehicle, controllable electric loads, lead acid battery pack to the micro-grid. MPPT mode is set for the PV system. The load is 500 W. Due to the sufficiency of the PV output power and the ability of storage modules to absorb more power; the supercapacitors and the electric vehicle are charged. Charging current is set with regard to the current generated power and power consumption. At about 9:47, charging current of the electric vehicle is increased to lower the charging current of supercapacitors. Ultimately, the control coordination of the DC micro-grid is accomplished. The power curve is shown in Figure 4.

6 CONCLUSION

Industrial control computer works as the host computer of the monitoring system of the DC micro-grid and communicates with subordinate computers for data exchange, realizing the monitoring and control of the micro-grid system. Each unit of the micro-grid can be combined with flexibility to push forward the research on wind and light complement, wind, light and energy storage complement, and complement of different kinds of energy storage. Convenience and expandability of the LabVIEW program paves the way for future research related to energy management.

REFERENCES

Ahmed T. Elsayed, Ahmed A. Mohamed & Osama A. Mohammed. 2015. DC micro-grids and distribution systems: An overview [J]. Electric Power Systems Research 119 (2015) 407–417.

Kumars Rouzbehi etc. 2014. Intelligent Voltage Control in a DC Micro-Grid Containing PV Generation and Energy Storage [C]. 2014 IEEE.

Kai Strunz, Ehsan Abbasi & Duc Nguyen Huu. 2014. DC Micro-grid for Wind and Solar Power Integration [J]. IEEE Journal of Emerging and Selected Topics In Power Electronics, Vol. 2, NO. 1.

Manuela Sechilariu etc. 2014. DC micro-grid power flow optimization by multi-layer supervision control. Design and experimental validation [J]. Energy Conversion and Management 82 (2014) 1–10.

Noroozian R. & M. Abedi. 2010. Distributed resources and DC distribution system combination for high power quality [J]. Electrical Power and Energy Systems 32 (2010) 769–781.

Salomonsson D., Soder L. & Sannino A. 2009. Protection of low-voltage DC micro-grids [J]. IEEE Transactions on Power Delivery, 2009, 24 (3):1045–1053.

Xiangyang Zhao & Shiyang Liu. 2013. Design of a Monitoring System of Micro-Grid [J]. Smart Grid and Renewable Energy, 2013, 4, 198–201.

Yuwu Liao, Jingjing Zhang & Guiping Zhao. 2012. Power Micro-grid Monitoring System Based on Configuration Software [J] 2012 International Conference on Intelligent Control and Information Processing Lecture Notes in Information Technology, Vols. 28–29.

Jian Z.H. & Z.Y. He. 2013. A Review of Control Strategies for DC Micro-grid [C]. 2013 Fourth International Conference on Intelligent Control and Information Processing (ICICIP) June 9–11, 2013, Beijing, China.

Advances in Power and Energy Engineering – Sun (Ed.)
© 2016 Taylor & Francis Group, London, ISBN 978-1-138-02846-3

The research and implementation of a new grounding resistance on-line monitoring method of tower transmission system

B. Zhang
College of Energy and Electrical Engineering, Hohai University Nanjing, Jiangsu, China

J.J. Song, J.B. Zhang & W. Li
College of Internet of Things Engineering, Hohai University Changzhou, Jiangsu, China

M. Xu
Pacific Automation Technology Co. Ltd., Changzhou, Jiangsu, China

ABSTRACT: In order to make real-time monitoring on grounding resistance of transmission system tower and find bad grounding resistance status in time, this paper puts forward a new grounding resistance on-line monitoring method of tower transmission system to avoid electrical equipment damage when lightning striking. The method mainly takes advantage of non-contact coupling with double iron core coils, and uses high frequency excitation signal to measure the grounding resistance value. After field application, the measurement precision of grounding resistance online monitoring instrument of tower transmission system can meet the requirements of on-line monitoring with comparison experiment.

1 INTRODUCTION

Grounding resistance value is an important technical indicator of transmission line grounding system (J. He et al. 2003, S. Furukawa et al. 1989), because it can affect the lightning overvoltage (A. Ametani & T. Kawamura 2005, I.M. Dudurych et al. 2010). If the grounding resistance of tower equipment is beyond the normal range, it will do serious damage to power system equipment when electric shock occurs. In order to ensure safe and reliable operation of all kinds of electrical equipment and electrical equipment grounding device, it must make regular measurement and monitoring for grounding resistance value of grounded system in accordance with the provisions. (H. Shu et al. 2010). However, because the transmission system poles are numerous, it becomes a heavy work to measure grounding resistance of grounded system. (Zhou Wenjun et al. 2015, Ma Hongjiang et al. 2005). The traditional methods for grounding resistance measurement are 3-pole method and clamp meter method. The 3-pole method needs to be decorated dozens of meters of electrode wires at the field, which costs a lot of work, while the error of clamp meter is great, which can't be corrected. (Jamnbal et al. 2001, Cai Chengliang 1997, Zhang Xiaodong & Ding Feng 2002).

This paper analyzes the source and mechanism of transmission tower grounding resistance measurement, and proposes a high frequency excitation signal method as a means of monitoring the resistance (Gan Yan et al. 2004).

2 GROUNDING RESISTANCE MEASUREMENT PRINCIPLE

2.1 Grounding resistance definition

Grounding resistance is defined as the comprehensive value of A point grounding resistance value, earth connecting cable metal conductor resistance, metal overhead lines resistance, the connection resistance value between earth connecting cable and metal overhead lines and B point grounding resistance value, as is shown in Figure 1 (Zhang Xiaodong & Ding Feng 2002).

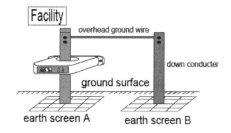

Figure 1. Principle diagram of the resistors circuit.

2.2 Grounding resistance measurement principle

Grounding resistance detection takes two groups of iron coils, a voltage coil N_t and a current coil N_r, and they all go through grounded line. Firstly, it adds a pulse signal to an iron coil in the measured loop, then it inducts a electromotive force in another iron coil by which the measured grounded loop generates induced current I. I outflows from R_x and flows back to meter from the grounding branch R_g. Through the study of the measurement of E and I, and through the formula: R = E/I, it can get the measured grounding resistance. The principle diagram of grounding resistance with double coils method is shown in Figure 2.

From the point of measurement principle, grounding resistance measured with double coils method must have a grounded loop. As is shown in Figure 2, grounding resistance tester is clamped on a branch of measured grounding electrode. It makes full use of the structure and characteristics of the grounding system, compared with traditional measurement methods, there is no necessary to cut off the power supply equipment or set up the auxiliary grounding electrode, thus it can realize online monitoring. In the diagram, R_x is the resistance supposed to measured, R_{earth} is the earth resistance, R_0 is n parallel grounding resistance, and $R_0 = R_1 // R_2 // R_3 // ... // R_n$, so the measured total loop resistance is all resistance of the grounded circuit included grounding resistance, that is $R_{loop} = R_x + R_{earth} + R_0$. Due to R_{earth} and R_0 are very small, and are far less than R_{earth}, it can be considered that $R_{loop} = R_{earth}$.

In brief, the measuring principle of grounding resistance with double coils method is whole circuit ohm's law which measures the circuit resistance value of the system. For the grounded device without forming a loop grounding system or closed metal loop, it needs to add auxiliary grounding electrode to make loop circuit, then uses grounding resistance meter to monitor the grounding resistance.

Grounded loop resistance measurement device is composed of lithium battery, solar panels, power management, exception handling circuit, low power consumption management circuit, key input circuit, LCD circuit, sound and light alarm circuit, real-time clock circuit, core processing circuit, excitation signal circuit, the first set of current transformers, grounding resistance circuit, a second set of current transformer, signal processing circuit, signal sampling circuit, level conversion circuit, SMS management circuit and RS485 communication circuit.

The first set and the second set of current transformer have the same current transformer structure which is made up of metal shell, plastic protection inner shell that winded by copper wire and permalloy magnetic core.

Excitation signal circuit includes high frequency signal generator, signal following the circuit and power amplifier circuit.

Signal processing circuit includes the active signal level amplifier circuit, filter circuit, signal amplification circuit, signal modulation and precision rectifier filter circuit and filter circuit.

The core processing circuit is connected to the excitation signal circuit by modulus conversion interface. Excitation signal circuit generates certain voltage signal and outputs the first set of current transformer. Resistance to be measured goes through the loop the first group and second group of current transformer current transformer that coupling induction into the voltage signal and output to the signal processing circuit, then the signal is flittered and amplified by the processing circuit and outputs to signal sampling circuit which connects with core processing circuit by digital to analog conversion interface. Core processing circuit calculates the value of grounding resistance according to the output and input signal. The principle diagram of grounded loop resistance tester is shown in Figure 3.

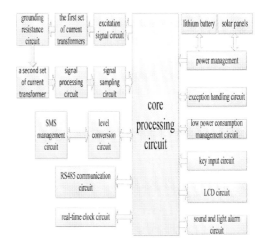

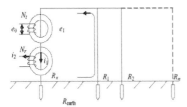

Figure 2. Principle diagram of grounding resistance with double coils method.

Figure 3. Principle diagram of grounded loop resistance tester.

Figure 4.   Diagram of iron core structure.

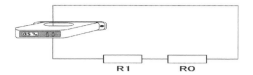

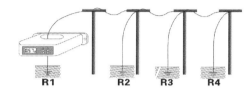

Figure 5.   Diagram of grounding resistance testing equivalent circuit.

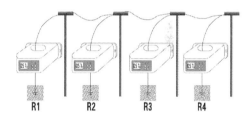

Figure 6.   Schematic diagram of grounding resistance in transmission system tower composition.

2.3   *The design of iron core coils*

Due to the grounding resistance minimum requirements can be measured $0.01\ \Omega$, in order to be able to accurately measure the grounding resistance value, two groups of iron core coils material uses permalloy. Permalloy is installed in a plastic case, and winding coil outside the case. One more group with 400–500 times around seems as drive source of grounding measurement loop, the other less group with 50–100 times around seems as sensors of grounding measurement loop. In order to reduce high voltage transmission interference to measurement and excitation coils, it uses pure iron as shielding case. Iron core structure is shown in Figure 4.

In the diagram, the number 3 indicates iron core; numbers 2 and 4 indicate plastic shells; numbers 1 and 5 indicate pure iron shields.

3   GROUNDING RESISTANCE MEASUREMENT ON MULTIPOINT OF MULTIPLE TOWERS

3.1   *Grounding resistance measurement principle of multiple towers multipoint*

Transmission system poles are commonly linked together by overhead earth wire to constitute multi point earth system. The detection principle of equivalent circuit is shown in Figure 5.

In the diagram, R1 is prediction of ground resistance, R0 is equivalent resistance of all the other tower grounding resistance in parallel, that is $R0 = R2//R3//R4//...//Rn$, the greater n is, the value of R0 is more close to 0, which is far less than R1, from the perspective of engineering we can see $R0 = 0$, so that the data obtained by detector is the value of R1, as can be seen in Figure 6.

As R1, R2, R3 and R4 are linked to the earth, although the soil resistivity of R1, R2, R3 and R4 point is different, the contact area of R1, R2, R3 and R4 is large, which can be thought as infinity, so that it is considered that the earth resistances among R1, R2, R3 and R4 are close to zero.

Figure 7.   Schematic diagram of many tower poles grounding resistance test.

3.2   *Installation requirements of earth loop resistance monitor site at multipoint on many poles*

Before installation, we should ensure that tower has a ground point. If there are two ground points, it has to dismantle the one for installation. In order to make measurement on each tower grounding resistance, it is necessary to install detector on each tower, thus it can test all tower grounding resistance at the same time. The schematic diagram of many tower poles grounding resistance test is displayed in Figure 7.

4   GROUNDING RESISTANCE TESTER LABORATORY AND FIELD TESTS

When making precise comparison test in the laboratory, we choose standard resistance to concatenate in the loop. As is shown in Figure 8, the comparison data of standard resistance and grounding resistance tester is as shown in Table 1.

In general, there are two wires on the tower connecting to the earth network with

Table 1. The comparison data of standard resistance and grounding resistance tester.

Number	Standard resistance/Ω	Test results/Ω	Error/%
1	0.01	0.01	0.0
2	0.5	0.49	2.0
3	1.0	0.99	1.0
4	5.1	5.09	1.9
5	10	9.99	1.0
6	200	199	0.5

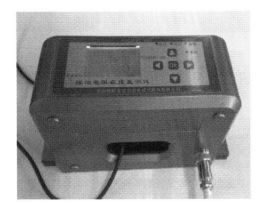

Figure 8. Ground loop resistance measuring instrument.

Figure 9. Installation drawing.

diagonal distribution. When installing a grounding resistance tester, in order to measure the value of grounding resistance rather than a ground loop, it is necessary to unfasten a grounding wire, and connect the other one to the tower through the middle of the grounding resistance tester, as in Figure 9. According to the above testing methods, we have installed 15 grounding resistance testers on 15 towers. At the field test, three tower grounding resistances were lightly larger in long sunny dry days. In the later, the users changed resistance grounding network, and the values are back to normal in less than 10 Ω.

5 THE NOTICE OF USING-GROUND RESISTANCE DETECTOR

SPECIAL INSTRUCTION: It only can be directly installed for loop grounding systems. For the systems without loop grounding, it must set auxiliary earth electrodes near the tested ones, and connect the tested and auxiliary earth electrodes in a loop, then install this detector.

5.1 Announcements

1. Install the tester to ground down line which needs to be tripped for late product maintenance. Pay attention to waterproof, preventing the rain, precautions against burglars and defending breakage.
2. Pay attention to the direction of the tester when installation. Keep vertical as far as possible, and downward the end of outgoing line.
3. Pay attention to the height to the ground when installation, and avoid water immersion equipment.
4. Prepare power cord according to the field distance that corresponding connection tester with solar panels.
5. Confirm that there is no polarity power supply anti, otherwise it is unable to work, and easy to damage the machine.
6. Place wireless com antenna on the outside, don't set in the shielding box.
7. Set the wireless receiving module beside the computer.
8. The installation of SIM communications card: please install SIM communications card before operation, don't install with charged. Due to the GSM module launches in stainless steel enclosure, please open the case to installed SIM communication card, press SIM communication card box on the right side of the small button with tweezers, pop-up the card box, and put in the SIM card.
9. Install the solar panels according to the actual situation. Try to make solar panels opposite the sun at noon which means the largest angle of the sun. Configure corrosion-resistant two cores.

6 CONCLUSIONS

The online transmission system tower grounding resistance tester according to the proposed principles can realize on-line monitoring circuit grounding resistance. The field tests show that the resistance measurement ranges 0.01 Ω–200 Ω; resistance resolution is 0.001 Ω; the measurement resolution is ±3% rdg ±3 dgt (20°C±5°C, below 70% RH), and it can satisfy the requirement of the tower grounding resistance test, then it can determine ground down line connection status, at the same time, it can measure the actual value of metal loop connecting resistance. Because it adopts non-contact measurement in real time and the ground wire goes hollow through without self-inspection, using GSM short message (or cable RS485), it can transmit data for remote communication to realize the remote on-line monitoring. There are sensors and circuit boards inside the completely closed detector, and it has the characteristics of raining proof, dustproof, anticorrosion, antifoaming. So it is especially suitable for long time on-line monitoring on field, mine and indoor. The measurement device has high measuring accuracy, stable performance and high reliability, etc. Monitoring software can and record data automatically stored data report, and it can also set alarm threshold, make real-time display and query history data. It is very suitable for independent installation.

ACKNOWLEDGEMENTS

This work was supported in part by the National Natural Science Foundation of China under Grant 61401147, the 2015 Special Funding for Science and Technology Achievements Transformation Projects in Jiangsu Province under Grant BA2015027, the Department of Jiangsu Province Policies Guiding plan (cooperative)—Prospective Joint Research Project under Grant BY2015030-1 and the 2015 Science and Technology Support Plan Project in Changzhou under Grant CE20150036.

REFERENCES

Cai Chengliang. 1997. The comparison of transmission line tower grounding resistance measurement methods. Electric measurement and instrument: 22–24.

Furukawa, S.O. Usuda, T. Isozaki, T. Irie. 1989. Development and applications of lightning arresters for transmission lines. IEEE Trans. Power Del.: 2121 Trans. A. Ametani and T. Kawamura. 2005. A method of a lightning surge analysis recommended in Japan using EMTP. IEEE Trans. Power Del.: 867E Tran I.M. Dudurych, T.J. Gallagher, J. Corbett, et al. 2010. Lightning withstand level calculation and anti-lightning method research for transmission lines across heavily icing region. IEEE Power Energy Eng: 150.

Gan Yan, RuanJiangjun, Chen Yunping. 2004. Application of unidimensional Finite Element Mehod (FEM) in Characteristics analysis of grounding mesh property. Power System Technology: 62–66.

He, J.R. Zeng, Y. Tu, J. Zou, S. Chen et al. 2003. Laboratory investigation of impulse characteristics of transmission tower grounding devices. IEEE Trans. Power Del.: 994.: Tr.

Jamnbal, Ahaad, Baker. 2001. Automatic maintenance of substation ground resistance. IEEE trans, Power engineering society summer meeting: 151–154.

Ma Hongjiang, Zeng Xiangjun, Wang Yuanyuan, et al. 2005. Grounding fault protection with fault resistance measuring for ineffectively earthed power systems. Transmission and distribution conference and exhibition: 1–3.

Shu H. et al. 2010. Lightning withstand level calculation and anti-lightning method research for transmission lines across heavily icing region. IEEE Power Energy Eng.

Zhang Xiaodong, Ding Feng. 2002. A practical method of grounding resistance measurement. Electric power automation equipment: 63–65.

Zhang Xiaodong, Ding Feng. 2002. A practical method of grounding resistance measurement. Electric power automation equipment: 63–65.

Zhou Wenjun, Liu Yushun, Li Gongxin, et al. 2015. Power Frequency Interference and Suppression in Measurement of Power Transmission Tower Grounding Resistance. Power Delivery, IEEE Transactions on: 1016–1023, April.

Advances in Power and Energy Engineering – Sun (Ed.)
© 2016 Taylor & Francis Group, London, ISBN 978-1-138-02846-3

An online coordinated optimal control strategy of a wind/photovoltaic/storage hybrid system

D. Yang
State Grid Shandong Electric Power Research Institute, Jinan, Shandong, China

P. Sun
State Grid Yantai Power Supply Company, Yantai, Shandong, China

W. Ci
Shandong Electric Power Dispatching and Control Center, Jinan, Shandong, China

L. Xing & T. Zheng
State Grid Shandong Electric Power Research Institute, Jinan, Shandong, China

ABSTRACT: An online coordinated optimal control strategy of a wind/photovoltaic/storage hybrid system is proposed. The power prediction of a wind turbine and photovoltaic generator is realized by the neural network. Considering the constraints of wind and PV power output and charge-discharge times of battery energy storage, the objective functions are established to maximize the residual capacity of the battery and minimize the exchange power between the microgrid and the distribution network. NSGA-II algorithm is used to optimize the operating reference point of the hybrid system. Finally, the operating reference point is adjusted based on the feedback of the real-time adjustable capacity of the wind turbine and photovoltaic generator at the time of control. The simulation results of the test system demonstrate that the proposed method can deal with uncertain wind and PV power outputs effectively, and the operating characteristics of the microgrid are also improved.

1 INTRODUCTION

Distributed Generation (DG) is resource-saving and environmental friendly, and the power supply is also flexible. The development of DG based on the centralized generation and large-scale power system has been the inevitable tendency of the smart grid both at home and abroad. According to the national energy development plan, until 2020, the installed capacity of DG will account up to 9% in China, becoming an indispensable component of the power supply system.

In order to resolve the problems by the integration of DG, the concept of the microgrid is presented. Microgrid integrates the dispersed distributed generation into a same physical network, and the uncertainties of wind and PV power output are solved by the coordinated control. In a study (Hossain, M.J. et al. 2015), a robust control design scheme of a multi-Distributed Energy Resource (DER) microgrid for power sharing was presented in both interconnected and islanded modes. Another study (Singaravel, M.M.R. & Daniel S.A. 2015) has proposed a new topology of a hybrid distributed generator based on the photovoltaic and wind-driven permanent magnet synchronous generator, and two low-cost controllers have also been proposed for the new hybrid scheme to separately trigger the DC-DC converter and inverter for tracking the maximum power from both sources.

Integration of wind and PV power increases the uncertainty of the operation of the microgrid. Therefore, a better control strategy is imperative to adapt to such a situation. Model Predictive Control (MPC) is an effective measure for solving such problems. MPC is widely applied in the industrial process control, and it is applicable to cope with the large uncertainties of system models (Camacho, E.F. & Bordons, C. 2004, Gallestey, E. et al. 2002, Camponogara, E. et al. 2002). MPC has strong adaptation and robustness. A MPC optimal scheme was adopted in one study (Xie, L. & Ilic, M.D. 2009) to deal with active power dispatch problems with large-scale wind power integration. So far, not much attention has been paid to the coordinated control in the microgrid by the MPC among the researchers.

Based on the MPC, an online coordinated optimal control strategy is proposed in this paper for a wind/photovoltaic/storage hybrid system. First, considering the constraints of wind and PV power output and charge-discharge times of an energy-storage battery, the objective function is established to minimize exchange power between the microgrid and the distribution network and maximize the residual capacity of the battery. Non-dominated Sorting Genetic Algorithm (NSGA-II) is adopted to optimize the operation reference point for the hybrid system. Finally, the reference point is adjusted based on the feedback of the real-time adjustable capacity of the wind turbine and photovoltaic generator at the time of control.

2 MODELING OF THE HYBRID SYSTEM

2.1 Modeling of the wind turbine

In the steady state, the relationship between the input and output of wind turbines is taken into account, while the dynamic process inside the wind turbine is not necessary to be considered. Therefore, the modeling of the wind turbine in this paper is mainly for modeling the wind power. Wind turbines transform wind energy to mechanical energy, and then mechanical energy is transformed to electric energy. The power extracted from the wind by wind turbines is given by

$$P_w = \frac{1}{2} C_p \rho A v_0^3 \tag{1}$$

where P_w is the wind power; C_p is the power coefficient; ρ is the air density; A is the rotor swept area; and v_0 is the wind speed.

The power coefficient C_p is closely related to the pitch angle of the blade and tip-speed ratio. According to Betz's law, theoretically, the maximum limitation of the power coefficient is given by

$$C_{p,\max} = \frac{16}{27} = 0.59 \tag{2}$$

Another important coefficient that influences the energy conversion is the tip-speed ratio, which can be defined as

$$P_w = \frac{1}{2} C_p \rho A v_0^3 \tag{3}$$

where λ is the tip-speed ratio; R is the rotor radius; Ω is the rotor rotational speed; and v_0 is the wind speed.

2.2 Modeling of the photovoltaic system

The modeling of the PV system is determined by the data fitting of the short circuit current, open circuit voltage, the voltage of the Maximum Power Point (MPP) and the current of MPP of the PV array. The specific expressions can be written as follows:

$$
\begin{aligned}
I &= I_{scc} \left[1 - C_1 e^{(V/C_2 V_{oc}) - 1} \right] \\
C_1 &= (1 - I_m/I_{scc}) e^{(-V_m/C_2 V_{oc})} \\
C_2 &= (V_m/V_{oc} - 1) \left[\ln (1 - I_m/I_{scc}) \right]^{-1} \\
I_{scc} &= I_{scc}^* (1 + \Delta S)(1 + a \cdot \Delta T) \\
V_{oc} &= V_{oc}^* (1 - c \cdot \Delta T) \ln (e + b \cdot \Delta S) \\
I_m &= I_m^* (1 + \Delta S)(1 + a \cdot \Delta T) \\
V_m &= V_m^* (1 - c \cdot \Delta T) \ln (e + b \cdot \Delta S) \\
\Delta T &= T - T_{\text{ref}} \\
\Delta S &= S/S_{\text{ref}} - 1
\end{aligned}
\tag{4}
$$

where I is the actual working current; V is the actual working voltage; I_{scc}, V_{oc}, I_m, V_m are the short circuit current, open circuit voltage, current of MPP and voltage of MPP at the present operation conditions, respectively; I_{scc}^*, V_{oc}^*, I_m^*, V_m^* are the short circuit current, open circuit voltage, current of MPP and voltage of MPP at standard test conditions; a, b, c are constants, whose values are related to concrete PV arrays; T is the actual working temperature; T_{ref} is the reference temperature, with $T_{\text{ref}} = 25°C$; S is the irradiance; and S_{ref} is the reference irradiance, with $S_{\text{ref}} = 1000$ W/m^2.

2.3 Modeling of battery energy storage

In the coordinated control process of the microgrid, battery energy storage attenuates the power fluctuations of wind turbines and photovoltaic generators. The power distribution reliability of the microgrid can also be improved based on the capacity effect and the power effect of batteries. The models of the capacity and the power of battery energy storage are as follows:

$$
\begin{aligned}
& SOC_{\min} < SOC(i) < SOC_{\max} \\
& \text{if } P_b(i) > 0, \\
& SOC(i) = SOC(i-1) \cdot (1 - \delta) + P_b(i) \cdot \gamma_C/S_{total} \quad (5) \\
& \text{if } P_b(i) < 0, \\
& SOC(i) = SOC(i-1) \cdot (1 - \delta) + P_b(i)/(\gamma_D \cdot S_{total})
\end{aligned}
$$

$$-P_{b\_dis\_\max} < P_b(i) < P_{b\_ch\_\max} \tag{6}$$

where $SOC(i)$, $SOC(i-1)$ are the State Of Charge (SOC) of battery energy storage at the time intervals i and $i-1$, respectively; $SOC_{\max}$, $SOC_{\min}$ are the upper and lower limits of SOC, respectively; δ is the self-discharge rate; γ_c is the charge efficiency; γ_D is the discharge efficiency; $P_b(i)$ is the charge/discharge power; $P_{b\_ch\_max}$ is the upper limit of charge power; and $P_{b\_dis\_max}$ is the upper limit of discharge power.

3 COORDINATED OPTIMAL CONTROL STRATEGY

Essentially, MPC is a finite-horizon closed-loop optimization algorithm based on models. It has been widely used in the industrial process control. In each sampling period, the current system state is taken as the initial state of the controller, and the prediction results of future time slots are obtained by prediction models. The current control actions are obtained by the rolling optimization of the finite time-horizon problem. The deviation between the future output and the reference trajectory should be minimized. Based on theories of the MPC, an online coordinated control strategy is proposed in this paper for the wind/photovoltaic/storage hybrid system, and the control model of prediction-online optimization feedback is established.

3.1 Prediction model

The prediction model forecasts future outputs based on the historical information of the process and future inputs, which provides prior knowledge for the optimization of the MPC. The prediction model focuses on the function, but not on the form of the model. Any model that can predict the future dynamic behaviors of the system can be used as the prediction model, regardless of their forms. In this paper, the wind and PV power output is predicted by the neural network in the prediction model.

3.2 Model of online optimization

In this model, the maximum wind and PV power output from the prediction model is regarded as the boundary conditions. The online optimization proceeds to obtain the power output of wind/PV/energy storage batteries. At a certain time period, the reference power output curves are generated to provide operation reference points for real-time control.

Objective function f_1 is to minimize exchange power between the microgrid and the distribution network, so that the microgrid can be regarded as a stable source or load, and the difficulty of control is decreased. If the microgrid is isolated, the exchange power is set to 0. The objective function f_2 is to minimize the change of SOC between the initial and end time intervals, so that the battery

energy storage can remain at a higher power level, which is given as follows:

$$\min f_1 = \frac{\sqrt{\sum_{i=1}^{N}[P_l(i)+P_{tie}(i)-P_w(i)-P_{PV}(i)+P_b(i)]^2}}{N}$$

(7)

$$\min f_2 = SOC(1) - SOC(N)$$

(8)

where N is the number of time intervals in the optimization time domain; $P_l(i)$ is the load power in the time interval I; $P_{tie}(i)$ is the exchange power between the microgrid and the distribution network in the time interval i (positive values mean that the microgrid transfers power to the distribution network, and negative values mean that the power is injected from the distribution network to the microgrid); $P_w(i)$ is the scheduled wind power output in the time interval i; $P_{PV}(i)$ is the scheduled PV power output in the time interval i; $P_b(i)$ is the scheduled power output of battery energy storage in the time interval i (positive values mean charge, while negative values mean discharge); and $SOC(1)$, $SOC(N)$ are the SOC of battery energy storage in time interval 1 and time interval N.

The constraints of online optimization include the wind and PV power output, and the charge/discharge cycles of batteries, which are as follows:

$$0 < P_w(i) < P_{w\_pre}(i)$$
$$0 < P_{PV}(i) < P_{PV\_pre}(i)$$
$$N_{ch\_dis} < N_{battery}$$

(9)

where $P_{w\_pre}(i)$ is the predicted wind power output in the time interval i; $P_{pv\_pre}(i)$ is the predicted PV power output in the time interval i; N_{ch_dis} is the charge-discharge times of the battery; and $N_{battery}$ is the maximum allowable charge-discharge times, with $N_{battery} = 1$.

There are three major performance evaluation indices for multi-objective optimization algorithms: 1) convergence to Pareto optimal set; 2) maintenance of dispersibility and diversity of the population; 3) avoiding the loss of Pareto optimal solutions. Accordingly, three key technologies make NSGA-II an excellent algorithm, namely fast non-dominated sorting, fast crowded distance estimation and simple crowded comparison operator. The procedure of NSGA-II has been discussed in detail previously (Deb, K. et al. 2002).

3.3 Model of feedback

Compared with the control time, the online optimization has a large advance in time, and prediction errors are also large. The obtained reference wind and PV power output curves may be higher than the maximum power output, so that the reference power output curves cannot be accomplished accurately, and the power compensation pressure of battery storage is increased.

In order to improve the completeness of the reference wind and PV power output curves, in the control instant, the reference wind and PV power output should be adjusted based on their real-time adjustable capacity. The energy storage battery should operate based on the online optimized curve as far as possible.

Real-time adjustable capacity of the wind turbine ΔP_w is calculated as follows:

$$\Delta P_w = P_{w\_est} - P_w \qquad (10)$$

where P_w is the scheduled wind power output and P_{w_est} is the real-time estimated wind power output (the estimated values can be rapidly calculated by the real-time wind speed and the running state of wind turbines).

Real-time adjustable capacity of the PV generator ΔP_{pv} is given by

$$\Delta P_{pv} = P_{pv\_est} - P_{pv} \qquad (11)$$

where P_{pv} is the scheduled PV power output and P_{pv_est} is the real-time estimated PV power output (the estimated values can be rapidly obtained by real-time illumination intensity and temperature).

The feedback module includes the following four circumstances:

1. The wind turbines and PV generators both have adjustable capacity, i.e. the adjustable capacity is positive and then the reference power output should be directly dispatched to wind turbines, PV generators and battery energy storage.
2. The wind turbines and PV generators have no adjustable capacity, i.e. the adjustable capacity is negative and then the reference power output should be dispatched to wind turbines and PV generators. The power shortage is compensated by battery energy storage.
3. The wind turbines have adjustable capacity, while PV generators have no adjustable capacity, and PV generators have the excess scheduling plan. The excess scheduling plan of PV generators should be transferred to wind turbines and the reference power output should be adjusted as follows:

$$P_{w\_sch} = P_w + \min (\Delta P_w, -\Delta P_{pv})$$
$$P_{pv\_sch} = P_{pv} - \min (\Delta P_w, -\Delta P_{pv}) \qquad (12)$$

where P_{w_sch} is the final scheduled wind power output and P_{pv_sch} is the final scheduled PV power output.

4. The wind turbines have no adjustable capacity, while PV generators have adjustable capacity, and wind turbines have the excess scheduling plan. The excess scheduling plan of wind turbines should be transferred to PV generators and the reference power output should be adjusted as follows:

$$P_{w\_sch} = P_w + \min (\Delta P_w, -\Delta P_{pv})$$
$$P_{pv\_sch} = P_{pv} - \min (\Delta P_w, -\Delta P_{pv}) \qquad (13)$$

4 CASE STUDY

The parameters of the test system are set as follows. The capacity of the wind turbine is 66 kW; the capacity of the PV generator is 200 kW; the battery energy storage is 90 kW/270 kWh; the load power is 120 kW; the exchange power between the distribution network and the microgrid is 60 kW; and the initial SOC of the battery is 0.5.

The predicted and actual maximum power output curves of the wind turbine and the PV generator are shown in Figures 1 and 2, respectively. The predicted wind power output is higher than the actual maximum power output during the whole time. The predicted PV power output is linked to

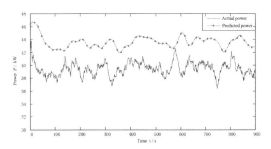

Figure 1. Predicted and actual maximum wind power output curves.

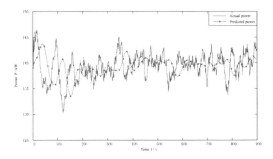

Figure 2. Predicted and actual maximum PV power output curves.

the actual maximum power output effectively, and the prediction precision is relatively high.

The time interval of online optimization is set to 15, i.e. each time interval takes 1 minute, and the total time of optimization is 15 minutes. NSGA-II is used for optimization. The population quantity is 400, and the number of generations is 200. It took 70 seconds for the optimization algorithm to satisfy the demands of online control. Typical Pareto optimal solutions are presented in Table 1.

If the objective function f_1 is taken as the main preference and f_2 is regarded as the secondary preference, scheme 1 is chosen as the optimal control strategy. The optimized reference power output curves of the wind turbine, PV generator and battery are shown in Figure 3.

The scheduled wind and PV power output is adjusted according to real-time conditions such as wind speed, illumination and temperature. The adjusted reference power output curves are shown in Figure 4. The excess scheduling plan of the wind turbine is transferred to the PV generator.

Table 1. Typical Pareto optimal solutions.

	f_1/kW	f_2
Scheme 1	0.1613	0.4907
Scheme 2	2.7910	0.6017

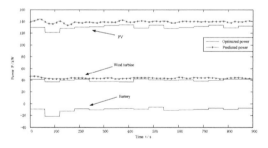

Figure 3. Optimized reference power output curves.

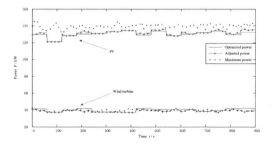

Figure 4. Reference power output curves after feedback.

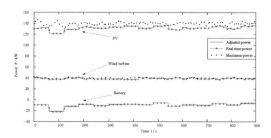

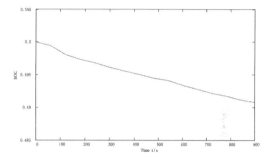

Figure 5. Real-time power output curves of the hybrid system.

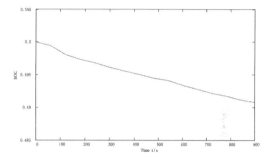

Figure 6. SOC curve of the battery energy storage.

As shown in Figure 5, the real-time wind and PV power output tracked adjusted the reference power output effectively. The battery energy storage operated according to the optimized reference power output curve fundamentally, and the power compensation pressure of the battery is reduced efficiently. The discharge process of the battery is shown in Figure 6.

5 CONCLUSIONS

An online coordinated optimal control strategy of a wind/photovoltaic/storage hybrid system is proposed in this paper. Based on theories of the MPC, the control model of prediction-online optimization feedback is established. The simulation results of the test system demonstrate the effectiveness of the control strategy. The influence of prediction errors can be corrected in time in the control process because the online optimization strategy is established based on the feedback of actual outputs. Compared with coordinated control in traditional microgrids, the control strategy reduces the demands on the precision of prediction models in the uncertain process. The drawbacks of the traditional coordinated control strategy including low precision of wind prediction models and highly uncertain outputs are curtailed by the proposed control strategy, and the operating characteristics of microgrids are improved effectively.

REFERENCES

Camacho, E.F. & Bordons, C. 2004. *Model Predictive Control.* New York: Springer-Verlag.

Camponogara, E. et al. 2002. Distributed model predictive control. *IEEE Control Systems Magazine* 22(1): 44–52.

Deb, K. et al. 2002. A fast and elitist multiobjective genetic algorithm: NSGA-II. *IEEE Transactions on Evolutionary Computation* 6(2): 182–197.

Gallestey, E. et al. 2002. Model predictive control and the optimization of power plant load while considering lifetime consumption. *IEEE Transactions on Power Systems* 17(1): 186–191.

Hossain, M.J. et al. 2015. Robust control for power sharing in microgrids with low-inertia wind and PV generators. *IEEE Transactions on Sustainable Energy* 6(3): 1067–1077.

Singaravel, M.M.R. & Daniel, S.A. 2015. MPPT with single DC-DC converter and inverter for grid-connected hybrid wind-driven PMSG-PV system. *IEEE Transactions on Industrial Electronics* 62(8): 4849–4857.

Xie, L. & Ilic, M.D. 2009. Model predictive economic/environmental dispatch of power systems with intermittent resources. *IEEE Power & Energy Society General Meeting* Calgary, AB, Canada: 1–6.

Advances in Power and Energy Engineering – Sun (Ed.)
© *2016 Taylor & Francis Group, London, ISBN 978-1-138-02846-3*

Dynamic Optimal Power Flow and analysis of the active distribution network

W.B. Li
State Grid Shandong Electric Power Research Institute, Jinan, Shandong Province, China

S. Yang
Economic and Technology Research Institute, State Grid Shandong Electric Power Company, Jinan, China

F.M. Meng
State Grid Laiwu Power Supply Company, Laiwu, Shandong Province, China

Z. Jiang & D.D. Zhang
State Grid Shandong Electric Power Research Institute, Jinan, Shandong Province, China

J.Y. Jiang
School of Electrical Engineering of Shandong University, Jinan, Shandong Province, China

ABSTRACT: With the amount of power sources increasing gradually in the distribution network, the issue of distribution network operation has caused wide concerns. The Optimal Power Flow (OPF) problem for the distribution network is studied in this paper. First, the regulating characteristics of different equipment are elaborated in detail, such as Distributed Generation (DG) and Energy Storage (ES). Then, by taking minimization of electricity purchasing cost as the objective function, an OPF model for the distribution network is established. The model can give full play to the regulation ability of adjustable equipment. As a result, optimal resource allocation can be achieved in the distribution network. Finally, examples illustrate the effectiveness and necessity of unified regulation in the distribution network, and the operation characteristics of the active distribution network are also demonstrated in detail.

1 INTRODUCTION

In the traditional power grid operation with centralized power supply, power grid regulation mainly relies on the transmission network. The corresponding OPF for the transmission network has been mature now with respect to concept, method and application, which mainly aims to reduce the cost of power generation and improve the security level (Carpentier 1962, Dommen & Tinney 1968, Sun et al. 1984, Torres & Quintana 1998). However, distribution network is mainly about power utilization and is in a passive position in terms of active power control. Therefore, the concept of OPF for the distribution network is mentioned rarely. Instead, it is dominated by concepts such as reactive power optimization and network reconfiguration with the reduction of network loss and improvement of the voltage level as the main purposes (Wang et al. 2008, Wang et al. 2009, Zhao et al. 2009).

Nowadays, under the dual influence of environment and resource crises, renewable DG connected to the distribution network has received a wide attention. It can be foreseen that DG will take a considerable proportion in the power grid. Thus, DG will coordinate with centralized generation in transmitting power into the grid, and the traditional power grid pattern will be broken. In this way, the distribution network will serve not only as a power user but also as a power provider. Furthermore, with the improvement of local network communication and intelligent regulation, the distribution network will have good ability to interact and coordinate with others. So, as an important sub-problem of entire power grid optimization, it is necessary to make further research on the concept and related issues of OPF for the distribution network.

Many scholars have carried out a study on the current development of the distribution network. For instance, the reactive power optimization of the distribution network with renewable DGs was analyzed by Chen et al. (2006). Many controllable regulation equipment, such as static compensator and diesel generator, were used to optimize for different objective functions. By taking transmission losses as the objective function, Ochoa & Harrison (2011) discussed the relationship between capacities of DGs and the losses of the distribution system, and proposed an optimization model to realize the optimal configuration of DGs. Under the background of "last-in first-out" admission control of DG, Dolan et al. (2012) proposed an online OPF model to diminish the branch overload. Gabash & Li (2012) considered the functions of ES in the operation of the distribution network, under which the wind curtailment and the total reactive power requirement were reduced effectively.

Starting with the market mechanism and taking the price of electricity as the measurement, this article establishes a dynamic OPF model of the distribution network to minimize the electricity purchasing cost. Examples are solved and analyzed using the GAM platform. The paper is organized as follows. Section 2 analyzes the regulating characteristics of DGs in detail. In Section 3, a dynamic OPF model of the distribution network is established and the features of the model are analyzed in depth. In Section 4, the mathematical examples are presented to prove that the application of OPF for the distribution network can give full play to the synergistic effect by various resources, realize DG absorption and reduce the electricity purchasing cost as far as possible.

2 CHARACTERISTICS OF DGS

2.1 Asynchronous wind power generator

Equivalent circuit of the asynchronous inductive wind generator is shown in Figure 1. R_s and X_s are stator resistance and reactance, respectively; X_r and R_r are rotor resistance and reactance; X_m is the magnetizing reactance; s is the slip; and V_W is the generator terminal voltage magnitude.

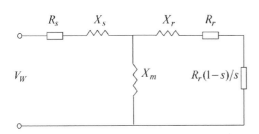

Figure 1. Equivalent circuit of the asynchronous wind power generator.

The maximum power gained by the asynchronous inductive wind generator from the wind turbine depends on the wind speed, which can be approximately expressed as follows:

$$P_{W\max} \approx P_m = \sum_{i=1}^{n} a_i v^i \tag{1}$$

where P_m is the maximum mechanical power of the wind generator; a_i is the power parameter; and v is the wind speed.

By using the wind turbine control system, the active power output of the asynchronous inductive wind generator can be adjusted within $P_{W\max}$ as follows:

$$P_{W\min} \leq P_W \leq P_{W\max} \tag{2}$$

where P_W is the actual active power of the asynchronous inductive wind generator.

Reactive power absorbed by the asynchronous inductive wind generator Q_W can be expressed as a function of active power and voltage (Feijoo & Cidras 2000) as follows:

$$Q_W = \frac{V_W^2}{X_m} + X \frac{V_W^2 + 2RP_m}{2(R^2 + X^2)}$$
$$-X \frac{\sqrt{(V_W^2 + 2RP_m)^2 - 4P_m^2(R^2 + X^2)}}{2(R^2 + X^2)} \tag{3}$$

where X is the sum of the leakage reactance of the rotor and stator, and R is the sum of the resistance of the rotor and stator.

Equations (1)~(3) together constitute active-reactive power features of the asynchronous inductive wind generation; that is, the absorbed reactive power increases with the increase in the active power output.

2.2 Doubly Fed Induction Generator (DFIG)

The operating limit range (Zhao et al. 2011) of the DFIG is shown in Figure 2. $P_{DFIGmax}$ is the maximum available active power of the DFIG under a certain wind speed. The area surrounded by ABCD is the safe operation area. The power characteristics of the DFIG can be described by Equations (4)~(6).

$$Q_{DFIG} \geq -\frac{U_{DFIG}^2}{X_s + X_m} \tag{4}$$

$$\left(\frac{P_{DFIG}}{1-s}\right)^2 + Q_{DFIG}^2 \leq (U_{DFIG}I_{s\max})^2 \tag{5}$$

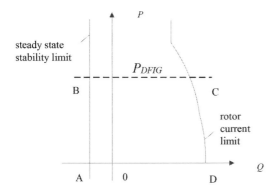

Figure 2. Safe operation region of the DFIG.

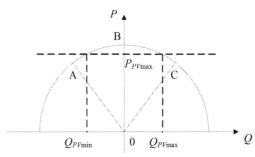

Figure 3. Operation limit of the PV system.

$$\left(\frac{P_{DFIG}}{1-s}\right)^2 + \left(Q_{DFIG} + \frac{V_{DFIG}^2}{X_s + X_m}\right)^2$$
$$\leq \left(\frac{X_m}{X_s + X_m} V_{DFIG} I_{r\max}\right)^2 \qquad (6)$$

where P_{DFIG} and Q_{DFIG} are active and reactive power of the DFIG; V_{DFIG} is the terminal voltage; I_s and I_r are the current amplitude of the rotor and stator; $I_{s\max}$ and $I_{r\max}$ are the maximum current amplitude of the rotor and stator.

2.3 Photovoltaic power generation

PV system is composed of a photovoltaic array and a SPWM inverter, which can convert solar energy into electrical energy and adjust reactive power (Wang et al. 2009) at the same time. Its active power output capability hinges on the sunshine condition and the outside environment. For a certain active power output, the capacity of reactive power is determined by the maximum capacity of the system. The operation limit of the PV system is shown in Figure 3, where OA and OC correspond to power factor constraints and the operation point must be located within OABC. The operation range can be described as follows:

$$0 \leq P_{PV} \leq P_{PV\max} \qquad (7)$$

$$Q_{PV}^2 + P_{PV}^2 \leq S_{PV\max}^2 \qquad (8)$$

$$P_{PV}/S_{PV} \geq \cos\phi_{\max} \qquad (9)$$

where $P_{PV\max}$ is the maximum available active power of the PV system; P_{PV} and Q_{PV} are, respectively, the actual output of active power and reactive power; $S_{PV\max}$ is the maximum capacity of the system; and $\cos\phi_{\max}$ is the minimum power factor allowed, which is set as 0.85 in this paper.

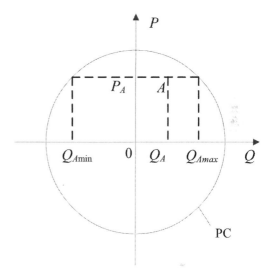

Figure 4. Operating range of the battery ESS.

2.4 Energy Storage Systems (ESS)

ESS can convert electric energy into other forms of energy for storage, which has its unique advantages in reducing the peak-to-valley difference and dealing with intermittent power generation. It is an essential component to maintain the power balance of the distribution network and achieve a flexible operation. As ESS are diverse, the storage battery is taken as a representative in this paper.

Battery ESS consists of a battery and a Power Convert System (PCS). PCS can not only meet the needs of the battery charging and discharging, but also adjust the reactive power. The operating range of the battery ESS is shown in Figure 4. Taking operating point A as an example, the active power P_A and reactive power Q_A at point A are both positive, which means that the system discharges active power and outputs inductive reactive power. In contrast, the adjustable range of reactive power under this active power model is between Q_{Amin} and Q_{Amax}. When leaving out the remaining energy of the battery, the operation range of the storage system can be expressed as follows:

$$Q_{ESS}^2 \leq S_{PCS\,max}^2 - P_{ESS}^2 \tag{10}$$

where P_{Ess} and Q_{Ess} are the active and reactive power of the storage system and S_{PCSmax} is the maximum apparent power of PCS.

3 THE DYNAMIC OPF MODEL OF THE DISTRIBUTION NETWORK

3.1 Objective function

Assuming that the time-of-use electricity price mechanism is adopted, the power distribution network will try to purchase electricity from the power transmission network at the lowest cost. So, the objective function is expressed as follows:

$$\min F = \sum_{i=1}^{NT} \lambda(t) \cdot P_{external}(t) \tag{11}$$

where NT is the number of time periods of the research cycle; $\lambda(t)$ is the electricity price during the period t; and $P_{external}$ is the active power exchanged between the distribution network and the external network.

3.2 Constraints

3.2.1 Power balance constraints

$$\begin{cases} P_{Gsum,i}(t) - P_{L,i}(t) = \sum_{j\in i} P_{ij} \\ Q_{Gsum,i}(t) - Q_{L,i}(t) = \sum_{j\in i} Q_{ij} \end{cases} i \in N; \ t = 1, 2, \ldots, NT \tag{12}$$

where N is the node set of the distribution network; $P_{Gsum,i}(t)$ and $Q_{Gsum,i}(t)$ are the total active power and reactive power of all units on the node i; $P_{L,i}(t)$ and $Q_{L,i}(t)$ are the active power load and reactive power load; and P_{ij} and Q_{ij} are active and reactive power flows of the branch i–j.

3.2.2 Network constraints

$$V_{\_i} \leq V_i(t) \leq \bar{V}_i \tag{13}$$

$$\underline{I}_l \leq I_l(t) \leq \bar{I}_l \quad l \in NL \tag{14}$$

where NL is the branch set and subscript l represents the branch corresponding to the variable.

3.2.3 Operating constraints of DGs

Operating range of DGs is constrained by both device characteristics and grid voltage. The specific forms of operating range constraints for various DGs discussed are described in Equations (1)~(10), which could be simplified as follows:

$$f_{G,i}(P_{G,i}(t), Q_{G,i}(t), V_{G,i}(t)) \leq 0 \tag{15}$$

3.2.4 Time correlation constraints of ESS

Energy balance equations are given as follows:

$$E_{Ess,i}(t+1) = E_{Ess,i}(t) + f_{in}(P_{Ess,i}(t)) \tag{16}$$

$$E_{Ess,i}(t+1) = E_{Ess,i}(t) + f_{out}(P_{Ess,i}(t)) \tag{17}$$

Equality constraint of the ESS status at the first and last time periods is given as follows:

$$E_{Ess,i}(1) - E_{Ess,i}(NT) = 0 \tag{18}$$

Top and bottom limitation constraint of ESS is given as follows:

$$E_{Ess,i,min} \leq E_{Ess,i}(t) \leq E_{Ess,i,max} \tag{19}$$

where $f_{in}()$ and $f_{out}()$ are the charging and discharging energy conversion functions, and $E_{Ess,i,max}$ and $E_{Ess,i,min}$ are minimum and maximum values of the ESS allowed.

3.3 Analyses of the dynamic OPF model of the distribution network

Equations (11)~(19) together constitute the dynamic OPF model of the distribution network. Driven by the objective function, it becomes an inevitable requirement for the distribution network to make full use of local renewable DGs in order to reduce the total power demand. However, subjected to the constraints of the network, the distribution network has different acceptance capacities to renewable DGs under different operation levels. Therefore, it is necessary to coordinate the storage equipment and the reactive power regulation device so as to achieve full acceptance to various DGs. In this regard, based on the research on developing adjustment ability of controllable equipment in the distribution network, this model considers active power, reactive power and transformer taps as decision variables, while considering all kinds of static constraints and dynamic correlation constraints of the equipment, so as to pursue the minimization of electricity cost of the distribution network under the precondition of ensuring the secure operation of the distribution network and DGs.

Although Equation (12) has the same form with that in the OPF model for the transmission network, the coupling of active power and reactive power is much closer and the degree of influence of active power on the voltage increases due to the large R/X ratio of the distribution branches. Equations (13) and (14) are static network operation constraints. Equation (15) represents constraints for DG operations. When the regulation capacity of DG is used fully, the opportunity cost of reactive power will reveal itself in a more obvious way. Furthermore, the effect can also be reflected in the whole distribution network, which may become an important feature of the distribution network. Equations (16)~(19) are correlated integral constraints and differential constraints of different time periods of the running state of the ESS. They widely link all time periods and enable the OPF to get hold of the coordination at different time periods. Under the above-stated operational features, transformer taps, as the decision content whose function goes beyond maintaining the voltage level of the distribution network, have a direct relation to the operation mode of active power and the acceptance level of DGs.

Since the transformers are discrete variables, it is apt to cause complications in solving. In this regard, the simple method of "treating them as the continuous quantity and then sorting out" is adopted, which simplifies the model into a nonlinear programming problem with only continuous variables. While there are already many mature solving methods for this, the model is solved on the basis of the GAM platform.

4 CASE ANALYSIS

4.1 *System data*

The 12-node distribution network is shown in Figure 5. The range of transformer tap positions is $(1.0 \pm 4) \times 0.0125$. The 24-hour load of each node, the wind power data and PV power all come from the actual data. Due to limited space, the detailed data are omitted, while the phenomenon is deeply described.

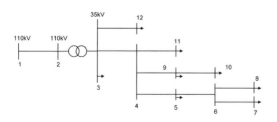

Figure 5. Structure of the distribution network.

When no DG is available, the total power demand of the distribution network for one day is shown in Figure 6. The total absorbed active power is 641.17 MWh and the cost is 10749 ¥.

Two access schemes of DGs are analyzed and the details are shown below.

Scheme I: Six DFIGs are connected to node 7 and node 8, respectively, five PV systems are connected to node 10, and ten sets of asynchronous wind power generators are connected to node 12.

Scheme II: Based on Scheme I, an ESS is connected to node 6.

4.2 *Optimization results and comparison*

4.2.1 *Total power demand of the distribution network*

The total active power demand of the distribution network under two schemes is shown in Figure 7. Massive access of DGs greatly reduces the active power demand. In some periods (periods 3 and 4), it even exports active power to the higher-level grid. In Scheme I, the total absorbed power is 281.72 MWh and the operation cost is 5945 ¥. In Scheme II, the total absorbed power is 277.32 MWh, meaning that the ESS can increase the acceptance level of DGs. In addition, it can reduce the cost to 5583 ¥.

The total active power output curve of DGs is shown in Figure 8, where the dashed line is the

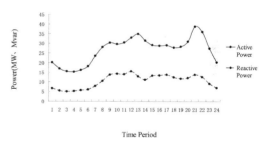

Figure 6. Total load curve of the distribution system without DG.

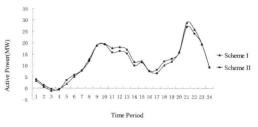

Figure 7. Comparison of the total active power demand under two schemes.

457

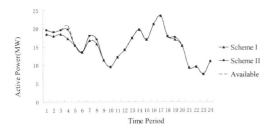

Figure 8. Comparison of the total active power output of DGs.

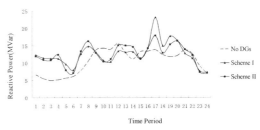

Figure 9. Comparison of the total reactive power of the distribution network.

available active power of DGs. In Scheme I, the total power reduction of DGs is 10.17 MWh, while in Scheme II, the reduction is 0.98 MWh, which means that the ESS has improved the capacity of the distribution network with DGs.

In the case of reactive power, the total reactive power demand of the distribution network under two schemes is shown in Figure 9, in which the dashed line shows the total reactive power demands of the distribution network without DGs. It can be seen that the total reactive power demand is not diminished. Instead, it increases dramatically in some periods, which means that more reactive power support from the higher-level grid is needed. However, reactive power demands under Scheme II are slightly lower than that under Scheme I, meaning that the ESS can ease the total reactive power demand.

Combining Figures 8 and 9, active power outputs of DGs have approximately the same change trend with the total reactive demand. At the time periods when there are higher active power outputs of DGs, the distribution network absorbs more reactive power. It means that, under the massive DG access, there are obvious interactive relations between the active power demand and the reactive power demand.

4.2.2 *Interactive relations of active power, reactive power and voltage*

The fourth time period of Scheme I is taken as an example to analyze the reason behind the reduction of the power form DGs and the increase in the overall reactive power demand. The voltage amplitude of each node is presented in Table 1, and the output power of DGs is given in Table 2.

As given in Table 1, the node connected with large capacity of DFIGs has a higher voltage (nodes 7 and 8), especially the node 7, at which the voltage reaches its upper limit. The main reason is that its output active power causes a backward power flow along the track of "node 6—node 5—node 4—node 3". Under the influence of a higher R/X ratio of the branch, the terminal voltage increases.

Table 1. Voltage magnitude of the nodes at the fourth time period under Scheme I.

Node	Voltage (pu)	Node	Voltage (pu)
1	1.050	7	1.050
2	1.043	8	1.045
3	0.995	9	0.961
4	0.975	10	0.953
5	1.000	11	0.950
6	1.030	12	0.965

Table 2. Output power of distributed generations at the fourth time period under Scheme I.

Node	Available power	Active power	Reactive power
7	9.367	7.515	−1.606
8	9.367	7.718	−1.589
9	2.011	2.011	−1.460

When the reverse active power is too high, the terminal voltage might cross the limit. At that time, in order to increase the output of active power of DGs, it is necessary to absorb more reactive power so as to ease the voltage increase. In this example, the DFIGs of nodes 7 and 8 use their reactive power regulation ability to absorb the reactive power (Table 2). However, when the devices absorb the maximum reactive power, the active power output of DGs will have to be diminished to avoid the terminal voltage reaching its ceiling. So, it is clear that when massive DGs are switched to the end of the distribution line, the only way to increase their active power output is to absorb more reactive power at its end, which obviously correlates the total active power demand with the total reactive power demand of the distribution network. That is why the active power output curve of DGs and the total reactive power demand curve of the distribution network have a similar trend.

4.2.3 *Functions of ESS*

Figure 10 shows the voltage curves of node 6 under two schemes. Figure 11 shows the output power of the ESS. Figure 12 shows a comparison of statistical active power for two schemes, where Difference Value I (DV I) stands for the extra active power absorbed by the distribution network under Scheme II compared with that under Scheme I, and DV II stands for the extra active power produced by DGs in Scheme II.

It can be seen from DV I, the distribution network increases the active power demand at the low-price periods and decreases the active power demand at peak periods, so as to reduce the operation costs through load transfer under the cooperation of the ESS.

Besides, to reduce the amount of purchased energy, power from DGs should be accepted as much as possible. In Scheme I, the active power

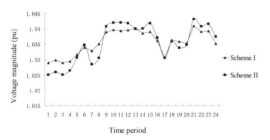

Figure 10. Comparison of the voltage magnitude at node 6 under two schemes.

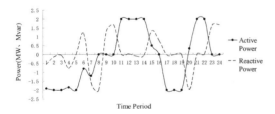

Figure 11. Active power and reactive power of the ESS.

Figure 12. Comparison of statistical active power for two schemes.

output of DGs is decreased at some time periods due to the reactive voltage issue. In Scheme II, this issue is dealt with by the regulation of the ESS. As can be seen from Figures 11 and 12, the active power output increases to some extent during periods 1 to 4, 7 to 8, 17 and 19 to 20, while the ESS absorbs the reactive power during these periods and the voltage level at node 6 is lower (Figure 10). So, it is clear that the ESS uses reactive power regulation ability to absorb the reactive power so as to reduce the voltage amplitude at the upstream node (node 6) of the access point (nodes 7 and 8) of DGs, which creates the conditions for the improvement of the terminal active power acceptance level of DGs. This phenomenon is quite obvious during time periods 4 and 8. While during time periods 5, 6 and 18, the ESS can absorb the active power from DGs, which, to some extent, inhibits the reverse active power flow, thus it is allowed to adjust the network through releasing the reactive power without diminishing the power from DGs caused by the voltage rise.

4.3 *Summary*

Under the influence of a high R/X ratio of the distribution line, the active power counter-flow produced by DG affects the power flow significantly. In order to achieve the maximum utilization of DG, the distribution network inevitably operates on the boundary constraints, resulting in interactive relations among the active power, reactive power and voltage. Under such circumstances, the ESS could transfer the load to reduce the cost, but can also exert a positive influence on reducing the network loss and easing the constraints of reactive resources under certain operation modes.

5 CONCLUSIONS

Dynamic OPF model of the active distribution network is established, and the interactive relationships among different variables are analyzed. The related conclusions can be summarized as follows:

The adjustment characteristics of several different kinds of DGs are analyzed in detail and the mathematical models to describe the operation ranges are established, which lay the foundation for the full utilization of various types of DGs.

In the distribution network with massive DGs, there is a coupling relationship among the active power, reactive power and voltage. In this regard, a dynamic OPF model of the active distribution network is established, which can coordinate various equipment to achieve the flexible adjustment of the distribution network.

As a flexible buffer device, the ESS can effectively ease safe operation constraints, such as active power constraint, reactive power constraint and voltage constraint. So, it plays an important role in improving the operation level and regulation ability of the distribution network.

REFERENCES

Carpentier, J. 1962. Contribution a l'etude du dispatching economique. *Bulletin de la Societe Francaise des Electriciens* 3: 431–447.

Chen, L. et al. 2006. Optimal reactive power planning of radial distribution systems with distributed generation. *Automation of Electric Power Systems* 30: 20–24.

Dolan, M.J. et al. 2012. Distribution power flow management utilizing an online optimal power flow technique. *IEEE Transactions on Power Systems* 27: 790–799.

Dommel, H.W. & Tinney, W.F. 1968. Optimal power flow solutions, *IEEE Transactions on Power Apparatus and Systems* PAS-87: 1866–1876.

Feijoo, A.E. & Cidras, J. 2000. Modeling of wind farms in the load flow analysis. *IEEE Transactions on Power Systems* 15: 110–115.

Gabash, A. & Li, P. 2012. Active-reactive optimal power flow in distribution networks with embedded generation and battery storage. *IEEE Transactions on Power Systems* 27: 2026–2035.

Ochoa, L.F. & Harrison, G.P. 2011. Minimizing energy losses: Optimal accommodation and smart operation of renewable distributed generation. *IEEE Transactions on Power Systems* 26: 198–205.

Sun, D.I. et al. 1984. Optimal power flow by Newton approach, *IEEE Transactions on Power Apparatus and Systems* PAS-103: 2864–2880.

Torres, G.L. & Quintana, V.H. 1998. An interior-point method for nonlinear optimal power flow using voltage rectangular coordinates. *IEEE Transactions on Power Systems* 13: 1211–1218.

Wang, W. et al. 2008. A distribution network optimal reconfiguration algorithm of reducing the number of spinning trees. *Proceedings of the CSEE* 28: 34–38.

Wang, W. et al. 2009. Action scope algorithm for optimal capacitor switching in distribution network. *Proceedings of the CSU-EPSA* 20: 36–40.

Wang, Y.B. et al. 2009. Steady-state power flow analyses of large-scale grid-connected photovoltaic generation system. *J Tsinghua Univ (Sci&Tech)* 49: 1093–1097.

Zhao, J.J. et al. 2009. A comprehensive optimization algorithm for injection power of distributed generation and distribution network reconfiguration based on particle swarm optimization. *Power System Technology* 33: 157–162.

Zhao, J.J. et al. 2011. Reactive power optimaization in distribution network considering reactive power regulation capability of DFIG wind farm. *Automation of Electric Power Systems* 35: 33–38.

Advances in Power and Energy Engineering – Sun (Ed.)
© *2016 Taylor & Francis Group, London, ISBN 978-1-138-02846-3*

Mixed measurement-based power system state estimation with measurement correlation

Z.G. Lu & S.H. Yang
State Grid Jiangsu Electric Power Research Institute, Nanjing, China

S. Yang
China Electric Power Research Institute, Beijing, China

J. Wang
Electric Power Planning and Engineering Institute, Beijing, China

ABSTRACT: Measurement error is the important factor affecting the precision of the power system state estimation. A power system state estimation method considering the measurement correlation based on the mixed measurement is proposed. The Phasor Measurement Unit (PMU) and Remote Terminal Unit (RTU) measurement error dependencies are taken into account in the proposed method. The actual mixed measurement configurations from Supervisory Control and Data Acquisition (SCADA) and Wide Area Measurement System (WAMS) are analyzed by taking into account measurement error transfer characteristics. The point estimation theory is used to calculate the modified measurement error covariance matrix, which replaces the traditional measurement error diagonal matrix. The Correlated Weighted Least Square (CWLS)-based state estimation method is proposed. Finally, the simulation results for the IEEE 14-bus and 57-bus system illustrate the validity of the proposed method, where better performance on the accuracy of statistical estimations can be obtained.

1 INTRODUCTION

Power system State Estimation (SE) is the core program of the Energy Management System (EMS) in the power system dispatching center. The role of the SE is to deal with the raw data as the real data. The raw measurement is of low precision including bad data occasionally, which is not useful for the power system analysis. The output of the SE is the state estimation (voltage magnitude and angle at each node) of the power system real value (Exposito, A.G. & Jaen, A.V. 2009 & Vanfretti, L. et al. 2011). By taking advantage of the SE program, the dispatch center could carry out the correct analysis and judgment for the security and economy of the power system. Therefore, the operating status is able to be understood preferably at all times (Abur, A. & Exposito, A.G. 2004 & Jazaeri, M. & Nasrabadi, M.T. 2015).

Nowadays, the Phasor Measurement Unit (PMU) is deployed at many nodes of the whole power grid widely, especially containing all the important nodes of 220 kV/500 kV (Kashyap, N. et al. 2014). While the PMU-based Wide Area Measurement System (WAMS) is available, a wealth of data of millisecond period would compensate

the traditional SE methods using the information only from Supervisory Control And Data Acquisition (SCADA) (Xing, W. et al. 2014). As the PMU being considered as a powerful tool for the power system SE, the mixed measurements from both the PMU and the Remote Terminal Units (RTU), which increase the measurement redundancy, are more practical and effective for the monitoring and forecasting of the power system operating status (Xianrong, C. & Rui, F. 2015).

At present, the practical method for the SE calculation in the EMS is the Weighted Least Square (WLS) method (Schweppe, F.C. & Wildes, J. 1970 & Monticelli, A. & Garcia, A. 1990. The improved algorithms have been further studied accordingly, which include the orthogonal transformation-based WLS (Kilokys, E. & Singh, N. 2000), the improved Givens method (Nagvajara, P. et al. 2007), the least-square Givens transformations method (Haibo, B. & Hua, W. 2012), the modified Newton method (normal equations with constraints, NE/C) (Ye, G. et al. 2012) and the optimal algorithm with inequality constraint (Yong, W. et al. 2008 & Yuan, Z. et al. 2012)., Moreover, many algorithms and improvements have also been proposed to deal with the mixed measurement

(Liu, W.Y. & Inseok, H. 2014, Sharma, A. et al. 2014 & Hongen, D. et al. 2014). A good many methods have been proposed by far, with a greater contribution to the accuracy and computational efficiency of the SE algorithm. However, the measurement error assumption for the traditional SE model is that each noise is the Gauss white noise, which is not correlated with each other. In fact, the measured data gathered at the same node is derived from the same source. In other words, there is a correlation between the relevant measurements. In one study (Morales, J.M. & Pérez-Ruiz, J. 2007 & Caro, E. et al. 2009), the practical relevance between the electrical digitals in each transformer substation is analyzed and the correlated measurement model is proposed. Therefore, the measurement error covariance matrix is updated in the WLS-based SE method. The proposed method can be treated as an effective one to establish an accurate measurement model in the power system SE.

Therefore, based on the above discussion, in this paper, the actual mixed measurement configurations are considered thoroughly. The RTU and PMU measurement error transfer characteristics are taken into account in the WLS algorithm. The point estimate method is used to compute the modified measurement error covariance matrix. Then, the Correlated WLS (CWLS)-based SE method is proposed. Finally, simulation results for the IEEE test system validate the effectiveness of the proposed methods.

The rest of this paper is organized as follows. In Section II, the WLS theory is introduced generally. In Section III, the mixed measurement configurations are revealed for the analysis of the error correlation. In Section IV, the point estimate approach is used to calculate the modified measurement error covariance matrix. In Section V, the CWLS model is verified by the simulations performed on the standard IEEE test system. Finally, the conclusion and some remarks are drawn in the last section.

2 WEIGHTED LEAST SQUARE METHOD

The aim of the SE is to determine the most appropriate operating status based on the measurement data. Under the conditions that the network topology and system parameters are known, the object function J is given as follows:

$$J(x) = \left[z - h(x)\right]^T R^{-1} \left[z - h(x)\right] \tag{1}$$

where x is the state vector of n dimensions; z is the measurement vector of m dimensions; h is the measurement function decided by Ohm's and

Kirchhoff's laws; R is the measurement error covariance matrix; and T is the calculation of the transposition matrix.

By using the Newton-Raphson method, Equation (1) can be calculated by using the following formula:

$$\begin{cases} \Delta x_k = \left[H^T(x_k)\, x R^{-1} H\,(x_k)\right]^{-1} \\ \qquad \times H^T(x_k)\, R^{-1} \left[z_k - h\,(x_k)\right] \\ x_{k+1} = x_k + \Delta x_k \end{cases} \tag{2}$$

where subscript k is the time sample and H is the Jacobian matrix calculated as follows:

$$H(x) = \frac{\partial h(x)}{\partial x} \tag{3}$$

The iterative computation would converge after the maximum correction value is below the given threshold. Then, the expectation of the measurement estimate error covariance is formulated as follows:

$$\begin{aligned} E\left[(z - z')(z - z')\right]^T \\ = E\left[\left(H(x')(x - x')\right)\left(H(x')(x - x')\right)^T\right] \\ = H(x')G^{-1}(x')H^T(x') \end{aligned} \tag{4}$$

where z' is the measurement estimate; x' is the state estimate; and G is the information matrix given as follows:

$$G(x) = H^T(x)R^{-1}H(x) \tag{5}$$

By using Equation (4), the information matrix is of significance, which characterizes the WLS estimation effect, i.e. the greater the information matrix, the smaller the estimation error, as well as the measurement estimate error covariance.

3 MIXED MEASUREMENT CORRELATION

3.1 RTU signals and SCADA measurements

The signals measured directly by the RTUs are voltage/current magnitudes and voltage-current angles at nodes or on lines. On the other hand, the SCADA measurements for the input data of the SE calculation are voltage magnitudes and active/reactive power. The correlation between the RTU signals and SCADA measurements is given by

$$\begin{cases} U_i = \dfrac{(U_i^{\mathrm{A}} + U_i^{\mathrm{B}} + U_i^{\mathrm{C}})}{3} \\ P_i = \displaystyle\sum_{f=\{A,B,C\}} U_i^f I_i^f \cos\left(\theta_i^f\right) \\ Q_i = \displaystyle\sum_{f=\{A,B,C\}} U_i^f I_i^f \sin\left(\theta_i^f\right) \\ P_{ij} = \displaystyle\sum_{f=\{A,B,C\}} U_i^f I_{ij}^f \cos\left(\theta_{ij}^f\right) \\ Q_{ij} = \displaystyle\sum_{f=\{A,B,C\}} U_i^f I_{ij}^f \sin\left(\theta_{ij}^f\right) \end{cases} \quad (6)$$

where U_i is the voltage magnitude measurement at node i; P_i is the injection active power flow measurement at node i; Q_i is the injection reactive power flow measurement at node i; P_{ij} is the line active power flow measurement on line ij at terminal i; Q_{ij} is the line reactive power flow measurement on line ij at terminal i; U_i^f is the voltage magnitude signal for phase f at node i; I_i^f is the current magnitude signal for phase f at node i; θ_i is the voltage-current angle signal for phase f at node i; I_{ij}^f is the current magnitude signal on line i–j at terminal i; and θ_{ij}^f is the voltage-current angle signal on line i–j at terminal i.

3.2 PMU signals and WAMS measurements

The PMU signals include voltage/current phasor signals. Meanwhile, the WAMS measurements include the voltage and current magnitudes. The corresponding correlation between the initial phasor signals and the measurements is given as follows:

$$\begin{cases} \dot{U}_i = \dot{U}_i^{\mathrm{A}} + \dot{U}_i^{\mathrm{B}} + \dot{U}_i^{\mathrm{C}} \\ \dot{I}_{ij} = \dot{I}_{ij}^{\mathrm{A}} + \dot{I}_{ij}^{\mathrm{B}} + \dot{I}_{ij}^{\mathrm{C}} \\ \dot{U}_j = \left[(\dot{Y}_{i0} + \dot{Y}_{ij}) \dot{U}_i - \dot{I}_{ij} \right] / \dot{Y}_{ij} \end{cases} \quad (7)$$

where U_i is the voltage phasor measurement at node i; U_j is the voltage phasor measurement at node j; I_{ij} is the current phasor measurement from node i to j on the line ij; U_i^f is the f-phase voltage phasor signal at node i $(f = \{A, B, C\})$, I_{ij}^f is the f-phase current phasor signal from node i to j on the line ij $(f = \{A, B, C\})$; Y_{i0} is the shunt admittance of node i; and Y_{ij} is the admittance of the line between node i and node j.

It is obvious that the SCADA and WAMS measurements are not Gauss distributions; that is to say, it is the error assumption that is no longer adequate now.

4 POINT ESTIMATION

From Equation (1), it can be seen that the measurement error covariance matrix R is not in accordance with the analysis of the mixed measurement correlation. Therefore, the modified measurement error covariance matrix is calculated by taking advantage of the point estimation method.

The signal vector p and measurement vector d are given as

$$\begin{cases} p = [p_1, \ldots, p_l, \ldots, p_\eta]^T \\ d = [d_1, \ldots, d_\rho, \ldots, d_{\eta'}]^T \end{cases} \quad (8)$$

where η is the dimension of p and η' is the dimension of d.

In terms of the mixed measurement configurations, RTU/SCADA and PMU/WAMS expressions are formulated by

$$\begin{cases} p_{RTU} = [U_i^f, I_i^f, \phi_i^f, I_{ij}^f, \ldots, I_{in'}^f, \phi_{in'}^f, \ldots, \phi_{in'}^f]^T \\ d_{SCADA} = [U_i, P_i, Q_j, P_{ij}, \ldots, P_{jn'}, Q_{ij}, \ldots, Q_{jn'}]^T \end{cases} \quad (9)$$

$$\begin{cases} p_{PMU} = [\dot{U}_i^f, \dot{I}_{ij}^f, \ldots, \dot{I}_{in'}^f]^T \\ d_{WAMS} = [U_i, \theta_i, U_j, \ldots, U_{jn'}, \theta_j, \ldots, \theta_{jn'}]^T \end{cases} \quad (10)$$

where θ_i is the angle measurement and n' is the number of nodes connected with node i directly.

The point estimation theory is used to calculate the modified measurement error covariance matrix. The key idea of the point estimation method is the aggregation of the effects of elements in variable sets with the scale factors. The derivative process is described in detail below.

The elements p_l are examined twice with the statistical characters, so the examination value pl, λ ($\lambda = 1, 2$) can be formulated by

$$\begin{cases} p_{l,1} = \mu_{p_l} + \sqrt{\eta}\sigma_{p_l} \\ p_{l,2} = \mu_{p_l} - \sqrt{\eta}\sigma_{p_l} \end{cases} \quad (11)$$

where μ_{pl} is the expectation of the element pl and σ_{pl} is the standard deviation of pl, which is determined by the accuracy of the telemetering device.

The elements in each variable set can be computed by using Equation (13), which are computed 2η times ($l = 1, \ldots, \eta$), and constitute the examination function vector $D(l,\lambda)$, whose component $D_\rho(l,\lambda)$ can be calculated by

$$D_\rho(l,\lambda) = F_\rho(\mu_{p_1}, \ldots, \mu_{p_{l-1}}, p_{l,\lambda}, \mu_{p_{l+1}}, \ldots, \mu_{p_\eta}) \quad (12)$$

where F_ρ is the functional form of measurement configuration expressed in Equations (5) and (6).

Then, the scale factors $s_{l,\lambda}$ are given by

$$s_{l,1} = \frac{1}{2\eta}, \quad s_{l,2} = \frac{1}{2\eta} \quad (13)$$

463

By using the values of $D_\rho(l,\lambda)$ and $s_{l,\lambda}$, the γth non-cross moments $\mu^\gamma_{D_\rho}$ of the measurement variable d_p in D can be evaluated by

$$\mu^\gamma_{D_\rho} = E[D^\gamma_\rho] \approx \sum_{l=1}^{\eta}\sum_{\lambda=1}^{2} s_{l,\lambda}\,(D_\rho(l,\lambda))^\gamma \qquad (14)$$

From Equation (12), the second centered non-cross moments $c_{d_p d_p}$ of the measurement element d_p can be easily obtained as follows:

$$c_{d_\rho d_\rho} = \sigma^2_{D_\rho} = E[D^2_q] - E[D_q]^2 \qquad (15)$$

where $c_{d_p d_p}$ is the variance $\sigma^2_{D_p}$ of D_p; $\mu^1_{D_p}$ is the calculation of $\mu^\gamma_{D_p}$, with $\gamma = 1$; and $\mu^2_{D_p}$ is the calculation of $\mu^\gamma_{D_p}$, with $\gamma = 2$.

Then, the second centered cross moment $c_{d_p d_{p'}}$ between d_p and $d_{p'}$ can be calculated by

$$c_{d_\rho d_{\rho'}} = E[D_\rho D_{\rho'}] \approx \sum_{l=1}^{\eta}\sum_{\lambda=1}^{2} s_{l,\lambda}(D_\rho(l,\lambda)D_{\rho'}(l,\lambda)) \quad (16)$$

Based on Equations (13) and (14), the elements, including both the diagonal and non-diagonal numbers, of the modified measurement error covariance submatrix R'_{z_i} can be formulated as follows:

$$R'_{D_\rho,D_\rho|z_i} = \sigma^2_{D_\rho} = E[D^2_\rho] - E[D_\rho]^2 \qquad (17)$$

$$R'_{D_\rho,D_{\rho'}|z_i} = E[D_\rho D_{\rho'}] - E[D_\rho]E[D_{\rho'}] \qquad (18)$$

where $R'_{D_p,D_p|z_i}$ is the diagonal element of R'_{z_i} and $R'_{D_p,D_p|z_i}$ is the non-diagonal element of R'_{z_i}.

For the whole network, all sets of mixed measurements at each node within all substations constitute the complete measurement data. Each submatrix R'_{z_i} ($i = 1, \ldots, m$) thus forms the modified measurement error covariance matrix R' as follows:

$$R' = \begin{bmatrix} R'_{z,1} & \cdots & 0 \\ \vdots & \ddots & \vdots \\ 0 & \cdots & R'_{z,m'} \end{bmatrix} \qquad (19)$$

where m' is the number of the measurement sets.

Conventionally, in the WLS, the squares of the expected errors between the measured and estimated values are weighted by R'^{-1}, which has always been taken as the diagonal form. Replacing by R^{-1}, the improvement in the WLS is already considered as the mixed measurement correlations. Then, the correlated WLS (Correlated Weighted Least Square, CWLS)-based SE technique is assessed by comparisons by using the two different measurement variance-covariance matrices with numerous tests.

5 CASE STUDY

To validate the effectiveness of the CWLS method, a lot of simulation tests are performed on the IEEE 14-bus and 39-bus systems. The comparisons of the results between the CWLS and WLS methods are analyzed and studied in detail.

Based on the basic data of power flow on the IEEE test system, the corresponding varieties within a day are simulated by using the 144 daily load changes obtained from the actual operating status (Fig. 1). Moreover, four sampling times are set up among each time interval of adjacent sampling points in the 144-point daily curve. A total of 720 load data are simulated at a 2-minute time interval per day. Finally, the 720 power flow snapshots are obtained. The Gauss white noises are added on the true values of power flows, which constitute the measurement data. The standard deviations of the measured signals are 0.01/0.02 for RTUs and 0.001/0.00 for PMUs.

To evaluate the effectiveness of the proposed method, some performance indices are provided as follows:

$$MAE_\kappa = \frac{\sum_{i=1}^{m_p}\left|\hat{z}^p_i - \overline{z}_i\right|}{m_p} \qquad (20)$$

where MAE_κ is the Mean Absolute Error (MAE) of the estimated value for each measurement category κ; z_i is the estimated value of z; z_i is the true value of z_i computed based on the power flow solution; and m_p is the number of measurement categories κ.

Moreover, the other performance indices of the state absolute error (ε_u and ε_θ) are given as follows:

$$\begin{cases} \varepsilon_{U_i} = \left|\hat{U}_i - \overline{U}_i\right| \\ \varepsilon_{\theta_i} = \left|\hat{\theta}_i - \overline{\theta}_i\right| \end{cases} \qquad (21)$$

where $\hat{U}_i$ is the measured value of the voltage magnitude at node i; $\overline{U}_i$ is the true value of the voltage

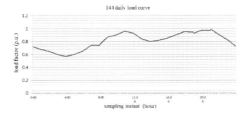

Figure 1. 144-daily load curve obtained from the actual operating status.

464

magnitude at node i; $\hat{\theta}_i$ is the measured value of the angle at node i; and $\overline{\theta}$ is the true value of the angle at node i.

5.1 IEEE 14-bus system

The modified measurement covariance matrix R' is adopted instead of the traditional diagonal matrix R in the WLS. A total of 720 simulation tests are carried out by both WLS and CWLS methods.

Figure 2 shows the distributions of the index differences between the WLS and CWLS results with respect to the estimated values. Each MAE for six performance indices is obtained by the traditional and proposed methods. Comparisons are represented by the histograms in each chart. These histograms represent 720 measurement snapshots for the correlated measurement configuration, and compare voltage magnitudes/angles and active/reactive power flows (lines and injections). The dotted line located in the "zero" position, vertically along with each histogram, depicts the same values of the two estimator results. In the histogram, each bar represents the number of tests where the corresponding index appears within the interval at the horizontal axis.

It should be noted that the dotted line is considered as a well-balanced benchmark for the average absolute estimator error of the two methods. It is on the right side of the dotted line where almost all the indices scatter over, as shown in Figure 2. In other words, $MAE^{WLS} > MAE^{CWLS}$. These representations indicate that the CWLS method, by contrast, has a much better performance.

It can be observed from Figure 2 that the performance on the accuracy of statistical estimations is relatively significant. It follows from these comparisons that the accuracies of state estimates obtained by using the CWLS are higher than the WLS method, which verifies the validity of the proposed method.

5.2 IEEE 57-bus system

The CWLS method is also tested on the IEEE 57-bus system. The comparisons between the CWLS and WLS are carried out. Table 1 gives a comparison of the absolute error statistics of the state estimates.

From Table 1, it can be seen that the average absolute error of the state estimate obtained by the CWLS method has an advantage over the WLS method. Meanwhile, the number of tests where the MAE of the CWLS method is lower than the WLS method is included in Table 2.

From Table 2, it can be seen that the error of power flows obtained from the CWLS method is lower than the WLS method in most or all of the tests.

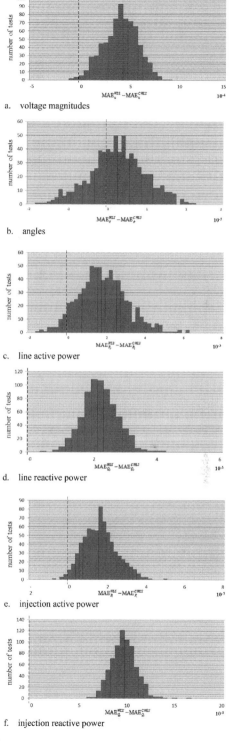

a. voltage magnitudes

b. angles

c. line active power

d. line reactive power

e. injection active power

f. injection reactive power

Figure 2. Comparison of estimated values on the IEEE 14-bus system.

Table 1. Comparison of the average absolute error performance index in the two methods.

| | State estimate (voltage magnitude and angle) | | | |
| | ε_u | | ε_θ | |
Methods	Average	Maximum	Average	Maximum
WLS	0.003725	0.008404	0.002988	0.007169
CWLS	0.001631	0.003167	0.001566	0.006326

Table 2. Total number of the tests with each performance index.

Number of tests (deviation: CWLS < WLS)

Voltage magnitude	Angles	Line active power	Line reactive power	Injection active power	Injection reactive power
720	594	579	632	542	720

Both from Tables 1 and 2 with the increase in network size, the proposed method is proved to be effective.

6 CONCLUSION

In this paper, based on the mixed measurements, the configurations between RTU/PMU signals and SCADA/WAMS measurements were expressed in detail. The correlated measurement error was taken into account in the WLS algorithm. The modified measurement error covariance matrix was calculated in the CWLS algorithm by using the point estimate method. Finally, the simulation test results on the IEEE test system demonstrated that the accuracy of state estimates could be improved obviously by using the proposed method.

REFERENCES

Abur, A. & Exposito, A.G. 2004. *Power System State Estimation. Theory and Implementations.* New York: Marcel Dekker, 2004.
Caro, E. et al. 2009. Power System state estimation considering measurement dependencies. *IEEE Transactions on Power Systems* 24(4): 1875–1885.
Exposito, A.G. & Jaen, A.V. 2009. Two-level state estimation with local measurement pre-processing. *IEEE Transactions on Power Systems* 24(2): 676–684.
Haibo, B. & Hua, W. 2012. A stochastic response surface method for probabilistic evaluation of the voltage stability considering wind power. *Proceedings of the CSEE* 32(13): 77–85.

Hongen, D. et al. 2014. Solution to bad data problem of phase angle reference bus for state estimation with hybrid measurement. *Automation of Electric Power System* 38(9): 132–136.
Jazaeri, M. & Nasrabadi, M.T. 2015. A new fast and efficient artificial neural network based state estimator incorporated into a linear optimal regulator for power system control enhancement. *Electric Power Components and Systems* 43(6): 644–655.
Kashyap, N. et al. 2014. Power system state estimation under incomplete PMU observability: a reduced-order approach. *IEEE Journal of Selected Topics in Signal Processing* 8(6): 1051–1062.
Kilokys, E. & Singh, N. 2000. Minimum correction method for enforcing limits and equality constraints in state estimation based on orthogonal transformations. *IEEE Transactions on Power Systems* 15(4): 1281–1286.
Liu, W.Y. & Inseok, H. 2014. On Hybrid State Estimation for Stochastic Hybrid Systems. *IEEE Transactions on Automatic Control* 59(10): 2615–2628.
Monticelli, A. & Garcia, A. 1990. Fast Decoupled State Estimator. *IEEE Transactions on Power System* 5(2): 556–564.
Morales, J.M. & Pérez-Ruiz, J. 2007. Point estimate schemes to solve the probabilistic power flow. *IEEE Transactions on Instrumentation and Measurement* 22(4): 1594–1601.
Nagvajara, P. et al. 2007. State estimation using sparse Givens rotation field programmable gate array. *39th North American Power Symposium (NAPS 2007) Boston, 12 February 2007.*
Sharma, A. et al. 2014. Multi Area State Estimation for Smart Grid Application Utilizing All SCADA and PMU Measurements. *2014 IEEE Innovative Smart Grid Technologies—Asia (ISGT Asia) Bangkok, Thailand 4-6 November* 2014: 525–530.
Schweppe, F.C. & Wildes, J. 1970. Power System Static State Estimation, Part I, II, III. *IEEE Transactions on Power System and Apparatus* 89(1): 120–135.
Vanfretti, L. et al. 2011. A phasor-data-based state estimator incorporating phasor bias correction. *IEEE Transactions on Power Systems* 26(1): 111–119.
Xianrong, C. & Rui, F. 2015. A mixed measurement partition state estimation method taking zero injection node constraint into account. *Power System Technology* 39(8): 2253–2357.
Xing, W. et al. 2014. Comparative research on three state estimation algorithms based on data compatibility of WAMS/SCADA. *East China Electric Power* 42(2): 240–246.
Ye, G. et al. 2012. Power system state estimation solution with zero injection constraints using modified Newton method. *Proceedings of the CSEE* 32(19): 96–100.
Yong, W. et al. 2008. A new decomposition and coordination algorithm for state estimation in interconnected power system. *Power System Technology* 32(10): 79–83.
Yuan, Z. et al. 2012. Parameter sensitivity and optimal allocation of UPFCs in bulk power systems reliability assessment. *Automation of Electric Power Systems* 36(1): 55–59.

Advances in Power and Energy Engineering – Sun (Ed.)
© 2016 Taylor & Francis Group, London, ISBN 978-1-138-02846-3

Research on undervoltage problems and coordination strategies of prevention and control in isolated receiving power grid

Z.H. Li, L.H. Wu, Q. Wang & X. Chen
China Electric Power Research Institute, Beijing, China

Y.X. Yu
North China Electric Power University, Baoding, Hebei, China

ABSTRACT: In a large interconnected power system, when a local receiving power grid is disconnected from the main power system and operates as an isolated grid, a serious undervoltage problem can occur if the active power or reactive power is unbalanced. It poses a serious threat to the security and stability operation of the isolated local power system. Based on RLC equivalent load model, it analyzes the voltage characteristics of the isolated grid. It deduces and calculates the range of the unbalanced active power or reactive power to meet the requirement of a stable operation of an isolated power system. It studies the influence factors of voltage characteristics by simulation, and analyzes the coordination control strategies of undervoltage load shedding and concentrated load shedding.

1 INTRODUCTION

In a large interconnected power system, serious faults, extreme weather events, geological disasters, or mis-operations may result due to the disconnection of weakly connected local power grids. For a locally receiving power system with a high ratio of received power, if the active power or reactive power is seriously unbalanced, then the undervoltage problem may occur. If the adjustment and control are not executed in time, voltage collapses and even blackouts may occur in the isolated grid (Huazhong Power System 7.1 Accident Investigation Team 2006, Huang et al. 2008 and NERC 2010). Therefore, it is required to monitor the operating status of the isolated grid online and take corresponding measures in time. In general, to increase the generator's reactive power and a shunt capacitor bank, SVC, or SVG should be put into use urgently, which is often used to solve the shortage of reactive power. Generator's reactive power is limited and the reactive power of shunt capacitor bank is lower under the condition of undervoltage than normal status, so it often leads to unsatisfactory results of voltage support and even worsens the voltage. Then, the distribution of SVC and SVG is limited. Therefore, other regulatory measures are needed to take into account, such as concentrated load shedding and undervoltage load shedding. Aiming at improving

Supported by the Research Fund of the Key Projects of the State Grid Corporation of China (*XT71-13-037*).

the security, stability, and reliability of the isolated grid, this paper focuses on the local receiving power system, which is disconnected from the major grid and analyzes its voltage characteristics. It deduces and calculates the range of the allowable unbalanced degree of active power or reactive power, studies the influence factors of voltage stability characteristics by the simulation, and analyzes the coordination strategies of undervoltage load shedding and concentrated load shedding.

2 VOLTAGE CHARACTERISTICS OF ISOLATED GRID

2.1 Change of voltage across load

The RLC equivalent load model is shown in Figure 1. The 'G_I' denotes the equivalent generator

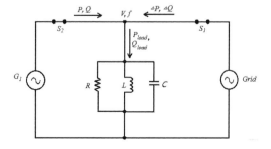

Figure 1. Isolated grid equivalent.

in the isolated grid and the load is described as the RLC parallel model. The isolated grid is connected with major grid '*Grid*' through the switch 'S_1'.

The load can be expressed by parameter '*R*,' '*L*,' and '*C*' or quality factor 'Q_f' and resonant frequency 'f_0' in formula (1). The 'ω' denotes the actual angular frequency.

$$Z_{load} = \frac{1}{\sqrt{\frac{1}{R^2} + \left(\frac{1}{\omega L} - \omega C\right)^2}} = \frac{R}{\sqrt{1 + Q_f^2\left(\frac{f_0}{f} - \frac{f}{f_0}\right)^2}}$$

(1)

The 'Q_f' and 'f_0' can be expressed as below:

$$f_0 = \frac{1}{2\pi\sqrt{LC}}$$

(2)

$$Q_f = \frac{R}{\omega_0 L} = \omega_0 CR$$

(3)

When the RLC load is connected to the major grid in which the voltage is '*V*' and the frequency is '*f*' ('*V*' equals to 1 pu and the '*f*' equals to 50 Hz when the major grid is operating normally), the active power and reactive power that the load of local power grid needed can be expressed in formulas (4) and (5) as below.

$$P_{load} = \frac{V^2}{R}$$

(4)

$$Q_{load} = P_{load} Q_f\left(\frac{f_0}{f} - \frac{f}{f_0}\right)$$

(5)

It can be seen from the Figure 1 that the 'G_1' and the RLC load form the isolated grid after the 'S_1' is switched off. The unbalanced power of isolated grid 'ΔP' and 'ΔQ' equals to the power received from major grid. If the active power and reactive power of '$G1$' is '*P*' and '*Q*,' 'ΔP' and 'ΔQ' can be expressed in formulas (6) and (7).

$$\Delta P = P_{load} - P$$

(6)

$$\Delta Q = Q_{load} - Q = Q_{load}$$

(7)

The worst situation to the stabilization of isolated grid is that the active power of generator in the grid remains unchanged and the reactive power equals to '0'. The applicability of the results is ensured through the consideration of the worst situation to analyze the voltage characteristics of load and range of power shortage that can ensure the isolated system's stable operation.

The active power and reactive power of load must be matched with the output of generator 'G_1'. It can be expressed in formulas (8) and (9).

$$P_{load} = P = \frac{V'^2}{R}$$

(8)

$$Q_{load} = Q = 0$$

(9)

Among above, '*V'*' denotes the voltage across the RLC load in the isolated grid.

The voltage '*V*' and '*V'*' across the load before and after '*S1*' is switched off can be deduced in formulas (10) and (11).

$$V = \sqrt{P_{load}R} = \sqrt{Q_{load} / (\frac{1}{\omega L} - \omega C)}$$

(10)

$$V' = \sqrt{PR} = \sqrt{Q / (\frac{1}{\omega L} - \omega C)}$$

(11)

It can be seen from formulas (10) and (11) that if $\Delta P > 0$, then $V' < V$, and if $\Delta P < 0$, then $V' > V$. And supposing the load frequency remains unchanged it can be seen that if $\Delta Q > 0$, then $V' < V$ and if $\Delta Q < 0$, then $V' > V$.

It can be summarized that the change of voltage across the load is related to the unbalanced degree of active power or unbalanced reactive power. When the active power or reactive power is insufficient in the isolated grid, the voltage across the load will decrease. When the active power and reactive power supplied is greater than the need of the load, the voltage across the load will increase.

2.2 Range of unbalanced active power

According to the voltage characteristics discussed above, the range of unbalanced active power to ensure the stable operation of isolated grid can be figured out through the allowable voltage range.

Before the 'S_1' is switched off, the '$\Delta P/P$' can be expressed in formula (12).

$$\frac{\Delta P}{P} = \frac{P_{load} - P}{P} = \frac{V^2}{RP} - 1$$

(12)

After the 'S_1' is switched off, the voltage across the load V' equals to $\sqrt{PR}$. Considering the range of '*V'*' which is between 'V_{min}' and 'V_{max}', the range of unbalanced active power can be deduced in formula (13).

$$\left(\frac{V}{V_{max}}\right)^2 - 1 \leq \frac{\Delta P}{P} \leq \left(\frac{V}{V_{min}}\right)^2 - 1 \qquad (13)$$

Among above, 'V' denotes the voltage across the load, which is 1 pu when the system is operating normally. 'V_{min}' and 'V_{max}', respectively, is the minimum and maximum voltage that the local power system can operate stably after the forming of the isolated grid. 'V_{min}' equals to 0.8 pu and 'V_{max}' equals to 1.2 pu. 'P' denotes the total active power output of the generator in the isolated grid. It can be calculated in formula (14).

$$-30.5\% \leq \frac{\Delta P}{P} \leq 56.25\% \qquad (14)$$

It's obvious that when the ratio of unbalanced active power and the total active power output satisfies formula (14), voltage instability will not arise in the isolated grid. In the actual power system, active power output of generators will change with the unbalanced active power. 'P' in formula (14) represents the actual active power output.

2.3 Range of unbalanced reactive power

Unbalanced reactive power can be expressed as:

$$\frac{\Delta Q}{P} = \frac{Q_{load}}{P} = \left(\frac{V}{V'}\right)^2 Q_f \left(\frac{f_0}{f} - \frac{f}{f_0}\right) \qquad (15)$$

After 'S_1' is switched off, considering the range of 'f_0' which is between 'f_{min}' and 'f_{max},' the range of unbalanced reactive power can be deduced in formula (16).

$$\left(\frac{V}{V_{max}}\right)^2 Q_f \left(\frac{f_{min}}{f} - \frac{f}{f_{min}}\right) \leq \frac{\Delta Q}{P}$$
$$\leq \left(\frac{V}{V_{min}}\right)^2 Q_f \left(\frac{f_{max}}{f} - \frac{f}{f_{max}}\right) \qquad (16)$$

Where $f = 50$ Hz–the frequency of normal power system, $f_{min} = 48$ Hz–the minimum frequency that the isolated grid can operate stably, and $f_{max} = 51.5$ Hz, maximum frequency that the isolated grid can operate stably.

It can be seen from formula (16) that the unbalanced reactive power is related to voltage and frequency. If the change of voltage and frequency exceeds the permitted range, high/low frequency protection and high/low voltage protection will operate and the isolated grid cannot operate stably. If the change of voltage and frequency is

within the allowed limits, isolated grid can operate at a new static operation point through their own regulatory mechanism.

Supposing 'Q_f' equals to 1.5, the unbalanced reactive power can be expressed in the formula (17).

$$-8.5\% \leq \frac{\Delta Q}{P} \leq 38.7\% \qquad (17)$$

It can be seen from formulas (14) and (17) that the effect of unbalanced reactive power to voltage is more sensitive than the effect of unbalanced active power. Generally, the unbalanced degree of reactive power to ensure the stable operation of the isolated grid is lower than the unbalanced degree of active power without regard to the power adjustment ability of grid itself. Hierarchical and regional balance of reactive power should be satisfied to the greatest extent in the arrangement of the operation mode. It can avoid the undervoltage problem of the local power system and improve the security, stability, and economy of the power system operation.

2.4 Influence factors to voltage stability

In the winter summit load operation mode of Sanhua (Center China, North China, and East China) synchronized power grid which has "two vertical and two horizontal" Ultra-High Voltage (UHV) structure, Hunan power system receives about 5800 MW active power and 220 Mvar reactive power, the receiving ratio of active power is about 27%. Simulation by the Power System Full Dynamic Simulation (PSD-FDS) shows that the risk of voltage collapse in the operation mode is large, though the receiving ratio is only 27%. So it's necessary to analyze other influencing factors. Using Hunan power system disconnected from the major system as an example, it analyzes the effect of the three factors (Load model, leap of load, and excitation system operation mode) to the voltage stability.

2.4.1 Load model

Two synthesis load models are studied in this paper (Walve 1986). Type I model is constituted by 65% dynamic load and 35% static load, Type II model is constituted by 50% dynamic load and 50% static load. Hunan power system acquiescently uses the Type I mode. In the simulation, Type 1 model is compared with Type II and static load model. The static load model is constituted of 30% constant reactance and 40% constant current and 30% constant power. The synthesis load model is constituted of dynamic load and static load (constant reactance and constant current). Effect of load model to the voltage stability is shown in Table 1.

Table 1. Effect of load model.

Load model	Minimum value of 500 kV steady-state voltage (pu)	Stable or not
Static load model	>0.93	Stable
Synthesis load model I	>0.88	Stable
Synthesis load model II	>0.89	Stable

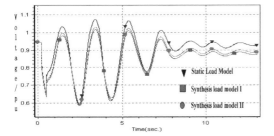

Figure 2. The minimum voltage with different load model.

Table 2. Effect of leap of load.

Increment of load (%)	Minimum value of 500 kV steady-state voltage (pu)	Stable or not
2	>0.87	Stable
4	>0.85	Stable

It can be seen that the higher the proportion of the dynamic load, the greater is the effect to the voltage stability. This is determined by starting characteristics of synthesis load induction motor and the undervoltage response characteristics (active power will decrease and reactive power will increase).

The minimum voltage is corresponding to different load model after the fault is shown in Figure 2.

2.4.2 Leap of load
Considering the load increasing in 2% or 4% in 30 s, the effect on voltage stability is shown in Table 2.

It can be seen from Table 2 that the greater the increment of load, the greater is the effect to voltage stability, and the voltage stability becomes worse.

2.4.3 Excitation system operation mode
Considering three modes including constant machine voltage, constant reactive power, and constant power factor, the effect of excitation sys-

Table 3. Effect of excitation system operation mode.

Excitation system operation mode	Minimum value of 500 kV steady-state voltage (pu)	Stable or not
Constant generator terminal voltage	>0.88	Stable
Constant reactive power	>0.87	Stable
Constant power factor	>0.88	Stable

tem operation mode to voltage stability is shown in Table 3.

It can be seen that after the disconnection fault, the effect of constant machine voltage mode and constant power factor mode to voltage stability is better than constant reactive power mode.

3 COORDINATION STRATEGIES OF UNDERVOLTAGE LOAD SHEDDING AND CONCENTRATED LOAD SHEDDING

Undervoltage load shedding (Balanathan 1998, Imai 2005, Liu Zhiyang 2009, Ma Shiying & Yi Jun 2009 & Sun Huadong et al. 2009) refers to the measure of load shedding through relevant undervoltage devices according to the undervoltage characteristics in the third defensive line of power system. The concentrated load shedding refers to the initiative measures of load shedding through stability control devices according to the serious fault feature in the second defensive line of power system. In china, the Standard *Guide on Security and Stability for Power System* stipulates that undervoltage load shedding should coordinate with other security control measures but without specific principle. In general, the range of load controlled by undervoltage load shedding is wider and it's easy to over shedding, and the load controlled by concentrated load shedding has higher pertinence. If the stability control devices can split load appropriately to keep the voltage in the normal level, the possible over shedding could be avoided and the range of blackout could be reduced.

The coordination strategies of undervoltage load shedding and concentrated load shedding mainly includes three aspects.

1. Non-overlapping load

Concentrated load shedding is more effective to solve the problem of insufficient active power and it aims at high voltage lines which are not tie-lined in the load center. And undervoltage load shedding is more effective to solve problem of insufficient reactive power. It aims to shed load in the region where generator output is deficient and reactive power support is weak, or the region where the transferred power flow has passed.

2. Action preset value and delay time.

Concentrated load shedding is a part of the second defensive line of power system and it should act antecedently. The third defensive line is initiated only if the second defensive line cannot control and reverse the emergency situation. Concentrated load shedding is commonly initiated by criterions of fault and undervoltage load shedding is commonly initiated by criterions of voltage. Shedding load in an isolated region may make the undervoltage load shedding act later to keep more loads connected. The range of unbalanced degree of power deduced above can be a guide to coordinate the two shedding ways.

3. Target of voltage stability

Concentrated load shedding aims at N-2 faults and the recuperative voltage is commonly above 0.95 pu. Undervoltage load shedding aims at the serious cascading faults and the recuperative voltage is commonly lower, such as 0.8 pu or 0.9 pu, because the voltage of some nodes in the power system may fail to a lower level due to the serious faults which is hard to recover.

4 CONCLUSION

When local receiving power system operates as an isolated grid, undervoltage problems may occur if active power or reactive power is unbalanced. Considering the allowable range of voltage and frequency, the range of unbalanced degree of active power calculated is between -30.5% and 56.25% and the range of unbalanced degree of reactive power calculated is between -8.5% and 38.7%. However, unbalanced degree in real power grid is smaller than the percent. Simulation verified that voltage stability is affected by load model, leap of load, and excitation system operation mode etc. In order to keep the voltage stability in an isolated grid, concentrated load shedding and undervoltage load shedding should be coordinated.

REFERENCES

Balanathan R. 1998. Under voltage load shedding to avoid voltage instability, Proceedings of IEEE T&D, 145(2):175–181

Huang He, Xu Guanghu & Yu Chang 2008. Operation Experience of Isolated Grid in CSG during Ice Disaster. Southern Power System Technology, 2(5):6–9

Huazhong Power System 7.1 Accident Investigation Team 2006. Huazhong '7.1' Accident Analysis Report.

Imai S. 2005. Under voltage load shedding improving security as reasonable measure for extreme contingencies. Proceedings of IEEE PES: Vol2, June 12–16: 1754–1759.

Liu Zhiyang 2009. Research on Undervoltage Load Shedding for Regional Isolated Grid. School of Electrical Engineering and Automation, Harbin Institute of Technology. 6.

Ma Shiying, Yi Jun & Sun Huadong et al. 2009. Study on Configuration of Undervoltage Load shedding in Power System. Automation of Electric Power Systems. 33(5):45–49.

North American Electric Reliability Corporation (NERC) 2010. Power plant and transmission system protection coordination.

Walve K. 1986. Modeling of power system components at severe disturbances. Proceedings of CIGRE: 18–38, August 27 ~ September 4, Paris, France.

Advances in Power and Energy Engineering – Sun (Ed.)
© 2016 Taylor & Francis Group, London, ISBN 978-1-138-02846-3

Impact of VSC–HVDC integration on the rotor angle stability of power systems

T.Z. Pan

Institute of Electrical Engineering, Chinese Academy of Sciences, Beijing, China
University of Chinese Academy of Sciences, Beijing, China

X.S. Tang

Institute of Electrical Engineering, Chinese Academy of Sciences, Beijing, China

ABSTRACT: With the integration of large-scale VSC–HVDC in power systems, the traditional synchronous generator will encounter a huge challenge in dynamics and transient characteristics, due to the fast electromagnetic process of VSC–HVDC. This paper presents an approach to analyze the impact of VSC–HVDC integration on the rotor angle stability of power systems. First, a simplified two-machine model is put forward according to the transient characteristics of VSC–HVDC. Second, the transient stability of power systems caused by VSC–HVDC is analyzed with extended equal area criterion. Additionally, the current angle instability) of VSC–HVDC and its influence on the power system's transient angle stability are discussed. Finally, the proposed method is verified through digital simulations of a two-machine power system.

1   INTRODUCTION

With the rapid growing of offshore Wind Farms (WFs) as well as the rapid evolution of power electronics technology, the VSC–HVDC system, also called HVDC light or HVDC plus, has gained wide development in recent years (Flourentzou et al. 2009). The world's first VSC–HVDC link was commissioned in Sweden in 1997 (Asplund et al. 1997), and the VSC–HVDC link "Tjaereborg" connecting a WF was built in 2000 (Eriksson 2001). Together with the existing LCC–HVDC links, a hybrid MIDC (HMIDC) system that consists of the VSC–HVDC and LCC–HVDC links can be envisioned in the near future (Feltes et al. 2011). Compared with the conventional MIDC systems with LCC–HVDC links, the VSC–HVDC integration in the power system will have a significant impact on various stability aspects of the power system (Guo & Zhao 2010, Zhao & Sun 2006, Zhong et al. 2008,).

A fully operational MTDC grid with offshore WFs can be regarded as a large (virtual) power plant capable of providing ancillary services to the mainland AC grid (Silva et al. 2012). In this sense, it is expected that MTDC grids can also provide fault ride through (FRT) capability for faults occurring in the mainland AC grid, in line with grid code requirements for onshore wind generators (Tsili & Papathanassiou 2009).

VSCs are based on valves that can be switched on and off by a control signal. By choice of the switching instant, it is possible to generate any desired wave shape. With higher switching frequency components, it is possible to use Pulse Width Modulation (PWM) technology to re-create the AC voltage with any phase angle or voltage amplitude (within a certain limit). Thus, PWM offers the possibility to control both active and reactive power independently. Despite the favorable controllability of VSC–HVDC (Latorre & Ghandhari 2011), the power electronic interface is vulnerable, especially during faults. This manifests itself by the FRT strategy, current limiting, and active power control actions taken to keep the VSC–HVDC scheme connected during and after a disturbance. These phenomena act on the rotor angle stability time frame of interest; thus, it is imperative to study the dynamic behavior of the hybrid AC/DC power system.

The analysis and performance evaluation of different control solutions for the provision of FRT requirements in point-to-point VSC–HVDC systems have been discussed elsewhere (Bianchi et al. 2012, Feltes et al. 2009, Vrionis et al. 2007). In any case, the rotor angle stability is ignored due to the favorable controllability of VSC–HVDC. At the same time, a lot of work has been done to analyze the effects of large-scale integration of wind generators on dynamic stability. The effect of wind power on the oscillations and damping was investigated by gradually replacing the power generated by the synchronous generator in the system with power from either constant or variable speed induction generators (Hagstrøm et al. 2005, Slootweg & Kling 2003).

In addition, a general explanation of the loss of synchronism caused by current angle instability (CAI) and the derivation of the associated stability limits in case of a remaining voltage at the fault location has been given elsewhere (Göksu 2012, Göksu et al. 2014). Different lock mechanisms have different effects on system stability (Wang 2014). Therefore, the rotor angle stability of VSC–HVDC cannot be ignored. In order to facilitate our understanding, a brief introduction about CAI is given as follows:

If the VSC was an uncontrolled AC voltage source, a current adequate to the impedance would flow automatically. However, as the VSC is a controlled voltage source behaving as a current source, the current controller tries to inject a current according to its current set points (for active and reactive currents). If the resulting current angle of the current set points fits with the impedance angle, a stable operation is ensured ('current angle stability'). If the resulting current angle of the current set points does not fit with the impedance angle, the current controller cannot find a steady-state operating point. Thus, the current angle moves and with it the local voltage angle, which is affected by the current injection.

As the stability of wind generators is always a concern of rotor angle stability (Hagstrøm et al. 2005, Slootweg & Kling 2003), this paper investigates the rotor angle stability of VSC–HVDC integration. A simplified two-machine model power system is developed based on the Extended Equal Area Criterion (EEAC) method to analyze its transient stability. For the first time, the impact on rotor angle stability during the conventional fault and the CAI fault in the AC system is put forward.

The paper is organized as follows: Section II develops a simplified two-machine model according to the transient characteristics of VSC–HVDC and describes the analysis methodology used in the study. Section III presents and discusses the results from the analysis. Section IV demonstrates the operation of the simplified two-machine model by time simulations from the electromagnetic transient simulation software PSCAD/EMTDC. Finally, Section V concludes this paper.

2 SYSTEM MODEL AND TRANSIENT STABILITY MECHANISM

2.1 *Simplified two-machine model*

This paper focuses on the rotor angle stability of the synchronous generator with VSC–MTDC integration in a power system. If the

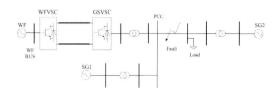

Figure 1. Simplified two-machine power system model with VSC–HVDC integration.

multi-machine system is a two-cluster swing mode when the fault occurs, it can be regarded as a two-machine system. In this way, the conclusion of the simplified two-machine model can be extended to the multi-machine power system. In order to demonstrate the impact of VSC–HVDC integration on the rotor angle stability of power systems based on the EEAC method, a simplified two-machine model was considered, as shown in Figure 1.

Power flow is sent from synchronous generator 1 (SG1) to Synchronous Generator 2 (SG2). VSC–HVDC connected to the AC grid at PCC between two synchronous generators and SG1 is relatively close. It also includes an offshore WF. The wind generator is connected to the WF VSC (WFVSC). The DC side of WFVSC is connected to the DC side of Grid-Side VSC (GSVSC) through DC lines. Then, the AC side of GSVSC is connected to the AC grid generator. The load bus is between SG1 and SG2. Furthermore, a three-phase-to-ground fault occurs on a transmission line between the load and SG1.

We consider the synchronous generator for the classic second-order model, and ignore the prime mover, speed control system and excitation system dynamic, while setting the load linear and the network linear. Fault point is set in the vicinity of the VSC–HVDC access point.

2.2 *Transient stability analysis method*

There is no rotor angle stability problem of VSC–HVDC due to its fast electromagnetic control process (Prabhu & Padiyar 2009). Therefore, this paper focuses on the synchronous generator's rotor angle stability with VSC–HVDC integration in power systems. Transient stability analysis can use the transient energy function method and the time-domain simulation method, and the two methods are combined in this paper. First, a qualitative analysis is carried out for a two-cluster swing mode based on the EEAC method.

For the power system without VSC–HVDC, when subjected to a large disturbance, we assume that the system unstable mode is the two-machine swing mode and the system Unstable Equilibrium

Point (UEP) is known. The seriously disturbed group is referred to as the S group; and the rest of the group is referred to as the Y group. Their inertia coefficient is M_S and M_Y, and the rotor angle is δ_S and δ_Y. The mechanical input power and the electrical output power are P_m and P_e, respectively.

For a single synchronous generator, the motion equation is given as follows:

$$M\ddot{\delta} = P_m - P_e \tag{1}$$

where M is the inertia coefficient; δ is the rotor angle; P_m is the mechanical input power; and P_e is the electrical output power. Then, the equations of motion inertia center S and Y are given by

$$M_S\ddot{\delta}_S = P_{mS} - P_{eS} \tag{2}$$

$$M_Y\ddot{\delta}_Y = P_{mY} - P_{eY} \tag{3}$$

For the above motion equations,

$$M = M_T^{-1} M_S M_Y \tag{4}$$

$$P_m = M_T^{-1}(M_Y P_{mS} - M_S P_{mY}) \tag{5}$$

$$P_e = M_T^{-1}(M_Y P_{eS} - M_S P_{eY}) \tag{6}$$

$$M_T = \sum_{i=1}^{n} M_i \tag{7}$$

When the multi-machine system stability is explored, it should be in accordance with the minimum criteria defining the most dangerous power angle image. The earliest arrival angle instability image is regarded as a simple dominant image. Stability dominant image determines the stability of the multi-machine system. With regard to energy, the dominant image is equivalent to the kinetic energy, which will alter the stability characteristics of the system. Therefore, the dominant image's equivalent kinetic energy P_{ac} is a key factor to determine the transient stability.

The equivalent kinetic energy of the dominant image is given by

$$P_{ac} = \frac{M_a \Delta P_S - M_S \Delta P_a}{M_S + M_a} \tag{8}$$

Since this paper explores the impact of large-scale VSC–HVDC integration on power systems, the synchronous machine and VSC–HVDC are regarded as disturbed groups.

3 EEAC ANALYSIS

In Figure 2, since the VSC–HVDC will not change the system inertia, P_m is assumed to be constant throughout the transient process. At the same time, VSC–HVDC tracks the phase and frequency of the power system with PLL, which has no influence on the accelerating power. Therefore, there is no effect of VSC–HVDC on the rotor angle stability of power systems. Moreover, VSC–HVDC can provide reactive power support during the fault period, and also fast power recovery after system voltage restoration. Therefore, system rotor angle stability can be significantly improved, as shown in Figure 2.

It is shown that when the reactive current, which is required by the grid codes, is injected, VSC–HVDC can lose its synchronism with the grid fundamental frequency and, thus, PLL frequency decreases. It simulates the drop in the equivalent generator speed and the electromagnetic power, which, in turn, increase the acceleration area that is detrimental to the transient stability of the system. Following the clearance of the fault, the VSC–HVDC recovers the normal frequency. In addition, electromagnetic power is restored. In the case of CAI, when only the first swing is considered, the deceleration area is reduced due to the fact that electromagnetic power cannot be restored immediately. It is detrimental to the rotor angle stability of the power system. However, when the case of multiple swings is considered, the deceleration area increases with the electromagnetic power recovery. It improves the rotor angle stability of the power system after swinging repeatedly, as shown in Figure 3.

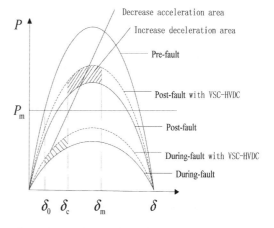

Figure 2. EEAC analysis of the VSC–HVDC-based power system.

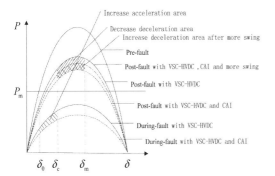

Figure 3. EEAC analysis of the VSC–HVDC-based power system when considering CAI.

Table 1. Main parameters used in the PSCAD model.

Main parameters	Value
Main grid size	150 MW
SG1 size	100 MW
SG2 size	30 WM
VSC–HVDC size	20 MW
Rated frequency of the main grid	50 HZ
Rated frequency of the WF	50 HZ
Fault duration	200 ms
DC voltage	20 kV

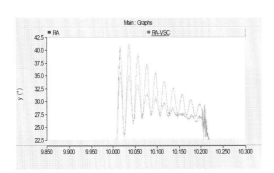

Figure 4. Rotor angle curve with and without VSC–HVDC when the fault occurs.

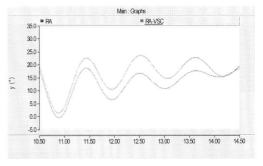

Figure 5. Rotor angle curve with and without VSC–HVDC after the fault is cleared.

4 SIMULATION AND VERIFICATION

To demonstrate the impact of VSC–HVDC integration on the rotor angle stability of power systems, the simplified two-machine power system model shown in Figure 4 is simulated in the time simulation software PSCAD/EMTDC.

A three phase-to-ground fault is imposed, as shown in Figure 1. The impact on rotor angle stability is observed for two kinds of fault situations, namely the conventional fault and the CAI fault, with the fault duration of 200 ms. Initially, the island system is connected to a VSC–HVDC system with constant frequency. In this mode, the impact of VSC–HVDC integration on the rotor angle stability of power systems is discussed. The initial condition is as follows. The total load in the system is 150 MW. The SG1 supplies 100 MW, while the VSC–HVDC supplies 20 MW with active and reactive power control actions. The rest of the loads are balanced by SG2, i.e. SG2 supplies 30 MW. The main parameters used in the PSCAD model are presented in Table 1.

In order to facilitate our understanding, we define the conventional power system case as the power system without VSC–HVDC; the VSC–HVDC case as the power system with VSC–HVDC; and the CAI case as the power system with VSC–HVDC when considering CAI.

In Figure 4, the power angle of the power system with and without VSC–HVDC is compared when the fault occurs. In the VSC–HVDC, the VSC provides reactive power to the system, and also provides the necessary dynamic support services. After the fault clearance, shown in Figure 5, the VSC–HVDC case is able to return to a stable operating state quickly. Therefore, it improves the rotor angle stability of the power system.

In Figure 6, the conventional power system case is compared with the CAI case when the fault occurs. The simulation results indicate that in the CAI case, VSC–HVDC integration has no influence on the rotor angle stability of the power system. Following the clearance of the fault, shown in Figure 7, when considering the first swing case, the swing angle of the CAI case is larger than that of the conventional power system case. It is

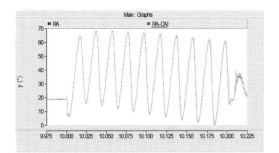

Figure 6. Rotor angle curve with and without VSC–HVDC when the fault occurs when considering CAI.

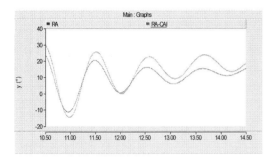

Figure 7. Rotor angle curve with and without VSC–HVDC after the fault is cleared when considering CAI.

detrimental to the rotor angle stability of the power system. However, when considering the multiple-swing case, the CAI case is able to return to a stable operating point quickly. Therefore, it has both detrimental and beneficial effects on the rotor angle stability of the power system.

5 CONCLUSION

In this paper, the impact of VSC–HVDC integration on the transient stability of the power system is explored, especially the rotor angle stability of the traditional synchronous generator. A simplified two-machine model power system is developed based on the EEAC method to analyze its transient stability. Two kinds of fault situations are considered in the AC system. The analysis and the simulation results based on the EEAC method demonstrate that VSC–HVDC can improve the rotor angle stability. However, in the CAI case, VSC–HVDC has no influence on rotor angle stability when the fault occurs, and when the fault is cleared, it has both detrimental and beneficial effects on rotor angle stability. It

is detrimental to the stability of the first swing. However, after multiple swings, the rotor angle stability is improved. Time-domain simulation results of the PSCAD using the two-machine model also demonstrate the impact of VSC–HVDC and CAI.

REFERENCES

Asplund, G. et al. 1997. DC transmission based on voltage source converters. *In CIGRE SC14 Colloquium*, South Africa: 1–7.

Bianchi, F. D, et al. 2012. Optimal control of voltage source converters under power system faults. *Control Engineering Practice*, 20(5), 539–546.

Eriksson, K. 2001. Operational experience of HVDC LightTM. *Seventh International Conference on AC and DC Transmission*:205–210.

Feltes, C. et al. 2009. Enhanced fault ride-through method for wind farms connected to the grid through VSC-based HVDC transmission. *Power Systems, IEEE Transactions on*, 24(3), 1537–1546.

Feltes, J.W. et al. 2011. From smart grid to super grid: Solutions with HVDC and FACTS for grid access of renewable energy sources. *In Power and Energy Society General Meeting*: 1–6. IEEE.

Flourentzou, N. et al. 2009. VSC-based HVDC power transmission systems: An overview. *Power Electronics, IEEE Transactions on*, 24(3), 592–602.

Göksu, Ö. 2012. Control of wind turbines during symmetrical and asymmetrical grid faults. Doctoral dissertation, *PhD thesis, Aalborg University, Department of Energy Technology*, Aalborg, Denmark.

Göksu, Ö. et al. 2014. Instability of wind turbine converters during current injection to low voltage grid faults and PLL frequency based stability solution. *Power Systems, IEEE Transactions on*, 29(4), 1683–1691.

Guo, C. & Zhao, C. 2010. Supply of an entirely passive AC network through a double-infeed HVDC system. *Power Electronics, IEEE Transactions on*, 25(11), 2835–2841.

Hagstrøm, E. et al. 2005. Large-scale wind power integration in Norway and impact on damping in the Nordic grid. *Wind Energy*, 8(3), 375–384.

Latorre, H. F, & Ghandhari, M. 2011. Improvement of power system stability by using a VSC-HVDC. *International Journal of Electrical Power & Energy Systems*, 33(2), 332–339.

Prabhu, N. & Padiyar, K.R. 2009. Investigation of sub-synchronous resonance with VSC-based HVDC transmission systems. *Power Delivery, IEEE Transactions on*, 24(1), 433–440.

Silva, B. et al. 2012. Provision of inertial and primary frequency control services using offshore multiterminal HVDC networks. *Sustainable Energy, IEEE Transactions on*, 3(4), 800–808.

Slootweg, J.G. & Kling, W.L. 2003. The impact of large scale wind power generation on power system oscillations. *Electric Power Systems Research*, 67(1), 9–20.

Tsili, M. & Papathanassiou, S. 2009. A review of grid code technical requirements for wind farms. *IET Renewable Power Generation*, 3(3), 308–332.

Vrionis, T.D. et al. 2007. Control of an HVDC link connecting a wind farm to the grid for fault ride-through enhancement. *Power Systems, IEEE Transactions on*, 22(4), 2039–2047.

Wang. Z. 2014. Analysis on Impact of Doubly Fed Induction Generations with Different Phase Lock Mechanism on Power System Small Signal Stability. *Proceedings of the CSEE*, 34(34), 6167–6176.

Zhao, C. & Sun, Y. 2006. Study on control strategies to improve the stability of multi-infeed HVDC systems applying VSC-HVDC. *In Electrical and Computer Engineering, CCECE'06. Canadian Conference on*, 2253–2257. IEEE.

Zhong, Q. et al. 2008. Study of HVDC Light for its enhancement of AC/DC interconnected transmission systems. *In Power and Energy Society General Meeting-Conversion and Delivery of Electrical Energy in the 21st Century*, 2008 IEEE: 1–6. IEEE.

Advances in Power and Energy Engineering – Sun (Ed.)
© 2016 Taylor & Francis Group, London, ISBN 978-1-138-02846-3

Analysis of the reliability of multi-circuit-on-same-tower power transmission systems

Y.J. Sun, Y.H. Zhang, Y.C. Zhang, X.H. Qin & Q.Y. Zhou
China Electric Power Research Institute, Beijing, China

ABSTRACT: Multi-circuit-on-same-tower power transmission technologies, which can help save transmission corridor sources, are often used in areas with scarce land sources or special geographical environments. The interferences between lines are serious and fault types are complex, which results in the particularity of system reliability and security in transmission systems. Since more power will be lost when multiple lines fail simultaneously, which may have an influence on the power systems, problems related to increasing voltages have received attention. Using the reliability assessment method, based on the separation model, this paper provides probability formulas for multi-circuit-on-same-tower systems, considering different numbers of element failure and various fault types. According to the outage frequency and repair time in practice, the reliability of two-circuit-on-same-tower, four-circuit-on-same-tower, and four-circuit-on-two-tower power transmission systems is compared. It is demonstrated that the reliability of multi-circuit-on-same-tower power transmission systems declines with the number of lines increasing and the reliability of four-circuit-on-same-tower power transmission lines is lower than that of four-circuit-on-two-tower power transmission lines.

1 INTRODUCTION

The multi-circuit-on-same-tower power transmission technology can save power corridor sources and improve the economy of power grid construction (Xu, 2001). In many countries, it has been often used to reduce the cost and satisfy the power transmission needs.

In China, it has been proposed that multi-circuit-on-same-tower power transmission technologies and the compact transmission technology (Liao & Li 2006, Meng 2003, Guo 2010, Shi 2012, Zhao et al. 2007, Zhao 2008, Zhang 2010,) can help save transmission corridor sources, reducing the investment cost and meeting the requirements of large-capacity power transmission systems. In addition, an upgrade of some of the existing lines in the process of power grid construction may also result in multi-circuit-on-same-tower power transmission systems with the same voltage grade or different voltage grades.

Studies on multi-circuit-on-same-tower power transmission technologies include the structure of towers and poles, the lightning-impulse withstand level, the differentiation of line protection faults, the selection of insulator types, the environmental impact, the characteristics of operation and maintenance (Zhang & Ge 2005) and insulation, lightning protection, and electromagnetic compatibility (Huang et al. 2006). At present, related

research carried out in China includes the study of arc current and recovery voltage (Cao et al. 2010), induced current and voltage (Ban et al. 2009), phase-order arrangement (Wang et al. 2009), the measurement to improve the transmission capacity (Zhou et al. 2001), fault calculation (Liu & Fan 2010, Suo et al. 1991), parameter correction method (Shu & Fan 2010), phase mode transformation (Zhang et al. 2009), protection (Jin et al. 2009), fault location (Yu et al. 2009), and electromagnetic environment simulation (Zhang et al. 2010). However, there is very little literature on the reliability of multi-circuit-on-same-tower power transmission systems. Gao et al. (2010) discussed the influence of the ratio of variable load effect to permanent load effect on the reliability of Chinese EHV/UHV multi-circuit-tower structure by structure reliability theories. Wang & Zhou (2003) deduced a reliability valuation unity model for two transmission lines in parallel by utilizing the theory of Markov stochastic process. Lin et al. (2008) proposed a reliability evaluation model for two transmission lines in parallel based on the separation model by considering the independent outage, the common cause outage, and the dependent outage. However, the general formulas and methods for analyzing the reliability of multiple-circuit-on-same-tower power transmission systems have not been proposed in the research mentioned above.

The multiple circuit power transmission lines on the same tower are used at the voltage of 500 kV or below at abroad. China has built 750 kV EHV and 1000 kV UHV power transmission lines. When the voltage grade is high, faults occurring in the multi-circuit-on-same-tower power transmission systems may produce an extremely serious influence. Therefore, the study on the reliability of multi-circuit-on-same-tower power transmission systems is of great significance for power grid planning and operation.

This paper focuses on the reliability of multi-circuit-on-same-tower power transmission systems. Using the reliability assessment method based on the separation model, the probability formulas of different types of faults in multi-circuit-on-same-tower power transmission systems are proposed. According to the statistical data, this paper evaluates and compares the probability of different types of faults in two-circuit-on-same-tower, four-circuit-on-same-tower, and four-circuit-on-two-tower power transmission systems. The conclusions obtained may be used as a reference for the practical power grid planning and construction.

2 RELIABILITY ANALYSIS OF MULTI-CIRCUIT-ON-SAME-TOWER POWER TRANSMISSION SYSTEMS

2.1 Outage model of multi-circuit-on-same-tower power transmission systems

2.1.1 The independent outage model of one single element

The independent outage model of different elements in N-circuit-on-same-tower power transmission systems is shown in Figure 1.

λ_i is the outage rate for the ith element and μ_i is the repair rate for the ith element, where $i = 1, 2, 3, 4, ..., N$.

2.1.2 The common cause outage model of multiple components

The common cause outage model of K different elements in N-circuit-on-same-tower power transmission systems is shown in Figure 2.

$\lambda_{KC1,2,... K}$ is the common cause outage rate of the 1st, 2nd,...Kth element and $\mu_{KC1,2,... K}$ is the repair

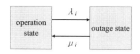

Figure 1. The independent outage model of different elements in N-circuit-on-same-tower power transmission systems.

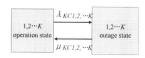

Figure 2. The common cause outage model of K different elements in N-circuit-on-same-tower power transmission systems.

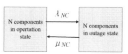

Figure 3. The common cause outage model of different elements in N-circuit-on-same-tower power transmission systems.

rate for the common cause outage of the 1st, 2nd,...Kth element.

2.1.3 The common cause outage model of N components

The common cause outage model of different elements in N-circuit-on-same-tower power transmission systems can be described in Figure 3.

λ_{NC} is the common cause outage rate of all the N elements and μ_{NC} is the repair rate for the common cause outage of all the N elements in N-circuit-on-same-tower power transmission systems.

2.2 The formulas for calculating the outage probability of multi-circuit-on-same-tower power transmission systems

2.2.1 The basic formulas for calculating the outage probability of different types of faults

The basic formulas for calculating the outage probability of different types of faults can be given based on the above outage model:

1. The independent outage probability of the ith element can be given by the following formula:

$$P_{1i}$$
$$= \left(\frac{\lambda_i}{\lambda_i + \mu_i} \right) \left(\prod_{\substack{k=1 \\ k \neq i}}^{N} \frac{\mu_i}{\lambda_i + \mu_i} \right) \left(\sum_{\substack{\forall k, j \in \{N\} \\ k \neq j}} \frac{\mu_{2Ckj}}{\lambda_{2Ckj} + \mu_{2Ckj}} \right)$$
$$\cdot \left(\sum_{\substack{\forall k, j, m \in \{N\} \\ k \neq j \neq m}} \frac{\mu_{3Ckjm}}{\lambda_{3Ckjm} + \mu_{3Ckjm}} \right) \cdots \left(\frac{\mu_{NC}}{\lambda_{NC} + \mu_{NC}} \right)$$
$$\tag{1}$$

where $\{N\}$ is a collection of all the N elements.

2. The concurrent independent outage probability of the 1st, 2nd, 3rd, ... Kth element can be given by the following formula:

$$
\begin{aligned}
P_{KI1,2,3\ldots K} &= \left(\prod_{i\in\{K\}} \frac{\lambda_i}{\lambda_i+\mu_i}\right)\left(\prod_{j\in\{N\}-\{K\}} \frac{\mu_j}{\lambda_j+\mu_j}\right) \\
&\cdot\left(\sum_{\substack{\forall k,j\in\{N\}\\k\neq j}} \frac{\mu_{2Ckj}}{\lambda_{2Ckj}+\mu_{2Ckj}}\right) \\
&\cdot\left(\sum_{\substack{\forall k,j,m\in\{N\}\\k\neq j\neq m}} \frac{\mu_{3Ckjm}}{\lambda_{3Ckjm}+\mu_{3Ckjm}}\right) \\
&\cdots\left(\frac{\mu_{NC}}{\lambda_{NC}+\mu_{NC}}\right)
\end{aligned}
$$

(2)

where $\{K\}$ is a collection of the *1*st, *2*nd, *3*rd, ..., Kth element in which independent outages occur at the same time; $\{N\}$ is a collection of the N elements; and $\{N\}-\{K\}$ is a collection of the remaining elements in which no outages occur.

3. The common cause outage probability of the *1*st, *2*nd, *3*rd, ..., Kth element can be given by the following formula:

$$
\begin{aligned}
P_{KC1,2,3,\cdots K} &= \left(\frac{\lambda_{KC1,2,3\cdots K}}{\lambda_{KC1,2,3\cdots K}+\mu_{KC1,2,3\cdots K}}\right)\left(\prod_{j=1}^{N} \frac{\mu_j}{\lambda_j+\mu_j}\right) \\
&\cdot\left(\sum_{\substack{\forall k,j\in\{N\}\\k\neq j}} \frac{\mu_{2Ckj}}{\lambda_{2Ckj}+\mu_{2Ckj}}\right) \\
&\cdot\left(\sum_{\substack{\forall k,j,m\in\{N\}\\k\neq j\neq m}} \frac{\mu_{3Ckjm}}{\lambda_{3Ckjm}+\mu_{3Ckjm}}\right) \\
&\cdots\left(\sum_{\substack{\forall i_1,i_2,\cdots i_K\in\{N\}\\i_1,i_2,\cdots i_K\notin\{K\}\\i_1\neq i_2\neq\cdots\neq i_K}} \frac{\mu_{KCi_1,i_2,\cdots i_K}}{\lambda_{KCi_1,i_2,\cdots i_K}+\mu_{KCi_1,i_2,\cdots i_K}}\right) \\
&\cdots\left(\frac{\mu_{NC}}{\lambda_{NC}+\mu_{NC}}\right)
\end{aligned}
$$

(3)

where $\{K\}$ is a collection of the *1*st, *2*nd, *3*rd, ... Kth element in which common cause outages occur; $\{N\}$ is a collection of the N elements; and $\{N\}-\{K\}$ is a collection of the remaining elements in which no outages occur.

4. The probability of independent outages of the *i1*st, *i2*nd,...i_Kth element and a common cause outage of the *j1*st, *j2*nd, ... j_Lth element at the same time can be given by the following formula:

$$
\begin{aligned}
P_{KIi_1 i_2\cdots i_K,LCj_1 j_2\cdots j_L} &= \left(\prod_{i\in\{K\}} \frac{\lambda_i}{\lambda_i+\mu_i}\right)\left(\frac{\lambda_{LCj_1 j_2\cdots j_L}}{\lambda_{LCj_1 j_2\cdots j_L}+\mu_{LCj_1 j_2\cdots j_L}}\right) \\
&\cdot\left(\prod_{j\in\{N\}-\{K\}} \frac{\mu_j}{\lambda_j+\mu_j}\right)\cdot\left(\sum_{\substack{\forall k,l\in\{N\}\\k\neq l}} \frac{\mu_{2Ckl}}{\lambda_{2Ckl}+\mu_{2Ckl}}\right) \\
&\cdot\left(\sum_{\substack{\forall k,l,m\in\{N\}\\k\neq l\neq m}} \frac{\mu_{3Cklm}}{\lambda_{3Cklm}+\mu_{3Cklm}}\right) \\
&\cdots\left(\sum_{\substack{\forall i_1,i_2,\cdots j_L\in\{N\}\\i_1,i_2,\cdots j_L\notin\{L\}\\i_1\neq i_2\neq\cdots\neq i_L}} \frac{\mu_{LCi_1 i_2\cdots i_L}}{\lambda_{LCi_1 i_2\cdots i_L}+\mu_{LCi_1 i_2\cdots i_L}}\right) \\
&\cdots\left(\frac{\mu_{NC}}{\lambda_{NC}+\mu_{NC}}\right)
\end{aligned}
$$

(4)

where $\{K\}$ is a collection of elements $i_1, i_2,\ldots i_K$ in which independent outages occur at the same time; and $\{L\}$ is a collection of elements $j_1, j_2,\ldots j_L$ in which a common cause outage occurs.

5. The independent outage probability of one arbitrary element can be given by the following formula:

$$
P_{1I} = \sum_{i=1}^{N} P_{1Ii}
$$

(5)

6. The independent outage probability of K arbitrary elements can be given by the following formula:

$$
P_{KI} = \sum_{\forall i_1, i_2, \cdots, i_K \in \{N\}} P_{KIi_1, i_2, \cdots, i_K}
$$

(6)

7. The common cause outage probability of K arbitrary elements can be given by the following formula:

$$
P_{KC} = \sum_{\forall i_1, i_2, \cdots, i_K \in \{N\}} P_{KCi_1, i_2, \cdots, i_K}
$$

(7)

8. The probability of independent outages of K arbitrary elements and a common cause outage of L arbitrary elements can be given by the following formula:

1. Suppose that an independent outage and a common cause outage will not occur in one

same element at the same time, then the probability can be given by the following formula:

$$P_{KILC} = \sum_{\substack{\forall i_1, i_2, \cdots, i_K \in \{N\} \\ \forall j_1, j_2, \cdots, j_L \in \{N\} - \{K\}}} P_{KIi_1, i_2, \cdots, i_K, LCj_1, j_2, \cdots, j_L}$$

(8)

2. Suppose that an independent outage and a common cause outage may occur in one same element, then the probability can be given by the following formula:

$$P_{KILC} = \sum_{\substack{\forall i_1, i_2, \cdots, i_K \in \{N\} \\ \forall j_1, j_2, \cdots, j_L \in \{N\}}} P_{KIi_1, i_2, \cdots, i_K, LCj_1, j_2, \cdots, j_L}$$

(9)

9. The independent outage probability of the N elements can be given by the following formula:

$$P_{NI}$$

$$= \left(\prod_{i=1}^{N} \frac{\lambda_i}{\lambda_i + \mu_i} \right) \left(\sum_{\substack{\forall k, j \in \{N\} \\ k \neq j}} \frac{\mu_{2Ckj}}{\lambda_{2Ckj} + \mu_{2Ckj}} \right)$$

$$\cdot \left(\sum_{\substack{\forall k, j, m \in \{N\} \\ k \neq j \neq m}} \frac{\mu_{3Ckjm}}{\lambda_{3Ckjm} + \mu_{3Ckjm}} \right) \cdots$$

$$\cdot \left(\sum_{k=1}^{N} \frac{\mu_{(N-1)C1,2,\cdots k-1,k+1,\cdots N}}{\lambda_{(N-1)C1,2,\cdots k-1,k+1,\cdots N} + \mu_{(N-1)C1,2,\cdots k-1,k+1,\cdots N}} \right)$$

$$\cdot \left(\frac{\lambda_{NC}}{\lambda_{NC} + \mu_{NC}} \right)$$

(10)

10. The common cause outage probability of the N elements can be given by the following formula:

$$P_{NC}$$

$$= \left(\frac{\lambda_{NC}}{\lambda_{NC} + \mu_{NC}} \right) \left(\prod_{i=1}^{N} \frac{\mu_i}{\lambda_i + \mu_i} \right)$$

$$\cdot \left(\sum_{\substack{\forall k, j \in \{N\} \\ k \neq j}} \frac{\mu_{2Ckj}}{\lambda_{2Ckj} + \mu_{2Ckj}} \right)$$

$$\cdot \left(\sum_{\substack{\forall k, j, m \in \{N\} \\ k \neq j \neq m}} \frac{\mu_{3Ckjm}}{\lambda_{3Ckjm} + \mu_{3Ckjm}} \right) \cdots$$

$$\cdot \left(\sum_{k=1}^{N} \frac{\mu_{(N-1)C1, 2, \cdots k-1, k+1, \cdots N}}{\lambda_{(N-1)C1, 2, \cdots k-1, k+1, \cdots N} + \mu_{(N-1)C1, 2, \cdots k-1, k+1, \cdots N}} \right)$$

(11)

2.2.2 Calculation formulas for the outage probability of different numbers of elements

The outage probability of one arbitrary element, P_1, is equal to the sum of the independent outage probability of any one element of the N ones.

$$P_1 = P_{1I}$$

(12)

The outage probability of two arbitrary elements, P_2, can be defined as the sum of the independent outage probability of any two elements, respectively, at the same time, and the common cause outage probability of any two elements:

$$P_2 = P_{2I} + P_{2C}$$

(13)

The outage probability of three arbitrary elements, P_3, can be defined as the sum of the probability of the following situations: independent outages occur in three elements, respectively, at the same time, a common cause outage occurs in three elements and an independent outage occurs in one element and a common cause outage occurs in two elements at the same time:

$$P_3 = P_{3I} + P_{3C} + P_{1I2C}$$

(14)

By analogy, the outage probability of K arbitrary elements, P_K, can be defined as the sum of the probability of the following situations: independent outages occur in K elements, respectively, at the same time, a common cause outage occurs in K elements, an independent outage occurs in one element and a common cause outage occurs in $(K-1)$ elements, independent outages occur in two elements, respectively, at the same time and a common cause outage occurs in $(K-2)$ elements, independent outages occur in $(K-2)$ elements, respectively, at the same time and a common cause outage occurs in two elements:

$$P_K = P_{KI} + P_{KC} + P_{1I(K-1)C} + P_{2I(K-2)C} + \cdots + P_{(K-2)I2C}$$

(15)

The outage probability of the N elements, P_N, can be defined as the sum of the probability of the following situations: independent outages occur in N elements, respectively, at the same time, a common cause outage occurs in the N elements, an independent outage occurs in one element and a common cause outage occurs in $(N-1)$ elements, independent outages occur in two elements, respectively, at the same time and a common cause outage occurs in $(N-2)$ elements, independent outages occur in $(N-2)$ elements, respectively, at the same time and a common cause outage occurs in two elements:

$$P_N = P_{NI} + P_{NC} + P_{1I(N-1)C} + P_{2I(N-2)C}$$
$$+ \cdots + P_{(N-2)I2C} \tag{16}$$

It should be noted that the calculation formulas given in this section have considered all the possible cases, and if there are some impossible situations in practice, the corresponding part should be excluded in the calculation formulas.

2.3 Reliability analysis of multi-circuit-on-same-tower power transmission systems

This section analyzes and compares the reliability of two-circuit-on-same-tower, four-circuit-on-same-tower, and four-circuit-on-two-tower power transmission systems.

Supposing that the one circuit failure frequency in two-circuit-on-same-tower, four-circuit-on-same-tower, and four-circuit-on-two-tower power transmission systems is 1.5 times per year, the repair time is 20 h per time, and then the average repair time T_{MTTRI} is equal to 30 h. The repair rate of one circuit can be calculated as follows:

$$\frac{8760}{T_{MTTR1}} = 292 \ times/year \tag{17}$$

Supposing that the concurrent failure frequency of two circuits in two-circuit-on-same-tower, four-circuit-on-same-tower, and four-circuit-on-two-tower power transmission systems is 1.5 times per

year, the repair time is 200 h per time, and then the average repair time T_{MTTR2C} is equal to 300 h. The repair rate of two circuits can be calculated as follows:

$$\frac{8760}{T_{MTTR2C}} = 29.2 \ times/year \tag{18}$$

Supposing that the concurrent failure frequency of four circuits in four-circuit-on-same-tower power transmission systems is 0.15 times per year, the repair time is 300 h per time, and then the average repair time T_{MTTR4C} is equal to 45 h. The repair rate of four circuits can be calculated as follows:

$$\frac{8760}{T_{MTTR4C}} = 194.7 \ times/year \tag{19}$$

Supposing that the concurrent failure frequency of four circuits in four-circuit-on-two-tower power transmission systems is 0.015 times per year, the repair time is 300 h per time, and then the average repair time T_{MTTR4C} is equal to 4.5 h. The repair rate of four circuits can be calculated as follows:

$$\frac{8760}{T'_{MTTR4C}} = 1947 \ times/year \tag{20}$$

Based on the above assumptions and the fault probability calculation formulas, the comparison result for the outage probability of different

Table 1. The comparison of the outage probability of different types of faults in two-circuit-on-same-tower, four-circuit-on-same-tower, and four-circuit-on-two-tower power transmission systems.

Fault type	Outage probability of two-circuit-on-same-tower power transmission systems	Outage probability of four-circuit-on-same-tower power transmission systems	Outage probability of four-circuit-on-two-tower power transmission systems
An independent outage of one line	$P_{1I} = 0.004731$	$P'_{1I} = 0.05341$	$P''_{1I} = 0.01780$
Independent outages of two lines	$P_{2I} = 0.00002378$	$P'_{2I} = 0.0004026$	$P''_{2I} = 0.0001342$
Independent outages of three lines	–	$P'_{3I} = 1.34878 \times 10^{-6}$	$P''_{3I} = 4.4959 \times 10^{-7}$
Independent outages of four lines	–	$P'_{4I} = 1.69445 \times 10^{-9}$	$P''_{4I} = 5.6482 \times 10^{-10}$
A common cause outage of two lines	$P_{2C} = 0.04851$	$P'_{2C} = 0.1369$	$P''_{2C} = 0.09128$
A common cause outage of two lines with an independent outage of one line	–	$P'_{2C1I} = 0.001376$	$P''_{2C1I} = 0.0004587$
A common cause outage of two lines with independent outages of two lines	–	$P'_{2C2I} = 6.9148 \times 10^{-6}$	$P''_{2C2I} = 1.1525 \times 10^{-6}$
A common cause outage of two lines with a common cause outage of two lines	–	$P'_{2C2C} = 0.007055$	$P''_{2C2C} = 0.002352$
Common cause outage of four lines	–	$P'_{4C} = 0.0020472$	$P''_{4C} = 0.0006824$

Table 2. The comparison of the outage probability of different quantitative element faults in two-circuit-on-same-tower, four-circuit-on-same-tower and four-circuit-on-two-tower power transmission systems.

Fault type	Outage probability of two-circuit-on-same-tower power transmission systems	Outage probability of four-circuit-on-same-tower power transmission systems	Outage probability of four-circuit-on-two-tower power transmission systems
Outage of one line	$P_1 = 0.004731$	$P'_1 = 0.05341$	$P''_1 = 0.01780$
Outage of two lines	$P_2 = 0.048535$	$P'_2 = 0.137319$	$P''_2 = 0.09141186$
Outage of three lines	–	$P'_3 = 0.00137739$	$P''_3 = 0.00045913$
Outage of four lines	–	$P'_4 = 0.00910866$	$P''_4 = 0.00303507$

types of faults in two-circuit-on-same-tower, four-circuit-on-same-tower, and four-circuit-on-two-tower power transmission systems is summarized in Table 1. The comparison result for the outage probability of different quantitative elements in two-circuit-on-same-tower, four-circuit-on-same-tower, and four-circuit-on-two-tower power transmission systems is summarized in Table 2.

The results indicate that the outage probability of a multi-circuit-on-same-tower power transmission system increases as the line numbers increase.

In fact, it is difficult to obtain the failure frequency accurately as there are only a few four-circuit-on-two-tower power transmission systems in operation. Thus, the deviation of specific assumptions may lead to an influence on the reliability results.

The method proposed in this paper can be used to calculate the outage probability of different numbers of element failure and different types of faults in multi-circuit-on-same-tower power transmission systems. Combined with a more accurate and reasonable element failure frequency and repair time, more detailed reliability analysis results of multi-circuit-on-same-tower power transmission systems can be further obtained.

3 CONCLUSION

By using the reliability assessment method based on separation model, this paper analyzes the reliability of multi-circuit-on-same-tower power transmission systems and provides general probability calculation formulas for different types of faults in multi-circuit-on-same-tower power transmission systems. Then, according to the outage frequency and the repair time in practice, the paper evaluates and compares the reliability of two-circuit-on-same-tower, four-circuit-on-same-tower, and four-circuit-on-two-tower power transmission lines. It is pointed out that the reliability of multi-circuit-on-same-tower power transmission systems declines with the number of power transmission lines increasing. Furthermore, the reliability of four-circuit-on-same-tower power transmission lines is lower than that of four-circuit-on-two-tower power transmission lines.

The method proposed in this paper can be used to calculate the outage probability of different numbers of element failure and different types of faults in multi-circuit-on-same-tower power transmission systems, which can also provide the guidance for the planning, construction, and operation of practical power grids.

REFERENCES

Ban, Liangeng. Wang, Xiaogang. Bai, Hongkun. et al. 2009. Simulative analysis of induced voltages and currents among multi circuit 220 kV and 500 kV transmission lines on same tower. *Power System Technology* 33(6): 45–49.
Cao, Huazhen. Cai, Guanglin. Wang, Jingyi. Wang, Xiaotong. 2010. Influence of Guangdong-500 kV-parallel-four-circuit-transmission lines on secondary-arc current and recovery voltage. *Electric Power* 43(11): 14–19.
Gao, Yan. Yang, Jingbo. & Han, Junke. 2010. Analysis on Structural Reliability of Multi-Circuit Tower for EHV and UHV AC Power Transmission Line. *Power System Technology* 34(9): 181–184.
Guo, Han. 2010. The design of the 500kV four-circuit-on-same-tower transmission lines from ShiYang to WuYi. *Coastal enterprises and science & technology* 7: 121–122.
Huang, Aihua. Zheng, Xu. Qian, Guangzhong. 2006. Application of technology of constructing multiple circuits on the same tower. *East China Electric Power* 34(8): 60–63.
Jin, Honghe. Ren, Mingzhu. Yuan, Cheng. & Tai, Nengling. 2009. Influence of mutual inductance to the distance protection of quadruple-lines on the single tower and the solution. *East China Electric Power* 37(8): 1346–1350.
Liao, Yi. & Li, Minsheng. 2006. Design and implementation of electric power transmission line with 500 kV/220 kV four circuits in the same tower. *Southern Power System Technology Research* 2(6): 34–53.
Lin, Zhimin. Lin, Han. & Wen, Buying. 2008. Reliability evaluation for parallel two-circuit transmission lines by using individual models. *East China Electric Power* 36(7): 34–37.

Liu, Ling. & Fan, Chunju,. 2010. Fault calculation for joint four transmission lines of different voltage grades on the same tower based on six-sequence-component method. *Power System Protection and Control* 38(9): 6–11.

Meng, Yu. 2003. The design and study of 220 kV 3-circuits-on-same-tower transmission lines. *Shanghai Electric Power* 4: 295–297.

Shi, Liuwu & Xi, Xiaoli. 2012. Transposition Tower Selection of 1000 kV UHV AC Double-circuit Transmission Lines on Same Tower. *Electric Power Construction* 33(1): 38–41.

Shu, Qiaojun. & Fan, Chunju. 2010. Parameters' Modification Method to Two-and Four-jointed Parallel Lines. *Modern Electric Power* 27(1): 45–48.

Suo, Nan. Ge, Yaozhong. & Tao, Huiliang. 1991. A new fault analysis method of the double circuit line on the same pole (DLSP) by use of the interconnection of six sequence networks. *Proceedings of the EPSA* 3(1): 92–106.

Wang, Xiaotong. Lin, Jiming. Ban, Liangeng. et al. 2009. Selection of phase sequence arrangement for Guangdong 500 kV power transmission line adopting structure of four circuits on the same tower. *Power System Technology* 33(19): 87–91.

Wang, Shao. & Zhou, Jia-qi. 2003. A Reliability evaluation model for two transmission lines in parallel. *Proceedings of the CSEE* 23(9): 53–56.

Xu, Jianguo. 2001. Investigation and Analysis on Transmission Line Technique of EHV Multiple-circuit on the Same Tower Abroad. *Electric Power Construction* 22(7): 15–18.

Yu, Sheng. Xu, Gang. Yu, Qiong. & Xu, Bin. 2009. Fault location for multiple circuit overhead lines of different voltage grade on the same tower. *Power System Protection and Control* 37(6): 44–47.

Zhang, Jiamin. & Ge, Rongliang. 2005. Features and application of power transmission technology of multi-circuit lines on the same tower. *East China of Electric Power* 33(7): 23–26.

Zhang, Qibing. Tai, Nengling, Yuan, Cheng. Jin, Honghe. LIN, Wei. 2009. Phase-mode Transformation of Four-parallel Lines on the Same Tower. *Proceedings of the CSEE* 29(34): 57–62.

Zhang, Xiao. Jia, Zhenhong. Wu, Suoping. Wu, Jianhong. Wang, Jianmin. Chen, Guang. & Zhou, Hao. 2010. Simulation Analysis on Electromagnetic Environment of Quadruple-Circuit Transmission Lines Belonging to Different Voltage Classes on the Same Tower. *Power System Technology* 34(5): 207–211.

Zhang, Yong. 2010. The application of multi-circuit-on-same-tower transmission lines in HuaiNan mining area, *Power and Energy* 17: 192–194.

Zhao, Quan-jiang. 2008. Discussion on 500 kV Triple-Circuit Transmission Line Design. *Electric Power Construction* 29(1): 13–16.

Zhao, Quan-jiang. XIE, Bang-hua. & Xu, Wei-yi. 2007. Study on Conductor Selection of the 750 kV Double-circuit Transmission Tower. *Electric Power Construction* 28(2): 11–15.

Zhou, Xiaoxin. Guo, Jianbo. Hu, Xuehao. & Tang, Yong. 2001. Engineering technologies and measures for improving the transmitting capability of 500 kV transmission lines. *Power System Technology* 25(3): 1–6.

Advances in Power and Energy Engineering – Sun (Ed.)
© 2016 Taylor & Francis Group, London, ISBN 978-1-138-02846-3

A bus voltage automatic optimization and adjustment method based on AIMMS software

P. Xu, F. Li, Y. Wang & F. Shi
China Electric Power Research Institute, Nanjing, China

A.A. Ni
Shanghai University of Electric Power, Shanghai, China

ABSTRACT: In the increasingly complicated power system, the bus voltage violation problem is becoming more and more remarkable, which makes it urgent to design an effective and economical approach to solve it. This paper proposes an easy and agile algorithm to optimize and adjust the generator bus voltage automatically on the basis of the AIMMS software. This algorithm computes the adjustment volume of generator bus voltages through the sensitivity and optimization theory and combines the available dispatching technical support system (D5000) with the AIMMS software by the Python language. The proposed algorithm is validated to be effective and efficient by tests on some provincial power system.

1   INTRODUCTION

In the operation of the power system, the bus voltage may exceed its limits for many reasons, such as dramatic load growth, insufficient power supply, environment changes, and so on, which in turn leads to other problems. The power system is highly non-linear and the relation among its state variables is quite complicated, and any irregular voltage adjustment could lead to more serious buses voltage violation or economic decline of the power grid. Meanwhile, to ensure the difference between the actual voltage of user equipment and its rated voltage remains within a permitted value is one basic task of power system operation (Chen 2007, He et al. 1985, Wu 1985). Therefore, the modern power system badly needs an effective and efficient method of voltage adjustment.

Based on the sensitivity index, the sensitivity method can help adjust the value of the independent variable input, and thus achieve the purpose of controlling the dependent variable output (Luo 1990, Wang & Zhao 1992). Its adjustment and control measures can improve system safety and stability, and therefore is widely used in power system (Begovic et al. 1992, Guo 2007, Liu 1996, Miao &, Zhou 1994, Yuan et al. 1997). However, there still exist many problems when applying the sensitivity index alone, that is, which variable and how much should be adjusted, whether or not these adjustments would cause new problems, which all together make it hard for operators to make a final decision. Keep the above in mind,

the paper progress the sensitivity based bus voltage automatic adjustment method by introducing the optimization theory, which taking both safety and economy into consideration. What's more, the proposed approach is developed on the basis of the existing net-modeling program, power flow calculation program, sensitivity analysis program in the D5000 system and the mature optimized Software (AIMMS), which guarantee the accuracy of results and compatibility with practical system.

2   REQUIREMENT OF SENSITIVITY BASED ON SENSITIVITY MATRIX

2.1   *Acquiring the sensitivity connection between generator bus voltage variation ΔV_G and load bus voltage variation ΔV_D*

The reactive balance formula of bus i:

$$Q_i - V_i \sum_{j \in i} V_j (G_{ij} \sin \theta_{ij} - B_{ij} \cos \theta_{ij})$$
$$\approx Q_i + \sum_{j \in i} V_j B_{ij} = 0 \qquad (1)$$

The above simplified formula is derived from the structure and operation of power system. According to the basic method of sensitivity analysis, we can get the relationship among the changes of various variables after they are disturbed by dropping the higher-order term in the form of Taylor expansion around the present state.

$$\Delta Q_i + \sum_{j \in i} B_{ij} \Delta V_j = 0$$

$$\Delta Q_i = -\sum_{j \in i} B_{ij} \Delta V_j \qquad (2)$$

The load and generator buses are separately arranged and can be written in a matrix form as below:

$$-\begin{bmatrix} B_{DD} & B_{DG} \\ B_{GD} & B_{GG} \end{bmatrix} \begin{bmatrix} \Delta V_D \\ \Delta V_G \end{bmatrix} = \begin{bmatrix} \Delta Q_D \\ \Delta Q_G \end{bmatrix} \qquad (3)$$

Formula 3 shares the same form with the V–Q iteration modification formula in P–Q decoupled load flow method. But we have to notice that ΔQ_D and ΔQ_G are the reactive power variation of loads and generators, namely, Formula 3 indicates a relationship between the changes of control variables and state variables based on the system's new steady and the old steady state.

Suppose that the load reactive power remains unchanged after V_G is adjusted, namely $\Delta Q_D = 0$, the first expression of Formula 3 would be:

$$B_{DD} \Delta V_D + B_{DG} \Delta V_G = 0 \qquad (4)$$

And then transformed it into:

$$\Delta V_D = -B_{DD}^{-1} B_{DG} \Delta V_G = S_{DG} \Delta V_G \qquad (5)$$

where S_{DG} = the sensitivity matrix between ΔV_D and ΔV_G. With the sensitivity matrix S_{DG}, we could know that which generator controls the load bus voltage most effectively.

2.2 Acquiring the sensitivity connection between voltage variation ΔV_G and reactive power variation ΔQ_G of generator bus

Transform Formula 3 into:

$$\begin{bmatrix} \Delta V_D \\ \Delta V_G \end{bmatrix} = -\begin{bmatrix} B_{DD} & B_{DG} \\ B_{GD} & B_{GG} \end{bmatrix}^{-1} \begin{bmatrix} \Delta Q_D \\ \Delta Q_G \end{bmatrix}$$

$$= \begin{bmatrix} R_{DD} & R_{DG} \\ R_{GD} & R_{GG} \end{bmatrix} \begin{bmatrix} \Delta Q_D \\ \Delta Q_G \end{bmatrix} \qquad (6)$$

Suppose that when the generator reactive power varies, the load reactive power could remain the same, namely $\Delta Q_D = 0$, then:

$$\Delta V_G = R_{GG} \Delta Q_G \qquad (7)$$

where R_{GG} = a sensitivity matrix between ΔV_G and ΔQ_G. With the sensitivity matrix R_{GG}, we could know that which generator is more sensitive between its reactive power and voltage.

3 OPTIMIZATION OF MATHEMATICAL MODEL

In the optimization mathematical model, the objection function is to keep the sum of all the generator buses voltage variation to the minimum, and the constraint condition is the generator bus voltage adjustment, generator reactive power and all bus voltages in the upper and lower limits. With the inequality constraints, direct application of the sensitivity results, namely "sensvq_gen" and "sens," avoids the tedious modeling procedure and improves the efficiency of optimization calculation by getting rid of the active and reactive power balance equations of the power grid. But it must be cautious that it would be certain deviation on the substitute of the simplified linear sensitivity results to the complicated nonlinear power flow equations. Plenty research and practical operation experience shows that, under the normal operation condition, the power flow equation is closely linear around the working point, since the deviation results are permitted to meet engineering demand (Wang & Zhao 1992). The optimizations of mathematical model are as below:

$$obj. \quad min\left(\sum_i dvg(i) \cdot dvg(i)\right)$$

$$s.t. \quad dvgmin(i) \le dvg(i) \le dvgmax(i)$$
$$rmn\_gen(i) \le r0\_gen(i)$$
$$+ dvg(i)/sensvq\_gen(i) \le rmx\_gen(i)$$
$$Vmin\_elc(e) \le v0\_elc(e)$$
$$+ \sum_i dvg(i) \cdot sens(i,e) \le Vmax\_elc(e)$$
$$(i = 1, 2, \ldots, m \quad e = 1, 2, \ldots, n) \qquad (8)$$

where $dvg(i)$, $dvgmax(i)$, $dvgmin(i)$ = the practical adjustment, the maximum and minimum of a generator bus voltage; $r0\_gen(i)$, $rmx\_gen(i)$, $rmn\_gen(i)$ = a current volume, the maximum and minimum limits of a generator reactive power; $v0\_elc(e)$, $Vmin\_elc(e)$, $Vmax\_elc(e)$ = the current volume and the maximum and the minimum limits of a bus voltage; $sensvq\_gen(i)$ = the sensitivity between a generator bus voltage variation and its reactive power variation; $sens(i,e)$ = the sensitivity between a generator bus voltage variation and a load voltage variation; m = the number of adjustable voltage generators; and n = the number of buses.

4 AIMMS SOFTWARE MODELING

1. Establish a text that could be directly read by AIMMS includes three composite tables: bus information table, generator information table, and sensitivity information table.

2. AIMMS software model instruction
   a. Establish main execution, a main item for controlling the inner optimized calculation procedure of AIMMS, which includes three aspects of data reading, optimization solving and result writing.
   b. Establish parameter vector, set vector, constraint vector and variable vector under the declaration of the optimization issue.

5 OPERATION PROCEDURE

This approach, relying on "glue language" Python which composes available net-modeling program, power flow program, sensitivity analysis program of D5000 system and commercial optimization software AIMMS into an integral whole, achieves an automatic adjustment of bus voltage which exceed its limit.

Calculation procedure:

1. Read the reference data of a network model, and establish network topology model by using net-modeling program.
2. Calculate the power flow by power flow program and judge whether the bus voltage exceeds its limit or not. If not, then stop any calculation or continue step 3.
3. Calculate the sensitivity connection between a generator bus voltage variation and a load bus voltage variation, the sensitivity connection between the reactive power variation and voltage variation of a generator by using sensitivity analysis program.
4. An optimized data text can be directly read by AIMMS relying on the calculation result of sensitivity of step 3 and the network parameter of step 1.
5. Execute the AIMMS interface function.
6. Adopt the AIMMS software to execute optimized calculation. (Before the first calculation, it is necessary to get an optimized modeling inside the AIMMS software.)
7. Adjust the generator voltages according to the result of the AIMMS optimized calculation.

The net-modeling program, the power flow calculation program, the sensitivity analysis program of D5000 system and the AIMMS interface function in the first five steps, are all based on C++, but step 6 is the optimization procedure based on AIMMS software; and step 7 is still established on the basis of C++. All the steps are coherently taken relying on Python language, as shown in Figure 1.

6 ACTUAL SAMPLE ANALYSES

An operating mode after fault under an extremely bad weather condition of ice disaster was simulated on the Zhejiang power grid in 2013. An ice covering caused a sequence of breaks of several 500 kV and 220 kV lines in a short time in some local districts, which caused several bus voltages exceed their lower limit as shown in Table 1.

Table 1 shows that the bus voltages of several power stations were lowered to different degrees, and among them, the 220 kV bus voltage of Tianyi power station was lowered by 8.8 kV and its 500 kV bus voltage lowered by 15 kV, both exceeding their lower limits, and therefore, it is the power station that needs a priority optimization adjustment.

Table 2 lists the generators, whose sensitivity to 500 kV.I bus voltage of Tianyi power station after fault is the top 14. It is obvious that No. 1 generator of Beilun station is the most sensitive to solving the 500 kV bus voltage violation problem of Tianyi power station, namely the sensitivity is 1.37, and the final optimization result also indicates that its generator terminal voltage has the highest magnitude adjust value of 0.52. The adjust values of other generator units have proportional counter relation to their sensitivity. Meanwhile, it can testify if the optimization could meet the constraints

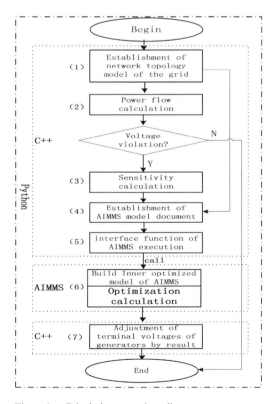

Figure 1. Calculation procedure diagram.

Table 1. Bus voltage contrast before and after fault.

Bus name	Voltage before fault	Voltage after fault	Limit	Off-limit percentage (%)
Tianyi power station 500 kV.I/II	514.7	499.4	505	1.1
Fuchunjiang power station 220 kV.I/II	225.1	223.4	225	0.7
Tianyi power station 220 kV.I/II	227.8	219	220	0.45
Wangsong power station 220 kV.I/II	226.4	224.4	225	0.22
......				

Table 2. Parameters of generators and its sensitivity to 500 kV.I bus voltage of Tianyi station after fault.

Generator	v0_gen	r0_gen	rmx_gen	rmn_gen	sensvq_gen	sens	dvg
Beicang station							
#1	18.519	88	250	0	0.01	1.37	0.52
#2	19.177	149	300	−80	0	0.81	0.31
#3	19.610	162	323	−80	0	0.93	0.36
#4	19.155	157	323	−80	0	0.93	0.35
#6	26.148	240	484	−130	0.01	0.55	0.21
#7	26.156	236	484	−130	0.01	0.55	0.21
Zhenghai station							
#3	13.948	72	133	0	0.01	0.77	0.29
#5	13.772	59	133	0	0.01	0.58	0.31
#6	13.682	50	133	0	0.01	0.58	0.31
Qiangjiao station							
#1	18.466	95	280	−110	0	0.73	0.28
#2	18.571	122	280	−110	0	0.73	0.28
#3	18.546	111	280	−110	0	0.73	0.28
#4	18.572	120	280	−110	0	0.73	0.28
Langxi station							
#3	19.455	43	180	−40	0.01	0.83	0.32

Table 3. Conditions of bus voltage violation before and after fault.

Bus name	Before the adjustment	After the adjustment	Limit
Shanxi power plant/10.5 kV.I	9.867	9.907	10
Shanxi power plant/10.5 kV.II	9.867	9.907	10
Jiaxin 2 power plant/500 kV.I	517.632	520.077	520
Jiaxin 2 power plant/500 kV.II	517.632	520.077	520

depending on the other information provided by the other generator units, and No. 1 generator of Beilun, for example, $r0\_gen(i) + dvg(i)/sensvq\_gen(i) = 88 + 0.52/0.01 = 140$, which could meet the needs for the upper and lower limits of the reactive power of the generator unit.

The calculation of the power flow by using the optimized generator bus voltages shows that all the bus voltages whose limits were exceeded in Table 1 have returned to the normal ranges, with the exception of four new bus voltage violation emerging, including two 10.5 kV buses and two 500 kV buses, such as the Table 3. However, the violation rate of the four buses is negligible, compared with the situation before. The bus violation problem was greatly improved.

7 CONCLUSION

This paper puts forward a simple and effective approach to adjust and optimize the bus voltage automatically based on the AIMMS software. The method combines the sensitivity analysis and the optimization theory, which realizes the combination of the D5000 program and the commercial software. Finally, the proposed algorithm is verified to be effective and efficient by the tests on Zhejiang Power grid.

REFERENCES

Begovic, M.M., et al. 1992. Control of voltage stability using sensitivity analysis. *IEEE Trans on Power Systems* 7(1): 114–123.

Chen H. 2007. *Power systems steady-state analysis*. Beijing: China Electric Power Press.

He, Y.Z. et al. 1985. *Power systems steady-state analysis*. Wuhan: Huazhong University of Science & Technology Press.

Liu Y.C. 1996. Research of direct corrective control of the abnormal voltage problem in power systems. *Marine Engineering College* 22(4): 101–104.

Luo, J. 1990. *Introduction to system sensitivity theory*. Xi'an: Northwestern Polytechnical University Press.

Miao, F.X. & Guo, Z.Z. 2007. A survey of sensitivity technique and its application in power systems analysis and control. *RELAY* 35(15): 72–76.

Wang, E.Z. & Zhao, Y.H. 1992. *Sensitivity analysis and power flow calculation of power network*. Beijing: China Machine Press.

Wu, J.S.1985. *Power system static* security *assessment*. Shanghai: Shanghai Jiao Tong University Press.

Yuan, J. et al. 1997. Summarization of the sensitivity analysis method of voltage stability in power systems. *Power System Technology* 21(9): 7–10.

Zhou, E.Z. 1994. Functional Sensitivity Concept and Its Application to Power System Damping Analysis. *IEEE Trans on PWRS* 9(1): 518–524.

Advances in Power and Energy Engineering – Sun (Ed.)
© 2016 Taylor & Francis Group, London, ISBN 978-1-138-02846-3

Research on Demand Side Management

M.L. Dong, D.H. You, J. Hu, G. Wang, C. Long, F. Zhang, Z. He & L. Dai
State Key Laboratory of Advanced Electromagnetic Engineering and Technology,
Huazhong University of Science and Technology, Wuhan, Hubei Province, China

ABSTRACT: Demand Side Management (DSM) is the development trend of power systems under new circumstances, such as contradiction between electricity demand growth and constraints of resources and environment, electricity market reform and large-scale renewable energy integration. Based on DSM researches and experiences around the world, a comprehensive research on DSM is conducted, including basic concepts of DSM, its influence on the power system, as well as the key points during its implementation process. Comprehensive benefits of DSM are analyzed. Suggestions and assumptions of developing DSM in China are finally proposed.

1 INTRODUCTION

With the rapid development of social economy, the increasing electricity demand will require further expansion of the power system. However, the expansion is faced with multiple constraints, such as energy, environment, land, capital, et al. Therefore, the demand side resource has become an alternative to relieve the bottleneck on supply side resources.

The pressure of fossil energy depletion, energy conservation and emissions reduction makes the permeability of wind power and other renewable energy power increase continually, but the uncertainty of power generation also increases as a result. It is not economical to simply rely on the generation side to ensure safe operation of the power system. Thus, demand side assistance is helpful to guarantee the system reliability and improve the power system's ability of accommodating renewable energy.

The electricity market reform enriches the electricity trading patterns, such as Time of Use (TOU) rate and Demand Side Bidding (DSB), to make it possible for demand side to participate in the market (Wang Xifan et al, 2014).

The concept of DSM arises in such a condition. The definitions and classifications of DSM, Efficiency Power Plant (EPP) and Demand Response (DR) are introduced in this paper. Based on extensive researches on DSM implementation at home and abroad, the influence of DSM on power system is analyzed, including participation in power system planning, scheduling and production simulation. And then key points during the implementation process of the DSM projects

are discussed. Finally the comprehensive benefits are summarized, and suggestions on DSM development in China are proposed.

2 DEFINITIONS AND EXPERIENCES OF DEMAND SIDE MANAGEMENT

2.1 DSM

DSM (Wang Mingjun 2005, Zeng Ming 2000), including energy efficiency management and load management, can be defined as methods to improve electricity efficiency or modify electricity consumption patterns of end-use customers, through technical means, economic means, guide means and administrative means. It can reduce peak demand and fuel consumption to achieve the purpose of saving resources and protecting the environment, while meeting the requirement of power supply reliability at the same time.

The main tasks of DSM include:

1. load management: adopting load adjustment measures to change the clients' electricity consumption patterns, like reducing peak electricity demand, or transferring part to off-peak time, in order to flatten the demand curve, improve the utilization rate of equipment during the off-peak time and the system operation reliability. Classic forms like: peak clipping and valley filling and load shifting are included.

2. electricity efficiency management: taking effective incentives to encourage users to take advantage of more advanced technologies to improve the efficiency of terminal equipment, reduce power consumption and save energy.

The main methods of DSM include technology means, economic means, administrative means, and guide means.

Not only is DSM beneficial for participants in the programs, it also plays an active role in the electricity market, system stability and social benefits. Customers participating in DR programs can expect savings in electricity bills and receive incentive payments if they participate in incentive programs. For power enterprises, the equipment utilization rate is improved, the short-term spare capacity is increased and high investments in expansion of generation, transmission and distribution networks can be avoided or deferred. For power system, due to the participation of customers, operators have more choice and resources to maintain system reliability without forced shutdown, so thus the risk of power outages is reduced, and the system reliability is improved. Electricity market can perform more fair and transparently by reducing the ability of main market players to exercise power in the market. What's more, an overall electricity price reduction is expected eventually because of a more efficient utilization of the available infrastructure. Environmental benefits of DSM programs are numerous, including better land utilization, air and water quality improvement and reduction of natural fuel depletion.

2.2 EPP

EPP has been in pilot phase in China since it was introduced as an innovation mode of DSM in 2005. EPP is a virtual power plant (Liu Jizhen 2014), which is to reduce the terminal electricity consumption by implementing packages of DSM programs. An EPP has lower construction cost than a conventional power plant, without land occupation and fuel consumption, so it is helpful to achieve the goal of energy conservation and emissions reduction. EPP concept vividly depicts the role of DSM project, simplifies the comparison and selection between supply side and demand side resources, and makes cost-effective DSM projects more likely to be taken into practice.

Niu Wenjuan (2014) built models of incentive-based and price-sensitive Demand Response based Virtual Power Plant (DR-VPP) were built separately based on the different mechanism. Based on the analysis on electricity-saving potential in the aspects of saved electricity quantity and peak load shifting, Wang Yanling (2014) established an electricity-saving potential optimization model.

2.3 DR

DR (Albadi M.H. 2008, Zhang Qin 2008) is the further extension of DSM, which can be seen as kind of load management in an open electricity market. DR refers to customers' response to change their inherent electricity consumption patterns according to the price signal or incentive mechanism.

DR measures can be classified into 2 types: Price-Based Demand Response (PBDR), and Incentive-Based Demand Response (IBDR), according to different response styles.

1. PBDR
   1. TOU: TOU is a kind of price mechanism to reflect the cost difference of power supply during different periods, including peak-valley price, seasonal price, and abundant-dry water price, etc. According to the load characteristics, periods in one day (1 year) can be divided into equal, peak, or valley hours (seasons). Then, use appropriate price signal to guide customers to consume electricity reasonably, like lowering the off-peak hours (seasons) tariff and raising the peak hours (seasons) tariff.
   2. Critical Peak Pricing (CPP): CPP rates include a pre-specified higher electricity usage price superimposed on TOU rates or normal flat rates. CPP prices are used during contingencies or high wholesale electricity prices for a limited number of days or hours per year. Customers are informed in advance (usually a day-head) to adjust their consumption plans.
   3. Real Time Pricing (RTP): RTP is a kind of dynamic pricing mechanism, which can be updated once an hour or even more short. It can accurately reflect the change of power supply cost in each period of one day and effectively convey electricity price signal to customers, through the linkage of electricity price between retail side and wholesale market. Many economists are convinced that RTP is the most optimal pricing mechanism. However, its implementation is faced with technology and cost limit.
2. IBDR
   1. Direct Load Control (DLC): Utilities have the ability to remotely shut down or control participant equipment (such as air conditioners and water heaters) during the peak time on a short notice (usually within 15 min). This kind of programs is of interest mainly to residential customers and small commercial customers.
   2. Interruptible Load (IL): Customers participating in Interruptible Load Programs will receive incentive payments or rate discounts when they are asked to reduce their load to predefined values during the peak time. Those who do not respond will face penalties, depending on the program terms.
   3. DSB: DSB (also called Buyback) programs are those in which consumers bid on specific load reductions in the electricity

whole sale market. A bid is accepted if it is lower than the market price. When a bid is accepted, the customer must curtail his load by the amount specified in the bid or face penalties. Power supply company, electricity retailers and large consumers can participate directly in DSB programs, while small scattered consumers can be involved in through a third party agent (Aggregator). Emergency Demand Response (EDR) and Capacity/Auxiliary Service Plan (CASP) are included.

2.4 *DSM experiences*

The United States is the earliest DR implementation country. Each state has established DR programs and energy saving potential is tremendous. The load quantity cut by DR programs is expected be up to 20% of peak load in 2019 (Commission F.E.R 2009). The Pacific gas and electric power company, and Southern California Edison electric company have been successively adopted interactive system to encourage consumers to actively participate in DR programs and reduce the peak load effectively.

In China, a document named Instructions on Power Demand Side Management Integrated Pilot Cities was issued by the Treasury Department and National Development and Reform Commission (NDRC) in July 2012. Since then, the trial of DSM projects have been undertaken in Beijing, Tangshan, Foshan, and Suzhou, while the trial of DR projects undertaken in Shanghai. In April 2015, the document Notification about Improving Emergency Mechanism and Promoting Power Demand Side Management Projects in Pilot Cities was issued to request strengthening the construction of DSM platform in the pilot cities, which can help encourage consumers to monitor their electricity consumption on-line, and promote energy services. In addition, the document also asked local authorities to improve the electricity pricing mechanism and guarantee the balance between power supply and demand in a more market-oriented form.

For the European Union, DSM projects are different in each member country, because the estimated penetration level of DR technologies, the amount of industrial manageable power, the household load curves differ across Europe. But thanks to policy support for smart meters and electricity market reform at national level, all countries have push forward DSM programs based on TOU or IL pricing mechanisms. A typical example of a program making use of different prices according to the weather is Electricite de France's Tempo tariff. Around 350,000 residential customers and more than 100,000 small business customers use the Tempo tariff. Days are distinguished according to price using a color system

(red, blue, and white), together with an indication of whether the hour is currently one of eight off-peak hours (Torriti J. et al, 2010).

When it comes to Japan, its experience in DSM projects is in the top in Asia, in addition to its high energy efficiency. Elaboration and strong maneuverability are its most important characteristics. The Japanese government has issued energy conservation laws, to formulate standards and measures about energy consumption for energy-inefficient factories, transportation, buildings construction, and other terminal consumption. An officer-research (learn)-enterprise cooperation pattern has been formed, to promote positive interaction between government, scientific research units, and enterprises, and ensure the maneuverability of DSM measures.

3 APPLICATION OF DSM IN POWER SYSTEM

3.1 *In power system planning*

With the steady development of social economy, the demand for electricity is growing day by day. The traditional way to solve the contradiction between electricity supply and demand is to expand power generation, which is expensive and time consuming, and restricted by factors such as resources, capital, and environment. The concept Integrated Resource Planning (IRP) (Joel N. 1997) arises, which utilizes the resources of demand side as an alternative of power supply in the planning optimization modeling, to maximize the power enterprise on the premise of keeping the energy service level (Zhou Jinghong et al, 2010).

Considering that electricity enterprises no longer monopolize generation, transmission and distribution, Hu Zhaoguang (2008) studied Integrated Resources Strategy Planning (IRSP) as the extension of IRP, which means demand side resources is introduced into resources integrated optimization of the planning process in the form of EPP, along with supply side resources across the country, according to the national energy development strategy. The ultimate goal of IRSP is to realize the maximum social benefit on the premise of meeting the demand electricity.

Zeng Ming (2007) & Tan Xiandong (2009) proposed a hybrid integer programming model, in which the EPP and supply side resources are comparably brought into generation expansion planning. Zhu Lan (2014) selected a typical micro grid consisting of a wind turbine, a photovoltaic, a storage battery, a diesel engine and controllable loads, as a case study to prove that the minimum total cost of system planning could be achieved if DR programs actively participated in the planning.

Demand side resources can be taken into consideration not only in the resource planning model, but also in the process of load forecast. Zeng Ming (2013) put forward a method of load forecasting considering impact of demand side resources.

Zheng Jing (2013) developed a two-level planning model of transmission system considering the optimal EPP placement, of which, the upper model took the minimum total cost of transmission system and EPP investment as its objective while the lower model took the minimum cost of EPP investment as its objective. With the interaction between two levels, the optimal transmission system planning scheme was finally determined by the upper model.

3.2 In optimal scheduling and dispatch modeling

Studies on DR programs involved in power system scheduling have become hot, due to DR roles in peak clipping, suppressing fluctuations caused by new energy integration, as well as energy saving and emission reduction.

Ai Xin (2014) proposed an optimal dispatch model considering interruptible load based on the scenario analysis, in which the interruptible loads was treated as virtual power plants compared to conventional plants. The objective function of the model is minimum cost expectation considering load surges and unit faults. The test results show that the optimal dispatch model can effectively handle the uncertainty in the system operation.

Considering subjective randomness of DR programs, Sun Yujun (2014) established a day-ahead scheduling model based on fuzzy chance constraints, in which the fuzzy parameters were used to fit the day-ahead load under the background of TOU, and the IL was regarded as the virtual generating unit to be integrated into the day-ahead scheduling, according to different response mechanisms of different demand side resource types.

With the ever-increasing penetration level of wind power, the impacts of its fluctuation, intermittence and randomness characteristics on power system reliability are becoming more significant, which leads to the increasing peak-shaving pressure of conventional units (Xue Yusheng 2014). Thus, demand side resources can be involved in the scheduling optimization modeling to improve wind utilization.

Wen Jing (2015) & Liu Wenying (2015) used the high energy load as a means of wind power accommodation thanks to its large adjustable capacity and fast response. A multi-objective optimal model based on source-load was established with the objective functions of both the maximum capacity of wind power accommodation and the minimum system operation cost. Song Yihang (2014) brought demand side response and energy storage technology into optimization of power generation scheduling. Based on that, the interval method was utilized to simulate the scene of wind farm, and Ju Liwei (2015) constructed a two-stage scheduling optimization model for wind farm and energy storage system considering DR, with the day-ahead and ultra-short term wind power forecast as random variables.

Recently, the smart grid has entered into an all-round construction period. The traditional distribution network is facing more challenges, due to the increasing amount of Distributed Generation (DG) integration and the popularity of controllable loads and electric vehicles. The C6.11 project team of International Conference on Large High Voltage Electric System (Conference International des Grands Reseaux Electriques, CIGRE) put forward the concept of Active power Distribution Network (ADN) in 2008, which could achieve active control and management of local DG by the use of flexible network topology to manage the power flow. The operation of ADN needs effective DSM programs to realize the integration of DG, renewable energy and mass concentrated generation effectively (Zhao Bo. 2014).

To deal with the uncertainties of intermittent DG and loads in ADS, Gao Yajing et al, (2015) discussed the corresponding probabilistic models of units in ADS, before constructing scenarios. Then, a two-step optimal dispatch model based on multiple scenarios technique was proposed, which consisted of day-ahead dispatch model and real-time dispatch model. The day-ahead dispatch model was responsible to determine purchasing energy for power supply companies, while the real-time dispatch model was used to develop IL programs and energy storage system charged or discharged plans, to ensure the balance of power.

3.3 To improve system reliability

The implement of DSM programs can improve energy efficiency, reduce the peak load demand, which is equivalent to increase the system capacity and improve the power system reliability as a result.

Guan Yibin (2003) adopted Equivalent Load Duration Curve (ELDC) to analyze the influence of DSM on power supply reliability quantitatively. However, ELDC belongs to the analytical method, the calculation process of which would be much more complicated when applied to complex system analysis. Therefore, the sequential Monte Carlo method was presented by Cai Dehua (2015) and tested with the IEEE Reliability Test System (RTS) as an example.

Zhou Jinghong (2010) introduced the approach and the calculation process of a variety of EPPs involved in power system production simulation. Based on the intelligent engineering and utilizing agent technology, Duan Wei (2014) constructed a stochastic production simulation model of EPP, and the Generation Agent (GA), EPP Agent (EPPA) and Coordination Agent (CA) were designed. According to the load condition and the quoted prices of power plants, the CA made generation scheduling and calculated the clearing price, and EPPA and GA adjusted the bidding strategy according to the clearing price.

4 IMPLEMENTATION AND EVALUATION OF DSM

4.1 Studies on consumers' response characteristics

Performance of DSM projects is significantly affected by participant rates and consumers' response. Thus, it is of great significance to fully grasp the load characteristics of consumers and forecast the response of consumers in different industries, in order to encourage users to participate in DSM programs, formulate appropriate DSM measures, and enhance DSM level.

Huang Yuteng (2013) proposed a new combinational load analysis method for DSM, by using the density based clustering to gain the representative load profile of a single customer and clustering customers into different clusters according to the cosine similarity function.

Among all the factors that affect the consumers' power consumption behavior, the impact of price is the most important. Generally, demand price elasticity is used to quantitatively reflect the influence of electricity price changes on consumers' response behavior. Wang Beibei (2014) applied three elasticity measures (price elasticity of demand, elasticity of substitution and arc elasticity) to estimate load impacts under an expected range of prices.

Economic incentive level affects the consumers' load curtailing capacity and response uncertainty. The higher the incentive level is, the higher the user's response enthusiasm (load curtailing ratio) will be, and the smaller the response randomness will be. Wang Beibei (2015) established a power-score incentive mechanism based on uncertainty demand response modeling, which gave feedback to target consumers according to their credit accumulation. The model could help electricity companies adjust the incentive level and target groups for different power shortage scenarios to realize the minimum cost.

4.2 Studies on TOU

TOU is a kind of effective PBDR strategy. Appropriate TOU price can provide consumers with adequate effective price signals, to guide consumers to arrange consumption plans, shift peak load; and reduce the power shortage loss, improve the power supply reliability at the same time.

According to the customers' demand response to multi interval electricity prices, a price elasticity matrix under peak-valley TOU price was derived (Chen Changyang et al, 2014), based on which, a peak-valley TOU price model was built, with the objective of realizing the maximized revenue of power corporation considering the constraint of reliability, power purchase risk and network loss. An optimal TOU electricity price model based on the demand price elasticity was established (Li Chunyan et al, 2015), with the objective to minimize difference between peak and valley peak. Tan Zhongfu (2009) established an optimization model consisting of TOU and IL price, based on the analysis of costs and benefits of power supply companies.

Considering the poor load transferring capability of heavy energy-consuming enterprises as well as their appeal to continually dividing the period of peak-valley TOU price, a stagger peak-valley TOU price mechanism was proposed (Cui Qiang et al, 2015).

4.3 Cost-benefit analysis

DSM projects usually need participation of power generation enterprises, power grid enterprises, consumers and government. Comprehensive analysis of costs and benefits is necessary to balance the interests of all parties, formulate reasonable DSM mechanism, as well as guide participants to make economic decision-making analysis. In other words, cost-benefit analysis is the key point to DSM projects implementation.

Considering uncertain factors and time value of capital, a set of cost-benefit analysis models with evaluation index system were established (Wang Mianbin et al, 2006). By introducing the theory of reliability benefits, the cost-benefit models of participants in interruptible price mechanism were proposed (Xiao Xin et al, 2014).

Li Li (2011) discussed the risk investment portfolio optimization of EPP under various electricity-saving measures in low carbon economy environment. Considering the fact that the income of EPP consists of the income from selling electricity and that from the trading in carbon market, so risks of electricity price fluctuation, carbon trading price fluctuation and electricity-saving fluctuation should be considered during the investment portfolio optimization of EPP. An analytic model for EPP participating in electricity market bidding transaction and the estimation index of its benefit in smart grid were constructed (Li Hongze et al, 2012).

The results showed that not only the participation of EPP in bidding transaction was favorable to electric power consumers and power supply enterprises, but also it contributed to economic and stable operation of power system.

4.4 DSM system design

The development of smart grid provides two-way communication equipment and power transmission technologies based on Advanced Metering Infrastructure (AMI) and Energy Management System (EMS) for DSM, making DR enter into the automatic age (Gao Ciwei 2014, Tian Shiming 2014, & Yan Huaguang 2015).

Tian Shiming (2014) made an investigation of the current situation of domestic and foreign DR from several aspects, such as benefits analysis, incentive mechanism design, support platform technologies, and integration of wind power. The wide-used DR support technologies such as advanced measurement system and power public service platform were introduced.

The research and design of demand side energy efficiency management and demand response system was discussed Yan Huaguang (2015), which helped monitor users' energy consumption data, control operating states of energy utilization system or devices and execute response instructions quickly. The system had been applied in the development and construction of national power demand side management platform and electric energy service management platform of China State Power Grid Corp.

Zeng Ming (2015) proposed a DR safeguard mechanism for ADN, based on the operating characteristics of ADN from the angles of government departments, power companies and consumers, to promote ADN from the whole society level.

5 SUMMARY AND OUTLOOK

DSM has become the solution to problems under new circumstances, such as contradiction between electricity demand growth and constraints of resources and environment, electricity market reform and large-scale renewable energy integration, with its comprehensive benefits as follows:

1. shift peak demand, improve the load rate and the efficiency of the equipment, defer the construction investment in power system;
2. conserve energy and reduce emissions, improve the renewable energy accommodation;
3. enrich the control and adjusting methods of power system operation, mitigate electrical system emergencies, reduce the number of blackouts and improve system reliability;

4. offer a variety of trading patterns, help market participants increase revenue, enhance the fairness and competitiveness of the electricity market by forming a reasonable price mechanism.

According to the development of power industry in China, suggestions are given to develop DSM programs as follows:

1. develop IRP on a broader scale to fully coordinate power supply side and demand side resources, to realize the optimal allocation of resources and produce maximum social benefits;
2. promote the electricity market reform, perfect the electricity price and incentive mechanism to improve the enthusiasm of the participants' response; and develop advanced DR support technologies to achieve the optimal automatic response;
3. make full use of DR projects to improve renewable energy utilization by providing system with auxiliary services, such as frequency modulation, peak shifting and spare capacity.

ACKNOWLEDGMENT

The research leading to this paper was conducted as part of the project Researches on UHVAC supported by National Energy Administration.

REFERENCES

Ai Xin, Zhou Shupeng, Zhao Yuequn. 2014. Research on optimal dispatch model considering interruptible loads based on scenario analysis. *Proceedings of the CSEE*, 34(S1): 25–31 (in Chinese).

Albadi M.H., El-Saadany E.F. 2008. A summary of demand response in electricity markets. *Electric Power Systems Research*, 78(11): 1989–1996.

Cai Dehua, Chen Baixi, Cheng Lefeng, Wang Liguo, Yu Tao. 2015. Effective study about the implementation of demand side management on improving reliability of generation system. *Power System Protection and Control*, 43(10): 51–56 (in Chinese).

Chen Changyang, Hu Bo, Xie Kaigui, Wan Lingyun, Xiang Bin. 2014. A peak-valley tou price model considering power system reliability and power purchase risk. *Power System Technology*, 38(8): 2141–2148 (in Chinese).

Commission F.E.R. 2009. A national assessment of demand response potential.

Cui Qiang, Wang Xiuli, Wang Weizhou. 2015. Stagger peak electricity price for heavy energy-consuming enterprises considering improvement of wind power accommodation. *Power System Technology*, 39(4): 946–952 (in Chinese).

Duan Wei, Hu Zhaoguang, Yao Mingtao, Zhang Jian, Zhou Yuhui. 2014. Power system production simulation considering demand response resource. *Power System Technology*, 38(6): 1523–1528 (in Chinese).

Gao Ciwei, Liang Tiantian, Li Yang. 2014. A survey on theory and practice of automated demand response. *Power System Technology*, 38(2): 352–359 (in Chinese).

Gao Yajing, Li Ruihuan, Liang Haifeng, Zhang Jiancheng, Ran Jinwen. 2015. Two step optimal dispatch based on multiple scenarios technique considering uncertainties of intermittent distributed generations and loads in the active distribution system. *Proceedings of the CSEE*, 35(7): 1657–1665 (in Chinese).

Guan Yibin, Zhao Cuiyu. 2003. Effect of demand side management technology on system reliability. *Automation of Electric Power systems*, 23(9): 63–66 (in Chinese).

Hu Zhaoguang. 2008. Brief introduction of integrated resources strategy planning. *Power DSM*, 10(2): 1–4 (in Chinese).

Huang Yuteng, Hou Fang, Zhou Qin, Fu Bo, Guo Chuangxin. 2013. A new combinational electrical load analysis method for demand side management. *Power System Protection and Control*, 41(13): 20–25 (in Chinese).

Joel N. Swisher G.D.M.J. 1997. *Tools and methods for integrated resource planning: improving energy efficiency and protecting the environment*. Danmark Riso National Laboratory.

Ju Liwei, Yu Chao, Tan Zhongfu. 2015. A two-stage scheduling optimization model and corresponding solving algorithm for power grid containing wind farm and energy storage system considering demand response. *Power System Technology*, 39(5): 1287–1293 (in Chinese).

Li Chunyan, Xu Zhong, Ma Zhiyuan. 2015. Optimal time-of-use electricity price model considering customer demand response. *Proceeding of the CSU-EPSA*, 27(3): 11–16 (in Chinese).

Li Hongze, Wang Bao, Guo Sen. 2012. Analytic model and benefits measurement of efficiency power plant participating in market bidding transaction in smart grid environment. *Power System Technology*, 36(7): 111–116 (in Chinese).

Li Li, Wang Jianjun, Li Ning, Tan Zhongfu, An Jianqiang. 2011. A risk investment portfolio optimization model of energy efficiency power plant based on mean semi-variance theory in low-carbon economy environment. *Power System Technology*, 35(8): 26–29 (in Chinese).

Liu Jizhen, Li Mingyang, Fang Fang, Niu Yuguang. 2014. Review on virtual power plants. *Proceedings of the CSEE*, 34(29): 5103–5111 (in Chinese).

Liu Wenying, Wen Jing, Xie Chang, Wang Weizhou, Liang Chen. 2015. Multi-objective optimal method considering wind power accommodation based on source-load coordination [J]. *Proceedings of the CSEE*, 35(5): 1079–1088 (in Chinese).

Niu Wenjuan, Li Yang, Wang Beibei. 2014. Demand response based virtual power plant modeling considering uncertainty. *Proceedings of the CSEE*, 34(22): 3630–3637 (in Chinese).

Song Yihang, Tan zhongfu, Li Huanhuan, Liu Wenyan. 2014. An optimization model combining generation side and energy storage system with demand side to promote accommodation of wind power. *Power System Technology*, 38(3): 610–615 (in Chinese).

Sun Yujun, Li Yang, Wang Beibei, Su Huiling, Liu Xiaocong. 2014. A day-ahead scheduling model considering demand response and its uncertainty. *Power System Technology*, 38(10): 2708–2714 (in Chinese).

Tan Xiandong, Hu Zhaoguang, Peng Qian. 2009. A Resource Combination Optimization Model Considering Efficiency Power Plan. *Power System Technology*, 33(20): 108–112 (in Chinese).

Tan Zhongfu, Xie Pinjie, Wang Mianbin, Zhang Rong, Qi Jianxun. 2009. The optimal design of integrating interruptible price with peak-valley time-of-use power price based on improving electricity efficiency. *Transactions of China Electro-technical Society*, 24(5): 161–168 (in Chinese).

Tian Shiming, Wang Beibei, Zhang Jing. 2014. Key Technologies for Demand Response in Smart Grid. *Proceedings of the CSEE*, 34(22): 3576–3589 (in Chinese).

Torriti J, Hassan M.G, Leach M. 2010. Demand response experience in Europe: Policies, programmes and implementation [J]. *Energy*, 35(4S1): 1575–1583.

Wang Beibei, Li Yang, Gao Ciwei. 2009. Demand side management outlook under smart grid infrastructure. *Automation of Electric Power Systems*, 33(20): 17–22 (in Chinese).

Wang Beibei, Sun Yujun, Li Yang. 2015. Application of uncertain demand response modeling in power-score incentive decision. *Automation of Electric Power Systems*, 39(10): 93–99 (in Chinese).

Wang Beibei. 2014. Research on consumers' response characterics and ability under smart grid: a Literatures Survey. *Proceedings of the CSEE*, 34(22): 3654–3663 (in Chinese).

Wang Mianbin, Tan Zhongfu, Cao Fucheng, Guan Yong, Li Yaqing. 2006. Cost-benefit analysis models for dsm project considering uncertain factors. *Power System Technology*, 30(14): 59–63 (in Chinese).

Wang Mingjun. 2005. Load management and demand side management in electricity market environment, *Power System Technology*, 29(5): 1–5 (in Chinese).

Wang Xifan, Xiao Yunpeng, Wang Xiuli. 2014. Study and analysis on supply-demand interaction of power systems under new circumstances, *Proceedings of the CSEE*, 34(29): 5018–5028 (in Chinese).

Wang Yanling, Wang Jingjing, Wu Jingkai, Huang Jie. 2014. An electricity-saving potential optimization model of efficiency power plant. *Power System Technology*, 38(4): 941–946 (in Chinese).

Wen Jing, Liu Wenying, Xie Chang, Wang Weizhou. 2015. Source-load coordination optimal model considering wind power consumptive benefits based on bi-level programming. *Transactions of China Electro-technical Society*, 30(8): 247–256 (in Chinese).

Xiao Xin, Zhou Yuhui, Zheng Kaizhong, Chen Jianfu, Zhang Ning, et al. 2014. Research on strategy of interruptible price and its cost-benefit model aimed at peak load shifting in Taiwan. *Proceedings of the CSEE*, 34(22): 3615–3622 (in Chinese).

Xue Yusheng, Lei Xing, Xue Feng, Yu chen, Dong Chaoyang, et al. 2014. A review on impacts of wind power uncertainties on power systems. *Proceedings of the CSEE*, 34(29): 5029–5040 (in Chinese).

Yan Huaguang, Chen Songsong, Zhong Ming, Jiang Limin. 2015. Research and design of demand side energy efficiency management and demand response system. *Power System Technology*, 39(1): 42–47 (in Chinese).

Zeng Ming, Han Xu, Li Bo. 2015. Study of demand response safeguard mechanism for active distribution networks. *Electric Power Construction*, 36(1): 110–114 (in Chinese).

Zeng Ming, Li Na, Wang Tao, Ouyang Shaojie, Wang Liang. 2013. Load forecasting compatible with demand-side resources. *Electric Power Automation Equipment*, 33(10): 59–62 (in Chinese).

Zeng Ming, Zhang Yanfu, Wang He, Jia Junguo. 2007. A linear programming on resource combination optimization based on integrated resources planning. *Automation of Electric Power Systems*, 31(3): 24–28 (in Chinese).

Zeng Ming. 2000. *Demand side management*. Beijing: China Electric Power Press. (in Chinese).

Zhang qin, Wang Xifan, Wang Jianxue, Feng Changyou, Liu Lin. 2008. Survey of demand response research in deregulated electricity markets. *Automatic of Electric Power Systems*, 32(3): 97–106 (in Chinese).

Zhao bo, Wang Chaisheng, Zhou Jinhui, Zhao Jinhui, Yang Yeqing, et al. 2014. Present and future development trend of active distribution network. *Automation of Electric Power systems*, 38(18): 125–135 (in Chinese).

Zheng Jing, Wen Fushuan, Li Li,Wang Ke, Gao Chao. 2013. Two-level planning of transmission system with optimal placement of efficiency power plants. *Electric Power Automation Equipment*, 33(6): 13–20 (in Chinese).

Zhou Jinghong, Hu Zhaoguang, Tian Jianwei, Xiaoxiao. 2010. Power system production simulation including efficiency power plant. *Automation of Electric Power Systems*, 34(18): 27–31 (in Chinese).

Zhou Jinghong, Hu Zhaoguang, Tian Jianwei, Xiao Xiao. 2010. A Power Integrated Resource Strategic Planning Model and Its Application. *Automation of Electric Power Systems*, 34(11): 19–22, 92 (in Chinese).

Zhu Lan, Yan Zheng, Yang Xiu, Fu Yang. 2014. Integrated resources planning in microgrid based on modeling demand response. *Proceedings of the CSEE*, 34(16): 2621–2628 (in Chinese).

Advances in Power and Energy Engineering – Sun (Ed.)
© *2016 Taylor & Francis Group, London, ISBN 978-1-138-02846-3*

An integrated decision-making method for condition assessment of power transformer

L.J. Sun, H.W. Yuan, B.Y. Zhang & W.J. Zuo
School of Automation Science and Electrical Engineering, Beihang University, Beijing, China

Z. Ma & Y.W. Shang
Department of Power Distribution, China Electric Power Research Institute, Beijing, China

ABSTRACT: Power transformer is a key equipment in the modern power system. With a view to eliminate the weakness and defects that exists in condition assessment for a power transformer, such as difficulty in tackling uncertain information, this paper proposes an integrated decision-making method for condition assessment of a power transformer using fuzzy set theory, Analytic Hierarchy Process (AHP), and evidence combination based on the distance of evidence. The integrated decision-making model relies on three factors: DGA data, oil testing data, and electrical testing data. The weights of characteristic quantity are obtained under the AHP method. In addition, the assessment results can be generated by the evidence combination based on the distance of evidence, which can avoid the error result facing conflict evidences. The case study shows that the integrated decision-making method is a reliable and effective tool to evaluate the condition of a power transformer.

1 INTRODUCTION

For decades, the judgment of the health condition for power transformer is mainly dependent on the regular maintenance. It will result in a huge waste of resources and reduce the reliability of power supply. Therefore, the development of condition based maintenance for power transformer is an inevitable trend (Wang & Srivastava 2002).

Due to the uncertain information of power transformer, the process for health condition assessment is very difficult. In this condition, the amount of information which needs to be considered is regarded as a multi-attribute decision-making problem. Over the years, many research achievements have been published. The typical method was proposed and applied in health diagnose for power transformer, such as fuzzy set theory (Chen & Lin 2001), fuzzy logic control, and others (Arsha & Khaliq 2014). Literature (Ahmed & Librahim 2012) analyzed the key operation factors of power transformer and adopted the classical mamdani fuzzy inference system. However, this approach did not give a good quantitative analysis and excessively depended on the expert rules. Literature (Liao et al. 2011) integrated the fuzzy set theory, group decision-making and evidence theory as a comprehensive evaluation model, but the classic evidence theory does not take into account the conflict evidence because of the complex system of power transformer (Liao et al. 2014).

Based on these findings, in order to calculate the health condition of power transformer effectively, this paper presents an intelligent model that integrates the fuzzy set theory, Analytic Hierarchy Process (AHP), and the evidence combination based on the distance of evidence. A novel multi-attribute decision-making algorithm for health condition of power transformer is demonstrated in detail to explain how to employ the intelligent model to obtain the health condition. The results obtained thus gave a reference to the condition based maintenance. The case study shows that this model is a reliable and effective tool to assess the health condition of a power transformer.

The paper is organized as follows: Section II shows how to select the key characteristic quantities of power transformer, and gives a condition assessment framework. Section III is made up of three subsection. The first subsection illustrates the health assessment model using the fuzzy set theory; the second subsection demonstrates the calculation of weights using the AHP; the third subsection obtains the final assessment result based on evidence theory. Section IV verifies the validity of the model by the case study. And section V gives the conclusion.

2 CONDITION ASSESSMENT SYSTEM

Because of the complex structure, various information of assessing indices for health diagnosis of

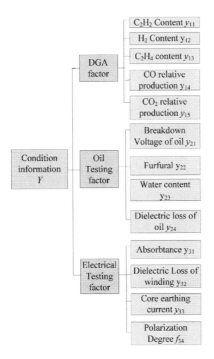

Figure 1. Condition information of power transformer.

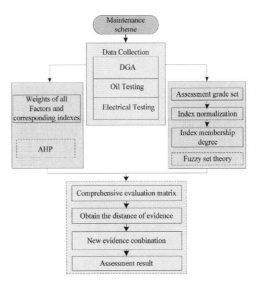

Figure 2. Condition evaluation model of power transformer.

power transformer is needed, which is regarded as a MADM problem. The essence of MADM problem (Yang & Singh 1994) is to integrate all information to achieve a uniform explanation and description of the recognized object. Therefore, finding out the power transformer condition indices in an effective and reasonable way is considered to be the primary task. On the basis of the aging mechanisms and fault characteristics of power transformer, the proposed hierarchical power transformer condition assessment index system is relied on three factors: the DGA factor, electrical testing factor, and oil testing factor.

As can be seen from Figure 1, the framework for power transformer condition assessment is made up of three-layer structure. The level 1 $F = \{f_1, f_2, f_3\}$ is named the objective layer of assessment system. The level 2 has three evaluation factors which illustrate the power transformer condition in different situation. Each factor is made up of several indices. For example, $f_2 = \{f_{21}, f_{22}, f_{23}, f_{24}\}$ represents the oil testing factor with eight indices.

3 CONDITION EVALUATION MODEL

An intelligent maintenance model is presented in this paper to evaluate the health condition of power transformer by using the fuzzy set theory,

AHP and the evidence combination based on the distance of evidence. The main idea of this model is simply described as below.

First, a popular method for evaluation is to construct an assessment grades set, then, the fuzzy set theory is adopted to tackle fuzziness of assessment indices and obtain an index membership degree of the assessment grades;

Second, by applying the AHP, the weights of the key characteristic quantities would be obtained (Chen 2010);

Third, the final assessment result would be given by the evidence combination based on the distance of evidence.

The flowchart of power transformer condition evaluation is shown in Figure 2 and the specific steps of modified condition evaluation model can be given as below.

3.1 Step 1: condition evaluation model based on the fuzzy theory

A general approach for evaluation is to build an assessment grades set, and the set in this paper is defined as:

$$H = \{H_1, H_2, H_3, H_4, H_5\}$$
$$= \{\text{excellent, good, moderate, poor, worst}\} \quad (1)$$

As each index value has different dimensions, various indices values can be not directly compared or calculated. Therefore, the actual measured data of indices need to be standardized.

Let y_{ij0} be the initial value which rely upon the specific power transformer, y_{ijl} be the limit value which is decided by actual tests and some operation norms, y_{ij} be the actual date. The normalized equation can be built as follows.

The values of indices if smaller the better, such as acetylene content, they can be defined as follows:

$$C_{ij} = \frac{y_{ij} - y_{ij0}}{y_{ijl} - y_{ij0}}, \quad y_{ij0} \leq y_{ij} \leq y_{ijl} \tag{2}$$

The values of indices if larger the better, such as absorptance, they can be defined to be:

$$C_{ij} = \frac{y_{ij0} - y_{ij}}{y_{ij0} - y_{ijl}}, \quad y_{ijl} \leq y_{ij} \leq y_{ij0} \tag{3}$$

where C_{ij} is the normalization value of the index y_{ij}, with its value ranges from 0 to 1.

According to the fuzzy set theory, intangibilities are contained in the evaluation process of the power transformer, and each assessment grade is an interval. One popular method is to employ the assigned membership function which is to get the border values by the extensive testing data and experts' opinions (Po-Chun & Ming-Ta 2014). The triangle membership function can be obtained as follows.

$$\begin{cases} f_{p^-}(x_{ij}) = \dfrac{m_{p+1} - x_{ij}}{m_{p+1} - m_p}, & p = 1, 2, 3, 4 \\[3mm] f_{p^-}(x_{ij}) = \dfrac{x_{ij} - m_{p-1}}{m_p - m_{p-1}}, & p = 2, 3, 4, 5 \end{cases} \tag{4}$$

$$\begin{cases} \mu_1(x_{ij}) = \begin{cases} 1, & x_{ij} \leq p_1 \\ f_{1^-}(x_{ij}), & p_1 < x_{ij} < p_2 \\ 0, & other \end{cases} \\[6mm] \mu_2(x_{ij}) = \begin{cases} f_{2^+}(x_{ij}), & p_1 < x_{ij} < p_2 \\ f_{2^-}(x_{ij}), & p_2 < x_{ij} < p_3 \\ 0, & other \end{cases} \\[6mm] \mu_3(x_{ij}) = \begin{cases} f_{3^+}(x_{ij}), & p_2 < x_{ij} < p_3 \\ f_{3^-}(x_{ij}), & p_3 < x_{ij} < p_4 \\ 0, & other \end{cases} \\[6mm] \mu_4(x_{ij}) = \begin{cases} f_{4^+}(x_{ij}), & p_3 < x_{ij} < p_4 \\ f_{4^-}(x_{ij}), & p_4 < x_{ij} < p_5 \\ 0, & other \end{cases} \\[6mm] \mu_5(x_{ij}) = \begin{cases} f_{5^+}(x_{ij}), & p_4 < x_{ij} < p_5 \\ 1, & p_5 < x_{ij} \\ 0, & other \end{cases} \end{cases} \tag{5}$$

where $p_1 = 0.1$, $p_2 = 0.3$, $p_3 = 0.5$, $p_4 = 0.7$, $p_5 = 0.9$, and x_{ij} is the ith index of the jth factor.

3.2 Step 2: calculation of weights based on AHP

The weight vector P can be obtained by the AHP method. In 1980, the AHP theory was initially proposed by Saaty. The algorithm principle was to structure a hierarchical framework, to resolve a complex problem and to compare the relative importance of two indices. Thus, the AHP method is widely adopted as the decision-making method due to its relative efficiency in tackling quantitative and qualitative criteria. The procedure of the method is described as follows:

1. A hierarchical model based upon the problem analysis and the assessment factors are built.
2. The judgment matrix is constructed. The elements in the judgments matrix stands for the decision-maker's knowledge about the relative importance of any pair of attributes. The judgment matrix for the n attributes f_1, f_2, ..., f_n is explained by the following n-by-n matrix F:

$$F = [a_{ij}] = \begin{bmatrix} 1 & a_{12} & \cdots & a_{1n} \\ \dfrac{1}{a_{12}} & 1 & \cdots & a_{2n} \\ \vdots & \vdots & \vdots & \vdots \\ \dfrac{1}{a_{in}} & \dfrac{1}{a_{2n}} & \cdots & 1 \end{bmatrix} \tag{6}$$

where $a_{ii} = 1$ and $a_{ij} = 1/a_{ji}$, i and $j = 1, 2, ..., n$, respectively.

3. The maximum eigenvalue and its corresponding eigenvector of the judgment matrix can be obtained as follows:

$$\lambda_{\max} = \frac{F^* G}{G} \tag{7}$$

where $\lambda_{\max}$ = maximum eigenvalue; and G = normalized eigenvector.

4. The index of Consistency Ratio (CR) is employed to verify the effectiveness of the comparison matrix.

$$CR = \frac{CI}{RI} \tag{8}$$

5. $CI = (\lambda_{\max} - n)/(n - 1)$ is the general consistency index of the judgment matrix, RI = random index of the judgment matrix, and the weight vector is considered acceptable if $CR \leq 0.1$.

3.3 Step 3: evidence combination based on the distance of evidence

Suppose that the distance between two evidences is computed by the Jousselme distance method and can be defined as follows:

$$d(m_1, m_2) = \sqrt{\frac{1}{2}(m_1 - m_2)^T \underline{D}(m_1 - m_2)} \qquad (9)$$

where $\underline{D} = 2^N \times 2^N$ matrix and its elements are defined as:

$$D(A, B) = \frac{|A \cap B|}{|A \cup B|}, \quad A, B \in 2^\Theta \qquad (10)$$

Assume that there are n BPAs, and the distance matrix can be obtained as:

$$D = \begin{bmatrix} 1 & d_{12} & \cdots & d_{1j} & \cdots & d_{1n} \\ \vdots & \vdots & \vdots & \vdots & \vdots & \vdots \\ d_{i1} & d_{i2} & \cdots & d_{ij} & \cdots & d_{in} \\ \vdots & \vdots & \vdots & \vdots & \vdots & \vdots \\ d_{n1} & d_{n2} & \cdots & d_{nj} & \cdots & 1 \end{bmatrix} \qquad (11)$$

Then, the similarity measure matrix is a matrix which expresses an agreement between bodies of evidence.

$$Sim(m_i, m_j) = 1 - d(m_i, m_j) \qquad (12)$$

Sum up each row of the similarity matrix, the support degree of BPA m_i can be computed as:

$$Sup(m_i) = \sum_{\substack{j=1 \\ j \neq i}}^n Sim(m_i, m_j) \qquad (13)$$

It is obvious that the support degree $Sup(m_i) = $ sum of each row, except itself.

Standardizing for (11), the weights of evidences can be calculated as:

$$w_{e_i} = \frac{Sup(m_i)}{\sum\limits_{i=1}^n Sup(m_i)} \qquad (14)$$

where

$$\sum_{i=1}^n w_{e_i} = 1 \qquad (15)$$

From (13), the new evidence can be expressed as below:

$$MAE(m) = \sum_{i=1}^n \left(w'_{e_i} \times m_i \right) \qquad (16)$$

Then, we can use the classical Dempster's combination rule $n - 1$ times if there are n BPAs in this framework (Deng et al. 2004).

4 CASE STUDY

The preventive test data of 110 KV transforms in 2013 from the Nanjing Electric Power Company in China is shown in Table 1.

Based on the AHP method, the weights of characteristic quantity for power transformer is obtained and shown in Table 2.

From Tables 1 and 2, the comprehensive assessment matrix, namely original BPA allocation, can be obtained based on step 1 as follows:

$$M(H) = \begin{bmatrix} H_1 & H_2 & H_3 & H_4 & H_5 \\ 0.115 & 0.0851 & 0.2 & 0.19 & 0.41 \\ 0.2 & 0.25 & 0.55 & 0 & 0 \\ 0.03 & 0.32 & 0.15 & 0 & 0.5 \end{bmatrix}$$

From the original BPA allocation, using (9)–(16), the modified evidence can be generated as below:

$$MAE(m_{new}) = \begin{bmatrix} 0.11 & 0.22 & 0.29 & 0.07 & 0.32 \end{bmatrix}$$

Table 1. Testing data of power transformer.

Indices	Testing data	Initial value	Limit value
H_2 (μL/L)	90	5.8	150
C_2H_2 (μL/L)	2.6	0	3
C_2H_4 (μL/L)	22.3	3.2	50
CO (% month)	15	0	100
CO_2 (% month)	21	0	200
Breakdown voltage (KV)	65.2	68	35
Furfural (mg/L)	0.07	0	0.2
Water content (mg/L)	13.1	3.5	25
Dielectric loss of oil (%)	1.404	0.5	4
Absorptance	1.73	2	1.3
Dielectric loss of winding (%)	0.294	0.17	0.8
Earth current of core (mA)	97	10	100
Polarization	2.12	2.5	1.5

Table 2. Weights of characteristic quantity.

Factor	Corresponding indices weights				
0.4956	0.232	0.491	0.138	0.092	0.046
0.1834	0.1767	0.2575	0.4034	0.1691	
0.3142	0.2071	0.121	0.5036	0.121	

504

According to the evidence combination based on the distance of evidence, the final assessment result can be calculated as follows:

$$M(H_1) = 0.04; \quad M(H_2) = 0.19; \quad M(H_3) = 0.31;$$
$$M(H_4) = 0.01; \quad M(H_5) = 0.45$$

Therefore, the final result of health condition of power transformer is regarded as the 'worst.' After overhaul of the power transformer, the condition of transformer is 'worst.' Thus, the integrated model is an effective tool to assess the condition of transformer.

5 CONCLUSION

This paper proposes a MADM model of condition evaluation for power transformer using the fuzzy set theory, AHP and the evidence combination based on the distance of evidence. The fuzzy set theory is applied in the index normalization and the condition assessment model of power transformer, the AHP is adopted to obtain weights of all factors and corresponding indexes, and the evidence combination based on the distance of evidence obtains the comprehensive assessing result of power transformer. In the case study, the presented model is an effective tool to assess the real health condition of power transformer. The contributions of this paper are summarized as following:

a. The proposed model has a strict logical reasoning and foundation of mathematics.
b. The presented assessment model of power transformer can decrease the unnecessary maintenance schedule and improve the reliability of equipment.
c. This intelligent model has a certain reference value for other evaluation field.

ACKNOWLEDGMENT

The research was supported by SGCC (State Grid Corporation of China) "Thousand Talents program" special support project (EPRIPDKJ (2014)2863).

REFERENCES

Ahmed E.B. Abu-Elanien, M.M.A. Salama, and M. Ibrahim. 2012. Calculation of a health index for oil-immersed transformers rated under 69 KV using fuzzy logic. *IEEE Transactions on Power Delivery* 27(4): 2029–2036.
Arshad, M.S. Islam, and A. Khaliq. 2014. Fuzzy logic approach in power transformers management and decision making. *IEEE Transactions on Dielectrics and Electrical Insulation* 21(5): 2343–2354.
Chen A.P. and C.C. Lin. 2001. Fuzzy approaches for fault diagnosis of transformers. *Fuzzy Sets Syst* 118(1): 139–151.
Chen. W. 2010. Quantitative decision-making model for distribution system restoration. IEEE Trans. Power Syst 25(1): 313–321.
Deng Yong, Shi WenKang, Zhu ZhenFu, and Liu Qi. 2004. Combining belief functions based on distance of evidence. *Decis. Support. Syst* 38(3): 489–493.
Liao et al. 2011. An Integrated Decision-Making Model for Condition Assessment of Power Transformers Using Fuzzy Approach and Evidential Reasoning. *IEEE Transactions on Power Delivery* 26(2): 1111–1118.
Liao et al. 2014. A cloud and evidential reasoning integrated model for insulation condition assessment of high voltahe transformers. *Int. Trans. Electr. Energ. Syst* 24: 913–926.
Po-Chun, L., G. Jyh-Cherng and Y. Ming-Ta. 2014. Intelligent maintenance model for condition assessment of circuit breakers using fuzzy set theory and evidential reasoning. *IET Generation, Transmission & Distribution* 8(7): 1244–1253.
Wang, M. A.J. Vandermaar and K.D. Srivastava. 2002. Review of condition assessment of power transformers in service. *IEEE Electrical Insulation Magazine* 18(6): 12–25.
Yang J.B. and M.G. Singh. 1994. An evidential reasoning approach for multiple-attrribute decision making with uncertainty. *IEEE Trans. Syst., Man, Cybern* 24(1): 1–18.

Advances in Power and Energy Engineering – Sun (Ed.)
© 2016 Taylor & Francis Group, London, ISBN 978-1-138-02846-3

Determination of harmonic responsibilities under the change of background harmonic impedance

J. Chen, T.L. Zang, L. Fu & Z.Y. He
School of Electrical Engineering, Southwest Jiaotong University, Chengdu, Sichuan Province, China

ABSTRACT: The change of background harmonic impedance caused by transformation of network structure or switching of loads reduces the accuracy of harmonic responsibilities determination. It is urgent to assess the harmonic responsibilities of customers in this case. This paper proposes a slope comparison method to segment the measurement data. The change time can be adaptively identified and the measurement data are segmented into several sections with different background harmonic impedances. Besides, the ranged major axis regression is introduced to determine the harmonic responsibility. The errors of independent variable are taken into account in the proposed regression method, so it will get higher accuracy in estimation. Furthermore, a total harmonic responsibility index is established to assess the total impact of customers. The simulations are conducted to verify the validity and correctness of the proposed method. Compared to current methods, the proposed method has higher effectiveness and accuracy.

1 INTRODUCTION

With the increasing use of non-linear loads, harmonic pollution has become an important issue. The injected harmonic currents by non-linear loads disturb the network by increasing power losses and damaging system equipment. Therefore, it is an urgent task to control and manage harmonic in the grid on the condition that modern electrical devices are strict to power quality. To limit the harmonic content in the network and assure the quality of electricity, the power grid corporation makes rules to punish the customers who inject harmonic currents into the system. And the most essential challenge before accomplishing the task is to identify the harmonic sources and determine their harmonic responsibilities (Bonavolontà, F. et al. 2016, Dan, A.M. 2009).

The primary problem in determination of harmonic responsibilities is to estimate the background harmonic impedance accurately (Xu, W. & Liu, Y. 2000). In current researches, the methods of calculating the background harmonic impedance can mainly be classified into two types, linear regression methods and fluctuation methods. Linear regression methods (Singh, S.K. et al. 2015, Xu, W. et al. 2009) are on the basis of Norton equivalent circuit or Thevenin equivalent circuit, and calculate the impedance by solving the regression coefficients according to the equations deduced from the circuit theory. Fluctuation methods (Wang, S.C. et al. 2012) make use of the fluctuations of harmonic voltage and harmonic current in the Point of Common Coupling (PCC) to calculate the background harmonic impedance. Among them linear regression methods are

commonly used and most of them are based on the least square regression. But the least square regression generates biased estimates of the regression coefficient as it ignores random errors of independent variables. Thus, the ranged major axis regression is introduced in this paper to avoid these problems.

In the field of harmonic responsibilities assessment, some achievements (De Andrade, G.V. et al. 2009, Hui, J. et al. 2010) get the result on constant background harmonic, while some others (Hua, H. et al. 2012, Mazin, H.E. et al. 2011, Shojaie, M. & Mokhtari, H. 2014) are got on condition of background harmonic fluctuation, but they attribute the fluctuation to the variation of background harmonic current, without consideration of background harmonic impedance change. In practical power system, the fluctuation of parameters in system side, especially background harmonic impedance, makes a great influence on the determination of harmonic responsibilities. The variation of background harmonic impedance, caused by some reasons such as switching of loads, reduces the accuracy of harmonic responsibilities assessment. The above mentioned methods are unable to determine the harmonic responsibilities precisely, thus a new method should be proposed to assess the harmonic responsibilities under the change of background harmonic impedance.

To provide economic incentives for encouraging the utility and customers to comply with the harmonic limits defined in the IEEE Std. 519-1992 (Duffey, C.K. & Stratford, R.P. 1989), it is essential to quantitatively assess the harmonic responsibilities of customers. Several researchers qualitatively study the influence of background harmonic impedance change on the determination of harmonic responsibilities. Ren, L.Z. & Chen, Y. (2014) discuss the harmonic distortion impact of background harmonic impedance change at PCC when the impedance rate in system (X/R) is constant or variable. Then, the major contributors, system or customers, are determined. But the exact value of harmonic responsibilities is not obtained. Considering the above problems, this paper quantitatively studies the determination of harmonic responsibilities under the change of background harmonic impedance. A total harmonic responsibility index is established in Section 2. Section 3 presents the principle of segmenting the measurement data with slope comparison method. And the ranged major axis regression is applied to estimate the background harmonic impedance and quantify the harmonic responsibilities in Section 4. To verify the effectiveness of the proposed method, the simulation is conducted on a Norton equivalent circuit in Section 5. Finally, conclusions are stated in Section 6.

2 ESTABLISHMENT OF TOTAL HARMONIC RESPONSIBILITY INDEX

In practical power system, harmonic voltage is often chosen as the evaluating indicator to assess the harmonic responsibilities. Traditionally, harmonic responsibility is defined as the modulus ratio of the projection of harmonic voltage generated by customer ($\dot{V}_c^h$) on harmonic voltage at PCC ($\dot{V}_{pcc}^h$) to $\dot{V}_{pcc}^h$. That is

$$\mu_c^h = \frac{|\dot{V}_c^h|\cos\alpha}{|\dot{V}_{pcc}^h|} \times 100\% \qquad (1)$$

But it does not apply in condition of background harmonic impedance change. Thus, the harmonic responsibilities under the change of background harmonic impedance should be redefined.

Taking h-th harmonic as an example, the Norton equivalent circuit is employed in the system side and customer side at PCC, as shown in Figure 1.

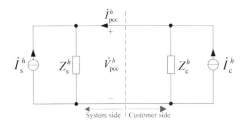

Figure 1. Norton equivalent circuit of system and customer.

The equivalent harmonic current source and background harmonic impedance are $\dot{I}_s^h$ and Z_s^h, respectively. In customer side, they are $\dot{I}_c^h$ and Z_c^h. Then $\dot{V}_c^h = Z_s^h \dot{I}_c^h$.

If the background harmonic impedance varies for N-1 times in a period of time, that is to say, the period can be segmented into N sections. The harmonic responsibilities of every section can be got according to Equation (1).

$$\mu_c^h(i) = \frac{|\dot{V}_c^h(i)|\cos\alpha(i)}{|\dot{V}_{pcc}^h(i)|} \times 100\%$$
$$\approx \frac{|Z_s^h(i)\dot{I}_{pcc}^h(i)|}{|\dot{V}_{pcc}^h(i)|} \times 100\% \quad (i=1, 2, ..., N) \quad (2)$$

In the sampling period, the measurement data are also segmented into N sections, and each section corresponds to different background harmonic impedance values. If time length of the i-th ($i = 1, 2, ..., N$) section is $h(i)$, the total harmonic responsibility in the period is determined by

$$HI = \frac{\sum_{i=1}^{N} \mu_c^h(i) \cdot h(i)}{\sum_{i=1}^{N} h(i)} \qquad (3)$$

3 SLOPE COMPARISON METHOD BASED MEASUREMENT DATA SEGMENTATION

In Equation (2), $\dot{I}_{pcc}^h$ and $\dot{V}_{pcc}^h$ are measurement data. Consequently, the first essential task is to estimate the value of background harmonic impedance. Most researches of harmonic impedance estimation are on the condition of constant background harmonic impedance. In this section, a slope comparison method that adaptively identifies the change of background harmonic impedance is proposed. The basic principle is to recognize the time of background harmonic impedance change so that the measurement data are segmented into sections with different background harmonic impedances.

In Figure 1, we can conclude that

$$\dot{V}_{pcc}^h = Z_s^h \dot{I}_{pcc}^h + \dot{V}_0^h \qquad (4)$$

where $\dot{V}_0^h$ = background harmonic voltage. Ignoring the impact of phase angle, $|\dot{I}_{pcc}^h|$ and $|\dot{V}_{pcc}^h|$ are taken as the independent variable and dependent variable respectively. The slope of equation can be considered as the modulus of background

508

harmonic impedance (Z_s^h). The slope comparison method is described as follows.

The change of background harmonic impedance can be simplified as shown in Equation 5.

$$\left|Z_s^k\right| = \begin{cases} A_1 & t_1 \leq t < t_2 \\ A_2 & t_2 \leq t < t_3 \\ \vdots & \vdots \\ A_{N-1} & t_{N-1} \leq t < t_N \end{cases} \tag{5}$$

where A_1, A_2, ..., A_{N-1} = constant background harmonic impedance values.

The change time are t_1, t_2, ... , t_{N-1}. During time t_i and t_{i+1}, the background harmonic impedance remains constant. The slope of any two adjacent sample points in the time section are similar to the value of the same background harmonic impedance. While on the change time t_{i+1}, the slopes of lines between points of the time and the previous or after time change largely. This feature is used to segment the measurement data.

Assuming that the number of sampling points is P, the two adjacent points are (x_k, y_k) and (x_{k+1}, y_{k+1}) ($k = 1, 2, ..., P-1$), as shown in Figure 2. The slope of the two points is

$$R_k = \frac{y_{k+1} - y_k}{x_{k+1} - x_k} \tag{6}$$

Then all the slopes can be calculated as R_1, R_2, ..., R_{P-1}. Set the threshold M and compare the difference between two adjacent slopes with the threshold. If the difference is no more than M, the corresponding sampling points are in the same section. Otherwise, they are in different sections. That is

$$\left|R_{k+1} - R_k\right| \begin{cases} > M & \text{in different sections} \\ \leq M & \text{in the same section} \end{cases} \tag{7}$$

Equation (6) is used to segment the measurement data.

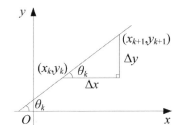

Figure 2. Slope of two adjacent points.

4 RANGED MAJOR AXIS REGRESSION BASED HARMONIC RESPONSIBILITIES DETERMINATION

After segmenting the measurement data, the background harmonic impedance of each section must be correctly estimated so that the harmonic responsibilities can be assessed properly.

Linear regression is a statistical analysis method which studies the interrelation of independent variables and dependent variables. In general, simple linear regression is modeled as

$$\hat{Y} = \varepsilon + \lambda \cdot X \tag{8}$$

where $\hat{Y}$ = estimates of dependent variable Y; X = independent variable; ε = regression intercept; and λ = regression coefficient.

Define the algebraic expressions Q_{XX}, Q_{YY}, Q_{XY} on X and Y as

$$Q_{XX} = \sum(X - \bar{X})^2 = \sum X^2 - (X)^2/n \tag{9}$$

$$Q_{YY} = \sum(Y - \bar{Y})^2 = \sum Y^2 - (Y)^2/n \tag{10}$$

$$Q_{XY} = \sum(X - \bar{X})(Y - \bar{Y}) = \sum XY - (XY)/n \tag{11}$$

where $\bar{X}$ = average value of X; and $\bar{Y}$ = average value of Y.

Simple linear regression works on the least square regression, and the two terms are often seen equivalent. The least square regression targets the minimum sum of squares of deviations of estimated Y-values and measurement Y-values. That is

$$\min \quad \sum(Y - \hat{Y})^2 \tag{12}$$

Then the estimated value of regression coefficient $\hat{\varepsilon}_L$ and $\hat{\lambda}_L$ can be got as

$$\hat{\lambda}_L = Q_{XY}/Q_{XX} \tag{13}$$

$$\hat{\varepsilon}_L = \bar{Y} - \hat{\lambda}_L \cdot \bar{X} \tag{14}$$

The result is obtained on the assumption that the independent variable X is measured without error. However, it may not satisfy the assumption in practical studies. Moreover, errors may be introduced into the estimation regression coefficient. Thus the ranged major axis regression (Legendre, P. 2013) is proposed to avoid the problem.

First transform the Y and the X variables into Y' and X', whose range is

$$Y_i' = \frac{Y_i - Y_{\min}}{Y_{\max} - Y_{\min}} \text{ or } X_i' = \frac{X_i - X_{\min}}{X_{\max} - X_{\min}} \tag{15}$$

The objective function of ranged major axis regression is to minimize the sum of the products of the respective deviations of X'-values and Y'-values. That is

$$\min \; \sum (X' - \hat{X}')(Y' - \hat{Y}') \qquad (16)$$

Based on the objective function, the estimated value of regression coefficient $\hat{\lambda}'_{G}$ are

$$\hat{\lambda}'_{G} = \text{sign}(\gamma) \cdot \sqrt{Q_{y'y'} / Q_{x'x'}} \qquad (17)$$

where sign (γ) = the sign of correlation coefficients γ, and

$$\gamma = Q_{x'y'} / \sqrt{Q_{x'x'} Q_{y'y'}} \qquad (18)$$

Then back-transform the estimated coefficient to the original units by multiplying it by the ratio of the ranges

$$\hat{\lambda}_{G} = \hat{\lambda}'_{G} \frac{Y_{\max} - Y_{\min}}{X_{\max} - X_{\min}} \qquad (19)$$

And the intercept can be calculated by

$$\hat{\varepsilon}_{G} = \overline{Y} - \hat{\lambda}_{G} \cdot \overline{X} \qquad (20)$$

The ranged major axis regression equation is

$$\hat{Y}_{G} = \hat{\varepsilon}_{G} + \hat{\lambda}_{G} \cdot X \qquad (21)$$

Comparing Equation (4) and Equation (21), the regression coefficient $\hat{\lambda}_{G}$ is the estimated value of modulus of background harmonic impedance $\left|Z_{s}^{h}\right|$.

The ranged major axis regression is applied in every section of measurement data to estimate the background harmonic impedance. Then Equation (2) is applied to calculate the harmonic responsibilities of each section. Finally, total harmonic responsibility is got based on Equation (3).

Based on the above, the procedure of harmonic responsibilities determination is listed on the following steps.

1. Collect the measurement data and analyze for $\dot{I}_{\text{pcc}}^{h}$ and $\dot{V}_{\text{pcc}}^{h}$;
2. Calculate the slopes of each two adjacent points based on Equation (6) and get slopes $R_{1}, R_{2}, \ldots, R_{P-1}$;
3. Set the threshold M and compare the difference between two adjacent slopes with M. Determine the change time and points in the same section according to Equation (7);
4. In the same section, take $|\dot{I}_{\text{pcc}}^{h}|$ as the independent variable and $|\dot{V}_{\text{pcc}}^{h}|$ as the dependent variable to complete ranged major axis regression analysis;

5. Take the result of regression coefficients as the estimation of background harmonic impedance;
6. Calculate the harmonic responsibilities of all sections based on Equation (2);
7. Calculate the total harmonic responsibility according to Equation (3).

5 SIMULATION STUDY

5.1 Segmentation of measurement data

The Norton equivalent circuit in Figure 1 is used to validate the proposed method. Taking 5-*th* harmonic as an example, the initial values of electric parameters are set as shown in Table 1.

A zero mean normally distributed random variable with 0.3 of center value as variance is multiplied to the customer harmonic current. The amplitudes of Z_{s}^{5} are set random variables between 0 and 1, and the angle varies randomly between 71 degree and 80 degree. The amplitude and angle of Z_{s}^{5} changes every 144 points. Data of harmonic voltage and harmonic current at PCC are got on the simulation as shown in Figure 3.

It can be seen from Figure 3 that sampling data show specific distribution. Set threshold $M = 0.01$, then by applying slope comparison method, these data are segmented into several sections. The simulation result reveals that the data are into 10 section with the starting points of 1, 145, 289, 433, 577, 721, 865, 1009, 1153, 1297. The segmentation result is highly consistent with the initial setting, which indicates the effectiveness of the slope comparison method.

Table 1. Initial value of parameters in Norton equivalent circuit.

Parameters	$\dot{I}_{c}^{5}$	$\dot{I}_{s}^{5}$	Z_{c}^{5}	Z_{s}^{5}
Initial values	14∠10°	12∠70°	7∠54°	0.1∠71°

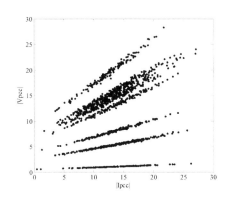

Figure 3. Scatter diagram of harmonic voltage and current at the bus.

5.2 Determination of harmonic responsibilities of multiple sections

In the same sections, background harmonic impedance is considered a constant. After the segmentation, the harmonic responsibilities of each section can be determined with the proposed ranged major axis regression. The results are compared with that of least square regression (method 1) (Xu, W. et al. 2009) and robust regression (method 2) (Che, Q. & Yang, H. G. 2004). The sections are numbered 1, 2,...,10 in chronological order. The estimation errors of background harmonic impedances are listed in Table 2.

Table 2 shows that the three methods all can be used to estimate background harmonic impedance. Compared with method 1 and method 2, the proposed method has the least errors in calculation. It indicates that the proposed method performs well in accuracy in estimating background harmonic impedance.

On the basis of estimation result of background harmonic impedance, the harmonic responsibilities of all sections are calculated on Equation (1). The estimation errors of harmonic responsibilities of all sections are shown in Table 3.

It can be seen from Table 3 that the calculation result of the proposed method gets the smallest errors, which means that the proposed method has the highest precision in determining the harmonic responsibilities. It demonstrates that the proposed method can be used to accurately assess the harmonic responsibilities.

5.3 Discussion on validity of segmentation

To test the validity of segmentation, a comparison between the calculation result of estimating without segmentation and estimating with segmentation is conducted, and the result of background harmonic impedance is shown in Figure 4.

Table 2. Estimation errors of background harmonic impedance in all sections.

Section	Errors of method 1/%	Errors of method 2/%	Errors of the proposed method/%
1	11.07	11.48	10.63
2	9.84	12.32	8.62
3	10.37	11.14	9.86
4	8.20	9.45	7.39
5	10.33	11.83	9.63
6	7.26	9.93	6.17
7	10.14	10.60	9.47
8	8.74	10.51	7.71
9	8.09	9.79	6.75
10	11.51	12.78	10.88
Average error/%	9.56	10.98	8.71

Table 3. Estimation errors of harmonic responsibilities in all segments.

Section	Errors of method 1/%	Errors of method 2/%	Errors of the proposed method/%
1	10.02	10.43	9.58
2	3.12	5.78	1.80
3	9.27	10.05	8.75
4	5.96	7.24	5.13
5	8.16	9.70	7.44
6	4.90	7.65	3.79
7	2.95	3.44	2.22
8	4.86	6.69	3.78
9	5.43	7.18	4.05
10	6.88	8.22	6.23
Average error/%	6.16	7.64	5.28

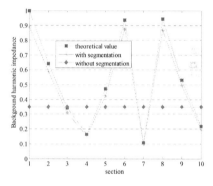

Figure 4. Comparison of estimation results of background harmonic impedances in two cases.

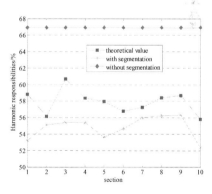

Figure 5. Comparison of estimation results of harmonic responsibilities in two cases.

The harmonic responsibilities are determined based on the background harmonic impedance shown in Figure 4. The calculation result with or without segmentation is shown in Figure 5. It can be seen in Figure 5 that he errors of estimating the background harmonic impedance without segmentation are so

511

Table 4. Estimation results of total harmonic responsibility.

Method	Theoretical value	Method 1	Method 2	The proposed method
Total harmonic responsibility/%	57.88	54.30	53.45	54.81

big in all sections that the calculation of harmonic responsibilities is inaccurate as a result of 66.91%.

In Figure 4 and Figure 5, it is evident that the estimation errors of background harmonic impedances and harmonic responsibilities in case of no segmentation are big. On the contrary, results with segmentation are close to the theoretical value. Thus the validity of segmentation is confirmed.

5.4 Calculation of total harmonic responsibility

According to Equation (3), total harmonic responsibility can be obtained with responsibilities of all sections. The result is shown in Table 4.

Table 4 shows that the above three methods are useful in assessing total harmonic responsibility, but the result of the proposed method is closest to the theoretical value. Therefore, the proposed method has higher accuracy compared to the others.

6 CONCLUSION

Considering the change of background harmonic impedance in practical power system, this paper proposes the slope comparison method to segment the measurement data into several sections with different background harmonic impedances. Besides, the ranged major axis regression is applied to estimating the background harmonic impedances. Finally, a total harmonic responsibility index is established to assess the total harmonic impact. By analyzing the simulation results in Norton-equivalent circuit, the following conclusions can be obtained: 1) In case of background harmonic impedance change, the slope comparison method can identify the change time adaptively, and the segmentation is valid in the circumstance; 2) Compared to current methods, ranged major axis regression performs precisely in the background harmonic impedance estimation; 3) The index of total harmonic responsibility is based on the change rule of background harmonic impedance, so it can be used to better measure the harmonic responsibility of customer.

REFERENCES

Bonavolontà, F. et al. 2016. Harmonic and interharmonic measurements through a compressed sampling approach. *Measurement* 77:1–15.

Che, Q. & Yang, H.G. 2004. Assessing the harmonic emission level based on robust regression method. *Proceedings of the CSEE* 24 (4):39–49.

Dan, A.M. 2009. Identification of Individual Harmonic sources and Evaluation their contribution in the Harmonic Distortion Level. *IEEE Power and Energy Society General Meeting-PESGM*. Calgary, CANADA.

De Andrade, G.V. et al. 2009. Estimation of the Utility's and Consumer's Contribution to Harmonic Distortion. *IEEE Transactions on Instrumentation and Measurement* 58 (11):3817–3823.

Duffey, C.K. & Stratford, R.P. 1989. Update of harmonic standard IEEE-519: IEEE recommended practices and requirements for harmonic control in electric power systems. *IEEE Trans. Ind. Appl.* 25 (6):1025–1034.

Hua, H. et al. 2012. Practical method to determine the harmonic contribution of a specific harmonic load. *Proceedings of 2012 IEEE 15th International Conference on Harmonics and Quality of Power, ICHQP 2012*. Hong Kong, China.

Hui, J. et al. 2010. Assessing Utility Harmonic Impedance Based on the Covariance Characteristic of Random Vectors. *IEEE Transactions on Power Delivery* 25 (3):1778–1786.

Legendre, P. 2013. Model II Regression User's Guide, R Edition, 1–14.

Mazin, H.E. et al. 2011. Determining the Harmonic Impacts of Multiple Harmonic-Producing Loads. *IEEE Transactions on Power Delivery* 26 (2):1187–1195.

Ren, L.Z. & Chen, Y. 2014. Influence of Harmonic Impedance Changes on PCC Harmonic Contribution. *Electric Power Construction* 35 (11):7–12.

Shojaie, M. & Mokhtari, H. 2014. A method for determination of harmonics responsibilities at the point of common coupling using data correlation analysis. *IET Generation Transmission & Distribution* 8 (1):142–150.

Singh, S.K. et al. 2015. Power system harmonic parameter estimation using Bilinear Recursive Least Square (BRLS) algorithm. *International Journal of Electrical Power & Energy Systems* 67:1–10.

Wang, S.C. et al. 2012. A Fluctuation Quantity Based Method to Evaluate Estimation Precision of Harmonic Impedance Amplitude at System Side. *Power System Technology* 36 (5):145–149.

Xu, W. et al. 2009. A Method to Determine the Harmonic Contributions of Multiple Loads. *IEEE Power and Energy Society General Meeting-PESGM*. Calgary, CANADA.

Xu, W. & Liu, Y. 2000. A method for determining customer and utility harmonic contributions at the point of common coupling. *IEEE Transactions on Power Delivery* 15 (2):804–811.

Advances in Power and Energy Engineering – Sun (Ed.)
© 2016 Taylor & Francis Group, London, ISBN 978-1-138-02846-3

Research and application of power flow diagram automatic generation method based on logical geographic information

W.C. Fang, S.Y. Liang, W. Zhu & R. Hu
China Southern Power Grid (CSG) Dispatching and Control Center, Guangzhou, China

T.F. Kang, Y.G. Li & R.P. Zhang
Beijing Sifang Automation Co. Ltd., Beijing, China

ABSTRACT: With the large scale development of interconnected power grid, the power flow diagram has become one of the most commonly used GUI tools to provide the system operators with a global perspective of the transmission network and auxiliary decision support. The manual maintenance of such diagram is very time consuming and error prone. In this paper, a novel method for the automatic generation of a power flow diagram, based on logical geographic information, is proposed to provide better system operation situation awareness and decision support. This method is based on the global model of the power transmission network and the logical geographic information of substations and control areas, optimizing the substations in the control area level and the control areas on a systematic level. Panorama dispatching and system situation awareness applications have been developed on basis of our proposed method in the China Southern Power Grid (CSG) dispatching and control center. An actual example is provided to prove the feasibility and advantage of our proposed method.

1 INTRODUCTION

As the toolkits for presenting system operation information and real-time status interactively and intuitively, Graphical User Interfaces (GUI) is one of the key components of smart grid and smart dispatching to enhance the ability of situation awareness, accuracy and efficiency for system management and real-time control. Power flow diagram is one of the most commonly used GUI tools for power grid dispatching and operation, which could provide the system operators with global perspective of the transmission network and auxiliary decision support (Manno & Dugan 1994, Ong et al. 2000, Protsko et al. 1991, Rochefort et al. 1996).

In the traditional Energy Management System (EMS), the power flow diagram is manually generated. However, with the development of smart grid, ultra high voltage and AC–DC hybrid power grid, the scale and complexity of power transmission network increases a lot. Therefore, manual creation and maintenance of such diagrams may not satisfy the requirements by system operators, and it is very time consuming and error prone. Automatic generation of such diagrams has become one of the emerging technologies of smart grid, which draws more and more attentions.

Generally, there are two types of power flow diagrams for large scale of power transmission

network, one is one-line format and the other is geographic information based. Diagrams of the former type layout the equipments of power grid in straightness, which are suitable for logical inference and system operation. On the contrary, diagrams of the latter type layout the equipments with reference to the geographical coordinate, which are more intuitive and better for perception of system operation status and decision support.

In this paper, we researched on diagram automatic generation method for automatic generation of geographic information based power flow diagram for large scale of power grid. In the proposed method, logical geographic information is taken into account for diagram automation generation to balance the layout of the diagram to prevent the crowding and disorder on the diagram due to high density of equipments in the area which increases the clarity, applicability, and elegancy. The examples of generated diagram in actual projects are provided to illustrate the effective of our proposed method.

2 DIAGRAM AUTOMATIC GENERATION

2.1 Diagram automatic generation and benefit

Diagram automatic generation is to layout the elements on the diagrams with the minimum number of overlapping nodes and number of crossings, etc.

The types of diagram include one-line formatted substation diagram, geographic information based power flow diagram and one-line formatted power flow diagram.

One-line formatted substation diagram focuses on the layout of the equipments in single substation, considering the connectivity between bus bars, transformers, circuit breakers, load, and so on. All the elements on the diagram are organized horizontally and vertically.

Geographic information based power flow diagram focuses on the layout of the power flow diagram of the whole transmission network with substations and transmission lines, according to the geographic information of substations, lines, etc.

One-line formatted power flow diagram also focuses on the layout of the power flow diagram with substations and transmission lines. However, the layout is horizontal and vertical.

Diagram automatic generation could ease the burden for creation and maintenance of power grid diagram and decrease the possibility of mistakes. Besides, geographic information based diagram satisfies the requirement for information investigation and situation awareness while one-line formatted diagram satisfies the requirement for detail information investigation and is suitable for system operation.

2.2 *Recent research*

As geographic power flow diagram could provide the global perspective of system operation, better situation awareness and decision support, automatic generation of such diagrams has drawn more and more attention. (Zhang et al. 2009) presented a method for generating power flow diagram of a provincial power grid based on Common Information Model (CIM) data source with geographical and relative location. However, the proposed method took more time for optimization which is effective just for small scale of power grid and unsuitable for online applications. (Zhang et al. 2013) improved the algorithm in (Zhang et al. 2009) with better solution efficiency. However, the new method takes average on the geographic location and degrades the systematic perspective by simply averaging the geographic information. Besides, no sub-area of the system is taken into consideration for layout optimization and sub-area based automatic generated diagram is unavailable.

2.3 *Challenges for diagram automatic generation based on geographic information*

Compared with the automatic generation of one-line formatted diagrams, there are some challenges for geographic information based diagram automatic generation methods.

First, the layout of substations and sub areas should satisfy the requirements on geographic information constraints. As there is a possibility of crowding in some special geographic area amongst substations with large scale and over spaciousness in some other areas, relative geographic relationship is alternative for such methods. Second, to obtain an elegant and balanced layout, partial release and compression of the equipments and sub areas in some limited regions is required. Third, automatic measurement mapping is required to provide the system operators with real-time situation awareness.

3 LOGICAL GEOGRAPHIC INFORMATION BASED DIAGRAM AUTOMATIC GENERATION FOR POWER TRANSMISSION NETWORK

3.1 *Logical geographic information and diagram automatic generation*

In this section, we researched on power flow diagram automatic generation method based on logical geographic information. In the method, CIM based model is used as the data source for diagram generation. Sub-control-area based logical geographic information is utilized for diagram generation. The overall flowchart of our proposed method is shown in Figure 1.

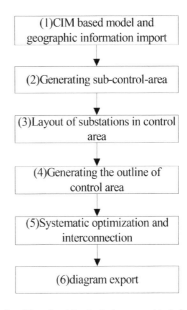

Figure 1. Flowchart for logical geographic information based power flow diagram automatic generation method.

3.2 CIM based model and geographic information import

IEC61970 CIM is the standard for information modeling and exchange between systems and applications. In our method, CIM based model is exported by EMS as XML type file. Then the file is imported into our method as the data source for diagram generation.

Geographic information related to the substations in the CIM model is also needed to as input parameters for our method.

3.3 Generating control areas

Usually, large scale of power transmission network is divided into different control areas for management and regulation. Different areas are interconnections by tie lines. In order to balance the generated diagram without the deficiency of clearance, we propose logical geographic information based on control areas. The generated diagram is optimized in and among control areas. Based on the CIM 61970 standard, the tag 'cim:SubControlArea' provides the relationship between control area and substations. The substations are grouped into separate control areas.

3.4 Layout of substations in control area

On the automatic generated diagram, the substations are represented by circles or squares. On the initial layout, the substations in control area may be imbalanced on geographic map. In such situations, we introduce a release/compress procedure to eliminate crowd or overlap due to such problems.

The aim of optimizing the layout is to keep the maximum density of all the substations in a limited space, and to ensure that the relative position of the substations is unchanged. With the geographic information of the substations, the position of each station is transformed into the relative position in the control area. First, the minimum and maximum distance between substations is assigned. Then sort the substations respectively in the X-axis and Y-axis direction according to the coordinates in an ascending order. If the distance between two substations is smaller than the minimum distance, these two substations need to be realized. Otherwise, they need to be compressed. The procedure for the 'release/compress' is shown in Figure 2.

1. Release procedure
   Throughout the substations in the same control area, if the distance between any two of stations is less than the minimum distance, the two substations need to be released. Take dL as the distance to be released. Decompose dL in both X and Y axis with dx and dy. All the substations in

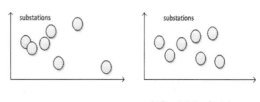

(a)Before optimization of control area   (b)After optimization of control area

Figure 2.   Procedure for release/compress of substations in control area.

the sequential array that is larger than the two substations in X axis would be increased by dx in X axis and correspondingly dy in Y axis.

2. Compress procedure
   Throughout the substations in the same control area, if the distance between any two of stations is larger than the maximum distance, the two substations need to be compressed. Take dL as the distance to be compressed. Decompose dL in both X and Y axis with dx and dy. All the substations in the sequential array that are larger than the two substations in X axis would be decreased by dx in X axis and correspondingly dy in Y axis.

3.5 Generating the outline of control area

After the layout of substations in control area is ready, all the control areas need to be automatically mapped together on the diagram of the interconnected power grid. Therefore, an appropriate outline needs to present the layout of the control area which covers all the substations with minimum area.

In our method, ellipse is proposed for representing the outline of control area. As shown in Figure 3, the algorithm for searching such ellipse is as follows:

1. Basing on standard coordinate system, calculate the coordinates of all the stations in the control area.
2. Through the minimum and maximum coordinate of the substations in the control area, find out the rectangle with one side across the horizontal coordinate and the other side perpendicular to the horizontal coordinate which covers all the substation in the control area and calculate the area as S (0);
3. Counterclockwise rotate the horizontal coordinate and the rotation angle 'I' is from 0 to 90 degrees with one degree as the single step for rotation. Then calculate the new coordinates of the substations and the new rectangle area. Its area is S (I).
4. Get the minimum of S (I) and relative ellipse to present the control area.

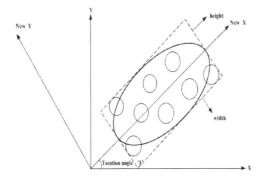

Figure 3. Procedure for generating the outline of control area.

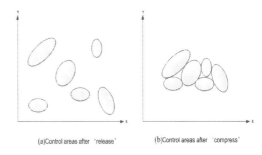

(a)Control areas after 'release'   (b)Control areas after 'compress'

Figure 4. Procedure for release/compress of control areas.

3.6 *Systematic optimization and interconnection*

The layout of control area should not only meet the constraints by the relative geographic location but also avoid collision or overlapping on the diagram. Therefore, according to the geographic information of different control areas, we will optimize the layout of the systematic diagram in the release and compress way similar to optimization of the layout of substations in the control area.

According to the initial gravity center of each control area ellipse, the actual distance between the gravity centers is released to ensure that there's no collision or overlapping on the ellipses. Then compress the distance between ellipses in the X and the Y axis direction respectively, with the constraint that no collision or overlapping occurs. In order to make the diagram more reasonable and convenient, the relative position of control area ellipse can be manually intervened and the layout optimized procedure is shown in Figure 4.

After systematic optimization of control areas, there are transmission lines to map on the diagram to realize interconnection of different control areas. The symbol that represents substation is typically a circle or square according to the type of substation. The transmission lines are mapped on the diagram across the direction that connects the centers of two substations.

3.7 *Diagram export*

The generated diagram could be exported in *.svg format and be utilized by EMS and other applications to available for advanced functions.

4 CASE STUDY

The actual application based on our proposed method has been established as panorama

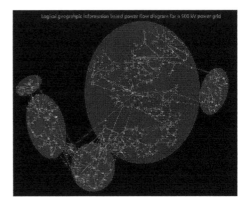

Figure 5. Example of automatically generated power flow diagram based on our proposed method.

dispatching and situation awareness system in the China Southern Power Grid (CSG) dispatching and control center. Figure 5 is one of the generated power flow diagrams for a 500 kV power grid. On the diagram, logical geographic information is consistent with the topological properties of the original transmission network model. The substations are automatically mapped with appropriate site density and sparse degree. The generated diagram is intuitive and better for perception of system operation status and decision support.

5 CONCLUSION

Geographic information based power flow diagram could provide the system operators with global perspective of the transmission network and auxiliary decision support. Automatic generation of such diagrams is one of the key technologies for smart grid and panorama dispatching.

In this paper, we researched on the logical geographic information based automatic generation method for power transmission network diagram. Based on the systematic model and geographic information, logical geographic information is utilized by optimization of the layout of substations and control areas. With logical geographic information, the generated diagram overcomes the problem of imbalance in small regions. The logical geographic information based power flow diagram is consistent with the topological properties of the original transmission network model, and maintains the relative geographical attributes with appropriate site density and sparse degree.

The example of actual generated diagram shows that our proposed method is compatible with both macroscopic and microcosmic vision of the power transmission network which satisfies the requirements by the system operators for situation perception and decision support.

REFERENCES

Manno, R. & Dugan, D. 1994. The dream and reality of automated wiring systems. *IEEE Computer Applications in Power* 7(1):20–23.

Ong, Y.S., Gooi, H.B. & Chan, C.K. 2000. Algorithms for Automatic Generation of One-line Diagrams, *IEEE Proceedings Generation, Transmission and Distribution* 147(5):292–298.

Protsko, L.B., Sorenson, P.G., Tremblay, J.P. & Schaefer, D.A. 1991. Toward the automatic generation of software diagrams. *IEEE Trans. Software Eng.* 17:10–21.

Rochefort, M., DeGuise, N. & Gingras, L. 1996. Development of a graphical user interface for a real-time power system simulator. *Elect. Power Syst. Res.* 36:203–210.

Zhang, J.M., Wang, Y. & Gu, W. 2009. Automatic generation of transmission network single-line power flow diagram for dispatching large screen display: Part one automatic layout. *Automation of Electric Power Systems* 33(24):43–47.

Zhang, J.M., Ye, Y., Chen, L.Y., Chen, J., Zhan, Z.B. & Zhuang, X.D. 2013. A Novel Force-direction Algorithm and Its Application in Automatic Generation of Uniformly-distributed Provincial Transmission Power Grid Diagram. *Automation of Electric Power Systems* 37(11):107–112.

Advances in Power and Energy Engineering – Sun (Ed.)
© 2016 Taylor & Francis Group, London, ISBN 978-1-138-02846-3

The design of an automatic batch test system for power quality monitoring equipment

R.S. Qin & Z. Xu
Yunnan Electric Power Research Institute of Yunnan Power Grid Co. Ltd., Kunming, Yunnan, China

X. Wang & J.W. Geng
CEIEC Shenzhen Electric Technology Inc., Shenzhen, Guangdong, China

ABSTRACT: Currently, the grid-accessing test of power quality monitoring equipment is mainly executed manually, which will cause some problems such as a heavy workload, low efficiency, data error, etc. Based on the deep analysis of the operation process of the grid-accessing test, an automatic batch test system for power quality monitoring equipment is proposed in this paper. The system supports the automatic control of the standard power source, data acquisition of different kinds of equipment, automatic generation of test report, etc., and can thus actualize a fully automatic test of power quality monitoring equipment and improve the efficiency of the grid-accessing test, which is beneficial for application and promotion.

1 INTRODUCTION

As the increasing use of power quality monitoring equipment in both, power grid and power consumers, power quality monitoring equipment of various factories and models are produced gradually more in the power quality equipment market. To ensure that the function, the performance and quality of power quality monitoring equipment are qualified to the national standard, all levels of regional management departments of power grid have established grid-accessing testing system of power quality monitoring equipment progressively, which is mainly for the accuracy and communication protocol test of power quality monitoring before accessing to the power grid. However, due to the limited technical development, currently the grid-accessing test of power quality monitoring equipment is mainly executed manually, which means the equipment wiring, standard source controlling, data recording, and report writing are all done by testers. This will result in some problems such as heavy workload, low efficiency, date error, etc.

To solve these problems, there are some researches that have discussed the issues. A calibration of harmonic electrical energy meter has been discussed in a research and a calibration scheme has been represented (Lu, C.G. and Yao, L. 2010). However, it is only a theory discussion. Another research has designed a calibration system of power quality monitoring equipment based on

virtual instrument technology, which has improved calibration speed and has decreased data error, while ensuring calibration accuracy (He, W. *et al.* 2010). However, the system hasn't achieved the automation of grid-accessing test. The testing report cannot be generated automatically. Thus, this situation needs to be improved.

Based on the detailed analysis of operation environment and operation process of grid-accessing test of power quality monitoring equipment, an automatic batch test solution of power quality monitoring equipment is proposed in this paper. Based on the technical solution, an automatic batch test system of power quality monitoring equipment is designed, which will reduce the manual workload of grid-accessing test of power quality monitoring equipment, and will improve the work efficiency.

2 RESEARCH OF AUTOMATIC BATCH TEST TECHNIQUE OF POWER QUALITY MONITORING EQUIPMENT

2.1 *Operation process of grid-accessing test*

The purpose of automatic test technique is to achieve the automation of operation process of power quality monitoring equipment's grid-accessing test and improve work efficiency. The operation processes of grid-accessing test are as follows, which are also shown in Figure 1.

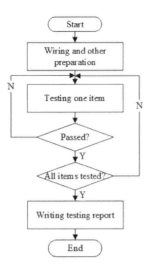

Figure 1. Workflow of grid-accessing test of power quality monitoring equipment.

1. Wiring preparation. Power quality standard source and power quality monitoring equipment are connected with a signal wire. Meanwhile, all equipment should be powered up;
2. Output signal control. Standard source is controlled manually to output power quality signals according to specified parameters;
3. Copy testing result. The testing result of power quality monitoring equipment will be copied manually or copied by supporting software. These two methods usually will only record one set of measurements;
4. Accuracy calculation. The accuracy of power quality monitoring equipment is usually calculated manually with Excel by tester;
5. Composing testing report. After all of the tests are done, the testing report is composed manually by tester.

According to the analysis of the operation process above, except wiring preparation, other steps can be finished automatically without manual involvement. That is, the automation of standard source controlling, testing result recording, accuracy calculating, and report generating can be finished in the automatic testing system.

2.2 Items of grid-accessing test

Currently, the grid-accessing test items of power quality monitoring equipment mainly cover two aspects: accuracy test and communication protocol test.

The accuracy test is mainly in accordance with "GB/T 19862-2005 General requirements for mon-

itoring equipment of power quality" (China Electric Power Press 2005), to test whether the accuracy index of power quality monitoring equipment meet the requirements. The testing items include steady-state power quality indexes such as voltage, frequency, voltage and current unbalance, harmonics, flicker, etc. Some regional power grid also requires accuracy test of transient-state power quality indexes such as voltage swell, voltage sag, voltage short interruption, and general electricity indexes such as current, power, etc. Items of accuracy test are shown in Tables 1 and 2. The actual test may differ from the contents of the tables.

Communication protocol test is to test whether the communication protocol or data format of power quality monitoring equipment meets the local communication requirement, and whether the monitoring equipment can be accessed to local power quality management platform. As the communication protocol requirement of power quality monitoring equipment hasn't been specified in "GB/T 19862-2005 General requirements for monitoring equipment of power quality", currently, the requirement of communication protocol of power quality equipment in different regional power grids has certain differentiation.

Table 1. Accuracy test of steady-state indexes.

Item	Testing reference and parameter settings	Error limit
Frequency	Based on "GB/T 19862-2005 General requirements for monitoring equipment of power quality"	± 0.01 Hz
Voltage deviation		$\pm 0.5\%$
Voltage unbalance		$\pm 0.2\%$
Current unbalance		$\pm 0.2\%$
Flicker		$\pm 5\%$

Table 2. Harmonic test (A-class equipment).

Measured item	Setting parameters*	Condition**	Error limit
Voltage	0.5%, 1%, 4%, 8%	$U_h \geq 1\%U_N$	$\pm 5\%U_h$
		$U_h < 1\%U_N$	$\pm 0.05\%U_N$
Current	1%, 3%, 20%	$I_h \geq 3\%I_N$	$\pm 5\%I_h$
		$I_h < 3\%I_N$	$\pm 0.15\%I_N$

*Based on the rated voltage and current signals, and fundamental frequency (50 Hz) of monitoring equipment, set the 3rd, 5th, 7th, 11th, 13th, 25th harmonic components as the table above required respectively. **U_N stands for rated voltage, I_N stands for rated current, U_h stands for harmonic voltage, I_h stands for harmonic current.

2.3 Standard source and its automatic control

Power quality standard source provides standard analogue signals to power quality monitoring equipment, which is for power quality monitoring equipment to monitor and calculate power quality indexes. The automatic test system is required to achieve the automation of the output power quality signals of standard source. The output signals should meet the requirement of section 2.2.

Currently, Fluke 6100 A power source is generally used as power quality standard source for the testing of power quality monitoring equipment in all regional levels of testing organizations. Research shows that Fluke 6100 A power source have the interface to communicate with external system with standard GPIB (General-Purpose Interface Bus) and control card (Fluke Corporation 2008). The communication protocol of Fluke 6100 A is standard IEEE-488 parallel bus interface standard. Meanwhile, Fluke 6100 A opens an interface list for external system to call for remote control, which includes various power quality index control interfaces. The list is shown in Table 3.

2.4 The access of various types of power quality monitoring equipment

To automatically record testing results of power quality monitoring equipment, the automatic test system should provide real-time communication with power quality monitoring equipment. Based on "GB/T 19862-2005 General requirements for monitoring equipment of power quality", power quality monitoring equipment should have communication function and communication protocol of Ethernet and RS-485. Thus, qualified power quality monitoring equipment can access to the automatic test system. Considering the inconsistency of communication protocol and data format of various types of power quality monitoring equipment in the market, the UAPI (Uniform Application Program Interface) technique is used to develop specified communication drive program for each kind of communication protocol of power quality monitoring equipment, which is to achieve the data recording of various types of power quality monitoring equipment (Li, P. et al. 2012).

2.5 Automatic generation of testing report

The writing of testing report is a heavy-workload step of grid-accessing test of power quality monitoring equipment. The tester needs to gather data from different equipment to a fixed formatted report, which is very cumbersome.

The research shows that the format of power quality monitoring equipment testing report of a certain region is fixed, while the format of testing report of different regions varies a lot. Hence, the format of testing report should not be fixed in the automatic test system. A flexible way of changing the format should be developed to satisfy the compatibility of testing report formats of different regions. Further research shows that the major differentiation of the format is the sequence of sections, the description of contents and the format of tables. The data of testing report, that is power quality index and its limit, testing results, etc., are similar.

Based on the analysis above, the Word-template method and anchor technique are proposed in this paper to support the automatic generation of different formats of testing reports. The template technique is for the testing report format differentiation of different region. The anchor technique is for the automatic content fill-in of the testing report.

Table 3. Control command of Fluke 6100A for power quality index.

Control command	Explanation
SOUR:FREQ	Set frequency of output signal
PHAS:VOLT:RANG	Set amplitude range of output voltage signal
PHAS:CURR:RANG	Set amplitude range of output current signal
PHAS:VOLT:MHAR: HARMn	Set amplitude and angel of output voltage harmonics
PHAS:CURR:MHAR: HARMn	Set amplitude and angel of output current harmonics
PHAS:VOLT:FLICK: FREQ	Set frequency of voltage variation
PHAS:VOLT:DIP:ENV	Set parameters of voltage dip

Conclusion: Fluke 6100A standard source can output various power quality indexes

3 THE DESIGN OF TEST SYSTEM OF POWER QUALITY MONITORING EQUIPMENT

3.1 Overall framework design

The main components of the automatic test system include power quality standard source, power quality monitoring equipment, and automatic test software. In order to achieve automatic test process, a closed-loop control method is used to connect the three components. The design of the functional framework of automatic test system is shown in Figure 2.

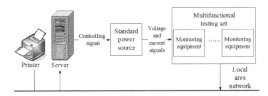

Figure 2. Functional framework of automatic test system.

The test software, which is deployed in the specific test server, mainly contains data collecting module, standard source controlling module, data processing module, and human-computer interaction module. Data collecting module is for the data collection from power quality monitoring equipment. Standard source controlling module is for the automatic control of the output signals of standard source. Data processing module is to calculate testing errors and assess the eligibility of test result. Human-computer interaction module is the key module of the test system, which is to set communication parameters, set testing items, implement testing scheme, edit, and generate testing report.

The major components of automatic test system are connected closed-looped by Ethernet. First, based on the default testing scheme, the standard source module controls power quality standard source to generate power quality signals (standard values). Then, power quality monitoring equipment receives power quality signals and outputs measured values. After that, automatic test software collects measured values and saves in local cache, then compare measured values with standard values, and calculate errors and measurement uncertainty. After the testing process, the testing data and results will be the output to the testing report automatically with VBA (Visual Basic for Application) technique (Walkenbach, J. 2010). If the external environment of the automatic test system is stable, the whole test process can be finished automatically without manual involvement.

3.2 Requirement analysis and function design

The power quality monitoring equipment automatic test system proposed in this paper is mainly for the grid accessing test of power quality monitoring equipment. After the requirement research and operation process analysis, the automatic test system should satisfy the following three explicit functional requirements:

1. Automatic test. Automatic test means that the whole testing process can be finished without manual involvement. The tester only needs to do the wiring and powering work of standard source and power quality monitoring equipment. After starting up the automatic test software and set the basic setting, the testing report (electronic version or printed version) can be generated after a period of time.
2. Batch test. Multiple power quality monitoring equipment can be tested in a single testing task.
3. Testing process control. During the testing process, the tester can pause/continue, and abort the process at any time, and retest the unqualified items. Meanwhile, tester can check testing data and progress at any time.

Actually, except the explicit requirements above, the design of automatic batch test system have many other implicit requirements. The main functional requirements of automatic batch test system are shown in Table 4.

Table 4. Main functional requirement of automatic batch test system.

No	Functional requirement	
1	Access function	Support the access of different types of power quality monitoring equipment, and support driver developing
2		Support IEC 61850 standard protocol system (International Electrotechnical Commission 2003)
3		Support the reading and analysis of PQDIF files (IEEE Standards Department 1995)
4		Support multiple power quality monitoring equipment testing tasks
5	Accuracy testing function	Satisfy the requirement of GB/T 19862 standard
6		Add/delete and adjust testing items and contents according to regional requirement
7	Automatic testing process	Support automatic test and manual interaction during testing process
8		Automatically control standard source output signals. Preferentially consider the support of Fluke 6100A, while consider the support of other standard source simultaneously
9		Automatically calculate data error based on the output data from testing equipment
10		The data display and warning function will automatically output unqualified items during testing process
11		Support automatically retest of unqualified items
12		Support the generation of regional report template, and reduce manual editing work.

3.3 *Human-computer interaction design*

The automatic batch test system is designed for improving working efficiency. Therefore, the interface of the system should be clear and concise. Tester can use the system without training or with simple tuition.

Based on the design idea above, the SDI-based (SDI, Single Document Interface) wizard-style interface prototype is used to design the automatic batch test system. Before starting the test, only three steps are needed: set basic parameters, set testing scheme, run the test program.

4 CONCLUSION

Based on the need of reducing labor cost and improving working efficiency of the grid-accessing test of power quality monitoring equipment, in this paper, an automatic batch test system is proposed. The closed-looped control theory is used to design the automatic batch test system. With the open-interface-based automatic control of standard source, UAPI-based accession of different types of power quality monitoring equipment, and template-and-anchor-technique-based automatic generation of testing report, the automation of grid-accessing test of power quality monitoring equipment can be achieved. With these features, the automatic batch test system can improve working efficiency. Besides, the automatic batch test system is compatible with various testing contents and testing reports of different regions, which possesses certain application value and promotion value.

REFERENCES

China Electric Power Press 2005. GB/T 19862-2005 General requirements for monitoring equipment of power quality.
Fluke Corporation 2008. Fluke 6100 A_User_Manual.
He, W. *et al.* 2010. Design and application of virtual instrument technology based calibration and detection system for power quality analyser. *Electrical Measurement & Instrumentation* 34:84–89.
International Electrotechnical Commission 2003. IEC 61850-10: Communication networks and systems in substations-part 10 Conformance testing.
IEEE Standards Department 1995. IEEE 1159.D3 Power quality data interchange format.
Li, P. *et al.* 2012. Research of the application of CAC-based substation unified communication platform in power quality monitoring system. *Electrotechnical Application* 22:26–29.
Lu, C.G. & Yao, L. 2010. Discussion on the calibration of harmonic electrical energy meter. *Electrical Measurement & Instrumentation* 47:35–42.
Walkenbach, J. 2010 *Excel 2010 power programming with VBA*. Beijing: Tsinghua University Press.
Wang, L. *et al.* 2014 Design and application of full-automatic calibration system for power quality monitoring equipment. *Electrical Measurement & Instrumentation* 51:95–100.

Advances in Power and Energy Engineering – Sun (Ed.)
© 2016 Taylor & Francis Group, London, ISBN 978-1-138-02846-3

Transmission line optimization on wind farm considering the effect of wind speed

Z.F. Jiang, C. Ma, X.G. He & Y.L. Zhong
Power Grid Planning Research Center, Guizhou Power Grid Corporation Ltd., CSG Guiyang, P.R. China

ABSTRACT: Avoidance of the wind curtailment often needs to use high capacity transmission lines and leads to a low utilization rate of such lines and an unsatisfactory economic effectiveness. By comparing the energy loss caused by wind curtailment with the investment on constructing new transmission lines, some optimization on the wind power delivery scheme is given with the consideration of wind speed, parameters of wind turbines and other factors involved. According to the analysis on the wind speed in two Canadian areas, named Regina and Swift, it is the distribution condition of wind speed and the rated wind speed of wind turbines that will largely influence the required capacity of transmission lines for the wind power delivery. If the wind curtailment is allowable to some extent, the joint delivery of wind power from different wind farms will have great effect on lowering the required capacity. Therefore, the effectiveness of the optimization scheme is proved by means of a case study.

1 INTRODUCTION

By the end of 2014, the installed wind power capacity soars up to 114609 MW in China, accounting for 31% of the global installed wind power capacity.

Based on the increasingly improvement on the awareness of environmental protection and progress of the wind power technology, wind power, as clean energy that can be exploited in large scale, has achieved strong growth in recent years and is supposed to afford more shares of total power supply.

However, the inherent randomness of wind speed results in the random output of wind turbines. Therefore, the output characteristics of a wind farm fully depend on the properties of wind speed, if large scale energy storage system is not considered.

Compared to conventional power sources, the annual utilization hour of wind farms is much less than that of coal-fired or hydraulic power plants. As a result, many power-transmission-paths will not be sufficiently utilized if these paths are planned and designed only on the basis of the wind installed power capacity. Additionally, since there are both correlation and complementarily among different wind farms in a same wind zone, it is reasonable to integrate these wind farms into a group and use the same paths to deliver power. Otherwise, it will lead to severely low path utilization and then bad economical efficiency.

Therefore, it is very necessary to study the joint output characteristics of different wind farms when it comes to planning and design for the wind power delivery.

To avoid Wind Energy Curtailment (WEC), power lines with larger cross-section have to be used, which will increase the cost. Otherwise, the curtailment is inevitable. So it is important to judge and weigh the loss caused by wind energy abandonment and the cost on improving and maintaining power paths, trying to optimize the wind power delivery lines.

Wind-speed forecasting (Bossanyi et al. 1985, Kamal et al. 1997, Alexiadis et al. 1998, Kariniotakis et al. 1996), the reliability evaluation of the power system including wind energy (Billinton et al. 2009, Ahmad et al. 2009, Billinton et al. 2008, Ribrant et al. 2007, Xie et al. 2012, Paul et al. 1983) and wind power control technology have recently been highlighted in the field of wind power, while some other research focuses on the way in estimating the credible capacity of wind farms (Hu et al. 2014, Voorspools et al. 2006, Wand Dee et al. 2006). However, few references refer to the factors, such as wind speed, wind turbine properties, which affect the selection of power lines for wind power delivery.

In this paper, based on the time-series wind speed model, it created the probability distribution models for the joint output power of wind farms; with the consideration on wind speed and wind turbines properties, the relationship between the amount and proportion of energy loss resulted from wind curtailment and the power delivery capacity of transmission paths is analyzed; according to the comprehensive comparison between curtailment loss and transmission paths construction cost, the optimization method for wind power transmission line are proposed for the purpose of obtaining maximal economical benefit, providing a new thought for the planning and designing of wind power transmission line.

2 TIME-SERIES MODELS FOR THE OUTPUT OF WIND FARMS AND WIND FARM GROUPS

When the parameters of wind turbine have been fixed, the output is determined by wind speed. The normal models for wind turbine output power can be expressed as follows (Ribrant et al. 2007), (Xie et al. 2012):

$$
P_t = \begin{cases}
0 & 0 \leq V_t < V_{ci} \\
(A + B \times V_t + C \times V_t^2) \times P_r & V_{ci} \leq V_t < V_r \\
P_r & V_r \leq V_t < V_{co} \\
0 & V_t \geq V_{co}
\end{cases} \tag{1}
$$

where P_t and P_r are the output power at time t and rated output power of wind turbine; V_{ci}, V_r and V_{co} are the cut-in, rated and cut-out wind speeds, respectively; the constants A, B, C can be calculated using V_{ci}, V_r and V_{co}, more details can be found in Reference 11.

As for a wind farm, to simplify the analysis, if ignoring the wake effects and equalizing the wind speed for each turbines, the corresponding output power of the farm can be indicated through equation (2),

$$
P_F(t) = \sum_{i=1}^{h} P_{i,t} \tag{2}
$$

where h = number of the wind turbines and $P_{i,t}$ = output of turbine i at the time of t.

Similarly, the models for the joint output of different wind farms at the time of t can be expressed as equation (3),

$$
P_G(t) = \sum_{F=1}^{z} P_F(t) \tag{3}
$$

where z = number of wind farms.

3 RELATIONSHIP BETWEEN DELIVERY CAPACITY AND WIND CURTAILMENT

Once the output of a wind farm has exceeded the delivery capacity of the transmission lines, wind curtailment becomes inevitable. Undoubtedly, different capacity corresponds to different curtailment, which can be expressed via the indicators such as annual power energy loss (APEL) of wind curtailment and wind curtailment proportion.

Suppose Pg is the delivery capacity of a power line, the APEL is shown as follows:

$$
APEL = \int_0^{8760} \left(P_G(t) - P_g \right) \times t dt \ \left[P_G(t) > P_g \right] \tag{4}
$$

The proportion of wind curtailment can be worked out through Equation (5),

$$
K_q = \frac{APEL}{W} \tag{5}
$$

where W = annual energy generated by wind farms group without delivery capacity limitation.

4 POWER TRANSMISSION LINES OPTIMIZATION MODELS POWER TRANSMISSION LINE COST ANALYSIS

4.1 Power transmission line cost analysis

In the process of power grid planning, selection of transmission lines is always determined by both technical and economical factors.

The cost of launching a power transmission line project is comprised of the initial investment of the project and the annual maintenance cost. In planning phase, the maintenance cost is often roughly estimated to be in proportion to the initial investment. Thus, the annual cost of a transmission line project is demonstrated in Equation (6),

$$
C_L = C_{L,i}(1 + x\%) \tag{6}
$$

where $x\%$ = percentile ratio between maintenance and initial investment; $C_{L,i}$ = equivalent annual worth of initial investment, which can be calculated via Equation (7) (Nilsson et al. 2007),

$$
C_{L,i} = I_k \times L \times \frac{(1+r)^{f_L} \times r}{(1+r)^{f_L} - 1} \tag{7}
$$

where I_k = initial investment of a transmission line for each kilometer, L = total length of the line, r = annual interest rate and f_L = operational life span of the line.

4.2 Commercial loss caused by wind curtailment

When selecting the transmission lines, high delivery capacity means high cost, while low delivery capacity often means poor transmission ability and then great extent of wind curtailment, imposing commercial loss on power generation utilities. The annual commercial loss caused by wind curtailment can be calculated via Equation (8),

$$C_q = APEL \times I_E \qquad (8)$$

where I_E = feed-in tariff of wind power.

4.3 The optimization models for transmission lines

In the process of planning the power transmission line of wind farms or wind farm groups, selection of power lines affects not only the cost but also the energy loss. Reducing the cost certainly means increasing the energy loss. These two factors are inversely correlated. Therefore, based on a basic scenario in which wind curtailment never exists, the saving investments ΔC_L and commercial loss ΔC_q is compared by gradually abating the investment of power transmission line projects. Here, the difference between ΔC_L and ΔC_q is defined as

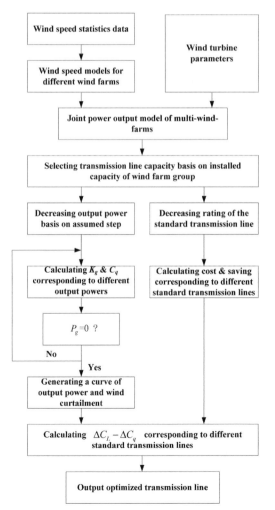

Figure 1. Flow Chart of Optimization.

annual comprehensive benefit C_T. To achieve the best C_T, the whole process of selecting the power lines should be optimized through Equation (9).

$$\max C_T = \Delta C_L\left(P_g, P_T\right) - \Delta C_q\left(P_g, P_T\right) \qquad (9)$$

By working out Equation (9), the optimal delivery capacity of transmission lines for wind power can be determined.

According to the types of actual delivery capacity of transmission lines used commonly, the optimization process is demonstrated in Figure 1.

5 CASE OF STUDIES

To make a better illustration of the optimized methods for wind power transmission line selection, based on the statistical data of wind speed per hour, which is gathered from the wind farms located in two sites of Regina and Swift in Canadian, a case of studies is analyzed under the assumption that each wind farm has 500 MW installed capacity.

5.1 Wind speed models

The wind speed ARMA time-series of the wind farms is given as follows (Ahmad et al. 2007):
Regina: ARMA(4,3):

$$y_t = 0.9336 y_{t-1} + 0.4506 y_{t-2} - 0.5545 y_{t-3} \\ + 0.1110 y_{t-4} + \alpha_t - 0.2033 \alpha_{t-1} - 0.4684 \alpha_{t-2} \\ + 0.2301 \alpha_{t-3}$$

$$\alpha_t \in NID\left(0, 0.409^2\right), \qquad (10)$$

Swift: ARMA(4,3):

$$y_t = 1.1772 y_{t-1} + 0.1001 y_{t-2} - 0.3572 y_{t-3} \\ + 0.0379 y_{t-4} + \alpha_t - 0.5030 \alpha_{t-1} - 0.2924 \alpha_{t-2} \\ + 0.1317 \alpha_{t-3}$$

$$\alpha_t \in NID\left(0, 0.525^2\right) \qquad (11)$$

The average wind speed of the above two wind farms in Regina and Swift is 5.405 and 5.795 m/s, with the standard deviation of 2.694 and 2.823 m/s.

5.2 The power output distribution of wind farms or wind farm groups

To comparatively analyze the power output difference between two single operational wind farms and the two wind farms in a joint operational group, the probabilistic distribution of these two scenarios is considered.

It is supposed that the cut-in speed, the rated speed the cut-out speed and the rated power of a wind turbine is respectively 3 m/s, 11 m/s, 25 m/s, and 2 MW. Based on the assumption above, by mathematical computation, the power output probabilistic distribution chart of each wind farm is respectively shown in Figures 2–3.

To analyze the joint scenarios effect on the requirement for delivery capacity of transmission paths, the power generated from the two wind farms, Regina and Swift are supposed to be delivered in bundle. Thus, based on the wind speed time-series models, the probabilistic distribution of joint power output is shown in Figure 4.

By comparison, joint operational wind farms are more likely to have steady power output than single operational wind farm. Obviously, the joint output is more concentrated.

5.3 The relationship curve of wind power delivery capacity and wind curtailment proportion

To clearly reflect the difference in the required wind power delivery capacity between the above two scenarios, the relationship curve of wind power delivery capacity and wind curtailment proportion is given in Figures 5–7.

As illustrated in Figures 5–6, through analyzing the time-series power output curve of each single wind farm, the power delivery capacity for Regina and Swift wind farm is respectively 173 MW and 334 MW, with the wind curtailment of 3%, which means the total delivery paths capacity should be over 507 MW. However, with the same curtailment, if joint delivered, such power requires for only 260 MW paths capacity (shown in Figure 7).

5.4 The wind turbine parameter effect analysis

Undoubtedly, different parameters of wind turbines correspond to different characteristics of wind power output. Taking Regina and Swift wind farms as an example, the effect on the required power delivery capacity caused by wind turbine

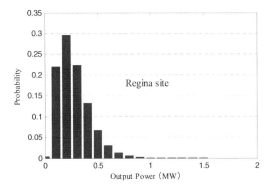

Figure 2. Output power probability distribution of a wind turbine for the Regina site.

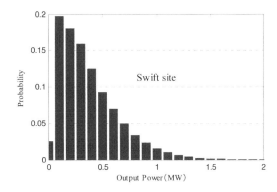

Figure 3. Output power probability distribution of a wind turbine for the Swift site.

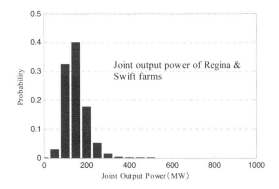

Figure 4. Output power probability distribution of wind farms Regina and Swift.

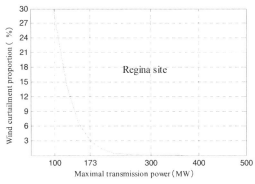

Figure 5. The curve of the transmission ability and the abandoned wind ratio for the Regina site.

parameters can be worked out by changing the cut-in, rated and cut-out wind speed of turbines and assuming the proportion of wind curtailment is no more than 3%. More detailed information is shown in Figures 8–10.

Based on Figures 8–10, the following issues can be seen. The impact on the requirement for wind power delivery capacity, caused by cut-in wind

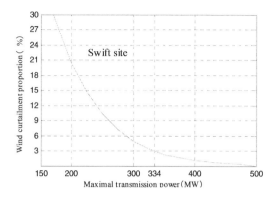

Figure 6. The curve of the transmission ability and the abandoned wind ratio for the Swift site.

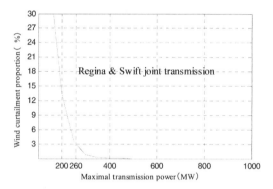

Figure 7. The curve of between transmission ability of wind farm group and the abandoned wind ratio for the Swift site.

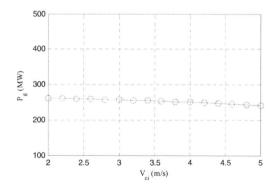

Figure 8. Relationship between cut-in speed and required path transmission ability for joint transmission.

speed, is tiny. Actually, with the increase in cut-in speed, the above requirement slightly decreases. It is the rated wind speed that largely impacts the power delivery capacity requirement. With the increase in rated speed, there will be a considerable decrease in such requirement. Furthermore,

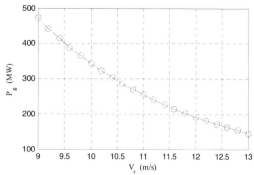

Figure 9. Relationship between rated speed and required path transmission ability for joint transmission.

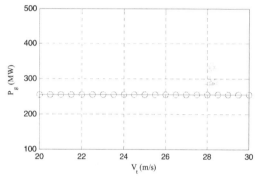

Figure 10. Relationship between cut-out speed and required path transmission ability for joint transmission.

cut-out wind speed scarcely has any effect on the power delivery requirement.

Wind curtailment due to insufficient power delivery capacity often occurs when the power output outnumbers the paths transmission capacity. Therefore, wind power generation units with different rated wind speed present in different levels of wind curtailment, when the wind speed is supposed to be the same. However, the probability that wind speed is more or less near the cut-out wind speed is comparatively low and the change in cut-out speed has little impact on wind curtailment. Cut-in wind speed affects the characteristics of power output of wind farms to some extent, but such effect mainly happens in the phase where the wind power output is low. In this situation, there will be mild modification in curves of wind power output, but it hardly affects the wind curtailment.

5.5 Optimization on power transmission lines

Based on certain wind turbine parameters, wind speed and the relationship between wind curtailment proportion and requirement for power delivery capacity, further analysis on annual loss caused by curtailment and the equivalent annual worth of power transmission projects investment, corresponding to different selection of power lines, can be made. Optimization on transmission lines for wind power can be achieved via Equation (9).

In accordance with Chinese national standard--technical rule for connecting wind farm to power system (General Administration of Quality Supervision et al. 2012), it is allowable that wind farms be connected to the power grid through a single line. Thus, in this article, the number of wind power delivery line for each wind farm is supposed to be one. Based on the assumed generation capacity of wind farms, it is considered to use 220 kV power lines to deliver wind power. The relevant parameters for the common types of 220 kV power lines are demonstrated in Table 1.

It is assumed that the length of transmission line is 100 km, the operational life span of the transmission line is 20 years, the annual interest rate is 6% and the feed-in tariff of wind energy is 0.6 yuan per kwh. Based on the above assumption, optimization on the selection of transmission lines for joint power delivery of Regina and Swift wind farms is made as follows, with the consideration that the total generation capacity of these two wind farms is 1000 MW.

According to Figure 11, it can be seen that by using line 9(LGJQ-600), whose continuous allowable transmission capacity is 374 MW, the maximal annual benefit can be achieved with the money value of 6.08 million yuan.

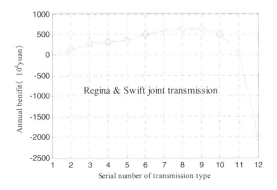

Figure 11. The annual benefits corresponding to different transmission lines (Regina Swift).

Apparently, by using transmission lines, whose delivery capacity is equal to or higher than the generation capacity of wind farms, the wind curtailment can be avoided, but it is not an optimal scheme with regards to economical benefit. In the process of optimized planning for wind power delivery lines, wind turbine parameters, wind speed, distance for power transmission and other relevant factors should be considered so that the reasonable and optimized selection of transmission lines as well as the rational selection of wind turbines can be made.

6 CONCLUSION

In actual operation, the high transmission capacity power lines are selected to avoid wind curtailment, which always leads to low utilization rate and economical compromise. With the consideration both on energy loss caused by wind curtailment and the investment on transmission line projects, an optimized planning scheme for wind power delivery line selection, aiming at maximal annual benefit, has been proposed. Based on the wind speed time-series model, it is analyzed the time-series power output of wind farms and wind farm groups, where the proportion of wind curtailment is used to demonstrate the levels of wind curtailment. The relationship between the proportion of wind curtailment and delivery capacity of transmission paths for wind power has been worked out. The impacts on the requirement for power delivery capacity due to wind turbine parameters have also been analyzed under certain level of wind curtailment. Therefore, the studies will provide a more economically reasonable method for wind power delivery planning.

Through the analysis on case of studies, it is demonstrated that the proposed method in this article

Table 1. Standard transmission line and its relative parameters (Power factor: 0.9).

Transmission lines type		Serial number	Continuous limit transmission capacity (MW)		Cost/km (10⁴ yuan)
			Allowable constant current (A)	Transmission capacity (MW)	
LG JQ	3 × 500	1	2898	994	210
	3 × 400	2	2535	869	185
	2 × 600	3	2180	748	170
	3 × 300	4	2130	730	160
	2 × 500	5	1932	663	155
	2 × 400	6	1690	580	145
	2 × 300	7	1420	487	135
	2 × 240	8	1220	418	125
	600	9	1090	374	110
	500	10	966	331	105
	400	11	845	290	100
	300	12	710	243	95

will make great saving in terms of transmission line projects, achieving optimized annual benefit compared to the traditional schemes in selecting types of transmission lines for wind power delivery. Considering certain extent of wind curtailment, there is a great difference in the requirement for power transmission capacity among different wind speed condition with the same generation capacity. With the same level of wind curtailment, the rated wind speed has a considerable effect on such requirement; while cut-in and cut-out wind speed have slight and even little effect on such requirement. By using the joint time-series power output models to analyze the required capacity of transmission lines, it can be concluded the requirement is much lower than that in the situation where the wind farm is working alone.

ACKNOWLEDGMENT

This work is supported by the China Southern Power Grid Program (No. K-GZ2013–468).

REFERENCES

Ahmad, S.D. & Mahmud, F.F. 2009. A Reliability Model of Large Wind Farms for Power System Adequacy Studies, *IEEE Trans. Energy Conversion* 24(3): 792–801.

Alexiadis, M. et al. 1998. Short term forecasting of wind speed and related electrical power, *Solar Energy* 63(1): 61–68.

Billinton, R. & Gao, Y. 2008. Multistate Wind Energy Conversion System Models for Adequacy Assessment of Generating Systems Incorporating Wind Energy, *IEEE Trans. Energy Conversion* 23(1): 163–170.

Billinton, R. et al. 2009. Composite System Adequacy Assessment Incorporating Large-Scale Wind Energy Conversion Systems Considering Wind Speed Correlation, *IEEE Trans. Power Systems* 24(3): 1375–1382.

Bossanyi, E.A. 1985. Short-term wind prediction using Kalman filters, *Wind Engineering* 9(2): 1–8.

General Administration of Quality Supervision et al. 2012. GB/T 19963–2011, Technical rule for connecting wind farm to power system, *Standards Press of China, Beijing.*

Hu, B. et al. 2014. Evaluation Model and Algorithm for Wind Farm Capacity Credit Considering Effect of Storage Systems Using the Bisection Method, *PES General Meeting | Conference & Exposition, 2014 IEEE, Washington.*

Kamal, L.Y. & Jafri, Z. 1997. Time series models to simulate and forecast hourly averaged wind speed in Wuetta, Pakistan, *Solar Energy* 61(1): 23–32.

Kariniotakis, G. et al. 1996. Wind power forecasting using advanced neural network models, *IEEE Trans. Energy Conversion* 11(4): 762–767.

Nilsson, J. & Bertling, L. 2007. Maintenance Management of Wind Power Systems Using Condition Monitoring Systems—Life Cycle Cost Analysis for Two Case Studies, *IEEE Trans. Energy Conversion* 22(1): 223–229.

Paul, G. & Kent, F. 1983. Development of a new procedure for reliability modeling of wind turbine generators, *IEEE Trans. Power Application Systems* 102(1): 134–143.

Ribrant, J. & Bertling, L.M. 2007. Survey of failures in wind power systems with focus on Swedish wind power plants during, *IEEE Trans. Energy Conversion* 22(1):167–173.

Voorspools, K.R. & D'Haeseleer, W.D. 2006. An analytical formula for the capacity credit of wind power, *Renewable Energy* 31(1): 45–54.

Wand Dee, W. & Billinton, R. 2006. Considering load-carrying capability and wind speed correlation of WECS in generation adequacy assessment, *IEEE Trans. Energy Conversion* 21(3): 734–741.

Xie, K. et al. 2012. Effect of Wind Speed on Wind Turbine Power Converter Reliability, *IEEE Trans. Energy Conversion* 27(1): 96–104.

Advances in Power and Energy Engineering – Sun (Ed.)
© 2016 Taylor & Francis Group, London, ISBN 978-1-138-02846-3

Optimal allocation of STATCOM using improved Harmony Search algorithm

X.Q. Xu, T. Zhang & Y.Q. Liu
College of Electrical Engineering and New Energy, China Three Gorges University, Yichang, China

ABSTRACT: The aim of this paper is to identify the optimal allocation of Static Synchronous Compensator (STATCOM) to enhance voltage stability and decrease the power loss in power systems. To overcome the Harmony Search (HS) algorithm, it is easy to fall into a local optimum problem when solving a high-dimensional multi-objective optimization problem, thus an improved Differential Evolution Harmony Search (DEHS) algorithm is proposed to solve the problem of optimal allocation of STATCOM. In order to validate the usefulness of the approach suggested here, a case study using the IEEE14 bus system is presented and discussed. Numerical results demonstrate that the proposed algorithm performs much better than the multi-objective HS in terms of the solution quality.

1 INTRODUCTION

One of the important operating tasks of power system is to keep the voltage within an allowable range for high power quality. When there is an increase in load demand, the generation and transmission units have to be operated at critical limits in power system. Voltage instability becomes one of the main problems in power system. When power system is operated close to the voltage stability limit, it becomes difficult to satisfy the reactive power demand. Flexible Alternating Current Transmission Systems (FACTS) devices have been proved to be effective for controlling power flow and regulating bus voltage in power systems (Dong 2011). The Static Synchronous Compensator (STATCOM) can adjust the injected reactive power to improve the voltage profile at the terminal bus (Bhattacharyya et al. 2014).

Recently, evolutionary algorithms have been employed to solve the optimal allocation of FACTS devices. A new Harmony Search (HS) algorithm, developed by Geem in 2001, was conceptualized using the musical process of searching for a perfect state of harmony. Harmony search algorithm with high solving speed has a simple structure, needs less parameters. With these characteristics, it has been successfully applied in dealing with various electrical engineering problems (Xu et al. 2013). In this paper, optimization problem of placement and sizing of STATCOM is solved by an improved Differential

Evolution Harmony Search (DEHS) algorithm. The proposed algorithm has been applied to the standard IEEE14 bus systems and the results are found to be more effective than the Multi-Objective Harmony Search (MOHS) algorithm.

2 MATHEMATICAL MODEL OF OPTIMAL PLACEMENT AND SIZING OF STATCOM

2.1 Objective function

In the paper, the mathematical model identifies the optimal allocation of STATCOM that was established by taking the minimum active power loss and voltage stability index L as target functions.

2.1.1 Minimization of active power loss
The active power loss (Seraj et al. 2013) calculated according to the formula are as follows:

$$\min P_{loss} = \sum_{i,j \in N_L} G_{ij}(U_i^2 + U_j^2 - 2U_iU_j \cos\theta_{ij}) \quad (1)$$

where N_L = number of branches in the power system; U_i = voltage magnitude at bus i; G_{ij} = real parts of nodal admittance matrix; θ_{ij} = phase difference between buses i and j.

2.1.2 Voltage stability index L
Kessel proposed a voltage stability index L (Liu et al. 2008) to correctly reflect the steady state of a system. Node in the system needs to be divided into two groups: one is characterized by the PQ node, defined as $\alpha_L = \{1, 2, ..., n_L\}$; the other comprises

This work was supported by National Natural Science Foundation of China (Grant 51307097) and the Graduate Scientific Research Foundation of China Three Gorges University (41040010314).

the PV node, defined as $\alpha_G = \{n_L + 1, n_L + 2, \ldots, n\}$. The transmission system can be represented in terms of a Hybrid (H) matrix.

$$\begin{bmatrix} U^L \\ I^G \end{bmatrix} = \begin{bmatrix} H_{11} & H_{12} \\ H_{21} & H_{22} \end{bmatrix} \begin{bmatrix} I^L \\ U^G \end{bmatrix} = \begin{bmatrix} Z^{LL} & F^{LG} \\ K^{GL} & Y^{GG} \end{bmatrix} \begin{bmatrix} I^L \\ U^G \end{bmatrix} \quad (2)$$

where Z^{LL}, F^{LG}, K^{GL} and Y^{GG} = sub matrices of the H-matrix; U^G and I^G = vectors of voltages and currents at PV nodes; U^L and I^L = vectors of voltages and currents at PQ nodes.

For each PQ node $j \in \alpha_L$, L_j is defined as follows:

$$L_j = \left| 1 - \frac{\sum\limits_{i \in \alpha_G} F_{ji} U_i}{U_j} \right| \quad (3)$$

where F_{ji} = ith row jth column element of the F^{LG} matrix.

The system voltage stability index L is defined as:

$$L = \max_{j \in \alpha_L} (L_j) \quad (4)$$

L value is between 0~1. The lower L means the better voltage stability degree of branches in system. The difference between L and 1.0 can be used as voltage stability margin of the system.

2.2 Constraints

2.2.1 Equality constraint
STATCOM can be modeled as an ideal reactive power generator which is capable of generating and absorbing reactive power. Thus, equality constraints include the active and reactive power flow equations (Zhao et al. 2012) of all nodes as follows:

$$\begin{cases} P_{Gi} - P_{Li} = U_i \sum\limits_{j=1}^{N} U_j (G_{ij} \cos\theta_{ij} + B_{ij} \sin\theta_{ij}) \\ Q_{Gi} + Q_{Ci} - Q_{Li} = U_i \sum\limits_{j=1}^{N} U_j (G_{ij} \sin\theta_{ij} - B_{ij} \cos\theta_{ij}) \end{cases} \quad (5)$$

where P_{Gi} and Q_{Gi} = active and reactive power of ith generator; P_{Li} and Q_{Li} = active and reactive loads at bus i; B_{ij} = imaginary parts of nodal admittance matrix; N = the number of buses.

2.2.2 Inequality constraints
The inequality constraints of this optimal problem can be divided into control variables and state variables constraints (Zhao et al. 2012).

Control variables include voltage magnitude generator, tap position of variable transformer and reactive power compensation capacity. State variables include the voltage magnitude load node and reactive power of generator. Inequality constraints are shown as follows:

$$\begin{cases} U_{Gi\min} \leq U_{Gi} \leq U_{Gi\max} & i = 1, 2, \cdots, N_G \\ T_{i\min} \leq T_i \leq T_{i\max} & i = 1, 2, \cdots, N_T \\ Q_{Ci\min} \leq Q_{Ci} \leq Q_{Ci\max} & i = 1, 2, \cdots, N_C \end{cases} \quad (6)$$

$$\begin{cases} U_{Li\min} \leq U_{Li} \leq U_{Li\max} & i = 1, 2, \cdots, N_L \\ Q_{Gi\min} \leq Q_{Gi} \leq Q_{Gi\max} & i = 1, 2, \cdots, N_G \end{cases} \quad (7)$$

where U_{Gi} = voltage magnitude of ith generator; T_i = tap position of ith variable transformer; Q_{Ci} = reactive power compensation capacity of ith compensation point; U_{Li} = voltage magnitude of ith load node; N_G, N_T, N_C and N_L = number of generators, variable transformers, reactive power compensation device and load nodes.

3 THE APPLICATION OF MULTI-OBJECTIVE HARMONY SEARCH ALGORITHM FOR OPTIMAL ALLOCATION OF STATCOM

3.1 Multi-objective harmony search algorithm

The details of Multi-Objective Harmony Search (MOHS) algorithm are described as follows (Wang 2012):

Step 1: Initialize the optimization algorithm parameters.
Initializing parameters include Harmony Memory Size (HMS), Harmony Memory Considering Rate ($HMCR$), Pitch Adjusting Rate (PAR) and arbitrary distance Band Width (BW).
Step 2: Initialize the harmony memory (HM) with randomly generated solution vectors.

$$HM = \{X_1, X_2, \cdots, X_{HMS}\}$$

$$X = \{x_1, x_2, \cdots, x_m\}$$

where m = pitch of the harmony.
Step 3: Generate Non-dominated solutions
Calculate objective function values of each harmony in HM. Rank the harmonies through non-dominated sorting approach. Harmonies with the first rank are conserved in solutions.
Step 4: Improvise a new harmony from the HM.
The new harmony vector x^{new} is generated by three rules: memory consideration, pitch adjustment and random selection. The procedure works as follows:

$$x_{i,j}^{new} = \begin{cases} x_{r,j} & rnd < HMCR \\ x_{j,\min} + rand & else \\ \times (x_{j,\max} - x_{j,\min}) \end{cases} \qquad (8)$$

where r = uniformly distributed random integer between 1~HMS; $x_{j,\max}$ and $x_{j,\min}$ = upper and lower boundaries of jth pitch; rnd, $rand$ = uniformly distributed random number between 0~1.

If any new vector generated from HM, pitch adjusting is carried out with PAR parameter, pitch adjustment and random selection is given as follows:

$$x_{i,j}^{new} = \begin{cases} x_{i,j} \pm rand \times BW & rand < PAR \\ x_{i,j} & else \end{cases} \qquad (9)$$

where PAR = pitch adjusting rate; BW = arbitrary distance band width.

Step 5: Update harmony memory.

Gather the new generated harmonies and HM, and then sort the harmonies with fast non-dominated sorting approach. Choose the solution vector from 1 to HMS remained in HM according to the rank.

Step 6: Check the stopping criterion.

If the maximal iteration number is satisfied, computation is terminated. Otherwise, turn to Step 4.

Step 7: Obtain the best compromised solution.

3.2 Code

The control variables are generator bus voltages, transformer tap positions, location and value of injected MVar of STATCOM. The harmonies coded as follows:

$$x_i = \left[U_{G1}, \cdots, U_{GN_G}, T_1, \cdots, T_{N_T}, L_1, \cdots, \right. \\ \left. L_{N_C}, Q_{C1}, \cdots, Q_{CN_C} \right] \qquad (10)$$

where U_G = voltage magnitude of generator; T = tap position of variable transformer; L = location point of compensate device; Q_C = reactive power compensation capacity.

3.3 Constant handling

In this paper, introducing a penalty function to construct the objective function to handling the solution vector of the state variable threshold crossings, the penalty function (He et al. 2013) works as follows:

$$f = \lambda_U \sum_{i=1}^{N_L} \left(\frac{\Delta U_i}{U_{i\max} - U_{i\min}} \right)^2 \\ + \lambda_G \sum_{i=1}^{N_G} \left(\frac{\Delta Q_{Gi}}{Q_{Gi\max} - Q_{Gi\min}} \right)^2 \qquad (11)$$

where

$$\Delta U_i = \begin{cases} U_{i\min} - U_i & U_i < U_{i\min} \\ 0 & U_{i\min} \leq U_i \leq U_{i\max} \\ U_i - U_{i\max} & U_{i\max} < U_i \end{cases}$$

$$\Delta Q_{Gi} = \begin{cases} Q_{Gi\min} - Q_{Gi} & Q_{Gi} < Q_{Gi\min} \\ 0 & Q_{Gi\min} \leq Q_{Gi} \leq Q_{Gi\max} \\ Q_{Gi} - Q_{Gi\max} & Q_{Gi\max} < Q_{Gi} \end{cases}$$

where λ_G and λ_U = penalty coefficient of the reactive power over limit of generators and the voltage violations of the load node; sets $\lambda_G = 5$, $\lambda_U = 10$; U_i = the voltage magnitude of ith load node.

4 IMPROVED HARMONY SEARCH ALGORITHM

To improve the optimization performance of harmony search algorithm, a Differential Evolution Harmony Search (DEHS) algorithm is presented in this paper. In the DEHS algorithm, mutation and crossover operation are adopted instead of harmony memory consideration and pitch adjustment operation (Cui et al. 2013). The DEHS and MOHS algorithm have differences in two aspects as follows:

In step 1, Harmony Memory Considering Rate ($HMCR$) and Pitch Adjusting Rate (PAR) are excluded from the DEHS, and mutagenic factor F and crossover rate CR are included in the DEHS.

In Step 4, the DEHS modifies improvisation step of the HS, The new harmony vector x^{new} is determined by two rules: Mutation and Crossover operation.

In order to take full advantage of the best individual information, accelerate the convergence rate, the variation works as formula (12) with global optimal search ability. To increase the diversity of harmony, harmonies randomly generated as formula (13) to take advantage of HM.

$$v_{ij} = x_{gbest,j} + F \times rand(x_{r1,j} - x_{r2,j}) \qquad (12)$$

$$v'_{ij} = x_{r1,j} + (2 * rand - 1) \times rand(x_{r2,j} - x_{r3,j}) \qquad (13)$$

where r_1, r_2 and r_3 = distinct integers uniformly chosen from the set$\{1, 2, \ldots, HMS\}$, $r_1 \neq r_2 \neq r_3 \neq best$; x_{gbest} = the best harmony vector in the current iteration; F = mutagenic factor.

The new harmony vector x_i^{new} generates by a binomial crossover operation as follows:

$$x_{ij}^{new} = \begin{cases} v_{ij} & rand < CR \\ v'_{ij} & else \end{cases} \qquad (14)$$

where CR = crossover rate.

The F and CR value increases dynamically with increasing generation as follows:

$$F_k = F_{min} + \frac{F_{max} - F_{min}}{\pi/2} \arctan k \qquad (15)$$

$$CR_k = CR_{max} - \frac{CR_{max} - CR_{min}}{K} k \qquad (16)$$

where K = maximum number of iterations; k = current number of iterations; F_{min} and F_{max} = minimum and maximum mutagenic factor; CR_{min} and CR_{max} = minimum and maximum crossover rate.

5 CASE STUDY AND SIMULATION RESULT

The proposed DEHS algorithm is applied to IEEE14 bus system (Zhang & Chen 1996). Take reference power of 100 MVA; use Newton–Raphson method to calculate the power flow. Without installation STATCOM, the total power loss is 17.91 MW, L index is equal to 0.076 for the IEEE14 bus system.

Parameter settings of simulation work as follows: the size of HM is 100; maximum iteration K equals 300. The upper and lower voltage limits for generators are 1.1 p.u. and 0.9 p.u., and that of load buses are 1.05 p.u. and 0.95 p.u.; the tap position of variable transformer ranges from 1 to 5, that is 1 ± 2.5%, the reactive power compensation capacity ranges from 100 MVar to 100 MVar.

In this article, two STATCOM are connected to the system. The optimal results of DEHS and MOHS are shown in Figure 1 and Figure 2 respectively.

Comparing the results, which are present in Figure 1 and Figure 2, the Pareto solution set of DEHS is better than the MOHS algorithm and more evenly distributed.

As optimal solution set gives a variety of options, out solutions of the two algorithms listed in Table 1. Figure 3 shows the bus voltage profile of the L-index minimum for the IEEE14 bus system.

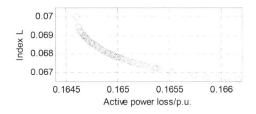

Figure 1. Pareto optimal solution set of DEHS algorithm.

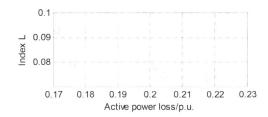

Figure 2. Pareto optimal solution set of MOHS algorithm.

Table 1. STATCOM location and sizes results for out solutions.

Performance	Ploss-Min		L index-Min	
	DEHS	MOHS	DEHS	MOHS
Location	9, 13	7, 12	9, 13	7, 12
Size (MVar)	25.60,	−39.34,	48.22,	59,
	10.2	−2	10.01	12.86
Ploss (MW)	16.461	17.113	16.616	21.760
L index	0.070	0.093	0.067	0.073

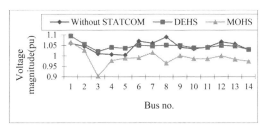

Figure 3. Bus voltage profile for different algorithm with minimum L index of IEEE14 bus system.

Comparing the out solutions of the two algorithms in Table 1, the objective functions of DEHS are lower than the MOHS algorithm and the compensation capacity is smaller. The out solutions of MOHS can only optimize one objective function.

6 CONCLUSION

The feasibility of the proposed DEHS algorithm for STATCOM allocation problem is demonstrated on the IEEE14 bus system. To avoid the problem of low efficiency and local optimum caused by HS, mutation and crossover operation are adopted instead of harmony memory consideration and pitch adjustment operation in the DEHS algorithm. Simulation results show that the DEHS algorithm is able to improve voltage profile along with loss minimization in the system. The results of the DEHS algorithm are better than MOHS. Take full advantage of the best individual information and current variables in HM, the proposed algorithm converges faster and get better solutions. Its ability of better convergence is utilized to find location and reactive power contribution of individual STATCOM.

REFERENCES

Bhattacharyya B., Gupta V.K. & Kumar S. 2014. Fuzzy-DE approach for the optimal placement of FACTS devices to relief congestion in a power system. International Conference on Control, Instrumentation, Energy and Communication, Calcutta, India: 291–295.

Cui Z.H., Gao L.Q., Ouyang H.B. & Li H.J. 2013. Hybrid differential evolution harmony search algorithm for numerical optimization problems, control and decision conference, Guiyang, China: 2930–2933.

Dong L. 2011. Research on optimal placement for FACTS devices based on reliability cost/benefits analysis. Chongqing: Chongqing University (in Chinese).

He X., Pang X., Zhu D.R. & Liu C.X. 2013. Multi-objective reactive power optimization based on chaos xparticle swarm optimization algorithm, Instrumentation and Measurement, Sensor Automation (IMSNA), Toronto, Canada: 1014–1017.

Liu J., Li D., Gao L.Q. & Song L.X. 2008. Vector evaluated adaptive particle swarm optimization algorithm for multi-objective reactive power optimization. Proceedings of the CSEE, 28(31): 22–28 (in Chinese).

Seraj A., Prof A.A., Prof M.A.K. & Sharad K.P. 2013. Optimal location of STATCOM using PSO in IEEE 30 bus system. International Journal Of Advance Research In Science And Engineering IJARSE, 2(5): 156–165.

Wang X.H. 2012. The modified Harmony Search Algorithm with Control Parameters Co-evolution and its Application, Shanghai: East China University.

Xu Y., Bo Z. & Zhang J.H. 2013. An improved self-adaptive harmony search algorithm for distribution system planning. IEEE conference on Power and Energy Society General Meeting, Vanvouver, BC: 1–5.

Zhang B.M. & Chen S.S. 1996. Advanced power network analysis. Beijing: Tsinghua University Press: 297–302 (in Chinese).

Zhao Y., Dong L. & Xie K.G. 2012. Research on optimal placement of FACTS devices based on reliability cost/benefits analysis, Power System Protection and Control, 40(1): 107–114 (in Chinese).

Advances in Power and Energy Engineering – Sun (Ed.)
© *2016 Taylor & Francis Group, London, ISBN 978-1-138-02846-3*

The harmonic model and its application to traction power supply system considering high-pass filter

D.D. Li, F.L. Zhou, Q. Liu, P. Zhu & H. Yu
College of Electrical Engineering, Southwest Jiaotong University, Chengdu, China

ABSTRACT: The high harmonic current which was produced by the AC-DC and AC-DC-AC locomotives of the high-speed railway may have caused traction resonance and harmonic current amplification in the traction network. Aiming at the harmonic resonance and harmonic current amplification in high speed railway systems, the amplification of the harmonic current was derived. The mathematical derivation for a harmonic model of the traction power supply system, considered with high-pass filter, was derived firstly. Based on this model the parameter for the harmonic filter in the traction network was optimized with the optimization objective of 10% of the total harmonic distortion with a filter in the head of the traction power supply system. An effective method to select the parameters of a high-passed filter was proposed. The results showed that: the smaller the resistance of a high-pass filter, the higher the capacity of the filter.

1 INTRODUCTION

With the development of high-speed railway, AC-DC-AC locomotives are widely adopted in the world. The content of high order harmonic such as 29th, 31th, 33th harmonic current increased, harmonic characteristics and reactive power characteristics have undergone significant changes.

The derivation and algorithm of harmonic model have been researched. Based on Carson theory the chain circuit analysis model was derived and has been proposed in (Lee Hanmin et al. 2006; Wang Qi 2009). Based on electromagnetic simulation software (such as MATLAB, PSCAD) some research had to be done about the influence factors of harmonic current magnification, including the length of traction network, locomotive position, the capacity of systems and traction transformers, the unit length impedance and admittance of traction network (Hu Haitao et al. 2012; Wu Mingli 2010; Zhang Yang & Liu Zhigang 2011). The suppression measures of harmonic current, including the compare of second-order high-pass filter and wave-trip high-pass filter. At present, most research has been carried out based on electromagnetic transient simulation analysis for research on harmonics. Relatively speaking the study on establishment of mathematical model was not perfect as it lacked the study and was mainly based

on the simulation model. Thus this paper derived a mathematical model of traction power supply system with wave-trap high-pass filter. The parameter optimization of filter was also presented based on the established mathematical model.

2 TRACTION POWER SUPPLY SYSTEM

2.1 Traction power supply system

The traction power supply system equivalent circuit is obtained in Figure 1 (LI Qunzhan 1991; MA Qixiao 2009). Z_{SS} is the impedance of the traction substation and system imputed to the secondary side of the transformer. Z_{hF} is the equivalent impedance of the filter. Z_a is the traction network impedance viewed from the locomotive to the

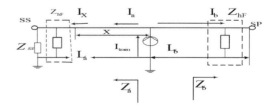

Figure 1. The equivalent diagram of traction network.

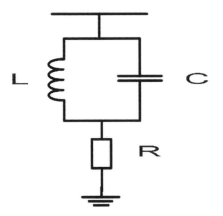

Figure 2.   The schematic diagram of high-pass filter.

direction of traction substation. Z_b is the traction network viewed from the locomotive to the direction of partition. I_a is the current flowing to the traction substation. I_b is the current flowing to the partition. I_x is the current away from the locomotive at the position of x (traction substation direction). L_a is the distance from locomotive to traction substation. L_b is the distance from locomotive to traction substation. The overall length of traction network $L_e = L_a + L_b$.

Different types of filters have different effects. This paper will take the wave-trap high-pass filter as an example. The wave-trap high-pass filter has advantages of high resistance at power frequency, low-pass at high-frequency, non-exchanged reactive power with the system and low loss.

2.2   The principle of high-pass filter

The wave-trap high-pass filter consists of L and C in parallel and R in series (Li Zihan et al. 2014). The structure is shown in Figure 2. The resonant frequency of high-pass filter $f_0 = \frac{1}{2\pi\sqrt{LC}}$. The parallel resonance occurs at power frequency. When $X_L = -X_C$, impedance of filter tends to infinity. When the frequency increases, the equivalent impedance decreases rapidly to provide a return path for the flow of high-order harmonics.

The equivalent impedance for high-pass filter is shown in follows.

$$Z_{hF} = \frac{X_{hL} \cdot X_{hC}}{X_{hL} + X_{hC}} + R = \frac{hX_C}{h^2 - 1} + R. \tag{1}$$

where X_{hL} = impedance value of reactor under h-order harmonic; X_{hC} = impedance value of capacitor under h-order harmonic; X_c = impedance value of capacitor under fundamental; And h = harmonic number.

3   THE DERIVATION OF TRACTION POWER SUPPLY SYSTEM MODEL

3.1   Traction network model

The π type equivalent circuit of transmission line is shown in Figure 3.

According to Figure 3 the parameter of π-type equivalent circuit can be obtained.

$$Z_\pi = Z_C(f)\sinh(\gamma(f)L_e) \tag{2}$$

$$\frac{Y\pi}{2} = \frac{\cos h(\gamma(f)L_e)1}{Z_C(f)\sinh(\gamma(f)L_e)} = \frac{1}{Z_C(f)}\tanh\left(\frac{\gamma(f)L_e}{2}\right) \tag{3}$$

where $Z_c(f)$ = characteristic impedance of transmission line; $\gamma(f)$ = propagation constant of transmission line; Z_π = equivalent impedance for transmission line of π type equivalent circuit; $\frac{Y\pi}{2}$ = equivalent admittance for transmission line of π type equivalent circuit; L_e = total length of transmission line.

3.2   The traction power supply system with filter in the head

The equivalent circuit of traction power supply system with filter in the head is shown as Figure 4.

It's known by the shunt relationship, the current flowing to traction substation is $I_a(f)$. The current away from the locomotive (traction substation direction) about x is $I_x(f)$. Then analyze the magnification of harmonic current in the head of the traction network when the locomotive is at the end of the traction network. The magnification of parallel current inflow of traction substation and

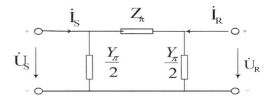

Figure 3.   The π equivalent circuit of transmission line.

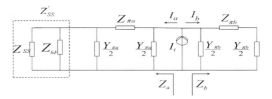

Figure 4.   The schematic diagram of traction network with high-pass filter in the head end.

540

filter is K_{magsb}. The current flowing into the traction substation is K_{mags}.

$$K_{magsb} = \cfrac{Z_C}{\cfrac{Z_{SS}\sin h(\gamma L_e)[hX_C + R(h^2-1)]}{(Z_{SS}+R)(h^2-1)+hX_C} + Z_C\cosh(\gamma L_e)} \tag{4}$$

$$K_{mags} = \frac{Z_C[hX_C + (h^2-1)R]}{Z_{SS}\sin h(\gamma L_e)[hX_C+(h^2-1)R]+Z_C\cosh(\gamma L_e)][(Z)_{SS}+R)(h^2-1)+hX_C]} \tag{5}$$

The magnification of harmonic current away from the locomotive (traction substation direction) about x without filter is K_{mag}.

$$K_{mag} = \frac{I_x}{I_t} = \cosh(\gamma(f)L_b) \times \frac{[Z_{SS}(f)\sin h(\gamma(f)(L_a-x))+Z_C(f)\cosh(\gamma(f)(L_a-x))]}{Z_{SS}(f)\sinh(\gamma(f)L_e)+Z_C(f)\cosh(\gamma(f)L_e)} \tag{6}$$

From Equation (6), it can be seen that the magnification of harmonic current at the traction substation is the most serious. It's expressed as follows.

$$K'_{mag} = \frac{Z_C(f)\cosh(\gamma L_b)}{Z_{SS}(f)\sin h(\gamma(f)L_e)+Z_C(f)\cos h(\gamma(f)L_e)} \tag{7}$$

4 CASE ANALYSIS

4.1 System parameters

There is one electrified railway that the power supply voltage level is 110 kV and the system short circuit capacity is 800 MVA. The traction transformer is single phase wiring with the capacity is 31.5 MVA and the percentage of short-circuit voltage is 10.5%. It's calculated that the equivalent impedance of traction substation and system is $Z_{SSh} = 1.18 + i9.75$ Ω/km. The length of traction power supply system arm is 30 km. The parameters of line are shown in Table 1.

Table 1. The line parameters.

Linear	Type	Material	DC resistance Ω/km
Contact line	CTSH-150	Copper and Tin 150	0.1580
Messenger line	JTMH-120	Cu Mg 120	0.2420
Positive feeder	LBGLJ-240/30	Aluminium clad steel	0.1133
Rail	P50		

The impedance and admittance matrix for per unit is obtained by Carson theory and phase-model transformation. The parallel multi-conductor system of traction network is equivalent to a wire by reduced order processing. The equivalent impedance and admittance for per unit length are $Z_h = 0.114 + i0.593$ Ω/k mand

$Y_h = i2.4765 \times 10^{-6}$ Ω/km. The contents of each harmonic for traction load are shown in Figure 5.

4.2 The system characteristic with filter in the head

There are differences in the filtering effect of each harmonic when the parameters of high-pass filter R and X_C are changing. When the harmonic frequency varies from 2 to 10 the relationship between maximum amplification multiple of harmonic current to R and X_C is shown in Table 2.

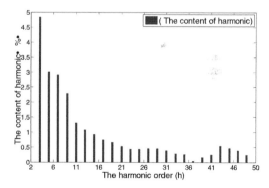

Figure 5. The harmonic content of traction load.

Table 2. Harmonic current enlargement times.

h	2	3	4	5
R	2	2	2	2
X_C	40	80	140	240
K	3.451	9.385	9.933	15.982
h	6	7	8	9
R	2	2	2	2
X_C	360	480	640	840
K	17.903	22.261	26.482	30.298

Thus when R and X_C change the amplification of harmonic current will be different at different times. Therefore it's necessary to analysis the filtering effect when R and X_C have different values and optimize the values of R and X_C.

Definitions: $K_h (h = 1, 2, 3...)$ is the amplification of the h-th harmonic current after filtering. When $h = 1$ it represents the magnification of fundamental current. $K_h' (h = 1, 2, 3...)$ is the amplification of the h-th harmonic current before filtering. When $h = 1$ it represents the magnification of fundamental current. $I_h (h = 1, 2, 3 ...)$ is the h-th harmonic current of locomotive. $h = 1$ represents the fundamental current. It's measured by the total harmonic distortion $\varepsilon_1 = \sqrt{\dfrac{\sum (I_h * K_h)^2}{(I_1 * K_1)^2}}$ after filtering. At the same time it's measured by the total harmonic distortion $\varepsilon_0 = \sqrt{\dfrac{\sum (I_h * K_h')^2}{(I_1 * K_1)^2}}$ before filtering. The compare waveform of total harmonic distortion with filter in the head or at the end is shown in Figure 8.

The total harmonic distortion is closely related to the filtering effect of filter. The filtering effect of filter is measured by the total harmonic distortion. The optimization goal is less than 10% of the

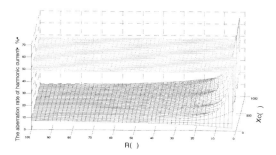

Figure 8. The harmonic current distortion with high-pass filter in the head end.

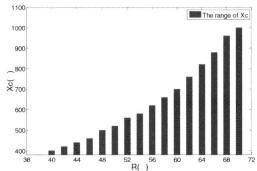

Figure 9. The parameter set for high-passed filter.

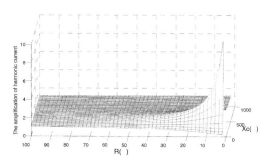

Figure 6. The harmonic current enlargement of the 4th harmonics.

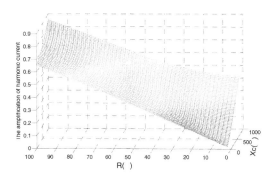

Figure 7. The harmonic current enlargement of the 20th harmonics.

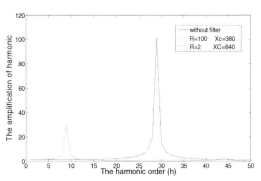

Figure 10. The contrast of optimization in the head end.

harmonic distortion. The cost is too high with higher capacity. The maximum capacity of filter is 2 MVA. The relationship between X_C and R is shown in Figure 9.

The relationship after optimization between R and X_C can be obtained from Figure 9 with filter in the head of the traction network. Take a set of parameter within the scope of optimization $R = 100\ \Omega$, $X_C = 380\ \Omega$. Take a set of parameter outside the scope of optimization $R = 2\ \Omega$, $X_C = 840\ \Omega$.

542

The relationship between harmonic number and the magnification of harmonic current is shown in Figure 10 without filter.

It's easy to get from Figure 10 that the magnification of harmonic is small after parameter optimization. It may lead to low harmonic resonance after filtering if parameters without optimization are selected.

5 CONCLUSIONS

It can be obtained by considering the derivation of the mathematical model for traction power supply system with filter.

1. The high-pass filter has some suppression characteristics of harmonic current resonance such as making the resonance point move later, decreasing magnification and it's obvious of harmonic suppression.
2. The filter parameter is relevant to the amplification of low harmonic. And the improper filter parameter can lead to low harmonic amplification significantly.
3. The selection of filter resistor. The capacity of filter is greater when the filter resistance is small. It has practical engineering value to provide a basis of high-pass filter parameter.

REFERENCES

Hu Haitao, He Zhengyou, Wang Jiangfeng, et al. Power flow calculation of high-speed railway traction network based on train-network compling systems [J]. Proceedings of the CSEE, 2012, 19:101–108+192.

Lee Hanmin, Lee changmu, JANGG ilsoo. Harmonic analysis of the Korean high-speed railway using the eight-port representation model [J]. IEEE Transactions on Power Delivery, 2006, 21(2):979–986.

Li Qunzhan. On the calculation and effect of harmonics [J] Journal of the China Railway Society, 199,1 (S1):59–69.

Li Zihan, Zhao Yuanzhe, Zhou Fulin, et al. Study on wave-trap high-pass filter in high-speed electrified railway [J]. Electric Railway, 2014, (1):13–17.

Ma Qixiao. Study of the harmonic resonance fault of highspeed electrified railway [D]. Chengdu: Southwest Jiaotong University, 2009.

Wang Qi. Simulation study on high-order harmonic resonance in traction power supply system of high-speed railway [D]. Chengdu: Southwest Jiaotong University, 2009.

Wu Mingli. Uniform chain circuit model for traction networks of electric railways [J]. Proceedings of the CSEE, 2010, 28:52–58.

Zhang Yang, Liuz higang. Modeling and characteristic analysis of harmonic in high-speed railway traction network based on PSCAD/EMTDC platform [J]. Power System Technology, 2011, 35(5):70–75.

Advances in Power and Energy Engineering – Sun (Ed.)

A comprehensive accessibility evaluation method for power network planning

J.Y. Xu, J. Yan, C. Wei & L. Qiao
State Grid Hubei Economic Research Institute, Wuhan, China

Y.X. Zhuo, Z.C. Wang, X. Zhou, C. Chen & X.N. Lin
State Key Laboratory of Advanced Electromagnetic Engineering and Technology,
Huazhong University of Science and Technology, Wuhan, China

ABSTRACT: To evaluate the optimum power network design, traditionally, the decision is made mainly depending on the expert's experience, which is somehow subjective. To solve this problem, an accessibility index based on power flow tracing for network evaluation is proposed, which is able to present the inner connectivity and power transfer ability of a power network. Then, a comprehensive evaluation including traditional short circuit, load flow, N-1, economic calculations, and the accessibility index is conducted on the power grid planning of a high-tech industrial park in China. The effectiveness of the proposed accessibility index is validated and network design comparisons are made from the viewpoint of reliability, accessibility and economy.

1 INTRODUCTION

The rapidly growing electricity power demand and the consumption of large-scale renewable energy require a more efficient power supply network (Ding M. et al. 2014). In addition to the network scheme design, determining the optimum alternative is also important, and, traditionally, the decision is made mainly depending on the expert's experience, which is somehow subjective (Xu G. et al. 2006, Zhang R. et al. 2010, Luo F. et al. 2012, Wang Y. et al. 2010). The improper network design may result in lower stability margin, extra power loss, and higher outage probability (Henderson M. et al. 2009, Tang Y. et al. 2012, Cha J. et al. 2009, Dobson I. et al. 2001, Yu Y. et al. 2012). Therefore, this paper proposes an accessibility index based on the power transfer ability of a network. Based on the power flow tracing method (Bialek J. 1996), the accessibility index analyzes the transmission paths from generators to loads, and explores the inner characteristics from the surface power flow level to the deep electric stream level, which provides an objective evaluation on the reasonableness of the power network scheme (Xu D. & Tang W. 2011, Razi Kazemi A.A. et al. 2011).

Therefore, in this paper, a comprehensive accessibility evaluation method is proposed, based on which three power supply network designs for load centers are analyzed from both electric and economic points of view.

2 THE ACCESSIBILITY INDEX

The accessibility of power transmission within a grid is reflected in many aspects: more transmission paths suggest more alternative options during power transmission from generator nodes to load nodes; shorter distance of transmission paths suggests lower resistance, smaller loss, and higher transmission efficiency; larger amounts of the total load suggest higher capability of power transmission. Based on these considerations, the definition of an integrated power grid accessibility index is proposed herein as follows:

$$S_{grid} = \frac{N_{path} \sum_{j=1}^{N_{path}} P_j}{\sum_{j=1}^{N_{path}} P_j d_j} \tag{1}$$

where S_{grid} is the integrated power grid accessibility index; N_{path} is the total count of power transmission paths in the grid; j is the serial number of a transmission path; P_j is the power value transferred by the transmission path j; and d_j is the electrical distance of the transmission path j, namely the reactance sum of all its constituent branches (Li Y. et al. 2010).

To calculate the above accessibility index, this paper uses an upstream flow tracing algorithm to analyze the power transmission path between generation–load node pairs. The upstream flow tracing uses the input power flow balance principle to track the input power flow of a given node, namely to track the power transmission path. The algorithm is based on the following assumption and principles:

1. Proportional distribution assumption: every MW of active power that flows out from a node contains proportional power from each input line into the node.
2. Lossless network equivalent principle: power flow transferred in a line is assumed as the average value of power flow into and out of the line, and power flow loss is deemed as the load evenly allocated on both ends of the line.
3. Upstream flow tracing principle: For the upstream tracking input power flow, the input power flow sum Pi of node i can be formulated as follows:

$$P_i = \sum_{j \in \alpha_i^{(u)}} \left| P_{i-j} \right| + P_{Gi} \qquad (2)$$

where $\alpha_i^{(u)}$ is the node collection that supplies power to node i (the corresponding lines must transfer power towards node i); P_{i-j} is the power flow into node i from line $j-i$; and P_{Gi} is the generated power of node i.

According to the lossless principle, $P_{i-j} = P_{j-i}$.

Line power P_{i-j} can be associated with the input power of node j by the relational expression, where $c_{ij} = P_{i-j}/P_j$. Therefore,

$$P_i = \sum_{j \in \alpha_i^{(u)}} c_{ij} P_j + P_{Gi} \qquad (3)$$

Thus, (3) can be transformed into

$$P_i - \sum_{j \in \alpha_i^{(u)}} c_{ij} P_j = P_{Gi} \quad \text{or} \quad A_u P = P_G \qquad (4)$$

where A_u is an n × n upstream power flow distribution matrix; P is the power flow vector through each node; and P_G is the generator node vector.

The element (i, j) of matrix A_u equals to

$$A_u(i,j) = \begin{cases} 1 & \text{for } i = j \\ -c_{ji} = P_{j-i}/P_j & \text{for } j \in \alpha_i^{(u)} \\ 0 & \text{otherwise} \end{cases} \qquad (5)$$

In (5), A_u is an asymmetric sparse matrix. From the definition of matrix A_u, it can be deduced that each row of the matrix marks the nodes supplying the power to the corresponding node, and the

power flow amount of each transmission path. By applying the flow tracing algorithm, the following factors can be calculated: which transmission paths transfer power to a certain lode node, and exactly how much power is supplied by these paths.

The power transmission path analysis of the IEEE57 bus system is shown in Figure 1(a). The power flow tracing of node 38 is shown in Figure 1(b).

In Figure 1(b), the number on each branch is the proportion of the load on node 38 through each

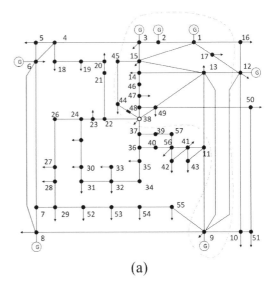

(a)

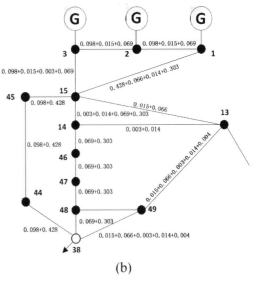

(b)

Figure 1. The flow tracing of node 38 in the IEEE57 system.

546

Table 1. Comparison of accessibility of different systems.

System	Accessibility	Total load (MW)	Power transmission paths	Total weighted distance
IEEE 9	48.07	315	6	39.3
IEEE 14	174.97	259	56	82.9
IEEE 24	635.20	2850	101	453.2
IEEE 30	442.43	189.2	90	38.5
IEEE 39	661.76	6254.2	57	538.7
IEEE 57	1427.70	1250.8	179	156.8
IEEE 118	2530.52	4242	431	722.5

transmission path. The addition of the number indicates that there are multiple transmission paths passing through that branch. Taking branch 44–38 as an example, "0.098 + 0.428" indicates that there are two transmission paths, searching "0.098" and "0.428", respectively, on upstream branches, and yields transmission path I 1-2-3-15-45-44-38 (0.098) and path II 1-15-45-44-38 (0.428), which means that path I and path II provide 9.8% and 42.8% power of load on node 38, respectively. Similarly, we can find other paths including 1-15-14-46-47-48-38, which accounts for 30.3% of the load. Therefore, generator 1 provides up to 73.1% power of load on node 38, while other generators account for only a small fraction.

The integrated power grid accessibility indices of different systems are compared in Table 1.

In general, the accessibility increases with grid size, as a larger system usually has a larger load and more transmission paths. Briefly, taking IEEE 24 and IEEE 39 as examples, although IEEE 39 has more nodes than and twice as many loads as the former system, IEEE 24 almost shares the same accessibility since it has a more complex structure that contains multi-circuit lines and more power transmission paths. For similar network candidate schemes, the one with more transmission paths and less weighted distance will have higher accessibility, as described later in Section 4.

3 COMPREHENSIVE ACCESSIBILITY EVALUATION PROCESS

The process of the comprehensive accessibility evaluation method is shown in Figure 2. Those proposed designs are subjected to basic electrical calculations in order to verify whether they satisfy the requirement of power flow, voltage level, short circuit current, and reliability. Lastly, the accessibility and economy calculation will be performed for further comparison of alternative solutions.

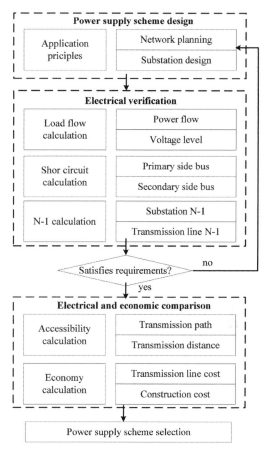

Figure 2. The process of comprehensive evaluation for the network.

4 CASE STUDY

4.1 Power supply network design

4.1.1 The original 500kV/220kV power grid design

The DHGX industrial park is a planned high-tech industrial development area of 175 square kilometers with a future load of 3214 MW, located in the city of Wuhan, China. Through integration with the planning power network, a 500 kV/110 kV direct transformation scheme is proposed and compared with the original 500 kV/220 kV network.

Based on the existing 500 kV/220 kV/110 kV power grid planning, a new 500 kV/220 kV and 8 220 kV/110 kV substation will be built. Among them, the DHGX 500 kV/220 kV substation will supply power for 220 kV JF, LQ1#, LQ2#, and HGW substations with a total load of 1625 MW and a power supply area of 83 square kilometers. The capacity-to-load ratio is 1.83 for the 220 kV network.

547

4.1.2 The 500 kV/110 kV direct transformation design

Substituting the original DHGX 500 kV/220 kV substation with 500 kV/110 kV direct transformation design, 220 kV QS, MJS, LQ1#, and HGW substations, shown in Figure 3, are cancelled, so that thirteen 110 kV substations can be supplied directly by the new DHGX 500 kV/110 kV substation. Three aspects are taken into consideration for the replacement of those four 220 kV substations:

1. 500 kV/110 kV substation should not substitute those 220 kV substations that are too far away and beyond the 110 kV transmission limit of 15 km.
2. The capacity of the 500 kV/110 kV substation is 4 × 500 MVA and approximately matches the total capacity of 220 kV QS, MJS, LQ1#, and HGW substations.
3. The 110 kV substations supplied by the 500 kV/110 kV substation can be connected to 220 kV/110 kV substations from other regions to ensure reliability.

The 500 kV/110 kV direct transformation can be designed as a new scheme with two independent 500 kV/110 kV substations, as shown in Figure 4, that is, DHGX 1# and DHGX 2#, with each substation capacity of 2 × 500 MVA, and supplies six and seven 110 kV substations, respectively.

Based on PSASP software, electrical calculations including load flow, short-circuit, and N-1 calculations are performed for the comparison and validation of the original 500 kV/220 kV design and the new 500 kV/110 kV scheme.

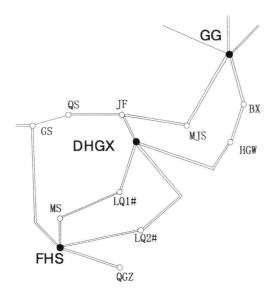

Figure 3. Original scheme—500 kV/220 kV power network design.

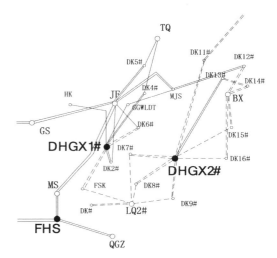

Figure 4. New scheme—500 kV/110 kV network design.

4.2 Short-circuit calculation

Three-phase short-circuit calculation result for buses of 500 kV/110 kV, and adjacent 500 kV/220 kV, 220 kV/110 kV, and 110 kV/35 kV substations are presented in Table 2.

As Table 2 shows, short-circuit current levels of the original and new schemes are roughly the same. The short circuit current levels are below 55 kA for the 500 kV bus, below 44 kA for the 220 kV bus, and below 35 kA for the 110 kV bus. Therefore, 63 kA, 50 kA, and 40 kA circuit breakers should be adopted for the 500 kV, 220 kV, and 110 kV buses, respectively, to avoid excessive short-circuit current.

4.3 Load flow and N-1 calculations

Load flow calculation graphs taking compensation solution 4 as an example are shown in Figure 5 and 6, which indicates that load flow distributions of the original and new schemes are reasonable, and transferring power is below 500 MVA on the 500 kV line, below 300 MVA on the 220 kV line, and below 50 MVA on the 110 kV line, which satisfies the requirements of the continuing operation current limitation of those transmission lines. Besides, N-1 calculations are performed for 500 kV lines, 500 kV/110 kV substations, 110 kV lines, and 110 kV substations, respectively. N-1 criterion can also be satisfied.

4.4 Accessibility evaluation

The accessibility evaluation results of the DHGX region based on the original and new schemes are given in Table 3.

Table 2. Comparison result of short circuit calculation.

| Substation | Voltage (kV) | Short circuit current (kA) | | Current limitations |
		Original scheme	New scheme	
DHGX	500	54.64	/	63
	220	32.88	/	50
DHGX 1#	500	/	53.72	63
	110	/	34.34	40
DHGX 2#	500	/	53.65	63
	110	/	34.24	40
FHS	500	49.88	49.27	63
	220	43.22	43.24	50
LQ2#	220	42.73	27.70	50
DK7#	110	24.47	28.99	40
DK8#	110	23.69	28.09	40
DK9#	110	20.63	27.68	40

Table 3. Accessibility of network schemes.

System	Accessibility	Total load (MW)	Power transmission paths	Total weighted distance
Original scheme	821.8	3214	121	473.2
New scheme	831.9	3214	114	440.4

Table 4. Construction comparison of the original and new schemes.

Category	Original scheme	New scheme
500 kV substation (billion $)	0.08	0.11
220 kV substation (billion $)	0.22	0.11
Transmission lines (billion $)	0.32	0.28
Total cost (billion $)	**0.62**	**0.50**
New site area (ha.)	**18.2**	**10.3**
500 kV site area (ha.)	3.5	3.6
220 kV site area (ha.)	14.7	6.7
Accessibility	**821.8**	**831.9**

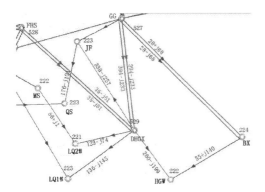

Figure 5. Load flow calculations for the original scheme.

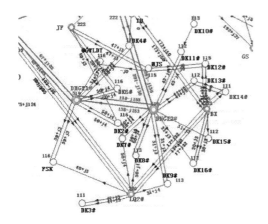

Figure 6. Load flow calculations for the new scheme.

Since the new scheme replaces four 220 kV substations that results in less power transmission paths, with a longer 500 kV transmission line towards the load center, the total weighted distance

is also reduced. Besides, the new scheme contains one more 500 kV/110 kV substation compared with the original scheme, decreasing the length of 110 kV transmission lines. Therefore, the comparison of the accessibility of the three networks is as follows: new scheme > original scheme, which indicates a better power transfer ability of the new scheme.

4.5 *Economic analysis*

As Table 4 shows, the total construction cost and the site area of the original scheme are higher than those of the new scheme, mainly because the original scheme needs to build more number of 220 kV substations than the new scheme. Since the new scheme has two independent 500 kV/110 kV substations, the investment of circuit breakers and the length of 110 kV lines are decreased, correspondingly the cost of the line also decreases. The results of the total investment comparison are as follows: original scheme > new scheme.

In summary, based on the above analysis, the new scheme is the optimum design in terms of total investment and accessibility while satisfying electrical requirements.

5 CONCLUSION

The proposed accessibility index is able to analyze the number, power, and electrical distance on transmission paths from generators to loads, and give an insight into the inner characteristics of the electric stream, which is validated by the IEEE benchmark test systems and the actual power supply network in Wuhan, and provide an objective evaluation on the reasonableness of the power network scheme.

A new 500 kV/110 kV scheme is proposed for the DHGX region in Wuhan, and demonstrated through the short circuit, load flow and N-1, accessibility index, and economic calculations. The results obtained indicate that the new scheme not only satisfies the electrical requirements but also has advantages on the substation-occupied areas and economy. Taking accessibility into account, the new scheme is the optimum design in terms of total investment and accessibility with a moderate site area occupation.

ACKNOWLEDGMENT

This work was supported in part by State Grid Hubei Electric Power Company and by the National Natural Science Foundation of China (51277082).

REFERENCES

Bialek, J. 1996. Tracing the flow of electricity. *IEEE Transaction On Power Delivery*, 143(4): 313–320.

Cha, J et al. 2009. Determination of a deterministic reliability criterion for composite power system expansion planning. *In Power & Energy Society General Meeting, 2009. PES 2009, 26–30 July 2009*: 1–6.

Ding, M. et al. 2014. A review on the effect of large-scale PV generation on power systems. *Proceedings of the CSEE*, 2014, 34(1): 1–14.

Dobson, I. et al. 2001. An initial model for complex dynamics in electric power system blackouts. *Proceedings of the Annual Hawaii International Conference on System Sciences*: 51–51.

Henderson, M. et al. 2009. Power system planning process and issues. *in Power & Energy Society General Meeting, 2009. PES 2009, 26–30 July 2009*: 1–6.

Li, Y. et al. 2010. The combination method which based on the shortest electric distance. *Power system protection and control*, 2010, 38(15): 24–27.

Luo, F. et al. 2012. Study and Application of Integrated Power Network Planning Information System. *In Power and Energy Engineering Conference, APPEEC 2012, 27–29 March 2012*: 1–5.

Razi Kazemi, A.A. et al. 2011. A probabilistic approach for remote terminal unit placement in power distribution systems, *in Telecommunications Energy Conference, INTELEC 2011, 9–13 October 2011*: 1–7.

Tang, Y. et al. 2012. Analysis and lessons of the blackout in Indian power grid on July 30 and 31, 2012. *Proceedings of the CSEE*, 2012, 32(25): 167–174.

Wang, Y. et al. 2010. Coordination Assessment of Power Plants and Power Network Based on Entropy and TOPSIS in Power System Planning. *In Electrical and Control Engineering, ICECE 2010, 25–27 June 2010*: 3503–3506.

Xu, D & Tang, W. 2011. Reliability evaluation of complex distribution networks based on regional accessibility analysis. Transactions of China Electrotechnical Society, 2011, 26(6): 172–178.

Xu, G. et al. 2006. A Novel Flexibility Evaluating Approach for Power System Planning under Deregulated Environment. *In Power System Technology, PowerCon 2006, 22–26 October 2006*: 1–6.

Yu, Y. et al. 2012. Detailed operation simulation based planning alternative evaluation of power system with large-scale wind power. *In Innovative Smart Grid Technologies—Asia, ISGT Asia 2012, 21–24 May 2012*: 1–5.

Zhang, R. et al. 2006. New challenges to power system planning and operation of smart grid development in China. *In Power System Technology, PowerCon 2010, 24–28 October 2006*: 1–8.

Advances in Power and Energy Engineering – Sun (Ed.)
© *2016 Taylor & Francis Group, London, ISBN 978-1-138-02846-3*

A novel ADP-based method for SCUC with power flow constraints

D.L. Long

Guilin Power Supply Bureau (Guangxi Power Grid Corporation), Guilin, Guangxi, China

ABSTRACT: This paper presents a novel method based on Approximate Dynamic Programming (ADP) for solving the Security-Constrained Unit Commitment (SCUC) problem with power flow constraints. A forward calculation based recursive approximate value function of SCUC is proposed for making decisions, which eliminates the accurate calculation of value function in dynamic programming. To cope with the SCUC constraints, a procedure for getting the contribution in the approximate value function is developed, which includes solving a simplified one-stage optimal power flow problem and calculating the start-up cost of performing action. In the procedure, the dynamic limits of active power are adopted to consider the ramp rate constraints. The proposed method performs effectively due to the short calculation time and substantial production cost savings, as compared with the existing methods. The results of test systems that range in size from 30 to 2737 buses over a 24 h horizon fully validate the effectiveness of the method.

1 INTRODUCTION

The Security-Constrained Unit Commitment (SCUC) in a power system provides a safeguarded and economic generation scheduling program on an hourly basis (Shafie-khah M. et al. 2014). Since it involves a large amount of 0–1 variables that represent the on/off status of all units, the SCUC problem is a typical large-scale nonlinear mixed-integer programming problem and it is hard to get the optimal solution. Moreover, the power flow equations, transmission flow constraints, and limitations of the bus voltage in the SCUC problem make the results suitable for the practical operation situation in power system. But these constraints increase the nonlinear degree and size of the optimization model, which makes the difficulties and calculations of solving the problem increase significantly.

Due to the complication of SCUC problem, several techniques have been proposed in order solve the problem effectively. An outer-approximation method is presented in (Ruiz, J.P. et al. 2013) for SCUC problems that can be modelled with DC power flow constraints. The method can achieved more accurate solutions in comparison with traditional piecewise linear approximation methodologies. In (Fu, Y. et al. 2005), the Benders decomposition is applied for separating the unit commitment in the master problem from the network security check in sub-problems. Capabilities and performances of core methodologies for the SCUC master problem and sub-problems are

evaluated in (Fu, Y. et al. 2013) through technical discussion and numerical testing. A fast bounding technique is proposed in (Wang, Y. et al. 2012) to solve the SCUC problem by improving the branch-and-cut algorithm.

Among the existed methods, the Dynamic Programming (DP) method (Guy, J.D. 1971) can consider variety constraint conditions easily, address the 0–1 variables directly and get global optimum solution (Lie, W.N. et al. 2014). However, DP suffers from the "curse of dimensionality". The search range must be cut to be suit to the application of DP, resulting in more cost for the generation scheduling. Approximate Dynamic Programming (ADP) is an iterative technique for searching optimal decisions in forward time so as to overcome the "curse of dimensionality" (Lee, J.Y. et al. 2012). Recently ADP algorithms (Powell, W.B. 2011) have been successfully applied to solve intractably large optimization problems (Powell, W.B. et al. 2012, Kim, J.H. & Powell, W.B. 2011, Anderson, R.N. et al. 2011). These algorithms develop approximations of the values of being in states and update the approximate values iteratively to reduce the calculation of value functions in DP.

This paper applies ADP to solve the SCUC problem with power flow constraints and proposes a novel ADP-based SCUC method (ADP-SCUC) in which the unit-scheduling problem is handled by ADP and the power flow constraints are handled by solving a simplified one-stage optimal power flow problem. A forward calculation based recursive approximate value function is proposed for making decisions, which eliminates the accurate calculation of value function in DP. By a mechanism of iteration, ADP-SCUC overcomes the bad effect caused by errors of approximate value estimation. The proposed method is tested on systems with the number of buses from 30 to 2737. The calculating results show that ADP-SCUC can get high-quality solutions for large-scale systems within a reasonable time. The "curse of dimensionality" problem has been alleviated. It is a very promising alternative for solving SCUC problems.

2 SCUC PROBLEM FORMULATION

In this section, we formulate the SCUC considered in this paper. The objective of SCUC is to determine a day-ahead UC to minimize the total cost of supplying the load, while meeting the operational and power flow constraints

$$\text{min. } F_{\text{Cost}} = \sum_{t \in T} \sum_{i \in S_G} \left[d_i^t F_i\left(P_{Gi}^t\right) + d_i^t \left(1 - d_i^{t-1}\right) C_{Ui}^t \right.$$
$$\left. + d_i^{t-1}\left(1 - d_i^t\right) C_{Di}^t \right] \tag{1}$$

subject to:

1. Power balance constraints (power flow equations)

$$\begin{cases} P_{Gi}^t - P_{Di}^t - \sum_{j \in S_B} [e_i^t(e_j^t G_{ij} - f_j^t B_{ij}) \\ \quad + f_i^t (f_j^t G_{ij} + e_j^t B_{ij})] = 0 \\ Q_{Gi}^t + Q_{cri}^t - Q_{Di}^t - \sum_{j \in S_B} [f_i^t(e_j^t G_{ij} - f_j^t B_{ij}) \\ \quad - e_i^t (f_j^t G_{ij} + e_j^t B_{ij})] = 0 \\ i \in S_B, t \in T \end{cases} \tag{2}$$

2. Transmission flow constraints

$$\underline{P}_{Lij} \leq P_{Lij}^t \leq \bar{P}_{Lij} \tag{3}$$

where

$$P_{Lij}^t = \left[(e_i^t)^2 + (f_i^t)^2 - e_i^t e_j^t - f_i^t f_j^t \right] G_{ij}$$
$$+ (e_i^t f_j^t - e_j^t f_i^t) B_{ij}, \quad ij \in S_L, t \in T$$

3. Spinning reserve requirements

$$\sum_{i \in S_G} d_i^t (\bar{P}_{Gi} - P_{Gi}^t) \geq R^t, \quad t \in T \tag{4}$$

4. Ramp rate limitations

$$\begin{cases} P_{Gi}^t - P_{Gi}^{t-1} \leq d_i^t P_{Gi}^{up} + (d_i^t - d_i^{t-1}) P_{Gi}^{start} + (1 - d_i^t) \bar{P}_{Gi} \\ P_{Gi}^{t-1} - P_{Gi}^t \leq d_i^t P_{Gi}^{down} + (d_i^{t-1} - d_i^t) P_{Gi}^{shut} + (1 - d_i^{t-1}) \bar{P}_{Gi} \end{cases} \tag{5}$$

5. Startup and shutdown characteristics of units

$$\begin{cases} (d_i^{t-1} - d_i^t)(T_i^{t-1} - \underline{T}_i^{on}) \geq 0 \\ (d_i^t - d_i^{t-1})(-T_i^{t-1} - \underline{T}_i^{off}) \geq 0 \end{cases} \tag{6}$$

6. Limits of active power

$$d_i^t \underline{P}_{Gi} \leq P_{Gi}^t \leq d_i^t \bar{P}_{Gi}, \quad i \in S_G \tag{7}$$

7. Limits of reactive power

$$\begin{cases} d_i^t \underline{Q}_{Gi} \leq Q_{Gi}^t \leq d_i^t \bar{Q}_{Gi}, \quad i \in S_G \\ \underline{Q}_{cri} \leq Q_{cri}^t \leq \bar{Q}_{cri}, \quad i \in S_{cr} \end{cases} \tag{8}$$

8. Limits of voltage at each bus

$$\bar{U}_i^2 \geq (e_i^t)^2 + (f_i^t)^2 \geq \underline{U}_i^2, \quad i \in S_B, t \in T \tag{9}$$

where T = total scheduling period; t = index for time; S_B = set of buses; S_G = set of generators; S_{cr} = set of reactive power sources containing no generators; S_L = set of transmission lines; $G_{ij} + jB_{ij}$ = transfer admittance between buses i and j; $d_i^t \in \{0,1\}$ = up/down status of unit $i \in S_G$; P_{Gi}^t = schedulable active power output of bus $i \in S_G$; Q_{Gi}^t = schedulable reactive power output of bus $i \in S_G$; Q_{cri}^t = schedulable reactive power output of bus $i \in S_{cr}$; P_{Lij}^t = active power of transmission line; $\bar{*}$ = upper limit of $*$; $\underline{*}$ = lower limit of $*$; $U_i^t = e_i^t + jf_i^t$ = node voltage; P_{Di}^t = active power demand; Q_{Di}^t = reactive power demand; R^t = system spinning reserve requirement; P_i^{up} = ramp up limit of unit $i \in S_G$; P_i^{down} = ramp down limit of unit $i \in S_G$; P_i^{start} startup ramp limit of unit $i \in S_G$; P_i^{shut} = shutdown ramp limit of unit $i \in S_G$; C_{Di}^t = shutdown cost; $F_i(P_{Gi}^t)$ = fuel cost; and C_{Ui}^t = startup cost. We define $F_i(P_{Gi}^t)$ and C_{Ui}^t as

$$F_i(P_{Gi}^t) = a_i + b_i P_{Gi}^t + c_i (P_{Gi}^t)^2 \tag{10}$$

$$C_{Ui}^t = \begin{cases} C_i^{hot} : \underline{T}_i^{off} \leq -T_i^t \leq \underline{T}_i^{off} + T_i^{cold} \\ C_i^{cold} : -T_i^t > \underline{T}_i^{off} + T_i^{cold} \end{cases} \tag{11}$$

a_i = constant term coefficient; b_i = monomial term coefficient; c_i = quadratic term coefficient; C_i^{cold} = cold start cost; C_i^{hot} = hot start cost; $\underline{T}_i^{on}$ = minimum up time; T_i^{off} = minimum down time; T_i^{cold} = cold start time; and T_i^t = continuously on (positive)/down (negative) time of unit i up to time t, namely

$$T_i^t = \begin{cases} > 0 & \text{the continuously on time} \\ < 0 & \text{the continuously down time} \end{cases} \quad (12)$$

3 IMPLEMENTATION OF ADP FOR SCUC PROBLEM

3.1 Reformulation in ADP process

In this section, we reformulate the SCUC problem in ADP process using pre-decision state variables and post-decision state variables.

Let $\hat{v}_k^t$ and $\tilde{V}_k^t$ be approximations of the value of being in pre-decision state S_k^t and post-decision state S_{k}^t. To reduce the calculation of value functions in DP, we construct an ADP solution process for updating the approximate value of pre-decision state variables and post-decision state variables by using iterations. Figure 1 illustrates this process.

In Figure 1, the pre-decision state S_k^t is the vector of up/down commitment statuses at time t in the kth iteration, in which we make a decision; the post-decision state $S_{a,k}^t$ is the vector of up/down commitment statuses at time t in the kth iteration after we have made a decision but before the next time step has arrived; $C^t(S_k^t, a_k^t)$ is the contribution from being in state S_k^t and taking the best action a_k^t.

The approximate value function $\hat{v}_k^t$ and $\tilde{V}_k^t$ can be calculated from the following equations (Powell, W.B. 2011):

$$\hat{v}_k^t = C^t(S_k^t, a_k^t) + \tilde{V}_{k-1}^t(S_{a,k}^t) \quad (13)$$

$$\tilde{V}_k^{t-1} = (1-\alpha)\tilde{V}_{k-1}^{t-1} + \alpha\hat{v}_k^t \quad (14)$$

where α = step length; and C^t = sum of fuel cost, startup cost and shutdown cost of a_k^t:

$$C^t(S_k^t, a_k^t) = \sum_{i=1}^{n} \Big[F_{\text{fuel}}^t + d_i^{t+1}\big(1-d_i^t\big) C_{\text{U}i}^t \\ + d_i^{t-1}\big(1-d_i^t\big) C_{\text{D}i}^t \Big] \quad (15)$$

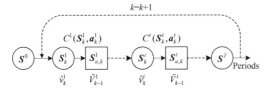

Figure 1. ADP solution process.

Here, F_{fuel}^t represents the fuel cost at time t, which is determined as the result of a simplified one-stage optimal power flow problem in this study.

The procedure of the proposed ADP-SCUC method is presented in the following steps.

Step 0: Read in system data. Set $k = 1$, $K_{\text{max}} = 50$, where k, K_{max}: iteration count and its maximum.
Step 1: Choose the initial state $S_{a,0}^t, t \in [1, ..., T]$ by Priority List method (PL) (Senjyu, T. et al. 2003), and initialize the approximate value function $\tilde{V}_0^t(S_{a,0}^t)$ of $S_{a,0}^t$.
WHILE ($k < K_{\text{max}}$) DO:
Step 2: Set $t = 1$, and choose the pre-decision state S_k^t.
Step 3: Solve the simplified one-stage optimal power flow problem to obtain the fuel cost F_{fuel}^t.
Step 4: Let a_k^t be the value of a^t that solves the decision function (16).

$$a_k^t = \{a^t \mid \forall a \in \mathbb{R}_k^t : F_S^t(S_k^t, a) \geq F_S^t(S_k^t, a^t)\} \quad (16)$$

where a^t is a decision at time t; $\mathbb{R}_k^t$ is the feasible decision space; $F_S^t(S_k^t, a^t)$ is the observation cost of choose the decision a^t in the pre-decision state S_k^t defined by

$$F_S^t(S_k^t, a^t) = C^t(S_k^t, a^t) + \tilde{V}_{k-1}^t(S_{a,k}^t) \quad (17)$$

Step 5: Evaluate the value at post-decision state approximately by (13) and (14).
Step 6: If $t < T$, set $t = t + 1$ and go to step 3; else proceed to next step.
Step 7: If $V_k^t = V_{k-1}^t$, let $a_k^t, t \in [1, ..., T]$ be the resulting decisions of the SCUC problem and stop; else set $k = k + 1$, go to step 2.
END DO.

3.2 Handling of SCUC constraints

We check the spinning reserve requirements and the startup and shutdown characteristics of units when we choose the pre-decision state. To cope with the power balance constraints, transmission flow constraints, and node voltage limits, a procedure for getting the contribution in the approximate value function is developed, which includes solving a simplified one-stage optimal power flow problem and calculating the start-up cost of performing action.

To be more specific, at a given pre-decision state, the up/down status of unit d_i^t is known. The fuel cost F_{fuel}^t in (15) is calculated by solving the simplified one-stage optimal power flow problem:

$$F_{\text{fuel}}^t = \min \sum_{i \in S_G} d_i^t f_i(P_{Gi}^t) \quad (18)$$

subject to: the system power flow equations (2), transmission flow constraints (3), limits of reactive power (8), limits of the voltage at each bus (9) and ramp rate constraints. We use dynamic limits of active power to cope with the ramp rate constraints. In Section 3.3, we will show the corresponding formulation of the dynamic limits of active power.

The aforementioned optimal power flow problem is a typical nonlinear optimization problem, the mathematical model of which was shown in the following.

$$\min \ f(\boldsymbol{x})$$
$$\text{s.t.} \ \boldsymbol{h}(\boldsymbol{x}) = 0 \qquad (19)$$
$$\underline{\boldsymbol{g}} \le \boldsymbol{g}(\boldsymbol{x}) \le \bar{\boldsymbol{g}}$$

where $\boldsymbol{x}$ = set of independent variables; $f(\boldsymbol{x})$ = objective function; $\boldsymbol{h}(\boldsymbol{x})$ = set of equality constraints; and $\boldsymbol{g}(\boldsymbol{x})$ = set of inequality constraints. Therefore, we can solve the simplified one-stage optimal power flow problem directly by using the interior point method (Wei, H. et al. 1998).

3.3 Dynamic limits of active power

During the course of updating of approximate value function, we get the contribution C^t by solving a simplified one-stage optimal power flow problem and calculating the start-up cost of performing action. At this time, there is a coupling relationship between each time period as a result of the existence of ramp rate limitations. In particular, we adopt the dynamic limits of active power so as to cope with the ramp rate limitations.

The dynamic limits of active power meet with the ramp rate limitations and the limits of active power simultaneously. When calculating C^t, we determine the dynamic limits of active power at time t by (20) based on the schedulable active power output of generators at time $t-1$, the upper and lower limits of active power output and the ramp rate limitations.

$$\begin{cases} P_{\text{G}i}^t \le \min\{\bar{P}_{\text{G}i}, P_{\text{G}i}^{t-1} + P_{\text{G}i}^{\text{up}}\} \\ P_{\text{G}i}^t \ge \max\{\underline{P}_{\text{G}i}, P_{\text{G}i}^{t-1} - P_{\text{G}i}^{\text{down}}\} \end{cases}, \ i \in S_{\text{G}} \qquad (20)$$

Assuming that unit $i \in S_{\text{G}}$ is under continuous operation at time t. Obviously, the up/down status $d_i^t = d_i^{t-1} = 1$ and the active power output of unit i at time $t-1$, namely $P_{\text{G}i}^{t-1}$, is known. We can determine the ramp rate limitations of unit i based on $P_{\text{G}i}^{t-1}$ (Fig. 2). Meanwhile the active power output of unit i at time t would be in the area of the upper limit $\bar{P}_{\text{G}i}$ and lower limit $\underline{P}_{\text{G}i}$. Then $P_{\text{G}i}^t$ should be in the overlaps of the range of ramp rate limitations

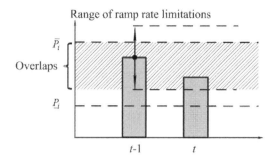

Figure 2. Dynamic limits of active power.

and the range of active power output, as shown by the shaded area.

When the unit $i \in S_{\text{G}}$ is starting up or shutting down, the active power output of thermal power unit is gradually increased or decreased, rather than fully loaded or zero. Therefore, it is necessary to take the startup ramp limit P_i^{start} and shutdown ramp limit P_i^{shut} into account. For the units that start up at time t ($d_i^t = 1, d_i^{t-1} = 0$), the dynamic limits of active power are presented as follows:

$$\underline{P}_{\text{G}i} \le P_{\text{G}i}^t \le \min\{\bar{P}_{\text{G}i}, P_{\text{G}i}^{\text{start}}\}, \ i \in S_{\text{G}} \qquad (21)$$

For the units that shut down at time $t + 1$ ($d_i^t = 1, d_i^{t+1} = 0$), the dynamic limits of active power at time t are presented as follows:

$$\begin{cases} P_{\text{G}i}^t \le \min\{\bar{P}_{\text{G}i}, P_{\text{G}i}^{\text{shut}}\} \\ P_{\text{G}i}^t \ge \max\{\underline{P}_{\text{G}i}, P_{\text{G}i}^{t-1} - P_{\text{G}i}^{\text{down}}\} \end{cases}, \ i \in S_{\text{G}} \qquad (22)$$

In the actual production, the unit the power output follows a predefined power trajectory when the unit is starting up or shutting down (Morales-Espana, G. et al. 2013). For the sake of simplicity, the predefined power trajectories of units are not considered; however, they can be easily introduced by limiting the scope of dynamic limits of active power to corresponding power output of the start-up ramp process or shut-down ramp process. And as the units in up period or down period do not provide spinning reserve capacity, the spinning reserve requirements should be checked again.

4 CASE STUDIES

The proposed ADP-SCUC method is tested on systems with the number of buses in the range of 30 to 2737, considering a 24 h scheduling horizon. The value parameters for system scales are depicted in Table 1, in which Polish-2737 is a simulation system. Let the spinning reserve requirements be 10% of hourly load demand. The ramp rate of each unit is selected as $P_{Gi}^{up} = P_{Gi}^{down} = 0.2\bar{P}_{Gi}, P_{Gi}^{start} = P_{Gi}^{shut} = 2\underline{P}_{Gi}$ (Han, D. et al. 2014). The hourly load distribution over the 24 h horizon in IEEE-30 bus, IEEE-300 bus and Polish-2737 bus systems are listed in Tables 2–4. The load data in the IEEE 118-bus system is the same with data in (Bai, X. & Wei, H. 2009).

Table 5 presents the comparison of the computing time and daily cost between UC and SCUC. Obviously, the proposed method can converge to the optimal solution reliably within a relatively

Table 1. Value parameters for system scales.

System	Buses	Units	Lines
IEEE-30	30	6	41
IEEE-118	118	54	186
IEEE-300	300	69	411
Polish-2737	2,737	117	3,506

Table 2. Hourly load in IEEE-30 bus system.

Hour	P_D (MW)	Q_D (MW)	Hour	P_D (MW)	Q_D (MW)
1	163.01	63.01	13	170.10	65.75
2	160.18	61.92	14	167.27	64.65
3	154.51	59.72	15	175.77	67.94
4	141.75	54.79	16	177.19	68.49
5	148.84	57.53	17	173.64	67.12
6	155.93	60.27	18	176.48	68.22
7	163.01	63.01	19	180.02	69.59
8	168.68	65.20	20	182.86	70.68
9	171.52	66.30	21	184.28	71.23
10	175.77	67.94	22	177.19	68.49
11	176.48	68.22	23	175.06	67.67
12	172.94	66.85	24	171.52	66.30

Table 3. Hourly load in IEEE-300 bus system.

Hour	P_D (MW)	Q_D (MW)	Hour	P_D (MW)	Q_D (MW)
1	28,314	10,945	13	28,611	11,059
2	28,195	10,899	14	28,492	11,013
3	27,958	10,807	15	28,849	11,151
4	27,423	10,600	16	28,908	11,174
5	27,720	10,715	17	28,760	11,117
6	28,017	10,830	18	28,878	11,163
7	28,314	10,945	19	29,027	11,220
8	28,552	11,036	20	29,146	11,266
9	28,670	11,082	21	29,205	11,289
10	28,849	11,151	22	28,908	11,174
11	28,878	11,163	23	28,819	11,140
12	28,730	11,105	24	28,670	11,082

Table 4. Hourly load in Polish-2737 bus system.

Hour	P_D (MW)	Q_D (MW)	Hour	P_D (MW)	Q_D (MW)
1	12,474	4,821.71	13	12,771	5,641.71
2	12,355.2	4,775.78	14	12,652.2	5,631.57
3	12,117.6	4,683.94	15	13,008.6	7,629.21
4	11,583	4,477.31	16	13,068	11,365.40
5	11,880	4,592.10	17	12,919.5	8,489.62
6	12,177	4,706.88	18	13,038.3	7,475.73
7	12,474	4,821.71	19	13,186.8	6,844.83
8	12,711.6	4,913.55	20	13,305.6	6,461.89
9	12,830.4	4,959.47	21	13,365	6,300.13
10	13,008.6	5,028.34	22	13,068	5,166.09
11	13,038.3	5,039.82	23	12,978.9	4,904.13
12	12,889.8	4,982.42	24	12,830.4	4,841.39

Table 5. Comparison of SCUC with UC.

System	Computing time (s)		Daily cost ($)	
	UC	SCUC	UC	SCUC
IEEE-30	0.3	6.5	12,388	12,489
IEEE-118	2.9	204.4	1,643,263	1,707,114
IEEE-300	3.1	252.3	22,070,885	22,114,781
Polish-2737	7.1	630.8	28,110,971	29,062,462

Table 6. Search range.

System	State space		Feasible states		Post-decision states	
	UC	SCUC	UC	SCUC	UC	SCUC
IEEE-30	186	186	172	172	83	83
IEEE-118	1,653	2,917	1,405	2,139	702	831
IEEE-300	1,491	1,751	1,423	1,519	243	358
Polish-2737	12,256	12,663	11,462	11,714	2,341	2,457

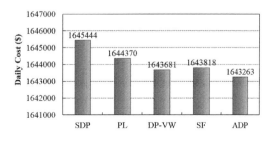

Figure 3. Results of different methods.

short time. The computing time and daily cost for each SCUC problem are higher than these for UC problem, since the SCUC problem considers the network security constraints. Table 6 shows the search range. The number of post-decision states evaluated in ADP is 40.65% of the number of feasible states on average.

Figure 3 compares the results of SDP (Bai, X. & Wei, H. 2009), PL (Senjyu, T. et al. 2003), DP-VW (Ouyang, Z. & Shahidehpour, S.M. 1991), SF (Hosseini, S.H. et al. 2007) and the proposed ADP-SCUC method. As seen from Figure 3, the solution given by ADP-SCUC is better than other methods.

5 CONCLUSIONS

In response to the computational difficulties brought by SCUC problem with optimal power flow constraints, this paper has established a novel ADP-based method, which can handle the power flow constraints and 0–1 variables directly. It makes use of approximating, instead of calculating exactly, the value functions for SCUC to alleviate the "curse of dimensionality" problem in dynamic programming. The effectiveness and validity of the proposed method have been demonstrated through its applications to test systems with the number of buses from 30 to 2737. It has been found that the proposed method is very powerful and efficient and it outperforms many other methods. It can alleviate the "curse of dimensionality" problem by estimating the values of about 40.65% of the feasible states. The method can be used to find good-quality solution in relatively short time for large scale application.

REFERENCES

Anderson, R.N. et al. 2011. Adaptive stochastic control for the smart grid. *Proceedings of the IEEE* 99(6): 1098–1115.
Bai, X. & Wei, H. 2009. Semi-definite programming-based method for security-constrained unit commitment with operational and optimal power flow constraints. *IET Gener Transm Distrib* 3(2): 182–197.
Fu, Y. et al. 2005. Security-constrained unit commitment with AC constraints. *IEEE Trans Power Syst* 20(3): 1538–1550.
Fu, Y. et al. 2013. Modeling and solution of the large-scale security-constrained unit commitment. *IEEE Trans Power Syst* 28(4): 3524–3533.

Guy, J.D. 1971. Security constrained unit commitment. *IEEE Transactions on Power Apparatus and Systems* PAS90(3): 1385–1390.
Hosseini, S.H. et al. 2007. A novel straighforward unit commitment method for large-scale power systems. *IEEE Trans Power Syst* 22(4): 2134–2143.
Han, D. et al. 2014. Outer approximation and outer-inner approximation approaches for unit commitment problem. *IEEE Trans Power Syst* 29(2): 505–513.
Kim, J.H. & Powell, W.B. 2011. Optimal energy commitments with storage and intermittent supply. *Operations Research* 59(6): 1347–1360.
Lee, J.Y. et al. 2012. Approximate dynamic programming for continuous-time linear quadratic regulator problems: relaxation of known input-coupling matrix assumption. *IET Control Theory Appl* 6(13): 2063–2075.
Lie, W.N. et al. 2014. Motion vector recovery for video error concealment by using iterative dynamic programming optimization. *IEEE Trans Power Syst* 16(1): 216–227.
Morales-Espana, G. et al. 2013. Tight and compact MILP formulation of start-up and shut-down ramping in unit commitment. *IEEE Trans Power Syst* 28(2): 1288–1296.
Ouyang, Z. & Shahidehpour, S.M. 1991. An intelligent dynamic programming for unit commitment application. *IEEE Trans Power Syst* 6(3): 1203–1209.
Powell, W.B. 2011. Approximate dynamic programming: Solving the curses of dimensionality (2nd ed.): 139–141. Hoboken: Wiley.
Powell, W.B. et al. 2012. SMART: a stochastic multiscale model for the analysis of energy resources, technology, and policy. *INFORMS Journal on Computing* 24(4): 665–682.
Ruiz, J.P. et al. 2013. Out-approximation method for security constrained unit commitment. *IET Gener Transm Distrib* 7(11): 1210–1218.
Senjyu, T. et al. 2003. A fast technique for unit commitment problem by extended priority list. *IEEE Trans Power Syst* 18(2): 882–888.
Shafie-khah, M. et al. 2014. Fast and accurate solution for the SCUC problem in large-scale power systems using adapted binary programming and enhanced dual neural network. *Energy Conversion and Mangement* 78: 477–485.
Wei, H. et al. 1998. An interior point nonlinear programming for optimal power flow problems with a novel data structure. *IEEE Trans Power Syst* 13(3): 870–877.
Wang, Y. et al. 2012. Fast bounding technique for branch-and-cut algorithm based monthly SCUC. In: *IEEE Power and Energy Society General Meeting*, Beijing, China.

Advances in Power and Energy Engineering – Sun (Ed.)
© *2016 Taylor & Francis Group, London, ISBN 978-1-138-02846-3*

Multi-source generation optimal dispatching strategy with significant renewable energy penetration under non-market condition

S. Ma, X.F. Li, Y.P. Xu & Y.F. Wang
China Electric Power Research Institute, Beijing, China

J.T. Zhang
Qinghai Electric Power Research Institute, Qinghai, China

ABSTRACT: With the rapid development of renewable energy generation in China, great challenges of renewable energy operation and dispatching are met by the system operators of the power grid with large-scale renewable energy generation. In a non-market power system, the traditional dispatching strategy usually focuses on conventional power sources and could not fully apply the flexibility of the system regulation ability. Therefore, it is difficult to properly cope with the increasing curtailment of renewable energy generation. In this paper, the bottlenecks of large-scale renewable energy generation dispatching and operation in China are presented, and a new dispatching strategy with multi-source generation coordinated optimization is proposed for more renewable energy generation accommodation. The optimal strategy is based on the Mixed-Integer Linear Programming (MILP) model, considering actual system operation constraints in a non-market situation. Moreover, validation is provided using the actual data from Qinghai Province. The time-series production simulation is also provided. The results indicate that the proposed strategy can effectively decrease the curtailment of renewable energy generation.

1 INTRODUCTION

The renewable generation accommodation capability of the grid has relevance to scale, area, and time horizon. Due to the natural randomness of renewable generation, this capability is basically about a coordinated dispatching between renewable energy consumption and power system operation with different types of power sources. The main issues that restrict renewable generation integration are insufficient system regulation capacity and transmission line limitation. In a non-market system, the common dispatching strategy usually focuses on conventional units and has no (or limited) coordinated consideration of renewables with other power sources. Recently, with large-scale renewable integration, this has resulted in a significant limitation of renewable energy utilization. Meanwhile, the renewable energy consumption capability of the system is difficult to be at a maximum. Several papers have discussed the power system operation with renewable energy generation integration, including long-term planning (Chen et al. 2010, Francois et al. 2008), system accommodation capability evaluation (Barth et al. 2006, Zhi et al. 2010), unit commitment model (Hetzer et al. 2008, Tuohy et al. 2009), optimal dispatching and scheduling (Zhaosui et al. 2011), coordinated dispatching and control (Guoqiang et al. 2011), and renewable energy curtailment (Fink et al. 2012, Qian-kun 2012).

However, most research mainly concentrates on the electricity market theory, without considering multi-source coordinated optimal dispatching; therefore, it is quite impractical for the system operator to schedule renewable generation in an optimal way in a non-market situation. This paper discusses the difficulties of renewable generation operation and problems of the conventional scheduling strategy in the Chinese power system. Based on the flexibility of hydro and thermal units, from the complementary and coordinated angle of wind–PV–hydro–thermal generation, a multi-source generation optimal dispatching strategy is proposed, taking into consideration the non-market situation of China. Validation is presented by real operation data from the Qinghai grid. Some of the conclusions may give directions for decreasing renewable generation curtailment.

2 RENEWABLE ENERGY GENERATION CURTAILMENT

By the end of 2014, China's wind power installed capacity was 96.4 GW, accounting for 7% of gross installed capacity. Wind power generation was 153 TWh, accounting for 2.8% of the total demand. In addition, the PV power installed capacity reached 28 GW, while the electricity generation was 25 TWh. The curtailed wind power generation was 13.3 TWh in 2014 according to NEA's data, and the average curtailment percentage of wind power was 8% in China.

2.1 *Insufficient flexibility of conventional generation units*

The wind is extremely strong in the winter in northern China, especially during midnight and early morning. However, at that time, the demand is always the lowest of the day and a number of thermal units have to maintain a certain level of output to ensure enough heat supply during the heating period. The regulation capacity of the system decreases further since it is impossible for those heat supply thermal units to stop running or operate at a lower level.

Moreover, compared with the EU and the US, the power source structure of China is much less flexible. For example, the ratios of flexible power capacity in the north and northeast grid are only 7.8% and 3.4%, respectively. Over 70% of the power capacity is thermal units in the northern area, and more than 60% of the thermal units must supply heat in the heating period. The regulation capability of conventional thermal units is quite inflexible during the heating period; therefore, curtailment is unavoidable, as shown in Figure 1.

2.2 *Transmission limitation*

Wind energy is useful only when the wind power can be transmitted from the wind turbine to terminal loads via the grid. However, with the rapid development of renewable energy, an appropriate construction plan for both renewable plants and transmission facilities has been lacking. The construction period of renewable plants is shorter than that of trans-

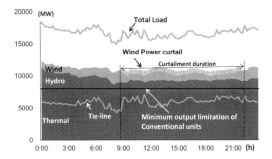

Figure 1. Curtailment due to inflexible regulation capacity.

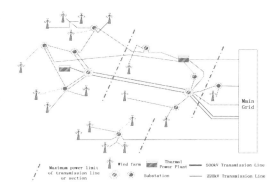

Figure 2. Curtailment due to the transmission limitation of the grid.

mission lines. Besides, unlike the distributed integration of EU, renewable plants in China are often located at remote areas where the resource is perfect but the demand is low; thus, the transmission capability shows a significant impact.

In the wind power base of Gansu and Jilin, wind farms are integrated intensively, and local transmission capacity is obviously insufficient, transmission system operates at its stability limitation all the time, and some substations also meet its capacity limitation. Once the weather is good, the output of wind farms/PV stations could not be fully transmitted to the load center due to transmission limitation and the curtailment is unavoidable, as shown in Figure 2.

3 MODELING AND OPTIMIZATION

In a non-market environment, it is unable to use the market tool to ensure the balance between supply and demand. System operators are more relying on the plan to maintain supply reliability. So, a more optimal dispatching strategy is needed to help the system operator to manage renewable, especially in the situation of renewable energy generation curtailment.

3.1 *Power grid*

The provincial power grid of China is actually complex as lots of substations and transmission lines are included. Considering the complexity of the radial physical model of the grid and that renewable generation plants are usually connected at the end of the grid, all transmission limitations can be equivalent to only a few main regional limitations after re-aggregation of the grid. The proposed regional aggregation method is based on the actual transmission limitation of the grid, and

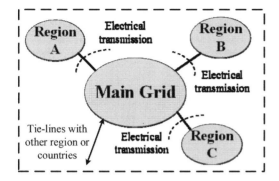

Figure 3. Aggregation model of the power grid.

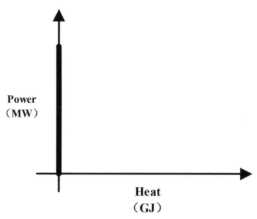

Figure 4. Relationship between power and heat of the condensing unit.

the aggregation mode of the power grid is created after classifying load, renewable energy, and conventional unit and transmission lines in different regions. Then, the complexity of the power grid is reduced, the model computing speed is improved, and it also gets closer to the practical requirement of the simulation. The aggregation model of the power grid is shown in Figure 3.

3.2 Coal-fired unit

Coal-fired units can be divided into condensing, back pressure, and extraction by different technologies. Those units that produce both electricity and heat have more impact on peak load following ability and renewable energy consumption in a certain power grid.

3.2.1 Condensing unit
Condensing unit supplies no heat; therefore, there is no relation between its power and heat outputs, and its characteristics are shown in Figure 4.

3.2.2 Back pressure unit
The operation feature of the back pressure unit is shown in Equation (1) and Figure 5:

$$P_{i,t} = H_{i,t} \cdot C_b \tag{1}$$

where C_b is the ratio of the unit power output and heat output; $H_{i,t}$ is the heat output; and $P_{i,t}$ is the power output.

3.2.3 Extraction unit
The operation feature of the extraction unit is shown in Equation (2) and Figure 6. Power output may vary in a range when the heat supply is solid, which is determined by C_b and C_v as follows:

$$\begin{cases} P_{i,t} \geq S_i^{Mx} + H_{i,t} \cdot C_b \\ P_{i,t} \leq S_i^{Ex} - H_{i,t} \cdot C_v \end{cases} \tag{2}$$

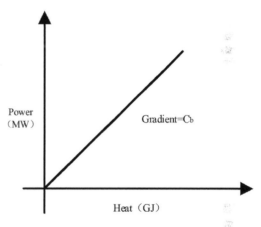

Figure 5. Relationship of power and heat for the back pressure unit.

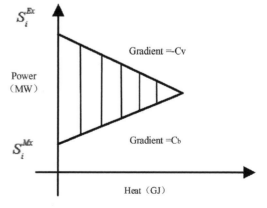

Figure 6. Relationship of power and heat for the extraction unit.

559

where S_i^{Mx} and S_i^{Ex} are minimum and maximum power outputs of the unit.

3.3 Hydro power unit

Hydro power output is closely related to water resource. The power of the run-off hydro unit depends on river flow, and its operation characteristics is similar to renewable generation. Hydro power plant with reservoir has no minimum output limitation, but the capacity of the reservoir and the flow have limitations of their outputs, as shown in Equation (3):

$$\begin{cases} W_s + W_{in} - \sum P_t^{rese} \geq W_{s+1} \\ W_{min} \leq W_s \leq W_{max} \end{cases} \tag{3}$$

where W_s is the initial available electricity generation of the reservoir; W_{in} is the incoming water available electricity generation; $\sum P_t^{rese}$ is the actual electricity generation; W_{s+1} is the initial available electricity generation next week; W_{min} is the minimum available electricity; and W_{max} is the maximum available electricity.

3.4 Renewable energy generation

Basically, the output of wind turbines and PV modules is variable and directly related to the local wind and solar resources. The model of the wind power and the PV power is shown in Equations (4) and (5):

$$P_{i,t}^{wind} \leq S^{wind} \cdot FLH \cdot \frac{P_{i,t}^{wh}}{\sum_{i,t}(P_{i,t}^{wh} \cdot 1)} \tag{4}$$

$$P_{i,t}^{PV} \leq S^{PV} \cdot FLH \cdot \frac{P_{i,t}^{ph}}{\sum_{i,t}(P_{i,t}^{ph} \cdot 1)} \tag{5}$$

where FLH is the full load hour of the wind or PV; S^{wind} and S^{PV} are the installed capacity of the wind and PV; P^{wh} and P^{ph} are the hourly average output of the wind and PV, calculated based on the capacity, wind speed, solar radiation, and power curve of the wind turbine and PV modules. The output series of renewable generation can be varied by changing capacity and FLH. In this paper, renewable generation is regarded as the time-series output.

3.5 Gas-fired unit

The operation feature of the gas-fired unit is similar to that of the back pressure unit. In northern China, the gas-fired unit usually operates as the spinning reserve due to its high operation cost and ancillary service compensation deficiency. Therefore, it can be treated as an extra reserve of the system for more flexible regulation.

3.6 Coordinated optimal dispatching model

3.6.1 Objective function
The optimal object of the model is to maximize the renewable energy generation electricity, as shown in Equation (6):

$$\max \sum_{t=1}^{T} \sum_{n=1}^{N} (P_w(t,n) + P_{pv}(t,n)) \tag{6}$$

where N is the number of regional power grids; n is one of the power grids; T is the dispatch duration; t is the optimization step; $P_w(t,n)$ and $P_{pv}(t,n)$ are the wind power and the PV power outputs of grid n in duration t.

3.6.2 Constraints
All the considered constraints in the model are shown in Equations (7)–(15), where Equation (7) is the power system reserve constraint, Equation (8) is the power balance constraint, Equations (9)–(12) are the output constraints of the conventional unit, Equation (13) is the startup and shutdown constraint of coal-fired units, Equation (14) is the FLH constraint of generating units, and Equation (15) is the transmission limitation constraint:

$$\begin{cases} \sum_{n=1}^{N} \sum_{j=1}^{J} (-P_{j,max}(t,n) \cdot S_j(t,n) - P_w(t,n) \\ \quad - P_{pv}(t,n)) \leq -\sum_{n=1}^{N} P_l(t,n) - P_{re} \\ \sum_{n=1}^{N} \sum_{j=1}^{J} (P_{j,min}(t,n) \cdot S_j(t,n) + P_w(t,n) \\ \quad + P_{pv}(t,n)) \leq \sum_{n=1}^{N} P_l(t,n) - N_{re} \end{cases} \tag{7}$$

$$\sum_{j=1}^{J} P_j(t,n) \cdot S_j(t,n) + P_w(t,n) \\ + P_{pv}(t,n) + L_i(t) = P_l(t,n) \tag{8}$$

$$0 \leq \Delta P_j(t,n) \leq \left[P_{j,max}(t,n) - P_{j,min}(t,n) \right] \cdot S_j(t,n) \tag{9}$$

$$P_j(t,n) = P_{j,min}(t,n) \cdot S_j(t,n) + \Delta P_j(t,n) \tag{10}$$

$$P_j(t+1,n) - P_j(t,n) \leq \Delta P_{j,up}(n) \tag{11}$$

$$P_j(t,n) - P_j(t+1,n) \leq \Delta P_{j,down}(n) \tag{12}$$

$$0 \leq S_j(t,n) \leq S_{j,max}(n) \tag{13}$$

$$E_{j,\min} \le \sum_{t=1}^{T} P_j(t,n)\Delta T \le E_{j,\max} \qquad (14)$$

$$-L_{i,\max} \le L_i(t) \le L_{i,\max} \qquad (15)$$

where $P_j(t,n)$ is the output of unit j in the power grid n at time t; $\Delta P_j(t,n)$ is the power output variation of unit j in the power grid n at time after the calculation; $P_{j,\max}$ and $P_{j,\min}$ are the proportion of the unit's maximum output over capacity and the unit's minimum output over capacity, respectively; S_j is the number of units; $\Delta P_{j,up}$ and $\Delta P_{j,down}$ are up- and down-ramping rates of unit j; and $L_i(t)$ is the power of the transmission line i that has transmission limitation.

4 MULTI-SOURCE GENERATION OPTIMAL DISPATCHING STRATEGY

4.1 Wind–PV–hydro coordinated optimal dispatch

The fundamental of wind–PV–hydro coordinated optimal dispatching is the regulation ability of the reservoir. The output of the hydro unit is reduced when the renewable energy generation output increases, otherwise the output of the hydro unit is increased. The total output of the renewable and hydro units has a relative constant value under the regulation of the reservoir. The short-term regulation ability of the hydro unit is the difference between its expected output and minimum output. However, for large-scale renewable energy generation, electricity balance should also be considered in dispatching (Lidong et al. 2010).

Intraday electricity generation of the hydro unit includes regulative electricity E_r and compulsory electricity E_c. E_r is for the regulation of renewable energy variation, and E_c is generated by must-run units. As the renewable generation electricity is E_{re} and the maximum output of renewable generation is F, the electricity of hydro power regulation for variation E_h can be calculated by using Equation (16):

$$E_h = 24F - E_{re} \qquad (16)$$

According to the relationship between E_h and E_r, the balance of hydro power regulation for renewable energy can be evaluated. There could be three scenarios:

1. When $E_h = E_r$, hydro power just balances the renewable generation electricity, and supplies no extra electricity for the other system regulation, as shown in Figure 7.

where Y, P, and Q represent the expected output, the average output, and the compulsory output of hydro

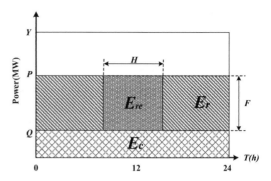

Figure 7. Hydro power just balances the renewable generation electricity while $E_h = E_r$.

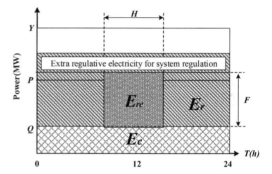

Figure 8. Hydro power fully balances the renewable generation electricity and supplies extra regulation ability while $E_h < E_r$.

power. H is the full load hour of renewable generation, which can be calculated by using Equation (17):

$$H = E_{re}/F \qquad (17)$$

2. When $E_h < E_r$, hydro power fully balances the renewable electricity and supplies extra regulative electricity for the system regulation, as shown in Figure 8.

The extra regulative electricity of hydro power for the system regulation is calculated by using Equation (18):

$$E_{h-extra} = 24(P - Q) - F(24 - H) \qquad (18)$$

3. When $E_h < E_r$, hydro power could not balance the renewable generation electricity; therefore, coal-fired and gas-fired units have to supply the remaining regulative electricity for renewable energy generation, as shown in Figure 9.

Regulative power supplied by hydro units can be calculated by using Equation (19):

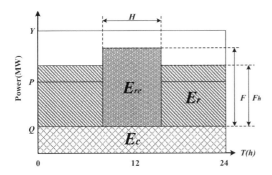

Figure 9. Hydro power could not balance the renewable generation electricity while $E_h < E_r$.

$$F_h = (24(P - Q))/(24 - H) \quad (19)$$

The remaining regulative electricity supplied by thermal units for renewable energy can be calculated by using Equation (20):

$$E_{remain} = F(24 - H) - 24(P - Q) \quad (20)$$

However, the above scenarios are all under the assumption of $Y - Q \geq F$. If $Y - Q < F$, no matter how much is the renewable generation electricity, hydro power could not balance the renewable generation electricity completely and other power sources need to participate for balance.

4.2 *Multi-source generation optimal dispatch*

After the calculation of required regulation capability for renewable energy generation regulation and available hydro power supplied regulation capability, the power generation schedule of all units can be dispatched. The basic dispatching strategy is given as follows: 1) renewable generation, must-run units and run-off hydro power, 2) regulation capacity of hydro power with the reservoir, 3) remaining part of the load should be covered by thermal units. When unit commitment is determined, the regulation capability of thermal unit supply for regulating renewable energy is the difference between its maximum and minimum output limitations.

5 VALIDATION AND ANALYSIS

The effectiveness of the proposed strategy is evaluated by using the time-series production simulation. The simulation is carried out with the actual operation data of the Qinghai power grid in 2015. Computation conditions of the simulation are as follows: PV power installed capacity is 5144 MW,

wind power capacity is 368 MW, and other conditions are the same as in real operation conditions.

Under the non-market condition, it is difficult for the system operators to evaluate the accuracy of renewable accommodation ability of the system, and for ensuring reliability, the general practice is to schedule the generation plan of traditional units ahead of time, and then to dispatch renewable generation. However it is too rough that the maximum regulation capability of traditional units is not considered and usually causes unnecessary curtailment of renewable energy during the load valley. The simulation result comparison between with and without proposed optimal strategies is presented in Table 1.

If the system operator performs dispatching with coordinated optimal consideration, for example, no pre-determined generation plan for conventional units and no solid electricity plan limitation of hydro power, the final curtailment of renewable generation will be less than the scenario of without proposed strategy. Besides, the proposed strategy presents a significant reduction in curtailment due to the insufficient regulation capability of the system to the greatest extent, and hydro and thermal units can present their maximum regulation capability coordinately for renewable energy.

A typical week's production simulations of this month without and with proposed strategies are shown in Figures 10 and 11. When implementing optimal dispatching, the total renewable curtailment is significantly reduced. Conventional units regulate more frequently for renewable energy and tie-line variation. As shown in Table 2, thermal units regulate 11 times in a week under optimization, whereas they regulate once in the same dispatching duration in the non-optimization scenario. Moreover, curtailment due to insufficient regulation capability in the non-optimization scenario is totally removed under optimization. Again, the optimal strategy has two curtailment days less than the non-optimal

Table 1. Simulation result comparison between with and without proposed optimal strategies.

Qinghai grid	Without proposed strategy	With proposed strategy
	MWh	MWh
RE generation	543,957	563,723
Curtailment	37,340	17,574
Curtailment (%)	6.4	3.0
Curtail due to transmission limitation	17,574	17,574
Curtail due to insufficient regulation cap	19,766	0

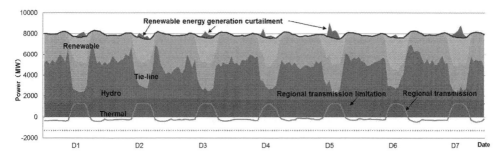

Figure 10. Time-series production simulation in a typical week of Qinghai grid without proposed dispatching strategy.

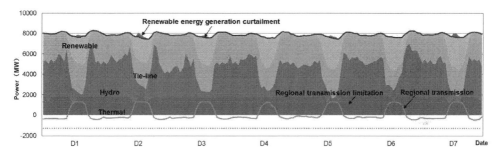

Figure 11. Time-series production simulation in a typical week of Qinghai under proposed dispatching strategy.

Table 2. Dispatching simulation result between with and without optimization in a typical week.

Qinghai grid	Non-optimization	Optimization
Curtailment days	7	5
Thermal units regulating times	1	11
Min output of hydro (MW)	600	0
Curtailment (%)	9.5	2.3
Curtailment due to insufficient regulation cap (MWh)	19,766	0

scenario. After optimization, renewable generation curtailment only has regional transmission limitation curtailment that could not be reduced by the proposed strategy of system operators, and the curtailment percentage is reduced by 7.2%.

6 CONCLUSION

The bottlenecks of China's renewable energy generation development and operation are discussed in this paper, and a multi-source generation coordinated optimal dispatching strategy in the non-market power system operation is presented.

Considering the flexibility of hydro units and coordination between hydro, thermal, and renewable energy, the total output of all generating units will keep a relative constant value; therefore, the impact of renewable's variability on the system is reduced. The effectiveness of the proposed strategy is validated via the time-series production simulation of the Qinghai power grid in a month. The results indicate that it could help the system operator to significantly decrease renewable energy generation curtailment due to the insufficient system regulation capability of the hydro, thermal, and renewable generation units by coordinated optimal dispatching.

ACKNOWLEDGMENT

This work was supported by the Key Projects in the National Science & Technology Program of China (2013BAA02B01), the Science and Technology Project of SGCC, and the PV Power Integration Technology Key Lab of Qinghai Province (2014-Z-Y34 A).

REFERENCES

Barth, R., Brand, H. & Meibom, P. 2006. A stochastic unit commitment model for the evaluation of the impacts of the integration of large amounts of wind power, in Proc. 9th Int. Conf. *Probabilistic Methods Applied to Power Systems*. Stockholm: Sweden.

Chen, F. Qing, X & Xin, S. 2010. Generation Maintenance Scheduling with Significant Wind Power Penetration, *Automation of Electric Power Systems* 19(33):20–25.

Fink, S. & Mudd, C. 2012. Wind Energy Curtailment Case Studies. *Report of National Renewable Energy Laboratory*. Washington, D.C: USA.

Francois, B & Francisco, D.G. 2008. Stochastic security for operations planning with significant wind power generation, *IEEE Transactions on Power Systems* 23(2):306–316.

Guoqiang, Z. Boming, Z & Wenchuan, W. 2011. Coordinated Roll Generation Scheduling Considering Wind Power Integration. *Automation of Electric Power Systems* 19(35):18–23.

Hetzer, J. Yu, D & Bhattarai, K. 2008. An economic dispatch model incorporating wind power. *IEEE Trans on Energy Conversion* 23(2):603–611.

Lidong, Y., Minyi, Z & Lei, W. 2010. A Computing Method for Peak Load Regulation Ability of Northwest China Power Grid Connected With Large-Scale Wind Farms. *Power System Technology* 2(34):129–132.

Qian-kun, W. 2012. Update and Empirical Analysis of Domestic and Foreign Wind Energy Curtailment. *East China Electric Power* 3(40):78–82.

Tuohy, A., Meibom, P. & O'Malley, M. 2009. Unit Commitment for Systems with Significant Wind Penetration. *IEEE Transactions on Power Systems* 2(24): 592–601.

Zhaosui, Z & Yuanzhang, S. 2011. A Solution of Economic Dispatch Problem Considering Wind Power Uncertainty. *Automation of Electric Power Systems* 22(35):125–131.

Zhi, L. Xueshan, H & Ming, Y. 2010. Power system dispatch considering wind power grid integration, *Automation of electric power systems* 34(19): 15–19.

Advances in Power and Energy Engineering – Sun (Ed.)
© 2016 Taylor & Francis Group, London, ISBN 978-1-138-02846-3

Research on the differentiated construction mode of smart distribution network

Z. Huang & S.G. Liu
State Grid Corporation of China, Beijing, China

Y.H. Liu, W. Zhang, W. Liu, L.M. Zhou & Y.M. Hou
China Electric Power Research Institute, Beijing, China

ABSTRACT: In order to improve the technology and economy of a smart distribution network, it is necessary to choose the suitable construction mode in the process of planning and construction. This paper investigates the differentiated construction mode of a smart distribution network and establishes a set of typical modes. Then, taking full consideration of the differences in construction regions based on political, economic, functional, and loads, a differentiated construction mode selection model is built. Based on the comprehensive evaluation technology, quantitative selection criteria are obtained by mixing clustering analysis, hierarchy analysis, and expert decision analysis methods. Applications of the results would effectively promote the smart distribution network construction development scientifically and orderly.

1 INTRODUCTION

At present, the development of smart grid has entered into the stage of comprehensive construction, and the intelligent construction projects of urban and rural power distribution networks also have been fully carried out. As an important part of smart grid, the smart distribution network is a key point directly related to the implement of smart grid (Qing & Ma 2010).

Smart distribution network construction has a characteristic rich in contents, projects, and wide range, mainly concerned with smart substations, distribution automation, distributed generations/microgrid interconnection, intelligent electricity, electric communication network, and integration of distribution and electricity (He & Sun 2010). So far, the construction of the smart distribution network depends mainly on the standards or specifications in accordance with types and projects; apparently, it lacks a global unified planning and design. In China, there are a lot of differences among regions, such as development level, network structure types, construction ideas, smart demands, and objectives. Thus, the construction scheme of the smart distribution network should be closely combined with the specific situation (Liu 2012). Otherwise, it will easily result in low equipment asset utilization rate and even serious waste, and the technology and economy of construction is not reasonable (Song et al. 2011).

In order to improve the lean planning of the smart distribution network, this paper establishes a set of typical and differentiated modes for smart distribution network construction based on the early experience. The modes involve different levels, covering electric primary systems such as capacity, wires and switches, and electric secondary systems including automation, measurement, and communication. Taking full consideration of the differences in construction areas based on political, economic, functional, and loads, a differentiated construction mode selection model is built. Based on the comprehensive evaluation technology, the selection criteria for the quantitative selection of differentiated construction modes are obtained by the mixing of the clustering analysis, the hierarchy analysis, and the expert decision analysis methods.

2 EASTABILISHMENT OF DIFFERENTIATED CONSTRUCTION MODES

Smart distribution network construction possesses the characteristic of informatization, automation, and interactive. It should be designed overall, especially the connection of each system, and the integration of power flow, business flow, and information flow (Lv 2006). A typical smart distribution network construction should take the power flow and the information flow as the two main lines. The power flow involved voltage levels from 110 kV to 0.4 kV. The flow of information refers to the application systems, collecting related information when the power is flowing. All the information is transmitted to the data center. It is able to interact with the next higher data center, and give feedback to the application systems for real-time controlling and visual displaying. More attention should be paid to the interaction with the power grid and the users.

Based on the experience of early demonstration projects, this paper puts forward a multi-level architecture of smart distribution network construction, as shown in Figure 1. Taking the architecture as the goal, this paper establishes a typical smart distribution network construction mode from all levels, including primary systems such as structures, lines, and switches, and secondary systems such as automation, measurement, and communication (see the attached table for details).

2.1 Base layer

The base layer mainly involves the network structures of different voltage grades. This article puts forward the range of different structures according to the reliability.

2.2 Equipment layer

The equipment layer involves the contents of new generation of intelligent substations, power distribution automation, smart distribution areas, and intelligent electricity construction. For the new generation of intelligent substations, the main differentiation is in the capacity of transformers and construction types of the primary system. In the secondary system, two layers of a single network hierarchical and distributed system structure are used jointly (Song et al. 2013). For the distribution automation, there are a lot of differences in the configuration, the coordination with relay protection, master stations, and terminal and communication system. Three typical modes in the construction of smart power distribution areas are as follows: indirect type, standard type, and extension type. Each type has the corresponding

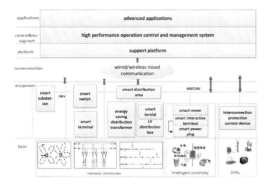

Figure 1. A typical architecture diagram of smart distribution network construction.

construction standards and the scope of applications. For the intelligent electricity, the modes mainly consider the differentiation in intelligent community, intelligent household, and demand side management of industrial users.

2.3 Communication layer

Communication layer is made up of a medium-voltage network and a low-voltage network. It takes all kinds of remote terminals in the distribution automation and electricity information collection system as the downward ending, and 110 kV or 35 kV substation's backbone communication network nodes as the ascending node, which provide the access channel for information systems (Xu 2014). Construction of the backbone gives priority to optical fiber. In the field of 10 kV and below, the differentiated construction modes are proposed according to the safety and reliability based on the business requirements.

2.4 Support platform layer

The realization of the smart distribution network depends on the intelligent systems, and the implementation of intelligent systems cannot exclude a mature, safe, reliable, and perfect communication network platform (Su & Li 2014). The support platform requires the ability of various terminal interconnections in distribution and electricity fields, including all kinds of substations, power distribution terminals, electric meters, and users' terminals. Moreover, the platform needs to manage the equipment and realizes the integration of all kinds of data to provide a common data acquisition and information interaction platform for advanced applications.

2.5 System and application layer

Based on the support platform, it is easy to achieve the applications of the efficiency of optimization maintenance, asset operation monitoring, comprehensive management, power quality monitoring and analysis, electric vehicle optimization scheduling, and multistage energy coordination. Differentiated function configurations are put forward according to the modes.

3 DIFFERENTIATED CONSTRUCTION MODE SELECTION MODEL

3.1 Evaluation index system establishment

Taking the influence factors of smart distribution network construction into fully consideration, the construction division criterion is determined by four aspects, including administrative category, economic, functional, and the power loads (Zhang et al. 2011). Based on these differences, an evaluation index system of the smart distribution network construction area division is set up, as presented in Table 1.

1. Administrative rank refers to the administrative level and is a national division for hierarchical management. It is an important characteristic of construction areas, so the paper takes it as one of the main criteria of region division.
2. Economic level refers to the level of regional economic and social development. It mainly includes indices of annual GDP per capita, annual household electricity consumption per capita, and annual social electricity consumption per capita.
3. Function difference means that the construction areas can be divided into commercial zone, government affairs zone, development zone, residential zone, mixed zone, and suburban area.
4. Load differences are referred to the load density and the load important level of construction areas.

The evaluation index system of smart distribution network construction includes four first-grade indices, namely political, economic, functional, and power loads, and seven second-grade indices.

3.2 Evaluation index classification

3.2.1 Clustering analysis of evaluation indices
Based on the data of multiple cities, the clustering analysis method is used to obtain the clustering results such as indices of annual GDP per capita, annual household electricity consumption per capita, annual social electricity consumption per capita, and load density (He et al. 2011). The clustering results obtained with statistical software are shown in Figure 2. It shows that the indices basically obey the normal distribution.

3.2.2 Functional index division
With the high speed and large-scale development of modern economic society, urban development pattern is changed from traditional that is industry, commerce, and residence concentrated and mixed distribution to

Table 1. The evaluation index system of the smart distribution network construction area division.

Classification	Evaluation index
Political	Administrative rank
Economic	Annual GDP per capita
	Annual household electricity consumption per capita
	Annual social electricity consumption per capita
Functional	Function division
Power loads	Load density
	Load important level

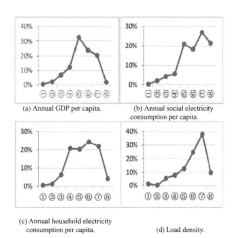

(a) Annual GDP per capita.

(b) Annual social electricity consumption per capita.

(c) Annual household electricity consumption per capita.

(d) Load density.

Figure 2. Clustering analysis results of the evaluation indices.

Table 2. The scheme of the functional index division.

Function	Type 1	Type 2	Type 3
Commercial zone	Center	Common	Market
	$\sigma \geq 40$	$30 \leq \sigma < 40$	$\sigma < 30$
Government affairs zone	Core	Common	Scattered offices
	$\sigma \geq 50$	$30 \leq \sigma < 50$	$\sigma < 30$
Development zone	National/ provincial and above	City/district	Rest
	$\sigma \geq 35$	$25 \leq \sigma < 35$	$\sigma < 25$
Residential zone	City/town	Suburban	Rural
	$\sigma \geq 30$	$15 \leq \sigma < 30$	$\sigma < 15$
Mixed zone	City	Town	Suburban
	$\sigma \geq 35$	$15 \leq \sigma < 35$	$\sigma < 15$
Suburban	City suburban	Town suburban	City/town remote area
	$\sigma \geq 1$	$0.1 \leq \sigma < 1$	$\sigma < 0.1$

functional and centralized distribution. According to the nature of the loads, combined with the load density (σ/MW per km²), modern urban regions can be divided into commercial zone, government affairs zone, development zone, residential zone, mixed zone, and suburban types. Each functional zone could be subdivided again, as given in Table 2.

4 EXAMPLE

4.1 Typical region overview

A typical region covering 2.89 square kilometers is selected as an example. The planning of the area contains industry, research, teaching, and living functions, and focuses on the field of biomedical engineering, financial, trade, commerce, service, and leisure activities, forming functions for residential, industrial, and public facilities, municipal facilities, and considering comprehensive supporting facilities and commercial facilities such as hotels, businesses, and conferences.

4.2 Weight calculation and grade standard setting

The example uses the Analytic Hierarchy Process (AHP) method to calculate the weights of evaluation indices (Guo 2003). The evaluation indices use the centesimal system, through normalized processing, to convert all kinds of data into the standardized one to make a direct comparison (Li et al. 2014). The comprehensive evaluation results are summarized in Table 3.

4.3 Construction mode selection

The quantitative selection criterion of the smart distribution network construction mode is presented in Table 4.

Table 3. Results of the comprehensive evaluation.

First-grade index	Weight	Score	Second-grade index	Weight	Score
Political	0.3106	18.6	Administrative rank	1.0000	60.0
Economic	0.2122	15.6	Annual GDP per capita	0.3108	75.2
			Annual household electricity consumption per capita	0.1958	78.8
			Annual social electricity consumption per capita	0.4934	70.5
Functional	0.1967	15.7	Function division	1.0000	80.0
Power load	0.2805	22.7	Load density	0.6812	86.2
			Load important level	0.3188	70.0

Table 4. Construction mode selection of the smart distribution network.

Comprehensive score	Construction mode selection
≥90	Type I
80–90	Type I/II
70–80	Type II
60–70	Type II/III
50–60	Type III
≤50	Type IV

The simple weighted method is used to calculate the comprehensive evaluation score. In this example, the region comprehensive evaluation score is 72.6 points. Thus, type II is the most suitable choice for the typical region of smart distribution construction.

5 CONCLUSION

It is a key point to choose the right mode of smart distribution network construction according to the characteristics of the regional types and differences. This paper proposed a set of differentiated and typical smart distribution network construction modes that cover all levels of power distribution. The comprehensive evaluation system is established and used in choosing the right construction mode based on the comprehensive evaluation method. Technology and economy of smart distribution network construction are improved apparently.

REFERENCES

Guo, Junhua 2003. The Research of Clustering Analysis in Data Mining. *Wuhan University of Technology master thesis*.

He, Guangyu & Sun, Yingyun 2010. *The Fundamentals of Smart Grid*. Beijing: China electric power press.

He, Yongxiu et al. 2011. *Comprehensive Evaluation Method and Its Applications of Power Grid*. Beijing: China electric power press.

Li, Rui et al. 2014. Assessment on typical Power Supply Mode for Important Power Consumers Based on Analytical Hierarchy Process and Expert Experience. *Power System Technology* 38(9):2336–2340.

Liu, Weibin 2012. Research of Rural Smart Distribution District Typical Construction Scheme. *North China electric power university master thesis*.

Lv, Yongbo 2006. *System Engineering*. Beijing: Beijing jiaotong university press.

Qing, Lijun & Ma Qiyan 2010. *Distribution Smart Grid and Its' Key Technologies*. Beijing: China electric power press.

Song, Xiaohui et al. 2011. A Modular Method for Planning and Design of Power Networks. *Power System Technology* 35(7):123–128.

Song, Xuankun et al. 2013. Overall design scheme for new generation intelligent substation. *Electric power construction* 34(6):3–8.

Su, Dawei et al. 2014. Architecture Design of Unified Data Collection and Information Support Platform for Power Distribution and Utilization. *Electric power automation equipment* 34(9):166–172.

Xu, Jiani 2014. Research on the Smart Distribution Grid Communication Network Mode. *South china university of technology master thesis*.

Zhang, Zuping et al. 2011. Exploration on Technical Issues in "The Twelfth Five-Year-Plan" Distribution Network Planning. *Distribution & Utilization* 28(4):1–4,40.

Attached table. Typical differentiated construction mode of the smart distribution network.

Level	Setting		Type I	Type II	Type III
High voltage	Network structure		Double chain, double loop, double radiation	Double/single chain, double loop, double radiation	Single radiation
	New generation of intelligent substation	Primary	Indoor; Smart devices and GIS	Indoor/half indoor	Outdoor
		Secondary	Large/medium capacity; Integration of intelligent components and primary equipment; Integrated monitoring and control system of substation, with functions of background monitoring, fault wave record, and network analysis; Two layers, single network	Medium/small capacity	Small capacity
Medium voltage	Network structure		Double loop; Single loop, multiple segments	Single loop, multiple segments, radiation	Multiple segments, radiation
	Primary	Line	Cable; Public/private	Overhead, cable; Public/private	Overhead; Public
		Segmentation	Looped network/switch station	Switch station	
		Switch	Vacuum circuit breaker, load switch; Energy-saving transformer	Load/column switch	Column switch
	Secondary	Terminal	Breaker; Three remotes	Load switch and fuse; Two remotes/single remote	Drop-fuse; Single remote
		Feeder automation	Automatic and centralized; Intelligent distributed/half automatic and centralized	Local reclosing/fault indicator	Fault indicator
		Station	Intelligent; Integrated/standard	Practical	

(Continued)

Level	Setting		Type I	Type II	Type III
Users	Interconnection		Users with characteristics access to the same transformer, and use ES, EV, and DSM to realize load operation smoothly		
	Meter		Bidirectional, smart meter		
	Intelligent electricity	Interactive	Big users	Optional	
		Configuration	Electricity and energy management/intelligent electricity community/smart home	Electricity and energy management	
	DG/microgrid	DG	PV/comprehensive utilization of energy	Wind/PV/comprehensive utilization of energy	
		Microgrid	Substation/feeder/LV	Feeder/LV	LV
	Multi-energy complementary		Wind/PV/ES/gas/biomass	Wind/PV/biomass	Wind/PV/biomass
Communication	Bone		Fiber		
	Interconnect		Fiber/public grid	Public grid	Public grid
Platform	Function configurations		Integration of distribution and utilization/maintenance optimization/quick fault handling/energy efficiency management	Maintenance optimization/quick fault handling/energy efficiency management	Quick fault handling

Advances in Power and Energy Engineering – Sun (Ed.)
© 2016 Taylor & Francis Group, London, ISBN 978-1-138-02846-3

Effect analysis of Distributed Generations in the Active Distribution Network

H. Fan
China Electric Power Research Institute, Beijing, China
State Grid Corporation of China, Beijing, China

H. Hui, Y.M. Hou & W. Liu
China Electric Power Research Institute, Beijing, China

Z. Huang
State Grid Corporation of China, Beijing, China

J. Su
China Electric Power Research Institute, Beijing, China

ABSTRACT: The Active Distribution Network (ADN) requires that Distributed Generations (DGs) be flexible in its access and that the power flow be controlled flexibly. After DG integration, the effect of DGs on the distribution network contains many aspects; in addition, the relevant influencing factors are also very complex. In this paper, the distribution network is divided into three layers: user layer, feeder layer, and network layer. The influence factors of these three levels are analyzed by means of survey data and simulation experiments, respectively. The results indicate that the capacity, type, and access modes of DGs affect the three layers of the distribution network.

1 INTRODUCTION

With a large scale of Distributed Generation (DG) integration, the power flow changes from one direction to multi-direction, and the distribution network is changed from passive to active. To maximize the acceptance to DGs, the distribution network is required to be changed from the traditional distribution network to the Active Distribution Network (ADN), which integrates the information, communication, and controlling technology. Through effective monitoring and optimization regulation of DGs, the ADN could consume DGs as much as possible (Hui et al. 2015).

To transform into the ADN, the planning and operation modes of the traditional distribution network need to be designed for adapting to DGs, which is a big challenge, so that it becomes necessary to research on the effect of DGs on the distribution network. The ADN could significantly improve the absorptive capacity and active management level of DGs, and improve the system's controllability, security, reliability, and economy under the new boundary conditions (Fan et al. 2015).

In the ADN, power flow is allowed to be controlled flexibly, thus the power flow direction will be changed obviously. The network loss will be reduced through controlling the power flow properly, thus the corporation will achieve the maximum benefit. In addition to this, the short circuit current will be also changed; in general, it will be bigger. Moreover, due to a large amount of power electronic equipment introduced, the harmonic will also be injected into the system. Taking into account the above analysis, it is necessary to analyze the effect of DG integration (D'Adamo et al. 2009).

This paper contains the following aspects: first, it presents the study ideal, in which the distribution network will be divided into some layers and the influencing factors of DG integration on different layers will be listed, respectively; second, it gives a deeper insight into the main influencing factors on these three layers; and finally, through some simulation experiments and survey materials, it analyzes the effect of the main factors and gives the analytical conclusion.

2 STUDY IDEAL

In view of the energy transmission direction of the distribution network, the system could be divided into three layers: user layer, feeder layer, and network layer.

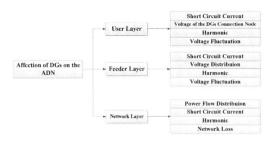

Figure 1. Affecting factors of DGs on the ADN.

If the power output of DGs cannot be consumed in one layer, the excess energy will be transferred to the upper layer. Particularly, when DGs are connected to the user layer, and the power output of DGs could not be consumed completely, the excess energy could be transferred to the feeder layer. If the total power output of DGs could not be consumed in the feeder layer, the redundant energy will be transferred to other feeders in the network layer. If the network layer could not consume all the energy from DGs, the surplus energy will be transferred to the transmission network. As the paper focuses on the distribution network, the relevant contents about the transmission network will not be considered.

As shown in Figure 1, for the user layer, the effect aspects include the short circuit current, the voltage of the DG connection node, harmonic, and voltage fluctuation. For the feeder layer, the impacts of DG integration are mainly the short circuit current, voltage distribution, harmonic, and voltage fluctuation. For the network layer, the impacts are the power flow distribution, short circuit current, harmonic, and network loss. This paper analyzes the effect of DG integration on these three layers, respectively, in the following sections (Hui et al. 2013).

3 EFFECT ON THE USER LAYER

If the power output of DGs could be balanced in the local area, the power users will completely consume the energy from DGs, so the effect of DGs on the distribution network is not obvious, and it mainly affects the DG connection node.

3.1 Short circuit current

The short circuit current of the DG connection node will increase after DG integration, which is related to the type and capacity of DGs. The inverter-interfaced DGs could provide 1–1.5 times less than the rated current of DGs, the synchronous motor-interfaced DGs could provide 5–10 times, and the asynchronous interfaced DGs could provide 3–6 times. Moreover, with the increasing

capacity of DGs, the short circuit current supplied by DGs is also increasing (Liu et al. 2013).

In addition, if the short circuit current of the DG connection node increases to a large extent, it is possible to make the equipment to withstand high current so that equipment will be damaged. Therefore, thermal stability verification should be done on the bus, line, and breaker of the system side.

3.2 Voltage of the DG connection node

The voltage of the DG connection node will increase after DG integration, and the increasing range is relevant to the ratio of DGs' capacity and the short circuit capacity of the system. Based on the standard that is already applied, the ratio should not be less than 10 (Li et al. 2012).

Based on Li's study, the closer the feeder on which DGs are installed, the higher the voltage increases. Therefore, the boundary condition of the problem is that if the load on the feeder is light, the voltage deviation should not exceed the regulations of GB/T 12325-2008 when DGs are connected to the end of the feeder.

3.3 Harmonic

After the inverter-interfaced DG integration, the Direct Current (DC) from DGs will be converted into Alternating Current (AC) before DGs are connected to the distribution network, and it will inject the harmonic current into the distribution network.

If the capacity of the inverter-interfaced DGs is not very high, the harmonic pollution can be controlled within the safe range. However, if the capacity is high enough to make the harmonic exceed the specified limit in IEEE519-1992 or GB/T14549-93, the capacity of DGs should be restricted to no more than a certain value (Xie et al. 2012).

3.4 Voltage fluctuation

When the power output of DGs and the local load are not coordinated, the voltage fluctuation of the DG connection node will increase because of the fluctuating power output and start–stop of DGs. In the distribution network, there are few dynamic reactive power regulating devices, so the voltage regulation depends only on the capacitors and reactors; thus, the difficulty of voltage regulation for the DG connection node will increase, and cause voltage fluctuation if the voltage is not regulated properly (Pei et al. 2008).

4 EFFECT ON THE FEEDER LAYER

If the energy could not be consumed in the user layer, the excess energy will be transferred to the feeder layer, and the impact aspects of DG integration include the short circuit current, voltage distribution, voltage fluctuation, and harmonic.

To analyze the effect of DGs on the feeder layer, this paper designs a simulation model of radial feeder, shown in Figure 2, to perform some simulation experiments. The model uses a common radiation type in China, whose voltage is 10 kV, node number is 10, length of the line is 5 km, type of the conductor is JKLYJ0.6/1-185 mm², maximum load is 6 MW, and short circuit current limit is 20 kA.

4.1 Short circuit current

The effect trend analysis of the capacity and type of DGs on the short circuit of the feeder layer is similar to that described in Section 3.1. So, this section mainly analyzes the effect of different DG access modes on the feeder layer.

In the simulation experiments, four common DG access modes are presented in Table 1. Four DGs are connected to the feeder and the capacity of each one is 1 MW.

In the simulation experiments, the fault occurs at the head of the feeder, and the experiments are carried out in these four different DG access modes. The simulation results are summarized in Table 2.

The simulation results indicate that the 'gathered at the head' mode affects the short circuit of the feeder most, and the least affected is the 'gathered at the end' mode. Therefore, considering the effect of the short circuit on the feeder, it is better to avoid installing too many DGs in the head of the feeder.

4.2 Voltage distribution

The effect trend analysis of the capacity and type of DGs on the voltage distribution of the feeder layer is similar to that described in Section 3.2. So, this section also focuses on the DG access modes. The DG access modes are the same as those given in Table 1 in the simulation, and the simulation results are shown in Figure 3.

From Figure 3, it can be seen that the effect of the 'distributed along the line' mode on the voltage of the feeder is average, and the effect of the 'gathered at the end' mode is the most. Therefore, installing too many DGs at the end of the feeder is not a good idea.

4.3 Harmonic

The harmonic effect trend analysis of the capacity and type of the feeder is similar to that described

Figure 2. A schematic diagram of the radial feeder.

Table 1. Access modes of DGs.

Access modes	DG1	DG2	DG3	DG4
Distributed along the line	3	5	7	9
Gathered at the head	1	1	1	1
Gathered in the middle	5	5	6	7
Gathered at the end	9	9	10	10

Table 2. Effect of different DG access modes on the short circuit current of the distribution network.

Access modes	Faulted node	DG1	DG2	DG3	DG4
Distributed along the line	14.283	0.082	0.078	1.177	1.103
Gathered at the head	15.557	0.087	0.087	1.8	1.72
Gathered in the middle	14.434	0.077	0.077	1.24	1.193
Gathered at the end	13.949	0.071	0.071	0.977	1.005

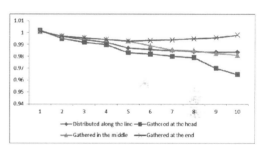

Figure 3. Effect of different DG access modes on the voltage distribution of the feeder.

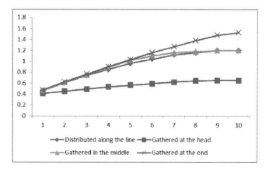

Figure 4. Total Harmonic Distortion (THD) of different DG access modes (%).

Table 3. Voltage fluctuation value of different DG access modes (%).

Access modes	1	2	3	4	5	6	7	8	9	10
Distributed along the line	0.063	0.097	0.131	0.159	0.185	0.203	0.219	0.231	0.238	0.238
Gathered at the head	0.063	0.063	0.063	0.063	0.062	0.062	0.062	0.062	0.061	0.061
Gathered in the middle	0.063	0.063	0.063	0.063	0.062	0.062	0.062	0.062	0.061	0.061
Gathered at the end	0.066	0.102	0.15	0.177	0.214	0.252	0.29	0.323	0.366	0.396

in Section 3.3. So, this paper mainly analyzes the DG access modes. The DG access modes are the same as those given in Table 1 in the simulation, and the simulation results are shown in Figure 4.

From Figure 4, we can see that the 'gathered at the end' mode is the most affected. In view of harmonic reduction, it is better to avoid installing too many DGs at the end of the feeder.

4.4 Voltage fluctuation

This section mainly analyzes the effect of different DG access modes on the feeder layer. In the simulation, the DG access modes are the same as those given in Table 1. Given that there is some kind of disturbance to DGs, the experiments were carried out in four different DG access modes, and the simulation results are summarized in Table 3.

From the simulation results, we can see that the effect of the 'gathered at the end' mode is the most and the least affected is the 'gathered at the head' mode.

5 EFFECT ON THE NETWORK LAYER

With a large scale of DG integration, the power flow reversion occurs in the local area; thus, it changes the power flow distribution of the whole distribution network. In the distribution network, the reverse power flow will bring changes to the network loss and the short circuit current. Moreover, the superposition of DGs may lead to the harmonic and the short circuit current, exceeding the standard limit. Furthermore, DG integration will affect the protection parameter setting, fault recovery, the distribution network automation design, system stability, and even the network planning. This section focuses on the basis analysis, and the results are discussed below.

5.1 Power flow distribution

With DG integration, if the power output of DGs could not be consumed completely in the feeder layer, the extra energy will be transferred to other feeders through a 10 kV bus. The changes are shown in Figure 5.

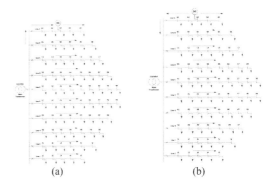

(a)   (b)

Figure 5. Power flow distribution with DG integration in the network layer. (a) Power flow with small DG integration and (b) power flow reserve with a large scale of DG integration.

The change in the power flow direction will bring changes to the regulatory policy and control modes, so that it becomes necessary to monitor the power flow direction of the feeder on which DGs are installed.

5.2 Short circuit current

For DG integration with the low voltage and the distribution mode, because the access mode is decentralized, the capacity of each DG is small and many DGs are photovoltaic generations; thus, the short circuit current of DGs is small, so that this type of DGs has a small impact on the short circuit current of the low voltage distribution network.

DG access to the medium-voltage feeder will have a great influence on the short circuit current of the distribution network. In recent years, with the rapid development of DGs (the capacity is large), access to the city distribution network through private lines, and the gas turbine, diesel engines, and other rotating motor types of DGs are developing rapidly, and the short-circuit current of the distribution network is increasing year by year, especially in some eastern cities with dense loading. It can be expected that a 10 kV short circuit current will be closer to the planning level in some area for the next 2 or 3 years (Yang et al. 2015).

If many rotary motor-interfaced DGs are connected to the distribution network, the short circuit current will exceed the interrupting capacity of the system circuit breaker, and the company will have to invest a lot to exchange the breakers. According to the statistics, if the interrupting capacity of the breaker increases by a grade, the equipment price will increase by 15%–20%.

With a large number of asynchronous motor and synchronous motor-interfaced DG integration, when a fault occurs in the system, the short circuit current from each DG will be collected to the fault point so as to make the short circuit current of the system much bigger, and it will put forward to a higher demand for the whole system.

In order to analyze the effect of DGs on the short circuit current of the distribution network deeply, the simulation experiments are carried out with the model, as shown in Figure 5. As analyzed in the above sections, the main effect factors of the short circuit current are as follows: capacity, type, and access modes of DGs. Through setting different parameters of different factors, we give the short circuit current range of key nodes (see Table 4), in which the maximum capacity of DGs is 10 MW with 27.8% penetration.

From the simulation results, we can see that the substation bus withstands the maximum short circuit current, so it is very important to do the thermal check on the substation bus before planning and designing the distribution network.

5.3 Harmonic

More and more power electronic devices will be applied in the distribution network after DG integration. Because the voltage regulation and the control mode of the distribution network are different from the conventional way, the frequent opening and closing will easily produce the harmonic and affect the network. Moreover, if many DGs are connected to the system, the harmonic superposition effect becomes more obvious.

Table 4. Short circuit current range of key nodes in the system after DG integration.

Key nodes	Short circuit current (kA)
Substation bus	(3.61, 5.19)
A1	(0.29, 4.16)
B1	(0.44, 4.12)
C1	(0.73, 4.06)
D1	(0.72, 4.13)
E1	(0.43, 4.25)

The simulation model uses the system, as shown in Figure 5. The main influencing factors of harmonic are as follows: capacity, type, and access modes of DGs, and the simulation results are summarized in Table 5.

From the simulation results, we can conclude that the node at the end of the feeder has the most limiting possibilities. Although existing DGs in the market have a relatively perfect harmonic control device, if the harmonic exceeds the standard regulation (GB14549-1993), DGs should be strictly controlled to be connected to the distribution network because of the obvious superposition effect.

5.4 Network loss

The power flow direction of each feeder in the whole distribution system may change after DG integration. According to the relationship between the load and DGs, three cases can be divided as follows:

1. The load of each node in the system is greater than or equal to the power output of DGs of the node.
2. At least one node on which the power output of DGs is greater than the load of the node, but the total power output of DGs is less than the total load of the system.
3. At least one node on which the power output of DGs is greater than the load of the node, but the total power output of DGs is larger than the total load of the system (Kou 2011).

In case (1), it will reduce the network loss after DG integration. In case (2), DG integration may increase the loss of some feeders in the distribution network, but the total loss will reduce for the whole system. In case (3), the system loss will increase a lot. In conclusion, the network loss is related to the position of DGs, the relative value of DGs and load, and some other factors.

Table 5. THD of key nodes in the system after DG integration.

Key nodes	THD
A1	(0.40%, 1.03%)
A5	(1.13%, 1.55%)
B1	(0.41%, 0.89%)
B6	(1.10%, 1.58%)
C1	(0.42%, 0.70%)
C7	(1.02%, 1.61%)
D1	(0.44%, 0.65%)
D8	(0.76%, 1.60%)
E1	(0.45%, 0.62%)
E9	(0.65%, 1.68%)

6 CONCLUSION

In this paper, the distribution network is divided into three layers: user layer, feeder layer, and network layer. Analyzing the effect of DGs on different layers, respectively, the results can be summarized as follows:

1. If the power output of DGs could be consumed completely in the user layer, the main impact aspects are the short circuit current, the voltage of DG connection node, harmonic, and voltage fluctuation. DGs mainly influence the connection node, and the type and capacity are the main influencing factors.
2. If the user layer could not consume the energy of DGs completely, the extra energy will be transferred to the feeder layer, and it mainly influences the short circuit current, voltage distribution, voltage fluctuation, and harmonic of the feeder. In addition to the type and capacity of DGs, the access mode is also the key factor.
3. If the power output of DGs could not be consumed completely, the excess energy needs to be transferred to the network layer. DGs also influence other feeders in the network layer, and the main impact aspects are as follows: power flow distribution, power quality, short circuit current, and network loss. With the flexible power flow, the network loss, short circuit current, and harmonic also occur with the related changes.

REFERENCES

D'Adamo C., Jupe S. & Abbey C. 2009. Global survey on planning and operation of active distribution networks update of CIGRE C6.11 working group activities. *CIRED2009 (20th International Conference on Electricity Distribution): IET Services Ltd: 1–4.*

Fan, M.T., Hui H., & Zhang, Z.P. 2015. Main impacts on Active Distribution System Planning. *Electric Power Construction 36(1):60–64.*

Hui H., Fan, M.T. & Zhang, Z.P. 2015. Research on the typical systems for active distribution network in China. *2015 international symposium on smart Electric Distribution Systems and Technologies (EDST 2015):206–211.*

Hui, H., Zhao, M.X., et al. 2013. Study on interactions between high penetration of distributed generation and the operation characteristics of distribution network. *Asia-Oceania CIGRE Technical Meeting 2013(AORC-CIGRE):60–61.*

Kou, F.H. 2011. Impacts of Distributed Photovoltaic Power Sources on Distribution Network Loss. *Power System and Clean Energy 27(11):62–68.*

Li, P. & Zhang H.E. 2012. Probe into Influence of Distributed Generation on Distribution Network Voltage. *Shanxi Electric Power 12:30–35.*

Liu, J., Lin, T., et al. 2013. Simulation Analysis on Influences of Distributed Photovoltaic Generation on Short-Circuit Current in Distribution Network. *Power System Technology 37(8):2080–2085.*

Pei, W., Sheng, K., et al. 2008. Impact and Improvement of Distributed Generation on Distribution Network Voltage Quality. *Proceedings of the CSEE 28(13):152–157.*

Xie, B.X., Wang, Z. & Fan, S.L. 2012. Influence on harmonic characteristics of distributed network with distributed generations. *Power System Protection and Control 40(2):116–119.*

Yang, S., Tong, X.Q., et al. 2015. Short-Circuit Current Calculation of Distribution Network With Distributed Generation. *Power System Technology 39(7):1977–1982.*

Advances in Power and Energy Engineering – Sun (Ed.)
© *2016 Taylor & Francis Group, London, ISBN 978-1-138-02846-3*

Constructive Heuristic Algorithm for integrated Generation and Transmission Network Expansion Planning

E.B. Cedeno
Business School, Nanjing Tech University, Nanjing, China

ABSTRACT: The integrated planning of transmission and generation capacities has become more relevant to researchers in recent years. However, empirical evidence suggests joint efforts from energy planners and practitioners to determine coordinated or integrated expansion plans for these sectors. This Generation and Transmission Network Expansion Planning problem is a complex non-convex Mixed Integer Non-Linear Programming. The solution methods for this problem cannot guarantee achieving global optimality. Primal solution methods attempt to identify satisfactory solutions with minimum effort. This paper presents a Constructive Heuristic Algorithm (CHA) to solve the GTNEP problem. The method expands simultaneously capacities of several transmission lines to avoid being trapped in local minima. The starting feasible solution considers all new generation nodes and all candidate transmission lines connected to the grid with zero capacities. Iteratively, the algorithm searches for better solutions until all unneeded generation has been removed. A 9-nodes test network illustrated a successful application of the method.

1 INTRODUCTION

In recent years integrated planning of transmission and generation capacities has become more relevant to researchers (Cedeno 2014; Bertsch & Fichtner 2015; Jin & Ryan 2014; Krishnan et al. 2015; Seddighi & Ahmadi-Javid 2015; Cedeño & Arora 2013). After electricity deregulation, with the unbundling of vertically integrated utilities coordinating or integrated planning for generation and transmission expansion is considered by many to go against competition and deregulation (Cedeño & Arora 2013; Rocha & Kuhn 2012). However, there is empirical evidence from different markets indicating joint efforts from energy planners and practitioners seeking to coordinate the planning of generation and transmission capacities involving several different owners (Minnesota Department of Commerce 2007; Upper Midwest Transmission Development Initiative 2010; The Minnesota Transmission Owners 2009; Unsihuay-Vila et al. 2011; UPME 2013; The CapX2020 Utilities. 2007). Complexities in the sector, especially regarding integration of renewable sources to the generation portfolio maximizing the use of existing transmission lines have increased the interest in integrated models for transmission and generation planning (Minnesota Department of Commerce 2007; The Minnesota Transmission Owners 2009). This research presents a model to determine optimal expansion plans considering restrictions in terms of the minimum capacity that must be added to the network from renewable sources while at the

same time limiting the amount of new transmission line additions.

The Generation and Transmission Network Expansion Planning (GTNEP) adds to the Generation Expansion Problem (GEP) complexities of the Transmission Network Expansion Planning problem (TNEP) resulting in Mixed Integer Non-Linear Programming (MINLP) problem. The TNEP problem is a complex non-convex MINLP problem with multiple local minima. Solution methods for this problem (Hemmati et al. 2013) in general cannot guarantee achieving global optimality and quality of the solution. Primal solution methods attempt to identify satisfactory solutions with minimum effort. This paper presents a Constructive Heuristic Algorithm (CHA) to solve the GTNEP problem. This method extends the research presented in (Bustamante-Cedeño & Arora 2009; Cedeño & Arora 2013).

The discussion continues in the following manner: Section 2 formulates the integrated model for the GTNEP problem as a MINLP. Section 3 describes the steps involved in the CHA for solving this complex problem. Section 4 presents a test case for a 9-nodes network. Section 5 gives some conclusions.

2 INTEGRATED GENERATION AND TRANSMISSION EXPANSION PLANNING MODEL

In this section a macro level model for integrated generation and transmission planning is presented

The optimization problem is presented below (Cedeño & Arora 2013):

Min TC

$$TC = \sum_{(i,j)\in E} w_{ii}\left(e^1_{ij} - e^0_{ij}\right) + \sum_{i\in Ns} v_i m_i + \sum_{i\in Nd} r_i s_i + \sum_{i\in Ns} g_i p_i \tag{1}$$

Power flow equations:

$$B(E)\ \theta + G(m) + r = D \tag{2}$$

Generation capacity constraints:

$$g_i \leq g_i^{\max} \forall i \in N_s \tag{3}$$

Generation capacity constraints for candidate generation nodes:

$$m_i g_i \leq m_i g_i^{\max} \forall i \in N_s^a \tag{4}$$

Line flow capacity constrains:

$$\left|f_{ij} e_{ij}\right| \leq c_{ij} e_{ij} \tag{5}$$

Line flow capacity constrains for new lines connecting candidate new generation nodes:

$$\left|m_i f_{ij} e_{ij}\right| \leq m_i c_{ij} e_{ij}, i \in N_s^a, j \in N \tag{6}$$

where $f_{ij} = (\theta_i - \theta_j)B_{ij}$ \hfill (7)

Initial line configuration constraints:

$$e^0_{ij} \leq e_{ij} \tag{8}$$

Maximum number of parallel lines along lines:

$$e_{ij} \leq e_{ij}^{\max} \tag{9}$$

Phase angle Constraints:

$$\theta_i \leq \theta_i^{\max} \forall i \tag{10}$$

$$\theta_i \geq \theta_i^{\min} \forall i \tag{11}$$

where $B(E) = $ is the corresponding susceptance matrix for the configuration being studied; $B_{ij} = $ susceptance of transmission line along line (i,j); $D = $ demand vector, a vector with element d_i at demand node i and 0 elsewhere; $e_{ij} = $ number of circuits added along transmission line connecting node i to node j given by $e_{ij} = e^1_{ij} - e^0_{ij}$; $e^1_{ij} = $ number of circuits along transmission line connecting node

i to node j in the final network; $e^0_{ij} = $ number of circuits along transmission line connecting node i to node j in the initial network; $f_{ij} = $ flow along line (i,j); $G(m) = $ Generation vector, a vector with element g_i at generation node i and 0 elsewhere; $m_i = $ binary variable representing the addition of new generation at node i; $N = $ total set of nodes; $N^0 = $ set of existing nodes; $N^a = $ set of alternative new nodes; $N_s = $ set of supply nodes; $N_d = $ set of demand nodes; $N_t = $ set of transshipment nodes; $p_i = $ generation production cost at generation node i; $R = $ artificial generation vector, a vector with element r_i at demand node i with load curtailed and 0 elsewhere; $s_i = $ load curtailment cost per unit at demand node i; $\theta = $ vector of phase angles at the nodes, its elements are real numbers within certain constraints for the purpose of maintaining stability; $v_i = $ investment cost of adding new generation i; $w_{ij} = $ cost of adding a circuit along transmission line connecting node i to node j.

Decision variables in the model are of three types:

Continuous decision variables: θ_i, g_i, r_i.
Discrete decision variables: e_{ij}
Binary decision variables: m_i

The problem formulated in (1) is a Mixed Integer Non-Linear Programming Problem (MINLP) since it involves cross product of the decision vectors E and θ in the objective function (1) and in the following constraints (2), (5), and (6). This is an inherent complexity to the TNEP problem to which it is added the complexity of the GEP problem.

3 CONSTRUCTIVE HEURISTIC ALGORITHM FOR GTNEP

A Constructive Heuristic Algorithm (CHA) is an iterative process that searches for a good quality solution, having followed a series of iterative steps. The algorithm stops when no further additions are required to meet the demand and a good quality solution has been found (Romero et al. 2005). As contrasted with the CHA presented in this research, most CHA algorithms follow an approach of single transmission line additions, then followed by a second phase decreasing capacities (one at a time) to reduce the expansion cost while maintaining feasibility.

Notation:
$\alpha = $ parameter to control selection of good solutions; $\delta = $ congested or saturated transmission line parameter; h = counter for the number of consecutive non-improving iterations, designed to get out of the local minimas; H = maximum number of non-improving iterations; IC = Incumbent cost; LC = Load curtailment cost; L = List of congested

transmission lines; L2 = List with a congested transmission line from M being studied; L3 = List of solutions within distance α of the incumbent TC; M = List of congested transmission lines from L being studied; u = counter for the number of iterations; U = maximum number of iterations; U.B. = upper bound on capacity additions; TC = Total Cost.

Elements of the feasible space for the proposed CHA are congested or saturated transmission lines, new lines and new generation nodes. Parameter δ determines congested transmission lines (δ is close to zero).

3.1. *CHA algorithm for the GTNEP problem (Bustamante-Cedeño & Arora 2009)*

Numbers in the steps below refer to numbers in the algorithm presented in Figure 1.

1. Select as a starting feasible solution, the existing transmission grid requiring capacity expansion

(i.e. LC ≠ 0) to meet a future given demand. Consider all the already connected circuits and proposed non-connected transmission lines as well as generator nodes as part of the grid, assuming the proposed new circuits to be connected to the grid with zero capacities. Evaluate Total Cost (TC) and Load Curtailment Cost (LC). Set TC as Incumbent Total Cost (IC), LC as incumbent load curtailment cost and E as incumbent solution vector. This incumbent TC includes costs for all possible generation nodes.

2. From the incumbent solution vector E, create a list L with transmission lines whose slack capacities are less than δ value.

3. If the list L is not exhausted, pick any transmission line (i,j) from the list, not previously studied. Otherwise go to step 8.

4. Select any node in the chosen transmission line (i,j) not previously studied. If both nodes of the chosen transmission line (i,j) have been studied go to step 3.

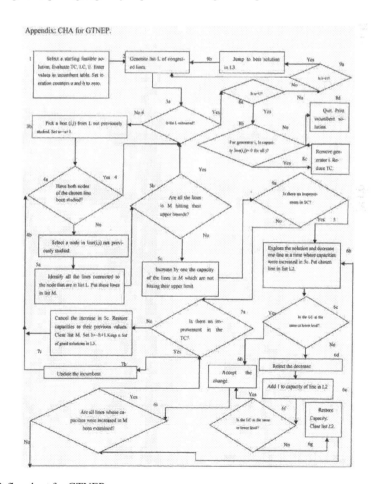

Figure 1.   CHA flowchart for GTNEP.

5. Identify all the transmission lines connected to the chosen node that belong to the list L. Increase the capacities of these transmission lines by one, which will not violate their upper bounds (UB's).
6. If LC decreases by the changes made in 5, explore reducing the capacities of these lines one by one. Identify the lines that decrease LC and retain these changes. Reject the decreases in transmission capacity, which increase LC, and explore again increasing their capacities and retain the favorable increases. If there is no reduction in LC go to step 7. 7. If there is a reduction in TC, update the incumbents and go to step 4. Otherwise, cancel the capacity increases done in step 5 and go to step 4. Create a list L3 of solutions within distance α of the incumbent TC.
8. If the maximum number of iterations U has been reached, remove unneeded generation capacity connected to transmission lines with zero capacities. Remove corresponding generation nodes from total cost. Then, quit with the incumbent solution as the "best" solution. Otherwise, go to step 9.
9. In case the algorithm has gone through a complete set of H iterations with no improvement, and it has not reached the maximum number of allowed iterations U, jump to best solution in list L3 and go to step 2.

The initial version of this algorithm has been used successfully to solve TNEP problems for standard test networks (Bustamante-Cedeño & Arora 2009). The CHA applied to a six node Garver's network outperforms results reported in (Gallego et al. 1998) requiring only 81 iterations. The CHA applied to a larger real life network consisting of 46 nodes requires 3790 iterations. A hybrid algorithm presented in (Gallego et al. 1998) requires 4835 iterations using random initialization. These iterations can be reduced by half using a modified Garver's initialization. The CHA applied to the 9 nodes test case presented in this paper requires only 93 iterations for the integrated GTNEP problem.

4 TEST CASE

Consider a power system is integrated by three different regions as depicted in Figure 2 (Cedeño & Arora 2013; Cedeno 2014). Generation and transmission assets are owned by several different companies, as in a deregulated system. Transmission lines are managed centrally by an Independent System Operator (ISO). The analysis presented in this example constitutes a joint effort to integrate all previous studies. Region A has one generation

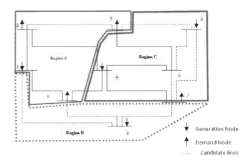

Figure 2. Network representation for 9-nodes test case.

node (node 1) and three demand nodes (nodes 2, 3 and 5). It is assumed that there is no potential for expansion of renewable generation capacities in this region. Initial demand in Region A corresponds to 10 MW; whereas initial generation capacity in Region A is 5 MW. Region B represents a remote area further away from load pockets but rich in renewable energy. A new generation facility is proposed to be added in this region, represented by node 8. Region C has one generation and one demand node, node 4 and node 7 respectively. This region also has potential for adding more expensive renewable generation (node 9). Initial demand in Region C corresponds to 1 MW; whereas initial generation capacity in Region C is 6 MW.

The initial installed capacity in the system is 11 MW. The future increment in capacity corresponds to 1 MW in each region. There are two new 3 MW generation projects that could be connected to the network to meet this future demand, as previously mentioned. In the initial configuration of this 9 nodes example, there are 7 nodes connected to the grid: 4 demand nodes, 2 generation nodes and one transshipment node (a node that has zero generation and zero loads).

There are two new proposed generation nodes not connected to the grid. There are 8 existing transmission lines and 7 new potential transmission lines to connect these new generation nodes to the existing network. The number of parallel circuits along any transmission line connecting nodes i and j is limited to 3. Relevant costs for these new investments are as follows: $v_8 = 315$ \$, $v_9 = 350$ \$, $g_{8\max} = 3$ MW, $g_{9\max} = 3$ MW. It is assumed that these investment costs have been annualized including the same set of externalities. The load curtailment cost is assumed to be constant at all demand nodes and is set equal to 100 \$/MW. The generation conditions are presented in Table 1. Demands in MW are given as follows: D2 = 4, D3 = 3, D5 = 4, D6 = 0, D7 = 2. Table 2 present transmission lines data (Cedeño & Arora 2013).

Table 1. Generation data for test case.

Generation nodes	Production/ dispatch cost ($/MW)	Maximum generating capacity (MW)
1	3	5
4	5	6
8	2.5	3
9	3.5	3

Table 2. Transmission line data for test case.

Transmission line	e^0_{ij}	$w_{ij\,(\$\,per\,line)}$	$f^{max}_{i,i}$ (MW)	B_{ij}
01,02	2	20	1	2.222222
01,03	1	19	1.7	1.666667
02,04	1	25	1.1	2.857143
02,05	1	17	1.1	2.325581
03,04	1	19	1	1.612903
04,06	1	21	1.3	3.333333
04,07	1	23	2.2	2.500000
05,06	2	22	1.2	2.857143
01,08	0	25	1.3	1.666667
03,08	0	37	1.6	1.754386
07,08	0	35	1.2	2.040816
05,09	0	25	1.1	1.960784
06,07	0	35	1.2	2.702703
06,09	0	40	1.6	2.127660
07,09	0	38	1.3	3.030303

The initial feasible solution includes both generation projects connected to the grid with zero capacities. This generates an initial high load curtailment cost plus a high investment cost. The optimization process would seek to minimize the total cost. The best design identified by the optimization model selects generation project at node 9 and adds the following transmission lines to the initial configuration: $e_{01,02} = 1$, $e_{01,03} = 1$, $e_{04,07} = 1$, $e_{05,09} = 3$. The total expansion cost is equal to \$487. The load curtailment cost is equal to zero as it is customary to expand capacities till there is no load curtailed. Solution requires adding the most expensive generation but transmission network requires only minor upgrades, then the total cost is the smallest to meet future demand. Solution conforms to the condition that the transmission network should only require minor upgrades or no upgrades.

5 CONCLUSIONS

Integrated planning of transmission and generation capacities has become more relevant to researchers in recent years. However, empirical evidence suggests joint efforts from energy planners and practitioners to determine coordinated or integrated expansion plans for these sectors. This GTNEP problem is a complex non-convex Mixed Integer Non-Linear Programming. In this research a CHA has been proposed to solve this problem. The method expands simultaneously capacities of several transmission lines to avoid being trapped in local minima (Bustamante-Cedeño & Arora 2009). The starting feasible solution considers all new generation nodes and all candidate transmission lines connected to the grid with zero capacities. Iteratively, the algorithm searches for better solutions until all unneeded generation has been removed. A 9-nodes test network illustrated a successful application of the method. Solution for this problem requires adding the most expensive generation but transmission network requires only minor upgrades, then the total cost is the smallest to meet future demand. The logic for the proposed CHA for the GTNEP is easy to explain to policy makers and energy planners. Then, perhaps there is scope for its application. However, test in larger networks are needed to evaluate the performance of the algorithm.

REFERENCES

Bertsch, V. & Fichtner, W., 2015. A participatory multi-criteria approach for power generation and transmission planning. *Annals of Operations Research*, pp.1–31.

Bustamante-Cedeño, E. & Arora, S., 2009. Multi-step simultaneous changes Constructive Heuristic Algorithm for Transmission Network Expansion Planning. *Electric Power Systems Research*, 79(4), pp.586–594.

Cedeno, E.B., 2014. Preemptive Goal Formulation for the Multi-Objective Generation and Transmission Planning Model in a Deregulated Environment. *International Journal of Electrical Energy*, 2(4), pp.308–313.

Cedeño, E.B. & Arora, S., 2013. Integrated transmission and generation planning model in a deregulated environment. *Frontiers in Energy*, 7(2), pp.182–190.

Gallego, R.A., Monticelli, A. & Romero, R., 1998. Comparative studies on non-convex optimization methods for transmission network expansion planning. *IEEE Trans. Power Syst*, 13, pp.822–828.

Hemmati, R., Hooshmand, R.A. & Khodabakhshian, A., 2013. Comprehensive review of generation and transmission expansion planning. *IET Generation, Transmission and Distribution*, 7(9), pp.955–964.

Jin, S. & Ryan, S.M., 2014. A tri-level model of centralized transmission and decentralized generation expansion planning for an electricity market-Part II. *IEEE Transactions on Power Systems*, 29(1), pp.142–148.

Krishnan, V. et al., 2015. *Co-optimization of electricity transmission and generation resources for planning and policy analysis: review of concepts and modeling approaches*, Springer Berlin Heidelberg.

Minnesota Department of Commerce, 2007. *The next generation renewable energy objective: Minnesota's smart renewable energy standard*.

Rocha, P. & Kuhn, D., 2012. Multistage stochastic portfolio optimisation in deregulated electricity markets using linear decision rules. *European Journal of Operational Research*, 216(2), pp.397–408.

Romero, R. et al., 2005. Constructive heuristic algorithm for the DC model in network transmission expansion planning. In *IEE Proceedings: Generation, Transmission and Distribution*. pp.277–282.

Seddighi, A.H. & Ahmadi-Javid, A., 2015. Integrated multiperiod power generation and transmission expansion planning with sustainability aspects in a stochastic environment. *Energy*, 86, pp.9–18.

The CapX2020 Utilities., 2007. *CapX C-BED Transmission Study Report*.

The Minnesota Transmission Owners, 2009. *Capacity Validation Study*.

Unsihuay-Vila, C. et al., 2011. Multistage expansion planning of generation and interconnections with sustainable energy development criteria: A multiobjective model. *International Journal of Electrical Power & Energy Systems*, 33(2), pp.258–270.

UPME, U.D.P.M.E., 2013. *Plan Indicativo De Expansión De Cobertura De Energía Eléctrica 2013–2017*.

Upper Midwest Transmission Development Initiative, 2010. *Executive Committee Final Report*.

Advances in Power and Energy Engineering – Sun (Ed.)
© 2016 Taylor & Francis Group, London, ISBN 978-1-138-02846-3

The electrical properties of polyester materials as insulators

K.B. Ewiss
Electrical Design Engineering Department, Shaker Consultant Group (SCG), Cairo

L.S. Nasrat
Electrical Power and Machines Department, Faculty of Engineering, Aswan University, Aswan

R.M. Sharkawy
Head of Electrical and Control Department, Faculty of Engineering and Technology, Arab Academy for Science, Technology and Maritime Transport, Cairo

ABSTRACT: This work presents a study of flashover voltage for polyester and composite insulators under dry, ultra-violet, and contaminated conditions. Cylindrical of polyester composite samples (with different lengths) have been prepared after doping with different concentrations of inorganic filler e.g., Magnesium hydroxide ($Mg(OH)_2$). The aim is to improve the electrical and thermal properties in addition to maximize surface flashover voltage and decrease tracking phenomena. The relationship between the flashover voltage and percentage doing is studied. A comparison between different concentrations of filler under various environmental conditions (dry and contaminated conditions) showed higher flashover voltage values for samples containing filler with higher concentrations of $Mg(OH)_2$. On the other hand, during exposure to Ultra Violet (UV) radiation, flashover voltage decreases by adding $Mg(OH)_2$ filler for polyester samples In this study, the effect of thermal performance with respect to surface of the sample under test have been investigated in details.

1 INTRODUCTION

High voltage insulators used in transmission lines and substations are being subjected to various operating conditions and environments. Contamination on the surface of the insulators enhances of leakage current develops which may lead to flashover and power system outage and it's the essential element in power system outages with contaminated insulators on the systems which are located in desert, costal sites, and manufacturing cities (Khan et al. 2006). However, since early sixties, alternative materials namely polymers have emerged and presently are being used extensively for a variety of insulator applications. Polymeric materials used for insulation have been classified to various types such as composite, non-ceramic, plastic, and synthetic insulation (Bernstorf et al. 2004).

Polymers have been preferred over porcelain and glass by many utilities for the housing of high voltage outdoor materials. Polymer has many advantages over ceramic such as; cheap, light weight, easy handling, and installation, shorter manufacturing time, high impact resistance, high mechanical strength, greater flexibility in product design, reduced breakage (non brittle characteristic), and resistance to vandalism.

These advantages have driven the utility to prefer polymer insulators over conventional porcelain or glass insulators. The polymeric materials, particularly silicon rubber, epoxy, Ethylene Propylene Diene Monomer (EPDM), and polyesters are used as insulators for transmission, distribution, termination of underground cables, bushings, and surge arrester housings (Ersoy et al. 2007).

High voltage transmission lines are often subjected to the deposition of contaminated substances, high temperature, humidity, and sandstorm. In the desert, sandstorm—that has particles—increases surface roughness of composite polymer insulators. The high degree of UV radiation can cause physical as well as chemical changes on the surface of composite insulators (Zedan & Akbar 2006).

Cylindrical rod specimens chemically prepared from polyester have been tested to examine the effect of desert weather such as, UV on the flashover voltage performance of composites. The ac (50 Hz) flashover voltage of polyester specimens has been investigated in this study, $Mg(OH)_2$ is used for increasing electrical performance of polyester specimens.

2 EXPERIMENTAL SETUP AND TECHNIQUES

2.1 *Material specimen*

Specimens have been prepared from unsaturated polyester resin [2121P-1]. The resin supplied already dissolved in styrene and physical properties are listed in Table 1.

Polyester specimens were fabricated as cylindrical rods have 10 mm diameter and 100 mm length cut into various lengths by lathe machine according to standard procedure. The powder inorganic filler was used such as $Mg(OH)_2$. It is an effective flame retardant and smoke suppressant for polymers with particle size 2 μm. The powder filler is mixed with the base polymer to improve the electrical surface tracking and erosion resistance as well as reinforced the thermal stability. Polymer composites were prepared by mixing different ratios of each of $Mg(OH)_2$ (20, 30, and 40 wt%) with unsaturated polyester/styrene mixture in the presence of 1% by weight of methyl ethyl ketone peroxide MEK-P. The prepared composites were left at room temperature ($25°C \pm 1$) until curing occur. The material compositions of polymeric samples at different concentration of filler are show in Table 2.

2.2 *Test apparatus*

2.2.1 *Electrical test supply and electrodes*

The schematic diagram of flashover test arrangement is shown in Figure 1. The AC (50 Hz) high voltage was obtained from a single-phase high

Table 1. Technical data and specification for polyester.

Item	Specification	Technical data
1	Appearance	Faint blue-transparency
2	Styrene monomer content (%)	36–39
3	Viscosity at 25°C	300–400
4	Specific gravity at 25°C	1.11

Table 2. Sample code and composition (weight%) of composite samples.

Samples (%)	Polyester resin (%)	Type of filer Magnesium hydroxide $Mg(OH)_2$ (%)
(Blank) 0	100	—
$[(Mg(OH)_2]_{20}$	80	20
$[(Mg(OH)_2]_{30}$	70	30
$[(Mg(OH)_2]_{40}$	60	40

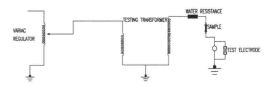

Figure 1. Schematic diagram of flashover test arrangement.

voltage transformer (150 kV–15 kVA). The electrodes were made of copper with 5 mm thickness and 10 mm diameter. The electrodes were fixed into the specimens, one at the top and the other at the bottom.

2.2.2 *UV radiation apparatus*

UV light is one of the major factors responsible for degradation of polymeric insulators. The main sources of UV rays are sun, corona formation, and dry band arcing activities on insulation surface. Simulate the UV light and observe the local discharge phenomenon on the composites surface, all test specimens had been exposed to UV radiation from different locations by using UV lamps. Three various times of UV radiation exposures have been recorded (720, 1440, and 2160 h) according to ASTM D 7238, the surfaces of all specimens have been monitored during UV exposure.

2.2.3 *Dry and contaminated conditions*

To simulate these effects and hence gain a better understanding in the tracking phenomenon and determined flashover voltage on insulation surface, polyester samples have been tested under two different environmental conditions; dry condition and contaminant condition. Two identical sets of samples have been prepared and tested in atmospheric air using ac (50 Hz) voltage. The different testing conditions are performed to study effect of flashover voltage on each set as follows:

1. The first set has been tested in dry condition according to ASTM D 2303.
2. The second set (contaminant condition) has been immersed in water, 10% solution of NaCl 10% solution of HCl for 7 days, then taken out, dried and tested using AC voltages according to ASTM D 2303.

2.2.4 *Flammability test*

Fire hazard of materials is one of major factors, which determines their acceptance or rejection for specific applications such as the use of materials in electrical insulation. For acceptance or rejection, flammability test is performed to assess the relative fire hazards of material according to ASTM D 635.

3   RESULTS AND DISCUSSION

3.1   Electrical tests

3.1.1   Flashover voltage of virgin specimens

To study the flashover voltages of polyester specimens with different concentrations (20, 30, and 40 wt%) of filler $Mg(OH)_2$ at dry weather, the voltage was applied and raised gradually until flashover occurs; all tests were applied at high voltage laboratory and the results are shown in the Figure 2.

From Figure 2, it can be observed that flashover voltage sharply increased for all samples (polyester bank 20%, 30%, and 40% $Mg(OH)_2$ filler).

3.1.2   Effect of UV on the flashover voltage of composite samples loaded with different percentage of filler

The simulation of UV effect and observe the local discharge phenomenon on the surface of the polyester specimens. A comparison between different percentages of filler $Mg(OH)_2$ (20, 30, and 40 wt%) have been undertaken for three different time durations (720, 1440, and 2160 h) to measure the ability of filler to withstand UV rays.

It was clear from Figure 3, that, increasing the percentage of $Mg(OH)_2$ filler decreases the flashover voltage of polyester composite under UV conditions. At (720, 1440 and 2160 hrs), the unfilled polyester composite have flashover voltages (20.26, 16.9 and 14.7 kV) respectively, these values drop to 19.22, 18.38 and 16.62 kV), (16.01, 15.08 and 14.05 kV) and (13.88, 12.25 and 10.9 kV) with addition (20, 30 and 40 wt%) of $Mg(OH)_2$ filler respectively. Lack of improvement in flashover voltage value for polyester sample loaded with 40 wt% has been found compared with polyester sample loaded with 30 wt% and 20 wt% of $Mg(OH)_2$ filler is 11.02% and 21.47% respectively when exposed to UV rays for 2160 hrs.

3.1.3   Comparison between dry and UV conditions

The first set is consisted of four samples with different concentration of filler that has not been

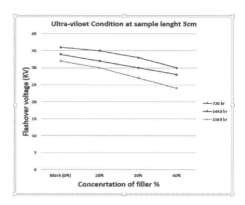

Figure 3.   Flashover voltage (kV) vs concentration of filler (%) for different UV radiation periods.

exposed to UV radiations as shown in Figure 4. The second set also contains 12 samples with different concentration of filler that was exposed to the UV radiation with different periods (720, 1440, and 2160 h) as shown in Figure 4.

The above graph shows the difference between dry condition and UV condition for the same sample with different concentration of filler:

1. Flashover voltage of blank sample is (45.39 kV) at dry condition while under UV condition, it is 36, 34, and 32 kV at 720, 1440, and 2160 h respectively.
2. At $[Mg(OH)_2]_{40\%}$ the flashover voltage for the dry condition increased to (52.64 kV) while the flashover voltage of UV condition decrease to (30, 28, and 24 kV) at (720, 1440, and 2160 h) respectively.
3. At dry condition by increasing the ratio of filler the flashover voltage increased by 13.7% and at UV condition the flashover voltage decreased by (16.7%, 17.6%, and 25%) at (720, 1440, and 2160 h) respectively by increasing the ratio of filler.

3.1.4   Wet condition

Figure 5 represents the flashover voltage (kV) for composite polyester samples at different concentration of filler $Mg(OH)_2$ after being immersed in water reagent for 7 days.

Figure 5 shows the effect of $Mg(OH)_2$ filler concentration on the flashover voltage for different samples of polyester after immersed in water. An improvement in flashover voltage value for polyester sample loaded with 40% $Mg(OH)_2$ filler compared with blank sample.

3.1.5   Salt–wet condition

Figure 6 shows flashover voltage (kV) of polyester composites with different concentration of filler after being immersed in NaCl solution.

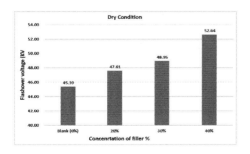

Figure 2.   Flashover voltage (kV) vs concentration of filler (%).

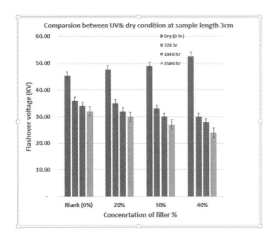

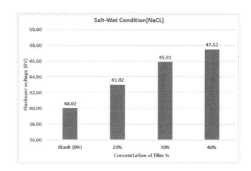

Figure 4. Flashover voltage (kV) for dry and UV conditions vs concentration of filler (%) with different interval of UV radiation.

Figure 6. Flashover voltage (kV) vs concentration of filler (%) after immersed in NaCl solution for 7 days.

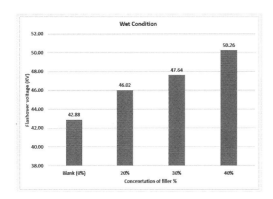

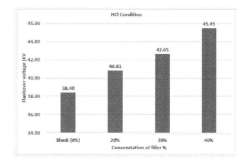

Figure 5. Flashover voltage (kV) vs concentration of filler (%) after immersed in water for 7 days.

Figure 7. Flashover voltage (kV) vs concentration of filler (%) after immersed in HCl solution for 7 days.

38.4 kV & 45.45 kV respectively and this meaning that the ratio of increasing between these two points is 15.5%.

3.1.7 Comparison between contaminated conditions (wet, salt–wet and hydrochloride)

This comparison shows that which condition has worse effect on the polymers that used as outdoor insulators and located at these conditions.

From Figure 8 we can note that the samples which immersed in water have the greatest values of flashover voltage at different concentrations of filler. The lowest values of flashover voltage were at the HCl conditions at different concentrations of filler So we can conclude from this comparison, the flashover voltage for composite polyester at wet area is greater than the composite polyester samples which located in coastal area (NaCl condition) and industrial regions (HCl condition).

Figure 6 illustrates the flashover voltage versus concentration of filler (blank, 20%, 30%, and 40%) which immersed in 10% NaCl solution. From this figure it can be noticed that, by increasing the filler concentration, the value of flashover voltage increases. Flashover voltage of 40% $Mg(OH)_2$ composite reaches to 47.52 kV with increasing percentage 35.5%.

3.2 Thermal test

3.2.1 Flammability test

Polyester, being essentially organic in nature, cannot resist the action of flame. They are all

3.1.6 Hydrochloride condition

Figure 7 illustrates flashover voltage (kV) of polyester composites with different concentration of filler after immersed in HCl solution for 7 days.

Figure 7 illustrates the relationship between flashover voltages against concentration of filler after being immersed in 10% HCl solution.

At the blank sample and $[Mg(OH)_2]_{40\%}$ sample, the values of flashover voltage at blank sample is

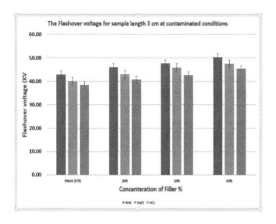

Figure 8. The flashover voltage (kV) against the concentration of filler for three sets of samples first one immersed in water for 7 days, second immersed in NaCl solution for 7 days, and the last one immersed in HCl solution for 7 days.

Table 3. The average burning time of styrenated polyester and composites.

Concentration of filler (%)	Average burning time(s)	Note
Blank	163	Burned with smoke
$[Mg(OH)_2]_{20}$	212	Burned with smoke after glow
$[Mg(OH)_2]_{30}$	233	Burned with slight smoke
$[Mg(OH)_2]_{40}$	280	Did not pass the 100 mm Marks. Smoke after ignition for 7s.

combustible, whether or not they are classified as "self-extinguishing." Addition of inorganic fillers such as $Mg(OH)_2$ reduces the burning rate of polyester. The average burning time of different concentration of $Mg(OH)_2$ in polymer composites was determined according to ASTM D635. The results are compared with the average burning time of styrenated polyester resin in Table 3.

The rate of burning time increases sharply and the density of smoke decreases as the concentration of $Mg(OH)_2$ increases. This is due to the fact that, the water released during the combustion of $Mg(OH)_2$ has the effect of diluting the combustion gases at the point of combustion and acts as a vapor barrier over the condensed phase to prevent oxygen from reaching the flame. The blank sample has the quickest burning time with 163 s. while the $[Mg(OH)_2]_{40\%}$ sample has the least burning time with 280 s.

4 CONCLUSION

This paper has focused on the effect of environmental conditions on the electrical and thermal performance of composite polymer insulators. The finding can be summarizing as following:

1. The analytical and experimental investigation for the flashover phenomena on the behaviour Adding some of inorganic fillers to original material (polyester) lead to improve the electrical and thermal properties of this material (composites material).
2. By increasing the concentration of filler, the value of flashover voltage increases the value of the flashover at all environmental condition except under UV radiation. This makes the doping with such concentrations more suitable for indoor purposes.
3. There is a critical percentage of filler that can be added with respect to the quantity of polyester, the most suitable percentage is 40% of $Mg(OH)_2$ filler.
4. From the comparison between the different contaminated conditions, the HCl condition has the lowest value of flashover voltage and the water condition has the largest value of flashover voltage.
5. $Mg(OH)_2$ filler prevent or delay ignition and retard combustion of the polymeric material (Polyester).

REFERENCES

Amin M., and M. Salman, "Aging of Polymeric Insulators (An overview)", Rev.Adv.Mater.Sci. 13, pp. 93–116, Pakistan, 2006.

Awad, M.M., H.M. Said, B.A. Arafa and A. Sadeek, "Effect of Sandstorms with Charged Particles on the Flashover and Breakdown of Transmission Lines", CIGRE 15–306 Paris, 2002.

Bernstorf, R.A., R.K. Niedermier and D.S. Winkler, "Polymer Compounds Used in High Voltage Insulators", Hubbell Power Systems, Ohio Brass Company, pp. 2–8, 2004.

Ersoy, A., M. Uğur, I. Güneş and A. Kuntman," A Study on the Insulation Capacity of Polymeric Composite Materials Blended with Boron", Journal of Electrical & Electronics Engineering, Vol. 7, No. 1, pp. 367–371, Turkey, 2007.

Ersoy, A., Y. Özcelep and A. Kuntman," A Study on Reliability of Polyester Insulators Blended with Borax", Journal of Electrical & Electronics Engineering, Vol. 16, No. 3, pp. 257–265, Turkey, 2008.

Gorur, R.S., R. Olsen, J. Crane, T. Adams, E. Jim Gurney and Jim Duxbury,"Prediction of Flashover Voltage of Insulators Using Low Voltage Surface Resistance Measurement", PSERC Publication 06–42, pp.1–10, Arizona, 2006.

Khan, Y.Z., A.A. Al-Arainy, N.H. Malik and M.I. Qureshi,"Effect of Thermo-Electrical Stresses and Ultra Violet Radiation on Polymeric Insulators", KSA University, final research report No. EE-18/26/27, pp. 1–7, Saudi Arabia, 2006.

Muniraj C., and S. Chandrasekar," Adaptive Neurofuzzy Inference System-based Pollution Severity Prediction of Polymeric Insulators in Power Transmission Lines", Article ID: 431357, pp.1–9, India, 2011.

Natural and Applied Sciences, "The Comparison of Ceramic and Non-Ceramic Insulators", E-Journal of new world sciences academy, Vol. 2, No. 4, Article no.: A0038, pp. 275–2888, Turkey, 2007.

Nasrat L.S., and A. Nosseir, "Evaluation of Polyester Insulation Materials under Polluted Conditions" CIGRE 15–402, Paris, 2002.

Zedan, F., M. Akbar, "Performance of H.V Transmission Line Insulators in Desert Conditions", IEEE Trans. on power delivery, Vol. 6, pp: 429–438, 1989.

Power transmission and distribution

Advances in Power and Energy Engineering – Sun (Ed.)
© 2016 Taylor & Francis Group, London, ISBN 978-1-138-02846-3

Testing and analysis of 220 kV transformer's abnormal noise

W. Jiang, X.S. Lan & Y.Q. Zhou
Sichuan Power Company, Electric Power Research Institute of State Grid, Chengdu, China

ABSTRACT: Increasing transformer noise can be caused by an internal fault and abnormal operating conditions. In this paper, the reason for the abnormal noise of a transformer in the Yulin station is determined by analyzing the operating conditions, testing for noise and vibration, measuring the DC of the transformer neutral point, and comparing it with the time between the monopole ground operation mode of the DC transmission system and the increasing noise of the transformer. The results from the testing and analysis found that the abnormal noise of the transformer is due to its DC bias phenomena, which are caused by the monopole ground return operation style of the Deyang converter station. Moreover, the recorded distance of the DC bias phenomena is up to 300 km.

Keywords: hysteresis scaling; transformer neutral point; DC bias; monopole operation

1 INTRODUCTION

Transformer noise comes from the iron core because of its silicon steel magnetic hysteresis stretching and electromagnetic coil wire. Increasing transformer noise may be caused by its internal faults or abnormal operating conditions, such as harmonics, DC bias, and abnormal increasing load (Alshawabkeh 2004, Hock 1985 & Shu Lu 1993).

At 07:30, on October 10, 2013, the sound of the 2# main transformer in the Yulin station was louder than usual. At the same time, its vibration was enhanced. The oil temperature, load, and bodies of the main transformer were not abnormal.

There are two main transformers in the Yulin station. Their neutral point grounding style is

that the 1# main transformer neutral point is non-grounding and the 2# main transformer neutral point is grounding. Both transformers are made by the same transformer company, which are a three-phase transformer with an independent iron core and connected oil circuit (see Figure 1). Their main parameters are as follows:

Model: SFSZ10-H-180000/220;
Rated capacity: 180000/180000/90000 kVA;
Rated voltage: (230 ± 8 × 1.25%)/115/38.5 kV;
Connection group label: YNyn0D11.

2 OPERATING CONDITIONS OF INCREASING TRANSFORMER NOISE

Does the load contribute to the reason for the increasing noise of the transformer with neutral point grounding? To find out this issue, its maximum load and maximum current were recorded, respectively, before its increasing noise (7:00) and after its increasing noise (8:00) on October 10 and the next day.

2.1 *The operating conditions of the 2# transformer at 7:00*

Power flow and load
220 kV bus II:
    Uab: 230.18 kV, Ua: 133.07 kV, Ub: 132.86 kV, Uc: 132.79 kV.
110 kV bus II:
    Uab: 111.36 kV, Ua: 64.352 kV, Ub: 64.346 kV, Uc: 64.368 kV.

Figure 1. 1# and 2# transformers of the Yulin station.

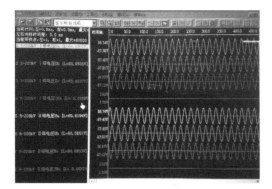

Figure 2. Recorded diagram of the 2# transformer at 7:40.

Current and active power
High voltage side:
    Ia: 185.820 A, P: −75.724 MW.
Middle voltage side:
    Ia: 390.791 A, P: −75.756 MW.

2.2 The operating conditions of the 2# transformer at 8:00

Power flow and load
220 kV bus II:
    Uab: 230.69 kV, Ua: 133.2 kV, Ub: 133.1 kV, Uc: 133.098 kV.
110 kV bus II:
    Uab: 111.405 kV, Ua: 64.365 kV, Ub: 64.297 kV, Uc: 64.273 kV.
Current and active power
High voltage side:
    Ia: 170.176 A, P: −69.250 MW.
Middle voltage side:
    Ia: 358.945 A, P: −69.684 MW.

From the above electrical data of the operating conditions, we can see that there were no abnormal data recorded for power flow, load, current and active power at the moment of increasing transformer noise.

3 OIL TEMPERATURE AND INFRARED TESTING

The noise of the 2# transformer in the Yulin station was increasing from October 10, and was not reduced until the next day. So, the operators performed the infrared temperature measurement and the oil temperature test for the 2# transformer. The testing results on October 12 are as follows:

Figure 3 shows the transformer body and a, b, and c phases. The highest temperature recorded is 46.3°C.

The oil temperature of the 2# transformer is detailed in Table 1.

On October 12, the oil and winding temperatures of the transformer were in the normal range.

4 TEST AND ANALYSIS OF NOISE AND VIBRATION

4.1 Test and analysis of noise

Is the transformer's abnormal noise related to its neutral point grounding style? To find out this issue, we tested the noise of the two transformers in accordance with their neutral point grounding styles. The one style is that the 2# transformer neutral point is grounding and the 1# transformer's is not. The other style is that the 1# transformer neutral point is grounding and the 2# transformer's is not.

After the transformer noise testing, we set the noise measurement instrument at the pool edge and its vertical elevation at 1.2 m. For each trans-

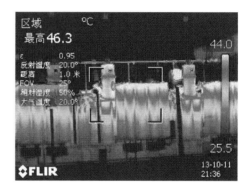

Figure 3. Infrared temperature measurement diagram of the 2# transformer.

Table 1. Testing results of the oil temperature of the 2# transformer (°C).

A phase	B phase	C phase	Winding	Environment
50.1	49.6	51.3	52.6	21.3

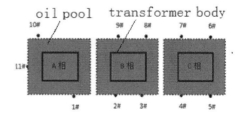

Figure 4. Noise measurement location of the 2# transformer.

Table 2. Noise testing results of the Yulin station transformer (equivalent sound level dB (A)).

Testing point	1	2	3	4	5	6	7	8
1# transformer's neutral point grounding	70.5	69.5	68	69	61.8	64.6	66.3	67
1# transformer's neutral point without grounding	57.9	64.1	63.9	62.9	61.1	57.1	59.9	58
Difference	**12.6**	**5.4**	**4.1**	**6.1**	**0.7**	**7.5**	**6.4**	**8.2**
2# transformer's neutral point grounding	71.4	76.6	72.9	70.7	70	71.1	73.7	73
2# transformer's neutral point without grounding	64.1	66.2	64.4	67.9	67.3	67	68.9	68
Difference	**7.3**	**10.4**	**8.5**	**2.8**	**2.7**	**4.1**	**4.8**	**5.9**

Note: A is the net weight.

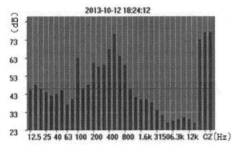

(A) When transformer's neutral point grounding

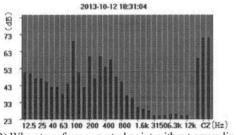

(B) When transformer neutral point without grounding

Figure 5. Distribution map of the transformer neutral point noise spectrum.

former, 11 testing points were measured, the location of which is shown in Figure 4.

The noise measurement results are summarized in Table 2.

According to the testing data given in Table 2, we can draw a conclusion that the noise of the transformer with neutral point grounding is higher than that of the other transformer without neutral point grounding. The difference in maximum noise is up to 12.6 dB.

From Figure 5, we can see that the noise of the transformer with neutral point grounding is significantly higher than that of the other transformer, when the noise spectrum ranges from 200 Hz to 800 Hz. In order to prove the relationship between the increasing noise of the transformer and its grounding style, we tested the noise of the 2# transformer with neutral point grounding, and found it to be about 10 dB higher than the noise of the 1# transformer without neutral point grounding. Then, we changed the grounding styles of those two transformers. In this case, the noise of the 1# transformer with neutral point grounding, was also found to be about 10 dB higher than that of the 2# transformer without neutral point grounding.

Therefore, we can draw a conclusion that the increasing noise of the transformer in the Yulin station is not caused by its inner fault but by its neutral point grounding style. Only the transformer with neutral point grounding had increasing noise for some time.

4.2 *Test and analysis of vibration*

For the vibration test of the transformer, a sensor was placed on the tank wall of the transformer during increasing noise. The test results are given as follows:

As shown in Figure 6, the vibration acceleration increases mainly at 100 Hz and increases concentrations from 200 Hz to 500 Hz for the transformer without neutral point grounding. At the same time, the vibration acceleration increases mainly at 500 Hz and obviously from 250 Hz to 700 Hz for the transformer with neutral point grounding.

5 TEST AND ANALYSIS OF TRANSFORMER NEUTRAL CURRENT

In order to analyze the effect of the neutral point current on the transformer, we changed the

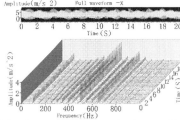

(A) When transformer's neutral point grounding.

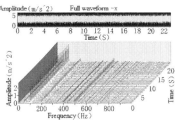

(B) When transformer's neutral point without grounding.

Figure 6. Distribution map of the transformer neutral point vibration spectrum.

Table 3. Neutral point current of the 1# transformer. (A).

Neutral point current	220 kV	110 kV
DC	0.9	0.5
AC	6.6	6.2

transformer's grounding styles and found the rule of the current and noise.

For one style of transformer grounding: 1# transformer with neutral point grounding and 2# transformer without neutral point grounding.

Then, we tested the neutral point current of the 1# transformer. The results are summarized in Table 3.

The other style of transformer grounding: for the 1# transformer, there is 220 kV side neutral point grounding and 110 kV side neutral point non-grounding, and for the 2# transformer, there is 220 kV side neutral point non-grounding and 110 kV side neutral point grounding.

From the above test styles, we can judge the main flowing path of DC. The testing results are summarized in Table 4.

The AC of the transformer neutral point is caused by unbalanced three phases. The maximum value is 6.6 A, which does not exceed the transformer's zero-sequence current protection setting value.

Table 4. Results of the neutral point current (A).

		110 kV	220 kV	Noise
1# transformer	DC	/	0.8	Louder
	AC	/	6.6	
2# transformer	DC	0.3	/	Slightly
	AC	5.9	/	louder

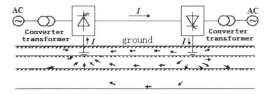

Figure 7. Monopole operation of the HVDC transmission system.

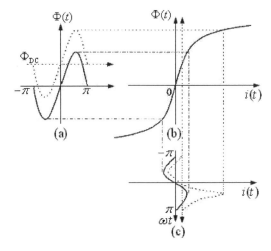

Figure 8. Principles of DC bias.

There is a 0.9 A DC current flow into the windings of the transformer in the Yulin station, which is due to its DC bias phenomena. As a result, the transformer's noise is increased.

6 INFLUENCE OF DC GROUND ELECTRODE CURRENT ON TRANSFORMERS

6.1 Introduction of transformer DC bias

When the HVDC transmission system works in monopole operation mode, DC flows through the neutral point by ground and causes DC bias of transformers around the HVDC grounding

Table 5. The time of Deyang switch converter station operation styles and the time of the 2# transformer changing noise.

Time	6:54 on Oct 10	18:52 on Oct 15	7:11 on Oct 16	9:50 on Oct 21
Deyang converter station	Monopole (II) ground return style	Restoration of bipolar operation style	Monopole (I) ground return style	Restoration of bipolar operation style
Time	7:30 on Oct 10	19:20 on Oct 15	7:36 on Oct 16	10:23 on Oct 21
2# transformer	Noise increasing	Noise back to normal	Noise increasing	Noise back to normal

electrode, which will pose a threat to their operation security (Zeng 2005, Villas 2003).

The principles of DC bias are shown in Figure 8. There is DC magnetic flux when DC flows through the transformer neutral point. If the total magnetic flux, including AC magnetic flux and DC magnetic flux, reaches transformer magnetic core saturation, excitation current half waveforms are distorted. Consequently, DC bias occurs (Bolduc 2000). At this time, there can be some exception, such as transformer local overheating, increasing vibration, noise, and other adverse reactions.

6.2 Transformer DC bias in the Yulin station

The noise changing time of the 2# transformer in the Yulin station is almost consistent with the time of adopting the monopole grounding operation style of the Deyang switch converter station (see Table 5), which is caused by the DC bias of the transformer.

7 CONCLUSION

To find the reason for the increasing noise of the transformer in the Yulin station, the infrared temperature measurement, vibration, and noise of neutral point flow test were carried out. In addition, we compared the time of Deyang switch converter station operation styles shifting with 2# transformer noise changing time. Finally, we found that the increasing noise of the transformer is caused by the DC grounding electrode current due to the Deyang switch converter station's adaptation of the monopole ground return operation style.

REFERENCES

Alshawabkeh, A.N., Sheahan, T.C., Wu X. 2004. Coupling of electrochemical and mechanical processes in soils under DC fields. Mechanics of Materials, 36(5):453–465.

Bolduc, L., Gaudreau, A., Dutil. A. 2000. Saturation time of transformers under dc excitation. Electric Power Systems Research, 56(2):95–102.

Hock, C.T., Glenn, W. 1985. On the Problem of Transformer overheating Due to Geomagnetically Induced Currents. IEEE Transactions on PAS, 104(1):212–219.

Shu Lu, Yilu Liu, Jaime De La Ree. 1993. Harmonics Generated From A DC Biased Transformer. IEEE Transactions on Power Delivery, l8 (2):725–731.

Villas, J.E.T., Portela, C.M. 2003. Soil heating around the ground electrode of an HVDC system by interaction of electrical, thermal, and electroosmotic phenomena. IEEE Trans on Power Delivery, 18(3):874–881.

Zeng, L.S., 2005. Impact of HVDC ground electrode current on the adjacent power transformer. High Voltage Engineering, 31(4):57–58.

Advances in Power and Energy Engineering – Sun (Ed.)
© 2016 Taylor & Francis Group, London, ISBN 978-1-138-02846-3

Critical line identification of the power system considering network structure and operating state

Z.Y. Qu & Q. Li
School of Information Engineering, Northeast Dianli University, Jilin, China

J.H. Du
Jilin Fengman Power Plant, Jilin, China

Y.X. Ren
State Grid Power Company in Jilin Province, Jilin Electric Power Company, China

ABSTRACT: Some critical lines in the large-scale power system have a great influence on the large-scale power outages and chain collapse accidents. Therefore, a practical method of critical line identification that conforms to the physical reality is need. First, a PTDF matrix is made based on PTDF (Power Transfer Distribution Factors) and the sensitivity of each branch at different load nodes is quantized from the point of network structure. Then, the model-based multi-machine power system transient energy function is used to solve the transient potential energy on each branch, and the vulnerability of branches is quantized from the perspective of running state based on the ratio of critical line transient potential energy and the share of transient potential energy. At last, based on the above processes, the validity and rationality of this method are verified by the simulating computation of IEEE39 nodes system and the comparison between this result and that of existing methods.

Keywords: power transfer distribution factors; transient potential energy; critical line identification; sensitivity of branches; integrated vulnerability

1 INTRODUCTION

The recent large-scale power outages world-wide have brought great troubles to the economy and people's daily life. The vulnerable parts in the power system structure has become a new theory in this field and many scholars are doing in-depth study on it but there aren't acknowledged definition and unified standards at the moment.

In Reference (Cao, Y.J.), the author quantizes the importance of all nodes and branch structures by using their betweenness and points out that the key part of the power system is the high betweenness of the nodes or branch structures. The author of Reference (Barbulescu) suggests that the electrical distance between nodes be improved to the strength of coupling of the voltage and the available rate of the branch structures.

In Reference (Ding. 2006) and (Han. 2008), the authors make a research from a new perspective-the mechanism of chain breakdown with the reactance as the circuit weight. They introduce two factors-the running capacity and running limitation and make a comparison of the relationship

between load and the running capacity to make out the location of the key part. The above methods are worked out based on the proposition that the generatrix power only spreads according to the shortest path, without regard to the Kirchhoff's law. That is, the influence of other possible transmission paths has been ignored in the calculation of nodes and circuit betweenness.

The author of Reference (Wei. 2010) comes up with a way of key route identification based on the highest contribution degree of transmission, revealing the key part of power system from the perspective of overall precaution. In Reference (Qu. 2012), the concept of electric betweenness is proposed, that is, unit current is used in some branch and the cumulative sum of the absolute value of the current is taken as the critical standard of branches.

Reference (Gross. 2004) is on the comprehensive evaluation model based on the running condition of the power system and network topology, but the topology factor still uses the evaluation index of the worldlet theory. To some extent, this method overcomes the defect of betweenness but it neglects the directivity of the flow of power transmission line.

Meanwhile, it doesn't take into consideration the fact that the circuital vulnerability is related not only to the structural vulnerability but also the condition vulnerability.

In this article, direct power transmission distribution factor is used to compute the sensitivity of power loss of different load node on each branch whose importance is quantized based on the network structure. Meanwhile, the model-based multi-machine power system transient energy function is used to solve the transient potential energy on each branch so as to evaluate the vulnerability of these branches according to their running condition.

By combining the two operations, the author scientifically quantizes the important role of each transmission line in the whole transmission to make its physical background closer to the actual power system.

2 ASSESSMENT OF THE STRUCTURAL SENSIBILITY BASED ON PTDF

2.1 Improved PTDF

In the power system, there is a transmission between the generator node and load node at every moment and the change of generator node output and that of load node's absorbing power will both cause the redistribution of power on each branch. Power Transfer Distribution Factor (PTDF) refers to the relative change in the power flow on a particular branch due to a change in injection and corresponding with-draw at the system swing or slack bus.

According to China's voltage class and requirement on power transmission line, the voltage is over 35 kV and the line whose voltage is 35 kV to 220 kV is called high-voltage power transmission line while that between 330 kV and 500 kV is called super-high-voltage power transmission line. In the study, the power transmission network is supposed to be super- high- voltage line whose cross sectional area and the geometric mean distance between circuits are big due to the big current. Thus, the resistance is far smaller than the reactance with r/x approximates to zero and the impedance angle approximates to pi/2 neglecting the admittance of parallel branches. As the voltage phase difference of two adjacent nodes in super-high-voltage power transmission line is almost zero, the trigonometric function in the expression of pure injected power on each node can be approximated to $\cos\theta_{ij} \approx 1, \sin\theta_{ij} \approx \theta_{ij}$. If the voltage amplitude of each node is nominal voltage, then $U_i \approx U_j = 1.0$ based on the per unit value.

According to the classic flow analysis theory, the active power and reactive power of a certain

generator node and its relevant load node on all branches included in-between are as followed:

$$P_{Gi} - P_{Li} = U_i \sum_{j \in i} U_j(G_{ij}\cos\theta_{ij} + B_{ij}\sin\theta_{ij})$$
$$Q_{Gi} - Q_{Li} = U_i \sum_{j \in i} U_j(G_{ij}\sin\theta_{ij} - B_{ij}\cos\theta_{ij})$$
(1)

In equation (1), P_{Gi} and Q_{Gi} represent the active power and reactive power given out by the generator node while P_{Li} and Q_{Li} represent the the flow those absorbed by the load node. U_i and U_j represent the per unit value of the voltage of node i and node j while θ_{ij} represents the voltage phase angle difference between voltage U_i and U_j.

The equation extracted from equation (1) about the power on the i-j branch is:

$$P_{ij} = U_i^2 g_{ij} - U_i U_j(g_{ij}\cos\theta_{ij} + b_{ij}\sin\theta_{ij})$$
$$Q_{ij} = -U_i^2(b_{ij} + b_{i0}) + U_i U_j(b_{ij}\cos\theta_{ij} - g_{ij}\sin\theta_{ij})$$
(2)

In equation (2), P_{ij} and Q_{ij} represent the active power and reactive power on the branch i-j, while g_{ij}, b_{ij}, b_{i0} represent the resistance, reactance and admittance. The circuit diagram corresponding to equation (2) is as followed:

Equation (2) can be simplified like this under the simplification conditions of super-high-voltage power transmission line:

$$P_{ij} = -b_{ij}(\theta_i - \theta_j) = \frac{\theta_i - \theta_j}{x_{ij}}$$
$$Q_{ij} = 0$$
(3)

In equation (3), x_{ij} represents the reactance value on the branch i-j.

It is obvious that by the simplification of direct flow computing, the power of branches in super-high-voltage power transmission line depends only on the branch reactance and the voltage angle phase difference between two nodes. Let us suppose

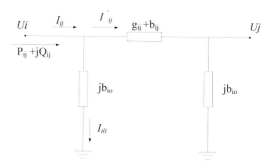

Figure 1.   Equivalent circuit diagram on the branch i-j.

the total number of load nodes is T and a certain load node is t, and the total number of branches is L and a certain branch is l, and we use equation (4) to compute the sensibility of any branch's reaction to the power change of load nodes.

$$\rho_{l,t} = \frac{dP_l}{dP_t} \tag{4}$$

$\rho_{l,t}$ represent the PTDF of branch l, P_l represents the commutative active power on branch l and P_t represents the active power cost by load node t.

2.2 Assessment of the sensibility of branches

We get equation (5) by substituting equation (3) into equation (4) to compute the sensibility of each branch.

$$\rho_{l,t} = \frac{d}{dP_t}\left[\frac{1}{x_{ij}}(\theta_i - \theta_j)\right]$$
$$= \frac{1}{x_{ij}}\left(\frac{d\theta_i}{dP_t} - \frac{d\theta_j}{dP_t}\right) = \frac{1}{x_{ij}}\left(\frac{1}{\frac{dP_t}{d\theta_i}} - \frac{1}{\frac{dP_t}{d\theta_j}}\right) \tag{5}$$

It can be simplified to equation (6):

$$\rho_{l,t} = \frac{1}{x_{ij}}(X_{it} - X_{jt}) \tag{6}$$

In equation (6), x_{ij} represents the reactance of the branch i-j while X_{it} and X_{jt} represent the corresponding resistance value in the impedance matrix in power system. Therefore, the PTDF matrix H can be introduced with the number of branches as the line.

Number and the number of load nodes as the column number. As what should be taken into consideration is only the degree of sensibility whether positive or negative, we only use the module value of PTDF to pick out the branch with high sensibility.

$$H_{PTDF} = \begin{bmatrix} |\rho_{11}| & |\rho_{12}| & \cdot & \cdot & |\rho_{1T}| \\ |\rho_{21}| & |\rho_{22}| & \cdot & \cdot & |\rho_{2T}| \\ \cdot & & & & \cdot \\ \cdot & & & \cdot & \cdot \\ |\rho_{L1}| & |\rho_{L2}| & \cdot & \cdot & |\rho_{LT}| \end{bmatrix} \tag{7}$$

The matrix of PTDF shows the sensibility of all branches in their generator-load node pairs and quantizes the usage of any branch of the power

transmission between generator node and load node. Different from the previous methods, PTDF matrix cuts down tow variables (X_{is}, X_{js}), which represents respectively elements of line i/column s and line j/column s in the reactance matrix, which makes it more quickly and more effectively compute the parameters of large-scale power system. As in the same branch, the PTDFs of load nodes supplied by different powers are not the same, we have the following equation.

$$K(i,j) = \sum_{l\in L,\, t\in T} |\rho_{l,t}(i,j)| \tag{8}$$

In equation (8), L represents the set of all branches, T represents the set of load nodes, and K (i, j) represents the overall efficiency of load nodes reacting to power commutation on the branch. The bigger the K(i, j), the higher the sensibility of the branch, seen from the computing process, the sensibility is only related to the node reactance matrix. As a result, the rapid assessment of the importance of each branch is made by the use of the node reactance matrix, which overcomes the previous disadvantage of the trend that only spreads along the shortest path. Thus, the assessment of the structural vulnerability of branches is more in accordance with the physical background of the power system.

3 THE ASSESSMENT OF THE IMPORTANCE OF BRANCHES BASED ON THE TRANSIENT ENERGY FUNCTION

3.1 The equation of the potential energy on a branch

Consider the voltage phase angles of node i and node j at the two ends of the branch l are θ_i, θ_j and their difference is $\theta_L = \theta_i - \theta_j$, then the equation of the multi-machine system transient energy function is as shown:

$$V = V_{PE} + V_{LE}$$
$$= \sum_{l=1}^{L}\int_{\theta_l^s}^{\theta_l}[P_l(u) - P_u^s]\,du + \frac{1}{2}\sum_{g=1}^{n}M_g\omega_g^2 \tag{9}$$

In equation (9), V represents the total energy of the system, V_{PE} represents the potential energy of the system, V_{LE} represents the kinetic energy of the system, ω_g represents the rotational angular frequency of generator g, M_g represents the rotational inertia of generator g, $P_l(\theta_l)$ represents the active power of the first branch, $P_l(\theta_l^s)$ represents the active power of the first branch in equilibrium state, and θ_l^s represents the phase angle difference of the first branch in equilibrium state [10]. Seen from the equation (9), the total transient energy includes the sum of kinetic energy of all generators and the sum of potential energy shared by each branch in the augmented network in the system.

The equation of the transient potential energy of any branch l in the network is as shown:

$$V_{PEl}(\theta_l) = \int_{\theta_l^s}^{\theta_l} \left[P_L(u) - P_l^s \right] du \qquad (10)$$

Taking a derivative of time of equation (10), we get equation (11):

$$V_{PEl}'(t) = [P_l(\theta_l(t)) - P_l^s]\omega_l(t) \qquad (11)$$

In equation (11), $\omega_l(t)$ represents the time derivative of the phase angle difference $\theta_l(t)$ of the branch.

3.2 The assessment of transient stability of branches

When the power system has been disturbed, a large amount of transient energy is injected into the system and the kinetic energy of the generator's rotor transforms into the potential energy of the augmented network. The power system is stable if it can absorb the injected kinetic energy, otherwise it will fail to maintain the stability. After being disturbed, the system's transient energy assembles on some branches instead of scattering equally on each branch in the network. Thus, we can evaluate the vulnerability of a branch by computing the transient potential energy and critical share on each line.

Based on the characteristic of the function change in the equation of the transient potential energy on a branch, equation (10) is taken a derivative of time with t_{bl} being the first moment of reaching the maximum. If $p_l(t_{bl}) - p_l^s$ is not zero, then $\omega_l(t_{bl}) = 0$ and the branch is stable when the phase angle difference reaches the constant at t_{bl}. If $[p_l(t_{bl}) - p_l^s] = 0$ and the slopes of both sides above and below the arrest point are positive, then $\omega_l(t_{bl}) > 0$. The phase angle difference of the branch will still enlarge so the branch is not stable.

Defining the transient potential energy $V_{PEl}(t_{al}, t_{bl})$ shared by the branch after the breakdown as the differentials between the potential t_{bl} at the first moment of reaching maximum and minimum and the transient potential energy of the critical branch $p_l(t_{bl}) - p_l^s$ as the differentials between the first maximum of the transient

potential energy after the breakdown and the potential energy in equilibrium state, then we get the function of the conditional vulnerability of the branch is as followed:

$$I_{Vl} = \left| \frac{\left[P_l(t_{bl}) - P_l^s \right]}{P_l^s} \cdot \frac{1}{V_{PEl}(t_{al}, t_{bl})} \right| \qquad (12)$$

The above equation shows that when I_{Vl} approximates to zero, the branch is very vulnerable and can be easily effected by the disturbance so it is the weak link as well as the key line that needs key monitoring in the system.

4 THE WAY OF INTEGRATED CRITICAL LINE IDENTIFICATION

The importance of lines is not only reflected in its structural characteristic but also related to the running state. The two perspectives of critical line identification-assessment of the structural sensibility of the power system based on PTDF and conditional vulnerability based on the transient potential energy function of branches are complementary with different focuses and are included in each other. In the actual running of the power system, the change of conditions and structural attribute coexist.

The way of integrated critical line identification also takes into consideration the structural sensibility of the power system and the conditional vulnerability, and it is defined as:

$$F(i, j) = \frac{1}{I_{Vl}(i, j)} \cdot K(i, j) \qquad (13)$$

In equation (13), i represents the starting nodes, j represents the ending node, j represents the ending node, and $F(i, j)$ represents the integrated importance. In the critical line identification, the value of $F(i, j)$ is found by the collection and computing of relevant electrical parameter and it is arranged in descending order with some lines in front as the critical lines of the network system.

5 THE ANALYSIS OF EXAMPLES

In this article, IEEE39 node system is used to verify the efficiency of the way of integrated critical line identification and Figure 1 is the IEEE39 node system. The node reactance matrix makes it easy to get the level of vulnerability of all branches at different load nodes. By Matlab simulation, the level of vulnerability of all 46 branches at 17 different load nodes. Then, by using equation (8), we get the overall efficiency of load nodes reacting to power commutation on the branch.

Table 1 shows the value of sensitivity of branches computed by improved PTDF function. The top ten branches is arranged according to the level of sensibility in descending order and compared with the methods of previous scholars.

According to Table 1, in Reference (Huang. 2014), branches 15–16, 16–17 and 2–25 in top ten branches are defined as the critical lines based on the weighted branch betweenness. In Reference (Ding. 2008), branches 14–15, 15–16 and 16–17 are defined as the critical lines according to the distribution degree index based on original PTDF. This conclusion is reasonable in the network diagram of the system because branches 15–16 and 16–17 are in the important transmission channels and if they break down the power on nodes 33, 34, 35 and 36 can't be delivered which will cause partial power shortage in the system. And this conclusion is also drawn by the use of this method.

In the bar diagram of the sensitivity level of each branch, the branches with high degree of sensitivity are obviously shown. However, the identification result which takes into consideration only the network structure factors regardless of the running state of lines is not convincing in the actual power system. Therefore, the importance of branches can be further modified according to the transient potential energy function of branches.

In the IEEE39 node system, three short-circuit faults are set based on the given load flow with the fault clearing time as o.18 seconds and the energy function algorithm is used to analyze the transient stability index of each branch.

First, the transient potential energy fluctuation conditions of each branch after fault clearing are

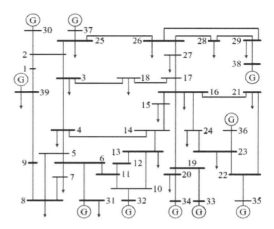

Figure 2. Wiring scheme of IEEE39 node system.

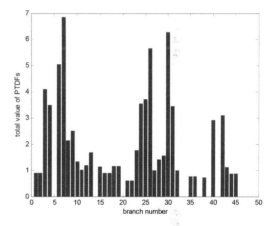

Figure 3. Assessment of the sensibility of branches based on PTDF.

made out by simulation as shown in Figure 3, but it is difficult to identify the level of importance through the transient potential energy fluctuation curves of each branch.

Therefore, the assessment of transient stability of branches is needed to quantize the stability of each branch. A large amount of transient energy is injected into the system and the kinetic energy of the generator's rotor transforms into the potential energy of the network. Then we should examine the ability of each branch to absorb the injected energy. The branch is stable if its critical transient potential energy is big, it is vulnerable if the critical transient potential energy is small.

Then, based on the theory of the assessment of transient stability of branches, equation (13) of the conditional vulnerability of branches is used

Table 1. Critical line identification result by branch sensitivity function.

Nr.	Sensitivity value of branches $K(i,j)$	Line	Method (Me.) of this article	Me. of Ref. [11]	Me. of Ref. [5]
1	6.8434	7	3–18	2–3	16–17
2	6.2826	30	17–18	16–19	2–3
3	5.6623	26	16–17	16–17	14–15
4	5.0519	6	3–4	15–16	15–16
5	4.1047	3	2–3	4–14	16–19
6	3.7192	25	15–16	2–25	4–14
7	3.5541	24	14–15	3–4	4–5
8	3.4918	4	2–25	25–26	17–27
9	3.4586	31	17–27	14–15	3–4
10	3.1080	40	25–26	16–24	3–18

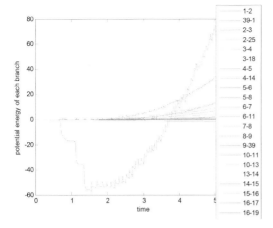

Figure 4. Transient potential energy fluctuation curves of each branch after fault clearing.

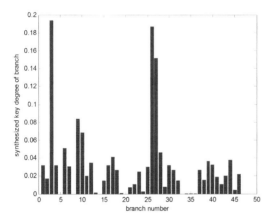

Figure 5. The integrated vulnerability of each branch.

Table 2. Result of integrated critical line identification.

Nr.	$F(i, j)$	Line	Branch
1	0.1939	3	2–3
2	0.1923	26	16–17
3	0.1520	27	16–19
4	0.0869	25	15–16
5	0.0845	4	2–25
6	0.0690	9	4–14
7	0.0513	10	5–6
8	0.0463	6	3–4
9	0.0376	28	16–21
10	0.0326	44	26–29

integrated importance assessment index. Seen from the tables, the critical branches are mostly near the generators. Besides, based on the integrated influence of structural sensitivity factors and branch transient potential energy, the analysis is made that the reason why branch 2–3 is of high importance is that it is not only the access of the power transmission of generator nodes 30 and 37 but also is closely related to the load nodes 3 and 4 which means it is responsible for the power transmission in the system.

Branch 2–3 will switch off if the protection of it removes, which will affect the power transmission of generator 30 and 37 causing the load flow transfer in some areas. Many important connecting lines will switch off if the protection of branch 2–3 fails to remove, which will isolate generator 30 and load node 3 causing the range fluctuation of load flow in the system. In conclusion, branch 2–3 is of high importance.

6 CONCLUSION

This article proposes the method of evaluating the sensitivity of branches based on the improved power transmission distribution function. Though the author ignores the reactance between the generator's node and nodes of the two ends of referenced branch, the result is almost the same as that found by previous methods. This improvement makes it more efficient in computing the mass data of large-scale power system. Meanwhile, the author puts forward the branch conditional vulnerability function on the basis of the previous method of assessment of branch transient stability. Combining the two methods, the network structural factors and running state factors are taken into consideration at the same time in critical line identification. The difference of indexes of critical lines and non-critical lines is enlarged,

to identify the function with high degree of vulnerability. At last, we get Figure 5 according to the way of integrated critical line identification. The diagram shows that the branch whose vulnerability degree is high tends to break down when the system is greatly disturbed so it is the critical line.

In Table 2 the integrated vulnerability degree indexes are arranged in descending order.

The comparison between Figure 2 and Figure 4 shows obviously that the previous critical line index is strengthened after the introduction of conditional factors while the indexes of branches with comparative small importance value decrease sharply, which makes the identification clearer. At the same time, the top ten branches are found based on the function of the which makes the result clearer and easy to be identified to some extent.

REFERENCES

Cao, Y.J. & Chen, X.G. 2006. Large-scale power systems on complex network theory based on fragile line identification [J]. Electric Power Automation Equipment, 2006, 26(12): 1–5, 31.

Ding, M. & Han, P.P. 2006. Large grid-based Small World Topological Model Based Vulnerability Assessment Algorithm [J]. Automation of Electric Power Systems, 2006, 30(8): 7–11.

Ding, M. 2008. Small World Network Vulnerability Assessment Weighted Topological Model [J]. China Electrical Engineering, 2008, 28(10): 20–25.

Qu, W.Y. & Li, Y.H. 2012. Identification based on the maximum contribution streaming gateway key power lines and nodes [J]. Automation of Electric Power Systems, 2012, 36(9): 6–12.

Xu, L. & Wang, X.L. 2011. Electrical betweenness and its application in power system critical path identification [J]. China Electrical Engineering, 2010, 30(1): 33–39.

Wei, Z.B. & Liu, J.Y. 2010. Guojun Zhu. Network vulnerability assessment model reliability Weighted Topological Model Based [J]. Electrical Technology, 2010, 25(8): 132–136.

Liu, M.H. & Gross, G. 2004. Role of distribution factors in congestion revenue rights applications, IEEE Transaction on Power Systems, Volume 19, Issue 2, May (2004). p. 802–810.

Barbulescu, C. 1990. Software Tool For PTDF Computing within the power systems.

Huang, Y. & Li, H.Q. 2014. Vulnerability assessment is based on the transient complex branch network and transient energy function [J]. Power system protection and control system, 2014, 42(20): 1–6.

Zhang, H.X. & Lv, F.P. 2014. Power Grid Vulnerability Assessment Based on the protection of vulnerable Weighted Topological Model [J]. China Electrical Engineering, 2014, 34(4): 613–619.

Advances in Power and Energy Engineering – Sun (Ed.)
© 2016 Taylor & Francis Group, London, ISBN 978-1-138-02846-3

Simulation and experimental study of a state-of-the-art M-STATCOM

S.A. Kamran & Javier Muñoz
Department of Industrial Technologies, Universidad de Talca, Talca, Chile

Yaojun Chen
Department of Electronic Countermeasure, Air Force Radar Academy, Wuhan, China

ABSTRACT: In this paper, simulation and experimental properties of a state-of-the-art STATCOM based on modular multilevel converter (M-STATCOM) is presented. The modulation scheme is not explanation in this paper while control strategy modifies the traditional capacitor balance strategy. In the control strategy, there are three loops in parallel that is used to control three variables respectively. Negative sequence double fundamental frequency synchronous rotating coordinate system NDdq0 coordinate system is used to convert three-phase circulating currents. A 14 KV DC input MMC is simulated under both the stable and dynamic condition MATLAB/SIMULINK environment with total 14 numbers of sub-modules per lag. To further verify these simulation results, a downscaled platform is established based on DSP and FPGA controller. The effectiveness and feasibility of the proposed control strategy is validated by both the simulation and experimental results.

1 INTRODUCTION

The STATCOM is a key element of the FACTS technology (Acha E. 2013). On the basis of voltage sourced converter, it can effectively improve the power quality and adjust reactive power smoothly from inductive to capacitive load (Jing Song Zhu et al. 2012). Supported by the fast development of high-power semiconductors over recent years, multilevel converters have emerged and enable better line-side behavior than current converter topologies (Martin Glinka et al. 2015). The multilevel converter based STATCOM can easily reach up to medium or high-voltage high-power applications without transformers, which make the whole system heavy and bulky (Allebrod et al. 2008).

As the disadvantages of traditional multilevel converters such as Neutral Point Clamped (NPC), the Flying Capacitor (FC), Cascade H-Bridge (CHB) in STATCOM are obvious so the topology of modular multilevel converter based on cascaded structure is examined. Besides its modular characteristic, MMC has public DC bus. It can realize the AC-DC and DC-AC transform. Meanwhile, MMC can be used in the high-voltage and high-power field without any other equipments (Wei Zhang et al. 2014).

MMC provides a viable approach to construct a reliable and cost effective STATCOM (called M-STATCOM) with increased number of levels, capable of eliminating interface transformers (Bina et al. 2011). The active power redistributed in the internal loops can be used for negative sequence balancing purpose. Therefore, M-STATCOM can

work continuously under three-phase unbalance conditions, capable of overcome symmetrical and asymmetrical faults without increasing the risk of a system collapse and furthermore it has intrinsic fault management capability (Adam et al. 2012).

The rest of the paper is organized as follows: the topology of M-STATCOM is explained in section two. Then, section three describes the control scheme; in section four M-STATCOM is simulated in Matlab/Simulink environment for its stable and dynamic responses; in section five a downscaled platform is established. Finally, section six concluded out the paper.

2 TOPOLOGY OF M-STATCOM

Figure 1 shows the topology structure of a three phase MMC. In the converter, each phase consists of two legs that are named as positive leg and negative leg, respectively. In this work, each leg contains 14 identical, evenly and serially connected Sub-Modules (SMs) along with an inductor l (that limits the fault current). Each sub-module consists of a floatingdc capacitor C and two insulated-gate bipolar transistors (T_1 and T_2) that form a bidirectional chopper (Junhua Guo et al. 2014).

As shown in (Casaei et al. 2012 & Rohner et al. 2010), the voltage of each capacitor is Vd/N, where Vd is the DC bus voltage and N is the number of sub-modules in each leg. In a sub-module, when the upper switch is on (and the lower switch is off), C is inserted in the circuit, in which the state of the sub-module

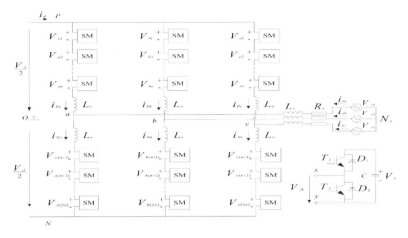

Figure 1.   The topology structure of three-phase MMC.

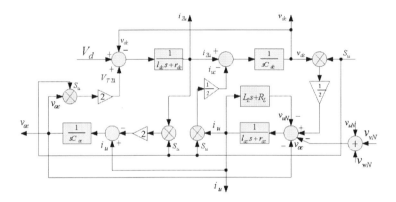

Figure 2.   Control diagram for MMC.

is defined "on" or "1"; when the upper switch is off (and the lower switch is on), C is by-passed, here the sub-module is "off" or "0". Then, by controlling the states of these sub-modules, the levels in the legs can be changed. Therefore, the terminal voltage of each sub-module can be either its capacitor voltage or zero, depending on the switching states.

3   CONTROL STRATEGY

The system control block with one phase is shown in Figure 2, in which there are three controllers to control the d-, q-, and 0-axis components of the circulating current respectively (Junhua Guo et al. 2014). The LPFs are used to filter the sixth harmonic components in the d- and q-axis, and because the i_{Zd} is coupled with i_{Zq}, there has a decoupled control to be sure to the control for i_{Zd} and i_{Zq} being respective. The zero sequence current i_{Z0} is the DC component of the circulating current

which affects the capacitor voltage and the output power.

The variability of the average capacitor voltage reflects the demand of the circulating current, so the output of the regulator of the capacitor voltage can be taken as the command of i_{Z0}. There has a forward control of i_{p0} (a dc component which indicates the magnitude of the output power), which is used to improve the response speed of the system. It should be worth attention that i_{p0} is also a DC current in the stable state and decided by the output active power. The output of the three controller are summed together and then been normalized, whose output plus the modulating signal in the positive leg and the negative leg respectively are compared with two groups that compose with four Level-shift triangle waves. The output PWM signals of these comparators are ordered by the voltage balance algorithm and then sent to every sub-module (Antonopoulos et al. 2009, Chen Yaojun et al. 2014, K. Jafri et al. 2012, Baichao Chen et al. 2015 & Chen Yaojun 2015).

4 SIMULATION RESULTS

Based on the aforementioned control strategy, a 14 KV DC input MMC is simulated with Matlab/Simulink which is further verified by a downscaled experimental platform.

4.1 *Simulation in the stable state*

The simulation parameters are shown in Table 1.

Table 1. Simulation parameters.

Capacitance in a module	3.7 nF
Inductance in the leg	4 mH
DC equivalent resistance	0.5 Ω
Load inductance	8 H
Load resistance	20 Ω
Number of modules per lag	14
Amplitude modulation rate m	1
Carrier wave frequency	5 KHZ
Input DC voltage	14 KV

Figure 3 simulation wave forms are shown without control, in which (a) and (b) are the three phase output voltage and current waveforms respectively. Both the three phase output voltage and current wave forms can achieve super waveform quality with THD values about 3.6% and 0.70%.

The waveforms of one phase (u-phase) of both the output current and circulating current are shown in Figure 3(c). It is clear that the second harmonic component is leading and amplitude of the harmonic current is about half as the output current. The waveforms of the circulating current in the NDdq0 coordinate system are shown in Figure 3(d), from which it can be found that there is sixth harmonic component in the d-, q-, and 0-axis, however comparing to the DC component these harmonic components are small, especially in the 0-axis.

In Figure 3(e) the waveforms of the three-phase v_{dc} (capacitor voltage) is shown, which reflects the sum of the voltages of the positive and the negative leg per phase and whose DC component

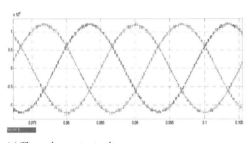

(a) Three-phase output voltage

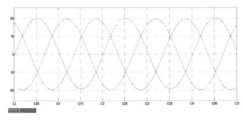

(b)Three-phase output current

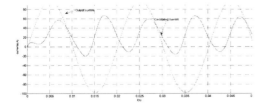

(c)Output and circulating current of one phase

Figure 3. Simulation waveforms without control.

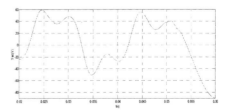

(d)Circulating current in NDdq0 coordinate system

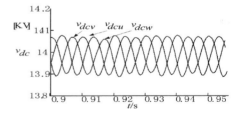

(e) The waveform of three-phase v_{dc}

(f) The waveform of three-phase v_{ac}

is equal to the input DC voltage (ignoring the equivalent resistance) while AC components is dominant by the second harmonic component. The waveform of v_{ac} is shown in Figure 3(f), which reflects the capacitor voltage difference between the positive leg and the negative leg, and compose with a fundamental component and 3rd odd order harmonic components.

4.2 *Dynamic simulation*

In order to verify the dynamic characteristic of the system, the system suddenly changed from no load to full load at 0.2 s, and then the system load abruptly changed from full load to no load at 0.6 S, the simulation waveforms of one phase (u-phase) of both the output current and circulating current are shown in Figure 4(a). When system is at no load, the output currents and circulation current are zero. At 0.2 s the system load abruptly changed from no load to full load, the output currents and circulation current reached to steady quickly with a steady and rapid transient. Now, the system load

abruptly changed from full load to no load at 0.6 s, the output currents change to zero immediately. But the circulating current is fluctuated slightly in the near zero after a short-term oscillation. The Figure 4(b) shows the waveform of circulation current in the NDdqo coordinate system, which is controlled dq components to near zero, indicating that the second harmonic has been eliminated, and the zero-sequence component is controlled to a desired size. At 0.6 s, there is a short time fluctuation in component of d- and q- axes.

Figure 4(c) and (d) are the dynamic and static simulation waveforms of v_{dc}. In, Figure 4(c) at 0.2 s, load suddenly added, its average value is still maintained at 14 kV, but the three-phase fluctuations are uneven, because the initial state of the circuit is not the same, resulting sometime is required in charging while the rest of the time for discharging. Hence after a certain period of time it restores the balance. After removal of load at 0.6 s, capacitor voltage on the DC side charging and discharging, the instability slowly decay to zero. In Figure 4(d), the steady-state waveform of v_{dc} is shown.

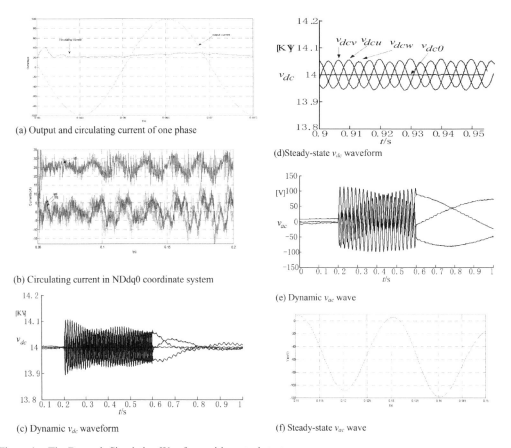

(a) Output and circulating current of one phase

(b) Circulating current in NDdq0 coordinate system

(c) Dynamic v_{dc} waveform

(d) Steady-state v_{dc} waveform

(e) Dynamic v_{ac} wave

(f) Steady-state v_{ac} wave

Figure 4.   The Dynamic Simulation Waveform with control strategy.

The average of the three phases is still maintained at 14 kV, the second harmonic of any naturally occurring fluctuations are also eliminated.

Figure 4(e) and (f) are the dynamic and static simulation waveforms of v_{ac} respectively. After sudden addition and removal of load at 0.2S and 0.6S respectively, the three-phase v_{ac} go through a period of adjustment and obtain the final balance. Actually, v_{ac} reflects the difference between positive and negative voltages of capacitors in a bridge arm, if the difference is too sweeping, change is likely to affect the stability of the system. Therefore, must be concerned in the control process of change v_{ac}. In Figure 4(f), the steady-state waveform of v_{ac} is shown, where the third harmonic is eliminated, leaving only the fundamental.

5 EXPERIMENTAL RESULTS

For final confirmation of these simulation results based on the control method mentioned above, a downscaled platform is designed whose parameters are listed in the Table 2. Controller is consists of DSP and FPGA. The function of FPGA is generating, measuring and controlling the capacitor voltage waveform, while DSP implements the relevant algorithm.

Figure 5 experimental waveforms are shown. In Figure (a) and (b) three-phase output current and voltage waveforms are shown which are purely sinusoidal. As there are 4 sub-modules per lag so the output voltage waveform achieved 5 levels.

In Figure 5(c) and (d), the output currents and the circulating currents of a phase are shown without and with control action. Figure 5(c), waveforms are under uncontrolled condition so the circulation current significantly contains DC component and the second order harmonic component. As, in Figure 5(d), waveforms are under controlled condition where low-order harmonics in circulation current have been suppressed.

In Figure 5(e) and (f), the sum of all the capacitor voltages v_{dc} between the positive and the

Table 2. Parameters of the experiment.

Capacitance in a module	2.9 nF
Inductance in the lag	6 mH
IGBT module	1.2 kV/50 A
Load inductance	8 mH
Load resistance	20 Ω
Number of modules per lag	4
Amplitude modulation rate m	0.90
Carrier wave frequency	5 kHz
Input DC voltage	500 V

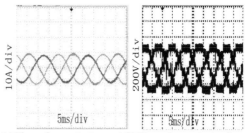

(a) Three-phase output current (b) Three-phase output voltage

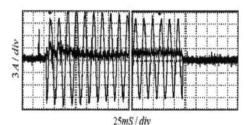

25mS / div

Output currents and circulation currents
(c) without control   (d) with control

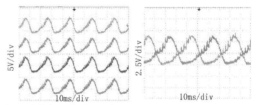

Sum of all the capacitor voltages v_{dc}
(e) withcontrol   (f) withoutcontrol

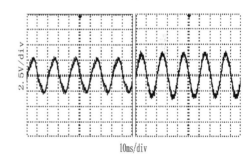

10ms/div

The difference between positive and negative voltages of capacitors in a bridge armvac
(g)   with control   (h) without control

Figure 5. Experimental waveforms.

negative legs are shown under controlled and uncontrolled condition. In Figure 5(f), the amplitude fluctuation is increased, due to two factors: one is the circulation of the second order harmonics, and another factor is the load current. In Figure 5(e), the second order harmonics are suppressed by the controller.

609

The difference between positive and negative voltages of capacitors in a bridge arm v_{ac} under controlled and un-controlled condition are shown in Figure 5(g) and (h). In Figure 5(g), third order harmonics are removed by controller as they are increasing the amplitude in Figure 5(h).

6 CONCLUSIONS

The computer simulation in the MATLAB/SIMULINK environment and experimental set-up verified the viability and effectiveness of the three-phase M-STATCOM. The novel control strategy modified the traditional capacitor balance strategy. The system remained stable in whole process. It can be seen the output system has good dynamic characteristics. Especially the elimination of sixth harmonic components in the d- and q-axes maintained voltage drop and reactive power. MMC is showing considerable promise as a power converter for medium-voltage motor-drives, High-Voltage Direct-Current (HVDC) systems, self-commutated STATCOMs, and Back-To-Back (BTB) systems.

ACKNOWLEDGMENT

The authors wish to thank the financial support provided by the Chilean Government through Project FONDECYT-CONICYT 3150034.

REFERENCES

Acha E. A New STATCOM Model for Power Flows Using the Newton–Raphson Method Power Systems. IEEE Transactions on Power Electronics, 2013, 28(3):2455–2465.

Adam, G.P., Giddani O. Anaya-Lara et al. STATCOM Based On Modular Multilevel Converter: Dynamic Performance And Transient Response During Ac Network Disturbance, 6th IET International Conference on Power Electronics, Machines and Drives, 2012:1–5.

Allebrod, S., R. Hamerski and R. Marquardt. New transformerless, scalable modular multilevel converters for HVDC-transmission. IEEE Power Electronics Specialists Conference (PESC), Rhodes, June 15–19, 2008:174–179.

Antonopoulos, A., L. Angquist, and H.P. Nee. On dynamics and voltage control of the modular multilevel converter, 13th European Conference on Power Electronics and Applications, 2009:1–10.

Baichao Chen, Yaojun Chen, etc. Analysis and Suppression of Circulating Harmonic Currents in a Modular Multilevel Converter Considering the Impact of Dead Time. IEEE Transactions on Power Electronics, 2015: 30(7):3542–3552.

Casadei G., Teodorescu R., Vlad C., et al. Analysis of dynamic behavior of Modular Multilevel Converters: Modeling and control. 16th International Conference on System Theory, Control and Computing (ICSTCC), Sinaia: 2012:1–6.

Chenyao Jun, Chen Baichao, Songzhong You. A voltage converter circuit and method for pulse width. Patent, 2012103419658.

Jafri, K., Chen Baichao, Chen Yaojun, et al. Direct current control and experiment of a novel 9-level modular multilevel converter, 2012 IET International Conference on Information Science and Control Engineering (ICISCE 2012), Shenzhen, China: 2012:1526–1530.

Jing Song Zhu, Lei Li, Min Pan. Study of a novel STATCOM based on modular multilevel inverter. IECON 2012–38th Annual Conference on IEEE Industrial Electronics Society, 2012:1428–1432.

Junhua Guo, Yaojun Chen, Baicao Chen et al. Decoupling control for the capacitor voltage and the circulating current harmonics of the modular multilevel inverter, 9th IEEE Conference on Industrial Electronics and Applications (ICIEA), 2014:173–178.

Martin Glinka, Rainer Marquardt. A New AC/AC Multilevel Converter Family. IEEE Transactions on Industrial Electronics, 2005, 52(3):662–669.

Mohammadi, H.P., Bina, M.T.A. Transformerless Medium-Voltage STATCOM Topology Based on Extended Modular Multilevel Converters. IEEE Transactions on Power Electronics, 2011, 26(5):534.

Ph.D thesis of Chen Yaojun. Research on Operation Characteristics and Control of the Modular Multilevel Converter, submitted to Department of Electric Power Engineering Wuhan University, China, October 2014.

Rohner, S., S. Bernet, Modulation, Losses, and Semiconductor Requirements of Modular Multilevel Converters. IEEE Transactions on Industrial Electronics, 2010, 57(8):2633–2642.

Wei Zhang, Qiang Gao, Bonan Su et al. Research on the Control Strategy of STATCOM Based on Modular Multilevel Converter. IEEE Power Electronics Conference (IPEC), Japan, 2014:614–618.

Advances in Power and Energy Engineering – Sun (Ed.)
© *2016 Taylor & Francis Group, London, ISBN 978-1-138-02846-3*

DC voltage characteristic analysis of MTDC distribution network

Z.Y. Zhao, B.S. Su & D. Xie
Department of Electrical Engineering, Shanghai Jiaotong University, Shanghai, China

ABSTRACT: A novel DC distribution network with advantages over the traditional AC network is of increasing concern. This paper tends to find DC voltage characters based on a Multi-Terminal DC (MTDC) distribution network, operating with various sources and loads. Considering changes of loads and probabilistic fluctuation of renewable energy, the voltage of a DC bus has to be robust enough to stay at an acceptable range. The conditions of distributed generation changes and load changes are analyzed and simulated in this paper. To maintain the voltage of a bus, the power flow in the grid must be balanced. A seven-terminal prototype grid based on Voltage Source Converter (VSC), comprising loads, renewable energy, and Energy Storage System (ESS) including batteries and super-capacitors, is established to illustrate the DC voltage characters of the system during load changing conditions. Simulation results are obtained and presented to illustrate the stability of voltage under acceptable load changes and potential problems existing in the system.

1 INTRODUCTION

In the late 19th century, Nikola Tesla and Thomas Edison came head-to-head in *"The War of the Currents."* Finally, Tesla's "Alternating Current" is generally considered the winner over Edison's "Direct Current" (Williamson et al. 2011). Today, with the development of renewable energy technology and power semiconductor technology, there is an increasing interest in DC distribution power network. Compared with AC networks, DC networks have the following advantages: 1) The cost of transmission line and bus of DC network are lower than that of AC network, 2) Decreased loss of transmission, 3) Increasing efficiency, reliability, and stability of the system, and 4) DC network benefits the environment. This is why more and more people study DC distribution power networks (Lee et al. 2010, Chen & Xu, 2011, Gan & Low, 2013, Cole et al. 2010).

The rapid development of power electronics manages power of a distribution power network in a more intelligent and flexible way.

Stability assessment of DC network is simpler than that of AC network because we can ignore the reactive power and frequency synchronization of the system (Eghtedarpour & Farjah, 2012). To ensure system stability, the DC voltage on the bus must be limited within an acceptable range at all time. Such is a DC system's nature that a constant DC voltage indicates the power balance within a DC grid. In other words, power flow in the network affects the DC voltage of the bus (Mohamed & Mohammed, 2010). Thus, generally we use control strategies to maintain the power balance to keep the DC voltage within an acceptable range. The control strategy of DC bus voltage can be categorized into two main types: centrally controlled and droop controlled (Jian et al. 2013). For a centrally controlled grid, there is a central controller in the system which receives information from all convertors connected with the DC network via communication paths; then it sends indications to every existing terminal to control its power in order to maintain the power balance within the grid. However, the reliability of the communication path cannot be guaranteed all the time. Serious and undesirable consequences will happen when unexpected error or time delay occurs (Xu et al. 2012). On the other hand, droop controlled grid is based on the droop characteristic of the converter which every terminal operates with all feedbacks detected locally (Beerten et al. 2011, Haileselassie & Uhlen, 2012).

According to characters of loads, the load change may be step up or step down type leading to rapid power changing being needed. Renewable energy, such as solar power system and wind power system, produces probabilistic power depending on the weather.

In this paper, a proposed seven-terminal DC distribution network model is studied to identify characters of DC voltage of the network and to find out potential problems that might exist in operation of the grid. We use detailed models of renewable energy generation and loads based on MATLAB/SIMULINK to analyze the operational characteristic of MTDC distribution network.

In order to make the simulation easier and more convenient, we simplify Wind Turbine (WT) into a controlled AC source and simplify Photovoltaic generator (PV) into a DC source with controllable power (Shen et al. 2012). The work in this paper is supported by the National High Technology Research and Development Program 863 of China under Grant 2015AA050103.

This paper is organized as follows. Section II defines DC network under investigation. Control strategies under different operations are introduced in section III. Simulation results are presented in section IV and section V draws the conclusion.

2 DC NETWORK CONFIGURATION

2.1 DC network configuration

In 2014, China Southern Power Grid and several universities cooperated to set up a seven-terminal real-time simulation lab for the purpose of further research on control and protection of MTDC distribution networks, as showed in Fig. 1. This seven-terminal grid is modeled as shown in Fig. 1, connected with seven kinds of terminals, including G-VSC, PV units, WT generation, super-capacitors, batteries, and AC and DC loads. The voltage level of this model is 500 V, which is not of the real distribution network. The PV is connected to the grid through a DC-DC boost converter whose duty ratio is adjusted by MPPT. Similarly, the WT is connected through an AC-DC converter with an algorithm for MPPT. The battery system is connected to the network through a bi-directional DC-DC converter for discharging and charging, respectively.

Generally, super-capacitors and batteries are considered as ESS. These seven elements, as shown

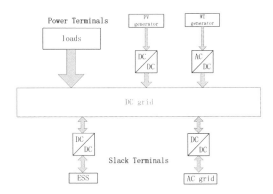

Figure 2. DC network configuration.

in Fig.2, are divided into two types: slack terminal and power terminal. Slack terminals, including AC grid, a battery, and a super capacitor, have flexible power to accommodate the power fluctuations caused by power terminals and to keep the system stable. The slack terminal ensures that total amount of power injecting into the grid equals the sum of the amount of power consumed by loads and the losses in the DC system. It accomplishes this task by maintaining DC voltage at its reference value. These generators and loads connected with the grid are considered as power terminals because they only produce or consume inconstant power to/from the DC grid but contribute nothing to the system stability. In addition, the change of power terminals usually causes instability of the system.

2.2 ESS

Although super-capacitors and battery system both accommodate DC voltage, there are some differences existing. Super-capacitor with smaller capacitance related to a battery has quick response to changes of DC voltage. Compared with super-capacitor, a battery system with large capacitance shows a relatively low response to supply or consume power for a long term (Gavriluta et al. 2013).

2.3 Operating modes

DC network should operate in both grid types, i.e., connected and islanded states. In each of these two states, there are different operating modes and the control strategy should regulate the DC bus voltage. Different operating modes are considered for the DC network, which are summarized in Table 1.

In Mode 1 and 2, the DC network is connected to an AC grid. When PV and WT generators produce enough power to satisfy the need of loads, the

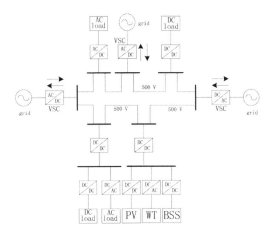

Figure 1. Seven-terminal DC distribution network.

Table 1. DC network operating modes.

Mode	Grid state	PV & WT state	BESS state	Loads
1	Grid connected	MPPT	Charging/off	Normal
2	Grid connected	MPPT	Charging/off	Normal
3	Islanded	MPPT	Discharging/off	Normal
4	Islanded	MPPT	Charging/off	Normal
5	Islanded	off- MPPT	off	Normal
6	Islanded	MPPT	Limited	Shedding

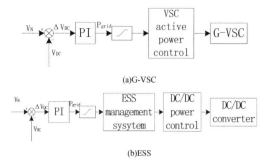

(a)G-VSC

(b)ESS

Figure 3. Control block diagram.

battery will charge to store the energy and maintain the bus voltage, and DC network will invert extra power into AC grid in Mode 1. In Mode 2, on the contrary, the battery will discharge and an AC grid will rectify the bus voltage. When the AC grid is in high peak or has some faults, the DC network has to operate in islanded state in Mode 3 to 6. Mode 3 and 4 correspond to the islanded state of the network where the PV and WT are operating at MPPT and the insufficient/surplus power is balanced by discharging or charging of the battery. In these two modes, the ESS is responsible to regulate the DC bus voltage. Mode 5 and 6 refer to the battery is fully charged or empty, using operating PV and WT at off-MPPT or load shedding to balance the power.

3 CONTROL STRATEGIES

3.1 G-VSC control

As long as the power needed to keep the DC network in a stable operating point does not exceed the power rating of the slack terminal and the DC voltage control is fast enough to handle severe transients, DC overvoltage can be avoided. Thus, it is assumed in this example that the AC grid is big enough to accommodate the power fluctuation of DC network caused by power generation fluctuation or load variation (Carlos et al. 2013, Pinto et al. 2012, Wilson et al. 2014).

The G-VSC accommodates such instant power surplus or deficit by regulating external exchanging power. There is:

$$P_{grid} = P_E + P_L - P_G \tag{1}$$

P_{grid} is the power flowing into a DC network from an AC grid, P_E is power of battery charging, P_L is power of loads, and P_G is power of power generators. As is shown in Fig.3 (a), a close loop

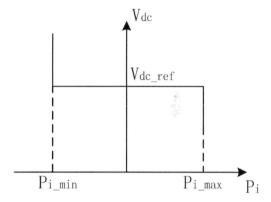

Figure 4. P_i-V_{dc} characteristic for DC bus.

DC regulator is designed to control the DC bus voltage.

Active power injection P_i, which indicates P_{grid} in this case, affects voltage value of DC bus. The P_i-V_{dc} characteristic of DC bus under voltage control is presented in Fig. 4. As shown in this figure, the bus will act as a voltage regulator as long as P_i is not reaching the power limits.

A PI controller is a straightforward choice to accommodate V_{dc} due to its ability to control DC voltage to its reference value.

3.2 ESS control

The energy storage system, including a battery and a super capacitor in this example, becomes the main regulator of DC voltage when the AC grid is disconnected. In the ESS, the battery is considered to accommodate power for a long time, and the super capacitor is designed to regulate transient fluctuation of power. Neglecting power of super capacitor, there is:

$$P_E = P_L - P_G \tag{2}$$

When P_E is negative, the battery is discharging. As is shown in Fig.3 (b), a closed loop control system for ESS is designed to maintain system stability.

3.3 PV and WT generator control and load control

In the case of a fully charged or an empty battery in islanded operations, control of PV and WT generators or loads is used to regulate DC voltage. The power is produced from distributed generators and consumed by loads in the network. Thus, there is:

$$P_G = P_L \qquad (3)$$

When the battery system is fully charged, DC voltage rises above upper limit V_{max} due to renewable energy producing more power than loads needed. PV and WT must operate at off-MPPT operation to balance the power injection and need. Similarly, some unnecessary loads should be cut off to reduce the power needed in case that battery system is empty and distributed generators connected with the network operates at MPPT operation.

4 SIMULATION AND RESULTS

The performance of DC voltage within the distribution network is evaluated and analyzed through simulation studies performed on the DC network. An MATLAB/SIMULINK simulation platform has been used to build a seven-terminal DC distribution network simulation model. The grid simulation uses a 500-V DC bus.

4.1 Test models

The test model used is shown in Fig. 5. This model based on the topological structure shown in Fig. 1 is simplified for a faster and more convenient simulation.

The PV system is modeled as a controlled AC source with a boost circuit and the WT system is modeled as a controlled AC source with an AC/DC converter, for simplicity. ESS, including a super-capacitor and a battery, is connected to the grid as a slack terminal to stabilize the voltage (Musolino et al. 2010, Henry et al. 2010). The capacitance of super-capacitor is 1 F and the battery with a bi-directional DC/DC converter is a nickel-metal-hydride battery with 240 V nominal voltage and 3000 Ah capacity. Load units include DC loads with DC/DC converters and AC loads with DC/AC converters. Three G-VSCs are modeled as

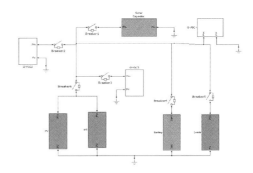

Figure 5. Simplified model of a DC network.

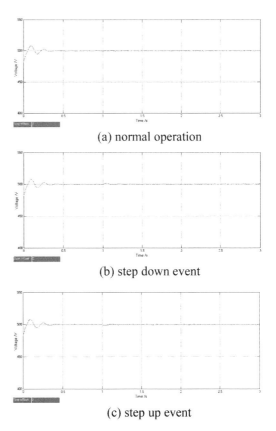

(a) normal operation

(b) step down event

(c) step up event

Figure 6. Simulation results of grid connected state.

infinite AC sources with VSC to supply sufficient power to the network.

4.2 Grid connected state

In the connected state, three G-VSCs can supply sufficient power to the grid to maintain DC voltage at 500 V. As is shown in Fig. 6 (a), when all generators and loads are operational at normal state the DC bus voltage is 500 V (The voltage oscillation at beginning is due to PI controller to control injection power).

Load units can control their power and is used for load demand variation study. The results of system operation with variations of load demand are plotted in Fig. 6(b) and Fig. 6(c), respectively. At the start of simulation, all elements operate at a normal state like Fig. 6(a). Simulation of a 100-KW step down event at 3 s was performed by load units by cutting off 100-KW loads. G-VSCs perform a good ability to maintain bus voltage at nominal level, as shown in Fig. 6(b). Similar simulation of a 100-KW step up event at 1s, in Fig. 6(c), was also performed by load units. G-VSCs also show good performance.

4.3 Islanded state

In an islanded state, G-VSC is cut off from the network. There are only distributed generations and the battery to supply the power needed by loads.

Compared with connected state, the power injected into the grid to accommodate the stability of the bus voltage is limited due to the battery's characters. Thus, overload events may happen, though it is an extreme condition in the real world.

With normal operation of the system, the voltage of the DC bus supplied by the battery is stable. Though, the voltage is not at 500 V accurately, the error is acceptable, as shown in Fig. 7(a). It is not hard to find that the response time of battery is obviously longer than that of G-VSC. Fig. 7(b–c) shows the results of the simulation with acceptable variation in load demand, and Fig. 7(d) shows the simulation of the overload event. According to the longer response time of battery, simulation time is extended to 20 s of these simulations. As shown in Fig. 7(b) and Fig. 7(c), the battery can keep the voltage stable with a 100-KW step down and up at 3 s, respectively. Comparing these two figures, we can find that the step up events have larger deviations than that of step down events.

The simulation of a 10-MW step up is considered as an overload event. As shown in Fig. 7(d), the voltage control collapsed after overload connecting to the grid at 3 s due to the absence of sufficient power injected into the network. Generally, load shedding is used to solve this problem, and the network operates at mode 6, according to section II in this paper. Although, mode 6 can prevent collapse of voltage at an overload event, a high quality battery system is expected to stabilize the voltage in a larger range.

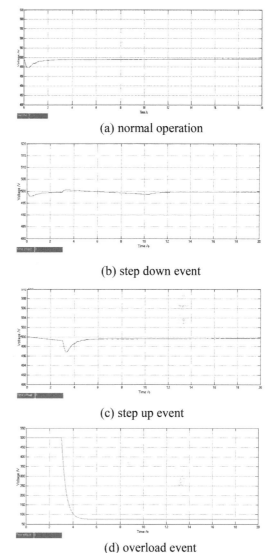

(a) normal operation

(b) step down event

(c) step up event

(d) overload event

Figure 7.　Simulation results of islanded state.

5 CONCLUSION

This paper proposes a seven-terminal DC distribution network model to study and identify characters of DC voltage within the network and potential problems that might exist in operation of the grid. Simulations corresponding to loads and renewable energy variations are performed. The results show that although the G-VSC and ESS can maintain the voltage stability during those events, voltage oscillations can occur on the bus. In extra case, voltage may go out of the range acceptable.

Therefore, it is suggested to consider enhancing the co-control of converters to balance power flow. Also, a high-quality ESS is required in this study.

REFERENCES

Beerten et al. 2011, VSC MTDC systems with a distributed DC voltage control-A power flow approach.

Carlos et al. 2013, A distributed DC voltage control method for VSC MTDC systems.

Chen & Xu, 2011, DC microgrid with variable generations and energy storage.

Cole et al. 2010, Generalized dynamic VSC MTDC model for power system stability studies.

Eghtedarpour & Farjah, 2012, Control strategy for distributed integration of photovoltaic and energy storage systems in DC micro-grids.

Gan & Low, 2013, Optimal power flow in direct current networks.

Gavriluta et al. 2013, Decentralized control of MTDC networks with energy storage and distributed generation.

Haileselassie & Uhlen, 2012, Impact of DC line voltage drops on power flow of MTDC using droop control.

Henry et al. 2010, Feasibility study of off-shore HVDC grids.

Jian et al. 2013, A review of control strategies for DC micro-grid.

Lee et al. 2010, Operational characteristic analysis of DC micro-grid using detailed model of distributed generation.

Mohamed & Mohammed, 2010, Smart power flow control in DC distribution systems involving sustainable energy sources.

Musolino et al. 2010, Simulations and field test results for potential applications of LV DC distribution network to reduce flicker effect.

Pinto et al. 2012, A novel distributed direct-voltage control strategy for grid integration of offshore wind energy systems through MTDC network.

Shen et al. 2012, Control of VSC HVDC system integrated with AC network.

Williamson et al. 2011, Project Edison: SMART-DC. ISGT Europe.

Wilson et al. 2014, Hamiltonian control design for DC microgrids with stochastic sources and loads with applications.

Xu et al. 2012, Energy management and control strategy for DC micro-grid in data center.

Advances in Power and Energy Engineering – Sun (Ed.)
© *2016 Taylor & Francis Group, London, ISBN 978-1-138-02846-3*

Synchronverter to damp multiple electromechanical oscillations

E.L. van Emmerik, B.W. França, A.R. Castro, G.F. Gontijo, D.S. Oliveira & M. Aredes
Federal University of Rio de Janeiro, Rio de Janeiro, Brazil

ABSTRACT: This paper presents a deep investigation on the synchronverter performance as a Power Oscillation Damping (POD) device. It is applied to damp multiple electromechanical oscillations in the Brazilian Transmission Grid. The electromechanical oscillations are identified through a filter that receives the measured active power and provides the reference to the synchronverter controller. In this way, the proper designing of the filter is substantial to ensure the effectiveness of the damping performance. This work is an extended investigation of the synchronverter operation in substitution of the SVC connected at the Bom Jesus da Lapa substation. Here the filter is designed to keep the former performance on damping the 0.31 Hz mode, plus to attenuate the 0.62 Hz mode. The filter designs are compared theoretically and a simulation of the investigated scenario shows the better performance of the proposed filter to damp both 0.31 and 0.62 Hz modes.

1 INTRODUCTION

The robustness and resilience of an Electric Power System (EPS) are important to ensure suitable power delivery to the load centers. In this manner, several researches have been conducted in order to increase the margin of stability of EPSs (Olulope et al. 2013), (Amarasekara et al. 2013). As the EPS's dimensions increase, its operation complexity rises as well.

Centralized Power Systems (CPSs) with continental dimensions, like the Brazilian electric system (also named as National Interconnected System—SIN), usually have long transmission systems to interconnect the power plants to the load centers. These systems may have Flexible AC Transmission Systems (FACTS) to ensure and optimize the power delivery such as Thyristor Controlled Series Capacitor (TCSC) and Static VAR Compensator (SVC) (Sadikovic 2006).

Recently, the synchronverter control method is being studied as a mean to add the benefits of the synchronous operation in power-electronic based devices (Zhong & Weiss 2009, Beck & Hesse 2007, Aouini et al. 2015). Its main applications are found in controlling renewable generation (wind and photovoltaic generation) like synchronous machines, through the front-end power electronic converters, and in storage systems with virtual inertia for frequency regulation (Zhong & Weiss 2011).

In (van Emmerik et al. 2015) the synchronverter operation as a FACTS with Power Oscillation Damper (POD) (Assis & Taranto 2008) functionality was evaluated in substitution of the SVC

connected at the Bom Jesus da Lapa substation. PODs are FACTS devices that perform the same function as a PSS. The advantages of using the synchronverter, instead of the SVC, rely on the possibility of future operational enhancements, such as integration with storage or generated energy for frequency regulation, active power control, providing virtual inertia etc.

This paper proposes an extended investigation of the scenario presented in (van Emmerik et al. 2015). Principally, the objectives here are:

- Provide a brief description of the scenario and the main conclusions obtained from the previous researches;
- Remove the current PSS that is being used in the system and investigate the system's robustness with only the synchronverter operation;
- Attenuate the 0.62 Hz oscillation observed in the last analysis.

The description of the simulated scenario is presented in Section 2. Section 3 presents the filter analysis with a comparison between the proposed filter and the filter presented in (van Emmerik et al. 2015). Section 4 presents the simulation results and section 5 the conclusions.

2 DESCRIPTION OF THE STUDIED SCENARIO

The main objectives of (van Emmerik et al. 2015) are to show the technical possibility to substitute

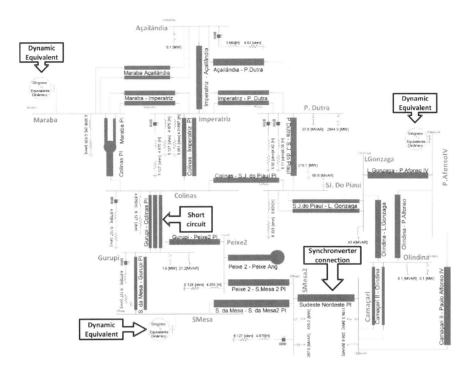

Figure 1. Modeled diagram of the simulated network; adapted from (van Emmerik et al. 2015).

the SVC by a synchronverter at the Bom Jesus da Lapa substation (250 MVAr), and to show that this synchronverter is capable to mitigate the North/South electromechanical oscillation mode in the range of 0.2 until 0.4 Hz. Therefore, dynamic equivalents of the SIN were developed to represent the electromechanical oscillations inherent at the SVC's area of operation (Fig. 1). There are four dynamic equivalents in this model: one at the Marabá substation to represent the North region, one at Paulo Afonso IV substation to represent the Northeast region, one at Serra da Mesa substation to represent Southeast/Mid-west regions, and one to represent a local generation at Lajeado region. It is important to highlight that in (van Emmerik et al. 2015) a Power System Stabilizer (PSS) was provided in the dynamic equivalent at Paulo Afonso IV substation to support the system's stability.

The SVC is located approximately at the middle of the Southeast/Northeast interconnection, which is far from power generators that contribute to the North/South mode. Consequently, the damping through reactive power control is more suitable than through active power control (Beza et al. 2011). Hence, the measured active powers from the transmission lines S. Mesa—Gurupi, at the Serra da Mesa substation, were employed to feed the synchronverter controller with information.

3 FILTER ANALYSIS

Filter parameters adjustments lead to the development of a filter with superior frequency response and performance. The filter design was based on the filter used on (van Emmerik et al. 2015), which showed good performance at damping the 0.31 Hz mode.

The filter used on a POD can be regarded as a three-stage filter:

- a Low-Pass stage (LP);
- a Washout Stage (WO);
- a delay stage.

The role of each one of these stages in composing the final filter's frequency response is depicted in the Bode diagram in Figure 2.

This filter has a gain difference ΔG between the modes of 0.62 Hz and 0.31 Hz of −8.9 dB, with a −66.1 dB gain at 0.31 Hz. The filter's gain peaks at ~0.09 Hz, over 6 dB higher than at 0.31 Hz. It introduces phase delay through the series association of three lag compensators to give a resulting phase of −90° at 0.3 Hz.

The objective of the POD is to damp low frequency, electromechanical oscillations from the power system. In this paper, it's achieved through the injection of reactive power and subsequent grid voltage modulation by the synchronverter. Hence, a filter is designed to receive measured active power

618

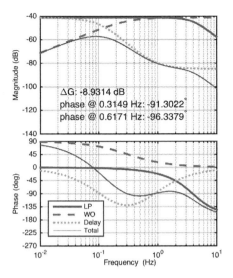

Figure 2. Bode diagram of the filter used in (van Emmerik et al. 2015) and its different stages.

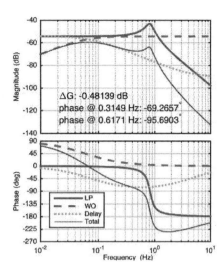

Figure 3. Bode diagram of the new filter and its different stages.

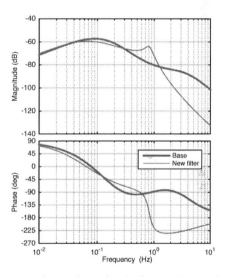

Figure 4. Comparison of Bode diagrams between base filter and new filter.

input and its output modulates the grid voltage (van Emmerik et al. 2015, Assis & Taranto 2008).

The LP reduces the influence of higher frequency components on the filter's output, as the objective of the POD is to act on lower frequencies. The WO eliminates de DC level present in the measured power, as the POD must act only on the oscillations of the modes being considered, ignoring the steady-state active power. The delay stage serves to introduce the phase difference between measured active power and grid voltage which best acts on damping the electromechanical oscillations through a negative feedback effect. The optimum delay, considering R<<X, is −90°. In this case, the modulated voltage is in phase with the angular speed of the dynamic equivalent, since the active power is measured from Serra da Mesa to Gurupi, which is the opposite direction of the power flow from all the dynamic equivalents with the exception of Serra da Mesa.

The new filter design aimed at improving the POD's performance at damping the 0.62 Hz mode, while keeping its good results on the 0.31 Hz mode. Hence, the gain at 0.31 Hz was kept constant and the gain difference ΔG from 0.62 Hz was decreased. This could be achieved by shifting the LP's cutoff frequency from 3.86 Hz to 0.82 Hz and decreasing its damping ratio from 0.74 to 0.14, creating a peak at this frequency, as seen in Figure 3.

In this design, the WO works together with the LP to form a band-pass filter. Its cutoff frequency was changed from 0.32 Hz to 0.06 Hz, which means a time constant of 2.65 s.

The delay stage's central frequency was shifted from 0.3 Hz to 0.57 Hz and only one delay transfer function was used in this stage. This makes for

a wider, better distributed among the frequency band of interest, phase delay response.

The new design improved the response at the 0.62 Hz mode and increased the system's gain margin. The frequency responses from both new and original filters are directly compared in Figure 4. This comparison shows how the magnitude response decreased at lower frequencies, which increased the gain margin as observed in simulations, and increased at the higher side of the frequency band of interest. The new filter shows better performance at higher frequencies as well.

4 SIMULATION RESULTS

All the shown figures in this section are based on the same scenario presented in (van Emmerik et al. 2015). Briefly recapping, in the first seconds the grid and equivalents are initialized, the synchronverter is connected and immediately starts controlling the voltage at the 500 kV transformer side. With a reference voltage of 1.04 p.u. and a droop D_q (van Emmerik et al. 2015) of 33 in the reactive loop of the synchronverter, the voltage at the 500 kV side is maintained in 1.05 p.u., which is the nominal voltage nowadays, to reduce transmission losses. At t = 5 s, a 20 ms short circuit is introduced at the middle point of the North-South connection to provoke electromechanical oscillations. In all figures, the POD signal is calculated since the moment in which the synchronverter is connected. However, it's only enabled a few seconds after short-circuit is applied, in order to better show the capacity of the POD signal. All the figures start 5 s after short-circuit.

In the first subsection, results are shown for the system without any PSS. This makes the system unstable because of the lowest inter-area oscillation mode. The reason for showing this situation is to illustrate the power of the POD on the synchronverter. In the last subsection, the results are shown with a PSS on the dynamic equivalent in Paulo Afonso to illustrate better the effect of the new filter in comparison with the former published one (van Emmerik et al. 2015), since the lowest mode is far more dominant than the 0.62 Hz mode.

4.1 *System without PSS*

In Figure 5, the angular speeds of the three biggest dynamic equivalents are shown. The POD signal is enabled at 25 s. The speed of the Lajeado equivalent is omitted in this figure since its mode of around 2 Hz naturally damps quickly. This equivalent is utilized to maintain the normal load flow on the North-South connection.

The equivalent of Paulo Afonso, ω_{pa}, oscillates in opposite phase with Serra de Mesa, ω_{sm}, with a frequency of 0.39 Hz. In the next-section and as

mentioned in (van Emmerik et al. 2015), with the PSS in Paulo Afonso, this inter-area oscillation mode shifts to 0.31 Hz. The angular speed of Marabá, ω_{mb}, is shown to be a result of its own mode of 0.6 Hz and the mode of Paulo Afonso, which Marabá is following. There may be no doubt about the effectiveness of the enabled POD. Right after t = 25 s de 0.39 Hz mode is damped instead of unstable. Since the utilized filter is the base one, the same as in (van Emmerik et al. 2015), the 0.6 Hz mode of Marabá is still poorly damped, though, not unstable.

4.2 *System with PSS*

One PSS has been connected again at the Paulo Afonso equivalent for all figures in this section. The motives are that the Brazilian electric system operates normally with PSS's in order to achieve stability and to better observe the effect of the new filter in comparison with the base one. The vertical scale in Figure 5 (without PSS) is around 10 times greater than the ones in Figures 8–10 (with PSS). With exception of the last figure in this section, all figures compare the effect of the base and new filter for the main quantities. For all figures the POD signal, Pref, has been enabled at t = 30 s.

The effect of the old and new filter on the measured active power, Pmesa, which is the input signal for the POD, is shown in Figure 6. Before enabling the POD signal, the active peak to peak power oscillation is nearly 150 MW and hardly damped. Right after 30 s the signal is significantly damped as was shown in (van Emmerik et al. 2015). As mentioned there, after 10 s, the 0.31 Hz mode is completely damped, leaving a poorly damped 0.62 Hz mode. The new filter shows a significant superior damping also for the 0.62 Hz mode.

The POD signal, Pref, derived from the power measurement Pmesa, as described in the former section on Filter Analysis, is shown in Figure 7. Initially, the new filter provokes a larger oscillation. However, after a few seconds, the signal's amplitude reduces below the level of that using the base filter, showing a better oscillation damping.

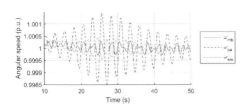

Figure 5. Angular speeds ω_{mb}, ω_{pa} and ω_{sm} (p.u.) of the equivalents in Marabá, Paulo Afonso and S. Mesa, respectively, for the system without PSS.

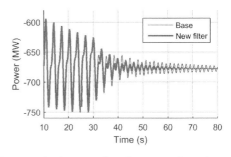

Figure 6. Active power from S. Mesa to Gurupi, Pmesa (MW), for the original and the new filter.

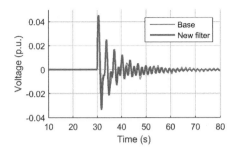

Figure 7. Additional voltage signal, Pref (p.u.), for the original and the new filter.

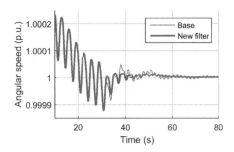

Figure 8. Angular speed of the equivalent in S. Mesa, ω_{sm} (p.u.), for the original and the new filter.

Further on, the base filter causes a visible oscillation in ~0.09 Hz, while it's unnoticeable for the new filter, as mentioned in Section 3. Besides the mentioned advantages of the new filter, it operates far within the operation limits, not even reaching 5% of the nominal operation limit in the initial moment.

The ~0.09 Hz mode can be seen as well with the base filter for the angular speed of the equivalent in Serra de Mesa, ω_{sm}, as shown in Figure 8. Again, the new filter has damped this mode. The damping on the 0.62 Hz mode by the new filter is evident.

The opposition of the angular speed for the 0.31 Hz mode in Paulo Afonso, ω_{pa}, with Serra de Mesa can be seen in Figure 9. The ~0.09 Hz mode with the old filter seems in phase with the one found in Serra de Mesa indicating that this mode is from the total system.

The 0.31 Hz mode is hardly found in the equivalent of Marabá as shown in Figure 10. Such result contrasts with the one in Figure 5, in which the angular speed is dominated by the unstable 0.31 Hz mode. As seen in the other two equivalents, the base filter introduces the ~0.09 Hz, while the new filter eliminates the mode. The 0.62 Hz mode, related to Marabá, is significantly better damped by the new filter.

The equivalent of Lajeado, the smallest one, has a relating mode of around 2 Hz. Its own mode is no longer observed within 5 s after the short-circuit.

In Figure 11, the reactive power Q of the synchronverter is shown in p.u. Like the POD signal, shown in Figure 7, the new filter causes slightly more work initially, though, rapidly after it shows less oscillation then the base one. The operation of the synchronverter is anyway far between the operation limits.

A similar curve can be found in Figure 12 for the controlled voltage at the 500 kV side of the transformer after the synchronverter, V_col. With initial peaks of around 3% of the steady state value, no stress on the grid is expected from the POD signal.

While the base filter provokes instability for gains above 150% of the original one presented in (van Emmerik et al. 2015), the new filter can

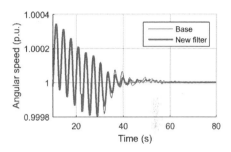

Figure 9. Angular speed of the equivalent in Paulo Afonso, ω_{pa} (p.u.), for the original and the new filter.

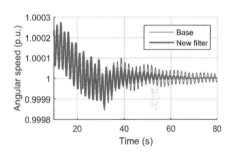

Figure 10. Angular speed of the equivalent in Marabá, ω_{mb} (p.u.), for the original and the new filter.

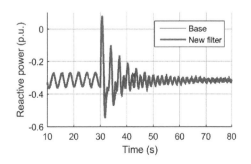

Figure 11. Reactive power of the synchronverter, Q (p.u.), for the original and the new filter.

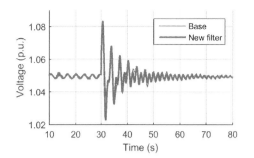

Figure 12. Voltage at the 500 kV side of the transformer in B.J. de Lapa, V_col (p.u.), for the original and the new filter.

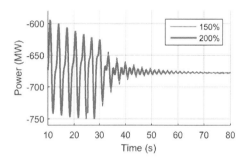

Figure 13. Active power from S. Mesa to Gurupi, Pmesa (MW), for the new filter with higher gain.

have gains even above 200% of the original one. Figure 13 shows the measured active power, Pmesa, for the new filter with gains of 150%, like all the figures before in this sub-section, and 200% for which the base filter is unstable. The higher gain shows even better behavior for the new filter.

5 CONCLUSIONS

This paper has firstly given a brief description of the former related research on the oscillation damping with the synchronverter (van Emmerik et al. 2015). While the utilized filter to generate the POD signal was effective for the 0.3 Hz inter-area oscillation mode, it hardly had any effect on the mode with double frequency.

A new simulation was produced, utilizing the filter from the previous paper, without any PSS in the system. Without PSS, the system is unstable. Nevertheless, the base filter has shown its capability to stabilize the electric system, even with the synchronverter far away from the active power measurement and from the equivalent of Paulo Afonso, which is related to the unstable mode.

The new filter design has been described in all three stages and it has been shown by an analysis in the frequency domain that the new filter is superior for lower frequencies and for inter-area modes until 0.8 Hz. The reference gain for 0.31 Hz has been kept constant while designing, to avoid an inferior behavior for this unstable mode.

The behavior of all important quantities has been shown in the time domain for the base and new filter and confirmed the superiority of the new filter. The system mode of ~0.09 Hz has been eliminated and the 0.62 Hz of Marabá has been damped significantly. With the new filter, the synchronverter with POD and measurement in Serra de Mesa at a distance of about 600 km has shown that it is capable to damp multiple inter-area modes, between 0.2 Hz and 0.8 Hz, coming from generator units at distances of over 1000 km.

REFERENCES

Amarasekara, H.W.K.M. et al. 2013. Impact of renewable power integration on VQ stability margin. *Power Engineering Conference (AUPEC), 2013 Australasian Universities*: 1–6.

Aouini, R. et al. 2015. Synchronverter-Based Emulation and Control of HVDC Transmission. *Power Systems, IEEE Transactions on* PP (99): 1–9.

Assis, T.M.L. & Taranto G.N. 2008. Increase of Transfer Limit in the Large Interconnected Brazilian System Constrained by Small-Signal Stability.

Beck, H.P. & Hesse, R. 2007. Virtual Synchronous Machine. *Electrical Power Quality and Utilisation, 2007. EPQU 2007. 9th International Conference on*: 1–6.

Beza, M. & Bongiorno, M. 2011. Power Oscillation Damping Controller by Static Synchronous Compensator with Energy Storage. *Energy Conversion Congress and Exposition (ECCE) 2011 IEEE*: 2977–2984.

Olulope, P.K. et al. 2013. Modeling and simulation of hybrid distributed generation and its impact on transient stability of power system. *Industrial Technology (ICIT), 2013 IEEE International Conference on*: 1757–1762.

Sadikovic, R.U.S.E.J.L.A. 2006. *Use of FACTS devices for power flow control and damping of oscillations in power systems*. (Doctoral dissertation, Faculty of Electrical Engineering, University of Tuzla).

van Emmerik, E.L. et al. 2015. A synchronverter to damp electromechanical oscillations in the Brazilian transmission grid. *Industrial Electronics (ISIE), 2015 IEEE 24th International Symposium on*: 221–226.

Zhong, Q.C. & Weiss, G. 2009. Static Synchronous Generators for Distributed Generation and Renewable Energy. *Power Systems Conference and Exposition, 2009. PSCE'09. IEEE/PES*: 1–6.

Zhong, Q.C. & Weiss, G. 2011. Synchronverters: Inverters that Mimic Synchronous Generators. *Industrial Electronics, IEEE Transactions on* 58(4): 1259–1267.

Advances in Power and Energy Engineering – Sun (Ed.)
© 2016 Taylor & Francis Group, London, ISBN 978-1-138-02846-3

Research on power stability of hybrid system with UHVDC Hierarchical Connection to AC grid

Y. Tang, B. Chen & L.L. Zhu
School of Electrical Engineering, Southeast University, Nanjing, China

J.C. Pi
State Grid Electric Power Research Institute, Nanjing, China

C.G. Wang
Jiangsu Electric Power Company Research Institute, Nanjing, China

ABSTRACT: In order to absorb the huge capacity of power transmitted by Ultra High Voltage Direct Current (UHVDC) from long distance renewable generation bases, a new transmission pattern called Hierarchical Connection (HC) is proposed to connect UHVDC with UHVAC (Ultra High Voltage Alternate Current) grid of two different voltage classes. In order to analyze the stability of hybrid system, the complex interaction mechanism should be taken into consideration. Based on a practical HC project in which the receiving end of UHVDC transmission line is connected with both 500 kV and 1000 kV UHVAC systems, the stability evaluation index and its affecting factors are studied to assess the power stability of hybrid system with UHVDC hierarchical connection to AC Grid. Furthermore, four important factors are utilized to analyze the power stability, including the operation mode of DC system, converter transformer tap position, the coupling degree between hierarchical system and the values of Short Circuit Ratio (SCR) of different layers. Analysis results show that the power transmission capability, power stability margin can be effectively improved by reducing coupling impedance and increasing the short circuit ratio of receiving end system under HC mode.

1 INTRODUCTION

With the characteristics of long transmission distance and large transmission capacity, more and more UHVDC projects are applied to transmit power energy from remote renewable bases to load centers in China. The existing UHVDC transmission projects (Zhao, W.J. 2011; Xu, Z. 2004; Liu, Z.Y. 2013; Chen, X.L. et al. 2011) in East and South China mainly utilize multi-infeed Monolayer Connection (MC) mode. Security and stability characteristics, reactive power control, DC terminal location selection and AC-DC interactive features of DC power transmission technology have been studied widely to analyze the multi-infeed mode (Gao, Y. & Han, M.X. 2014; Guo, X.J. et al. 2013; Zhou, Q.Y. et al. 2013; Xiao, J. & Li, X.Y. 2014; Li, C.S. et al. 2014; Liu, Z.Y. et al. 2013; Li, S.H. et al. 2015). In order to absorb the power energy transmitted by DC system, where the DC terminal's

location is and what the DC connection mode is become more and more intensive. Due to the limitation of existing MC mode, a new transmission pattern called Hierarchical Connection (HC) is proposed to connect UHVDC with UHVAC grid of two different voltage classes. State Grid Corporation of China (SGCC) is building two UHVDC transmission projects which will utilize DC hierarchical connection mode, one is a ±800 kV project from Ximeng of Inter-Mongolia to Taizhou of Jiangsu province, while another one is a ±1100 kV UHVDC transmission project from Hulunbuir of Inter-Mongolia to Wannan of Anhui province. The research on UHVDC hierarchical connection mode is still in the preliminary stage by now, mainly focus on the applicability of the Multi-Infeed Short Circuit Ratio (MISCR) under the HC mode (Guo, L. et al. 2015), the overall design of DC control system (Nayak, O.B. et al. 1995), and the coordinated control of power (Thallam, R.S. et al. 1992).

Under UHVDC hierarchical connection mode, topology of DC system becomes more complex, and leads to different interaction mechanism between AC and DC systems. Therefore, power stability of hybrid system under UHVDC hierarchical

Project supported by State Gird Corporation of China (XT71-15-045); Project supported by JiangSu Electric Power Company (5210EF14001P).

connection mode should be studied far more. Based on the practical HC project, HC model and described mathematic equations of UHVDC system were put forward, then this paper improved the definition and application of Short Circuit Ratio (SCR) and Voltage Stability Factor (VSF) of hybrid system more perfect while taking reactive power compensation devices into consideration, and the maximum power curve method was generalized and improved to analyze the power stability characteristics of hybrid system while considering different converter transformer taps, coupled impedance and condition of AC systems.

2 MODEL OF UHVDC HIERARCHICAL CONNECTION

Taken Ximeng-Taizhou UHVDC project as an example, the features of AC-DC system under HC mode will be introduced in detail. Ximeng-Taizhou UHVDC transmission project was planned to operate in 2017, the voltage class is ±800 kV, the transmission capacity is 10000 MW. The transmission lines of the UHVDC project start at Ximeng converter station, end at Taizhou converter station, the total transmission length is about 1620 km. The development of Ximeng clean energy base will be promoted and more electricity can be sent to east of China after the project is put into operation. Since the transmission capacity is huge, a new transmission pattern called hierarchical connection mode is taken to discharge the power flow better. Schematic of the project with HC mode is shown in Figure 1.

The double low-end converters (TZ converter station I) are connected to the 1000 kV bus, while the double high-end converters (TZ converter station II) are connected to 500 kV bus, and there is no electrical connection between the two TZ converter stations. The 500 kV bus of TZ converter

transformers is connected to 500 kV grid of Jiangsu province by sextuple-circuit 500 kV lines. The 1000 kV bus of TZ converter transformers is connected to 1000 kV grid of East China directly due to the Taizhou converter station and UHVAC station will be built jointly.

Under HC mode, AC filters and reactive compensation devices are respectively configured in different layers, and converter buses of different layers are connected by transformers. Converter valves are connected in series in each pole. In order to achieve the decoupling control to the voltage of same pole valve, DC control systems include double-pole control, double-layer control, pole control and valve control under HC mode. A simplified model of the AC-DC system under UHVDC HC mode is shown in Figure 2.

Where U_d and I_d are DC voltage and current, $U_i \angle \delta_i$ is the voltage of AC bus of converter station, P_{dn} and Q_{dn} are active and reactive power of converter valve, P_{ij} and Q_{ij} are active and reactive power between two buses, P_{aci} and Q_{aci} are active and reactive power of AC system, Q_{Cn} is reactive compensation capacity, $Z_i \angle \theta_i$ is equivalent impedance of AC system, $Z_{ij} \angle \theta_{ij}$ is equivalent coupling impedance, $E_i \angle \Psi_i$ is equivalent emf.

Figure 3 is the equivalent simplified model network of the receiving end AC system. Where Z_T is equivalent impedance of transformer, k is ratio of transformation.

The admittance matrix of node i, j can be obtained by matrix transformation as follows if AC filters and reactive compensation devices are considered.

$$Y = \begin{bmatrix} \dfrac{1}{Z_i} + \dfrac{1}{k^2 Z_T + Z_{ij}} + jB_{Ci} & -\dfrac{k}{k^2 Z_T + Z_{ij}} \\ -\dfrac{k}{k^2 Z_T + Z_{ij}} & \dfrac{1}{Z_j} + \dfrac{k^2}{k^2 Z_T + Z_{ij}} + jB_{Cj} \end{bmatrix} \quad (1)$$

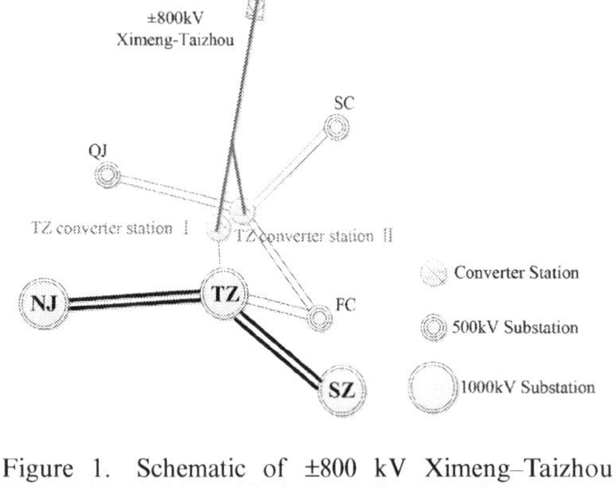

Figure 1. Schematic of ±800 kV Ximeng–Taizhou UHVDC with hierarchical connection to AC system.

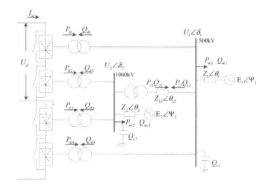

Figure 2. Simplified model of AC-DC system under UHVDC HC mode.

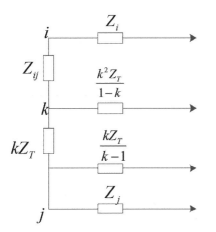

Figure 3. Simplified model network.

$$P_{aci} = [U_i^2 \cos\theta_i - E_iU_i\cos(\delta_i + \theta_i - \psi_i)]/|Z_i| \quad (10)$$

$$P_{ij} = [U_i^2\cos\theta_{ij} - U_iU_j\cos(\delta_i + \theta_{ij} - \delta_j)]/|Z_{ij}| \quad (11)$$

$$Q_{aci} = [U_i^2\sin\theta_i - E_iU_i\sin(\delta_i + \theta_i - \psi_i)]/|Z_i| \quad (12)$$

$$Q_{ij} = \left[U_i^2\sin\theta_{ij} - U_iU_j\sin(\delta_i + \theta_{ij} - \delta_j)\right]/|Z_{ij}| \quad (13)$$

$$Q_{Ci} = B_{Ci}U_i^2 \quad (14)$$

$$P_{d1} + P_{d4} = P_{ac1} + P_{12} \quad (15)$$

$$P_{d2} + P_{d3} = P_{ac2} + P_{21} \quad (16)$$

$$Q_{d1} + Q_{d4} + Q_{ac1} + Q_{12} = Q_{C1} \quad (17)$$

$$Q_{d2} + Q_{d3} + Q_{ac2} + Q_{21} = Q_{C2} \quad (18)$$

$$I_{d1} = I_{d2}, I_{d3} = I_{d4} \quad (19)$$

$$P_{12} + P_{21} = 0, Q_{12} + Q_{21} = 0 \quad (20)$$

The impedance matrix of node i, j can be obtained by inversing the admittance matrix.

$$Z = Y^{-1} = \begin{bmatrix} Z_{eqii} & Z_{eqij} \\ Z_{eqji} & Z_{eqjj} \end{bmatrix} \quad (2)$$

The equivalent impedance of the receiving end AC system i is:

$$|Z_{eqi}| = |Z_{eqii}| + |Z_{eqij}|\frac{P_{dNj}}{P_{dNi}} \quad (3)$$

The equivalent impedance of the receiving end AC system j is:

$$|Z_{eqj}| = |Z_{eqjj}| + |Z_{eqji}|\frac{P_{dNi}}{P_{dNj}} \quad (4)$$

The equivalent impedance between the receiving end AC systems i and j is:

$$Z_{eqij} = Z_{eqji} \quad (5)$$

The characteristics of the simplified model showed in Figure 2 and Figure 3 can be described by the following equations.

$$P_{dn} = C_nU_i^2[\cos 2\gamma_n - \cos(2\gamma_n + 2\mu_n)] \quad (6)$$

$$P_{dn} = C_nU_i^2[\cos 2\gamma_n - \cos(2\gamma_n + 2\mu_n)] \quad (7)$$

$$Q_{dn} = C_nU_i^2[2\mu_n + \sin 2\gamma_n - \sin(2\gamma_n + 2\mu_n)] \quad (8)$$

$$U_{dn} = P_{dn}/I_{dn} \quad (9)$$

Here, $i = 1$ while $n = 1, 4$ and $i = 2$ while $n = 2, 3$.

where n ($n = 1, 2, 3, 4$) is the number of the converter valves, i, j respectively represent the low-voltage layer and the high-voltage layer. γ_n is the extinction angle, μ_n is the commutation angle, C_n and K_n respectively represent two constants related to the parameters of inverter and the reference value of DC system. The expression of C_n is

$$C_n = \frac{3S_{Tn}}{4\pi P_{dNn}u_{kn}\%\tau_n^2} \quad (n = 1, 2, 3, 4) \quad (21)$$

where S_{Tn} = capacity of the converter transformers; and $u_{kn}\%$ ($n = 1\sim4$) = short-circuit ratio of the converter transformers.

It can be assumed that the parameters of converter station equipment are fixed without considering the changes of transformer taps and the switches of compensation capacitors while the system is running. The parameters of receiving end system are also fixed while the system operating mode is determined. The equivalent emf E_i is also assumed to be fixed. Then the remaining parameters of operational state in above equations are: γ_n, μ_n, U_{dn}, I_{dn}, P_{dn}, Q_{dn}, U_i, δ_i, P_{ij}, Q_{ij}, Q_{Ci} ($n = 1, 2, 3, 4, i, j = 1, 2, i \neq j$), the total number is 38. The equations from (6) to (20) contain 34 constraint equations while $i, j = 1, 2$ and $i \neq j$. Therefore, the remaining variables can be calculated once 4 of the 38 state variables are identified. The values of parameters of converter station are set as follows: $S_{Tn} = 1.15 P_{dNn}$, $u_{kn}\% = 0.18$, $\tau_n = 1$. The value of C_n can be obtained by formula (21): $C_n = 1.525$. Assumed that the system operated at rated conditions, $\gamma_n = \gamma_N = 18°$, $U_i = 1$, $P_{dn} = 1$, $I_{dn} = 1$.

3 STABILITY EVALUATION INDEX

3.1 *Short Circuit Ratio (SCR)*

The calculation method of *SCR* under HC mode has been put forward (Guo, L. et al. 2015). Each of the layers respectively transmits half of DC power as well as provides the power supportability for each other while the DC system operates at rated condition, which will affect the value of *SCR* greatly. Both of the factors are considered when analyzing the *SCR*.

a *Long electrical distance between layers*
When the electrical distance is farther between 1000 kV and 500 kV converter buses, electrical coupling degree between layers is weak and as well as the electrical contact. It can be thought that $Z_{12} = +\infty$. Then the *SCR* of receiving end system is:

$$SCR_i = \frac{S_{aci}}{P_{deqi}} = \frac{U_{Ni}}{P_{dNi}|Z_i|} = \frac{1}{|Z_i|}(i = 1, 2) \quad (22)$$

The DC transmission power will be halved in each layer when the transmission power in 1000 kV side and 500 kV side is equal, meanwhile, the *SCR* of receiving end AC system respectively doubles. It can greatly improve the intensity of hybrid system.

b *Short electrical distance between layers*
When the electrical distance is closer between 1000 kV and 500 kV converter buses, electrical coupling degree between layers is strong and as well as the electrical contact. Then the short circuit ratio of receiving-end system is:

$$SCR_i = \frac{S_{aci}}{P_{deqi}} = \frac{U_{Ni}^2/Z_i}{P_{dNi} + \left|\frac{Z_{ij}}{Z_j}\right| P_{dNj}}$$

$$= \frac{1}{|Z_i| P_{dNi} + \left|\frac{Z_{ij}}{Z_j}\right| P_{dNj}} \quad (23)$$

If $P_{dNi}/P_{dNj} = 1$ and $P_{dNi} + P_{dNj} = 1$, then the formula (23) can be transformed as:

$$SCR_i = \frac{1}{|Z_i| P_{dN}/2 + \left|\frac{Z_{ij}}{Z_j}\right| P_{dN}/2} \quad (24)$$

Table 1 presents the values of *SCR* under HC mode and MC mode while utilizing different system parameters. It can be seen that the value of *SCR* of the hybrid system can be efficiently improved

Table 1. The value of *SCR* under different connection mode.

Connection mode	Z_{12}	Z_i	U/kV	SCR
MC mode	–	1/3	500	3
HC mode	1	1/3	500	6.3333
		1/4	1000	7.6000
HC mode	1	4/5	500	3.4167
		1/4	1000	6.3077
HC mode	$+\infty$	1/3	500	6
		1/4	1000	8

while appropriately reducing the equivalent impedance of receiving end system under HC mode, and the value increases to nearly twice the original value compared to single-infeed MC mode.

3.2 *Voltage Stability Factor (VSF)*

The voltage of converter bus will change with the fluctuation of reactive power. The hybrid system operates at a stable voltage region while $VSF > 0$. The smaller the *VSF* is, the changes of AC converter bus voltage following the system disturbances are less, which represents that the voltage of converter station bus is stable. In contrast, the voltage of converter station bus is unstable while $VSF < 0$ (Xu, Z. 2004; Aik, D.L.H. & Andersson, G. 1997).

The total reactive power of the converter bus i is Q_i, thus the voltage stability factor index is defined as follows.

$$VSF = \left.\frac{dU/U_b}{dQ_i/Q_b}\right|_{P_{d0}} = \frac{Q_b/U_b}{dQ_i/dU}$$

$$= Q_b/U_b \times (dQ_{aci}/dU + dQ_{ij}/dU \quad (25)$$

$$+ dQ_{Li}/dU + dQ_{di}/dU - dQ_{Ci}/dU)^{-1}$$

where U_b and Q_b are respectively the reference voltage and the reference capacity.

The various derivative of *VSF* can be represented by difference quotient while conducting calculation:

$$VSF \approx \left(\frac{\Delta Q_{aci}}{\Delta U} + \frac{\Delta Q_{ij}}{\Delta U} + \frac{\Delta Q_{Li}}{\Delta U} + \frac{\Delta Q_{di}}{\Delta U} - \frac{\Delta Q_{Ci}}{\Delta U}\right)^{-1} \quad (26)$$

The mathematical model of HC mode is shown in Figure 2. The incremental equations of physical quantities can be obtained corresponding to the changing variable ΔU of converter buses voltage U. It is assumed that these three parameters of equivalent AC system, $|Z|(|Z_{eqi}|, |Z_{eqj}|, |Z_{eqij}|)$, θ (θ_i, θ_j, θ_{ij}) and E (E_i, E_j), are constant while analyzing the voltage static stability of AC bus, P_{di}, P_{aci} and P_{ij} keep balance regardless of changes in U,

$$\Delta P_{di} = \Delta P_{aci} + \Delta P_{ij} \qquad (27)$$

Thus there will be 10 incremental equations, and two more incremental equations if considering the DC control system. There are totally 12 incremental equations when the voltage U at the converter AC bus obtains increment ΔU. The matrix expression is shown in equation (28).

Where, J is a Jacobian matrix with the order of 12*10. The value of each element in J shall be calculated at the operating point ($P_d = P_{d0}$). The elements at 11th and 12th row in J can be determined by control mode of DC system. Matrix J will be converted into the following form through elimination calculation.

$$\begin{bmatrix} \Delta Q_{di} \\ \Delta Q_{aci} \\ \Delta Q_{Ci} \\ \Delta Q_{ij} \\ 0 \\ 0 \\ 0 \\ 0 \\ 0 \\ 0 \\ 0 \\ 0 \end{bmatrix} = [J] \begin{bmatrix} \Delta U_i \\ \Delta U_j \\ \Delta \delta_i \\ \Delta \mu_i \\ \Delta \gamma_i \\ \Delta I_{di} \\ \Delta U_{di} \\ \Delta P_{di} \\ \Delta P_{aci} \\ \Delta P_{ij} \end{bmatrix}$$

$$= \begin{bmatrix} a_{11} & \vdots & & & \\ a_{21} & \vdots & & 0 & \\ a_{31} & \vdots & & & \\ a_{41} & \vdots & & & \\ \dots & \dots & \dots & \dots & \dots \\ & & & \ddots & 0 \\ & & X & & \ddots \\ & & & & \ddots \end{bmatrix} \begin{bmatrix} \Delta U_i \\ \Delta U_j \\ \Delta \delta_i \\ \Delta \mu_i \\ \Delta \gamma_i \\ \Delta I_{di} \\ \Delta U_{di} \\ \Delta P_{di} \\ \Delta P_{aci} \\ \Delta P_{ij} \end{bmatrix}$$

$$(28)$$

Thus:

$$\frac{\Delta Q_{di}}{\Delta U_i} = a_{11}, \frac{\Delta Q_{aci}}{\Delta U_i} = a_{21}, \frac{\Delta Q_{Ci}}{\Delta U_i} = a_{31}, \frac{\Delta Q_{ij}}{\Delta U_i} = a_{41} \quad (29)$$

$$VSF \approx \left(a_{11} + a_{21} + a_{41} - a_{31} \right)^{-1} \qquad (30)$$

Given an example of typical control mode for constant current to the rectifier side and constant extinction angle to the inverter side, the values of VSF index were calculated. Two control equations of the control mode is

$$\Delta I_{di} = 0, \Delta \gamma_i = 0 \qquad (31)$$

Let γ be the constant angle to the inverter, set $\gamma = \gamma_N = 18°$. Figure 4 shows the curve of VSF index along with I_d changing under different equivalent impedances angle θ_i of AC system and it can be seen that the larger θ_i is, the sooner the system voltage becomes instable. Figure 5 is the curve of VSF index along with I_d changing under different control target of extinction angle. As can be seen, the larger the DC current is, the greater the VSF index

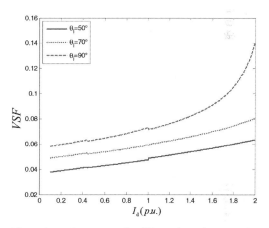

Figure 4. VSF curve under different impedances angle.

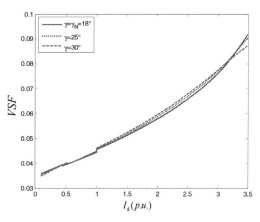

Figure 5. VSF curve under different extinction angle.

and the faster the growth rate are, but the overall value remains within a small positive number in the range. It presents that this control mode has a good voltage stable characteristics and different values of γ have little influence on the *VSF* Index.

4   POWER STABILITY ANALYSIS

The inverter side of DC system worked at the control mode of constant extinction angle in the simplified model showed in Figure 2. Assume that the rated value of the extinction angle is 18° and the four converter valves have equal currents. P_{di} ($i = 1, 2, 3, 4$) will increase with I_d increasing, until reaching a certain value, where P_{di} begins to decline since that the extent of U_{di} declining is greater than the increasing of I_d. Therefore, there will exist a maximum DC power point under the operating mode. The maximum DC power P_{di} should satisfy the following equation.

$$\frac{\mathrm{d}P_{di}}{\mathrm{d}I_d} = 0, (i = 1, 2, 3, 4) \tag{32}$$

Figure 6 shows the maximum power curve of HC mode and single-infeed MC mode. Under HC mode, the values of the equivalent impedance of receiving end system are $Z_1 = 1/2$ in 500 kV side and $Z_2 = 1/3$ in 1000 kV side, the values of the interconnection impedance is $Z_{12} = 1$. The values of the equivalent impedance of receiving end system are 1/2 in 500 kV side and 1/3 in 1000 kV side under the single-infeed MC mode. It can be seen from the maximum power curve that the system is stable when the system operates on the left of the maximum DC power point, where $\mathrm{d}P_d/\mathrm{d}I_d > 0$. In contrast, the system is unstable when the system operates on the right of the maximum DC power point. The rated operating point is on the right side of the maximum power point for 500 kV single-infeed DC system and the system is unstable for rated operating. Using HC mode can efficiently improve the power transmission capacity of DC system. The smaller the equivalent impedance of receiving end AC system is, the higher the intensity of AC system and the greater the DC transmission power are.

Figure 7 shows the curves of DC transmission power for each layer and four converter valves. The maximum power curve of each layer is not obviously different since that the equivalent impedances of receiving end systems are similar. The maximum power curve of two converter valves in the same layer coincide entirely as is showed in Figure 7 since two converter valves in the same layer adopt same control mode. With different values of converter transformer taps in the same layer: $\tau_1 = 1$, $\tau_2 = 0.95$, $\tau_3 = 1$, $\tau_4 = 1.05$, the maximum power curve of four converter valves changing along with I_d will be different as is shown in Figure 8. The four converter valves respond independently and the maximum power curve in the same layer no longer coincide. It is illustrated that the power coordinate distribution can be achieved between two layers by independent control.

Figure 9 shows the DC transmission power in 1000 kV side changing along with I_d with different coupling impedances Z_{12}. It can be seen that reducing the electrical distance between layers can effectively improve the power stability margin and power transmission capacity of DC system under HC mode.

Analyzing the transmission capability of single DC system under MC mode, the rated operating point perhaps is on the right side of the maximum

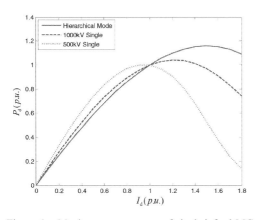

Figure 6.   Maximum power curve of single-infeed MC and HC mode.

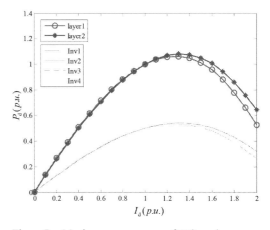

Figure 7.   Maximum power curve of HC mode.

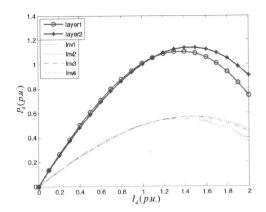

Figure 8. Maximum power curve with different control mode in the same layer converter.

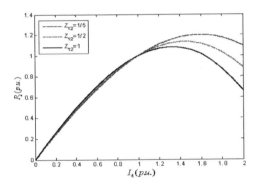

Figure 9. Maximum power curve of HC mode with different Z_{12}.

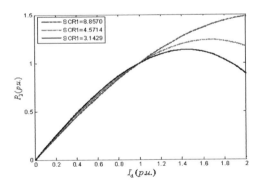

Figure 10. Maximum power curve with different SCR.

power point and system is unstable when SCR of hybrid system is relatively small. Improving the SCR can effectively improve power stability margin. In order to study how changes in SCR of 500 kV side affect the power stability of DC system

in 1000 kV side, the parameters Z_2 of receiving end system in 1000 kV side are set to 1, 1/3, 1/5, then the SCR in 500 kV side calculated is respectively 3.1429, 4.5714, 8.8570 and the power in 1000 kV side changing along with I_d is showed in Figure 10.

Comparison of the above curves shows that properly increasing the SCR of hybrid system in 500 kV side can improve the maximum power and power stability margin of DC system in 1000 kV side effectively.

5 CONCLUSIONS

The power stability of hybrid AC & DC power system under hierarchical connection mode is studied in this paper. Assessment of the power stability by using evaluation indexes and affection factors are applied, and then four important factors are utilized to analyze the power stability, including the operation mode of DC system, converter transformer tap position, the coupling degree between hierarchical system and the values of Short Circuit Ratio (SCR) of different layers. The simulation results demonstrated that: 1) Higher value of SCR stands for the AC system is more strong and the hybrid system is more stable; 2) VSF index can effectively reflecting the voltage stability under different control method under the new HC mode; 3) Compared to single-infeed DC system, using HC mode can effectively improve the power stability margin and increase the transmission capacity.

REFERENCES

Aik, D.L.H. & Andersson, G. 1997. Voltage stability analysis of multi-infeed HVDC systems. IEEE Transactions on Power Delivery. 12(3): 1308–1318.

Chen, X.L. et al. 2011. Study and design of main circuit parameters for ±800 kV/7500 MW DC power transmission project from Xiluodu to Zhexi. Power System Technology. 35(12): 26–32.

Gao, Y. & Han, M.X. 2014. Steady State Characteristic of Multi Infeed UHVDC Power Transmission. Power System Technology. 38(12): 3447–3452.

Guo, X.J. et al. 2013. A Method for Multi DC Terminal Location Selection Based on Multi-infeed Short Circuit Ratio. Proceedings of the CSEE. 33(10): 36–42.

Guo, L. et al. 2015. Power coordination control of ±1100 kV UHVDC system with hierarchical connection mode. Automation of Electric Power System. 39(11): 24–30.

Li, S.H. et al. 2015. Control system design for UHVDC hierarchical connection to AC grid. Proceedings of the CSEE. 38(10): 2409–2416.

Li, C.S. et al. 2014. Research on security and stability control strategy for weak sending-end system of large capacity HVDC power transmission system. Power System Technology. 38(1): 28–32.

Liu, Z.Y. 2013. Ultra high voltage AC/DC grid. Beijing: China Electric Power Press.

Liu, Z.Y. et al. 2013. Study on the application of UHVDC hierarchical connection mode to multi-infeed HVDC system. Proceedings of the CSEE. 33(10): 1–7.

Nayak, O.B. et al. 1995. Control Sensitivity Indices for Stability Analysis of HVDC Systems. IEEE Transactions on Power Delivery. 10(4): 2054–2060.

Thallam, R.S. 1992. Review of the design and performance features of HVDC systems connected to low short circuit ratio AC systems. IEEE Transactions on Power Delivery. 7(4): 2065–2073.

Xu, Z. 2004. Dynamic performance analysis for AC/DC systems. Beijing: Mechanical industry press.

Xiao, J. & Li, X.Y. 2014. Analysis on multi-infeed interaction factor of multi-infeed AC/DC system and its influencing factor. Power System Technology. 38(1): 1–7.

Zhao, W.J. 2011. HVDC transmission technology. Beijing: China Electric Power Press.

Zhou, Q.Y. et al. 2013. A method to select terminal locations in multi-infeed HVDC power transmission system considering weights of HVDC transmission lines. Power System Technology. 37(12): 3336–3341.

Advances in Power and Energy Engineering – Sun (Ed.)
© 2016 Taylor & Francis Group, London, ISBN 978-1-138-02846-3

Electrical breakdown of air gap between conductor and tower covered by insulation sheath

J. Chen, Z. Fan, Z.C. Zhou, Y. Liu, Y.L. Lu & W. Liang
State Grid Jiangsu Electric Power Research Institute, Nanjing, China

ABSTRACT: In order to solve transmission line trip-out accidents caused by windage yaw, quad-bundle conductors were covered by insulation sheath and the electrical breakdown characteristic of air gap between conductor and tower was studied in this paper. During experiments, the conductors were applied to AC 318 kV voltage for 1 min, and the distance between conductor and tower were changed. Experiment results showed that electric breakdown voltage increased with gap distance for bare conductor. While applied on AC 318 kV voltage, the distance of electric breakdown decreased 9.3% for quad-bundle conductor by covered insulator sheath. The breakdown points were located in different sheath connection space which filled by room temperature vulcanized silicone rubber. This paper indicates that the electric breakdown strength of air gap between the wires and the tower for conductors covered by insulation sheath has been effectively improved. However, the construction process should be strictly controlled in order to improve the actual results.

Keywords: transmission line; windage yaw; insulation sheath; electrical breakdown; room temperature vulcanized silicone rubber

1 INTRODUCTION

In recent years, transmission line trip-out accidents caused by windage yaw occurred frequently (Long Lihong et al. 2006, Zhu Kuanjun et al. 2010, Fu Guanjun et al. 2013, Chen Hao et al. 2006, Hu Yi 2004, Li Li et al. 2011). The existing researches indicate that strain and tangent towers are two common types of tower where the windage yaw discharge accidents occurred. Strong winds accompanied by heavy rain or hail will appear in the discharge area when the accident occurs (Liu Yousheng et al. 2007, Liu Mingliang et al. 2009, Geng Cuiying et al. 2009, Zhang Yufang 2005, Chen Hao et al. 2007, Xu Haining et al. 2006, Wang Shengxue et al. 2008, Huang Xinbo et al. 2011). The discharge point has obvious arc burn mark and clear discharge path. What's more, the windage yaw discharge accidents tend to occur on the operating voltage and generally can't reclose, which will lead the transmission lines to stop working and cause great economic losses and social influence (Chen Haibo et al. 2009, Li Na et al. 2010, Jia Yuzhuo et al. 2010, Mei Hongwei et al. 2011, Yang Qing et al. 2013, Wan Qifa et al. 2012, Chen Weijiang et al. 2013, Yu Zhanqing et al. 2013). The inadequate insulation distance caused by strong winds and breakdown of air gap between conductor and tower is the most direct reason lending the accident of windage yaw discharge. In this

paper, quad-bundle conductors were covered by insulation sheath and the electrical breakdown characteristic of air gap between conductor and tower was studied for 500 kV transmission line tower.

2 ELECTRIC FIELD SIMULATION OF THE CONDUCTOR COVERED BY INSULATION SHEATH

In order to simulate the effect of insulation sheath on avoiding windage yaw discharge accidents, electromagnetic field simulation software was used to calculate the electric field distribution between the conductors covered by insulation sheath and ground line. The calculation model is two LGJ—400/35 conductors with a cross-section diameter of 26.82 mm. One of the conductors is applied to 500 kV with a peak phase voltage of 408 kV and the other has a potential of 0 V. The thickness of the insulation sheath is set to 5 mm and the distance between the conductors are 0.8 m, 1.0 m, 1.2 m and 1.5 m respectively.

Simulation results are shown in Figure 1. It could be concluded that the maximum electric field of conductors covered by insulation sheath decreased with air gap lengthen. When the conductor distance increases from 0.8 m to 1.5 m, the maximum electric field of the high potential conductor

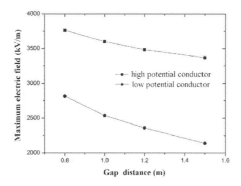

Figure 1. Maximum electric field distribution of conductors changed with gap distance.

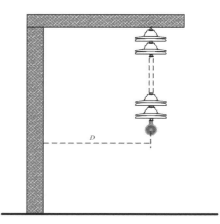

Figure 2. Schematic of experiment.

decreases from 3757 kV/m to 3371 kV/m, and the maximum electric field of the low potential conductor decreases from 2806 kV/m to 2138 kV/m.

3 EXPERIMENTAL EQUIPMENTS AND METHODS

3.1 *Experimental platform building*

A simulate side edge phase model of transmission tower is shown in Figure 2, where the bare and covered conductors are suspend by the insulators.

3.2 *Method of covering the conductor with insulation sheath*

Insulation sheath are made of fluorine silicone rubber. The insulation box covered the hardware fittings compose of multiple parts and are adhesive into a complete box body. Then the insulation box and sheath are pasted together to cover the conductors and hardware fittings completely which was shown in Figure 3.

The length of insulation sheath is 6 m (3 meters on either side) and the thickness of insulation sheath is 10 mm. The sheath was kept for 48 hours after construction, in order to make the room temperature vulcanized silicone rubber solidify completely and form enough mechanical and electrical strength.

3.3 *Experimental method*

In this paper, the gap distance D is measured the vertical distance between the center of conductor and the tower material.

3.3.1 *Breakdown experiment study with different air gap for bare conductor*

The gap distance D is changed from 1.2 m to 1.0 m, 0.8 m and 0.6 m respectively. The applied voltage rise from zero to air gap breakdown voltage in a

Figure 3. Photos of insulation sheath for quad-bundle conductor.

minute linearly. All experiment was made twice and average voltage value was acquired.

3.3.2 *Experimental method of air gap breakdown before and after covered by insulation sheath*

Set the initial gap distance to 0.9 m. The voltage rise from zero to 318 kV within 1 minute linearly and maintain for 1 minute. If the gap electric breakdown has not occurred, take 2 cm as step length to shorten the gap distance and another experiments was carried out until electrical breakdown happens and take the bigger voltage value of 2 times repeated tests. The gap breakdown distance of the bare conductors and conductors covered by insulation sheath respectively were recorded as $D1$ and $D2$. The reduced degree of electric breakdown distance is described as formula 1:

$$\Delta D/D1 = (D1 - D2)/D1 \qquad (1)$$

Table 1. Electric breakdown test results of bare conductor air gap.

Gap distance/m	Breakdown voltage/kV
0.6	
1st time	231.65
2nd time	231.86
0.8	
1st time	315.08
2nd time	313.32
1.0	
1st time	385.44
2nd time	384.03
1.2	
1st time	447.35
2nd time	433.89

Table 2. Experiment results of bare quad-bundle conductor.

Gap distance/m	Results	
	1st time	2nd time
0.90	Pass	Pass
0.88	Pass	Pass
0.86	Fail	Fail

Table 3. Experiment results of quad-bundle conductor covered by composite insulation sheath.

Gap distance/m	Results	
	1st time	2nd time
0.86	Pass	Pass
0.84	Pass	Pass
0.82	Pass	Pass
0.80	Pass	Pass
0.78	Fail	Pass
0.76	–	Fail

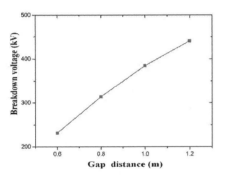

Figure 4. Breakdown voltage with different air gap for bare conductor.

4 EXPERIMENTAL RESULTS OF AIR GAP BREAKDOWN FOR BARE CONDUCTOR

Electric breakdown experimental results of bare conductor air gap are shown in Table 1.

According to the data in Table 1, the breakdown voltages under different air gap distance could be illustrated as in Figure 4. It could be concluded that the breakdown voltage and gap distance has a linear relationship. The breakdown position was located in insulator metal ends where the conductor is closest to the tower.

5 DISCHARGE EXPERIMENT OF CONDUCTORS COVERED BY INSULATION SHEATH

Results of experiment between quad-bundle conductors covered by composite insulation sheath and quad-bundle bare conductors are as shown in Table 2 and Table 3 respectively.

From Table 2 and Table 3, it could be concluded that the breakdown gap distance $D1$ of bare conductor is 0.86 m and the breakdown gap distance $D2$ of conductors covered by insulation sheath is 0.78 m, so the reduced breakdown gap distance is shown as below:

$$\Delta D/D1 = (D1 - D2)/D1 = 9.3\% \qquad (2)$$

The breakdown points were located in sheath connection space which was filled by room temperature vulcanized silicone rubber for both quad-bundle conductors.

Pores will exist in the bonded area when the construction process is undeserved and insulation vulnerabilities will form when the thickness of the seal is too thin. Under the effect of high electric field strength, the weak point would firstly cause electrical insulation breakdown, resulting in overall gap electrical breakdown.

6 CONCLUSIONS

Simulations results show that the maximum electric field of conductors covered by insulation sheath decreased with air gap lengthen.

For bare conductor, the breakdown voltage and gap distance has a linear relationship. In the gap distance of 0.6, 0.8, 1.0, 1.2 m, the breakdown voltage are 231, 314, 385, 314 kV respectively and the breakdown position are located in insulator metal ends where the conductor is closest to the tower.

While applied on 318 kV, the distance of electric breakdown decreased 9.3% for quad-bundle conductor covered by insulator sheath. The breakdown points were located in sheath connection space which is filled by room temperature vulcanized silicone rubber.

This paper indicates that conductors covered by insulation sheath could effectively improve the electric breakdown characteristic of air gap between the wire and the tower. However, the construction process should be strictly controlled in order to improve the actual results.

REFERENCES

Chen Haibo, Wang Cheng, Li Jun-feng, 2009, Application of on-line monitoring technologies for UHV AC transmission lines [J]. Power System Technology, (10): 55–58.

Chen Hao, Chen Chao, Ma Zhigang, 2007, Accident analysis for the transmission line trip-out because of wind [J]. Central China Electric Power, 20(05): 63–65.

Chen Hao, Hao Fuzhong, Shi Xinqing, 2006, Analysis and application of anti-windage yaw technology for HV overhead transmission line [J]. Electric Power, 39(05): 45–48.

Chen Weijiang, Zeng Rong, He Hengxin, 2013, Research progress of long air gap discharge [J]. High Voltage Engineering, 39(6): 1281–1295.

Ding Yujian, Li Qingfeng, Liao Weiming, 2013, Switching impulse discharge characteristics and altitude corrections for typical long air gaps at high altitude areas [J]. High Voltage Engineering, 39(6): 1441–1446.

Fu Guanjun, Wang Liming, Guan Zhicheng, 2013, Torsional Stiffness and Galloping Mechanism Analysis on Bundle Conductors for Overhead Transmission Lines. [J]. High Voltage Engineering, 39(05): 1273–1280.

Geng Cuiying, Chen Shouju, Liu Shenyu, 2009, Experimental study of raining effect on frequency breakdown voltage of air-gap [J]. High Voltage Apparatus, 45(01): 36–39.

Hu Yi, 2004, Study on Trip Caused by windage Yaw of 500 kV transmission line [J]. High Voltage Engineering, 30(8): 9–10.

Huang Xinbo, Tao Baozhen, Zhao Long, 2011, Design of transmission lines new wind deviation on-line monitoring system [J]. High Voltage Engineering, 37(10): 2350–2355.

Jia Yuzhuo, Xiao Maoxiang, You Bo, 2010, The windage yaw numerical simulation of 500 KV overhead transmission lines [C]. International Conference on Power System Technology. Hangzhou, China: IEEE, 1–5.

Li Li, Cao Huajin, Luo Xianguo, 2011, Galloping andwind-induced vibration control on transmission tower-line system [J]. High Voltage Engineering, 37(05): 1253–1260.

Li Na, Lv Yuxiang, Ma Fei, 2010, Research on overhead transmission line windage yaw online monitoring system and key technology [C]. International Conference on Computer and Information Application (ICCIA). Tianjin, China: IEEE, 71–74.

Li Xin-min, Zhu Kuan-jun, Li Jun-hui, 2011, Review on analysis and prevention measures of galloping for transmission line [J]. High Voltage Engineering, 37(02): 484–490.

Liu Mingliang, Li Jinzhu, Xu Kuanhong, 2009, Analysis of windage yaw discharge accident occurred on tangent tower [J]. Electrical Safety Technology, (07): 32–34.

Liu You-sheng, Fan Peilun, 2007, Analysis of windage yaw discharge occurred on jumping wire and pilot current wire on strain tower of 110 kV power line [J]. Power System Technology, S1: 227–228.

Long Lihong, Hu Yi, Li Jinglu, 2006, Study on windage yaw discharge of transmission line [J]. High Voltage Engineering, 32(04): 19–21.

Mei Hongwei, Chen Jinjun, Peng Gongmao, 2011, Research on insulation jacket put on transmission line conductor hung [J]. Proceedings of the CSEE, (01): 109–116.

Wan Qifa, Huo Feng, Xie Liang, 2012, Summary of research on flashover characteristics of long air-gaps [J]. High Voltage Engineering, 38(10): 2499–2505.

Wang Shengxue, Wu Guangning, Fan Jianbin, 2008, Study on flashover of suspension insulator string caused by windage yaw in 500 kV transmission line [J]. Power System Technology, 32(09): 65–69.

Xu Haining, Xiang Wenxiang, Liu Dengyuan, 2006, Study on the flashover due to windage yaw of 500 kV transmission line [J] Hubei Electric Power, (06): 46–48.

Yang Qing, Dong Yue, YE Xuan, 2013, Insulating Coordination Between Composite Insulator and Parallel Gap. [J]. High Voltage Engineering, 39(02): 407–414.

Yu Zhanqing, Yu Junjie, Zeng Rong, 2013, Resistance characteristics of arc in long air gap [J]. High Voltage Engineering, 39(08): 1881–1885.

Zhang Yufang, 2005, Study of the flashover due to windage yaw of 500 kV transmission line [J]. Power System Technology, 29(07): 65–67, 73.

Zhu Kuanjun, Di Yuxian, Li Xinmin, 2010, Analysis of overhead transmission line for asynchronous swaying by the finite element method [J]. High Voltage Engineering, 36(04): 1038–1043.

Advances in Power and Energy Engineering – Sun (Ed.)
© 2016 Taylor & Francis Group, London, ISBN 978-1-138-02846-3

Impulse characteristics of grounding devices of transmission tower based on a simulation experiment in a two-layer soil

L.S. Xiao, Q. Li & Z.Q. Rao
Guangdong Electric Power Research Institute of China Southern Power Grid, Guangzhou, China

X. Yang & J.R. Huang
School of Electrical Engineering, Wuhan University, Wuhan, China

ABSTRACT: Reducing the impulse grounding resistance of grounding devices of a transmission tower is a major measure to improve the lightning impulse, withstand the level, and to ensure power system stability. As the tower mainly adopts a box with radiation grounding devices, this paper uses a simulation experiment method based on similarity theory and researches the influence of radiation length, radiation angle and the proportion of the box and radiation, and the branch length on the box with radiation grounding devices' impulse characteristics in the layer soil model. Results of the experiments show that the branches have an effective use of length; the box and radiation have an optimal proportion improper length radiation or proportion will render the resistance reduction ineffective. According to the results of the experiments, we have got optimal-length rays and put forward an impulse characteristic featuring a better tower grounding structure which is named T2-15-25 & H2-15.

1 INTRODUCTION

Reducing the impulse grounding resistance of grounding devices of transmission towers is an effective measure to improve lightning impulse withstanding level (Guan 2003). The impulse grounding resistance will directly influence the lightning protection of the transmission line (Almeida & Correia 1996). Lightning current flows into the earth through the grounding conductor. Then, the electrical field strength around the conductor will increase and create the ionization in the soil when the field strength exceeds the soil's breakdown electric field strength (Bellaschi 1942). The spark discharge will increase the equivalent grounding radius and the soil conductivity (Bellaschi 1942). Meanwhile, lightning electric wave frequency components are mainly concentrated in 10~200 kHz; the conductor inductance and the inductance to the ground will prevent the current from flowing to the distance. Due to the inductive effect, the conductor will not be used sufficiently (Nor 2000). The spark discharge and inductive effect will make a large difference between the power frequency grounding resistance and the impulse grounding resistance. Therefore, the method to decrease the power frequency grounding resistance is not applicable to the impulse grounding resistance.

The existing resistance reduction methods mainly include using the resistance reducing agent,
replacing the local soil around the grounding conductor, and changing the grounding network structure. The resistance reducing agent could improve the electrical conductivity of the soil around the grounding electrodes directly. But its low adhesion, easy solubility, corrosion to the conductor, and the pollution to the soil limit its application (Liu et al. 1981). Replacement of soil to reduce the grounding resistance in engineering was also used in several situations (Tang et al. 2006, Li et al. 2004, Yao et al. 2007), namely to replace the soil of high resistivity around the grounding conductor with a soil of low resistivity. This paper (Liu et al. 2010) has established a "replacement of soil" model to calculate the grounding resistance, and it indicates that the resistance reduction rate is lower than 5%. The ways to change grounding structure mainly include extending the radiation length, adding some branches to the radiation, and so on. In a high resistivity area, increasing the radiation length can effectively reduce the power frequency grounding resistance, but the effect is not obvious to the impulse grounding resistance. In some special regions, the radiation length was longer than 200 m and the power frequency grounding resistance is under 2 Ω, but the tripping operation caused by lightning flashover still happens (Liew & Darveniza 1974). Simply increasing the length of electrode is not an effective method to reducing the impulse grounding resistance. There are also many

other applications of adding conductors on the radiation branch. In those papers (Sima et al. 2011, Yuan et al. 2012), the authors studied the way to reduce the impulse resistance by adding branches in the perspective of improving the flowing current distribution. But they didn't take the size of conductors into account or get the optimal length of radiation and branches of the tower grounding devices directly.

In order to optimize the impulse characteristics of the grounding system in a two-layer soil (Yu et al. 2005) and reduce the impulse grounding resistance effectively, this paper adopted the method of a simulation test (He et al. 2003) and tested the commonly used box with a radiation grounding system in the transmission tower. The influences of the radiation length, radiation angle, the proportion of box, and radiation length to the impulse characteristics were studied. At the same time, this paper researches the influence of branch length to the horizontal grounding conductor and that of the total length on the box grounding conductor. By comparing the impulse features of different structures of grounding systems, an optimized arrangement of grounding network has been proposed.

2 EXPERIMENT METHOD

The testing field is the grounding laboratory located in the Chinese Electric Power Research Institute UHV AC test site. The test connection is shown in Fig. 1. The grounding conductor is buried in a hemispherical simulation slot (Fig. 2) whose diameter is 8 m. The slot is filled with soil and the wall is designed for reflowing. The testing power is an impulse current generator which can produce 60kV/10kA current. Through changing the adjustable resistance and inductance, the needed current is exported and injected into the grounding device. The current flows through the soil and is collected by the slot wall, then comes back into the low-voltage terminal of the impulse current generator. The voltage divider used for measuring the voltage waveform in the feed point is a weak damping resistance voltage divider whose voltage divider ratio is 115.7:1. The shunt used for measuring the loop current waveform is a Pearson coil whose changing ratio is 0.01V/A. The soil resistivity in the hemispherical slot is 30 Ω·m. We have used 35-cm-thick silver sand with the resistivity of 2180 Ω·m on the soil, so a two-layer soil model is ready. To simulate the impulse features in high soil resistivity area, the simulated grounding electrode is buried in the sand.

According to similarity theory, the paper takes the main simulation scale as follows (Gao 2003):

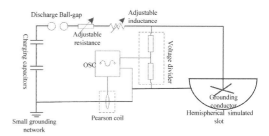

Figure 1. Connecting diagram of experiment.

Figure 2. Physical map of the experiments site.

Soil resistivity: $\rho_1 = \rho_2$
Soil breakdown electric field strength: $E_{01} = E_{02}$
Length: $l_1 = nl_2$
Time: $t_1 = nt_2$
Amplitude of impulse current: $I_{m1} = n^2 I_{m2}$
Wave front time of impulse current: $\tau_1 = n\tau_2$
Impulse grounding resistance: $R_1 = R_2/n$
Impulse coefficient: $\alpha_1 = \alpha_2$

In each equation, the physical quantities with the subscript 1 represent the true type of grounding device and those with the subscript 2 represent the simulated type of grounding device. Also, n is the simulation ratio. The wave front time of impulse current will greatly influence the impulse characteristics of the grounding device. In order to make the wave front time of the testing impulse current close to the standard lightning current and reduce the loop inductance, the current will be injected into the earth by three copper-braided straps. The length of simulation test electrode shall not be longer than 1/4 times the diameter of the simulated slot (He & Chen 1994). Considering the test requirement for wave front time and the size of electrode, the simulation ratio has been set as n = 50:1. As shown in Fig. 3, the wave front time of impulse current in the test is between 0.08 μs

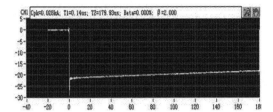

Figure 3. Current waveform of experiment.

Table 1. Structure of the experiments grounding devices.

Name	Shape	Real size（m）
T2-a-L		a=5m, 10m, 15m, 20m, 25m, L=15m, 25m, 35m, 45m, 55m
T3-15-55		a=15m L=55m
H2-L-L1		L=60-2·L1 L1=10m, 20m, 30m
T2-a-L& H2-L1		a=15m L=55-2·L1 L1=10m, 15m, 20m, 27.5m

and 0.14 μs. After the conversion, the wave front time is between 4 μs and 7 μs. It is relatively close to the wave front time of standard lightning current which is 1~4 μs. In a real tower grounding device whose structure is the box with radiation, the length of the box side is commonly 8~15 m, the radiation length is 0~60 m. The shapes and sizes of grounding device are shown in Table 1. The influences of radiation length, radiation angle and the proportion of box and radiation, and branch length on the box with radiation grounding devices' impulse characteristics in layer soil model have been discussed.

3 ANALYSIS OF TEST RESULTS

3.1 *Influence of radiation length*

In order to reduce the grounding resistance in areas with high soil resistivity, the radiation length is commonly more than 60 m. But as the radiation length had increased, the inductive effect of grounding device also increased, which reduces the availability of the grounding device. When the length reaches a certain value, the impulse grounding resistance will decrease slowly. To explore the effective length of the radiation length in the box with a radiation grounding device under the condition of high soil resistivity, the paper will study the impulse characteristics on radiation lengths of 15 m, 25 m, 35 m, 45 m, and 55 m, respectively with the length of the box side as 15 m. A ZC-8 grounding resistance measuring instrument is used to measure the power frequency grounding resistance of the above five kinds of grounding devices. The grounding resistances transformed by the simulation result are shown in Table 2.

The impulse characteristics of different grounding devices with high impulse current injected are shown in Fig. 4. The high current makes the effect of spark stronger than the inductive effect, so the impulse coefficient is lower than 1. The test results show that the impulse grounding resistances of different grounding devices reduced with the increase of current amplitude. Due to the increase

Table 2. Power frequency grounding resistance of different grounding electrodes.

Name	T2-15-15	T2-15-25	T2-15-35	T2-15-45	T2-15-55
Grounding resistance\ Ω	23.9	17.5	13.6	11.1	9.4

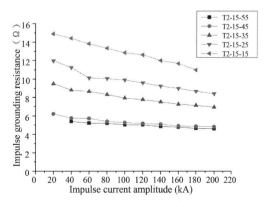

Figure 4. Impulse characteristics of box grounding devices at different radiation lengths.

of impulse current amplitude, the partial field strength around the grounding device will increase. When the value is higher than the soil breakdown electric field strength, the spark discharge phenomenon will occur. Ignoring the residual resistivity of the spark discharge area and assuming that the potential of the partial discharge area is the same as the potential of grounding device; it is equivalent to increasing the diameter of grounding device or increasing the area of the current flow. It will also reduce the impulse grounding resistance. However, when the amplitude increases to a certain value, the spark discharge tends to be saturated. It makes the falling trend of the impulse grounding resistance more and more unobvious. At the same current amplitude, Fig. 5 shows the trend of the impulse grounding resistance with different radiation lengths. When the length of the rays reaches 55 m, the inductive effect will be more obvious and the impulse grounding resistance will not decrease at all. The above tells us that effective length of the radiation length exists. It's meaningless to set a very long radiation length.

3.2 Influence of ray angle

To study the effect of ray angle, this paper had compared the impulse characteristic of T2-15-55 and T3-15-55 in Table 1. The two kinds of grounding devices have the same length of box side and radiation length. The test result is shown in Fig. 6. They are almost the same. So the ray angle has little influence when the radiation length is comparatively long.

3.3 Influence of the proportion of box and radiation

When the total length is a certain value, how to get the optimal effect by controlling the box length

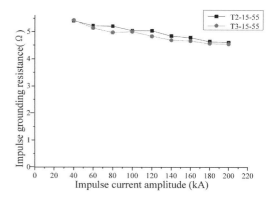

Figure 6. Impulse characteristics of box grounding devices in different radiation angle.

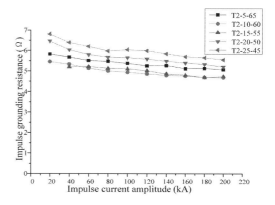

Figure 7. Impulse characteristics of grounding devices in different proportion of box side and radiation length.

and the radiation length? The paper had tested five kinds of boxes with radiation grounding devices: T2-5-65, T2-10-60, T2-15-55, T2-20-50, and T2-25-45. The total lengths of these grounding devices are 70 m. The proportion of box and radiation are, respectively, 1/13, 1/6, 3/11, 2/5, and 5/9; the results are shown in Fig. 7. As the proportion increases from 1/13 to 5/9, the impulse grounding resistance decreases first and then increases. With the increase of the length of box side, the conductor spacing increases. It makes the mutual impedance decrease and the grounding resistance decrease. But the field strength generated by the impulse current also decreases, the spark discharge weakens and the grounding resistance increases. When the proportion is less than 1/6, the mutual impedance plays a leading role; when the proportion is greater than 3/11, the spark effect plays a leading role. So it is appropriate to set the proportion between 1/6 and 3/11.

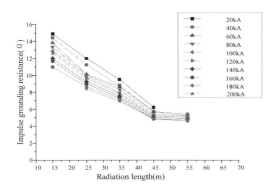

Figure 5. Effects of the length of radiation on the impulse grounding resistance.

3.4 *Influence of branch length on the radiation*

In the actual grounding transformation, adding branches on the radiation is often used to reduce the impulse grounding resistance. So the paper studied its rules. The result of horizontal grounding device is shown in Fig. 8. When the total length is a certain value (60 m), adding branches could reduce the impulse grounding resistance of horizontal grounding device. With the increase of the branches, the impulse grounding resistance decreases first and increases later. It is similar to the proportion of the box and radiation. The appropriate proportions of the branch and radiation need to be explored.

In order to explore the influence of the branch length on the impulse characteristic, the paper selects five kinds of grounding devices: T2-15-55, T2-15-35 & H2-10, T2-15-25 & H2-15, T2-15-15 & H2-20, and T2-15-0 & H27.5 in Table 1(d). In these grounding devices, the length of the box side is a, equal to 15 m; the total length of radiations and branches is 55 m; the lengths of branches L1 are 0 m, 10 m, 15 m, 20 m, and 27.5 m. The special one T2-15-0 & H27.5 represents that the radiation length is 0 m and the branches are connected to the four vertices of the grounding system. The test result is shown in Fig.9. It is similar to the horizontal grounding electrode. In the same condition, with the increase of branch length, the impulse grounding resistance decreases first and increases later. Among these grounding devices, the T2-15-25 & H2-the 15-m one is the best. To a certain extent, the branches can reduce the impulse grounding resistance. But on the contrary, the branches will also influence the efficiency of the current flow into the earth. So it is necessary to choose the optimal branch length on the radiation.

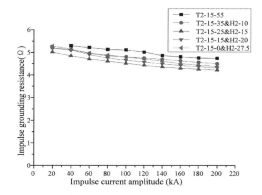

Figure 9. The impulse characteristics of box grounding devices in different branch lengths.

4 CONCLUSION

1. In the two-layer soil model, a simulation test has been used to study the impulse characteristics of the box with radiations, which is commonly used in the grounding devices of transmission tower. Also, we have studied its influences on the impulse characteristics, including the radiation length, the radiation angle, the proportion of box, and radiation and the branch.
2. The radiation angle has little influence on the impulse characteristics. There is a certain value for a special radiation line; once exceeding the value, the impulse grounding resistance decreases slowly. There is an optimal proportion of box and radiation to get the best grounding resistance reducing effect. All of this could be the guidance in the design or transformation of the grounding system.
3. There will be some effect to add the branch on the radiation. When the total length of the radiations and branches is of a certain value, there will be an optimal branch length for the best resistance reducing effect. The simulation test shows that in the two-layer soil model (the upper soil resistivity is about 2180 $\Omega{\cdot}$m; the lower soil resistivity is about 30 $\Omega{\cdot}$m), the T2-15-25 & H2-15 kind of grounding structure has the minimum impulse grounding resistance among all of the tested grounding devices. The structure and the proportion can offer some references for future engineering papers.

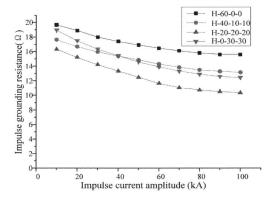

Figure 8. The impulse characteristics of horizontal grounding devices in different branch lengths.

REFERENCES

Almeida M.E. & Correia De Barros M.T. 1996. Accurate modeling of rod driven tower footing. *Ieee Trans. Power Delivery*. 11: 1606–1609.

Bellaschi P.L. 1942. Impulse and 60-cycle characteristics of driven ground—II. *Amer. Inst. Elect. Eng. Trans.* 61(3): 349–363.

Gao Yanqing 2003. Research on Soil breakdown mechanism and grounding system transient characteristics. Beijing: Tsinghua University.

Guan Genzhi 2003. *High Voltage Engineering Fundamentals*. Beijing: China Electric Power Press.

He Jinliang et al. 2003. Laboratory investigation of impulse characteristics of transmission tower grounding devices. *IEEE Transaction on Power Delivery*. 18(3): 994–1000.

He Jin-liang & Chen Xianlu 1994. Simulation experiments theory of the impulse resistance characteristics of transmission line ground devices. *Journal of Tsinghua University(Science and Technology)*. 34(4): 38–43.

Li Jinglu et al. 2001 Discussion on defining the values of related parameters about grounding project. *High Voltage Apparatus*. 40(4): 264–266.

Liew A.C. & Darveniza M. 1974. Dynamic model of impulse characteristics of concentrated earths. *Proc. Inst. Elect. Eng.* 121(2): 123–135.

Liu Ji et al. 1981. Research on heavy current impulse characteristics of long-acting chemical resistance reducer grounding electrode. *High Voltage Engineering*. 45(4): 1–8.

Liu Yugen et al. 2010. Calculation and analysis of grounding parameters of grounding grid with artificial improvement soil surrounding the grounding grid's conductors. *High Voltage Apparatus*. 46(10): 16–19.

Nor, N.M. et al. 2000 Characterization of soil ionization under fast impulses. *Int. Conf. Lightning Protection, Rhodes, Greece*. 2000: 417–422 A.

Sima Wenxia et al. 2011. Experimental study on grounding resistance reduction based on improved grounding electric field distribution induced by the diffuser of impulse current. *High Voltage Engineering*. 37(9): 2294–2301.

Tang Shiyu et al. 2006. Grounding treatment in substation in high soil resistivity area. *High Voltage Engineering*. 32(3): 121–122.

Yao Tianzhong & Wang Jufeng 2007. Grounding resistance calculation of soil-exchanging method based on resistance reduction technology of backfilled soil. *Guangxi Electric Power*. 12(3): 81–82.

Yu Gang et al. 2005. Optimal design rules of grounding system considering seasonal influence. *High Voltage Apparatus*. 41(2): 104–106.

Yuan Tao et al. 2012. Analysis of grounding resistance reduction effect based on enhancing impulse current leakage efficiency. *Transactions of China Electrotechnical Society*. 27(11): 276–284.

Advances in Power and Energy Engineering – Sun (Ed.)
© 2016 Taylor & Francis Group, London, ISBN 978-1-138-02846-3

Investigation on fault diagnosis of substation grounding grid using distribution of ground potential

W.D. Qu, D. Huang & M. Yang
State Grid Shanghai Changxin Supply Company, Shanghai, China

W.R. Si
State Grid Shanghai Electric Power Research Institute, Shanghai, China

Z.B. Xu & X. Guo
Red Phase Electric Power Equipment Co. Ltd., Xiamen, China

F.H. Wang
Shanghai Jiaotong Univeristy, Shanghai, China

ABSTRACT: In order to effectively improve the accuracy of the conductors' fault diagnosis of the grounding grid in the substation, this thesis took regular- and irregular-shaped grounding grid models as research objects and we measured the distribution of the ground potential under typical faults. The experimental results indicate that the analysis of the ground potential distribution could relatively accurately determine the fault location and fault type. If the grounding grid has void-welding or sealing-off, the potential of the surface ground above fault point will significantly increase while the ground potential above its adjacent node is slightly lower. If the grounding grid has fracture faults, the potential of the surface ground above fault point will significantly increase while the ground potential above adjacent parallel conductors is slightly lower. If the grounding grid has corrosion faults, the potential of surface ground above fault point will significantly increase while the ground potential above the adjacent parallel conductors is slightly lower. The results could be an important supplement of grounding grid's fault diagnosis according to ground potential distribution.

1 INTRODUCTION

The function of grounding grids in substations is to ensure the electric equipment is well-grounded and to keep the equipment at the same ground potential. When a grounding fault occurs in the substation, the grounding grid could improve the potential distribution and quickly disperse the flow of current, which would ensure the safety of the equipment and operation personnel as well as maintain the safe and reliable operation of power systems (Zhang & Huang 2002; Zhao et al. 2006). But the faults of grounding conductors like void-welding, sealing-off, corrosion, or conductor fracture may cause the power system faults, which may result in accidents and bring huge economic losses. Therefore, it is necessary to analyze and research the fault diagnosis method of substation grounding grids which could be easily realized and has high accuracy to determine fault location and fault types without excavation, thus, for repairing the grounding system and ensuring the safety and stability of power systems.

Scholars in China and abroad have conducted many studies on fault diagnosis of grounding grids without excavation. F. Dawalibi has proposed a diagnosis method based on the magnetic field measurement near the ground (Dawalibi 1986; Hyung et al. 1998), which diagnoses and locates conductor breakpoints by applying different frequency AC excitation signals to the grounding grid directly and measuring the magnetic induction distribution near the ground surface (Liu et al. 2008). However, it is difficult to measure the electromagnetic signal in substations with large background electromagnetic interferences, and it would directly affect the accuracy of diagnosis. Electrochemical detection is one of the most effective ways to characterize the corrosion state and corrosion mechanism of the grounding material (Zhang et al. 2008). But as it is difficult for the traditional electrochemical measurement system to accurately measure the corrosion state of the grounding conductors and there exists much interference signal and stray current, this method lacks diagnosis accuracy. Resistance change analysis of grounding conductors to diagnosing grounding fault is mostly studied. Since grounding grids' topology are determined after completion of power plant or substation construction, the grid could be deemed to be composed of resistance grid (Huang 2007). We can consider grounding wires as the nodes of the grid and compare the resistance change between nodes to diagnose faults (Xiao et al. 2001). But the resistance is of milliohm level if the grounding grids were in normal condition or the conductor corrosion was slight. Thus, it is difficult to eliminate measurement error and fault diagnosis has low accuracy. Xu Lan has calculated potential distribution around grounding grid and proposed a diagnosis method of grounding faults based on the surface potential distribution (Xu 2011). Although there are simple simulating test results of grounding grid, that paper is mainly about simulating calculation which is quite different from actual working conditions, and it only discussed one fault type, that is, conductor fracture.

Aiming at these problems, this thesis designed two types of grounding grid models based on the actual grounding grids and carried out experimental research on ground potential distribution under typical faults to get more accurate laws of ground surface potential distribution and to improve the accuracy of fault location.

2 THEORY ANALYSIS OF FAULT DIAGNOSIS OF GROUNDING GRIDS

When a DC or AC power supply is connected to the earth through grounding conductors, a constant current field will be established in the substation ground. Similar to electrostatic field, the constant current field is produced by electric charge. The charge distribution is not time-varying and it meets the Laplace equation which can be calculated as a constant field.

According to the uniqueness of the theorem of electrostatic field, if the source and the boundary conditions are determined and any point in the field meets the Poisson equation or Laplace equation, the potential of the point will be uniquely determined, namely the static electric field has a unique solution.

For a single point source, a point potential in space can be solved by the image method. For a linear electrode, the point potential generated by a leakage current is generally calculated by the Green's function. If we define infinity point as the reference point and there is a current flowing through grounding device, the potential of any point generated by leakage current can be calculated by the following formula:

$$V_P = \iint_S G(P, Q) J(Q) dS \qquad (1)$$

In the formula, $J(Q)$ is for the leakage current density of point Q in the electrode surface S; $G(P, Q)$ is the Green's function corresponding to the electrode geometry which is for the potential of P in space generated by leakage current of point Q at the electrode surface.

Potential of points in the field varies with the change of source. Solving the point potential distribution by point source and boundary conditions is the only solution to problems in electromagnetic movements. On the contrary, to solve the field source distribution by point potential distribution and boundary conditions will get non-uniqueness solutions, namely that the electromagnetic inverse problem may have multiple solutions. For a practical substation grounding grid, its topology and spatial location are determined after construction is completed. When fault current flows into grounding grid, the potential distribution of the space field is generated. If the grid has faults, such as sealing-off, fracture, or corrosion, the source of field will be changed; thus, electric field and potential distribution will be changed correspondingly. Therefore, change of ground potential distribution in space field around the grounding grid can reflect the source change accordingly. Since the potential distribution at the ground surface could be easily obtained, this thesis selects grounding grid as the study object and puts forward the grounding grid's fault diagnosis method based on potential distribution of ground surface.

3 SIMULATION TEST OF A GROUNDING GRID'S FAULT DIAGNOSIS

In order to improve the validity and accuracy of a grounding grid's fault diagnosis based on potential distribution of ground surface, we carried out simulation experiments on two types of grounding grids (regular-shaped grid and irregular-shaped grid) under normal state and three kinds of faults.

3.1 Design of grounding grid models

Figure 1 shows two types of grounding grid models designed on the basis of dimensional similarity principle. Among them, figure (a) is for regular-shaped grounding grid and figure (b) is for irregular-shaped grounding grid. Their specific size can be found in the figures and the unit is in meters. The regular-shaped grid has an area of $7\,m \times 7\,m$. Its horizontal conductors are arranged by nonuniform design optimization and length of the vertical grounding electrodes is $0.3\,m$. The irregular-shaped grid has an area of $7\,m \times 9\,m$ and

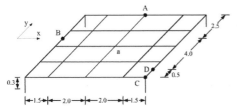

(a) Regular-shaped grounding grid

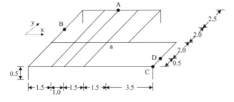

(b) Irregular-shaped grounding grid

Figure 1.   Grounding grid models.

the vertical grounding electrode is 0.5 m. Horizontal conductors of two types of grids above are galvanized flat steel of 25 × 6 mm. The vertical grounding electrodes are 38-mm round steel in diameter and buried depth of grounding grids is 0.2 m.

In addition, A, B, C, and D are analog fault locations. In consideration of method of the faults' set and reset, we adopted the means of rivet to make grid models. In the following figures, a is for current injection point.

3.2   Experiment description

In the experiment, ground potential distributions under normal state and different fault conditions are measured after injecting excitation current to the grounding grid. Signal source used in experiment is Xiamen Hongxiang-8100 type system of grounding device parameter measurement and management. The frequency of input current is set to 65 Hz to avoid interference from 50 Hz power frequency. Considering the limit of the source's output power and the test circuit impedance, the maximum output current of signal source is adjusted to 4 A within bounds of power output. The voltage reference point and current reference electrode are set according to the 0.618 method. Exciting current is injected to point a and flown out from current reference electrode. We first tested the grounding grid under normal state and then measured ground potential distribution of grid under fault conditions by excavating and setting faults.

3.3   Fault settings

When listing facts use either the style tag List signs or the style tag List numbers.

Table 1.   Simulated fault settings.

Faults	Fault settings
Desoldering at A	Disconnect riveting at A
Fracture at B	Disconnect riveting at B
Conductor corrosion segment	Replace C-D segment at C-D with thin conductor

Faults of grounding grid are usually caused by void-welding or sealing-off in construction, soil corrosion, and electrodynamics of grounding current. These processes could make the horizontal conductors diminution or cause fracture. Therefore, we mainly simulated the following three typical faults: desoldering, conductors fracture, and corrosion. Specific fault settings are shown in Table 1.

4   ANALYSIS OF FAULT DIAGNOSIS

4.1   Ground surface potential distribution under normal conditions

Figure 2 shows surface potential distribution of two types of grounding grids calculated by equipotential algorithm based on electromagnetic field theory and circuit theory. Coordinates x and y are corresponding to Figure 1. As shown in the figure, when current injection point is in the center of the grid, surface potential of regular-shape grounding grid is symmetrical about the grid's center. Potential of conductors' ends is slightly higher than the middle period and the closer the injection point, the higher the potential. Surface potentials of four corners is higher because of the influence of vertical grounding electrodes. The largest potential appeared in the current injection point and the minimum surface potential appeared on the edge of the ground grid, are 5.36 V and 1.67 V, respectively. For the irregular-shaped grounding grid, the surface potential distribution also suits the law that potential of conductors' ends are lightly higher than middle period; the potential close to injection point is higher and the potential of four corners is higher. The largest potential appeared in the current injection point and the minimum surface potential appeared in the center of the mesh at the lower right corner are 7.52 V and 1.50 V, respectively.

4.2   Ground surface potential distribution under fault conditions

Figure 3 shows the ground surface potential distribution of two types of grids near A under fault 1 (desoldering at A). Potential distribution under normal conditions is also shown in the diagram.

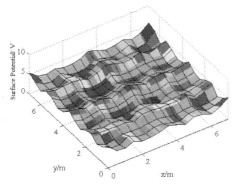

(a) Regular-shaped grounding grid

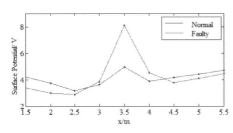

(b) Irregular-shaped grounding grid

Figure 2.   Ground surface potential distribution under normal conditions.

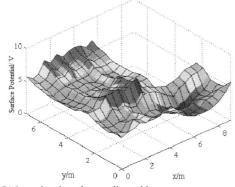

(a) Regular-shaped grounding grid

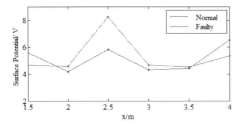

(b) Irregular-shaped grounding grid

Figure 3.   Surface potential above conductor at A under fault 1.

Direction of coordinates x is the same as what is shown in Figure 1, indicating location of measuring points. We can indicate from the figure that the potential at ground surface above point A increased significantly under fault conditions while potential above its adjacent nodes slightly decreased.

Figure 4 shows the potential distribution at ground surface above conductor at B of two types of grid under fault 2 (fracture at B).

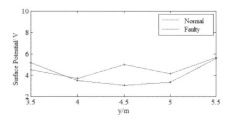

(a) Potential at ground surface above point B (regular-shaped grounding grid)

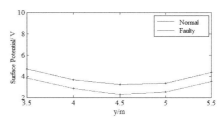

(b) Potential at ground surface above parallel conductors of B (regular-shaped grounding grid)

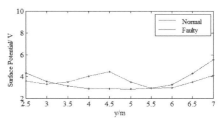

(c) Potential at ground surface above point B (irregular-shaped grounding grid)

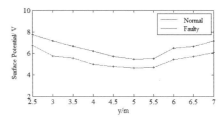

(d) Potential at ground surface above parallel conductors of B (irregular-shaped grounding grid)

Figure 4.   Ground surface potential above conductor near B under fault 2.

Potential distribution under normal conditions is also shown in the diagram. Direction of coordinates y is the same as what is shown in Figure 1, indicating location of measuring points. We can indicate from the figure that the potential at ground surface above point B increased significantly under fault conditions while the potential above its parallel conductors slightly decreased.

Figure 5 shows the potential distribution at ground surface above C-D conductor segment of two types of grid under fault 3

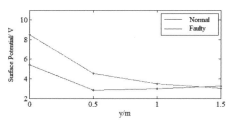

(a) Potential at ground surface above C-D segment (regular-shaped grounding grid)

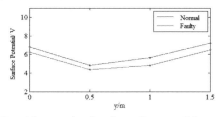

(b) Potential at ground surface above adjacent parallel conductors of C-D segment (regular-shaped grounding grid)

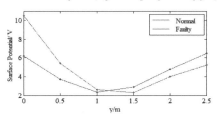

(c) Potential at ground surface above C-D segment (irregular-shaped grounding grid)

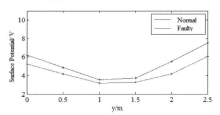

(d) Potential at ground surface above adjacent parallel conductors of C-D segment (irregular-shaped grounding grid)

Figure 5. Ground surface potential above C-D conductor segment under fault 3.

(Conductor corrosion at C-D segment). Potential distribution under normal condition is also shown in the diagram. Direction of coordinates y is the same as what is shown in Figure 1, indicating location of measuring points. We can indicate from the figure that the potential at ground surface above C-D segment increased significantly under fault conditions while the potential above its adjacent parallel conductors slightly decreased.

From the above analysis, for regular- and irregular-shaped grounding grids, when there is void-welding or sealing-off at the nodes of level conductors, the ground surface potential above fault point will increase by about 55% compared with normal conditions and surface potential above its adjacent nodes slightly decrease. When there is a fracture of level conductors, the ground surface potential above fault point will increase by about 35% while surface potential above its adjacent conductors decrease slightly. Similarly, when there is conductor corrosion, the ground surface potential above fault conductor will increase by about 50% while surface potential above its parallel adjacent conductors slightly decrease.

5 CONCLUSION

This thesis took two types of grounding grid model as research objects and carried out tests around ground surface potential distribution under typical faults. The test results are as follows:

1. Under normal conditions, surface ground potential of substation has maximum value in the current injection point and the surface potential is on the decline with the current scattered flows along a conductor. In addition, surface potential above conductor's middle period is lower than the conductor's node on both ends and surface potential above mesh center is lower than conductor section.

2. When there is void-welding or sealing-off at node of level conductors, the ground surface potential above fault point will greatly increase and surface potential above its adjacent nodes slightly decrease. When there is fracture of level conductors, the ground surface potential above fault point will greatly increase while surface potential above its adjacent conductors slightly decrease. When there is conductor corrosion, the ground surface potential above fault conductor will greatly increase while surface potential above its parallel adjacent conductors slightly decrease.

3. Experimental results further illustrated that the grounding grid fault diagnosis method based on surface potential distribution of the substation is effective.

4. Experimental results of two types of grounding grid indicate that when the grounding grid is under fault conditions like void-welding, sealing-off or corrosion, ground potential above fault point, and its adjacent conductor will present different degree of rising or decrease, to which we can determine the fault position and type according. But the universality of this conclusion and the quantitative standards still need to be further field tested, and this is also our future work.

REFERENCES

Dawalibi F., Electromagnetic fields generated by overhead and buried short conductors Part 2—Ground conductor [J]. IEEE Trans. on Power Delivery, 1986: 112–119.

Hyung S.L., Jung H., Dawalibi F., Efficient ground grids designs in layered soils [J]. IEEE Trans. on PWRD, 1998, 13(3):745–751.

Huang Lihong, Measurement of Resistance between Down-lead Nodes of Earth Network and Its Fault Diagnosis [J]. Guangdong Electric Power, 2007, 7: 13–15.

Liu Yang, Cui Xiang, Lu Tiebin, Diagnosis Method of Breaking Points in Grounding Grid for Substations [J]. Power System Technology, 2008, 2:21–25.

Xiao Xinhua, Liu Hua, Chen Xianlu, Analysis of Theory and Method about the Corrosion as well as the Broken Point of the Grounding Grid [J]. Journal of Chongqing University (Nature Science Edition), 2001, 3:72–75.

Xu Lan, Grounding Grid's Potential Parameter Calculation and its Fault Diagnosis in substation [D]. Hu Bei: Huazhong University of Science and Technology, 2011.

Zhang Xiaolin, Huang Qingquan. Fault diagnosis of grounding grid of electric power plants and substations [J]. Proceedings of the EPSA, 2002, 1(3):49–51.

Zhao Zhibin, Cui Xiang, Li Lin. Analysis of grounding grids buried in complex soil [J]. Journal of North China Electric Power University, 2006, 3(4):33–36.

Zhang Xiuli, Luo Ping, Mo Ni, Development and Application of Electrochemical Detection System for Grounding Grid Corrosion State [J]. Proceedings of the CSEE, 2008, 19:152–156.

Advances in Power and Energy Engineering – Sun (Ed.)
© *2016 Taylor & Francis Group, London, ISBN 978-1-138-02846-3*

Study on impulse characteristics of substation grounding grid with experimental analysis

Z.X. Lu, R. Zhou & M.B. Wu
State Grid Shanghai Changxing Power Supply Company, Shanghai, China

W.R. Si
State Grid Shanghai Electric Power Research Institute, Shanghai, China

X. Guo & Z.B. Xu
Red Phase Electric Power Equipment Co. Ltd., Xiamen, China

F.H. Wang
Shanghai Jiaotong University, Shanghai, China

ABSTRACT: In order to analyze the impact characteristics of grounding system in substation and improve the operation reliability of the electrical equipment to guarantee personal safety, this thesis adopted dimensional similarity principle to design and customize two types of grounding system model with regular shape and irregular shape based on typical grounding grid in 35 kv substation. We carried out experiments on impulse characteristics of grounding grid under different impulse current into different injection points by means of simulating typical faults including conductor diminution, fracture and desoldering. The results show that the transient dispersing characteristics of regular-shape grounding grid is better than grounding grid with irregular-shape. For the same current injection point, the shorter the wave-head time, the greater impulse grounding resistance. And when the injection point is at the center of grounding grid, it has the least grounding resistance. When there are typical faults exist in grounding grid, the impulse grounding resistance is greater than the normal working condition and the growth rate is associated with wave-head time of impulse current.

1 INTRODUCTION

Grounding performance of substations' grounding grid has always been highly concerned by design, production and operation department. It is of great significance to ensure the safety of substation equipment and person as well as reliable operation of the power system, and the increase of precision equipment used for protection and control in substation puts forward higher request for current dispersing and voltage sharing function of the grounding grid (Li. 2008). In addition, when lightning accidents occurred in or near the substation, there might be a shock current injection with high frequency and large amplitude into substation grounding system. It will arouse complex transient characteristics of the grounding grid. So it is necessary to analyze the impulse characteristics of the grounding grid in substation.

At present, many scholars in China and abroad did a lot of research on the impulse characteristics of the substation grounding grid and the basic methods adopted includes transmission line theory, analysis method based on circuit theory and finite element analysis method, etc. (Zhang et al. 2002; Zhang et al. 2011; Wang et al. 2010). Some researchers also did improving efforts to these methods from the perspective of improving calculation precision and calculation efficiency of the algorithm (Wang et al. 2003; Zou et al. 2005). At the same time, the grounding model also developed from equal-potential model to transient unequal-potential model with conductors' potential drop considered for which the more accurate ground surface potential distribution is obtained (Cidras et al. 2000; Zhang et al. 2000). Besides, CDEGS, which is an international famous grounding analysis software, is used in research (IEEE Std 80-2000). The existing calculation method or software have guiding significance to the design and optimization of grounding systems in most cases. When analyzing large grounding grid or grounding performance with higher frequency impulse current injection, calculation process is more time-consuming and accuracy is not satisfactory. In addition, for practical grounding grid, numerical calculation method is difficult to get accurate substation grounding physical model due to construction, renovation, corrosion and other factors. There are some researchers trying to carry out simulation experiment on customized model to obtain impulse characteristics and impulse dispersing characteristics of grounding grid (Tu et al. 2006). Results of grounding grid impulse characteristics from experimental analysis could be reference to verify simulating calculation results, to optimize the design of the grounding system and to help operating personnel understand the impulse characteristics. However, grounding model used in experiment is different to some extent from actual grounding grid and analysis of impulse characteristics under typical faults was less involved, still needs further research.

Aiming at these problems, this thesis designed and customized two types of grounding system model based on typical grounding grid in 35 kv substation. We carried out experiments to analyze impulse characteristics at normal condition and under typical faults to obtain more accurate and all-sided impulse characteristics and provide the basis to optimum design and condition assessment.

2 EXPERIMENT DESCRIPTION OF GROUNDING IMPULSE CHARACTERISTICS

2.1 Two types of grounding grid model

Figure 1 shows two types of grounding model for 35 kV substation based on the principle of dimensional similarity and the size unit is m here. Figure (a) shows the $7\,m \times 7\,m$ model with regular shape, of which horizontal conductors made from galvanized flat steel are arranged by non-uniform optimized design and vertical earthing electrode made from round steel in 38 mm diameter is 0.3 m long. m and n are current injection points. Figure (b) shows $7\,m \times 9\,m$ model with irregular shape of which the vertical earthing electrode is 0.5 m long. m, n, p and q are current injection points. In addition, the A, B, C, D for typical faults position adopts riveting method.

2.2 Experiment description

Lightning current generator used for obtaining impulse characteristics of grounding grid is the type-4051 lightning impulse resistance tester produced by Red Phase Electric Power Equipment Company. This tester could automatically adjust the injection current to obtain the maximum power output. Voltage reference point in the test circuit and current reference electrode are set according to the 0.618-method.

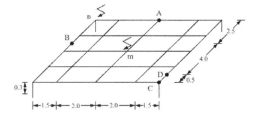

(a) Regular-shape grounding grid

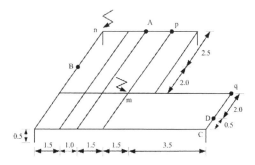

(b) Irregular-shape grounding grid

Figure 1. Two types of grounding model.

In the experiment, we tested impulse resistance of two types of grounding grid in normal state at first step. Then we set up faults at simulation fault points and tested impulse resistance after landfill. Grounding resistance are measured and analyzed in experiment under different lightning current waveform injected to different injection points respectively. The impulse current waveform are 0.25/100 μs, 1/20 μs, 8/20 μs and 10/350 μs specifically in test. The injection points are m, n for regular-shape grid while m, n, p and q for irregular-shape grid.

Considering the grounding grid's defects are often caused by void-welding or sealing-off, soil erosion, electric force of grounding current and other factors, which would make horizontal homogeneous conductors thinner or fracture. Therefore, this thesis carried out tests mainly under the three typical faults which are void-welding, conductors' fracture and conductor corrosion. The specific setting mode of faults is as shown in Table 1.

3 RESULTS ANALYSIS

3.1 Impulse characteristics under different lightning current waveform

Data shown in Table 2 is impulse resistance of two types of grounding model in normal state under

Table 1. Settings of simulated faults.

Fault	Fault description	Setting way	Fault recovery
1	Desoldering at A	Disconnect riveting at A	Connect riveting at A
2	Fracture at B	Disconnect riveting at B	Connect riveting at B
3	Conductor corrosion at C-D segment	Replace C-D segment with thin conductor	None

Table 2. Impulse grounding resistance under different lightning current waveform.

Lightning current waveform/μs	Impulse grounding resistance/Ω	
	Regular-shape grid	Irregular-shape grid
0.25/100	9.983	15.480
1/20	4.425	6.767
8/20	2.612	2.975
10/350	2.560	2.813

Table 3. Impulse resistance of regular-shape grounding grid with different injection points.

Lightning current waveform/μs	Impulse grounding resistance/Ω	
	Injection point m	Injection point n
0.25/100	9.983	11.320
1/20	4.425	7.097
8/20	2.612	3.230
10/350	2.560	3.054

Table 4. Impulse resistance of irregular-shape grounding grid with different injection points.

Lightning current waveform/μs	Impulse grounding resistance /Ω			
	Injection point m	Injection point n	Injection point p	Injection point q
0.25/100	15.480	18.016	16.972	18.804
1/20	6.767	11.236	10.987	12.542
8/20	2.975	4.059	3.737	3.948
10/350	2.813	3.132	2.993	3.225

different lightning current waveform injecting to the same point m.

We can infer from Table 2 that for the two types of grounding grid, the shorter front time, the greater impulse resistance. And impulse resistance of two types of grounding grid is close under current waveform with relatively long front time. Besides, the resistance of irregular-shape model is always greater than regular-shape model's, which means that the dispersing characteristics of regular-shape grounding grid are better. Therefore, we should take the influence of grounding performance under different lightning current into consider and try to adopt regular-shape grounding model when analyzing the impulse characteristics of substation grounding grid. According to that, we suggest to analyze the lightning activities when designing grounding system.

3.2 Influence of injection point to impulse characteristics

Data shown in Table 3 and Table 4 is impulse resistance of two types of grounding model in normal state under different lightning current waveform injected to different point. We can conclude from the table that the impulse resistance is the least when injection point is at the center of grounding grid for two types of grid with same impulse current. Besides, this conclusion is suitable for situation of different lightning current waveform.

For the regular-shape grounding grid, the shorter front time of impulse waveform, the greater impulse resistance. And the difference between impulse resistances of different injection points will be small when the front time of lightning current waveform is relatively long.

By contrast, for the irregular-shape grounding grid, there are obvious differences between impulse resistances of different injection points. The impulse resistance has maximum value when the injection point is at the corner of grid and the minimum value when the injection point is at the center of grid oppositely. And when the impulse current injected into frame of grid, the resistance value is between the above two kinds of situations. In addition, the above conclusion is also applicable to conditions of different lightning current waveform.

3.3 Grounding grid's impulse characteristics under typical faults

Data in Table 5 and Table 6 are respectively impulse resistance of regular-shape grid and irregular-shape grid under different lightning current injected to the same point m. The tables show us that the impulse grounding resistances of two types of grounding grid have increased relative to the normal working condition. The shorter the front-time of lightning current waveform, the greater increase of impulse grounding resistance is.

Specifically, the impulse resistance of regular-shape grounding grid increased 21.07% to normal

Table 5. Impulse resistance of regular-shape grounding grid under typical faults.

Lightning current waveform/μs	Impulse grounding resistance/Ω			
	Fault 1	Fault 2	Fault 3	Fault 4
0.25/100	9.983	11.037	12.396	12.827
1/20	4.425	4.704	5.224	4.966
8/20	2.612	2.911	2.671	2.818
10/350	2.560	2.800	2.619	2.756

Table 6. Impulse resistance of irregular-shape grounding grid under typical faults.

Lightning current waveform/μs	Impulse grounding resistance/Ω			
	Fault 1	Fault 2	Fault 3	Fault 4
0.25/100	15.480	17.403	18.123	20.988
1/20	6.767	7.688	7.432	8.682
8/20	2.975	3.329	3.186	3.221
10/350	2.813	2.888	2.957	2.992

condition when the injection current is c lightning waveform and increased about 6.45% when the injection current is 10/350 μs lightning waveform. By contrast, the impulse resistance of irregular-shape grounding grid increased 21.69% to normal condition when the injection current is 0.25/100 μs lightning waveform and increased about 4.72% when the injection current is 10/350 μs lightning waveform.

4 CONCLUSION

The analysis results of impulse characteristics of regular-shape and irregular-shape grounding grid are as follows.

1. The impulse grounding resistance of regular-shape model is always less than irregular-shape model's under different lightning waveforms, which means that the dispersing characteristic of regular-shape grounding grid is better. Therefore we should adopt regular-shape grounding model in research.
2. For the two types of grounding grid with the same injection point, the shorter front time of lightning waveform, the greater impulse resistance. So we suggest taking lightning waveform into account based on analysis of lightning activities to determine appropriate parameters of grounding impulse characteristics.
3. The impulse resistance has maximum value when the injection point is at the corner of grid and the resistance has minimum value when the

injection point is at the center of grid oppositely. And when the impulse current injected into frame of grid, the resistance value is between the above two kinds of situations. As a result, we should set the earthing point at the center of grid rather than the corner when designing the grounding grid.

4. The impulse resistances of grounding grid under faults have increased relative to the normal working condition. And the shorter the front-time of lightning current waveform, the greater increase of impulse resistance is. Therefore we can assess grounding grid's condition according to impulse resistance change. In addition, it is necessary to point out that the analysis should also base on impulse current waveform injected into the substation.

The research results could provide important references to grounding grid's optimal design and condition evaluation as well as knowing substation grounding grid's impulse characteristics well.

REFERENCES

Cidras, J., A.F. Otero, C. Garrido. Nodal Frequency Analysis of Grounding Systems Considering the Soil Ionization Effect. IEEE Transaction on Power Delivery, 2000, 15(1): 103–107.

IEEE-SA Standards Board. IEEE Guide for Safety in AC Substation Grounding [S]. IEEE Std 80-2000.

Li Xiaoli, Study on impulse characteristics of grounding system in substation [D]. Chongqing: Chongqing University, 2008.

Lu Zhiwei, Wen Xishan, Shi Yanlin, etc. Numerical calculation of large substation grounding grids in industry frequency [J]. Proceedings of the CSEE, 2003, 23(12): 89–92.

Tu Youping, He Jinliang, Zeng Rong. Lightning impulse performances of grounding devices covered with low resistivity materials [J]. IEEE Transactions on Power Delivery, 2006, 21(3): 1706–1713.

Wang Yihua, Wang Fenghua, Jin Zhijian, etc. Application of Finite Element Method to the Grounding Resistance Measurement with Short Leads [J]. High voltage apparatus, 2010, 46(7): 81–85.

Zhang Bo, Cui Xiang, Zhao Zhibin. Analysis of grounding grids at large scale substations in frequency domain [J]. Proceedings of the CSEE, 2002, 22(9): 59–63.

Zhang Haifeng, Wang Fenghua, Jin Zhijian, etc. Impacts of underground metal pipelines on grounding resistance of substation grounding grid and its measurement [J]. Power System Technology, 2011, 35(1): 134–140.

Zhang Liping, Yuan Jiansheng, Li Zhongxin. Calculation of substation grounding grids with unequal-potential model [J]. Proceedings of the CSEE, 2000, 20(1): 1–3.

Zou Jun, Guo Jian, Zhang Bo, etc. Fast approach grounding transient calculation based on MBPE in both frequency domain and spatial domain [J]. Proceedings of the CSE.E, 2005, 25(9): 121–125.

Advances in Power and Energy Engineering – Sun (Ed.)
© 2016 Taylor & Francis Group, London, ISBN 978-1-138-02846-3

One high sensitive residual current protection apparatus

Feng Du, Weigang Chen & Yue Zhuo
Siemens SLC, China

M. Anheuser
Siemens AG, Germany

ABSTRACT: In this paper, one high sensitive residual current protection apparatus down to 10 mA with AC/DC (alternating current, direct current) sensing method had been proposed. To realize the residual current protection even under such a small residual current from DC to several kilohertz (kHz), two key problems had to be solved. One is how to realize small alternate residual current sensing and energy harvesting at the same time in the voltage-independent scenario. The other is how to realize the small direct residual current sensing under the voltage-dependent condition. The above two challenges had been solved by three proposed innovative solutions, 1) adaptive structure for current transformers in the protection device. 2) A unique energy harvesting circuit based on resonant circuit. 3) An accurate DC sensing based on magnetic modulation technology with double feedback properties. Per to the function test results according to the standard of IEC60947-2 for type B RCD (residual current device), the high sensitive residual current protection apparatus had be realized successfully.

1 BACKGROUND

The residual current protection is widely required in power distribution system for our personnel and asset safety. Among the three types of residual current protection products: type AC, type A and type B, type B residual current protection is the most versatile. It can protect all kinds of residual currents, i.e., AC, pulsated DC and DC. In the state of art, the type B residual current protection device can only provide the protection down to 30 mA. However, per to the standard of IEC 60947-2 (IEC, 2013) for better safety, 10 mA or even 6 mA is the preferred minimum protected residual current. Especially, in the application scenario like humid environment and hospital, there is an urgent market demand.

To protect extremely small type B residual current, the difficulty is how to detect this residual current reliably and act correspondingly. Especially this kind of product should work at both powered (voltage dependent) and un-powered (voltage independent) modes. Any false protection action is risky (Czapp, S, 2008). It will result in unnecessary power off. Thus, the downtime will lead to asset loss and even threaten the personnel safety, e.g., in hospital. In other words, two critical problems exist. One is how to realize small alternate residual current sensing (up to several kilohertz) and energy harvesting at the same time in a voltage-independent scenario.

The other is how to realize the small direct residual current sensing under the voltage-dependent condition. And all these requirements should be satisfied in the small dimension and low cost for such kind of protection device.

In this paper, the type B residual current protection apparatus which is aiming to extend its protection range down to 10 mA is explored. The paper is organized in the following way. In the section two, the working principle including the details of the challenges, theoretical analysis and corresponding proposed solutions has been gone through. The verification based on the test has been introduced in the section three. At last, the summary and possible future work has been pointed out.

2 WORKING PRINCIPLE

2.1 *Adaptive structure*

Now there are mature technologies to realize the type B residual current protection device down to 30 mA. e.g., manufacturers like Siemens have this kind of products. Wherein, double magnetic core structure is applied in (Siemens, AG, 2012). The typical structure is shown in Figure 1. The summation current transformer W1 monitors the AC and pulse current-type residual currents. The summation current transformer W2 detects the smooth DC residual currents and, in the event of a fault,

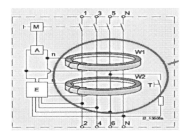

Figure 1. Typical structure diagram of existing RCCB.

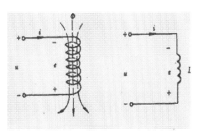

Figure 2. Equivalent circuit.

relays the tripping command through electronic unit E to release A, which uses mechanics to disconnect the circuit.

Because of the size limit of magnetic core and only usage of W1, very small residual current, e.g., 10 mA cannot provide enough energy for residual current measurement and trip energy at the same time at voltage independent working mode.

The key point is the power which can be got from the current transformer. Theoretically, as known in Figure 2, the possible energy is from the inductor of the primary side,

$$\psi = N_1 \phi \qquad (1)$$

$$L = \frac{\psi}{i} \qquad (2)$$

$$e = -N_1 \frac{d\phi}{dt} = -\frac{d\psi}{dt} \qquad (3)$$

$$e = -L \frac{di}{dt} \qquad (4)$$

$$u = -e \qquad (5)$$

$$P_1 = u \cdot i = 4.44 N_1 \cdot f \cdot B \cdot S \cdot I_1 \qquad (6)$$

where Ψ = flux linkage; Φ = magnetic flux; B = magnetic strength in the core; f = frequency of the primary current; N_1 = turns of the winding; and S = cross-section area.

With formula (6), the possible energy that can be got from the current transformer can be calculated.

Considering the practical dimension of the existing RCCB, one adaptive structure is proposed to enlarge the possible energy from the primary side by fully taking advantage of the two cores, as shown in the Figure 3.

The high sensitive residual current protection apparatus is possible to be realized based on the structure,

1. One voltage independent first residual current detection part and one voltage dependent second residual current detection part. The first residual current transformer and the second residual current transformer. The first electronic control switch and the second electronic control switch.
2. When working at voltage independent mode, i.e., the line voltage from phase A, B, C is not introduced into the apparatus, Under control of the electronic control switch 1 and 2, the first and second residual current transformers work in serial mode. The induced residual current signal by the two residual current transformers is injected to the first residual current detection part. The first residual current detection part processes the signal and sends out trip signal to trip component.
3. When working at voltage dependent mode, i.e., the line voltage from phase A, B, C is introduced into the apparatus. Under control of the electronic control switch 1 and switch 2, the first and second residual current transformers work respectively. The signal from the first residual current transformer is injected to the magnetic modulation circuit in the second residual current detection part. Then the DC and low frequency residual current can be detected. The signal from the second residual current transformer is injected to the filter and amplifier circuit in the second residual current detection part. Then the high frequency residual current can be detected. The judgment circuit in the second residual current detection part processes the signals from the magnetic modulation circuit and the filter & amplifier circuit. Then sending a trip signal to the trip component.

With above adaptive structure, the required functionality in residual current protection device

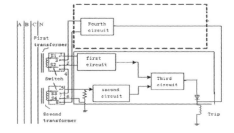

Figure 3. Proposed adaptive structure.

can be assured with providing maximum energy output at the same time, even under the voltage-independent condition.

2.2 Energy harvesting based on the resonant circuit in voltage independent condition

To realize the energy harvesting in the condition of voltage independent, the Figure 4 shows the conventional circuit in the existing protection device to trip maglatch. The induced secondary current acts as a current source, whose energy will be conducted to a storage capacitor. Obviously, when the leakage current is small, e.g.,10 mA, the induced current will be small which can't provide enough required energy.

The energy in the storage is used to trip the maglatch, thus, the high voltage is required. To ensure the enough voltage on the capacitor, a smart solution based on resonant circuit is applied, as shown in Figure 5.

For a current transformer, the magnetic potential in the core keeps the below relationship of formula (7)

$$N_1 I_1 = N_1 I_m + N_2 I_2 \tag{7}$$

where I_m = magnetizing current in the core; N_2 = turn number of the secondary winding; I_1 = leakage current as the primary current; and I_2 = induced secondary current.

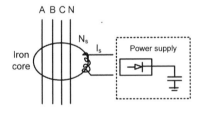

Figure 4.   Schematic of existing power supply.

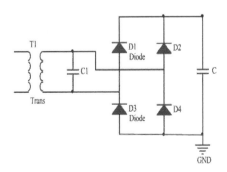

Figure 5.   Energy harvesting based on resonant circuit.

The Figure 6(a) shows the theoretical analysis of the above vectorial relationships when there is no the resonant capacitor C_1, and the Figure 6(b) shows the condition when there is a resonant capacitor. It can be found when the capacitor is added to parallel with the secondary winding, the phase of the induced current I_2 will be ahead a step due to the capacitive reactance of the capacitor. In this way, the induced E2 will be increased due to the increased magnetizing current.

For a resonant circuit, the value of the capacitor is critical. To determine the optimum value of the capacitor, both simulation and test had been conducted. The Figure 7 shows the simulation results about the output voltage when varying the input residual current and the value of paralleled capacitor.

The test configuration with two cores is shown in Figure 8, which is based on the serial connection of the two cores according to the proposed adaptive structure.

Table 1 shows the test result when the capacitor is 330 nF.

Both the simulation results and test results comply with the resonant formula (8)

$$\omega^2 = \frac{1}{L \cdot C} \tag{8}$$

where ω = frequency of the residual current; L = inductance of the secondary winding; and C = paralleled capacitance in the circuit.

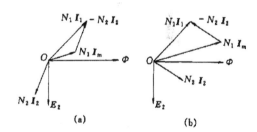

(a)   (b)

Figure 6.   Vectorial analysis wo/w resonant capacitor.

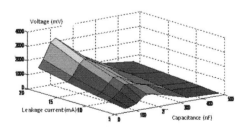

Figure 7.   Simulation for optimizing the capacitance.

Figure 8. Serial connections of the two cores.

Table 1. Test results with 330 nF.

Leakage current (mA)	5	10	15	20	25		30	100
Induced voltage (V)	1.42	3.42	4.4	5.08	5.75 V	6.8	5.9

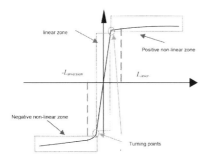

Figure 9. Control of the turning point in the B-H curve.

2.3 An accurate DC sensing based on the magnetic modulation with double feedbacks

There are a lot of methods to detect direct residual current, such as technology based on hall sensor, fiber, and so on (Xu, 2009). However, considering the reliability and cost, the magnetic modulation method is applied in the residual current protection device. As noted in the paper (Min, 2011), the proposed magnetic modulation method with double feedbacks can have better sensing accuracy than the classical magnetic modulation, which can strictly control the inversion points as shown in the Figure 9. Thus, the technology is applied to sense the small direct current when the protection device is in the voltage-dependent condition.

By smartly taking advantage of the feedbacks of the induced current and its derivative, the solution can ensure low power consumption and measurement accuracy at the same time. More theoretical analysis can be found in the paper (Min, 2011).

3 VERIFICATION

For a type B residual current protection device, the function test items in the IEC60947-2 should be

Table 2. Test result of alternate residual current.

Trip value (mA)				
Frequency	50 Hz		1 kHz	
Waveform	Test results	Required in standard	Test results	Required in standard
Full wave	7.93	5–10	82.5	10–140
Half wave	7.75	3.5–20	No requirement	
90° wave	9.21	2.5–20	No requirement	
135° wave	12.19	1.1–20		

Table 3. Test result of direct residual current.

Trip value (mA)		
	Test results	Required in standard
DC	7.65	5–20

passed. The proposed solutions had been implemented in the evaluator. Per to the below test results in Table 2 and Table 3, the high sensitive protection apparatus can realize the protection even under a residual current of 10 mA.

4 SUMMARY

With the proposed three solutions, the two critical challenges had been overcome successfully to realize a high sensitive type B residual current protection device, which can fully pass the function test in the related standard. The future work is on the research of extending the protection range down to 6 mA, which is also mentioned in the standard.

REFERENCES

Czapp, S, 2008. The impact of higher-order harmonics on tripping of *residual current* devices. The13th International Power Electronics and Motion Control Conference.
IEC, 2013. IEC. Low-voltage switchgear and controlgear Circuit breakers (IEC 60947-2).
Min, Y.Z. 2011. An AC/DC sensing method based on adaptive magnetic modulation technology with double feedback properties. IEEE International Workshop on Applied Measurements for Power Systems.
Siemens AG, 2012. Residual Current Protective Devices Technology Primer.
Xu, Z.L, 2009. Type B RCD with a simplified magnetic modulation/demodulation method, The 6th IEEE Power Electronics and Motion Control Conference.

Advances in Power and Energy Engineering – Sun (Ed.)
© 2016 Taylor & Francis Group, London, ISBN 978-1-138-02846-3

The arc model of single phase grounding fault and EMTP/ATP simulation research

Y. Yang
Electric Power College, Inner Mongolia University of Technology, Hohhot, Inner Mongolia, China

C.M. Li
College of Information Engineering, Inner Mongolia University of Technology, Hohhot, Inner Mongolia, China

ABSTRACT: In this paper, a new algorithm of arc modeling of single phase grounding fault and simulation research is proposed. The logic judgments transfer function models in the ATP/EMTP are used to establish an accurate control blocks model of dynamic characteristics of primary and secondary arc. And to apply the DB2 discrete wavelet transform to arc fault signal strangeness detection. A large number of simulation experiments show that this method can be used to extract high-frequency component of fault arc signal effectively, detect all the abnormal information of arc signal quickly and identify the arc quenching time accurately.

1   INTRODUCTION

Transmission lines within electrical power systems tend to be the weak link, accounting for most faults within the system. According to the operating experience, more than 90% of the EHV transmission line faults are single-phase grounding fault, and among them, more than 80% are transient faults.

Generally, the reclosing action is the most commonly used method to solve line faults. However, a blind reclosing action to permanent faults or reclosing the line before the secondary arc quenched would have a huge impact on power system stability, and cause serious damage to operating equipment (Song Guobing et al.2006). Therefore, prior identification of arc quenching time and fault properties can reduce the impact on the system, as well as providing insight to the operating system safety and stability.

Recently, scholars locally and abroad have done a great deal of researches on adaptive auto-reclosing, and the adaptive single-phase reclosing has been improved.

The typical voltage criterion method, utilizes capacitive coupling voltage characteristics to distinguish the fault types in the recovery voltage of the transient fault of the, although its sensitivity and resistance to transition resistance ability at the cost of the line for length. Additionally, it is not suitable for long EHV transmission line with shunt reactor. In modern EHV large capacity power system, it is imperative to install a certain number of shunt reactors. Installing shunt reactor in the EHV transmission line enables the compensation of the line capacitance in normal operation as well as the installation of neutral small reactance is mainly to compensate capacitive coupling, resulting in the power arc and avoiding resonance overvoltage (Nagy I. EIkalashy & Hatem A. Darwish 2007). The application of the reactor greatly modifies many characteristics of the electric parameters of the faulty transmission line, it makes them wrong what are application of capacitive coupling is used to identify the fault voltage of typical voltage criterion method (Nian Peixin & Luo Shihuang 2000), on this basis, a variety of voltage compensation criterion is proposed (Cheng Ling & Xu Yuqing 2007) and the fault phase recovery voltage amplitude phase will be based on the fault phase voltage amplitude phase recovery (Zhou Niancheng 2008) and so on.

The following paper will propose a series of measures to solve the aforementioned issues, first by choosing the best neutral small reactor and simulating the 500 kV EHV transmission lines with shunt reactors by establishing arc models rely on the ATP/EMTP software. Through the analysis of the simulation results, the arc extinguishing time will be confirmed, so as to provide theoretical support to further improvement for the Single-phase adaptive re-close EHV transmission line performance.

2   THE CHOICE OF BEST NEUTRAL REACTOR

A variety of solutions have been proposed in related literatures locally and abroad, naming accelerating arc extinguishing as the solution. The most common way to increase the probability of success on reclosing is to install neutral point reactors in the lines with shun reactors (Liu Hao-fang & Wang Zengping 2006). By taking advantage of neutral reactor optimal choice, arc current can be inhibited and arc will be extinguished immediately.

Figure 1 is the image for double side EHV transmission equivalent circuit with shunt reactor. In this situation, shunt reactor and neutral reactor respectively were converted to the compensation of capacitive coupling C_m and ground capacitance C_0 as L_m and L_0. After transformation, after the transient fault disappears, before reclosing, the fault phase (assuming A phase as the fault phase) of equivalent circuit is shown in Figure 2.

Figure 2: C_0 = line ground capacitance, and C_m = the capacitive coupling, The inductances of shunt reactors are L_1, L_2 & $L_1 = L_2$. The neutral small reactor L_{n1}, L_{n2} & $L_{n1} = L_{n2}$. Take A phase for instance, in the case of being no paralleling reactance compensation, $X_C = X_{C0}$, $X_m = X_{Cm}$. For lines with shunt reactor compensation,

$$L = L_0 + 3L_n \tag{1}$$

$$L_m = \frac{L_1}{L_{n1}}(L_1 + 3L_{n1}) \tag{2}$$

$$X_0 = \frac{X_{L0}X_{C0}}{X_{L0} - X_{C0}} \tag{3}$$

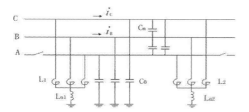

Figure 1. The equivalent circuit of transmission line double-phase operation.

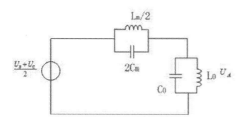

Figure 2. Laplacian equivalent circuit of recovery voltage.

$$X_m = \frac{X_{Lm}X_{Cm}}{X_{Lm} - X_{Cm}} \tag{4}$$

X_{Lm}, X_{L0} respectively L_m, L_0 are the respective corresponding impedance for X_{Lm}, X_{L0}. where L = three phase parallel compensation inductance; L_n = a neutral small reactor; F = compensation degree; C_1 = a positive sequence capacitance and, C_0 = zero sequence capacitance; L_m = equivalent mutual inductance; L_0 = to equivalent capacitance to earth.

What we can know after analyzing Figure 2 is that, when the value of $L_n = L_m/2$, it will form a parallel resonance with $2C_m$. There is no current in the circuit, and the capacitive coupling voltage $U_A = 0$. This is the best time to extinguish the arc quickly.

If the reactance value were designed according to the principle of capacitive coupling full compensation (Liu Haofang 2007)

$$L_m = 1/(\omega^2 C_m) \tag{5}$$

The proper value for neutral reactor can be calculated through the formula above as follows:

$$L_m = \frac{C_1 - C_0}{3F\omega^2 C_1[C_0 - (1 - F)C_1]} \tag{6}$$

Currently, EHV lines installed with shunt reactors are usually loading in owed compensation mode, which means that, commonly, the shunt compensation degrees F will be 60%~80% in 500 kV EHV transmission lines.

3 TRANSIENT FAULT ARC MODEL

When the transient fault appears, the arc fault can be divided into two stages: the primary arc stage according to the size of the arc current: the primary arc stage starts the moment when fault appears to the moment when circuit breaker is broken. The secondary arc stage will continue to the time when arc fault is extinguished.

3.1 Primary arc stage

At the primary arc stage, the circuit breaker is still connected on both ends of the line, so the arc will maintain power and the current is larger. When such a large current goes through the fault point, the arc column would not be prolonged significantly, but there will be a larger cross section. Therefore, the arc will not be easily affected by outside conditions, such as wind etc. (Han Xiao et al 2005).

The primary arc characteristics simulation expression is Formula 7:

$$\frac{dg_p}{dt} = (G_p - g_p)/T_p \quad (7)$$

where G_p = the primary arc Time-varying conductance. When the external conditions remain in force, and the arc current could keep for a relative long time, g_p = the conductance of the arc; T_p = the primary arc time constant, which is the voltage rising in the process of reaching a steady-state voltage, usually it is a fixed value.

The computational expressions: $T_P = \alpha I_P/l_p$; where I_P = a primary arc value; l_p = the primary arc length of primary arc, a fixed value; scale coefficient $\alpha \approx 2.85 \times 10^5$.

The computational expressions: $G_P = |i|/V_p l_P$; where V_P = a voltage per unit length of primary arc, It's size is associated with the arc current in the cycle; and the value is approximately equal in the same period.

3.2 Secondary arc stage

Compared with the primary arc current, the secondary arc current and the arc cross-sectional area are much smaller. The biggest difference between the primary arc and the secondary arc is the combustion power. The secondary arc is greatly influenced by external factors, such as atmospheric environment, magnetic field force etc. The secondary arc is a process of burns, burns out, restrict, and extinction, which is repeated until the arc voltage is no longer greater than the arc voltage to reignite, then the secondary arc is extinguished (He Bona & Zhao Yunwei 2013).

Through a lot of experimental analysis, the secondary arc dynamic characteristics are shown as following constraints:

$$G_S = \frac{|i|}{V_S L_S(t_r)} \quad (8)$$

$$T_S = \frac{\beta I_S^{1.4}}{V_S L_S(t_r)} \quad (9)$$

$$V_S = 0.75 I_S^{0.4} \quad (10)$$

where g_s = secondary arc conductivity changing with time, G_S = the secondary arc steady conductance, T_S = time constant, L_S = arc length, I_S = the secondary arc current peak, β = a proportionality coefficient, V_S = a voltage per unit length of secondary arc t_r = secondary arc time, i = the absolute value of the arc current.

Through a variety of simulations with different voltage grades, length of lines, and different sizes of arc current, we can get:

$$V_{r(t_r)} = \left[5 + \frac{1620 T_e}{2.15 + I_S}\right](t_r - T_e)h(t_r - T_e) \times 10^3 \quad (11)$$

$$h(t_r - T_e) = \begin{cases} 0 & (t_r \langle T_e) \\ 1 & (t_r \rangle T_e) \end{cases} \quad (12)$$

$$V_r = |V_r(t_r)| \times l_S(t_r \quad (13)$$

where $V_{r(tr)}$: Potential gradient of arc re-breakdown; T_e: Arc produced to extinguished of time; V_r: the voltage of restrike arc.

With the time goes and the arc length increases, the arc restrict voltage rises. The arc will be completely extinguished when the arc voltage is no longer higher than the restrict voltage.

Identifying the characteristics of the secondary arc can help to diagnosis the character of fault and detect the arc extinguishing time, which is important when deciding the optimal reclining time of reclosing device.

4 ATP SIMULATION

By taking the line model from related literature for reference, the line model reference related literature (Jeong-Yong Heo 2012 & Shang Liqun 2008). The model selects a 500 kV, 360 km EHV line system. As shown in Figure 3. The equivalent parameters for double power supply system:

$X_{1m} = 92.72\ \Omega$, $X_{0m} = 47.49\ \Omega$;
$X_{1n} = 148.09\ \Omega$, $X_{0n} = 45.95\ \Omega$.

Line phase conductors: $4 \times$ JL/G1A-400/50-54/7, Line parameter:

$R_1 = 0.027\ \Omega/km$, $R_0 = 0.1974\ \Omega/km$;
$L_1 = 0.965\ mH/km$, $L_0 = 2.21\ mH/km$;
$C_1 = 0.013\ \mu F/km$, $C_0 = 0.0083\ \mu F/km$.

So the compensation degree $F = 86.52\%$, Paralleling reactance: $X_L = 2000\ \Omega$, Neutral small reactor $X_{Ln} = 650\Omega$.

In this paper, EMTP/ATP are used to carry out the digital simulation, as well as, MODELS and TACS model being used to set up an arc model.

Experimentation: A phase ground fault occurs on line, as $t_1 = 50\ ms$ line down, into the secondary

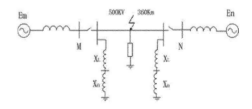

Figure 3.   500 kV transmission line system.

657

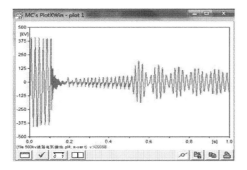

a: transient fault grounding fault phase voltage

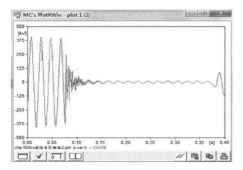

b: permanent fault grounding fault phase voltage

Figure 4. The phase voltage difference between transient fault and permanent faults.

arc stage, as $t_2 = 100$ *ms*, the circuit breakers open on both ends of the line, under the logical switch control, the primary arc model disconnect, into secondary arc stage, as $t_3 = 600$ *ms* the arc completely extinguished, the secondary arc model is broken.

Through comparison, it is clear that the fault phase voltage is affected by arc characteristics, and also it can be clearly identified. While the arc is burning, the nonlinear characteristic of arc causes the fault phase voltage distortion. Fault phase voltage will not be stabilizing until the arc is extinguished.

As Figure 4a,b show: in primary arc stage and secondary arc stage, the arc voltage was not decreased, but increased gradually, until the arc extinguishing. When the actual voltage is not larger than the voltage to renew, the arc will be completely extinguished.

5 DETECTION OF ARC FAULT SIGNAL SINGULARITY RESEARCH BASED ON WAVELET TRANSFORMATION

In this paper, Daubechies wavelet analysis was used to extract the fault phase of high frequency signals. By using the Db wavelet, signals can be divided

into a low frequency signal and a high frequency signal, which fully reflects the characteristics of high frequency information and low fault information we need.

As the EHV line system shows in the figure. Terminal voltage signal is sampled at a transient fault and permanent fault of the fault phase.

Sampling period is 0~1.5s, sampling frequency is 20 kHz, and sampling 1000 points in one cycle. Db2 divides detail coefficients, and the results of the 2^1, 2^2, 2^3, 2^4 are show in Figure 5, 6.

When the transient faults appear, the fault phase voltage is impacted by the non-linear resistance of arc, where the waveform will show as much uninterruptible mutation point existing in wavelet decomposition. It is can be seen from the results of the first layer that the waveform mutation is clear when the fault occurs, the time when fault phase breaker open and the moment when arc extinguish. However, a permanent fault appears, at two

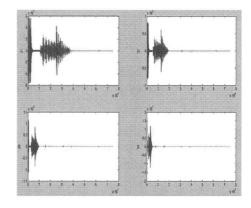

Figure 5. Db2 2^1, 2^2, 2^3, 2^4 scale detail coefficients on transient fault.

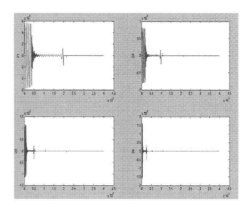

Figure 6. Db2 2^1, 2^2, 2^3, 2^4 scale detail coefficients on permanent fault.

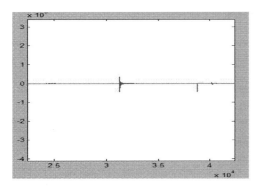

Figure 7. Positioning arc quenching time.

distinct point mutations when fault appears and the moment when fault-phase circuit breaker open.

After the wavelet transform, wavelet coefficients fully reflect the mutation information of the original signal, and the variation of original signal can be obtained by analyzing the wavelet coefficients. Therefore, this method can clearly detect distortion the corresponding arc characteristics.

As is shown in Figure 7, when the arc extinguished, there would be a large frequency distortion of the fault phase voltage. We identified three larger singular points, which the time can roughly determine the arc is extinguished in about 620 ms.

6 CONCLUSION

In this paper, we combined the research results of arc fault locally and overseas. According to the experience of the arc mathematical equations, we built a primary arc model and a secondary arc model to analyze their dynamic characteristics.

In the process of setting up the arc model, by using the ATP/EMTP logic judgment, we established the expression model of primary and secondary arc dynamic control block accurately, which vividly simulated the transient process when single-phase transient fault was happened. The simulation result is consistent with the results in reference (Shang Liqun 2008).

All the useful information of fault arc voltage signal can be reflected by the wavelet transformation through the wavelet coefficients. By Db2 wavelet analyze the voltage signal of the fault phase, to extract the high frequency component. Through detecting the singularities of arc fault signal, arc-extinguishing times can be identified more accurately. We were able to increase reliability as well as improve the reaction time of the reclosing action, which played a very important basis in adaptive reclose technology in EHV transmission line.

REFERENCES

Cheng Ling & Xu Yunqing, 2007, Single-phase Adaptive Reclosure of EHV Transmission Line Based on the Arc Characteristic, *Relay*, 35(22):18–21.

Han Xiao etc, 2005, Simulation Research on Protection Criterion for High-voltage Transmission Line Based on Theory of Singularity Detection, *Relay*, 33(11):26–30.

He bo-na & Zhao Yunwei, 2013, Single-phase adaptive reclose for shunt compensated UHV transmission lines fault identification method, *High Voltage Electric*, 49(12),13–18.

Jeong-Yong Heo, 2012, Development of Adaptive Auto-reclosure Algorithm in Transmission Lines, *power Systems Protection*, April 2012.

Liu Haofang & Wang Zengping, 2006, Criterion for Determining Fault Nature in Adaptive Single-phase Reclosing for Shunt Compensated EHV/UHV transmission lines. *Power System Technology*, 30(18):29–34.

Liu Haofang, 2007, New UHV Transmission Line Protection Principle and the Adaptive Reclosing Technology Research, *North China Electric Power University*.

Nagy I. Elkalashy & Hatewn A. Darwish, 2007, An Adaptive Single pole Auto-reclosure Based on Zero Sequence Power, *Electric Power Systems Research*, 77:438–446.

Nian Peixin, 2000, Protection of Fault arc in The Fields of Low Voltage Power Distribution, *Low Voltage Apparatus*: 22–26.

Shang Liqun.2008, Fault Nature Identification for Single-phase Adaptive Reclosure on Transmission Lines with Shunt Reactors, *Automation of Electric Power Systems*, 32(6):81–84.

Song Guobing & Suonanjiale, 2006, A Survey on Methods Distinguish Permanent Faults from Instantaneous Faults in Transmission Lines, *Power System Technology*, 30(18):75–80.

Zhou Niancheng, 2008, Power Spectrum of Fault Phase Voltage Based Single-phase Adaptive Reclosure, *DRPT 2008 Nanjing China*, 1287–1292.

Advances in Power and Energy Engineering – Sun (Ed.)
© 2016 Taylor & Francis Group, London, ISBN 978-1-138-02846-3

Study on electric field distribution of 1000 kV AC double-circuit transmission line Y-type insulator string

F. Huo & T. Xu
China Electric Power Research Institute, Wuhan, China

X.H. Fan, M.T. Wei & C.P. Huang
School of Electrical Engineering, Wuhan University, Wuhan, China

ABSTRACT: The Y-type insulator string has the advantages of external insulation and reduction of corridor width. This paper presents calculation results on the electric field distribution of 1000 kV AC double-circuit transmission line Y-type composite insulator string, using 3D simulations based on the Finite Element Method (FEM). The impact of length of each part, angle of V-part, single and double units on electric field distribution has been analyzed. It is found that the effect is closely related to the distance from conductor to the upper cross arm. The results can give a reference for design of UHV AC Y-type insulator string.

1 INTRODUCTION

In order to meet the development of economy and solve the problem of long distance and bulk capacity power transmission, UHV power transmission was developed quickly in China (Huang, D.C. & Shu, Y.B. 2009). In 2012, a program obtained the National Science and Technology Progress Award grand prize in China, which researched the key technology, the complete set of equipment and engineering application of UHV AC transmission (Liu, Z.Y. 2013). That same year, the first commercial transmission project of the world officially was put into operation, which is the UHV AC double-circuit transmission line on the same tower (from Huainan to Shanghai). Nowadays, how to improve the economy of transmission project while ensuring the reliable operation of line has become an important research direction of the UHV engineering.

The insulator string is an important component of UHV AC transmission line. It plays an important role in the electrical insulation and mechanical support. In the transmission project (from Huainan to Shanghai), a new type insulator string (Y-type) has been used. The Y-type insulator string is a combination of I-type and V-type insulator string, and it has more advantages, for example, it can prevent wind yaw, improve flashover voltages under wet and pollution conditions, improve the height of conductor, and compress the length of cross arm and the corridor width. Hence, it has advantages of external insulation and reduction of corridor width (Liu, Z.Y. 2013 & Fan, J.B. 2007). So far, a lot of studies have been made for conventional I-type

and V-type insulator strings at home and abroad, about the electromagnetic environment, external insulation characteristics, structural properties and so on (IIhan, S., 2011, Huang, D.C. 2013). The study of Y-type insulator string mainly focuses on external insulation test and mechanical performance (Liu, Z.Y. 2013, Fan, J.B. 2007). China Electric Power Research Institute (EPRI) carried out a research about flashover characteristics of Y-type insulator string under polluted condition on DC transmission line. The result shows that the DC polluted flashover voltage of Y-type insulator string is about 9% higher than that of I-type. When the angle of V-part is 110°, the salt of this part loses fast in contamination test. The voltage distribution of Y-type insulator strings for AC/DC 500 kV transmission line is measured by sphere gap method. Hence, it is necessary to carry out researches systematically on grading rings' optimization, string type, external insulation characteristics and insulation configuration.

With the method of FEM, this paper studies the electric field distribution of UHV Y-type insulator string, and analyzed the impact of length of each part, angle of V-part, single and double units on electric field distribution. The results can give a reference on the design of UHV AC Y-type insulator string.

2 MODEL PARAMETERS AND SIMULATION

According to the original design parameters of the Huainan-Shanghai UHV AC transmission line,

UHV AC double-circuit transmission line 1/2 3D electro-static field model was built, including the tower, insulator, grading rings, phase conductors, and other hardware fittings. Figure 1 shows the simulation model.

The tower type is SZY323 which is 115.4 m high. The conductor type is an eight-bundled conductor, which diameter is 33.75 mm and bundled spacing is 400 mm. And the length of the line is 100 m. The total length of the insulator is 9 m, and the I-part is 2 m, V-part is 7 m, and the angle of V-part is 105°. The sheds of composite insulator include a big one with diameter of 180 mm and a small one with diameter of 150 mm. Figure 2 shows the parameters of grading rings, for the racetrack type grading ring R = 800 mm, r = 120 mm, H = 140 mm, and coupling distance is 600 mm, for the small one, h = 20 mm.

All entities are surrounded by two volumes of air. The first layer is rectangular 60 m long, 100 m wide, and 120 m high; the second one is a

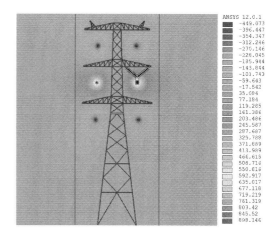

(a) Whole model

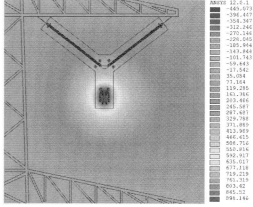

(b) Middle phase

Figure 3. Potential distribution of this model.

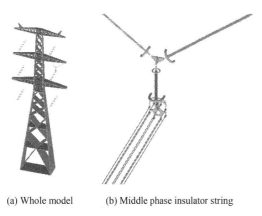

(a) Whole model   (b) Middle phase insulator string

Figure 1. Simulation model.

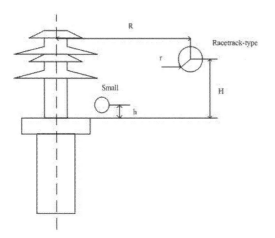

Figure 2. Parameters of grading rings.

1/2 cylinder, 100 m thick, and radius of 200 m. Considering the characteristic of analysis object, two simplifications are made for improving the calculation efficiency as follows: (1) ground wire is ignored for having little impact on electric field distribution and (2) the adjacent phases only build conductors and no insulator for reducing element. The total number of elements are nine million.

According to the actual situation, the middle phase loads maximum peak value Um = 898.146 kV of 1000 kV transmission line phase voltage, which contains conductors and grading rings close to the high voltage end, the other two phase load—Um/2. The connection fittings of I-part and V-part need potential coupling. The tower, ground, the outside air boundary, and the low voltage end load zero.

Figure 3 shows potential distribution of this model. Figure 4 shows the electric field distribution.

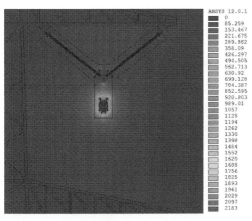

(a) Whole model

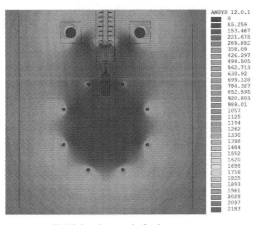

(b) High voltage end of string

Figure 4. Electric field distributions.

The maximum electric field on racetrack grading ring surface is 18.81 kV/cm, while that on composite insulator surface is 5.70 kV/cm.

3 EFFECT OF DIFFERENT FACTORS

3.1 Length of each part

The total length of composite insulator string is 9 m. When the length of each part is changed, the maximum of surface electric field (Emax) on racetrack grading ring and composite insulator are shown in Table 1. As the length of I-part increases, the distance from conductor to the upper cross arm increases, but Emax on racetrack grading ring decreases, Emax on composite insulator decreases too.

The impact of different length of each part to potential and electric field distribution along the insulator string in I-part is shown in Figure 5.

Table 1. Impact of the length of each part.

Y-type insulator string		Distance from conductor to the upper cross arm [m]	The maximum of surface electric field [kV/cm]	
Length of I-part [m]	Length of V-part [m]		Racetrack grading ring	Composite insulator
2	7	8.91	18.81	5.70
3	6	9.61	18.55	5.59
4	5	10.01	18.47	5.50

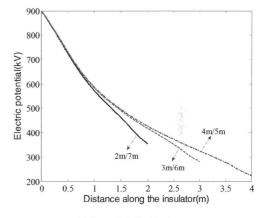

(a) Potential distribution curve

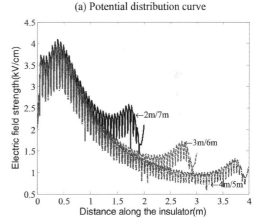

(b) Electric field distribution curve

Figure 5. Impact of the length of each part in I-part.

While length of I-part is shorter, potential distribution curve is steeper than others, and electric field distribution curve is above others. Both potential and electric field distribution curve close to the end of high voltage are similar for different length of each part.

Through the analysis we can infer that, because different length of each part changes the distance from conductor to the upper cross arm, the potential and electric field distribution is changed. The farther the distance from conductor to the upper cross arm is, the lower electric field on the Y-type insulator string is.

3.2 Angle between the two branches of V-part

When the angle between the two branches of V-part is changed, the maximum of surface electric field (Emax) on racetrack grading ring and composite insulator are shown in Table 2. With the increase of between-branches angle, the distance from conductor to the upper cross arm decreases, but Emax on racetrack grading ring increases. Emax on composite insulator increases too.

The impact of different between-branches angle to potential and electric field distribution along the insulator string in I-part is shown in Figure 6. While angle of V-part is bigger, potential distribution curve is above others, and electric field distribution curve is lower than others. Both potential and electric field distribution curve close to the end of high voltage are similar for between-branches angle.

Through the analysis we can infer that because between-branches angle changes the distance from conductor to the upper cross arm, the potential and electric field distribution is changed. The impact of distance from conductor to the upper cross arm is same as above.

3.3 Single and double units

The model of double units is similar to single units, only grading ring of I-part is different, the big grading ring is a circle one with R = 1400 mm, r = 120 mm, and H = 140 mm. Figure 7 shows the model.

Figure 8 shows the electric field distribution curve along the middle of insulator string in I-part

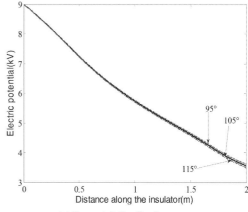

(a) Potential distribution curve

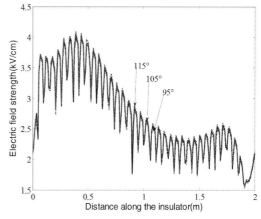

(b) Electric field distribution curve

Figure 6. Potential distribution curve by the impact of branches angle.

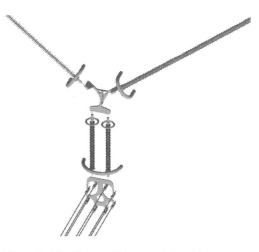

Figure 7. Double units Y-type insulator string.

Table 2. Result under different between-branches angle.

Angle between-branches	Distance from conductor to the upper cross arm [m]	The maximum of surface electric field [kV/cm]	
		Racetrack grading ring	Composite insulator
95°	9.51	18.72	5.65
105°	8.91	18.81	5.70
115°	8.51	18.97	5.75

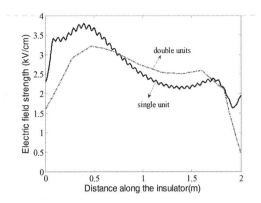

Figure 8. Impact of double and single unit on electric field distribution in I-part.

of single and double units. Compared with double units, the electric field distribution curve of single unit is higher in both ends of I-part, and lower in the middle of I-part. Double units can make electric field distribute evenly.

4 CONCLUSION

A 3D simulation model of UHV Y-type insulator was modeled by the author, and with the use of FEM method, the results of the electric field distribution were presented in this paper. The impact of length of each part, angle of V-part, single and double units on electric field distribution has been analyzed.

Both lengths of each part and the angle between-branches have an influence on the distribution of electric field by changing the distance from conductor to the upper cross arm. The farther the distance from conductor to the upper cross arm is, the lower electric field on the Y-type insulator string is. The original parameter is a better choice, I-part of 2 m length, between branches angle of 105°. And the double units can make electric field distribute more evenly.

REFERENCES

Fan, J.B. & Wu, X. 2007. Research on voltage distribution of Y-type insulator strings under AC and DC condition, Chinese Power System Technology, vol. 31, no. 14, pp. 6–9.

Huang, D.C., & Shu, Y.B. 2009. Ultra high voltage transmission in China: developments, current status and future prospects, Proceedings of the IEEE, vol. 97, no. 3, pp. 555–583, 2009.

Huang, D.C. & Xie, X.G. 2013. Grading ring parameter design and corona characteristic test arrangement of 1000 kV AC compact transmission line, Chinese High Voltage Engineering, vol. 39, no. 12, pp. 2933–2942.

IIhan, S. & Ozdemir, A. 2011. 380 kV Corona ring optimization for ac voltage. IEEE Transactions on Dielectrics and Electrical Insulation, vol. 18, no. 2, pp. 410–417.

Liu, Z.Y. 2013. Innovation of UHVAC transmission technology in China, Chinese Power System Technology, vol. 37, no. 3, pp. T1–T8.

Liu, B. & Liu, C.L. 2013. A study on swing and buckling of Y-type insulator strings under strong wind action for ±800 kV UHDV transmission lines. 2013 IEEE Power and Energy Engineering Conference, pp. 1–4.

Advances in Power and Energy Engineering – Sun (Ed.)
© 2016 Taylor & Francis Group, London, ISBN 978-1-138-02846-3

Study on the power grid structural scheme of the Ultra-High Voltage power delivery system used in large-scale thermal power bases

P. Ji
China Electric Power Research Institute, Beijing, China

J.X. Liu
State Grid Shanxi Electric Power Company, Taiyuan, China

H.T. Yang
China Electric Power Research Institute, Beijing, China

W. Tang
State Grid Anhui Electric Power Company, Hefei, China

Y.T. Song & C. Zheng
China Electric Power Research Institute, Beijing, China

L.N. Zhang
State Grid Shanxi Electric Power Company, Taiyuan, China

X.F. Hu
State Grid Anhui Electric Power Company, Hefei, China

A. Wang
State Grid Shanxi Electric Power Company, Taiyuan, China

J.J. Wang
State Grid Anhui Electric Power Company, Hefei, China

ABSTRACT: China's coal resources are relatively abundant, and the large coal bases have a strong demand for power delivery. The transmission capacity of the 500-KV lines may not meet the requirements of the power delivery due to the stability of the power grid, the load level, and the transmission distance. Through research, a kind of structure of the power delivery system was determined. In the structure, generators closer to the load center of the sending power system can be pooled to one 500-KV substation and then transformed to a 1000-KV voltage level, generators farther from the load center can be transformed to a 1000-KV voltage level, and the midway 1000-KV stations have direct access power generating units to provide adequate voltage support. The results show that the studied power grid structural scheme has obvious application advantages, sending the output of newly installed generators to the receiving power system by Ultra-High Voltage (UHV) transmission lines, reducing the power supply pressure of 500-KV transmission channel, and not significantly increasing short-circuit current levels in 500-KV power grids.

1 INTRODUCTION

China's energy resource distribution is very uneven. Nearly two-thirds of exploitable hydropower is distributed in Sichuan, Yunnan, and Tibet provinces (area) in western China. Nearly two-thirds of the coal reserves are concentrated in Shanxi, Inner Mongolia, and Shaanxi provinces (area) in northern China. Large hydropower bases and coal bases are formed (Bai Jianhua 2007, Song Yunting 2009 & Zhou Jianping 2011). Distribution of energy resources and economic development in China is in reverse, which determines the necessity of a large scale flow of energy and power.

As the large-scale power base is a long distance away from the load center, all countries have adopted the Extra-High Voltage (EHV) or Ultra-High Voltage (UHV) transmission mode in order to improve the transmission efficiency, reduce investment, reduce the transmission corridor, and improve the transmission reliability. For example, in addition to a small part of the power provided by the Itaipu hydropower station generators in Paraguay supplying Paraguay, the rest of the power is transferred to Brazil by a DC transmission system. The power provided by the generators in Brazil is transferred to the load center in the East and South of Brazil by three transmission lines on 765-kV voltage level.

China's Three Gorges Power station is accessed to DC Converter Station through the 500-KV AC lines and complete power transmission by the DC transmission system. The two-beach power plant is connected to the Sichuan AC system by three transmission lines of 500 KV. Huainan coal base power plants are connected to the power grid with 500-KV and 220-KV voltage levels. The Yangcheng power plant is in the mode of specialized power plant, specialized transmission lines, and specialized power supply to transfer electricity to the Huaian side of Jiangsu Province (Song Yunting 2009).

UHV transmission line has the characteristics of high voltage, long distance transmission, and large transmission capacity. These advantages are accompanied by the UHV transmission lines putting into operation that increased transmission channel, strengthened power grid structure, greatly improved stability of transmission section and transmission capability, ensured regional security power supply, and higher ability to optimize the allocation of resources in power grid. Moreover, when the power grid fails, the UHV power grid can enhance the transfer capability of power flow, and the ability to resist multiple serious failures of the power grid will be significantly enhanced. The UHV power grid is an important and effective way to use clean energy to prevent air pollution. It is helpful for the country to change energy utilization mode and transfer to the strategy of "clean energy as the main and fossil energy as supplement" (Chen Hao 2014).

2 LARGE-SCALE COAL BASE DEVELOPMENT AND PROBLEMS IN ACCESSING THE POWER SYSTEM

2.1 Amount of development in large-scale coal bases

In literature (Bai Jianhua 2007), the optimization model was built by using the minimum total cost of power supply in planning period as the goal, using coal supply capacity, water supply, environmental loading, transport capacity, and market space as the constraint conditions, and carrying out optimized matching according to the principle

Table 1. The power transmission scale and target market of each coal—powerbase in 2020 (MW).

Coal bases	Receiving system			
	Beijing Tianjin Hebei and Shandong provinces	Four provinces in Central and Eastern China	Eastern China	Total
Northern Shanxi	10250	—	—	10250
Central Shanxi	900	—	5800	6700
Southeastern Shanxi	1200	3000	22100	26300
Northern Shaanxi	3600	—	13600	17200
Eastern Ningxia	1500	9500	4000	15000
West Inner Mongolia	28000	7000	6600	41600
Xilin Gol League	21600	—	—	21600
Hulun Buir League	14000	—	—	14000
Hami	—	7000	—	7000
Total	81050	26500	52100	159650

of maximum transmission efficiency. Based on this model, a reasonable coal transmission scale solved by optimal solution for nine coal bases in 2020 was determined, namely Northern Shanxi, Central Shanxi, Southeastern Shanxi, Northern Shaanxi, Eastern Ningxia, West Inner Mongolia, Xilin Gol League, Hulun Buir League, and Hami, as shown in Table 1 above. In 2020, power supply size is about 160,000 MW.

2.2 Problems in accessing the power system

Due to the distribution constraints of primary energy, the conventional power supply is developing in the direction of layout centralization for the power plant and large capacity of single units. At the same time, with the rapid growth of electricity demand, the scale of power grids is increasing rapidly. Intensive high voltage transmission power grid makes the power grid electrical distance shorter, decreases the system impedance, and the short circuit current is increased year by year, which have caused bus short-circuit current to exceed standards in some substations. These problems have become a bottleneck restricting the development of the power system. Large capacity power supply accessing the power system will increase the short circuit current level of the buses near the accessing bus (Luo Tao 2015, Wang Hai-qian 2014, Wang Hua 2009, Yao Yingbei 2014 & Zhou Ji-an 2012), which will inevitably lead to the risk of short-circuit current exceeding the switch rupturing capacity in the sending systems.

The transmission capacity of the 500-KV lines will be limited by the stability level of the network, load level, and transportation distance (Kang Hongqiu 2008). The transmission capacity of the 500-KV transmission line is mainly determined by the stability limit of the line. The lack of adequate power supply support usually makes the stability limit of the long distance transmission line low and it will also restrict the transmission capacity of the 500-KV line. Therefore, for large-scale coal bases, large capacity, long-distance transportation, and the use of 500-KV lines may not meet the needs of transportation.

3 SCHEME OF LARGE CAPACITY UNITS CONNECTING TO THE UHV POWER GRID

3.1 All large capacity generators directly connected to the 1000-KV power grid

In the power industry standard of the People's Republic of China "Guide on Security and Stability for Power System," published in 2001, the main power plants should be directly connected to the highest level of voltage power grid (Power industry standard of the People's Republic of China 2001). But in the UHV power grid, the scheme of connecting the main power plants to power grids according to the actual situation is needed. The cost of each single-phase 1000-KV transformer is relatively high. If all of the new-installed large capacity units are directly connected to the 1000-KV voltage level, the investment only for transformer equipment will be considerable, which is not reasonable from the aspect of the invest economy. Therefore, after UHV power grids are completed and put into operation, the scheme of new main power plants accessed to power system should be determined combined with the actual situation, and not have all generators directly connected to the 1000-KV power grid.

3.2 Scheme of partial large capacity generating units having directly access to UHV

As described in the Section 3.1, after UHV power grid was completed and put into operation, the scheme of new main power plants accessed to power system should be determined combined with the actual situation. This section will discuss the determination principle of power plants connected to power grid to achieve economy, reliability, and operability.

In this paper, the principle of power plants connected to the power grid is as follows:

1. Generators closer to the load center of the sending power system can be pooled to one 500-KV substation and then transformed to the 1000-KV voltage level, which can reduce the network loss, as shown in Figure 1. A, B, C, and D, which are 1000-KV substations where A and B are the sending system power stations, C is the midway substation, D is the receiving system power station, A_1 and A_2 for 500 KV substations, G_1-G_9 for the power plants, T_1-T_9 for the transformers, and L for the load in receiving system. Apparently, G_1-G_5 are pooled to the 500-KV substations A_1 and A_2 and then transformed to 1000-KV substation A.

2. Generators further from the load center can be directly transformed to 1000-KV voltage level, and the midway 1000-KV stations have direct-access power generating units to provide adequate voltage support, as shown in Figure 2. Apparently, G_1-G_5 directly access the 1000-KV voltage level, and in the midway stations B and C, there are direct-access power generating units.

3. The proportion of units connected to power grid in (1) and (2) is determined according to the following principles: the demand of the receiving system and the power situation of the sending system.

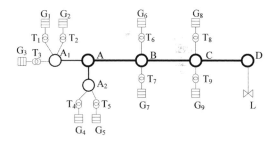

Figure 1. Generators pooled to one 500-KV substation and then transformed to the 1000-KV voltage level.

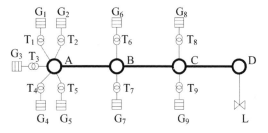

Figure 2. Generators directly transformed to the 1000-KV voltage level.

Based on the analysis of the Section 2 and the Section 3, this section presents power grid structural scheme of the ultra-high voltage power delivery system used in the large scale thermal power bases, as shown in Figure 3. In this structure, A-H are 1000-KV substations, and both A and B, as well as E in sending systems, C and F midway substations, D and H in receiving systems; A_1, B_1 and C_1 are 500-KV substations, G_1-G_8 are generators, T_1-T_8 are transformers, and L_1 and L_2 are load in receiving systems.

The structure has the following characteristics:

1. G_1 and G_2 of the power supply system are directly connected to the 1000-KV network, and the other units are pooled to 500-KV substations and then transformed to the 1000-KV voltage level, which can not only meet the demand of power supply, but also ensure that the short-circuit current of the 500-KV power grid in the sending system is maintained at a reasonable level, and there will not be a problem of short circuit current.

2. B and E are interconnected in the system, and the stability and reliability of the power system are significantly improved compared with the two substations not interconnected. This connecting scheme will also have a beneficial effect on the transmission capability of the 500-KV transmission channel in the sending system.

5.1 *Case introduction*

In this paper, the power transmission system of a large-scale coal power base is studied. The thermal base is expected to increase new installed generators with a total capacity of 3000 MW in 2017. All of this new increased power output will be transmitted to the receiving systems. Currently, there are two transmission channels at the 500-KV voltage level in this sending system.

5.2 *All new-installed capacity connected to the 500-KV power grid scheme*

Figure 4 shows power grid structure diagram that all the new installed capacities are connected to the 500-KV network, and then complete the power transmission through the 500-KV sending channels. Among these generators, G_1, G_2, G_4, G_5, G_6, G_8, and G_{10} are the new units, A_2 and E_1 are new 500-KV stations.

As described in Section 5.1, the current system has two sending channels at the 500-KV voltage level. It is clear that, when adding new power transmission, only using 500-KV power transmission channels may not meet the power delivery demand. Moreover, G_1, G_2 connected to A_1, A_1 will lead to short-circuit current exceeding the rated interrupting capacity of switch (63 kA). The bus

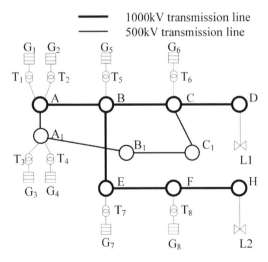

Figure 3. A power grid structural scheme of the ultra-high voltage power delivery system used in the large scale thermal power bases.

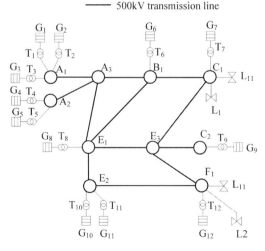

Figure 4. All new-installed capacities connected to the 500-KV power grid scheme.

short-circuit current of the 500-KV electric power grid is listed in Table 2.

5.3 *Scheme of partial large capacity generating units directly accessible to UHV*

In accordance with the large scale coal base UHV power transmission system power grid structure in the Section 4.1, the units G_1, G_2, G_6, and G_{11} have direct access to the UHV substation, and the sending power system structure as shown in Figure 5. On the one hand, this scheme sends the output of new newly installed generators to the receiving power system by ultra-high voltage transmission lines, reducing the power supply pressure of 500-KV transmission channels; on the other hand, it has not significantly increased the short-circuit

Table 2. The bus short-circuit current of the 500-KV electric power grid with all new-installed capacity connected to the 500-KV network.

	Short-circuit current (kA)	
Bus name	No new-installed capacity	All new-installed capacity connected to the 500-KV power grid scheme
A1	59.73	63.34
A3	53.66	59.27
E2	48.39	50.16

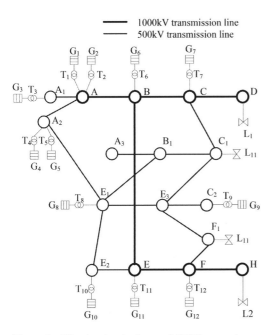

—— 1000kV transmission line
— 500kV transmission line

Figure 5. The structural scheme of UHV power transmission network used in the studied case.

Table 3. The bus short-circuit current of the 500-KV electric power grid with scheme of partial large capacity generating units directly accessed to UHV.

	Short-circuit current (kA)	
Bus name	No new-installed capacity	Scheme of partial large capacity generating units directly accessed to UHV
A1	59.73	59.75
A3	53.66	53.66
E2	48.39	48.41

current level of 500-KV power grids, as shown in Table 3.

5.4 *Contrast in results*

Comparing the calculation results of Sections 5.2 and 5.3, it is found that with the scheme of the newly installed capacities connected to the 500-kV power grid, using only the 500-kV transmission channel may be not able to satisfy the power delivery requirements and will lead to short-circuit current levels over the rated power, interrupting the capacities of switches. By comparison, with the scheme of partial large capacity generating units directly accessed to UHV, the newly installed power supply is mostly transmitted through the UHV transmission line to the receiving system, which reduces the power supply pressure on the 500-kV transmission channel and has no obvious increase in the short-circuit current level of the 500-kV power grid. It is clear that the scheme of partial large capacity generating units directly accessed to UHV has obvious application advantages.

6 CONCLUSION

China's large coal bases have strong demand for power delivery. The transmission capacity of the 500-KV lines may not meet the requirements due to various factors. The results in this paper show: the studied power grid structure has obvious application advantages, in which generators closer to the load center of the sending power system can be pooled to one 500-KV substation and then transformed to the 1000-KV voltage level; generators further from the load center can be transformed to the 1000-KV voltage level, and the midway 1000-KV stations have direct-access power generating units to provide adequate voltage support. The structure can send the output of newly installed generators to the receiving power system by ultrahigh voltage transmission lines, reduce the power supply pressure of 500 KV transmission channel, and not significantly increase short-circuit current levels in the 500-KV power grid.

REFERENCES

Bai Jianhua, Wang Yaohua, Zhang Feng-ying, et al. 2007. Optimizing research on power transmission scale and target market of main coal-power bases. *Electric Power*, 40(12): 10–13 (in Chinese).

Chen Hao, Shen Tugang, Cao Lu, et al. 2014. Construction of the second and third defense lines of UHV AC grid. *East China Electric Power*, 42(12): 2770–2775 (in Chinese).

Kang Hongqiu 2008. *Research on improving transfer capability of 500 KV lines in North China Power Grid by adopting compact transmission technology*. Beijing: Beijing, North China Electric Power University for Master Degree of Engineering: 5–7 (in Chinese).

Luo Tao, Liu Lixia, Wang Kui, et al 2015. Influence of UHV Grid on Short-Circuit Current of Tianjin 500 kV Grid and Its Restrictive Measures. *Electric Power Construction*, 36(8): 79–83 (in Chinese).

Power industry standard of the People's Republic of China 2001. *Guide on Security and Stability for Power System*: 6 (in Chinese).

Wang Hai-qian, Wang Xu, Qi Wan-chun, et al. 2014. Integration of 1 000 MW Coal-fired Generation Unit to 220k V Grid. *East China Electric Power*, 42(7): 1305–1309 (in Chinese).

Song Yunting, Zhao Liang, Qin Xiaohui, et al. 2009. *Research on security and stability, control measures and access patterns of large-scale hydropower or thermal power bases connecting to power systems* [R]. China Electric Power Research Institute: 3–20.(in Chinese).

Wang Hua 2009. *The Study Of Million Units Of Capacity Access To 220k V Power Grid*. Shanghai: Shanghai Jiao Tong University for Master Degree of Engineering: 2–10(in Chinese).

Yao Ying-bei, Miao Yuan-cheng, Zhuang Kan-qin, et al. 2014. 500 KV Short-Circuit Current Control Strategy in the Initial Operation Stage of East China Grid UHV AC Transmission Project. *East China Electric Power*, 42(12): 2775–2778 (in Chinese).

Zhou Ji-an, Jin Dan, Wang Weizhou, et al. 2012. Analysis of influence of huge power supplies grid-connection modes on short-circuit current. *Advanced Technology of Electrical Engineering and Energy*, 31(1): 48–51 (in Chinese).

Zhou Jianping, Qian Gangliang 2011. Planning and development status of the thirteen hydropower bases. *Water Conservancy and Hydropower Construction*, (1): 1–7 (in Chinese).

Advances in Power and Energy Engineering – Sun (Ed.)
© 2016 Taylor & Francis Group, London, ISBN 978-1-138-02846-3

Test research of live working of 1000 kV UHV substation

B. Xiao, K. Liu, T. Wu, T. Liu, Y. Peng, Z.M. Su, P. Tang & X.L. Lei
China Electric Power Research Institute, Wuhan, China

ABSTRACT: Live line maintenance is of importance to maintain the safety and reliability of UHV substations. In order to investigate the discharge characteristics of gaps and safety distances, typical working condition of live maintenance in 1000 kV UHV substation was analyzed in this article initially. Switching impulse discharge tests were conducted on air gaps that between the workers and the ground as well as the conductors and the discharge characteristic curve of typical live working gaps were acquired with analyzing typical working condition. Minimum approach distance of live working, the minimum complex distance of entering and leaving the split busbar and the minimum complex gap distance of entering and leaving hard tube bus by insulation lifting platform were calculated, according to switching overvoltage level of the general designation for substation. Finally, this article checked the risk rate of live working, which provides technical support for the safety of live working in UHV substation.

1 INTRODUCTION

The first China 1000 kV UHV AC demonstration project "Jindongnan-Nanyang-Jingmen" was completed in 2008. Then, "Huai Nan-Shanghai" and "Northern Zhejiang-Fuzhou", etc. were put into operation successively. UHV becomes the backbone of our power grid gradually. UHV grid has the characters of huge transmission capacity and remote transmission distance etc. Since the stability of the whole power grid will be impacted in case of faults, therefore, the safety and reliability are quite important.

Live working is an important technological means for the maintenance of power grid, which has obvious advantages comparing with outage repair. Studying of live working technology is interested all the time in China, and achieves many results in the field of EHV/UHV transmission lines. Reference (Wang et al. 2006, Hu et al. 2006, Shu & Hu 2007) conducts researches on minimum approach distance of live working and safety protection measures in 1000 kV transmission lines, and illuminates the methods for operation and maintenance on UHV AC transmission lines. Spacer replacement and defects elimination of UHV transmission lines from equal potential have been done many times. However, the structures of equipment in UHV substation are much more complex. The gap structure between workers and equipment during maintenance is different from line, so technical parameters of live working of transmission line can't be followed completely.

Developed countries launched a number of tests in live working of substation with voltage class below 500 kV, such as America and French, etc., and formed complete operation methods (Assad & Braga 2000, Assad & Rosa 2003). The main substation live working projects launched in our country currently are overheat elimination of 110 kV and 220 kV disconnect switch clips, disconnecting and reconnecting of conductor, and live cleaning and live line water washing of tube post insulators with voltage class below 500 kV (An & Wang 2006, Sun & Hua 2013). There are no relevant research results reports about critical parameters of live working in 1000 kV UHV substation at home and abroad until now.

2 TYPICAL WORKING CONDITIONS

Different operation methods are adopted for different structures of buses in live maintenance of UHV substation. Generally speaking, buses in substation are divided into two types, four split conductor buses and single hard tube bus. Flexible buses are hanged by insulators, while hard tube buses are braced by post insulators. There are three live working methods including ground potential operation, intermediate potential operation and equal potential operation according to different potentials of workers. Due to high voltage class of 1000 kV and large phase to ground gap distance, operation method of equal potential is generally adopted, and the typical operation working condition is as shown in Figures 1 and 2.

Gaps under typical working condition of split flexible bus mainly have four kinds as shown in Figure 1: gap d_1 between workers heads and upper structure while workers are locating in split bus with equal

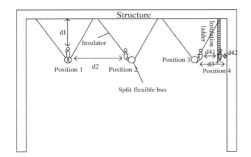

Figure 1. Typical working conditions diagram of split flexible bus.

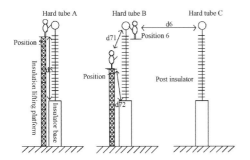

Figure 2. Typical working condition schematic diagram of hard tube bus.

potential, gap d_2 between human body and adjacent bus while workers are located in bus with equal potential, gap d_3 between workers and lateral structure while workers are located in equal potential with side phase, and complex gap $d_{41} + d_{42}$ among charged bus, human body and lateral structure during workers entering and leaving bus with equal potential.

Gaps under typical working condition of hard tube bus mainly include three kinds as shown in Figure 2: gap d_5 between human body and post insulator base while workers are located in tube bus with equal potential; d_6 between workers in equal potential and adjacent tube bus; complex gap $d_{71} + d_{72}$ among charged tube bus, human body and insulator base during workers entering and leaving equal potential with insulator lifting platform.

3 GAPS DISCHARGE TESTS

In general design of transmission and substation engineering, distance of tube bus to ground is 17.5 m, and to four split overhead bus is 19.5 m. Phase to ground distance of structure bus span is 11.3 m, and phase to phase distance is 14.2 m (Liu 2014). Test sample is set according to typical

working condition with the scale of 1:1. Diameter of tube bus for test is 0.25 m with length of 15 m. Height of mental base is 10 m. Four split simulated conductor split radius is 0.5 m with length of 18 m. Diameter of equalizing ring in ends is 3.5 m. Height of simulating workers for test is 1.8 m with width of 0.5 m. Test equipment contains 7200 kV and 4800 kV impulse generators and the measurement equipment. Uncertainty of the whole set of measurement system is smaller than 3% through verification of national high voltage metering station.

Up and down method were adopted for gap discharge tests. Switching impulse discharges were conducted on the gap for many times. U_{50} (50% discharge voltage) of the gaps were calculated according the test results under different gap length, then the discharge characteristic of gaps were obtained. Since discharge voltage of imposing positive polarity switching wave in the same gap is lower than that of negative polarity switching wave, positive polarity standard switching wave (250/2500 μs) was imposed on the gaps of phase to ground in the test. Positive polarity standard switching wave was imposed on buses of phase to phase test, and negative polarity standard switching wave was imposed on adjacent phase. The electric conductor connecting was as shown in Figures 3 and 4, respectively.

Standing posture was adopted for simulating workers for test in operation position 1, which was in the same potential as four split simulated conductor. Upper simulated structure was grounded. Distances of simulating worker and simulated structure were adjusted, discharge tests were conducted and the results were in Table 1.

Holding posture was adopted for simulating workers for test in operation position 2. Four split simulated conductor side was in the same potential with joint. Split bus in adjacent phase was in back of simulating workers. Distances of simulating

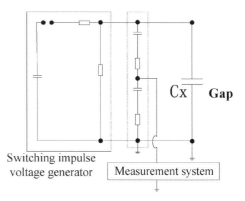

Figure 3. Electric wiring diagram of phase to ground test.

674

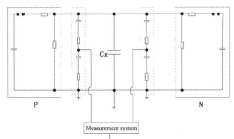

P is positive polarity switching impulse generator, N is negative polarity switching impulse generator, and Cx is test gap.

Figure 4. Electric wiring diagram of phase to phase test.

Table 1. Test results in position 1.

d_1/m	U_{50}/kV	Standard deviation δ
6.5	1654	4.7
7.5	1867	3.5
8.5	1975	4.1
9.5	2103	5.3

Table 2. Test results in position 2.

d_2/m	U_{50}/kV	Standard deviation δ
7.5	2254	5.5
8.5	2368	4.9
9.5	2587	3.2
10.5	2714	4.6

Table 3. Test results in position 3.

d_3/m	U_{50}/kV	Standard deviation δ
6.5	1947	3.9
7.5	2094	4.7
8.5	2308	4.3
9.5	2445	5.2

Table 4. Test results in position 4.

D_{41}/m	D_{42}/m	U_{50}/kV	Standard deviation δ
0.5	6.0	1873	3.8
0.5	7.0	2059	5.3
0.5	8.0	2198	4.4
0.5	9.0	2335	4.9

Table 5. Test results in position 5.

d_5/m	U_{50}/kV	Standard deviation δ
6.5	1894	5.1
7.5	2045	5.7
8.5	2211	4.9
9.5	2345	5.3

workers and adjacent phase were adjusted and the test results were in Table 2.

Holding posture was adopted for simulating workers for test in operation position 3. Four split simulated conductor side was in the same potential with joint. Vertical simulated structure was located in back of simulating workers and grounded. Distances of simulating workers and vertical structure were adjusted and the test results were in Table 3.

Test in operation position 4 belongs to complex gap test, which was used to determine safety that workers enter and leave busbar with equal potential from lateral structure. According to test results of live working of UHV transmission line, the U_{50} is the lowest in the whole horizontal access path when simulating workers are 0.5 m away from bus. Therefore, the lowest discharge position test is not conducted in the test, and to determine discharge characteristic of path only. Insulation ropes were first used to fix simulating workers in 0.5 m from split bus in test. Distances of vertical structure behind simulating workers and simulating workers were adjusted and the test results were results are shown in Table 4.

Standing posture was adopted for simulating workers for test with support of insulation platform in operation position 5. Hands were located in the same potential as simulated busbar. Lower post insulator bases are grounded. Distances of surface under simulating workers foot and insulator base, were adjusted and the test results were shown in Table 5.

The posture of simulating workers for test in operation position 6 was the same as that in position 5. Distance of simulating workers and adjacent bus were adjusted and the test results were shown in Table 6.

Operation position 7 and position 4 both belong to complex gap test. Standing posture of simulating workers in the test is in insulation platform. Distance of 0.5 m is kept between heads and tubes. Distances of surface under simulating workers foot and insulator base were adjusted and the test results were shown in Table 7.

Discharge characteristic curves of live working minimum approach distance and complex gap of ultra high voltage substation were obtained

Table 6. Test results in position 6.

d_6/m	U_{50}/kV	Standard deviation δ
7.5	2278	5.0
8.5	2411	5.4
9.5	2651	5.3
10.5	2803	4.2

Table 7. Test results in position 7.

d_{71}/m	D_{72}/m	U_{50}/kV	Standard deviation δ
0.5	6.0	1814	4.7
0.5	7.0	1998	4.3
0.5	8.0	2173	5.2
0.5	9.0	2288	5.1

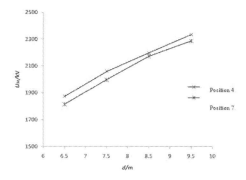

Figure 6. Discharge characteristics curve of combined gap.

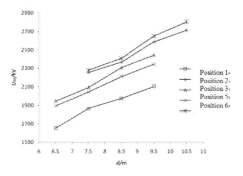

Figure 5. Discharge characteristic curves of minimum approach distance.

according to test results of operation gap of typical working conditions above in different positions, as shown in Figures 5 and 6, respectively.

From discharge characteristic curve in Figure 5, discharge voltages of phase to ground gaps in position 1, position 3 and position 5 are obviously lower than those of phase to phase gaps in position 2 and position 6. Discharge voltage in position 1 is the lowest among three phases to ground gaps. In other words, when workers operate in split bus through riding, air gap between head and upper structure is most likely to discharge. Controlling posture and occupation area of human body shall be paid attention to while operation is conducted in the position.

From discharge characteristic curves of two kinds of complex gaps of entering and leaving different equal potentials in Figure 6, discharge voltage of complex gap during entering and leaving tube bus with equal potential from ground with help of insulation platform method is lower than that of entering and leaving split bus complex gap from structure side with insulation flexible ladder or hanging basket method. In other words, entering and leaving equal potential from structure side is safer, but the difference of two kinds of discharge characteristics is not obvious.

4 CALCULATION OF LIVE WORKING RISK RATE AND MINIMUM APPROACH DISTANCE

In previous live working researches, the most important index of measuring the operation safety is risk rate. Risk rate of live working refers to certain insulation breakdown probability during operation. Generally, acceptable hazard rate criteria of live working is believed lower than 10^{-5}, namely discharge probability of live working gap shall be lower than one of hundred thousand, when maximum system switching overvoltage occurs once each time (Ding & Tan 1998).

It is assumed that probability distribution of system switching overvoltage and probability of air gap breakdown follow normal distribution, then Risk rate of live working can be calculated and obtained through the following equation:

$$R_0 = \frac{1}{2}\int_0^\infty P_0(U)P_d(U)dU \qquad (1)$$

where $P_0(U)$ = probability density distribution function of switching overvoltage amplitude.

$$P_0(U) = \frac{1}{\sigma_0\sqrt{2\pi}}e^{-\frac{1}{2}\left(\frac{U-U_{av}}{\sigma_0}\right)^2} \qquad (2)$$

where $P_0(U)$ = probability distribution function that air gap discharges under switching overvoltage with amplitude of U.

$$P_d(U) = \int_0^U \frac{1}{\sigma_d \sqrt{2\pi}} e^{-\frac{1}{2}\left(\frac{U-U_{50}}{\sigma_d}\right)^2} dU \qquad (3)$$

where U_{av} = mean value of switching overvoltage; σ_0 = standard aviation of switching overvoltage; U_{50} = 50% discharge voltage of air gap; σ_d = standard aviation of discharge voltage of air gap.

Design *Code for* 1,000 kV *Substation* stipulates: switching overvoltage level at ultra high voltage substation side is 1.6 p.u. (1.0 p.u. = $\sqrt{2} \times 1100/\sqrt{3}$ kV) for phase to ground, and 2.8 p.u. for phase to phase (China electricity council 2011). According to discharge characteristic under typical working condition in each operation position, minimum approach distance and complex gap of live working are calculated with switching overvoltage level of substation, and then modified to elevation 1000 m. See the results in Tables 8 and 9.

Minimum safety clearance distance of ultra high voltage substation in transmission and substation project general design: phase to phase of

Table 8. Calculation results of Minimum Approach Distance (MAD).

Position	MAD/m	MAD in altitude 1,000 m/m
1	6.4	6.8
2	8.4	9.1
3	5.6	6.0
5	5.8	6.2
6	8.3	9.0

Table 9. Calculation results of Complex Gap (CG).

Position	CG/m	CG in altitude 1,000 m/m
4	5.8	6.2
7	5.9	6.3

Table 10. Rate of live working in minimum designed clearance distance.

	MAD/m	Risk rate
Phase-ground	6.8	6.34×10^{-6}
Phase-phase	9.2	4.61×10^{-6}
	11.3	1.96×10^{-8}

four split conductor is 9.2, phase to phase of tube is 11.3, and all phase to phase is 6.8 (Sun, Y. & Hua, Y. 2013). Risk rate of live working is checked and design of minimum safety clearance distance as shown in Table 10.

5 CONCLUSIONS

1. According to operation gap test results of typical working condition of 1,000 kV substation live working, discharge voltage of workers in operation gap of split bus with equal potential is about 10% lower than that in tube operation with equal potential under same gap length, and safety of tube live working is higher.
2. Minimum minimum approach distance of phase to ground of 1,000 kV substation live working in areas with altitude 1,000 m and below is 6.8 m, minimum minimum approach distance of phase to phase is 9.1 m; minimum complex gap of entering and leaving split bus from structure side is 6.2 m, and minimum complex gap of entering and leaving pipe type bus bar with insulation lifting platform method is 6.3 m, which comply to relevant requirements of *Power Safety Working Regulation*.
3. Live working risk rates of minimum safety clearance distance of phase to ground and phase to phase of ultra high voltage substation in transmission and substation project general design are smaller than 10^{-5}, and live working can be safe conducted.

ACKNOWLEDGMENT

This work was funded from scientific and technological project of State Grid Corporation of China in 2013 with project number of GY71-13-048 "Technological Research and Application of Live Working of 1000 kV Substation /±800 kV Converter Station".

REFERENCES

Assad, L.A.X., Braga, A.E. and Martins, M.D. et al. 2000. Mounting of by-pass disconnect switches on energized 138 kV and 230 kV substation with barehand technique. *IEEE 9th International Conference on Transmission and Distribution Construction, Operation and Live-Line Maintenance* 280–286.
Assad, L.A.X. & Rosa, J.E. 2003. Improvement in the 500 kV substations using barehand methods without outage of power flow in Brazil's north-south interconnection. *IEEE 10th International Conference on Transmission and Distribution Construction, Operation and Live-Line Maintenance* 111–117.

An, J, Wang, Y. 2006. Dealing with the problem of conductor clip overheating in 110 kV isolating switch with power. *Power System Technology* 30(S1): 228–229

China electricity council. 2011. GB50697-2011 Design Code for 1,000 kV Substation. Beijing: *China plans press.*

Ding, Y.& Tan K. 1998. Live working technology foundation. Beijing: *China Electric Power Press.*

Hu, Y., Wang L. & Liu, K. et a1. 2006. Research of safety protection for live working on 1000 kV ultra high voltage transmission line. *High Voltage Engineering.* 32(12):74–77.

Liu Z. 2014. Transmission and substation project of State Grid Corporation of China general design (1,000 kV substation book). Beijing: China Electric Power Press.

Shu, Y. and Hu Y. 2007. Maintenance and live working technology for ultra high voltage transmission line. *High Voltage Engineering* 33(6):1–5.

Sun, Y, Hua, Y. 2013. Application of polymer live cleaning technology in 220 kV substation. *East China Electric Powe*r 41(1):238–240.

Wang, L., Hu, Y. & Shao, G. et al. 2006. Research on minimum approach distance for live working on 1000 kV AC transmission line. *High Voltage Engineering* 32(12):78–84.

Smart grid technologies

Advances in Power and Energy Engineering – Sun (Ed.)
© 2016 Taylor & Francis Group, London, ISBN 978-1-138-02846-3

Application of disconnecting circuit breaker in new generation smart substation

J.B. Li, R. Hu & J. Chen
Hubei Electric Power Research Institute, Wuhan, China

Y.X. Wen
State Grid Ezhou Electric Power Supply Company, China

J.G. Yang
Jiangsu Electric Power Research Institute, China

ABSTRACT: With the development of new generation smart substations in China, disconnecting circuit breaker, a new kind of primary equipment is applied. Comparing with traditional circuit breakers and open-air disconnecting switches, this paper studied the design concept, structural and technical features of disconnecting circuit breaker. The functional integration of disconnecting switch, earthing switch, electronic current transformer, and blocking system are analyzed. Through the research on the special requirements of the type test of disconnecting circuit breaker, this paper introduces its special combination functional requirements and analyzes the differences from traditional high voltage switchgear. The research result would provide experience and references for similar engineering application of disconnecting circuit breaker in new generation smart substations in the future.

1 INTRODUCTION

In traditional substations, the design principle was to surround circuit breakers with disconnecting switches. It was based on the fact that circuit breakers needed a lot of maintenance and were therefore surrounded by disconnecting switches to enable maintenance without disturbing nearby circuits. Nowadays, development in circuit breakers technology has led to a significant decrease in maintenance and an increase in reliability. Reliability of circuit breakers has increased due to evolution of primary breaking technology, from air blast, oil minimum, SF_6 dual pressure to today's SF_6 single pressure type. The number of series interrupters has been reduced. And removal of grading capacitors for live tank circuit breakers with two interrupters has further simplified the primary circuit and thus increased the availability. Today live tank circuit breakers up to 550 kV are available with one interrupter per pole without grading capacitors (Larsson et al. 2010). At the same time, the operating mechanisms have also been improved from pneumatic or hydraulic to spring type, which leads to more reliable designs and lower maintenance. However, no significant improvements in maintenance requirements and reliability have been made with open air disconnecting switches, which focus on cost reductions

by optimizing production materials during the same period (Faxå 2006).

To adapt the unmatched reliability of disconnecting switches, a new kind of primary equipment named Disconnecting Circuit Breaker (DCB) is designed and introduced to the grid (Song et al. 2013). The DCB combines the switching and disconnecting functions into one device. The change of design principle has enabled the integration of the disconnecting function with the circuit breaker. Since the primary contact of DCB is in an SF_6 protected environment, its disconnecting function is highly reliable. Therefore, the maintenance interval is increased, providing greater overall availability for the substations (Song et al. 2012).

The first installation of the DCB was in Sweden, 2000, and today DCBs are available from 72.5 kV to 550 kV voltage level with already more than 1370 sets in 25 countries all over the world. The China State Grid Corporation first applied DCBs in 2012 in new generation smart substations, and has achieved a good application effect.

2 TECHNICAL CHARACTERISTICS OF DISCONNECTING CIRCUIT BREAKER

A DCB combines the function of a traditional circuit breaker and adjacent disconnecting switches, performing both the breaking function and the disconnecting function. The DCB is based on the well proven traditional SF$_6$ circuit breakers, and its interrupter contacts also provide the disconnecting switch's function when in an open position. The disconnecting requirements are fulfilled by an increased insulation level between the breaker contacts. The DCB at least consists of a breaker and a blocking device. In the new generation smart substations of China State Grid, the DCB was also configured with electronic current transformer and earthing switch on the same structure. This integration makes it possible to build stations without any disconnecting switches, earthing switches or current transformers. And the application effect is obvious, having all primary contacts encapsulated in SF$_6$ decreases the maintenance work and increases the reliability of the primary system. Therefore, the main reasons for selecting the DCB instead of traditional solutions are system reliability, availability, maintenance aspects, space requirements, easier rehabilitation work and life cycle costs. The structure of DCB is shown in Figure 1.

2.1 Electronic current transformer of DCB

The electronic current transformer is installed between the extinguish chamber and the supporting insulator. The collector is in the top of the optical fiber insulator, with its data transmission optical fiber inside the insulator. The electronic current transformer is fitted outside the frame of the DCB, and utilizes Rogowski coil principle to measure the current. It has the advantages of higher measurement accuracy, larger dynamic range and better transient characteristic. The structure of the electronic current transformer is displayed in Figure 2.

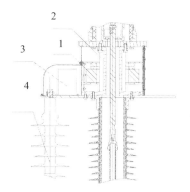

1- current coil 2- primary conductor 3- collector 4- fibre optic isolators

Figure 2. Structure of DCB's electronic current transformer.

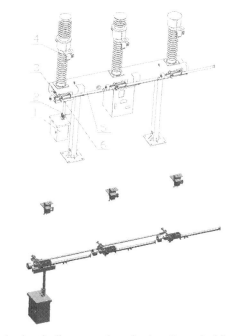

1- electrically operated mechanism 2- vertical links 3- three-phase linkage system 4- fixed contact of the earthing switch 5- moving contact of the earthing switch 6- frame

Figure 3. Structure of DCB's earthing switch.

2.2 Earthing switch of DCB

The earthing switch of DCB is placed outside the breaking chamber and the position of the earthing blades can clearly be seen from distance. The earthing switch is of single arm type, sharing the

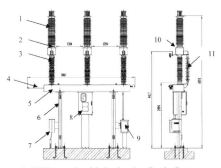

1- DCB's arc extinguish chamber 2- coil and collector of electronic current transformer 3- supporting insulator 4- moving contact of earthing switch 5- frame 6- outrigger 7- fiber splice tray 8- interrupter operating mechanism 9- earthing switch operating mechanism 10- fixed contact of earthing switch 11- fibre optic isolators

Figure 1. Structure of DCB.

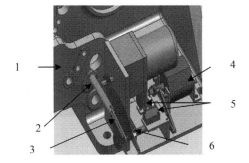

1- platform of the operating mechanism 2- locking lever 3- pushrod of the locking lever 4- auxiliary switch 5- travel switch 6- manually locked position

Figure 4. Structure of the DCB's interlocking and blocking system.

same underframe with the poles. Its direction of motion is perpendicular to the line terminals' direction. The three-phase linkage assembly system is integrated within the frame. The structure of the earthing switch is shown in Figure 3.

2.3 *Interlocking and bloking system of DCB*

The blocking device is integrated into the circuit breaker drive cabinet. DCBs need to be locked in the open position in a failsafe way. The locking consists of electrical blocking of the operating mechanism, as well as mechanical locking of the linkage system to the main contacts. Thereafter the adjacent earthing switch is closed, the visible closed earthing switch verifies that the part of the system is de-energized and safe for maintenance. The structure of the interlocking and blocking system of DCB is shown in Figure 4.

Table 1. Combined function test requirements of DCB.

Type test item	Type test technical requirements			Type test specification
	Operation sequence	Breaking current percentage/%	Operation number	
Short-circuit combined function test	O–0.3 s–CO–180 s–CO	100	2	(1) The 2 cycles of operation should be performed at the beginning and end of the short-circuit combined function test, the TRV should be 100% without any interference with the DCB between the tests. (2) The insulating characteristics of DCB after the short-circuit combined function test should satisfy the requirements of the insulation test.
	O	100	4	
	CO	100	2	
	12 times rated short circuit current interrupt			
	Operation sequence	Control voltage	Operation number	
Mechanical operations combined function test	C–30 s–O–30 s–C	110% C 120% O	500	(1) It is allowed to check the fasteners and lubrication parts after every operation cycle. (2) The insulating characteristics of DCB after the mechanical operations combined function test should satisfy the requirements of the insulation test.
		100% C 100% O	500	
		85% C 65% O	500	
	O–0.3 s–CO–180 s–C–30 s	100% C 100% O	250	
	5 000 times operations			
	Insulation test project		Voltage/kV	
Insulation test	Rated short-time withstand voltage test with working frequency (effective value)	Phase to ground Contact to contact	230 230 (+73)	(1) Rated short-time withstand voltage test with working frequency: Fixed contact and moving contact 1 times respectively, 1 min. (2) Lighting impulse withstand voltage test: Fixed contact and moving contact 30 times respectively, with positive and negative polarity each applied 15 times.
	Lighting impulse withstand voltage test (peak value)	Phase to ground Contact to contact	550 550 (+103)	

3 TYPE TEST REQUIREMENTS OF DCB

Since a DCB has to fulfill both applicable circuit breaker's standards and disconnecting switch's standards, a specific standard for disconnecting circuit breakers was issued by IEC in 2005, named IEC 62271-108 "High-voltage alternating current disconnecting circuit-breakers for rated voltage of 72.5 kV and above". In China, a similar standard was issued in 2011. Additional tests are required to demonstrate that the device complies with the relevant requirements of a disconnecting switch. The DCB shall fulfill the dielectric requirements for the isolating distance not only in new conditions but also in service, therefore the dielectric withstand across the isolating distance shall be demonstrated after a mechanical operation test as well as after the specified short-circuit test-duty. These important parts of this standard are the combined function tests, including the short-circuit combined function test and mechanical operations combined function test. These tests verify that the disconnecting properties of the DCB are fulfilled during its service life, despite contact wear and any decomposition by-products generated by arc interruption (Li et al. 2014). This is ensured by making all the breaking and mechanical tests first and thereafter confirming the disconnecting dielectric properties.

The separate test requirements are detailed in Table 1.

4 CONCLUSIONS

The construction objectives of State Grid Corporation's new generation of smart substation are "highly integrated system, reasonable structure and layout, advanced and applicable equipment". The switch gear is the most important primary equipment, and it is necessary to make it become more intelligent and integrated. This paper studied the design concept, structural and technical features of DCB. The functional integration of circuit breaker, disconnecting switch, earthing switch, current transformer, and blocking system are analyzed. The paper also studied the special type test requirements of DCB. The conclusions of this paper are beneficial for the application and maintenance management of DCB in the new smart substations.

REFERENCES

Faxå M. 2006. Applications of disconnecting circuit breakers [C]. *2006 IEEE PES Transmission and Distribution Conference and Exposition, Latin America, Venezuela.*

Larsson J.R., Sölver C.-E., Haglund L. 2010. Disconnecting circuit breaker enables smarter substation design [C]. *2010 IEEE PES Transmission and Distribution Conference and Exposition: Smart Solutions for a Changing World.*

Li Jinbin, Ruan Ling, Chen Jun 2014. Application of disconnecting circuit breaker in new generation smart substation [J]. *Electric Power Construction* 35(1): 30–34.

Song Xuankun, Li Jinru, Xiao Zhihong 2012. Overall design scheme for new generation intelligent sustation [J]. *Electric Power Construction* 33(11): 1–6.

Song Xuankun, Yan Peili, WuLei 2013. Review of key technology for smart substation pilot projects [J]. *Electric Power Construction* 34(7): 11–15.

Advances in Power and Energy Engineering – Sun (Ed.)
© 2016 Taylor & Francis Group, London, ISBN 978-1-138-02846-3

Research of medium voltage flexible DC device locating

Y.Y. Cui, T. Wei & W. Liu
China Electric Power Research Institute, Beijing, China

F.X. Hui
North China Electric Power University, Beijing, China

ABSTRACT: Flexible DC technology has been widely used in the field of power transmission in recent years, but lags behind in the distribution development. With the demand of the urban distribution network to improve the reliability and line capacity, it is necessary to study the flexible DC device to upgrade the existing distribution. The AC/DC hybrid power grid structure and typical networking using flexible DC devices are introduced in the paper. Then the problem of medium voltage flexible DC device site selection is researched. The evaluation indices of flexible DC device location are established. And the method of flexible DC ring network control device locating based on entropy and TOPSIS is proposed. Then the operability of the method is analyzed through the case. This method can promptly and accurately evaluate and sequence the alternative schemes, and the optimal location can be got.

1 INTRODUCTION

With the development of power electronic technology, the flexible HVDC technology has been developed greatly because of its strong controllability and convenient access to the distributed power supply. However, the application in the field of urban distribution network is lagging behind, especially in the medium voltage distribution network. Nowadays, the development of distribution network is facing the challenge of improving the power supply reliability and receiving the distributed energy sources at the global level (Hertem & Ghandhari 2010). In order to solve this problem, domestic and foreign relevant research institutions and well-known scholars all believe that the use of flexible direct current technology has a strong advantage in the island power supply, urban distribution network capacity expansion, AC systems interconnection, large-scale wind farm grid-connection and other aspects. Strengthening the communication and distribution network with the flexible HVDC technology (Huang et al. 2011) is an effective way to incorporate the distributed power supply into the power grid, it can also solve the problems such as large short circuit current and dynamic reactive power compensation, and it is the developing direction and strategic choice of the power system in the future (Jiang & Zheng 2012).

At present, the typical network structure of middle voltage distribution network in China mainly has double ring type, single ring type, multi segment moderate connection and radial structure.

Most of the urban medium voltage distribution network is ring design and open loop operation, which is not capable of the power flow adjustment, load balance and continuous load transfer. In view of the above problems, the scholars put forward ring network power controller based on DC back-to-back interconnection (Zhou et al. 2013), to achieve closed-loop operation and flow control (Yang et al. 2012). In view of the further requirement of the power supply reliability improvement, capacity increase and large number of DG access to distribution system in domestic urban distribution network, it is necessary to develop a flexible ring network control device, which can upgrade the existing urban distribution network. The closed-loop operation of the distribution network can be realized by installing a flexible DC control device at the key nodes, it can also carry out the power flow distribution and reactive power compensation to improve the power supply capacity of the existing distribution network and the stability of the voltage. So it is important to determine the installation position and capacity of the control device of the flexible DC link network.

With regard to the problem of location selection, it has been widely used in the substation construction. (Liu & Zhang 2007) proposed a particle swarm optimization algorithm to solve the problem of location selection in distribution network substation and improve the particle swarm optimization algorithm. (Wang et al. 2005) proposed a method for the two stage optimization planning of substation. (Tao et al. 2010) introduced the geographic information system, and putted forward the model and algorithm of the substation location based on the geographic information system. The characteristic of the location problems is to establish a mathematical model, using intelligent algorithms, to obtain the optimal solution and solution space is continuous. But for some discrete solution of problems, some literatures use a comprehensive evaluation method to treat site selection, such as analytic hierarchy process method and the Delphi method and fuzzy comprehensive evaluation (Gao et al. 2004), in order to get the best site. Different location should be considered by different situation, the solution space of the location of flexible DC device is discrete, and the location to be selected of different network is fixed, so the comprehensive evaluation method is appropriate.

2 HY DISTRIBUTION NETWORK

The existing distribution network is upgraded and reconstructed using 10 kV Flexible DC Ring Network Control Device (FDNCD) to form a new network structure of AC and DC distribution network. The FDNCD performance indicators are as follows: (1) AC port number ≥2, installed capacity 50 MW, AC side power factor [−1, 1], devices efficiency > 97% devices availability > 99%. Medium voltage FDNCD may achieve the following functions: optimized flow control and power distribution; reactive power and voltage control; emergency power support; multi-contact start-stop control; more communication and coordination control; distribution network loop operation.

2.1 Interconnection of AC line and DC load through FDNCD

Figure 1 shows the wiring pattern of interconnection of AC line and DC load through flexible device. In the mode of connection, the AC network is interconnected through the three-terminal devices communication networks and the DC load can be connected to the grid. The entire network can meet the N-1 criterion, and the power supply reliability of the network is higher.

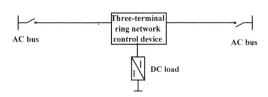

Figure 1. Interconnection of AC line and DC load through flexible device.

2.2 Interconnection of AC line and DC line through FDNCD

Figure 2 shows the wiring pattern of interconnection of AC line and DC line through flexible device. In the mode of connection, the AC networks are interconnected through flexible devices, and directly connect with DC network through the switch. The mode is relatively complex, the location and capacity of flexible DC devices will affect the technical and economic indicators of the entire system. The three-terminal flexible DC device is used to achieve interoperability of AC and DC systems, for any part of system failure or maintenance, it is able to restore the power supply promptly through the flexible device and interconnection switches, the system power supply reliability is greatly improved.

2.3 Interconnection of AC lines through FDNCD

Figure 3 shows the wiring pattern of interconnection of AC lines through flexible devices. This wiring pattern is the type of the project focused on in the study. The looped operation of different AC power distribution network is realized by using the flexible device, which is significant for the repowering of urban distribution network. The control mode is diverse. The entire network meets the N-1 criterion, it is intelligent and controllable. The flexible device can not only provide dynamic

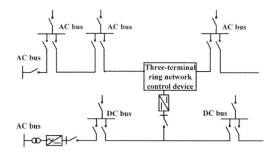

Figure 2. Interconnection of AC line and DC line through flexible device.

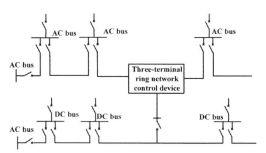

Figure 3. AC lines interconnected through FDNCD.

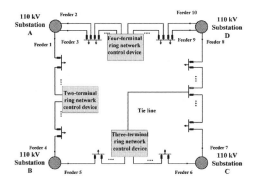

Figure 4. MV network configuration with FDNCD.

Table 1. Evaluation index system of FDNCD locating.

Target layer	Criteria layer	Base layer
FDNCD locating	Economic factors	FDNCD cost
		Annual operating costs
	Technical factors	Transformer load rate
		Emergency ability of reactive power compensation
	Geographical factors	Flood protection and water logging prevention conditions
		Construction conditions
	Environmental factors	Noise pollution

reactive power support, but also achieve load transfer, thus reduce short-term outage of the transfer process. Meanwhile, by means of the ring flexible control device, we can control the system flow flexible, realize the load balance among the more feeders, enlarge the power grid capacity, improve the reliability of the distribution network, and optimize the equipment utilization.

2.4 MV network configuration with FDNCD

Generally, Urban MV distribution network is designed as ring network and operates in open-loop. For the multi-feeders, it is possible to achieve loop operation in the medium voltage distribution network (10 kV) by using flexible DC device, as shown in Figure 4. It not only provides dynamic reactive power support, but also transfers the load economically and safely, as a result, reduce short-term outage significantly in the transfer process when meet overload of equipment or troubleshooting. Flexible DC ring network device can be set in form of two-terminal or three-terminal according to the actual situation.

3 FDNCD LOCATING BASED ON ENTROPY AND TOPSIS

3.1 Index System of FDNCD locating

In accordance with scientific, comprehensive, operability and important principle, the evaluation index system of FDNCD locating is established. It includes seven basic indicators, economic factors, technical factors, social factors and geographic factor, as shown in Table 1.

3.2 Indicators of entropy weight

The Entropy first appeared in the thermodynamic entropy, and introduced by Shannon to information theory late. Entropy is a state function, the system status remains unchanged, its entropy

is determined. Increase in entropy, showing loss of information. The higher the disorder degree of a system, the greater its entropy, the less amount of information it has.

The steps of index weight calculating with entropy method are as follows:

$$Y = (y_{ij})_{mn} \tag{1}$$

$$d_i^+ = \sqrt{\sum_{j=1}^{n}(r_j^+ - r_{ij})^2} \tag{2}$$

where Y = dimensionless matrix; y_{ij} = dimensionless data; and x_{ij} = raw data.

$$p_{ij} = y_{ij} \Big/ \sum_{j=1}^{n} y_{ij} \tag{3}$$

where P_{ij} = the proportion of y_{ij}.

$$e_j = e(y_{ij})\Big/ \ln n \tag{4}$$

$$e(y_{ij}) = \begin{cases} -\sum_{i=1}^{m} y_{ij} \ln y_{ij} & y_{ij} > 0 \\ 0 & y_{ij} = 0 \end{cases} \tag{5}$$

where $e(y_{ij})$ = information entropy; and e_j = output entropy of i-th index.

$$g_j = 1 - e_j g_j \in [0, 1] \tag{6}$$

where g_i = the difference coefficient of j-th index.

$$w_j = g_j \Big/ n - E_j \tag{7}$$

$$E_j = \sum_{j=1}^{n} e_j \tag{8}$$

where w_j = the weight of j-th index.

$$R = \left[r_{ij}\right]_{mn} = \left[w_j y_{ij}\right]_{mn}$$

$$= \begin{bmatrix} w_1 r_{11} & w_2 r_{12} & \cdots & \cdots & w_n r_{1n} \\ w_1 r_{21} & w_2 r_{22} & \cdots & \cdots & w_n r_{2n} \\ \vdots & \vdots & \vdots & \vdots & \vdots \\ w_1 r_{m1} & w_2 r_{m2} & \cdots & \cdots & w_n r_{mn} \end{bmatrix} \quad (9)$$

where R = weighting matrix; m = the number of candidates; and n = the number of evaluation index.

3.3 *Technique for Order Preference by Similarity to an Ideal Solution (TOPSIS)*

Technique for Order Preference by Similarity to an Ideal Solution (Hereinafter referred to as TOPSIS), is a multi-index case in a comprehensive evaluation of multiple program.

3.3.1 *The Positive and negative ideal solutions*

Let's assume efficiency indicators belong to Ω^+, cost indicators belong to Ω^-.

The positive ideal solutions:

$$R^+ = \left\{ \max r_{ij} \middle| 1 \le i \le m, j \in \Omega^+; \min r_{ij} \middle| 1 \le i \le m, \right.$$
$$\left. j \in \Omega^- \right\} \quad (10)$$

The negative ideal solutions:

$$R^- = \left\{ \max r_{ij} \middle| 1 \le i \le m, j \in \Omega^-; \min r_{ij} \middle| 1 \le i \le m, \right.$$
$$\left. j \in \Omega^+ \right\} \quad (11)$$

3.3.2 *Euclidean distance*
Distance from the positive ideal solution:

$$d_i^+ = \sqrt{\sum_{j=1}^{n} \left(r_j^+ - r_{ij}\right)^2} \quad (12)$$

Distance from the negative ideal solution:

$$d_i^- = \sqrt{\sum_{j=1}^{n} \left(r_j^- - r_{ij}\right)^2} \quad (13)$$

3.3.3 *Relative proximity*
Calculate the relative proximity between ideal solution and the indicators of each solution.

$$c_i = \frac{d_i^-}{d_i^+ + d_i^-} = 1 - \frac{d_i^+}{d_i^+ + d_i^-} \quad (14)$$

where $i = 1, 2, 4 \ldots m$

The corresponding alternative is depend on c_i, the greater c_i is, the more excellent, on the contrary the worse.

4 CASE ANALYSIS

According to the above calculation process, the typical 10 kV network in a demonstration area is analyzed bellow. The line is single connect-line with three segmentations. Five candidates are formed according to the number of segments points and connection points. The candidates are named as scheme 1 ... 5 respectively, as shown in Figure 5.

The initial data of the evaluation indicators are determined by the analysis of the load of the network and main transformer, as shown in Table 2.

Where, the value of A+ range from 95 to 100; the value of A range from 90 to 95 (not including 95); the value of B+ range from 85 to 90 (not including 90); the value of B range from 80 to 85 (not including 85).

The pre-indicators matrix is obtained by the data preprocessing:

$$Y = \begin{pmatrix} 0.5499 & 0.5314 & 0.4691 & 0.4709 & 0.4553 & 0.4364 & 0.4459 \\ 0.4124 & 0.4251 & 0.4356 & 0.4163 & 0.4409 & 0.4606 & 0.4162 \\ 0.2749 & 0.3543 & 0.4021 & 0.3822 & 0.4457 & 0.4388 & 0.4682 \\ 0.4055 & 0.4110 & 0.4423 & 0.4436 & 0.4409 & 0.4461 & 0.4385 \\ 0.5361 & 0.4924 & 0.4825 & 0.5119 & 0.4529 & 0.4538 & 0.4652 \end{pmatrix}$$

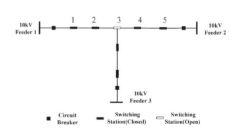

Figure 5. Typical 10 kV distribution network.

Table 2. Indicators raw data of 5 candidates.

Number of candidates	1	2	3	4	5
FDNCD cost (¥10,000)	700	600	350	550	650
Annual operating cost (¥10000)	20	18	16	17.5	18.5
Transformer load rate	0.61	0.57	0.5	0.58	0.65
Emergency ability of reactive power compensation	A+	A−	B+	A−	A+
Flood and water proofing	A	A−	A−	A−	A
Implementation	B+	A−	B+	B+	A−
Noise pollution (dB)	110	95	120	105	115

688

1. Calculating indicator entropy
   Firstly, calculate the output entropy of index,

   $$g_j = \begin{pmatrix} 0.4492 & 0.4382 & 0.4392 & 0.4417 & 0.2827 & 0.4389 & 0.4379 \end{pmatrix}$$

   Then, calculate the index difference coefficient,

   $$e_j = \begin{pmatrix} 0.5508 & 0.5617 & 0.5607 & 0.5583 & 0.7173 & 0.5611 & 0.5621 \end{pmatrix}$$

   Finally, calculate index Entropy,

   $$w_j = \begin{pmatrix} 0.1534 & 0.1497 & 0.1500 & 0.1509 & 0.0965 & 0.1499 & 0.1496 \end{pmatrix}$$

2. Weighted norm matrix
   Weighted normal matrix R is,

   $$R = \begin{pmatrix} 0.0844 & 0.0795 & 0.0704 & 0.0710 & 0.0440 & 0.0654 & 0.0667 \\ 0.0601 & 0.0604 & 0.0621 & 0.0596 & 0.0626 & 0.0656 & 0.0591 \\ 0.0401 & 0.0504 & 0.0573 & 0.0548 & 0.0633 & 0.0625 & 0.0665 \\ 0.0591 & 0.0584 & 0.0630 & 0.0636 & 0.0626 & 0.0635 & 0.0623 \\ 0.0781 & 0.0700 & 0.0688 & 0.0733 & 0.0643 & 0.0646 & 0.0661 \end{pmatrix}$$

3. Positive and negative ideal solution calculation

   $$R^- = \begin{pmatrix} 0.0801 & 0.0755 & 0.0688 & 0.0548 & 0.0626 & 0.0621 & 0.0661 \end{pmatrix}$$

   $$R^+ = \begin{pmatrix} 0.0401 & 0.0504 & 0.0573 & 0.0733 & 0.0647 & 0.0656 & 0.0591 \end{pmatrix}$$

4. Distance from ideal solution

   $$d_i^+ = \begin{pmatrix} 0.0590 & 0.0268 & 0.0203 & 0.0240 & 0.0449 \end{pmatrix}$$
   $$d_i^- = \begin{pmatrix} 0.0257 & 0.0275 & 0.0487 & 0.0294 & 0.0197 \end{pmatrix}$$

   where d_i^+ = the distance from positive ideal solution; and d_i^- = the distance from negative ideal solution.

5. Relative proximity calculation

   $$c_i = \begin{pmatrix} 0.3036 & 0.5070 & 0.7060 & 0.5507 & 0.3052 \end{pmatrix}$$

   The relative proximity of the third scheme is 0.7060. This scheme is the best solution followed by scheme 4, 2, scheme 1, 5 are poor.

5 CONCLUSION

The typical grid structures of upgraded AC power distribution network by use of the flexible DC devices are proposed in this paper. The method of the device locating is also researched. Firstly, the candidate sites are identified according to the locating index system and the characteristics of the network structure. And then the candidates are evaluated and ordered based on the entropy theory and TOPSIS. The case shows that the method is simple and clear, it makes a comprehensive evaluation on the candidates. And the evaluation result is more reliable. The comparison and pre-assessment of medium voltage flexible DC device locating schemes is of great significance for the construction of demonstration projects.

ACKNOWLEDGEMENT

I would like to express my gratitude to all those who helped me during the writing of this thesis. I wish to extend my thanks to the library colleagues who gave me kind encouragement and useful instructions all through my writing.

The project is supported by The National High Technology Research and Development of China 863 Program (2015AA050102).

REFERENCES

Gao, Ciwei et al 2004. The application of fuzzy evaluation of blind information in electric network planning. *Proceeding of CSEE* 24(9): 24–29.

Hertem, D.V. & Ghandhari, M. 2010. Multi-terminal VSC HVDC for the European supergrid: Obstacles. *Renewable and Sustainable Energy Reviews* 14(9): 3156–3163.

Huang, A.Q. et al 2011. The Future Renewable Electric Energy Delivery and Management (FREEDM) system: The Energy Internet. *Proceedings of the IEEE* 99(1): 133–148.

Jiang, Daozhuo & Zheng, Huan 2012. Research status and developing prospect of dc distribution network. *Automation of Electric Power Systems* 36(8): 98–104.

Liu, Zifa & Zhang, Jianhua 2007. Optimal planning of substation locating and sizing based on refined multi-team pso algorithm. *Proceedings of the CSEE* 27(1): 105–111.

Tao, Qingsong et al 2010. Geographic information based substation locating and sizing model and algorithm. *Proceedings of the CSU-EPSA* 22(6): 32–37.

Wang, Chengshan et al 2005. Two phase optimization planning approach to substation locating and sizing. *Automation of Electric Power Systems* 29(4): 62–66.

Yang, Yuefeng et al 2012. Research on control and protection system for Shanghai Nanhui MMCVSC-HVDC demonstration project. *The 10th IET International Conference on AC and DC Power Transmission, 2012.* Birmingham, United Kingdom: 1–6.

Zhou, Hao et al 2013. Research on insulation coordination for converter stations of Zhoushan Multi-Terminal VSC-HVDC transmission project. *Power System Technology* 37(4): 879–890.

Advances in Power and Energy Engineering – Sun (Ed.)
© *2016 Taylor & Francis Group, London, ISBN 978-1-138-02846-3*

Comparative analysis on the economy of coal transportation and electricity transmission

K. Zheng & F.Q. Zhang
State Grid Energy Research Institute, Beijing, China

Y. Wang
State Grid Xingtai Electric Power Supply Company, Hebei, China

ABSTRACT: The economic comparison of coal transportation and electricity transmission is a sensitive issue with a lot of external conditions. In this paper, a mathematical model to compare the economy of the coal transportation and electricity transmission based on the method of difference in coal prices is proposed. On the basis of the investigation of the current external conditions, including coal price, shipping costs, environmental external costs, investment of power plants, utilization hours, parameters of wire size, static investment, transmission power and power loss of different UHV transmission modes, etc., the comparative analysis on the economy of coal transportation and electricity transmission is implemented. The results show that, in the 12 UHV transmissions listed in the "13th five-year plan" investigated in this paper, electricity transmission is more economical than coal transportation for most transmissions, but for some transmission when the difference in coal prices is not large enough between sending and ending, coal transportation is more economical. Meanwhile, by comparing the current results and the historical data, when coal prices are at record levels, the economic advantages of electricity transmission reduced from 2008 to 2015, the main reason being that current coal price differences between the coal bases and load centers narrowed sharply since the supply and demand situation of the coal market reversed from 2008 to 2015.

1 INTRODUCTION

Coal occupies the dominant proportion of the energy supply and consumption structure in China; thus, coal transportation and electricity transmission become the two main modes of energy transmission to resolve the problem of imbalance in China's resource distribution and energy demand. The comparison of the coal transportation and electricity transmission is a comprehensive and complicated study which includes economy, efficiency, environment effects, land occupation, coordination of the regional economy, etc., and there has been a lot of controversy about which way is better. Though more and more people support developing electricity transmission preferentially, recently, because the effects of electrical transmission on environment being less than that caused by coal transportation, the economic comparison of these two energy transmission modes has always been a controversial issue and of great concern to the engineers and researchers (Development Research Center of the State Council (2009), Shao et al. 1985, Yang et al. 1987).

Considering the economic comparison of coal transportation and electricity transmission is a sensitive issue with a lot of external conditions (Deng et al. 2005, Ji. 2011, Wang et al. 2007), so the results would change with different external conditions in different periods of time. In recent years, the capacity of the coal transportation and electricity transmission have both been promoted dramatically thanks to progress in technologies, especially the electrification construction of the railway and the operation of heavy-load coal lines contribute a lot to the transportation ability of coal (Liang. 2011, He et al. 2014), and the rapid development of the Ultrahigh Voltage (UHV) transmission technology also improves the electricity transmission capacity (Zhang. 2007, Yin et al. 2010); on the other hand, the shipping charges and the coal prices have changed a lot: the average railway shipping charge was 0.1551 rmb/(ton•km) in 2015, which has increased 50% more than that in 2012. Compared to the peak value of the historical coal price (August of 2008), now the steam-coal (5500 kcal/kg) price has decreased almost 45% from the peak.

Thus, the previous opinion that it is economical to use coal transportation for long distances and use the electricity transmission for short distances (Zhang et al. 2014) is not appropriate for the new situation. In this paper, a mathematical model to calculate the difference in the price of electricity between the sending-end and receiving-end based on the difference in the price of coal of these two ends is proposed, and by comparing the electricity price difference and the electricity transmission price, the economy of coal transportation and UHV transmission listed in the "13th five-year plan" is analyzed. Meanwhile, the comparison with the current and historic results is implemented. The results of this paper are not only a choice criterion of the ways of energy transmission, but also of significant reference value for the energy and power planning.

2 METHODLOGY

2.1 *Method based on the coal price difference*

Two ways of energy transmission, i.e., the Electricity Transmission mode (ET) and Coal Transportation mode (CT) are selected to be compared. The schematic diagram of these two modes is shown in Figure 1. ET is transmitting the electricity through the transmission lines from the pithead power plant first to the load center; from there, the ground electricity price is calculated (Price 1); CT is transporting the coal by rail or a combination of various transportation modes (e.g., rail plus shipping) from the coal base to the load center. The coal is converted into electricity by the local power plant, then the electricity price of the local grid is calculated (Price 2). If Price 1 is less than Price 2, ET is more economical than CT; if Price 2 is less than Price 1, CT is more economical than ET.

2.2 *Mathematical model*

In CT, set x_s is the price of the coal in the coal base, the electricity price calculated based on x_s is $y_s = f_s(x_s)$ $y_s = f_s(x_s)$; set x_r is the price of the coal in the load center, the electricity price calculated based on x_r is $y_r^c = f_r(x_r)$; the set coal price difference between coal base and load center is x_t; so Price 2 is:

$$y_r^c = f_r(x_s + x_t) \tag{1}$$

In ET, set y_t is the electricity transmission price for one certain transmission type, and η is the transmission loss so Price 1 is:

$$y_r^e = y_s + y_t + \frac{\eta}{1-\eta} y_s = \frac{1}{1-\eta} f_s(x_s) + y_t \tag{2}$$

If ET is the same as CT, i.e., Price 1 = Price 2, then:

$$f_r(x_s + x_{ct}) = \frac{1}{1-\eta} f_s(x_s) + y_t \tag{3}$$

where X_{ct} = threshold of the coal price difference once ET is the same as CT, rewording equation (3) as:

$$x_{ct} = f_r^{-1}\left(\frac{1}{1-\eta} f_s(x_s) + y_t \right) - x_s \tag{4}$$

It can be considered approximately that $f_s(x_r) = a_s x_s + b_s$ and $f_r(x_r) = a_r x_r + b_r$, where a_s and a_r are the conversion coefficients of the electricity price calculated by the coal price in the coal base and load center, respectively, b_s is the non-fuel costs such as the depreciation of the plant, b_r is the sum of the profits and taxes. In view of $f_r^{-1}(y) = (y - b_r)/a_r$ and $f_r^{-1}(y_1 + y_2) = f_r^{-1}(y_1) + f_r^{-1}(y_2) - f_r^{-1}(0)$, the equation (4) can be reworded as:

$$x_{ct} = \left(f_r^{-1}(y_t) - f_r^{-1}(0) \right)$$
$$+ \left(f_r^{-1}\left(\frac{1}{1-\eta} f_s(x_s) \right) - x_s \right)$$

which is:

$$x_{ct} = x_{ct1} + x_{ct2} + x_{ct3}$$
$$= \frac{y_t}{a_r} + \frac{\eta}{1-\eta} f_r^{-1}(f_s(x_s)) + \left(f_r^{-1}(f_s(x_s)) - x_s \right) \tag{5}$$

where X_{ct1} = price of the electricity transmission which is not including the transmission loss; X_{ct2} = price of the coal transportation calculated based on the transmission loss; X_{ct3} = price of the coal transportation calculated based on the coal price in the coal base and the cost difference of the construction and operation of the power plants in the coal base and load center.

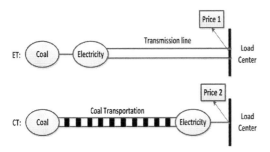

Figure 1. The schematic diagram of Electricity Transmission mode (ET) and Coal Transportation mode (CT).

The comparative analysis of the economy of CT and ET can be determined by the relation of the coal price difference between sending end and receiving end (x_t) and the threshold X_{ct}, as shown in equation (6):

$$\begin{cases} x_t > x_{ct}, & \text{ET is economical} \\ x_t = x_{ct}, & \text{CT is the equally economical as ET} \quad (6) \\ x_t < x_{ct}, & \text{CT is economical} \end{cases}$$

3 BOUNDARY CONDITIONS

3.1 Price of coal and transportation

Considering that in the current coal market supply exceeds demand, and the transportation is highly commercialized, the coal price difference from the coal base to the load center can represent the charge of coal transportation between the two places approximately. In this paper, we choose Inner Mongolia, C6, etc., six provinces as the coal bases and L3, L4, L5, etc., 10 provinces as the load centers. For consideration of the confidentiality agreements of the "13th five-year power plan of State Grid," we renamed the six coal bases as C1-C6 and the 10 load centers as L1-L10. Based on the coal prices in August 2015, the coal prices of all the places above and their price differences are shown in Table 1.

3.2 Environmental external costs caused by burning coal

The environmental external costs mainly include the damage to the environment, human health, and agricultural production caused by burning coal. According to the pollutant emission performance and the average coal consumption of different regions, and considering the environmental external costs of

Table 1. Coal prices of the coal bases and load centers based on the data in August, 2015 (The coal chosen is steam-coal with a heat value of 5500 kcal/kg; the units of the prices are in rmb/ton).

Coal bases					
C1	C2	C3	C4	C5	C6
252	259	160	157	253	144

Load centers				
L1	L2	L3	L4	L5
405	416	422	475	520
L6	L7	L8	L9	L10
495	530	470	465	520

*The price data is collected from http://www.cctd.com.cn/.

Table 2. Environmental external costs caused by power industry coal burning (unit: rmb/ton).

North China	East China	Central China	Northeast	Northwest
27	55	53	27	10

different pollutant emissions (i.e., soot, SO_2, NO_x) caused by burning coal, the total environmental external costs caused by power industry coal burning is calculated, the results are shown in Table 2.

3.3 Calculation of electricity price

The desulfurization and denitrification devices are installed in the power plants both in the coal base and the load center. The investment of the power plant refers to the typical investment to build a new power plant locally, and the capital Internal Rate of Return (IRR) of the investment is set to 8%. According to the thermal power industry benchmarking of China in 2015, for the power plants in the coal base, we use the air-cooling units, and the average coal consumption is 327 grams of standard coal per kWh; for the power plants in the load center, we use the water-cooling units, and the average coal consumption is 313 grams of standard coal per kWh.

3.4 Calculation of electricity transmission price

The three transmission modes, i.e., 500 kV HVAC, 1000 kV UHVDC, ±800 kV UHVDC, are investigated in this paper. The HVAC and UHVAC investigated use the double-circuit transmission lines on one tower, and the HVDC and UHVDC use the bipolar transmission lines. The utilization hours of HVAC and UHVAC investigated are set to 5000 hours, and of HVDC and UHVDC investigated are set to 6000 hours. The parameters of wire size, static investment, transmission power, and power loss of the different transmission modes investigated are shown in Table 3 (Electric Power Planning & Engineering Institute. 2015).

The static investments of the electricity transformation for the transmission modes investigated are shown in Table 4 (Electric Power Planning & Engineering Institute. 2015). For the HVAC and UHVAC transmission projects, the compensation rate of the high resistance is set to 0.8, series compensation is set to 40%.

The IRR of the investment of all the transmission projects investigated is set to 8%. The transmission distance from the coal base to the load center is set to 1.2 times of the straight line distance between these two points.

Table 3. Parameter settings of the different transmission modes.

Transmission modes	Wire size	Static investment (10,000 rmb/km)	Transmission power (MW)	Power loss/ 100 km
500 kV AC	4 × 630	309	2140	0.340%
1000 kV AC	8 × 630	905	9540	0.356%
±800 kV DC	6 × 1000	307	8000	0.280%

Table 4. Static investments of electricity transformation for the different transmission modes.

Station	Transforming project	Static investment (10,000 rmb)
500-kV substation	Construction of 1 × 750 MVA	16815
	Expansion of 1 × 750 MVA	5272
1000-kV substation	Construction of 2 × 3000 MVA	193835
	Expansion of 1 × 3000 MVA	30014
±800-kV convertor station	8000-MW convertor station (2 terminals)	899612

Table 5. Coal price difference matrix.

Load centers	Coal bases					
	C1	C2	C3	C4	C5	C6
L1	*153*	*147*	245	248	*152*	261
L2	*163*	*157*	256	258	*163*	271
L3	*170*	*163*	262	265	*169*	278
L4	223	217	**315**	**318**	222	**331**
L5	268	262	**360**	**363**	267	**376**
L6	243	237	**335**	**338**	242	**351**
L7	278	272	**370**	**373**	277	**386**
L8	218	212	**310**	**313**	217	**326**
L9	213	207	**305**	**308**	212	**321**
L10	268	262	**360**	**363**	267	**376**

Table 6. Transmission prices calculation results.

Project	Voltage level (kV)	Price (rmb/kWh)
C3-L3	±800 DC	0.063
C4-L3	±800 DC	0.076
C3-L7	±800 DC	0.065
C2-L9	±800 DC	0.075
C4-L4	±800 DC	0.076
C5-L5	±800 DC	0.074
C3-L4	±800 DC	0.073
C1-L4	±800 DC	0.066
C3-L3	1000 AC	0.044
C3-L1	1000 AC	0.039
C2-L3	1000 AC	0.06
C3-L8	1000 AC	0.072

4 RESULTS ANALYSIS

4.1 Coal price differences

Based on the coal prices shown in Table 1, the matrix of the coal price differences for each coal base to each load center is established as shown in Table 5. It can been found from Table 5 that the coal price differences are from 147 to 386 rmb/ton, and the larger differences are from C3, C4, and C6 to L4-L10 which are above 300 rmb/ton underlined in bold; the less differences are from C1, C2, and C5 to L2 and L3 which are below 200 rmb/ton underlined in bold and italic.

4.2 Transmission prices

According to the 13th five-year power plan of State Grid, 12 UHVAC and UHVDC transmission lines are selected to calculate the transmission prices based on the boundary conditions introduced in Section 3, the results are shown in Table 6.

4.3 Economic analysis

On the basis of the coal price differences and the prices of electricity transmission above, using the method introduced in Section 2, the comparative analysis of the economy of CT and ET for the 13 UHVAC and UHVDC above is made. There are 10 transmissions that ET is more economical than CT, the price of ET is less than CT from 0.01 to 0.07 rmb/kWh. The transmissions from C3, C4, and C6 have greater economic advantages (the price of ET is less than CT over 0.04 rmb/kWh). The main reason for that is that the coal price differences of these transmissions are larger (see the bold part of Table 5). On the other hand, there are two transmissions (C2-L9, C2-L3) where CT is more economical than ET; the main reason for that is the coal price of C2 is relatively higher in all the coal bases, which leads to the coal price differences of those two transmissions above being lesser, but the economic advantages of CT are not obvious in that the price of CT is less than ET from 0.007 to 0.01 rmb/kWh.

4.4 Comparison with historical results

In order to highlight the changes of the current analysis and the historical results, we use the same method to calculate the price items introduced in Section 2 with the data of August 2008 when the coal prices were at record levels. The prices of steam coal with heat value of 5500 kcal/kg in the main coal bases (C1-C6) are 220–400 rmb/ton, the prices for the main load centers (L1-L10) are 720–930 rmb/ton, and the coal price difference between the coal bases and load centers is 281–711 rmb/ton.

The calculation results with the 2008 data show that the price of ET is less than CT from 0.03 to 0.13 rmb/kWh, which is much larger than the current calculation result. The main reason for this is that the current coal price differences between the coal bases and load centers narrowed sharply compared to the situation in 2008. Specifically, the supply and demand situation of the coal market has been reversed from 2008 to 2015, and the costs of the intermediate link, such as the inventory and shipping fee, the railway wagon fee, etc., are basically eliminated when the supply is exceeding demand. However, when the demand is exceeding supply, such as in 2008, the costs mentioned above accounted for half of the total intermediate-link transmission costs and for the one-third of the factory coal price of the load center.

5 CONCLUSION

The economic comparison of the coal transportation and electricity transmission is a comprehensive study, so the comparing results would change once the external conditions change in different periods of time. Considering the current external conditions including the coal price, the shipping costs, the environmental external costs, the investment of the power plants, utilization hours, the parameters of wire size, static investment, transmission power and power loss of different UHV transmission modes, etc., with the mathematical model proposed in this paper, the economy of 12 UHV transmission listed in the "13th five-year plan" is calculated. By comparing the economy of the coal transportation between the relative coal bases and load centers, it is found that for most transmissions, electricity transmission is more economical than coal transportation, but for some transmissions where the coal price difference is not large enough between the sending and ending, coal transportation is more economical. Meanwhile, by

comparing the current results and the historical data when the coal price is at a record level, the economic advantages of electricity transmission reduced from 2008 to 2015. The main reason is that current coal price differences between the coal bases and load centers narrowed sharply since the supply and demand situation of the coal market reversed from 2008 to 2015.

REFERENCES

Deng Jianli, Ge Zhengxiang, Jiang Liping. 2005. A preliminary judgment on general balance of coal, electric power and transportation in the year of 2020 and constraint analysis on coal transport capacity to eastern power development. *Electric Power Technologic Economics* 17(3): 5–8.

Development Research Center of the State Council. 2009. Develop power transmission and coal transportation simultaneously and give the power transmission priority. *State Grid* 3: 29–31. Beijing: China Electric Power Press.

Electric Power Planning & Engineering Institute. 2015. *Power engineering design control limit indicators (2014)*. Beijing: China Electric Power Press.

He Yuqiang, Li Zhichen, Yin Chengfei. 2014. Study on the optimization of the 20000 tons train plan for the optimization of the new moon yellow railway. *Railway Transportation and Economy* 5: 25–29.

Ji Yuhua. 2011. Research on economic comparison of intra-regional coal transportation and power transmission in southern power grid. *Technological Development of Enterprise* 30(21): 184–185.

Liang Chenggu. 2011. China railway heavy haul transportation to create the wonders of the world-Datong Qinhuangdao railway volume exceeded 4 tons. *Railway Transportation and Economy* 1: 14.

Shao Hua, Zhou Shuren, Meng Xianshen. 1985. Rational utilization of coal resources: comparison between coal transportation, electricity transmission and gas transportation. *Coal Chemical Industry Design* 4:1–8.

Wang Yaohua, Zhang Fengying, Bai Jianhua. 2007. Economic comparison between coal transportation and power transmission. *Electric Power* 40(12): 6–9.

Yang Dezhuang, Wen Xun, Zhu Gang. 1987. The problem of coal transportation and electricity transmission. *Chinese Journal of Management Science* 2: 37–44.

Yin Shugang, Miao Peiqing, Bai Keming, et al. 2010. To improve utilization efficiency of global energy resources through UHV power transmission technology. *Electric Power* 43(2): 1–5.

Zhang Lei, Lu Xiaoqian, Wang Jing. 2014. Optimization on the structure of coal transportation and electricity transmission in China. *Journal of China University of geosciences* 14(6):13–22.

Zhang Yunzhou. 2007. The economic performance analysis of the over-voltage AC power transmission. *Electrical Equipment Industry* 4: 34–39.

Advances in Power and Energy Engineering – Sun (Ed.)
© 2016 Taylor & Francis Group, London, ISBN 978-1-138-02846-3

Comprehensive evaluation of regional grid smart level based on key technology

Q.M. Zhao, W.J. Qi, X.J. Li, H.J. Jia & Y.L. Liu
Key Laboratory of Smart Grid of Ministry of Education, Tianjin University, Tianjin, China

H. Huang & L. Liu
State Grid Energy Research Institute, Beijing, China

ABSTRACT: According to the current development features of the smart grid, a new idea to construct evaluation system of regional power grid intelligent level is presented in this paper which introduces the concept of "technology applicable degree". This paper also constructs extrapolation model in time dimension and spatial dimension. The extrapolation in time dimension means future evaluation outcome can be predicted by current and past evaluation outcome; the extrapolation in spatial dimension means the evaluation outcome of region ready for evaluation can be gotten quickly by the outcome of typical regions which we have evaluated. Furthermore, this paper introduces a method to decouple the index with strong relevance. Finally, By means of Gram-Charlier series, uncertain analysis of underlying index is considered. Case study reveals index system can provide reference to power grid plan. Index decoupling makes the result more accurate. The outcome of extrapolation in time and spatial dimension is credible. Gram-Charlier series is used to analysis uncertain index, which will deduce the confidence interval and confidence of final index, increasing the credibility of the evaluation outcome.

1 INTRODUCTION

Practices of smart grid is still in its infancy, a comprehensive and scientific evaluation to overall efficiency of smart grid is essential, which is meaningful for guidance of planning, practicing, operating smart grid.

At present, domestic and foreign research in related areas of smart grid construction and smart grid technology innovation has set off a new wave. Various methods to construct smart grid evaluation system have been developed. Gao & Yan (2013) proposed a comprehensive evaluation system of the smart grid. Li (2012) proposed a evaluation system to self-healing ability of smart distribution grid network. Ni, He, Shen, Deng, Deng & Huang (2010) sorted the relevant index out in the field of low-carbon electricity from generation, transmission, distribution and use. Tan, He, Liu, Huang, Deng & Deng (2010) analyzed the smart grid evaluation of US. While these evaluation systems are different from each other, the basic idea is consistent: index selecting always start from features of smart grid itself. Development of smart grid,

however, involves many smart grid technologies, especially for key technologies. Development of key technologies can always determine the development of smart grid. In addition, data getting of evaluation of smart grid is time-consuming, in normal case, we always want to get the outcome of evaluation system quickly, so we need to find a way, to get the evaluation result of the region ready for evaluation quickly. The correlation of the underlying index is strong; the accuracy of outcome can be influenced by the correlation without finding a way to eradicate it. What's more, there exist many uncertain underlying indexes, if we don't analysis uncertainty index, the credibility of the results will be greatly reduced.

This paper considers the key smart grid technologies; establish a new evaluation system with smart grid key technologies. For index with strong correlation, this paper eliminate the correlation with the way of weight revision; Gram-Charlier series is used to analysis the uncertain of underlying index; Extrapolation model in time and space dimension is built, deduction from point to plane is realized. Totally evaluation result can be quickly get in some degree. Finally, case of Henan, Jiangsu is used to verify the evaluation system, extrapolation model, and uncertain analysis.

This paper is supported by Project of Comprehensive evaluation of strong smart grid policy.

2 REGIONAL GRID COMPREHENSIVE EVALUATION SYSTEM

2.1 *Smart grid key technology applicable degree*

This paper considers the influence of smart grid technologies to the development of smart grid. For smart grid technologies, while some technologies are advanced, they are not applicable to this region; as a result, they can't support the smart grid development of the region. So we propose the concept of "key technology applicable degree". "Key technology applicable degree" is index which cover not only the development of technology, but also applicable degree of the technology in this region.

Grade of "key technology applicable degree" is the comprehensive evaluation result of underlying technology grade. To get the grade of underlying technology, we need to use smart grid technology maturity index model to calculate the maturity grade, and then multiply the grade with an "applicable degree" parameter.

For example, in index system, subordinate index of "strong and reliable" include not only index reflecting "strong and reliable", but also involve technologies related to "strong and reliable", which is shown in Table 1.

2.2 *Correlation and decoupling of index*

In the index system of regional grid evaluation, some index reflect the higher level index in different aspects, which results in the correlation between indexes (Zeng et al. 2013). As shown in Figure 1, some index point to two upper indexes, thus the calculation is inevitably repetitive, so a method is needed to eradicate the adverse effects brought by correlation, to make the result complete and correct.

In this paper, following method is used to decouple the correlation between indexes.

2.2.1 *Structure decoupling—pretreatment*

Firstly, decouple index with strong correlation in structure, as shown in following picture, in picture 1, the middle index in the second line is correlated with two upper indexes. Structure decouple in needed to make the underlying index point to only one upper index, as shown in Figure 2. It is to know, the final result is irrelevant with which upper index we point to, which will be showed below. The example in Figure 2 points the left index. Pretreatment in index system will be showed in section 1.3.

2.2.2 *Weight revision*

Structure decoupling just eradicates correlation on the surface, furthermore, correlation should be considered to revise weight of index system, to make the result accurate.

Given X_1 and X_2 as 2 indexes, while $X_1 \cap X_2 \neq \Phi$, in other words, if set X_1 and X_2 contain common element, we say X_1 and X_2 are overlapping. Define λ_{ij} as the correlation part in X_i of X_i and X_j, λ_{ij} is the correlation part, in other words, weight. Obviously:

$$\omega_i \lambda_{ij} = \omega_j \lambda_{ji} \qquad (1)$$

ω_i, ω_j is the weight of X_i and X_j, respectively.

Assume: $X_1 = \{A, B, C\}$, $X_2 = \{A, B, D\}$, weight in set X_1 is distributed as $\{\omega_{1A}, \omega_{1B}, \omega_{1C}\}$, weight in set X_2 is distributed as $\{\omega_{2A}, \omega_{2B}, \omega_{2D}\}$ (Weight has been regarded as the normalized).

In index system, set X_1 and X_2 the only two index below Z index, $X_1 = \{A, B, C\}$, $X_2 = \{A, B, D\}$, obviously, $X_1 \cap X_2 = \{A, B\}$, $X_1 - X_1 \cap X_2 = \{C\}$,

Table 1. Key technologies in index system.

First layer	Second layer	Third layer
Strong and reliable index	Power quality	Average total harmonic distortion
		Average phase imbalance
		User frequency pass rate
		Users voltage pass rate
	Level of safe operation	Transmission N − 1 pass rate
		Static stability reserve pass rate
		Dynamic stability pass rate
		N − 2 pass rate
	Supply reliability	Customer average interruption frequency
		Customer average interruption time
		Power supply reliability rate (RS-3)
	Key technology applicable degree	WAMS technology
		FACTS technology
		UHV AC and DC transmission technology
		Reactive power compensation technology

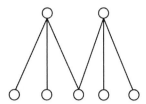

Figure 1. Schematic diagram before structure decoupling.

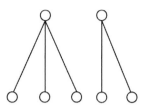

Figure 2. Schematic diagram after structure decoupling.

$X_2 - X_1 \cap X_2 = \{D\}$. For $X_1 \cap X_2$ is not empty, so take $\omega_1 X_1 + \omega_2 X_2$ as the final value is not suitable, adjustment is essential. Assume the value of index A, B, C, D is a, b, c, d. Therefore:

$$X_1 = \omega_{1A} \cdot a + \omega_{1B} \cdot b + \omega_{1C} \cdot c \qquad (2)$$

$$X_2 = \omega_{2A} \cdot a + \omega_{2B} \cdot b + \omega_{2D} \cdot d \qquad (3)$$

At this time there are two ways can be used to eliminate overlaps

1. Reserve part X_1 in $X_1 \cap X_2$. At this time, we have X_1 and $X_2 - X_1 \cap X_2$ two indexes.
2. Reserve part X_2 in $X_1 \cap X_2$. At this time, we have X_2 and $X_2 - X_1 \cap X_2$ two indexes.
   Define: $X_1' = X_1 - X_1 \cap X_2$, $X_2' = X_2 - X_1 \cap X_2$.
According to the first way:

$$Z_1 = \frac{\omega_1 \cdot X_1 + \omega'_2 \cdot X'_2}{K}$$
$$= \frac{\omega_1 (\omega_{1A} \cdot a + \omega_{1B} \cdot b + \omega_{1C} \cdot c) + \omega_2 \cdot \omega_{2D} \cdot d}{K} \qquad (4)$$

In the formula, ω_2' is the weight of X_2'. $K = \omega_1 + \omega_2' = \omega_1' + \omega_2$ the reason we divide K is to normalize the weight which is modified. For this question:

$$\omega'_2 = (1 - \lambda_{21})\omega_2 = \omega_{2D} \cdot \omega_2 \qquad (5)$$

$$X'_2 = \frac{\omega_{2D} \cdot d}{\omega_{2D}} = d \qquad (6)$$

Assume weight of index X_1 and X_2, which has been modified, is Y_1 and Y_2 known by the above explanation.

$$Y_1 = \frac{\omega_1}{K} \qquad (7)$$

$$Y_2 = \frac{\omega'_2}{K} \qquad (8)$$

The same way, process as the second method:

$$Z_1 = \frac{\omega'_1 X'_1 + \omega_2 X_2}{K}$$
$$= \frac{\omega_2 (\omega_{2A} a + \omega_{2B} b + \omega_{2D} d) + \omega_1 \omega_{1C} c}{K} \qquad (9)$$

The result is the same.

2.2.3 Correlation and decoupling of index

According to China's smart grid features, key technology and its application in electric power system of power generation, transmission, substation, power distribution and utilization, dispatching and communication, combining the five attributes of strong smart grid, strong and reliable, economic and efficient, transparent and open, clean and environmental protection, friendly interactive (Wang et al. 2011), and with the five attributes as the primary index, under each level indicators through the key technology for degrees, construct a comprehensive evaluation index system.

1. Strong and reliable
   Connotation of Strong and reliable refers to the power transmission capacity, ability to maintain the safe and stable operation of power grid, the abundance of the power and ability to safeguard the power quality. Strong and reliable index is shown in Table 2.
2. Economic and efficiency
   Economic and efficiency mainly refers to the optimization of the smart grid investments, reduction of line losses, giving full play to the power grid operations and social benefits. Through the application of smart technology, optimization of the whole power system operation and assets are deployed (Zhang et al. 2013). Index with strong relevance is "line loss percent", which is linked to both "Use efficiency" and "Operational level" index, in accordance to the decoupling method mentioned in section 1.2, we put "line loss" index under "efficiency" index, and adjust the weights through calculation. Economic and efficient index system is shown in Table 3.
3. Clean and environmental protection
   Transparent and open of smart grid mainly refers to the promotion of development and absorption of clean energy, development of low-carbon economy, energy conservation and environmental

Table 2. Strong and reliable index.

First layer	Second layer	Third layer
Strong and reliable index	Power quality	Average total harmonic distortion rate
		Average three-phase imbalance rate
		Frequency qualification rate of the user
		Voltage qualification rate of the user
	Safe operation level	N − 1 pass rate
		Static stability reserve check pass rate
		Large disturbance stability check pass rate
		N − 2 pass rate
	The power supply reliability	Average interruption times of customer
		Average interruption hours of customer
		Power supply reliability (RS-3)
	Key technology applicable degree	WAMS technology
		FATCS
		Uhv ac transmission technology
		Reactive compensation tech

Table 3. Strong and reliable index.

First layer	Second layer	Third layer
Economic and efficiency index	Use efficiency	The number of hours using annual maximum load of power grid
		The main transformer average load rate
		Average load rate
		Line loss percent
	Operational level	FACTS equipment application proportion
		Power transmission and transformation equipment condition monitoring rate online
		Distribution automation terminal coverage
		Auxiliary decision system percent
		Intelligent inspection rate
	Economic efficiency	Value-added service income scale
		Power supply per unit investment
		Input-output ratio of the power grid operation
	Key technology applicable degree	Load forecasting technology
		Intelligent substation technology
		Device status online monitoring technology

Table 4. Clean and environmental protection index.

First layer	Second layer	Third layer
Clean and environmental protection index	Configuration of absorption of clean energy	Newly increased grid synchronize capacity
		Clean energy permeability
	Clean energy dispatching level	Peak load regulating capacity
		Electric car emission reduction
		Difference between peak and valley
	Key technology applicable degree	Large scale photovoltaic power generation & synchronize technology
		Large scale wind power generation large scale wind power & synchronize technology
		Distributed power dispatching technology

protection (Tang 2011). Index with strong relevance is "Clean energy reductions", in accordance to the decoupling method mentioned in section I.B, we put "Clean energy reductions" index under "efficiency" index, and adjust the weights through calculation. Clean environmental protection layer index as shown in Table 4.

4. Transparent and open

Transparent and open of smart grid mainly refers to information sharing between power

Table 5. Transparent and open index.

First layer	Second layer	Third layer
Transparent and open index	Newly users percent	Users percent of distributed power
		Electric car users percent
	Information sharing	Information released percent
		Proportion of users participated in information sharing
	Key technology applicable degree	Big data related technology
		AMI technology
		Active distribution network technology

Table 6. Friendly interactive index.

First layer	Second layer	Third layer
Friendly interactive index	Load management	Users power information utilization percent
		Demand response load
	Service level	Customer satisfaction rate
		The third party evaluation satisfaction rate
	Intelligent level of communication	Smart meter coverage
		Two-way communication coverage
		Permeability of mobile terminal applications
	Key technology applicable degree	Demand response technology
		Electric car charge and discharge technology
		User electricity information acquisition technology

grid, power source and users. Grid is freely open without discrimination. Transparent and open index system is shown in Table 5.

5. Friendly interactive

Smart grid friendly interaction refers to realize the actively participation in power grid operation between power enterprises and user, friendly interactive equipment such as smart meters is wide used, demand response management is widely achieved (Wang et al. 2009). Friendly interactive index system is shown in Table 6.

3 EXPLORATION OF REGIONAL GRID SMART LEVEL

In actual assessment, evaluation results in time and space dimensions can be got by extrapolation.

3.1 Time exploration

Time extrapolation, means future evaluation outcome can be deduced by current evaluation result. Firstly, evaluate grid smart level of past years by comprehensive evaluation method, then, construct time extrapolation model by logistic model, finally, use the model to predict evaluation result in future years.

Logistic model can describe some increasing phenomenon which has boundary, describe the dependent variable changes with the time trend (Yu 2003), which calculation implication is simple, and now is used in many areas. Logistic First order differential equation of prediction model is shown below:

$$\frac{dF}{dt} = bF(1-F) \qquad (10)$$

In the equation, $F = \frac{y(t)}{m}$, $y(t)$ is the index level at time t; y_m is maximum of regional gird smart level. F is the ratio of index grade at a given time and index maximum grade; b is constant. By variable separation method, we get:

$$F(t) = \frac{1}{1 + a\exp(-bt)} \qquad (11)$$

In the equation, a is constant, the fastest growing time of regional smart level is $T_{\max} = \ln(a)/b$.

For Logistic model is nonlinear, so we can't get parameter estimated value by linear least square method. This paper use Nonlinear Least Square to get parameter of Logistic model. Criterion of NLS is to minimize the residual sum of squares, the objective function is:

$$\min \sum_{t=1}^{N} \left[Y(t) - y(t) \right]^2 \qquad (12)$$

In the equation, $y(t)$ is the actual grade of t; index, $y(t)$ is the predicted value of Logistic model. The schematic diagram is showed in Figure 3.

3.2 Space extrapolation

3.2.1 Basic information of space extrapolation

In actual evaluation, we can also get extrapolation evaluation result in space dimension. Space extrapolation means we can get other regions' evaluation result much more quickly by calculating some typical region in detail. For in normal case, we always want to grasp the evaluation result of region ready for evaluation quickly, but there exists many underlying indexes, which will spend lots of manpower and resources, which will lead to a long evaluation period. So, space extrapolation is essential to lead us to grasp the grid smart level quickly.

Space extrapolation need a bridge, which is region basic information, as shown below: when we precede space extrapolation, we should use basic information and grid smart level of the typical region we have evaluated, constructing mapping relationship between region basic information and region grid smart level. Region basic information is independent variable, and region grid smart level is dependent variable. After getting the mapping relationship, substitute basic information of extrapolation as independent variable, we can calculate grid smart level of extrapolation region out. The schematic diagram is showed in Figure 4.

Basic information of region is shown in Table 7.

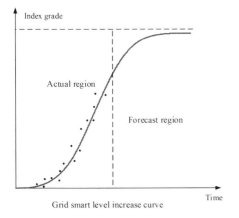

Figure 3. Regional grid smart level time extrapolation Logistic model.

Figure 4. Schematic diagram of smart level space extrapolation.

Table 7. Regional basic information.

Basic information of region
Population in the region
Total power sale quantity
Total residents power sale
Total industrial users power sale
Total number of motor vehicles
Number of substations above 35 kv
Line mileage above 110 kv
Gdp per capita
Per capita disposable income

3.2.2 Space extrapolation method analysis

Essence of spatial extrapolation is mapping relationship between two variables; this question can be solved by multiple regression analysis theory. Least square regression analysis method is a common method of multiple regression analysis. Simple method and little computational complexity is a big advantage of this method. However, when strong multiple relevance exist in independent system, Model deduced by least square regression is not stable and accurate; and when there are not much samples and independent variables, we have to use forward selection variables method, back to delete variables method and stepwise regression method and other methods of independent variable screening, which is no doubt increase the burden of forecasting work.

Partial least squares regression analysis is a new kind of multivariate statistics analysis method, which combines multiple linear regression analysis, canonical correlation analysis and principal component analysis into an organic whole, which can construct a model in condition of little samples and independent variables with strong relevance (Mao et al. 2008). Therefore, using partial least squares regression analysis method to analyze spatial extrapolation, the conclusion will be more reliable.

3.2.3 Principle of partial least squares algorithm

Partial least squares regression is an organic combination of principle components analysis, canonical correlation analysis and multivariate linear regression analysis. In principle components analysis principle, first principle component t_1 and u_1 extracted from independent variable X and dependent Y should take variation information of original data as much as possible, making the variance of data extracted largest (He 2009). That is:

$$D(t_1) \to \max \qquad (13)$$

$$D(u_1) \to \max \qquad (14)$$

In the equation, $D(t_1)$, $D(u_1)$ express variance of t_1 and u_1 respectively.

In canonical correlation analysis, in order to ensure the relevance between independent variable and dependent variable, in the process of extracting principle component t_1 and u_1, relevance parameter between principle components should be largest, that is:

$$r(t_1, u_1) \to \max \qquad (15)$$

In the equation, $r(t_1, u_1)$ express relevance parameter of t_1 and u_1, so, component t_1 and u_1 extracted from X and Y can not only take information from X and Y the greatest, they but also ensure strongest interpret ability of t_1 and u_1. After extracting first principle component t_1, regression analysis to t_1 of X and Y should be deployed. If precision meet requirements, calculation can be ceased; Otherwise, Use residual information of X after be explained by t_1, residual information of Y after be explained by u_1 to extract the second principle component. Recycle, until meet the precision requirement. Finally, if m principle components t_1, t_2, ..., t_m are extracted from X, regression analysis Y to X is transformed to multiple linear regression analysis Y to t_1, t_2, ..., t_m, and principle component t_1, t_2, ..., t_m can be expressed by X, so, regression equation of Y to X can be also get.

This paper take 9 basic information in Table 6 as the independent variable X, take region grid smart level as variable Y, dealing partial least squared analysis.

3.2.4 Basic flow of region smart grid smart level evaluation

Step 1: Collect data. Follow underlying indexes in index system, collecting the data corresponding to underlying indexes. The score of technology is calculated by technology maturity model evaluation system, and then multiply by an applicable parameter. Index dimensionless processing is deal by normalization method.

Step 2: Index weight design. Using order relation method to calculate comprehensive weight every layer.

Step 3: Index decoupling. According to the weight calculated, using weight revision is to decouple the coupling index.

Step 4: Calculate the comprehensive evaluation system result. Using linear weighting method is to calculate quantitative evaluation result.

Step 5: Uncertainty analysis. Using Gram-Charlier series (Zhang & Lee 2005), processing uncertain analysis to underlying uncertain index.

4 CASE STUDY

4.1 Comprehensive evaluation of Henan smart grid

Choosing Henan regions' smart grid as a research object, using multilevel analysis model, utilizing order relation method is to evaluate Henan's grid smart level in "twelfth five-year" period.

1. Fill in basic index data
2. Using order relation method to calculate weight of every basic index, weight of every basic index is shown in Table 8.
3. With reference to weight of every layer, index value of every layer can be calculated. Table 9 list out evaluation value of every basic property.
4. Using weight and index value to calculate final result by year, we can get evaluation result of Henan in "Twelfth five-year" period. As shown in Table 10:
5. Analysis of evaluation result: Figure 5 is radar map of evaluation result of every basic property of Henan in "Twelfth five-year" period. From picture we can get: 1) In "Twelfth five-year" period, Henan power grid smart level has always maintained a stable development trend, development speed is relatively stable. 2) Economic and efficient, Transparent and open, Friendly interactive have a high growth, which growing fast. 3) For strong and reliable, development is more gently, growth is limited,

Table 8. Weight of every basic index.

Strong and reliable	Economic and efficient	Clean and environmental protection	Transparent and open	Friendly interactive
0.2457	0.1653	0.1575	0.1833	0.2479

703

Table 9. Comprehensive evaluation result of every property in Henan.

Henan	2009	2010	2011	2012	2013
Strong and reliable	0.4917	0.4839	0.533	0.5228	0.5067
Economic and efficient	0.3305	0.3401	0.3653	0.4209	0.4451
Clean and environmental protection	0.2131	0.2191	0.2348	0.2434	0.2761
Transparent and open	0.2707	0.343	0.3729	0.4045	0.7154
Friendly interactive	0.3139	0.3429	0.4132	0.4895	0.5593

Table 10. Comprehensive evaluation result of Henan.

Year	2009	2010	2011	2012	2013
Henan	0.3691	0.3831	0.4229	0.4449	0.4862

Table 11. Grid smart level result of jiangsu—evaluation result of basic property.

Jiangsu	2009	2010	2011	2012	2013
Strong and reliable	0.4783	0.5057	0.5847	0.6054	0.6387
Economic and efficient	0.4932	0.5712	0.6668	0.7529	0.8120
Clean and environmental protection	0.2929	0.3699	0.4579	0.5967	0.7711
Transparent and open	0.3026	0.3345	0.3005	0.3126	0.3984
Friendly interactive	0.3731	0.4635	0.5736	0.6604	0.7363

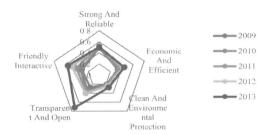

Figure 5. Grid smart level evaluation result of Jiangsu—Radar map.

Table 12. Comprehensive evaluation result of jiangsu per year.

Year	2009	2010	2011	2012	2013
Result	0.3691	0.3831	0.4229	0.4449	0.4862

the overall level of change is not big 4) Growth of clean and environmental protection is slow, which is related to Henan's power structure—thermal power plants take a large proportion.

4.2 Comprehensive evaluation of Jiangsu smart grid

Using the same steps with Henan region, calculate the result of grid smart level of Jiangsu.

1. Evaluation result of every property is shown Table 11:
2. Comprehensive evaluation result is shown in Table 12:
3. Analysis of evaluation result: Figure 6 is radar map of evaluation result of every basic property of Jiangsu in "Twelfth five-year" period. From picture we can get: 1) In "Twelfth five-year" period, Jiangsu power grid smart level has always maintained a rapid growth. 2) Friendly interaction, Economic and efficient, Clean and environmental increase fast. 3) For strong and reliable, development is more gently, growth is limited, the overall level of change is not big.

4) Transparent and open has a limited growth, which should attach more importance in future planning.

4.3 Time extrapolation case study

Take evaluation result of 2009–2012 in Henan as basic data to deal with the time extrapolation, erect logistic model to predict evaluation result of 2013, and verify the result with actual result of 2013. Fitting curve is shown in the Figure 7.

Extrapolation result of 2013 is 0.4731, actual data is 0.48626, error is 2.6%, which is acceptable, and extrapolation result is conceivable.

4.4 Space extrapolation case study

1. Space extrapolation error analysis

At the present stage, for little data of typical region is got, we can't compare the error between

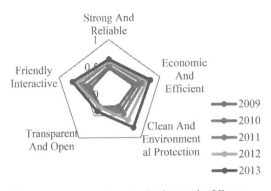

Figure 6. Grid smart level evaluation result of Jiangsu—Radar map.

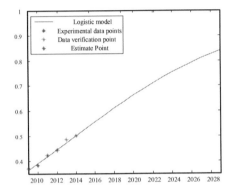

Figure 7. Grid smart level of Henan—time extrapolation fitting curve.

Table 13. Regional grid smart level—Space extrapolation error analysis table.

Anti-derivative value	0.5177	0.6272
Actual value	0.4863	0.6296
Error	6%	0.30%

actual result and extrapolation result, so we take "Forward and backward count" to verify the error.

Step 1: Take region basic data and grid smart level of Henan and Jiangsu to erect regression function.

Step 2: Extrapolate grid smart level of Beijing, Tianjin, Shandong, Neimenggu Province by regression function.

Step 3: Take region basic data and grid smart level of Henan and Jiangsu to erect regression function, and anti-derivative grid smart level of Jiangsu and Henan.

Step 4: Compare actual smart level of Jiangsu and Henan to anti-derivative smart level of Jiangsu and Henan.

Table 13 shows the outcome of space extrapolation by comparison, error between anti-derivative value and actual value is little, which is acceptable, this method is accurate, and extrapolation is conceivable.

2. Result analysis of Space extrapolation

Based on error analysis above, we take Henan and Jiangsu in 2013 as typical region, erect partial least square regression relationship between region basic information and region smart level, extrapolate region grid smart level of Huadong region.

Step 1: Fill in region basic data

Step 2: Deal partial least square regression, calculate regression equation.

Step 3: Take in region basic data of Huadong, calculate the final result: Grid smart level of Huadong in 2013 is 0.6013.

4.5 Uncertain analysis

Precondition of uncertain analysis: assume index A is uncertain, its value is μ, we assume this index follow normal distribution with expectation μ and variance $0.01 \times \mu$.

Table 14. Region grid smart level—uncertain index list.

Uncertain indexes	Data after normalization				
	2009	2010	2011	2012	2013
Intelligent inspection rate	0.1213	0.1312	0.1413	0.1423	0.1489
Power supply per unit investment	0.3221	0.3543	0.3656	0.3787	0.3790
Input-output ratio of the power grid operation	0.5665	0.5778	0.5898	0.5990	0.6012
Clean energy permeability	0.3104	0.3208	0.3098	0.2890	0.3421
Electric car emission reduction	0.3513	0.3718	0.3898	0.3990	0.4001
Peak load regulating capacity	0.2456	0.2667	0.2781	0.2891	0.3001
Clean energy emission reduction	0.1123	0.2331	0.2442	0.3112	0.4553
Users percent of distributed power	0.3221	0.3555	0.5231	0.6352	0.6821
Demand response load	0.2513	0.3526	0.3627	0.4362	0.5637

Table 15. Region grid smart level—uncertain analysis result.

Henan	Uncertain result		
	Lower limit	Upper limit	Confidence coefficient
2009	0.35	0.39	0.9116
2010	0.36	0.4	0.8801
2011	0.4	0.44	0.8410
2012	0.43	0.47	0.7955
2013	0.47	0.51	0.7285

Take results of Henan from 2009 to 2013 as a case to analysis the uncertain indexes.

Analytically, there exist 9 uncertain indexes in region smart grid index system, as shown in Table 14, the indexes value in the table are data of Henan after normalization.

Calculate result is shown in Table 15, which is uncertain analysis result of Henan from 2009~2013.

From Table 15, we can get the influence of 9 uncertain indexes to final result, which shows a decreased confidence coefficient, which is because the variance of uncertain data increased year by year, so lead to the growth of total variance.

5 CONCLUSION

According to features of smart grid construction of our country, this paper proposes take key technology into the construction of smart grid smart level. Case application shows, method proposed could reflect all kinds of properties of smart grid and the overall development. The overall construction effect could supply advice to the planning of grid construction, uncertain analysis could reveal effect of underlying index to final result. Time extrapolation provide prediction to future state of smart grid, space extrapolation utilize partial least square regression, achieve evolvement from point to plane, which all with a certain accuracy, result is conceivable.

REFERENCES

Gao X.H. & Yan Z. 2013. Smart grid comprehensive evaluation based on principle component-cluster analysis. *Power System Technology* 37(8): 2239–2243.

He Y.X 2011. *Comprehensive Evaluation of Power and Its Application*. Beijing: China Electric Power Press.

Li T.Y. 2012. Research on smart grid distribution network self-healing function and benefit evaluation model. *North China Electric Power University*.

Mao L.F, Jiang Y.H, Long R, Li N, Huang H, & Huang S. 2008. Long term power load forecasting based on partial least-squares regression analysis. *Power System Technology* 32(19): 72–77.

Ni J.M, He G.Y, Shen C, Deng Y, Deng Z.Y. & Huang W.Y. 2010. America smart grid evaluation review. *Automation of Electric Power Systems* 34(8): 9–13.

Tan W, He G.Y, Liu F, Huang W.Y, Deng Z.Y, & Deng Y. 2010. Low carbon index system of smart grid. *Automation of Electric Power Systems* 34(17): 51–54.

Tang H. 2011. Comprehensive evaluation research on region grid smart development level. *North China Electric Power University*.

Wang B, Li Y & Gao C.W. 2009. Demand side management under the framework of the smart grid vision and thinking. *Automation of Electric Power Systems* 33(20): 17–21.

Wang L.Q, Ruan Y.J, & Niu G.X. 2011. Research on relevance of evaluation index. *Computer & Digital Engineering* 39(7): 51–54.

Yu A.H. 2003. Research on Logistic model. *Nanjing Forestry University,*

Zeng M, Li L.Y, Ma M.J, & Li N. 2013. Smart grid economic benefit research based on interval number. *East China Electric Power*. 41(2): 0250–0252.

Zhang D.X, Yao L.Z, & Ma W.Y. 2013. Smart grid development strategy of China and foreign countries. *Proceedings of the CSEE* 33(31): 1–14.

Zhang P, & Lee S T. 2005. Probabilistic load flow computation using the method of combined cumulant and Gram-Charlier expansion. *IEEE Trans on Power Systems* 20(4): 1843–1850.

Advances in Power and Energy Engineering – Sun (Ed.)
© 2016 Taylor & Francis Group, London, ISBN 978-1-138-02846-3

Author index

An, H.H. 205
Anheuser, M. 651
Aredes, M. 251, 617
Aredes, M.A. 251
Arifullah, M. 171

Bai, S.Q. 411
Braslavsky, J.H. 217

Castro, A.R. 617
Cedeno, E.B. 577
Chang, F.Y. 199
Chen, B. 623
Chen, C. 289, 545
Chen, D.H. 75
Chen, J. 507
Chen, J. 631
Chen, J. 681
Chen, J.G. 159
Chen, K.M. 107
Chen, N.S. 335
Chen, W. 651
Chen, X. 401
Chen, X. 467
Chen, X.M. 141
Chen, Y. 605
Chen, Y.Y. 159
Cheng, Y.G. 185
Ci, W. 447
Cui, Y.Y. 685

Dai, L. 493
Deng, S.C. 401
Deng, W. 69
Dong, M.L. 493
Dong, T. 429
Dong, W. 411
Dong, Y.W. 385
Du , T. 257
Du, F. 651
Du, J.H. 597
Du, W. 317
Du, Y. 47
Du, Y. 91

Ewiss, K.B. 583

Fan, H. 101
Fan, H. 571
Fan, Q. 159
Fan, X.H. 661
Fan, X.K. 237, 241
Fan, Z. 631
Fang, W.C. 513
Feng, B.C. 395
Feng, H. 283
França, B.W. 617
Fu, L. 185
Fu, L. 507

Gao, J. 245
Gao, Q. 163
Gao, Y.L. 257
Ge, X.J. 335
Geng, J.W. 519
Ghosh, S. 7
Gontijo, G.F. 617
Gu, W. 159
Gui, J.T. 309
Guo, B.W. 411, 417
Guo, T.F. 263
Guo, X. 225, 405
Guo, X. 641, 647

Han, B. 91
Hao, X.D. 245
Hassan Ibraheem, M.A. 193
He, C.Y. 153
He, R.K. 115
He, X.G. 525
He, Y. 159
He, Y.F. 269
He, Y.F. 289
He, Z. 493
He, Z.Y. 507
Hou, Y.M. 565, 571
Hu, J. 493
Hu, R. 513
Hu, R. 681
Hu, T.S. 429
Hu, X.F. 667
Hu, Y.N. 163
Huang, C.P. 661

Huang, D. 641
Huang, H. 697
Huang, J.R. 635
Huang, Y. 107, 135
Huang, Y.H. 263
Huang, Y.M. 107
Huang, Z. 565, 571
Hui, F.X. 685
Hui, H. 571
Huo, F. 661
Huo, Y.H. 309

Ibrahim, N.A. 171
Inyang, H.I. 177

Ji, P. 667
Jia, H.J. 697
Jiang, J.Y. 453
Jiang, K. 303
Jiang, L. 91
Jiang, W. 591
Jiang, Z. 363, 453
Jiang, Z.F. 525
Ju, X.B. 153

Kaanagbara, L.L. 177
Kamran, S.A. 605
Kang, T.F. 513
Kong, W.Y. 211
Kou, P.G. 185

Lan, X.S. 591
Lei, B.L. 391
Lei, X.L. 673
Li, C. 231
Li, C.M. 655
Li, D.D. 539
Li, F. 435
Li, F. 487
Li, H.T. 275
Li, J. 217
Li, J.B. 681
Li, L. 231
Li, L.F. 401
Li, M.Y. 141
Li, P.F. 411

Li, Q. 597
Li, Q. 635
Li, S.Q. 391
Li, S.Y. 309
Li, W. 211
Li, W. 441
Li, W.B. 453
Li, X.F. 557
Li, X.J. 697
Li, X.R. 429
Li, X.S. 55
Li, X.Y. 129
Li, Y.G. 35
Li, Y.G. 513
Li, Y.Y. 141
Li, Z.H. 467
Li, Z.J. 275, 347
Li, Z.Y. 17
Liang, S.Y. 513
Liang, W. 631
Liang, X.F. 35
Liang, Y. 327
Lin, C.H. 159
Lin, X.F. 435
Lin, X.N. 55, 75,
    269, 545
Ling, M. 405
Littler, T. 317
Liu, B. 435
Liu, C. 263
Liu, D. 355
Liu, J.X. 667
Liu, K. 673
Liu, L. 163
Liu, L. 295
Liu, L. 697
Liu, M.X. 335
Liu, Q. 163
Liu, Q. 539
Liu, Q.F. 355
Liu, Q.Y. 355
Liu, R.H. 355
Liu, S. 327
Liu, S.G. 565
Liu, T. 673
Liu, W. 565, 571, 685
Liu, Y. 327
Liu, Y. 631
Liu, Y.H. 565
Liu, Y.L. 697
Liu, Y.Q. 533
Liu, Y.X. 29
Liu, Y.Y. 245
Liu, Y.Y. 385
Liu, Z.B. 63, 87
Liu, Z.F. 115
Long, C. 493

Long, D.L. 551
Lu, B.F. 303
Lu, S.F. 91
Lu, Y. 283
Lu, Y.L. 631
Lu, Z.G. 461
Lu, Z.X. 647
Lukic, V.P. 177
Lv, C. 317
Lv, J.X. 225, 405

Ma, C. 525
Ma, L.C. 199
Ma, M. 115
Ma, S. 557
Ma, Z. 225, 405, 501
Mao, C.X. 327
Meng, F.M. 453
Meng, Z.J. 211
Mero Hussin, A.S. 171
Muñoz, J. 605

Nasrat, L.S. 583
Ni, A.A. 487
Ni, W. 371
Nie, W.S. 205
Niu, Q.C. 237, 241
Niu, Y. 257

Oliveira, D.S. 617

Pan, L. 23
Pan, T.Z. 473
Pan, X.J. 377
Pei, W. 47, 69
Peng, C. 141
Peng, L. 55
Peng, Y. 673
Pi, J.C. 623
Pu, T.J. 69

Qi, S.P. 295
Qi, W.J. 697
Qi, Y. 335
Qiao, L. 545
Qin, R.S. 519
Qin, X.H. 479
Qiu, S. 7
Qu, W.D. 641
Qu, Z.Y. 597
Qu, Z.Z. 63, 87

Rao, Z.Q. 635
Ren, Y.X. 597

Shalsh, F.J. 171
Shang, Y.W. 501

Shangguan, M.X. 3
Sharkawy, R.M. 583
Shen, J. 29, 35
Shen, S. 275
Shi, F. 487
Shi, L.X. 385
Shi, T. 411
Shi, W. 97
Si, W.R. 641, 647
Song, C.N. 435
Song, G.Z. 97
Song, J.J. 441
Song, Y.T. 667
Su, B.S. 611
Su, J. 571
Su, S. 55, 75
Su, Z.M. 673
Sun, H.H. 153
Sun, H.S. 377
Sun, L.J. 501
Sun, P. 447
Sun, Y.J. 479

Tan, W.P. 423
Tang, P. 673
Tang, Q. 283
Tang, W. 395
Tang, W. 667
Tang, X.S. 473
Tang, Y. 623
Teng, Z.F. 163
Tu, C.G. 269
Tu, L.M. 327

van Emmerik, E.L. 617

Wang, A. 667
Wang, C. 153
Wang, C.G. 623
Wang, D.F. 101
Wang, F.H. 641, 647
Wang, G. 493
Wang, G.L. 107
Wang, J. 263, 461
Wang, J. 401
Wang, J.J. 667
Wang, J.Q. 327
Wang, L. 303
Wang, M.S. 335
Wang, Q. 467
Wang, S.R. 121
Wang, S.T. 41
Wang, W. 121
Wang, W.S. 263
Wang, W.X. 129
Wang, X. 519
Wang, X.D. 335

Wang, X.G. 75
Wang, X.R. 283
Wang, X.T. 107, 135
Wang, X.Y. 75, 83
Wang, Y. 377
Wang, Y. 487
Wang, Y. 691
Wang, Y.C. 327
Wang, Y.F. 557
Wang, Z.B. 135
Wang, Z.C. 545
Wang, Z.X. 269, 341
Wang, Z.X. 303
Wei, C. 545
Wei, M.T. 661
Wei, R.F. 417
Wei, T. 685
Wei, Y.G. 341
Wen, S.F. 41
Wen, X.K. 159
Wen, Y.X. 681
Wu, D.M. 237, 241
Wu, F. 211
Wu, Jy. 177
Wu, L.H. 467
Wu, L.X. 115
Wu, M. 245
Wu, M.B. 647
Wu, N. 363
Wu, T. 673
Wu, W.J. 129
Wu, X. 35

Xi, J.H. 377
Xia, J.R. 153
Xia, X. 371
Xiao, B. 673
Xiao, J. 327
Xiao, J. 423
Xiao, L.S. 635
Xiao, Y. 159
Xie, D. 611
Xin, E.C. 225
Xing, L. 363, 447
Xu, B.Y. 115
Xu, D.Z. 429
Xu, J.Y. 545
Xu, L.Y. 289
Xu, M. 441
Xu, M.M. 159
Xu, P. 487
Xu, T. 661
Xu, X.Q. 417
Xu, X.Q. 533

Xu, Y.P. 377
Xu, Y.P. 557
Xu, Y.T. 159
Xu, Z. 519
Xu, Z.B. 641, 647

Yan, B. 237, 241
Yan, H.R. 347
Yan, J. 545
Yan, T. 63, 87
Yang, D. 363, 447
Yang, H.T. 667
Yang, J.G. 681
Yang, L.B. 257
Yang, M. 641
Yang, S. 263, 461
Yang, S. 453
Yang, S.H. 461
Yang, X. 635
Yang, Y. 655
Yang, Y.B. 245
Ye, P. 163
Ye, S.Y. 283
You, D.H. 493
Young, D. 177
Yu, A.W. 395
Yu, F.Z. 17
Yu, G.K. 429
Yu, G.M. 141
Yu, H. 539
Yu, H.J. 107
Yu, R.Y. 83
Yu, T. 69
Yu, Y.X. 467
Yuan, B. 323
Yuan, H.B. 405
Yuan, H.W. 225, 501
Yuan, J. 417
Yuan, Y. 237, 241

Zang, T.L. 507
Zhang, B. 429
Zhang, B. 441
Zhang, B.Y. 501
Zhang, C. 153
Zhang, C. 303
Zhang, C.K. 17
Zhang, D.D. 453
Zhang, F. 29
Zhang, F. 257
Zhang, F. 493
Zhang, F.Q. 691
Zhang, G.M. 163
Zhang, H. 429

Zhang, J.B. 441
Zhang, J.L. 23, 29, 35
Zhang, J.T. 557
Zhang, L. 83
Zhang, L.N. 667
Zhang, L.X. 309
Zhang, Q.L. 17
Zhang, Q.L. 309
Zhang, R.P. 513
Zhang, T. 533
Zhang, W. 289
Zhang, W. 565
Zhang, X. 69
Zhang, X.W. 153
Zhang, Y. 135
Zhang, Y.C. 479
Zhang, Y.H. 479
Zhang, Y.J. 231
Zhang, Y.Q. 97
Zhang, Z.Y. 55
Zhao, D.C. 23
Zhao, H. 347
Zhao, J.X. 3
Zhao, Q.M. 697
Zhao, R.G. 355
Zhao, S.Y. 417
Zhao, X. 363
Zhao, Z.Y. 611
Zheng, C. 667
Zheng, K. 691
Zheng, T. 447
Zhong, Y.L. 525
Zhou, F.L. 539
Zhou, L.M. 225,
   405, 565
Zhou, Q.Y. 479
Zhou, R. 647
Zhou, W.H. 17
Zhou, W.J. 411, 417
Zhou, X. 29
Zhou, X. 289, 545
Zhou, Y.Q. 591
Zhou, Z.C. 631
Zhu, G.W. 63, 87
Zhu, L.L. 623
Zhu, L.Z. 83
Zhu, P. 539
Zhu, W. 513
Zhuo, Y. 651
Zhuo, Y.X. 269, 289, 545
Zong, J. 323
Zou, Q. 385
Zuo, L.H. 101
Zuo, W.J. 501